NEUROLOGIC COMPLICATIONS OF CANCER

Second Edition

CONTEMPORARY NEUROLOGY SERIES

NEUROLOGIC COMPLICATIONS OF CANCER

Second Edition

LISA M. DeANGELIS, MD
Chair, Department of Neurology
Co-Executive Director, Brain Tumor Center
Lillian Rojtman Berkman Chair in Honor
   of Jerome B. Posner
Memorial Sloan-Kettering Cancer Center
New York, NY

JEROME B. POSNER, MD
Evelyn Frew American Cancer Society Clinical
Research Professor
George C. Cotzias Chair of Neuro-oncology
Memorial Sloan-Kettering Cancer Center
New York, NY

OXFORD
UNIVERSITY PRESS

2009

OXFORD
UNIVERSITY PRESS

Oxford University Press, Inc., publishes works that further
Oxford University's objective of excellence
in research, scholarship, and education.

Oxford New York
Auckland Cape Town Dar es Salaam Hong Kong Karachi
Kuala Lumpur Madrid Melbourne Mexico City Nairobi
New Delhi Shanghai Taipei Toronto

With offices in
Argentina Austria Brazil Chile Czech Republic France Greece
Guatemala Hungary Italy Japan Poland Portugal Singapore
South Korea Switzerland Thailand Turkey Ukraine Vietnam

Published by Oxford University Press, Inc.
198 Madison Avenue, New York, New York 10016
www.oup.com

Oxford is a registered trademark of Oxford University Press

Library of Congress Cataloging-in-Publication Data
DeAngelis, Lisa M.
Neurologic complications of cancer/Lisa M. DeAngelis, Jerome B. Posner.—2nd ed.
p. cm.—(Contemporary neurology series; 73)
Rev. ed. of: Neurologic complications of cancer/Jerome B. Posner. c1995.
Includes bibliographical references and index.
ISBN 978-0-19-536674-7
1. Neurologic manifestations of general diseases. 2. Cancer–Complications.
3. Neurophysiology. 4. Nervous system–Pathophysiology.
[DNLM: 1. Neoplasms–complications. 2. Neoplasms–therapy. 3. Central Nervous System
Neoplasms–secondary. 4. Nervous System Diseases–physiopathology. QZ 202 D281n 2008]
I. Posner, Jerome B., 1932- II. Posner, Jerome B., 1932- Neurologic complications
of cancer. III. Title. IV. Series.
RC347.P67 2008
616.8–dc22
2008000669

9 8 7 6 5 4 3 2 1

Printed in China
on acid-free paper

Preface

Much has changed in the field of neuro-oncology since the publication of the first edition of this book in 1995. As the first edition was being written in the early 1990s, few academic centers had physicians who considered themselves neuro-oncologists. Those who were neuro-oncologists generally dealt with primary brain tumors rather than neurologic complications of cancers that arise outside of the nervous system. Now, most academic centers have at least one neuro-oncologist and many have neuro-oncology units. However, most of these units still deal primarily or exclusively with primary CNS tumors. Thus, there remains a need for a book aimed at clinicians who do not necessarily specialize in neuro-oncology, but who care for patients with cancer. This book is intended for them. We hope that it will be useful to neurologists who are asked to assist in the diagnosis and treatment of patients with cancer affecting the nervous system either directly or indirectly. We also hope that the book will be equally useful for oncologists and oncology nurses who encounter neurologic complications while treating patients for cancer.

Since the first edition, there have been many new developments in neuro-oncology. The treatment of brain metastases and to a lesser degree epidural spinal cord compression has been revolutionized by advances in stereotactic radiation. New therapeutic agents, including new chemotherapy, monoclonal antibodies, and small molecules, have produced new neurologic complications, involving both the peripheral and central nervous system. Other neurologic complications of chemotherapy and radiation therapy, although clinically described for many years, are often sufficiently uncommon that they are not recognized.

These advances, and many others, have required extensive rewriting of the second edition. Every chapter has been revised and thoroughly updated. Of particular note, new developments in the biology of metastases required a complete rewrite of Chapter 2. Although the clinical signs and symptoms of metastases have not changed, new treatments have required extensive revision of Chapters 5 through 8. New therapeutic agents and new techniques in radiation therapy have required revisions of Chapters 12 and 13. The references have been updated, emphasizing the most recent papers and reviews.

Because the book is intended to help both oncologists and neurologists, oncologists may find some of the neurology more arcane than necessary (after all we are neurologists), and neurologists may find some of the oncology too detailed. But don't give up; the information contained within is of equal use to both specialists. The reader will find redundancies in the text. This is deliberate, because we hope the book will be both read and used as a reference. The redundancies have decreased the need to keep referring to other sections of the book when reading a specific topic.

The addition of Dr. Lisa DeAngelis, Chair of the Department of Neurology at Memorial Sloan-Kettering Cancer Center, as an author brings new insights into some of the problems discussed in this book. She was extremely helpful in preparing the first edition; many of the ideas expressed in that edition were hers. Also added to this edition are new illustrations, beautifully drawn by Ms. Terry Helms.

We are extremely grateful to colleagues who have read portions of the book and made extremely valuable suggestions. These include Dr. Anne Chiang, who read Chapter 2 on the biology of metastasis and made many helpful suggestions; Dr. Fabio Iwamoto, who read Chapter 4; Dr. Edward Avila, who reviewed the section on seizures and made helpful suggestions; Dr. Andrew Lassman, who read Chapter 5; Dr. Vivian Tabar, who assisted in the preparation of the surgical section of Chapter 5; Dr. Mark Bilsky, who read Chapter 6; Dr. Yoshiya (Josh) Yamada, who read the radiation oncology section of Chapter 6; Dr. Kent Sepkowitz, who read Chapter 10; and Dr. Robert Darnell, whose paper with Dr. Posner in the *Seminars in Oncology* formed the basis of Chapter 15. All of these readers made suggestions and corrections, but they are not responsible for any errors that may be contained in those chapters. Ms. Judy Lampron and Ms. Carol D'Anella read the manuscript for spelling and syntax and certainly improved its readability. Finally, we are extremely grateful to Craig Panner of Oxford University Press for his patience, good advice, and easy availability during the preparation of the edition.

We dedicate this book not only to our spouses but also to our patients, whose illnesses and suffering have taught us much about clinical diagnosis, treatment, and the humane care of patients. We hope that by imparting the knowledge we have gained from these patients, the readers of this book will be able to improve their patients' quality of life.

Lisa M. DeAngelis
Jerome B. Posner

Contents

Part 2.   Metastases

Part 3.   Nonmetastatic Complications of Cancer

INTRODUCTION   325

FREQUENCY   326

PATHOPHYSIOLOGY OF CEREBROVASCULAR DISEASE   327

    Hypocoagulation   327
    Hypercoagulation   328

CENTRAL NERVOUS SYSTEM HEMORRHAGE   330

    Hemorrhage into Brain Metastases   330
    Subdural Hemorrhage   334
    Coagulopathic Hemorrhage   335
    Leukostasis   338
    Hyperviscosity   339
    Hypertension   339

CENTRAL NERVOUS SYSTEM INFARCTION   340

    Atherosclerosis   340
    Disseminated Intravascular Coagulation   341
    Arterial Occlusion by Extrinsic Tumor   346
    Cerebral Emboli   346
    Thrombotic Microangiopathy   354
    Cerebral Vasculitis   355
    Intravascular Lymphoma   356
    Thrombocytosis   356

VENOUS OCCLUSIONS   357

    Compressive Venous Sinus Occlusion   357
    Venous Sinus Thrombosis   358
    Tumor Emboli   360

OTHER DISORDERS   360

    Systemic Thrombophlebitis   360
    Episodic Neurologic Dysfunction in Patients with Hodgkin Disease   360
    Systemic Hypotension   360
    Air Embolism   362

APPROACH TO THE PATIENT   362

REFERENCES   362

10.   CENTRAL NERVOUS SYSTEM INFECTIONS   369

INTRODUCTION   369

PATHOPHYSIOLOGY OF CENTRAL NERVOUS SYSTEM
    INFECTION   371

    Host Defenses   371
    Infection Sites within the Central Nervous System   374

PART **1**

GENERAL PRINCIPLES

Chapter 1

Overview

INTRODUCTION

Neuro-oncology is a medical discipline that deals with the diagnosis and treatment of (1) primary central nervous system (CNS) neoplasms, (2) metastatic and nonmetastatic neurologic complications of cancers that originate outside of the nervous system (called here systemic cancers), and (3) symptoms associated with cancer that have nervous system implications, for example, pain and fatigue. This book primarily addresses the second aspect, that is, neurologic complications of systemic cancer. Where appropriate, some comments, figures, and case illustrations that apply to

primary nervous system tumors are provided, and a chapter is devoted to symptoms, of cancer such as pain, fatigue, seizures, and other symptoms that require care to improve quality of life (see Chapter 4).

The American Cancer Society has estimated that approximately 213,380 persons developed lung cancer and approximately 160,390 individuals have died of that disease in 2007.[1] However, it is not the lung lesion that kills most of those patients. They, like most patients with cancer, die not from the primary lesion but from metastases to vital organs or from nonmetastatic complications, such as infection associated with the immune suppression, that can be caused either by the cancer or its treatment. Metastases and other potentially lethal complications often affect the nervous system. Patients with brain and spinal metastases, if not treated, often die within weeks of the development of CNS symptoms. Systemic infections and treatment side effects can also cause severe nervous system symptoms. Furthermore, nervous system complications, including metastases, may occur in patients not known to have cancer. A significant number of patients with lung cancer do not present with pulmonary symptoms but present with nervous system symptoms; correct identification of a brain or spinal lesion as a metastasis, rather than a primary CNS tumor, leads to discovery of the underlying lung cancer. Paraneoplastic syndromes (see Chapter 15) affecting the nervous system are rare but important, because approximately two-thirds of patients present to physicians with neurologic symptoms, the cancer being identified only when the physician recognizes the neurologic disorder as paraneoplastic. Even long-term survivors of cancer may not escape neurologic problems. A recent study of adverse health outcomes in 1362 survivors of childhood cancer, who were assessed a median of 17 years after treatment, indicated at least one adverse event in 75% of them. Neurologic problems, including seizures, motor dysfunction, and sensory loss, affected 262 patients (19%).[2] For these reasons and those listed in the following paragraphs, neurologists, oncologists, and general physicians must be familiar with neurologic symptoms as a manifestation of a systemic cancer.

This chapter has three sections: The first describes the rationale for a discipline called *neuro-oncology*. The second establishes a working classification of neurologic complications of systemic cancer that serves as an outline of subsequent chapters. The third describes the diagnostic approach to a patient with nervous system symptoms possibly due to cancer and includes tables that help assure uniformity among caregivers assessing such patients.

RATIONALE FOR NEURO-ONCOLOGY

Neurologists, oncologists, and other physicians must give special attention to neurologic complications of cancer for several reasons (Table 1–1).

Nervous System Complications of Cancer Are Common

Cancer is the nation's second leading cause of death, and when age-adjusted death rates are considered, for individuals less than 85 years, it is the leading cause.[1] Data indicate that in 2007, more than 1.4 million persons developed a new cancer (excluding carcinoma in situ and non-melanotic skin cancer) and more than half a million died of cancer[1] (Table 1–2). Although both the incidence and mortality from most cancers is decreasing, the increase in population and the number of older people has led to an overall increase in prevalence. However, the statistics are encouraging. Mortality rates from cancer have decreased by 1.6% per year in men and by 0.8% in women between 1993 and 2003.[1] Even more encouraging are recent data that overall cancer death rates decreased by 2.1% a year from 2002 to 2004, almost twice the

Table 1–1. **Importance of Neurologic Complications of Cancer**

Neurologic complications of cancer are common.
Neurologic complications of cancer are increasing.
Neurologic complications are serious.
Diagnosis is often difficult.
Treatment helps.
Neurologic complications are unique.
Research is essential.
The brain and cancer have biologically important relationships.

Table 1–2. **Frequency of Intracranial Metastases from Systemic Cancers**

Primary Tumors	New Cases (United States 2007)	No. of Deaths (United States 2007)	Percentage with Intracranial Tumor at Autopsy at MSKCC[6]	Estimated Total No. of Deaths with Intracranial Tumor
Lung	213,380	160,390	34	54,533
Breast	180,510	40,910	30	12,273
Colon	112,340	52,180	7	3653
Urinary organs	120,400	27,340	23	6288
Melanoma	59,940	8110	72	5839
Prostate	218,890	27,050	31[*]	8386
Pancreas	37,170	33,370	7	2336
Leukemia	44,240	21,790	23[†]	5012
Lymphoma (non–Hodgkin)	63,190	18,660	16[†]	2986
Female genital tract	78,290	28,020	7	1961
Brain and CNS	20,500	12,740	100	12,740
All Sites	1,444,920	559,650	24	134,316

[*] Largely skull and dura.
[†] Largely leptomeningeal.
MSKCC, Memorial Sloan-Kettering Cancer Center; CNS, central nervous system.

annual decrease of that from 1993 to 2003.[3] The lesser decrease in women results from the fact that lung cancer mortality increased by 0.3% per year between 1995 and 2003,[1] a result of cigarette smoking peaking approximately 20 years later in women than it did in men. However, lung cancer incidence in women is no longer increasing, and death rates are increasing at slower rates than in the past.[3] Also encouraging is the decrease in both incidence and mortality from breast cancer in women[4,5] that may partially result from decreased use of hormone replacement therapy.[5] The most recent data indicate a decrease in breast cancer incidence of 3.5% a year from 2001 to 2004.[3] Even mortality rates from brain cancers decreased between 1990 and 2003,[1] although, as indicated in the following text, it is unlikely that the same is true for either the incidence or mortality from brain metastases. Despite these encouraging trends, the lifetime probability of developing cancer is 45% for men and 38% for women.[1]

Autopsy data from Memorial Sloan-Kettering Cancer Center (MSKCC)[6] identified intracranial metastases in 24% (572) of 2375 patients with cancer. If the prevalence of intracranial metastases in patients dying from specific cancers at MSKCC is similar to that in the rest of the nation, and one multiplies by the American Cancer Society national figures, one obtains a rough estimate of the number of patients dying with intracranial metastases in the United States each year. The 2007 figure is over 130,000.

Unfortunately, given the current low autopsy rates (now below 10%[7]), these data from the 1970s are unlikely to be replicated with more up-to-date information. Despite modern imaging techniques, substantial errors that might alter patient treatment are still encountered in modern autopsy studies[8,9]; this includes errors in diagnosing and treating malignancies[10,11] (Fig. 1–1). A clinical analysis of patients dying with brain metastases[12] suggests that two-thirds or 80,000 will have had significant neurologic symptoms during life. Even assuming that these numbers are overestimated by 100%, more than 40,000 patients die with symptomatic intracranial metastases each year, making

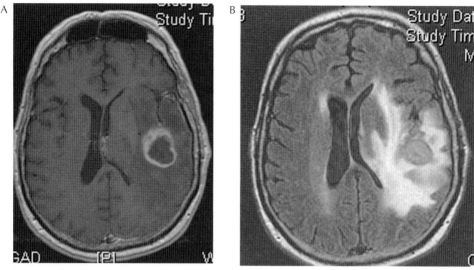

Figure 1–1. Neurologic complications are common. A 65-year-old man was found to have a right upper lobe tumor on a screening chest X-ray. After local resection of an adenocarcinoma, he was treated with chemotherapy. He then developed headache and confusion. **A**: A contrast-enhanced MRI revealed a ring-enhancing mass, with a somewhat thickened area anteriorly, suggesting tumor. **B**: The mass was surrounded by edema (hyperintensity in the white matter, sparing the gray matter). Although the neuroradiologist, noting restricted diffusion, suggested the possibility of infection (see p 386) because brain metastases from lung cancer are so common, the lesion was assumed to be a metastasis. A decision was made to remove the tumor in order to relieve symptoms. Pathological examination of the brain lesion revealed "necrotizing granulomatous inflammation" with associated pigmented hyphae, consistent with a fungal infection. Had this patient not had surgery and died of the lesion, in the absence of an autopsy, his death would almost certainly have been classified as death from a brain metastasis.

Table 1–3. **Frequency of Brain Metastases (Clinical)**

Primary Tumor	No. of Brain Metastases/Total No. of Tumors[*]	Percentage of Brain Metastases[*]	No. of Brain Metastases/Total No. of Tumors[†]	Percentage of Brain Metastases[†]
Lung	11,763/59,038	19.9	156/938	16.3
Breast	2635/51,898	5.1	42/802	5.0
Renal	467/7205	6.5	12/114	9.8
Melanoma	566/8229	6.9	12/150	7.4
Colon	779/42,817	1.8	10/720	1.2

[*] Data from Ref. 13.
[†] Data from Ref. 14.

the disorder over three times as common as primary CNS malignancies (12,740) and five times as common as Hodgkin disease (8190).

Clinical series identify a lower incidence of brain metastases (Table 1–3). A population-based study of the metropolitan Detroit area reported an incidence of brain metastases from lung cancer of approximately 20%,[13] with a lower incidence for other primary tumors (Table 1–3).[13] The results of other population-based studies are similar.[14] The likelihood of a patient developing a brain metastasis varies substantially, depending on the primary tumor. For example, one report describes the incidence of brain metastases from over 1000 patients with esophageal cancer as 1.5%.[15]

Table 1–4. **Brain Lesions in 1905 Autopsies at Memorial Sloan-Kettering Cancer Center (1970–1973)**

Brain Lesions	Number	Percentage Total	Percentage Brain Lesions
Total	687	36	100
Metastases	310	16	45
Vascular lesions	151	8	22
Infections	80	4	12
Other lesions°	146	8	21

° Primary brain tumors (benign and malignant) and lesions not related to cancer, for example, aneurysms, vascular malformations.

Conversely, 24% of patients with small-cell lung cancer have brain metastases (11% asymptomatic)[16] as do 30% of patients with HER2-positive breast cancer.[16a]

Intracranial metastases, however, are only one neurologic complication of systemic cancer; other brain lesions are also common. When Posner and Chernik[6] (Table 1–4) examined the brains of patients dying of cancer over a 4-year period (1970–1973), they found lesions within the brain of potential clinical significance in 36% of patients; 16% had brain metastases (lower than the 24% intracranial metastases because lesions of the dura and leptomeninges were not considered); 8% had vascular lesions (see Chapter 9); and 4% had CNS infections (see Chapter 10). In addition, clinicians estimate that 5% to 10% of patients with metastatic cancer suffer clinically significant spinal epidural metastases.[17] In an old autopsy study, epidural spinal cord compression, a major cause of disability in patients with cancer, was found in approximately 5% (37) of 704 patients[18] (see Chapter 6). A recent population-based study found the frequency of malignant spinal cord compression in Ontario was 2.5% of cancer patients.[19] The likelihood ranged from almost 8% in patients with myeloma, more than 7% in patients with prostate cancer, and 5% in patients with breast cancer to less than 1% in patients with bladder, neck, stomach, ovarian, or pancreatic cancer.[19] The authors of this study believed that these figures probably underestimate the true incidence.

Neurologic complications of systemic cancer are as common in the clinic as they are at autopsy (Tables 1–5 and 1–6). Of 2137 inpatients for whom neurologic consultation was requested, the most common complaint was pain, followed by mental status changes and muscle weakness (Table 1–5). The frequency of neurologic complications is similar in children, although the types of systemic cancers differ. Antunes conducted a 3-year prospective study (1997–2001) of neurologic consultations requested by the Pediatric Department of MSKCC.[20] Five hundred and twenty-eight consultations were requested on 372 patients. Pain, as in adults, was the most common symptom, but headache was more common than back pain. Altered mental state was the second most common complaint, but it was far less common in children than in adults, whereas seizures were far more common in children.[21–23]

When all causes of neurologic disability are considered, approximately 15% of patients with cancer suffer a symptomatic neurologic complication during the course of the disease. For many patients, the neurologic disorder may occur late, when the cancer is already widespread, but for others, the neurologic symptom may be the first evidence of cancer. A survey from a Netherlands cancer hospital[24] reports that among 7004 new adult patients examined during a 2-year period, 1105 (16%) were referred for neurologic evaluation. Breast cancer was the most frequent primary tumor, followed by lung, ovarian, head and neck, and non-Hodgkin lymphoma. Pain was the most common complaint with nerve root, plexus, and spinal cord problems being the most common final diagnosis. Other studies yield even higher figures. Gilbert and Grossman[25] from the Johns Hopkins Cancer Center report that, with the exception of planned admissions for

Table 1–5. Neurological Complaints in 2137 Adult Inpatients with Cancer Referred to the Memorial Sloan-Kettering Cancer Center Neurology Department

Complaint	No. of Patients	Percentage of Patients
Pain		
Back pain	385	18
Headache	192	9
Pain in a limb	62	2
Neck pain	50	2
Other	48	2
Other Neurologic Complaints		
Altered mental status	521	24
Leg weakness	179	8
Ataxia or gait disturbance	156	7
Sensory disturbance	173	8
Visual disturbance or diplopia	54	2
Arm weakness	120	5
Seizures	156	7
Speech or language disturbance	52	2
Hemiparesis	96	4
Total	2244°	

° Some patients had more than one complaint.

chemotherapy, neurologic problems were the most common reason for admission to the Solid Tumor Service at their center. The major problems were changes in mental status, epidural spinal cord compression, and brain metastases. Collectively, neurologic problems represented more than 50% of the admissions. Sculier et al.[26] report that of 641 patients with small cell lung cancer, 29.5% (189) had at least one symptomatic neurologic disorder during the course of the disease. A prospective evaluation of 432 consecutive patients with small cell lung cancer found a neurologic disorder in more than half the patients; in approximately 50%, the neurologic symptoms were present at diagnosis of the cancer.[27] Brain metastases were the most common complication; 10% of patients had symptomatic brain metastases at diagnosis and the remainder (25%) developed a brain metastasis during the course of their illness.[27]

Table 1–6 lists a wide variety of final diagnoses in patients for whom neurologic consultation was sought at MSKCC.

Neurologic Complications of Cancer Are Increasing

Autopsy data from MSKCC showed a steady increase in the postmortem incidence of intracranial, brain, and leptomeningeal metastases from 1970 through 1976.[6] Although these data may have some percentage of error because patients with neurologic disorders are likely to be referred to this hospital and no epidemiologic data are available for comparison, other autopsy reports support our findings. Pickren et al.[28] report a steady increase in brain metastases encountered at autopsy between 1959 and 1979. Unfortunately, more recent prevalence data are not available because of the low autopsy rate (in our hospital <10%). In addition, the brains of asymptomatic patients are not routinely imaged. Imaging techniques help, but many patients with widely metastatic disease who have neurologic symptoms are not imaged because further treatment is not planned. Lung cancer is an exception. Because

Table 1–6. Neurologic Diagnoses in 2137* Adult Inpatients with Cancer Referred for Neurologic Consultation at Memorial Sloan-Kettering Cancer Center (1999–2000)

Diagnoses	No. of Patients	Percentage of Patients
Common Cancer-Related Neurologic Diagnoses		
Metastatic		
Brain metastasis	407	19
Pain associated with bone metastasis only	282	13
Epidural extension or metastasis	298	14
Tumor plexopathy	41	2
Leptomeningeal metastasis	224	10
Paravertebral radiculopathy	36	1
Base-of-skull metastasis	48	2
Nonmetastatic		
Metabolic or drug-related encephalopathy	275	12
Paraneoplastic syndromes	7	0.3
Related to cancer treatment		
Chemotherapy-related peripheral neuropathy	40	1
Intracranial hemorrhage associated with thrombocytopenia	12	0.6
Radiation plexopathy	3	0.1
Common Non–Cancer-Related Neurologic Diagnoses		
Cerebrovascular disease	169	8
Headache associated with systemic illness or unclassifiable	67	3
Degenerative disease of the spine	57	2
Syncope	45	2
Epilepsy	34	1
Other Neurologic Diagnoses or Unknown	174	8
Total	2219[†]	

* Does not include 1474 patients admitted directly to the neurologic inpatient unit.
[†] Some patients had more than one diagnosis.

brain metastases from lung cancer are so common, asymptomatic patients often undergo brain imaging. In a study of 112 patients with normal neurologic findings, a magnetic resonance image (MRI) of the brain demonstrated asymptomatic brain metastases in 17 (15%).[29] In another study, MRI of the brain revealed asymptomatic brain metastasis in 3% of 177 patients with potentially operable non–small cell lung cancer.[30] In one study of 809 patients with non–small cell lung cancer and brain metastases, the metastasis was present at initial staging in 181 (22%); of these, 61 (7.5%) patients were asymptomatic.[31]

The reason that brain metastases appear to be increasing is that the CNS serves as a sanctuary for neoplastic cells when systemic tumor is controlled by chemotherapy or immunotherapy. The increased incidence of CNS complications recapitulates previous experience with acute lymphoblastic leukemia: leptomeningeal leukemia was an uncommon clinical problem before effective chemotherapy for the systemic illness was developed.[32] Beginning in the 1950s, as systemic treatment improved survival, the incidence of symptomatic meningeal leukemia increased rapidly until approximately 50% of children developed that complication during the course of their illness.[33] Leptomeningeal leukemia developed because leukemic cells found sanctuary behind the blood–brain barrier (see Chapter 3) and were inaccessible to chemotherapy with parenteral water-soluble chemotherapeutic agents.

The solution was to treat the CNS with either radiation therapy (RT) or agents delivered directly into the CNS (intrathecally).[34] A situation similar to that in the early days of leukemia is now occurring in patients with breast cancer. In patients receiving adjuvant systemic chemotherapy, not only with trastuzumab,[35,36] but also with taxanes[37] (both of which cross the blood–brain barrier poorly), the first site of relapse is more likely to be the brain than in those not receiving systemic chemotherapy (Fig. 1–2).[38] When chemotherapy was added to radiation for the treatment of breast cancer, distant metastases were significantly decreased, but brain metastasis remained unchanged. Other tumors, including testicular cancer[39] and both small cell and non–small cell lung cancer may also find sanctuary in the brain. The increasing incidence of brain metastasis from several cancers has raised the question of prophylactic brain irradiation, which is part of the routine therapy of some patients with small cell lung cancer; some investigators have also considered its use in patients with breast cancer[40] and non–small cell lung cancer.[41,42]

Nonmetastatic complications of systemic cancer may also be increasing. Although no epidemiologic data are available, nervous system complications of RT and chemotherapy (see Chapters 13 and 12, respectively) seem to be more frequent as these therapies have become more vigorous and have increased survival.

Neurologic Complications of Systemic Cancer Are Serious

Many patients suffering from systemic cancer, even some with widespread metastases, can, with appropriate therapy, function normally for prolonged periods. For example, a woman with breast cancer and widespread bone metastases may, with pain medication and perhaps a back brace, lead a virtually normal life. However, if a vertebral metastasis compresses the spinal cord, then pain increases, neurologic disability develops, and her functional state worsens dramatically.[42a] Disorders of the brain and spinal cord, with their attendant symptoms of paralysis, incontinence, dementia, and seizures, often render a previously functional patient bedridden or hospitalized for the remainder of his or her life (Fig. 1–3).

A recent editorial describes a patient's perspective on brain metastases in breast cancer[43]:

> Your brain controls your independence, your quality of life, your entire existence. Brain mets (*sic*) can bring on a loss of hope and a fear of loss of self. It's not just a body part that's at risk, it's our life as who we are. My first thought was not my brain! To me, that meant I would lose me. (p 1623). (See also Ref. 42a.)

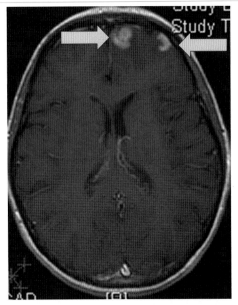

Figure 1–2. Neurologic complications are increasing. A 50-year-old woman with breast cancer received adjuvant treatment with doxorubicin and cyclophosphamide followed by paclitaxel and trastuzumab. An MRI of the brain before chemotherapy was negative. When she developed headache, a new MRI revealed brain metastases (*arrows*) that had developed during the course of successful treatment for her systemic tumor. As will be discussed in Chapter 5, there appears to be an increasing incidence of brain metastases in patients with breast cancer who respond systemically to trastuzumab, a molecule that does not cross the blood–brain barrier.

Diagnosis Is Often Difficult

The neuro-oncologist is equipped through training and experience to handle difficult neurologic problems, because the discipline combines familiarity of neurologic disease with knowledge of the common causes of neurologic disability in patients with cancer. Different disorders often present similar clinical symptoms and signs, requiring meticulous and

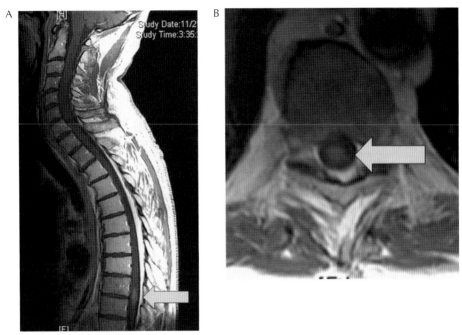

Figure 1–3. Neurologic complications are serious. A 53-year-old man with small cell lung cancer and known brain metastases was functioning well until he began to note gait unsteadiness and leg weakness. The symptoms progressed to the point that he was unable to walk and became incontinent. **A**: A sagittal MRI revealed an enhancing mass in the lower thoracic cord (*arrow*). **B**: The mass occupied almost the entire spinal cord (*arrow*).

sophisticated clinical and laboratory evaluation to reach a definitive diagnosis (Fig. 1–4). Many oncologists may be unaware, for example, that painless epidural spinal cord compression can clinically mimic a midline cerebellar syndrome[44] (see Chapter 6), or that syncope can be a sign of a nasopharyngeal[45] or neck neoplasm[46] (see Chapter 8). Conversely, common neurologic syndromes often have uncommon causes when they occur in patients with systemic cancer, so that neurologists, unfamiliar with the spectrum or complications of cancer, may miss the diagnosis. Examples are strokes due to non-bacterial thrombotic endocarditis or disseminated intravascular coagulation (see Chapter 9). CNS infections are usually caused by organisms different from those that the neurologist encounters in a general hospital population (see Chapter 10). To further complicate matters, approximately 20% of patients with cancer referred for neurologic consultation suffer from a neurologic disorder unrelated to their cancer, such as migraine, neurocardiogenic syncope, and herniated disc, all potentially treatable.[47,48] Furthermore, cancer and other systemic

diseases acting together may produce unusual neurologic pictures. An example is diabetic neuropathy exacerbated by pelvic radiation (see Chapter 13). Thus, an accurate diagnosis, required to prescribe appropriate therapy, demands knowledge not only of the nervous system, but also of the cancer from which the patient is suffering, the treatment the patient is receiving, and the patient's likely neurologic complications. Failure to make the correct diagnosis can have tragic consequences.

Treatment Helps

Patients with neurologic complications of systemic cancer can suffer persistent or progressive problems that require special management. Once a correct diagnosis is made, appropriate therapy directed at the nervous system complication frequently relieves symptoms and either prolongs life or improves its quality (Fig. 1–5). Evidence presented in Chapter 6 indicates that, in patients suffering from epidural spinal cord compression, early diagnosis and vigorous

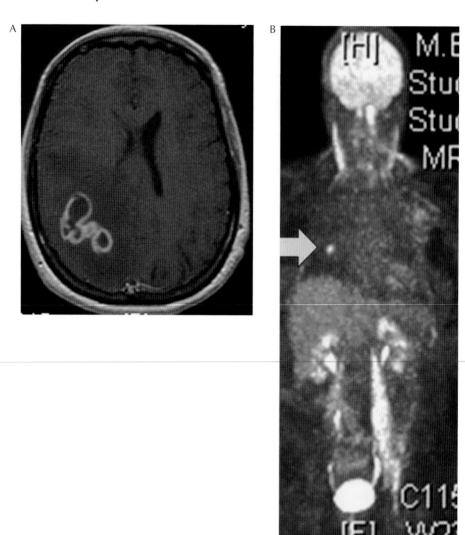

Figure 1–4. Diagnosis is often difficult. A 47-year-old woman, previously healthy, developed weakness of her left arm followed by focal motor seizures. **A**: An MRI of the brain showed a multilobulated right parietal mass surrounded by edema (edema is dark on T_1-weighted images). **B**: There was a lesion in the right lung that was hypermetabolic on PET scan (*arrow*), suggestive of malignant disease (the hypermetabolic areas seen in the neck and left pectoralis muscle were believed to be physiologic). A tentative diagnosis of lung cancer with brain metastasis was made. At craniotomy, the lesion was found to be an abscess caused by Nocardia. The lung lesion was the same. The patient had no cancer. The lesions cleared with antibiotic therapy.

treatment maintain the patient's ability to walk, often for the remainder of his or her life. Patients who are not treated inevitably become paraplegic or quadriplegic and spend the remainder of their life bedridden and incontinent. The same is true for patients with metastatic brain tumors (see Chapter 5). With appropriate treatment, many patients remain free of neurologic symptoms for the remainder of their life, with death occurring from systemic disease. Prevention or effective treatment of paraplegia or hemiplegia considerably enhances the quality of a patient's remaining life and makes him or her less of a burden on loved ones. Similar therapeutic considerations also apply to nonmetastatic complications of cancer.

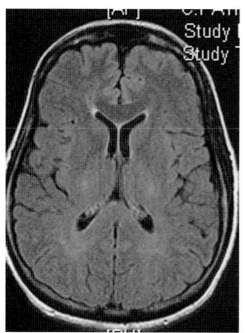

Figure 1–5. Treatment helps. A 54-year-old woman with known breast cancer developed facial numbness. She then developed increasingly severe headaches. The headaches occurred suddenly, usually on arising in the morning. The headaches were sometimes associated with vomiting, a rushing noise in her head, and weakness of her legs. These episodes lasted 5 to 10 minutes and then cleared. The symptoms became increasingly frequent and occurred with almost any change in posture. The cerebrospinal fluid was positive for malignant cells, establishing the diagnosis of leptomeningeal tumor. Her symptoms were believed to be from increased intracranial pressure causing plateau waves (see Chapter 3) despite a normal magnetic resonance image that failed to show the expected hydrocephalus or periventricular white matter hyperintensity suggesting transependymal obstruction of the obstructed cerebrospinal fluid. A ventriculoperitoneal shunt was placed and her symptoms immediately resolved. (At first the neurosurgeon was reluctant to shunt, believing that in the absence of hydrocephalus, the pressure would be normal. After agreeing to do the procedure, the neurosurgeon bet the neurologist $50 that the pressure would be normal. When the ventricle was entered, fluid spurted out under greatly increased pressure.)

Problems in Neuro-Oncology Are Unique

Several neurologic complications of cancer occur because of the unique anatomy and physiology of the nervous system; similar complications do not occur when cancer affects other organ systems (Table 1–7).

Table 1–7. Unique Aspects of Neurologic Complications of Cancer

The brain and spinal cord are enclosed in bone.

Small lesions can cause "large" symptoms.

The blood–brain (blood–nerve) barrier excludes many chemotherapeutic agents.

The CNS lacks lymphatics, making removal of edema and detritus difficult.

The CNS does not possess the capacity for clinically significant regeneration.

Occult cancers can destroy nervous tissue ("remote effects") without direct contact.

CNS, central nervous system.

For example, the brain and spinal cord are enclosed in nondistensible bone. Small tumors that grow in the abdomen might simply move the normal organs aside and remain asymptomatic. In the brain or spinal cord, these tumors often cause severe and sometimes irreversible or lethal symptoms by compression and distortion of tissue that cannot move aside. Also, the CNS is heterogeneous; different areas have different functions. A tumor must destroy a large fraction of liver, lung, or kidney to cause organ failure, but in some areas of the nervous system such as the brainstem or spinal cord, severe disability may be caused by very small tumors. Anatomic and physiologic barriers (see Chapter 3) separate the nervous system from the rest of the body. These barriers often exclude water-soluble chemotherapeutic agents that might otherwise kill tumor cells sequestered behind them. On the other hand, when the blood–brain barrier is disrupted, as often occurs with brain metastases and other forms of brain injury, water-soluble molecules, some of which may be neurotoxic, can enter and spread into surrounding normal brain. The barrier disruption also causes edema that, by its mass effect, may produce more brain dysfunction than the tumor itself (Fig. 1–6).

Another problem is that damaged CNS lacks the ability to regenerate meaningfully. The clinical implication is that nervous system damage by metastatic cancer or its treatment (see Chapters 12–14) may be irreversible, even if the treatment destroys the tumor effectively. This failure of the damaged nervous system to regenerate effectively puts the onus on the physician to make the diagnosis early and to treat

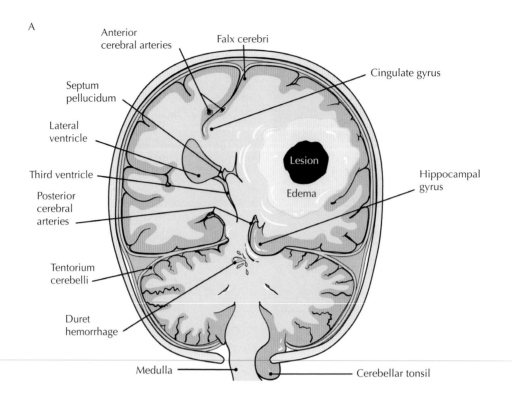

A

Anterior
cerebral arteries

Falx cerebri

Cingulate gyrus

Septum
pellucidum

Lateral
ventricle

Third ventricle

Posterior
cerebral
arteries

Tentorium
cerebelli

Duret
hemorrhage

Lesion

Edema

Hippocampal
gyrus

Medulla

Cerebellar tonsil

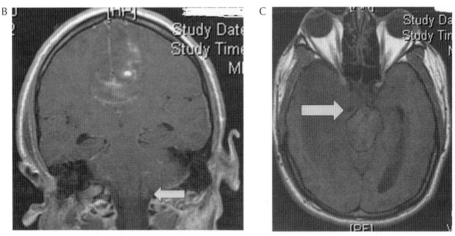

B

C

Figure 1–6. Neuro-oncology problems are unique. **A**: This schematic coronal section of a brain, enclosed in the skull, illustrates the changes caused by a brain metastasis and why neuro-oncologic problems are unique. The lesion itself (black sphere) might cause only minimal symptoms, but the additional mass of the edema (yellow) surrounding the metastasis increases neurologic dysfunction. Note that the edema involves only white matter, sparing the nearby cerebral cortex (gray matter). Because of the combined effect of the tumor and the surrounding edema, the brain shifts, compressing normal brain structures locally and remotely. Herniation of the cingulate gyrus under the falx cerebri compresses not only the contralateral frontal lobe, but also the anterior cerebral arteries. Such herniation can cause bilateral frontal ischemia, with weak legs, urinary incontinence, and mental changes. The diencephalon shifts toward the contralateral side, compressing itself, the third ventricle, and the opposite diencephalon. The resulting diencephalic dysfunction causes diminished consciousness. Herniation of the uncus and hippocampal gyrus of the temporal lobe into the tentorial notch compresses the posterior cerebral artery, leading to infarction in the ipsilateral and sometimes bilateral occipital lobe(s) (see Fig. 3–6C). Uncal herniation also stretches the ipsilateral oculomotor nerve and compresses the

vigorously, when the nervous system can still be salvaged.

Research Is Essential

As we indicate in this book, even though diagnostic and therapeutic endeavors in neuro-oncology have advanced (Fig. 1–7), treatment is still mostly unsatisfactory. In every area of neuro-oncology, treatment needs major improvements that are likely to come only through research. Research areas particularly relevant to neurologic complications of cancer include gene therapy and nervous system regeneration. The hope is that in the future, one can both abolish cancer in the nervous system and repair the damage that cancer or nonmetastatic complications of cancer have inflicted on the nervous system. Most oncologists believe that patients who enroll in clinical trials generally have better outcomes than those who do not. Although this has been disputed,[49] certainly such patients do at least as well as those treated by conventional means.

This book is concerned primarily with the diagnosis and treatment of neurologic complications of cancer and is addressed to the clinician, not the scientist. However, clinicians must play an important role in fostering research. The whole concept of translational research is based on the ability of the clinically trained investigator to move observations in the laboratory to the clinic. Controlled trials of promising treatments are still the major hope for progress. These trials should be coupled with laboratory programs that investigate the biology of neuro-oncologic problems at the cellular and molecular levels. Such therapeutic research is difficult and often frustrating. Unfortunately, too few patients enter into investigative treatment protocols even when they are available. In a recent study, only 30% of adult cancer patients eligible for entry into The National Cancer Institute (NCI)–sponsored clinical trials actually enrolled.[50] Patients with fee-for-service insurance coverage were more likely to enroll than those with managed care despite evidence that clinical trials do not substantially increase cost.[51] Mahaley et al.[52] estimated that only 8% of patients with malignant gliomas are enrolled in protocol studies, despite the fact that 138 protocols were active in North America in 1993.[53] These percentages are low despite the evidence that patients who are enrolled into protocol treatments generally live longer and do better than patients who are not.[54] While doing better, they also help advance the field and provide therapeutic answers. We physicians make an ethical judgment about clinical research every time we see a patient, even when we choose not to offer a research program. As Eisenberg[55] points out, "not to act is to act."

Relationships between the Brain and Systemic Cancer Are Biologically Important

Evidence, although controversial, suggests a reciprocal relationship between systemic cancer and brain function. The field of psychoneuroimmunology[56] provides evidence for the effect of the nervous system on the cancer, its metastases, and its treatment. Potential mechanisms include stress-induced immune suppression,[57] inhibition of DNA repair, induction of sister chromatid exchange, and inhibition of apoptosis.[56] However, most studies of patients suggest that stress alone does not cause cancer,[58,59] but that psychological factors may potentiate the effects of physical factors in causing cancer.[60,61] A recent study indicates that while there was no evidence to support an independent association between personality variables and the development of breast cancer,[62] stress and lack of social support appeared to increase the incidence of breast cancer.[63] In humans, no substantial evidence is available yet to support the

---

midbrain, causing changes in the state of consciousness. Compression of the opposite cerebral peduncle against the tentorium cerebelli causes a hemiparesis that is ipsilateral to the side of the lesion (a false-localizing sign). Downward displacement of the brainstem alters consciousness and may cause midbrain and pontine hemorrhages (Duret hemorrhages). Herniation of the cerebellar tonsils through the foramen magnum compresses the lower brainstem and may cause respiratory arrest. Hydrocephalus occurs in the contralateral lateral ventricle as a result of obstruction of the third ventricle and Sylvian aqueduct by compression. Any combination of the mechanisms illustrated in this figure may play a role in causing the symptoms of a brain metastasis (see Chapter 5). **B**: Magnetic resonance image (MRI) of a patient with an intracerebral hemorrhage from promyelocytic leukemia (see Chapter 9), demonstrating tonsillar herniation (*arrow*). **C**: MRI of a patient with a brain tumor, demonstrating uncal herniation (*arrow*) compressing the midbrain.

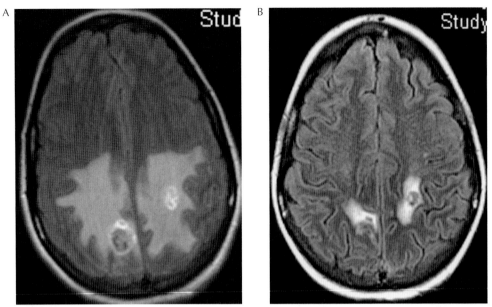

Figure 1–7. Research is essential. A 40-year-old woman developed occipital headaches that were worse in the morning. Magnetic resonance image revealed two contrast-enhancing lesions, one in each parietal lobe, surrounded by edema (**A**). Brain metastases were diagnosed. A search of the body revealed a renal adenocarcinoma. Corticosteroids relieved the headache (Chapter 3) and she was given whole-brain radiation therapy. The radiation did not shrink the tumors, and when the corticosteroids were tapered, the headaches returned. She was begun on therapy with sunitinib (see Chapter 12). After one course, her headaches resolved. Both the lesions and the edema had improved substantially (**B**). She was able to discontinue her corticosteroids. For many years, chemotherapy with water-soluble agents was believed to be useless for metastatic brain tumors. Basic research on the blood–brain barrier (Chapter 3) and clinical experience has shown that once the tumor has disrupted the barrier, as indicated by gadolinium enhancement, if the primary tumor is sensitive so are the brain metastases (Chapter 5). This figure illustrates a dramatic response of a metastatic neoplasm to chemotherapy. It also illustrates the effectiveness of newly developed small molecules in treating cancer.

belief that patients can either prevent cancer or treat an established cancer effectively by the psychological or behavioral techniques of relaxation, biofeedback, or imagery. Reports that such techniques prolong life in patients with cancer are conflicting. Alterations of behavior, such as cessation of smoking and improvement of diet, are another matter. The widespread belief among laypersons that depressive symptoms predispose to cancer appears to be without merit.[64]

The brain also may interfere with the treatment of cancer. Anticipatory nausea and vomiting is an obvious example of classical conditioning in human beings.[65] The phenomenon, which affects approximately 20% of patients on chemotherapy, can be severe enough that the patient becomes nauseated and may begin to vomit on entering the hospital, smelling alcohol, or beginning an intravenous infusion, even before the chemotherapeutic agent is added. Some patients are so severely affected that they abandon therapy.

An opposite relationship, the effect of cancer on the nervous system, is a focus of this book. The literature abounds with examples of small and occult cancers that cause major changes in behavior and neurologic function by mechanisms that are not clearly established (Fig. 1–8). Limbic encephalitis (see Chapter 15) associated with small cell lung cancer causes severe memory loss when the cancer is small, restricted to the lung and mediastinum and at times undetectable by diagnostic techniques, including computed tomography (CT).

CLASSIFICATION OF NEURO-ONCOLOGIC DISORDERS

The rationale given above dictates that neurologic complications of systemic cancer deserve the serious attention of physicians, both neurologists and non-neurologists, who manage patients with cancer, and of neuroscientists interested

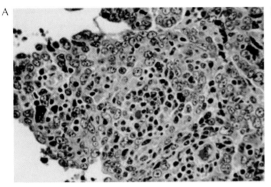
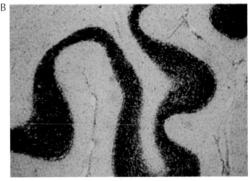

Figure 1–8. Relationship between cancer and the nervous system. A 56-year-old woman, previously in excellent health, noted that her gait became unsteady. Within a few days, she had difficulty walking, her arm movements became uncoordinated, and her speech became slurred. Within 3 weeks, she was unable to walk, could not sit unsupported, and her speech was difficult to understand. Examination revealed a pure cerebellar syndrome. Her serum was positive for the anti-Yo paraneoplastic antibody (see Chapter 15). Although imaging (not shown) failed to reveal an underlying neoplasm (this was before positron emission tomography scanning was available), an oophorectomy was performed, revealing a small cancer in one ovary (**A**). Pathologic examination of patients with paraneoplastic cerebellar degeneration has revealed total loss of Purkinje cells in the cerebellum (**B**) (see also Purkinje cells, Fig. 15–3A).

in the effects of the body on the brain and the brain on the body. The first step in studying neurologic complications of cancer is to classify the ways in which a cancer may affect the nervous system (Table 1–8). This classification also serves to outline the detailed presentation of neurologic complications of systemic cancer in the subsequent chapters of this book.

Nervous system complications can be divided into two groups: (1) those caused by direct spread of cancer to the nervous system (metastasis), and (2) those in which the symptoms are an indirect effect of the cancer (nonmetastatic or paraneoplastic).

Spread of Cancer to the Nervous System (Metastatic or Direct Effects of Cancer)

The direct neurologic complications of systemic cancer can be subdivided by the anatomic areas affected: intracranial (see Chapter 5), spinal (see Chapter 6), leptomeningeal (see Chapter 7), nerve root or plexus and muscle (see Chapter 8). A given patient can, of course, have metastases to more than one region of the nervous system, often complicating both the diagnosis and the therapy.

The classification by anatomic area is useful because the diagnostic evaluation, the therapeutic approach, and even the pathophysiology

of signs and symptoms are different for each area. For example, most intracranial metastases occupy the brain parenchyma, causing symptoms partly by replacing brain tissue and partly by shifts of brain structures caused by the tumor and its surrounding edema. By contrast, most spinal metastases are extradural and do not invade the substance of the spinal cord. Instead, they cause symptoms primarily by compression. Leptomeningeal metastases often appear as a microscopic sheet of tumor cells within the leptomeninges, neither invading nor compressing the parenchyma of the brain or spinal cord. Symptoms may result only from obstruction of cerebrospinal fluid (CSF) pathways, leading to hydrocephalus. Cranial and peripheral nerve dysfunction caused by extradural tumors is usually compressive, whereas intradurally, the nerves are more likely to be invaded.

Laboratory evaluation also differs among the groups. For example, lumbar puncture is unhelpful and sometimes dangerous in the evaluation of intracranial or spinal metastases (see Chapter 14), whereas it is vital to the evaluation of leptomeningeal disease (see Chapter 7) and CNS infection (see Chapter 10). Conversely, an MRI is essential for the diagnosis of intracranial and spinal metastases.

The therapy of metastatic disease depends on an accurate diagnosis of the anatomic area involved. Not only must the radiation oncologist know where to draw the radiation portals, but

Table 1–8. **Classification of Neurologic Complications of Cancer**

Spread of Cancer to the Nervous System (Metastatic or Direct Effects of Cancer)
　　Intracranial (usually to brain parenchyma)
　　Spinal (usually epidural)
　　Leptomeningeal (diffuse or multifocal—micro or macroscopic)
　　Nerves (cranial nerves, peripheral nerves, nerve plexuses, or nerve roots)
　　Muscle (direct spread and metastasis)

Indirect Neurologic Complicatons of Cancer (Nonmetastatic or Paraneoplastic)
　　Vascular disorders (hemorrhage or infarction)
　　Infections (meningitis, abscess)
　　Nutritional disorders
　　Metabolic disorders
　　Side effects of therapy
　　　Chemotherapy
　　　Radiation therapy
　　　Surgery and other diagnostic or therapeutic procedures
　　　Biologic therapy
　　"Remote" or paraneoplastic syndromes

also that the principles of treatment vary with the locus of the lesion. For example, delaying RT of brain metastases for 48 to 72 hours after diagnosis, to allow steroids to effect dramatic improvement, will make the patient better able to tolerate the radiation. On the other hand, except for pain relief, steroids are only modestly effective in ameliorating the symptoms of spinal cord compression, and radiation must begin promptly as an urgent measure (see Chapter 6).

Indirect Neurologic (Nonmetastatic or Paraneoplastic) Complications of Cancer

Indirect neurologic complications of systemic cancer can be divided into five subgroups, based on etiology. The pathophysiology, diagnosis, and management of these subgroups will be considered consecutively in Chapters 9 through 15.

VASCULAR DISORDERS

Cerebrovascular disease frequently complicates systemic cancer. Most patients with cancer have measurable coagulation disorders (see Chapter 9). Intracranial or spinal hemorrhage can occur when patients become thrombocytopenic or if other clotting disorders develop. An intracranial subdural hematoma represents a potentially lethal but treatable complication of thrombocytopenia found occasionally in patients with cancer. Hemorrhage into a metastatic brain tumor may be the first symptom of cancer or the first symptom of brain disease in a patient known to have cancer. Certain metastatic tumors, such as melanoma and choriocarcinoma, are frequent causes of intracerebral hemorrhage, but any metastatic tumor may bleed. Cerebral infarction occurring in patients with cancer may result from the common causes of stroke in the general population, such as atherosclerotic cerebral thrombosis or emboli from atherosclerotic plaques in the carotid, or may also be caused by cancer-specific causes, such as nonbacterial thrombotic endocarditis, tumor emboli, or occlusion of the dural sinuses.

INFECTIONS

The organisms that cause CNS infections differ from those encountered in a general hospital and occur in patients with cancer whose immune responses are abnormal as a consequence of the malignancy or its treatment (see Chapter 10). *Listeria monocytogenes* and *Cryptococcus neoformans*, rare in the general population, are more common in those immunocompromised by cancer or its treatment. Pneumococcal, meningococcal, and (before immunization) *Haemophilus influenzae* meningitis, the common infections in the general population, are rare in the cancer population. Brain abscesses are usually caused by fungi, such as mucormycosis or aspergillosis; by parasites, such as toxoplasmosis; or by unusual bacterial organisms, such as nocardia, rather than by the bacterial organisms that cause infection in the general population. Viral encephalitis may be caused by herpes zoster or JC virus (progressive multifocal leukoencephalopathy) as well as by herpes simplex encephalitis,[66] the most common cause of sporadic encephalitis in the general population. In patients with CNS infections related to cancer, knowledge of the underlying primary tumor and the patient's immune status often allows the physician to predict not only whether CNS infection is causing the symptoms, but also which organism is most likely to be involved.

METABOLIC AND NUTRITIONAL DISORDERS

Metabolic encephalopathy is a common cause of neurologic symptoms in patients with widespread cancer (see Chapter 11). Many individual causes of metabolic encephalopathy in patients with cancer, including hypoxia, vital organ failure, and electrolyte abnormalities (particularly hypercalcemia), are similar to those found on a general hospital ward. Other causes, such as iatrogenically induced opioid overdose, are not usually encountered on general hospital wards. Some unusual causes of metabolic brain disease, such as radiation-induced thyroid failure, are unique to cancer patients and must always be considered, to avoid overlooking a treatable illness.

SIDE EFFECTS OF THERAPY

The increase in CNS metastatic disease, which some observers believe has occurred as a result of effective systemic chemotherapy and immunotherapy, can be considered a complication of cancer treatment. More direct complications include those that arise from surgery around the head, neck, or spine (see Chapter 14); those that are caused by irradiation of the brain, spinal cord, or peripheral nerves (see Chapter 13); and those that result from chemotherapy, whether given systemically or intrathecally (see Chapter 12). New chemotherapeutic agents, with new and differing nervous system toxicity, seem to appear almost every month. Radiation therapy combined with certain chemotherapeutic agents may lead to synergistic toxicity even when the doses of the individual agents are delivered within a safe range.

"REMOTE EFFECTS"— PARANEOPLASTIC SYNDROMES

Although all nonmetastatic complications of cancer can be considered "paraneoplastic," the term *paraneoplastic syndrome* is usually reserved for a rare group of disorders that probably have an immune-mediated pathogenesis (see Chapter 15). Although they are rare, paraneoplastic syndromes are important for two reasons: (1) In approximately two-thirds of affected patients, neurologic symptoms develop before the cancer is apparent and, if recognized, the neurologic syndrome may lead to a fruitful search for an occult and potentially curable cancer. (2) Although their etiology is presently unknown, they hint at a relationship between the brain and cancer, which, if explored, may enhance knowledge of the biochemistry of each.

APPROACH TO THE PATIENT

Generally, the physician encounters a patient with a possible neurologic complication of cancer in two different settings (Table 1–9): (1) The patient is known to have or has had cancer. (2) The patient develops neurologic symptoms but is not known to have or ever had cancer. In the first setting, the cancer may be newly discovered, may have already been staged and under active treatment, or may have occurred in the recent or distant past and is believed to be either in remission or cured.

Many patients, with different neurologic complications, present with similar clinical symptoms. For example, parenchymal brain metastases, leptomeningeal metastases, and

Table 1–9. Approach to the Patient with Suspected Neurologic Complications of Cancer

Known Cancer	No Known Cancer
1. Search for metastases: MRI of involved site, CSF cytology	1. Search for cancer: Body CT and/or PET, mammograms, serum cancer markers (e.g., CEA)
2. Search for nonmetastatic disorder (see Chapters 9–14)	2. CSF for cells, IgG, oligoclonal bands, cytologic examination
3. CSF for inflammatory cells and IgG	3. Serum and CSF for paraneoplastic antibodies
4. Serum and CSF for paraneoplastic antibodies (see Chapter 15)	4. Follow and search again for cancer if numbers 2 and 3 are positive

MRI, magnetic resonance image; CSF, cerebrospinal fluid; CEA, carcinoembryonic antigen; IgG, immunoglobulin G; CT, computed tomography; PET, positron emission tomography.

metabolic encephalopathy can all appear as delirium, pointing to the brain as the site of the lesion but giving no clue as to the cause. Similarly, patients with epidural spinal cord compression, radiation-induced myelopathy, and paraneoplastic myelopathy can all present with similar symptoms, pointing to spinal cord dysfunction but not to its cause. Because such situations arise commonly in neuro-oncology, it is often useful for the physician to determine the site of the lesion first and then to consider the various neuro-oncologic disorders that may cause such a lesion. This approach is outlined in Table 1–10, which lists the common causes of neuro-oncologic disorders localized to specific portions of the central and peripheral nervous system.[67]

Known Cancer

No matter what the current situation, in almost any patient with a history of cancer who develops a neurologic disorder, metastatic disease affecting the nervous system must be considered the most likely diagnosis. Accordingly, the first step should be a high-resolution MRI of the area of the nervous system identified as abnormal by the clinical history and examination. If the scans suggest a brain metastasis in a patient whose primary cancer was not in the lung, a CT of the chest may reveal metastasis to the lung, strongly suggesting that the brain lesion seen on MRI is part of the metastatic

cascade (see Chapter 2). Even if a non-cancer diagnosis seems obvious, imaging should be strongly considered. For example, a patient with a history of cancer presenting with apparently clear-cut psychological depression may instead be suffering from bilateral frontal lobe metastases or hydrocephalus from a posterior fossa metastasis. A patient with diabetes and an apparently obvious diabetic peripheral neuropathy may instead be suffering from leptomeningeal metastasis. An MRI should always be obtained both without and with contrast. Contrast-enhanced images may reveal brain or intraspinal metastases too small to be seen on other sequences. Enhancement of nerve roots in the cauda equina may define leptomeningeal tumor not visible on the unenhanced scan.

If the MRI does not reveal metastatic disease and the neurologic signs point to a CNS or intraspinal nerve root process, a lumbar puncture should be performed (see Chapter 10). Measurement of opening pressure is mandatory. A patient suffering from cancer with increased intracranial pressure because of compression or occlusion of venous sinuses or leptomeningeal tumor may not have papilledema or other obvious symptoms or signs of increased intracranial pressure. In addition to the pressure, routine cell count, protein, and glucose concentration, fluid should be collected for cytologic examination; at least 4 cc is required. Unless the fluid can be taken to the laboratory immediately, preservatives should

Table 1–10. **Neurologic Complications in Cancer Patients by Site**

Site	Usual Causes	Typical Symptoms and Signs
Brain	Metastasis Leptomeningeal metastasis Metabolic/toxic encephalopathy Infection (meningitis, brain abscess) Radiation encephalopathy Cerebral hemorrhage or infarction Paraneoplastic (limbic encephalopathy)	Headache, confusion, hemiparesis, seizures, ataxia
Spinal cord and cauda equina	Epidural spinal cord compression Leptomeningeal metastasis Intramedullary metastasis Epidural abscess or hematoma Radiation myelopathy Myelopathy following intrathecal chemotherapy Paraneoplastic myelopathy	Back pain, paraparesis, sensory level, incontinence
Cranial and peripheral nerves	Extrinsic compression by tumor or other mass (e.g., hematoma) Direct infiltration by tumor Drug toxicity Varicella-zoster infection Radiation plexopathy Paraneoplastic neuropathy	Focal pain, sensory loss, motor weakness, decreased reflexes in nerve distribution (focal lesion) or distally in hands and feet (polyneuropathy)
Muscle	Metastasis Steroid myopathy Cachectic myopathy Paraneoplastic polymyositis or dermatomyositis	Proximal weakness without sensory loss

From Ref. 67 with permission.

be added to maintain cellular morphology. In addition, if the underlying cancer is associated with a tumor marker (e.g., prostate-specific antigen [PSA] in prostate cancer), serum and CSF should be sent to assay that marker. Relative concentrations of the marker in serum and CSF will help determine whether there is tumor in the CNS.

If a paraneoplastic syndrome is suspected (see Chapter 15), assaying for paraneoplastic antibodies may establish the diagnosis. Although most paraneoplastic syndromes occur in patients not known to have cancer, development of a paraneoplastic syndrome may occur after the cancer has appeared and been treated and may, in some instances, presage relapse.

If metastatic disease cannot be identified, one must consider the non-neurologic complications of cancer described in detail in Chapters 9 to 15. Vascular disorders of the CNS are usually detected by imaging. However, imaging sequences may be different from those best for tumor. If the physician suspects infarction, diffusion and perfusion images are required. If hemorrhage is suspected, gradient echo sequences may be necessary.

Infections can be identified by appropriate examination of the CSF if the leptomeninges are affected. Brain abscesses, on the other hand, may be difficult to distinguish from either primary brain tumors or metastases, although diffusion-weighted images and magnetic resonance spectroscopy (MRS) may help (see Chapter 5); biopsy may be required to establish the diagnosis.

In patients under active treatment for cancer or those who have had extensive treatment in the past, always consider the side effects of therapy. Neurologic side effects of radiation and chemotherapy may occur either during the course of treatment or months to years after treatment is completed (see p 536). In many instances, the side effects are so well known that they are immediately apparent to the examiner. In others, the patient may have been treated with either relatively new drugs or unusual combinations or dosages of agents, so that it may not be apparent that the neurologic symptoms are related to treatment. Furthermore, some neurologic symptoms (e.g., neck weakness in patients radiated for Hodgkin disease [see Chapter 13]) are not widely known, and the physician must search the literature to identify them. In addition to treatments administered by the physician, one must consider side effects of alternative therapy. Many (perhaps most) patients suffering from advanced cancer try "alternative therapy." The physician may be unaware that the patient is using these agents unless he or she specifically inquires. For example, one of our patients with colon cancer, whose treatment included chemotherapeutic agents not known to be neurotoxic, developed a peripheral neuropathy. A compulsive neurology resident ordered a heavy metal screen over the objections of the attending physician (J.B.P.). The final diagnosis of arsenic intoxication was established when the Chinese herb that the patient was taking, unbeknownst to his doctors, contained a high level of arsenic.

Often, the nature of the underlying cancer or its treatment suggests a nutritional or metabolic disorder associated with neurologic symptomatology. For example, hypercalcemia, a common complication of multiple myeloma, may cause either neuromuscular disturbances or encephalopathy. Hypercortisolism from lung cancer may cause muscle weakness or depression. Some patients with advanced cancer are malnourished and a few may develop specific vitamin deficiencies such as Wernicke encephalopathy (see Chapter 11).

Finally, the physician must always consider the possibility that a neurologic disorder occurring in a patient with cancer may be unrelated to the cancer. Co-morbid diseases such as diabetes, hypertension, or atherosclerosis can cause nervous system symptoms unrelated to the underlying cancer and must always be considered in the differential diagnosis.

No Known Cancer

The approach in the patient not known to have cancer is often more difficult than in the patient known to have cancer. Two different scenarios are encountered. The first is the patient who develops a neurologic syndrome and is found on imaging to have an abnormality consistent with a neoplasm in the nervous system. For example, approximately 10% of patients with carcinoma of the lung present with neurologic symptoms and a brain MRI is compatible with metastatic disease. As many as 20% of patients with epidural spinal cord compression are not known to have cancer until neurologic symptoms lead to an image of the spine revealing tumor in vertebral bodies. In such patients, a CT of the chest, abdomen, and pelvis or a body positron emission tomography (PET) scan may reveal the primary site, and a biopsy of that site may establish the diagnosis. In some instances, however, a primary site is not immediately evident and biopsy or removal of the lesion causing the nervous system symptoms may establish the diagnosis (and treat the patient). Biochemical markers of cancer in the serum such as PSA may help establish the source of the primary tumor.

In the second setting, a patient develops neurologic symptoms not immediately identifiable on imaging as being neoplastic. In most instances, the patient is suffering from an undiagnosed neurologic disease that the physician thinks may be a paraneoplastic syndrome. The nature of the neurologic disease may help direct the search for an underlying cancer. For example, if the patient suffers from proximal lower extremity weakness and has the electrophysiologic abnormality that defines the Lambert–Eaton myasthenic syndrome (see Chapter 15), the underlying cancer is likely small cell lung cancer. If the patient is a woman with subacutely developing cerebellar signs, the likely site is breast or ovary. Examination of the serum for paraneoplastic antibodies will help identify the neurologic syndrome as paraneoplastic and also suggest the possible underlying cancer.

The search for cancer can include a CT of the chest, abdomen and pelvis, mammograms, serum cancer markers and, if these do not reveal a source, a body PET using fluorodeoxyglucose to identify occult lesions.[68] The finding of white cells in the CSF along with an elevated immunoglobulin G (IgG) suggests, in the appropriate clinical setting, the possibility of a paraneoplastic syndrome even in the absence of positive paraneoplastic antibodies.

EVALUATING THE PATIENT

A number of scales are used to evaluate a patient's overall performance status and standardize the assessment of the patient's overall level of function (Table 1–11). Other scales are used to assess pain (Table 1–12), weakness from cord compression (Table 1–13), or severity of delirium (Tables 1–14 and 1–15). These scales are used in many clinical trials in an effort to normalize eligibility criteria or uniformly assess response to therapy. There is also a scale to classify patients with brain metastases into prognostic categories (Table 1–16) and a scale to grade peripheral neuropathy (Table 1–17). These scales appear throughout the general and neuro-oncology literature and are fundamental tools to quantify the patient's symptoms, signs, and functional capacity. They are included here and will be referred to in future chapters.[69–74]

Table 1–11. **Performance Scales: Karnofsky and ECOG**

Karnofsky Scale	Description	ECOG	Description
100	Normal; no complaints; no evidence of disease	0	Fully active, able to carry on all pre-disease activities without restriction
90	Able to carry on normal activity; minor signs or symptoms of disease	1	Restricted in physically strenuous activity, but ambulating and able to carry out work of a light or sedentary nature, for example light housework/office work
80	Normal activity with effort; some signs or symptoms of disease		
70	Cares for self; unable to carry on normal activity or to do active work	2	Ambulatory and capable of all self-care, but unable to carry out any work activities; up and about more than 50% of waking hours
60	Requires occasional assistance but is able to care for most personal needs		
50	Requires considerable assistance and frequent medical care	3	Capable of limited self-care; confined to bed or chair 50% or more of waking hours
40	Disabled; requires special care and assistance		
30	Severely disabled; hospitalization is indicated, although death not imminent	4	Completely disabled; cannot carry on any self-care; totally confined to bed or chair
20	Very sick; hospitalization necessary; active support treatment is necessary		
10	Moribund; fatal processes progressing rapidly		
0	Dead	5	Dead

ECOG, Eastern Cooperative Oncology Group.

Table 1–12. **Comprehensive Pain Assessment**

History

Pain
  Intensity
    At rest
    With movement
    Interference with activities
  Location
  Pathophysiology
  Somatic: pain in skin, muscle, bone described as aching, stabbing, throbbing, pressure
  Visceral: pain in organs or viscera described as gnawing, cramping, aching, sharp
  Neuropathic: pain caused by nerve damage described as sharp, tingling, burning, shooting
  If patient is unable to communicate, consider alternative method to obtain pain rating and response
  History: onset, duration, course, aggravating factors, associated symptoms, alleviating factors, response to current and prior treatment including reasons for discontinuing
  Etiology
    Cancer
    Cancer therapy or procedures
    Coincidental or non-cancer
  Response to current therapy
    Pain relief and side effects
    Patient adherence to medication plan

Medical
  Current medications including prescribed, over-the-counter, complementary, and alternative therapies
  Oncologic
  Other significant medical illnesses
Psychosocial
  Patient distress
  Family and other support
  Psychiatric history including current or prior history of substance abuse
  Special issues relating to pain
  Meaning of pain for patient/family
  Patient/family knowledge and beliefs surrounding pain
  Cultural belief toward pain
  Spiritual or religious considerations
Risk Factors for Under-Treatment of Pain
  Pediatric, geriatric, communication barriers, history of substance abuse, neuropathic pain, minorities, female, cultural factors
Risk Factors for Aberrant Use or Diversion of Pain Medication
  Patient factors
  Environmental factors
Physical Examination
Relevant Laboratory and Imaging Studies

Modified from NCCN Practice Guidelines in Oncology, v.1.2007.

Table 1–13. **The ASIA Impairment Scale**

Grade	Description
A complete	No motor or sensory function is preserved at S4–5.
B incomplete	Sensory but not motor function is preserved below the neurological level and extends through S4–5.
C incomplete	Motor function is preserved below the neurological level and the majority of key muscles below the neurological level have a muscle grade < 3.
D incomplete	Motor function is preserved below the neurological level and the majority of key muscles below the neurological level have a muscle grade ≥ 3.
E normal	Motor and sensory function are normal.

From Ref. 69 with permission.

Table 1–14. **Mini-Mental State Examination**

Maximum Score	Patient's Score	
		Orientation
5	( )	What is the (year) (season) (date) (day) (month)?
5	( )	Where are we: (state) (county) (town) (hospital) (floor)?
		Registration
3	( )	Name three objects, one second to say each, then ask the patient to repeat all three after you have said them. Give 1 point for each correct answer. Continue repeating all three objects until the patient learns all three. Count trials and record.
		Attention and Calculation
5	( )	Serial 7's. One point for each correct response. Stop after five answers. Alternatively, spell "world" backward.
		Recall
3	( )	Ask for the three objects named in *Registration*. Give one point for each correct answer
		Language
2	( )	Name a pencil and watch.
1	( )	Repeat the following: "No ifs, ands, or buts."
3	( )	Follow a three-stage command: "Take paper in your right hand, fold it in half, and put it on the floor."
1	( )	Read and obey the following: "CLOSE YOUR EYES."
1	( )	Write a sentence.
1	( )	Copy a design.
30	( )	

Assess level of consciousness along a continuum

Alert	Drowsy	Stupor	Coma

From Ref. 70 with permission.

Table 1–15. **The Confusion Assessment Method for the Intensive Care Unit (CAM-ICU)***

Feature 1: Acute Onset of Mental Status Changes or Fluctuating Course
- Is there evidence of an acute change in mental status from the baseline?
- Did the (abnormal) behavior fluctuate during the past 24 hours, that is, tend to come and go or increase and decrease in severity?

Sources of Information: Serial Glasgow Coma Scale or sedation score ratings over 24 hours as well as readily available input from the patient's bedside critical care nurse or family.

Feature 2: Inattention
- Did the patient have difficulty focusing attention?
- Is there a reduced ability to maintain and shift attention?

Sources of Information: Attention screening examinations by using either picture recognition or Vigilance A random letter test (see Methods and Appendix 2 for description of Attention Screening Examinations[71]). Neither of these tests requires verbal response, and thus they are ideally suited for mechanically ventilated patients.

Continued on following page

Table 1–15.—*continued*

Feature 3: Disorganized Thinking
- Was the patient's thinking disorganized or incoherent, such as rambling or irrelevant conversation, unclear or illogical flow of ideas, or unpredictable switching from subject to subject?
- Was the patient able to follow questions and commands throughout the assessment?
    1. "Are you having any unclear thinking?"
    2. "Hold up this many fingers." (Examiner holds two fingers in front of the patient)
    3. "Now, do the same thing with the other hand." (Not repeating the number of fingers)

Feature 4: Altered Level of Consciousness
- Any level of consciousness other than "alert."
- Alert—normal, spontaneously fully aware of environment and interacts appropriately
- Vigilant—hyperalert
- Lethargic—drowsy but easily aroused, unaware of some elements in the environment, or not spontaneously interacting appropriately with the interviewer; becomes fully aware and appropriately interactive when prodded minimally
- Stupor—difficult to arouse, unaware of some or all elements in the environment, or not spontaneously interacting with the interviewer; becomes incompletely aware and inappropriately interactive when prodded strongly
- Coma—unarousable, unaware of all elements in the environment, with no spontaneous interaction or awareness of the interviewer, so that the interview is difficult or impossible even with maximum prodding

From Ref. 71 with permission.
° Delirium is diagnosed when both Features 1 and 2 are positive, along with either Feature 3 or Feature 4.

Table 1–16. **Prognostic Factors for Brain Metastases by RPA Class***

	Median Survival (months)
RPA Class 1	
KPS ≥ 70, age < 65 years, controlled primary tumor, no extracranial disease	7.1
Single metastasis	13.5
Multiple metastases	6.0
RPA Class 2	
All other situations	4.2
Single metastasis	8.1
Multiple metastases	4.1
RPA Class 3	
KPS < 70	2.3

From Ref. 74 with permission.
° Survival results for overall RPA classes are from Radiation Therapy Oncology Group trials,[72] and those for the single and multiple metastases subdivisions of Class 1 and 2 are from Lutterbach et al.[73]
RPA, recursive partitioning analysis; KPS, Karnofsky Performance Status (see Table 1–11).

Table 1–17. **Peripheral Neuropathy Grading Systems**

Type	Grade 1	Grade 2	Grade 3	Grade 4
NCI-CTC Version 2				
Motor	Subjective weakness but no objective findings	Mild objective weakness interfering with function but not interfering with activities of daily living	Objective weakness interfering with activities of daily living	Paralysis
Sensory	Loss of DTRs or paresthesia (including tingling) but not interfering with function	Objective sensory loss or paresthesia interfering with function but not interfering with activities of daily living	Sensory loss or paresthesia interfering with activities of daily living	Permanent sensory loss interfering with function
NCI-CTC Version 3				
Motor	Asymptomatic, weakness on examination/ testing only	Symptomatic weakness interfering with function but not interfering with activities of daily living	Weakness interfering with activities of daily living, bracing, or assistance to walk (e.g., cane or walker) indicated	Life threatening, disabling (e.g., paralysis)
Sensory	Asymptomatic, loss of DTRs or paresthesia but not interfering with function	Sensory alteration or paresthesia interfering with function but not interfering with activities of daily living	Sensory alteration or paresthesia interfering with activities of daily living	Disabling
WHO	Paresthesias and/or decreased DTRs	Severe paresthesias and/or mild weakness	Intolerable paresthesias and/or marked motor loss	Paralysis
ECOG	Decreased DTRs, mild paresthesias, mild constipation	Absent DTRs, severe paresthesias, severe constipation, mild weakness	Disabling sensory loss, severe peripheral nerve pain, obstipation; severe weakness, bladder dysfunction	Respiratory dysfunction secondary to weakness, obstipation requiring surgery, paralysis confining patient to bed/ wheelchair

From Ref. 75 with permission.
DTR, deep tendon reflex; NCI-CTC, National Cancer Institute Common Toxicity Criteria; ECOG, Eastern Cooperative Oncology Group.

REFERENCES

1. Jemal A, Siegel R, Ward E, et al. Cancer statistics, 2007. *CA Cancer J Clin* 2007;57:43–66.
2. Geenen MM, Cardous-Ubbink MC, Kremer LC, et al. Medical assessment of adverse health outcomes in long-term survivors of childhood cancer. *JAMA* 2007;297:2705–2715.
3. Espey DK, Wu XC, Swan J, et al. Annual report to the nation on the status of cancer, 1975–2004, featuring cancer in American Indians and Alaska Natives. *Cancer* 2007;110:2119–2152.
4. Stewart SL, Sabatino SA, Foster SL, et al. Decline in breast cancer incidence—United States, 1999–2003. *Morb Mortal Wkly Rep* 2007;56:549–553.
5. Jemal A, Ward E, Thun MJ. Recent trends in breast cancer incidence rates by age and tumor characteristics among U.S. women. *Breast Cancer Res* 2007;9:R28.
6. Posner JB, Chernik NL. Intracranial metastases from systemic cancer. *Adv Neurol* 1978;19:575–587.
7. Lundberg GD. Low-tech autopsies in the era of high-tech medicine: continued value for quality assurance and patient safety. *JAMA* 1998;280:1273–1274.
8. Sonderegger-Iseli K, Burger S, Muntwyler J, et al. Diagnostic errors in three medical eras: a necropsy study. *Lancet* 2000;355:2027–2031.
9. Aalten CM, Samson MM, Jansen PA. Diagnostic errors: the need to have autopsies. *Neth J Med* 2006;64:186–190.
10. Combes A, Mokhtari M, Couvelard A, et al. Clinical and autopsy diagnoses in the intensive care unit: a prospective study. *Arch Intern Med* 2004;164:389–392.
11. Burton EC, Troxclair DA, Newman WP, III. Autopsy diagnoses of malignant neoplasms: how often are clinical diagnoses incorrect? *JAMA* 1998;280:1245–1248.
12. Cairncross JG, Kim J-H, Posner JB. Radiation therapy for brain metastases. *Ann Neurol* 1980; 7:529–541.
13. Barnholtz-Sloan JS, Sloan AE, Davis FG, et al. Incidence proportions of brain metastases in patients diagnosed (1973 to 2001) in the metropolitan Detroit cancer surveillance system. *J Clin Oncol* 2004;22:2865–2872.
14. Schouten LJ, Rutten J, Huveneers HAM, et al. Incidence of brain metastases in a cohort of patients with carcinoma of the breast, colon, kidney, and lung and melanoma. *Cancer* 2002;94:2698–2705.
15. Yoshida S. Brain metastasis in patients with esophageal carcinoma. *Surg Neurol* 2007;67:288–290.
16. Seute T, Leffers P, ten Velde GPM, et al. Detection of brain metastases from small cell lung cancer. *Cancer* 2008; 112:1827–1834.
16a. Yardley DA. *2007 San Antonio breast cancer symposium (abstract 6049).*
17. Mut M, Schiff D, Shaffrey ME. Metastasis to nervous system: spinal epidural and intramedullary metastases. *J Neuro-Oncol* 2005;75:43–56.
18. Barron KD, Hirano A, Araski S, et al. Experiences with metastatic neoplasms involving the spinal cord. *Neurology* 1959;9:91–106.
19. Loblaw DA, Laperriere NJ, Mackillop WJ. A population-based study of malignant spinal cord compression in Ontario. *Clin Oncol (R Coll Radiol)* 2003;15:211–217.
20. Antunes NL. The spectrum of neurologic disease in children with systemic cancer. *Pediatr Neurol* 2001;25:227–235.
21. Singh G, Rees JH, Sander JW. Seizures and epilepsy in oncological practice: causes, course, mechanisms and treatment. *J Neurol Neurosurg Psychiatry* 2007;78:342–349.
22. Antunes NL. Seizures in children with systemic cancer. *Pediatr Neurol* 2003;28:190–193.
23. Ochs JJ, Bowman WP, Pui CH, et al. Seizures in childhood lymphoblastic leukaemia patients. *Lancet* 1984;2:1422–1424.
24. Hovestadt A, van Woerkom CM, Vecht CJ. Frequency of neurological disease in a cancer hospital. *Eur J Cancer* 1990;26:765–766.
25. Gilbert MR, Grossman SA. Incidence and nature of neurologic problems in patients with solid tumors. *Am J Med* 1986;81:951–954.
26. Sculier J-P, Feld R, Evans WK, et al. Neurologic disorders in patients with small cell lung cancer. *Cancer* 1987;60:2275–2283.
27. Seute T, Leffers P, Ten Velde GP, et al. Neurologic disorders in 432 consecutive patients with small cell lung carcinoma. *Cancer* 2004;100:801–806.
28. Pickren JW, Lopez G, Tsukada Y, et al. Brain metastases: an autopsy study. *Cancer Treat Symp* 1983;2:295–313.
29. Hochstenbag MMH, Twijnstra A, Wilmink JT, et al. Asymptomatic brain metastases (BM) in small cell lung cancer (SCLC): MR-imaging is useful at initial diagnosis. *J Neuro-Oncol* 2000;48:243–248.
30. Yokoi K, Kamiya N, Matsuguma H, et al. Detection of brain metastasis in potentially operable non-small cell lung cancer—a comparison of CT and MRI. *Chest* 1999;115:714–719.
31. Shi AA, Digumarthy SR, Temel JS, et al. Does initial staging or tumor histology better identify asymptomatic brain metastases in patients with non-small cell lung cancer? *J Thorac Oncol* 2006;1:205–210.
32. Moore EW, Thomas LB, Shaw RK, et al. The central nervous system in acute leukemia. A postmortem study of 117 consecutive cases, with particular reference to hemorrhages, leukemic infiltrations, and the syndrome of meningeal leukemia. *Arch Intern Med* 1960;105:141–158.
33. Nesbit ME, Sather H, Robison LL, et al. Sanctuary therapy: a randomized trial of 724 children with previously untreated acute lymphoblastic leukemia. A report from Children's Cancer Study Group. *Cancer Res* 1982;42:674–680.
34. Gokbuget N, Hoelzer D. Meningeosis leukaemica in adult acute lymphoblastic leukaemia. *J Neuro-Oncol* 1998;38:167–180.
35. Lai R, Dang CT, Malkin MG, et al. The risk of central nervous system metastases after trastuzumab therapy in patients with breast carcinoma. *Cancer* 2004;101:810–816.
36. Yau T, Swanton C, Chua S, et al. Incidence, pattern and timing of brain metastases among patients with advanced breast cancer treated with trastuzumab. *Acta Oncol* 2006;45:196–201.
37. Souglakos J, Vamvakas L, Apostolaki S, et al. Central nervous system relapse in patients with breast cancer is associated with advanced stages, with the presence

of circulating occult tumor cells and with the HER2/neu status. *Breast Cancer Res* 2006;8:R36.

38. Boogerd W, Hart AAM, Tjahja IS. Treatment and outcome of brain metastasis as first site of distant metastasis from breast cancer. *J Neuro-Oncol* 1997;35:161–167.

39. Gerl A, Clemm C, Kohl P, et al. Central nervous system as sanctuary site of relapse in patients treated with chemotherapy for metastatic testicular cancer. *Clin Exp Metastasis* 1994;12:226–230.

40. Gabos Z, Sinha R, Hanson J, et al. Prognostic significance of human epidermal growth factor receptor positivity for the development of brain metastasis after newly diagnosed breast cancer. *J Clin Oncol* 2006;24:5658–5663.

41. Chen AM, Jahan TM, Jablons DM, et al. Risk of cerebral metastases and neurological death after pathological complete response to neoadjuvant therapy for locally advanced nonsmall-cell lung cancer: clinical implications for the subsequent management of the brain. *Cancer* 2007; 109:1668–1675.

42. Gore E. Prophylactic cranial irradiation versus observation in stage III non-small-cell lung cancer. *Clin Lung Cancer* 2006;7:276–278.

42a. Abrahm JL, Banffy MB, Harris MB. Spinal cord compression in patients with advanced metastatic cancer: "All I care about is walking and living my life". *JAMA* 2008;229:937–946.

43. Mayer M. A patient perspective on brain metastases in breast cancer. *Clin Cancer Res* 2007;13:1623–1624.

44. Hainline B, Tuzynski MH, Posner JB. Ataxia in epidural spinal cord compression. *Neurology* 1992; 42:2193–2195.

45. Wang C-H, Ng S-H. Syncope as the initial presentation of nasopharyngeal carcinoma. *J Neuro-Oncol* 1995;25:73–75.

46. Bauer CA, Redleaf MI, Gartlan MG, et al. Carotid sinus syncope in head and neck cancer. *Laryngoscope* 1994;104:497–503.

47. Clouston PD, DeAngelis LM, Posner JB. The spectrum of neurologic disease in patients with systemic cancer. *Ann Neurol* 1992;31:268–273.

48. Antunes NL. The spectrum of neurologic disease in children with systemic cancer. *Pediatr Neurol* 2001;25:735.

49. Peppercorn JM, Weeks JC, Cook EF, et al. Comparison of outcomes in cancer patients treated within and outside clinical trials: conceptual framework and structured review. *Lancet* 2004;363:263–270.

50. Klabunde CN, Springer BC, Butler B, et al. Factors influencing enrollment in clinical trials for cancer treatment. *South Med J* 1999;92:1189–1193.

51. Reynolds T. Costs studies show clinical trials, standard therapy may be equal. *J Natl Cancer Inst* 2000;92:1116–1118.

52. Mahaley MS, Jr, Mettlin C, Natarajan N, et al. National survey of patterns of care for brain-tumor patients. *J Neurosurg* 1989;71:826–836.

53. Bernstein M, Rutka J. Brain tumor protocols in North America. *J Neuro-Oncol* 1993;17:231–251.

54. Stiller CA. Centralised treatment, entry to trials and survival. *Br J Cancer* 1994;70:352–362.

55. Eisenberg L. The social imperatives of medical research. Impeding medical research, no less than performing it, has ethical consequences. Not to act is to act. *Science* 1977;198:1105–1110.

56. Kiecolt-Glaser JK, Glaser R. Psychoneuroimmunology and cancer: fact or fiction? *Eur J Cancer* 1999;35:1603–1607.

57. Cleeland CS, Bennett GJ, Dantzer R, et al. Are the symptoms of cancer and cancer treatment due to a shared biologic mechanism? A cytokine-immunologic model of cancer symptoms. *Cancer* 2003;97:2919–2925.

58. McGee R. Does stress cause cancer? There's no good evidence of a relation between stressful events and cancer. *BMJ* 1999;319:1015–1016.

59. Protheroe D, Turvey K, Horgan K, et al. Stressful life events and difficulties and onset of breast cancer: case-control study. *BMJ* 1999;319:1027–1030.

60. Jung W, Irwin M. Reduction of natural killer cytotoxic activity in major depression: interaction between depression and cigarette smoking. *Psychosom Med* 1999;61:263–270.

61. Grossarth-Maticek R, Eysenck HJ, Boyle GJ, et al. Interaction of psychosocial and physical risk factors in the causation of mammary cancer, and its prevention through psychological methods of treatment. *J Clin Psychol* 2000;56:33–50.

62. Price MA, Tennant CC, Smith RC, et al. The role of psychosocial factors in the development of breast carcinoma. Part I: The cancer prone personality. *Cancer* 2001;91:679–685.

63. Price MA, Tennant CC, Butow PN, et al. The role of psychosocial factors in the development of breast carcinoma. Part II: Life event stressors, social support, defense style, and emotional control and their interactions. *Cancer* 2001;91:686–697.

64. Garssen B, Goodkin K. On the role of immunological factors as mediators between psychosocial factors and cancer progression. *Psychiatry Res* 1999;85:51–61.

65. Marchioro G, Azzarello G, Viviani F, et al. Hypnosis in the treatment of anticipatory nausea and vomiting in patients receiving cancer chemotherapy. *Oncology* 2000;59:100–104.

66. Schiff D, Rosenblum MK. Herpes simplex encephalitis (HSE) and the immunocompromised: a clinical and autopsy study of HSE in the settings of cancer and human immunodeficiency virus-type 1 infection. *Hum Pathol* 1998;29:215–222.

67. DeAngelis LM. Neurologic complications. In: Kufe DW, Bast RC, Hait WN, Hong WK, Pollack RE, Weichselbaum RR, Holland JF, Frei E eds. *Cancer Medicine*. 7th ed. Hamilton, Ontario: BC Decker, 2006. pp. 2061–2076.

68. Antoine JC, Cinotti L, Tilikete C, et al. [18F]fluoro-deoxyglucose positron emission tomography in the diagnosis of cancer in patients with paraneoplastic neurological syndrome and anti-Hu antibodies. *Ann Neurol* 2000;48:105–108.

69. Savic G. Inter-rater reliability of motor and sensory examinations performed according to American Spinal Injury Association standards. *Spinal Cord* 2007;45:444–451.

70. Folstein MF, Folstein SE, McHugh PR. "Mini-mental state." A practical method for grading the cognitive state of patients for the clinician. *J Psychiatr Res* 1975;12:189–198.

71. Ely EW, Margolin R, Francis J, et al. Evaluation of delirium in critically ill patients: validation of the Confusion Assessment Method for the Intensive Care Unit (CAM-ICU). *Crit Care Med* 2001;29:1370–1379.

72. Gaspar LE, Scott C, Murray K, et al. Validation of the RTOG recursive partitioning analysis (RPA) classification for brain metastases. *Int J Radiat Oncol Biol Phys* 2000;47:1001–1006.

73. Lutterbach J, Bartelt S, Ostertag C. Long-term survival in patients with brain metastases. *J Cancer Res Clin Oncol* 2002;128:417–425.

74. Kaal EC, Niel CG, Vecht CJ. Therapeutic management of brain metastasis. *Lancet Neurol* 2005;4:289–298.

75. Walker M, Ni O. Neuroprotection during chemotherapy. *Am J Clin Oncol* 2007;30:82–90.

Chapter 2

Pathophysiology of Nervous System Metastases

INTRODUCTION

Unlike primary brain tumors (e.g., glioblastoma) that cause death as a direct result of local tumor growth, most other cancers can be controlled locally; death occurs only if the tumor metastasizes. Unfortunately, as many as 60% of patients with cancer have overt or occult metastases at the time of diagnosis,[1] rendering them virtually incurable. Metastases to the nervous system are a major cause of neurologic disability in patients with cancer and are often the direct cause of death. Metastases to the nervous system cause approximately one-half

of the significant neurologic complications of cancer.[2,3]

As Table 2–1 indicates, cancer can reach the nervous system in contiguity with the primary tumor or one of its metastases (direct extension) or arise not in contiguity with the primary tumor (indirect extension) through the metastatic process. Examples of direct extension include a tumor arising in the apex of the lung (Pancoast tumor) that may compress the brachial plexus or upper thoracic roots.[4] Pain and neurologic dysfunction of the arm may be the only signs that a lung tumor is present. Tumors may invade nerves and grow within the

31

Table 2–1. **How Cancer Spreads to the Nervous System**

Mode	Example
Metastasis	
Hematogenous	Brain metastasis
Lymphatics	Perineural lymphatics → leptomeninges
Body fluids	Leptomeningeal metastasis
	"Drop metastases" from brain tumor
Direct Extension	
From tumor itself	Pancoast tumor → brachial plexus
From lymph node	Breast cancer nodes → brachial plexus
From metastasis	Skull metastases → brain
Perineural growth	Neurotropic tumor → leptomeninges

perineural sheath to affect peripheral or cranial nerves and ultimately the central nervous system (CNS) (see Chapter 8).[5,6]

Metastasis has three mechanisms: (1) hematogenous dissemination (the usual mechanism of brain metastases), (2) lymphatic dissemination (because most of the CNS does not contain lymphatics, this mechanism applies primarily to peripheral nervous structures) (see Chapter 8), and (3) dissemination through body fluids such as the cerebrospinal fluid (CSF) (the mechanism for leptomeningeal dissemination of tumor) (see Chapter 7).

PATHOPHYSIOLOGY OF THE METASTATIC PROCESS

Metastasis is a complicated pathophysiologic process that is not yet completely understood.[7] As Figure 2–1 and Table 2–2 indicate, many steps are involved in metastasis, so many that only a small number of tumor cells ever complete this difficult and arduous process. One estimate is that of the many millions of cancer cells that enter the circulation, less than 0.01% are able to form metastases; cancer cells are often found in the circulation or microscopically in distant organs in patients in whom overt metastases never develop. Indeed, it is surprising that any cell possesses the necessary multifaceted capacity for metastasis, each step of which is controlled by multiple genes.

This chapter describes certain aspects of the metastatic pathogenesis relevant to neuro-oncology. Recent reviews detail the complex processes involved in metastatic spread.[1,8–10]

Transformation and Growth

The development of a cancer is a complicated process, involving a series of genetic and epigenetic steps in which tumor suppressor genes are downregulated or deleted and oncogenes are activated to render cells independent of external growth factors. The genes involved in this process, the so-called tumorigenic genes, promote proliferation, genetic instability, self-renewal, evasion of apoptosis, evasion of immunity, and resistance to hypoxia (Fig. 2–1).[10] Although arising from a single cell, the cells become heterogeneous as they proliferate, in part because some of the early events promote genetic instability. Increasing evidence suggests that the cell of origin of cancer, the tumor-initiating cell, is a stem cell or one of its immediate progeny (Fig. 2–2).[11–13] A stem cell is a slowly reproducing cell that has the capacity of self-renewal and the ability to form differentiated progeny. When a stem cell divides, it produces another stem cell and a transient amplifying cell. The stem cell can reproduce indefinitely, albeit usually quite slowly. Stem cells reside in specific anatomic locations surrounded by other cells (niche

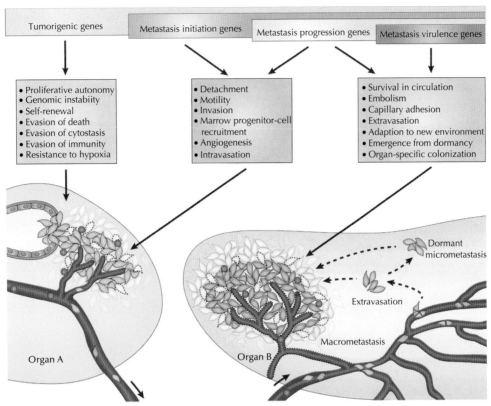

Figure 2–1. Classes of genes participating in the metastasis process. On the basis of their level of participation in the metastatic process, one can distinguish four general classes of cancer genes. Tumorigenic genes initiate the process that causes a tumor that may or may not become metastatic. Metastasis initiation genes promote mobility, invasion, and intravasation into blood vessels that pave the way for tumor cells to enter the circulation. Metastasis progression genes promote survival of metastatic cells in the circulation, adhesion to the capillaries of the organ that will harbor the metastasis, extravasation into the parenchyma, and growth of cells in the new organ. Some of these genes are also important in initiating metastases. Metastasis virulence genes are those that provide a selective advantage in secondary sites but not in the primary tumor, thus participating in metastatic colonization but not in primary tumor development. These genes are important in determining whether a metastasis will grow within the new organ and if so how fast. For the neuro-oncologist, Organ A is any systemic organ; Organ B is the brain or other nervous system structure. (From Ref. 10 with permission.)

cells) that not only serve to save the stem cell from depletion but also to prevent too rapid proliferation.[14] Niche cells may play a role in transformation to cancer. Transient amplifying cells can also reproduce but eventually differentiate and become postmitotic. Cancer stem cells, when injected into another host, reproduce the tumor. However, in any cancer, there may be only a few stem cells; the other tumor cells are not tumorigenic, but can continue to grow and proliferate.[11] Evidence suggests that a subset of cancer stem cells possess genes that promote metastases.[11]

Years or sometimes decades before a CNS metastasis appears, one or more cells in a non-CNS organ undergoes a series of genetic or epigenetic[15,16] alterations that allow proliferation in an uncontrolled manner. A cell destined to become a cancer undergoes two types of genetic alterations.[15] The first type involves activation of gatekeeper genes that control cellular proliferation. The second type involves silencing[16] of caretaker genes responsible for genetic stability, leading to the so-called mutator phenotype.[17,18] Thus, the cancer cell loses control of cell division and becomes genetically unstable, leading to additional mutations. Genetic changes include point mutations, deletions, overexpression, and translocations. Epigenetic changes include methylation of

Table 2–2. **Sequence of Events Leading to Nervous System Metastases**

In Primary Organ
Transformation and growth
Angiogenesis
Invasion
Intravasation

In Circulation
Transport

In First Metastatic Organ
Arrest (first capillary bed, e.g., lung)
Dormancy or growth
Passage to arterial circulation

In Nervous System
Arrest (final capillary bed, e.g., brain)
Extravasation
Dormancy or growth

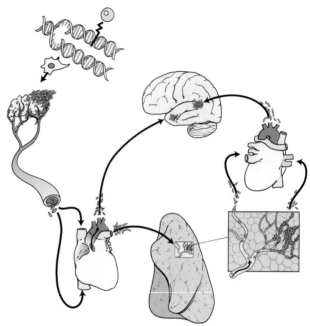

Figure 2–2. Pathophysiology of the metastatic process. A normal cell (upper left), possibly a stem cell, undergoes a genetic mutation (in this case a double-stranded DNA break) and is transformed into a cancer cell. As it grows uncontrollably, it invades its host organ, develops its own vascular supply, and eventually enters either the lymphatic or the venous circulation. When it reaches the right side of the heart, it can enter the arterial circulation if there is a patent foramen ovale or it can exit the right ventricle to enter the pulmonary circulation. In the lung, it can either grow as a metastasis or bypass the pulmonary circulation. In either event, tumor cells can enter the left side of the heart and reach the arterial circulation. Once tumor cells have entered the arterial circulation, they have access to the brain, where a metastasis will develop if the tumor cell and the brain are a match.

DNA within the promoter region of tumor suppressor genes, leading to inactivation of those genes and thus failure of tumor suppression.[16] Epigenetic changes can also silence caretaker genes.

Because cancer cells are characterized by genetic instability, as the tumor grows it becomes increasingly heterogeneous.[15] Some, but not all, cells in a given cancer have the capacity to metastasize and some can select the brain as their preferential metastatic site.[19–21] Whether metastatic potential resides in only a few cells within the primary tumor[22] or in most of the cells[23] is in dispute, as is whether the capacity for metastases arises early, in very small tumors, or later as the tumor grows to a larger size.[24]

Uncontrolled proliferation (i.e., tumor growth) is necessary but not sufficient for metastases. Current evidence suggests that the tendency to metastasize is acquired early, caused by one or more genetic changes, perhaps involving mutations similar to or the same as those that support proliferation.[25] Although, in general, larger cancers are more likely to metastasize than smaller ones (e.g., each millimeter increase in a breast cancer decreases survival, that is, increases metastasis by 1%[26]), some very small primary tumors may already have metastasized when first detected.[27] As already noted, up to 60% of cancers are metastatic at the time

of diagnosis.[1] A brain metastasis occasionally causes death at a time when the primary tumor is undetectable, sometimes even at autopsy.[28]

Some genetic alterations predispose a patient to systemic metastases. Genetic alterations that predict brain metastases from lung[29] and breast[30] cancer have been identified. Genes that are required for breast cancer to metastasize to bone (including vertebrae, causing spinal cord compression) have also been identified.[31,32] Two of those genes, interleukin-11 and *CTGF*, encode osteolytic and angiogenic factors whose expression is further increased by the pro-metastatic cytokine, transforming growth factor β (TGFβ). Overexpression of this bone metastasis gene set is superimposed on a poor prognosis gene expression signature already present in the parental breast cancer population, suggesting that metastasis requires a set of functions beyond those underlying the emergence of the primary tumor. Both of these mechanisms are probably required: Cells destined to metastasize contain a specific genetic signature found in poor-prognosis primary tumors. However, additional genetic changes are probably required for a metastasis to actually occur. Metastatic tumors utilize a different set of so-called metastasis initiation genes that, as indicated in the following paragraphs, allows cells to develop a blood supply, detach from the primary site, migrate to blood vessels, and finally enter blood vessels or lymph channels at the primary site.[10] In addition, a specific microRNA induced by a specific transcription factor can increase expression of a prometastatic gene,[33] causing a previously nonmetastatic breast tumor to invade and metastasize.

Vascularization (Angiogenesis)

Metastasis depends on growth of the primary tumor. Tumors smaller than 1 mm in diameter receive oxygen and other nutrients by diffusion from capillaries of the host organ. To grow beyond 1 to 2 mm ($\sim 10^6$ cells), a tumor requires an adequate blood supply. Neoplasms insure themselves of this blood supply by promoting new vessel formation facilitated by tumor angiogenesis factors. (Tumors can also co-opt normal host blood vessels,[34] although this is usually not sufficient to support tumor growth in most organs; however, see Ref. 35.) Thus, angiogenesis is important for growth of both the primary tumor and its metastases. For example, tumors implanted in the anterior chamber of the eye grow only to 1 mm³ unless they contact the cornea, ciliary body, or other tissue from which they can grow a capillary bed.[36] Tumors grow slowly at a linear rate before neovascularization but switch to exponential growth after vascularization.

Vascular endothelial growth factors (VEGFs) are among several agents that promote angiogenesis. Some members of the VEGF family also promote lymphogenesis. Angiogenesis can occur at any stage of tumor progression, depending both on the type of tumor and the nature of the microenvironment.[37] Newly formed blood and lymph vessels are probably more susceptible to tumor cell invasion than

are normal vessels.[38] Tumor vessels differ from normal vessels in several ways: the vascular structure is often aberrant, and endothelial cell–pericyte interactions are abnormal, as are blood flow and vascular permeability. A recent experimental model suggests that angiogenesis may be associated with metastasis independent of either growth of the primary tumor or the enhanced permeability that facilitates tumor cell intravasation into blood or lymph vessels.[39] In this model, mammary tumor nests are enveloped by endothelial cells and enter the circulation as tumor emboli within vessels. The tumor may then grow in the lung vasculature, without extravasation into lung parenchyma.[39]

The development of a new capillary bed is particularly important in the CNS because the newly formed capillaries differ from normal CNS vessels in that these neovessels have a fenestrated rather than continuous endothelial surface and the cells lack tight junctions. Thus, newly vascularized metastases have an absent or deficient blood–brain barrier, the primary cause of brain edema (see Chapter 3). Their abnormal vascular permeability also accounts for the contrast enhancement seen on computed tomography (CT) or magnetic resonance imaging (MRI).

Agents promoting angiogenesis, in addition to VEGF, include basic fibroblast growth factors (FGFs), matrix metalloproteinase,[40] macrophages,[41] and others.[42,43] These factors also serve to break down the basement membrane of the host organ capillaries, promote migration and proliferation of capillary endothelial cells into the tumor, and organize the endothelial cells into capillaries. Metastatic tumor cells secrete heparinase, which may serve to release FGFs. The brain is enriched in FGFs and other growth factors, but whether they play a role in CNS metastasis is unknown. Heparin inhibits angiogenesis and decreases metastasis;[44] it may also augment the antimetastatic activity of other substances such as interferon and tumor necrosis factor.

The degree of angiogenesis in a primary tumor, as measured by vessel count and density, correlates with the likelihood of metastasis.[45] The critical importance of angiogenesis in maintaining growth of both the primary tumor and any metastases explains the increasing use of anti-angiogenic agents, such as VEFG inhibitors, in cancer therapy.[46,47]

VEGFs promote not only angiogenesis but also lymphangiogenesis.[48] VEGFs induce the formation of lymphatics in tumors and promote the spread of tumor cells in experimental animals.[49,50] Conversely, a soluble form of VEGF receptor inhibits lymphangiogenesis.[51] What role these factors play in the lymphatic spread of cancer in humans is not known.

Invasion

The vascularized primary tumor invades the normal host tissue from which it arose in the several steps delineated in the preceding text. Many of the same genes that promote angiogenesis also promote invasion.[52] First, tumor cells must loosen their cell–cell and cell–stromal attachments.[1] This is generally done by downregulation of adhesion molecules, particularly the cadherin–catenin complex.[53] E-cadherin, an adhesion molecule, is downregulated in most cancers that have the potential for metastasizing. E-cadherin suppresses the metastatic potential of several cell lines. Next, the cancer cells must attach to the extracellular matrix of the host tissue. Integrins are cell adhesion molecules that promote such binding and play an important role in invasion and metastasis.[54] Tumors must also degrade the extracellular matrix in order to migrate to blood vessels and lymph channels. Matrix metalloproteinases play an important role in this function and are upregulated in many metastatic cancers.[55] Migration also requires local proteolysis and sequential attachment and detachment of cells to underlying matrix. This is stimulated by a number of factors, some of which are secreted by the tumor. Examples include FGFs, TGFβ,[56] and others.[43] Some factors necessary for invasion are produced by stromal cells around the tumor as well as cancer cells.[57] Cancer cells also attract inflammatory cells that promote the ability of the cancer cells to invade and metastasize.[58] Bone marrow–derived mesenchymal stem cells within the tumor supply paracrine signals that increase the metastatic potential of weakly metastatic breast cancer cells.[59]

The tumor cell then moves through the extracellular matrix to reach blood or lymph vessels. Tumor cells with a high propensity for metastasis have a greater orientation toward blood vessels and are less likely to fragment than nonmetastatic tumor cells. Many of the motile cells in the tumor mass are not neoplastic cells, but are probably derived from the immune system. These cells may either help orient the tumor cells toward blood vessels or generate pores in blood or lymph vessels so that tumor cells can more easily intravasate.[60]

Intravasation

Intravasation is a critical step in the development of metastatic tumors. Once the cells invade and detach, they must find a route to other tissues. Two important routes of metastatic spread are via the lymphatics and the vascular system. (Direct spread can also occur via body fluids or along nerves; see Table 2–1.) The cancer cells may enter venules, capillaries, or lymph channels. To do so, the tumor cells produce enzymes such as heparinase that degrade the basement membrane of the vessel. Entry into either capillaries or lymph channels leads to a venous circulation: the vena cava, the portal vein, or Batson vertebral veins. A tumor can also directly invade a large vein, allowing a large tumor embolus to enter the circulation. However, most metastases arise from single cells that have entered the circulation.

Current evidence suggests that individual cells isolated from metastases differ both genetically and phenotypically from the cells in the primary tumor.[24,60a] Moreover, cells that enter lymphatics differ from those that enter the bloodstream. However, metastases to lymph nodes can cause distant tumors secondarily by first growing in the lymph node and then entering the bloodstream to spread to other organs.

Tumor blood vessels differ from normal vessels in the host organ in that the tumor-induced endothelial cells are more porous, allowing cancer cells to enter the vessel lumen more easily.[61] This allows over 1 million cells per gram of tumor tissue to be shed into the circulation each day.[62] (If this figure is correct, and if Liotta and Kohn's estimate[63] of 0.01% of shed cells becoming metastases is also correct, each gram of tissue would produce 100 metastases each day. Obviously, one figure is not accurate.)

The importance of vascular and lymphatic invasion to the development of metastases is emphasized by Freedman et al.[64] Their work indicates that both vascular and lymphatic invasion are independently and significantly associated with risk of relapse, but their importance varies with the biology of the primary tumor. For example, in head and neck cancers, distant metastases are rare in the absence of lymph node metastases, whereas they are common in patients with breast cancer.[24]

The lymphatic mode of spread is probably important in the peripheral nervous system (e.g., when the brachial plexus is involved by metastatic tumor from the lung or breast) but, because the CNS has no lymphatics, hematogenous spread is far more important to the development of brain metastases. Separation of these two modes of spread is rather arbitrary because the blood and lymph systems are linked; lymphatic invasion can lead to the vascular system via venous–lymphatic anastomoses or via the thoracic duct. Batson venous plexus probably is important in disseminating tumor cells to the vertebral bodies and skull (Fig. 2–3).[65]

Circulation

Several mechanisms conspire to destroy tumor cells once they reach the circulation. The immune system, especially natural killer cells, attacks and lyses tumor cells in the blood stream.[66] Cells that have lost anchorage may develop detachment-induced apoptosis, called anoikis (Greek for homelessness).[67] Cancer cells that express anti-apoptotic molecules resist anoikis and are more likely to survive in the circulation.[67] A substantial number of tumor cells are probably also destroyed by mechanical damage caused by shear forces as they circulate. In one study, only 1.5% of intravenously injected cancer cells survived more than 24 hours.[68] However, in another study, destruction of cells by hemodynamic processes such as shear forces and anoikis was not an important cause of metastatic inefficiency.[69] Of cells arresting in the lungs after intravenous injection, 74% were still alive at 3 days. However, only 16% were still alive at 2 weeks, either remaining as solitary cells or developing into metastases.

Shear forces occur both as the tumor circulates and when it arrests in the capillary bed, partially or completely obstructing the vessel. Those tumor cells that can attach firmly to the vessel wall overcome shear forces and may live to form metastases.[70] Arrest in a small

capillary probably also deforms the tumor cell.[71] However, tumor cells may form complexes with platelets and leukocytes, protecting the cells against mechanical damage and attack by the immune system.[72] The complexes also promote arrest in larger capillary beds and protect the tumor there as well. In some tumors, P-selectin, an adhesion molecule, plays an important role in formation of these complexes[72]; heparin inhibits P-selectin and decreases metastases in some experimental models.[72] Tumor cells that express surface integrin receptors can bind integrins on the surface of platelets.[73]

Many circulating tumor cells survive the bloodstream and even extravasate into organs.[75] Nevertheless, studies of patients undergoing peritoneal–venous shunts to alleviate abdominal pain and distention from malignant ascites indicate that, although large numbers of malignant cells are introduced directly into the circulation, metastatic spread to other organs is uncommon and when metastasis does develop, it is often not in the first capillary network encountered.[75] A similar situation occurs when cerebroventricular–peritoneal shunts are performed for the alleviation of hydrocephalus caused by either brain or leptomeningeal tumors. Seeding of the peritoneal cavity by the malignancy rarely occurs after ventriculoperitoneal shunting for leptomeningeal metastases and is almost never a clinical problem. Only rarely do clinically symptomatic metastases to the peritoneal cavity from medulloblastoma follow shunting when the CNS tumor is quiescent (interestingly, we have had one patient who developed a large peritoneal metastasis without evidence of recurrence of the primary medulloblastoma or of leptomeningeal seeding of tumor). Shunted tumor cells fail to seed the peritoneum because host cell–to–tumor cell recognition factors are lacking. Furthermore, the presence of tumor cells in the circulation does not appear to indicate a worse prognosis than the absence of tumor cells. Although the first recorded identification of circulating tumor cells dates to 1869, new technologies allow both identification and characterization of circulating tumor cells.[76,76a]

A few circulating tumor cells survive both physical destruction as well as identification and destruction by the host immune system,[77] in part because defective expression of MHC (major histocompatibility complex) proteins on their surface prevents their recognition by the immune system. Tumors with metastatic

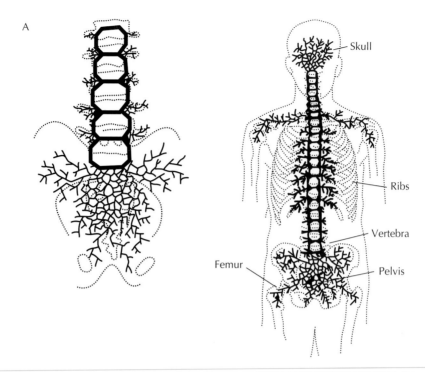

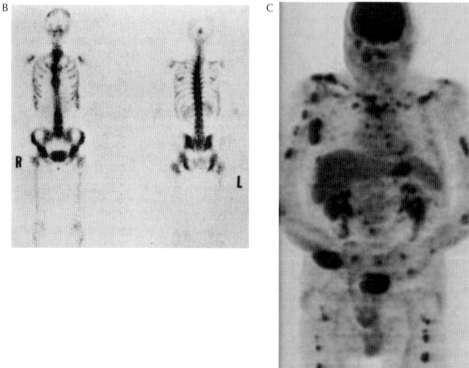

Figure 2–3. A: The vertebral venous system is a two-way thoroughfare frequently used by carcinoma of the prostate, breast, adrenal glands, and other types of cancer. **B**: This bone scan of a patient with prostate carcinoma illustrates widespread metastases in vertebral bodies and other areas supplied by Batson plexus. Virtually no metastases are observed in the axial skeleton and only a few in the skull. Parenchymal metastases were absent. **C**: A PET scan from a patient with lung cancer showing diffuse metastases in the axial skeleton, presumably arterial in origin. (A, B: From Ref. 138 with permission.)

potential are less likely to express class I MHC proteins than are nonmetastatic tumor cells. Likewise, metastases are usually MHC class I negative, whereas the primary tumor is usually positive. Metastasis virulence genes promote survival in the circulation, adhesion to capillaries in the bed of another organ, extravasation into that organ, adaptation to the new organ, and finally growth there to form a metastasis.[10]

Arrest: First Capillary Bed

Because tumor cells are generally larger than capillary vessels, once they are in the systemic circulation, they usually arrest in the first capillary bed they encounter[78]; capillaries usually are 3 to 8 µm as opposed to the diameter of 20 µm or more of tumor cells.[78] Accordingly, one would expect that cancers of organs other than lung would arrest first in lung, liver, or vertebral bodies, depending on whether the venous drainage is into vena cava, portal vein, or Batson plexus.[79] However, some tumor cells pass through the first capillary bed to reach the arterial circulation and then arrest in the capillary bed of other organs. In one study, less than 10% of tumor cells that were shed into the venous circulation reached the arterial circulation.[62] Arrest is not only related to the size of tumor cells, but also to the recognition by tumor cells of surface molecules on the endothelium of the host organ.[80] These so-called endothelial addressins are unique to specific organs[81] such as the brain. In addition, tumor cells may be attracted to specific organs by chemo-attractants produced by that organ. Some circulating tumor cells may find their way back to the original primary site and re-seed the primary tumor.[82]

Experimental evidence indicates that anticoagulation with heparin or warfarin inhibits the formation of metastasis by injected tumor cells[44,83] and that glucocorticoids given as a single agent usually but not always facilitate metastases[84] (but see Ref. 85). The effect of heparin on cancer progression and metastasis is complex and involves effects on proliferation, invasion, and immunity as well as the vascular system.[86] Prostacyclin may also inhibit metastases by anti-aggregatory effects on platelets.[87] Aspirin also inhibits metastases, but probably more through its effect on tumor cell invasiveness and angiogenesis than its effects on platelets.[88] Other nonsteroidal anti-inflammatory agents also appear to inhibit metastases, probably through their effect on cyclooxygenase 2 overexpression, which enhances lymphatic invasion and metastases.[89,90] Anticoagulants appear to have only modest effects on established human cancer metastases.[91] Warfarin[91] and heparin[44] may enhance survival in lung cancer, but aspirin does not.[92]

Although the clinical implications of these findings are not entirely clear, many patients with cancer are known to be hypercoagulable (see Chapter 9); this state may enhance the tendency for a tumor to metastasize. In addition, some reports indicate that tumors in patients who receive corticosteroids tend to metastasize to sites rarely selected in patients not receiving steroids.[93] It is unknown whether the extensive use of glucocorticoids at the relatively high doses that are commonly used to treat patients with CNS metastases promotes further spread of the cancer.

Passage to Arterial Circulation

Lung cancer is the most common source of brain metastasis, at least in part because lung tumors have direct access to pulmonary veins that lead to the arterial circulation via the left heart. Cancer from other organs first enters the venous circulation to reach the right side of the heart and then, in the absence of a right-to-left cardiac shunt, exits the heart via the pulmonary artery to reach the capillary bed of the lungs. Tumor cells may lodge in the lung to form pulmonary metastases, a frequent finding in patients whose brain metastasis comes from tumors other than lung cancer. Brain metastases usually occur late in the course of a patient's cancer, when the tumor is already widely metastatic to other organs. In those instances where the primary site of the cancer has been successfully treated, a brain metastasis may be a result of secondary metastasis (e.g., a lung metastasis from a breast cancer metastasizes to the brain). To reach the arterial circulation and seed the brain, the tumor must (1) either originate or grow in the lung to form a tumor that invades the pulmonary venous circulation and enters the left side of the heart and then the systemic circulation; (2) traverse the pulmonary capillary bed to enter the left side of the heart;[68] capillaries vary in size and tumor cells vary in their ability to deform and squeeze through them; or (3) cross a patent foramen ovale. Tumors that directly invade veins at the primary site can embolize to the brain through a patent foramen ovale, causing an infarct.[94] Smaller tumor emboli may be responsible for the so-called three-stage course of clinical symptoms: (1) The tumor embolus occludes a cerebral vessel, causing focal ischemia and a clinical course suggesting a stroke. (2) After causing neurologic symptoms, the embolus breaks up, enters the capillary bed, and the symptoms resolve. (3) The tumor cells in the capillary bed extravasate and grow, recapitulating, several months later, the same symptoms caused by the original embolus.[95]

Arrest: Final Capillary Bed

When tumor cells reach the arterial circulation, their distribution should be dependent on the blood flow to a given organ. The brain receives approximately 20% of cardiac output in the resting state and thus is expected to receive that proportion of circulating tumor cells. Experimental studies demonstrate that circulating tumor cells arrest in brain in the distribution of the blood flow and some clinical evidence is also supportive. For example, brain metastases usually occur at the gray–white junction and in the watershed areas of brain at the same sites as cerebral emboli.[96] The usual distribution of brain metastases is approximately 80% cerebral hemisphere, 15% cerebellum, and 3% brainstem, similar to blood flow. If that were the whole story, one would expect the distribution of metastases to be based entirely on the relative blood flow to an organ, and brain metastases might be considered to be so common because the brain receives so much of the cardiac output. The distribution of metastases based on blood flow was proposed by James Ewing and was called the *mechanical hypothesis*.[97] However, blood supply does not account for the propensity for some organs to be common sites of metastases and others to be rare sites, or for specific cancers to metastasize to specific organs. For example, liver metastases are common but spleen metastases are rare, despite comparable blood flow; muscle metastases are uncommon despite their high blood flow during exercise (see Chapter 8). Bone receives a much lesser blood supply than

does muscle, but is a more common site of metastasis. Breast cancer has a propensity to metastasize to the ovaries, whereas lung cancer has a propensity to metastasize to the adrenal glands.

In experimental animals, after intravenous injection of clonal tumor cells, the cells are uniformly distributed throughout an organ (lung); the distribution of dormant cells, those that do not produce metastases, is also uniform.[69] However, metastases are not uniform, but are dependent on location even within a relatively homogeneous organ such as the lung. The metastases were more likely to develop along the lung surface and around arterial and venous vessels than elsewhere in the lung.[69]

Brain metastases from lung cancer are common, possibly because the pulmonary capillary bed serves as a filter that prevents tumors elsewhere in the body from entering the arterial circulation. However, this does not explain why breast cancer and melanoma commonly metastasize to the brain whereas prostate cancer rarely does. Likewise, specific lung cancers such as small cell lung cancer are more likely to metastasize to brain than non–small cell lung cancer, and adenocarcinomas metastasize to brain more commonly than squamous carcinomas. Such observations led to the *seed and soil hypothesis* proposed by Paget[98] more than 100 years ago. As currently conceived, the hypothesis postulates that genetic changes in some cancer cells (the seed) allow them to find the biochemical environment of a particular host organ (the soil) an especially favorable place in which to grow. Ample clinical evidence supports this hypothesis for the CNS. Fidler et al.[99] have been able to derive clones of the B16 melanoma that have a strong predilection to grow only in certain organs such as the ovary or brain, even when injected into the venous circulation. Furthermore, some clones metastasize preferentially to specific areas of the brain (e.g., parenchyma, meninges, ventricle).

Both the mechanical and the seed and soil hypotheses play roles in the distribution of CNS metastases. One estimate suggests that 66% of metastases can be explained on blood flow alone, but the remainder cannot.[100] When the likelihood of metastasis to an organ based on autopsy data was divided by its blood flow (metastatic efficiency index [MEI]), no primary tumor had a high MEI for brain (melanoma, a tumor that has a high propensity to metastasize to the brain,[101] was not included in this analysis); carcinoma of the ovary, prostate, stomach, bladder, and osteosarcoma had a low MEI for brain.[100]

Tumor Emboli

Most tumor emboli are small and do not obstruct vessels that are large enough to produce clinical symptoms. They distribute themselves much like small emboli and, in the brain, are likely to lodge in the watershed areas, that is, at the terminations of major end arteries (watersheds), and also in the gray–white junction, where penetrating arterioles first separate into capillary beds (Fig. 2–4).

An occasional tumor embolus is large enough to cause clinical symptoms that resemble a transient ischemic attack or, rarely, a full-blown stroke. If the symptoms clear and the tumor embolus invades the brain, the same symptoms may recur months later from the growing brain metastasis (see Chapter 5). Tumor emboli may be large enough to cause symptoms of peripheral arterial insufficiency.[102]

Extravasation

Tumor cells that arrest in brain capillaries must cross into brain parenchyma to form metastases. However, they may begin to reproduce while still in the capillary bed.[70] Tumor cells that lodge in the brain or other organs appear to extravasate easily across the capillary membrane into the parenchyma of the organ. The blood–brain barrier is no hindrance to tumor cell extravasation, although it does provide sanctuary to those cells against many therapeutic agents, for example, water-soluble chemotherapeutic agents and large molecules such as antibodies. Only after the cells form a mass that disrupts the blood–brain barrier does the brain metastasis become susceptible to the same agents that kill its cousins in the periphery. The time course of extravasation and the factors that promote extravasation are not well known, but they differ from factors that promote the development of a cancer cell into a metastasis because many extravasated cells remain dormant.

Using the technique of in vivo videomicroscopy, Chambers et al.[74] have studied extravasation of cancer cells into organs of experimental animals. They found no evidence that tumors that arrested in vessels replicated in those vessels and then extravasated by proteolytic destruction of blood vessel basement membranes. Instead, what they observed were single cells extending themselves along one wall of the blood vessel and then through the vessel wall into the surrounding tissue. The cells are first found in the tissue a few hours after initial arrest and, by 4 days, virtually all arrested cells are found extravasated. At least in this model, there is no evidence to indicate degradation of the blood vessel basement membrane and the extracellular matrix. Furthermore, the previously held belief that highly metastatic cells extravasate more rapidly than cells of lesser metastatic potential also does not appear to be true. Chambers et al. found that most blood-borne cancer cells, whether metastatic or not, can successfully extravasate into surrounding tissue after arrest in the microcirculation. They suggest that extravasation is not a rate-limiting step in metastasis; rather, the rate-limiting step appears to be the ability of the extravasated cells to grow within the new host tissue and form mass lesions. In the study, virtually all B16 melanoma cells extravasated by 3 days, but only 2% had started to divide to reach the size of 4 to 16 cells. By the 13th day, only 1% of these micrometastases had grown to form tumors; the rest had disappeared. Some single extravasated cells became dormant. The dormant cells could explain the late occurrence of clinically identifiable metastasis in organs such as brain many years after successful treatment of the primary tumor.

Dormancy

Tumors that leave the CNS vasculature to enter the brain, spinal cord, and leptomeninges do not necessarily grow.[78] If the soil is not propitious, the tumor cells may die or lie dormant for months or even years. There is ample evidence for a solitary brain metastasis developing in patients years (and sometimes decades) after apparent cure of the primary cancer. This phenomenon is particularly prominent in melanoma and breast cancer but occurs with other tumors as well. Because such patients do not have recurrence of the tumor elsewhere in the body, these cancer cells must have been in the brain at the time the original tumor was successfully treated. Surgery may have eradicated the primary tumor, and chemotherapy with water-soluble agents may have effectively eradicated micrometastases elsewhere in the body, but the blood–brain barrier may have prevented CNS micrometastases from being eradicated by the chemotherapy. This is one of the reasons that prophylactic whole brain radiation successfully reduces the incidence of subsequent brain metastases in patients with small cell lung cancer.[103]

One concept of dormancy is that extravasated tumor cells either remain quiescent as single cells or reproduce enough to develop into micrometastases, but grow no further. The concept is that a balance is reached between proliferation and apoptosis such that the metastasis remains microscopic. These microscopic metastases are viable, as indicated by the fact that if removed from experimental animals they will reproduce in vitro and make tumors in vivo.[104]

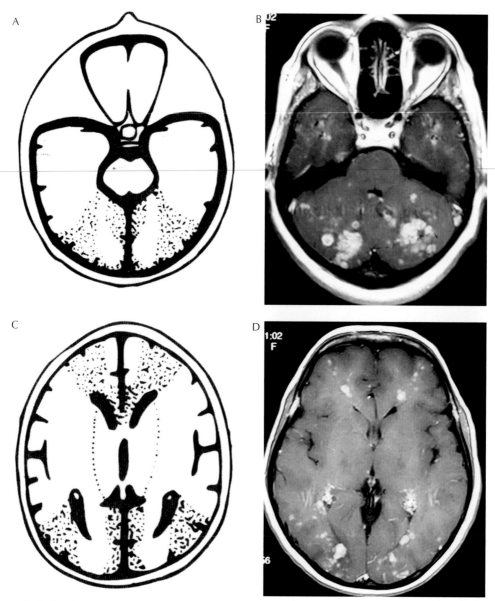

Figure 2–4. (cont.)

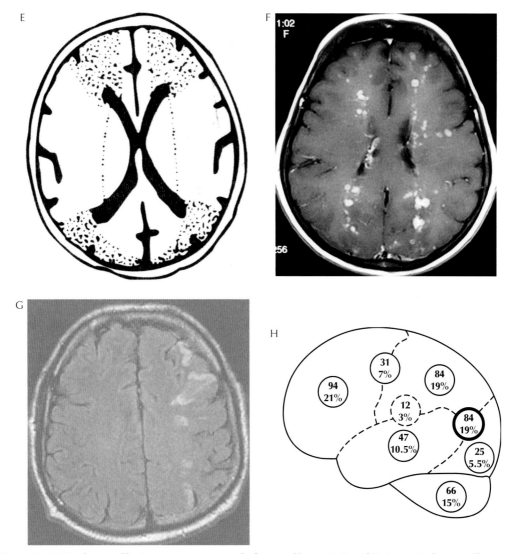

Figure 2–4. Distribution of brain metastases to watershed areas of brain. **A**, **C**, and **E**: Anatomic diagrams illustrating the boundary zones between major cerebral arteries in the brain (watershed areas). **B**, **D**, and **F**: Scans of a patient with multiple cerebral metastases from breast cancer demonstrating the distribution of these metastases within the watershed areas of the brain. Compare with (**G**), which shows carotid occlusion in a patient who had a hypotensive episode resulting in infarction in the watershed area. **H**: Distribution of metastases noted on CT scans. There was a total of 443 metastases in 256 patients, each with fewer than 5 metastases. The bold circle indicates the over-representation of metastases in the watershed area between the middle and posterior cerebral arteries. There was also over-representation in the anterior border zone between the anterior and middle cerebral arteries. (H: From Ref. 96 with permission.)

Growth in the CNS

In order for a brain metastasis to form and become clinically or radiographically apparent, the brain must provide a hospitable environment for the tumor cell and the tumor cell must express molecules that support its growth in the brain. For example, the presence of the cell surface receptor P75 NTR on tumor cells strongly correlates with the brain colonization potential of melanoma cells.[21,105] The ligand of this low affinity neurotropic receptor, nerve growth factor (NGF), is made by astrocytes and may be induced by the presence of the receptor on the metastatic cell. NGF, and perhaps other neurotropic factors, may induce astrocytes to produce heparinase, allowing the metastatic cells to invade the brain.[105] Likewise, VEGF production is necessary, but not sufficient, for cancer cells to grow in the brain.[106] Various isoforms of VEGF either promote angiogenesis within the metastasis or dilate and make permeable normal cerebral vessels.[107] This co-option of normal vessels permits growth of a metastasis without angiogenesis.[35] Evidence that the brain microenvironment actually alters the nature of the tumor cells comes from experiments with melanoma: when an amelanotic melanoma is injected into the carotid artery, it forms a brain metastasis that is melanotic. If the melanotic metastasis is then removed and implanted subcutaneously, it again becomes amelanotic.[99] Although tumor cells may easily reach the brain by taking advantage of the brain's blood supply, their ability to grow and form a metastasis depends on many factors present in the tumor and the brain. Clinical implications of this are enormous, allowing one to consider therapeutic interventions that not only affect the cancer cell but also affect the brain's microenvironment, rendering it less attractive for tumor growth. Bone marrow–derived hematopoietic and endothelial progenitors that are activated by the primary tumor travel to other organs, where they promote growth and angiogenesis of metastases.[107a,107b]

Despite the fact that each metastasis has a clonal origin, heterogeneity rapidly develops as the metastatic tumor grows, owing to mutations of the rapidly growing and genetically unstable tumor cells. Thus, in a patient with multiple metastases, each metastatic tumor may have arisen from a different clone within the primary tumor and may become even more heterogeneous as it grows, causing differential sensitivity to chemotherapeutic agents.[108]

The host tissue may influence the responsivity of a metastasis to chemotherapeutic agents. In CNS metastases, there are at least three factors that determine sensitivity. The first is the intrinsic chemosensitivity of the clone that metastasized. The second is the blood–brain barrier, which often diminishes the amount of water-soluble chemotherapeutic agents that a metastasis may encounter. The third is the organ environment, which may induce relative resistance or sensitivity of cancer cells to chemotherapeutic agents.[75] For example, the host organ may also promote or suppress angiogenesis by metastatic tumor cells.[75]

Organ-specific or tissue-specific growth or repair factors that are stimulated by tissue damage (e.g., trauma or ischemia) may promote tumor growth, as may the normal organ-specific matrix, possibly explaining the observations that brain or spinal metastases are sometimes found at the site of prior CNS damage, considerably complicating the clinical diagnosis. Furthermore, organ-specific growth factors may promote the growth of individual tumor cell clones that possess those specific growth factor receptors. The abundance of growth factors in brain may be one reason for the frequency of metastasis to that organ.

CLINICAL CONSIDERATIONS RELEVANT TO NERVOUS SYSTEM METASTASIS

Tumor Size and Lymph Node Involvement

Although there are many exceptions, in general, the larger the primary tumor, the more likely it is to metastasize and cause death.[109–111] In patients with breast cancer, as the size of the tumor increases, the risk of metastasis increases, whether axillary nodes are positive or negative. Axillary metastases occur in fewer than 25% of women with breast tumors of 1 cm or less in diameter, but occur in 80% of those with tumors larger than 5 cm.[112] Only 27% of patients with primary breast tumors less than 2.5 cm in diameter develop distant metastases at a median time of 42 months, whereas 92% of patients with primary tumors larger than 8.5 cm in diameter develop such metastases at a median time of four months.[113] The situation is similar in other cancers. For example, in patients with malignant melanoma, only 4% of patients with tumors less than 1 mm in thickness develop distant metastases, compared with 72% of patients whose tumors are greater than 4 mm in thickness. If a patient with head and neck cancer has three lymph node metastases in the neck, the likelihood of distant metastases is 50%, as opposed to 7% in patients without lymph node metastases.[114] Furthermore, strong evidence indicates that local control of a tumor decreases the likelihood of distant metastasis if the local lymph nodes are negative, but local recurrence increases the likelihood, suggesting that the recurrent tumor cells that grow more rapidly and are genetically unstable develop the metastatic phenotype.[115] A recent study reports improved survival in patients with metastatic breast cancer who undergo complete excision of the primary tumor,[116] perhaps by decreasing the overall tumor burden. Similar results have been found in other primary tumors.[117] However, some very small primary tumors (at times too small to be detected by clinical or laboratory tests[118]) do metastasize, and some very large ones or even recurrent ones do not.[113]

Even more important than the size of the tumor is whether it has metastasized to local lymph nodes. Patients with small breast tumors (<2 cm) are likely to be node negative, but if there are four or more positive nodes (<10%), the 5-year survival falls from 96% (negative nodes) to 66%.[119] The presence of lymphatic vessel invasion within the primary tumor is predictive of node positivity.[120] In melanoma, although thickness of the primary tumor is important, it is less important than whether nodes are positive or negative and how many nodes are positive.[121] Treating positive lymph nodes may also improve survival.[122] Because local spread to lymph nodes usually occurs in a predictable manner, there has been an increasing emphasis on carefully studying the first lymph node to which the cancer is likely to drain (sentinel lymph node), particularly for breast cancer and melanoma.[123] Study of these lymph nodes by immunohistochemistry or polymerase chain reaction (PCR) appears to be substantially more sensitive than histologic evaluation of that lymph node or multiple lymph nodes.[124] False-negative sentinel node biopsies are uncommon and vary from study to study (0%–29%), but overall amount to only 7.3%.[125]

Properties of Metastases

The proliferative potential of a metastatic brain tumor, as measured by the 5-bromodeoxyuridine labeling index, is often greater than that of the primary tumor from which it arose,[75] perhaps accounting for a large brain metastasis from a small lung cancer. The effect of resection of the primary tumor on the growth of metastases and survival is controversial. Some studies suggest that even in the presence of metastases, resection of the primary tumor promotes survival, perhaps by lessening the overall tumor burden.[116,126] Alternatively, other studies suggest the contrary,[127,128] perhaps because surgery induces angiogenesis in the metastases.[128,129]

Differences between a Primary Tumor and Its Metastases

The genetic composition of a metastasis may be similar to or substantially different from that of the primary tumor,[20,29,130] although both have some common genetic signatures. Moreover, the genetic structure of a metastasis in one organ differs from that of a metastasis in another organ.[20] Furthermore, two different metastases to the same organ may have different cellular characteristics, such as the presence or absence of estrogen receptors in breast cancer or the immunophenotype of melanoma metastases.[135] The primary tumor of the patients in whom early metastases have occurred usually contains a large percentage of cells with metastatic potential, whereas if the metastasis occurs late in the course of the cancer, only a small percentage of primary tumor cells have metastatic potential.[24]

Because the metastatic cells may differ from the majority of cells in the primary tumor[132] from which they arose, their biologic behavior, including their response to therapy, may differ as well. Factors in the host environment may also determine the response of metastases to therapy.[75] Similar clones of tumor cells growing in different organs have different chemosensitivities. In one experiment, mouse colon cancer cells growing subcutaneously responded to doxorubicin treatment, whereas the same clone of cells in the lung did not.[75] These results imply that radiation or chemotherapy that is effective for the primary tumor may not be similarly effective for a brain metastasis, or vice versa.[133]

Site of Metastases

As indicated in the preceding text, four major factors determine the site to which a tumor will metastasize: (1) the anatomic distribution of the blood and lymph drainage from the organ harboring the primary tumor (mechanical or hemodynamic hypothesis),[97] (2) the receptivity of the organ receiving the metastasis (seed and soil hypothesis, or the molecular recognition hypothesis[98]), (3) the genetic nature of the tumor cell that reaches the organ, and (4) the genetic nature of the primary tumor itself. These factors are illustrated in Figure 2–5.[10]

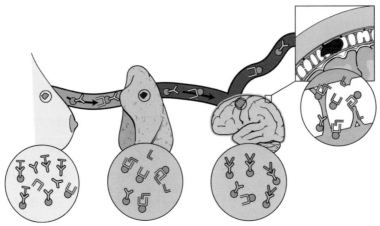

Figure 2–5. Adaptation of Nguyen and Massague's[10] model integrating four metastasis theories. In this illustration, tumor cells arising in the breast mostly express factor Y, with a rare cell expressing factor U. The microenvironment of the breast provides a factor T that promotes growth of the Y-expressing cells. Both the Y-expressing and the rarer U-expressing cells invade the venous circulation and reach the first capillary bed they encounter, in this case the lung. The lung provides a factor L that promotes growth of the U-expressing but not the T-expressing cells, so that only those cells expressing the rare factor U will grow, illustrating the seed and soil hypothesis (see text). If tumor cells reach the arterial circulation either from the lung metastasis or by bypassing the lung (see Fig. 2–2), they can seed the brain or leptomeninges. However, in this illustration, only those tumor cells expressing factor Y will grow in the brain because of the presence of a promoting factor V, whereas only those tumor cells expressing U will grow in the meninges because of the presence of factor L.

HEMODYNAMIC HYPOTHESIS: REVELANCE TO NERVOUS SYSTEM METASTASES

Some nervous system metastases[134] can be explained by the anatomy of regional venous and lymphatic drainage.[135,136] For example, the anatomy of the breast dictates that the first site of metastasis is often an axillary lymph node draining lymph channels from the breast with potential compression of the brachial plexus. Another common site is the thoracic spine, reached via venous channels of Batson plexus,[137–139] explaining the frequency of thoracic spinal cord compression from breast cancer and lumbar vertebral metastasis in prostate cancer[139] (Fig. 2–3). Cancer of the right side of the colon spreads via the portal circulation to the liver first, but cancer of the left side spreads via the systemic circulation to reach the lungs and subsequently, at times, the brain. Because primary tumors of organs other than lung are likely to metastasize to the lung first, most patients with brain metastases that arise from a nonpulmonary primary cancer have lung metastases as well. Thus, in general, a patient with breast or colon cancer and no lung metastasis is less likely to have a brain metastasis.

Anatomic factors, however, do not explain the predilection for breast cancers to select the brain as the first site of relapse,[140] or why cutaneous malignant melanoma and bladder cancer spread to the brain, whereas more than 80% of bone metastases with potential spinal cord compression are caused by breast,[141] lung, and prostate cancers, which together account for only 42% of all cancers.[142]

SEED AND SOIL HYPOTHESIS: RELEVANCE TO NERVOUS SYSTEM METASTSES

The predilection of certain tumors to spread to certain organs involves interactions between the tumor cell itself (seed) and the organ receiving the metastasis (soil).[98] Fidler[135,143] has restated the definition of the seed and soil hypothesis as encompassing three principles: (1) The primary neoplasm is biologically heterogeneous. (2) The process of metastasis is highly selective, favoring the survival and growth of a small subpopulation of cells that preexist in the heterogeneous parent neoplasm. (3) The outcome of metastasis depends on multiple interactions of metastatic cells (seed) with homeostatic mechanisms (soil). The host organ may synthesize and secrete factors such as hormones or growth factors that attract certain clones of circulating tumor cells and promote their growth. Clones of tumor cells themselves may have factors on their surface that react with the endothelium of particular organs and allow preferential growth in a particular organ environment. For example, B16 melanoma will spread to an implanted artificial organ that releases lung extract (the organ of preference for that tumor) but not liver extract.[99]

Subtypes

Illustrating the relevance of the seed and soil hypothesis to nervous system metastases is the fact that specific cancer subtypes tend to spread to different portions of the nervous system.[20,74,75,100,144] For example, infiltrating ductal carcinoma of the breast has a predilection to cause parenchymal brain metastasis, whereas infiltrating lobular carcinoma is more likely to affect the leptomeninges.[145] Metastatic non–Hodgkin lymphoma has a predilection to spread to the leptomeninges, but primary lymphomas of the CNS usually cause mass lesions within the brain. Breast cancer often metastasizes either to bone (vertebral bodies) or to the brain, but usually not to both. One breast cancer cell line produces intramedullary spinal metastases after intracardiac injection whereas another metastasizes to vertebral bodies, causing spinal cord compression.[146]

Within the brain itself, selectivity is evident. Delattre et al.[96] found that although most brain metastases are distributed roughly in proportion to cerebral blood flow, certain pelvic tumors tend to metastasize to the cerebellum rather than to supratentorial structures (Fig. 2–6). Most brain metastases are in white matter, but melanoma often chooses gray matter. Also, mouse melanoma cell lines have been developed that have specific predilection either for brain parenchyma or leptomeninges.[99]

Timing of Metastasis

The soil (brain) may also account for the time when metastases become apparent. Lung cancer often, but not always, metastasizes early, so that a significant percentage (~10%) of patients with lung cancer develop symptoms of brain metastasis before the pulmonary lesion is identified. Of 2682 patients in whom brain metastases occurred within 2 months of discovery of the primary tumor, 75% were from the lung;[147] 11% were from cancers where the primary was not identified, most probably also from the lung.[28]

Carcinomas of the breast and malignant melanoma are other tumors that commonly metastasize to the brain, but they rarely cause neurologic symptoms before the cancer has been identified (only 1 in 137 patients with brain metastases from breast cancer); they may not cause neurologic symptoms for many years after the primary cancer has been discovered and effectively treated and may be the only site of relapse.[140]

THE CASCADE PROCESS

Although experimental evidence indicates that tumor cells can pass through the pulmonary capillary bed to reach organs such as the brain,[148] clinical evidence suggests that this process occurs in only a minority of patients with CNS metastases. In animal experiments, only pulmonary metastases were encountered after foot-pad implantation of tumor, but widespread metastatic lesions developed after the lungs were bypassed by intra-arterial injection.[99] In most patients (~70% in one series[96]), the CNS appears to be reached by tumor cells that have entered the arterial circulation from the lung, either because the primary tumor originated in the lung or because the primary tumor has

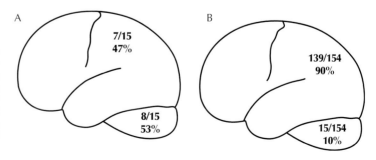

Figure 2–6. This schematic shows the distribution of single metastases to the brain. The posterior fossa is significantly overrepresented in patients with pelvic (prostate or uterus) tumors and gastrointestinal primary tumors (**A**), compared with patients with other primary tumors (**B**). (From Ref. 96 with permission.)

A

7/15
47%

8/15
53%

B

139/154
90%

15/154
10%

metastasized to the lung via the systemic venous or lymphatic circulation. Thus, when the primary tumor arises outside the lung and metastasizes to the nervous system, it usually does so via a secondary metastasis from a metastatic lung tumor, the so-called cascade process.[24,139,149]

The issue of whether the cascade hypothesis is valid is important clinically because, if secondary spread of a lung metastasis goes to the brain rather than a primary spread to both organs, more aggressive treatment of the lung metastasis (e.g., metastasectomy)[150] is warranted.

CNS metastases do occur, however, in the absence of obvious pulmonary metastases. Factors that might contribute to such metastases include the following.

Large paradoxical emboli from tumors sometimes occlude brain vessels: In one patient,[94] the middle cerebral artery was occluded by tumor cells from a testicular carcinoma; a second tumor embolus was seen at autopsy lying astride a patent foramen ovale. It is unknown how often paradoxic embolization causes brain metastasis.

Batson plexus: Batson plexus is a valvular system of veins lying in the epidural space. It drains vertebral bodies, the skull, and parts of the pelvis. Increases in intra-abdominal or intrathoracic pressure, such as coughing or straining, usually shift blood from the systemic venous system into this vertebral venous system. Batson[137] originally proposed that this system was a potential source of metastases to the CNS, including the brain. It is clearly a source of metastases to vertebral bodies and skull, according to the clinical studies of del Regato[138] and others[139,151] and the experimental studies of Coman and DeLong[152] and Harada et al.[153] In both studies, tumor cells were injected into the tail vein of rodents with or without abdominal compression. Those animals without abdominal compression developed liver and lung metastases, while those with abdominal compression developed vertebral metastases. The unanswered question is whether tumor cells traveling via Batson plexus can seed the parenchyma of the brain or spinal cord. Veins of Batson plexus do communicate with the cerebral venous sinus system and through that system with veins draining the brain, so that a potential communication with the brain exists. Such a system may account for the small number of

patients who have only a solitary brain metastasis in whom, even at autopsy, no metastases are found in the lungs. Delattre et al.,[96] however, found no patient with a brain metastasis who did not, after a careful search at autopsy, have metastatic disease in the lungs. Therefore, if this system plays any role in generating brain metastases, it must be a rare one; its importance in spinal metastasis is unequivocal.[138]

Passage through pulmonary vessels: Zeidman and Buss[148] demonstrated in experimental animals that tumor cells could traverse the capillary bed of the lungs without causing pulmonary metastases. The ability to do so was different for different tumors, and independent of the size of the tumor cell. They concluded that some tumor cells may be more pliable or may produce substances that dilate pulmonary vessels.

Small pulmonary metastases: Occasionally, a large metastatic brain tumor is encountered and only later is it found that the patient has a very small pulmonary metastasis. As previously mentioned, growth of metastatic brain tumors may exceed growth of the primary or of other metastatic foci.

Other factors relevant to growth of metastatic tumor: Once tumor cells reach the nervous system, what determines whether they will grow? The nervous system itself may play a role in metastatic spread. Recent evidence suggests that destruction of the sympathetic nervous system by chemical axotomy increases the number of metastases without affecting normal cell activity or the immune reaction to prior tumor exposure.[154] As indicated earlier, cell adhesion molecules probably play an important role, as does the expression of MHC class 1 antigens on the tumor cell surface, allowing an immune reaction that may protect against metastases. In some human tumors, human lymphocyte antigen expression is lost on metastatic but not nonmetastatic cells.[77] Organ damage is a factor. Systemic cancers have a predilection to metastasize to vertebral bodies that have been the site of previous injury or Paget disease.[155] This situation can cause considerable diagnostic confusion, both clinically because the patient may have had back pain for a long time with only recent worsening, and radiographically because the vertebral collapse could be due to either the old trauma or the more recent metastatic lesion. Similarly, in the CNS, tumor cells occasionally may seed an ischemic area created by cerebral vascular disease but spare normal brain.[156]

PATHOPHYSIOLOGIC FACTORS RELEVANT TO SPREAD AND GROWTH OF CANCER TO SPECIFIC NEURAL STRUCTURES

Brain

Most metastatic brain tumors arise by embolization of tumor cells to the brain that arrest in the white matter at the gray–white matter junction.[96] As already mentioned, brain metastases generally lodge in the terminal ends of the arterial supply, the so-called watershed areas[96,157] (Fig. 2–4A). Most tumors distribute themselves between the supratentorial and infratentorial compartments in proportion to the weight and blood supply of those structures, that is, 85% hemispheres, 15% posterior fossa (Fig. 2–4D). Certain tumors, however, particularly those arising in the pelvis (prostate or uterus) or gastrointestinal tract,[96,158] have a predilection to metastasize to the cerebellum. Thus, approximately 50% of single brain metastases from these organs are found in the cerebellum, whereas only 10% of single brain metastases from other organs involve the cerebellum (Fig. 2–6). The reason for the cerebellar predilection of some tumors is unknown. It does not apply to all pelvic or abdominal tumors; renal cell cancer, for instance, does not appear to select the cerebellum.[159] Experimental evidence is currently unavailable to support the Batson plexus hypothesis of how these tumors reach the brain, but evidence supports the Batson hypothesis for skull and intracranial dural metastases.

Because growth of the metastasis disrupts the blood–brain barrier, edema of the white matter, wet-appearing and somewhat grayish, surrounds all metastatic brain tumors, although its amount varies. Occasionally, small tumors are surrounded by massive amounts of edema whereas some large tumors by lesser amounts (see Fig. 3–4). Some investigators believe that adenocarcinomas, particularly from the lung, are associated with more brain edema than other metastatic tumors, but we have not been able to confirm such observations.

Aquaforin-4, normally expressed in astrocyte foot processes, is important for transporting water across the blood–brain barrier. It is reported to be upregulated in some high-grade gliomas and also in reactive astrocytes that surround some metastatic adenocarcinomas. Upregulated aquaforin-4 expression correlated with blood–brain barrier disruption. Upregulation of this protein may explain why some metastatic tumors are surrounded by a great deal of edema and others by little edema.[160]

Most metastases in the brain grow as smooth, spherical masses, displacing rather than destroying brain tissue and creating edema in the surrounding white matter. Other intracranial metastases (e.g., dural metastases) are more irregular (see Chapter 5). Brain metastases have a granular, fleshy appearance and are soft to the touch. They are usually well demarcated from surrounding brain, both grossly and microscopically, and thus often can be completely excised surgically, although recurrence is common (40%) unless patients are radiated postoperatively.[161] Certain cancers, particularly poorly differentiated lymphomas, renal carcinoma, and melanoma, may invade surrounding brain tissue microscopically, although they appear grossly to be well demarcated. Brain metastases are usually solid but, because they grow rapidly, they often undergo central necrosis. Cystic lesions occasionally occur, particularly from primary breast cancer. Metastatic brain tumors are usually highly vascular. Metastases from melanomas, choriocarcinomas, and thyroid and testicular carcinomas may be hemorrhagic,[162–164] because these tumors either invade the walls of blood vessels[165,166] or promote extensive neovascularization.[167] Metastases from melanomas and lymphomas often invade gray matter. Occasionally, widespread miliary metastases invade the cerebral cortex as small or microscopic foci of tumor, which may or may not be identifiable on CT or MRI, giving rise to the clinical syndrome of carcinomatous encephalitis.[168–170]

The microscopic appearance of a brain metastasis resembles the primary tumor from which it arises, even though it may differ genetically. Therefore, in patients with an unknown primary cancer, extirpation of the cerebral metastasis with microscopic examination often provides a significant clue as to the original site of the primary tumor, although a metastatic brain tumor may have a different appearance (and growth rate) from the primary; even when the routine histology is uninformative, immunohistochemical analysis may point to the primary.[171] In mixed epidermoid and adenocarcinomas of the lung, it is the adenocarcinoma component that is more likely to metastasize to the brain; in mixed testicular tumors, choriocarcinoma elements are more likely to metastasize than the other portions.[172] Primary tumors vary in their propensity to metastasize to the brain. Even different tumors arising from the same primary site differ in that propensity. For example, in breast cancer, infiltrating ductal carcinoma is more likely to metastasize to the brain than other histologies.[173] Although it is likely that overexpression of HER-2 selects for brain metastases, treatment with trastuzumab, by suppressing systemic metastases, also increases the likelihood of brain metastases, often as the only site of relapse.[174] As indicated, adenocarcinoma of the lung is more likely to metastasize than squamous carcinoma, though less likely than small-cell lung cancer. Even among adenocarcinomas, specific genetic patterns increase the likelihood of metastases.[175]

Spinal Cord

Unlike the brain, in which most metastases are parenchymal, metastases affecting the spinal cord usually act by compressing the cord from the epidural space. Metastatic tumor reaches the epidural space in one of three ways (Fig. 2–7): (1) extension from a vertebral metastasis, (2) invasion through the intervertebral foramina by a paravertebral mass, or (3) direct hematogenous spread to the epidural space.

VERTEBRAL METASTASES

Metastases to bone are much more common than might be predicted from bone's low overall blood flow (11 mg/kg per minute) or from the fact that skeletal blood flow accounts for only 4% to 10% of cardiac output.[142,176] Bone metastases occur in 65% to 75% of patients with advanced breast and prostate cancer and in 30% to 40% of patients with advanced lung cancer.[177] The most common site of bone metastasis is the vertebral column. In many patients, it is the pain caused by a vertebral metastasis that first leads to the diagnosis of cancer. The high incidence of bone metastases results from the following: (1) The large capacity of the bone marrow's capillary bed, six to eight times that of the arterial system, allows

the circulation to come to a virtual standstill. (2) The walls of sinusoids have discontinuities that allow tumor cells to escape blood vessels. (3) The intersinusoidal cords of hematopoietic tissue form culs-de-sac in which tumor cells can lodge and proliferate. (4) Production of prostaglandin E_2, osteoclast-activating factor(s), parathormone-like substances, and TGFβ by the tumor serves to stimulate osteoclastic activity. (5) Products of bone resorption stimulate tumor growth and attract monocytes that produce interleukin-1, which promotes bone resorption.[178,179]

The usual mechanism for spinal cord compression is hematogenous metastasis to the vertebral body or, less commonly, the vertebral lamina, pedicle, or spinous process. On the basis of the hypothesis that the vertebral bodies would be affected in different locations if the tumor reached them via Batson plexus rather than by arterial spread, Yuh et al.[180] determined the pattern of metastases in 433 lesions identified in 218 vertebral bodies. The distribution of lesions did not support either hypothesis. A histologic study of metastatic vertebral tumors suggests that the vertebral column is invaded in the bone marrow of the dorsal region of the vertebral body.[181] The ligaments surrounding the vertebral bodies serve as barriers to tumor progression, although the posterior longitudinal ligament is the weakest and is sometimes destroyed by the tumor at the point of perforating vessels.[181] Posterior vertebral elements are involved approximately 15% as often as the vertebral body. Metastatic clones appear to have a predilection to invade bone already damaged by trauma or Paget disease,[155] complicating the diagnosis. Tumor invasion of the vertebral body often leads to collapse followed by progressive kyphosis. The tumor, collapsed bone, and occasionally intervertebral disc herniate posteriorly into the spinal canal to compress the cord. The posterior longitudinal ligament, which separates the vertebral body from the epidural space, may be buckled backward or actually invaded and breached by the tumor to promote direct entry of tumor into the anterior or anterolateral portion of the epidural space (Fig. 2–7). If the posterior elements are also destroyed either by tumor or by prior laminectomy (see Chapter 6), a forward-flexion deformity results. Less commonly, but especially in children, the pedicles, lamina, or posterior spinous process

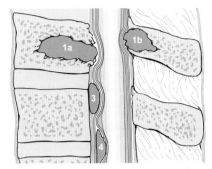

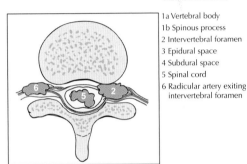

METASTATIC SITES

1a Vertebral body
1b Spinous process
2 Intervertebral foramen
3 Epidural space
4 Subdural space
5 Spinal cord
6 Radicular artery exiting
   intervertebral foramen

Figure 2–7. Most spinal cord dysfunction is caused by a metastasis to a vertebral body that invades the epidural space and compresses the anterior portion of the spinal cord. (1a) The vertebral body may also collapse, causing bone to herniate into the epidural space. Less common are metastases to the spinous process (1b) or vertebral lamina that compress the posterior or lateral portion of the spinal cord. Somewhat less common than vertebral metastases are those that compress the lateral portion of the cord by growing from the paravertebral space through the intervertebral foramen (2). The vertebral body may or may not be involved. Even less common are hematogenous metastases directly to the epidural space (3), subdural metastases (4), or parenchymal metastases (5). Metastatic lesions may rarely cause symptoms by compressing a radicular artery (6) or cause radicular symptoms, but not spinal cord symptoms, by invading or compressing the nerve root as it exits the intervertebral foramen.

may compress the posterior or posterolateral portion of the cord.

Vertebral invasion is the mechanism of epidural spinal compression in 85% to 90% of patients; as a result, most patients with spinal cord compression have identifiable abnormalities of the vertebral body at the compression site by plain X-ray, radionuclide bone scan, or CT and especially by MRI[182,183] (see Chapter 6). In most patients with acute spinal cord compression due to metastases both anterior and posterior elements are involved.[182]

The tumor may reach the vertebral body or the epidural space hematogenously, either arterially or by venous dissemination via Batson plexus. Other modes include spread along the lymphatics and direct extension from retroperitoneal or posterior mediastinal tumor.

Individual tumors drain differently into Batson plexus and thus reach different portions of the vertebral column. The breast drains principally by the azygos vein into the thoracic level of the spine. Cervical cancers metastasize to lumbar spine and sacrum, usually to the left side, probably representing direct drainage from tumor-bearing lymph nodes. The prostate drains through the pelvic venous plexus, reaching Batson plexus in the pelvis and lower spine. Some tumors, by contrast, drain principally via the pulmonary vein into the left heart and thus, reach the vertebral body via the arterial circulation. In humans, particularly those with carcinomas of the breast and prostate, bony metastases are often distributed in a pattern outlining Batson plexus, that is, they involve the axial skeleton and spare the appendicular skeleton (Fig. 2–3). Such patients are likely to have multiple vertebral metastases and often skull metastases without involvement of long bones.

The anatomy of Batson plexus is not the complete explanation for the very high incidence of vertebral body and epidural metastases. The sinusoidal pattern of the vasculature of vertebral bone marrow makes that structure particularly vulnerable to metastases. Furthermore, certain tumors have a predilection to invade and grow in bone, probably because they produce osteoclastic substances, including osteoclast-activating factor and prostaglandins that allow them to grow more easily within vertebral structures. Bone marrow stromal cells and hematopoietic growth factors in vertebral bodies also promote growth of the metastatic cancer. Conversely, many tumors promote bone growth and cause blastic rather than lytic lesions. Arguello et al.[184] constructed an experimental model of arterial spread to the epidural space. This model, produced by injection of tumor into the heart, led to invasion of the vertebral bodies near the cartilaginous plate at the site where end arterioles terminate as tortuous loops. The growing tumor then compressed and thrombosed vertebral veins and entered the spinal canal where the thrombosed veins traversed the intervertebral foramina. In the spinal canal, the tumor either can grow as an anterior mass compressing the spinal cord or can work its way around the cord, causing a more posterior mass. In this experimental model, the site of the mass depended on the nature of the tumor.

FORAMINAL INVASION

Less common than vertebral metastasis is invasion (10%–15%) of the epidural space through the intervertebral foramen by paravertebral tumor. Lymphomas and neuroblastomas are particular offenders. The tumor usually remains in the extradural space and compresses the lateral portion of the spinal cord. The clinical implication is that spinal cord compression can occur in the absence of involvement of the vertebral bodies, so that the results of X-rays and nuclide scans of the vertebral bodies may be normal. An MRI (see Chapter 6) usually identifies both the paravertebral mass and the intraspinal extension.

EPIDURAL METASTASES

Direct hematogenous spread of tumor to the epidural space is uncommon. Most such exclusively epidural tumors are leukemias or lymphomas. Approximately 5% of spinal lymphomas originate in the epidural space rather than extend there from adjacent bone or paravertebral lesions.[185] The same phenomenon occasionally is encountered in leukemia,[186] and isolated cases of primary epidural Ewing sarcoma[187,188] and osteosarcoma[189] have been reported. In this situation, an MRI may fail to distinguish epidural tumor from other epidural masses such as a herniated disk[190] or extramedullary hematopoiesis.[191]

CONSEQUENCES OF SPINAL CORD COMPRESSION

Whatever the mode of spread, once the tumor reaches the epidural space, it not only compresses the cord at its entry site (anterior for vertebral metastases, lateral for those that enter through the intervertebral foramen) but also may take the path of least resistance to spread in a circumferential manner around the cord and compress it from all sides (Fig. 6–2C). The physiological and pathological changes in the spinal cord that cause the clinical symptomatology of spinal cord compression are discussed in Chapter 6.

Leptomeninges

Malignant cells can reach the CSF by several routes[192–195] (Fig. 2–8) (see also Table 7–4):

Direct extension: Tumor metastases to brain or spinal cord that abut either the subarachnoid or ventricular surface can directly seed cells into the CSF.[196,197] Cerebellar lesions are particularly likely to cause meningeal metastases because of the large subarachnoid surface of the cisterna magna. Approximately one-third of our patients whose cerebellar metastases were surgically resected subsequently developed leptomeningeal metastases.[198] In another series, 6 of 9 patients with cerebellar metastases and 3 of 18 patients with cerebral metastases developed leptomeningeal metastases 2 to 3 months after surgery.[199] Direct spread from brain or spinal cord in the absence of surgery is not a frequent mode of CSF entry for two reasons. First, metastatic tumor of the brain tends to originate in white matter (at the gray–white matter junction); the lesions grow largely below the subarachnoid surface but do not reach the ventricular surface. Second, even when tumor grows through the cortex to the subarachnoid surface, a reactive response develops in the leptomeninges that often seals off that metastasis, leading to a focal leptomeningeal deposit but not the diffuse or widespread multifocal leptomeningeal process encountered with tumor disseminated throughout the CSF.[200] Because tumors rarely metastasize to the spinal cord parenchyma (see Chapter 6), this is a rare source of leptomeningeal tumor. Primary CNS lymphomas in the brain usually grow periventricularly and thus, more frequently seed the ventricles and leptomeninges.[201]

Hematogenous: Tumor spread via the arterial circulation to the choroid plexus and from the choroid plexus into the cerebral ventricles is also uncommon. When leptomeningeal tumors develop from a choroid plexus metastasis, the cells are carried by CSF flow into the subarachnoid space.[194] Because the ventricles have a smooth ependymal surface without areas of stasis, tumor cells generally do not adhere to the ventricular wall (although exceptions have been observed), but instead find their way into the subarachnoid space, in which regions of stasis permit growth. Choroid plexus metastases are, in our experience, a rare source of leptomeningeal tumor.

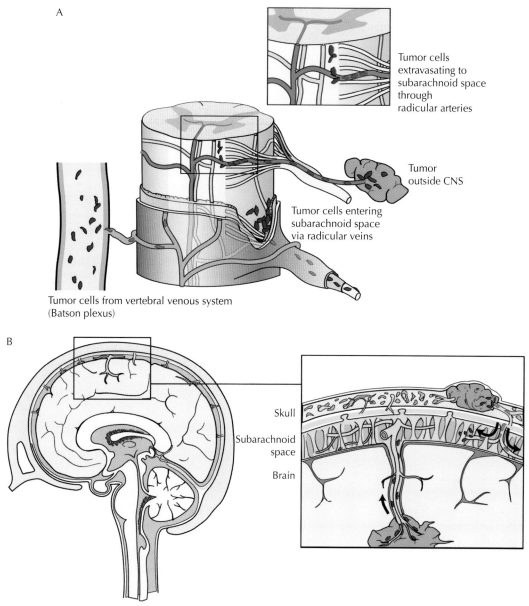

Figure 2–8. Pathophysiology of leptomeningeal metastases. **A:** This schematic of the spinal canal depicts the mechanisms of tumor cell entry into the spinal subarachnoid space. The tumor may invade the vertebral body and grow along vertebral veins into the subarachnoid space. The tumor may invade peripheral nerves or nerve roots outside the vertebral canal and grow along the nerve sheath into the spinal canal to seed the leptomeninges. The tumor can invade blood vessels outside the CNS and traverse subarachnoid veins into the subarachnoid space. **B:** This schematic of the skull illustrates the possible mechanisms of tumor entry into the cerebral subarachnoid space. Tumor may enter the cranial subarachnoid space via metastases either to the skull or the brain, diploic veins of the skull, or directly from subarachnoid veins. Tumor in the leptomeninges can grow down the perivascular space (Virchow–Robin) and exit that space to invade the brain. The choroid plexus is also an occasional site for the formation of leptomeningeal tumor. CNS, central nervous system.

Venous access: Tumor cells may metastasize directly to the subarachnoid space, entering through leptomeningeal veins.[202] This probably represents a common site of entry of leukemia cells.

Venous drainage from bone marrow: Tumor cells involving the bone marrow of vertebrae or skull may grow along veins exiting the marrow to reach the dura and then along perivenous adventitial tissue, connecting the dura mater with the subarachnoid space. A high incidence of bone marrow infiltration in patients who develop leptomeningeal metastases, and a high incidence of leptomeningeal metastases in those tumors that either arise in or commonly affect the marrow, indicate that this is a common source of spread to the subarachnoid space.[203] A study contrasting the different histology of breast cancers metastatic to brain with those metastasizing to the leptomeninges also supports this mode of spread. Parenchymal cerebral metastases were associated almost exclusively with infiltrating ductal carcinoma and meningeal infiltration with infiltrating lobular carcinoma, the histologic variety that tends to metastasize to bone marrow.[145] However, prostate cancer that spreads to vertebral bodies via Batson plexus[65,139] is a rare cause of leptomeningeal metastases, indicating that the molecular nature of the tumor is as important as anatomy in determining whether a given tumor will spread to the leptomeninges.[99]

Cranial and peripheral nerves: Tumors may grow along cranial or peripheral nerves (Fig. 2–8), usually along the surface of the epineurium, sometimes along the perineurium or even within the nerve fascicle or along lymphatic and venous channels that travel with the nerve, to enter the subarachnoid space with those nerves and then spread cells into the CSF.[193,194,204,205] The high frequency of paravertebral metastases in patients with leptomeningeal metastases, as well as pathologic evidence of growth along spinal and cranial nerves that lead to the subarachnoid space, supports this mode of spread as a common one.[194] Breast, lung, and especially skin cancers of the face are common offenders.

Once the tumor has reached the CSF, it can seed any portion of the nervous system that has contact with the CSF. The seeding, usually widespread with a particular predilection for the base of the brain and the cauda equina, includes the depths of the sulci (rather than the surface of the gyri); the lateral surface of the hemispheres; and the quadrigeminal, ambient, and peduncular cisterns surrounding the brainstem.[206] Tumor tissue is often found in the lateral recesses of the fourth ventricle and in the Sylvian fissure between the frontal and temporal lobes. In the cauda equina, it tends to be concentrated over the dorsal portions of that structure rather than the ventral surface. The dorsal surface of the brain usually is seeded more sparsely than the ventral surface, and the ependyma usually is spared. At times, tumor cells are found in the arachnoid villi. Because the villi are the site of CSF resorption, plugging of the arachnoid villi is one potential cause of hydrocephalus and/or increased intracranial pressure.

In many instances, even with extensive seeding of the leptomeninges, no abnormalities are apparent to the naked eye. Sometimes, the arachnoid membrane has a slight milky appearance, which may indicate to the pathologist that leptomeningeal metastases are present, particularly in younger patients whose arachnoid membranes usually are transparent. At times, the tumor forms macroscopic masses, particularly along nerve roots of the cauda equina, making it evident in life to the radiologist on a gadolinium-enhanced MRI.

Microscopically, the metastatic tumor grows in sheets along the surface of the brain, spinal cord, cranial nerves, and spinal roots. Several different patterns of meningeal reaction are seen commonly but do not appear to characterize individual tumors:

1. Dense infiltration of the leptomeninges by tumor entirely obliterates normal meningeal structures but with little inflammatory or fibrotic reaction.
2. More sparse infiltration with tumor cells is organized in a linear manner, following the outline of the pia-arachnoid and interconnecting trabeculae; the subarachnoid space may be relatively preserved, with little evidence of reactive fibrosis.
3. Focally dense reactions of the leptomeninges result when a few tumor cells cause meningeal fibrosis with a moderate number of reactive meningothelial cells.
4. A thick exudative reaction is caused by a large number of tumor cells accompanied

by a marked inflammatory reaction; many inflammatory and cancer cells infiltrate the leptomeninges and seed the CSF, producing an inflammatory picture on lumbar puncture and giving rise to the commonly used term *carcinomatous meningitis*.

Leptomeningeal metastases may remain restricted to the subarachnoid space or may infiltrate the parenchyma of the brain and spinal cord by growing down the Virchow–Robin spaces. These potential spaces are a tunnel-like extension of the subarachnoid space formed by blood vessels that penetrate from the surface of the brain into its depths. Virchow–Robin spaces, strictly speaking, lie outside the brain, separated from it by pia and astrocytic foot processes. As the tumor grows in this space, however, it may penetrate the space to invade brain or spinal cord parenchyma itself. This tendency for tumor to settle deep in the sulci of brains and to invade the Virchow–Robin spaces has led some investigators to believe that intrathecally injected drugs would not reach all tumor and that more penetrating treatments, such as radiation therapy, are required for adequate prophylaxis or treatment.

Leptomeningeal metastases also invade cranial and peripheral nerves as they pass through the subarachnoid space, a more common invasion than those to the brain or spinal cord. The first pathologic change appears to be interruption of the myelin sheath with relative preservation of axis cylinders, but if the tumor continues to grow, the nerve is disrupted entirely.

Cranial and Peripheral Nerves

Metastatic cancers affect peripheral nerves by either compression or direct invasion (Fig. 2–9). Nerves can be compressed by the growing primary tumor or by metastases to bone or lymph nodes. Primary tumors that result in compression include Pancoast tumors that compress the brachial plexus, nasopharyngeal tumors that compress cranial nerves exiting the base of the skull, and pelvic tumors, including colonic and cervical carcinomas that compress the sacral plexus with direct posterior growth. In addition, prostate and breast cancers often metastasize to the bones of the skull base, compressing cranial nerves; tumors metastasizing to pelvic bones can compress the sciatic nerve; lymph node metastases compress

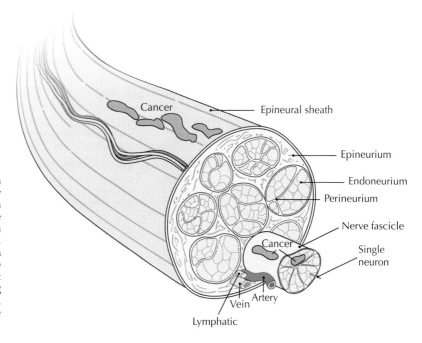

Figure 2–9. Cross-section of a cranial nerve. The only lymphatics associated with the peripheral nerve are found in the epineurium and perineural sheath. Neurotropic carcinoma may spread along these lymphatic channels but more often spreads along the epineural sheath or, less frequently, a nerve fascicle.

Cancer
Epineural sheath
Epineurium
Endoneurium
Perineurium
Nerve fascicle
Cancer
Single neuron
Vein
Artery
Lymphatic

the brachial plexus in some patients with metastatic breast cancer and the lumbar or sacral plexus in patients with metastatic spread of colon, prostate, or bladder cancer. Rarely, peripheral nerves, such as the ulnar nerve, may be compressed by enlargement of nearby lymph nodes because of metastatic tumor.

An infrequent cause of peripheral nerve dysfunction is direct invasion of the nerve by the tumor either from hematogenous spread of tumor to peripheral nerve or dorsal root ganglion[207–209] or, more commonly, by direct extension to the nerve from a surrounding structure. An example of the latter is nerve invasion from nerve root involvement in the subarachnoid space in a patient with leptomeningeal metastases (see Chapter 7). The epineurium and perineurium generally form an effective barrier against tumor penetration. At times, metastases entirely encompass the nerve and sometimes grow down its perineural septa, causing severe compressive damage but not actually invading the nerve.

Whether neurologic dysfunction is caused by compression or invasion of the nerve is clinically important. Nerve compression lesions tend to be more painful but are associated with less neurologic dysfunction than nerve invasion. Compressive lesions of the nerve usually cause radicular pain accompanied by severe local aching. Invasive lesions are more likely to cause dysesthetic, tingling, or burning pain, or, occasionally, even neuralgia-like pains (sharp, shooting, short-lived, lightning-like pains resembling trigeminal neuralgia or the lightning pains of tabes dorsalis). Invasive lesions of mixed sensorimotor nerves often cause selective motor weakness, sometimes without pain and with relative sparing of sensory fibers. A similar situation is seen in the vascular disorder of nerves that causes diabetic amyotrophy.[210]

Because compressive lesions of the nerve come either from tumor in bone or from enlarging lymph nodes or organs, they are more easily detected by imaging techniques than are invasive lesions, which are usually microscopic.

Although the mechanism by which tumor compression causes nerve dysfunction is not entirely known, probably both direct pressure and ischemia participate. Compressive lesions may lead to ischemia, inflammation, or physical distortion that in turn may cause demyelination or axonal loss. Invasive lesions

often affect all compartments of the nerve, particularly the endoneurium, to cause demyelination, axonal damage, or both.[211] Vascular lesions such as intravascular lymphomatosis may infarct the nerve (see Chapter 9).[212] The first pathologic change in compressive lesions appears to be demyelination with relative preservation of the axon. Later, axon loss occurs as well, suggesting ischemia of the nerve. How invasive lesions of the nerve cause their symptoms is not known entirely. At some point in the invasion of a nerve by tumor, the entire nerve is destroyed, but before that, changes may be produced by interference with vascular supply or by competition for essential metabolites. Early destruction of myelin sheaths often appears with relative preservation of axons, similar to the effects of compressive lesions.

The CNS and peripheral nervous systems differ from each other in several respects that influence their response to treatment of metastases. Peripheral-nerve myelin sheaths are formed and nourished by Schwann cells rather than by oligodendroglial cells as in the CNS. Peripheral nerves have a better regenerative capacity and thus, treatment is likely to be more effective if the primary process can be relieved. If the peripheral-nerve lesion arises near the cell body, however, there may be enough dying back to destroy the cell body, thus eliminating the possibility of regeneration.

The CNS and peripheral nervous systems are similar in two important and relevant aspects: (1) A blood–nerve barrier similar to the blood–brain barrier (see Chapter 3) serves to exclude water-soluble chemotherapeutic agents, providing a sanctuary for cells.[213] (2) The peripheral nerve is encased in a relatively unyielding perineural sheath and like the CNS, is susceptible to edema, which may, in and of itself, produce neurologic disability. These physiologic aspects of the peripheral nervous system have not been studied as carefully as they have been in the CNS, however; thus, their exact relevance to the pathophysiology of neural dysfunction and its treatment has not been established.

Once a tumor invades a peripheral nerve, it can grow along the nerve, sometimes for a considerable distance, actually entering the CNS via that route. The opposite situation, in which leptomeningeal tumor invades nerve roots and follows them outside the CNS, has been shown in experimental animals,[214] and probably occurs in humans as well.

The pathogenesis of the neurotropism and perineural invasion of some cancers is uncertain. Some tumors express adhesion molecules that may allow them to react with nerves.[215–217] Perineural invasion in bile duct cancers appears to be associated with expression of glial-derived neurotrophic factor (GDNF)[218]; the co-expression of GDNF in tumor and its receptor in peripheral nerve may attract tumor cells to the nerve. The tumors appear to infiltrate the perineural space, but not blood vessels or lymph channels associated with the nerves. One proposal suggests that less physical resistance in the perineural space facilitates tumor growth in this compartment.[219]

Whatever the mechanism(s), the clinical findings of nerve damage may precede other evidence that a tumor is present: Tumors of the lung apex frequently involve the brachial plexus, causing pain and sometimes other neurologic signs long before the initial tumor is discovered.[220] Breast tumors metastatic to axillary lymph nodes can also affect the brachial plexus as a presenting complaint. Abdominal and pelvic tumors may involve the lumbar or sacral plexus; pain and sacral plexus dysfunction (e.g., incontinence) may be the presenting complaint or herald recurrence of a previously treated prostate or colon cancer. We have seen a few patients with cranial neuropathies, usually trigeminal but occasionally facial or even oculomotor, in whom only a scar on the face from a forgotten resection of a benign lesion years ago gave a clue to the diagnosis (see Chapter 8).[221] Imaging often establishes the diagnosis.[221,222]

REFERENCES

1. Kohn EC. Invasion and metastases. In: Holland JF, Frei EI, eds. *Cancer Medicine*. 7th ed. London: BC Decker, 2006. pp. 148–156.
2. Clouston PD, DeAngelis LM, Posner JB. The spectrum of neurologic disease in patients with systemic cancer. *Ann Neurol* 1992;31:268–273.
3. Rees J. Neurological manifestations of malignant disease. *Hosp Med* 2000;61:319–325.
4. Arcasoy SM, Jett JR. Superior pulmonary sulcus tumors and Pancoast syndrome. *N Engl J Med* 1997;337:1370–1377
5. Fowler BZ, Crocker IR, Johnstone PA. Perineural spread of cutaneous malignancy to the brain: a review of the literature and five patients treated with stereotactic radiotherapy. *Cancer* 2005;103:2143–2153.
6. Ginsberg LE. MR imaging of perineural tumor spread. *Magn Reson Imaging Clin N Am* 2002;10:511–525.
7. Gupta GP, Massague J. Cancer metastasis: building a framework. *Cell* 2006;127:679–695.
8. Fidler IJ. Critical determinants of metastasis. *Semin Cancer Biol* 2002;12:89–96.
9. Fidler IJ, Yano S, Zhang RD, et al. The seed and soil hypothesis: vascularisation and brain metastases. *Lancet Oncol* 2002;3:53–57.
10. Nguyen DX, Massague J. Genetic determinants of cancer metastasis. *Nat Rev Genet* 2007;8:341–352.
11. Li F, Tiede B, Massague J, Kang Y. Beyond tumorigenesis: cancer stem cells in metastasis. *Cell Res* 2007;17:3–14.
12. Houghton J, Morozov A, Smirnova I, et al. Stem cells and cancer. *Semin Cancer Biol* 2006;17:191–203.
13. Polyak K, Hahn WC. Roots and stems: stem cells in cancer. *Nat Med* 2006;12:296–300.
14. Scadden DT. The stem-cell niche as an entity of action. *Nature* 2006;441:1075–1079.
15. Macaluso M, Paggi MG, Giordano A. Genetic and epigenetic alterations as hallmarks of the intricate road to cancer. *Oncogene* 2003;22:6472–6478.
16. Herman JG, Baylin SB. Gene silencing in cancer in association with promoter hypermethylation. *N Engl J Med* 2003;349:2042–2054.
17. Jefford CE, Irminger-Finger I. Mechanisms of chromosome instability in cancers. *Crit Rev Oncol Hematol* 2006;59:1–14.
18. Bignold LP. Chaotic genes make chaotic cells: the mutator phenotype theory of carcinogenesis applied to clinicopathological relationships of solid tumors. *Cancer Invest* 2004;22:338–342.
19. Nathoo N, Chahlavi A, Barnett GH, et al. Pathobiology of brain metastases. *J Clin Pathol* 2005; 58:237–242.
20. Minn AJ, Kang Y, Serganova I, et al. Distinct organ-specific metastatic potential of individual breast cancer cells and primary tumors. *J Clin Invest* 2005;115:44–55.
21. Marchetti D, Denkins Y, Reiland J, et al. Brain-metastatic melanoma: a neurotrophic perspective. *Pathol Oncol Res* 2003;9:147–158.
22. Fidler IJ, Kripke ML. Metastasis results from preexisting variant cells within a malignant tumor. *Science* 1977;197:893–895.
23. Ramaswamy S, Ross KN, Lander ES, et al. A molecular signature of metastasis in primary solid tumors. *Nat Genet* 2003;33:49–54.
24. Pantel K, Brakenhoff RH. Dissecting the metastatic cascade. *Nat Rev Cancer* 2004;4:448–456.
25. Bernards R, Weinberg RA. A progression puzzle. *Nature* 2002;418:823.
26. Michaelson JS, Silverstein M, Sgroi D, et al. The effect of tumor size and lymph node status on breast carcinoma lethality. *Cancer* 2003;98:2133–2143.
27. Engel J, Eckel R, Kerr J, et al. The process of metastasisation for breast cancer. *Eur J Cancer* 2003;39:1794–1806.

28. Van de Pol M, Van Aalst VC, Wilmink JT, et al. Brain metastases from an unknown primary tumour: which diagnostic procedures are indicated? *J Neurol Neurosurg Psychiatry* 1996;61:321–323.

29. Kikuchi T, Daigo Y, Ishikawa N, et al. Expression profiles of metastatic brain tumor from lung adenocarcinomas on cDNA microarray. *Int J Oncol* 2006;28:799–805.

30. Albiges L, Andre F, Balleyguier C, et al. Spectrum of breast cancer metastasis in BRCA1 mutation carriers: highly increased incidence of brain metastases. *Ann Oncol* 2005;16:1846–1847.

31. Kang Y, He W, Tulley S, et al. Breast cancer bone metastasis mediated by the Smad tumor suppressor pathway. *Proc Natl Acad Sci USA* 2005;102:13909–13914.

32. Kang Y, Siegel PM, Shu W, et al. A multigenic program mediating breast cancer metastasis to bone. *Cancer Cell* 2003;3:537–549.

33. Ma L, Teruya-Feldstein J, Weinberg RA. Tumour invasion and metastasis initiated by microRNA-10b in breast cancer. *Nature* 2007;449:682–688.

34. Holash J, Maisonpierre PC, Compton D, et al. Vessel cooption, regression, and growth in tumors mediated by angiopoietins and VEGF. *Science* 1999;284:1994–1998.

35. Leenders W, Kusters B, Pikkemaat J, et al. Vascular endothelial growth factor-A determines detectability of experimental melanoma brain metastasis in GD-DTPA-enhanced MRI. *Int J Cancer* 2003;105:437–443.

36. Folkman J. How is blood vessel growth regulated in normal and neoplastic tissue? G.H.A. Clower Memorial Award Lecture. *Cancer Res* 1986;46:467–473.

37. Bergers G, Benjamin LE. Tumorigenesis and the angiogenic switch. *Nat Rev Cancer* 2003;3:401–410.

38. Laferriere J, Houle F, Huot J. Regulation of the metastatic process by E-selectin and stress-activated protein kinase-2/p38. *Ann N Y Acad Sci* 2002;973:562–572.

39. Sugino T, Kusakabe T, Hoshi N, et al. An invasion-independent pathway of blood-borne metastasis—a new murine mammary tumor model. *Am J Pathol* 2002;160:1973–1980.

40. Deryugina EI, Quigley JP. Matrix metalloproteinases and tumor metastasis. *Cancer Metastasis Rev* 2006;25:9–34.

41. Condeelis J, Pollard JW. Macrophages: obligate partners for tumor cell migration, invasion, and metastasis. *Cell* 2006;124:263–266.

42. Carmeliet P. Angiogenesis in life, disease and medicine. *Nature* 2005;438:932–936.

43. Fidler IJ, Singh RK, Yoneda J, et al. Critical determinants of neoplastic angiogenesis. *Cancer J Sci Am* 2000;6 (Suppl 3):S225–S236.

44. Engelberg H. Actions of heparin that may affect the malignant process. *Cancer* 1999;85:257–272.

45. Fidler IJ. Angiogenesis and cancer metastasis. *Cancer J Sci Am* 2000;6 (Suppl 2):S134–S141.

46. Cardones AR, Banez LL. VEGF inhibitors in cancer therapy. *Curr Pharm Des* 2006;12:387–394.

47. Purow B, Fine HA. Progress report on the potential of angiogenesis inhibitors for neuro-oncology. *Cancer Invest* 2004;22:577–587.

48. Achen MG, Mann GB, Stacker SA. Targeting lymphangiogenesis to prevent tumour metastasis. *Br J Cancer* 2006;94:1355–1360.

49. Stacker SA, Caesar C, Baldwin ME, et al. VEGF-D promotes the metastatic spread of tumor cells via the lymphatics. *Nat Med* 2001;7:186–191.

50. Skobe M, Hawighorst T, Jackson DG, et al. Induction of tumor lymphangiogenesis by VEGF-C promotes breast cancer metastasis. *Nat Med* 2001;7:192–198.

51. Makinen T, Jussila L, Veikkola T, et al. Inhibition of lymphangiogenesis with resulting lymphedema in transgenic mice expressing soluble VEGF receptor-3. *Nat Med* 2001;7:199–205.

52. Liotta LA, Kohn EC. Invasion and Metastases. In: Bast RC, Kufe DW, Pollock RE, Weichselbaum RR, Holland JC, Frei EI, eds. *Cancer Medicine.* Hamilton, BC: Decker, 2000. pp. 121–131.

53. Birchmeier W. Cell adhesion and signal transduction in cancer. Conference on cadherins, catenins and cancer. *EMBO Rep* 2005;6:413–417.

54. Guo W, Giancotti FG. Integrin signalling during tumour progression. *Nat Rev Mol Cell Biol* 2004;5:816–826.

55. Curran S, Murray GI. Matrix metalloproteinases: molecular aspects of their roles in tumour invasion and metastasis. *Eur J Cancer [A]* 2000;36:1621–1630.

56. Pasche B. Role of transforming growth factor beta in cancer. *J Cell Physiol* 2001;186:153–168.

57. Egeblad M, Werb Z. New functions for the matrix metalloproteinases in cancer progression. *Nat Rev Cancer* 2002;2:161–174.

58. Coussens LM, Werb Z. Inflammation and cancer. *Nature* 2002;420:860–867.

59. Karnoub AE, Dash AB, Vo AP, et al. Mesenchymal stem cells within tumour stroma promote breast cancer metastasis. *Nature* 2007;449:557–563.

60. Wyckoff JB, Jones JG, Condeelis JS, et al. A critical step in metastasis: *in vivo* analysis of intravasation at the primary tumor. *Cancer Res* 2000; 60:2504–2511.

60a. Wu JM, Fackler MJ, Halushka MK, et al. Heterogeneity of breast cancer metastases: comparison of therapeutic target expression and promoter methylation between primary tumors and their multifocal metastases. *Clin Cancer Res* 2008;14:1938–1946.

61. Chang YS, di Tomaso E, McDonald DM, et al. Mosaic blood vessels in tumors: frequency of cancer cells in contact with flowing blood. *Proc Natl Acad Sci USA* 2000;97:14608–14613.

62. Butler TP, Gullino PM. Quantitation of cell shedding into efferent blood of mammary adenocarcinoma. *Cancer Res* 1975;35:512–516.

63. Liotta LA, Kohn EC. Cancer's deadly signature. *Nat Genet* 2003;33:10–11.

64. Freedman LS, Parkinson MC, Jones WG, et al. Histopathology in the prediction of relapse of patients with stage I testicular teratoma treated by orchiectomy alone. *Lancet* 1987;2:294–298.

65. Geldof AA. Models for cancer skeletal metastasis: a reappraisal of Batson's plexus. *Anticancer Res* 1997;17:1535–1539.

66. Nieswandt B, Hafner M, Echtenacher B, et al. Lysis of tumor cells by natural killer cells in mice is impeded by platelets. *Cancer Res* 1999;59:1295–1300.

67. Grossmann J. Molecular mechanisms of "detachment-induced apoptosis–Anoikis". *Apoptosis* 2002;7:247–260.

68. Fidler IJ. Metastasis: quantitative analysis of distribution and fate of tumor embolilabeled with

125 I-5-iodo-2'-deoxyuridine. *J Natl Cancer Inst* 1970;45:773–782.

69. Cameron MD, Schmidt EE, Kerkvliet N, et al. Temporal progression of metastasis in lung: cell survival, dormancy, and location dependence of metastatic inefficiency. *Cancer Res* 2000; 60:2541–2546.

70. Ito S, Nakanishi H, Ikehara Y, et al. Real-time observation of micrometastasis formation in the living mouse liver using a green fluorescent protein gene-tagged rat tongue carcinoma cell line. *Int J Cancer* 2001;93:212–217.

71. MacDonald IC, Groom AC, Chambers AF. Cancer spread and micrometastasis development: quantitative approaches for in vivo models. *Bioessays* 2002;24:885–893.

72. Borsig L, Wong R, Feramisco J, et al. Heparin and cancer revisited: mechanistic connections involving platelets, P-selectin, carcinoma mucins, and tumor metastasis. *Proc Natl Acad Sci USA* 2001;98:3352–3357.

73. Felding-Habermann B, Habermann R, Saldivar E, et al. Role of beta3 integrins in melanoma cell adhesion to activated platelets under flow. *J Biol Chem* 1996;271:5892–5900.

74. Chambers AF, MacDonald IC, Schmidt EE, et al. Clinical targets for anti-metastasis therapy. *Adv Cancer Res* 2000;79:91–121.

75. Fidler IJ. Critical determinants of cancer metastasis: rationale for therapy. *Cancer Chemother Pharmacol* 1999;43 (Suppl):S3–S10.

76. Loberg RD, Fridman Y, Pienta BA, et al. Detection and isolation of circulating tumor cells in urologic cancers: a review. *Neoplasia* 2004;6:302–309.

76a. Weissleder R, Pittet MJ. Imaging in the era of molecular oncology. *Nature* 2008;452:580–589.

77. Marincola FM, Jaffee EM, Hicklin DJ, et al. Escape of human solid tumors from T-cell recognition: molecular mechanisms and functional significance. *Adv Immunol* 2000;74:181–273.

78. Chambers AF, Groom AC, MacDonald IC. Dissemination and growth of cancer cells in metastatic sites. *Nat Rev Cancer* 2002;2:563–572.

79. Oge HK, Aydin S, Cagavi F, et al. Migration of pacemaker lead into the spinal venous plexus: case report with special reference to Batson's theory of spinal metastasis. *Acta Neurochir (Wien)* 2001;143:413–416.

80. Gassmann P, Enns A, Haier J. Role of tumor cell adhesion and migration in organ-specific metastasis formation. *Onkologie* 2004;27:577–582.

81. Pasqualini R, Arap W. Profiling the molecular diversity of blood vessels. *Cold Spring Harb Symp Quant Biol* 2002;67:223–225.

82. Norton L, Massague J. Is cancer a disease of self-seeding? *Nat Med* 2006;12:875–878.

83. Culp LA, Lin WC, Kleinman NR. Tagged tumor cells reveal regulatory steps during earliest stages of tumor progression and micrometastasis. *Histol Histopathol* 1999;14:879–886.

84. Fidler IJ, Lieber S. Quantitative analysis of the mechanism of glucocorticoid enhancement of experimental metastasis. *Res Comm Chem Pathol Pharmacol* 1972;4:607–613.

85. Folkman J, Langer R, Linhardt RJ, et al. Angiogenesis inhibition and tumor regression caused by heparin or a heparin fragment in the presence of cortisone. *Science* 1983;221:719–725.

86. Smorenburg SM, Van Noorden CJF. The complex effects of heparins on cancer progression and metastasis in experimental studies. *Pharmacol Rev* 2001;53:93–105.

87. Schirner M, Lichtner RB, Schneider MR. The stable prostacyclin analogue Cicaprost inhibits metastasis to lungs and lymph nodes in the 13762NF MTLn3 rat mammary carcinoma. *Clin Exp Metastasis* 1994;12:24–30.

88. Murono S, Yoshizaki T, Sato H, et al. Aspirin inhibits tumor cell invasiveness induced by Epstein-Barr virus latent membrane protein 1 through suppression of matrix metalloproteinase-9 expression. *Cancer Res* 2000;60:2555–2561.

89. Jones MK, Wang H, Peskar BM, et al. Inhibition of angiogenesis by nonsteroidal anti-inflammatory drugs: insight into mechanisms and implications for cancer growth and ulcer healing. *Nat Med* 1999; 5:1418–1423.

90. Murata H, Kawano S, Tsuji S, et al. Cyclooxygenase-2 overexpression enhances lymphatic invasion and metastasis in human gastric carcinoma. *Am J Gastroenterol* 1999;94:451–455.

91. Zacharski LR, Henderson WG, Rickles FR, et al. Effect of warfarin anticoagulation on survival in carcinoma of the lung, colon, head and neck and prostate: final report of VA Cooperative Study #75. *Cancer* 1984;53:2046–2052.

92. Lebeau B, Chastang C, Muir JF, et al. [The "Petites Cellules" group]. No effect of an antiaggregant treatment with aspirin in small cell lung cancer treated with CCAVP16 chemotherapy. Results from a randomized clinical trial of 303 patients. *Cancer* 1993; 71:1741–1745.

93. Sherlock P, Hartmann WH. Adrenal steroids and the pattern of metastases of breast cancer. *JAMA* 1962;181:313–317.

94. Thompson T, Evans W. Paradoxical embolism. *Q J Med* 1930;23:135–150.

95. Paillas JE, Pellet W. Brain metastases. In: Vinken PJ, Bruyn GW, eds. *Handbook of Clinical Neurology, Vol. 18.* New York: Elsevier, 1975. pp. 201–232.

96. Delattre J-Y, Krol G, Thaler HT, et al. Distribution of brain metastases. *Arch Neurol* 1988;45:741–744.

97. Ewing J. Metastasis. In: Ewing J, ed. *Neoplastic Diseases: A Treatise on Tumours.* 4 ed. Philadelphia: WB Saunders, 1940. pp. 62–74.

98. Paget S. The distribution of secondary growths in cancer of the breast. *Lancet* 1889;i:571–573.

99. Fidler IJ, Schackert G, Zhang R, et al. The biology of melanoma brain metastasis. *Cancer Metastasis Rev* 1999;18:387–400.

100. Weiss L. Comments on hematogenous metastatic patterns in humans as revealed by autopsy. *Clin Exp Metastasis* 1992;10:191–199.

101. Posner JB, Chernik NL. Intracranial metastases from systemic cancer. *Adv Neurol* 1978;19:575–587.

102. Xiromeritis N, Klonaris C, Papas S, et al. Recurrent peripheral arterial embolism from pulmonary cancer. Case report and review of the literature. *Int Angiol* 2000;19:79–83.

103. Auperin A, Arriagada R, Pignon JP, et al. Prophylactic cranial irradiation for patients with small-cell lung

cancer in complete remission. Prophylactic Cranial Irradiation Overview Collaborative Group. *N Engl J Med* 1999;341:476–484.

104. Goodison S, Kawai K, Hihara J, et al. Prolonged dormancy and site-specific growth potential of cancer cells spontaneously disseminated from non-metastatic breast tumors as revealed by labeling with green fluorescent protein. *Clin Cancer Res* 2003;9:3808–3814.

105. Marchetti D, Li J, Shen R. Astrocytes contribute to the brain-metastatic specificity of melanoma cells by producing heparanase. *Cancer Res* 2000;60:4767–4770.

106. Yano S, Shinohara H, Herbst RS, et al. Expression of vascular endothelial growth factor is necessary but not sufficient for production and growth of brain metastasis. *Cancer Res* 2000;60:4959–4967.

107. Kusters B, de Waal RM, Wesseling P, et al. Differential effects of vascular endothelial growth factor A isoforms in a mouse brain metastasis model of human melanoma. *Cancer Res* 2003;63:5408–5413.

107a. Kaplan RN, Rafii S, Lyden D. Preparing the "soil": the pre-metastatic niche. *Cancer Res* 2006;66:11089–11093.

107b. Rafii S, Lyden D. A few to flip the angiogenic switch. *Science* 2008;319:163–164.

108. Fidler IJ, Poste G. The cellular heterogeneity of malignant neoplasms: implications for adjuvant chemotherapy. *Semin Oncol* 1985;12:207–221.

109. Shah C, Johnson EB, Everett E, et al. Does size matter? Tumor size and morphology as predictors of nodal status and recurrence in endometrial cancer. *Gynecol Oncol* 2005;99:564–570.

110. Konishi T, Watanabe T, Kishimoto J, et al. Prognosis and risk factors of metastasis in colorectal carcinoids: results of a nationwide registry over 15 years. *Gut* 2007;56:863–868.

111. Tresserra F, Rodriguez I, Garcia-Yuste M, et al. Tumor size and lymph node status in multifocal breast cancer. *Breast J* 2007;13:68–71.

112. Price JE. The biology of metastatic breast cancer. *Cancer* 1990; 66:1313–1320.

113. Peckham M. Clinical aspects of metastases. In: Bock G, Whelan J, eds. *Metastasis. Ciba Foundation Symposium 141.* Chichester, UK: John Wiley, 1988. pp. 223–243.

114. Leemans CR, Tiwari R, Nauta JJ, et al. Regional lymph node involvement and its significance in the development of distant metastases in head and neck carcinoma. *Cancer* 1993;71:452–456.

115. Yorke ED, Fuks Z, Norton L, et al. Modeling the development of metastases from primary and locally recurrent tumors: comparison with a clinical database for prostatic cancer. *Cancer Res* 1993;53:2987–2993.

116. Rapiti E, Verkooijen HM, Vlastos G, et al. Complete excision of primary breast tumor improves survival of patients with metastatic breast cancer at diagnosis. *J Clin Oncol* 2006;24:2743–2749.

117. Swanson G, Thompson I, Basler J, et al. Metastatic prostate cancer-does treatment of the primary tumor matter? *J Urol* 2006;176:1292–1298.

118. Maesawa S, Kondziolka D, Thompson TP, et al. Brain metastases in patients with no known primary tumor—the role of stereotactic radiosurgery. *Cancer* 2000;89:1095–1101.

119. Carter CL, Allen C, Henson DE. Relation of tumor size, lymph node status, and survival in 24,740 breast cancer cases. *Cancer* 1989;63:181–187.

120. Wong JS, O'Neill A, Recht A, et al. The relationship between lymphatic vessel invasion, tumor size, and pathologic nodal status: can we predict who can avoid a third field in the absence of axillary dissection? *Int J Radiat Oncol Biol Phys* 2000;48:133–137.

121. Morton DL, Wanek L, Nizze JA, et al. Improved long-term survival after lymphadenectomy of melanoma metastatic to regional nodes: Analysis of prognostic factors in 1134 patients from the John Wayne Cancer Clinic. *Ann Surg* 1991; 214:491–499.

122. Khan SA, Stewart AK, Morrow M. Does aggressive local therapy improve survival in metastatic breast cancer? *Surgery* 2002;132:620–626.

123. Scoggins CR, Chagpar AB, Martin RC, et al. Should sentinel lymph-node biopsy be used routinely for staging melanoma and breast cancers? *Nat Clin Pract Oncol* 2005;2:448–455.

124. Sakorafas GH, Tsiotou AG. Sentinel lymph node biopsy in breast cancer. *Am Surg* 2000;66:667–674.

125. Kim T, Giuliano AE, Lyman GH. Lymphatic mapping and sentinel lymph node biopsy in early-stage breast carcinoma: a metaanalysis. *Cancer* 2006;106:4–16.

126. Flanigan RC, Salmon SE, Blumenstein BA, et al. Nephrectomy followed by interferon alfa-2b compared with interferon alfa-2b alone for metastatic renal-cell cancer. *N Engl J Med* 2001;345:1655–1659.

127. Coffey JC, Wang JH, Smith MJ, et al. Excisional surgery for cancer cure: therapy at a cost. *Lancet Oncol* 2003;4:760–768.

128. Baum M, Demicheli R, Hrushesky W, et al. Does surgery unfavourably perturb the "natural history" of early breast cancer by accelerating the appearance of distant metastases? *Eur J Cancer* 2005;41:508–515.

129. O'Reilly MS, Holmgren L, Shing Y, et al. Angiostatin: a novel angiogenesis inhibitor that mediates the suppression of metastases by a Lewis lung carcinoma. *Cell* 1994;79:315–328.

130. Kuukasjarvi T, Karhu R, Tanner M, et al. Genetic heterogeneity and clonal evolution underlying development of asynchronous metastasis in human breast cancer. *Cancer Res* 1997;57:1597–1604.

131. Bystryn J-C, Bernstein P, Liu P, et al. Immunophenotype of human melanoma cells in different metastases. *Cancer Res* 1985;45:5603–5607.

132. Morse HG, Moore GE, Ortiz LM, et al. Malignant melanoma: from subcutaneous nodule to brain metastasis. *Cancer Genet Cytogenet* 1994;72:16–23.

133. Dexter DL, Leith JT. Tumor heterogeneity and drug resistance. *J Clin Oncol* 1986;4:244–257.

134. Sugarbaker EV. Patterns of metastasis in human malignancies. *Cancer Biol Rev* 1981;2:235–278.

135. Fidler IJ. The pathogenesis of cancer metastasis: the 'seed and soil' hypothesis revisited. *Nat Rev Cancer* 2003;3:453–458.

136. Weiss L. Patterns of metastasis. *Cancer Metastasis Rev* 2000;19:281–301.

137. Batson CV. Function of vertebral veins and their role in spread of metastases. *Ann Surg* 1940;112:138–149.

138. del Regato JA. Pathways of metastatic spread of malignant tumors. *Semin Oncol* 1977;4:33–38.

139. Bubendorf L, Schöpfer A, Wagner U, et al. Metastatic patterns of prostate cancer: an autopsy study of 1,589 patients. *Hum Pathol* 2000;31:578–583.

140. Higashi H, Fukutomi T, Watanabe T, et al. Seven cases of breast cancer recurrence limited to the central nervous system without other visceral metastases. *Breast Cancer* 2000;7:153–156.

141. Yoneda T. Cellular and molecular basis of preferential metastasis of breast cancer to bone. *J Orthop Sci* 2000;5:75–81.

142. Woodhouse EC, Chuaqui RF, Liotta LA. General mechanisms of metastasis. *Cancer* 1997;80:1529–1537.

143. Fidler IJ. Regulation of neoplastic angiogenesis. *J Natl Cancer Inst Monogr* 2000;2000:10–14.

144. Radinsky R. Modulation of tumor cell gene expression and phenotype by the organ specific metastatic environment. *Cancer Metastasis Rev* 1995;14:323–338.

145. Smith DB, Howell A, Harris M, et al. Carcinomatous meningitis associated with infiltrating lobular carcinoma of the breast. *Eur J Surg Oncol* 1985;11:33–36.

146. Engebraaten O, Fodstad O. Site-specific experimental metastasis patterns of two human breast cancer cell lines in nude rats. *Int J Cancer* 1999;82:219–225.

147. Thomas AJ, Rock JP, Johnson CC, et al. Survival of patients with synchronous brain metastases: an epidemiological study in southeastern Michigan. *J Neurosurg* 2000;93:927–931.

148. Zeidman I, Buss JAM. Transpulmonary passage of tumor cell emboli. *Cancer Res* 1952;12:731–733.

149. Viadana E, Bross ID, Pickren JW. An autopsy study of some routes of dissemination of cancer of the breast. *Br J Cancer* 1973;27:336–340.

150. Kondo H, Okumura T, Ohde Y, et al. Surgical treatment for metastatic malignancies. Pulmonary metastasis: indications and outcomes. *Int J Clin Oncol* 2005;10:81–85.

151. Nishijima Y, Uchida K, Koiso K, et al. Clinical significance of the vertebral vein in prostate cancer metastasis. *Adv Exp Med Biol* 1992;324:93–100.

152. Coman D, DeLong RP. The role of the vertebral venous system in the metastasis of cancer to the spinal column. *Cancer* 1951;4:610–618.

153. Harada M, Shimizu A, Nakamura Y, et al. Role of the vertebral venous system in metastatic spread of cancer cells to the bone. In: Karr JP, Yamanaka H, eds. *Prostate Cancer and Bone Metastasis.* New York: Plenum, 1992. pp. 83–92.

154. Brenner GJ, Felten SY, Felten DL, et al. Sympathetic nervous system modulation of tumor metastases and host defense mechanisms. *J Neuroimmunol* 1992;37:191–202.

155. Powell N. Metastatic carcinoma in association with Paget's disease of bone. *Br J Radiol* 1983;56:582–585.

156. Nielson SL, Posner JB. Brain metastasis localized to an area of infarction. *J Neuro-Oncol* 1983;1:191–195.

157. Hwang TL, Close TP, Grego JM, et al. Predilection of brain metastasis in gray and white matter junction and vascular border zones. *Cancer* 1996;77:1551–1555.

158. Cascino TL, Leavengood JM, Kemeny N, et al. Brain metastases from colon cancer. *J Neuro-Oncol* 1983;1:203–209.

159. Postler E, Meyermann R. Brain metastasis in renal cell carcinoma: clinical data and neuropathological differential diagnoses. *Anticancer Res* 1999; 19:1579–1581.

160. Saadoun S, Papadopoulos MC, Davies DC, et al. Aquaporin-4 expression is increased in oedematous human brain tumours. *J Neurol Neurosurg Psychiatry* 2002;72:262–265.

161. Patchell RA, Tibbs PA, Regine WF, et al. Postoperative radiotherapy in the treatment of single metastases to the brain: a randomized trial. *JAMA* 1998;280:1485–1489.

162. Rogers LR. Cerebrovascular complications in cancer patients. *Oncology* 1994;8:23–30.

163. Lieu AS, Hwang SL, Howng SL, et al. Brain tumors with hemorrhage. *J Formos Med Assoc* 1999;98:365–367.

164. Nutt SH, Patchell RA. Intracranial hemorrhage associated with primary and secondary tumors. *Neurosurg Clin N Am* 1992;3:591–599.

165. Kondziolka D, Bernstein M, Resch L, et al. Significance of hemorrhage into brain tumors: clinicopathological study. *J Neurosurg* 1987;67:852–857.

166. Pullar M, Blumbergs PC, Phillips GE, et al. Neoplastic cerebral aneurysm from metastatic gestational choriocarcinoma. *J Neurosurg* 1985;63:644–647.

167. Schackert G, Price JE, Bucana CD, et al. Unique patterns of brain metastasis produced by different human carcinomas in athymic nude mice. *Int J Cancer* 1989;44:892–897.

168. Floeter MK, So YT, Ross DA, et al. Miliary metastasis to the brain: clinical and radiologic features. *Neurology* 1987;37:1817–1818.

169. Madow L, Alpers BJ. Encephalitic form of metastatic carcinoma. *Arch Neurol Psychiatry* 1951;65:161–173.

170. Bhushan C. "Miliary" metastatic tumors in the brain. *J Neurosurg* 1997;86:564–566.

171. Becher MW, Abel TW, Thompson RC, et al. Immunohistochemical analysis of metastatic neoplasms of the central nervous system. *J Neuropathol Exp Neurol* 2006;65:935–944.

172. Takakura K, Sano K, Hojo S, et al. *Metastatic Tumors of the Central Nervous System.* Tokyo: Igaku-Shoin, 1982.

173. Tham YL, Sexton K, Kramer R, et al. Primary breast cancer phenotypes associated with propensity for central nervous system metastases. *Cancer* 2006;107:696–704.

174. Yau T, Swanton C, Chua S, et al. Incidence, pattern and timing of brain metastases among patients with advanced breast cancer treated with trastuzumab. *Acta Oncol* 2006;45:196–201.

175. Hayes DN, Monti S, Parmigiani G, et al. Gene expression profiling reveals reproducible human lung adenocarcinoma subtypes in multiple independent patient cohorts. *J Clin Oncol* 2006;24:5079–5090.

176. Orr FW, Lee J, Duivenvoorden WC, et al. Pathophysiologic interactions in skeletal metastasis. *Cancer* 2000;88:2912–2918.

177. Lipton A. Pathophysiology of bone metastases: how this knowledge may lead to therapeutic intervention. *J Support Oncol* 2004;2:205–213.

178. Layman R, Olson K, Van PC. Bisphosphonates for breast cancer: questions answered, questions remaining. *Hematol Oncol Clin North Am* 2007;21:341–367.

179. Cotta CV. Metastatic tumors in bone marrow: histopathology and advances in the biology of the tumor cells and bone marrow environment. *Ann Diagn Pathol* 2006;10:169–192.

180. Yuh WT, Quets JP, Lee HJ, et al. Anatomic distribution of metastases in the vertebral body and modes of hematogenous spread. *Spine* 1996;21:2243–2250.

181. Fujita T, Ueda Y, Kawahara N, et al. Local spread of metastatic vertebral tumors. A histologic study. *Spine* 1997;22:1905–1912.

182. Khaw FM, Worthy SA, Gibson MJ, et al. The appearance on MRI of vertebrae in acute compression of the spinal cord due to metastases. *J Bone Joint Surg [Br]* 1999;81:830–834.

183. Kim JK, Learch TJ, Colletti PM, et al. Diagnosis of vertebral metastasis, epidural metastasis, and malignant spinal cord compression: are T_1-weighted sagittal images sufficient? *Magn Reson Imaging* 2000;18:819–824.

184. Arguello F, Baggs RB, Duerst RE, et al. Pathogenesis of vertebral metastasis and epidural spinal cord compression. *Cancer* 1990;65:98–106.

185. Lyons MK, O'Neill BP, Marsh WR, et al. Primary spinal epidural non-Hodgkin's lymphoma: report of eight patients and review of the literature. *Neurosurgery* 1992;30:675–680.

186. Mostafavi H, Lennarson PJ, Traynelis VC. Granulocytic sarcoma of the spine. *Neurosurgery* 2000;46:78–83.

187. Allam K, Sze G. MR of primary extraosseous Ewing sarcoma. *AJNR Am J Neuroradiol* 1994;15:305–307.

188. Kaspers G-J, Kamphorst W, van de Graaff M, et al. Primary spinal epidural extraosseous Ewing's sarcoma. *Cancer* 1991;68:648–654.

189. Schimandle JH, Levine AM. An isolated nonosseous metastasis to the epidural space from an osteogenic sarcoma. *Cancer* 1992;69:103–107.

190. Goodkin R, Carr BI, Perrin RG. Herniated lumbar disc disease in patients with malignancy. *J Clin Oncol* 1987;5:667–671.

191. Alorainy IA, Al Asmi AR, Del Carpio R. MRI features of epidural extramedullary hematopoiesis. *Eur J Radiol* 2000;35:8–11.

192. Grossman SA, Krabak MJ. Leptomeningeal carcinomatosis. *Cancer Treat Rev* 1999;25:103–119.

193. Maroldi R, Ambrosi C, Farina D. Metastatic disease of the brain: extra-axial metastases (skull, dura, leptomeningeal) and tumour spread. *Eur Radiol* 2005;15:617–626.

194. Kokkoris CP. Leptomeningeal carcinomatosis. How does cancer reach the pia-arachnoid? *Cancer* 1983;51:154–160.

195. Mareel M, Leroy A, Bracke M. Cellular and molecular mechanisms of metastasis as applied to carcinomatous meningitis. *J Neuro-Oncol* 1998; 38:97–102.

196. Mirimanoff RO, Choi NC. Intradural spinal metastases in patients with posterior fossa brain metastases from various primary cancers. *Oncology* 1987;44:232–236.

197. Mirimanoff RO, Choi NC. The risk of intradural spinal metastases in patients with brain metastases from bronchogenic carcinomas. *Int J Radiat Oncol Biol Phys* 1986;12:2131–2136.

198. Patchell RA, Cirrincoine C, Thaler HT, et al. Single brain metastases: surgery plus radiation or radiation alone. *Neurology* 1986;36:447–453.

199. Van der Ree TC, Dippel DWJ, Avezaat CJJ, et al. Leptomeningeal metastasis after surgical resection of brain metastases. *J Neurol Neurosurg Psychiatry* 1999;66:225–227.

200. Boyle R, Thomas M, Adams JH. Diffuse involvement of the leptomeninges by tumour—a clinical and pathological study of 63 cases. *Postgrad Med J* 1980;56:149–158.

201. Nasir S, DeAngelis LM. Update on the management of primary CNS lymphoma. *Oncology (Huntingt)* 2000;14:228–234.

202. Azzarelli B, Mirkin DL, Goheen M, et al. The leptomeningeal vein. A site of re-entry of leukemic cells into the systemic circulation. *Cancer* 1984;54:1333–1343.

203. Azzarelli B, Roessmann U. Pathogenesis of central nervous system infiltration in acute leukemia. *Arch Pathol Lab Med* 1977;101:203–205.

204. Herman DL, Courville CB. Pathogenesis of meningeal carcinomatosis. Report of a case secondary to carcinoma of cecum via perineural extension; with a review of 146 cases. *Bull Los Angeles Neurol Soc* 1965;30:107–117.

205. Hayat G, Ehsan T, Selhorst JB, et al. Magnetic resonance evidence of perineural metastasis. *J Neuroimaging* 1995;5:122–125.

206. Olson ME, Chernik NL, Posner JB. Infiltration of the leptomeninges by systemic cancer. A clinical and pathologic study. *Arch Neurol* 1974;30:122–137.

207. Dickenman RC, Chason JL. Alterations in the dorsal root ganglia and adjacent nerves in the leukemias, the lymphomas and multiple myeloma. *Am J Pathol* 1958;34:349–357.

208. Johnson PC. Hematogenous metastases of carcinoma to dorsal root ganglia. *Acta Neuropathol (Berl)* 1977;38:171–172.

209. Wigfield CC, Hilton DA, Coleman MG, et al. Metastatic adenocarcinoma masquerading as a solitary nerve sheath tumour. *Br J Neurosurg* 2003;17:459–461.

210. Kelkar P. Diabetic neuropathy. *Semin Neurol* 2005;25:168–173.

211. Odabasi Z, Parrott JH, Reddy VV, et al. Neurolymphomatosis associated with muscle and cerebral involvement caused by natural killer cell lymphoma: a case report and review of literature. *J Peripher Nerv Syst* 2001;6:197–203.

212. Beristain X, Azzarelli B. The neurological masquerade of intravascular lymphomatosis. *Arch Neurol* 2002;59:439–443.

213. Allt G, Lawrenson JG. The blood-nerve barrier: enzymes, transporters and receptors—a comparison with the blood-brain barrier. *Brain Res Bull* 2000;52:1–12.

214. Ushio Y, Chernik NL, Posner JB, et al. Meningeal carcinomatosis: development of an experimental model. *J Neuropathol Exp Neurol* 1977;36:228–244.

215. Ayala GE, Dai H, Ittmann M, et al. Growth and survival mechanisms associated with perineural invasion in prostate cancer. *Cancer Res* 2004; 64:6082–6090.

216. Ayala GE, Dai H, Li R, et al. Bystin in perineural invasion of prostate cancer. *Prostate* 2006;66:266–272.

217. McLaughlin RB, Jr, Montone KT, Wall SJ, et al. Nerve cell adhesion molecule expression in squamous cell carcinoma of the head and neck: a predictor of propensity toward perineural spread. *Laryngoscope* 1999;109:821–826.

218. Iwahashi N, Nagasaka T, Tezel G, et al. Expression of glial cell line-derived neurotrophic factor correlates

with perineural invasion of bile duct carcinoma. *Cancer* 2002;94:167–174.

219. Hassan MO, Maksem J. The prostatic perineural space and its relation to tumor spread: an ultrastructural study. *Am J Surg Pathol* 1980; 4:143–148.

220. Ichinohe K, Takahashi M, Tooyama N. Delay by patients and doctors in treatment of Pancoast tumor. *Wien Klin Wochenschr* 2006;118:405–410.

221. Hughes TA, McQueen IN, Anstey A, et al. Neurotropic malignant melanoma presenting as a trigeminal sensory neuropathy. *J Neurol Neurosurg Psychiatry* 1995;58:381–382.

222. Gandhi D, Gujar S, Mukherji SK. Magnetic resonance imaging of perineural spread of head and neck malignancies. *Top Magn Reson Imaging* 2004;15:79–85.

Chapter 3

Blood–Nervous System Barrier Dysfunction: Pathophysiology and Treatment

INTRODUCTION

Three aspects of the anatomy and physiology of the nervous system make it particularly vulnerable to the effects of metastases.

First, the central nervous system (CNS) is enclosed by bone. The skull protectively encloses the brain, and the vertebral column surrounds the spinal cord. The result is that when tumors grow, they not only distort and compress the normal tissue nearby, but also raise tissue pressure both in the immediate area and at a distance (i.e., overall intracranial pressure), thus interfering with functions of neural tissue that is not directly invaded or compressed by the tumor (see Fig. 1–6). Once peripheral and cranial nerves leave the spinal canal or skull, they are not encased in bone but in an unyielding

connective tissue sheath, the perineurium. The clinical consequence is the same as in the CNS, that is, compression and distortion increase tissue pressure, often causing serious interference with neural function before the tumor causes significant direct destruction.

Second, physiologic barriers termed the *blood–brain barrier* (BBB), *blood–spinal fluid barrier* (blood–CSF), and *blood–nerve barrier* separate the systemic circulation from the neuraxis[1–4] (Fig. 3–1). These barriers reside largely in the endothelium of the vessels feeding the nervous system supported by perivascular astrocytes. They also exist in choroid plexus and pia mater (Fig. 3–2). They retard entry of most water-soluble substances into the nervous system. One clinical consequence of these barriers is that many chemotherapeutic agents

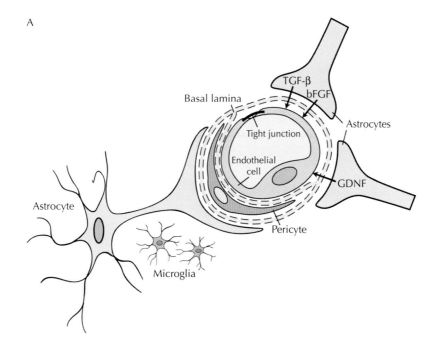

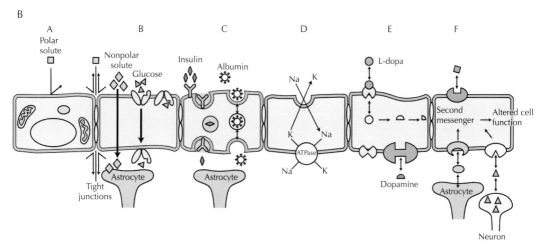

Figure 3–1. The blood–brain barrier. **A**: The figure shows normal brain capillary endothelial cells connected by a tight junction and surrounded by a basal lamina that encompasses a pericyte. Most of the capillary is covered by astrocytic foot processes. The astrocytes secrete transforming growth factor β (TGF-β), basic fibroblast growth factor (bFGF), and glial derived neurotrophic factor (GDNF) as well as other substances. Microglial cells are also in the area. All of these elements contribute to the blood–brain barrier. **B**: A schematic representation of the normal blood–brain barrier. The endothelial cell lipid membrane and the tight junctions connecting endothelial cells prevent the movement of polar solutes (green squares) between the blood and the brain. Lipid-soluble solutes (yellow diamonds) and solutes for which the endothelial cell has transporters, such as glucose (blue triangles), some amino acids, and some micronutrients such as thiamine, cross from blood to brain by two-way transport systems. Insulin and transferrin (blue diamonds) as well as other large molecules such as albumin (black circles) are transported by endocytosis. Substances, such as potassium and sodium, are transported asymmetrically from either the blood to the brain or the brain to the blood. Some molecules that cross the luminal endothelial membrane, such as L-dopa, are metabolized within the endothelial cell so that a metabolic product such as dopamine reaches the brain. Receptors on the endothelial surface, including selectins and integrins, react with circulating messengers that affect the brain via a second messenger system.

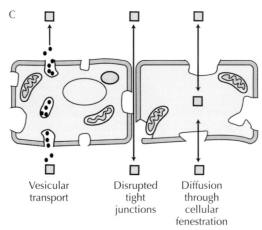

Vesicular transport

Disrupted tight junctions

Diffusion through cellular fenestration

Figure 3–1. (*cont.*) **C**: When a brain metastasis promotes angiogenesis, the newly formed vessels do not have a normal blood–brain barrier. Substances kept out by the normal barrier are now allowed to enter because the tight junctions are disrupted, fenestrations occur in the endothelial cell and vesicular transport is increased.

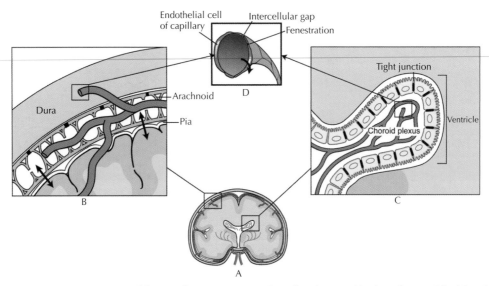

Figure 3–2. Major compartments of the central nervous system and interfaces between blood, cerebrospinal fluid (CSF), and brain tissue. A barrier similar to the blood–brain barrier exists between the blood and CSF, both in the brain (**A**) and the spinal cord (not shown). There is no barrier to passage of materials in the dura (**B**). There is relatively free passage of material from CSF into brain and spinal cord and vice versa (*arrows* in **B**). In the choroid plexus (**C**), the endothelial cells do not form a barrier, but allow easy entry of substances because of intercellular gaps and fenestrations within the endothelial cell (**D**). However, tight junctions between ependymal cells form a barrier, almost, but not quite, as strong as the barrier in brain capillaries.

do not have access to tumor cells sequestered behind the BBB, perhaps causing the increased incidence of brain metastases from certain primary tumors that have been controlled by systemic chemotherapy (see Fig. 1–2).

Tumors in the brain that grow beyond a few millimeters promote angiogenesis. The resulting capillaries lack tight junctions and other qualities of normal brain capillaries (see Fig. 3–1).[5–7] The process at least partially disrupts the BBB.[7] Water-soluble agents can then gain some (but variable) access to tumor cells behind the disrupted barrier, although not always in concentrations equal to that achieved in systemic organs. Disruption of the BBB in and immediately around metastatic tumors exposes the tumor and, by diffusion, exposes the normal brain surrounding the tumor to potentially neurotoxic chemotherapeutic agents. The disruption also allows the entry of normal substances, such as electrolytes and proteins that usually have only limited access to the brain. In addition to their own potential toxicity, these osmolar particles bring water with them and cause edema, thus adding mass to the already expanding tumor.[8,9]

Third, the brain parenchyma lacks lymphatics. Lymph channels that are positioned along cranial nerves, especially the olfactory nerve, and spinal roots may absorb some cerebrospinal fluid (CSF) and its macromolecules[10,11] (Fig. 3–3). The cranial nerve–associated lymphatics drain to the lymphatic network of the nasal cavity and nasopharynx as well as to cervical lymph nodes[12] and may play a role in reducing elevated intracranial pressure.[13,14] Lymphatic drainage, however, is rarely sufficient to absorb

enough brain edema to reduce increased intracranial pressure. Instead, absorption occurs by a long, tortuous route of convection or bulk flow from the tumor through the brain substance to the ventricular or subarachnoid CSF, where it is absorbed into the systemic circulation.[15,16] Unfortunately, the tumor and edema fluid, by causing increased intracranial pressure, often disrupt the normal CSF absorptive pathways, compounding the problem of brain edema.[17]

The ultimate result of these three anatomic–physiologic factors is that when the BBB is disrupted, plasma-derived fluid enters the brain and increases brain mass,[18] further raising already increased tissue pressure in the tumor,[19–21] first locally, and then at a distance from the tumor (see Fig. 1–6). The increasing mass distorts surrounding normal tissue and interferes with its blood flow,[19] often leading to ischemia, more edema, and more neurologic dysfunction. Furthermore, substances in the edema fluid (e.g., potassium and glutamate) may be neurotoxic, adding to the dysfunction (Fig. 4–3). Distant neural structures physiologically linked to the area of tumor and edema also fail to function normally, presumably because of remote trans-synaptic impairment (diaschisis), and may contribute to neurologic dysfunction.[22,23] The sum of these physiologic abnormalities often causes more symptoms than does the tumor itself, a principle proved by the rapid clinical response to corticosteroids, which restore the integrity of the BBB and with time decrease edema.[24–26]

For reasons that are not always clear, the degree of edema varies substantially among tumors that are seemingly of the same size (Fig. 3–4). Some small metastases are accompanied by massive edema, whereas some larger ones have little or no edema.

ANATOMY AND PHYSIOLOGY

Normal brain capillaries of the BBB differ from most capillaries elsewhere in the body (Table 3–1). Exceptions are testis[27,28] (and perhaps ovary)[29,30] and skin,[31] which have similar barriers. Even in the brain, there are areas where barrier function is normally defective or absent. These include the median eminence, neurohypophysis, the pineal gland, the organum vasculosum of the lamina terminalis, the subfornical organ, the subcommissural organ,

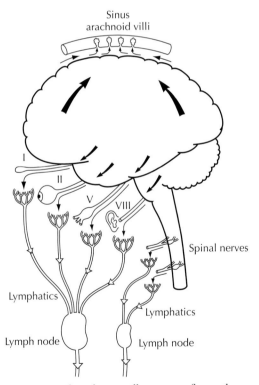

Figure 3–3. This schematic illustrates outflow pathways from the cranial and spinal subarachnoid space across the arachnoid villi to dural sinus blood or along certain cranial nerves and spinal nerve roots to lymphatics: olfactory nerve (I), optic nerve (II), trigeminal nerve (V), and acoustic nerve (VIII). (From Ref. 11 with permission.)

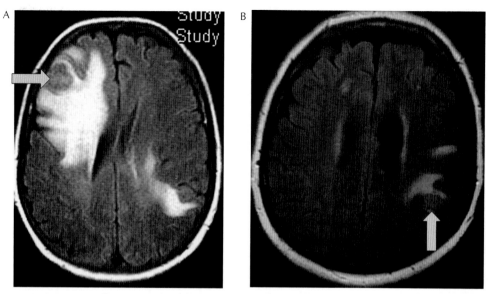

Figure 3–4. Brain scans of two patients with lung cancer and multiple metastases to the brain. **A:** In one patient the tumor (*arrow*) was surrounded by a massive amount of edema with midline shift. The patient had a dramatic response to corticosteroids. **B:** In the second patient, a tumor of the same size (*arrow*) produced little edema. In both patients, contrast enhancement with gadolinium indicated that the blood–brain barrier was disrupted.

Table 3–1. **Properties of Endothelial Cells**

Brain	Periphery
Tight intercellular junctions	Loose intercellular junctions
Few pinocytotic vesicles	Many pinocytotic vesicles
No fenestrae	Fenestrae
High electrical resistance	Low electrical resistance
Pericytes	Pericytes
Astrocytic foot processes	No astrocytes
Many transporters	Few transporters
Influx, e.g., Glut-1	
Efflux, e.g., P-glycoprotein	

and, perhaps most importantly for neuro-oncology, the area postrema, which is probably responsible for much of the nausea and vomiting induced by chemotherapy. Taken together, these structures comprise the circumventricular organs (CVO).[4]

Blood–Brain Barrier

Endothelial cells in capillaries in most of the body are connected one to another by relatively loose junctions, normally 6 to 7 nm in size, slightly smaller than an albumin molecule.[3] These endothelial cells contain abundant pinocytotic vesicles and have the capacity to form fenestrae that allow substances to pass through the endothelial cell (see Fig. 3–1, Table 3–2). On the other hand, as shown in Figure 3–1 and Table 3–2, several anatomic features of the BBB determine which substances enter the brain:

1. Capillary endothelial cells are connected by a tight rather than a gap junction.[3,4] Tight junctions are perforated by aqueous channels with diameters no larger

Table 3–2. **Physical Characteristics of the Blood–Brain Barrier as a Modified Tight Epithelium**

- It is a sheet of cells on a basement membrane.
- The cells are connected by tight junctions.
- Permeability to hydrophilic nonelectrolytes is very low in the absence of a membrane carrier.
- Transport of certain organic solutes is by facilitated transport with saturation kinetics as well as stereospecificity and competitive interactions between related compounds.
- It has features of a tight epithelium, including low hydraulic conductance, low ionic permeability, and high electrical resistance.
- These characteristics hold for endothelial cells (blood–brain barrier) and the choroidal epithelium (blood–CSF barrier).

CSF, cerebrospinal fluid.

than 0.6 to 0.8 nm, restricting free diffusion of most substances. Even water does not move as freely across the BBB as it does into other organs.[32–34] Several proteins are important in tight junction physiology. These include occludin, claudin-5, and junction adhesion molecules that are trans-membrane proteins. The zonula occludens protein ZO1, p120 catenins, claudin, and cadherin are cytoplasmic proteins[4] that also contribute to tight junctions.

2. Large molecules, including proteins, find their way into the interstitial space of most organs through capillary fenestrations. Brain capillaries lack these fenestrations, substantially retarding movement of these molecules.

3. Pinocytotic vesicles transport plasma or interstitial fluid contents bidirectionally across the endothelium. The density of pinocytotic vesicles in the brain is approximately 5% of that in other organs, substantially diminishing the opportunity for this kind of transport.

4. Mitochondria that supply the energy for active transport across the endothelium are 10 times more abundant in brain endothelial cells than in other organs.[35] The increased density of mitochondria in the brain fuels several pumps that permit the entry of water-soluble substances that would not otherwise cross the BBB. These pumps also promote the exit of potentially toxic substances that might accumulate as a result of brain activity (e.g., potassium).[35]

5. P-glycoprotein, a unique membrane glycoprotein with a molecular weight of 170

to 180 kDa, is encoded by a multidrug resistance gene. Increased expression of P-glycoprotein has been associated with resistance of tumors to several chemotherapeutic agents, including doxorubicin, etoposide, and vincristine. P-glycoprotein is expressed normally in the endothelial cells of blood–tissue barriers, including brain, skin, and testis. It is not expressed in those parts of the brain without a BBB, except the pineal gland. P-glycoprotein appears to play a role in the BBB and in keeping water-soluble chemotherapeutic agents out of the brain.[3] Primary and metastatic tumors of the brain express little or no P-glycoprotein, although other normal capillary markers such as factor VIII are present.[3] Other unique proteins are also expressed in brain endothelial cells[3] (Table 3–3).

Many water-soluble agents that are necessary for brain function but are not synthesized by the brain, including proteins and amino acids, must be conveyed by either active or carrier-facilitated transport across the BBB (see Fig. 3–1). The brain is an obligate user of glucose, and the density of the glucose transporter in its capillaries is 10 to 20 times higher than in the endothelial membranes of other tissues.[36] Transporters exist for neutral, acidic, and basic amino acids,[37] monocarboxylic acids such as lactic acid (primarily for efflux of lactic acid produced by the brain), ketones, and electrolytes. A sodium–potassium adenosine triphosphatase pump regulates the concentration of potassium in the brain's extracellular space at approximately 2.8 mEq/L, compared with the concentration of 3 to 5 mEq/L in the

Table 3–3. Drug Transporters Putatively Involved in Forming the Blood–Brain Barrier, as well as Chemotherapeutic Agents that Are Substrates for Each Transporter, and Compounds Used as Potential Inhibitors of Transporter Function

Transporter	HUGO Name	Substrates	Inhibitors
P-glycoprotein	ABCB1	Doxorubicin, daunorubicin, docetaxel, paclitaxel, epirubicin, idarubicin, vinblastine, vincristine, etoposide	Verapamil, cyclosporin A, quinidine, PSC 833 (valspodar), GF120918 (elacridar), VX-710 (biricodar), LY 335979 (zosuquidar), XR9576 (tariquidar)
MRP1	ABCC1	Etoposide, teniposide, daunorubicin, doxorubicin, epirubicin, melphalan, vincristine, vinblastine	Probenecid, sulfinpyrazone, MK-571, some P-glycoprotein inhibitors (e.g., cyclosporin A, verapamil, PSC 833)
MRP2	ABCC2	Similar to MRP1	Probenecid, MK-571, leukotriene C6
MRP3	ABCC3	Similar to MRP1	Sulfinpyrazone, indomethacin, probenecid
MRP4	ABCC4	Methotrexate, 6-mercaptopurine, thioguanine	Probenecid
MRP5	ABCC5	6-Mercaptopurine, thioguanine	Probenecid, sildenafil
MRP6	ABCC6	Actinomycin D, cisplatin, daunorubicin, doxorubicin, etoposide	Probenecid, indomethacin
BCRP	ABCG2	Mitoxantrone, methotrexate, SN-38, topotecan, imatinib, erlotinib, gefitinib	GF120918, fumitremorgin C

From Ref. 3 with permission.
HUGO, human genome.

plasma. Higher levels of potassium that follow BBB breakdown may lead to seizures and other abnormalities of neurologic function.

The capillary endothelium of the brain is surrounded by a basement membrane in which pericytes are embedded that help determine capillary diameter and thus, control cerebral blood flow. There is some evidence that pericytes play an important role in regulating BBB function.[38–40] These cells, particularly their actin filaments, are disrupted in brain tumor vessels, contributing to their increased capillary permeability.[41] In addition, over 90% of the surface of cerebral capillaries is ensheathed by astrocytic foot processes; they, and perhaps neurons and perivascular macrophages, play a role in maintaining the structure and function of the BBB.[1] Cerebral capillaries in vitro maintain tight junctions when co-cultured with astrocytes or astrocytic-conditioned media but, if cultured alone, lose their tight junctions.[1]

Astrocytic and pericytic damage caused by primary and metastatic tumors may contribute to BBB disruption.[41]

Three of the cerebral CVOs, the subfornical organ, the organum vasculosum of the lamina terminalis, and the area postrema, contain neurons that have the capacity to monitor substances circulating in the blood; they are referred to as sensory CVOs. Four of the areas, the median eminence, the neurohypophysis, the intermediate lobe of the pituitary gland, and the pineal gland are secretory organs; secretory CVOs contain terminals, axons, and glial and epithelial cells. The subcommissural organ is less well understood but may play both sensory and secretory roles.[42,43] The area postrema contains an emetic chemoreceptor trigger zone that may explain, in part, the nausea and vomiting that occur when patients are given chemotherapeutic agents that do not cross the BBB (Fig. 4–8).

Blood–Peripheral Nerve Barrier

Less is known about the blood–peripheral nerve barrier than the BBB.[44] Like the BBB, the blood–peripheral nerve barrier restricts movement of water, proteins, ions, and certain other water-soluble agents between the blood and peripheral nerve. However, it is slightly less restrictive than the BBB. The barrier site appears to be at endoneurial capillaries; epineurial vessel permeability does not differ from that of vessels elsewhere in the body, but the perineurium also forms a partial barrier.[44] Endoneurial capillaries, like brain capillaries, possess tight junctions, no fenestrations, and few pinocytotic vessels.[45] Enzymes, transporters, and receptors are slightly different at the endothelial and perineurial blood–nerve barriers from those at the BBB. The functional meaning of these differences is not clear.

The blood–peripheral nerve barrier is absent in the cellular, but not the nerve fiber portion of the dorsal root ganglia[46] (the cell body of sensory nerves), and in the terminal branches of both sensory and motor nerves. The absence of a blood–nerve barrier at the dorsal root ganglion may explain in part the sensory neuronopathy that sometimes follows cisplatin and other chemotherapeutic agents (see Chapter 12). The absence of a barrier in terminal branches of the nerve may permit ingress and retrograde transport of chemotherapeutic agents such as vincristine, which are neurotoxic. Tumor cells, particularly leukemia and lymphoma cells that traverse the endoneurium, are isolated from most systemic chemotherapeutic agents. This sequestration may explain some episodes of peripheral neuropathy from tumor invasion in patients after apparently successful systemic treatment for leukemia or lymphoma (see Chapter 8).

Blood–Cerebrospinal Fluid Barrier

The blood–CSF barrier differs from the BBB in several respects (Fig. 3–2).[3,47] Like the BBB, the blood–CSF barrier at the choroid plexus possesses tight junctions, but the junctional proteins differ slightly, making the barrier more leaky than the BBB.[48] The tight junctions are located between the epithelial cells of the choroid and not on endothelial cells as in the BBB. Thus, water-soluble agents can cross capillaries between capillary endothelial cells, but are restricted by choroid cells. As a result of their slightly increased leakiness, water-soluble agents, including some drugs that do not cross the BBB, can be found at low levels in the CSF. Choroid plexus cells possess enzymes and cofactors that can metabolize (detoxify) several drugs in a manner similar to liver detoxification.[48] The choroid plexus has a substantially smaller surface area than that of brain capillaries (5000-fold less), although on a weight basis, the reverse is true.[47]

The major sites of transport across the blood–CSF barrier are at the choroid plexus (where a substantial portion of the spinal fluid is secreted and the composition of newly formed CSF is determined) and at the arachnoid villi, where CSF is absorbed by bulk flow. To a lesser extent, transport also occurs along the spinal nerve root sheaths. Certain trace elements essential for brain nutrition, such as folic acid and vitamin B_{12}, that do not cross the BBB are transported into the nervous system via the choroid plexus and reach the brain by diffusion.[48] Certain acidic substances, such as penicillin and methotrexate, as well other neurotoxic metabolites, are transported from the CSF to the blood by the choroid plexus[47,48] (Fig. 3–2), particularly the choroid in the fourth ventricle.

The CSF is reabsorbed by bulk flow through the arachnoid villi, along spinal nerves exiting the subarachnoid space and through the cribriform plate into cervical lymphatics.[10,11] The rate of spinal fluid formation and absorption is approximately 0.35 mL/min. The CSF volume in the adult is approximately 150 mL, an amount that is achieved by the age of 4. Thus, the intrathecal dose of methotrexate or other drugs should be the same in children older than 4 years as in adults.[49] The total CSF volume turns over approximately four times a day. A barrier does not exist between the CSF and the brain because the ependyma lining the ventricles and the pia-arachnoid surrounding the brain and spinal cord allow free diffusion of substances, including proteins such as albumin. However, the diffusion rate into and through brain parenchyma is considerably slower than that of bulk flow. Furthermore, many substances are reabsorbed by the brain capillary bed once they have diffused a short distance into the parenchyma. Thus, most substances introduced into the CSF do not penetrate very far into the brain or spinal cord. The result is that drugs injected intrathecally, while exposing leptomeningeal tumor to high concentrations, usually fail to reach parenchymal lesions in significant concentration. Consequently, the method is unreliable for treating intraparenchymal tumors.[50] In experimental animals, appreciable concentrations of methotrexate have been shown to reach approximately 40% of the brain an hour after intraventricular administration, but this cannot be achieved in humans because of the much larger mass of the human brain. In both experimental animals and humans, white matter adjacent to CSF contains the highest drug concentration, possibly explaining the tendency for drug-induced leukoencephalopathy to be periventricular.

The CSF also serves as the brain's lymphatic system. Substances that either cross the BBB to enter brain interstitial space or are secreted or excreted by neurons and glia leave the nervous system by diffusing into the CSF and then are carried to the arachnoid granulations by bulk flow. In addition to diffusion, hydrostatic pressure may drive substances either toward or away from the spinal fluid. Thus, in the presence of a brain tumor, the tumor and the surrounding plasma-derived edema cause increased tissue pressure; the pressure drives substances from the tumor through the normal brain toward the subarachnoid space. Conversely, when subarachnoid pathways are blocked, pressure may drive substances from the CSF into the brain. Intrathecal methotrexate appears to be more toxic in patients with leptomeningeal tumor (with partial blockage of subarachnoid absorptive pathways) than when given prophylactically to individuals with normal subarachnoid pathways. Toxicity may be severe with intraventricular injection in patients with obstructive hydrocephalus.[51]

Most lipid-soluble substances enter the brain with little difficulty. Some examples common in neuro-oncologic practice are phenytoin, codeine, methadone, and carmustine. Morphine and aspirin enter the brain poorly. Most water-soluble chemotherapeutic agents, such as cytarabine and methotrexate, also enter the brain poorly, although the physician can substantially increase the concentration of these substances in the brain by raising the parenteral dose. For penetration of chemotherapeutic agents discussed later in this chapter, see Table 3–11.

DISRUPTION OF BLOOD–CENTRAL NERVOUS SYSTEM BARRIERS

Metastatic tumors in the nervous system can grow up to 1 to 2 mm without acquiring their own blood supply.[7,52] Thus, these small tumors are behind the BBB and find sanctuary from chemotherapeutic agents. Most water-soluble chemotherapeutic agents do not reach them in adequate concentrations, giving the tumors an opportunity to grow even if they are sensitive to the parenteral drugs received by the patient. This sequestration may explain the increasing frequency of brain tumors in patients whose systemic tumor is well controlled with chemotherapeutic treatment.

As tumors enlarge, they produce angiogenic factors[52] that promote the emergence of new capillaries to supply the growing tumors. These newly formed capillaries in metastatic brain tumors are fenestrated, lack tight junctions, and have an increased number of pinocytotic vesicles and an irregular basal lamina. As a result, they do not possess a normal blood–CNS barrier, so that they are more accessible to water-soluble chemotherapeutic agents than normal cells. Paradoxically, chemotherapy with water-soluble agents of established brain tumors is more effective than treatment of microscopic tumors, because the former have a disrupted BBB, whereas the latter have an intact barrier. Experiments by Ushio and coworkers[53,54] have demonstrated in laboratory animals with brain or leptomeningeal tumors that the tumors are less susceptible to systemically administered chemotherapeutic agents early in their growth, when the BBB is relatively intact, than they are later, when the barrier has been disrupted.

Even with an altered BBB, some experimental studies indicate that a water-soluble chemotherapeutic agent is less likely to penetrate a metastatic brain tumor than the same tumor found in other sites in the body.[55] However, other data suggest that in metastatic brain tumors with disrupted BBB, therapeutic levels of water-soluble chemotherapeutic agents can be achieved[56] (see Chapter 5).

Blood vessels in "normal" brain surrounding the tumor probably do not have a normal BBB[6,57,58] either. In some instances, this is due to an invasion of tumor cells into the parenchyma beyond the bulk of the tumor, a phenomenon uncommon in brain metastases except for some melanomas, renal cell carcinomas, and lymphomas. However, the disruption of the barrier does not necessarily depend on direct contact with tumor cells.[6] The degree of macrophage infiltration seen on immunoperoxidase staining correlates with the amount of peritumoral edema. Macrophagic infiltration is particularly prominent in metastatic brain tumors, perhaps partially explaining the marked edema that even small metastatic tumors often cause.[59] Tumor disruption of the BBB is rarely complete; it varies from tumor to tumor and within different regions of a single tumor.

The clinical observation is that most metastatic tumors show substantial contrast enhancement, which corresponds to disruption of the normal BBB. The commonly found enhancing rim and nonenhancing center does not indicate an intact BBB within the tumor center, but represents low blood flow to the poorly vascularized and sometimes necrotic tumor center. Delayed scans, after an injection of contrast, usually yield enhancement in the tumor center. If invaded by tumor, the leptomeninges, peripheral nerves, and nerve roots can also enhance on gadolinium magnetic resonance imaging (MRI) after gadolinium. The BBB in peritumoral edema may be abnormal, but is not sufficiently disrupted to enhance.

Brain Edema

Brain edema is an abnormal accumulation of fluid within the brain parenchyma. The increased fluid may be either in the extracellular space of the brain (vasogenic edema) or within cells, usually astrocytes (cytotoxic edema) (Table 3–4).[60] *Hydrocephalic* or interstitial edema is really a form of vasogenic edema. Hydrocephalic edema is particularly important in neuro-oncology because when tumors obstruct the ventricular system, the resulting hydrocephalus may drive CSF and substances injected into it into the brain parenchyma. As previously mentioned, when the BBB is focally disrupted, water-soluble substances and large molecules such as proteins can enter the brain more easily. Because the brain parenchyma has no lymphatics, these substances are not easily eliminated but are often driven into surrounding normal brain by increased hydrostatic pressure from the tumor, eventually to be eliminated by filtering through the white matter and then into the cerebral ventricles to be reabsorbed by bulk flow. Experimental evidence suggests that some protein can be reabsorbed by reverse vesicular transport across endothelial cells in the edematous brain,[61] perhaps aided by the presence of aquaporin-4 in astrocytic foot processes,[62] but few data are available to determine whether this mechanism is significant clinically.

Brain tumors cause vasogenic edema (Table 3–4). Because of arborization and interdigitation of dendritic spines, the edema develops primarily in white matter rather than in the more closely packed and flow-resistant gray matter. Vasogenic edema is extracellular and tends to follow the path of least resistance along white matter fiber tracts rather than diffuse in a spherical pattern. Thus, a small occipital tumor may cause edema infiltrating along fiber tracts all the way to the tip of the temporal lobe, leading to a characteristic MRI scan, with the hyperintense edema fluid extending like fingers into the white matter, outlining the cerebral cortex.

Vasogenic edema fluid has an elevated protein content including albumin, and the fluid volume of the extracellular space is increased. Most studies suggest that vasogenic edema, no matter how it is produced, is similar in content. One study of human tumors suggested, however, that the mean serum protein content in tissues adjacent to the tumor varied considerably, depending on the nature of the tumor: it was high in glioblastomas and low in peritumoral edema surrounding metastases.[64] Dexamethasone has similar salutary effects on water content in either malignant gliomas or metastases.[26,64,65] Because the edema fluid has more water and is less dense than normal brain, it can be readily imaged by MRI or computed tomography (CT). Characteristically, edema fluid is hypodense on CT scan and hyperintense on the T2-weighted or FLAIR MRI. As indicated earlier, edema fluid does not enhance with contrast, making it easy to differentiate metastasis from edema (see Fig. 3–5). However, in some primary brain tumors, infiltrating cells cannot easily be

Table 3–4. Classification of Brain Edema

	Vasogenic	Cellular (Cytotoxic)	Hydrocephalic (Interstitial)
Characteristics			
Pathogenesis	Increased capillary permeability	Cellular swelling; glial, neuronal, endothelial	Increased brain fluid from blockage of CSF absorption
Location of edema	Chiefly white matter	Gray and white matter	Chiefly periventricular white matter, hydrocephalus
Edema fluid composition	Plasma filtrate including plasma proteins	Increased intracellular water and sodium levels	CSF
Extracellular fluid volume	Increased	Decreased	Increased
Capillary permeability to large molecules (RIHSA, insulin)	Increased	Normal	Normal
Clinical Disorders			
Syndrome	Brain tumor, abscess, infarction, hemorrhage, lead encephalopathy Ischemia Purulent meningitis (granulocytic edema)	Hypoxia, ischemia hypo-osmolality (e.g., water intoxication) Dysequilibrium syndromes Ischemia Infarction Purulent meningitis (granulocytic edema) Reye syndrome	Obstructive hydrocephalus Purulent meningitis (granulocytic edema)
EEG changes	Focal slowing common	Generalized slowing	EEG often normal
Therapeutic Effects			
Steroids	Beneficial in brain tumor, abscess	Not effective (possibly Reye syndrome)	Uncertain effectiveness (possibly pseudotumor or meningitis)
Osmotherapy	Reduces volume of normal brain tissue only, acutely	Reduces brain volume acutely	Rarely useful Improves compliance
Acetazolamide	May be useful[109]	No direct effect	Minor usefulness
Furosemide	May be useful[122]	No direct effect	Minor usefulness

Adapted from Ref. 63.
CSF, cerebrospinal fluid; RIHSA, radioiodinated human serum albumin; EEG, electroencephalogram.

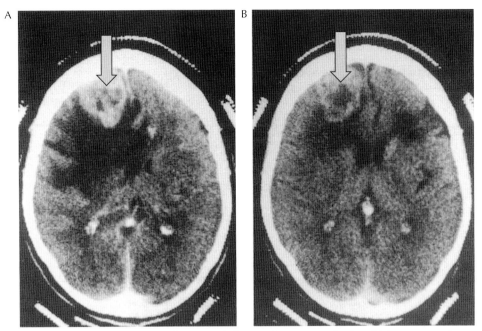

Figure 3–5. A contrast-enhanced computed tomograph showing the response of a frontal-lobe metastatic brain tumor (*arrows*) to dexamethasone. **A**: Massive edema with deformity and compression of lateral ventricles as well as herniation across the midline was present before dexamethasone treatment. **B**: Remarkable decrease in edema volume after the administration of dexamethasone. Note the decreased enhancement of the tumor without a change in size.

distinguished from edema on routine scans, although perfusion-weighted MRI may demonstrate the increased blood flow associated with tumor infiltration.[66]

In metastatic brain tumors, vasogenic edema occurs when angiogenesis produces tumor vessels that do not have a normal BBB. Albumin and other intravascular substances leak into the tumor and are driven by hydrostatic forces into the extracellular space of the surrounding white matter.[67,68] The formation within the tumors of new vessels that do not have a BBB is certainly important to the pathogenesis of brain edema, but new vessel formation (i.e., angiogenesis) may not be the only way that edema is formed in brain tumors. Substantial evidence indicates that, in addition to angiogenic substances, tumors secrete substances that not only promote angiogenesis of new vessels lacking a BBB, but also promote leakiness of normal brain capillaries in the brain immediately surrounding the tumor.[68,69] Among other substances that play a role in producing vasogenic edema, vascular endothelial growth factor (VEGF) is probably the most important. Other factors include arachidonic acid metabolites, nitrous oxide, leukotrienes, and

prostaglandins.[68] Aquaporin-4 may play a role in producing cytotoxic brain edema and in resolving vasogenic brain edema.[62]

Other Sources of Disruption

Tumor growth is not the only mechanism that can disrupt the blood–CSF barrier. An increase in hydrostatic pressure, such as that caused by severe arterial hypertension, can disrupt the barrier[70] and lead to brain edema (see also posterior reversible encephalopathy syndrome [Chapter 12, p 451]). Compression of nervous system tissue can also disrupt the barrier, probably by interfering with venous drainage and increasing capillary hydrostatic pressure; this is possibly the mechanism of tissue edema caused by tumors compressing the brain or spinal cord but not directly invading them.[71] Thus, mechanisms of edema formation in metastatic brain tumors and epidural spinal cord compression appear different. In metastatic brain tumors, leaky neovessels within the substance of the tumor and altered normal vessels in the brain surrounding the tumor lead to the formation of brain edema. Whether increased tissue pressure disrupts the BBB in the brain surrounding a tumor is not entirely clear. On the other hand, in the patient with epidural spinal cord compression, the tumor is not in direct contact with the spinal cord, and neovascularity is absent; the edema is a result of normal vessels leaking within the spinal cord. It has not been fully established whether hydrostatic pressure disrupts the blood–spinal cord barrier or whether some other damaging effect of spinal cord compression occurs. Edema caused by either intrinsic or extrinsic tumors responds to treatment with corticosteroids (see following text). Other forms of edema are less amenable to corticosteroid therapy (Table 3–4).

Whether edema fluid is toxic to the brain in and of itself (see Fig. 4–3), or whether all of its symptoms are a consequence of mass effect distorting normal tissue, is not clear. Edema fluid has a higher potassium concentration than normal brain extracellular fluid, possibly promoting depolarization of neurons, which could lead to seizures. Patients who have prolonged seizures may develop cytotoxic edema as well.[72] The pragmatic issue is that cerebral edema unquestionably is instrumental in causing symptoms in patients with metastatic tumors. Amelioration of the edema with corticosteroids

or other substances (see following text) often substantially improves the symptoms even before the tumor itself is treated.

Other Substances Causing Blood–Brain Barrier Disruption

Some chemotherapeutic agents may cause BBB disruption. In experimental animals, the intracarotid infusion of etoposide or vinorelbine disrupts the BBB.[73,74] Barrier disruption also appeared to occur in a patient given high-dose intravenous methotrexate.[75] Other substances that may disrupt the barrier when given in the carotid artery include protamine sulfate,[76] and leukotriene C4[77] and sodium caprate (C10).[78] These are probably of no clinical significance (Table 3–5). Therapeutic barrier disruption with hyperosmolar mannitol or Cereport (RMP-7), in order to enhance entry of chemotherapeutic agents, is considered in the paragraphs that follow.

Seizures can cause BBB disruption, probably related to seizure-induced arterial hypertension.[72,79] The sudden accumulation of brain edema following a seizure may worsen neurologic symptoms or even cause herniation in patients with metastatic brain tumors—a reason to control seizures in such patients.

Table 3–5. Agents Modifying Brain Endothelial Function and Blood–Brain Barrier Tightness*

Agents that Impair BBB Function

Bradykinin, histamine, serotonin, glutamate

Purine nucleotides: ATP, ADP, AMP

Adenosine, platelet-activating factor

Phospholipase A2, arachidonic acid, prostaglandins, leukotrienes

Interleukins: IL-1α, IL-1β, IL-6

Tumor necrosis factor α, macrophage-inhibitory proteins MIP1 and MIP2

Complement-derived polypeptide C3a-desArg

Free radicals, nitric oxide

Cerebral arterial infusion of hypertonic agents (mannitol)

Agents that Cause Barrier Tightening and Improve Function

Steroids, elevated intracellular cyclic AMP, adrenomedullin, noradrenergic agents, corticotropin-releasing factor, bevacizumab (Avastin)

Modified from Ref. 1 with permission.
* A number of chemical agents circulating in the plasma or secreted from cells associated with the BBB are capable of increasing brain endothelial permeability and impairing endothelial transport and metabolic functions. Other agents have the opposite effect, improving tightness and BBB function.

Therapeutic irradiation also increases permeability across the blood–CNS barrier (see Chapter 13). Levin et al.[80] observed a significantly increased permeability of the BBB during the first 24 hours after cranial radiation with 200 to 400 cGy. Permeability was reduced after 2000 cGy in 10 fractions. The disruption of the BBB may explain acute radiation reactions sometimes encountered clinically (see Chapter 13, p 515). Delattre et al.[81] reported that, in an experimental brain tumor model, a single dose of 3000 cGy or a total dose of 3000 cGy delivered in 10 fractions (the usual schedule for treatment of brain metastases) increased the blood-to-brain transport of small molecules and partially disrupted the BBB in normal brain, although no change was seen in the already partially disrupted barrier of the brain tumor. The clinical implication is that concurrent radiation and chemotherapy may promote the entry of chemotherapeutic agents into the brain without increasing entry into the tumor. However, a recent report using MRI in patients being treated with conformal radiation for malignant gliomas found increased BBB opening in the tumor but not in normal brain.[82] In addition, one report found that delivery of 200 cGy 5 days a week to experimental animals resulted in no increase in BBB opening until 90 days after the treatment.[83] A recent study in experimental animals indicates that radiation of the brain reduces P-glycoprotein expression allowing entry of substances that it normally excludes.[84] Unfortunately, the clinical importance of BBB disruption in patients receiving standard doses of radiation remains unresolved.

Transient "opening" of the BBB by the intracarotid infusion of hyperosmolar agents such as mannitol (BBB disruption) has been used therapeutically to treat patients with brain or leptomeningeal metastases.[85] The barrier is widely open for only several minutes, but may not return to baseline for several hours.[86] The increase in barrier permeability is both greater and longer lasting in normal brain than in the brain tumor.[87] During the time the barrier is open, water-soluble agents administered systemically or intra-arterially can enter the brain at a substantially greater rate. The procedure may cause complications[88–90]; whether it is efficacious is controversial.

The mechanism by which hyperosmolar agents open the BBB is controversial. The best evidence indicates that shrinkage of endothelial cells widens inter-endothelial cell junctions, allowing substances to cross capillaries[91]; other studies suggest increased trans-endothelial vesicular transport.[92–94] Experimentally, several studies have shown that modification of the BBB has a far greater effect on drug entry into normal brain than into the tumor, thus exposing normal brain to greater concentrations of potentially neurotoxic drugs.[95]

A bradykinin analog called Cereport can open the BBB when given intravenously[88,96]; the barrier remains open for 20 minutes. However, its clinical usefulness has not yet been established (Table 3–6).

Table 3–6. **Some Mediators of Brain Edema**

Mediator	Proposed Mechanism
Oxygen-free radicals damage endothelial cells	Peroxidation of lipid membranes
	DNA strand breaks
	Depletion of cellular energy stores
	? Reaction with nitric oxide to form toxic peroxynitrite
Bradykinin	Dilation of cerebral vessels
Arachidonic acid	Dilation of cerebral vessels
	? Production of oxygen-free radicals
Glutamate	Cellular energy depletion
	Dilation of cerebral vessels
Hormones: arginine vasopressin, atrial natriuretic peptide	Disruption of water and electrolyte homeostasis
Nitric oxide?	Vasodilation
Histamine	Vasodilation

CONSEQUENCES OF BLOOD–BRAIN BARRIER DISRUPTION: INCREASED INTRACRANIAL PRESSURE, PLATEAU WAVES, AND CEREBRAL HERNIATION

In patients with mass lesions involving or compressing nervous system structures, the breakdown of blood–nervous system barriers and the production of brain edema often lead to increased intracranial pressure. The pressure may be distributed evenly throughout the brain (e.g., obstruction of the superior vena cava or sagittal sinus, pseudotumor cerebri or sometimes multiple metastases affecting both hemispheres and cerebellum evenly raise the pressure diffusely) or it may be distributed unevenly, as exemplified in patients with focal mass lesions.[97,98] Evenly distributed increased intracranial pressure may cause headache, but no focal neurologic symptoms until the intracranial pressure approaches the arterial blood pressure, resulting in cerebral ischemia. Focally increased tissue pressure causes herniation of portions of normal brain into areas of lower pressure. The most common sites of herniation, as illustrated in Figures 1–6 and 3–6, include herniation of the medial frontal lobe under the falx cerebri, herniation of the medial temporal lobe through the tentorium cerebelli, and herniation of the cerebellar tonsils through the foramen magnum. The clinical consequences of these herniations are described in Chapter 5. Focally increased tissue pressure can also cause symptoms by interfering with the local blood supply. Both mechanisms are probably responsible for many of the symptoms caused by brain and spinal cord tumors.

Some patients with normal or slightly elevated intracranial pressure at rest, or with chronically elevated intracranial pressure from mass lesions or compensated hydrocephalus from leptomeningeal tumors, suffer from intermittent episodes of neurologic dysfunction that result from sudden rises in intracranial pressure called *plateau waves*.[100–102] These episodes last 5 to 20 minutes and then cease (Table 3–7).

Plateau waves, first described by Lundberg[101] in 1960 (Fig. 3–7), appear to result from an increase in cerebral blood volume because of a sudden decrease in cerebral vascular resistance[102–105]; blood flow may actually decrease as the blood volume rises. Active vasoconstriction is believed to terminate the pressure rise.[102] Patients in whom an intracranial mass and elevated pressure have caused diminished CSF absorption are more likely to develop plateau waves than those with normal CSF absorption.[103] Conversely, patients who have plateau waves develop signs of cerebral herniation at a higher pressure than those who do not. One suggestion is that CSF pooling resulting from decreased CSF absorption forms a buffer to protect against herniation.[103]

Plateau waves can occur spontaneously or be precipitated in patients with increased intracranial pressure or hydrocephalus. Common precipitating causes include tracheal suctioning, coughing or sneezing, and particularly rising from a lying or sitting position,[106] especially in the morning after a night's sleep. Upon arising, both CSF and venous volume in the brain decrease, leading to a fall in intracranial pressure. Vasodilatation compensates for the decrease in venous and CSF volume in the normal individual, but there appears to be an overshoot in patients with increased intracranial pressure, leading to the plateau wave.[107,108] Plateau waves are often asymptomatic even when they cause a dramatic increase in intracranial pressure. Symptoms, when they occur, can be quite variable (Table 3–8); the most common are headache, altered consciousness, and sudden weakness of both lower extremities with collapse but preservation of consciousness. Plateau waves respond dramatically to relief of raised intracranial pressure by either corticosteroids or other mechanisms. Although corticosteroids constitute the most widely used treatment, both indomethacin, a cerebral vasoconstrictor,[105] and acetazolamide, a carbonic anhydrase inhibitor, have proved effective in some reports.[109] Occasionally, ventriculoperitoneal shunting is necessary. Because of their dramatic symptomatology, plateau waves can be mistaken for seizures, transient attacks of cerebral ischemia, hemorrhage into a metastatic tumor, or migraine equivalents. A correct diagnosis of plateau waves depends on the physician's suspecting that plateau waves are responsible for the patient's symptoms. Intracranial pressure monitoring may also establish the diagnosis but is not always feasible or necessary. The diagnosis is often established on the basis of a therapeutic trial with corticosteroids. Plateau

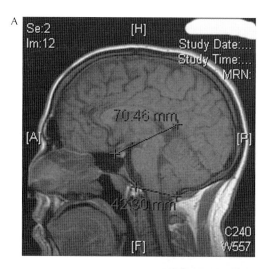
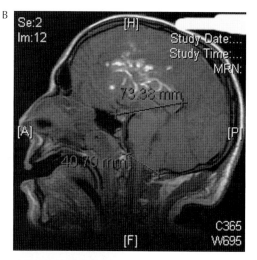
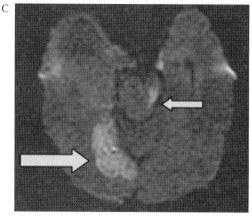

Figure 3–6. Cerebral herniation from a supratentorial hemorrhage. A young man with promyelocytic leukemia developed headache during the course of treatment. Magnetic resonance imaging (MRI) scan (**A**) was essentially normal. A line drawn from the anterior tuberculum sellae to a point marking the confluence of the straight sinus, great cerebral vein, and the inferior sagittal sinus identified the upper end of the aqueduct to be in normal position (upper line).[99] A line drawn from the inferior lip of the clivus to the bony base of the posterior lip of the foramen magnum (lower line) identified the tonsils of the cerebellum also to be in normal position. His headache grew worse and his state of consciousness deteriorated. A repeat MRI (**B**) identified transtentorial herniation. Both the cerebral aqueduct and the cerebellar tonsils are below their normal position. The patient subsequently died from uncontrollable hemorrhage. (**C**) A diffusion-weighted MRI scan demonstrating two of the consequences of cerebral herniation: (1) an ipsilateral occipital lobe infarct from compression of the posterior cerebral artery (*thick arrow*) and (2) a contralateral hemorrhagic infarct in the midbrain (Kernohan notch) from compression of the midbrain against the tentorium cerebelli (*thin arrow*) (see also Fig. 1–6).

Table 3–7. **Characteristics of Plateau Waves**

Usually associated with intracranial hypertension

May occur without any appreciable increase in blood pressure

Usually not observed in hydrocephalic infants with an open fontanelle

Often preceded by hypercarbia, painful stimuli, activity, increase in blood pressure, or change in body position

Temporarily controlled by ventricular drainage, administration of hypertonic solution, hyperventilation

Wider arterial vessels during the plateau wave than in the interval between waves shown by angiograms during spontaneous plateau waves, but venous phase unaffected

Increased cerebral blood volume and reduced cerebral blood flow reported during plateau waves

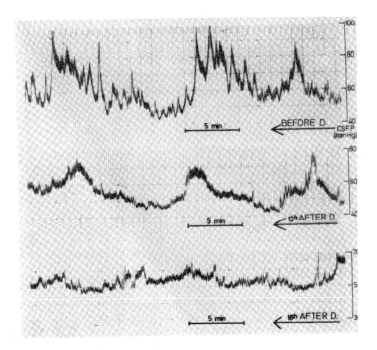

Figure 3–7. Continuous registration of the cerebrospinal fluid pressure (CSFP) in a patient with increased intracranial pressure before and after being treated with dexamethasone (D). The curves read from right to left. The amplitude and frequency of the plateau waves diminish rapidly; the overall intracranial pressure diminishes more slowly. (From Ref. 24 with permission.)

Table 3–8. Paroxysmal Symptoms and Signs from Plateau Waves in Patients with Intracranial Space-Occupying Lesions

Impairment of consciousness	Opisthotonus, trismus
Trance-like state	Rigidity and tonic extension/flexion of the arms and legs
Unreality/warmth	
Confusion, disorientation	Bilateral extensor plantar responses
Restlessness, agitation	Sluggish/absent deep tendon reflexes
Disorganized motor activity, carphologia	Generalized muscular weakness
Sense of suffocation, air hunger	Facial twitching
Cardiovascular/respiratory disturbances	Clonic movements of the arms and legs
Headache	Facial/limb paresthesias
Pain in the neck and shoulders	Rise in temperature
Nasal itch	Nausea, vomiting
Blurring of vision, amaurosis	Facial flushing
Mydriasis, pupillary areflexia	Pallor, cyanosis
Nystagmus	Sweating
Oculomotor/abducens paresis	Shivering and "goose flesh"
Conjugate deviation of the eyes	Thirst
External ophthalmoplegia	Salivation
Dysphasia, dysarthria	Yawning, hiccoughing
Nuchal rigidity	Urinary and fecal urgency/incontinence
Retroflexion of the neck	

waves can lead to cerebral herniation, respiratory arrest, and death.[103]

Treatment of Cerebral Herniation

Immediate treatment of increased intracranial pressure is required to reverse or prevent cerebral herniation and prevent death. The treatment of increased intracranial pressure and cerebral herniation is discussed extensively elsewhere[97,110,111] and includes hyperventilation, hyperosmolar agents, and adrenocorticosteroids (Table 3–9).

Hyperventilation is the most rapid technique for lowering intracranial pressure and reversing herniation. Hyperventilation lowers pCO_2, causing cerebral vasoconstriction and decreasing cerebral blood volume. If the patient is unconscious, an endotracheal tube should be inserted and the patient ventilated to a pCO_2 level between 25 and 30 mm Hg, which lowers intracranial pressure rapidly in most patients but only transiently. Many patients with cerebral herniation spontaneously hyperventilate to these levels and require no additional respiratory intervention from the physician. However, intubation is advisable to protect the airway. Respiratory function must be monitored carefully because brain herniation can cause respiratory failure. Mechanical ventilation can raise as well as lower intracranial pressure; patients who have brain lesions and require ventilation are at risk for exacerbation of their elevated intracranial pressure.[112] Controversy exists as to whether hyperventilation decreases blood flow enough to cause cerebral ischemia and thus make the situation worse. Most studies have been done on head-injured patients, a situation analogous to patients with increased intracranial pressure from a brain tumor. One study suggests that even moderate hyperventilation (PCO_2, 27–32 mm Hg) causes hypoxia in normal brain near the injured area.[113] Another suggests that despite marked decreases in blood flow, oxygen metabolism is preserved.[114] Clinical observations of brain tumor patients suggest that brief periods (20 minutes) of moderate hyperventilation ($\sim$30 mm Hg) are safe and effective.[114a]

Hyperosmolar agents decrease the water content of the brain by creating an osmolar gradient between the blood and that portion of the brain with an intact BBB (Table 3–10). The agent of choice had been mannitol given IV over 10 to 20 minutes as a 20% solution at a dose of 0.5 to 2 g/kg[115,116]; however, there has been a recent shift toward the use of hypertonic saline in preference to mannitol.[97,117] Mannitol may also function to decrease CSF formation and volume.[118] In a few patients with severe and sustained increased intracranial pressure, pressure can be monitored and the dose tailored to maintain decreased pressure.[111] In patients with normal intracranial pressure, mannitol actually may increase the pressure briefly, probably because of a transient increase in cerebral blood flow and volume. This does not appear to occur in patients with increased intracranial pressure.[115] Mannitol effects are rapid and last several hours. Mannitol injections may be repeated in smaller doses if the patient responds and subsequently relapses. Repeated doses of hyperosmolar agents also can cause a reverse osmotic

---

Table 3–9. **Emergency Treatment of Impending Herniation in Acutely Decompensating Patients**

Therapy	Dosage or Procedure	Onset (Duration) of Action
Hyperventilation	Lower pCO_2 to 25–30 mm Hg	Seconds (minutes)
Osmotherapy	Mannitol, 0.5–2.0 g/kg (IV) over 15 minutes, followed by 25 g (IV "boosters") as needed or hypertonic saline, 300 mL of a 3% solution given over 20 minutes, or 7.5% saline 2 mL/kg	Minutes (hours)
Corticosteroids	Dexamethasone, 100 mg IV push, followed by 40–100 mg/24 hr, depending on symptoms	Hours (days)

Table 3–10. **Principles of Osmotherapy**

1. Brain volume and elastance decrease in the presence of an osmotic gradient.
2. Osmotic gradients are short lived; solutes reach equilibrium in the brain after a delay of a few hours.
3. With vasogenic edema, the normal brain areas "shrink"; regions with a disrupted blood–brain barrier do not.
4. Osmotherapy is most effective in reducing intracranial pressure in normal subjects and least effective when the blood–brain barrier is impaired.
5. Rebound may follow the administration of any hypertonic solution.
6. Brain adapts to sustained hyperosmolality with the equilibration of the administered solute and the appearance of "idiogenic" osmoles.

From Ref. 63 with permission.

effect. Some of the agent finds its way into the brain, particularly in areas of BBB disruption. Those areas of the brain become hyperosmolar and when the serum osmolality returns to normal, fluid is drawn into the brain, exacerbating cerebral edema. The addition of a diuretic (furosemide) enhances and prolongs the hyperosmolar effect.[119]

Many physicians prefer hypertonic saline to mannitol.[120] The dose is in the range of 7 to 10 g of sodium chloride in concentrations from 3% to 23.4% administered by rapid injection or, in some cases, by continuous intravenous infusion titrated to the intracranial pressure[117,120,121] or to achieve a serum sodium concentration of 145 to 155 mEq/L. Although no controlled trials comparing mannitol with hypertonic saline in patients with brain tumors have been reported, the results with saline appear to be at least comparable to those with mannitol.[120] Brain dehydration can be maintained for several days without complications.

Dexamethasone (100 mg, IV) is given immediately, followed by doses of 40 to 100 mg/24 hr, depending on the patient's response to treatment. Some physicians add furosemide (40 to 120 mg, IV) to the corticosteroid, and believe that the combination is better than corticosteroids alone.[122]

Barbiturate anesthesia proved successful in one patient with intractable increased intracranial pressure from choriocarcinoma metastases.[123] She subsequently responded to chemotherapy. Barbituates are rarely used to treat herniation from brain tumors because other methods are so effective.

In patients with intractable brain edema from traumatic brain injury or massive stroke, *decompressive craniectomy* is sometimes used to relieve what would otherwise be lethal intracranial pressure.[110] In patients with a large single brain metastasis, surgical resection can, under appropriate circumstances, both relieve pressure and eliminate its cause (see Chapter 5).

With such vigorous treatment, most patients who herniate from a brain metastasis can be stabilized and many have complete amelioration of their symptoms within a few days. Fortunately, most patients with brain metastases, even with extensive peritumoral edema, do not require more than anti-edema therapy with corticosteroids (see following text).

TRANSPORT OF CHEMOTHERAPEUTIC DRUGS ACROSS THE BLOOD–BRAIN BARRIER

Lipid-soluble drugs penetrate the BBB with ease and are carried to the brain and tumors in direct proportion to blood flow. Blood flow is heterogeneous in most primary and metastatic tumors and, in many areas, the flow may be quite low.[124] Thus, in general, when lipid-soluble chemotherapeutic agents are used, the normal brain receives greater exposure than does the tumor. A few substances, such as melphalan, are transported across capillary endothelium by facilitated diffusion.[125]

Several factors determine how much chemotherapeutic agent, lipid- or water-soluble, will enter the brain[126,127]:

1. One factor is the concentration of free drug in the plasma, unbound to plasma

proteins. Binding of a drug to a plasma protein usually limits its entry into the brain, but some plasma protein–bound substances may become more available for transport through the barrier because of specific interactions between plasma proteins and the endothelial surface, interactions called endothelial-enhanced absorption of ligand-protein. The degree to which a given ligand is transported through the BBB depends on the site the ligand binds to. Albumin has at least six different endothelial binding sites.

2. Another factor determining entry of a drug into the brain is the drug's ionization. An ionized drug crosses the BBB with much more difficulty (10,000 times) than a non-ionized form.

3. Plasma concentration level and the drug's duration in the circulation (the "concentration × time" product) are both important. Therefore, clearance of the drug also determines exposure, and the area under a concentration × time curve expresses the total exposure of tissue to a chemotherapeutic agent.

4. For water-soluble drugs, the permeability of the capillaries and their total surface area are also important factors. The increased capillary permeability in tumors leads to the entry of a greater amount of water-soluble chemotherapeutic agents into the tumor than into normal brain.

5. The route of administration can also be important, especially for a drug with a very short plasma half-life. Intra-arterial injections lead to substantially greater entry of drug into both tumor and normal brain than the same dose administered intravenously.[128] The "concentration × time" advantage of intra-arterial chemotherapy is achieved during the first transcapillary passage of the drug since there is no difference in drug exposure between intra-arterial and intravenous administration for the second and subsequent passages of the drug through tumor capillaries. The first capillary passage advantage can be calculated from the ratio of the cardiac output over the blood flow. ([Net drug flux] BBB = [Drug concentration] Arterial blood × Blood flow × [BBB permeability] Extraction fraction − [Drug concentration] Tumor or Brain × [BBB permeability] Efflux constant).

The technique of intra-arterial injection allows one, depending on the above factors, to expose the tumor to more chemotherapeutic agent than the bone marrow or other systemic organs. However, the distribution of intra-arterially injected drugs is variable and not entirely predictable because streaming within the arterial lumen affects drug distribution.

The result of all the factors described in the preceding text is that certain chemotherapeutic agents cross the normal BBB and enter both normal brain and brain tumor; others cross only a disrupted BBB, and still others rarely reach either the brain or the tumor in clinically significant concentrations. The effects of chemotherapy on the treatment of brain metastases are described in Chapter 5. The ability of chemotherapeutic agents to penetrate normal brain is listed in Table 3–11.

Table 3–11. Brain Capillary Permeability of Chemotherapeutic Agents

High		Intermediate	Low
Carmustine	Cytarabine (high dose)	Cisplatin/carboplatin	Vincristine
Lomustine	Topotecan	Irinotecan	Taxanes
Procarbazine	Hydroxyurea	Bleomycin	Gemcitabine
Thiotepa	Methotrexate (high dose)	Proteins (e.g., interferon α, trastuzumab)	
Nimustine	Temozolomide	Etoposide/teniposide	Doxorubicin

From Ref. 129 with permission.

RESTORATION OF BLOOD–CENTRAL NERVOUS SYSTEM BARRIERS

Lowering Intracranial Pressure

The breakdown of blood–CNS barriers leads to the formation of edema. Several agents and methods are used to decrease nervous system edema and lower tissue pressure. As indicated earlier, physical methods include hyperventilation and hyperosmolar agents such as mannitol or saline.[111,119,130]

Corticosteroids are almost miraculous in ameliorating the clinical symptoms of metastatic brain tumors, diminishing pressure waves, lowering intracranial pressure, and decreasing peritumoral edema. However, because of their side effects (see Chapter 4) attempts have been made to find less toxic substitutes. A wide variety of drugs have been proposed, based on either experimental data or case reports. Some of these drugs include nonsteroidal anti-inflammatory agents such as indomethacin,[131–133] Cox-2 inhibitors,[8,134,135] lazaroids,[136] Boswellic acids,[8,137] glutamate inhibitors,[138] thrombin inhibitors,[139] and estrogen and progesterone.[140,141] Glycerophosphoinositol improves the trans-endothelial electrical resistance of the BBB in vitro in a manner similar to that of dexamethasone.[142] The drug appears to be substantially less toxic than dexamethasone, which similarly restores the BBB. The clinical efficacy of these drugs is unproved.

Corticotropin-releasing factor (CRF) has proved effective in ameliorating tumor-induced brain edema in experimental animals.[143] Preliminary evidence suggests that it may be effective in humans as well.[144] The mechanism of action is not known. The edema-ameliorating effects are not a result of CRF-induced cortisol release because CRF is effective in adrenalectomized brain tumor–bearing animals. Clinical trials are now under way. VEGF inhibitors such as bevacizumab are also promising in the treatment of brain edema,[145] although more clinical data are necessary to delineate their role fully.

Corticosteroids

The most widely used drugs in neuro-oncology are the synthetic glucocorticoids, commonly referred to simply as steroids or corticosteroids.[146–148] They are the mainstay of treatment for nervous system edema and increased pressure. The history of brain edema and corticosteroids is interesting. Prados et al. in 1945 noted that when the brain was exposed to air, the BBB failed and brain edema developed.[149] If the brain was sprayed with adrenal extract, the BBB remained intact and the brain did not swell. In 1952, Ingraham et al.[150] recognized that cortisone and adrenocorticotropic hormone (ACTH) "smoothed the postoperative course" of patients after surgery for craniopharyngioma. Kofman et al.[151] first demonstrated, in 1957, the effectiveness of corticosteroids in relieving symptoms of breast cancer metastatic to the brain (an effect that may have, in part, been oncolytic). In 1961, Galicich and French[152] demonstrated amelioration of symptoms of primary brain tumors by dexamethasone (see following text). Since that time, these drugs have been used in virtually all patients suffering from metastatic brain and spinal cord tumors. Corticosteroids have an oncolytic effect on a few tumors such as lymphomas and breast cancer, but are used by the neuro-oncologist primarily to relieve the symptoms of edema affecting the brain, spinal cord, and possibly nerves and nerve roots. They are also used to treat CNS edema that occurs as a side effect of chemotherapy or radiation therapy.

Buttgereit and coworkers[153,154] have postulated that, when used clinically, corticosteroids have three different mechanisms of action. The best established mechanism is genomic; steroids pass through the cell membrane and bind to a receptor in the cytoplasm. The steroid-receptor complex moves to the nucleus where it binds to specific DNA sites and serves as a transcription factor. Only physiologic doses are required to activate this pathway, but at least 30 minutes are required to see an effect. The second mechanism is a specific nongenomic one that requires the steroid to bind to receptors that are probably in the cell membrane. Higher doses are necessary, but the action occurs within minutes. The third mechanism is nonspecific and nongenomic. The mechanism probably results from direct effects of the steroid on cell membranes. As a result, clinical improvement begins rapidly,[155] but does not peak for 24 to 72 hours, probably because time is required to clear the edema. Pharmacologic doses are required, but the effect occurs in seconds.[153,154] Corticosteroids probably exert their major effect in a nonspecific, nongenomic manner by partially reversing the disruption of the BBB, blood–spinal cord barrier, or blood–nerve barrier caused by metastases to those organs.[25,65] Some other mechanisms are considered in the final section of this chapter.

A one-day prevalence study in 1986 indicated that 33% (13 of 40) of patients on the Neurology Service at The New York Hospital and 69% (22 of 32) of the patients on the Neuro-Oncology Service at Memorial Sloan-Kettering Cancer Center (MSKCC) were receiving corticosteroids. Overall, approximately 15% of all patients hospitalized at MSKCC receive corticosteroids at some time.

The optimal dose and best preparation of corticosteroids are not established.[156] The dosage probably should differ with both the nature of the problem and its severity.[157] Equivalent doses of the corticosteroids commonly used by neuro-oncologists are indicated in Table 3–12. These doses apply to the genomic effects of steroids. For nonspecific, nongenomic effects, if methylprednisolone is 1.0, then prednisolone is 0.4 and dexamethasone is 1.2.[154] Because the exact mechanism of corticosteroid effect on brain edema is not known, most physicians use genomic equivalents to switch from one corticosteroid to another.

Dexamethasone is the most widely used corticosteroid in neuro-oncology, largely because it was the drug used by Galicich and French[152] to treat brain tumors. Dexamethasone may be preferred to other synthetic glucocorticoids for several theoretical reasons. First, it has no mineralocorticoid effect, so that it is the least likely corticosteroid to cause salt retention and systemic edema formation. Some investigators believe that it is also less likely than other synthetic corticosteroids to cause cognitive and behavioral dysfunction[67] (see Chapter 4). Dexamethasone is said to inhibit leukocyte migration to a lesser extent than other synthetic glucocorticoids, perhaps decreasing the risk of infection.[158] On the other hand, experimental evidence indicates that dexamethasone and other fluorinated corticosteroids are more likely to cause steroid myopathy. For many disorders, prednisone often is given every other day to minimize side effects; however, every-other-day prednisone is ineffective in controlling brain edema and its symptoms. Because dexamethasone has a long half-life, an every-other-day dosage does not lessen side effects. On balance, dexamethasone appears to be the best drug to treat metastatic CNS tumors.

Only a few experiments address a dose–response curve for CNS effects of corticosteroids. One such experiment in an animal model of spinal cord compression suggests that doses equivalent to 100 mg/24 hr of dexamethasone may be superior to lesser doses both in decreasing edema and in ameliorating clinical symptomatology.[159] Because reports thus far give no indication that these higher doses given for a few days are deleterious, our practice is to use high doses in seriously symptomatic or deteriorating patients, tapering the dose after

Table 3–12. **Relative Potency of Corticosteroids**

Steroid	Glucocorticoid Activity[*]	Mineralocorticoid Activity[*]	Equivalent Dose (mg)
Short acting (8–12 hr[†])			
Cortisol (hydrocortisone)	1.0	1.0	20.0
Cortisone	0.8	1.0	25.0
Intermediate (12–36 hr[†])			
Prednisolone	4.0	0.8	5.0
Prednisone	4.0	0.8	5.0
Methylprednisolone	5.0	0.5	4.0
Long acting (>36 hr[†])			
Dexamethasone	25.0	0.0	0.75

[*] Relative to cortisol.
[†] Biologic half-life.

a few days to the lowest dose consistent with symptomatic control.

Corticosteroid treatment begins with the physician's selecting a dose appropriate for the neurologic disorder and its severity[146] (see following text). Because of its long half-life, dexamethasone can be given twice daily, although most physicians give four divided doses. The drug is well absorbed from the gastrointestinal tract, but first-pass hepatic metabolism may decrease the effectiveness of an oral dose, especially in patients taking phenytoin (see Chapter 4). Because side effects are numerous and often serious, patients should be maintained on the lowest dose of corticosteroids that affords relief of symptoms. Thus, once symptomatic control is established and more definitive therapy (e.g., surgery, radiation, or chemotherapy) is under way, the corticosteroid should be tapered to the lowest possible dose (see following text).

BRAIN METASTASES

Symptomatic but stable patients can begin on 16 mg of dexamethasone daily (one report suggests that 4 or 8 mg is as effective as 16 mg).[160] Approximately 70% to 80% of patients with brain metastases achieve symptomatic improvement. The improvement is more dramatic in patients with symptoms and signs of *generalized brain dysfunction*, that is, headache, lethargy, papilledema, and cerebral herniation caused by distortion of the intracranial contents by tumor and edema. The drug is less effective in patients who have only focal signs, for example, hemiparesis, aphasia, or visual field defects without headache or lethargy. If the standard dose fails to produce a clinical response within 48 hours, the dose should be doubled each 48 hours until a response is achieved. Up to 100 mg/24 hr of dexamethasone may be required.[157,161] For patients with signs of increased intracranial pressure, cerebral herniation, or plateau waves, an IV bolus of 40 to 100 mg may be given, followed by a daily dose equal to the size of the effective bolus. If the initial dose controls symptoms, consider beginning a taper (see section on corticosteroid taper) to achieve the lowest dose for symptom control. This may require tapering until symptoms reappear and then raising to the next higher dose.

Patients with metastatic tumors of the brain, and sometimes the spinal cord, begin to improve within an hour after an IV injection of dexamethasone. Maximal clinical improvement on continued treatment occurs within 24 to 72 hours. The positron emission tomography (PET) scans of humans with brain tumors show an effect on the BBB as early as 6 hours after a 100 mg IV bolus of dexamethasone. The transfer constant of rubidium, a water-soluble marker resembling potassium, decreased after steroid administration.[25] The apparent blood and tissue volumes in and around the tumor also decreased at a later time.

Patients with brain tumor headache, lethargy, and sometimes weakness are demonstrably better within a few hours. Alberti et al.[24] have shown that the first change is a decrease in plateau waves (see Fig. 3–7) followed by gradual decline of the initially increased intracranial pressure over a period of 48 to 72 hours. CT and MRI often fail to show a substantial amelioration of edema in patients who have shown dramatic clinical improvement, but many scans do show decreased contrast enhancement within the tumor, suggesting partial restoration of the integrity of the BBB.[65] On rare occasions, the tumor, as defined by contrast enhancement, disappears after the corticosteroid treatment. The brain tumor whose MRI responds most dramatically to corticosteroids is lymphoma. Approximately 60% of contrast-enhancing masses that are due to lymphoma in the brain resolve partially or completely with corticosteroid treatment, probably reflecting a direct oncolytic effect. With gliomas and metastatic brain tumors, dramatic changes in the size or intensity of contrast enhancement occur rarely, usually with small metastases. Thus, in most patients suspected of harboring neoplasms in the CNS, scans should be performed before corticosteroids are administered unless the drugs are necessary to control serious symptoms.

SPINAL CORD COMPRESSION

Corticosteroid effects seem to be less dramatic in epidural spinal cord compression than in metastatic brain tumor. This difference may be because most patients with spinal cord compression receive other therapy (i.e., surgical decompression, radiation therapy, or cytotoxic drugs) almost immediately, without an opportunity for the physician to assess corticosteroid effects. By contrast, in patients with metastatic brain tumors, other therapy is often deferred 2 or 3 days until the full effects of steroids have occurred. In addition, spinal cord edema arises from a mechanism different from that of brain metastases. The edema results from compression of the spinal cord vasculature rather than from direct contact of tumor with the CNS. Nevertheless, animal experiments that demonstrate the presence of vasogenic edema when the spinal cord is compressed by epidural tumor have shown that the edema can be reversed (and neurologic symptoms improved) by corticosteroids[162]; higher doses are more effective than lower ones.[159]

Corticosteroids alone have a salutary effect on the neurologic signs and symptoms of epidural cord compression in some patients and are effective in relieving the pain in most, often within an hour of administration.[163–165] For patients with clinical or MRI evidence of cord compression but without neurologic symptoms except for moderate pain, one can begin with the standard dose (16 mg/24 hr of dexamethasone), increasing the dose if pain persists or new symptoms develop. For patients with severe pain or evidence of myelopathy, an IV bolus of 100 mg of dexamethasone should be administered, followed by 100 mg/24 hr in divided doses orally. The drug should be tapered as the patient is treated with more definitive modalities (see Chapter 6). IV dexamethasone is infused slowly over a period of 5 to 10 minutes; some patients complain of severe genital burning as the drug is infused, but this lasts only a few minutes and is easily tolerated by the patient, particularly if forewarned.[166] One study suggests that no additional benefit is obtained from these higher doses,[165] but animal experiments,[159] along with a great deal of anecdotal human experience, suggest an advantage to the higher dose. Experimental evidence in animals with epidural tumor-induced spinal cord compression suggests that within 1 hour after a single large dose of steroids, the transfer constant of water-soluble substances from blood to spinal cord decreases; this effect is greater at 12 hours.[167] A recent randomized trial of high-dose corticosteroids versus no steroids indicated improved function in the steroid group 6 months after radiation.[168]

CORTICOSTEROIDS FOR OTHER NEUROLOGIC COMPLICATIONS OF CANCER

The indications for corticosteroids in the treatment of leptomeningeal metastases involving nerve roots, or of tumor involving peripheral nerves or nerve plexuses, are not established. Many patients are relieved of pain at standard or slightly higher doses; few appear to have improved neurologic function. The drugs are worth trying in the initial treatment of patients with these conditions, but should be abandoned if a clinical response is not achieved within 48 hours.

Corticosteroids may also help treat chemotherapy-induced leukoencephalopathy (see Chapter 12) and some of the delayed effects of radiation on the nervous system (see Chapter 13). Although the dosage for these effects is not established, it is probably best to start with large doses and either taper rapidly or abandon the treatment if a response is not observed in a few days.

CORTICOSTEROID TAPER

Because of deleterious effects of corticosteroids (see Chapter 4), patients should be treated with the lowest effective dose for the shortest time possible. Virtually all patients begun on corticosteroid therapy for brain metastases or spinal cord compression are treated subsequently with surgery, radiation therapy, or chemotherapy. During or after this more definitive treatment, the patient should be weaned from the steroids entirely, if possible. The steroid taper begins 3 to 4 days after surgery or during week 2 of radiation therapy and should be gradual enough to prevent the development of steroid withdrawal symptoms, but rapid enough so the patient is not taking the drugs for an extended period. For patients receiving 16 mg of dexamethasone, the drug can be tapered by 2 to 4 mg every fifth day. If at any time during the taper the patient develops either recurrent symptoms or the corticosteroid withdrawal syndrome (see Table 4–8), the drug is increased to the prior dose for 4 or 5 days before tapering again. If, after drug withdrawal, the patient develops brain tumor symptoms it is probably wise to start the full regimen of 16 mg/24 hr of dexamethasone. Patients started on steroids with minimal or no symptoms (e.g., an MRI reveals an asymptomatic brain metastasis) can often be tapered more rapidly. Some asymptomatic patients do not require steroids at all.

For patients who have been on corticosteroids for many months and fail the usual taper schedule or have large amounts of residual tumor, the drug is tapered more slowly (e.g., 1–2 mg/wk) to the lowest dose tolerable. For patients taking large doses of corticosteroids (e.g., 100 mg/24 hr dexamethasone) who have stabilized and are receiving more definitive treatment, the dose can be halved every 4 to 5 days, depending on the patient's clinical state. The dose should be raised again if the clinical state deteriorates.

Weissman et al.[169] propose a more rapid taper schedule, beginning at 16 mg/day (8 mg bid) for 4 days followed by 8 mg/24 hr for 4 days and 4 mg/24 hr until completion of radiation therapy. This procedure was well tolerated by their patients with brain metastases.

Patients receiving corticosteroids, particularly those receiving the drug for more than a few months in relatively high doses, are at risk for developing pneumocystis pneumonia.[170,171] The infection is more likely to occur in those whose CD4 counts fall below 200 mm^3.[172] The infection often occurs during tapering of the corticosteroids. For those patients with metastatic brain or spinal tumors in whom corticosteroids are likely to be required for an extended period (>6 weeks), we recommend prophylaxis with trimethoprim-sulfamethoxazole, one double strength tablet twice a day, 3 days a week. The drug can be stopped after the patient has discontinued the corticosteroids for a few weeks.

MECHANISMS OF CORTICOSTEROID ACTION ON BLOOD–BRAIN BARRIER

The mechanisms by which corticosteroids stabilize the BBB, blood–spinal cord barrier, and probably blood–CSF barrier are not entirely known, although several have been proposed. Corticosteroids inhibit the production or release of a number of biochemical substances shown to increase vascular permeability and to induce vasodilatation (an effect that, by increased hydrostatic pressure, also increases permeability). Corticosteroids induce formation of lipocortins, which inhibit phosphorylase A2, thus preventing the release of arachidonic acid.[173] Arachidonic acid and its metabolites increase vascular permeability; thus, the reduction of arachidonic acid by corticosteroids may reduce brain edema. Nonetheless, other inhibitors of this pathway, such as indomethacin, do not

stabilize the BBB or ameliorate the symptoms of brain tumors as effectively as corticosteroids. Also, corticosteroids inhibit the release of interleukin 1; whether the interleukins play a role in BBB breakdown is not known. Corticosteroids also appear to have a direct effect on endothelial cells in several organisms, inhibiting the increased permeability that results from their interaction with a number of chemical agents.[174] Experimental evidence suggests that corticosteroids can induce the synthesis of a protein that inhibits microvascular permeability, a direct action on the endothelial cell. The inhibitory protein appears to be distinct from lipocortin, and thus, the effect is independent from the inhibition of phosphorylase A2. Whatever the mechanism, corticosteroids appear to be unique in their ability to ameliorate clinical symptoms and stabilize the BBB.

In addition to restoring capillary impermeability,[175] dexamethasone also decreases cerebral blood flow and volume[25,176] and increases the fractional extraction of oxygen throughout the brain without affecting oxygen utilization. Reports suggest that the drug probably has a direct vasoconstrictive effect on cerebral blood vessels.[177] Dexamethasone treatment reduces or eliminates the filtration of plasma-derived fluid across tumor capillaries and also reduces the movement of albumin through the extracellular space by solvent drag. These effects may be mediated by reducing the size of the extracellular space or decreasing the pore size of tumor capillaries, probably representing an important mechanism for corticosteroid control of tumor in peritumoral brain edema.[178] Although most studies have not demonstrated an effect of dexamethasone on the BBB of normal brain, at least one study[179] suggests that dexamethasone may decrease the permeability to macromolecules of even normal cerebral vasculature, possibly by interfering with vesicular transport.[179]

MRI studies of primary brain tumors demonstrate that corticosteroids reduce the permeability of cerebral blood vessels and decrease the water content in the edematous tissue surrounding the tumor.[26,65] Because so few patients with metastatic brain tumors were studied, these changes, although suggestive, were not statistically significant.

REFERENCES

1. Abbott NJ, Ronnback L, Hansson E. Astrocyte-endothelial interactions at the blood-brain barrier. *Nat Rev Neurosci* 2006;7:41–53.
2. Abbott NJ. Dynamics of CNS barriers: evolution, differentiation, and modulation. *Cell Mol Neurobiol* 2005;25:5–23.
3. Deeken JF, Loscher W. The blood-brain barrier and cancer: transporters, treatment, and Trojan horses. *Clin Cancer Res* 2007;13:1663–1674.
4. Ueno M. Molecular anatomy of the brain endothelial barrier: an overview of the distributional features. *Curr Med Chem* 2007;14:1199–1206.
5. Folkman J. Angiogenesis. *Annu Rev Med* 2006;57:1–18.
6. Ohnishi T, Sher PB, Posner JB, et al. Increased capillary permeability in rat brain induced by factors secreted by cultured C6 glioma cells: role in peritumoral brain edema. *J Neuro-Oncol* 1991;10:13–25.
7. Zhang RD, Price JE, Fujimaki T, et al. Differential permeability of the blood-brain barrier in experimental brain metastases produced by human neoplasms implanted into nude mice. *Am J Pathol* 1992;141:1115–1124.
8. Wick W, Küker W. Brain edema in neurooncology: radiological assessment and management. *Onkologie* 2004;27:261–266.
9. Grabb PA, Gilbert MR. Neoplastic and pharmacological influence on the permeability of an in vitro blood-brain barrier. *J Neurosurg* 1995;82:1053–1058.
10. Knopf PM, Cserr HF, Nolan SC, et al. Physiology and immunology of lymphatic drainage of interstitial and cerebrospinal fluid from the brain. *Neuropathol Appl Neurobiol* 1995;21:175–180.
11. Koh L, Zakharov A, Johnston M. Integration of the subarachnoid space and lymphatics: is it time to embrace a new concept of cerebrospinal fluid absorption? *Cerebrospinal Fluid Res* 2005;2:6.
12. Zakharov A, Papaiconomou C, Djenic J, et al. Lymphatic cerebrospinal fluid absorption pathways in neonatal sheep revealed by subarachnoid injection of Microfil. *Neuropathol Appl Neurobiol* 2003;29:563–573.
13. Boulton M, Armstrong D, Flessner M, et al. Raised intracranial pressure increases CSF drainage through arachnoid villi and extracranial lymphatics. *Am J Physiol Regul Integr Comp Physiol* 1998;275:R889–R896.
14. Mollanji R, Bozanovic-Sosic R, Li ISB, et al. Intracranial pressure accommodation is impaired by blocking pathways leading to extracranial lymphatics. *Am J Physiol Regul Integr Comp Physiol* 2001;280:R1573–R1581.
15. Abbott NJ. Evidence for bulk flow of brain interstitial fluid: significance for physiology and pathology. *Neurochem Int* 2004;45:545–552.
16. Marmarou A, Hochwald G, Nakamura T, et al. Brain edema resolution by CSF pathways and brain vasculature in cats. *Am J Physiol Heart Circ Physiol* 1994;267:H514–H520.
17. Van Crevel H. Pathogenesis of raised cerebrospinal fluid pressure. *Doc Ophthalmol* 1982;52:251–257.
18. Del Maestro RF, Megyesi JF, Farrell CL. Mechanisms of tumor-associated edema: a review. *Can J Neurol Sci* 1990;17:177–183.
19. Jain RK, Tong RT, Munn LL. Effect of vascular normalization by antiangiogenic therapy on interstitial hypertension, peritumor edema, and lymphatic metastasis: insights from a mathematical model. *Cancer Res* 2007;67:2729–2735.
20. Huber PE, Bischof M, Jenne J, et al. Trimodal cancer treatment: beneficial effects of combined antiangiogenesis, radiation, and chemotherapy. *Cancer Res* 2005;65:3643–3655.
21. DiResta GR, Lee J, Larson SM, et al. Characterization of neuroblastoma xenograft in rat flank. I. Growth, interstitial fluid pressure, and interstitial fluid velocity distribution profiles. *Microvasc Res* 1993;46:158–177.
22. Rozental JM, Levine RL, Nickles RJ, et al. Cerebral diaschisis in patients with malignant glioma. *J Neuro-Oncol* 1990;8:153–161.
23. Kajimoto K, Oku N, Kimura Y, et al. Crossed cerebellar diaschisis: a positron emission tomography study with l-[methyl-11C]methionine and 2-deoxy-2-[18F]fluoro-d-glucose. *Ann Nucl Med* 2007;21:109–113.
24. Alberti E, Hartmann A, Schutz HJ, et al. The effect of large doses of dexamethasone on the cerebrospinal fluid pressure in patients with supratentorial tumors. *J Neurol* 1978;217:173–181.
25. Jarden JO, Dhawan V, Poltorak A, et al. Positron emission tomographic measurement of blood-to-brain and blood-to-tumor transport of 82Rb: the effect of dexamethasone and whole-brain radiation therapy. *Ann Neurol* 1985;18:636–646.
26. Sinha S, Bastin ME, Wardlaw JM, et al. Effects of dexamethasone on peritumoural oedematous brain: a DT-MRI study. *J Neurol Neurosurg Psychiatry* 2004;75:1632–1635.
27. Fijak M, Meinhardt A. The testis in immune privilege. *Immunol Rev* 2006;213:66–81.
28. Wong CH, Cheng CY. The blood-testis barrier: its biology, regulation, and physiological role in spermatogenesis. *Curr Top Dev Biol* 2005;71:263–296.
29. Zhu Y, Brannstrom M, Janson PO, et al. Differences in expression patterns of the tight junction proteins, claudin 1, 3, 4 and 5, in human ovarian surface epithelium as compared to epithelia in inclusion cysts and epithelial ovarian tumours. *Int J Cancer* 2006;118:1884–1891.
30. Zhu Y, Maric J, Nilsson M, et al. Formation and barrier function of tight junctions in human ovarian surface epithelium. *Biol Reprod* 2004;71:53–59.
31. Madison KC. Barrier function of the skin: "la raison d'etre" of the epidermis. *J Invest Dermatol* 2003; 121:231–241.
32. Kimelberg HK. Water homeostasis in the brain: basic concepts. *Neuroscience* 2004;129:851–860.
33. Zador Z, Bloch O, Yao X, et al. Aquaporins: role in cerebral edema and brain water balance. *Prog Brain Res* 2007;161:185–194.
34. Manley GT, Binder DK, Papadopoulos MC, et al. New insights into water transport and edema in the central nervous system from phenotype analysis of aquaporin-4 null mice. *Neuroscience* 2004;129:983–991.

35. Oldendorf WH, Cornford ME, Brown WJ. The large apparent work capability of the blood-brain barrier: a study of the mitochondrial content of capillary endothelial cells in brain and other tissues of the rat. *Ann Neurol* 1977;1:409–417.

36. Qutub AA, Hunt CA. Glucose transport to the brain: a systems model. *Brain Res Brain Res Rev* 2005;49:595–617.

37. Hawkins RA, O'Kane RL, Simpson IA, et al. Structure of the blood-brain barrier and its role in the transport of amino acids. *J Nutr* 2006;136:218S–226S.

38. Kim JA, Tran ND, Li Z, et al. Brain endothelial hemostasis regulation by pericytes. *J Cereb Blood Flow Metab* 2006;26:209–217.

39. Lai CH, Kuo KH. The critical component to establish in vitro BBB model: pericyte. *Brain Res Brain Res Rev* 2005;50:258–265.

40. Shepro D, Morel NM. Pericyte physiology. *FASEB J* 1993;7:1031–1038.

41. Arismendi-Morillo G, Castellano A. Tumoral micro-blood vessels and vascular microenvironment in human astrocytic tumors. A transmission electron microscopy study. *J Neuro-Oncol* 2005;73:211–217.

42. Fry M, Ferguson AV. The sensory circumventricular organs: brain targets for circulating signals controlling ingestive behavior. *Physiol Behav* 2007;91:413–423.

43. Fry M, Hoyda TD, Ferguson AV. Making sense of it: roles of the sensory circumventricular organs in feeding and regulation of energy homeostasis. *Exp Biol Med (Maywood)* 2007;232:14–26.

44. Allt G, Lawrenson JG. The blood-nerve barrier: enzymes, transporters and receptors—a comparison with the blood-brain barrier. *Brain Res Bull* 2000;52:1–12.

45. Beamish NG, Stolinski C, Thomas PK, et al. Freeze-fracture observations on normal and abnormal human perineurial tight junctions: alterations in diabetic polyneuropathy. *Acta Neuropathol* 1991;81:269–279.

46. Abram SE, Yi J, Fuchs A, et al. Permeability of injured and intact peripheral nerves and dorsal root ganglia. *Anesthesiology* 2006;105:146–153.

47. Kusuhara H, Sugiyama Y. Efflux transport systems for organic anions and cations at the blood-CSF barrier. *Adv Drug Deliv Rev* 2004;56:1741–1763.

48. Strazielle N, Khuth ST, Ghersi-Egea JF. Detoxification systems, passive and specific transport for drugs at the blood-CSF barrier in normal and pathological situations. *Adv Drug Deliv Rev* 2004;56:1717–1740.

49. Bleyer WA. Clinical pharmacology of intrathecal methotrexate. II. An improved dosage regimen derived from age-related pharmacokinetics. *Cancer Treat Rep* 1977;61:1419–1425.

50. Grossman SA, Reinhard CS, Loats HL. The intracerebral penetration of intraventricularly administered methotrexate: a quantitative autoradiographic study. *J Neuro-Oncol* 1989;7:319–328.

51. Shapiro WR, Chernik NL, Posner JB. Necrotizing encephalopathy following intraventricular instillation of methotrexate. *Arch Neurol* 1973;28:96–102.

52. Folkman J. How is blood vessel growth regulated in normal and neoplastic tissue? G.H.A. Clower Memorial Award Lecture. *Cancer Res* 1986;46:467–473.

53. Ushio Y, Chernik NL, Shapiro WR, et al. Metastatic tumor of the brain: development of an experimental model. *Ann Neurol* 1977;2:20–29.

54. Ushio Y, Shimizu K, Aragaki Y, et al. Alteration of blood-CSF barrier by tumor invasion into the meninges. *J Neurosurg* 1981;55:445–449.

55. Gerstner ER, Fine RL. Increased permeability of the blood-brain barrier to chemotherapy in metastatic brain tumors: establishing a treatment paradigm. *J Clin Oncol* 2007;25:2306–2312.

56. Fine RL, Chen J, Balmaceda C, et al. Randomized study of paclitaxel and tamoxifen deposition into human brain tumors: implications for the treatment of metastatic brain tumors. *Clin Cancer Res* 2006;12:5770–5776.

57. Ohnishi T, Posner JB, Shapiro WR. Vasogenic brain edema induced by arachidonic acid: role of extracellular arachidonic acid in blood-brain barrier dysfunction. *Neurosurgery* 1992;30:545–551.

58. Stewart PA, Hayakawa K, Farrell CL, et al. Quantitative study of microvessel ultrastructure in human peritumoral brain tissue. Evidence for a blood-brain barrier defect. *J Neurosurg* 1987;67:697–705.

59. Shinonaga M, Chang CC, Suzuki N, et al. Immunohistological evaluation of macrophage infiltrates in brain tumors. Correlation with peritumoral edema. *J Neurosurg* 1988;68:259–265.

60. Marmarou A. A review of progress in understanding the pathophysiology and treatment of brain edema. *Neurosurg Focus* 2007;22:E1.

61. Vorbrodt AW, Lossinsky AS, Wisniewski HM, et al. Ultrastructural observations on the transvascular route of protein removal in vasogenic brain edema. *Acta Neuropathol [Berl]* 1985;66:265–273.

62. Bloch O, Manley GT. The role of aquaporin-4 in cerebral water transport and edema. *Neurosurg Focus* 2007;22:E3.

63. Fishman RA. *Cerebrospinal Fluid in Diseases of the Nervous System*. 2nd ed. Philadelphia: WB Saunders, 1992.

64. Bodsch W, Rommel T, Ophoff BG, et al. Factors responsible for the retention of fluid in human tumor edema and the effect of dexamethasone. *J Neurosurg* 1987;67:250–257.

65. Armitage PA, Schwindack C, Bastin ME, et al. Quantitative assessment of intracranial tumor response to dexamethasone using diffusion, perfusion and permeability magnetic resonance imaging. *Magn Reson Imaging* 2007;25:303–310.

66. Cha S. Differentiation of glioblastoma multiforme and single brain metastasis by peak height and percentage of signal intensity recovery derived from dynamic susceptibility-weighted contrast-enhanced perfusion MR imaging. *Am J Neuroradiol* 2007;28:1078–1084.

67. Fishman RA. Brain edema. *N Engl J Med* 1975;293:706–711.

68. Stummer W. Mechanisms of tumor-related brain edema. *Neurosurg Focus* 2007;22:E8.

69. Strugar J, Rothbart D, Harrington W, et al. Vascular permeability factor in brain metastases: correlation with vasogenic brain edema and tumor angiogenesis. *J Neurosurg* 1994;81:560–566.

70. Suzuki N, Sako K, Yonemasu Y. Effects of induced hypertension on blood flow and capillary permeability in rats with experimental brain tumors. *J Neuro-Oncol* 1991;10:213–218.

71. Ushio Y, Posner R, Posner JB, et al. Experimental spinal cord compression by epidural neoplasm. *Neurology* 1977;27:422–429.

72. Kim JA, Chung JI, Yoon PH, et al. Transient MR signal changes in patients with generalized toni-coclonic seizure or status epilepticus: periictal diffusion-weighted imaging. *AJNR Am J Neuroradiol* 2001;22:1149–1160.

73. Spigelman MK, Zappulla RA, Strauchen JA, et al. Etoposide induced blood-brain barrier disruption in rats: duration of opening and histological sequelae. *Cancer Res* 1986;46:1453–1457.

74. Mouchard-Delmas C, Gourdier B, Vistelle R, et al. Modification of the blood-brain barrier permeability following intracarotid infusion of vinorelbine. *Anticancer Res* 1995;15:2593–2596.

75. Phillips PC, Dhawan V, Strother SC, et al. Reduced cerebral glucose metabolism and increased brain capillary permeability following high-dose methotrexate chemotherapy: a positron emission tomographic study. *Ann Neurol* 1987;21:59–63.

76. Strausbaugh LJ. Intracarotid infusions of protamine sulfate disrupt the blood-brain barrier of rabbits. *Brain Res* 1987;409:221–226.

77. Black KL, King WA, Ikezaki K. Selective opening of the blood-tumor barrier by intracarotid infusion of leukotriene C4. *Acta Neurochir* 1990;51 (Suppl):140–141.

78. Preston E, Slinn J, Vinokourov I, et al. Graded reversible opening of the rat blood-brain barrier by intracarotid infusion of sodium caprate. *J Neurosci Methods* 2008;168:443–449.

79. Nitsch C, Klatzo I. Regional patterns of blood-brain barrier breakdown during epileptiform seizures induced by various convulsive agents. *J Neurol Sci* 1983;59:305–322.

80. Levin VA, Edwards MS, Byrd A. Quantitative observations of the acute effects of x-irradiation on brain capillary permeability, Part I. *Int J Radiat Oncol Biol Phys* 1979;5:1627–1631.

81. Delattre J-Y, Shapiro WR, Posner JB. Acute effects of low-dose cranial irradiation on regional capillary permeability in experimental brain tumors. *J Neurol Sci* 1989;90:147–153.

82. Cao Y, Tsien CI, Shen Z, et al. Use of magnetic resonance imaging to assess blood-brain/blood-glioma barrier opening during conformal radiotherapy. *J Clin Oncol* 2005;23:4127–4136.

83. Yuan H, Gaber MW, Boyd K, et al. Effects of fractionated radiation on the brain vasculature in a murine model: blood-brain barrier permeability, astrocyte proliferation, and ultrastructural changes. *Int J Radiat Oncol Biol Phys* 2006;66:860–866.

84. Bart J, Nagengast WB, Coppes RP, et al. Irradiation of rat brain reduces P-glycoprotein expression and function. *Br J Cancer* 2007;97:322–326.

85. Fortin D, Gendron C, Boudrias M, et al. Enhanced chemotherapy delivery by intraarterial infusion and blood-brain barrier disruption in the treatment of cerebral metastasis. *Cancer* 2007;109:751–760.

86. Siegal T, Rubinstein R, Bokstein F, et al. In vivo assessment of the window of barrier opening after osmotic blood-brain barrier disruption in humans. *J Neurosurg* 2000;92:599–605.

87. Carson RE, Zunkeler B, Blasberg RG, et al. Quantitative measurement of hyperosmotic blood-brain barrier disruption with ^{82}Rb and PET. *J Nucl Med* 1990;31:810. Abstract no. 429.

88. Kemper EM, Boogerd W, Thuis I, et al. Modulation of the blood-brain barrier in oncology: therapeutic opportunities for the treatment of brain tumours? *Cancer Treat Rev* 2004;30:415–423.

89. Galor A, Ference SJ, Singh AD, et al. Maculopathy as a complication of blood-brain barrier disruption in patients with central nervous system lymphoma. *Am J Ophthalmol* 2007;144:45–49.

90. Marchi N, Angelov L, Masaryk T, et al. Seizure-promoting effect of blood-brain barrier disruption. *Epilepsia* 2007;48:732–742.

91. Rapoport SI, Robinson PJ. Blood-tumor barrier disruption controversies. Letter to the editor. *J Cereb Blood Flow Metab* 1991;11:165–168.

92. Cosolo WC, Martinello P, Louis WJ, et al. Blood-brain barrier disruption using mannitol: time course and electron microscopy studies. *Am J Physiol* 1989;256 (2 Pt 2):R443–R447.

93. Farrell CL, Shivers RR. Capillary junctions of the rat are not affected by osmotic opening of the blood-brain barrier. *Acta Neuropathol (Berl)* 1984;63:179–189.

94. Groothuis DR, Warnke PC, Molnar P, et al. Effect of hyperosmotic blood-brain barrier disruption on transcapillary transport in canine brain tumors. *J Neurosurg* 1990;72:441–449.

95. Shapiro WR, Voorhies RM, Hiesiger EM, et al. Pharmacokinetics of tumor cell exposure to [14C] methotrexate after intracarotid administration without and with hyperosmotic opening of the blood-brain and blood-tumor barriers in rat brain tumors: a quantitative autoradiographic study. *Cancer Res* 1988;48:694–701.

96. Borlongan CV, Emerich DF. Facilitation of drug entry into the CNS via transient permeation of blood brain barrier: laboratory and preliminary clinical evidence from bradykinin receptor agonist, Cereport. *Brain Res Bull* 2003;60:297–306.

97. Posner JB, Saper CB, Schiff ND, et al. *Plum and Posner's Diagnosis of Stupor and Coma.* 4 ed. New York: Oxford University Press, 2007.

98. Weaver DD, Winn HR, Jane JA. Differential intracranial pressure in patients with unilateral mass lesions. *J Neurosurg* 1982;56:660–665.

99. Panullo SC, Reich JB, Krol G, et al. MRI changes in intracranial hypotension. *Neurology* 1993;43:919–926.

100. Hayashi M, Kobayashi H, Handa Y, et al. Plateau-wave phenomenon (II). Occurrence of brain herniation in patients with and without plateau waves. *Brain* 1991;114:2693–2699.

101. Lundberg N. Continuous recording and control of ventricular fluid pressure in neurosurgical practice. *Acta Neurol Scand* 1960;36 (Supp 149):1–193.

102. Daley ML, Leffler CW, Czosnyka M, et al. Plateau waves: changes of cerebrovascular pressure transmission. *Acta Neurochir Suppl* 2005; 95:327–332.

103. Hayashi M, Kobayashi H, Handa Y, et al. Brain blood volume and blood flow in patients with plateau waves. *J Neurosurg* 1985;63:556–561.

104. Matsuda M, Yoneda S, Handa H, et al. Cerebral hemodynamic changes during plateau waves in brain-tumor patients. *J Neurosurg* 1979; 50:483–488.

105. Imberti R, Fuardo M, Bellinzona G, et al. The use of indomethacin in the treatment of plateau waves: effects on cerebral perfusion and oxygenation. *J Neurosurg* 2005;102:455–459.

106. Magnaes B. Body position and cerebrospinal fluid pressure. Part I: Clinical studies on the effect of rapid postural changes. *J Neurosurg* 1976;44:687–697.

107. Magnaes B. Movement of cerebral spinal fluid within the craniospinal space when sitting up and lying down. *Surg Neurol* 1978;10:45–49.

108. Alperin N, Hushek SG, Lee SH, et al. MRI study of cerebral blood flow and CSF flow dynamics in an upright posture: the effect of posture on the intracranial compliance and pressure. *Acta Neurosurg* 2005;95 (Suppl):177–181.

109. Watling CJ, Cairncross JG. Acetazolamide therapy for symptomatic plateau waves in patients with brain tumors—report of three cases. *J Neurosurg* 2002;97:224–226.

110. Hutchinson P, Timofeev I, Kirkpatrick P. Surgery for brain edema. *Neurosurg Focus* 2007;22:E14.

111. Raslan A, Bhardwaj A. Medical management of cerebral edema. *Neurosurg Focus* 2007;22:E12.

112. Colice GL. How to ventilate patients when ICP elevation is a risk: monitor pressure, consider hyperventilation therapy. *J Crit Illness* 1993;8:1003–1020.

113. Imberti R, Bellinzona G, Langer M. Cerebral tissue PO_2 and $SjvO_2$ changes during moderate hyperventilation in patients with severe traumatic brain injury. *J Neurosurg* 2002;96:97–102.

114. Diringer MN, Videen TO, Yundt K, et al. Regional cerebrovascular and metabolic effects of hyperventilation after severe traumatic brain injury. *J Neurosurg* 2002;96:103–108.

114a. Gelb AW, Craen RA, Rao GS, et al. Does hyperventilation improve operating condition during supratentorial craniotomy? A multicenter randomized crossover trial. *Anesth Analg* 2008;106:585–594.

115. Ravussin P, Abou-Madi M, Archer D, et al. Changes in CSF pressure after mannitol in patients with and without elevated CSF pressure. *J Neurosurg* 1988;69:869–876.

116. Videen TO, Zazulia AR, Manno EM, et al. Mannitol bolus preferentially shrinks non-infarcted brain in patients with ischemic stroke. *Neurology* 2001;57:2120–2122.

117. Koenig MA, Bryan M, Lewin JL 3rd, et al. Reversal of transtentorial herniation with hypertonic saline. *Neurology* 2008;70:1023–1029.

118. Donato T, Shapria Y, Artru A, et al. Effect of mannitol on cerebrospinal fluid dynamics and brain tissue edema. *Anesth Analg* 1994;78:58–66.

119. Pollay M, Fullenwider C, Roberts PA, et al. Effect of mannitol and furosemide on blood-brain osmotic gradient and intracranial pressure. *J Neurosurg* 1983;59:945–950.

120. Rabinstein AA. Treatment of cerebral edema. *Neurologist* 2006;12:59–73.

121. Larive LL, Rhoney DH, Parker D, Jr, et al. Introducing hypertonic saline for cerebral edema: an academic center experience. *Neurocrit Care* 2004;1:435–440.

122. Rottenberg DA, Hurwitz BJ, Posner JB. The effect of oral glycerol on intraventricular pressure in man. *Neurology* 1977;27:600–608.

123. Wolf AL, Adcock LL, Hachiya JT, et al. Choriocarcinoma with brain metastases. Successful management of increased intracranial pressure with barbiturates. *Cancer* 1986;57:1432–1436.

124. Groothuis D, Blasberg RG, Molnar P, et al. Regional blood flow in avian sarcoma virus (ASV)-induced brain tumors. *Neurology* 1983;33:686–696.

125. Groothuis DR, Blasberg RG. Rational brain tumor chemotherapy. The interaction of drug and tumor. *Neurol Clin* 1985;3:801–816.

126. Fenstermacher JD. Pharmacology of the blood-brain barrier. In: Neuwelt EA, ed. *Implications of the Blood-Brain Barrier and Its Manipulation, Vol. 1: Basic Science Aspects.* New York: Plenum Medical Book Company, 1989. pp. 137–155.

127. Frankenheim J, Brown RME. *Bioavailablity of Drugs to the Brain and the Blood-Brain Barrier* (National Institute on Drug Abuse Research Monograph Series, No. 120). Washington, DC: United States Department of Health and Human Services, 1992.

128. Hiesiger EM, Voorhies RM, Basler GA, et al. Opening the blood-brain and blood-tumor barriers in experimental rat brain tumors: the effect of intracarotid hyperosmolar mannitol on capillary permeability and blood flow. *Ann Neurol* 1986;19:50–59.

129. Peereboom DM. Chemotherapy in brain metastases. *Neurosurgery* 2005;57:S54–S65.

130. Hartwell RC, Sutton LN. Mannitol, intracranial pressure, and vasogenic edema. *Neurosurgery* 1993;32:444–450.

131. Ackerman NB, Jacobs R. The effects of steroidal and nonsteroidal anti-inflammatory agents on uptake of Evans blue in experimental metastasis. *Microvasc Res* 1988;35:1–7.

132. Reichman HR, Farrell CL, Del Maestro RF. Effects of steroids and nonsteroidal anti-inflammatory agents on vascular permeability in a rat glioma model. *J Neurosurg* 1986;65:233–237.

133. Siegal T, Siegel T, Shapira Y, et al. Indomethacin and dexamethasone treatment in experimental neoplastic spinal cord compression. Part 1. Effect on water content and specific gravity. *Neurosurgery* 1988;22:328–333.

134. Rutz HP, Hofer S, Peghini PE, et al. Avoiding glucocorticoid administration in a neurooncological case. *Cancer Biol Ther* 2005;4:1186–1189.

135. Kaal EC, Vecht CJ. The management of brain edema in brain tumors. *Curr Opin Oncol* 2004;16:593–600.

136. King WA, Black KL, Ikezaki K, et al. Tumor-associated neurological dysfunction prevented by lazaroids in rats. *J Neurosurg* 1991;74:112–115.

137. Streffer JR, Bitzer M, Schabet M, et al. Response of radiochemotherapy-associated cerebral edema to a phytotherapeutic agent, H15. *Neurology* 2001;56:1219–1221.

138. Atsumi T, Hoshino S, Furukawa T, et al. The glutamate AMPA receptor antagonist, YM872, attenuates regional cerebral edema and IgG immunoreactivity following experimental brain injury in rats. *Acta Neurochir Suppl* 2003;86:305–307.

139. Ohyama H, Hosomi N, Takahashi T, et al. Thrombin inhibition attenuates neurodegeneration and cerebral edema formation following transient forebrain ischemia. *Brain Res* 2001;902:264–271.

140. O'Connor CA, Cernak I, Vink R. Both estrogen and progesterone attenuate edema formation following diffuse traumatic brain injury in rats. *Brain Res* 2005;1062:171–174.

141. Jones NC, Constantin D, Prior MJ, et al. The neuroprotective effect of progesterone after traumatic brain injury in male mice is independent of both the inflammatory response and growth factor expression. *Eur J Neurosci* 2005;21:1547–1554.

142. Cuculo L, Hallene K, Dini G, et al. Glycerophosphoinositol and dexamethasone improve transendothelial electrical resistance in an in vitro study of the blood-brain barrier. *Brain Res* 2004;997:147–151.

143. Tjuvajev J, Uehara H, Desai R, et al. Corticotropin-releasing factor decreases vasogenic brain edema. *Cancer Res* 1996;56:1352–1360.

144. Mechtler L, Alksne J, Wong E, et al. Interim report of the phase III open label study of xerecept (corticorelin acetate injection) for treatment of peritumoral brain edema in patients with primary or secondary brain tumors. *Neuro-Oncology* 2006;8:446–447. Abstract.

145. Gonzalez J, Kumar AJ, Conrad CA, et al. Effect of bevacizumab on radiation necrosis of the brain. *Int J Radiat Oncol Biol Phys* 2007;67:323–326.

146. Weissman DE. Glucocorticoid treatment for brain metastases and epidural spinal cord compression: a review. *J Clin Oncol* 1988;6:543–551.

147. Koehler PJ. Use of corticosteroids in neuro-oncology. *Anticancer Drugs* 1995;6:19–33.

148. Gomes JA, Stevens RD, Lewin JJ, III, et al. Glucocorticoid therapy in neurologic critical care. *Crit Care Med* 2005;33:1214–1224.

149. Prados M, Strowger B, Feindel WH. Studies on cerebral edema. II. Reaction of the brain to exposure to air; physiologic changes. *Arch Neurol Psychiatry* 1945;54:290–300.

150. Ingraham FD, Matson DD, McLaurin RL. Cortisone and ACTH as an adjunct to surgery of craniopharyngiomas. *N Engl J Med* 1952;246:568–571.

151. Kofman S, Garvin JS, Nagamani D, et al. Treatment of cerebral metastases from breast carcinoma with prednisolone. *JAMA* 1957;163:1473–1476.

152. Galicich JH, French LA. Use of dexamethasone in the treatment of cerebral edema resulting from brain tumors and brain surgery. *Am Practit-Dig Treat* 1961;12:169–174.

153. Gold R, Buttgereit F, Toyka KV. Mechanism of action of glucocorticosteroid hormones: possible implications for therapy of neuroimmunological disorders. *J Neuroimmunol* 2001;117:1–8.

154. Buttgereit F, Brand MD, Burmester GR. Equivalent doses and relative drug potencies for non-genomic glucocorticoid effects: a novel glucocorticoid hierarchy. *Biochem Pharmacol* 1999;58:363–368.

155. Jarden JO, Dhawan V, Moeller JR, et al. The time course of steroid action on blood-to-brain and blood-to-tumor transport of 82Rb: a positron emission tomographic study. *Ann Neurol* 1989;25:239–245.

156. Oka K, Shimodaira H. Telepharmacodynamics to predict therapeutic effects of glucocorticoids. Letter to the editor. *Lancet* 1991;338:385.

157. Renaudin J, Fewer D, Wilson CB, et al. Dose dependency of decadron in patients with partially excised brain tumors. *J Neurosurg* 1973; 39:302–305.

158. Peters WP, Holland JF, Senn H, et al. Corticosteroid administration and localized leukocyte mobilization in man. *N Engl J Med* 1972;282:342–345.

159. Delattre J-Y, Arbit E, Thaler HT, et al. A dose-response study of dexamethasone in a model of spinal cord compression caused by epidural tumor. *J Neurosurg* 1989;70:920–925.

160. Vecht CJ, Hovestadt A, Verbiest HBC, et al. Dose-effect relationship of dexamethasone on Karnofsky performance in metastatic brain tumors: a randomized study of doses of 4, 8, and 16 mg per day. *Neurology* 1994;44:675–680.

161. Lieberman A, LeBrun Y, Glass P, et al. Use of high-dose corticosteroids in patients with inoperable brain tumors. *J Neurol Neurosurg Psychiatry* 1977;40:678–682.

162. Ushio Y, Posner R, Kim J-H, et al. Treatment of experimental spinal cord compression by extradural neoplasms. *J Neurosurg* 1977;47:380–390.

163. Cantu RC. Corticosteroids for spinal metastases. Letter to the editor. *Lancet* 1968;2:912.

164. Greenberg HS, Kim J-H, Posner JB. Epidural spinal cord compression from metastatic tumor: results with a new treatment protocol. *Ann Neurol* 1980;8:361–366.

165. Vecht CJ, Haaxma-Reiche H, van Putten WLJ, et al. Initial bolus of conventional versus high-dose dexamethasone in metastatic spinal cord compression. *Neurology* 1989;39:1255–1257.

166. Czerwinski AW, Czerwinski AB, Whitsett TL, et al. Effects of a single, large intravenous injection of dexamethasone. *Clin Pharmacol Ther* 1972;13:638–642.

167. Shapiro WR, Hiesiger EM, Cooney GA, et al. Temporal effects of dexamethasone on blood-to-brain and blood-to-tumor transport of 14C-α-aminoisobutyric acid in rat C6 glioma. *J Neuro-Oncol* 1990;8:197–204.

168. Sorensen PS, Helweg-Larsen S, Mouridsen H, et al. Effect of high-dose dexamethasone in carcinomatous metastatic spinal cord compression treated with radiotherapy: a randomised trial. *Eur J Cancer* 1994;30A:22–27.

169. Weissman DE, Janjan NA, Erickson B, et al. Twice-daily tapering dexamethasone treatment during cranial radiation for newly diagnosed brain metastases. *J Neuro-Oncol* 1991;11:235–239.

170. Wen PY, Schiff D, Kesari S, et al. Medical management of patients with brain tumors. *J Neuro-Oncol* 2006;80:313–332.

171. Bollee G, Sarfati C, Thiery G, et al. Clinical picture of pneumocystis jiroveci pneumonia in cancer patients. *Chest* 2007;132:1305–1310.

172. Hughes MA, Parisi M, Grossman S, et al. Primary brain tumors treated with steroids and radiotherapy: low CD4 counts and risk of infection. *Int J Radiat Oncol Biol Phys* 2005;62:1423–1426.

173. Chan PH, Fishman RA. The role of arachidonic acid in vasogenic brain edema. *Fed Proc* 1984;43:210–213.

174. Weissman DE, Stewart C. Experimental drug therapy of peritumoral brain edema. *J Neuro-Oncol* 1988;6:339–342.

175. Williams TJ, Yarwood H. Effect of glucocorticosteroids on microvascular permeability. *Am Rev Respir Dis* 1990;141:S39–S43.

176. Tajima A, Yen M-H, Nakata H, et al. Effects of dexamethasone on blood flow and volume of perfused microvessels in traumatic brain edema. *Adv Neurol* 1990;52:343–350.

177. Leenders KL, Beaney RP, Brooks DJ, et al. Dexamethasone treatment of brain tumor patients: effects on regional cerebral blood flow, blood volume and oxygen utilization. *Neurology* 1985;35:1610–1616.

178. Nakagawa H, Groothuis DR, Owens ES, et al. Dexamethasone effects on [125I] albumin distribution in experimental RG-2 gliomas and adjacent brain. *J Cerebral Blood Flow Metab* 1987;7:687–701.

179. Hedley-Whyte ET, Hsu DW. Effect of dexamethasone on blood-brain barrier in the normal mouse. *Ann Neurol* 1986;19:373–377.

Chapter 4

Supportive Care and Its Complications

INTRODUCTION

Patients with neuro-oncologic disorders often suffer symptoms that are not strictly neurologic but result from the cancer, its treatment, or from comorbid illness[1,2] (Table 4–1). Some of these non-neurologic abnormalities, such as cachexia,[2–4] generalized weakness,[5,6] fatigue[7,8] (often associated with anemia[9]), sleep disturbances,[8,10] depression,[10] cognitive impairment,[10] gastrointestinal (GI) ulcerations, deep vein thromboses,[11] and side effects of the many drugs used for supportive care, are particularly common in patients with neuro-oncologic disorders. Effectively managing these disorders and their attendant symptoms can improve the quality of the patient's life and may even increase survival.[12–14] As patients with cancer, even those not cured, are living longer and often receiving more toxic treatments, the control of symptoms that impact on quality-of-life has become increasingly important. As a result, more and more attention is being paid to these problems, both in standard oncologic journals and in specialized supportive care journals.

Although most of the disorders listed in Table 4–1 are not considered to be neurologic illnesses, all must, in order to be perceived by the patient, affect the brain. Cleeland et al.[16] have suggested that many of the symptoms related to cancer and its treatment, including pain, anorexia, cachexia, fatigue, cognitive impairment, anxiety, and depression, can be explained as resulting from a shared cytokine–immunologic model that has its ultimate expression in the brain. Figure 4–1A illustrates this model. The model suggests certain therapeutic interventions that are illustrated in Figure 4–1B. Recent reviews have also addressed the mechanisms of behavioral abnormalities in patients with cancer.[10,17] Further details describing individual symptoms can be found in the paragraphs that follow.

Supportive care and symptom management is a vast field. Since the publication of the first edition of this book, several journals, including the *Journal of the National Comprehensive Cancer* *Network* and the *Journal of Supportive Oncology and Palliative and Supportive Care* (the *Journal of Supportive Care and Cancer* began publication in 1993 as the first edition of this book was being written), have appeared. Periodic updates of guidelines for the management of

Table 4–1. Symptom Prevalence among 922 Patients with Advanced Cancer

Symptom	Number	Percentage	Symptom	Number	Percentage
Pain	775	84	Confusion	192	21
Easy fatigability	633	69	Dizzy spells	175	19
Weakness	604	66	Dyspepsia	173	19
Anorexia	602	65	Belching	170	18
Lack of energy	552	60	Dysphagia	165	18
Dry mouth	519	56	Bloating	163	18
Constipation	475	52	Wheezing	124	13
Early satiety	473	51	Memory problems	108	12
Dyspnea	457	50	Headache	103	11
Sleep problems	456	50	Hiccup	87	9
Weight loss	447	49	Sedation	86	9
Depression	376	41	Aches/pains	84	9
Cough	341	37	Itch	80	9
Nausea	329	36	Diarrhea	77	8
Edema	262	28	Dreams	62	7
Taste change	255	28	Hallucinations	52	6
Hoarseness	220	24	Mucositis	47	5
Anxiety	218	24	Tremors	42	5
Vomiting	206	22	Blackout	32	4

From Ref. 15 with permission.

Figure 4–1. A: Biologic/physiologic mechanistic framework for cytokine-induced sickness behavior. In the afferent arm (solid lines), proinflammatory cytokines and chemokines (interleukin 1, tumor necrosis factor α, IL-6, interferon α, and interferon γ) are released in the periphery by activated immunocytes. They exert their effects on peripheral nerves and directly on the brain to induce various aspects of the sickness response. These behavioral/physiologic changes are elicited by mediators acting downstream from the cytokines. Glutamate, nitric oxide, prostaglandins, and substance P act on brain regions including the paraventricular nucleus of the hypothalamus and the amygdala. Turnover of monoamines (serotonin, dopamine, norepinephrine) in these brain regions is affected. Availability of monoamine precursors (e.g., tryptophan) may be decreased. The hypothalamic–pituitary–adrenal axis is activated with upregulation of plasma concentrations of corticosteroids, which in turn can provide feedback (dotted lines) to limit cytokine production. Other mediators, such as the anti-inflammatory cytokine IL-10, also have roles in activation and regulation of responses. ACTH, adrenocortical trophic hormone; CRH, corticotropin releasing hormone; CVO, circumventricular organ; NO, nitric oxide; PG, prostaglandin. **B**: Treatment strategies for cancer-related symptoms. The flow diagram in the center is a modification of the sickness response illustrated in more detail in A. Sites for the implementation of treatments directed at the sickness response circuit are shown. (1) Immunologic treatments (such as soluble receptors of TNF-α and IL-1 receptor antagonists) that are designed to inhibit cytokine signaling directly, or treatments that block downstream mediators of inflammation, including prostaglandins, nitric oxide, and substance P; (2) Neurobiologic treatments that target CNS mediators of behavioral alterations including the monoamines and CRH; (3) Symptomatic treatments (such as narcotics for alleviation of pain, stimulants to combat fatigue, antidepressants for relief from depression) that address the ultimate manifestations of upstream mediators; and (4) Treatments designed to take advantage of the normal endogenous feedback circuits that limit sickness responses in settings such as viral illness. CNS, central nervous system; TNF, tumor necrosis factor; COX, cyclooxygenase; NOS, nitric oxide synthase. (From Ref. 16 with permission.)

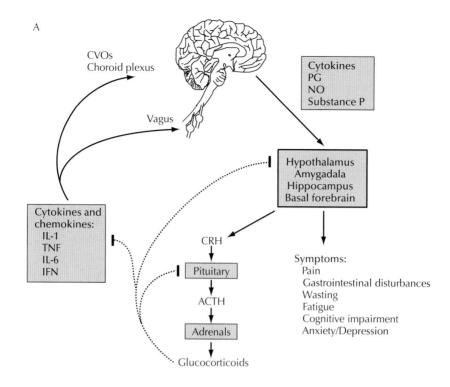

A

CVOs
Choroid plexus

Cytokines
PG
NO
Substance P

Vagus

Hypothalamus
Amygadala
Hippocampus
Basal forebrain

Cytokines and
chemokines:
IL-1
TNF
IL-6
IFN

CRH

Pituitary

ACTH

Adrenals

Glucocorticoids

Symptoms:
Pain
Gastrointestinal disturbances
Wasting
Fatigue
Cognitive impairment
Anxiety/Depression

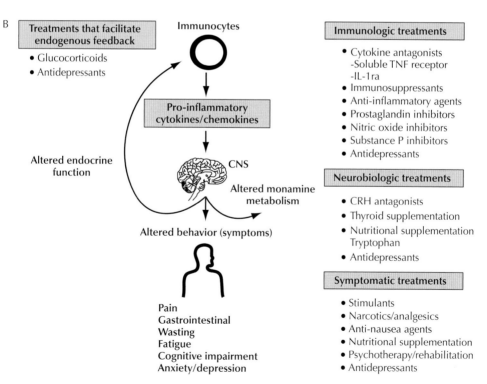

B

**Treatments that facilitate
endogenous feedback**

- Glucocorticoids
- Antidepressants

Immunocytes

Altered endocrine
function

Pro-inflammatory
cytokines/chemokines

CNS

Altered monamine
metabolism

Altered behavior (symptoms)

Pain
Gastrointestinal
Wasting
Fatigue
Cognitive impairment
Anxiety/depression

Immunologic treatments

- Cytokine antagonists
  -Soluble TNF receptor
  -IL-1ra
- Immunosuppressants
- Anti-inflammatory agents
- Prostaglandin inhibitors
- Nitric oxide inhibitors
- Substance P inhibitors
- Antidepressants

Neurobiologic treatments

- CRH antagonists
- Thyroid supplementation
- Nutritional supplementation
  Tryptophan
- Antidepressants

Symptomatic treatments

- Stimulants
- Narcotics/analgesics
- Anti-nausea agents
- Nutritional supplementation
- Psychotherapy/rehabilitation
- Antidepressants

several supportive care problems (Table 4–2) can be found on the Internet (www.nccn.org).

Only a few problems of particular relevance to neuro-oncology are addressed in this chapter. These include agents used for supportive care of patients with cancer and their neurologic side effects (Table 4–3).

EDEMA/CORTICOSTEROIDS

Corticosteroids are among the most frequently used drugs in both general oncology and

Table 4–2. Guidelines for Supportive Care of the Cancer Patient

Antiemesis
Cancer and treatment-related anemia
Cancer-related fatigue
Distress management
Fever and neutropenia
Myeloid growth factors
Palliative care
Senior adult oncology
Venous thromboembolic disease

Adapted from www.nccn.org National Comprehensive Cancer Network

neuro-oncology,[18,19] and they are widely used in nononcologic illnesses. In one study, 0.9% of the general population and 2.5% of the elderly were receiving the drugs.[20] In one survey of patients with cancer admitted to hospice, one-third were receiving glucocorticoids.[21] More than two-thirds of the patients on the inpatient Neuro-oncology Ward at Memorial Sloan-Kettering Cancer Center (MSKCC) are receiving corticosteroids.

The major use of corticosteroids in neuro-oncology is to control brain and spinal cord edema (see Chapter 3). However, these drugs also have oncolytic effects on breast cancer, lymphoma, and some other tumors. Other important salutary effects include improved appetite, elevated mood, decreased pain, and restoration of the integrity of the blood–brain barrier. Unfortunately, because corticosteroids are so useful, the drugs (usually dexamethasone) are often used reflexly in any patient with a central nervous system (CNS) metastasis. A typical patient who is found to have a brain metastasis (even an asymptomatic one found on a screening scan) is usually begun on dexamethasone, antiseizure medication, and sometimes several other drugs, all of which could have side effects and none of which may be required in that particular patient (Table 4–4). The physician

Table 4–3. Some Agents Used for Supportive Care of the Neuro-Oncology Patient

Problem	Agent(s)
Brain, spinal cord edema	Corticosteroids
Seizures	Anticonvulsants
Venous thromboembolism	Anticoagulants
GI problems	
Anorexia/cachexia	Megestrol, cannabinoids, etc.
Nausea/vomiting	Antiemetics
GI bleeding/perforation	H-2 blockers, proton pump inhibitors
Constipation	Laxatives
Pain	Opioids, anticonvulsants, antidepressants, corticosteroids
Cognitive/behavior problems	
Depression	SSRIs, tricyclics
Fatigue	Methylphenidate, modafinil
Delirium	Haloperidol
Infection	Antibiotics (? prophylactic)

GI, gastrointestinal; SSRI, selective serotonin reuptake inhibitor.

Table 4–4. Reflexive Treatment of a Typical Patient before Referral to Neuro-Oncology*

Dexamethasone, 4 mg q 6 hr

Phenytoin, 100 mg tid

Omeprazole, 20 mg qd

Trimethoprim/sulfamethoxazole, 1 DS tab bid tiw

Chemotherapy

Antihypertensives

Hypoglycemics

Antidepressants

*Whether indicated or not (see text).

Table 4–5. Salutary Effects of Glucocorticoids in Neuro-Oncology

Effect	Reference
Reverses CNS edema	28*
Improves overall sense of well being	29
Antiemesis	24
Relieves pain	27
Increases appetite	28
Reduces fatigue	26

*See also Chapters 5 and 6.
CNS, central nervous system.

should determine whether the edema associated with a CNS metastasis is symptomatic or not before beginning corticosteroids. Prophylactic antiseizure medications are not indicated (see p 112).

Salutary Effects of Corticosteroids

Corticosteroids increase the quality of life in patients with preterminal or terminal cancer.[12,14] They improve appetite,[22,23] decrease nausea and vomiting induced by radiation or chemotherapy,[24,25] may relieve fatigue,[26] enhance the feeling of well-being in many patients, and appear to have substantial pain-relieving qualities (see Table 4–5).[16,27] The exact mechanism of pain relief is unclear, although several corticosteroid properties, including their anti-inflammatory and anti-edema effects, their ability to reduce ectopic firing of damaged peripheral nerves,[30] and their inhibition of nociceptive cytokine release,[16] may play a role. In patients with spinal cord compression, pain relief is often dramatic and is clinically independent of its effect on other spinal cord dysfunction (see Chapter 6). Many patients with severe pain can decrease their intake of opioids when they begin taking corticosteroids, and in some, the pain relief from corticosteroids is more complete than that from opioids.

The specific corticosteroids that achieve the best salutary effects, the dose, and the timing of divided doses have not been established. Prednisolone, methylprednisolone, and dexamethasone have all been used in various studies and have yielded approximately comparable results. Relative potencies of individual preparations are discussed in Chapter 3 and can be found in Table 3–12 and in Ref. 154. Relatively low doses of steroids (e.g., 32 mg/24 hr of methylprednisolone) appear to suffice for most indications, but our experience at MSKCC suggests that in spinal cord compression, higher doses are necessary to control pain (see Chapter 3). The side effects of high-dose steroids (e.g., 100 mg/24 hr of dexamethasone) appear to us to be no greater than those from lower doses (16 mg/24 hr dexamethasone) if given for only a short period. However, at least one study suggests a higher frequency of serious side effects with high-dose steroids.[31]

Dexamethasone is the corticosteroid preparation most used by neurologists, partially for historical reasons (see p 84), but partially because its lack of mineralocorticoid effect obviates the salt retention sometimes induced by other corticosteroids. The absence of mineralocorticoid effect is so complete that the physician must remember that in patients with adrenal failure associated with cancer, for example, adrenal metastases, dexamethasone cannot be substituted for hydrocortisone without risking volume depletion and vascular collapse (adrenal crisis).[32,33] (Adrenalectomy was once used to treat metastatic breast cancer. Neurologists at MSKCC would encounter about one such patient a year in whom hypotension, encephalopathy, and hyperkalemia developed after dexamethasone was prescribed to treat a CNS complication and the hydrocortisone was discontinued as being unnecessary. The result was acute Addisonian crisis.)

Protein binding differs. At standard doses, dexamethasone is 70% protein bound, whereas prednisone may be as much as 95% protein bound, so that relatively more dexamethasone than prednisone can be found in the brain and cerebrospinal fluid (CSF) after systemic administration.[34] This difference may be the reason that patients with acute leukemia receiving dexamethasone suffer fewer CNS relapses than those receiving prednisone.[34]

One report suggests that dexamethasone causes less inhibition of the migration of white cells into injured tissue than do other corticosteroids, and thus it may be less immunosuppressive.[35] However, substituting dexamethasone for prednisone resulted in an increased rate of septic episodes and toxic deaths in children being treated with an intensive conventional induction regimen for acute lymphoblastic leukemia.[36] Some investigators believe that the fluorinated corticosteroids (e.g., dexamethasone) are more likely to produce steroid myopathy than non-fluorinated corticosteroids (e.g., prednisone).[37]

Corticosteroids are useful in treating community-acquired bacterial meningitis[38] as well as certain other infections such as pneumonia resulting from *Pneumocystis carinii* (*jiroveci*) in the immunosuppressed patient (note also that patients chronically immunosuppressed by corticosteroids are susceptible to *Pneumocystis* pneumonia) (see p 106). Corticosteroids may also suppress aseptic meningitis caused by blood in the subarachnoid space that sometimes follows neurosurgical procedures for primary or metastatic brain or spinal tumors[39] (see Chapter 14).

Unwanted Effects of Corticosteroids

Table 4–6 summarizes the common side effects of corticosteroids. Details of these and other effects may be found in a number of reviews[19,40,41] and most textbooks of pharmacology.

In one study[42] of patients treated with steroids for spinal cord compression, 50% developed at least one side effect and 20% required hospital admission. The side effects included hyperglycemia (19%), infections (22%), GI disturbances (14%), myopathy (19%), and psychiatric problems (3%). These developments and other important, relevant effects are discussed in the following paragraphs.

Table 4–6. **Some Side Effects of Corticosteroids**

Common but Usually Mild	Non-Neurologic but Serious	Neurologic (Common)	Neurologic (Uncommon)
Insomnia	Osteoporosis	Myopathy	Psychosis
Urinary frequency (nocturia)	Osteonecrosis (hip)	Behavioral alterations	Delirium
Increased appetite (weight gain)	GI bleeding	Hallucinations (high-dose)	Seizures
Abdominal bloating	Bowel perforation	Hiccups	Memory loss
Moon facies	Diabetes	Tremor	
Visual blurring	Opportunistic infections (Pneumocystis)	Brain atrophy	
Acne	Glaucoma		
Edema (legs)	Kaposi sarcoma		
Lipomatosis (spinal cord compression)	Pancreatitis		
Genital burning (IV push)			
Candidiasis			
Cataracts			

GI, gastrointestinal; IV, intravenous.

MYOPATHY

Corticosteroids weaken muscles. Corticosteroid myopathy occurs in two settings. In the first, a patient with cancer, a rheumatologic disorder, or other illness requiring chronic corticosteroids develops, after a few weeks or months of therapy, weakness of proximal muscles (chronic steroid myopathy).[43–45] In the second setting, a critically ill patient in an intensive care unit receives corticosteroids with or without neuromuscular blocking agents and develops muscle weakness that outlasts the use of the neuromuscular blocking agents.[46]

Chronic steroid myopathy occurs in most patients who receive conventional doses (16 mg/24 hr of dexamethasone or equivalent) for more than 2 or 3 weeks.[43] The disorder begins with proximal muscle weakness in the hip girdle and wasting of thigh muscles.[37] Affected patients first note difficulty rising from the toilet seat unless they push off with their hands. As the disorder becomes more severe, it affects the ability to climb stairs. Weakness may spread to the proximal arm muscles (Fig. 4–2), preventing the lifting of heavy objects above the head. At its most florid, severe weakness is noted in the shoulder and pelvic girdles and in the neck muscles, particularly neck flexors. The onset is usually slow but may be sudden with accompanying myalgia. In some instances, respiratory function is compromised.[43,47,48] (Disuse atrophy of the diaphragm begins within 18 hours of mechanical ventilation, whether or not the patient is on steroids.)[48a] Recovery generally occurs over weeks or months after the corticosteroids are tapered or discontinued. Most patients who develop clinically significant steroid myopathy also have the typical cushingoid appearance of the face, trunk, and extremities. The diagnosis is made by clinical examination. Serum creatine kinase is generally normal, and the EMG may show nonspecific small motor units, polyphasic potentials, rare fibrillations or positive sharp waves,[47] or be normal. The muscle biopsy also may be normal but usually shows abnormal variation in fiber diameter, atrophic and necrotic fibers, particularly affecting type 2 (fast) muscle fibers.[47]

The cause of chronic steroid myopathy is not known. Corticosteroids are catabolic, probably inhibiting protein synthesis.[44] Changes have been reported in glycogen metabolism,[50] expression of insulin growth factors 1 and 2,[48,51] and in messenger RNA levels of sarcoplasmic endoplasmic reticulum calcium pumps. Apoptotic pathways are activated.[52]

Acute corticosteroid myopathy,[46,53] also called acute quadriplegic myopathy or critical illness myopathy, occurs in critically ill patients treated in intensive care units, most of whom have been exposed to neuromuscular blocking drugs. One group of patients develops necrotizing myopathy involving type 2 muscle fibers usually associated with high levels of serum creatine kinase. Another group develops similar weakness with a normal or only mildly elevated creatine kinase, a low compound muscle action potential, and loss of myelin filaments on biopsy. Like the chronic steroid myopathy,

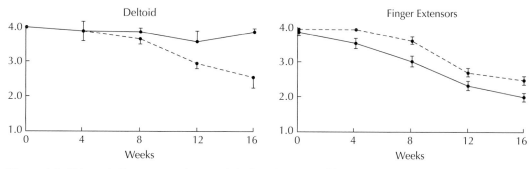

Figure 4–2. This graph illustrates muscle strength (ordinate) over time (abscissa) in patients receiving steroids for systemic lymphoma. Vertical bars indicate the standard deviation. The low-dose dexamethasone group (solid line) receives 25 mg three times a week and the high-dose group (broken line) 50 mg three times a week. Proximal weakness of the deltoids was evident by week 12 and more pronounced by week 16 in all patients receiving high-dose corticosteroids. Distal muscles most affected by vincristine, such as the finger extensors, were not affected by corticosteroid dose; patients receiving a higher dose of dexamethasone had less weakness in distal muscles than those patients receiving the lower dose. An identical pattern of weakness was seen in the leg muscles (not shown). (From Ref. 49 with permission.)

weakness is usually most marked proximally. Distal weakness is rare. Sensation remains normal, and deep tendon reflexes are preserved. The exact mechanism is unknown, but there is impaired excitation–contraction coupling at the level of the sarcolemma, increased muscle protein breakdown, necrosis, and apoptosis. Inflammatory cytokines in these critically ill patients probably also play an important role.[46]

The differential diagnosis of proximal muscle weakness in patients with cancer who are taking corticosteroids includes metabolic and nutritional myopathies, leptomeningeal tumor, spinal cord compression causing predominantly proximal weakness, Lambert–Eaton myasthenic syndrome, and paraneoplastic polymyositis (see Chapter 15). It is particularly important to consider polymyositis because that disorder is treated with steroids. Laboratory abnormalities in polymyositis include electromyographic changes of increased insertional activity of muscle, elevated muscle enzyme concentrations in serum, and histologic findings of muscle necrosis or inflammation. These laboratory findings are usually minor or absent in patients with steroid myopathy. Progressive spinal cord compression in patients being treated with steroids may cause increasing proximal leg weakness that should be distinguished from steroid myopathy because the compression may require an increase in the corticosteroid dose.

Steroid myopathy probably develops more rapidly in patients who are inactive and may be ameliorated by exercise[54] and perhaps by supplying essential amino acids.[55] Experimentally, insulin-like growth factor-1[51] and growth hormone reverse the myopathy.[56] One report suggests that patients with brain tumors treated with phenytoin are at lower risk (8%) for developing steroid myopathy than those not taking the anticonvulsant (16%), probably because of drug interactions, discussed in a later section of this chapter.[57] Because many elderly patients are vitamin D–deficient, an abnormality that also leads to muscle weakness, it is probably worth measuring vitamin D and replacing it if it is low.[58]

BEHAVIORAL AND/OR COGNITIVE ABNORMALITIES (STEROID PSYCHOSIS)

The term *steroid psychosis* is used to refer not only to psychotic behavior but also to delirium, mood disorders, cognitive changes, and other psychiatric syndromes that may be induced by the ingestion of corticosteroids (Table 4–7).

Mild behavioral changes, including euphoria and depression, are common side effects of corticosteroids.[59,60] Most of our patients on 8 to 16 mg of dexamethasone a day describe themselves as feeling "wired." Some, particularly those who have complained of fatigue, enjoy the feeling; others complain of agitation and irritability.

More severe mental and behavioral abnormalities are uncommon.[59] In one series of ophthalmologic patients free of psychiatric illness, an 8-day course of prednisone (mean dose 119 mg/day) resulted in the development of elevated mood in 26% of patients and depression in 10%. The affective symptoms usually occurred early during treatment.[61] Other studies have found either lesser or greater changes in mood, with depression being about as common as elevation of mood. Recurrent episodes of mood disorder, usually manic but sometimes depressive, may occur in patients given recurrent courses of corticosteroids.[62] Cognitive changes, particularly abnormalities of working memory, are common and may reflect the effects of corticosteroids on the hippocampus.[63–66]

Severe psychiatric reactions are of three general types—affective, schizophrenic-like, and delirious[60,67]:

1. Affective disorders cannot be distinguished from non–steroid-associated psychiatric illness. Mania is more common with exogenous steroid therapy; depression is more common with Cushing syndrome (endogenous). The disorder usually begins early in the course of therapy, is more likely to affect women, and is probably dose related. It resolves

Table 4–7. **Behavioral/Cognitive Effects of Corticosteroids**

Affective
Euphoria → Mania
Depression
Both (cycling)
Psychosis
Delirium
Dementia

when the corticosteroids are withdrawn and can be treated with neuroleptic drugs including lithium, valproic acid, haloperidol, or olanzapine. Tricyclic agents are not effective and may worsen symptoms. Affective disorders occasionally begin during the tapering of steroids but resolve once the corticosteroids are discontinued entirely. A few patients taking alternate-day corticosteroids cycle with mood elevation on the on-days and depression on the off-days; the phenomenon appears to be dose related. Lithium and valproate have been used prophylactically and may prevent the affective form of steroid psychosis. A history of steroid psychosis does not predict recurrent psychosis with a second course of corticosteroids.

2. Schizophrenia may be mimicked by corticosteroids. The disorder usually begins acutely. The patient becomes withdrawn, paranoid, or both, and may experience auditory or visual hallucinations. The disorder cannot be distinguished from the primary psychiatric illness, but always responds to either corticosteroid withdrawal or neuroleptic treatment.

3. Delirium may be caused by corticosteroids. Affected patients become distractible, confused, and unable to attend appropriately to environmental stimuli; visual hallucinations are common. One of our patients, a young woman with metastatic melanoma to the brain, developed a florid delirium when dexamethasone was started. She was wide awake, terrified, and believed she had died and gone to hell. She vividly described the fires that surrounded her. As the steroids were tapered, the delirium cleared and the patient denied any memory of the episode. Mild hallucinations are often not reported because patients attribute them to the drug and are neither surprised nor concerned about them.

The affective and schizoid steroid psychoses occurring in a patient under the stress of cancer treatment must be differentiated from a psychiatric disorder. Steroid-induced delirium must be distinguished from similar symptoms that are due to other metabolic disorders that may complicate the cancer or its treatment (see Chapter 11).

INSOMNIA

Insomnia and other sleep disturbances are common in patients with cancer[68] and are particularly prominent if the patients are on steroids as well. The insomnia is often exacerbated by steroid-induced nocturia. Patients should be warned about this side effect and hypnotics used if necessary.

HYPERGLYCEMIA

As many as 5% of patients with apparently normal glucose tolerance may develop steroid-induced diabetes. Blood glucose levels should be monitored and appropriate treatment applied. Oral hypoglycemics often suffice. Because of the known deleterious effects of hyperglycemia on brain function,[69] particularly cognition, all patients on corticosteroids should have glucose levels carefully measured. The physician should attempt to keep the serum glucose level below 150 mg/dL.

BONE DISORDERS

Osteoporosis is a common side effect of steroid use[70] and begins after only a few months of treatment. It can cause vertebral and other fractures. Pathogenetic factors include both direct and indirect effects of the drugs on bone, including reduced calcium absorption, increased calcium excretion, reduced osteocyte function, and inhibition of sex hormones.[70,71] The back pain from an osteoporotic fracture in a patient with cancer may be difficult to distinguish from a vertebral metastasis. Steroid-induced osteoporosis reverses itself, at least in young persons, when the drug is discontinued. Bisphosphonates,[72] calcium (1200 mg daily if renal function is normal), and vitamin D (400–800 IU daily) may help prevent the disorder.[73] Oral bisphosphonates are cumbersome to give and can cause esophageal ulceration. Zoledronic acid and pamidronate can be given intravenously. Rarely, osteonecrosis of the jaw complicates bisphosphonate treatment, especially if given intravenously to patients with myeloma.[74]

For the patient with neuro-oncologic disorders, avascular necrosis of the hips[75] or occasionally of the shoulders, wrist, clavicle, or vertebral body is a more important problem. The disorder may be confused with spinal cord compression or peripheral neuropathy. Affected patients have usually been on steroids for months, although occasional patients develop osteonecrosis after only a few weeks of treatment. In a series of 71 patients with severe acute respiratory distress given high-dose steroids, 10% developed osteonecrosis.[76]

The hip disorder is characterized by pain, often radiating down the anterior aspect of the thigh to the knee, resembling, in some respects, a femoral neuropathy or lumbar radiculopathy. Sometimes, the pain may be localized to the buttock or groin, hindering the diagnosis. The pain is exacerbated by walking. The diagnosis is suggested by reproducing the pain on rotation of the hip. In some patients, a "click" can be heard as the hip is passively flexed and extended. The results of plain X-rays, nuclide bone scans, and computed tomography (CT) may be normal early in the course of the disorder, but eventually they will reveal the necrosis. The most sensitive diagnostic test is magnetic resonance imaging (MRI).[77]

LIPOMATOSIS

Corticosteroids cause redistribution of fat to the face, causing the characteristic *moon facies*; to the abdomen, causing a *pot belly*; and to the posterior lower neck, causing the characteristic *buffalo hump*. Many patients suffer a sensation of chronic abdominal bloating. Additionally, fat redistributes to the retro-orbital space (causing exophthalmos—unilateral after a retro-orbital steroid injection)[78,79] to the mediastinum, and to the epidural space. Excess fat in the epidural space can cause spinal cord compression.[80–82] In some instances, surgical decompression of the cord has been necessary to relieve neurologic symptoms. Both MRI and CT easily differentiate the typical fat density of the steroid-induced lesions in the epidural and mediastinal spaces from other lesions such as metastases.

VISUAL PROBLEMS

The visual system can be affected by steroids in several ways.[83] Many patients receiving corticosteroids complain of visual blurring, which appears to be due to a change in refraction associated with the corticosteroids. The symptoms may be precipitated by either an increase or a decrease in corticosteroid dose but generally disappear after the agents are discontinued. Cataracts are common in patients on long-term corticosteroids and have been reported to be caused by intermittent dexamethasone used as an antiemetic. Corticosteroids increase intraocular pressure, sometimes leading to glaucoma.[84]

GASTROINTESTINAL DISORDERS

Ulceration

Most patients with neuro-oncologic disorders (at least in our institution) who are prescribed corticosteroids are also given *gastro-protective* agents despite the fact that the evidence that steroids alone cause peptic ulceration is weak and the evidence that gastro-protective agents prevent so-called steroid-induced peptic ulceration is even weaker.[85] Most investigators believe that corticosteroids in combination with either aspirin or nonsteroidal anti-inflammatory agents promote GI bleeding, but even here the evidence is controversial, at least one investigator believing that corticosteroids protect against non-steroidal anti-inflammatory drug (NSAID)-induced GI bleeding.[86] Thus, we do not recommend routine use of either histamine (H2) blockers or proton pump inhibitors in patients receiving corticosteroids. Furthermore, gastro-protective agents can have side effects: Misoprostol can cause diarrhea, and abdominal pain and, rarely, confusion. Omeprazole can cause abdominal pain, constipation, or diarrhea. CNS effects include headache, dizziness, somnolence, confusion, and hallucinations. Myalgias may also occur. Omeprazole is metabolized by the cytochrome P-450 system and thus can inhibit hepatic metabolism of several drugs including some chemotherapeutic agents.

H2 blockers have been associated with encephalopathy and occasionally coma in the elderly. Whether H2 blockers react synergistically with other drugs to promote sedation needs clarification. They can cause thrombocytopenia and interact with chemotherapeutic agents.[87] Sucralfate is not absorbed and thus should be safe, but it binds a number of drugs in the stomach, preventing adequate absorption. It also can cause constipation, dry mouth and, rarely, abdominal pain. Thus, if it is used, it should be given 2 hours before or after other oral agents that require absorption to be efficacious.

Bowel Perforation

Bowel perforation generally affects the sigmoid colon.[88–90] Corticosteroids are a major risk factor for bowel perforation. NSAIDs and opioids are lesser risk factors.[88] Affected patients are usually constipated because of bowel dysfunction due to spinal cord compression, drugs, or inactivity. Clinical symptoms usually begin with moderate-to-severe pain in the abdomen, but the physician is often misled because bowel sounds persist and the rebound tenderness characteristic of peritonitis may not be present, being masked by the corticosteroids or abdominal muscle weakness resulting from spinal cord compression. The diagnosis can be confirmed by finding free air in the abdomen. The patient usually must be treated surgically, although a few bowel perforations heal spontaneously. The risk is particularly high in patients with spinal cord compression because they are more prone to constipation. Prevention or early treatment of constipation may avert this serious complication.

Pneumatosis Cystoides Intestinalis

Pneumatosis cystoides intestinalis is a benign syndrome related to steroid therapy.[91] Within the intestinal wall, air appears as cysts or linear streaks.[92] The cysts may rupture, leading the physician to suggest surgery,[93] but the ruptures heal spontaneously and do not require surgical repair.

Colonic Ileus

Colonic ileus, or pseudo-obstruction of the colon (Ogilvie syndrome), is a disorder of bowel motility that affects severely ill patients, including those with spinal cord compression. Patients present with abdominal pain and marked distention. They are usually constipated, but a few have diarrhea; bowel sounds remain. Plain films of the abdomen show a grossly distended cecum that may, if untreated, rupture causing peritonitis and in some cases death.[94] Rupture is rare unless the cecum is greater than 12 cm. The pathophysiology of the disorder is not fully understood. It is believed to be neurogenic resulting from excessive sympathetic stimulation via the lumbar sympathetic chain and/or decreased parasympathetic tone via the vagus nerve and S2–4 sacral roots.[95,96] Conservative treatment, consisting of rectal tube placement, enemas, and discontinuing medications that decrease bowel motility is effective in most patients. When that fails, pharmacologic treatment with neostigmine[97] may be effective. Decompression colonoscopy and even surgery may be necessary in some cases.[96]

Constipation/Diarrhea

Whether taking corticosteroids or not, many patients with neuro-oncologic disorders are plagued by either constipation or diarrhea. Overall, about 10% of patients with advanced cancer suffer one of these two symptoms.[98] Diarrhea can be caused by chemotherapeutic agents, including 5-fluorouracil, methotrexate, irinotecan, taxanes, monoclonal antibodies, and hormonal agents[98] or by infection; at its worst, the disorder can be life threatening, leading to hemodynamic collapse. Occasionally, diarrhea develops around a fecal impaction; the symptoms should not be missed because treating the diarrhea exacerbates the impaction. Diarrhea is usually treated with loperamide and sometimes with octreotide.[98]

For most patients, constipation is the more distressing symptom.[99,100] One of our patients, who underwent a craniotomy and a thoracotomy during the same hospitalization to treat lung cancer, reported that the pain and suffering caused by the surgeries were far less than those caused by the constipation. Constipation can result from immobility, opioids, antiemetics, paraneoplastic neuropathy,[101] and chemotherapeutic agents, including thalidomide, cisplatin, vinca alkaloids, and temozolomide.

Laxatives are particularly important drugs for patients with neurologic complications of cancer,[102] especially for those taking glucocorticoids. Strong efforts should be made to maintain bowel function in any patient taking corticosteroids. Virtually all patients on opioids, and patients with advanced cancer not taking steroids, should take a stool softener[102]; laxatives should be administered, as needed, to maintain normal bowel function. Macrogol, which is an osmotic laxative also known as polyethylene glycol, is probably the drug of choice.[102] A laxative escalation ladder has been published.[102]

Patients should be encouraged to increase bulk in their diet and to drink adequate liquids to maintain stool softening. In some patients in whom opioids have produced severe constipation, oral naloxone or methylnaltrexone[103,104] appears to relieve constipation without altering the analgesic effect of the opioids.

OTHER SIDE EFFECTS

Several other side effects of corticosteroids are occasionally encountered in neuro-oncologic practice. A particularly unpleasant one is *hiccup*,[105] a dose-related but idiosyncratic corticosteroid effect that afflicts a few patients; corticosteroids may stimulate the hiccup reflex.[106,107] (Several other drugs of interest to neuro-oncologists, including benzodiazepines, antidepressants, antibiotics, anabolic steroids, and opioids have been reported to cause hiccups.[106,108] Treatment of steroid hiccup is often difficult unless the dose is decreased. Phenothiazine, nifedipine, valproate, metoclopramide, and carbamazepine sometimes suppress the symptom.[109]

Hypersensitivity and even *anaphylaxis* to corticosteroids have been reported,[110,111] but the sensitivity to one agent may not apply to all corticosteroids; a careful challenge can identify a safe drug.[110] Occasional patients develop psychological dependence on corticosteroids.[112] A peculiar side effect of IV bolus injection of dexamethasone is a severe anogenital itching or tingling that is extremely distressing but generally subsides within a few minutes. The patient who is warned of the possibility in advance finds it tolerable.[113] Corticosteroid-enhanced free-water clearance may cause *nocturia* and sometimes may awaken the patient hourly. Several drugs, including corticosteroids, alter the senses of *taste* and *smell* and occasionally cause anorexia.[114] Intrathecal administration of corticosteroids can cause a severe *pachymeningitis*.[115]

Because of their known immune-suppressing capacities, corticosteroids, whether exogenous or endogenous, can lead to opportunistic *infections* (see Chapter 10). Oropharyngeal candidiasis (thrush) is common, occurring in about one-third of patients in palliative care units who are receiving corticosteroids.[116] The infection may be asymptomatic or may cause oral burning or dysphagia. The mouth and throat of all patients receiving corticosteroids should be inspected at each visit. The disorder usually presents with white patches but may be erythematosus.[117] Culture of the lesions will establish the diagnosis, but patients often are treated simply on the clinical appearance. For mild cases, nystatin as a troche or oral rinse may be effective, but the material should be kept in the mouth for as long as possible before swallowing. Prophylactic therapy is unnecessary.[118]

In the neuro-oncologic setting, an occasional problem is *P. carinii* in patients who have been on corticosteroids and are being tapered to lower doses.[119,120] Some physicians recommend prophylaxis with trimethoprim-sulfamethoxazole (cotrimoxazole 160/800 mg three times a week).[121–123] We suggest that the drug be given to those patients who will be receiving corticosteroids for more than 6 to 8 weeks; the drug is discontinued a month after corticosteroids are stopped. Some reports indicate that for children with acute lymphoblastic leukemia, using dexamethasone rather than prednisone resulted in more bacterial sepsis and toxic deaths,[36,124] despite the fact that dexamethasone may be more effective than prednisone in treatment of the leukemia.[124,126] This has led some authors to question the use of dexamethasone during maintenance therapy for acute lymphoblastic leukemia.[127] Corticosteroids also may increase the severity of varicella infections in children being treated for leukemia.[128]

DRUG INTERACTIONS

Because patients with neuro-oncologic problems treated with corticosteroids often also take other drugs, it is important to be aware of untoward drug interactions, especially interactions with drugs that induce hepatic microsomal enzymes (cytochrome P-450 system). Drugs such as phenytoin, phenobarbital, and perhaps carbamazepine increase the metabolic clearance of corticosteroids and may decrease their therapeutic effect.[129] In one study, an oral dose of dexamethasone was decreased to 20% of its previous bioavailability after the addition of phenytoin.[130] As a result, some patients who have brain metastases and are on stable doses of corticosteroids develop increased symptoms when they are started on anticonvulsants. Conversely, one report[131] suggests that administration of dexamethasone may increase phenytoin levels, but our own experience has been the opposite. We have found that some patients on stable doses of phenytoin develop toxicity as the corticosteroid dose is decreased, or they develop seizures because phenytoin levels become subtherapeutic when the corticosteroid dose is increased.[132]

A recent study of 405 patients being treated for cancer identified at least one potential deleterious drug interaction in 109 (27%) patients. Twenty-five (9%) were classified as major and 211 (77%) as moderate. Most of the interactions involved noncancer agents, including corticosteroids and anticonvulsants (see following text).[133]

CORTICOSTEROID WITHDRAWAL

Withdrawal of corticosteroids from patients also causes neurologic disability (Table 4–8).[134,135] Adrenal insufficiency is the most dangerous complication of corticosteroid withdrawal, but *steroid pseudorheumatism*[136] is the most painful. Patients develop acute myalgias, arthralgias, or both, sometimes severe. One such patient was admitted to MSKCC with a presumptive diagnosis of spinal cord compression because severe pain in his legs prevented him from walking. In others, the disorder is milder and can be ameliorated by increasing the dose of corticosteroids for a few days and then tapering more slowly. Steroid pseudorheumatism usually follows a rapid taper of corticosteroids, but occasionally occurs when the drug is decreased by as little as 2 mg of dexamethasone a week. In affected patients, each reduction seems to produce an exacerbation of the arthralgia, which then disappears over a period of 36 to 48 hours.

Headache, lethargy, and sometimes low-grade fever are also common symptoms of corticosteroid withdrawal. The corticosteroid withdrawal syndrome was first described by Amatruda

Table 4–8. **Corticosteroid Withdrawal Syndrome**

Myalgia, arthralgia (steroid pseudorheumatism)
Headache
Lethargy
Fever
Nausea, vomiting, anorexia
Postural hypotension
Papilledema
Pneumocystis pneumonia

et al.[134] in patients who had no underlying neurologic disease. When it occurs in patients with CNS disease, the physician may believe that the symptoms are due to recurrent brain tumor rather than corticosteroid withdrawal.

In children, pseudotumor cerebri can complicate withdrawal from prolonged corticosteroid treatment.[137] This finding may be caused by decreased CSF absorption associated with corticosteroid withdrawal.[138] One report describes withdrawal symptoms in 50% (prednisone) to 75% (dexamethasone) of patients after induction therapy for leukemia.[139]

SEIZURES/ANTICONVULSANTS

Incidence/Epidemiology

Seizures are a common problem in neuro-oncology, occurring in patients with brain tumors (primary or metastatic), leptomeningeal metastases, CNS infections, after stem cell transplantation, and associated with several other neurologic complications of cancer (Table 4–9). Figure 4–3 identifies some of the causative mechanisms producing seizures in patients with primary or metastatic brain tumors.

In one retrospective analysis of 470 patients with brain metastases in a neuro-oncology database, seizures occurred either at presentation or during the course of illness in 113 patients (24%).[141] The likelihood of seizures was highest with brain metastases from melanoma (67%), but lung (29%), breast (16%), and unknown primary tumors (25%) were also common causes. Seizures occurred in 4% of 1256 cancer patients without structural brain lesions. (In patients with primary brain tumors, seizures occurred in 69% of patients with astrocytomas, 50% with oligodendrogliomas, and 56% with mixed gliomas. Seizures occurred in 45% of patients with meningiomas. Seizures were less common in patients with anaplastic tumors, occurring in 44% of patients with anaplastic gliomas and 48% of patients with glioblastoma.)

Seizures in patients with brain metastases may be focal or generalized. If generalized, they have a focal onset, although the focal discharge may be asymptomatic. Conversely, the presence of an aura does not necessarily indicate a focal onset; a recent report suggests that as many as 70% of patients with idiopathic

Table 4–9. **Some Causes of Seizures in the Patient with Cancer**

Metastatic Central Nervous System Neoplasms	*Treatment-Related Causes*
Parenchymal metastases	Radiation Therapy
Dural-based metastases	Acute
Leptomeningeal metastases	Early–delayed
	Late–delayed
Metabolic Conditions	Chemotherapy
Hyponatremia	Antimetabolites
Hypoglycemia	Methotrexate
Hypoxia	Cytarabine
Hypocalcemia	L-asparaginase
Hypomagnesemia	Vinca alkaloids
	Topoisomerase inhibitors
Cerebral Infarction	Alkylators
	Ifosfamide
Cerebral Hemorrhage	Nitrosoureas
	Cisplatin
Infections	Biologic response modifiers
Bacterial	Interferon
Listeria monocytogenes	Interleukin-2
Viral	Opioids
Cytomegalovirus	Meperidine
Herpes simplex	Antiemetics
Fungal	Phenothiazines
Cryptococcus neoformans	Butyrophenones
Aspergillus fumigatus	Antibiotics
Candida species	Penicillins
Parasites	Imipenem-cilastatin
Toxoplasma gondii	

generalized epilepsy report an aura during video monitoring.[142] Generalized seizures can be lethal if they induce status epilepticus (see subsequent text). The treatment of seizures is considered in the paragraphs that follow.

When a seizure occurs, anticonvulsants are indicated. However, we know of no published studies that address whether anticonvulsants prevent further seizures in patients with brain metastases or other neurologic complications of cancer, even when blood levels are within the *therapeutic range*. Therapeutic range is a statistical term that may not necessarily apply to an individual patient. Seizure control may be achieved in some patients whose blood levels are below the therapeutic range; others may require doses within the so-called

toxic range. Conversely, some patients demonstrate evidence of toxicity with blood levels within the therapeutic range, whereas others are quite comfortable with their blood levels within the toxic range. Evidence suggests a poor correlation between *therapeutic* serum phenytoin levels and control of partial (i.e., focal) seizures, despite a good correlation between such concentrations and control of generalized seizures.[143] Several of our patients with low-grade primary brain tumors suffer brief focal seizures as often as several times a day despite trying the gamut of treatments offered by experienced epileptologists. One of our patients even received a vagal nerve stimulator, without success. The seizures resolved after the tumor was treated. These

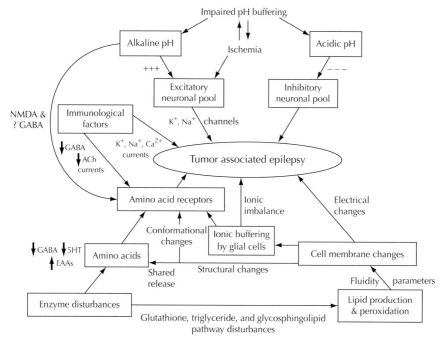

Figure 4–3. Summary of possible causative and influencing mechanisms on tumor-associated epilepsy. The rich interplay of the varied factors and many plausible routes for seizure causation is highlighted. (From Ref. 140 with permission.)

patients often prefer having the seizures to the side effects of what appear to be useless anticonvulsants.

Therapy

Generalized seizures can be lethal if they induce status epilepticus (see following text). Seizures can exhaust brain metabolic reserves, induce excitotoxic death of neurons,[144,145] and raise cerebral blood flow, blood volume, and intracranial pressure, potentially causing cerebral herniation (see Chapter 3). Focal seizures, if repetitive, can also cause permanent neurologic damage in patients with brain metastases. Accordingly, patients actively having seizures should be treated with anticonvulsant drugs on an urgent basis.[146,147] Seizures should be stopped first with IV diazepam (5 mg over 1–2 minutes, repeated once if necessary) or, probably better,[148] lorazepam (2 mg over 1–2 minutes, repeated once if necessary—up to 0.1 mg/kg) followed by 15–20 mg/kg by slow IV injection (no more than 50 mg/min) or more safely with fosphenytoin.[147] Intravenous midazolam also appears effective and safe.[149,150] A recent study indicates that sodium valproate (40 mg per kilogram IV bolus) may be more effective than phenytoin as initial therapy.[151]

Once the seizures are under control, the patient is switched to oral medication. A number of antiepileptic drugs are available (Table 4–10). One group of anticonvulsants induces liver microsomal enzymes, thus interfering with the metabolism of several chemotherapeutic agents (Table 4–11). Patients receiving such anticonvulsants require dose adjustment for many chemotherapeutic agents. Anticonvulsants can also be classified by their mechanism of action and by the nature of their adverse effects (Table 4–12).

Although approved by the FDA only for *add-on* therapy, levetiracetam is probably the best agent for initiating antiseizure treatment in patients with neuro-oncologic disorders, being as efficacious as carbamazepine for newly diagnosed epilepsy.[154,155] It has no known interactions with other agents, little or no protein binding, and a unique mechanism of action. Furthermore, in experimental animals, it appears to have long-lasting antiseizure effects after the drug is discontinued.[156] The most common side effect is somnolence, noted in about one-third of patients, which usually wears off with continued use. Behavioral

Table 4–10. Antiepileptic Drugs

	Average Dose (Serum Therapeutic Range)	Metabolism	Mechanism of Action	Some Adverse Effects
Enzyme-Inducing AEDs				
Phenytoin	20 mg/kg load; then 3–5 mg/kg, daily or twice daily (10–20 µg/mL)	Hepatic	Sodium channel	Rash, osteomalacia, Stevens–Johnson syndrome
Carbamazepine	800–2400 mg, two to four times a day (8 to 12 µg/mL)	Hepatic	Sodium channel	Drowsiness, diplopia, rash, Stevens–Johnson syndrome, leukopenia, hyponatremia
Phenobarbital	10 mg/kg load, then 1–3 mg/kg/d (15–40 µg/mL)	75% hepatic; 25% renal	GABA	Drowsiness, Stevens–Johnson syndrome, frozen shoulder
Oxcarbazepine	900–2400 mg two to four times a day	80% hepatic	Sodium channel	Hyponatremia, diplopia
Non–Enzyme Inducing AEDs				
Valproic acid	10–60 mg/kg three to four times a day (60–100 µg/mL); intravenous infusion rate is 20 mg/min, same dose as oral	Hepatic	GABA, sodium channel	Hair loss, weight gain, pancreatitis, thrombocytopenia, platelet dysfunction, tremor, parkinsonism, extrapyramidal syndrome
Gabapentin	900–4800 mg daily in three to four doses	Renal	GABA	Drowsiness, rapid titration, ataxia, weight gain
Topiramate	100–400 mg twice a day	30%–50% hepatic; 50%–70% renal	Sodium channel, GABA, AMPA/kainate	Cognitive impairment, paresthesias, slow titration, weight loss, renal calculi
Levetiracetam	500–3000 mg twice a day	Enzymatic hydrolysis	Synaptic vesicle protein binding	Agitation, psychosis, drowsiness, glaucoma
Lamotrigine	300–500 mg twice a day (3–20 µg/mL)	85% hepatic	Sodium channel	Drowsiness, rash, particularly with concurrent valproate, slow titration
Tiagabine	24–56 mg daily in two to four doses	90% hepatic	GABA	Drowsiness, tremor, slow titration
Zonisamide	200–600 mg once or twice a day (10–30 µg/mL)	>90% hepatic	Calcium, sodium channel	Drowsiness, headache, weight loss, renal calculi, slow titration

Adapted from Refs. 152 and 153.
AEDs, antiepileptic drugs; GABA, gamma-aminobutyric acid; AMPA, alpha-amino-3-hydroxy-5-methyl-4-isoxazoleprionate.

Table 4–11. **Protocol for Treatment of Convulsive Status Epilepticus at Memorial Sloan-Kettering Cancer Center***

First 5 Minutes

  ABCs
  Diagnose status epilepticus
  Obtain IV access
  Begin ECG monitoring
  Fingerstick for glucose—correct if necessary
  Draw blood for BMP, Mg, Ca, Ph, CBC, LFT, AED levels (PHB, PHT, VPA, CBZ), toxicology screen
  Call Neurology consult

6–10 Minutes

  Thiamine 100 mg IV; 50 mL of D50 IV in appropriate clinical setting
  Lorazepam 4 mg IV over 2 minutes; if necessary repeat once in 5 minutes. If no IV access give diazepam 20 mg rectally or midazolam 10 mg intranasally, or intramuscularly.

10–20 Minutes

  Add fosphenytoin 20 mg/kg IV at 50 mg/min with BP and ECG monitoring. Can re-bolus fosphenytoin 10 mg/kg if seizures persist. Maintain level 20–30 µg/mL

20–60 Minutes

  If seizures persist, *intubate* and start phenobarbital IV 20 mg/kg at 50–100 mg/min
  If still seizing can add or switch (PHB) to midazolam: load 0.2 mg/kg; repeat 0.2–0.4 mg/kg boluses every 5 minutes until seizures stop, up to a maximum total loading dose of 2 mg/kg. Initial rate 0.1 mg/kg/hour. Continuous IV range 0.05–2 mg/kg/hr.
  Or
  Propofol: Load 1 mg/kg; repeat 1–2 mg/kg boluses every 3–5 minutes until seizures stop, up to a maximum total loading dose of 10 mg/kg. Initial rate 2 mg/kg/hour. Dose range 1–15 mg/kg/hour

After 60 Minutes

  If seizures persist, use anesthetics
  Continuous IV propofol: load 1 mg/kg; initial rate 2 mg/kg/hr. Titrate until burst suppression.
  Will need to arrange continuous EEG monitoring (preferably as soon as the patient does not awaken rapidly)

Another possible consideration for fourth line treatment is valproate 40 mg/kg over 10 minutes. Can re-bolus 20 mg/kg over 5 minutes.

If bacterial meningitis is suspected start ceftriaxone, vancomycin, and ampicillin (can start along with treatment for SE). Start acyclovir if HSV encephalitis is suspected. Perform LP when stable.

* Protocol devised by Drs. Edward K. Avila and Lisa M. DeAngelis.

changes, including hostility and aggression, have also been reported.[157] Other drugs recommended by neuro-oncologists include gabapentin,[158] oxcarbazepine, and topiramate.[159]

STATUS EPILEPTICUS

Status epilepticus is a term used to describe patients with either continuous or intermittent seizures without recovery of consciousness between them.[146] Status epilepticus can be either convulsive or nonconvulsive.[160,161] Generalized convulsive status epilepticus is a medical emergency requiring vigorous treatment often to the point of general anesthesia[146]; one approach is illustrated in Figure 4–4. (The figure refers to all varieties of status, mostly in patients with epilepsy, and differs somewhat from the MSKCC

Table 4–12. Interactions between Antiepileptic Drugs and Cytotoxic Drugs

AED	CTDs that Decrease AED Concentration	CTDs that Increase AED Concentration	CTDs Whose Concentrations Are Lowered by AED	CTDs Whose Toxicity Is Increased by AED
Phenytoin	Nitrosourea	5-FU	Busulfan	
	Cisplatin	Tegafur	Vincristine	
	Etoposide	Tamoxifen	Methotrexate	
	Dacarbazine	Dexamethasone	Paclitaxel	
	Doxorubicin		Topotecan	
	Carboplatin		Irinotecan	
	Vinblastine		9-Aminocampthotecin	
	Methotrexate		Teniposide	
	Bleomycin		Dexamethasone	
	Dexamethasone			
Carbamazepine	Cisplatin		Vincristine	
	Doxorubicin		Methotrexate	
			Paclitaxel	
			9-Aminocampthotecin	
			Teniposide	
Valproic acid	Methotrexate			Nitrosoureas
	Cisplatin			Cisplatin
	Doxorubicin			Etoposide
Phenobarbital			Thiotepa	
			Nitrosoureas	
			Vincristine	
			Methotrexate	
			Paclitaxel	
			9-Aminocampthotecin	
			Teniposide	
			Procarbazine	
			Prednisone	

Data from 169.
AED, antiepileptic drugs; CTD, cytotoxic drugs.

approach, detailed in Table 4–11. However, the general principles of both apply to patients with neuro-oncologic disorders.)

Nonconvulsive status epilepticus is a disorder characterized by alterations of consciousness, including, at its most severe, coma. In one series, 8% of comatose patients without overt signs of seizure activity were found to be in electrographic status epilepticus.[161] The diagnosis should be suspected in comatose patients with risk factors for seizures who show subtle motor or oculomotor movements.[160] The diagnosis is made by the electroencephalogram[162,163]; however, not all patients with electrographic abnormalities suggesting seizures are in nonconvulsive status and, conversely, nonconvulsive status may occur in comatose patients whose electroencephalograms do not suggest seizure activity. When suspicion for the disorder is high, intravenous injection of lorazepam may normalize the electroencephalogram (EEG) and awaken the patient.

Rarely, absence status can paradoxically be caused by *therapeutic* levels of carbamazepine or phenytoin and perhaps other anticonvulsants as well.[164,165]

PROPHYLATIC ANTICONVULSANTS

As indicated in Table 4–4, most patients who present with a brain metastasis or primary brain

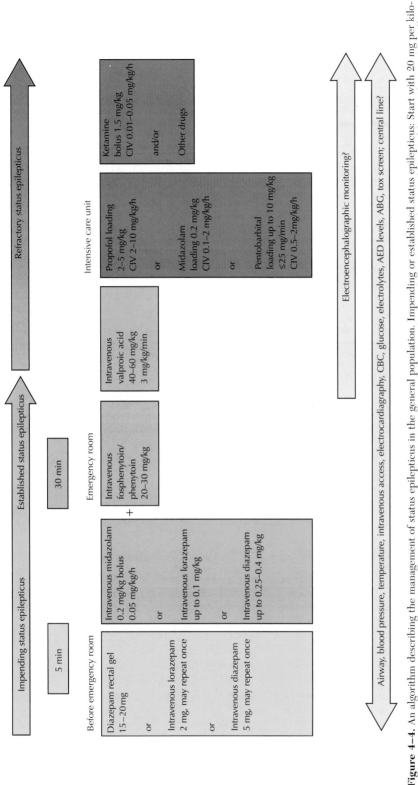

Figure 4–4. An algorithm describing the management of status epilepticus in the general population. Impending or established status epilepticus: Start with 20 mg per kilogram of fosphenytoin or phenytoin, and if status epilepticus persists, given additional 10 mg/kg. Follow the flow chart UNLESS there is a history of drug intolerance (e.g., allergy to phenytoin or benzodiazepine) then replace by intravenous valproic acid 40 to 60 mg per kilogram or IV phenobarbital 20 mg/kg; UNLESS treatment-induced hypotension slows rate of delivery; UNLESS history of progressive or juvenile myoclonic epilepsy (phenytoin/fosphenytoin harmful in the progressive myoclonic epilepsy and ineffective in juvenile myoclonic epilepsy), replace with IV valproic acid or IV phenobarbital; UNLESS tonic status epilepticus with Lennox–Gastaut syndrome (might be worsened by benzodiazepines), replace with IV valproic acid; UNLESS focal status epilepticus without impairment of consciousness, IV treatment not indicated, load anticonvulsant orally or rectally. Refractory status epilepticus: IV valproic acid—start with 40 mg/kg and if status epilepticus persists, give an additional 20 mg/kg. Continuous intravenous infusion usually starts with the lower dose which is titrated to achieve seizure suppression and is increased as tolerated if tachyphylaxis develops. Ketamine: Rule out increased intracranial pressure before administration. Other drugs: Felbamate, topiramate, levetiracetam, lidocaine, inhalation anesthetics, etc. Dosage and pharmacokinetics of most anticonvulsants must be adjusted appropriately in patients with hepatic or renal failure or with drug interactions. Some patients in refractory status epilepticus will need systemic and pulmonary artery catheterization, with fluid and vasopressors as indicated to maintain blood pressure. CBC, complete blood count; AED, antiepileptic drugs; ABG, arterial blood gas. (From Ref. 146 with permission.)

tumor receive anticonvulsants whether or not they have had a seizure. A number of studies indicate that prophylactic anticonvulsants for patients with brain metastases or primary brain tumor are ineffective.[166,167] An American Academy of Neurology Practice Parameter recommends that "prophylactic anticonvulsants should not be used routinely in patients with newly diagnosed brain tumors (standard)."[167]

Current evidence suggests that prophylactic anticonvulsants in patients undergoing neurosurgery reduce the risk of seizures in the first postoperative week by 40% to 50%, but probably have no effect on seizures occurring after that.[168] This evidence supports the guideline of the American Academy of Neurology suggesting "tapering and discontinuing anticonvulsants after the first postoperative week, particularly in patients who are medically stable and who are experiencing anticonvulsant-related side effects (guideline)."[167]

The issue would be moot were it not for both the interactions of anticonvulsants with chemotherapeutic agents and other drugs and the side effects of many anticonvulsants. These considerations are addressed in the following paragraphs.

Untoward Effects of Anticonvulsants

DRUG INTERACTIONS

Many anticonvulsants interact with cytotoxic drugs used to treat the underlying cancer. There are four types of interaction[169]: (1) The cytotoxic drug decreases the concentration of the anticonvulsant, leading to poor control of seizures. Phenytoin, carbamazepine, and valproate may be affected. (2) Cytotoxic drugs may increase the concentration of anticonvulsants, causing toxicity. Interactions between dexamethasone and phenytoin are particularly difficult. In patients on phenytoin, toxicity may develop as dexamethasone is tapered; the toxic side effect of ataxia may mimic tumor recurrence. Conversely, beginning phenytoin may decrease the effective concentration of dexamethasone, leading to increased intracranial pressure. (3) Anticonvulsants may lower the concentration of cytotoxic drugs. This was fully recognized when commonly used cytotoxic drugs were given to brain tumor patients. Not only were they ineffective, but the recommended doses failed

to produce the usual toxicity because of the anticonvulsants that most patients were receiving. (4) Valproic acid uniquely can increase the toxicity of some cytotoxic drugs, particularly nitrosoureas, cisplatin, and etoposide. These four types of interactions are listed in Table 4–12.

In occasional patients, a drug interaction that causes a decrease in anticonvulsant serum concentration is clinically significant and leads to seizures in patients previously controlled. To compensate for the lower anticonvulsant levels, the physician may decide to increase the dose, only to have the level overshoot to toxic ranges. To compound the problem, elevated levels of several anticonvulsants cause toxicity that may mimic the symptoms of a brain tumor or may even increase seizure activity. Conversely, anticonvulsants such as phenytoin, phenobarbital, and others[170] lower the serum concentration of antineoplastic agents,[171–174] decreasing their effectiveness and, at least in the case of leukemia, decreasing survival.[173]

In some patients with asymptomatic brain metastases, iodinated contrast given for a CT scan can provoke seizures within minutes of its administration. An oral prophylactic dose of a benzodiazepine (e.g., 5–10 mg of diazepam or 1–2 mg of lorazepam given 15–30 minutes before contrast injection) substantially decreases the likelihood of seizures.[175] Intravenous gadolinium, the contrast agent for MRI, is not associated with seizures.

SIDE EFFECTS

All anticonvulsant agents have side effects (Table 4–13). Lethargy and cognitive dysfunction can be caused by any of the agents listed in Table 4–12, even when blood levels are within the *therapeutic* range. However, some drugs are more likely to cause lethargy or cognitive abnormalities than are others and a few cause little or no such problems (Table 4–14). In one study of patients with low-grade gliomas, anticonvulsants were associated with cognitive dysfunction. Whether this was caused by the drugs or by seizures is unclear.[175a]

Individual patients may tolerate one anticonvulsant better than another. Carbamazepine and primidone, if given initially at full doses, cause profound drowsiness and may not be tolerated. These drugs must be started at a low dose and gradually increased to reach appropriate levels. Phenytoin, carbamazepine, lamotrigine, and other types of drugs (sulfamethoxazole,

Table 4–13. **Side Effects of Anticonvulsants**

Idiosyncratic

    Stevens–Johnson syndrome (phenytoin, carbamazepine, lamotrigine)
    Vasculitis (phenytoin)
    Arthritis, shoulder–hand syndrome (phenobarbital)
    Stupor (valproate)
    Aseptic meningitis (carbamazepine)
    Agranulocytosis (carbamazepine)
    Pseudolymphoma (phenytoin)

Dose-Related

    Seizures (phenytoin, carbamazepine)
    Cognitive dysfunction (all)
    Diplopia (carbamazepine, phenytoin)
    Ataxia-nystagmus (phenytoin)
    Asterixis (all)

Metabolic

    Microsomal enzyme inducer (phenytoin, carbamazepine)
    Osteomalacia (phenytoin)
    Increased liver enzymes (phenytoin, valproate)
    Metabolic acidosis (topirmate)

Table 4–14. **Effects of Anticonvulsants on Cognition and Behavior**

Drug	Cognitive Effects	Behavioral Effects
Phenobarbital	++	++
Phenytoin	+	0
Carbamazepine	+	0
Sodium valproate	+?	0
Clobazam	+	+
Clonazepam	++	+
Lamotrigine	+	0
Vigabatrin	0	+
Gabapentin	+ (transient)	0
Topiramate	+	+
Tiagabine	0	0
Oxcarbazepine	+?	0
Zonisamide	0	+?
Levetiracetam	+	+ to ++

Modified from Ref. 176 with permission.
0, no effect; +?, possible effect; +, mild effect; ++, marked effect.

oxicam, allopurinol) have been reported to cause the Stevens–Johnson syndrome. The syndrome may particularly affect patients who receive whole-brain radiation while on a decreasing dose of steroids.[177–179] If the patient has taken the anticonvulsant drug for more than a month without toxicity, radiation treatment to the brain while steroids are being tapered appears to be safe. Asian patients who are HLA-B° 1502 positive are particularly at risk for carbamazepine-induced Stevens–Johnson; in such patients, the drug should be avoided.[180]

The US Food and Drug Administration (FDA) has analyzed several placebo-controlled studies of various antiepileptic drugs and concluded that the risk of suicide was approximately twice that of the patient's taking placebo. The analysis encompassed all of the drugs described in the paragraphs that follow.

Phenytoin has been reported to cause granulomatous vasculitis[181] and has, on occasion, caused pulmonary failure; it and other anticonvulsants have also been associated with osteomalacia.[182] The levels of alkaline phosphatase and other liver enzymes in the serum may increase in patients receiving phenytoin.[183] Involuntary movements, particularly choreoathetosis, may be a sign of phenytoin intoxication, substituting for the more common and recognized toxic signs of ataxia and nystagmus.[184] A severe but reversible myopathy has been related to phenytoin therapy[185] despite its reported protective effect on steroid myopathy. Myelotoxicity from phenytoin is rare but serious, particularly in the patient with cancer who might receive other myelosuppressive agents.

Many patients taking *carbamazepine* complain of intermittent diplopia as well as drowsiness even at therapeutic doses. Carbamazepine lowers the white cell count, which may cause concern in patients being treated with myelosuppressive chemotherapeutic agents, but it very seldom leads to true agranulocytosis, and it is not clear that it potentiates the myelosuppression from chemotherapeutic agents. The syndrome of inappropriate antidiuretic hormone secretion with resultant hyponatremia sometimes occurs as a carbamazepine side effect.[186] The drug has been reported to cause aseptic meningitis.[187] Prolonged *absence status epilepticus* mimicking stupor has been reported to be caused by carbamazepine in one patient.[188] Reversible abnormalities of musical pitch perception have been reported.[189] A parenteral preparation is not available, but the drug can be given rectally.

About 20% of patients with a brain tumor who take *phenobarbital* as a therapeutic agent develop pain and dysfunction in the shoulder, and sometimes in the entire upper extremity (shoulder–hand syndrome).[147,190] This syndrome is usually contralateral to the tumor site

and may occur even in patients without motor deficits.

Valproate is an effective anticonvulsant for many seizure types and is preferred by some neurologists. An intravenous preparation has been used to treat status epilepticus.[191] The drug occasionally causes hepatic dysfunction in young children receiving multiple anticonvulsant agents.[192] It also occasionally causes thrombocytopenia[193] and interferes with the synthesis of coagulation factors such as fibrinogen. Patients taking valproate sometimes develop stupor or coma that clears when the valproate is discontinued.[192,193] This syndrome may be a result of valproate-induced hyperammonemia, although the pathogenesis is not entirely understood. Postural tremor is frequent; reversible extrapyramidal signs resembling Parkinson disease may also complicate valproate therapy.[194] Dose-related tremor, alopecia, and weight gain are also complications, but the last may be helpful in patients with cancer.[195] Sedation and cognitive change are uncommon.

Lamotrigine[196] is a drug that causes little cognitive impairment or sedation compared to most other anticonvulsants, perhaps because it has an arousing or alerting effect that may cause agitation in some elderly patients. Dizziness, ataxia, somnolence, diplopia, especially when the drug is added to carbamazepine, and nausea have been reported but may be avoided by increasing the dose slowly. A major problem is skin rash that can sometimes lead to Stevens–Johnson syndrome. Whether this occurs more often in patients being treated for brain metastasis than in the general population is unknown. The drug interacts with several anticonvulsant drugs. Its interactions with chemotherapeutic and other drugs used in cancer patients have not been fully studied.

Topiramate[196] appears to be an effective anticonvulsant but produces somnolence, fatigue, psychomotor slowing, confusion, and in some patients weight loss. The drug is a weak carbonic anhydrase inhibitor causing metabolic acidosis, mostly in infants and young children.[197] Acute myopia and glaucoma have been reported. The drug is a weak inducer of cytochrome-C 450 enzymes. Its interactions with anticancer agents are not known.

Tiagabine also appears to be relatively safe but does cause dizziness, somnolence, difficulty concentrating, and sometimes tremor.

Nonconvulsive status epilepticus has been reported with this drug.[196]

Gabapentin appears quite safe.[196] The drug has been widely used in cancer patients for the treatment of neuropathic pain as well as for seizures. The drug is widely used because it has little or no interaction with other agents. It can, however, cause cognitive dysfunction, including somnolence, dizziness, ataxia, and fatigue, which often disappears with continued therapy. Some patients gain weight, which may be an advantage in patients with cancer. *Pregabalin*, closely related to gabapentin, has an improved pharmacokinetic profile[198] and is probably more effective.[199]

Felbamate is no longer widely used because of the side effects of aplastic anemia and hepatic failure.

Zonisamide is generally used as add-on therapy in patients who fail monotherapy with other agents. It has been reported to produce convulsions. CNS toxicity includes depression, psychomotor retardation, and somnolence or fatigue. The drug does not appear to interfere with the metabolism of other drugs that utilize the cytochrome P-450 enzyme system.

Oxcarbazepine[196] has the same side effect profile as carbamazepine but at lesser incidence and severity. The drug is somewhat less sedating than carbamazepine. It can cause somnolence and hyponatremia. It is less likely to cause rash than carbamazepine.

Levetiracetam[200] is generally used as an add-on anticonvulsant. The drug can cause somnolence, fatigue, and ataxia. It does not appear to have significant interactions with other drugs and, as indicated in the preceding text, is considered by many neuro-oncologists to be the drug of choice for neuro-oncological patients.[155,201,202] An intravenous preparation has recently become available.[203]

Vigabatrin (not available in the United States), like gabapentin, does not bind to plasma proteins and is excreted largely unchanged in the urine; thus, it should not have significant interactions with other drugs. Side effects are usually mild and transient and include drowsiness, dizziness, fatigue, and weight gain. However, severe drowsiness leading to stupor and significant changes in behavior including depression have been reported. The drug has also been reported to induce myoclonus as well as tremor, athetosis, and dystonia. A unique side effect is irreversible constriction of visual fields, apparently due to damage to retinal neurons. Skin reactions do not appear to occur.[196]

Some anticonvulsants (e.g., valproate, topiramate, and 2-pyrrolidinone-*n*-butyric acid, a metabolite of levetiracetam) inhibit histone deacetylases and thus could serve as anticancer agents.[204]

VENOUS THROMBOSIS/ ANTICOAGULANTS

That patients with cancer are at increased risk for venous thromboembolism has been known since 1865, when Trousseau described a relationship between GI cancer and venous thromboses in patients, including himself. Approximately 10% of patients with idiopathic venous thromboembolism have cancer as the underlying cause.[205]

Patients with pulmonary emboli[206,207] do not always present with pulmonary symptoms. One of our patients who had a primary intraventricular brain tumor, but had not had previous seizures, had a generalized convulsion while at home. She recovered consciousness but had a persistent tachycardia beyond the usual postictal period. An astute physician ordered a CT scan of the chest, which revealed pulmonary emboli (Fig. 4–5). She responded well to treatment. Seizures and other neurologic symptoms have been reported as the presenting complaint in patients with pulmonary embolization.[208,209] Conversely, dyspnea may be

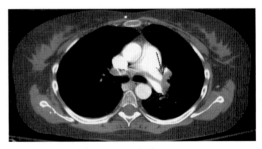

Figure 4–5. Pulmonary embolus presenting as a seizure. As described in the text, this patient with a brain tumor but no prior history of seizures had a generalized convulsion. A persistent tachycardia led the physician to suspect a pulmonary embolus. The chest computed tomography revealed a massive pulmonary embolus (*arrows*). That pulmonary emboli can present with neurologic symptoms is well established in the literature (see text).

a manifestation of partial epilepsy rather than pulmonary disease.[210]

Most patients with cancer have coagulation disorders (see Chapter 9); circulating fibrinogen degradation products may be found in as many as 90%.[211] Thrombophlebitis, with or without pulmonary embolism, is a major problem in patients with both primary and metastatic brain tumors, particularly in those who undergo surgical treatment.[212–214] In addition to these general risk factors for thrombophlebitis, patients with brain lesions often suffer immobility of the extremities, predisposing to stagnation of venous blood and clotting. Even in patients without motor deficits, when thrombophlebitis develops, it is twice as likely to occur in the extremity contralateral to a primary or metastatic brain tumor than in the ipsilateral limb.[215]

Precautions may help decrease the frequency of thrombophlebitis in patients with cancer, particularly in the postoperative period. Graduated elastic compression stockings and external pneumatic calf compression devices applied in the operating and recovery room, and continued until the patient is fully ambulatory, decrease but do not eliminate the incidence of thrombophlebitis and pulmonary embolism.[216] In one study, 19% of patients who underwent craniotomy for brain metastases developed overt venous thrombosis despite prophylaxis.[217] However, another study of 413 patients undergoing craniotomy for brain metastases gave an incidence of 2.6% for deep vein thrombosis (DVT) and 3.3% for pulmonary embolism.[218]

Venous compression of the extremities does not alter intracranial pressure even in brain-injured patients.[219] The mechanism of pneumatic intermittent compression boots is still debated. Both a hemodynamic action by increasing blood flow velocity and a biochemical action by stimulating endogenous fibrinolytic activity have been proposed. In one study, hemodynamic mechanisms appeared to be dominant.[220] Thus, placing the device on one leg when cortical mapping is necessary during surgery is as effective as placing the device on both legs.[216] Even placing the device on an arm may be helpful.

In cancer patients who do not have brain tumors but who require prophylaxis against venous thromboembolism, the consensus is that low molecular weight heparins (LMWHs) are the drugs of choice (Fig. 4–6).[221] These drugs have better bioavailability, simpler dosing with longer dosing intervals, a more predictable anticoagulant response with less need for laboratory monitoring, and are safer than unfractionated heparin. However, they are harder to reverse in an emergency such as an intracranial hemorrhage.

The situation is less clear in patients with brain tumors, whether primary or metastatic, undergoing craniotomy. Some suggest that the risk of intracranial hemorrhage outweighs the benefit of preventing venous thromboembolism.[222] In one randomized study of a LMWH initiated before the induction of anesthesia and continued throughout the hospital stay, postoperative intracranial hemorrhage occurred in 5 of 46 patients on the drug and none of those using compression boots,[223] without a significant difference in thrombosis between the patients treated with the drug and those treated with pneumatic compression boots. Conversely, a meta-analysis of the use of low molecular weight or unfractionated heparin in neurosurgery indicates that the drugs were effective for prophylaxis of venous thromboembolism without excessive bleeding risk.[224] Most of the studies began heparin 18 to 24 hours postoperatively. This analysis supports that author's previous study that LMWHs begun 24 hours after craniotomy and combined with compression stockings is more effective than compression stockings alone for the prevention of venous thromboembolism after elective neurosurgery and does not cause excessive bleeding.[225]

On balance, the evidence would appear to support the view that for most patients undergoing neurosurgery, LMWH begun the day after surgery is safe and effective.

For neurosurgical patients at MSKCC, 4500 units of a LMWH (tinzaparin) are given subcutaneously once a day beginning the morning after surgery. Patients wear compression stockings and pneumatic compression boots in the operating room. The boots are continued postoperatively while the patient is in bed. However, patients are aggressively mobilized the morning after surgery.

New anticoagulants that selectively inhibit specific coagulation factors such as fondaparinux, which inhibits factor Xa, or dabigatran, which inhibits thrombin, are useful for

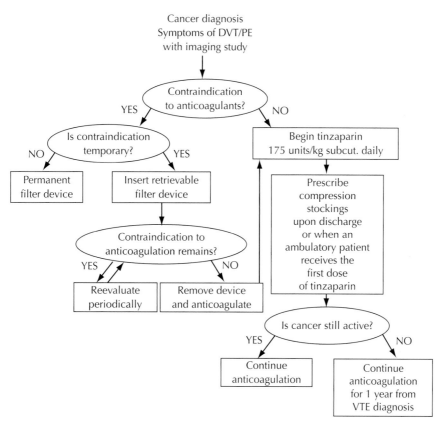

Figure 4–6. The Memorial Sloan-Kettering algorithm for the treatment of deep vein thrombosis. Note that low molecular weight heparin (tinzaparin is the one in our formulary) is preferred over warfarin, permanently in patients with active cancer. Note also that filters are placed only when anticoagulation is contraindicated; if possible a retrievable filter is placed.

prophylaxis of venous thromboembolism after orthopedic surgery,[226–228] but their role in neuro-oncology, if any, is still unclear.

For nonsurgical patients with brain tumors and brain metastasis suffering from established thromboembolic disease, substantial evidence exists to indicate that anticoagulation is both safe and effective.[214,229] Patients with proximal DVT do better if they are mobile during the acute treatment, and wearing compression stockings, than if they are kept at bed rest. Pain and swelling resolved significantly faster in the ambulatory patients and pulmonary embolism was not significantly increased.[11,230,231]

The use of anticoagulants in patients with or without cancer is not without other risks.

Heparin can induce thrombocytopenia and thrombosis,[232] and warfarin occasionally can induce necrosis of the skin and sometimes of the breast, penis, or toes.[233,234]

Because of the concern over intracranial bleeding and other anticoagulant complications, many physicians place inferior vena cava filters to prevent pulmonary emboli in patients with cancer and lower extremity thrombophlebitis. These devices have been placed prophylactically in high-risk patients undergoing extensive spinal surgery. The filters reduced the incidence of DVT, but one of the 22 patients developed bilateral deep vein thromboses requiring thrombolytic therapy.[235] Although these filters are usually effective in preventing 90% of emboli, a significant percentage of patients still suffer complications including pulmonary embolism (12%), progressive thrombosis of vena cava, leg, or pelvic veins, or massive edema of the lower extremities.[214,236] We prefer, when possible, to anticoagulate with LMWHs, and continue for a year rather than switching to warfarin,[211,237] reserving filters for those patients in whom anticoagulation is clearly contraindicated. In those patients in whom the need for thromboembolism prophylaxis is likely to be for a limited time, for example, a postoperative patient in whom anticoagulation is temporarily contraindicated, a retrievable filter might be preferable to a permanent one. Retrievable filters appear to be as efficacious and have as few acute complications as permanent filters and may have a lower long-term complication rate.[238,239]

Table 4–15 summarizes the current therapy of venous thromboembolism as suggested by NCCN.

PAIN/ANALGESIA

About 75% of patients with advanced cancer experience pain. Pain is a particular problem in patients with epidural spinal cord compression or with peripheral nerve invasion or compression where it can occur in newly diagnosed cancer patients as well as those with end-stage disease. A vast literature addresses the problem of pain relief in patients with cancer[240–243] and the National Comprehensive Cancer Network (NCCN) has recently issued detailed guidelines for assessment and management of cancer pain in adults and children[244,245]; only a few aspects are addressed here.

A careful assessment of the nature and severity of the pain is essential (Table 1–12). In general, cancer pain may be divided into two broad categories: nociceptive (pain resulting from stimulation of pain receptors called nociceptors in skin, bones, viscera, etc.) and neuropathic (pain initiated or caused by a lesion within the nervous system). The importance of distinguishing between the two is that while both can be treated with conventional analgesic agents, nociceptive pain responds better to opioids and neural blockade than neuropathic pain, which may respond to anticonvulsant and psychotropic agents.

Cancer pain is best treated by attacking the cause. Bone pain caused by cancer may respond to radiation therapy or to bisphosphonates[246,247] (see Chapter 6). Headache related to brain or leptomeningeal metastasis may respond to corticosteroids, radiation, or chemotherapy (see Chapters 5 and 7). When the pain does not respond to treatment of its cause, analgesic agents should be used in sufficient quantity and frequency to control the pain. Using the World Health Organization (WHO) 3 step ladder (Fig. 4–7) and the NCCN guidelines for pain control, the vast majority of patients will respond without developing unacceptable side effects.

For mild nociceptive pain, such as pain caused by bony metastases, the clinician should begin with a peripherally acting analgesic such as acetaminophen, aspirin, or an NSAID. In patients suffering gastric distress or those also taking corticosteroids, a cox-2 inhibitor is probably better.[244] Although aspirin and the NSAIDs can cause GI bleeding (but see p 104), they rarely cause neurotoxicity. Rarely, NSAIDs cause acute aseptic meningitis with headache, fever, stiff neck, and pleocytosis.[248] As CSF eosinophilia is common in drug-induced aseptic meningitis, it should prompt one to suspect a drug side effect before considering infectious meningitis or meningeal metastases.

Unfortunately, most cancer pain will not respond to these drugs. For those whose pain persists, one should quickly begin opioids, starting with codeine or oxycodone and, if ineffective, escalating promptly to morphine or morphine-like agents. Large doses of opioid drugs are sometimes required to control pain and should be used if they do not produce substantial side effects.[245] Individual responses to different opioids, both with respect to pain relief and side effects, are quite variable. Genetic, pharmacokinetic, pharmacodynamic, and environmental factors all play a role in causing individual variability. For example, genetic differences in P-glycoprotein may decrease entry of opioids into the brain, and genetic differences in opioid receptors may increase or decrease the effectiveness of individual agents. Accordingly, opioid rotation[249] and opioid switching[250] may both increase efficacy and decrease side effects. A table of equivalent doses of individual opioids can be used as a guide to switching agents, but again there is a great deal of individual variability[250] (Table 4–16).

Table 4–15. **Prophylaxis and Treatment of Venous Thromboembolism**

Inpatient Prophylactic Therapy

- LMWH

  Dalteparin 5000 units subcutaneous daily

  Enoxaparin 40 mg subcutaneous daily

  Tinzaparin 4500 units (fixed dose) subcutaneous daily or 75 units/kg subcutaneous daily
- Pentasaccharide fondaparinux 2.5 mg subcutaneous daily
- Unfractionated heparin: 5000 units subcutaneous twice daily
- Graded compression stockings
- Pneumatic compression boots
- Ambulation

Therapeutic Anticoagulation Treatment for DVT, PE, and Catheter-Associated Thrombosis

Immediate

- LMWH

  Dalteparin (100 units/kg every 12 hours)

  Enoxaparin (1 mg/kg every 12 hours)

  Tinzaparin (175 units/kg daily)
- Penasaccharide fondaparinux (5.0–7.5–10 mg subcutaneous daily)

  Unfractionated heparin (IV) (80 units/kg load, then 18 units/kg per hour, target aPTT
  to 2.0x–2.9x control)

Long-term

- LMWH is preferred as monotherapy without warfarin in patients with proximal DVT or PE and for prevention of recurrent VTE in patients with advanced or metastatic cancer
- Warfarin (2.5–5 mg every day initially, subsequent dosing based on INR value; target INR, 2.0–3.0)

Duration of long-term therapy:

- Minimum time of 3–6 months for DVT and 6–12 months for PE
- Consider indefinite anticoagulation if active cancer or persistent risk factors
- For catheter associated thrombosis, anticoagulate as long as catheter is in place

Clinical Scenarios Warranting Consideration of Filter Placement

- Contraindications to anticoagulation
- Pulmonary embolism while on adequate anticoagulation for DVT
- New pulmonary embolism while on adequate anticoagulation for PE
- Patient noncompliance with prescribed anticoagulation
- Baseline pulmonary dysfunction severe enough to make any new or recurrent PE life threatening
- Patient with documented multiple PE and chronic pulmonary hypertension
- Recent CNS bleed, intracranial or spinal lesion at high risk for bleeding
- Active bleeding (major): more than 2 units transfused in 24 hours
- Chronic, clinically significant measurable bleeding >48 hours
- Thrombocytopenia (platelets <50,000 per mcL)
- Severe platelet dysfunction (uremia, medications, dysplastic hematopoiesis)
- Recent major operation at high risk for bleeding
- Underlying coagulopathy

  Clotting factor abnormalities

  Elevated PT or aPTT (excluding lupus inhibitors)
- Spinal anesthesia/lumbar puncture
- High risk for falls

Modified from NCCN Practice Guidelines in Oncology, v.1.2006.
LMWH, low molecular weight heparin; DVT, deep vein thrombosis; PE, pulmonary embolism; VTE, venous thromboembolism;
CNS, central nervous system.

Opioids may add to existing neurologic dysfunction. Their side effects, including sedation, may augment or mask the symptoms caused by a growing mass lesion in the brain. Furthermore, opioid overdose can decrease respiration, thus increasing pCO_2 and causing a rise in intracranial pressure that may exacerbate brain tumor symptoms. Multifocal myoclonus and seizures may complicate the use of any opioid, especially meperidine (see Chapter 11). The same dose of opioid that was previously well tolerated may, if pain is relieved by other means, cause excessive sedation and even respiratory insufficiency. Also, if the pain is relieved by other means, the patient may cease taking the opioids abruptly and, if physically dependent, may suffer withdrawal unless instructed to taper the drugs even in the absence of pain. Genetic variation in the genes for multidrug resistance (MDR1) and catechol amines (COMT) affect the likelihood of side-effects.[250a]

Neuropathic pain may be exceedingly difficult to treat and often responds poorly to opioids.[251] Gabapentin and pregabalin are efficacious agents with few side effects. Tricyclic antidepressants, sometimes at surprisingly low doses such as 10 to 25 mg of amitriptyline at bedtime, may be effective.[245] Other drugs, including anticonvulsants,[252,253] antidepressants, baclofen, corticosteroids, oral local anesthetics, and clonidine, are usually combined with opioids. The effect of antidepressants on neuropathic pain is independent of their effect on mood. The response to high-dose IV corticosteroids in patients with epidural spinal cord compression is particularly dramatic (see Chapter 6).[254] For patients who do not respond to pharmacologic treatment or those who cannot tolerate its side effects, a number of non-pharmacological approaches should be considered[255] (Table 4–17).

Physical and cognitive modalities can be used in conjunction with pharmacologic agents

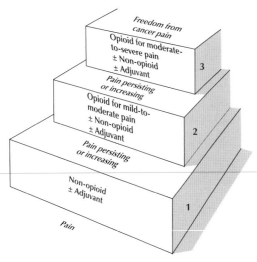

Figure 4–7. World Health Organization (WHO) analgesic ladder. (From cancer pain relief and palliative care: Report of a WHO expert committee. Technical reports series, number 804. Geneva, World Health Organization, 1990. With permission from WHO.)

Table 4–16. **Approximate Oral and Parenteral Dose Equivalents of Opioids Based on Single Dose Data**

Opioid Analgesic	Oral Dose (mg)	Parenteral Dose	Duration of Action (hr)	Half Life (hr)	Not Recommended
Codeine	100	50 mg	q3–4	2.9	Propoxyphene
Hydrocodone	15	N/A	q3–4	3.8 ± 0.3	Meperidine
Oxycodone	7.5–10	N/A	q3–4	3.2	Mixed agonist–antagonist
Morphine	15	5 mg	q3–4	1.5–2.0	
Hydromorphone	4	0.75–1.5 mg	q3–4	2.5	Partial agonists
Levorphanol	1	1 mg	q6–8	11–30	Placebos
Methadone	10	5 mg	q6–8	759	
Fentanyl	N/A	50 µg	1–3		
Transdermal fentanyl	N/A	50 µg/hr	q48–72	1–3	

Table 4–17. **Some Non-Pharmacological and Anesthesia Treatments for Pain**

Physical Modalities

    Bed, bath, and walking supports

    Positioning instruction

    Physical therapy

    Massage

    Heat and/or ice

    TENS

    Acupuncture or acupressure

    Ultrasonic stimulation

Cognitive Modalities

    Imagery/hypnosis

    Distraction training

    Relaxation training

    Active coping training

    Graded task assignments, setting goals, pacing, and prioritizing

Anesthesia Procedures

    Epidural—easy to place, requires large volumes and an externalized catheter; for infusion of opioids, local anesthetic preparations, clonidine

    Intrathecal—easy to internalize to implanted pump; consider for >3 months' life expectancy; for infusions of opioids, local anesthetics, clonidine

Surgical Procedures

    Head and neck: peripheral nerve neurolysis

    Upper extremity: brachial plexus neurolysis

    Thoracic wall: epidural neurolysis

    Upper abdominal pain (visceral): celiac plexus block; thoracic splanchnicectomy

    Midline pelvic pain: superior hypogastric plexus block

    Rectal pain: intrathecal neurolysis, midline myelotomy

    Unilateral pain syndromes: cordotomy

Modified from Ref. 245.
TENS, transcutaneous electrical nerve stimulation.

and often enhance their effectiveness. Nerve blocks and neurodestructive procedures represent a last resort for treatment of intractable pain. Their success is limited at best.[244]

PSYCHOLOGICAL DISTRESS/ PSYCHOTROPIC AGENTS

Psychological distress, especially anxiety and depression,[256] are frequent and significant complications of cancer, particularly in patients with neurologic complications.[257] The cancer subspecialty of psycho-oncology, with its own journal, has been developed to address some of these problems.[256,258] Nevertheless, psychological distress is often undiagnosed and undertreated,[256] perhaps because many physicians are reluctant to use antidepressants to treat psychological distress in patients with cancer, falsely believing that the anxiety or depression, which the physician feels is appropriate for the patient's situation, will not respond to drug therapy. In fact, antidepressant drugs quite effectively treat depression associated with cancer, just as anxiolytic drugs are effective in treating anxiety.[259]

The so-called psychostimulants include dextroamphetamine, modafinil, methylphenidate, and pemoline.[260] These drugs are reported to improve concentration and attention and thus enhance cognitive function. They have some effect in reversing sedation from opioids. In some settings, the drugs appear to relieve fatigue (see following text).[260]

In order to treat depression, the physician must know whether the patient is depressed. The best way to find out is to ask directly about mood changes. Many of the somatic symptoms of depression, including weight loss, insomnia, and fatigue, may be a result of the cancer and not associated with a depression of mood, so that a patient who denies depression is probably not depressed. However, in patients with somatoform disorders (conversion reactions, *hysteria*), the neurologic symptoms may mask the feelings of depression. In any depressed patient, specific causes such as brain metastasis, hypercalcemia or drugs, including corticosteroids, beta blockers, antiemetics, opioids, benzodiazepines, and some antibiotics, such as macrolides and fluoroquinolones, must be considered. In patients who are taking drugs that can cause depression, the drugs should be withdrawn, if possible, before beginning antidepressive therapy.

Many patients with cancer not only are anxious and depressed but also suffer insomnia and anorexia. Tricyclic antidepressants with sedative properties, such as amitriptyline, are often extremely effective in such patients. The patient receives both the sedative and the antidepressant effects of the drug administered as a single dose at night and gradually increased from 10–25 to 150 mg or more. The sedative effects take hold immediately, and patients often feel substantially better after their first good night's sleep. Antidepressant effects take somewhat longer. The selective serotonin reuptake inhibitors,[261] effective for treating depression may cause anorexia. For anxious patients, a short course of benzodiazepine may be useful[259]; however, these drugs can cause depression. At low doses, most of these drugs appear to have minimal side effects in the cancer population.

Cognitive impairment, also common in patients with cancer,[262] may be caused by psychological distress or may have another cause but be assumed by the physician to be related to psychological distress. Because cognitive impairment is so commonly related to treatment[263] rather than the cancer itself, it is discussed in detail in Chapter 12.

FATIGUE

Fatigue is a major problem in patients suffering from cancer.[7,242,264,265,265a] The NCCN defines cancer-related fatigue as "a distressing persistent, subjective sense of tiredness or exhaustion related to cancer or cancer treatment that is not proportional to recent activity and interferes with usual functioning." Fatigue occurs as a direct result of the cancer itself, probably via production of cytokines,[16] or anemia, with the severity of the anemia predictive of the degree of fatigue,[265] or of cancer treatments including chemotherapy with cytotoxic or biological agents and radiation therapy (Table 4–18). The fatigue can disrupt a patient's life both during the treatment of the cancer and for months or years after treatment ends.

Table 4–18. **Potential Contributing Factors to Cancer-Related Fatigue**

Physiologic
    Underlying disease
    Treatment for the disease
        Chemotherapy
        Radiotherapy
        Surgery
        Biologic response modifiers
    Intercurrent systemic disorders
        Anemia
        Infection
        Pulmonary disorders
        Congestive heart failure
        Hepatic failure
        Renal insufficiency
        Malnutrition
        Neuromuscular disorders
        Dehydration or electrolyte disturbances
    Sleep disorders
    Immobility and lack of exercise
    Chronic pain
    Use of centrally acting drugs (e.g., opioids)

Psychosocial
    Anxiety disorders
    Depressive disorders
        Stress related
        Environmental reinforces

From Ref. 266 with permission.

The treatment of fatigue begins with careful assessment of the patient's complaint. In many patients, fatigue is associated with pain, emotional distress, insomnia, anemia, or hypothyroidism. Treating these disorders, especially anemia, may ameliorate the fatigue. Curiously, when anemic patients are treated with erythropoietin, the fatigue sometimes improves before the anemia responds. Erythropoietin receptors are present in the brain and may play a role in alleviating fatigue.[267] Other factors that should be addressed include the patient's nutritional and metabolic state.

The fatigue itself can be treated by a variety of non-pharmacologic interventions including education, exercise, and cognitive therapy.[7,264] Pharmacologic therapy includes treatment of some of the associated factors listed above, including analgesics, antidepressants, hypnotics for sleep, erythropoietin for anemia,[268] and hormonal replacement for hypothyroidism or other endocrine disorders. Methylphenidate[269] and modafinil[270] are both effective in some patients. Selective serotonin reuptake inhibitors are ineffective.[271]

NAUSEA, VOMITING/ ANTIEMETIC AGENTS

Nausea and vomiting are distressing symptoms in patients with cancer. The symptoms usually result from treatment, particularly chemotherapy, but other causes must always be considered. These include posterior fossa brain metastases (see Chapter 5), vestibular dysfunction, bowel obstruction, gastroparesis, drugs other than chemotherapeutic agents (e.g., opioids), metabolic disturbances, and psychological factors. As illustrated in Figure 4–8, a number of neural pathways can mediate nausea and vomiting. These include the chemoreceptor trigger zone, an area in the brainstem that lacks a blood–brain barrier and thus is exposed to water-soluble chemotherapeutic agents.

Powerful antiemetic agents have revolutionized the treatment of chemotherapy-induced nausea and vomiting. These include the neurokinin-1 receptor antagonist, aprepitant, the serotonin 5-HT3 receptor antagonists, vestibular suppressants such as lorazepam as well as corticosteroids, particularly dexamethasone.

NCCN guidelines detail use of these agents. Table 4–19 outlines the approach to preventing emesis in patients receiving highly emetogenic chemotherapy.

Other antiemetic agents such as metoclopramide, prochlorperazine, and haloperidol have dopamine-blocking activity that can cause acute extrapyramidal reactions or respiratory dyskinesias. Extrapyramidal signs generally occur in the young and are characterized by dystonic posturing associated with akathisia and often severe agitation, much greater than that appropriate for any fright caused by the physical symptoms. The disorder can usually be reversed by IV diphenhydramine (Benadryl). In rare instances, however, diphenhydramine itself causes dystonia.[273] Extrapyramidal reactions usually begin while the patient is taking the drug, but occasionally may begin as long as 48 hours after the offending agent has been discontinued. The reaction may follow a single dose. The diagnosis may be difficult, particularly if the physician is unaware that the patient has received an antiemetic.

Akathisia (generalized restlessness often associated with anxiety) in the absence of other extrapyramidal symptoms is a particularly common and distressing side effect of many anti-emetics.[274,275] The symptom is frequently not mentioned to the physician, but simply tolerated by the patient. It should not be confused with the restless leg syndrome that may accompany a peripheral neuropathy of chemotherapy.[276] Akathisia usually responds to discontinuing or changing the antiemetic agent.

INFECTION/ANTIBIOTICS

Both cancer and its treatment can immunosuppress patients, thus increasing susceptibility to a variety of opportunistic infective agents. At times, the infection affects the nervous system. CNS infectious complications of cancer are discussed in Chapter 10. All require antibiotic treatment, but many of the antibiotics exhibit neurotoxicity.[277] Neurotoxicity can affect either the peripheral nervous system or the CNS (Table 4–20).

Penicillin, cephalosporins, quinolones, and carbapenems block the effect of γ-aminobutyric acid, increasing CNS excitability. Multifocal

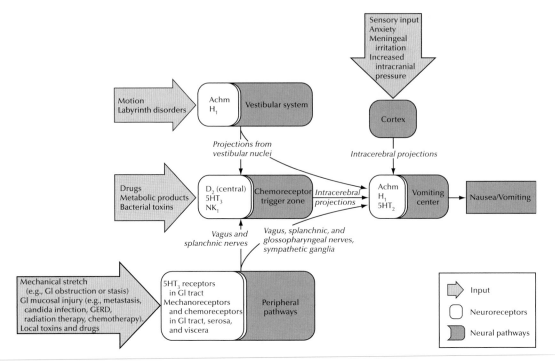

Figure 4–8. Inter-relationships between neural pathways that mediate nausea and vomiting. (From Ref. 272 with permission.)

Table 4–19. Prevention of Emesis in Patients Receiving High Emetic Risk Chemotherapy

Start Before Chemotherapy

Aprepitant, 125 mg PO day 1, 80 mg PO daily days 2–3

Dexamethasone, 12 mg PO or IV day 1, 8 mg PO or IV daily days 2–4, and

5-HT3 antagonist:

　　Ondansetron, 16–24 mg PO or 8–12 mg (maximum 32 mg) IV day 1, or

　　Granisetron, 2 mg PO or 1 mg PO bid or 0.01 mg/kg (maximum 1 mg) IV day 1, or

　　Dolasetron, 100 mg PO or 1.8 mg/kg IV or 100 mg IV day 1, or

　　Palonosetron, 0.25 mg IV day 1, and

±Lorazepam, 0.5–2 mg PO or IV or sublingual either every 4 or every 6 hr, days 1–4

Modified from NCCN guidelines. www.NCCN.org

myoclonus, encephalopathy and seizures can occur in patients receiving other antibiotics that lower seizure threshold, especially those with renal failure, or those in whom the blood–brain barrier has been disrupted.[278] Seizures usually occur in patients receiving high-dose IV therapy and are often preceded by multifocal myoclonus. When seizures or myoclonus develop, the dose should be decreased;

anticonvulsants are unnecessary. Penicillin can also cause aseptic meningitis.[279]

　Aminoglycosides such as gentamicin can cause ototoxicity and vestibulotoxicity.[280] Patients may develop tinnitus and hearing loss as the first symptom, but vestibulotoxicity may be the only problem. An ill, bedridden patient may complain only of mild dizziness while receiving the medication. After the course of

Table 4–20. **Neurotoxicity of Antibacterial Agents**

Side Effects	Common	Uncommon	Comments
Headache	TMP-SMX Cycloserine	Erythromycin Clarithromycin Itraconazole Azithromycin Griseofulvin Delavirdine Saquinavir Abacavir Efavirenz Foscarnet Polymyxin B	TMP-SMX may cause headache or aseptic meningitis
Aseptic meningitis	TMP-SMX	Trovafloxacin	TMP-SMX often is overlooked as a cause of aseptic meningitis Trovafloxacin may cause muscle spasms of the neck (meningismus) mimicking meningitis
Encephalopathy	Trovafloxacin Clarithromycin	Cycloserine Foscarnet Ethambutol Amantadine Ethionamide Ganciclovir Polymyxin B	Trovafloxacin central nervous system side effects include mental confusion and encephalopathy, which clear when trovafloxacin is discontinued
Neuroexcitatory symptoms	Ciprofloxacin	Cycloserine Ofloxacin	Neuroexcitatory effects may be minimized by taking antibiotics with neuroexcitatory effects in the morning rather than in the evening
Seizures	Penicillin Ciprofloxacin Imipenem Trovafloxacin Cycloserine Acyclovir Valcyclovir Famciclovir	Amantadine Rifampin Foscarnet Ganciclovir Metronidazole TMP-SMX Nalidixic acid Erythromycin	Most common causes of antibiotic-related seizures are ciprofloxacin, imipenem, and trovafloxacin Levofloxacin and meropenem do not cause seizures
Cerebellar ataxia	Metronidazole Amantadine	TMP-SMX Nitrofurantoin Nalidixic acid Polymyxin B	Disappears after drug withdrawal

Continued on following page

Table 4–20.—*continued*

Side Effects	Common	Uncommon	Comments
Myasthenic syndrome, neuromuscular blockade	Aminoglycosides Capreomycin	Polymyxin B Clindamycin Erythromycin	May occur with aminoglycosides/ polymyxin B administered to the lungs by aerolizer or with aminoglycoside/ clindamycin by peritoneal lavage Disappears after drug withdrawal. Neuromuscular blockage may follow massively absorbed quantities of drug interfering with neuromuscular transmission.
Depression	Cycloserine	Ethionamide	Disappears after drug withdrawal
Psychosis	Foscarnet Ethionamide Efavirenz	Trovafloxacin Amprenavir	Disappears after drug withdrawal
Pseudotumor cerebri (benign intracranial hypertension)	Tetracycline		
Peripheral neuropathy	INH Griseofulvin Cycloserine Trovafloxacin	Nitrofurantoin Ethionamide Polymyxin B Ethambutol	INH-induced peripheral neuropathy may be prevented by concomitant administration of pyridoxine Peripheral neuropathy owing to nitrofurantoin is associated with prolonged high-dose therapy in patients with renal insufficiency and may be permanent
		Metronidazole Foscarnet d4T Lamivudine (3TC) ddC ddI AZT	
Muscular tremors, spasticity	Amantadine Trovafloxacin Foscarnet	Ganciclovir	Parkinsonian symptoms reversible when drug is discontinued
Severe myalgias	Quinupristin/ dalfopristin		Although severe myalgias are uncommon, they are prolonged/very painful when they occur
Ototoxicity: Deafness (cochlear)	Aminoglycosides	Vancomycin	Deafness from erythromycin may follow rapid intravenous infusion

Continued on following page

Table 4–20.—*continued*

Side Effects	Common	Uncommon	Comments
	Erythromycin	Capreomycin	Aminoglycoside-associated cochlear deafness is usually irreversible
			Aminoglycoside deafness may occur after prolonged, highly elevated peak serum concentrations
Ototoxicity: Dizziness (vestibular)	Minocycline Streptomycin	Aminoglycoside Efavirenz Abacavir	Vestibular toxicity from minocycline is due to its high lipid solubility and concentration in vestibular cells
			Among the aminoglycosides, streptomycin has the greatest otologic potential
Blindness	Ethambutol	Chloroquine Ganciclovir	Ganciclovir may cause retinal detachment
			Ethambutol causes red–green color blindness
			Ethambutol causes dose-dependent ↓visual acuity in high doses; does not occur with daily doses of <15 mg/kg
			Ethambutol may cause central scotoma
Dysphagia	Indinavir	Amprenavir	
Circumoral paresthesias	Amprenavir		Reversible when drug is discontinued

From Ref. 290 with permission.
NSAIDs, Non-steroidal anti-inflammatory drugs.

the antibiotic has been completed, when the patient begins to ambulate, he or she discovers striking gait ataxia, which impairs walking. A characteristic complaint is oscillopsia, that is, when the patient attempts to walk, the environment does not appear to be stationary but instead bounces. Examination reveals absent vestibular responses to caloric stimulation and sometimes high-frequency hearing loss. Patients receiving aminoglycoside antibiotics should be monitored carefully and questioned about dizziness or vertigo; the dose should be decreased or the drug discontinued if these occur. The toxicity is usually associated with a prolonged antibiotic course, high serum levels of the drug, and renal failure. The aminoglycosides also cause synergistic ototoxicity with cisplatin even if the drugs are not given simultaneously. The vestibular failure rarely reverses, but younger patients can compensate. Physical therapy, sometimes with the use of a neck collar to stabilize the head, helps with walking.

Amphotericin B occasionally causes brain white matter damage. The disorder occurs after bone marrow or stem cell transplantation, when the amphotericin has been given to treat an opportunistic fungal infection. The disorder is characterized by encephalopathy with confusion and disorientation progressing to akinetic mutism or coma. Corticospinal tract signs include hyperactive reflexes, upgoing toes, and sometimes diffuse extensor rigidity. Parkinsonism features may be present.[281,282] Stupor may begin abruptly sometime after the course of therapy is completed, even after the patient has fully recovered from the marrow transplant and returned home. The illness may be reversible,[281,283] but patients who lapse into coma usually die. White matter changes are identified on the MRI as nonenhancing lesions best seen on T2 or fluid attenuated inversion recovery (FLAIR) images. Pathologic changes are typical of leukoencephalopathy and are indistinguishable from encephalopathy caused by methotrexate (see Chapter 12) or radiation (see Chapter 13). Lipid formulations may be less toxic.[284]

Trimethoprim-sulfamethoxazole has been reported to be responsible for aseptic

Table 4–21. Antibiotic Drug Interaction

Interaction	Consequence	Syndrome
Chloramphenicol–phenytoin	Phenytoin toxicity	Ataxia
Erythromycin–theophylline	Theophylline toxicity	Seizures
Erythromycin–lovastatin	Sarcolemmal damage	Rhabdomyolysis
Erythromycin–cyclosporin	Cyclosporin toxicity	Hypertensive encephalopathy
		Seizures
Isoniazid–carbamazepine	Vestibulocerebellar toxicity	Ataxia
Isoniazid–cycloserine	Unknown mechanism	Ataxia, dizziness
Quinolones–theophylline	Theophylline toxicity	Seizures
Linezolid	SSRIs	Serotonin syndrome

Modified from Ref. 290.
SSRIs, selective serotonin reuptake inhibitors.

meningitis[285,286] as well as a painful sensory and autonomic neuropathy.[287,288]

Quinolones can cause seizures, hallucinations, tremors, encephalopathy, and psychosis. Elevated glutamate levels may be the culprit.[289] Some of the drugs interact with theophylline and produce theophylline toxicity.[290,291]

Pentamidine used to treat *P. carinii* pneumonia may cause severe hypoglycemia with stupor or coma.[292] The hypoglycemia usually is caused by systemically administered pentamidine but occasionally has been associated with the aerosolized drug. When hypoglycemic coma occurs in a patient with a known brain lesion, the diagnosis may not be apparent.

Acyclovir and other antiviral drugs are not usually neurotoxic. They occasionally cause tremor or myoclonus and delirium,[293,294] rarely with focal symptoms. The neurotoxicity may occur at peak serum levels 24 to 48 hours after the drug is initially administered. Reversible changes on MRI with multiple enhancing lesions have been reported.[295] Antibiotics also have drug interactions that may be neurotoxic.

INTENSIVE CARE/COMMON AGENTS

Neurologic problems are common in patients ill enough to require intensive care for non-neurologic reasons. Table 4–21 indicates some of the agents commonly used in intensive care units that may cause neurotoxicity.

REFERENCES

1. Corcoran ME. Polypharmacy in the older patient with cancer. *Cancer Control* 1997;4:419–428.
2. Davis MP, Dickerson D. Cachexia and anorexia: cancer's covert killer. *Support Care Cancer* 2000;8:180–187.
3. Zajicek G. Pernicious cachexia: a different view of cancer. *Anticancer Res* 1999;19:4907–4912.
4. Nelson KA. The cancer anorexia-cachexia syndrome. *Semin Oncol* 2000;27:64–68.
5. Erlington GM, Murray NM, Spiro SG, et al. Neurological paraneoplastic syndromes in patients with small cell lung cancer. A prospective survey of 150 patients. *J Neurol Neurosurg Psychiatry* 1991;54:764–767.
6. Hovi L, Era P, Rautonen J, et al. Impaired muscle strength in female adolescents and young adults surviving leukemia in childhood. *Cancer* 1993;72:276–281.
7. Atkinson A, Barsevick A, Cella D, et al. NCCN practice guidelines for cancer-related fatigue. *Oncology (Huntingt)* 2000;14:151–161.
8. Mulrooney DA, Ness KK, Neglia JP, et al. Fatigue and sleep disturbance in adult survivors of childhood cancer: report from the childhood cancer survivor study (CCSS). *Sleep* 2008;31:271–281.
9. Ryan JL, Carroll JK, Ryan EP, et al. Mechanisms of cancer-related fatigue. *Oncologist* 2007;12 (Suppl 1): 22–34.
10. Bower JE. Behavioral symptoms in patients with breast cancer and survivors. *J Clin Oncol* 2008;26:768–777.
11. Lee AY, Levine MN. Management of venous thromboembolism in cancer patients. *Oncology (Huntingt)* 2000;14:409–417, 421.
12. Osoba D. Health-related quality-of-life assessment in clinical trials of supportive care in oncology. *Support Care Cancer* 2000;8:84–88.
13. Passik SD, Theobald D, Holtsclaw E, et al. The Oncology Symptom Control and Research (OSCR) Program. A psycho-oncology and symptom management program for community-based ambulatory oncology. *Oncology (Huntingt)* 2000;15:14–18.

14. Walsh D, Doona M, Molnar M, et al. Symptom control in advanced cancer: important drugs and routes of administration. *Semin Oncol* 2000;27:69–83.

15. Walsh D, Rybicki L. Symptom clustering in advanced cancer. *Support Care Cancer* 2006;14:831–836.

16. Cleeland CS, Bennett GJ, Dantzer R, et al. Are the symptoms of cancer and cancer treatment due to a shared biologic mechanism? A cytokine-immunologic model of cancer symptoms. *Cancer* 2003;97:2919–2925.

17. Miller AH, Ancoli-Israel S, Bower JE, et al. Neuroendocrine-immune mechanisms of behavioral comorbidities in patients with cancer. *J Clin Oncol* 2008;26:971–982.

18. Twycross R. The risks and benefits of corticosteroids in advanced cancer. *Drug Saf* 1994;11:163–178.

19. Koehler PJ. Use of corticosteroids in neuro-oncology. *Anticancer Drugs* 1995;6:19–33.

20. Van Staa TP, Leufkens HG, Abenhaim L, et al. Use of oral corticosteroids in the United Kingdom. *Q J Med* 2000;93:105–111.

21. Needham PR, Daley AG, Lennard RF. Steroids in advanced cancer: survey of current practice. *BMJ* 1992;305:999.

22. Hardy JR, Rees E, Ling J, et al. A prospective survey of the use of dexamethasone on a palliative care unit. *Palliat Med* 2001;15:3–8.

23. Loprinzi CL, Kugler JW, Sloan JA, et al. Randomized comparison of megestrol acetate versus dexamethasone versus fluoxymesterone for the treatment of cancer anorexia/cachexia. *J Clin Oncol* 1999;17:3299–3306.

24. Aapro MS, Alberts DS. High-dose dexamethasone for prevention of *cis*-platin-induced vomiting. *Cancer Chemother Pharmacol* 1981;7:11–14.

25. Kris MG, Pendergrass KB, Navari RM, et al. Prevention of acute emesis in cancer patients following high-dose cisplatin with the combination of oral dolasetron and dexamethasone. *J Clin Oncol* 1997; 15:2135–2138.

26. Bruera E, Roca E, Cedaro L, et al. Action of oral methylprednisolone in terminal cancer patients: a prospective randomized double-blind study. *Cancer Treat Rep* 1985;69:751–754.

27. Salerno A, Hermann R. Efficacy and safety of steroid use for postoperative pain relief. Update and review of the medical literature. *J Bone Joint Surg [Am]* 2006;88:1361–1372.

28. Devor M, Govrin-Lippmann R, Raber P. Corticosteroids suppress ectopic neural discharge originating in experimental neuromas. *Pain* 1985;22:127–137.

29. Helfer EL, Rose LI. Corticosteroids and adrenal suppression. Characterising and avoiding the problem. *Prac Therap* 1989;38(5):838–845.

30. Heimdal K, Hirschberg H, Slettebo H, et al. High incidence of serious side effects of high-dose dexamethasone treatment in patients with epidural spinal cord compression. *J Neurooncol* 1992;12:141–144.

31. Bonnette P, Puyo P, Gabriel C, et al. Surgical management of non-small cell lung cancer with synchronous brain metastases. *Chest* 2001;119:1469–1475.

32. Yavuzsen T, Davis MP, Walsh D, et al. Systematic review of the treatment of cancer-associated anorexia and weight loss. *J Clin Oncol* 2005; 23:8500–8511.

33. Jacobs TP, Whitlock RT, Edsall J, et al. Addisonian crisis while taking high-dose glucocorticoids. An unusual presentation of primary adrenal failure in two patients with underlying inflammatory diseases. *JAMA* 1988;260:2082–2084.

34. Balis FM, Lester CM, Chrousos GP, et al. Differences in cerebrospinal fluid penetration of corticosteroids: possible relationship to the prevention of meningeal leukemia. *J Clin Oncol* 1987;5:202–207.

35. Peters WP, Holland JF, Senn H, et al. Corticosteroid administration and localized leukocyte mobilization in man. *N Engl J Med* 1972;282:342–345.

36. Hurwitz CA, Silverman LB, Schorin MA, et al. Substituting dexamethasone for prednisone complicates remission induction in children with acute lymphoblastic leukemia. *Cancer* 2000; 88:1964–1969.

37. Pascuzzi RM. Drugs and toxins associated with myopathies. *Curr Opin Rheumatol* 1998;10:511–520.

38. van de Beek D, de Gans J. Dexamethasone in adults with community-acquired bacterial meningitis. *Drugs* 2006;66:415–427.

39. Carmel PW, Greif LK. The aseptic meningitis syndrome: a complication of posterior fossa surgery. *Pediatr Neurosurg* 1993;19(5):276–280.

40. Twycross R. The risks and benefits of corticosteroids in advanced cancer. *Drug Saf* 1994;11:163–178.

41. Hempen C, Weiss E, Hess CF. Dexamethasone treatment in patients with brain metastases and primary brain tumors: do the benefits outweigh the side-effects? *Support Care Cancer* 2002;10:322–328.

42. Weissman DE, Dufer D, Vogel V, et al. Corticosteroid toxicity in neuro-oncology patients. *J Neurooncol* 1987;5:125–128.

43. Batchelor TT, Taylor LP, Thaler HT, et al. Steroid myopathy in cancer patients. *Neurology* 1997;48:1234–1238.

44. Owczarek J, Jasinska M, Orszulak-Michalak D. Drug-induced myopathies. An overview of the possible mechanisms. *Pharmacol Rep* 2005;57:23–34.

45. Ruff RL, Weissmann J. Endocrine myopathies. *Neurol Clin* 1988;6:575–592.

46. Friedrich O. Critical illness myopathy: what is happening? *Curr Opin Clin Nutr Metab Care* 2006;9:403–409.

47. Decramer M, de Bock, V, Dom R. Functional and histologic picture of steroid-induced myopathy in chronic obstructive pulmonary disease. *Am J Respir Crit Care Med* 1996;153:1958–1964.

48. Gayan-Ramirez G, Vanderhoydonc F, Verhoeven G, et al. Acute treatment with corticosteroids decreases IGF-1 and IGF-2 expression in the rat diaphragm and gastrocnemius. *Am J Respir Crit Care Med* 1999;159:283–289.

48a. Levine S, Nguyen T, Taylor N, et al. Rapid disuse atrophy of diaphragm fibers in mechanically ventilated humans. *N Engl J Med* 2008;358:1392–1394.

49. DeAngelis LM, Gnecco C, Taylor L, et al. Evolution of neuropathy and myopathy during intensive vincristine/corticosteroid chemotherapy for non-Hodgkin's lymphoma. *Cancer* 1991;67:2241–2246.

50. Fernandez-Sola J, Cusso R, Picado C, et al. Patients with chronic glucocorticoid treatment develop changes in muscle glycogen metabolism. *J Neurol Sci* 1993;117:103–106.

51. Schakman O, Gilson H, de Coninck V, et al. Insulin-like growth factor-I gene transfer by electroporation prevents skeletal muscle atrophy in glucocorticoid-treated rats. *Endocrinology* 2005;146:1789–1797.

52. Lee MC, Wee GR, Kim JH. Apoptosis of skeletal muscle on steroid-induced myopathy in rats. *J Nutr* 2005;135:1806S–1808S.

53. Hanson P, Dive A, Brucher JM, et al. Acute corticosteroid myopathy in intensive care patients. *Muscle Nerve* 1997;20:1371–1380.

54. LaPier TK. Glucocorticoid-induced muscle atrophy. The role of exercise in treatment and prevention. *J Cardiopulm Rehabil* 1997;17:76–84.

55. Paddon-Jones D, Wolfe RR, Ferrando AA. Amino acid supplementation for reversing bed rest and steroid myopathies. *J Nutr* 2005;135:1809S–1812S.

56. Kanda F, Okuda S, Matsushita T, et al. Steroid myopathy: pathogenesis and effects of growth hormone and insulin-like growth factor-I administration. *Horm Res* 2001;56 (Suppl 1):24–28.

57. Dropcho EJ, Soong S-J. Steroid-induced weakness in patients with primary brain tumors. *Neurology* 1991;41:1235–1239.

58. Venning G. Recent developments in vitamin D deficiency and muscle weakness among elderly people. *BMJ* 2005;330:524–526.

59. Brown ES, Khan DA, Nejtek VA. The psychiatric side effects of corticosteroids. *Ann Allergy Asthma Immunol* 1999;83:495–503.

60. Patten SB, Neutel CI. Corticosteroid-induced adverse psychiatric effects: incidence, diagnosis and management. *Drug Saf* 2000;22:111–122.

61. Naber D, Sand P, Heigl B. Psychopathological and neuropsychological effects of 8-days of corticosteroid treatment. A prospective study. *Psychoneuroendocrinology* 1996;21:25–31.

62. Wada K, Yamada N, Suzuki H, et al. Recurrent cases of corticosteroid-induced mood disorder: clinical characteristics and treatment. *J Clin Psychiatry* 2000;61:261–267.

63. Wolkowitz OM, Lupien SJ, Bigler E, et al. The "steroid dementia syndrome": an unrecognized complication of glucocorticoid treatment. *Ann N Y Acad Sci* 2004;1032:191–194.

64. Wolkowitz OM, Reus VI, Canick J, et al. Glucocorticoid medication, memory and steroid psychosis in medical illness. *Ann N Y Acad Sci* 1997;823:81–96.

65. Lupien SJ, McEwen BS. The acute effects of corticosteroids on cognition: integration of animal and human model studies. *Brain Res Rev* 1997;24:1–27.

66. Lupien SJ, Gillin CJ, Hauger RL. Working memory is more sensitive than declarative memory to the acute effects of corticosteroids: a dose-response study in humans. *Behav Neurosci* 1999;113:420–430.

67. Stiefel FC, Breitbart WS, Holland JC. Corticosteroids in cancer: neuropsychiatric complications. *Cancer Invest* 1989;7:479–491.

68. Savard J, Morin CM. Insomnia in the context of cancer: a review of a neglected problem. *J Clin Oncol* 2001;19:895–908.

69. Sima AA, Kamiya H, Li ZG. Insulin, C-peptide, hyperglycemia, and central nervous system complications in diabetes. *Eur J Pharmacol* 2004;490:187–197.

70. Cooper MS. Sensitivity of bone to glucocorticoids. *Clin Sci (Lond)* 2004;107:111–123.

71. Lloyd ME, Spector TD, Howard R. Osteoporosis in neurological disorders. *J Neurol Neurosurg Psychiatry* 2000;68:543–547.

72. Homik JE, Cranney A, Shea B, et al. A metaanalysis on the use of bisphosphonates in corticosteroid induced osteoporosis. *J Rheumatol* 1999;26:1148–1157.

73. Sambrook PN. How to prevent steroid induced osteoporosis. *Ann Rheum Dis* 2005;64:176–178.

74. Migliorati CA, Siegel MA, Elting LS. Bisphosphonate-associated osteonecrosis: a long-term complication of bisphosphonate treatment. *Lancet Oncol* 2006;7:508–514.

75. Mirzai R, Chang C, Greenspan A, et al. The pathogenesis of osteonecrosis and the relationships to corticosteroids. *J Asthma* 1999;36:77–95.

76. Chan MH, Chan PK, Griffith JF, et al. Steroid-induced osteonecrosis in severe acute respiratory syndrome: a retrospective analysis of biochemical markers of bone metabolism and corticosteroid therapy. *Pathology* 2006;38:229–235.

77. Beltran J, Opsha O. MR imaging of the hip: osseous lesions. *Magn Reson Imaging Clin N Am* 2005;13:665–676, vi.

78. Gupta OP, Boynton JR, Sabini P, et al. Proptosis after retrobulbar corticosteroid injections. *Ophthalmology* 2003;110:443–447.

79. Van Dalen JT, Sherman MD. Corticosteroid-induced exophthalmos. *Doc Ophthalmol* 1989;72:273–277.

80. Fassett DR, Schmidt MH. Spinal epidural lipomatosis: a review of its causes and recommendations for treatment. *Neurosurg Focus* 2004;16:E11.

81. Fogel GR, Cunningham PY, III, Esses SI. Spinal epidural lipomatosis: case reports, literature review and meta-analysis. *Spine J* 2005;5:202–211.

82. Zentner J, Buchbender K, Vahlensieck M. Spinal epidural lipomatosis as a complication of prolonged corticosteroid therapy. *J Neurosurg Sci* 1995;39:81–85.

83. Carnahan MC, Goldstein DA. Ocular complications of topical, peri-ocular, and systemic corticosteroids. *Curr Opin Ophthalmol* 2000;11:478–483.

84. Tripathi RC, Parapuram SK, Tripathi BJ, et al. Corticosteroids and glaucoma risk. *Drugs Aging* 1999;15:439–450.

85. Marcus P, McCauley DL. Steroid therapy and H2-receptor antagonists: pharmacoeconomic implications. *Clin Pharmacol Ther* 1997;61:503–508.

86. Filaretova LP, Podvigina TT, Bagaeva TR, et al. Gastroprotective action of glucocorticoid hormones during NSAID treatment. *Inflammopharmacology* 2005;13:27–43.

87. Dorr RT, Soble MJ. H2-antagonists and carmustine. *J Cancer Res Clin Oncol* 1989;115:41–46.

88. Morris CR, Harvey IM, Stebbings WS, et al. Anti-inflammatory drugs, analgesics and the risk of perforated colonic diverticular disease. *Br J Surg* 2003;90:1267–1272.

89. Mpofu S, Mpofu CM, Hutchinson D, et al. Steroids, non-steroidal anti-inflammatory drugs, and sigmoid diverticular abscess perforation in rheumatic conditions. *Ann Rheum Dis* 2004;63:588–590.

90. Fadul CE, Lemann W, Thaler HT, et al. Perforation of the gastrointestinal tract in patients receiving steroids for neurological disease. *Neurology* 1988;38:348–352.

91. Zulke C, Ulbrich S, Graeb C, et al. Acute pneumatosis cystoides intestinalis following allogeneic transplantation—the surgeon's dilemma. *Bone Marrow Transplant* 2002;29:795–798.

92. Hoer J, Truong S, Virnich N, et al. Pneumatosis cystoides intestinalis: confirmation of diagnosis by endoscopic puncture, a review of pathogenesis, associated disease and therapy and a new theory of cyst formation. *Endoscopy* 1998;30:793–799.

93. Braumann C, Menenakos C, Jacobi CA. Pneumatosis intestinalis—a pitfall for surgeons? *Scand J Surg* 2005;94:47–50.

94. Saunders MD, Kimmey MB. Systematic review: acute colonic pseudo-obstruction. *Aliment Pharmacol Ther* 2005;22:917–925.

95. Brading AF, Ramalingam T. Mechanisms controlling normal defecation and the potential effects of spinal cord injury. *Prog Brain Res* 2006;152:345–358.

96. Fazel A, Verne GN. New solutions to an old problem: acute colonic pseudo-obstruction. *J Clin Gastroenterol* 2005;39:17–20.

97. Paran H, Silverberg D, Mayo A, et al. Treatment of acute colonic pseudo-obstruction with neostigmine. *J Am Coll Surg* 2000;190:315–318.

98. Gibson RJ, Keefe DM. Cancer chemotherapy-induced diarrhoea and constipation: mechanisms of damage and prevention strategies. *Support Care Cancer* 2006;14:890–900.

99. Sykes NP. The pathogenesis of constipation. *J Support Oncol* 2006;4:213–218.

100. Mancini I, Bruera E. Constipation in advanced cancer patients. *Support Care Cancer* 1998;6:356–364.

101. Jun S, Dimyan M, Jones KD, et al. Obstipation as a paraneoplastic presentation of small cell lung cancer: case report and literature review. *Neurogastroenterol Motil* 2005;17:16–22.

102. Klaschik E, Nauck F, Ostgathe C. Constipation—modern laxative therapy. *Support Care Cancer* 2003;11:679–685.

103. Friedman JD, Dello Buono FA. Opioid antagonists in the treatment of opioid-induced constipation and pruritus. *Ann Pharmacother* 2001;35:85–91.

104. Yuan CS. Clinical status of methylnaltrexone, a new agent to prevent and manage opioid-induced side effects. *J Support Oncol* 2004;2:111–117.

105. LeWitt PA, Barton NW, Posner JB. Hiccup with dexamethasone therapy. Letter to the editor. *Ann Neurol* 1982;12:405–406.

106. Dickerman RD, Overby C, Eisenberg M, et al. The steroid-responsive hiccup reflex arc: competitive binding to the corticosteroid-receptor? *Neuro Endocrinol Lett* 2003;24:167–169.

107. Dickerman RD, Jaikumar S. The hiccup reflex arc and persistent hiccups with high-dose anabolic steroids: is the brainstem the steroid-responsive locus? *Clin Neuropharmacol* 2001;24:62–64.

108. Thompson DF, Landry JP. Drug-induced hiccups. *Ann Pharmacother* 1997;31:367–369.

109. Friedman NL. Hiccups: a treatment review. *Pharmacotherapy* 1996;16:986–995.

110. Erdmann SM, Abuzahra F, Merk HF, et al. Anaphylaxis induced by glucocorticoids. *J Am Board Fam Pract* 2005;18:143–146.

111. Schonwald S. Methylprednisolone anaphylaxis. *Am J Emerg Med* 1999;17:583–585.

112. Flavin DK, Frederickson PA, Richardson JW, et al. Corticosteroid abuse—an unusual manifestation of drug dependence. *Mayo Clin Proc* 1983;58:764–766.

113. Czerwinski AW, Czerwinski AB, Whitsett TL, et al. Effects of a single, large intravenous injection of dexamethasone. *Clin Pharmacol Ther* 1972;13:638–642.

114. Bromley SM. Smell and taste disorders: a primary care approach. *Am Fam Physician* 2000;61:427–436, 438.

115. Latham JM, Fraser RD, Moore RJ, et al. The pathologic effects of intrathecal betamethasone. *Spine* 1997;22:1558–1562.

116. Hardy JR, Rees E, Ling J, et al. A prospective survey of the use of dexamethasone on a palliative care unit. *Palliat Med* 2001;15:3–8.

117. Muzyka BC. Oral fungal infections. *Dent Clin North Am* 2005;49:49–65.

118. Pankhurst C. Candidiasis (oropharyngeal). *Clin Evid* 2005;13:1701–1716.

119. Gluck T, Geerdes-Fenge HF, Straub RH, et al. Pneumocystis carinii pneumonia as a complication of immunosuppressive therapy. *Infection* 2000;28:227–230.

120. Kim DS, Park SK, Choi WH, et al. Pneumocystis carinii pneumonia associated with a rapid reduction of cortisol level in a patient with ectopic ACTH syndrome treated by octreotide and ketoconazole. *Exp Clin Endocrinol Diabetes* 2000;108:146–150.

121. Mathew BS, Grossman SA. Pneumocystis carinii pneumonia prophylaxis in HIV negative patients with primary CNS lymphoma. *Cancer Treat Rev* 2003;29:105–119.

122. Sowden E, Carmichael AJ. Autoimmune inflammatory disorders, systemic corticosteroids and pneumocystis pneumonia: a strategy for prevention. *BMC Infect Dis* 2004;4:42.

123. Castro M. Treatment and prophylaxis of Pneumocystis carinii pneumonia. *Semin Respir Infect* 1998;13:296–303.

124. Mantadakis E, Danilatou V, Stiakaki E, et al. Infectious toxicity of dexamethasone during ALL remission-induction chemotherapy: report of two cases and literature review. *Pediatr Hematol Oncol* 2004;21:27–35.

125. Bostrom BC, Sensel MR, Sather HN, et al. Dexamethasone versus prednisone and daily oral versus weekly intravenous mercaptopurine for patients with standard-risk acute lymphoblastic leukemia: a report from the Children's Cancer Group. *Blood* 2003;101:3809–3817.

126. Silverman LB, Gelber RD, Dalton VK, et al. Improved outcome for children with acute lymphoblastic leukemia: results of Dana-Farber Consortium Protocol 91–01. *Blood* 2001;97:1211–1218.

127. Te Poele EM, de Bont ES, Marike BH, et al. Dexamethasone in the maintenance phase of acute lymphoblastic leukaemia treatment: is the risk of lethal infections too high? *Eur J Cancer* 2007;43:2532–2536.

128. Hill G, Chauvenet AR, Lovato J, et al. Recent steroid therapy increases severity of varicella infections in children with acute lymphoblastic leukemia. *Pediatrics* 2005;116:e525–e529.

129. Putignano P, Kaltsas GA, Satta MA, et al. The effects of anti-convulsant drugs on adrenal function. *Horm Metab Res* 1998;30:389–397.

130. Chalk JB, Ridgeway K, Tro'r B, et al. Phenytoin impairs the bioavailability of dexamethasone in neurological and neurosurgical patients. *J Neurol Neurosurg Psychiatry* 1984;47:1087–1090.

131. Lawson LA, Blouin RA, Smith RB, et al. Phenytoin-dexamethasone interaction: a previously unreported observation. *Surg Neurol* 1981;16:23–24.

132. Lackner TE. Interaction of dexamethasone with phenytoin. *Pharmacotherapy* 1991;11:344–347.

133. Riechelmann RP. Potential drug interactions and duplicate prescriptions among cancer patients. *J Natl Cancer Inst* 2007;99:592–600.

134. Amatruda TT, Jr, Hurst MH, D'Esopo ND. Certain endocrine and metabolic facets of the steroid withdrawal syndrome. *J Clin Endocrinol* 1965;25:1207–1217.

135. Byyny RL. Withdrawal from glucocorticoid therapy. *N Engl J Med* 1976;295:30–32.

136. Dixon RA, Christy NP. On the various forms of corticosteroid withdrawal syndrome. *Am J Med* 1980;68:224–230.

137. Rimsza ME. Complications of corticosteroid therapy. *Am J Dis Child* 1978;132:806–810.

138. Johnston I, Gilday DL, Hendrick EB. Experimental effects of steroids and steroid withdrawal on cerebrospinal fluid absorption. *J Neurosurg* 1975;42:690–695.

139. Saracco P, Bertorello N, Farinasso L, et al. Steroid withdrawal syndrome during steroid tapering in childhood acute lymphoblastic leukemia: a controlled study comparing prednisone versus dexamethasone in induction phase. *J Pediatr Hematol Oncol* 2005;27:141–144.

140. Beaumont A, Whittle IR. The pathogenesis of tumour associated epilepsy. *Acta Neurochir (Wien)* 2000;142:1–15.

141. Oberndorfer S, Schmal T, Lahrmann H, et al. [The frequency of seizures in patients with primary brain tumors or cerebral metastases. An evaluation from the Ludwig Boltzmann Institute of Neuro-Oncology and the Department of Neurology, Kaiser Franz Josef Hospital, Vienna]. *Wien Klin Wochenschr* 2002;114:911–916.

142. Boylan LS, Labovitz DL, Jackson SC, et al. Auras are frequent in idiopathic generalized epilepsy. *Neurology* 2006;67:343–345.

143. Turnbull DM, Rawlins MD, Weightman D, et al. "Therapeutic" serum concentration of phenytoin: the influence of seizure type. *J Neurol Neurosurg Psychiatry* 1984;47:231–234.

144. Ingvar M. Cerebral blood flow and metabolic rate during seizures. Relationship to epileptic brain damage. *Ann N Y Acad Sci* 1986; 462:194–206.

145. Meldrum BS. Excitotoxicity and selective neuronal loss in epilepsy. *Brain Pathol* 1993;3:405–412.

146. Chen JW, Wasterlain CG. Status epilepticus: pathophysiology and management in adults. *Lancet Neurol* 2006;5:246–256.

147. Bleck TP. Management approaches to prolonged seizures and status epilepticus. *Epilepsia* 1999;40 (Suppl 1):S59–S63.

148. Prasad K, Al-Roomi K, Krishnan PR, et al. Anticonvulsant therapy for status epilepticus. *Cochrane Database Syst Rev* 2005:CD003723.

149. Hanley DF, Pozo M. Treatment of status epilepticus with midazolam in the critical care setting. *Int J Clin Pract* 2000;54:30–35.

150. Prasad A, Worrall BB, Bertram EH, et al. Propofol and midazolam in the treatment of refractory status epilepticus. *Epilepsia* 2001;42:380–386.

151. Misra UK, Kalita J, Patel R. Sodium valproate vs phenytoin in status epilepticus: a pilot study. *Neurology* 2006;67:340–342.

152. Pruitt AA. Treatment of medical complications in patients with brain tumors. *Curr Treat Options Neurol* 2005;7:323–336.

153. Schaller B, Ruegg SJ. Brain tumor and seizures: pathophysiology and its implications for treatment revisited. *Epilepsia* 2003;44:1223–1232.

154. Brodie MJ, Perucca P, Ben-Menachem E, et al Comparison of levetiracetam and controlled-release carbamazepine in newly diagnosed epilepsy. *Neurology* 2007;68:402–408.

155. van Breemen MS, Wilms EB, Vecht VJ. Epilepsy in patients with brain tumours: epidemiology, mechanisms, and management. *Lancet Neurol* 2007;6:421–430.

156. Ji-qun C, Ishihara K, Nagayama T, et al. Long-lasting antiepileptic effects of levetiracetam against epileptic seizures in the spontaneously epileptic rat (SER): differentiation of levetiracetam from conventional antiepileptic drugs. *Epilepsia* 2005;46:1362–1370.

157. Correno M. Levetiracetam. *Drugs Today* 2007;43:769–794.

158. van Breemen MS, Vecht CJ. Optimal seizure management in brain tumor patients. *Curr Neurol Neurosci Rep* 2005;5:207–213.

159. Maschio M, Dinapoli L, Zarabia A, et al. Issues related to the pharmacological management of patients with brain tumours and epilepsy. *Funct Neurol* 2006;21:15–19.

160. Husain AM, Horn GJ, Jacobson MP. Non-convulsive status epilepticus: usefulness of clinical features in selecting patients for urgent EEG. *J Neurol Neurosurg Psychiatry* 2003;74:189–191.

161. Towne AR, Waterhouse EJ, Boggs JG, et al. Prevalence of nonconvulsive status epilepticus in comatose patients. *Neurology* 2000;54:340–345.

162. Kaplan PW. Assessing the outcomes in patients with nonconvulsive status epilepticus: nonconvulsive status epilepticus is underdiagnosed, potentially overtreated, and confounded by comorbidity. *J Clin Neurophysiol* 1999;16:341–352.

163. Claassen J, Mayer SA, Kowalski RG, et al. Detection of electrographic seizures with continuous EEG monitoring in critically ill patients. *Neurology* 2004;62:1743–1748.

164. Gayatri NA, Livingston JH. Aggravation of epilepsy by anti-epileptic drugs. *Dev Med Child Neurol* 2006;48:394–398.

165. Osorio I, Reed RC, Peltzer JN. Refractory idiopathic absence status epilepticus: a probable paradoxical effect of phenytoin and carbamazepine. *Epilepsia* 2000;41:887–894.

166. Sirven JI, Wingerchuk DA, Drazkowsici JF, et al. Seizure prophylaxis in patients with brain tumors: a meta-analysis. *Mayo Clin Proc* 2004;79:1489–1494.

167. Glantz MJ, Cole BF, Forsyth PA, et al. Practice parameter: anticonvulsant prophylaxis in patients with newly diagnosed brain tumors—Report of the Quality Standards Subcommittee of the American Academy of Neurology. *Neurology* 2000;54:1886–1893.

168. Temkin NR. Prophylactic anticonvulsants after neurosurgery. *Epilepsy Curr* 2002;2:105–107.

169. Vecht CJ, Wagner GL, Wilms EB. Interactions between antiepileptic drugs and chemotherapeutic agents. *Lancet Neurol* 2003;2:404–409.

170. Benedetti MS. Enzyme induction and inhibition by new antiepileptic drugs: a review of human studies. *Fundam Clin Pharmacol* 2000;14:301–319.

171. Ghosh C, Lazarus HM, Hewlett JS, et al. Fluctuation of serum phenytoin concentrations during autologous bone marrow transplant for primary central nervous system tumors. *J Neurooncol* 1992;12:25–32.

172. Grossman SA, Hochberg F, Fisher J, et al. Increased 9-aminocamptothecin dose requirements in patients on anticonvulsants. *Cancer Chemother Pharmacol* 1998;42:118–126.

173. Relling MV, Pui CH, Sandlund JT, et al. Adverse effect of anticonvulsants on efficacy of chemotherapy for acute lymphoblastic leukaemia. *Lancet* 2000;356:285–290.

174. Lehmann DF, Hurteau TE, Newman N, et al. Anticonvulsant usage is associated with an increased risk of procarbazine hypersensitivity reactions in patients with brain tumors. *Clin Pharmacol Ther* 1997;62:225–229.

175. Pagani JJ, Hayman LA, Bigelow RH, et al. Diazepam prophylaxis of contrast media-induced seizures during computed tomography of patients with brain metastases. *AJNR Am J Neuroradiol* 1983;140:787–792.

175a. Klein M, Heimans JJ, Aaronson NK, et al. Effect of radiotherapy and other treatment-related factors on mid-term to long-term cognitive sequelae in low-grade gliomas. a comparative study. *Lancet* 2002;360:1361–1368.

176. Kwan P, Brodie MJ. Neuropsychological effects of epilepsy and antiepileptic drugs. *Lancet* 2001;357:216–222.

177. Khafaga YM, Jamshed A, Allam AAK, et al. Stevens-Johnson syndrome in patients on phenytoin and cranial radiotherapy. *Acta Oncologica* 1999;38:111–116.

178. Duncan KO, Tigelaar RE, Bolognia JL. Stevens-Johnson syndrome limited to multiple sites of radiation therapy in a patient receiving phenobarbital. *J Am Acad Dermatol* 1999;40:493–496.

179. Micali G, Linthicum K, Han N, et al. Increased risk of erythema multiforme major with combination anticonvulsant and radiation therapies. *Pharmacotherapy* 1999;19:223–227.

180. Lonjou C, Borot N, Sekula P, et al. A European study of HLA-B in Stevens-Johnson syndrome and toxic epidermal necrolysis related to five high-risk drugs. *Pharmacogenet Genom* 2009;18:99–107.

181. Gaffey CM, Chun B, Harvey JC, et al. Phenytoin-induced systemic granulomatous vasculitis. *Arch Pathol Lab Med* 1986;110:131–135.

182. Wolinsky-Friedland M. Drug-induced metabolic bone disease. *Endocrinol Metab Clin North Am* 1995;24:395–420.

183. Aldenhovel HG. The influence of long-term anticonvulsant therapy with diphenylhydantoin and carbamazepine on serum γ-glutamyltransferase, aspartate aminotransferase, alanine aminotransferase and alkaline phosphatase. *Eur Arch Psychiatry Neurol Sci* 1988;237:312–316.

184. Ahmad S, Laidlaw J, Houghton GW, et al. Involuntary movements caused by phenytoin intoxication in epileptic patients. *J Neurol Neurosurg Psychiatry* 1975;38:225–231.

185. Barclay CL, McLean M, Hagen N, et al. Severe phenytoin hypersensitivity with myopathy: a case report. *Neurology* 1992;42:2303.

186. van Amelsvoort T, Bakshi R, Devaux CB, et al. Hyponatremia associated with carbamazepine and oxcarbazepine therapy: a review. *Epilepsia* 1994;35:181–188.

187. Dang CT, Riley DK. Aseptic meningitis secondary to carbamazepine therapy. *Clin Infect Dis* 1996;22:729–730.

188. Callahan DJ, Noetzel MJ. Prolonged absence status epilepticus associated with carbamazepine therapy, increased intracranial pressure, and transient MRI abnormalities. *Neurology* 1992;42:2198–2201.

189. Miyaoka T, Seno H, Itoga M, et al. Reversible pitch perception deficit caused by carbamazepine. *Clin Neuropharmacol* 2000;23:219–221.

190. Taylor LP, Posner JB. Phenobarbital rheumatism in patients with brain tumor. *Ann Neurol* 1989;25:92–94.

191. Yamamoto LG, Yim GK. The role of intravenous valproic acid in status epilepticus. *Pediatr Emerg Care* 2000;16:296–298.

192. Gram L, Bentsen KD. Valproate: an updated review. *Acta Neurol Scand* 1985;72:129–139.

193. Gidal B, Spencer N, Maly M, et al. Valproate-mediated disturbances of hemostasis: relationship to dose and plasma concentration. *Neurology* 1994;44:1418–1422.

194. Sasso E, Delsoldato S, Negrotti A, et al. Reversible valproate-induced extrapyramidal disorders. *Epilepsia* 1994;35:391–393.

195. Mattson RH. Efficacy and adverse effects of established and new antiepileptic drugs. *Epilepsia* 1995;36 (Suppl 2):S13–S26.

196. Sabers A, Gram L. Newer anticonvulsants: comparative review of drug interactions and adverse effects. *Drugs* 2000;60:23–33.

197. Tebb Z, Tobias JD. New anticonvulsants—new adverse effects. *South Med J* 2006;99:375–379.

198. Guay DR. Pregabalin in neuropathic pain: a more "pharmaceutically elegant" gabapentin? *Am J Geriatr Pharmacother* 2005;3:274–287.

199. Bialer M. New antiepileptic drugs that are second generation to existing antiepileptic drugs. *Expert Opin Investig Drugs* 2006;15:637–647.

200. Willmore LJ. Clinical pharmacology of new antiepileptic drugs. *Neurology* 2000;55:S17–S24.

201. Newton HB, Goldlust SA, Pearl D. Retrospective analysis of the efficacy and tolerability of levetiracetam in brain tumor patients. *J Neurooncol* 2006;78:99–102.

202. Maschio M, Albani F, Baruzzi A, et al. Levetiracetam therapy in patients with brain tumour and epilepsy. *J Neurooncol* 2006;80:97–100.

203. Ruegg S, Naegelin Y, Hardmeier M, et al. Intravenous levetiracetam: treatment experience with the first 50 critically ill patients. *Epilepsy Behav*, in press.

204. Zain J. Valproic acid monotherapy leads to CR in a patient with refractory diffuse large B cell lymphoma. *Leuk Lymphoma* 2007;48:1216–1218.

205. Prandoni P, Piccioli A. Thrombosis as a harbinger of cancer. *Curr Opin Hematol* 2006;13:362–365.

206. Marine JE, Goldhaber SZ. Pulmonary embolism presenting as seizures. *Chest* 1997;112:840–842.

207. Tapson VF. Acute pulmonary embolism. N Engl J Med 2008;358:1037–1052

208. Kupnik D, Grmec S. Pulmonary thromboembolism presenting as epileptiform generalized seizure. *Eur J Emerg Med* 2004;11:346–347.

209. Fred HL, Willerson JT, Alexander JK. Neurological manifestations of pulmonary thromboembolism. *Arch Intern Med* 1967;120:33–37.

210. Cohen-Gadol AA, DiLuna ML, Spencer DD. Partial epilepsy presenting as episodic dyspnea: a specific network involved in limbic seizure propagation. Case report. *J Neurosurg* 2004;100:565–567.

211. Mandala M, Falanga A, Piccioli A, et al. Venous thromboembolism and cancer: guidelines of the Italian Association of Medical Oncology (AIOM). *Crit Rev Oncol Hematol* 2006;59:194–204.

212. Brandes AA, Scelzi E, Salmistraro G, et al. Incidence of risk of thromboembolism during treatment of high-grade gliomas: a prospective study. *Eur J Cancer* 1997;33:1592–1596.

213. Marras LC, Geerts WH, Perry JR. The risk of venous thromboembolism is increased throughout the course of malignant glioma—an evidence-based review. *Cancer* 2000;89:640–646.

214. Schiff D, DeAngelis LM. Therapy of venous thromboembolism in patients with brain metastases. *Cancer* 1994;73:493–498.

215. Ruff RL, Posner JB. The incidence of systemic venous thrombosis and the risk of anticoagulation in patients with malignant gliomas. *Trans Am Neurol Assoc* 1981;106:223–226.

216. Auguste KI, Quinones-Hinojosa A, Berger MS. Efficacy of mechanical prophylaxis for venous thromboembolism in patients with brain tumors. *Neurosurg Focus* 2004;17:E3.

217. Chan AT, Atiemo A, Diran LK, et al. Venous thromboembolism occurs frequently in patients undergoing brain tumor surgery despite prophylaxis. *J Thromb Thrombolysis* 1999;8:139–142.

218. Smith SF, Simpson JM, Sekhon LH. Prophylaxis for deep venous thrombosis in neurosurgical oncology: review of 2779 admissions over a 9-year period. *Neurosurg Focus* 2004;17:E4.

219. Davidson JE, Willms DC, Hoffman MS. Effect of intermittent pneumatic leg compression on intracranial pressure in brain-injured patients. *Crit Care Med* 1993;21:224–227.

220. Christen Y, Wutschert R, Weimer D, et al. Effects of intermittent pneumatic compression on venous haemodynamics and fibrinolytic activity. *Blood Coagul Fibrinolysis* 1997;8:185–190.

221. Pruemer J. Treatment of cancer-associated thrombosis: distinguishing among antithrombotic agents. *Semin Oncol* 2006;33:S26–S39.

222. Danish SF, Burnett MG, Ong JG, et al. Prophylaxis for deep venous thrombosis in craniotomy patients: a decision analysis. *Neurosurgery* 2005;56:1286–1292.

223. Dickinson LD, Miller LD, Patel CP, et al. Enoxaparin increases the incidence of postoperative intracranial hemorrhage when initiated preoperatively for deep venous thrombosis prophylaxis in patients with brain tumors. *Neurosurgery* 1998;43:1074–1081.

224. Iorio A, Agnelli G. Low-molecular-weight and unfractionated heparin for prevention of venous thromboembolism in neurosurgery: a meta-analysis. *Arch Intern Med* 2000;160:2327–2332.

225. Agnelli G, Piovella F, Buoncristiani P, et al. Enoxaparin plus compression stockings compared with compression stockings alone in the prevention of venous thromboembolism after elective neurosurgery. *N Engl J Med* 1998;339:80–85.

226. Kubitza D, Haas S. Novel factor Xa inhibitors for prevention and treatment of thromboembolic diseases. *Expert Opin Investig Drugs* 2006;15:843–855.

227. Bauer KA. New anticoagulants: anti IIa vs anti Xa—is one better? *J Thromb Thrombolysis* 2006;21:67–72.

228. Eriksson BI, Dahl OE, Rosencher N, et al. Dabigatran etexilate versus enoxaparin for prevention of venous thromboembolism after total hip replacement: a randomised, double-blind, non-inferiority trial. *Lancet* 2007;370:949–956.

229. Cohen AT. Venous thromboembolic disease management of the nonsurgical moderate- and high-risk patient. *Semin Hematol* 2000;37:19–22.

230. Partsch H, Blattler W. Compression and walking versus bed rest in the treatment of proximal deep venous thrombosis with low molecular weight heparin. *J Vasc Surg* 2000;32:861–869.

231. von Depka PM, Karthaus M, Ganser A, et al. Anticoagulant prophylaxis and therapy in patients with cancer. *Antibiot Chemother* 2000;50:149–158.

232. Atkinson JL, Sundt TM, Jr, Kazmier FJ, et al. Heparin-induced thrombocytopenia and thrombosis in ischemic stroke. *Mayo Clin Proc* 1988;63:353–361.

233. McGehee WG, Klotz TA, Epstein DJ, et al. Coumadin necrosis associated with hereditary protein C deficiency. *Ann Intern Med* 1984;100:59–60.

234. Essex DW. Late-onset warfarin-induced skin necrosis: case report and review of the literature. *Am J Hematol* 1998;57:233–237.

235. Rosner MK, Kuklo TR, Tawk R, et al. Prophylactic placement of an inferior vena cava filter in high-risk patients undergoing spinal reconstruction. *Neurosurg Focus* 2004;17:E6.

236. Decousus H, Leizorovicz A, Parent F, et al. A clinical trial of vena caval filters in the prevention of pulmonary embolism in patients with proximal deep-vein thrombosis. Prevention du Risque d'Embolie Pulmonaire par Interruption Cave Study Group. *N Engl J Med* 1998;338:409–415.

237. Lee AY, Levine MN, Baker RI, et al. Low-molecular-weight heparin versus a coumarin for the prevention of recurrent venous thromboembolism in patients with cancer. *N Engl J Med* 2003;349:146–153.

238. Getzen TM, Rectenwald JF. Inferior vena cava filters in the cancer patient: current use and indications. *J Natl Compr Canc Netw* 2006;4:881–888.

239. Kim HS, Young MJ, Narayan AK, et al. A comparison of clinical outcomes with retrievable and permanent inferior vena cava filters. *J Vasc Interv Radiol* 2008;19:393–299.

240. Levy MH, Samuel TA. Management of cancer pain. *Semin Oncol* 2005;32:179–193.

241. Rainone F. Treating adult cancer pain in primary care. *J Am Board Fam Pract* 2004;17 (Suppl):S48–S56.

242. Rao A, Cohen HJ. Symptom management in the elderly cancer patient: fatigue, pain, and depression. *J Natl Cancer Inst Monogr* 2004;150–157.

243. Balducci L. Management of cancer pain in geriatric patients. *J Support Oncol* 2003;1:175–191.

244. Cherny NI. The management of cancer pain. *CA Cancer J Clin* 2000;50:70–116.

245. Benedetti C, Brock C, Cleeland C, et al. NCCN practice guidelines for cancer pain. *Oncology (Huntingt)* 2000;14:135–150.

246. Michaelson MD, Smith MR. Bisphosphonates for treatment and prevention of bone metastases. *J Clin Oncol* 2005;23:8219–8224.

247. Bender T, Donath J, Barna I, et al. The analgesic effect of pamidronate is not caused by the elevation of beta endorphin level in Paget's disease—a controlled pilot study. *Neuro Endocrinol Lett* 2006;27:513–515.

248. Hopkins S, Jolles S. Drug-induced aseptic meningitis. *Expert Opin Drug Saf* 2005;4:285–297.

249. Estfan B, LeGrand SB, Walsh D, et al. Opioid rotation in cancer patients: pros and cons. *Oncology (Williston Park)* 2005;19:511–516.

250. Ross JR, Riley J, Quigley C, et al. Clinical pharmacology and pharmacotherapy of opioid switching in cancer patients. *Oncologist* 2006;11:765–773.

250a. Ross JR, Taegetmeyer AB, Sato H, et al. Genetic variation and response to morphine in cancer patients: catechol-O-methyltransferase and multidrug resistance-1 gene polymorphisms are associated with central side effects. *Cancer* 2008;112:1390–1403.

251. Gilron I, Watson CP, Cahill CM, et al. Neuropathic pain: a practical guide for the clinician. *CMAJ* 2006;175:265–275.

252. Tremont-Lukats IW, Megeff C, Backonja MM. Anticonvulsants for neuropathic pain syndromes: mechanisms of action and place in therapy. *Drugs* 2000;60:1029–1052.

253. MacPherson RD. The pharmacological basis of contemporary pain management. *Pharmacol Ther* 2000;88:163–185.

254. Greenberg HS, Kim J-H, Posner JB. Epidural spinal cord compression from metastatic tumor: results with a new treatment protocol. *Ann Neurol* 1980;8:361–366.

255. Kim PS. Interventional cancer pain therapies. *Semin Oncol* 2005;32:194–199.

256. Sharpe M, Strong V, Allen K, et al. Major depression in outpatients attending a regional cancer centre: screening and unmet treatment needs. *Br J Cancer* 2004;90:314–320.

257. Berney A, Stiefel F, Mazzocato C, et al. Psychopharmacology in supportive care of cancer: a review for the clinician. III. Antidepressants. *Support Care Cancer* 2000;8:278–286.

258. Kash KM, Mago R, Kunkel EJ. Psychosocial oncology: supportive care for the cancer patient. *Semin Oncol* 2005;32:211–218.

259. Stiefel F, Berney A, Mazzocato C. Psychopharmacology in supportive care in cancer: a review for the clinician. I. Benzodiazepines. *Support Care Cancer* 1999;7:379–385.

260. Homsi J, Walsh D, Nelson KA. Psychostimulants in supportive care. *Support Care Cancer* 2000;8:385–397.

261. Cheer SM, Goa KL. Fluoxetine: a review of its therapeutic potential in the treatment of depression associated with physical illness. *Drugs* 2001;61:81–110.

262. Poppelreuter M, Weis J, Kulz AK, et al. Cognitive dysfunction and subjective complaints of cancer patients: a cross-sectional study in a cancer rehabilitation centre. *Eur J Cancer* 2004;40:43–49.

263. Tannock IF, Ahles TA, Ganz PA, et al. Cognitive impairment associated with chemotherapy for cancer: report of a workshop. *J Clin Oncol* 2004;22:2233–2239.

264. Ahlberg K, Ekman T, Gaston-Johansson F, et al. Assessment and management of cancer-related fatigue in adults. *Lancet* 2003;362:640–650.

265. Cella D, Lai JS, Chang CH, et al. Fatigue in cancer patients compared with fatigue in the general United States population. *Cancer* 2002;94:528–538.

265a. Stone PC, Minton O. Cancer-related fatigue. *Eur J Cancer* 2008; Mar 30 [Epub ahead of print].

266. Portenoy RK, Itri LM. Cancer-related fatigue: guidelines for evaluation and management. *Oncologist* 1999;4:1–10.

267. Siren AL, Fratelli M, Brines M, et al. Erythropoietin prevents neuronal apoptosis after cerebral ischemia and metabolic stress. *Proc Natl Acad Sci USA* 2001;98:4044–4049.

268. Sabbatini P. Contribution of anemia to fatigue in the cancer patient. *Oncology (Huntingt)* 2000;14:69–71.

269. Bruera E, Valero V, Driver L, et al. Patient-controlled methylphenidate for cancer fatigue: a double-blind, randomized, placebo-controlled trial. *J Clin Oncol* 2006;24:2073–2078.

270. Morrow GR, Shelke AR, Roscoe JA, et al. Management of cancer-related fatigue. *Cancer Invest* 2005;23:229–239.

271. Morrow GR, Hickok JT, Roscoe JA, et al. Differential effects of paroxetine on fatigue and depression: a randomized, double-blind trial from the university of Rochester cancer center community clinical oncology program. *J Clin Oncol* 2003;21:4635–4641.

272. Wood GJ. Management of intractable nausea and vomiting in patients at the end of life: "I was feeling nauseous all of the time…nothing was working." *JAMA* 2007;298:1196–1207.

273. Lavenstein BL, Cantor FK. Acute dystonia. An unusual reaction to diphenhydramine. *JAMA* 1976;236:291.

274. Kawanishi C, Onishi H, Kato D, et al. Unexpectedly high prevalence of akathisia in cancer patients. *Palliat Support Care.* 2007;5:351–354.

275. Iqbal N, Lambert T, Masand P. Akathisia: problem of concern or concern of today. *CNS Spectr* 2007;9 (Suppl 14):1–13.

276. Kushida CA. Clinical presentation, diagnosis, and quality of life issues in restless legs syndrome. *Am J Med* 2007;120 (1 Suppl 1):S4–S12.

277. Cunha BA. Antibiotic side effects. *Med Clin North Am* 2001;85:149–185.

278. Grondahl TO, Langmoen IA. Epileptogenic effect of antibiotic drugs. *J Neurosurg* 1993;78:938–943.

279. River Y, Averbuch-Heller L, Weinberger M, et al. Antibiotic induced meningitis. *J Neurol Neurosurg Psychiatry* 1994;57:705–708.

280. Segal JA, Harris BD, Kustova Y, et al. Aminoglycoside neurotoxicity involves NMDA receptor activation. *Brain Res* 1999;815:270–277.

281. Manley TJ, Chusid MJ, Rand SD, et al. Reversible parkinsonism in a child after bone marrow transplantation and lipid-based amphotericin B therapy. *Pediatr Infect Dis J* 1998;17:433–434.

282. Mott SH, Packer RJ, Vezina LG, et al. Encephalopathy with parkinsonian features in children following bone marrow transplantations and high-dose amphotericin B. *Ann Neurol* 1995; 37:810–814.

283. Antonini G, Morino S, Fiorelli M, et al. Reversal of encephalopathy during treatment with Amphotericin-B. *J Neurol Sci* 1996;144:212–213.

284. Robinson RF, Nahata MC. A comparative review of conventional and lipid formulations of amphotericin B. *J Clin Pharm Ther* 1999;24:249–257.

285. Muller MP, Richardson DC, Walmsley SL. Trimethoprim-sulfamethoxazole induced aseptic meningitis in a renal transplant patient. *Clin Nephrol* 2001;55:80–84.

286. Moris G, Garcia-Monco JC. The challenge of drug-induced aseptic meningitis. *Arch Intern Med* 1999;159:1185–1194.

287. Craven W, Donofrio P. Sensory and autonomic polyneuropathy associated with trimethoprim-sulfamethoxazole. *Ann Neurol* 1992;32:281–282. Abstract.

288. Vincent FM. Acute polyneuropathy possibly associated with co-trimoxazole. *Lancet* 1977;ii:980.

289. Smolders I, Gousseau C, Marchand S, et al. Convulsant and subconvulsant doses of norfloxacin in the presence and absence of biphenylacetic acid alter extracellular hippocampal glutamate but not γ-aminobutyric acid levels in conscious rats. *Antimicrob Agents Chemother* 2002;46:471–477.

290. Thomas RJ. Neurotoxicity of antibacterial therapy. *South Med J* 1994;87:869–874.

291. Stahlmann R, Lode H. Toxicity of quinolones. *Drugs* 1999;58 (Suppl 2):37–42.

292. O'Brien JG, Dong BJ, Coleman RL, et al. A 5-year retrospective review of adverse drug reactions and their risk factors in human immunodeficiency virus-infected patients who were receiving intravenous pentamidine therapy for Pneumocystis carinii pneumonia. *Clin Infect Dis* 1997;24:854–859.

293. Rajan GR, Cobb JP, Reiss CK. Acyclovir induced coma in the intensive care unit. *Anaesth Intensive Care* 2000;28:305–307.

294. Ernst ME, Franey RJ. Acyclovir- and gangciclovir-induced neurotoxicity. *Ann Pharmacother* 1998;32:111–113.

295. Blohm MEG, Nürnberger W, Aulich A, et al. Reversible brain MRI changes in acyclovir neurotoxicity. *Bone Marrow Transplant* 1997;19:1049–1051.

PART 2

METASTASES

Chapter 5

Intracranial Metastases

GENERAL PRINCIPLES

Introduction

Metastasis to the cranium or its intracranial contents is a common neurologic complication of cancer. Although skull metastases are more common than parenchymal brain metastases, the latter are much more likely to be symptomatic (Fig. 5–1). Symptoms of brain metastases may be the presenting complaint in patients not known to have cancer. In one series from a general hospital, 42% of patients with

autopsy-confirmed lung cancer presented with symptoms of a brain metastasis. Asymptomatic brain metastases affect as many as 11% of patients with newly discovered lung cancer,[1,2] and they are most common in patients with adenocarcinomas and those younger than 70 years.[3] Overall, approximately 10% of patients with lung cancer present with neurologic symptoms. Cerebral symptoms such as seizures, dementia, hemiparesis, aphasia, and headache are feared by both patient and physician[4] and often lead physicians to abandon treatment. With early diagnosis and vigorous treatment, however,

141

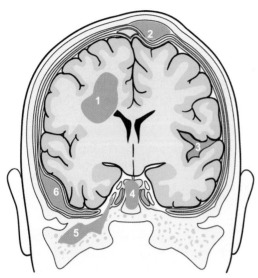

Figure 5–1. Cranial metastases. Most metastases affect the brain directly by hematogenous spread to the white matter of the cerebral hemispheres (1) (see also Figs. 5–2 and 1–6). The brain may be affected secondarily by a skull metastasis that invades the epidural space and compresses the brain. The skull metastasis may also compress the sagittal sinus (2) (Fig. 5–9). The tumor may involve the cranial leptomeninges and invade the brain by growing down the Virchow–Robin spaces (3) (Fig. 2–8B). A metastasis to the base of the skull may affect the pituitary gland (4) (Fig. 5–13) or cranial nerves (5) as they exit from the skull. Subdural metastases may cause effusions (6) that compress the brain (Fig. 5–10).

the symptoms can usually be reversed, often returning the patient to a useful life, at least for some time. Long-term survival, although not common, is seen in some patients.[5,6] In one series of 740 patients, the actuarial survival was 8.1% at 2 years, 4.8% at 3 years, and 2.4% at 5 years.[7] In another series of 1288 patients, 32 (2.5%) survived 5 years[8]; longer survivals have been reported.[9] Approximately 15% of patients with intracranial metastases from breast cancer survive more than 1 year.[10–12] Good pretreatment performance status (RPA class 1)[13] (see section on prognosis and Table 1–16) and primary tumor type (in one series, ovarian cancer had the best prognosis) are prognostic factors. Other good prognostic factors include response to steroid treatment, absence of extracranial metastases, normal serum lactic dehydrogenase, age less than 60 years, and fewer than three brain metastases.[14] In a series of surgically resected single brain metastases, significant prognostic factors included age less than 65 years, lack of extracranial metastasis,

control of primary tumor, histology, and treatment with stereotactic radiosurgery. Those in RPA class 1 had a median survival of over 21 months.[15] However, even considering prognostic factors, outcome is not predictable in a given patient; thus, all patients deserve consideration for aggressive treatment.

Certain tumors have a predilection to metastasize to the brain (e.g., lung and breast cancer and melanoma); others are uncommon or rare (e.g., myeloma,[16] nasopharyngeal carcinoma) but can spread directly to leptomeninges or cranial nerves (see Chapter 8).[17] Individual tumors are considered in the section that follows the general discussion of brain metastases.

The biology of brain metastases is discussed in Chapter 2.

Classification

Metastatic lesions may affect the cranium or any of the several intracranial structures (Fig. 5–2). The most important metastatic site, the brain parenchyma, may be affected in several different ways (Fig. 5–2). Table 5–1 classifies intracranial metastases by their anatomic sites.

Metastases to the skull are particularly common in patients with carcinoma of the breast and prostate. Skull metastases can be subdivided into those affecting the calvarium and those affecting the base of the skull. Calvarial metastases are usually asymptomatic (symptomatic calvarial metastases are discussed in this chapter, whereas those to the base, which often cause pain and cranial nerve dysfunction, are discussed in Chapter 8). Pachymeningeal (dural) metastases include those to the epidural space, usually by direct extension from skull metastases, and those to the subdural space, either by direct extension from an epidural metastasis or by hematogenous spread. The extensive neovascularization of dural metastases may cause subdural hematomas or effusions that are larger and more symptomatic than the tumor itself[18–20] and can occasionally obscure the metastases. Leptomeningeal metastases are discussed in Chapter 7.

Several other intracranial sites are sometimes affected by metastatic spread. Pituitary metastases are common, particularly in patients with breast cancer[21]; these usually involve the posterior lobe and are typically asymptomatic but may cause diabetes insipidus[21–23] or cranial nerve abnormalities by extension to the leptomeninges or the cavernous sinus (see Chapter 8). The pineal[24,25] and choroid plexus[26,27] are less common sites. Even rarer are metastases from a systemic primary cancer to an already extant intracranial brain tumor[28] such as a meningioma[29] (often breast cancer to a meningioma),[30] a glioma,[28] an acoustic neurinoma,[31] an ependymoma, a pituitary adenoma,[32] or even to a vascular malformation.[33,34] Brain metastases may arise in areas of cerebral damage from ischemic infarction,[35] or in chronic subdural hematoma membranes.[36] The metastasis may be the first symptom of a cancer or may appear many years after apparent cure of a tumor.

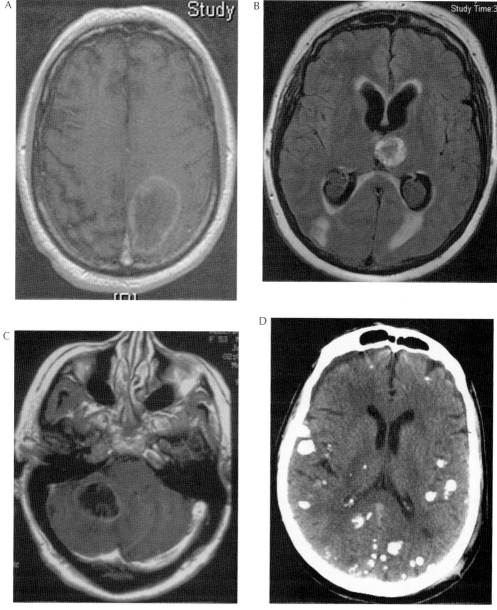

Figure 5–2. (*cont.*)

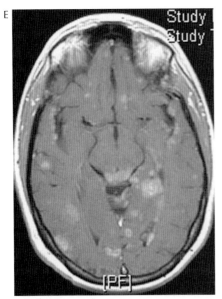
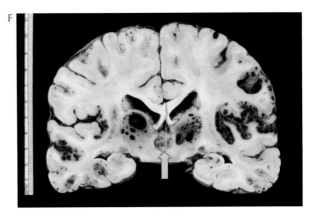

Figure 5–2. The wide range of brain metastases. **A**: A large single metastasis from non–small cell lung cancer (NSCLC). The lesion is surgically accessible; it has a necrotic center. There was little or no edema surrounding it. Compare this with Figure 3–4A, where a lesion slightly smaller causes massive edema (see discussion in Chapter 3). **B**: A large, surgically inaccessible lesion in a patient with small cell lung cancer. The lesion is solid and compresses the third ventricle, causing hydrocephalus. Note the hyperintensity surrounding the ventricles, suggesting transependymal absorption of cerebrospinal fluid. The patient required a shunt. **C**: A cerebellar metastasis from NSCLC. The lesion is cystic with an enhancing mural nodule of tumor, resembling a hemangioblastoma, a common primary cerebellar tumor. **D**: A computed tomogram showing multiple calcified metastases in a patient with adenocarcinoma of the lung. **E**: Multiple metastases in a patient with breast cancer. The lesions were numerous but the patient's only symptoms were slight confusion and an unsteady gait. **F**: A patient with metastatic melanoma. Note that unlike the previous patient where the lesions were mostly in the white matter, these lesions are in gray matter, both cortical and basal ganglia (and the pineal gland [*arrow*]). The gray matter distribution makes seizures more likely.

Incidence

The term *incidence* is used here in a colloquial sense to describe not only true incidence and prevalence, but also overall frequency in various studies. Autopsy studies find intracranial metastases in approximately 25% of patients who die of cancer. As shown in Table 5–2, the brain parenchyma is involved in approximately 15% of patients and is the exclusive site of intracranial metastases in approximately 10%.

One study of 15,000 autopsies (number of patients with systemic cancer not specified) found 237 patients (1.6%) with brain metastases[40]; single brain metastases were found in 42%. There was an equal likelihood that lung cancer and melanoma caused multiple metastases, although the exact number is not specified. Interestingly, the cerebellum was overrepresented, containing 24.5% of all metastatic tumors (the cerebellum represents ~15% of brain blood flow). Breast cancers and non–small cell lung cancers (NSCLC) were overrepresented in the cerebellum, findings not apparent in other studies where pelvic tumors usually have a greater propensity to metastasize to the posterior fossa.

As indicated in Chapter 1 (Table 1–3), the clinical incidence of brain metastases from lung cancer is approximately 20%, with other cancers occurring slightly less frequently. Brain metastases from solid tumors occur even less commonly in children (1.45% of 1100 patients).[41] The primary sources are different: in children sarcomas are the most common source.

Although there is anecdotal evidence to suggest that as systemic disease is better controlled the likelihood of developing a brain metastasis increases (HER-2 positive breast cancer responding to trastuzumab being an example),[42] one study identifying the yearly rates of detection of brain metastasis from breast

Table 5–1. **Classification of Intracranial Metastases by Anatomic Site**

Skull
   Calvarium
   Base (see Chapter 8)

Cerebral Meninges
   Epidural
      Extension from skull metastases
      Hematogenous to dura
         Compressing brain
         Compressing venous sinus(es)
Subdural
      Causing mass lesion
      Causing effusion
   Leptomeningeal (see Chapter 7)
      Focal
      Diffuse

Brain
   Cerebral hemisphere(s)
   Brainstem
   Cerebellum

Other Intracranial Sites
   Pituitary
   Pineal
   Choroid plexus
   Preexisting brain lesions (e.g., tumor, infarct, vascular malformation)

Table 5–2. **Incidence of Intracranial Metastases in Autopsy Studies**

	Posner and Chernik[37] (1978)	Takakura et al.[38] (1982)	Pickren et al.[39] (1983)
Total no. of autopsies	2375	3359	10,916
Site of metastases			
Intracranial	572 (24%)	860 (26%)	
Dural	467 (20%)	645 (19%)	
Leptomeningeal	184 (8%)	90 (3%)	
Brain	361 (15%)	555 (17%)	954 (8.7%)
Brain only	225 (9%)	–	–
Single*	106 (47%)**	–	378 (39%)
Solitary*			32 (0.3%)

* Single—one brain metastasis, primary tumor active, or other metastases present outside the brain.
Solitary—primary controlled, no other metastases.
** Percentage of total brain metastases that are single.

and lung cancer in Olmsted county Minnesota between January 1, 1988, and December 31, 2001, did not find an overall difference. Interestingly, women were twice as likely as men to have brain metastases once they developed lung cancer.[43]

BRAIN METASTASES (PARENCHYMAL)

Source

Table 5–3 lists the primary tumors from a group of 210 patients treated for brain metastases

Table 5–3. Primary Tumors in 210 Patients with Brain Metastases

Organ	Number	Histology
Lung	85	
Non, small cell		73
Small cell		12
Breast	39	
Melanoma	20	
Genitourinary	15	
Renal		7
Testis		4
Bladder		2
Prostate		2
Gastrointestinal	14	
Colon		8
Esophagus		4
Gastric		1
Pancreas		1
Gynecologic	10	
Ovary		3
Choriocarcinoma		3
Cervix		3
Endometrium		1
Sarcoma	7	
Unknown Primary	5	
Head and Neck	4	
Thyroid	4	
Miscellaneous	7	
Lymphoma		2
Thymoma		1
Neuroendocrine		2
Adrenal		1
Mediastinal germ cell		1

at Memorial Sloan-Kettering Cancer Center (MSKCC) during the first 10 months of 1994. Table 5–4 lists the histological types of tumor in 729 patients with brain metastases treated at the University of Minnesota between 1973 and 1993.[44] Both lists indicate that two-thirds of brain metastases are due to lung cancer, breast cancer, and melanoma. However, brain metastases can arise from any primary systemic cancer, including prostate and ovarian cancer, sarcomas (particularly in children) and Hodgkin disease, all of which rarely cause brain metastases.

Table 5–5 lists the likelihood of single or multiple (greater than one) metastases from different primary tumors, based on computed tomography (CT)[44,45] and autopsy data.[37,40] The figures from CT and autopsy data are about the same, suggesting that despite the ability of magnetic resonance imaging (MRI) to detect more and smaller lesions[46]; the CT figures are relatively accurate. About one-half of the patients have a single lesion and an additional 20% have only two; the clinical implication is that at least 70% of patients are potential candidates for focal therapy (see section on treatment).

Most brain metastases originate from cancers of the lung, breast, and melanoma.[40,44] Quite a few patients develop brain metastases from an unknown primary source.[47,48] In a series reported from general neurologic or neurosurgical services, the percentage of unknown primary tumors is greater than in series reported from a cancer hospital because of referral patterns. The primary site of such lesions, if eventually discovered, is usually the lung or, less often, other sites including the gastrointestinal (GI) tract or kidney.[47,48] Often the primary site is not found, even at autopsy.[47,49] A CT of the chest, abdomen, and pelvis and/or an FDG (2-fluorodeoxy-D-glucose) body PET (positron emission tomography)[50] will usually locate the primary site if it can be found.[51] If the brain metastasis is resected, immunohistochemical examination may help identify the primary site.[47,52] New molecular techniques are being developed and, may become useful in the future.[53,53a]

Most brain metastases, particularly those that arise from primary neoplasms other than lung cancer, occur late, when metastases are usually also present elsewhere in the body. The presence of more than one brain metastasis also suggests a broad dissemination of the primary tumor. In 201 patients with brain metastases treated

Table 5–4. **Median Survival Times for Primary Tumor Types in 729 Patients with Brain Metastases**

Primary Tumor Type	Total No.	% Single	Diagnosis to Metastasis (mos)	Metastasis to Death (mos)	Diagnosis to Death (mos)
NSCLC	178	50	3	4	10
Breast	121	49	40	4	53
SCLC	110	43	6	3	10
Melanoma	80	49	31	5	41
Renal cell	45	56	28	6	46
Gastrointestinal	45	67	14	3	22
Uterine/vulvar	38	53	23	3	27
Unknown	33	70	<1	7	8
Ovarian	14	57	23	8	32
Bladder	14	64	15	3	23
Prostate	11	82	22	3	31
Testicular	11	55	15	4	21
Miscellaneous	29	65	16	3	30
Total cohort	729	53	12	4	19

Modified from Ref. 44 with permission.
NSCLC, non–small cell lung carcinoma; SCLC, small cell lung carcinoma.

Table 5–5. **Number of Metastases and Primary Site of Tumors in 288 Patients with Brain Metastases**

No. of Metastases	No. of Patients (%)	Primary Site	Single %	Multiple % (Two or More)
1	141 (49)	Lung	46	54
2	60 (21)	Melanoma	41	59
3	38 (13)	Unknown	32	68
4	17 (6)	Breast	56	44
5+	32 (11)	Pelvis–abdomen	69	31

From Ref. 45 with permission.

at MSKCC,[54] only 38 (19%) were without evidence of other metastatic disease, and several of these developed clinical evidence of systemic metastases within a few weeks after the initial negative evaluation. In a large autopsy study of patients who had lung cancer, brain metastases were present in 25% but were solitary in only 3%.[55] Approximately 67% of patients with brain metastases from organs other than the lung also had pulmonary metastases; approximately 25% had bone metastases.[45]

At times, brain metastases may appear as isolated lesions years after "successful" treatment of the primary tumors. In patients with primary lung cancer, the brain may be the only site of metastasis.[56–58] Such lung lesions are more likely to be peripheral or apical rather than central, and the brain lesion is more likely to be single[3] than multiple, making both pulmonary and brain lesions more amenable to resection.[59] However, by the time of death, many patients have multiple metastases; only 20 of 100 patients who died of lung cancer were found to have a solitary brain metastasis.[3]

As a result of their late appearance when the systemic cancer has disseminated widely, brain metastases often are discovered only at autopsy, having been asymptomatic in 25% to 30% of patients.[54] MRI screening can detect asymptomatic metastases.[60] In one series of 809 patients with NSCLC, 120 presented with neurologic symptoms; another 61 (7.5%) patients had brain metastases that were asymptomatic.[60] Jacobs et al.[61] reported a 6% incidence of asymptomatic metastases detected by CT in patients scheduled for surgery for lung cancer. Salbeck et al.[62] found asymptomatic brain metastases on CT scans at the initial evaluation for lung cancer in 11% of 232 patients. Most often, these patients had limited small cell lung cancer (SCLC), stage III NSCLC, were greater than age 70, or had adenocarcinoma. Preoperative scans should be routine in patients at high risk for brain metastases in whom major surgery for the primary tumor is planned, because the presence of a silent brain metastasis may change the treatment; for example, the brain tumor may be treated first.

Pathophysiology of Neurologic Symptoms and Signs

Neurologic symptoms and signs can be classified as focal or generalized (Table 5–6). Focal signs may either identify the site of the brain metastasis or be false localizing[63] (see following text). Generalized signs are the result of increased intracranial pressure that either develops gradually and persists, causing sustained symptoms, or develops suddenly with transient increases (plateau waves) that can cause intermittent symptoms (Chapter 3). Increased intracranial pressure in and of itself commonly causes headache, nausea, vomiting, and lethargy but also can cause confusing symptoms such as tinnitus, stiff neck, paresthesias in the legs, and back pain, all of which can be relieved by lowering intracranial pressure. False-localizing signs result from distortion or compression of structures distant from the actual tumor.[63–71]

Neurologic symptoms and signs have multiple causes. Destruction or replacement of brain tissue by a metastatic tumor is unusual. The central necrosis often seen pathologically and on CT or MRI is because of tumor necrosis, not destruction of the underlying brain tissue. When a metastatic tumor is treated effectively by corticosteroids, radiation, or surgical extirpation, neurologic signs usually improve substantially or disappear altogether. However, when a substantial portion of brain tissue has been destroyed by the tumor, neurologic signs persist.

Displacement of brain tissue by the rapidly growing metastatic tumor is the common cause of focal neurologic symptoms. The growing mass compresses surrounding brain tissue, squeezes out blood and interstitial fluid, and distorts normal anatomy. In addition, hemorrhage, cyst formation, or necrosis within the tumor can

Table 5–6. **Presenting Symptoms and Signs of Brain Metastasis***

Symptoms	Percentage of 363 Patients	Signs	Percentage of 363 Patients
Headache	49	Impaired cognitive function	58
Focal weakness	30	Hemiparesis	59
Mental disturbances	32	Mild–moderate	27
Gait ataxia	21	Severe	31
Seizures	18	Hemisensory loss	21
Focal motor	4	Papilledema	20
Generalized	7	Gait ataxia	19
Other focal	7	Aphasia	18
Speech difficulty	12	Visual field cut	7
Visual disturbance	6	Limb ataxia	6
Sensory disturbance	6	Depressed level of consciousness	4
Limb ataxia	6		

Data from Refs. 54 and 72.

* Patients may present with more than one symptom or sign.

suddenly increase its size, rapidly compressing normal brain and causing acute focal symptoms.

"Irritation" by the tumor or the surrounding edema of the overlying gray matter probably accounts for the high incidence of focal seizures (Fig. 4–3).[73,74] Although the exact mechanism is unknown, it is likely that changes in the extracellular fluid composition caused by blood–brain barrier disruption are important. In patients who have melanoma that has invaded the gray matter directly, seizures are more frequent than they are with metastases from other primaries that remain confined to white matter.

Cerebral edema increases the size of the mass in the hemisphere, further compressing surrounding brain. Cerebral herniations (i.e., the shifting of cerebral structures from one compartment to another) may result from increased tissue pressure caused by the metastasis and its surrounding edema. Common sites and effects of herniation caused by supratentorial tumors (Fig. 1–6) are as follows:

1. Herniation of the cingulate gyrus under the falx compresses both anterior frontal lobes and leads to urgency incontinence and bilateral extensor plantar responses. The ipsilateral anterior cerebral artery also may be compressed, causing frontal lobe ischemia.

2. At the tentorium cerebelli, the uncus of the temporal lobe or the hippocampal gyrus, or both, herniate, displacing the brainstem toward the opposite side and compressing the opposite peduncle against the tentorium cerebelli (Kernohan notch), leading to an ipsilateral hemiparesis (Fig. 3–6C). The herniated uncus may also compress the third nerve and the posterior cerebral artery. Compression of the third nerve leads initially to ipsilateral pupillary dilatation and subsequently to total third nerve palsy. Compression of the posterior cerebral artery may cause ischemia or infarction of the ipsilateral occipital lobe, with a contralateral homonymous hemianopia (Fig. 3–6C). On rare occasions, both posterior cerebral arteries are compressed by massive herniation, causing bilateral occipital infarction and cortical blindness.

3. The enlarging supratentorial mass also shifts the diencephalon both contralaterally and caudally, resulting in progressive loss of consciousness, beginning with drowsiness and proceeding to stupor and finally to coma. Whether lateral or caudal shifts are primarily responsible for coma is disputed.[75] Lateral shifts also obstruct the third ventricle and foramen of Monro, leading to contralateral enlargement of the lateral ventricle(s) and further increasing tissue pressure in the supratentorial compartment. Rostral-caudal shifts compress the midbrain and pons, sometimes causing brainstem hemorrhage (Duret lesion).

4. Herniation of brain substance through the tentorial notch interferes with cerebrospinal fluid (CSF) absorptive pathways, leading to hydrocephalus. The same is true of tonsillar herniation through the foramen magnum.

5. A large supratentorial mass causes herniation of the cerebellar tonsils through the foramen magnum in about one-third of cases. Tonsillar herniation is even more common in patients with posterior fossa masses. Herniation of the cerebellar tonsils through the foramen magnum compresses the brainstem. When the compression is acute, it can cause apnea as a first sign, followed by loss of consciousness. Cough syncope may be a symptom of a cerebellar metastasis.[76] Tonsillar herniation probably causes the sudden death that (rarely) follows lumbar puncture in patients with brain tumors or meningitis.[75] A second form of herniation from posterior fossa tumors is that of the superior vermis of the cerebellum through the tentorial notch (upward herniation).[77–79] Upward herniation compresses the brainstem at the mesencephalic level, causing pupillary and eye movement abnormalities; it may interfere with subarachnoid absorptive pathways either by aqueductal stenosis (causing obstructive hydrocephalus) or by obliteration of perimesencephalic cisterns (causing communicating hydrocephalus). Although not a herniation, an enlarging cerebellar mass also pushes the brainstem forward, compressing it against the clivus anteriorly, sometimes obstructing the fourth ventricle, leading to hydrocephalus.

Compression of venous structures by the tumor mass leads to increased resistance to CSF absorption and elevated intracranial pressure.[80,81] This rarely causes hydrocephalus, but is often associated with intermittent vasoparalysis, leading to plateau waves[82,83] (Chapter 3).

Pathology

Brain metastases are generally well demarcated from surrounding normal but edematous brain. However, metastases do have the capacity to invade normal brain often along perivascular spaces, a phenomenon most commonly seen with metastatic lymphoma, SCLC and melanoma. One histologic study of 76 patients with brain metastases examined at autopsy found infiltration beyond the border of the metastasis in 48.[84] This included infiltration present in 70% of NSCLC and 83% of melanoma brain metastases. The maximum depth of infiltration was less than a millimeter in most patients save those with SCLC and melanoma. In another study, 26 of 66 metastatic anaplastic small cell carcinomas had poorly defined borders and a diffuse pattern of invasion. The authors referred to this as a *pseudogliomatous* growth pattern.[85] The clinical implication is that radiosurgical treatment should treat a margin beyond the MRI-defined border of the metastasis. However, metastatic tumors rarely invade brain as extensively as do primary tumors. Accordingly, unlike primary brain tumors, total resection of metastatic tumors is often possible.

In experimental animals, metastatic tumors that do not express vascular endothelial growth factor (VEGF) invade brain by co-opting the normal vasculature to allow growth. The co-opted blood vessels have a normal appearance and thus, an intact blood–brain barrier. In those tumors that express VEGF, the tumor grows by both angiogenesis and invasion, and the barrier is disrupted.[86] Such co-option of normal vessels by metastatic tumors raises the question of whether angiogenesis-inhibiting factors are likely to be effective in treating those brain metastases.

Clinical Findings

SYMPTOMS AND SIGNS

Presenting symptoms of brain metastases are summarized in Table 5–6. The percentages total more than 100 because many patients present with more than one symptom or sign.

In these as well as other series,[87–89] headache is the most common symptom, occurring in slightly less than one-half of patients. Focal weakness and impaired cognitive function are the most common signs, each occurring in slightly more than one-half of patients.

Headache
Headache is a common symptom, both in individuals who are otherwise well and in patients with cancer; the presence of headache in a cancer patient does not necessarily signify that the patient has a brain metastasis.[90] In a prospective study of 54 cancer patients with newly appearing headache or a change in the pattern of a preexisting headache syndrome, only 29 had brain metastases as the cause. In our own series of 97 patients in whom headache prompted a neurologic consultation, only 20 had intracranial metastases as the cause.[91] Are there clues that can lead the physician to suspect a brain tumor in a patient, either with or without cancer, who complains of headache? First consider individuals not known to have cancer.

In one study of 44,000 adult non-trauma patients evaluated in an emergency room, 4.5% had a primary complaint of headache. One hundred thirty-nine were hospitalized and 329 were discharged from the emergency department. Intracranial pathologic findings were subsequently found in 10% of the hospitalized patients and in 1% of the non-hospitalized patients. There were only four tumors, two in 139 hospitalized patients, and two in 329 non-hospitalized patients. The tumors were two meningiomas, one glioma, and one metastatic melanoma.[92] The authors concluded that patients older than 55 years with an acute onset of headache, and with an occipital, nuchal location of headache, as well as associated neurologic symptoms, required evaluation for structural disease.

In a study of 100 consecutive patients referred to a neurologist with headache of recent onset, that is, the last 12 months, or a change in character of a previous headache, 80 had a normal neurologic examination. An intracranial tumor was found in 21 patients, 13 of whom had a normal neurologic examination. The tumors included glioblastoma (4), anaplastic astrocytoma (4), low-grade astrocytoma (1), oligodendroglioma (1), metastasis (4), meningioma (4), vestibular schwannoma (2), and pituitary adenoma (1). Three of the four glioblastoma patients had normal neurologic examinations, as did three of the four with metastases.[93]

A study of 111 patients who underwent imaging to evaluate headache disclosed abnormal findings in only 39. The only statistically significant clinical findings predicting a structural lesion were papilledema, cognitive impairment, and paralysis.[94]

What about headache in the cancer patient? As indicated in the preceding text, headache as a presenting complaint, which may be focal or generalized, occurs in only 50% or fewer patients with brain metastases.[90,95–98] When the headache is focal, it has localizing value. Simonescu[99] reports that, in 72% of patients, the headache was localized at the site of the metastasis, but in most brain tumor patients examined at MSKCC, the headache has been diffuse or bilateral in the frontal or occipital regions, and thus without localizing value. The classic *brain tumor headache*, which occurs in only one-quarter to one-third of patients,[96,100] is mild at onset, begins when the patient awakens in the morning, disappears shortly after he or she arises, and recurs the following morning. The headaches gradually increase in frequency, duration, and severity until, in their later stages, they are almost constant and may be associated with other signs of increased intracranial pressure, including drowsiness, nausea, and vomiting. The headache may worsen with changes in intracranial pressure such as those precipitated by Valsalva maneuver, head shaking, or straightening from a bent position.

Headache as an isolated symptom is more common in patients with multiple metastases than in those with single lesions. It also is more common in patients with cerebellar metastases than in those with cerebral hemisphere metastases.[88] Rarely, patients with large frontal lesions will report severe and recurrent headache without other signs or symptoms, but most brain tumor patients with headache have other symptoms or signs. An otherwise occult brain tumor may occasionally present as a typical common or classic migraine.[101]

In one series, intracranial metastases were predicted by headaches of less than 10 weeks duration complicated by vomiting, and headaches not characteristic of tension-type headaches.[97] In another series, significant predictors of brain metastasis included (1) bilateral frontotemporal headaches more pronounced on the side of the metastasis, with a duration of more than 8 weeks and a pulsating quality of moderate-to-severe intensity, (2) vomiting, (3) gait instability, and (4) Babinski sign.

To complicate matters, brain tumor headache may mimic a variety of headache types not usually associated with brain tumors.[88] These include migraine with aura, cluster headaches, and several other types of headaches classified as trigeminal autonomic cephalgias (TAC).[88,102] These unusual headaches are more likely to occur with intracranial tumors involving cranial nerves (see Chapter 8). Furthermore, brain tumors may only increase a patient's preexisting chronic headaches.

The brain itself does not have pain receptors. Thus, brain tumor headaches result when the mass lesion causes traction on pain-sensitive intracranial structures such as the dura at the base of the brain, the cranial nerves, and the large venous sinuses. Such headaches imply an increase in intracranial pressure, either focal or diffuse. The clinical implication is that many patients, particularly older patients with atrophic brains that can accommodate the tumor mass without increasing pressure, are less likely to suffer headache. Nevertheless, one does encounter patients with small tumors who complain of headache that resolves after the tumor is treated. In these patients, there is no clear shift of structures and no raised intracranial pressure, so it is hard to understand their exact pathogenesis.

Because headaches caused by a metastasis are usually accompanied by increased intracranial pressure, one might expect a high frequency of papilledema in such patients. The contrary is true.[103] In the series reported by Young et al.,[72] 53% of patients reported headaches but only 26% had papilledema. In patients with cerebellar tumors, headaches were present in 70% and papilledema in only 25%. In another series, 10 of 29 patients (34%) with brain metastases had papilledema; one of 4 patients without brain metastasis was also reported to have papilledema.[90] In short, almost 50% of the patients with brain metastases did not have headache as a presenting complaint, and very few patients had papilledema. The absence of these two textbook hallmarks of brain tumor does not exclude a symptomatic brain metastasis.

Overall, it is impossible to predict in any given patient with cancer whether headache heralds the presence of brain metastases. All patients with a new headache or worsening previous headaches require careful neurologic examination and imaging.

Focal Weakness

Focal weakness is the second-ranking complaint reported in most series. The weakness may range from a gradually developing mild monoparesis or hemiparesis to acute hemiplegia. Focal weakness usually localizes the tumor to the contralateral cerebral hemisphere accurately, but because the effects of the tumor and its surrounding edema are so widespread, it does not necessarily localize the lesion to the motor area of that hemisphere. If only one extremity is weak, the tumor is more likely to be located in the white matter directly below the cortical area subserving that function. Motor disturbances are found more often than sensory disturbances even though more metastatic tumors are in the postcentral than in the precentral gyrus.

Cognitive/Behavioral Impairment

Cognitive/behavioral impairment is the initial complaint in approximately 30% of patients. A patient or family member reports a change in personality or memory. Mental and behavioral changes may be a result of either focal disturbances of brain function or more generalized cerebral abnormalities. Focal abnormalities of mental function include such disorders as aphasia, alexia, acalculia, agnosia, apraxia, or amnesia.[72] When present in pure form, these disorders have localizing value. Dressing apraxia that results from a right parietal lesion is a surprisingly common presenting complaint. One must take care to distinguish these focal neurologic disorders from confusion, the latter caused by more generalized brain dysfunction. It is particularly common for the physician to believe that a patient with aphasia is confused, when in fact all cognitive functions save language may be intact.

Affective changes,[104,105] usually depression, can also be the presenting complaints. These affective changes have no localizing value,[105] but may delay accurate diagnosis while the patient is treated for depression.

Much more common are mental and behavioral changes that result either from multiple cerebral metastases or from strategically placed lesions that cause increased intracranial pressure. In the early stages, patients typically are awake and alert, although they may find that they sleep more than usual and take an unaccustomed afternoon nap. They complain of difficulty with memory and concentration, or the family reports increased apathy, withdrawal, irritability, and forgetfulness. Occasionally, business colleagues complain that the patient's judgment has become poor. Despite the patient's denial of being depressed, psychiatric consultation is often sought before imaging is procured.

Although mental changes may occur in the absence of any other neurologic symptom, headache is a commonly associated symptom because both symptoms are related to increased intracranial pressure. If a diagnosis is not made and the patient is not treated, symptoms usually progress and intensify, with loss of alertness, increasing drowsiness, increasing memory impairment, and finally, loss of orientation (first for time, and later for place and person).

Seizures

Seizures, whether focal or generalized, are common; they can occur as the presenting complaint or develop later in a patient with known brain metastases.[106] In one recent series, a seizure was the presenting symptom in 7 of 35 patients (20%) with brain metastases[107]; 3 were not even known to have cancer. Five patients later developed seizures, making the total incidence of seizures 34%. In a study of 470 patients with brain metastases, 113 (24%) had seizures.[73] The likelihood of having a seizure varied with the nature of the primary tumor; seizures occurred in 67% of patients with melanoma (because so many melanoma metastases are cortical as well as subcortical and are often hemorrhagic), 29% of patients with lung cancer, and 16% of patients with breast cancer. In our patients (Table 5–6), 11 of 18 had seizures with a clear focal onset. In the other seven, the seizure was generalized; however, all generalized seizures in a patient with brain metastases must have a focal onset that may be silent or erased from the patient's memory by ictally induced amnesia.

Seizures are more likely to occur in patients with multiple metastases and in patients with combined brain and leptomeningeal metastases. The seizures that are clearly focal in onset have localizing value; rarely are they false-localizing.[64] Although seizures are common, status epilepticus is rare, but that disorder, although controllable, can be devastating in patients with brain metastases in whom it predicts early mortality.[108] Nonconvulsive status epilepticus can be the presenting complaint in patients with brain metastases.[109] Seizures, including nonconvulsive status epilepticus, can lead to reversible MRI abnormalities that can be confused with brain metastases. The enhancement, unlike most brain metastases, is usually cortical or leptomeningeal,[110] but transient corpus callosum lesions have also been reported.[111] The treatment of seizures and status epilepticus, as well as the use of prophylactic anticonvulsants, was discussed in Chapter 4.

The characteristic seizure is a focal motor seizure beginning with clonic jerking of the face or of one extremity. The seizure may progress to involve the other extremity on the same side (Jacksonian march) or may generalize to a grand mal convulsion. Following the seizure, the patient often suffers Todd paralysis of the extremities. The paralysis is usually transient, with recovery ranging from several minutes to hours, but on occasion, single or repetitive seizures lead to permanent paralysis of the involved extremities. Todd paralysis occurs more frequently in patients with brain metastases than in those with focal seizures without tumor; it probably results when seizures increase blood flow and blood volume, increasing edema, worsening previously elevated intracranial pressure and leading to focal ischemia and sometimes to infarction. Alternate explanations[112] include "neuronal exhaustion" from substrate depletion, inhibitory neuronal discharges, arteriovenous shunting causing ischemia,[113] or release of endogenous inhibitory substances. Postictal diffusion-weighted MRIs show transient, focally increased signal intensity, with a decreased apparent diffusion coefficient of cortical gray and subcortical white matter, suggesting that both cytotoxic and vasogenic edema may be a consequence of prolonged ictal activity.[114] Seizures sometimes cause transtentorial herniation.

Episodes of transient paralysis, speech arrest, or other loss of neurologic function may mimic transient ischemic attacks or complicated migraine, but probably represent either nonconvulsive seizures or plateau waves.[115]

The pathogenesis of seizures in patients with brain or leptomeningeal metastases (tumor-associated epilepsy)[116] is not fully understood, but is probably multifactorial (Fig. 4–3). The tumor itself cannot be responsible for the seizures; it must be the effect of the tumor on the surrounding brain tissue. Figure 4–3 illustrates possible causative and influencing mechanisms of the seizures. Among these changes, hypoxia and pH alterations related to tumor neovascularization seem particularly important.[117] In brain metastases, many of these changes are present in the peritumoral edema, which may explain why corticosteroids seem to decrease the frequency of seizures.

Other Symptoms

Other symptoms include *gait ataxia*, a prominent presenting complaint when the tumor involves the cerebellum or brainstem. Ataxia occasionally also occurs as a result of a large frontal lobe metastasis or hydrocephalus caused by obstruction of CSF pathways.[118] Ataxia from hydrocephalus can be relieved by shunting of the ventricular system. Rarely, parietal lesions cause a particular form of unilateral ataxia called optic ataxia, in which visually guided limb movements are poorly performed.[119,120]

Aphasia was a prominent presenting complaint in approximately 10% of patients in the series reported by Young et al.,[72] but was an isolated complaint in only 1% of Paillas and Pellet's[121] patients, and was seen in 4% of Hildebrand's[122] patients. Aphasia is an excellent localizing symptom when it occurs in the absence of other cognitive changes, but it may be confused with non-aphasic naming difficulties, a common symptom of more generalized brain dysfunction. Aphasia, particularly that resulting from parietotemporal lesions, where there may be no paralysis, is often labeled confusion by relatives and even some physicians. The physician should carefully inquire as to the nature of the "confusion," making sure that it is not a language difficulty rather than a more diffuse cognitive failure.

Sensory changes are an uncommon complaint in all series, despite the fact that the parietal lobe, the site of the sensory cortex, is involved in many patients. *Visual abnormalities*, usually a hemianopia, were a prominent presenting complaint in one surgical series.[123] Some patients with hemianopia are aware that they have visual difficulty, although they cannot specify its exact nature. Other patients are totally unaware of any visual deficit until, for example, they are involved in an automobile accident because they failed to see a car or other object in their hemianopic field.

Presenting Signs

In most series, signs and symptoms are lumped together. Young et al.,[72] and Cairncross et al.,[54] however, separately described the presenting signs and symptoms of brain metastases in 363 patients (Table 5–6). The major finding was that careful neurologic examination often revealed signs of which the patient was unaware. For example, although only 108 (30%) of the patients complained of focal weakness, examination revealed evidence of focal weakness in 214 (>50%).

More striking was the high incidence of impaired cognitive or behavioral function in patients at the time of diagnosis. Young et al.[72] found that 75% of the patients studied could not perform normally on the standard tests of mental status. This figure is higher than reported in most other series, but does suggest that careful testing of patients suspected of brain metastases often reveals subtle and unexpected changes in cognitive function. Paillas and Pellet[121] reported mental disturbances at diagnosis in 24% of their patients, and generalized asthenia, including clouding of consciousness and torpor, in an additional 24%. Elkington[124] reported mental changes in 75% of patients. Part of the discrepancy among these series may be attributed to unclear and inconsistent definitions of mental disturbances.

When usually safe doses of opioids or sedative agents provoke delirium or impaired cognition, the presence of previously asymptomatic brain metastases may be suspected. Early hydrocephalus from obstruction of the fourth ventricle may cause cognitive or behavioral changes mimicking depression.

Although neurologic findings on examination are often more prominent than those noted by the patient, ataxia in cerebellar tumors is an exception. Patient complaints of gait unsteadiness are often strikingly out of proportion to the little found on examination. Patients with a single metastasis in the cerebellum usually present with gait ataxia with or without limb ataxia. Headache, dizziness, diplopia, and vomiting are initial symptoms in fewer than 50% of patients

with cerebellar metastases, as are papilledema and nystagmus.[103,125] Vomiting with or without nausea is an occasional presentation of a posterior fossa metastasis. Particularly suggestive is vomiting on first awakening in the morning. If the patient is known to have disseminated cancer, especially in the liver or abdomen, the physician may fail to consider a brain lesion as a potential cause.

Unusual and perplexing focal signs are associated with tumors in unusual places. Examples are bilateral arm weakness sparing the legs (*man-in-a-barrel* syndrome) associated with symmetric tumors of the arm area of the motor strip,[126] hemichorea associated with basal ganglia tumors,[127] symptoms of a diffuse encephalopathy associated with miliary metastases to the brain,[128,129] cardiac arrhythmias leading to syncope,[130] and the syndrome of pseudobulbar palsy from bilateral, symmetric lesions of the frontal operculum. Brainstem metastases can also cause perplexing symptoms, including gaze and cranial nerve palsies, ptosis, and ataxia.[131–133] Other unusual signs of posterior fossa tumors include hypertension,[134] orthostatic hypotension,[135] hiccups,[136] achalasia,[137] night terrors,[138] inability to sneeze,[139] and neurogenic hyperventilation.[140] Because such brainstem metastases may be quite small, they may not be detected on CT but are invariably detected by MRI.

Onset of Symptoms

The signs and symptoms of a brain metastasis usually begin insidiously and evolve over a period of days or a few weeks or, more rarely, over a period of weeks or months. Sometimes the onset of the neurologic symptoms is sudden, occasionally because of hemorrhage into a tumor,[141,142] although the acute worsening of preexisting, slowly evolving symptoms is the more common presentation of intratumoral hemorrhage. Paillas and Pellet[121] report an acute onset of symptoms in 47% (83) of 178 cases: 19% (34) had seizures, 18% (32) had an acute stroke-like syndrome with hemiplegia, and the remainder had other acutely developing neurologic signs such as sensory loss, aphasia, and dementia. Van Eck et al.[143] indicate that 30 of their 104 patients with brain metastases had a "more or less acute onset of symptoms" but do not specify further. Our own experience has been that if seizures are excluded (15%), acute or apoplectic onset of neurologic symptoms occurs in fewer than 10% of patients. Metastatic tumor can mimic transient ischemic attacks, stroke-in-evolution, completed stroke, or multi-infarct dementia.[144]

Paillas and Pellet[121] report a three-stage course: The first stage is an acute onset of neurologic symptoms, which they believe represents a tumor embolus occluding a blood vessel.[145] Next is a phase of improvement that may last several weeks to months, followed by recurrent and progressive neurologic symptoms. The phase of improvement reflects the return of function in an ischemic but not infarcted area; the secondary increase in neurologic symptoms represents the growth of tumor from the tumor embolus. We have encountered such events rarely in patients at MSKCC. One of our patients with metastatic thyroid cancer developed an acute left hemiparesis resulting from a cerebral infarct. She recovered fully but 6 months later the hemiparesis reappeared and a metastasis was found at the site of prior infarct. Although this could have been an instance of a metastatic tumor arising in an area of ischemia unrelated to cancer,[35] it seems more likely that the original cerebral infarct was caused by a tumor embolus that subsequently extravasated and grew.

Metastases that present with apoplectic onset, the three-stage course, or transient attacks have all been grouped together under the term *pseudovascular syndromes*.[146] The clinical importance of these unusual modes of onset is that they are usually mistaken for stroke rather than tumor.

The pathogenesis of gradually developing symptoms is that of a slowly increasing, space-occupying mass with surrounding edema. The pathogenesis of the acutely developing symptoms other than seizures is more perplexing. In some instances, acute onset results from a hemorrhage into a previously silent metastatic tumor. This mode of onset is particularly common with choriocarcinomas and with malignant melanomas, but hemorrhage into a brain metastasis can occur with any primary tumor and is most common in metastases from lung cancer[142] because of the large number of brain metastases from lung cancer. Hemorrhage is suggested clinically by the acute onset of symptoms and can be confirmed by CT or MRI (gradient-ECHO sequences). In many instances, however, patients with brain metastasis suffer the apoplectic onset of hemiplegia or other neurologic signs without clinical or pathologic

evidence of hemorrhage. In these cases, the pathogenesis is not certain. In some instances, tumor may compromise arterial circulation, causing a cerebral infarct.

More perplexing are patients with brain metastases who suffer episodic neurologic symptoms often lasting a few hours, after which the symptoms clear.[82,115,147] These symptoms cause language or motor dysfunction, but can be confusing if they cause behavioral abnormalities, vaguely described episodes of dizziness, or difficulties in comprehending a situation or the environment. Several explanations are possible: (1) They may represent partial seizures originating in the limbic system. Although partial seizures (nonconvulsive status epilepticus) can last for a long period, they are usually shorter than the 30- to 90-minute episodes seen in patients with brain tumors. (2) The episodes may result from transient ischemia, although they last longer than the few minutes of most transient ischemic attacks. Furthermore, most transient ischemic attacks cause motor or sensory abnormalities rather than behavioral disturbances. (3) The episodic symptoms may result from plateau waves.[82,147] Most are probably seizures and should be treated as such.

Laboratory Findings

MAGNETIC RESONANCE IMAGING

An MRI is the best, and often the only, diagnostic test necessary for brain metastases (Figs. 5–3 and 5–4). A carefully done contrast-enhanced MRI, if negative, effectively rules out brain metastasis, except, at times, for miliary cortical metastases (see Chapter 7). The MRI detects lesions not seen on CT.[46,148] A standard 1.5 Tesla MRI using a standard dose of gadolinium (0.1 mmol/kg) is almost always sufficient for diagnostic purposes. However, increasing the dose of contrast and using a 3 Tesla magnet will identify some lesions smaller than 5 mm.[149] Post-contrast FLAIR (fluid attenuated inversion recovery) images are sometimes helpful in evaluating possible metastases.[150] MRI is also preferable to CT because the contrast material is safer; however, CT may be useful in identifying hemorrhage and calcium in metastases, as well as identifying bony lesions at the base of the skull.

Because almost any single symptom or constellation of neurologic symptoms can be

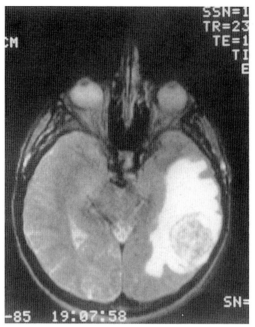

Figure 5–3. A T2-weighted magnetic resonance image of a patient with metastatic malignant melanoma. The tumor is visible as a heterogeneous mass surrounded by hyperintense edema in the white matter. Notice how the finger-like projections of the edema invade the white matter and spare the cortex of the temporal lobe.

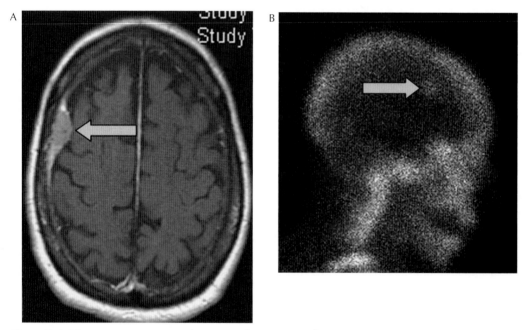

Figure 5–4. Meningioma in a patient with metastatic breast cancer. The patient was asymptomatic, but magnetic resonance imaging (**A**) revealed a dural-based lesion and multiple skull metastases (*arrow*). While the patient was being prepared for whole-brain radiation therapy, a neurologic consultant, believing that the lesion looked suspiciously benign, discovered a 3-year-old bone scan (**B**) that showed an identical lesion (*arrow*). The plan for radiation therapy was aborted and the patient remained asymptomatic.

caused by metastatic brain tumors, including no symptoms, an MRI should be performed on any patient with cancer who has an unexplained alteration in brain function, as well as in those patients who are at high risk for brain metastases and are about to undergo potentially curative systemic treatment. Examples include patients with lung cancer who are being considered for surgical resection of the primary tumor, and patients with metastatic melanoma or SCLC who are about to undergo systemic chemotherapy or immunotherapy.

The standard MRI includes T1, T2, and FLAIR images performed before the injection of contrast material (Fig. 5–3) (if one fails to do precontrast T1 images, one might mistake hemorrhage, apparent by hyperintensity on the precontrast T1 scan, as enhancement suggesting tumor). The unenhanced T1-weighted image often fails to show a metastatic lesion unless it is hemorrhagic, in which case the intensity is increased. Calcified lesions may be associated with areas of absent signal. The T2 and FLAIR images are almost always positive, revealing an area or areas of increased intensity in the white matter, usually encompassing both the tumor and the surrounding edema. Occasionally, the intensity of the tumor is less than that of the surrounding edema, permitting easy identification and localization (see Fig. 5–3). Almost all metastatic tumors are hyperintense on T1-weighted images after the injection of contrast (gadolinium), allowing for clear identification of the size, shape, and location of the metastasis. An enhanced brain metastasis usually is spherical and well demarcated from surrounding brain. The center is often devoid of contrast but surrounded by a rim of enhancement. If the images are performed in more than one plane, the tumor can be exquisitely localized by the area of enhancement, which is especially useful if focal therapy is contemplated. Acute cerebral infarcts are often identified by being increased on diffusion-weighted images (DWIs) with a decreased apparent diffusion coefficient. Similar changes can be seen in some brain metastases, although the brain metastases almost always contrast enhance.[151] Gradient echo sequences are best at identifying hemorrhage. Magnetic resonance spectroscopy (MRS) may help differentiate brain metastases from primary brain tumors.[152] Although elevated choline levels are found in both primary brain tumors and metastases, peritumoral areas show elevated choline levels only in primary tumors; myoinositol levels are present within the enhancing area of gliomas, but not in metastases.

The MRI is also useful in following the course of brain metastases after therapy. However, one must interpret the scans with great caution. Contrast enhancement first appearing 96 hours following surgical removal of a metastasis may be the result of the surgery rather than representing residual tumor.[153] No change or even a transient enlargement of the metastasis following radiosurgery is relatively common and does not necessarily represent treatment failure.[154] Apparent worsening in a few weeks following whole-brain radiation also must be interpreted with caution as it may represent *pseudoprogression*.[154a]

BIOPSY

Stereotactic needle biopsy is a relatively simple diagnostic procedure that may be performed under local anesthesia with a small scalp incision and requires only 1 day of hospitalization, yet a definitive diagnosis can be made more than 90% of the time.[155] A small sample size sometimes prevents an adequate diagnosis, although this usually is not a major issue when the question is one of metastasis versus non-metastasis. More important, there is a small risk of hemorrhage with worsening neurologic signs; neurologic worsening may also result from cerebral edema caused by trauma from the biopsy needle. Brain edema sometimes increases after the biopsy, leading to neurologic worsening even when corticosteroid doses are increased.

The major indication for stereotactic needle biopsy in patients with cancer is to distinguish among metastases, primary brain tumor, and non-neoplastic (usually inflammatory) lesions. Because most neoplastic lesions are better treated by surgical removal, stereotactic biopsy is indicated only when the physician suspects either a non-neoplastic lesion that would not benefit by direct surgical attack or a primary central nervous system (CNS) lymphoma.

OTHER LABORATORY TESTS

If an MRI cannot be performed (cardiac pacemaker, recent implantation of expanders for breast reconstruction), a CT should take place both before and after the injection of contrast (iodine). If a brain metastasis is strongly suspected, consider giving 5 to 10 mg of diazepam 30 minutes before the contrast injection to prevent a seizure induced by the contrast material[156] (see Chapter 4). Most metastatic lesions of the brain are hypodense or isodense on the precontrast CT except for melanoma, choriocarcinoma, colon cancer, and leukemia,[157] which are often hyperdense. Virtually all metastatic tumors increase their density after the injection of contrast. A CT without contrast can identify those rare metastases that calcify[158] (Fig. 5–2D).

If the diagnosis cannot be made with reasonable certainty by CT or MRI, no other test short of biopsy (or CSF evaluation for malignant cells when coexisting leptomeningeal tumor is present) can establish the diagnosis. FDG PET does not add to MRI in the diagnosis of brain metastasis,[159–161] but can help distinguish metastatic from primary tumors. Brain metastases are more hypermetabolic than gliomas; peritumoral hypermetabolism is found in gliomas but not metastases.[162]

A few reports suggest that an elevated plasma carcinoembryonic antigen (CEA) level may be an early sign of brain metastasis from lung or breast cancer, and that an increased CSF/serum ratio of beta human chorionic gonadotropin (βhCG) may be the first sign of brain metastasis from choriocarcinoma. Even with choriocarcinoma, however, MRI is more sensitive than βhCG. A caveat: One patient with a meningioma and a history of breast cancer developed an elevated plasma CEA caused by the meningioma, not by recurrent or metastatic breast cancer.[163]

Differential Diagnosis

Table 5–7 lists the many lesions that may mimic brain metastases. Usually, the clinical history,

Table 5–7. Brain Metastasis: Differential Diagnosis

Primary brain tumors
  Meningioma
  Glioma
  Pituitary adenoma
  Vestibular schwannoma
Vascular disease
  Cerebral hemorrhage
    Intratumoral (metastasis or primary brain tumor)
    Vascular anomaly
  Cerebral embolus
    Nonbacterial thrombotic endocarditis
    Tumor
    Cardiac thrombosis
  Cerebral thrombosis
    Arterial
    Venous
Infections
  Abscess—especially Hodgkin disease and lymphoma
  Viral infection—progressive multifocal leukoencephalopathy (PML)
Side effects of therapy
  Methotrexate leukoencephalopathy
  Radionecrosis
Other disorders
  Related to cancer—"paraneoplastic syndromes"
  Not related to cancer (e.g., multiple sclerosis)

combined with the results of an MRI, establishes the diagnosis of brain metastasis with reasonable certainty. However, a definitive diagnosis of metastatic brain tumor cannot be made on scan results alone; absolute certainty requires biopsy. Stereotactic needle biopsy establishes the diagnosis in 92% of patients, with a complication rate of 2.9% and a mortality rate of 1.2%.[155] The biopsy can also be done as part of an open craniotomy to remove the lesion. Even a typical MRI only suggests, but does not prove, that the lesion is a brain metastasis rather than a primary brain tumor or other lesion.[164] Multiple lesions increase the likelihood of brain metastasis, and a known systemic cancer makes the diagnosis even more likely. In one study, however, 6 of 54 patients with known cancer and a single brain lesion did not have a metastasis on biopsy; three did not even have a neoplastic lesion.[164]

Recognizing that the diagnosis by scan is never certain, particular clues suggest brain metastasis rather than another brain lesion. Brain metastases tend to be spherical, but primary gliomas are irregular, with finger-like extensions of contrast-enhancing tumor running along white matter bundles and fiber tracts. Primary brain lymphomas are often multiple, periventricular in distribution, uniform rather than ring enhanced, and have irregular margins. Meningiomas are distinguished from brain metastases because of their usual homogeneous contrast enhancement, relative lack of peritumoral edema, and their attachment to the dura of the calvarium, falx, or the base of the brain (Fig. 5–4). Metastatic breast, prostate, myeloma, and other cancers also arise from the calvarium or falx (dural metastasis) and can be supplied by branches of the external carotid artery, making the distinction between dural metastasis and meningioma impossible except by surgery.[165,166] Meningiomas are benign and potentially curable tumors that are reported to occur more frequently in women with breast carcinoma than in women without this malignancy.[167,168] Therefore, if the neurologic symptoms have developed very slowly or if the scan suggests a lesion abutting the falx or the inner table of the skull, serious consideration must be given to a meningioma. To further complicate the issue, breast cancer may metastasize to a meningioma.[167]

Other primary brain tumors, such as acoustic neurinomas or pituitary adenomas, are at times impossible to distinguish from metastases in the same area. Cerebral infarction may be confused with a brain metastasis, but the problem does not arise if the patient is seen during the acute stage because infarcts do not enhance; in fact, only DWI MRI may be abnormal in the first hours after an acute cerebral infarct. If the scan is delayed several days, however, contrast enhancement develops and may become intense between 10 days and 3 weeks after the stroke. Contrast enhancement of an infarct characteristically involves the pial surface of the overlying brain and has a serpiginous outline corresponding to the cortical gyri, unlike the usual spherical, ring-like contrast enhancement of a brain metastasis. After several weeks, the contrast enhancement in an infarct diminishes and finally disappears; the area of infarction changes to a low-density, fluid-filled cavity.

An acute cerebral hemorrhage is hyperdense on the non-contrast CT and hypointense on gradient echo MRI.[169,170] If the hemorrhage ruptures into the ventricles, blood can be identified in the ventricular cavities or in the subarachnoid space. Contrast enhancement is not seen early but may appear 3 to 6 weeks after the hemorrhage, at which time the clot is isodense and a ring of contrast enhancement develops, resembling a metastasis or abscess. Early enhancement or early extensive surrounding edema suggests that the hemorrhage has occurred in an underlying metastasis. In one study from a tertiary referral center, 5 of 93 metastatic brain tumors were initially misdiagnosed as stroke, despite CT or MRI.[171] The usual error is not to give contrast or to attribute the edema to the hematoma, but edema takes at least 24 hours to develop after acute hemorrhage. Cerebral hemorrhage is the most common wrong diagnosis, although incorrect diagnoses of cerebral infarct[171] or hemorrhagic infarct[172] have been described. In occasional patients, a *spontaneous* cerebral hemorrhage is removed and small amounts of tumor found in the resected clot.[173] In one series of 31 spontaneous cerebral hemorrhages examined histologically, there were two instances of metastatic cancer (both NSCLC); neither patient was known to have cancer at the time of the hemorrhage.[174] In one of our patients, who almost certainly had a hemorrhagic metastatic melanoma, no residual tumor was found on histologic examination of the blood clot.

Cerebral abscess cannot always be distinguished from brain metastasis by scan alone. An important clue is that cerebral abscesses have restricted diffusion (hyperintense on DWI with low ADC [apparent diffusion coefficient] values), whereas other cystic lesions such as metastases do not have restricted diffusion.[175] Cerebral abscesses usually occur in immunosuppressed patients (see Chapter 10) who suffer from conditions in which brain metastases are rare, such as Hodgkin disease and other lymphomas. In Hodgkin disease, a toxoplasma abscess may mimic a brain tumor. Biopsy may be required for correct diagnosis.

Radiation necrosis of the brain may mimic a metastatic tumor[176] (see Chapter 13). FDG PET,[177] MRS,[178] perfusion and DWI MRI[179] all help differentiate, but none is entirely accurate.[180]

Multiple sclerosis (MS) may be characterized by one or more contrast-enhancing lesions indistinguishable from a brain tumor.[181,182] One clue is that in MS the ring of enhancement is often incomplete.[183] With MS, however, the enhancement disappears in 6 to 8 weeks (rarely longer) and other nonenhancing lesions may be present, findings unlikely with brain metastases.

Methotrexate leukoencephalopathy causes hyperintensity of the white matter on T2-weighted MRI (decreased density on CT), usually bilaterally and associated with ventricular enlargement. The lack of contrast enhancement distinguishes it from a metastatic brain tumor.

Progressive multifocal leukoencephalopathy (PML) (see Chapter 10) causes white matter lesions of diminished density on CT or hyperintensity on T2-weighted MRI, which are irregular in outline and rarely contrast enhance (see Chapter 12).

Meningeal hematopoiesis in a patient with myelofibrosis can appear as multiple enhancing masses mimicking dural-based neoplasms.

As indicated on page 153 transient changes in scans sometimes follow focal or generalized epilepsy in the absence of an underlying primary or metastatic brain tumor. The lesions may occasionally mimic brain tumors, but they lack significant edema and disappear within a few weeks after seizures are controlled.

Approach to the Patient without Known Cancer

Figure 5–5 outlines an approach to diagnosing metastasis in patients with or without known cancer (see also Ref. 183a).

The first and best diagnostic test is an MRI. If a patient without known cancer presents with neurologic symptoms and the results of the MRI show a contrast-enhancing lesion suggesting metastatic brain disease, one should search for a malignant primary tumor before treating the brain metastasis.[184,185] Because most such brain metastases originate from the lung, either because the primary originated in the lung or because pulmonary metastases are present, careful attention should be directed to the chest. In one recent series, 35 of 49 (76%) patients with brain metastases from an unknown primary source were eventually discovered to have lung cancer.[47] The usual diagnostic approach is to obtain a CT of the chest, abdomen and pelvis, and/or an FDG body PET, to look for the primary or other metastatic lesions. At the same time, biochemical tumor markers, such as carcino-embryonic antigen, might help identify the lesion. If the primary is not discovered, surgical removal of the brain metastasis will sometimes give clues as to the primary site. Nevertheless, in a series of 99 patients who had resection of brain metastases from an unknown primary site, the primary was identified in only 49; 14 patients, of whom 9 underwent autopsy, died without the primary having been discovered.[47]

Treatment

The first step in treating brain metastases is stabilizing the patient neurologically, either by addressing increased intracranial pressure threatening cerebral herniation or repetitive seizures/status epilepticus in critically ill individuals. The treatment of increased intracranial pressure is considered in Chapter 3 and the treatment of seizures is discussed in Chapter 4. It is important to recognize that some patients do not require "stabilizing" treatment. If a brain metastasis is discovered in an asymptomatic patient, neither corticosteroids nor anticonvulsants are usually required. Even in a mildly symptomatic patient, corticosteroids may be deferred until the decision concerning definitive therapy has been reached. In patients who have not had seizures, prophylactic anticonvulsants are not indicated (see p 112).[106]

Once stabilized, unless the patient is eligible for a protocol study, the physician must decide which of the multiple therapeutic

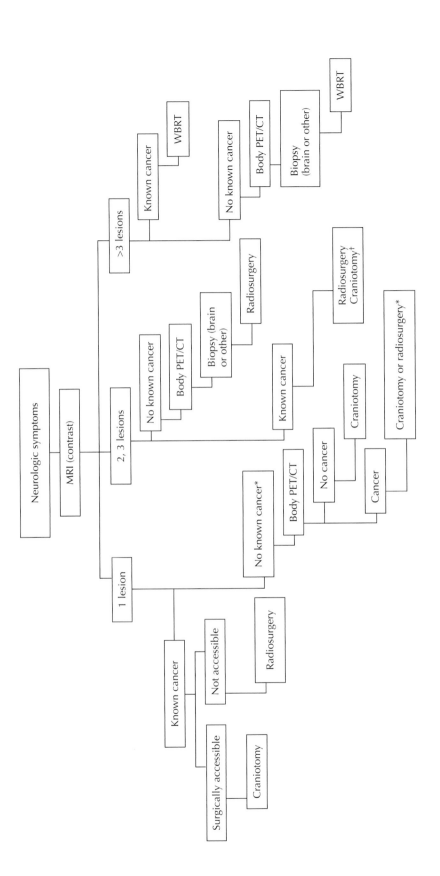

* If lesion suggests primary brain tumor, one can proceed directly to craniotomy.

† May be considered in selected patients.

Figure 5–5. Diagnostic approach to patients with possible brain metastasis(es). MRI, magnetic resonance imaging; PET, positron emission tomography; CT, computed tomography; WBRT, whole-brain radiation therapy.

options available is best for the individual patient. Treatments may include surgery, radiation therapy (RT) delivered either to the whole brain or focally to the metastasis (or both), chemotherapy delivered either systemically or directly to the nervous system,[186] or other agents. To select the appropriate therapy, the physician must first consider the brain metastasis(es), including the number of brain metastases, their size, location, and histology. The physician must also consider the extent of systemic disease. For example, a patient with a large right frontal lobe brain metastasis who is free of disease elsewhere should be considered for surgery, whereas the same patient with widespread active metastatic disease might be better treated with corticosteroids and radiation. The paragraphs that follow first describe the therapies available, and then describe potential approaches to specific tumors.

SYMPTOMATIC THERAPY

No controlled studies have compared the effects of managing metastatic brain tumors without treatment or with "symptomatic" therapy only. The available literature on the natural history of brain metastases suggests a median survival of fewer than 7 weeks if a symptomatic metastasis is not treated.[187,188] Symptoms almost never remit spontaneously, and patients develop increasing intracranial pressure, obtundation, stupor, and coma. Death results from cerebral herniation and brainstem compression or from intercurrent infection (e.g., pneumonia) in neurologically crippled patients. Spontaneous regressions are extremely rare.[189–193] All available evidence indicates that survival is longer and quality of life better if brain metastases are treated.

Corticosteroids as the sole treatment for brain metastases without subsequent surgery or radiation has been addressed in a few studies. Evidence that corticosteroids alone prolong life is sparse. In one study, the median duration of remission for patients treated with corticosteroids alone was approximately 2 months,[188] perhaps a month longer than no treatment at all. Nevertheless, as indicated in Chapter 3, corticosteroids rapidly ameliorate many symptoms of brain metastasis and should be used at the onset of therapy for most symptomatic patients.

RADIATION THERAPY

The vast majority of patients with brain metastases are treated with RT. Depending on the circumstances, radiation can be given to the entire brain, whole-brain radiation therapy (WBRT), or focused on the metastatic tumor, as in stereotactic radiosurgery (SRS) (Table 5–8).

Whole-Brain Radiation Therapy

For most patients with multiple metastases, WBRT is the preferred treatment.[148,194,195] The usual dose is 30 Gy delivered in 10 fractions.[194] For patients likely to have a short survival, some radiation oncologists deliver 20 Gy in five fractions.[195] Conversely, for patients likely to have a long survival, it might be wise to protract the radiation, delivering 40 Gy in 20 fractions or 45 Gy in 25 fractions.[196] Whatever the fractionation schedule, median survival varies from 2.3 months to 13.5 months depending on prognostic factors (see Table 1–16), with patients often succumbing to systemic disease rather than brain metastases. Long-term survival and even sterilization of the metastasis occurs occasionally.[197–199] Dose escalation beyond 30 Gy in 10 fractions does not increase survival or local control in patients with multiple brain metastases.[200] Other studies suggest that a short course of WBRT (4 Gy in 5 fractions over 5 days) yielded results similar to the standard 3 Gy × 10, at least for non–small cell lung[201] and breast cancer.[202]

Table 5–8. **Radiation Therapy of Brain Metastases: Options**

Whole-Brain—External Beam
  With focal boost
  High fraction—short course
  Low fraction—prolonged course
Focal Radiation Therapy
  External beam
  Stereotactic
  Interstitial radiation
    Heavy particles
    Photons
Enhancers
  Chemotherapy
  Hypoxic enhancers
Protectors

Patients about to undergo WBRT should be on corticosteroids for 48 to 72 hours before beginning therapy. The dose may vary depending upon the clinical situation. Some investigators use 4 to 8 mg a day of dexamethasone; we often begin with 8 mg twice daily. Tumors with a large amount of edema probably require larger doses and this is particularly true of metastases in the posterior fossa. The corticosteroids can be tapered to tolerance after the radiation is completed.

Using the above regimen, side effects are minimal (see Chapter 13). Neurocognitive functioning usually improves after the radiation, suggesting that neurocognitive dysfunction in patients with brain metastases is caused by the tumor rather than the radiation.[194] In one series, cognitive improvement was maintained in those who survived 15 months.[194] Nevertheless, as discussed in Chapter 13, radiation can cause cognitive damage, particularly in long-term survivors.

Three agents (radiation sensitizers) may be successful in sensitizing tumors to the effects of the radiation: Motexafin gadolinium is a paramagnetic (visible on MRI) agent that appears to enhance the effects of RT by reducing metabolites necessary for repairing oxidative damage. A randomized trial demonstrated

improvement in neurologic progression and quality of life, but no demonstrable effect on survival.[203,204] Studies are continuing. A phase III study of efaproxiral (RSR13), a drug that binds hemoglobin, decreasing its oxygen affinity and resulting in increased oxygenation of tissue, suggested that the drug may improve response rate and survival, particularly in breast cancer metastases.[205] The chemotherapeutic agent temozolomide increases survival in glioblastoma patients when the drug is used in conjunction with RT.[206] Preliminary studies suggest it might have a similar salutary effect on brain metastasis.[207,208]

Certain tumors are more sensitive to RT than others. The clone of cells that has metastasized from the primary tumor to the brain may vary greatly in sensitivity from the bulk of the primary tumor. Most series have reported no statistically significant differences in survival among patients by tumor type after treatment of brain metastases. Studies from MSKCC indicate that patients harboring metastases from breast or lung, especially small cell lung (Fig. 5–6), carcinoma are more likely to respond clinically and radiographically to RT and, if they respond, are less likely to die of their metastatic brain tumor.[54] In patients who have a single brain metastasis from lung cancer and whose

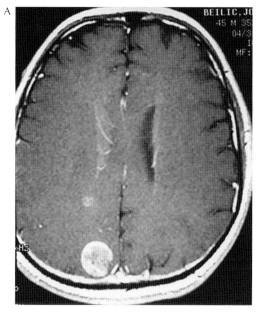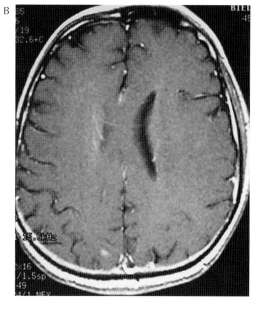

Figure 5–6. Response of brain metastases to whole-brain radiation therapy. **A**: A patient with small cell lung cancer and multiple brain metastases. **B**: Four months after a standard course of radiation (30 Gy in 10 fractions), the tumors had almost disappeared.

systemic disease is otherwise under control, the most common cause of death after whole-brain RT is recurrence of the brain metastasis.[164] Patients with melanoma,[209] colon cancer,[210] and renal cancer[211] are less likely to have a good response to WBRT, and are more likely to die of metastatic brain tumor than of systemic disease. These patients may respond better to radiosurgery (see following text). However, even so-called radioresistant metastases occasionally respond to WBRT.[210,212] Thus, if no other treatment is available, even radioresistant tumors should be treated with WBRT.

RT is effective palliative therapy. Depending on the series, between 66% and 75% of patients improve clinically. Patients whose neurologic symptoms are mildest at the time of treatment have the best improvement. Occasional patients who enter the hospital stuporous or comatose from herniation have a dramatic and very useful response to corticosteroids combined with RT, but even a failure to respond to corticosteroids does not indicate that RT will fail. By CT criteria, an objective response (50% or greater shrinkage) is seen in 50% to 60% of patients 6 weeks after RT[54,213] (Fig. 5–6). Smaller tumors respond better than larger tumors. The presence of more than three lesions is a poor prognostic sign for survival.[214] In occasional patients, RT sterilizes the tumor, and if there is no other disease, the patient is cured.[197] A residual contrast-enhancing mass on CT scan may mark the site of necrosis in the sterilized tumor.[198]

Several clinical issues remain unresolved: Does radiation tolerance of normal brain tissue differ from individual to individual because some are genetically predisposed to radiosensitivity? Patients with ataxia-telangiectasia, a syndrome associated with a homozygous mutation of the *ATM* gene, are extremely radiosensitive. Between 1% and 3% of the population is heterozygous for this mutation but their radiosensitivity is unknown. Heterozygous mice appear to be more radiosensitive than their litter mates.[215] Whether other genetic abnormalities confer unusual sensitivity to untoward effects of RT is presently unclear.

The still unresolved question of whether radiosurgery should be added to WBRT is discussed in the next section. Blood flow to tumors is not continuous. A number of investigators have demonstrated intermittent tumor blood flow.[216] Such intermittent flow can render an area of the tumor hypoxic at times, normoxic at other times; thus, the sensitivity of areas of tumor to RT and chemotherapy may vary from moment to moment or hour to hour.

Use of WBRT prophylactically to prevent the development of brain metastases, or to prevent their reappearance after surgery or stereotactic radiation, is discussed in paragraphs that follow.

In summary, WBRT is indicated as the primary treatment modality for brain metastases in all patients who do not meet appropriate criteria for either surgery or radiosurgery (Table 5–9). This group of patients probably constitutes 70% to 80% of patients with symptoms of metastatic brain disease. Symptoms are controlled with a combination of corticosteroids and whole-brain radiation in two-thirds to three-quarters of patients, with only mild side effects of fatigue and hair loss.[217]

Radiosurgery

Stereotactic radiosurgery (Fig. 5–7) delivered by multiple cobalt sources (gamma knife), or a linear accelerator, has become increasingly popular in the management of single and sometimes multiple, particularly small or surgically inaccessible metastases.[218–220] The various radiosurgical techniques achieve equivalent results (Fig. 5–7). In some respects, brain metastases are ideal targets for radiosurgery because most are spherically shaped with easily identifiable margins. The technique allows one to deliver high doses of radiation to the tumor, sparing normal tissue. The indications for radiosurgery are listed in Table 5–10.

Tumors amenable to SRS should be less than 4 cm in size (preferably 3 cm or less). The

Table 5–9. **Indication for Whole-Brain Radiation Therapy**

Multiple (>3) metastases

1–3 metastases with poorly controlled systemic disease

1–3 metastases too large for radiosurgery (>3–4 cm)

Reirradiation after late WBRT failure

Post-surgery or radiosurgery

Prophylaxis for small cell lung cancer

WBRT, whole-brain radiation therapy.

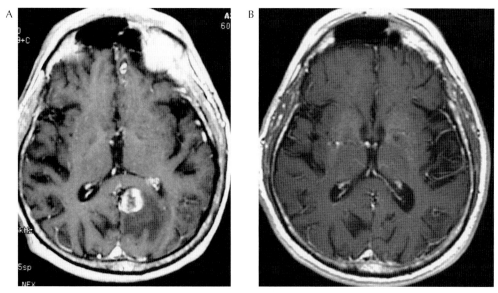

Figure 5–7. Response of a brain metastasis to stereotactic radiosurgery. **A**: A single occipital lobe metastasis from non–small cell lung cancer received 19 Gy via a linear accelerator. **B**: A year later, there was no evidence of tumor.

Table 5–10. **Indications for Radiosurgery**

Single or 2–3 metastases

Metastasis <3–4 cm

Post-surgery

Highly radioresistant tumor

Metastases inaccessible to surgery

local control rate was 85% when tumors were less than 2 cm, but fell to 45% when tumors were greater than 3 cm.[221] *Local control* is defined as stable disease or better, so not all patients have regression of their metastases following SRS. The technique appears to be particularly useful in the treatment of brain metastases resistant to standard radiation regimens, including melanomas and renal cell carcinomas. In fact, tumors highly resistant to conventional radiotherapy have a greater differential response to radiosurgery than tumors sensitive to conventional radiotherapy. How radiosurgery differs from conventional RT in causing tissue damage is not entirely clear, but it does appear to affect both normal and tumor tissues differently.[222–224]

Weltman et al.[225] devised a scoring system that they believe identifies patients who may not benefit from radiosurgery (Table 5–11). Giving a numerical score of 0 to 2 to age, Karnofsky Performance Status, systemic disease status, number of lesions, and volume of the largest lesion, they demonstrate that patients whose Score Index for Stereotactic Radiosurgery of Brain Metastases (SIR) is 5 or less have a median survival of only 4.5 months with a 6-month survival probability of only 36%. These patients are thought to be poor candidates for SRS (see section on prognosis). A recent study comparing WBRT with SRS in 186 patients (RPA classes 1 and 2) who had one to three brain metastases found that SRS alone improved local control, but did not significantly affect survival.[226]

The question of whether patients treated with radiosurgery should have WBRT as well is controversial. One survey found that complaints of adverse side effects including memory, concentration, and mood problems were substantially greater in patients who received WBRT (63%) in addition to stereotactic radiosurgery (34%). A recent randomized controlled trial[228] of SRS plus WBRT versus SRS alone for patients with one to four brain metastases found no difference in survival, but a substantial difference in 12-month CNS recurrence rate (47% vs. 76%). The report also found no

Table 5–11. **Prognostic Factor Categories for the Calculation of Score Index for Radiosurgery in Brain Metastases***

Category	Age (years)	Karnofsky Performance Score	Systemic Disease Status	Largest Lesion Volume (cm³)	Number of Lesions
0	≥60	≥50	Progressive disease	>13	≥3
1	51–59	51–70	Partial remission to stable disease	5–13	2
2	≤50	>70	Complete clinical remission to no evidence of disease	<5	1

From Ref. 227 with permission.

* Each independent prognostic factor is classified in categories 0, 1, and 2, in ascending order according to expected survival. The score index is the summation of these marks, ranging from 0 to 10: sum score of 1–3 indicates a median survival of 2.9 months, 4–7 of 7.0 months, and 8–10 of 31.4 months.

significant differences in neurologic functional preservation or toxicity of radiation. However, the Mini-Mental Status Examination that was used to assess cognitive function is insensitive to the cognitive changes produced by radiation (see Ref. 229).

Risk factors for development of new brain metastases after radiosurgery include more than three metastases, stable or poorly controlled extracranial disease, and melanoma. Patients having none of these factors had a 1-year actuarial freedom from new brain metastases of 83% versus 26% for all others.[230]

The converse question of whether patients receiving WBRT should have additional radiosurgery is also unresolved. In a randomized phase III trial examining the addition of an SRS boost to WBRT,[231] there was no difference in overall survival in patients with multiple metastases, by tumor size or dose. In a post-hoc subgroup analysis, patients with a single brain metastasis had longer survival (6.5 vs. 4.9, $p = 0.04$) when SRS was added to WBRT. However, the routine use of an SRS boost to WBRT was not recommended.[231] On the other hand, Kondziolka et al. terminated a comparable study early because of the superior effect on survival of the combination in treating patients with two to four brain metastases.[232]

One retrospective study of 132 patients, 46 of whom had four or more lesions, showed a better survival for the group with one to three lesions, but determined that the only significant prognostic factor was the RPA class (see Table 1–16).[233] Thus, the patient's age and performance status should play a role in any decision regarding radiosurgery. Long-term survival is possible; in one series, the actuarial survival was 6.9% at 5 years.[234] Local failure can be treated by reirradiation usually with standard fractionation schedules.[234]

In summary, radiosurgery should be the primary treatment in patients with one to three metastases who are not surgical candidates (see following text). In selected instances, particularly melanoma and renal carcinoma, one might choose to treat more than three metastases. As indicated subsequently, radiosurgery might prevent recurrence of tumor after successful surgical removal. Because the data are as yet inconclusive, the decision of whether to follow WBRT with radiosurgery to selected metastases or to give patients treated by radiosurgery additional WBRT should, in the absence of a controlled trial into which the patient could be entered, be made on an individual basis. Factors should include histology of the lesion, number and size of metastases, and estimated prognosis. The question of whether a metastasis should be surgically removed or treated with radiosurgery (or both) is addressed in paragraphs that follow.

Interstitial Radiation

Several techniques that deliver either photon radiation or radioactive substances directly into the tumor have been reported. These all require a surgical procedure. Among these are the photon radiosurgery system,[235] a miniature

X-ray generator that can stereotactically radiate intracranial tumors using low energy photons, and the GliaSite system,[236] an inflatable balloon filled with an aqueous solution of iodine 125 that delivers a low continuous dose of radiation to the resection cavity and stereotactic temporary placement of I-125.[237] The I-125 seeds can be either left in place[238,239] or removed after several days.[237] The radioactive iodine can be attached to antibody reactive tumor cells.[240] All of these techniques are experimental. Most of the studies have been in primary brain tumors, but each series contains a few metastatic tumors.

Prophylactic Cranial Irradiation

Prophylactic cranial irradiation is rarely used now in patients with leukemia because of toxicity (see Chapter 13) but, along with intrathecal chemotherapy, it was effective in eradicating sanctuaried cells and preventing the development of meningeal leukemia.[241] A similar approach is now routine for treating limited-stage SCLC, to prevent the development of brain metastases, and is being considered for non–small cell lung and breast cancer (see section on Specific Tumors).

Reirradiation after Relapse

Patients who initially respond to either WBRT or SRS and subsequently relapse are candidates for reirradiation with an initial or second course of WBRT.[242,243] Although the response rate (~40%) is lower than after initial RT, some patients do respond after a second or even a third course. Delayed radiation necrosis (see Chapter 13) is more likely, but most patients do not survive long enough to develop this complication. Those patients who have had a good clinical and MRI response to RT, and who survive more than 6 months after the initial RT course, are the best candidates for repeat irradiation. If the patient's systemic tumor is under control, surgical extirpation or chemotherapy should also be considered.

SRS can be administered if there is progression of one to three brain metastases following WBRT, or if patients develop new lesion(s) at a previously untreated site. Two courses of SRS are rarely administered to the same site. While radionecrosis is rarely an issue following two courses of WBRT, it is a common occurrence following SRS and WBRT given sequentially or when a patient receives two courses of SRS to the same site. Radionecrosis must be differentiated from tumor progression if subsequent MRIs demonstrate increased enhancement in the treated area.

SURGERY

MRI plays an important role in the preoperative and postoperative surgical management of patients with brain metastases, also allowing metastases to be removed successfully from language and sensorimotor areas of the cortex, the so-called eloquent areas.[244] Preoperative localization of intracranial lesions can mark the lesion, allowing for navigation under MRI visualization during surgery.[173] Functional MRI (fMRI) identifies language and sensorimotor areas that can be avoided during surgery. If necessary, surgery can be performed while the patient is awake, allowing monitoring of language during the procedure. Cortical mapping identifies sensorimotor areas during surgery even under general anesthesia. Intraoperative MRI allows the surgeon to inspect the tumor during the procedure and to confirm the degree of resection.[245] Postoperative MRI, although far less important after removal of the usually well-circumscribed brain metastasis than after removal of infiltrating gliomas, also plays a role,[246] allowing the surgeon to determine how complete the lesion removal has been. A postoperative, contrast-enhanced MRI should be performed within the first 24 to 72 hours after surgery (we usually do the MRI the day after surgery); after that time, contrast enhancement resulting from neovascularization from surgical trauma may mimic residual tumor even when surgical removal has been complete. The operatively induced contrast enhancement may not disappear for several months.[153,246,247] Scans performed after RT also must be interpreted with caution, because tumor progression may be mimicked by radiation changes weeks to months after completion of cranial RT (see Chapter 13).

Recent reviews describe in detail the technique and results of surgery.[248–250] Two of three randomized trials[164,251,252] support several noncontrolled studies that indicate patients with a single brain metastasis live longer, have fewer brain recurrences, and a better quality of life if treated by surgery followed by whole-brain irradiation than if they are treated by WBRT alone. The third study, which yielded no support for surgery,[252] appears to have treated patients with more widespread systemic disease, and many patients randomized to WBRT alone had surgery for their brain relapse. In one series of selected brain metastases treated by surgery, 22% of patients survived more than 4 years, with a median survival of more than 2 years.[253]

Surgery has several purposes: (1) It establishes the diagnosis. In a series by Patchell et al.,[164] 11% of patients with presumed metastatic brain tumors who underwent biopsy did not have metastatic cancer; two patients had benign brain lesions. Since the mortality and morbidity of removing a surgically accessible mass is about the same as a stereotactic biopsy, resection is preferred because it is both diagnostic and therapeutic. (2) Surgery improves symptoms. Even in patients with multiple metastases who would not ordinarily be candidates for surgery, a large symptomatic metastasis may be first treated by surgery, followed by treating the remaining metastases in another manner. (3) Surgery may provide a cure. In a patient free of other systemic disease who harbors only a single metastasis in the brain, surgical removal offers the only hope of cure. (4) Corticosteroids may become unnecessary. Because surgery removes the mass, surrounding edema resolves rapidly, allowing corticosteroids to be tapered. Other modalities are neither as quick nor as effective in decreasing brain edema. (5) Successful surgeries are increasingly common. Preoperative and intraoperative techniques that allow exquisite localization of the tumor and its relationship to important neurologic structures have substantially improved the ability of the neurosurgeon to remove tumors successfully, even in areas once thought untouchable.[244]

Metastatic tumors should be ideal surgical targets. They are generally well circumscribed, can easily be identified by the surgeon, and do not infiltrate surrounding tissue (but see p 150). Nevertheless, a significant percentage of patients who undergo surgical removal of a single metastasis will develop recurrence at the same area.[254] They are also, of course, at risk of developing metastatic disease elsewhere in the brain. In a randomized study comparing the effects of WBRT after removal of a single brain metastasis, Patchell et al.[254] identified recurrence at the surgical site in 46% of those who were not radiated, but only 10% of those who were. WBRT also decreased metastases at other sites in the brain (37% vs. 14%). Substantial evidence suggests that radiosurgery to the tumor bed also effectively prevents recurrence at that site, but of course not elsewhere in the brain.

Although surgery is usually reserved for patients with a single metastasis, one retrospective study suggests that removal of two to four metastases in a selected patient group yields an equivalent median survival.[255]

An unresolved issue is the comparison of surgery with radiosurgery.[256] Bindal et al.[257] compared 62 patients with brain metastases who were operated on with 13 patients treated by radiosurgery. The median survival was longer in the operated patients, the shorter survival in the radiosurgery patients being due to local recurrence. One attempt at a controlled trial failed because of lack of recruitment.[256] A recent retrospective analysis of over 200 patients in RPA classes 1 or 2 (Table 1–16) with one or two brain metastases found that stereotactic radiosurgery alone appeared to be as effective as resection plus WBRT in promoting survival.[258] A recent Cochrane Database review[259] found no randomized trials and thus could draw no meaningful conclusions.

At times, in a patient with multiple metastases, the larger one may be removed surgically and the others treated by radiosurgery.

Operative mortality in stable patients is almost nonexistent. Operative morbidity in the form of increased neurologic disability affects approximately 10% of patients; in most patients it is transient. The neurologic worsening may be present immediately after surgery or may begin 24 to 48 hours later. In most instances, this exacerbation of symptoms is due to brain swelling, and if corticosteroids are continued or increased without anything further being done, the patient recovers in a few days. In other instances, however, the symptoms may be due to an intraoperative ischemic insult or accumulation of subdural or epidural blood, which must be evacuated by a second operation. Before CT scans were available, 5% to 15% of patients undergoing operative intervention for brain metastases developed sufficiently severe secondary symptoms to necessitate a surgical re-exploration; usually only brain swelling was found, and the patient recovered without incident. Postoperative MRI permits physicians to distinguish between hemorrhage and swelling, so that reoperation is done only for patients who hemorrhage. In a series of 382 patients operated for brain metastasis by the neurosurgical service at MSKCC from 1999 to 2003, there was only one death. That patient had had a prior pneumonectomy and died a respiratory death 3 days after surgery for an occipital metastasis. There were postoperative complications in 27 other patients (7%). These included eight cerebral hemorrhages and three cerebral infarcts. Two patients developed hydrocephalus, five had postoperative seizures, three developed a CSF leak, and two developed a wound infection. Pulmonary embolus in one patient resulted in placement of an inferior vena cava filter (see Chapter 9). Two patients developed GI perforations, presumably steroid induced. One patient was treated for sepsis. All of the patients recovered without residual neurologic disability (data courtesy of Dr. Vivian Tabar).

The number of patients who are candidates for surgery is limited. In the series reported by Cairncross et al.,[54] of 201 consecutive patients, only 18 were considered to be candidates for surgical extirpation. Of these, four had unknown primary tumors and the surgery was done in part for diagnosis. One-half of the surgical candidates had no evidence of systemic disease, and in those who did, the disease was under good control.

Some unresolved questions persist concerning surgical therapy of brain metastases. One concerns leptomeningeal tumor, which may develop in as many as 50% of patients after surgery of a posterior fossa metastasis.[260] This complication occurs in a much smaller percentage of patients with supratentorial lesions.[261,262] In one series, 9 of 18 patients who received surgery for posterior fossa metastasis subsequently developed leptomeningeal spread, whereas only 4 of 62 with comparable lesions treated with SRS developed this complication.[260] Resection of tumor in the cerebellum likely spills tumor cells into the cisterna magna, seeding the meninges. There is substantially less contact with the subarachnoid space when supratentorial lesions are removed, unless the lateral ventricle is entered. Prophylactic intrathecal therapy with methotrexate might be considered in patients after successful extirpation of a single cerebellar metastasis.

A second unanswered question concerns postoperative radiation, either SRS, WBRT, or both. Postoperative radiation has two purposes: (1) to eradicate residual tumor at the surgical site and (2) to treat micrometastases elsewhere in the brain. As indicated in the preceding text, a randomized trial was conducted following resection of a single lesion,[254] and patients who received surgery alone had a significantly higher CNS relapse rate at the surgical site and elsewhere in the brain compared to those who received postoperative WBRT. However, survival of both groups was identical because the irradiated patients died of their systemic disease while those without RT died of uncontrolled CNS disease.

A recent uncontrolled study suggests that treatment of patients by surgery and WBRT, who received a boost after the WBRT (5 fractions of 3 Gy to patients who had previously received the standard 3 Gy $\times$ 10 or 5 fractions of 2 Gy in patients who had previously received 2 Gy $\times$ 20), improves outcome.[263]

Because strong evidence suggests that WBRT given postoperatively is effective, but does carry risk of cognitive dysfunction in long-term survivors, we believe the choice should be individualized. Patients who are at low risk for recurrence and are likely to have a long survival, or who have highly radioresistant primaries, might best be followed carefully, preserving radiation for recurrence. Radiosurgery to the tumor bed probably carries less risk of serious side effects, although radiation necrosis can occur.

Table 5–12 lists the indications for surgery of a brain metastasis.

Table 5–12. **Indications and Best Candidates for Surgery**

Indications
Establishes the diagnosis
Improves quality of life
Promotes survival
Decreases corticosteroids

Best Candidates
Solitary metastasis (no other disease)
Single metastasis >4 cm
Multiple metastases but one large
   and symptomatic

Uncertain diagnosis: In patients not known to have systemic cancer, but with a solitary, surgically accessible mass lesion in the brain, surgery achieves both diagnosis and therapy. If the lesion is not surgically accessible, stereotactic needle biopsy is probably indicated.

Poor control by RT: One or two accessible lesions that cannot be controlled by RT or chemotherapy should be removed. If there is one very large and several small lesions, resection of the large lesion should be considered before radiation of the smaller ones.

Failure of RT in a symptomatic patient: If a patient with a single brain metastasis fails radiation, the patient should undergo surgery as indicated, provided systemic disease is under control. If the patient has no symptoms, or if the symptoms can be controlled easily by corticosteroids (particularly in the presence of widespread systemic disease), surgery is probably not indicated. The point at which one should consider RT a failure is not clear but, if the tumor has not shrunk 8 weeks following RT and the patient remains symptomatic, radiation failure seems fairly certain. A patient who has been irradiated successfully and then relapses 6 months to a year later should be carefully assessed concerning the advisability of surgery, which may be the best option. A repeat course of RT is sometimes more appropriate.

Uncontrollable neurologic symptoms: In a few patients, symptoms of increased intracranial pressure that cannot be controlled by corticosteroids or repetitive and intractable seizures that cannot be controlled with anticonvulsant drugs represent an indication for surgery whether or not RT has been used previously. If increased intracranial pressure cannot be controlled with corticosteroids, RT may temporarily worsen the symptoms, so resection of surgically accessible lesions before RT is best. This may be applicable to patients with multiple brain metastases if one lesion is the predominant cause of symptoms. For example, in addition to having numerous supratentorial metastases, such a patient may have a large lesion in the posterior fossa obstructing the fourth ventricle.

Recurrence after successful surgery: Relapse after surgery can be followed by reoperation in those patients who remain good surgical candidates. In one series, median survival following a second craniotomy was 9 months, and the actuarial 2-year survival was 25%.[264] In another series, median survival was 10 months.[265]

CHEMOTHERAPY

Increasing evidence shows that chemotherapy (Table 5–13), either alone or in combination with immunotherapy and radiation, effectively treats some brain metastases[207] (Fig. 5–8).

Cytotoxic drugs can be administered orally, intravenously, into the carotid or vertebral artery supplying the brain metastases, directly into the tumor itself,[266] or intrathecally into the lumbar or ventricular CSF (see Chapter 7). The drugs can also be administered intravenously after opening the blood–brain barrier by intra-arterial hyperosmolar mannitol infusion[267] or by use of the drug Cereport (RMP-7).[268]

The two major factors that determine responsiveness of a brain metastasis to chemotherapy are blood–brain barrier permeability and the intrinsic sensitivity of the tumor to chemotherapy. Thus, chemotherapy of brain metastasis has been most effective in metastases from SCLC and breast cancer and least effective in melanoma, renal carcinoma, and NSCLC.

The physiology of the blood–brain and blood–CSF barriers is discussed in Chapter 3. The history of the relationship between these barriers and the treatment of CNS metastases has undergone many changes. When chemotherapy first became available to treat acute lymphoblastic leukemia, it was discovered that cells could find sanctuary within the CNS and lead to meningeal relapse after effective

Table 5–13. **Chemotherapy of Brain Metastases**

1. Nonspecific
   A. Anticonvulsants (see Chapter 4)
   B. Corticosteroids (see Chapter 3)
2. Specific
   A. Hormones
   B. Cytotoxic drugs
      1. Oral or intravenous
      2. Intra-arterial
      3. Intravenous with osmotic opening of blood–brain barrier
      4. As radiation therapy sensitizer
      5. Intrathecal

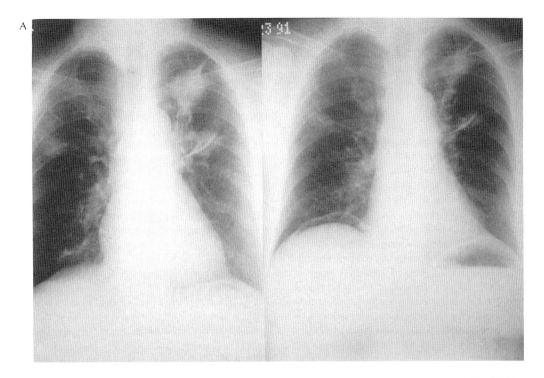

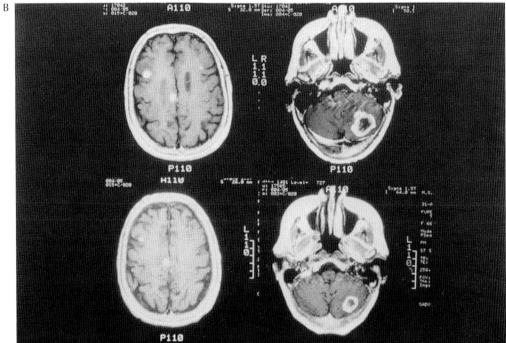

Figure 5–8. A: Chest film of a patient with breast carcinoma and pulmonary metastasis before (*left*) and after (*right*) chemotherapy. **B**: Contrast-enhanced T1-weighted magnetic resonance image of the brain in the same patient before (*top*) and after (*bottom*) chemotherapy. The brain metastases have decreased in size to about the same extent as the lung metastasis. The small lesion in the left frontal lobe has disappeared but the other brain metastases are only slightly smaller. The patient subsequently underwent whole-brain radiation therapy with further shrinkage of the tumor but died a year later of systemic disease.

171

therapy of the systemic disease. This led to the concept that the blood–brain barrier prevented the use of chemotherapy for treatment of any CNS metastasis despite the fact that the blood–brain barrier was clearly disrupted with brain metastases, as easily identified by enhancement of the lesions on MRI or CT.

As early as 1986, Rosner et al.[268] described the results of systemic chemotherapy in 100 patients with brain metastases from breast cancer; 10 were complete responders and 40 were partial responders. Several combinations of chemotherapy appeared equally efficacious. Similar responses were obtained by Cocconi and coworkers.[269,270] Using a combination of platinum and etoposide, they found responses in both breast cancer and NSCLC metastases.

Experimental studies supported the efficacy of systemic chemotherapy for brain metastases. Therapeutic levels of water-soluble chemotherapeutic agents such as paclitaxel reach metastatic brain tumors when given parenterally.[271] When compared with normal brain, levels of the drug are substantially higher, both in the center of the tumor and at its periphery.[271,272] This is in part due to permeable blood vessels within the tumor that lack a normal blood–brain barrier. Blood vessels in metastatic tumors also have low levels of P-glycoprotein, a protein that promotes egress of drugs[272] (see Chapter 3).

The clinical situation is not quite so clear. Earlier studies suggested that brain metastases were just as susceptible to systemic chemotherapy as were lesions elsewhere in the body, at least at initial treatment.[207,273,274] However, a recent study of the response of asymptomatic brain metastases from SCLC to first-line systemic chemotherapy indicates that the brain metastases are less susceptible to systemic chemotherapy than lesions elsewhere in the body.[275]

The other factor—some believe the only factor—is the intrinsic sensitivity of the tumor cells themselves to chemotherapy. Because this factor plays such an important role, the use of chemotherapy for brain metastasis is discussed in the following text where specific tumors that metastasize to the brain are described.

The sensitivity of cells in a brain metastasis is not necessarily identical to that of the primary tumor or metastases in other organs. Brain metastases arise from a clone of tumor cells in the primary tumor and this clone may be more or, more likely, less sensitive to chemotherapy than cells in the bulk of the primary tumor. It is even possible that two metastases within the brain, from the same primary, may differ from each other in response to chemotherapy. However, in general, agents that are effective against the primary tumor are used first to treat the metastasis.

Several questions remain unanswered: The first is how to predict drug resistance in brain metastases. In theory, if a brain metastasis develops after a patient has received chemotherapy, the metastasis is likely to be resistant to the previously used drugs and perhaps others as well. As indicated earlier, this concept is supported by established evidence of better control by chemotherapy of CNS metastases occurring at diagnosis than metastases developing later in the course of the disease. On the other hand, if microscopic tumor is sequestered behind the blood–brain barrier during the course of initial chemotherapy for the primary tumor, those cells may still be sensitive to the same drugs; breakdown of the blood–brain barrier as the tumor grows will permit cytocidal levels of the drug to reach the metastasis that previously had not been exposed.

Increasingly, chemotherapy is being used with good efficacy for the treatment of brain metastases, although it is usually reserved for those patients who have progressed following surgery and/or RT. The exceptions are patients in whom asymptomatic brain metastases are discovered during screening and are scheduled to receive chemotherapy for their systemic disease. In such patients, chemotherapy can be given and intracranial disease monitored with serial imaging; responses are frequent. The advent of small molecules that penetrate the blood–brain barrier readily and the increasing recognition that drugs or regimens tailored to the sensitivity of the primary tumor are effective suggest that systemic chemotherapy for the treatment of brain metastases is likely to grow in the coming years.

Prognosis

Once the patient develops a brain metastasis, the prognosis is poor. Most patients live only a few months, although with aggressive therapy, and appropriate circumstances, long-term survival[276] and even cure have been reported. Although the outcome of individual patients cannot be predicted, several prognostic factors play a role in predicting outcome in groups of patients. Table 1–16 outlines the use of recursive partitioning analysis (RPA) done by the Radiation Therapy Oncology Group (RTOG) to predict prognosis after RT. Median survival ranges from 2.3 to 13.5 months depending on the patient's RPA class and whether metastases are single or multiple.

SPECIFIC TUMORS

Small Cell Lung Cancer

SCLC represents approximately 20% of lung cancer cases, with that percentage decreasing as the prevalence of smoking decreases.[277] Once almost exclusively a disease of men, SCLC, in some centers, affects more women than men. In one series, brain metastases were present in 38 of 181 patients (21%) at diagnosis of the cancer; 24 were asymptomatic (63%).[275] More brain metastases developed during treatment, but in one autopsy series of 174 patients, brain metastases were present in only 17%.[278]

Surgery is almost never an option because tumors are usually multiple. Rare exceptions include very large and symptomatic metastases that have not responded to WBRT, which is the usual treatment. Patients respond initially, but survival is short. Median survival of 154 such patients was 8.6 months, 4.2 months, and 2.3 months for RPA classes 1, 2, and 3, respectively.[279] Among 127 patients with brain metastases from SCLC, none survived more than 5 years.[8]

SCLC is among the most chemosensitive of solid tumors and patients often have asymptomatic brain metastases discovered on screening MRI at initial diagnosis. Thus, these tumors were among the earliest treated with systemic chemotherapy. Lee et al.[280] reported objective responses in 82% of patients with brain metastases from SCLC. Kristensen et al.[281] reported a 76% response rate without radiation in patients with intracranial metastases at diagnosis, whereas the response rate to chemotherapy of brain relapses was only 43%. More recently, a review of several previous studies reported a 66% response rate to chemotherapy when a brain metastasis was discovered and treated at presentation of the cancer.[282] Later treatment resulted in a 52% response rate.

Likewise, a study of 80 patients, most of whom had received prior chemotherapy, yielded only a 33% response rate after treatment with teniposide.[283] A recent phase II trial of irinotecan and carboplatin yielded a 65% response rate after two cycles of chemotherapy and a median survival of 6 months.[284] Adding WBRT to chemotherapy increases the response rate but not survival.[285] In addition, as indicated in the preceding text, a recent and carefully done study of patients with asymptomatic brain metastases at diagnosis showed a poorer response of the brain metastases than the systemic metastases to chemotherapy.[286]

SCLC is the one solid tumor for which there is good evidence that prophylactic cranial irradiation prevents brain metastases and increases survival and quality of life.[287,288] Patients who do best are those with limited-stage SCLC who have a complete response to systemic chemotherapy. However, there is increasing evidence that patients with extensive disease who respond to chemotherapy also do better with either 20 Gy/5 fractions or 30 Gy/12 fractions.[289] Delayed cognitive dysfunction was present in only 2% (but see Ref. 289). A recent study of patients tested before and after prophylactic irradiation revealed that almost half had evidence of impaired cognitive function before treatment. Although there appeared to be an early decline in some cognitive functions, persistent decline in cognitive function did not occur.[291]

Non–Small Cell Lung Cancer

As indicated in Table 1–2, lung is the most common primary to cause brain metastasis, and most are NSCLCs. Adenocarcinoma is the most common type, representing in one series over half of all NSCLC brain metastases. The others were large cell carcinoma (18%) and squamous carcinoma (10%).[60] In that series, of 809 patients with both NSCLC and brain metastases, 181 (22%) had brain metastasis at initial staging and one-third were asymptomatic.[60] In another series of 211 patients with locally advanced NSCLC, 51 had a complete response of their systemic disease[56]; the most common site of recurrence in those 51 patients was the brain, occurring in 22 of the patients and representing 71% of all isolated metastases. In a series of 422 patients with stage III

NSCLC, 268 (64%) experienced disease progression; 54 of these relapses (20%) were in the brain only and another 17 were in the brain plus other sites.[292] Furthermore, the majority of patients who present with brain metastases from an unknown primary source eventually turn out to have lung cancer.

Single metastases that are surgically resectable should probably be resected, certainly if they are larger than 3 cm. Survivals of greater than 10 years have been reported.[57,293] For smaller lesions, whether radiosurgery is the equal of surgery is still unsettled.[256,259] If the patient has multiple lesions, but one is quite large and symptomatic, the large one should probably be resected. Patients with two or three metastases may respond to either surgery or radiosurgery. In the other patients, WBRT produces at least short-term palliation in most. When WBRT is combined with temozolomide, the combined treatment is more effective and improves quality of life.[294,295]

NSCLC is far less sensitive to chemotherapy than SCLC. Of 156 patients with NSCLC who presented with brain metastases, chemotherapy alone produced a brain response in 27% and WBRT followed by chemotherapy in 35%, a difference not statistically significant. Progression-free survival was 6 months and median survival 11 months.[279] A variety of chemotherapeutic agents, in addition to the platinum-based compounds, have been tried either alone or in combination. These include temozolomide,[296–298] topotecan,[299] high-dose methotrexate,[300] and others. Some responses are reported in most series, but most are short lived and not dramatic.

Small molecules that target the epidermal growth factor receptor (EGFR), including gefitinib and erlotinib, are effective against a subset of NSCLCs, especially adenocarcinomas in nonsmoking and Asian women. In a group of patients who responded to gefitinib, the CNS was the initial site of recurrence in 33%,[301] suggesting that the CNS may be a sanctuary that the agents could not penetrate. Once brain metastases develop, if they show the appropriate EGFR mutation, they appear responsive to the inhibitors.[302–304]

A recent review summarizes the current state of systemic therapy for brain metastases from lung cancer, both small cell and non–small cell.[305]

Prophylactic cranial irradiation has been reported to decrease the incidence or prolong the time to development of brain metastases in patients with locally advanced NSCLC. A review summarizing three randomized and six nonrandomized trials suggests that the treatment is effective and relatively safe.[306] It is still unclear exactly which patients benefit. Several trials are under way to further define the role of this treatment modality.[288]

Breast

Although the percentage of patients who develop brain metastases from breast cancer is smaller than that of lung, renal cancer or melanoma, the incidence of breast cancer is so great that brain metastases from breast cancer constitutes the second largest group of patients, after those with lung cancer. In population-based studies, the clinical frequency of brain metastases from breast cancer is approximately 5% (see Table 1–3). However, other studies report a larger incidence (10%–15%)[307]; as many as 30% of patients have CNS metastases at autopsy (see Table 1–2). Substantial evidence suggests that both the incidence of brain metastases from breast cancer and the likelihood of brain metastases being the first and often the only site of relapse are increasing. This phenomenon may be particularly true for patients whose systemic tumor is successfully treated with trastuzumab,[308,309] a large molecule that does not cross the blood–brain barrier.[310] However, this is complicated by the fact that HER-2 positivity, the reason for using the drug, is itself probably a risk factor for developing brain metastases. Other risk factors include young age, negative estrogen receptors, and positive lymph nodes.[311–313] Expression of cytokeratin and overexpression of epidermal growth factor receptor may also be risk factors.[314] The presence of a *BRCA1* mutation is an especially important risk factor for the development of brain metastases.[315] Some evidence suggests that infiltrating lobular carcinoma has a predilection to metastasize to the leptomeninges, whereas intraductal carcinoma is more likely to metastasize to the brain parenchyma.[316,317]

Brain metastases from breast cancer may appear many years after the cancer is apparently cured. One report encountered 11 patients over a 10-year period who developed brain metastases 11 to 30 years after the initial diagnosis of breast cancer.[318]

Experimental evidence suggests that breast cancer metastases to brain differ from the primary tumors. Although HER-2 expression does not affect the number of experimental brain metastases, HER-2 overexpression increases the outgrowth of metastatic tumor cells in the brain.[319] Furthermore, experimental evidence also suggests that breast cancer cells metastatic to brain alter their energy metabolism to adapt to the brain's metabolism.[320]

Breast cancer metastases to brain are both more radiosensitive and more chemosensitive that those from NSCLC and many other primaries. Large single metastases that are surgically accessible should probably be treated surgically and followed by RT.[321,322] Radiosurgery is also quite effective, particularly for single metastasis. In some series, local control rates are over 90%.[323,324] In a recent series of 84 patients, median survival was 19 months and 2-year survival 56% for those treated initially with SRS.[325] One- and two-year local control rates were 79% and 49%, respectively. Good performance status, being postmenopausal, and having positive estrogen receptors predicted better survival.[325]

Because the tumors are relatively radiosensitive, WBRT is reasonably effective.[10] An analysis of 174 patients so treated yielded survivals at 1, 2, and 3 years of 33.4, 16.7, and 8.8%, respectively. Surgically treated patients who subsequently received WBRT fared better, as did patients with good performance status and absent extracranial metastases.

Brain metastases from breast cancer are also sensitive to chemotherapy once they become large enough to disrupt the blood–brain barrier. Rosner et al.'s study of 100 women treated with several different regimens yielded a 50% response rate,[268] independent of whether they had received prior adjuvant chemotherapy. Several chemotherapy regimens have proved efficacious.[307] These include platinum compounds with etoposide,[269] capecitabine,[326] with or without temozolomide,[327] high-dose methotrexate,[300] and others.[328] The metastases also respond to hormonal therapy with tamoxifen,[329] megestrol, and[330] letrozol[331] in estrogen receptor–positive disease and tyrosine kinase inhibitors such as lapatinib in HER2-positive tumors.[332] Because a brain metastasis

may have been sequestered behind the blood–brain barrier at the time systemic disease was treated with chemotherapy, those cells may not have "seen" the water-soluble chemotherapeutic agents. Thus it is possible that repeating prior chemotherapy would be effective once the brain metastasis has become clinically apparent.

Melanoma

Although much less common than either breast or lung cancer, the incidence of melanoma is rising and the frequency of brain metastases is so high that melanoma is the third most common cause of brain metastasis.[333] The brain metastases are much more likely to be hyperintense on T1-weighted MRI, either from free radicals within the melanin or from hemorrhage into the tumor.[334] Hemorrhage is more common in melanoma than in most other brain metastases, with the exception of gestational trophoblastic lesions. Melanoma metastases are also more likely to be hypointense on T2 gradient echo images, representing small, often subclinical hemorrhage. Melanoma is probably the only brain metastasis that is occasionally identified only on gradient echo images.[335] Furthermore, metastases are more likely to invade the cortex than metastases from other tumors, perhaps accounting for the high incidence of seizures with this tumor.

Brain metastases from melanoma are more resistant to either radiation or chemotherapy than those from breast cancer. Melanoma cells are better able to repair themselves than many other cancer cells, thus requiring higher fractional doses of RT for cell kill. This factor makes these tumors especially good candidates for radiosurgery.

Because melanoma is resistant to most nonsurgical modalities, whenever possible, metastases should be removed, even if there are two or three lesions.[334,336] Whether WBRT in the postoperative period is helpful is unclear.[334,336] In one series, 3- and 5-year survivals were 9% and 5%, respectively.[336] Very long-term survivals occur occasionally.[192] WBRT yields some objective responses[337]; the addition of temozolomide did not change the response rate.

SRS of single or multiple lesions is the treatment of choice in patients who are not surgical candidates.[338–340] One report indicates that multimodality treatment including radiosurgery of up to 5 metastases is useful.[338] Long-term survivors are seen. In one series, 6.5% survived more than 4 years after radiosurgery.[340] Local control rates are more than 80%.[339] Favorable prognostic factors include single brain metastasis, controlled systemic disease, and good performance status.[339]

Two chemotherapeutic agents, fotemustine (a nitrosourea) and temozolomide, are the most widely used for brain metastases from melanoma. Given alone, or in combination with WBRT, fotemustine is generally ineffective.[341,342] However, the combination appears to delay time to cerebral progression without affecting objective response or overall survival.[342] The drug is also better than dacarbazine in delaying the development of brain metastases when used to treat extracranial metastases.[343]

In one series, temozolomide, alone or in combination with other chemotherapeutic agents, gave a partial response rate of 24% and stable disease in another 20%. The median time-to-progression, however, even in responders, was only 3.9 months.[344] Small asymptomatic metastases treated with temozolomide with or without immunotherapy gave responses or stable disease in 13 of 52 patients, with time-to-progression varying from 2 to 15 months.[345] This allowed physicians to defer WBRT. One other study shows a less than 5% objective response rate for asymptomatic brain metastases.[346]

A recent report describes the use of chemotherapy with either dacarbazine or temozolomide in combination with interleukin-2 and interferon-α-2B.[347] Those patients who had brain metastases had received SRS. The presence of a brain metastasis did not affect survival.

Renal

Metastases from renal cell carcinoma to the intracranial cavity have some unusual characteristics. They may occur many years after the patient has apparently been cured of the primary tumor.[348] They may seed areas of the brain that are relatively ischemic.[349] They often affect areas other than the brain parenchyma (e.g., choroid plexus, dura, or ventricle). They may even undergo spontaneous regression.[190] They also have a tendency to hemorrhage. In one series, the tumors were hemorrhagic in 46% of 50

patients; 4% required emergency craniotomy to evacuate the hematoma.[350]

Because the metastases are relatively resistant to both radiation and chemotherapy, and because they often occur in patients with a minimal systemic burden, surgery is the treatment of choice whenever possible, even for more than one metastasis.[350–353] WBRT after surgery may[352] or may not[350] increase survival. Favorable prognostic factors include more than a 1-year latency after nephrectomy and an excellent neurologic performance status.[350,353]

The results of WBRT are generally unsatisfactory,[354–356] although some patients derive benefit.[355,357] Several series demonstrate that SRS gives good local control, ranging from 92% to 100%.[355,358,359] Whether the addition of WBRT is necessary or desirable is unclear.

Neither chemotherapy nor immunotherapy has a clear role in the treatment of brain metastases from renal cell carcinoma. However, the multiple tyrosine kinase inhibitor, sunitinib, which is so effective for metastatic renal cell cancer, is under study for treatment of brain metastases.

Colon

Approximately 3% of patients with colon cancer develop brain metastases.[360] The metastases are usually associated with pulmonary metastases, which arise from the left side of the colon rather than the right side, which leads to hepatic metastases. Brain metastases affect the cerebellum disproportionately.[361]

Little data exist on the treatment of brain metastasis from colorectal cancer. Surgery, when feasible, is the treatment of choice,[362] although in one series cerebellar metastases fared less well than those above the tentorium. WBRT gives poor results, but radiosurgery usually provides good local control. No data exist concerning the effects of chemotherapy, although one might consider using the same chemotherapy that is sometimes effective in systemic colon cancer.

Testis

Testicular cancer metastatic to brain is curable (consider Lance Armstrong). The two major categories of testicular cancer are seminomas and nonseminomatous germ cell tumors; the latter can also arise in the mediastinum.

Seminomas rarely metastasize to the brain; in one series only 2 of 44 testicular cancers causing brain metastases were seminomas.[363] Brain metastases can present either at diagnosis, following complete systemic remission, or with extracranial progression.[364] When identified at presentation, chemotherapy should be the first treatment.[363] Surgery, WBRT, aggressive chemotherapy, and radiosurgery[365] are all effective modalities.[366] Aggressive therapy is appropriate as long-term survival is possible.[364]

Gynecologic Tumors

Gynecologic tumors include those of the uterine cervix, uterine endometrium, fallopian tube, ovary (epithelial), and gestational trophoblastic tumors. Brain metastases from trophoblastic tumors are now rare because the primary tumors are treated early, but at one time hemorrhage into brain metastases was a major cause of death. Brain metastases from trophoblastic disease can be successfully treated with a combination of chemotherapy and WBRT[367]; surgery is sometimes helpful.

Other gynecologic tumors do not metastasize commonly to the brain. Approximately 1% of patients with invasive cervical cancer develop brain metastases[368]; even at autopsy, the incidence is under 5%.[369] Surgery,[370] followed by WBRT, may occasionally lead to long-term survival.[371] WBRT improves symptoms in most patients,[372] albeit temporarily. Chemotherapy using a variety of agents appears to improve survival after WBRT.[372] One report describes spontaneous remission of a brain metastasis and other metastases after removal of the primary tumor.[373]

Less then 1% of endometrial cancers metastasize to brain.[374] Surgical therapy followed by WBRT can yield long-term survival[370,375]; WBRT temporarily improves symptoms.[374]

Ovarian and fallopian tube cancers metastasize to brain in approximately 1% to 3% of patients,[376,377] but the incidence may be increasing as a result of better systemic chemotherapy.[378] In one series, 6 of 52 patients (11.6%) developed brain metastases after chemotherapy for the primary tumor.[379] In about one-half, the brain was the only site of metastasis and in about two-thirds multiple brain metastases were present. Surgical treatment followed by WBRT has been successful.[370,380] WBRT and radiosurgery are also useful.[377] Chemotherapy may control small, asymptomatic metastases

(Fig. 5–8).[377,381] The presence of systemic metastases confers a worse prognosis. Patients who received supportive care only lived 1.6 months compared to those who received surgery followed by WBRT (23 months).

Sarcomas

Sarcomas, tumors that arise from mesenchymal tissue, are much less common than carcinomas. There are a large variety of sarcomas, most of which involve bone or soft tissue. These include osteogenic sarcoma, Ewing sarcoma, rhabdomyosarcoma, and alveolar soft part sarcoma. Many of these tumors are more common in children than adults and all can metastasize to brain. In one series of 480 patients with musculoskeletal sarcomas, 20 (4.2%) developed brain metastases.[382] These included 5/69 malignant fibrous histiocytomas, 1/54 liposarcomas, 1/27 malignant peripheral nerve sheath tumors, 2/13 rhabdomyosarcomas, 2/8 extra-skeletal Ewing sarcomas, 1/6 hemangiopericytomas, 3/103 osteosarcomas, and 2/18 bone Ewing sarcomas. Most of these patients were adults. In children, a review of over 2000 patients identified hematogenous brain metastases in 19/429 neuroblastomas (4.4%), 11/574 rhabdomyosarcomas(1.9%), 25/386 osteosarcomas (6.5%), 16/487 Ewing sarcoma (3.3%).[383] The metastases are usually single rather than multiple. In our experience, the incidence is increasing in patients with neuroblastoma as a result of more effective systemic therapy. Whenever possible, the tumor should be removed.[384] The role of postoperative radiation and radiosurgery is unclear.

Hematologic Cancers

Included in this category are the leukemias, non–Hodgkin lymphomas, Hodgkin disease, and the plasma cell dyscrasias, including myeloma. Collectively, when these tumors metastasize to the intracranial cavity, they usually affect the dura and the leptomeninges more than the brain parenchyma.[385] For example, primary CNS lymphoma (non–Hodgkin) is essentially a parenchymal brain tumor, whereas metastatic non–Hodgkin lymphoma usually involves the leptomeninges and spares the brain parenchyma.[386] Leptomeningeal metastases are discussed in detail in Chapter 7.

Plasma cell dyscrasias may metastasize to the dura and be mistaken for meningiomas.[387,388] Because most of these tumors are sensitive to chemotherapy, this is often the modality of choice, using drugs based on the chemosensitivity of the primary tumor.

Unknown Primary

A significant percentage of patients with metastatic brain tumors present to the neurologist and neurosurgeon without a prior diagnosis of cancer.[5] These brain metastases of unknown primary source represent as many as 15% of patients who undergo surgery for brain metastases.[44] Even in a cancer hospital, approximately 2% of metastatic brain tumors have no primary source (Table 5–3). For this reason, it is becoming increasingly common for patients who present with multiple brain lesions, and many who present with single brain lesions, to undergo a systemic search for primary cancer before biopsy or removal of the brain lesion. The search usually consists of a CT of the chest, abdomen, and pelvis and, if negative, an FDG PET scan. The search often proves negative and the tumor is discovered to be metastatic only when examined histologically. At times, histology or immunohistology suggests the diagnosis.[185]

Brain metastases of unknown primary source usually arise from lung, but a significant percentage (27% in one series)[47] remained undiagnosed after clinical evaluation and 9% were undiagnosed even at autopsy.

Treatment is much the same as for other metastatic brain tumors, except it is difficult to select appropriate chemotherapy when the primary remains unknown. The prognosis is similar to that of most patients in whom the primary tumor has been identified.[5] Identification of the primary tumor does not appear to affect overall survival.

LESS COMMON SITES OF INTRACRANIAL METASTASES

Calvarial Metastases

Metastases to the skull vault are usually asymptomatic and are discovered during routine imaging of the patient with cancer. They probably reach the skull via Batson plexus (see Chapter 2).

Occasionally, expansion of the bones or involvement of the periosteum may cause pain or a palpable mass. Sometimes a painless mass becomes quite large and may be disfiguring.[389] In some instances, a skull mass may be the first evidence that a known tumor has metastasized,[390] or more rarely the first evidence of cancer.[390] A skull metastasis may reach sufficient size to cause brain compression. Symptoms result from direct compression of the brain by the tumor or from secondary edema.

If the calvarial mass is situated over the superior sagittal or lateral sinus, it may compress that structure and cause increased intracranial pressure, usually with papilledema, and sometimes headache (Fig. 5–9). The neurologic examination is otherwise normal. Papilledema is also associated with compression of a dominant jugular vein at the base of the skull.[391] RT should be delivered to the lesion to decompress the venous structure. Hemorrhagic infarction of the brain involving one or both hemispheres, and associated with severe headache and focal signs including focal seizures, sometimes occurs with sagittal sinus obstruction caused by occlusion but is rare with metastatic compression (see Chapter 9).

In general, the treatment of symptomatic skull metastases is RT; surgical removal of a single large metastasis, particularly if it compresses the brain, may be indicated.[390] Asymptomatic metastases do not require specific treatment and usually respond to systemic therapy commensurate with any other site of disease.

A peculiar and unique phenomenon of skull metastasis is seen rarely in modern experience in children with neuroblastoma. The skull is widely seeded with metastatic tumor, the skull sutures may or may not split, and the skull expands as if the patient had hydrocephalus. The size of the brain and its ventricular system do not increase, however; the enlarging head is a direct result of tumor in both the skull and underlying epidural space. A CT scan easily establishes this diagnosis.[392]

Dural Metastases

In one autopsy series, 9% of patients had metastases to the dura; in 4% the dural lesion was the only intracranial metastasis.[37] Metastases to the dura (Fig. 5–10) may arise by extension from the skull, by direct hematogenous spread,

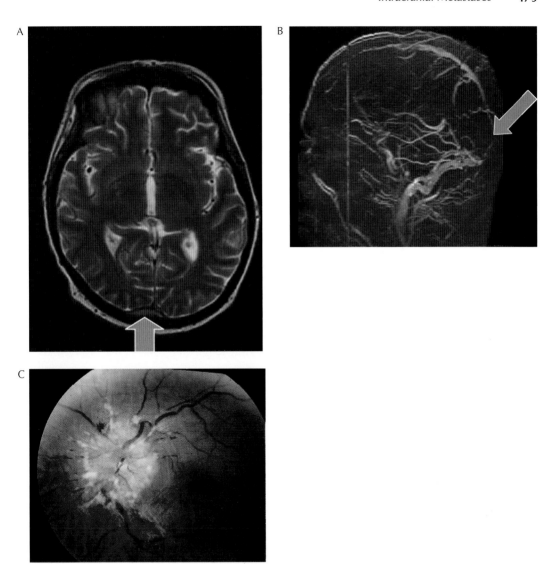

Figure 5–9. A skull metastasis compressing the sagittal sinus (*arrow*) and causing headache and papilledema but no focal neurologic signs (pseudo-pseudotumor). **A**: An axial magnetic resonance image of a patient with prostate cancer and multiple bony metastases. One metastasis, to the occipital bone, invaded the epidural space and compressed the sagittal sinus (*arrow*). **B**: His magnetic resonance venogram (MRV) showing severe compression of the sagittal sinus. **C**: His fundus photograph showing severe papilledema with hemorrhages over the disk and dilated venous vessels. The patient recovered after successful irradiation of the skull metastasis.

or more rarely by direct extension from a brain metastasis (most common in melanoma)[166,393] Carcinomas of the lung, prostate, and breast have a predilection for the dura,[393] but any tumor, including melanoma, leukemias and lymphomas, and even cancer of unknown primary site, may metastasize there.[393]

Two clinicopathologic syndromes result. In the first, the tumor compresses or invades the underlying brain. The symptoms resemble those of a parenchymal brain metastasis, although focal seizures at onset may be more common. The lesion is easily identified by MRI, but it may be difficult to distinguish dural-based metastases from meningiomas by radiologic means. Surgical extirpation, the treatment of choice for a single large lesion, prolongs survival.[393] In patients who are not appropriate candidates for surgery, RT has salutary effects on most metastases.

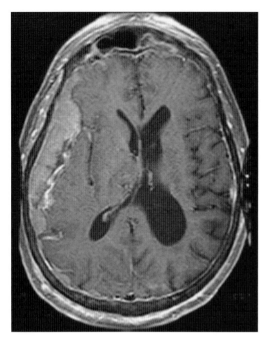

Figure 5–10. Dural metastases in a patient with prostate cancer.

The second syndrome results when subdural metastases exude fluid into the subdural space, producing a subdural hematoma or effusion.[394] The fluid presumably represents a leakage of plasma and blood from newly formed blood vessels by the tumor. The neurologic signs and symptoms are similar to those of a subdural hematoma or hygroma. Patients complain of headache, lethargy, and sometimes contralateral weakness. On examination, changes in mental state are out of proportion to focal neurologic signs, and localized headache and skull tenderness may be present. Unless there is clear evidence of nodular tumor on the MRI in addition to the subdural hemorrhage or effusion, a tumor effusion cannot be distinguished from a nontumorous chronic subdural hematoma. Drainage of the fluid with cytologic examination and/or biopsy of the dura is the only way of establishing the diagnosis.

If the metastasis is present in the midline, it may compress the posterior part of the sagittal sinus or the torcula, causing increased intracranial pressure, mimicking pseudotumor cerebri. Such patients have headache and papilledema, but no focal neurologic signs to suggest an intrinsic brain metastasis. A magnetic resonance venogram shows obstruction or significant narrowing of the sagittal sinus or sometimes a dominant lateral sinus. Treatment of the neoplasm ameliorates the symptoms.

A single tumor mass in the dura may be indistinguishable from a meningioma.[166,395] Metastatic tumor is more likely to cause edema in the underlying brain, and clinical symptoms are more likely to be present or to develop more rapidly. However, only a biopsy establishes the diagnosis. Also possible, as mentioned before, is a metastasis to a previously present meningioma.

Pineal

Pineal metastases are quite uncommon (Fig. 5–11). In one series of over 5000 intracranial metastatic tumors, only 17 (0.3%) were found in the pineal region.[24] Other studies have found that pineal metastases are slightly more common, occurring in 1.8% to 4% of patients with disseminated cancer.[396,397] Any tumor can metastasize there, but lung is probably the most common. Pineal metastases rarely cause symptoms unless they grow large enough to compress the cerebral aqueduct. A large pineal metastasis might cause symptoms similar to a primary pineal tumor, including hydrocephalus, large, fixed pupils, poor upgaze, convergence–retraction nystagmus (Parinaud syndrome), and hearing loss.[398] On rare occasions, symptomatic

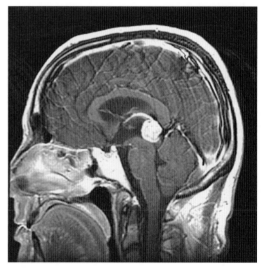

Figure 5–11. Melanoma metastatic to the pineal gland (see also Fig. 5–2F).

pineal region metastases may be the first evidence of cancer.[399] In one series of 191 patients referred for surgical management of pineal tumors of unknown etiology, 10 (5%) were metastases.[400] The MRI will demonstrate a pineal region tumor but does not have characteristics sufficient to differentiate it from other more common primary tumors that occur in the pineal region. The treatment is similar to that of an intracranial metastasis in other sites. Pineal metastases are particularly prone to seed the leptomeninges.[400] Interestingly, a pinealectomy prolongs life in cancer-prone mice,[401] and melatonin inhibits growth of mammary tumors in mice.[402]

Choroid

Choroid plexus metastases are encountered at autopsy in 2.6% of patients with cancer (Fig. 5–12).[396] Choroid metastases often occur in the setting of multiple other metastatic lesions[396] but can be solitary.[403] In one review of 18 cases of single metastasis to the choroid plexus, 8 were from kidney.[403] In some instances, these renal metastases were mistaken for benign tumors such as an intraventricular meningioma[404]; choroid plexus metastases may be followed for extended periods.[405] Lesions of the choroid plexus rarely produce symptoms unless they either seed cells into the CSF, causing leptomeningeal metastasis (see Chapter 7), or obstruct CSF outflow pathways at the foramen of Monro, leading to hydrocephalus. The

diagnosis and treatment resemble those of brain metastases.

Pituitary

Metastatic spread to the pituitary gland is more frequent than pineal or choroid metastases (Fig. 5–13). In one autopsy series of 737 patients with malignant tumors, pituitary metastases were found in 26 (3.5%).[396] In another series of 1000 autopsies, pituitary metastases were found in 1.8% of all cancer patients and in 9% with breast cancer.[406] In the past, when hypophysectomy was a treatment for breast cancer, metastases were found in approximately 10% of pituitary glands. Breast cancer is the most common cause[23] of pituitary metastasis. Other causes include SCLC, leukemia, or adenocarcinoma of unknown cause.[407] Only a minority of pituitary metastases are symptomatic.[407] It may be difficult to distinguish a pituitary metastasis from the more common pituitary adenoma (Fig. 5–13).[22,408] However, when metastases cause symptoms, the first is usually diabetes insipidus, distinguishing it from primary tumors,

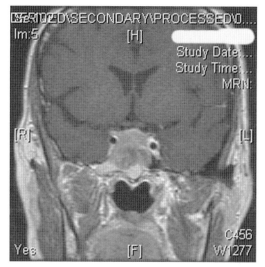

Figure 5–13. Colon cancer metastatic to the pituitary gland. This patient with colon cancer and liver metastases developed headache and diplopia. Magnetic resonance imaging revealed a large sellar mass compressing the abducens nerve on the right. Cortisol and thyroid levels were low, but other hormone levels were normal. There was no diabetes insipidus. The tumor was believed to be a pituitary adenoma but on transsphenoidal resection was discovered to be a metastasis.

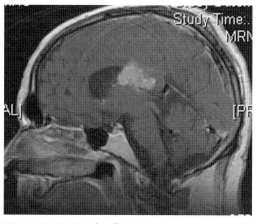

Figure 5–12. Renal cell carcinoma metastatic to the choroid plexus of the lateral ventricle.

which usually cause anterior pituitary failure or excessive production of a pituitary hormone. Most metastases to the pituitary gland involve the posterior lobe only, or the posterior lobe in combination with the anterior lobe. The reason is unclear, but the posterior lobe receives its blood supply directly from the systemic circulation, whereas the anterior lobe receives its blood supply via the portal circulation; tumor cells would first have to reach the hypothalamus before entering that circulation to involve the anterior lobe. In addition, many metastases, particularly those from carcinoma of the breast, involve the bony sella turcica immediately contiguous to the posterior lobe. Rarely, a metastasis may hemorrhage or infarct and present acutely as pituitary apoplexy.[409] The clinical symptoms include sudden onset of visual loss, ocular palsies, acute hypopituitarism, and alteration in consciousness varying from confusion to coma. Fever and nuchal rigidity may suggest the diagnosis of bacterial or viral meningitis. The CSF may contain blood or white cells, the latter in response to spilled necrotic material. Most patients respond to conservative treatment, replacing absent hormones and using corticosteroids to suppress inflammation. In some patients, emergency decompression of the enlarged pituitary is necessary to relieve symptoms. Untreated pituitary apoplexy can be fatal. If the tumor has grown into the suprasellar cistern, it may exude cells into the CSF; lumbar puncture may identify the malignant cells of a metastatic tumor.

REFERENCES

1. Sanchez de Cos EJ, Masjoans MD, Sojo Gonzalez MA, et al. Silent brain metastasis in the initial staging of lung cancer: evaluation by computed tomography and magnetic resonance imaging. *Arch Bronconeumol* 2007;43:386–391.

2. Yokoi K, Kamiya N, Matsuguma H, et al. Detection of brain metastasis in potentially operable non-small cell lung cancer—a comparison of CT and MRI. *Chest* 1999;115:714–719.

3. Tomlinson BE, Perry RH, Stewart-Wynne EG. Influence of site of origin of lung carcinomas on clinical presentation and central nervous system metastases. *J Neurol Neurosurg Psychiatry* 1979; 42:82–88.

4. Mayer M. A patient perspective on brain metastases in breast cancer. *Clin Cancer Res* 2007;13:1623–1624.

5. D'Ambrosio AL, Agazzi S. Prognosis in patients presenting with brain metastasis from an undiagnosed primary tumor. *Neurosurg Focus* 2007;22:E7.

6. Schueller P, Schroeder J, Micke O, et al. 9 years tumor free survival after resection, intraoperative radiotherapy (IORT) and whole brain radiotherapy of a solitary brain metastasis of non-small cell lung cancer. *Acta Oncol* 2006;45:224–225.

7. Hall WA, Djalilian HR, Nussbaum ES, et al. Long-term survival with metastatic cancer to the brain. *Med Oncol* 2000;17:279–286.

8. Chao ST, Barnett GH, Liu SW, et al. Five-year survivors of brain metastases: a single-institution report of 32 patients. *Int J Radiat Oncol Biol Phys* 2006;66:801–809.

9. Takeshima H, Kuratsu J, Nishi T, et al. Prognostic factors in patients who survived more than 10 years after undergoing surgery for metastatic brain tumors: report of 5 cases and review of the literature. *Surg Neurol* 2002;58:118–123.

10. Viani GA, Castilho MS, Salvajoli JV, et al. Whole brain radiotherapy for brain metastases from breast cancer: estimation of survival using two stratification systems. *BMC Cancer* 2007;7:53.

11. Boogerd W, Vos VW, Hart AAM, et al. Brain metastases in breast cancer: natural history, prognostic factors and outcome. *J Neuro-Oncol* 1993;15:165–174.

12. Kamby C, Soerensen PS. Characteristics of patients with short and long survivals after detection of intracranial metastases from breast cancer. *J Neuro-Oncol* 1988;6:37–45.

13. Patel RR, Mehta MP. Targeted therapy for brain metastases: improving the therapeutic ratio. *Clin Cancer Res* 2007;13:1675–1683.

14. Lagerwaard FJ, Levendag PC, Nowak PJ, et al. Identification of prognostic factors in patients with brain metastases: a review of 1292 patients. *Int J Radiat Oncol Biol Phys* 1999;43:795–803.

15. Tendulkar RD, Liu SW, Barnett GH, et al. RPA classification has prognostic significance for surgically resected single brain metastasis. *Int J Radiat Oncol Biol Phys* 2006;66:810–817.

16. Patriarca F, Zaja F, Silvestri F, et al. Meningeal and cerebral involvement in multiple myeloma patients. *Ann Hematol* 2001;80:758–762.

17. Ngan RKC, Yiu HHY, Cheng HKM, et al. Central nervous system metastasis from nasopharyngeal carcinoma—a report of two patients and a review of the literature. *Cancer* 2002;94:398–405.

18. Castaldo JE, Bernat JL, Meier FA, et al. Intracranial metastases due to prostatic carcinoma. *Cancer* 1983;52:1739–1747.

19. Cross PA, Ortega LG, Thedore N, et al. A subdural hematoma from metastatic esophageal cancer. *Acta Oncol* 1995;34:538–542.

20. Minette SE, Kimmel DW. Subdural hematoma in patients with systemic cancer. *Mayo Clin Proc* 1989;64:637–642.

21. Huinink DT, Veltman GA, Huizinga TW, et al. Diabetes insipidus in metastatic cancer: two case reports with review of the literature. *Ann Oncol* 2000;11:891–895.

22. Max MB, Deck MDF, Rottenberg DA. Pituitary metastasis: incidence in cancer patients and clinical differentiation from pituitary adenoma. *Neurology* 1981;31:998–1002.

23. Kimmel DW, O'Neill BP. Systemic cancer presenting as diabetes insipidus. Clinical and radiographic features of 11 patients with a review of metastatic-induced diabetes insipidus. *Cancer* 1983;52:2355–2358.

24. Hirato J, Nakazato Y. Pathology of pineal region tumors. *J Neuro-Oncol* 2001;54:239–249.
25. Vaquero J, Martinez R, Magallon R, et al. Intracranial metastases to the pineal region. Report of three cases. *J Neurosurg* 1991;35:55–57.
26. Kadrian D, Tan L. Single choroid plexus metastasis 16 years after nephrectomy for renal cell carcinoma: case report and review of the literature. *J Clin Neurosci* 2004;11:88–91.
27. Leach JC, Garrott H, King JA, et al. Solitary metastasis to the choroid plexus of the third ventricle mimicking a colloid cyst: a report of two cases. *J Clin Neurosci* 2004;11:521–523.
28. Tally PW, Laws ER, Jr, Scheithauer BW. Metastases of central nervous system neoplasms. Case report. *J Neurosurg* 1988;68:811–816.
29. Aghi M, Kiehl TR, Brisman JL. Breast adenocarcinoma metastatic to epidural cervical spine meningioma: case report and review of the literature. *J Neuro-Oncol* 2005;75:149–155.
30. Baratelli GM, Ciccaglioni B, Dainese E, et al. Metastasis of breast carcinoma to intracranial meningioma. *J Neurosurg Sci* 2004;48:71–73.
31. Leblanc RA. Metastasis of bronchogenic carcinoma to acoustic neurinoma. *J Neurosurg* 1974;41:614–617.
32. Molinatti PA, Scheithauer BW, Randall RV, et al. Metastasis to pituitary adenoma. *Arch Pathol Lab Med* 1985;109:287–289.
33. Chan CH, Fabinyi GC, Kalnins RM. An unusual case of tumor-to-cavernoma metastasis. A case report and literature review. *Surg Neurol* 2006;65:402–408.
34. Greene GM, Hart MN, Poor MM, Jr, et al. Carcinoma metastatic to a cerebellar vascular malformation: case report. *Neurosurgery* 1990;26:1054–1057.
35. Nielson SL, Posner JB. Brain metastasis localized to an area of infarction. *J Neuro-Oncol* 1983;1:191–195.
36. Cheng CL, Greenberg J, Hoover LA. Prostatic adenocarcinoma metastatic to chronic subdural hematoma membranes. Case report. *J Neurosurg* 1988;68:642–644.
37. Posner JB, Chernik NL. Intracranial metastases from systemic cancer. *Adv Neurol* 1978;19:575–587.
38. Takakura K, Sano K, Hojo S, et al. *Metastatic Tumors of the Central Nervous System.* Tokyo: Igaku-Shoin, 1982.
39. Pickren JW, Lopez G, Tsukada Y, et al. Brain metastases: an autopsy study. *Cancer Treat Symp* 1983;2:295–313.
40. Graf AH, Buchberger W, Langmayr H, et al. Site preference of metastatic tumours of the brain. *Virchows Arch A Pathol Anat Histopathol* 1988;412:493–498.
41. Kebudi R, Ayan I, Gorgun O, et al. Brain metastasis in pediatric extracranial solid tumors: survey and literature review. *J Neuro-Oncol* 2005;71:43–48.
42. Lin NU, Winer EP. Brain metastases: the HER2 paradigm. *Clin Cancer Res* 2007;13:1648–1655.
43. Yawn BP, Wollan PC, Schroeder C, et al. Temporal and gender-related trends in brain metastases from lung and breast cancer. *Minn Med* 2003;86:32–37.
44. Nussbaum ES, Djalilian HR, Cho KH, et al. Brain metastases—histology, multiplicity, surgery, and survival. *Cancer* 1996;78:1781–1788.
45. Delattre J-Y, Krol G, Thaler HT, et al. Distribution of brain metastases. *Arch Neurol* 1988;45:741–744.
46. Schellinger PD, Meinck HM, Thron A. Diagnostic accuracy of MRI compared to CCT in patients with brain metastases. *J Neuro-Oncol* 1999;44:275–281.
47. Giordana MT, Cordera S, Boghi A. Cerebral metastases as first symptom of cancer: a clinico-pathologic study. *J Neuro-Oncol* 2000;50:265–273.
48. Ruda R, Borgognone M, Benech F, et al. Brain metastases from unknown primary tumour: a prospective study. *J Neurol* 2001;248:394–398.
49. Eapen L, Vachet M, Catton G, et al. Brain metastases with an unknown primary: a clinical perspective. *J Neuro-Oncol* 1988;6:31–35.
50. Pelosi E, Pennone M, Deandreis D, et al. Role of whole body positron emission tomography/computed tomography scan with 18F-fluorodeoxyglucose in patients with biopsy proven tumor metastases from unknown primary site. *Q J Nucl Med Mol Imaging* 2006;50:15–22.
51. Van de Pol M, Van Aalst VC, Wilmink JT, et al. Brain metastases from an unknown primary tumour: which diagnostic procedures are indicated? *J Neurol Neurosurg Psychiatry* 1996;61:321–323.
52. Srodon M, Westra WH. Immunohistochemical staining for thyroid transcription factor-1: a helpful aid in discerning primary site of tumor origin in patients with brain metastases. *Hum Pathol* 2002;33:642–645.
53. Krafft C, Shapoval L, Sobottka SB, et al. Identification of primary tumors of brain metastases by infrared spectroscopic imaging and linear discriminant analysis. *Technol Cancer Res Treat* 2006;5:291–298.
53a. Weissleder R, Pittet MJ. Imaging in the era of molecular oncology. *Nature* 2008;452:580–589.
54. Cairncross JG, Kim J-H, Posner JB. Radiation therapy for brain metastases. *Ann Neurol* 1980;7:529–541.
55. Galluzzi S, Payne PM. Brain metastases from primary bronchial carcinoma: a statistical study of 741 necropsies. *Br J Cancer* 1956;10:408–414.
56. Chen AM, Jahan TM, Jablons DM, et al. Risk of cerebral metastases and neurological death after pathological complete response to neoadjuvant therapy for locally advanced nonsmall-cell lung cancer: clinical implications for the subsequent management of the brain. *Cancer* 2007;109:1668–1675.
57. Shahidi H, Kvale PA. Long-term survival following surgical treatment of solitary brain metastasis in non-small cell lung cancer. *Chest* 1996;109:271–276.
58. Figlin RA, Piantadosi S, Feld R, et al. Intracranial recurrence of carcinoma after complete surgical resection of stage I, II, and III non-small cell lung cancer. *N Engl J Med* 1988;318:1300–1305.
59. Hoseok I, Lee JI, Nam DH, et al. Surgical treatment of non-small cell lung cancer with isolated synchronous brain metastases. *J Kor Med Sci* 2006; 21:236–241.
60. Shi AA, Digumarthy SR, Temel JS, et al. Does initial staging or tumor histology better identify asymptomatic brain metastases in patients with non-small cell lung cancer? *J Thorac Oncol* 2006;1:205–210.
61. Jacobs L, Kinkel WR, Vincent RG. 'Silent' brain metastasis from lung carcinoma determined by computerized tomography. *Arch Neurol* 1977;34:690–693.
62. Salbeck R, Grau HC, Artmann H. Cerebral tumor staging in patients with bronchial carcinoma by computed tomography. *Cancer* 1990;66:2007–2011.

63. Gassel MM. False localizing signs. *Arch Neurol* 1961;4:526–554.

64. Arseni C, Maretsis M. Focal epileptic seizures ipsilateral to the tumour. *Acta Neurochir* 1979;49:47–60.

65. Collier J. The false localising signs of intracranial tumour. *Brain* 1904;27:490–508.

66. Jenkyn LR, Kinlaw WB, Bernat JL, et al. Falsely localizing third-nerve palsy. Letter to the editor. *N Engl J Med* 1980;303:161–162.

67. Klingele TG, Gado MH, Burde RM, et al. Compression of the anterior visual system by the gyrus rectus. Case report. *J Neurosurg* 1981;55:272–275.

68. Gelmers HJ. Relation of false localizing signs and remote hyperaemia in patients with intracranial mass lesions. *Acta Neurochir (Wien)* 1982;60:81–89.

69. Marshman LA, Polkey CE, Penney CC. Unilateral fixed dilation of the pupil as a false-localizing sign with intracranial hemorrhage: case report and literature review. *Neurosurgery* 2001;49:1251–1255.

70. Matsuura N, Akinori A. Trigeminal neuralgia and hemifacial spasm as false localizing signs in patients with a contralateral mass of the posterior cranial fossa Report of three cases. *J Neurosurg* 1996;84:1067–1071.

71. Lepore FE. False and non-localizing signs in neuro-ophthalmology. *Curr Opin Ophthalmol* 2002;13:371–374.

72. Young DF, Posner JB, Chu F, et al. Rapid-course radiation therapy of cerebral metastases: results and complications. *Cancer* 1974;34:1069–1076.

73. Oberndorfer S, Schmal T, Lahrmann H, et al. [The frequency of seizures in patients with primary brain tumors or cerebral metastases. An evaluation from the Ludwig Boltzmann Institute of Neuro-Oncology and the Department of Neurology, Kaiser Franz Josef Hospital, Vienna.] *Wien Klin Wochenschr* 2002;114:911–916.

74. Singh G, Rees JH, Sander JW. Seizures and epilepsy in oncological practice: causes, course, mechanisms and treatment. *J Neurol Neurosurg Psychiatry* 2007;78:342–349.

75. Posner JB, Saper CB, Schiff ND, et al. *Plum and Posner's Diagnosis of Stupor and Coma.* 4th ed. New York: Oxford University Press, 2007.

76. Larson SJ, Sances A, Jr, Baker JB, et al. Herniated cerebellar tonsils and cough syncope. *J Neurosurg* 1974;40:524–528.

77. Cuneo RA, Caronna JJ, Pitts L, et al. Upward transtentorial herniation: seven cases and a literature review. *Arch Neurol* 1979;36:618–623.

78. Osborn AG, Heaston DK, Wing SD. Diagnosis of ascending transtentorial herniation by cranial computed tomography. *AJR Am J Roentgenol* 1978;130:755–760.

79. Adamson DC, Dimitrov DF, Bronec PR. Upward transtentorial herniation, hydrocephalus, and cerebellar edema in hypertensive encephalopathy. *Neurologist* 2005;11:171–175.

80. Van Crevel H. Pathogenesis of raised cerebrospinal fluid pressure. *Doc Ophthalmol* 1982;52:251–257.

81. Van Crevel H. RIHSA cisternography in cerebral tumours. *Neuroradiology* 1979;18:133–138.

82. Ingvar DH, Lundberg N. Paroxysmal systems in intracranial hypertension, studied with ventricular fluid pressure recording and electroencephalography. *Brain* 1961;84:446–459.

83. Lundberg N. Continuous recording and control of ventricular fluid pressure in neurosurgical practice. *Acta Neurol Scand* 1960;36 (Suppl 149):1–193.

84. Baumert BG, Rutten I, Dehing-Oberije C, et al. A pathology-based substrate for target definition in radiosurgery of brain metastases. *Int J Radiat Oncol Biol Phys* 2006;66:187–194.

85. Neves S, Mazal PR, Wanschitz J, et al. Pseudoglioma-tous growth pattern of anaplastic small cell carcinomas metastatic to the brain. *Clin Neuropathol* 2001;20:38–42.

86. Küsters B, Leenders WPJ, Wesseling P, et al. Vascular endothelial growth factor-A$_{165}$ induces progression of melanoma brain metastases without induction of sprouting angiogenesis. *Cancer Res* 2002;62:341–345.

87. Arnold SM, Patchell RA. Diagnosis and management of brain metastases. *Hematol Oncol Clin North Am* 2001;15:1085–1107, vii.

88. Kirby S, Purdy RA. Headache and brain tumors. *Curr Neurol Neurosci Rep* 2007;7:110–116.

89. Klos KJ, O'Neill BP. Brain metastases. *Neurologist* 2004;10:31–46.

90. Argyriou AA, Chroni E, Polychronopoulos P, et al. Headache characteristics and brain metastases prediction in cancer patients. *Eur J Cancer Care (Engl)* 2006;15:90–95.

91. Clouston PD, DeAngelis LM, Posner JB. The spectrum of neurologic disease in patients with systemic cancer. *Ann Neurol* 1992;31:268–273.

92. Ramirez-Lassepas M, Espinosa CE, Cicero JJ, et al. Predictors of intracranial pathologic findings in patients who seek emergency care because of headache. *Arch Neurol* 1997;54:1506–1509.

93. Duarte J, Sempere AP, Delgado JA, et al. Headache of recent onset in adults: a prospective population-based study. *Acta Neurol Scand* 1996; 94:67–70.

94. Sobri M, Lamont AC, Alias NA, et al. Red flags in patients presenting with headache: clinical indications for neuroimaging. *Br J Radiol* 2003;76:532–535.

95. Ekbom K, Hornsten G, Johansson T. Posterior cranial fossa tumors. Headaches, oculostatic disorders and scintillation camera findings. *Headache* 1974;14:119–132.

96. Forsyth PA, Posner JB. Headaches in patients with brain tumors: a study of 111 patients. *Neurology* 1993;43:1678–1683.

97. Christiaans MH, Kelder JC, Arnoldus EPJ, et al. Prediction of intracranial metastases in cancer patients with headache. *Cancer* 2002;94:2063–2068.

98. Rushton JG, Rooke ED. Brain tumor headache. *Headache* 1962;2:147–152.

99. Simonescu ME. Metastatic tumors of the brain. A follow-up study of 195 patients with neurosurgical considerations. *J Neurosurg* 1960;17:361–373.

100. Pfund Z, Szapary L, Jaszberenyi O, et al. Headache in intracranial tumors. *Cephalalgia* 1999; 19:787–790.

101. Pepin EP. Cerebral metastasis presenting as migraine with aura. Letter to the editor. *Lancet* 1990;336:127–128.

102. Trucco M, Mainardi F, Maggioni F, et al. Chronic paroxysmal hemicrania, hemicrania continua and SUNCT syndrome in association with other pathologies: a review. *Cephalalgia* 2004;24:173–184.

103. van Crevel HV. Absence of papilloedema in cerebral tumours. *J Neurol Neurosurg Psychiatry* 1975;38:931–933.

104. Malamud N. Psychiatric disorder with intracranial tumors of limbic system. *Arch Neurol* 1967;17:113–123.

105. Madhusoodanan S, Danan D, Moise D. Psychiatric manifestations of brain tumors: diagnostic implications. *Expert Rev Neurother* 2007;7:343–349.

106. Sperling MR, Ko J. Seizures and brain tumors. *Semin Oncol* 2006;33:333–341.

107. Lynam LM, Lyons MK, Drazkowski JF, et al. Frequency of seizures in patients with newly diagnosed brain tumors: A retrospective review. *Clin Neurol Neurosurg* 2007;109:634–638.

108. Cavaliere R, Farace E, Schiff D. Clinical implications of status epilepticus in patients with neoplasms. *Arch Neurol* 2006;63:1746–1749.

109. Blitshteyn S, Jaeckle KA. Nonconvulsive status epilepticus in metastatic CNS disease. *Neurology* 2006;66:1261–1263.

110. Hormigo A, Liberato B, Lis E, et al. Nonconvulsive status epilepticus in patients with cancer: imaging abnormalities. *Arch Neurol* 2004;61:362–365.

111. Mirsattari SM, Lee DH, Jones MW, et al. Transient lesion in the splenium of the corpus callosum in an epileptic patient. *Neurology* 2003;60:1838–1841.

112. Rolak LA, Rutecki P, Ashizawa T, et al. Clinical features of Todd's post-epileptic paralysis. *J Neurol Neurosurg Psychiatry* 1992;55:63–64.

113. Yarnell PR. Todd's paralysis: a cerebrovascular phenomenon? *Stroke* 1975;6:301–303.

114. Kim JA, Chung JI, Yoon PH, et al. Transient MR signal changes in patients with generalized tonicoclonic seizure or status epilepticus: periictal diffusion-weighted imaging. *AJNR Am J Neuroradiol* 2001;22:1149–1160.

115. Fisher CM. Transient paralytic attacks of obscure nature: the question of non-convulsive seizure paralysis. *Can J Neurol Sci* 1978;5:267–273.

116. Beaumont A, Whittle IR. The pathogenesis of tumour associated epilepsy. *Acta Neurochir (Wien)* 2000;142:1–15.

117. Schaller B. Influences of brain tumor-associated pH changes and hypoxia on epileptogenesis. *Acta Neurol Scand* 2005;111:75–83.

118. Maurice-Williams RS. Mechanism of production of gait unsteadiness by tumours in the posterior fossa. *J Neurol Neurosurg Psychiatry* 1975;38:143–148.

119. Damasio AR, Benton AL. Impairment of hand movements under visual guidance. *Neurology* 1979;29:170–174.

120. Ando S, Moritake K. Pure optic ataxia associated with a right parieto-occipital tumour. *J Neurol Neurosurg Psychiatry* 1990;53:805–806.

121. Paillas JE, Pellet W. Brain metastases. In: Vinken PJ, Bruyn GW, eds. *Handbook of Clinical Neurology, Vol. 18.* New York: Elsevier, 1975. pp. 201–232.

122. Hildebrand J. *Lesions of the Nervous System in Cancer Patients.* New York: Raven, 1978. Monograph Series of the European Organization for Research on Treatment of Cancer, Vol. 5.

123. Gamache FW, Jr, Galicich JH, Posner JB. Treatment of brain metastases by surgical extirpation. In: Weiss L, Gilbert HA, Posner JB, eds. *Brain Metastasis.* Boston: GK Hall, 1980. pp. 394–414.

124. Elkington JSC. Metastatic tumors of the brain. *Proc Royal Soc Med* 1935;28:1080–1096.

125. Fadul C, Misulis KE, Wiley RG. Cerebellar metastases: diagnostic and management considerations. *J Clin Oncol* 1987;5:1107–1115.

126. Moore AP, Humphrey PR. Man-in-the-barrel syndrome caused by cerebral metastases. *Neurology* 1989;39:1134–1135.

127. Rudd A, McKenzie JG, Millard PH. Carcinoma of the bronchus presenting with hemichorea. Letter to the editor. *J Neurol Neurosurg Psychiatry* 1986;49:1210–1211.

128. Floeter MK, So YT, Ross DA, et al. Miliary metastasis to the brain: clinical and radiologic features. *Neurology* 1987;37:1817–1818.

129. Madow L, Alpers BJ. Encephalitic form of metastatic carcinoma. *Arch Neurol Psychiatry* 1951; 65:161–173.

130. Van der Sluijs BM, Renier WO, Kappelle AC. Brain tumour as a rare cause of cardiac syncope. *J Neuro-Oncol* 2004;67:241–244.

131. DeCarvalho C, Shuttleworth E, Knox D, et al. Bilateral gaze paralysis with positive computerized tomography findings. A clinicoanatomic correlation. *Arch Neurol* 1980;37:184–186.

132. Derby BM, Guiang RL. Spectrum of symptomatic brain-stem metastasis. *J Neurol Neurosurg Psychiatry* 1975;38:888–895.

133. Stevenson GC, Hoyt WF. Metastasis to midbrain from mammary carcinoma. *JAMA* 1963;186:514–516.

134. Evans CH, Westfall V, Atuk NO. Astrocytoma mimicking the features of pheochromocytoma. *N Engl J Med* 1972;286:1397–1399.

135. Hsu CY, Hogan EL, Wingfield W, Jr, et al. Orthostatic hypotension with brainstem tumors. *Neurology* 1984;34:1137–1143.

136. Stotka VL, Barcay SJ, Bell HS, et al. Intractable hiccough as the primary manifestation of brain stem tumor. *Am J Med* 1962;32:313–315.

137. Abello R, Yeakley JW, Goldman P. Secondary achalasia in a patient with brainstem metastases from lung carcinoma. Letter to the editor. *J Clin Gastroenterol* 1992;14:176–178.

138. Mendez MF. Pavor nocturnus from a brainstem glioma. Letter to the editor. *J Neurol Neurosurg Psychiatry* 1992;55:860.

139. Martin RA, Handel SF, Aldama AE. Inability to sneeze as a manifestation of medullary neoplasm. *Neurology* 1991;41:1675–1676.

140. Jaeckle KA, Digre KB, Jones CR, et al. Central neurogenic hyperventilation: pharmacologic intervention with morphine sulfate and correlative analysis of respiratory, sleep, and ocular motor dysfunction. *Neurology* 1990;40:1715–1720.

141. Kondziolka D, Bernstein M, Resch L, et al. Significance of hemorrhage into brain tumors: clinicopathological study. *J Neurosurg* 1987; 67:852–857.

142. Mandybur TI. Intracranial hemorrhage caused by metastatic tumors. *Neurology* 1977;27:650–655.

143. Van Eck JHM, Go KG, Ebels EJ. Metastatic tumours of the brain. *Psychiatr Neurol Neurochir* 1965;68:443–462.

144. Weisberg LA, Nice CN. Intracranial tumors simulating the presentation of cerebrovascular syndromes. Early detection with cerebral computed tomography (CCT). *Am J Med* 1977;63:517–524.

145. Limper AH, Prakash UB, Kokmen E, et al. Cardiopulmonary metastatic lesions of osteosarcoma and associated cerebral infarction. *Mayo Clin Proc* 1988;63:592–595.

146. de Divitiis E, Spaziante R, Stella L, et al. Le syndrome pseudo-vasculaire des metastases intracraniennes. *Neurochirurgie* 1978;24:235–238.

147. Seute T, Leffers P, Ten Velde GP, et al. Detection of brain metastases from small cell lung cancer: Consequences of changing imaging techniques (CT versus MRI). *Cancer* 2008;112:1827–1834.

148. Kaal EC, Taphoorn MJ, Vecht CJ. Symptomatic management and imaging of brain metastases. *J Neuro-Oncol* 2005;75:15–20.

149. Ba-Ssalamah A, Nobauer-Huhmann IM, Pinker K, et al. Effect of contrast dose and field strength in the magnetic resonance detection of brain metastases. *Invest Radiol* 2003;38:415–422.

150. Terae S, Yoshida D, Kudo K, et al. Contrast-enhanced FLAIR imaging in combination with pre- and postcontrast magnetization transfer T1-weighted imaging: usefulness in the evaluation of brain metastases. *J Magn Reson Imaging* 2007;25:479–487.

151. Geijer B, Holtas S. Diffusion-weighted imaging of brain metastases: their potential to be misinterpreted as focal ischaemic lesions. *Neuroradiology* 2002;44:568–573.

152. Opstad KS, Murphy MM, Wilkins PR, et al. Differentiation of metastases from high-grade gliomas using short echo time 1H spectroscopy. *J Magn Reson Imaging* 2004;20:187–192.

153. Cairncross JG, Pexman JHW, Rathbone MP, et al. Postoperative contrast enhancement in patients with brain tumor. *Ann Neurol* 1985;17:570–572.

154. Huber PE, Hawighorst H, Fuss M, et al. Transient enlargement of contrast uptake on MRI after linear accelerator (linac) stereotactic radiosurgery for brain metastases. *Int J Radiat Oncol Biol Phys* 2001;49:1339–1349.

154a. de Bruin HG, Bromberg JE, Swaak-Kragten AT, et al. The incidence of pseudo-progression in a cohort of malignant brain glioma patients treated with chemo-radiation with temozolomide. *J Clin Oncol* 2007;25:775 (abstract).

155. Ferreira MP, Ferreira NP, Pereira Filho AA, et al. Stereotactic computed tomography-guided brain biopsy: diagnostic yield based on a series of 170 patients. *Surg Neurol* 2006;65 (Suppl 1):S1.

156. Pagani JJ, Hayman LA, Bigelow RH, et al. Diazepam prophylaxis of contrast media-induced seizures during computed tomography of patients with brain metastases. *AJNR* 1983;140:787–792.

157. Wendling LR, Cromwell LD, Latchaw RE. Computed tomography of intracerebral leukemic masses. *Am J Roentgenol* 1979;132:217–220.

158. Yamazaki T, Harigaya T, Noguchi O, et al. Calcified miliary brain metastases with mitochondrial inclusion bodies. *J Neurol Neurosurg Psychiatry* 1993;56:110–111.

159. Pfannenberg C, Aschoff P, Schanz S, et al. Prospective comparison of 18F-fluorodeoxyglucose positron emission tomography/computed tomography and whole-body magnetic resonance imaging in staging of advanced malignant melanoma. *Eur J Cancer* 2007;43:557–564.

160. Posther KE, McCall LM, Harpole DH, Jr, et al. Yield of brain 18F-FDG PET in evaluating patients with potentially operable non-small cell lung cancer. *J Nucl Med* 2006;47:1607–1611.

161. Ludwig V, Komori T, Kolb D, et al. Cerebral lesions incidentally detected on 2-deoxy-2-[18F]fluoro-D-glucose positron emission tomography images of patients evaluated for body malignancies. *Mol Imaging Biol* 2002;4:359–362.

162. Kwee SA, Ko JP, Jiang CS, et al. Solitary brain lesions enhancing at MR imaging: evaluation with fluorine 18-fluorocholine PET. *Radiology* 2007;244(2):557–565.

163. Louis DN, Hamilton AJ, Sobel RA, et al. Pseudopsammomatous meningioma with elevated serum carcinoembryonic antigen: a true secretory meningioma. *J Neurosurg* 1991;74:129–132.

164. Patchell RA, Tibbs PA, Walsh JW. A randomized trial of surgery in the treatment of single metastases to the brain. *N Engl J Med* 1990;322:494–500.

165. Lath CO, Khanna PC, Gadewar S, et al. Intracranial metastasis from prostatic adenocarcinoma simulating a meningioma. *Australas Radiol* 2005;49:497–500.

166. Laigle-Donadey F, Taillibert S, Mokhtari K, et al. Dural metastases. *J Neuro-Oncol* 2005;75:57–61.

167. Caroli E, Salvati M, Giangaspero F, et al. Intrameningioma metastasis as first clinical manifestation of occult primary breast carcinoma. *Neurosurg Rev* 2006;29:49–54.

168. Custer BS, Koepsell TD, Mueller BA. The association between breast carcinoma and meningioma in women. *Cancer* 2002;94:1626–1635.

169. Linfante I, Llinas RH, Caplan LR, et al. MRI features of intracerebral hemorrhage within 2 hours from symptom onset. *Stroke* 1999;30:2263–2267.

170. Patel MR, Edelman RR, Warach S. Detection of hyperacute primary intraparenchymal hemorrhage by magnetic resonance imaging. *Stroke* 1996;27:2321–2324.

171. Morgenstern LB, Frankowski RF. Brain tumor masquerading as stroke. *J Neuro-Oncol* 1999;44:47–52.

172. Petzold GC, Valdueza JM, Zimmer C. Cerebral metastasis of renal carcinoma mimicking venous haemorrhagic infarction. *J Neurol Neurosurg Psychiatry* 2004;75:477.

173. Couce ME, Medina-Flores R, Wong M, et al. Occult germ cell tumour presenting as spontaneous intracerebral haemorrhage. *Histopathology* 2007;50:789–793.

174. Abrahams NA, Prayson RA. The role of histopathologic examination of intracranial blood clots removed for hemorrhage of unknown etiology: a clinical pathologic analysis of 31 cases. *Ann Diagn Pathol* 2000;4:361–366.

175. Mishra AM, Gupta RK, Jaggi RS, et al. Role of diffusion-weighted imaging and in vivo proton magnetic resonance spectroscopy in the differential diagnosis of ring-enhancing intracranial cystic mass lesions. *J Comput Assist Tomogr* 2004;28:540–547.

176. Omuro AM, Leite CC, Mokhtari K, et al. Pitfalls in the diagnosis of brain tumours. *Lancet Neurol* 2006;5:937–948.

177. Hustinx R, Pourdehnad M, Kaschten B, et al. PET imaging for differentiating recurrent brain tumor from radiation necrosis. *Radiol Clin North Am* 2005;43:35–47.

178. Chernov M, Hayashi M, Izawa M, et al. Differentiation of the radiation-induced necrosis and tumor recurrence after gamma knife radiosurgery for brain metastases: importance of multi-voxel proton MRS. *Minim Invasive Neurosurg* 2005;48:228–234.

179. Sugahara T, Korogi Y, Tomiguchi S, et al. Posttherapeutic intraaxial brain tumor: the value of perfusion-sensitive contrast-enhanced MR imaging for differentiating tumor recurrence from non-neoplastic contrast-enhancing tissue. *AJNR Am J Neuroradiol* 2000;21:901–909.

180. Weber MA. Diagnostic performance of spectroscopic and perfusion MRI for distinction of brain tumors. *Neurology* 2006;66:1899–1906.

181. Kepes JJ. Large focal tumor-like demyelinating lesions of the brain: intermediate entity between multiple sclerosis and acute disseminated encephalomyelitis? A study of 31 patients. *Ann Neurol* 1993;33:18–27.

182. Peterson K, Rosenblum MK, Powers JM, et al. Effect of brain irradiation on demyelinating lesions. *Neurology* 1993;43:2105–2112.

183. Masdeu JC, Moreira J, Trasi S, et al. The open ring. A new imaging sign in demyelinating disease. *J Neuroimaging* 1996;6:104–107.

183a. Ewend MG, Elbabaa S, Carey LA. Current treatment paradigms for the management of patients with brain metastases. *Neurosurgery* 2005;57 (5 suppl):S66–77.

184. Shaw PH, Adams R, Jordan C, et al. A clinical review of the investigation and management of carcinoma of unknown primary in a single cancer network. *Clin Oncol (R Coll Radiol)* 2007;19:87–95.

185. Polyzoidis KS, Miliaras G, Pavlidis N. Brain metastasis of unknown primary: a diagnostic and therapeutic dilemma. *Cancer Treat Rev* 2005;31:247–255.

186. Ewend MG, Brem S, Gilbert M, et al. Treatment of single brain metastasis with resection, intracavity carmustine polymer wafers, and radiation therapy is safe and provides excellent local control. *Clin Cancer Res* 2007;15(13):3637–3641.

187. Chang D-B, Yang P-C, Luh K-T, et al. Late survival of non-small cell lung cancer patients with brain metastases. Influence of treatment. *Chest* 1992;101:1293–1297.

188. Hazra T, Mullins GM, Lott S. Management of cerebral metastases from bronchogenic carcinoma. *Johns Hopkins Med J* 1972;130:377–383.

189. Papac RJ. Spontaneous regression of cancer. *Cancer Treat Rev* 1996;22:395–423.

190. Hensiek AE, Kellerman AJ, Hill JT. Spontaneous regression of a solitary cerebral metastasis in renal carcinoma followed by meningioma development under medroxyprogesterone acetate therapy. *Br J Neurosurg* 2000;14:354–356.

191. Guthbjartsson T, Gislason T. Spontaneous regression of brain metastasis secondary to renal cell carcinoma. *Scand J Urol Nephrol* 1995;29:215–217.

192. Bauman ML, Price TR. Intracranial metastatic malignant melanoma: long-term survival following subtotal resection. *South Med J* 1972; 65:344–346.

193. Omland H, Fossa SD. Spontaneous regression cerebral and pulmonary metastases in renal cell carcinoma. *Scand J Urol Nephrol* 1989; 23:159–160.

194. Li J, Bentzen SM, Renschler M, et al. Regression after whole-brain radiation therapy for brain metastases correlates with survival and improved neurocognitive function. *J Clin Oncol* 2007;25:1260–1266.

195. Tsao MN, Lloyd NS, Wong RK. Clinical practice guideline on the optimal radiotherapeutic management of brain metastases. *BMC Cancer* 2005;5:34.

196. DeAngelis LM, Delattre J-Y, Posner JB. Radiation-induced dementia in patients cured of brain metastases. *Neurology* 1989;39:789–796.

197. Cairncross JG, Chernik NL, Kim J-H, et al. Sterilization of cerebral metastases by radiation therapy. *Neurology* 1979;29:1195–1202.

198. Libshitz HI, Jing B-S, Wallace S, et al. Sterilized metastases: a diagnostic and therapeutic dilemma. *Am J Roentgenol* 1983;140:15–19.

199. Tsao MN. Radiotherapeutic management of brain metastases: a systematic review and meta-analysis. *Cancer Treat Rev* 2005;31:256–273.

200. Rades D. Dose escalation beyond 30 grays in 10 fractions for patients with multiple brain metastases. *Cancer* 2007;110:1345–1350.

201. Rades D. Two radiation regimens and prognostic factors for brain metastases in nonsmall cell lung cancer patients. *Cancer* 2007;110:1077–1082.

202. Rades D, Lohynska R, Veninga T, et al. Evaluation of 2 whole-brain radiotherapy schedules and prognostic factors for brain metastases in breast cancer patients. *Cancer* 2007;110(11):2587–2592.

203. Richards GM, Mehta MP. Motexafin gadolinium in the treatment of brain metastases. *Expert Opin Pharmacother* 2007;8:351–359.

204. Mehta MP, Rodrigus P, Terhaard CH, et al. Survival and neurologic outcomes in a randomized trial of motexafin gadolinium and whole-brain radiation therapy in brain metastases. *J Clin Oncol* 2003;21:2529–2536.

205. Suh JH, Stea B, Nabid A, et al. Phase III study of efaproxiral as an adjunct to whole-brain radiation therapy for brain metastases. *J Clin Oncol* 2006;24:106–114.

206. Stupp R, Mason WP, van den Bent MJ, et al. Radiotherapy plus concomitant and adjuvant temozolomide for glioblastoma. *N Engl J Med* 2005;352:987–996.

207. Chang JE, Robins HI, Mehta MP. Therapeutic advances in the treatment of brain metastases. *Clin Adv Hematol Oncol* 2007;5:54–64.

208. Kouvaris JR, Miliadou A, Kouloulias VE, et al. Phase II study of temozolomide and concomitant whole brain radiotherapy in patients with brain metastases from solid tumors. *Onkologie* 2007; 30:361–366.

209. Byrne TN, Cascino TL, Posner JB. Brain metastasis from melanoma. *J Neuro-Oncol* 1983;1:313–317.

210. Cascino TL, Leavengood JM, Kemeny N, et al. Brain metastases from colon cancer. *J Neuro-Oncol* 1983;1:203–209.

211. Gay PC, Litchy WJ, Cascino TL. Brain metastasis in hypernephroma. *J Neuro-Oncol* 1987;5:51–56.

212. Retsas S, Gershuny AR. Central nervous system involvement in malignant melanoma. *Cancer* 1988;61:1926–1934.

213. Van der Steen-Banasik E, Hermans J, Tjho-Heslinga R, et al. The objective response of brain metastases on radiotherapy. A prospective study using computer tomography. *Acta Oncol* 1992;31:777–780.

214. Swift PS, Phillips T, Martz K, et al. CT characteristics of patients with brain metastases treated in RTOG study 79–16. *Int J Radiat Oncol Biol Phys* 1993;25:209–214.

215. Hall EJ. Do no harm—normal tissue effects. *Acta Oncologica* 2001;40:913–916.

216. Durand RE, Aquino-Parsons C. Clinical relevance of intermittent tumour blood flow. *Acta Oncologica* 2001;40:929–936.

217. Khuntia D, Brown P, Li J, Mehta MP. Whole-brain radiotherapy in the management of brain metastasis. *J Clin Oncol* 2006;24:1295–1304.

218. Yamamoto M. Radiosurgery for metastatic brain tumors. *Prog Neurol Surg* 2007;20:106–128.

219. Yen CP, Sheehan J, Patterson G, et al. Gamma knife surgery for metastatic brainstem tumors. *J Neurosurg* 2006;105:213–219.

220. Kased N, Huang K, Nakamura JL, et al. Gamma knife radiosurgery for brainstem metastases: the UCSF experience. *J Neuro-Oncol* 2008;86:195–205.

221. Smith ML, Lee JY. Stereotactic radiosurgery in the management of brain metastasis. *Neurosurg Focus* 2007;22:E5.

222. Oh BC, Liu CY, Wang MY, et al. Stereotactic radiosurgery: adjacent tissue injury and response after high-dose single fraction radiation. Part II: Strategies for therapeutic enhancement, brain injury mitigation, and brain injury repair. *Neurosurgery* 2007;60:799–814.

223. Niranjan A, Gobbel GT, Kondziolka D, et al. Experimental radiobiological investigations into radiosurgery: present understanding and future directions. *Neurosurgery* 2004;55:495–504.

224. Oh BC, Pagnini PG, Wang MY, et al. Stereotactic radiosurgery: adjacent tissue injury and response after high-dose single fraction radiation: Part I: Histology, imaging, and molecular events. *Neurosurgery* 2007;60:31–44.

225. Weltman E, Salvajouli JV, Brandt RA, et al. Radiosurgery for brain metastases: Who may not benefit? *Int J Radiat Oncol Biol Phys* 2001; 51:1320–1327.

226. Rades D, Pluemer A, Veninga T, et al. Whole-brain radiotherapy versus stereotactic radiosurgery for patients in recursive partitioning analysis classes 1 and 2 with 1 to 3 brain metastases. *Cancer* 2007;110(10):2285–2292.

227. Weltman E, Salvajoli JV, Brandt RA, et al. Radiosurgery for brain metastases: a score index for predicting prognosis. *Int J Radiat Oncol Biol Phys* 2000;46:1155–1161.

228. Aoyama H, Shirato H, Tago M, et al. Stereotactic radiosurgery plus whole-brain radiation therapy vs stereotactic radiosurgery alone for treatment of brain metastases: a randomized controlled trial. *JAMA* 2006;295:2483–2491.

229. Correa DD, DeAngelis LM, Shi W, et al. Cognitive functions in survivors of primary central nervous system lymphoma. *Neurology* 2004;62:548–555.

230. Sawrie M, Guthrie BL, Spencer SA, et al. Predictors of distant brain recurrence for patients with newly diagnosed brain metastases treated with stereotactic radiosurgery alone. Int J Radiation Oncology Biol. Phys. 2007;70:181–186

231. Andrews DW, Scott CB, Sperduto PW, et al. Whole brain radiation therapy with or without stereotactic radiosurgery boost for patients with one to three brain metastases: phase III results of the RTOG 9508 randomised trial. *Lancet* 2004;363:1665–1672.

232. Kondziolka D, Patel A, Lunsford LD, et al. Stereotactic radiosurgery plus whole brain radiotherapy versus radiotherapy alone for patients with multiple brain metastases. *Int J Radiat Oncol Biol Phys* 1999;45:427–434.

233. Nam TK, Lee JI, Jung YJ, et al. Gamma knife surgery for brain metastases in patients harboring four or more lesions: survival and prognostic factors. *J Neurosurg* 2005;102 (Suppl):147–150.

234. Yu CP, Cheung JY, Chan JF, et al. Prolonged survival in a subgroup of patients with brain metastases treated by gamma knife surgery. *J Neurosurg* 2005;102 (Suppl):262–265.

235. Curry WT, Jr, Cosgrove GR, Hochberg FH, et al. Stereotactic interstitial radiosurgery for cerebral metastases. *J Neurosurg* 2005;103:630–635.

236. Rogers LR, Rock JP, Sills AK, et al. Results of a phase II trial of the GliaSite radiation therapy system for the treatment of newly diagnosed, resected single brain metastases. *J Neurosurg* 2006;105:375–384.

237. Mayr MT, Crocker IR, Butker EK, et al. Results of interstitial brachytherapy for malignant brain tumors. *Int J Oncol* 2002;21:817–823.

238. Bogart JA, Ungureanu C, Shihadeh E, et al. Resection and permanent I-125 brachytherapy without whole brain irradiation for solitary brain metastasis from non-small cell lung carcinoma. *J Neuro-Oncol* 1999;44:53–57.

239. Dagnew E, Kanski J, McDermott MW, et al. Management of newly diagnosed single brain metastasis using resection and permanent iodine-125 seeds without initial whole-brain radiotherapy: a two institution experience. *Neurosurg Focus* 2007;22:E3.

240. Reardon DA, Akabani G, Coleman RE, et al. Salvage radioimmunotherapy with murine iodine-131-labeled antitenascin monoclonal antibody 81C6 for patients with recurrent primary and metastatic malignant brain tumors: phase II study results. *J Clin Oncol* 2006;24:115–122.

241. Pui CH. Central nervous system disease in acute lymphoblastic leukemia: prophylaxis and treatment. *Hematol Am Soc Hematol Educ Prog* 2006;142–146.

242. Cooper JS, Steinfeld AD, Lerch IA. Cerebral metastases: value of reirradiation in selected patients. *Radiology* 1990;174(3 Pt 1):883–885.

243. Dritschilo A, Bruckman JE, Cassady JR, et al. Tolerance of brain to multiple courses of radiation therapy. I. Clinical experiences. *Br J Radiol* 1981;54:782–786.

244. Weil RJ, Lonser RR. Selective excision of metastatic brain tumors originating in the motor cortex with preservation of function. *J Clin Oncol* 2005;23(6):1209–1217.

245. Pamir MN, Peker S, Ozek MM, et al. Intraoperative MR imaging: preliminary results with 3 tesla MR system. *Acta Neurochir Suppl* 2006;98:97–100.

246. Élster AD, DiPersio DA. Cranial postoperative site: assessment with contrast-enhanced MR imaging. *Radiology* 1990;174:93–98.

247. Wakai S, Andoh Y, Ochiai C, et al. Postoperative contrast enhancement in brain tumors and intracerebral hematomas: CT study. *J Comput Assist Tomogr* 1990;14(2):267–271.

248. Rock JP, Haines S, Recht L, et al. Practice parameters for the management of single brain metastasis. *Neurosurg Focus* 2000;9:ecp2.

249. Sills AK. Current treatment approaches to surgery for brain metastases. *Neurosurgery* 2005;57:S24–S32.

250. Ranasinghe MG, Sheehan JM. Surgical management of brain metastases. *Neurosurg Focus* 2007;22:E2.

251. Vecht CJ, Haaxma-Reiche H, Noordijk EM, et al. Treatment of single brain metastasis: radiotherapy alone or combined with neurosurgery? *Ann Neurol* 1993;33:583–590.

252. Mintz AH, Kestle J, Rathbone MP, et al. A randomized trial to assess the efficacy of surgery in addition to radiotherapy in patients with a single cerebral metastasis. *Cancer* 1996;78:1470–1476.

253. Moser RP, Johnson ML. Surgical management of brain metastases: how aggressive should we be? *Oncology* 1989;3:123–127.

254. Patchell RA, Tibbs PA, Regine WF, et al. Postoperative radiotherapy in the treatment of single metastases to the brain: a randomized trial. *JAMA* 1998;280:1485–1489.

255. Bindal RK, Sawaya R, Leavens ME, et al. Surgical treatment of multiple brain metastases. *J Neurosurg* 1993;79:210–216.

256. Siker ML, Mehta MP. Resection versus radiosurgery for patients with brain metastases. *Future Oncol* 2007;3:95–102.

257. Bindal AK, Bindal RK, Hess KR, et al. Surgery versus radiosurgery in the treatment of brain metastasis. *J Neurosurg* 1996;84:748–754.

258. Rades D, Bohlen G, Pluemer A, et al. Stereotactic radiosurgery alone versus resection plus whole-brain radiotherapy for 1 or 2 brain metastases in recursive partitioning analysis class 1 and 2 patients. *Cancer* 2007;109:2515–2521.

259. Fuentes R, Bonfill X, Exposito J. Surgery versus radiosurgery for patients with a solitary brain metastasis from non-small cell lung cancer. *Cochrane Database Syst Rev* 2006:CD004840.

260. Siomin VE, Vogelbaum MA, Kanner AA, et al. Posterior fossa metastases: risk of leptomeningeal disease when treated with stereotactic radiosurgery compared to surgery. *J Neuro-Oncol* 2004;67:115–121.

261. Mirimanoff RO, Choi NC. Intradural spinal metastases in patients with posterior fossa brain metastases from various primary cancers. *Oncology* 1987;44:232–236.

262. Patchell RA, Cirrincione C, Thaler HT, et al. Single brain metastases: surgery plus radiation or radiation alone. *Neurology* 1986;36:447–453.

263. Rades D. A boost in addition to whole-brain radiotherapy improves patient outcome after resection of 1 or 2 brain metastases in recursive partitioning analysis class 1 and 2 patients. *Cancer* 2007;110:1551–1559.

264. Sundaresan N, Sachdev VP, DiGiacinto GV, et al. Reoperation for brain metastases. *J Clin Oncol* 1988;6:1625–1629.

265. Arbit E, Wronski M, Burt M, et al. The treatment of patients with recurrent brain metastases. A retrospective analysis of 109 patients with nonsmall cell lung cancer. *Cancer* 1995;76:765–773.

266. de Vries NA, Beijnen JH, Boogerd W, et al. Blood-brain barrier and chemotherapeutic treatment of brain tumors. *Expert Rev Neurother* 2006;6:1199–1209.

267. Fortin D, Gendron C, Boudrias M, et al. Enhanced chemotherapy delivery by intraarterial infusion and blood-brain barrier disruption in the treatment of cerebral metastasis. *Cancer* 2007;109:751–760.

268. Rosner D, Nemoto T, Lane WW. Chemotherapy induces regression of brain metastases in breast carcinoma. *Cancer* 1986;58:832–839.

269. Cocconi G, Lottici R, Bisagni G, et al. Combination therapy with platinum and etoposide of brain metastases from breast carcinoma. *Cancer Invest* 1990;8:327–334.

270. Franciosi V, Cocconi G, Michiara M, et al. Front-line chemotherapy with cisplatin and etoposide for patients with brain metastases from breast carcinoma, nonsmall cell lung carcinoma, or malignant melanoma. A prospective study. *Cancer* 1999;85:1599–1605.

271. Fine RL, Chen J, Balmaceda C, et al. Randomized study of paclitaxel and tamoxifen deposition into human brain tumors: implications for the treatment of metastatic brain tumors. *Clin Cancer Res* 2006;12:5770–5776.

272. Gerstner ER, Fine RL. Increased permeability of the blood-brain barrier to chemotherapy in metastatic brain tumors: establishing a treatment paradigm. *J Clin Oncol* 2007;25:2306–2312.

273. Tosoni A, Ermani M, Brandes AA. The pathogenesis and treatment of brain metastases: a comprehensive review. *Crit Rev Oncol Hematol* 2004;52:199–215.

274. Peereboom DM. Chemotherapy in brain metastases. *Neurosurgery* 2005;57:S54–S65.

275. Seute T, Leffers P, Wilmink JT, et al. Response of asymptomatic brain metastases from small-cell lung cancer to systemic first-line chemotherapy. *J Clin Oncol* 2006;24:2079–2083.

276. Pollock BE, Brown PD, Foote RL, et al. Properly selected patients with multiple brain metastases may benefit from aggressive treatment of their intracranial disease. *J Neuro-Oncol* 2003;61:73–80.

277. Murray N, Turrisi AT, III. A review of first-line treatment for small-cell lung cancer. *J Thorac Oncol* 2006;1:270–278.

278. Jereczek B, Jassem J, Karnicka-Mlodkowska H, et al. Autopsy findings in small cell lung cancer. *Neoplasma* 1996;43:133–137.

279. Videtic GM, Adelstein DJ, Mekhail TM, et al. Validation of the RTOG recursive partitioning analysis (RPA) classification for small-cell lung cancer-only brain metastases. *Int J Radiat Oncol Biol Phys* 2007;67:240–243.

280. Lee JS, Murphy WK, Glisson BS, et al. Primary chemotherapy of brain metastasis in small-cell lung cancer. *J Clin Oncol* 1989;7:916–922.

281. Kristensen CA, Kristjansen PE, Hansen HH. Systemic chemotherapy of brain metastases from

small-cell lung cancer: a review. *J Clin Oncol* 1992;10:1498–1502.

282. Grossi F, Scolaro T, Tixi L, et al. The role of systemic chemotherapy in the treatment of brain metastases from small-cell lung cancer. *Crit Rev Oncol Hematol* 2001;37:61–67.

283. Postmus PE, Smit EF, Haaxma-Reiche H, et al. Teniposide for brain metastases of small-cell lung cancer: a phase II study. *J Clin Oncol* 1995;13:660–665.

284. Chen G, Huynh M, Chen A, et al. Chemotherapy for brain metastases in small cell lung cancer. *Clin Lung Cancer* 2008;9:35–38.

285. Postmus PE, Haaxma-Reiche H, Smit EF, et al. Treatment of brain metastases of small-cell lung cancer: comparing teniposide and teniposide with whole-brain radiotherapy—a phase III study of the European Organization for the Research and Treatment of Cancer Lung Cancer Cooperative Group. *J Clin Oncol* 2000;18:3400–3408.

286. Seute T, Leffers P, Wilmink JT, et al. Response of asymptomatic brain metastases from small-cell lung cancer to systemic first-line chemotherapy. *J Clin Oncol* 2006;24:2079–2083

287. Lee JJ, Bekele BN, Zhou X, et al. Decision analysis for prophylactic cranial irradiation for patients with small-cell lung cancer. *J Clin Oncol* 2006;24:3597–3603.

288. Pugh TJ, Gaspar LE. Prophylactic cranial irradiation for patients with lung cancer. *Clin Lung Cancer* 2007;8:365–368.

289. Slotman B, Faivre-Finn C, Kramer G, et al. Prophylactic cranial irradiation in the extensive small cell lung cancer. *N Engl J Med* 2007;357:664–672.

290. D'Ambrosio DJ, Cohen RB, Glass J, et al. Unexpected dementia following prophylactic cranial irradiation for small cell lung cancer: Case report. *J Neurooncol* 2007; 85:77–79.

291. Grosshans DR, Meyers CA, Allen PK, et al. Neurocognitive function in patients with small cell lung cancer: effect of prophylactic cranial irradiation. Cancer 2008;112:589–595

292. Gaspar LE, Chansky K, Albain KS, et al. Time from treatment to subsequent diagnosis of brain metastases in stage III non-small-cell lung cancer: a retrospective review by the Southwest Oncology Group. *J Clin Oncol* 2005;23:2955–2961.

293. Chee RJ, Bydder S, Cameron F. Prolonged survival after resection and radiotherapy for solitary brain metastases from non-small-cell lung cancer. *Australas Radiol* 2007;51:186–189.

294. Athanassiou H, Synodinou M, Maragoudakis E, et al. Randomized phase II study of temozolomide and radiotherapy compared with radiotherapy alone in newly diagnosed glioblastoma multiforme. *J Clin Oncol* 2005;23:2372–2377.

295. Addeo R, Caraglia M, Faiola V, et al. Concomitant treatment of brain metastasis with whole brain radiotherapy [WBRT] and temozolomide [TMZ] is active and improves quality of life. *BMC Cancer* 2007;7:18.

296. Ebert BL, Niemierko E, Shaffer K, et al. Use of temozolomide with other cytotoxic chemotherapy in the treatment of patients with recurrent brain metastases from lung cancer. *Oncologist* 2003;8:69–75.

297. Omuro AM, Raizer JJ, Demopoulos A, et al. Vinorelbine combined with a protracted course of temozolomide for recurrent brain metastases: a phase I trial. *J Neuro-Oncol* 2006;78:277–280.

298. Giorgio CG, Giuffrida D, Pappalardo A, et al. Oral temozolomide in heavily pre-treated brain metastases from non-small cell lung cancer: phase II study. *Lung Cancer* 2005;50:247–254.

299. Wong ET, Berkenblit A. The role of topotecan in the treatment of brain metastases. *Oncologist* 2004;9:68–79.

300. Lassman AB, Abrey LE, Shah GD, et al. Systemic high-dose intravenous methotrexate for central nervous system metastases. *J Neuro-Oncol* 2006;78:255–260.

301. Omuro AM, Kris MG, Miller VA, et al. High incidence of disease recurrence in the brain and leptomeninges in patients with nonsmall cell lung carcinoma after response to gefitinib. *Cancer* 2005;103:2344–2348.

302. Shimato S, Mitsudomi T, Kosaka T, et al. EGFR mutations in patients with brain metastases from lung cancer: association with the efficacy of gefitinib. *Neuro Oncol* 2006;8:137–144.

303. Cappuzzo F, Ardizzoni A, Soto-Parra H, et al. Epidermal growth factor receptor targeted therapy by ZD 1839 (Iressa) in patients with brain metastases from non-small cell lung cancer (NSCLC). *Lung Cancer* 2003;41:227–231.

304. Chiu CH, Tsai CM, Chen YM, et al. Gefitinib is active in patients with brain metastases from non-small cell lung cancer and response is related to skin toxicity. *Lung Cancer* 2005;47:129–138.

305. Oh Y, Stewart DJ. Systemic therapy for lung cancer brain metastases: a rationale for clinical trials. *Oncology* 2008;22:168–188.

306. Gore E. Prophylactic cranial irradiation versus observation in stage III non-small-cell lung cancer. *Clin Lung Cancer* 2006;7:276–278.

307. Lin NU, Bellon JR, Winer EP. CNS metastases in breast cancer. *J Clin Oncol* 2004;22:3608–3617.

308. Lai R, Dang CT, Malkin MG, et al. The risk of central nervous system metastases after trastuzumab therapy in patients with breast carcinoma. *Cancer* 2004;101:810–816.

309. Burstein HJ, Lieberman G, Slamon DJ, et al. Isolated central nervous system metastases in patients with HER2-overexpressing advanced breast cancer treated with first-line trastuzumab-based therapy. *Ann Oncol* 2005;16:1772–1777.

310. Stemmler HJ, Schmitt M, Willems A, et al. Ratio of trastuzumab levels in serum and cerebral spinal fluid is altered in HER2-positive breast cancer patients with brain metastases and impairment of the blood brain barrier. *Anticancer Drugs* 2007;18:23–28.

311. Pestalozzi BC, Zahrieh D, Price KN, et al. Identifying breast cancer patients at risk for Central Nervous System (CNS) metastases in trials of the International Breast Cancer Study Group (IBCSG). *Ann Oncol* 2006;17:935–944.

312. Evans AJ, James JJ, Cornford EJ, et al. Brain metastases from breast cancer: identification of a high-risk group. *Clin Oncol (R Coll Radiol)* 2004;16:345–349.

313. Gabos Z, Sinha R, Hanson J, et al. Prognostic significance of human epidermal growth factor receptor

positivity for the development of brain metastasis after newly diagnosed breast cancer. *J Clin Oncol* 2006;24:5658–5663.

314. Hicks DG, Short SM, Prescott NL, et al. Breast cancers with brain metastases are more likely to be estrogen receptor negative, express the basal cytokeratin CK5/6, and overexpress HER2 or EGFR. *Am J Surg Pathol* 2006;30:1097–1104.

315. Albiges L, Andre F, Balleyguier C, et al. Spectrum of breast cancer metastasis in BRCA1 mutation carriers: highly increased incidence of brain metastases. *Ann Oncol* 2005;16:1846–1847.

316. Smith DB, Howell A, Harris M, et al. Carcinomatous meningitis associated with infiltrating lobularcarcinoma of the breast. *Eur J Surg Oncol* 1985;11:33–36.

317. Lamovec J, Zidar A. Association of leptomeningeal carcinomatosis in carcinoma of the breast with infiltrating lobular carcinoma. An autopsy study. *Arch Pathol Lab Med* 1991;115:507–510.

318. Piccirilli M, Sassun TE, Brogna C, et al. Late brain metastases from breast cancer: clinical remarks on 11 patients and review of the literature. *Tumori* 2007;93:150–154.

319. Palmieri D, Bronder JL, Herring JM, et al. Her-2 overexpression increases the metastatic outgrowth of breast cancer cells in the brain. *Cancer Res* 2007;67:4190–4198.

320. Chen EI, Hewel J, Krueger JS, et al. Adaptation of energy metabolism in breast cancer brain metastases. *Cancer Res* 2007;67:1472–1486.

321. Wronski M, Arbit E, McCormick B. Surgical treatment of 70 patients with brain metastases from breast carcinoma. *Cancer* 1997;80:1746–1754.

322. Pieper DR, Hess KR, Sawaya RE. Role of surgery in the treatment of brain metastases in patients with breast cancer. *Ann Surg Oncol* 1997;4:481–490.

323. Firlik KS, Kondziolka D, Flickinger JC, et al. Stereotactic radiosurgery for brain metastases from breast cancer. *Ann Surg Oncol* 2000;7:333–338.

324. Amendola BE, Wolf AL, Coy SR, et al. Gamma knife radiosurgery in the treatment of patients with single and multiple brain metastases from carcinoma of the breast. *Cancer J* 2000;6:88–92.

325. Akyurek S, Chang EL, Mahajan A, et al. Stereotactic radiosurgical treatment of cerebral metastases arising from breast cancer. *Am J Clin Oncol* 2007;30:310–314.

326. Ekenel M, Hormigo AM, Peak S, et al. Capecitabine therapy of central nervous system metastases from breast cancer. *J Neuro-Oncol* 2007;85(2):223–227.

327. Rivera E, Meyers C, Groves M, et al. Phase I study of capecitabine in combination with temozolomide in the treatment of patients with brain metastases from breast carcinoma. *Cancer* 2006;107:1348–1354.

328. Zulkowski K, Kath R, Semrau R, et al. Regression of brain metastases from breast cancer after chemotherapy with bendamustine. J Cancer Res Clin Oncol 2002;128:111–113

329. Salvati M, Cervoni L, Innocenzi G, et al. Prolonged stabilization of multiple and single brain metastases from breast cancer with tamoxifen. Report of three cases. *Tumori* 1993;79:359–362.

330. Stewart DJ, Dahrouge S. Response of brain metastases from breast cancer to megestrol acetate: a case report. *J Neuro-Oncol* 1995;24:299–301.

331. Madhup R, Kirti S, Bhatt ML, et al. Letrozole for brain and scalp metastases from breast cancer—a case report. *Breast* 2006;15:440–442.

332. Moy B, Goss PE. Lapatinib: current status and future directions in breast cancer. *Oncologist* 2006;11:1047–1057.

333. Majer M. Management of metastatic melanoma patients with brain metastases. *Curr Oncol Rep* 2007;9:411–416.

334. Wronski M, Arbit E. Surgical treatment of brain metastases from melanoma: a retrospective study of 91 patients. *J Neurosurg* 2000;93:9–18.

335. Gaviani P, Mullins ME, Braga TA, et al. Improved detection of metastatic melanoma by T2°-weighted imaging. *AJNR Am J Neuroradiol* 2006;27:605–608.

336. Zacest AC, Besser M, Stevens G, et al. Surgical management of cerebral metastases from melanoma: outcome in 147 patients treated at a single institution over two decades. *J Neurosurg* 2002;96:552–558.

337. Douglas JG, Margolin K. The treatment of brain metastases from malignant melanoma. *Sem Oncol* 2002;29:518–524.

338. Samlowski WE, Watson GA, Wang M, et al. Multimodality treatment of melanoma brain metastases incorporating stereotactic radiosurgery (SRS). *Cancer* 2007;109:1855–1862.

339. Mathieu D, Kondziolka D, Cooper PB, et al. Gamma knife radiosurgery in the management of malignant melanoma brain metastases. *Neurosurgery* 2007;60:471–481.

340. Kondziolka D, Martin JJ, Flickinger JC, et al. Long-term survivors after gamma knife radiosurgery for brain metastases. *Cancer* 2005;104:2784–2791.

341. Jacquillat C, Khayat D, Banzet P, et al. Final report of the French multicenter phase II study of the nitrosurea Fotemustine in 153 evaluable patients with disseminated malignant melanoma including patients with cerebral metastases. *Cancer* 1990;66:1873–1878.

342. Mornex F, Thomas L, Mohr P, et al. A prospective randomized multicentre phase III trial of fotemustine plus whole brain irradiation versus fotemustine alone in cerebral metastases of malignant melanoma. *Melanoma Res* 2003;13:97–103.

343. Avril MF, Aamdal S, Grob JJ, et al. Fotemustine compared with dacarbazine in patients with disseminated malignant melanoma: a phase III study. *J Clin Oncol* 2004;22:1118–1125.

344. Bafaloukos D, Tsoutsos D, Fountzilas G, et al. The effect of temozolomide-based chemotherapy in patients with cerebral metastases from melanoma. *Melanoma Res* 2004;14:289–294.

345. Boogerd W, De Gast GC, Dalesio O. Temozolomide in advanced malignant melanoma with small brain metastases: can we withhold cranial irradiation? *Cancer* 2007;109:306–312.

346. Schadendorf D, Hauschild A, Ugurel S, et al. Dose-intensified bi-weekly temozolomide in patients with asymptomatic brain metastases from malignant melanoma: a phase II DeCOG/ADO study. *Ann Oncol* 2006;17:1592–1597.

347. Majer M. Biochemotherapy of metastatic melanoma in patients with or without recently diagnosed brain metastases. *Cancer* 2007;110:1329–1337.

348. Cimatti M, Salvati M, Caroli E, et al. Extremely delayed cerebral metastasis from renal carcinoma: report of

four cases and critical analysis of the literature. *Tumori* 2004;90:342–344.

349. Erbay SH, O'Callaghan MG, Bhadelia RA, et al. Unusual case of 2 renal cell carcinoma metastases ipsilateral to an occluded internal carotid artery. *J Neuroimaging* 2006;16:176–178.

350. Wronski M, Arbit E, Russo P, et al. Surgical resection of brain metastases from renal cell carcinoma in 50 patients. *Urology* 1996;47:187–193.

351. Harada Y, Nonomura N, Kondo M, et al. Clinical study of brain metastasis of renal cell carcinoma. *Eur Urol* 1999;36:230–235.

352. Salvati M, Scarpinati M, Orlando ER, et al. Single brain metastasis from kidney tumors. Clinico-pathologic considerations on a series of 29 cases. *Tumori* 1992;78:392–394.

353. Badalament RA, Gluck RW, Wong GY, et al. Surgical treatment of brain metastases from renal cell carcinoma. *Urology* 1990;36:112–117.

354. Wronski M, Maor MH, Davis BJ, et al. External radiation of brain metastases from renal carcinoma: a retrospective study of 119 patients from the MD Anderson Cancer Center. *Int J Radiat Oncol Biol Phys* 1997;37:753–759.

355. Doh LS, Amato RJ, Paulino AC, et al. Radiation therapy in the management of brain metastases from renal cell carcinoma. *Oncology (Williston Park)* 2006;20:603–613.

356. Culine S, Bekradda M, Kramar A, et al. Prognostic factors for survival in patients with brain metastases from renal cell carcinoma. *Cancer* 1998;83:2548–2553.

357. Cannady SB, Cavanaugh KA, Lee SY, et al. Results of whole brain radiotherapy and recursive partitioning analysis in patients with brain metastases from renal cell carcinoma: a retrospective study. *Int J Radiat Oncol Biol Phys* 2004;58:253–258.

358. Noel G, Valery CA, Boisserie G, et al. LINAC radiosurgery for brain metastasis of renal cell carcinoma. *Urol Oncol* 2004;22:25–31.

359. Shuto T, Inomori S, Fujino H, et al. Gamma knife surgery for metastatic brain tumors from renal cell carcinoma. *J Neurosurg* 2006;105:555–560.

360. Sundermeyer ML, Meropol NJ, Rogatko A, et al. Changing patterns of bone and brain metastases in patients with colorectal cancer. *Clin Colorectal Cancer* 2005;5:108–113.

361. Hammoud MA, McCutcheon IE, ElSouki R, et al. Colorectal carcinoma and brain metastasis: distribution, treatment, and survival. *Ann Surg Oncol* 1996;3:453–463.

362. Wronski M, Arbit E. Resection of brain metastases from colorectal carcinoma in 73 patients. *Cancer* 1999;85:1677–1685.

363. Bokemeyer C, Nowak P, Haupt A, et al. Treatment of brain metastases in patients with testicular cancer. *J Clin Oncol* 1997;15:1449–1454.

364. Lutterbach J, Spetzger U, Bartelt S, et al. Malignant germ cell tumors metastatic to the brain: a model for a curable neoplasm? The Freiburg experience and a review of the literature. *J Neuro-Oncol* 2002;58:147–156.

365. Nicolato A, Ria A, Foroni R, et al. Gamma knife radiosurgery in brain metastases from testicular tumors. *Med Oncol* 2005;22(1):45–56.

366. Salvati M, Piccirilli M, Raco A, et al. Brain metastasis from non-seminomatous germ cell tumors of the testis: indications for aggressive treatment. *Neurosurg Rev* 2006;29:130–137.

367. Cagayan MS, Lu-Lasala LR. Management of gestational trophoblastic neoplasia with metastasis to the central nervous system: a 12-year review at the Philippine General Hospital. *J Reprod Med* 2006;51:785–792.

368. Cormio G, Pellegrino A, Landoni F, et al. Brain metastases from cervical carcinoma. *Tumori* 1996; 82:394–396.

369. Badib AO, Kurohara SS, Webster JH, et al. Metastasis to organs in carcinoma of the uterine cervix. Influence of treatment on incidence and distribution. *Cancer* 1968;21:434–439.

370. Tangjitgamol S, Levenback CF, Beller U, et al. Role of surgical resection for lung, liver, and central nervous system metastases in patients with gynecological cancer: a literature review. *Int J Gynecol Cancer* 2004;14:399–422.

371. Robinson JB, Morris M. Cervical carcinoma metastatic to the brain. *Gynecol Oncol* 1997;66:324–326.

372. Chura JC, Shukla K, Argenta PA. Brain metastasis from cervical carcinoma. *Int J Gynecol Cancer* 2007;17:141–146.

373. Gaussmann AB, Imhoff D, Lambrecht E, et al. Spontaneous remission of metastases of cancer of the uterine cervix. *Onkologie* 2006;29:159–161.

374. Gien LT, Kwon JS, D'Souza DP, et al. Brain metastases from endometrial carcinoma: a retrospective study. *Gynecol Oncol* 2004;93:524–528.

375. Orrru S, Lay G, Dessi M, et al. Brain metastases from endometrial carcinoma: report of three cases and review of the literature. *Tumori* 2007;93:112–117.

376. Güth U, Huang DJ, Bauer G, et al. Metastatic patterns at autopsy in patients with ovarian carcinoma. *Cancer* 2007;110:1272–1280.

377. Pectasides D, Pectasides M, Economopoulos T. Brain metastases from epithelial ovarian cancer: a review of the literature. *Oncologist* 2006;11:252–260.

378. Kolomainen DF, Larkin JM, Badran M, et al. Epithelial ovarian cancer metastasizing to the brain: a late manifestation of the disease with an increasing incidence. *J Clin Oncol* 2002; 20:982–986.

379. Hardy JR, Harvey VJ. Cerebral metastases in patients with ovarian cancer treated with chemotherapy. *Gynecol Oncol* 1989;33:296–300.

380. Cormio G, Maneo A, Colamaria A, et al. Surgical resection of solitary brain metastasis from ovarian carcinoma: an analysis of 22 cases. *Gynecol Oncol* 2003;89:116–119.

381. Tay SK, Rajesh H. Brain metastases from epithelial ovarian cancer. *Int J Gynecol Cancer* 2005;15:824–829.

382. Ogose A, Morita T, Hotta T, et al. Brain metastases in musculoskeletal sarcomas. *Jpn J Clin Oncol* 1999;29:245–247.

383. Curless RG, Toledano SR, Ragheb J, et al. Hematogenous brain metastasis in children. *Pediatr Neurol* 2002;26:219–221.

384. Wronski M, Arbit E, Burt M, et al. Resection of brain metastases from sarcoma. *Ann Surg Oncol* 1995;2:392–399.

385. Recht L, Mrugala M. Neurologic complications of hematologic neoplasms. *Neurol Clin* 2003; 21:87–105.

386. Recht L, Straus DJ, Cirrincione C, et al. Central nervous system metastases from non-Hodgkin's lymphoma: treatment and prophylaxis. *Am J Med* 1988;84:425–435.

387. Sahin F, Saydam G, Ertan Y, et al. Dural plasmacytoma mimicking meningioma in a patient with multiple myeloma. *J Clin Neurosci* 2006; 13:259–261.

388. Haegelen C, Riffaud L, Bernard M, et al. Dural plasmacytoma revealing multiple myeloma. Case report. *J Neurosurg* 2006;104:608–610.

389. Stark RJ, Henson RA. Cerebral compression by myeloma. *J Neurol Neurosurg Psychiatry* 1981;44:833–836.

390. Stark AM, Eichmann T, Mehdorn HM. Skull metastases: clinical features, differential diagnosis, and review of the literature. *Surg Neurol* 2003;60:219–225.

391. Graus F, Slatkin NE. Papilledema in the metastatic jugular foramen syndrome. *Arch Neurol* 1983;40:816–818.

392. Healy JF, Bishop J, Rosenkrantz H. Cranial computed tomography in the detection of dural, orbital, and skull involvement in metastatic neuroblastoma. *J Comput Tomogr* 1981;5:319–323.

393. Kleinschmidt-DeMasters BK. Dural metastases—a retrospective surgical and autopsy series. *Arch Pathol Lab Med* 2001;125:880–887.

394. Tseng SH, Liao CC, Lin SM, et al. Dural metastasis in patients with malignant neoplasm and chronic subdural hematoma. *Acta Neurol Scand* 2003;108:43–46.

395. Montano N, Puca A, Pierconti F, et al. Extremely delayed falx metastasis from renal cell carcinoma. *Neurology* 2007;68:1541–1542.

396. Schreiber D, Bernstein K, Schneider J. [Metastases of the central nervous system: a prospective study. 3rd Communication: metastases in the pituitary gland, pineal gland, and choroid plexus.] *Zentralbl Allg Pathol* 1982;126:64–73.

397. Lauro S, Trasatti L, Capalbo C, et al. Unique pineal gland metastasis of clear cell renal carcinoma: case report and review of the literature. *Anticancer Res* 2002;22:3077–3079.

398. Leigh RJ, Zee DS. *The Neurology of Eye Movements.* 4th ed. New York: Oxford, 2006.

399. Ahn JY, Chung YS, Kwon SO, et al. Isolated pineal region metastasis of small cell lung cancer. *J Clin Neurosci* 2005;12:691–693.

400. Lassman AB, Bruce JN, Fetell MR. Metastases to the pineal gland. *Neurology* 2006;67:1303–1304.

401. Bulian D, Pierpaoli W. The pineal gland and cancer. I. Pinealectomy corrects congenital hormonal dysfunctions and prolongs life of cancer-prone C3H/He mice. *J Neuroimmunol* 2000;108:131–135.

402. Cos S, Sanchez-Barcelo EJ. Melatonin, experimental basis for a possible application in breast cancer prevention and treatment. *Histol Histopathol* 2000;15:637–647.

403. Kitajima K, Morita M, Morikawa M, et al. Choroid plexus metastasis of colon cancer. *Magn Reson Med Sci* 2003;2:155–158.

404. Quinones-Hinojosa A, Chang EF, Khan SA, et al. Renal cell carcinoma metastatic to the choroid mimicking intraventricular meningioma. *Can J Neurol Sci* 2004;31:115–120.

405. Lauretti L, Fernandez E, Pallini R, et al. Long survival in an untreated solitary choroid plexus metastasis from renal cell carcinoma: case report and review of the literature. *J Neuro-Oncol* 2005;71:157–160.

406. Abrams HL, Spiro R, Goldstein N. Metastases in carcinoma. Analysis of 1000 autopsied cases. *Cancer* 1950;3:74–85.

407. Fassett DR, Couldwell WT. Metastases to the pituitary gland. *Neurosurg Focus* 2004;16:E8.

408. Verhelst J, Vanden Broucke P, Dua G, et al. Pituitary metastasis mimicking a pituitary adenoma. A description of two cases. *Acta Clin Belg* 1995;50:31–35.

409. Cantón A, Simó R, Gil L, et al. Headache, vomiting and diplopia—pituitary apoplexy caused by a solitary metastasis from breast carcinoma. *Postgrad Med J* 1997;73:357–359.

Chapter 6

Spinal Metastases

INTRODUCTION

Most symptomatic intracranial metastases involve the brain parenchyma (see Chapter 5). In the spinal canal, most symptomatic tumors compress the spinal cord or cauda equina from the epidural space and only rarely involve the cord directly. As is true in the brain, several other complications of cancer can affect spinal cord function, mimicking metastatic disease (Table 6–1).

Epidural lesions are usually caused by a tumor that has metastasized to vertebral bodies (Fig. 6–1); the neurologic symptoms result from compression rather than invasion of the spinal cord (Fig. 6–2; see also Fig. 2–6). Spinal cord compression requires that the mass obliterates the subarachnoid space and distorts the spinal cord or the nerve roots of the cauda equina. Epidural tumors rarely breach the dura; when

compressed by tumor, the dura may be more than a millimeter thick and quite resistant to penetration.[1] When the intradural space or the spinal cord is invaded directly, it is usually either by growth of the tumor along the spinal roots or by hematogenous spread directly to the spinal cord. Together, these invasions account for less than 5% of symptomatic spinal cord involvement.[2–7]

At autopsy, approximately 5% of patients who die from cancer exhibit spinal cord or cauda equina compression.[9] The vertebral body is the most common site of bone metastases.[10,11] In an autopsy study of 842 patients with known cancer, metastases from lung and breast cancers, lymphoma, and myeloma accounted for 65% of vertebral metastases.[11] In another study of 1589 patients with prostate cancer, distant metastases that developed as a consequence of hematogenous dissemination were found in

194

556 (35%). The bone was the most common site of metastatic disease, present in 501 patients (90%); of these, 90% were present in the spine.[12] Fortunately, most vertebral metastases are not large enough to compress the spinal cord.

Table 6–1. **Some Causes of Spinal Cord Dysfunction in Patients with Cancer**

Metastatic
  Epidural Metastases
    From vertebral bone
    From paravertebral structures
    Arising in epidural space
  Intradural metastases
    Single mass
    Leptomeningeal metastases (Chapter 7)
  Intramedullary metastases
Nonmetastatic
  Infections (Chapter 10)
    Epidural abscess
  Vascular disorders (Chapter 9)
    Hemorrhage
    Infarct
  Side effects of therapy
    Radiation (Chapter 13)
    Chemotherapy (Chapter 12)
  Paraneoplastic syndromes (Chapter 15)

A population-based study of spinal cord compression indicated that the cumulative probability of suffering at least one episode of spinal cord compression related to malignant disease in 5 years before death was 2.5%.[13] The likelihood ranged from 0.2% in pancreatic carcinoma to 7.9% in myeloma. The prevalence of spinal cord compression at diagnosis of the cancer was 0.2%. The cumulative incidence of spinal cord compression from prostate cancer was 7.2%, and for breast cancer 5.5%. The authors of this study recognized that their survey underestimated the true incidence. Table 6–2 lists the primary tumors causing spinal cord compression at Memorial Sloan-Kettering Cancer Center (MSKCC). Leukemia[14,15] and lymphoma[16] sometimes invade the epidural space without causing vertebral destruction.[17,18]

Symptoms and signs of spinal cord compression, other than pain, usually evolve rapidly. If untreated, weakness and ultimately paralysis invariably ensue.[19–21] When spinal cord compression is diagnosed and treated early, most patients either maintain or regain their ability to walk.[19,22] If diagnosis and treatment are delayed until the patient becomes paraplegic or quadriplegic, functional recovery is rare.[19,23] Therefore, early diagnosis of spinal cord compression is essential, and if the diagnosis is delayed until neurologic signs are rapidly worsening, treatment is urgent. In a study of 301 consecutive

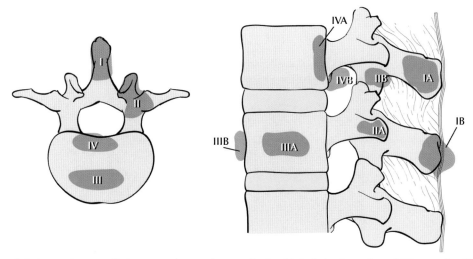

Figure 6–1. Anatomic extent of spine tumors by zone (see text for details). *Left:* Axial cuts through L1. zones I to IV: zone I is the spinous process, zone II the pedicle, zone III the central vertebral body, zone IV the posterior vertebral body. Lateral view of L1: In this view the zones labeled A are restricted to bone, and zones labeled B indicate areas where the tumor has broken through the bone to involve surrounding structures. Redrawn from Ref. 8.

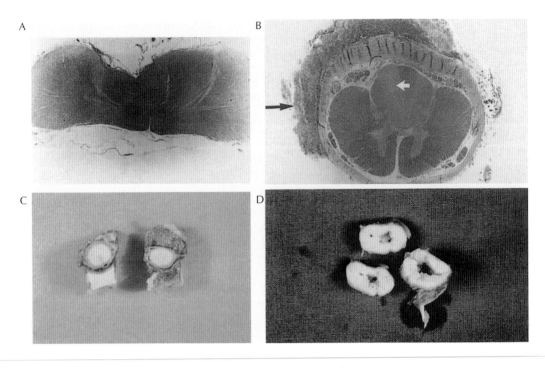

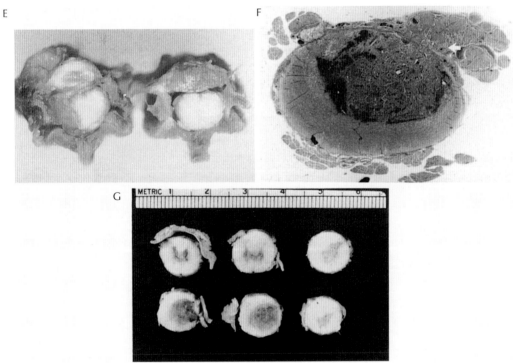

Figure 6–2. Anatomy and pathology of spinal cord dysfunction. **A**: Severe flattening of the spinal cord compressed by an anteriorly located epidural tumor. The lateral columns are demyelinated, but the anterior and posterior columns and the central gray area are relatively well preserved. **B**: An epidural tumor originating from a metastasis to the spinous process causing compression of the spinal cord posteriorly. Notice, in addition, that the tumor has partially surrounded the spinal cord on its lateral aspect (*black arrow*) and has caused demyelination in the dorsal columns (*white arrow*). **C**: A pathologic

patients presenting to a regional cancer center with epidural spinal cord compression, the median delay from onset of symptoms to treatment was 14 days, ranging from 0 to 840 days. During that time, many patients experienced deterioration of motor or bladder function.[24] The same considerations apply to compression of the nerve roots of the cauda equina, which occupy the spinal canal from vertebral body L1 or L2 to the sacrum. Because symptoms and prognosis are similar when either the cauda equina or the spinal cord is involved, they are discussed together in this chapter and are included under the general topic of spinal cord compression or invasion. Several recent reviews consider various aspects of the diagnosis and treatment of spinal metastases and spinal cord disease in general.[21,25–29] Specific tumors are considered in the following text.

EPIDURAL METASTASES

Frequency

The vertebral bodies, a common site of metastasis, are involved in 25% to 70% of patients with metastatic cancer.[30,31] Fortunately, most vertebral metastases are either asymptomatic or cause pain only; a small percentage initiate significant spinal cord or cauda equina compression (Fig. 6–3). The spine is the most common site of bone metastasis irrespective of the primary tumor responsible[1,32] (Table 6–3). So common are vertebral metastases that even a solitary vertebral tumor is more likely to be a metastasis than a primary bone tumor.[33] The pathophysiology and management of bone metastasis in general has been reviewed recently.[34]

Table 6–2. Primary Cancers Causing Symptomatic Spinal Cord Compression in 583 Patients at Memorial Sloan-Kettering Cancer Center

Primary Tumor	No. of Patients	Percentage
Breast	127	22
Lung	90	15
Prostate	58	10
Lymphoreticular system	56	10
Sarcomas	52	9
Kidney	39	7
Gastrointestinal tract	29	5
Melanoma	23	4
Unknown primary	21	4
Head and neck	19	3
Miscellaneous	69	12
Total	583	100

◄ ——————————————————————————————

specimen at two levels from a patient suffering with spinal compression by malignant neuroblastoma. Although the tumor originated in the vertebral body, once it entered the epidural space it grew circumferentially around the cord, compressing the spinal cord anteriorly and posteriorly. Shrinkage artifact because of fixation makes the compression less apparent than it actually was. **D**: Central necrosis of a compressed spinal cord at the cervical level caused by epidural tumor metastatic to the C5 vertebral body. Evidence of vascular occlusion was absent, although all of the radicular arteries were not carefully examined. **E**: Spinal cord compression in a rat. The section on the right is taken from a normal animal. The section on the left demonstrates cord compression from a tumor that grew through the intervertebral foramen to compress the cord laterally. The bones of the vertebra are not affected. **F**: Intramedullary metastasis caused by small cell lung cancer that invaded the spinal cord via the posterior root (*arrow*). Notice the infiltration of the posterior root and the apparent direct growth from the root into the spinal cord. **G**: Hematogenously spread intramedullary metastasis from malignant melanoma. The leptomeninges and the roots are not involved.

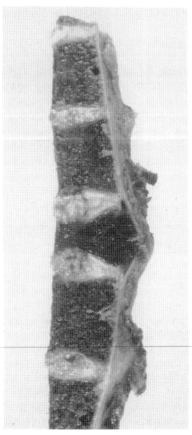

Figure 6–3. The thoracic spinal cord of a patient with a metastasis to a vertebral body. The vertebral body is collapsed anteriorly. Tumor and the posteriorly protruding bone compromise the spinal canal but do not breach the posterior longitudinal ligament.

The exact incidence of spinal cord compression is not well established for several reasons: (1) Epidural spinal cord compression usually occurs late during the course of metastatic cancer, often during its terminal stages. Before computed tomography (CT) and magnetic resonance imaging (MRI), the diagnosis of spinal cord compression required a myelogram; many patients and their physicians chose not to pursue spinal cord signs with that invasive procedure, particularly if a patient was very ill. (2) Most pathologists do not routinely examine the spinal cord at autopsy, thus underestimating the incidence of spinal cord disease. (3) Even when the spinal cord is examined at autopsy, pathologic changes may be minimal and not indicate that symptomatic spinal cord compression was present during life. Therefore, the population-based study giving an incidence of 2.5% in 5 years before death is almost certainly an underestimate.[13]

Epidural spinal cord compression, particularly from vertebral metastasis, is a relatively common neurologic complication of cancer. At MSKCC in 1994, a diagnosis of epidural metastasis was made in 136 patients, a little over one-half the frequency of brain metastasis.

Back pain is the most common reason for neurologic consultation at MSKCC.[35] Of those patients referred, approximately 33% have epidural metastases and another 33% have vertebral metastases without epidural invasion. The remainder suffered degenerative spine disease and paravertebral or meningeal metastases.

Table 6–3. **Vertebral Metastases in Patients with Cancer**

Primary Tumor	No. of Patients	Percentage Metastases at Autopsy	Percentage with Vertebral Metastases but Negative Radiograph (False Negative)	Percentage Vertebral Collapse on Radiograph but No Metastases Pathologically (False Positive)
Breast	113	74	25	8
Lung	138	45	24	3
GI	115	25	38	4
Lymphoma	107	29	64	3
Female GU	78	20	25	3
Head and neck	35	20	40	3
Kidney	24	29	0	8
Prostate	21	90	10	5
Total	832	36	26 (78/300)	22 (34/155)

Data from Ref. 31. GI, gastrointestinal tract; GU, genitourinary tract.

Table 6–4. **Site of Compression in 265 Patients with Epidural Spinal Cord Compression**

Primary Tumor	No. of Patients	Percentage	Cervical	T1 to T6	T7 to T12	Lumbosacral
Breast	58	21.9	7	25	26	11°
Lung	49	18.5	3	28	13	7
Prostate	25	9.4	1	7	10	9
Kidney	19	7.2	1	5	11	5
Lymphoreticular system	18	6.8	1	7	8	4
Melanoma	10	3.8	1	4	4	2
Gastrointestinal tract	14	5.3	3	1	5	6
Soft-tissue sarcoma	19	7.2	2	4	12	2
Unknown primary tumor	14	5.3	1	7	7	2
Miscellaneous†	39	14.7	3	17	15	8
Total	265	100	22	105	111	56

° Some patients had more than one site of compression.
† Thyroid, bladder, Ewing sarcoma, germ-cell tumors, neuroblastoma, and others.

Overall, approximately 20,000 people (5% of 400,000 dying of cancer) develop spinal cord compression each year.[36]

Spinal cord compression may be the first sign of cancer.[37] Schiff et al. estimated that epidural metastases are the initial manifestation of malignancy in 20% of patients with epidural spinal cord compression.[37] In a 10-year clinical study[38] at the London Hospital, 131 patients with cancers other than hematologic malignancies were found to have neurologic signs that were due to spinal cord or cauda equina compression; almost 50% were not known to have cancer when they presented with spinal symptoms. In 18 patients, the primary site was never discovered. In a similar study conducted over 15 years in Paris and Naples[39] of 600 patients with spinal metastases and neurologic signs, carcinomas of the breast (25%), lung (15%), and reproductive system (12%) were the common primary sites; in 12%, the primary site was unknown, fourth in the order of frequency. Hematologic tumors represented only 6%, perhaps because the report comes from a neurosurgical service; most patients with spinal cord compression from hematologic tumors are treated medically.

Table 6–4 lists the particular types of tumor and sites of spinal cord compression found in 265 patients with cancer at MSKCC. The thoracic spine is the most common location of epidural compression, followed by the lumbosacral and cervical spine in a ratio of approximately 4:2:1. Breast and lung cancer are distributed about equally through vertebral bodies, whereas prostate, renal, and gastrointestinal (GI) tumors are more likely to affect the lower thoracic or lumbar vertebrae.

Biology of Epidural Metastases

The biology of spinal metastasis is discussed in detail in Chapter 2. A few additional aspects are considered here.

VERTEBRAL METASTASES

Products produced by the tumor and the bone micro-environment present several potential therapeutic targets both to prevent and treat bone metastases (Fig. 6–4).[40] Some of these are listed in Table 6–5. Of particular note, the bisphosphonates have been used both in the adjuvant and therapeutic setting for the prevention and treatment of bone metastases.[41,42]

The usual mechanism for spinal cord compression is hematogenous metastasis to the

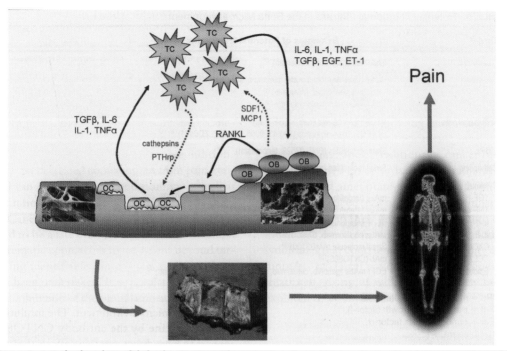

Figure 6–4. Pathophysiology of skeletal metastases. The activation of osteoblasts and osteoclasts by cancer cells results in a vicious cycle of bone destruction and increased tumor growth, resulting in pain, fractures, and spinal cord compression. TC, tumor cell; OC, osteoclast; OB, osteoblast; SDF-1, stromal-derived factor 1; MCP-1, monocyte chemoattractant protein 1; EGF, epidermal growth factor; IL-6, interleukin 6; IL-1, interleukin 1; TNFα, tumor necrosis factor α; TGFβ, transforming growth factor β; ET-1, endothelium 1; PTHrp, parathyroid hormone-related peptide; RANKL, receptor activator at nuclear factor κB ligand. (From Ref. 40 with permission.)

vertebral body or, less commonly, the vertebral lamina, pedicle, or spinous process. For purposes of staging vertebral tumors, Weinstein[43] divided the vertebral body into four zones (Fig. 6–1). Zone I includes the spinous process and pars interarticularis along with the inferior facets. Zone II includes the superior articular facet, the transverse process and pedicle from the level of the pars to its junction with the vertebral body. Zone III includes the anterior three fourths of the vertebral body. Zone IV includes the posterior one-fourth of a vertebral body. Each zone is given a letter, (A) indicating tumor restricted to bone, (B) indicating extraosseous spread, and (C) indicating more distant spread.

MRI is far more sensitive than radionuclide bone scan in detecting metastases of the axial skeleton, at least in high-risk patients with prostate cancer.[44] In most patients with acute spinal cord compression because of metastases, a portion of the lamina and both anterior and posterior elements are involved.[45]

Pathophysiology of Symptoms and Signs

The mechanism(s) by which neurologic symptoms and signs are caused by epidural spinal cord compression is only partially understood. Direct compression, ischemia, and edema are probably all important: (1) In experimental animals, compression of the spinal cord by epidural tumor[46] or abscess[47] causes weakness associated with white matter edema and axonal swelling at a time when spinal cord blood flow remains normal[46] and blood vessels are patent. The edema appears to be a result of either direct compression of the cord or venous congestion from compression of the epidural venous plexus,[47] or both. At this stage, removal of the compression usually leads to full functional recovery. (2) With progressive compression, spinal cord blood flow decreases. The normal dilatation of spinal cord vessels in response to carbon dioxide inhalation (autoregulation) fails

Table 6–5. **Potential Therapeutic Targets in the Bone Microenvironment**

Target	Examples of Potential Treatments
Osteoblast	Endothelin receptor (atrasentan)
Osteoclast	Hydroxyapatite (zoledronate, samarium, strontium)
	src tyrosine kinase (dasatinib)
	Receptor activator of nuclear factor (mAb denosumab)
Endothelial cells	Vascular endothelial growth factor (mAb bevcizumab, VEGF Trap)
	Vascular endothelial growth factor receptor tyrosine kinase (BAY 43-9006, PTK 787, ZD6474)
Matrix metalloproteinases	Small molecules (BMS-27529, tanosmastat)
Cathepsins	Small molecules (relacatib), AAE581
Cytokines	
Transforming growth factor β	mAbs (lerdelimumab, metelimumab), antisense (AP12009)
Tumor necrosis factor α	mAbs (etanercept, infliximab, adalimumab)
Interleukin 1	Recombinant interleukin 1 receptor antagonist (anakinra)
Interleukin 6	mAbs (tocilizumab, CNT0328)
CXCL12/CXCR4 (stromal-derived factor 1)	Small molecule (AMD3100)
CCL2/CCR2 (monocyte chemoattractant protein 1)	mAb (CNT0888)
Epidermal growth factor	Epidermal growth factor receptor mAbs and small molecules (gefitinib, cetuximab, erlotinib, lapatinib, trastuzumab)

Modified from Ref. 40 with permission.

at the site of compression and caudally even before blood flow itself begins to decrease.[46] Tumors located either posteriorly or anteriorly to the cord occlude the vasculature to the center of the cord, whereas lateral masses of larger size do not.[48] At this stage, if untreated, the cord eventually infarcts, causing irreversible neurologic damage. (3) The spinal cord edema caused by tumor compression is vasogenic[49,50] and can be partially ameliorated by corticosteroids[51,52] (Fig. 6–5), improving clinical symptoms.

The biochemical processes associated with spinal cord compression and edema are not fully understood. Pharmacologic evidence implicates an increase in the synthesis or release of prostaglandin E2 and 6-ketoprostaglandin F-1a.[53] Serotonin 5-hydroxytryptamine (5HT) metabolism may be altered as well; 5-hydroxy-indoleacetic acid (5-HIAA)/5-HT ratios are elevated in the compressed cord, suggesting that 5-HT utilization is increased; the type-2 serotonin receptor antagonists cyproheptadine or ketanserin reportedly delay the onset of paraplegia in experimental animals.[54,55] Release of the

excitatory transmitter, glutamate, may also play a role because the glutamate receptor antagonists ketamine and MK-801 appear to decrease the edema induced by experimental cord compression.[56] The role these substances play in clinical cord compression in humans is unknown.

The rate of spinal cord compression is important. Tarlov and Klinger,[57] using inflated epidural balloons in dogs, have shown that the more rapidly spinal cord compression develops, the less likely recovery is after decompression, and conversely, the more slowly compression develops, the longer the animal can be paralyzed and still recover. Experiments in cats gave similar results.[58] Data in humans[19,59] suggest that metastatic tumors causing neurologic dysfunction that develops over hours to days carry a worse prognosis, even after treatment, than do those that evolve more slowly. For example, Smith[59] found that fewer than 10% of patients with symptoms progressing over 2 months benefited from surgical decompression, whereas 33% of patients whose symptoms developed more slowly benefited. It is widely recognized that

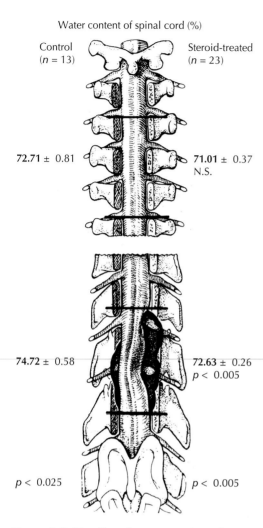

Water content of spinal cord (%)

Control
($n = 13$)

Steroid-treated
($n = 23$)

72.71 ± 0.81

71.01 ± 0.37
N.S.

74.72 ± 0.58

72.63 ± 0.26
$p < 0.005$

$p < 0.025$

$p < 0.005$

Figure 6–5. The effect of corticosteroids on the water content of the rat spinal cord. In this experimental model, the epidural tumor compressed the spinal cord anteriorly and laterally after growing through the intervertebral foramina at several levels (see Fig. 6–2E). A significantly higher ($p < 0.025$) water content is found in the compressed spinal cord than in the cord distant from the tumor. After steroid treatment, the water content of the compressed spinal cord drops significantly ($p < 0.001$) but is still greater than that of the normal cord. The animals improved symptomatically as well. (From Ref. 50 with permission.)

even paraplegic patients with slowly developing symptoms, such as caused by a meningioma, may regain neurologic function, a phenomenon much rarer with metastatic tumors. As indicated in the following text, one negative prognostic factor in patients radiated for epidural spinal cord compression is the development of weakness in fewer than 14 days.

In most patients with epidural spinal cord compression, neurologic symptoms other than pain evolve rapidly to paraplegia over several days to a few weeks. In some patients, with or without preceding pain, paraplegia begins suddenly, suggesting infarction of the cord. This sequence occurred in 28% of the patients reported by Costans et al.,[39] but much less frequently in our experience at MSKCC. Because the evolution of neurologic symptoms is usually rapid and one cannot predict when slowly evolving symptoms will suddenly accelerate, early diagnosis and prompt treatment are essential.

Pathology

Displacement and distortion of the spinal cord may be present on gross inspection of a fixed pathologic specimen (see Fig. 6–2A) but more often are absent (see Fig. 6–2C). When present, the degree and extent of gross distortion do not correlate with the degree of abnormalities noted microscopically.[60] Almost all symptomatic patients with spinal cord compression exhibit pathologic changes identified microscopically, although the changes are often minor and consist of only white matter edema and axonal swelling. Demyelination may not be present, infarction may not have occurred, and the axon cylinders may remain intact. The extent of spinal cord or nerve root pathology does not reflect the degree of neurologic impairment. The absence of major pathologic changes in symptomatic patients strongly suggests that physiologic changes predominate early in cord compression, and it is only later that the anatomic changes participate in the pathogenesis of symptoms.

Despite the usually anterior or anterolateral location of the epidural metastasis, microscopic changes at the level of compression are most common and severe in the dorsal and lateral funiculi, with relative sparing of the anterior funiculi and the central gray matter.[60,61] The earliest change in experimental cord compression appears to be extravasation of IV contrast into gray matter and small hemorrhagic areas in the dorsal columns.[46,47] Microscopic abnormalities in the white matter may be limited to tract demyelination; however, tissue necrosis may be diffuse and extensive. Even with severe white matter damage, the central gray matter is often fairly well preserved; the most common changes are degeneration of anterior horn cells and diffuse gliosis. Vacuolization of white matter indicating edema is most common at the cord periphery. Other changes include scattered small focal infarcts in white and gray matter, secondary tract degeneration away from the compression, and nerve root demyelination.

Pencil-shaped, longitudinal softening of the ventral portion of the posterior columns or posterior horns that extend over several segments is a relatively common complication of spinal cord compression,[62] either from vascular compression (arterial circulation or venous drainage) or directly from mechanical compression. This disorder tends to occur in the upper thoracic cord and, in one series, was present in 6 of 15 patients with epidural spinal cord compression[62] (Fig. 6–2D). The lesion is always associated with either transverse or patchy necrosis or with status spongiosis of the cord. Filled with necrotic debris and macrophages, the cavity can actually form a mass lesion compressing the surrounding cord. Rarely, epidural metastases may cause a visible syrinx on MRI that resolves after treatment.[63]

Infarction may occur as a result of vascular occlusion from compression of radicular arteries by tumor in the intervertebral foramen, as well as direct compression of the spinal arteries from an epidural metastasis.

Clinical Findings

PAIN

Pain is the earliest and most frequent presenting symptom of spinal cord compression, occurring in 80% to 95% of patients (Table 6–6).[19,64,65]

Pain from spinal metastasis is usually mild at first, but unless the diagnosis is established and treatment started, it becomes progressively more severe, often incapacitating the patient. Other signs and symptoms of spinal cord or cauda equina compression, including weakness, sensory loss, and autonomic dysfunction, develop later. The bone marrow is insensitive to pain, so that vertebral metastases cause pain only when the enlarging mass breaks through the bony cortex to involve the periosteum, paravertebral soft tissues, or other pain-sensitive structures such as dural nerves. Pain and other symptoms are also caused by compression or invasion of nerve roots, by pathologic fractures of vertebrae, by spinal instability[64–68] that occurs when posterior vertebral elements are destroyed, and by spinal cord compression itself. The tumor often secretes osteoclast-activating factor as well as prostaglandins E1 and E2, substances that assist the tumor in invading the bone and also cause hyperalgesic effects.[69] Factors produced by the injured spinal cord and its attached roots both mediate and modulate pain.[70] The clinical implication is that prostaglandin inhibitors may relieve pain and retard metastatic development. Other pain-producing substances, including acetylcholine, histamine, serotonin, bradykinin, and substance P, also play a role in producing pain by stimulating nociceptors.

Pain may begin from hours to months (median = 7 weeks) before other neurologic symptoms or signs develop.[19] The duration of pain before the development of other symptoms is shorter for lung cancer metastases than for breast cancer metastases.[38] Treatment is most effective when pain is the only symptom.

Table 6–6. **Signs and Symptoms of Spinal Cord Compression in 211 Patients at Memorial Sloan-Kettering Cancer Center**

	First Symptom		Present at Diagnosis	
	No. of Patients	Percentage of Patients	No. of Patients	Percentage of Patients
Pain	201	94	207	97
Weakness	7	3	157	74
Autonomic dysfunction	0	0	111	52
Sensory loss	1	0.5	112	53
Ataxia	2	1	17	8

The physician must be prepared to pursue vigorously the workup of patients with cancer and spinal pain, and he or she must caution patients, many of whom have widespread cancer with multiple aches and pains, that pain in the neck, back or radicular pain should be reported immediately and not ignored.[65]

Failure to evaluate the patient when pain is the only symptom may result in missed opportunities for effective therapy that preserves function. For example, Gilbert et al.[19] reported that although 125 of 130 adults with spinal cord compression had pain as the first symptom, by the time the diagnosis was eventually made, 99 (76%) complained of weakness, 74 (57%) of autonomic dysfunction, and 66 (51%) of sensory loss. Only five patients presented with painless weakness of the lower extremities. Other series of adults give a slightly different incidence of pain: Costans et al.,[39] 61%; Stark et al.,[38] 69%; and Bernat et al.,[71] 99% of patients. Of these, 43% had only local pain, 13% radicular pain, and 43% both. Although pain is absent in only approximately 5% of adults with myelopathy from metastatic spinal cord compression, as many as 20% of children may be pain-free at presentation.[64,65]

Back pain caused by cancer may be of several types: local pain without instability (also called biologic), pain caused by mechanical instability, pain from compression of nerve roots (radiculopathy), and pain perceived at a distance from a lesion (funicular and referred pain).

Local Pain

In most patients, the initial pain is local and perceived as steady and aching at the site of the involved vertebral body. The pain is nocturnal or morning pain that resolves over the course of the day. The pain is often exacerbated by coughing or straining (i.e., the Valsalva maneuver), but not by movement, unless the spine is unstable. Characteristically, the pain is more severe when the patient is supine.[72] One hypothesis for the etiology is the diurnal variation in endogenous secretion of steroids. Tumors secrete inflammatory mediators that result in pain. When endogenous corticosteroid secretion is reduced at night, patients experience flare inflammation that dramatically improves over the course of the day. Other possible etiologies include stretching of the vertebral body periosteum or epidural venous congestion. Local pain often responds to exogenous steroid administration or sometimes nonsteroidal anti-inflammatory mediators.

Whatever the pathogenesis, nocturnal spine pain that is relieved by arising and walking suggests a spinal cord tumor and helps to differentiate this pain from herniated disk pain, which is usually relieved by lying down.[73] At times, the positional pain requires the patient to sleep sitting up, particularly if there is spinal instability (see following text); however, some physicians have found this sign unreliable.[74] Local pain is usually associated with tenderness to percussion over the vertebral body, although some physicians question the usefulness of this sign. O'Rourke et al.[75] found percussion tenderness in only 43% of patients with CT-positive vertebral metastases and in only 65% with epidural disease. When present, however, the sign helps localize the metastasis. At times the pain is exacerbated by neck flexion if the involved vertebral body is cervical, upper, or midthoracic. Straight-leg raising often exacerbates the pain of lower thoracic or lumbar spinal metastases. Both signs suggest spinal cord compression rather than just vertebral metastasis.

Much local pain from spinal cord compression is probably not simply the result of vertebral body involvement, but originates from compression or inflammation of neural structures because it is usually dramatically relieved by administering corticosteroids, whereas other bone pain (such as hip metastasis) is not. The site of the local pain may help distinguish vertebral metastasis from the more common degenerative back lesion. Most degenerative back disease is either cervical or lumbar, causing neck or low back pain, whereas most epidural spinal cord compression is thoracic, yielding local thoracic pain.

Pain from Mechanical Instability

Mechanical instability is movement-related pain that is level dependent. As opposed to local pain, which usually responds to radiation therapy (RT), patients with mechanical instability often require operation or are dependent on an external orthosis for life. Clinical instability in the atlantoaxial spine is characterized by severe pain on flexion, extension, and rotation. Patients with subaxial cervical and thoracic spine instability are often worse with extension. These patients often give a history of sleeping in a recliner in an attempt to maintain the spinal lordosis resulting from the vertebral body fracture. Mechanical radiculopathy occurs with lumbar spine instability; the patient experiences severe radiculopathy on axial load (i.e., sitting or standing).

Radicular Pain

Compression of nerve roots within the spinal canal, or on exit through the intervertebral foramen, generates radicular pain; although it usually follows local pain, it sometimes precedes or is independent of it. Radicular pain is present in 80% of patients with cervical lesions. It generally radiates down one arm or sometimes both arms. Of patients with thoracic compression, 55% characteristically have radiating pain in a tight band around the chest or abdomen, almost always bilaterally; 90% of patients with lumbar spine involvement have pain radiating down one or both legs.[19] While most of these patients have local pain as well, radicular pain may be the only pain, and it may be felt in only one portion of the dermatome remote from the cord compression site.[76] One of our patients underwent extensive workup for isolated knee pain before careful examination revealed that the pain could be reproduced by straight-leg raising; an epidural metastasis was found.

Referred and Funicular Pain

Referred pain applies to pain perceived at a distance from the lesion but not dermatomal (radicular) in distribution. For example, L1 vertebral metastases can cause pain over the sacroiliac joint. In such cases, the physician may undertake a fruitless search for a lesion at the site of referred pain. Funicular pain is caused by compression of ascending spinal cord tracts; its presence can be misleading. For instance, cervical cord compression can cause funicular pain in the lower extremities, simulating sciatica,[77] and can also cause pain and paresthesias in a band-like distribution around the thorax or abdomen. Pseudoclaudication in the legs may be the only symptom of lumbar nerve root compression.[78]

WEAKNESS

Weakness is the second most common finding.[19,65,79] It usually results from corticospinal tract dysfunction and therefore is associated with spasticity, hyperactive deep tendon reflexes, and extensor plantar responses (Babinski sign).

Upper Motor Neuron Weakness

Spinal cord compression usually causes weakness that begins in the legs, regardless of the spinal compression site, and is more marked proximally than distally early in the course. Thus, the patient usually complains of difficulty walking and especially climbing stairs or rising from low chairs or toilet seats. The patient may notice that the knees buckle even when walking on level ground. As the illness progresses, the weakness becomes more profound, both proximally and distally, leading to increased difficulty in walking. With an upper thoracic or cervical cord lesion, the patient may next notice a weak cough or difficulty in sitting up from a recumbent position because of weak abdominal muscles. Only late in the development of cervical cord compression do the arms become substantially weak.

Although weakness is usually bilateral and symmetric, at times it may predominate in one leg or arm. Examination during the early stages of upper motor neuron weakness may fail to reveal spasticity or hyperactive reflexes; only modest weakness of iliopsoas and hamstring muscles may be observed, with apparently normal strength in distal muscles. Even at this time, the patellar and Achilles reflexes are usually slightly more active than the upper extremity reflexes; extensor plantar responses may also be elicited. As the illness progresses and the weakness becomes more profound, spasticity, hyperactive reflexes, and unequivocal extensor plantar responses develop. These signs are usually bilateral and symmetric, although sometimes one side may be more affected than the other. If the patient has received prior neurotoxic chemotherapy, reflexes may be diminished or absent from a toxic neuropathy reducing their localizing value. If the onset is sudden and leads to complete paraplegia, most patients are flaccid with areflexia, reflecting distal spinal reflex inhibition.

The distribution of the weakness and reflex changes do not identify the site of cord compression in the horizontal plane; that is, laterally placed lesions do not necessarily cause a Brown–Sequard syndrome[80] (dysfunction of the compressed site of the cord causing ipsilateral weakness and proprioceptive loss with contralateral temperature and pain loss). Such lesions may actually cause more contralateral than ipsilateral weakness, probably because of torsion and ischemia of the cord. Similarly, no clinical findings differentiate between anteriorly and posteriorly placed lesions.[60] However, slowly growing tumors are more likely to cause the anatomically expected localizing signs.

Lower Motor Neuron Weakness

When the cauda equina rather than the spinal cord is involved, the weakness reflects lower motor neuron dysfunction. Characterized by hypotonia, atrophy, fasciculations, and areflexia, lower motor neuron weakness can also occur with spinal cord compression as a result of dysfunction of anterior horn cells. This dysfunction is probably vascular rather than compressive. Lower motor neuron hand weakness, with weakness and sometimes atrophy of one or both hands, may be an early and isolated sign of lower cervical cord compression. With cauda equina compression, the weakness is usually more marked distally than proximally, although buttock and hamstring muscles are usually affected as well. The distal weakness begins with foot drop and proceeds to flail weakness below the knees. Initially, such patients may have more difficulty descending than climbing stairs; later, they catch their foot as they walk on an uneven surface or climb a curb. Achilles reflexes disappear early and the patellar reflexes follow. Plantar responses are absent or flexor. Fasciculations occasionally appear in the lower extremities. Severe atrophy is uncommon, probably because the disease evolves rapidly. In rare instances, fasciculations in the lower extremities have been reported with cervical cord compression[81] or injury.[82] We have seen one patient in whom fasciculations in leg muscles disappeared after a C7 epidural metastasis was treated successfully.

With lower motor neuron dysfunction from spinal cord compression, the segmental reflex disappears at the compression site, although reflexes below this site are hyperactive. For example, with C5 cord compression, the biceps reflex disappears and unilateral selective weakness may develop in the deltoid and biceps muscles. The triceps and finger-stretch reflexes are hyperactive, with better strength in those muscles. With T1 compression, pain in the upper back, radiating down the medial aspects of one or both arms or localized to the elbow(s), is accompanied by weakness of the small muscles of the hands and absent finger-stretch reflexes. The finger-stretch reflex (Hoffmann sign) may be absent in normal individuals but is present in most patients stimulated by the anxiety, tension, and pain associated with spinal metastases. Its unilateral absence is strongly suggestive of a T1 spinal lesion. Some of our patients have further developed severe weakness and atrophy of the small hand muscles without long tract signs. One caveat: Atrophy of the small hand muscles, usually without much weakness, may accompany upper cervical cord compression as a false-localizing sign, possibly from compression of descending arteries.[83]

SENSORY LOSS

Usually concurrent with the development of weakness or shortly afterward, patients notice either paresthesias or numbness (loss of sensation) or both in the lower extremities. Sensory complaints without pain are rare with spinal cord or cauda equina compression. Only 1 of 83 patients reported by Greenberg et al.[79] presented with sensory symptoms without pain, weakness, or ataxia. Sensory abnormalities usually begin in the toes and ascend in a stocking-like fashion, eventually reaching the level of the lesion. The patient may complain of either numbness ("as if I am wearing thick stockings") or paresthesias ("pins and needles" tingling of the lower extremities) and may state "My feet feel as if they are asleep." Although sensory changes evolve rapidly, the earliest sign appears to be a slight decrease of vibration and position sense, with pain and temperature loss ensuing. By the time most such patients present for diagnosis, sensory loss for both touch and pinprick are found to be one to five levels below the site of actual spinal cord compression. On occasion, the level of sensory loss is one or two segments *above* the site of spinal cord compression, presumably because of ischemia to the cord from compression of ascending vessels.

Sacral Sparing

Loss of sensation to pinprick in the legs and trunk with preservation in the perianal region and buttocks was originally believed to be characteristic of intramedullary disease, but actually commonly occurs with epidural spinal cord compression. Approximately 20% of our patients with extramedullary compression have exhibited some degree of sacral sparing. Usually limited to the buttocks, it may extend down the posterior aspect of the thighs (S2 dermatome) and, on rare occasions, down the posterior or even lateral aspect of the leg and foot (S1-L5 dermatome). Preserved sensation in the sacral area may explain why autonomic dysfunction is a late occurrence with spinal cord compression.

With cauda equina compression, the sensory loss is dermatomal and usually bilateral, involving the perianal area, the posterior thigh, and the lateral aspect of the leg. Substantial sensory loss is almost always associated with significant motor loss but sometimes occurs independently of motor change; therefore, a careful examination for a sensory level should be performed in all patients complaining of back pain.

Sensory loss is rarely as profound as weakness. Even in patients who are paraplegic, some appreciation of gross touch is usually present, that is, the patient perceives something happening when the leg is lightly squeezed. In some patients, however, loss of position sense may preclude independent walking even when strength is sufficient.

LHERMITTE SIGN

Originally believed to be a pathognomonic symptom of multiple sclerosis, Lhermitte sign is an electric shock-like sensation radiating into the back or legs when the neck is flexed.[84–86] Unpleasant but not painful, it sometimes complicates cervical or thoracic spinal cord compression. The sign is more common in patients with radiation myelopathy (see Chapter 13) and cisplatin neuropathy (see Chapter 12).

AUTONOMIC DYSFUNCTION

Isolated autonomic dysfunction is rarely the sole presenting complaint. Even when other neurologic signs are absent, the patient complains of back pain. Costans et al.[39] reported isolated sphincter disturbances in 2% of patients studied. We have found bladder and bowel dysfunction in more than 50% of patients by the time the diagnosis of spinal cord or cauda equina compression is made; most men are impotent. It is difficult to say whether impotence results from the cord compression or from pain and systemic illness. Constipation is usually present, but it may be a result of opioids taken to relieve pain. Occasional patients have

observed that enemas or laxatives induce fecal incontinence.

The most characteristic autonomic abnormality is bladder dysfunction. Urinary urgency with incontinence is an occasional complaint if the motor symptoms are evolving slowly. More commonly, patients simply develop painless urinary retention, usually associated with severe weakness and sensory loss of the lower extremities. Patients may or may not be aware that they have not voided recently, and the bladder may grow to enormous size. As the bladder rises above the pelvic brim, some patients suffer from diffuse abdominal pain, probably from peritoneal irritation, but many have no pain even when the bladder reaches the umbilicus. The physician suspecting cord compression is obligated to percuss the abdomen frequently and to ask the patient, if not hospitalized, to keep records of frequency and quantity of urinary output. If the bladder cannot be percussed because of a patient's obesity, one can measure residual urine by bladder ultrasound.[87] One woman referred because of weakness in the lower extremities was told that her cancer was rapidly progressing because of an enlarging abdominal mass; the mass disappeared when 2000 mL of urine were drained, gradually, from her painless bladder.

Other autonomic changes of lesser importance, but occasional clinical usefulness, include a definable sweat level (i.e., the absence of sweating below the level of a spinal block) and a Horner syndrome. A sweat level usually develops only when weakness is profound, but sometimes, in patients unable to cooperate for sensory examination, the level of the lesion can be discerned by running one's hand, a dry tuning fork, or a spoon up the back or abdomen. The tuning fork begins to stick or the finger perceives moisture at the level where sweating begins. In patients who are profoundly paraplegic, increased sweating above the level of the lesion is a normal mechanism for heat dissipation. Horner syndrome is more common in patients with paravertebral tumors than with cord compression, but when cord compression is at C7 to T1 vertebral level, Horner syndrome, either unilateral or bilateral, occasionally occurs.

An unusual autonomic presentation takes place when the cord is compressed at the T10 to T12 vertebral bodies at or just above the conus medullaris. Many patients, after experiencing pain for several days to weeks, suddenly develop urinary retention and severe constipation without weakness or sensory loss. The neurologic examination is usually normal save for a lax anal sphincter and an enlarged bladder. Vertebral tenderness is often present over the lower thoracic and upper lumbar area.

Examination of all patients with suspected spinal cord compression should include a rectal examination to assess sphincter tone, the patient's ability to contract the sphincter voluntarily, and the reflex contraction of the sphincter when the skin around the anus is scratched (anal wink). If any one of the above findings is abnormal, cauda equina or conus medullaris compression should be suspected.

ATAXIA

Gait ataxia is common, but *cerebellar signs* are usually obscured by motor weakness and sensory loss. In occasional patients, isolated ataxia, either with or without pain, can be a major complaint.[19,88,89] Ataxia without pain is probably the most confusing sign of cord compression the clinician encounters. An initial clinical diagnosis of either paraneoplastic cerebellar disease or hysteria has delayed proper diagnosis in several of our patients. One patient with pain and ataxia, but no other neurologic signs, was locked in a psychiatric ward after punching a house officer who told the patient he was "drunk and hysterical." The alcohol the patient drank was an attempt to control the severe thoracic pain resulting from an esophageal carcinoma metastatic to the T4 vertebral body. A woman with painless cord compression from Hodgkin disease was treated for several days for anxiety and *conversion symptoms* before a consultant suggested the correct diagnosis. Lack of dysarthria and upper extremity ataxia in the presence of severe gait and lower extremity ataxia should lead to a suspicion of spinal disease, although the same signs also can occur with lesions of the cerebellar vermis. The ataxia of spinal cord compression probably results from compression of the spinocerebellar tracts. Why these tracts should be selectively and symmetrically involved is not clear. Ataxia may persist after restoration of strength and sensation following treatment of spinal cord compression, making it impossible for the patient to walk; at times gait training helps.

UNUSUAL SYMPTOMS AND SIGNS

Unusual signs of spinal cord compression (Table 6–7) can lead to an incorrect diagnosis.

Clinically perplexing, rare signs include papilledema, hydrocephalus,[90,91] and nystagmus in patients with thoracic cord compression, as well as facial weakness, diplopia, and lower extremity fasciculations in patients with cervical or upper thoracic cord compression. More understandable are Brown–Sequard syndrome,[92] *Herpes zoster*, pseudoclaudication, pain in the neck and numbness of the tongue on head turning (the neck–tongue syndrome),[93] myoclonic contractions of abdomen and leg muscles, and the syndrome of painful legs and moving toes. Brown–Sequard syndrome is more often encountered in radiation myelopathy (see Chapter 13). Its presence with cord compression does not necessarily indicate that the tumor is positioned ipsilaterally. *Herpes zoster* at the cord compression site has been noted by several authors[19]; it is believed to be due to activation of the latent virus from tumor compression of the dorsal root ganglion.

Facial pain and numbness sometimes may delay accurate diagnosis of upper cervical cord compression. The upper cervical sensory fibers, particularly from the C2 and C3 roots, supply the back of the head and also a variable portion of the face. The extent of sensory loss on the face caused by a cervical lesion can vary with physiologic manipulation of the spinal cord.[94] Some patients with compression of C2 and C3 roots complain of pain, often tic-like,

Table 6–7. Unusual Symptoms and Signs in Patients with Spinal Cord Compression

"Painful legs and moving toes"
Lhermitte sign with cervical or thoracic tumor
Lower extremity fasciculations with cervical tumor
Sciatica with cervical tumor
Facial paresis with cervical tumor
Nystagmus with thoracic tumor
Pseudoclaudication
Hydrocephalus (papilledema)
Spinal myoclonus
Tongue numbness
Inverted knee jerk

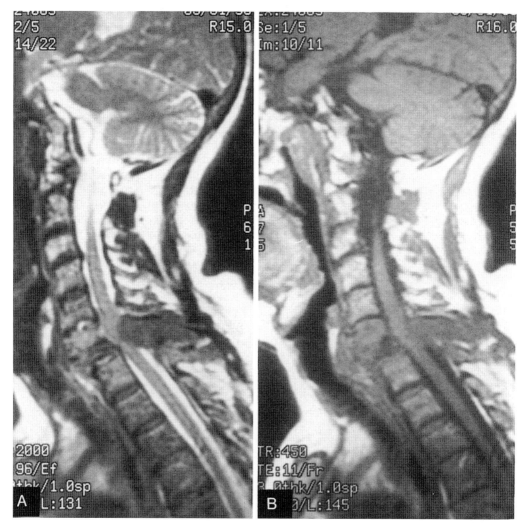

Figure 6–6. A magnetic resonance imaging scan in a patient with breast carcinoma reveals anterior and posterior epidural spinal cord compression. The posterior spinous process and the vertebral body were invaded by the tumor. Cord compression is more apparent in the T2-weighted image (**A**) than the T1-weighted image (**B**).

and numbness involving the lateral portion of the cheek and jaw. The pain and numbness never reach the medial portion of the face, but can mislead a physician into thinking that the symptoms originate with the trigeminal nerve rather than the upper cervical cord.

Laboratory Examination

APPROACH TO THE PATIENT

The most sensitive and specific diagnostic test for spinal lesions caused by cancer is an MRI[95,96] (Fig. 6–6). An MRI often detects compression of the spinal cord by metastatic tumor in patients without pain or other neurologic disability (16% of patients with metastatic prostate cancer in one series[97]). If the entire spine is imaged both before and after contrast enhancement, a normal study effectively rules out vertebral and paravertebral metastases, epidural spinal cord compression, intramedullary spinal metastases, and many leptomeningeal tumors. Those rare occasions when epidural mass lesions are missed by MRI, even after contrast enhancement, appear to be due to failure to examine the spine in its most lateral aspects and to obtain appropriate transverse as well as sagittal images. MRI also distinguishes malignant lesions from benign ones such as disk herniation. A complete MRI study is arduous for a patient in pain. Less expensive and time-consuming diagnostic tests often suffice, but all too often, failure to begin with an MRI results in the patient undergoing a series of increasingly sophisticated images (e.g., plain radiographs, a bone scan followed by CT), only to finally have the diagnosis established by MRI after a significant delay and increased exposure to radiation. In patients with severe pain, pain localized to the thoracic area (where disk disease is uncommon), or pain of recent onset, MRI is the place to start. Because multiple lesions may be present, the entire spine should be imaged.

Noncontiguous metastatic lesions occur with high frequency, either near each other or at a distance.[97–99] The presence of these other lesions may alter both the patient's symptoms and the physician's treatment plan. Thus, in most instances, it is important to image the entire spine. It is also important to try to determine if spinal cord compression associated with vertebral collapse is a result of tumor, bone in the epidural space, or (as is usual) both. If bone alone is the culprit, RT and chemotherapy will not work. Figure 6–7 shows a modified algorithm initially devised by Portenoy et al.[100] for the diagnosis and management of patients with back pain and cancer. The specific use of imaging techniques is detailed in the paragraphs that follow.

PLAIN X-RAYS

Depending on the series, in 85% to 95%[98,101] of patients with spinal cord compression, plain radiographs identify the responsible vertebral lesion.[19,98,101,102] In a few patients, radiographs will identify a paravertebral mass when vertebral bodies are normal. Vertebral body abnormalities include radiolucencies or radiodensities from direct involvement by tumor, erosion of pedicles, collapse, and sometimes subluxation of vertebral bodies. In one prospective study,[102] 42 of 45 patients with spinal blocks identified by myelography had abnormalities on spine radiographs. Plain radiographs are also both highly sensitive (91%) and highly specific (86%) for predicting epidural disease. Particularly useful predictive

features of epidural spinal cord compression are greater than 50% vertebral collapse and pedicle erosion. Fewer than 10% of patients with radiologic evidence of vertebral metastases without vertebral collapse suffer epidural spinal cord compression.[98] However, plain radiographs are not entirely reliable for several reasons: (1) Certain vertebral bodies, such as T1, may be obscured behind other bony shadows and not easily detected by routine spine radiographs. (2) Even when the vertebral body is seen, the tumor may not have destroyed enough bone to be identifiable. The estimate is that 30% to 50% of the vertebral body must be destroyed before noticeable changes are seen on radiographs. (3) If multiple vertebral bodies are involved by metastatic tumor, the radiographs will not identify which area is causing cord compression. (4) All too often, the oncologist or neurologist refers the patient to the radiologist with a note saying "suspected cord compression; please do cervical, lumbar, and thoracic spine films." In these instances, even a careful radiologist may miss a subtle lesion unless a localizing clue is available.

Autopsy studies suggest that approximately 25% of spinal lesions are not identifiable by plain radiographs[31] (see Table 6–3). The incidence is highest in lymphomas (63%), testicular tumors (50%), and nasopharyngeal tumors (40%). By contrast, a collapsed vertebral body in a patient with cancer is not absolute evidence of metastatic disease. In one study, 34 of 155 patients with vertebral collapse did not have bony metastases, a false-positive rate of over 20%.[31] Vertebral compression fractures with back and leg pain, but without epidural disease, may be a presenting symptom of acute lymphoblastic leukemia.[103,104]

RADIONUCLIDE BONE SCANS

Bone scans are as sensitive as radiographs (91%) but less specific in predicting the site of spinal cord compression.[98] They are more sensitive in identifying vertebral metastases that, even in asymptomatic patients, may be associated with a partial block. Bone scans, however, are often abnormal at sites in which vertebral bodies are diseased by processes other than cancer. In one prospective series, bone scans predicted with only 66% accuracy the presence and location of epidural metastases.[102] In the few cases in which spinal abnormalities were visualized

Diagnostic Approach to Neck/Back Pain

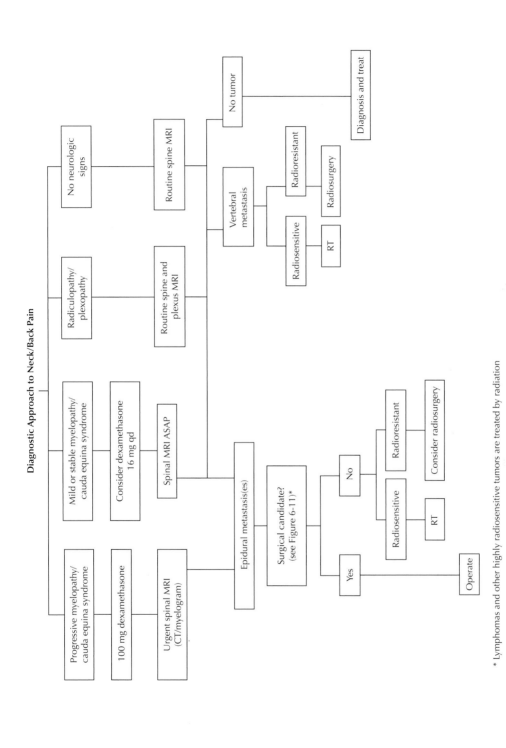

* Lymphomas and other highly radiosensitive tumors are treated by radiation

Figure 6–7. Management of back pain in the patient with cancer.

on bone scan, but not on plain radiography, the myelogram was normal. As with radiographs, bone scans may reveal multiple uptake areas in the vertebral bodies but may fail to identify the site of cord compression. Lesions of the vertebral body that do not involve bone cortex are likely to be negative on radionuclide bone scan, although positive on MRI.[44,105]

Increased radionuclide uptake, the usual abnormal finding on bone scan, is due to increased blood flow and sequestration of the technetium isotope at sites of new bone formation. When a destructive tumor neither substantially increases blood flow nor stimulates new bone formation, the scan may be negative despite extensive destruction. This occurs often with myeloma, occasionally with metastatic solid tumors and at the site of previous RT. In addition, the metastatic disease is sometimes so diffuse that the uptake by bone scan is uniform throughout, and the bone scan is interpreted as normal because there are no normal areas against which to compare those involved by metastatic tumor. Sometimes this is referred to as a *super scan* or a *super normal scan*. Single photon emission computed tomography (SPECT) images are occasionally positive when standard bone scans are negative.[106] Bone scans using technetium-99m methoxy isobutyl isonitrile (MIBI) can distinguish nonmalignant vertebral fractures from metastases in patients whose standard bone scan is positive.[107]

COMPUTED TOMOGRAPHY

When an MRI is not possible, for example a patient with a pacemaker, CT often identifies metastatic lesions in vertebral bodies when plain radiographs and bone scans are negative.[108–110] On modern scanners, reconstructed images can provide an excellent view of the complete spine in the sagittal plane, providing visualization comparable to that on an MRI. At times CT is helpful in identifying the nature of the vertebral lesions seen on MRI, particularly to assess pedicle and posterior element integrity before surgery.[108]

POSITRON EMISSION TOMOGRAPHY

Positron emission tomography (PET) scanning using ^{18}F-FDG can identify metabolically active lesions in the bone, making it a good screening test for vertebral metastases and helping distinguish benign from malignant bone lesions.[111,112] It is particularly effective when combined and fused with CT images. Of 242 spinal lesions detected on PET/CT, 220 were identified by PET alone, and only 159 by CT alone.[112]

MAGNETIC RESONANCE IMAGING

Because MRI is the procedure of choice to evaluate the vertebral column and the spinal cord, it usually eliminates the need for other diagnostic tests[96,97,113,114] (Fig. 6–6). Thus, the MRI has major clinical advantages both as a screening tool and as the ultimate diagnostic test to evaluate spinal cord compression. Despite its high cost, it is actually cost effective to start with an MRI in a patient at high risk for epidural spinal cord compression rather than begin with a less costly but also less definitive test.[115] The only disadvantage of MRI is that imaging the entire spine may take up to an hour and movement substantially degrades the MR image; many patients with severe back pain cannot lie quietly for that period. However, corticosteroids, as indicated in the paragraphs on treatment that follow, often give enough pain relief to allow the patient to lie still.

MYELOGRAPHY

At MSKCC, improvements in MRI, including faster imaging, have rendered myelography for diagnostic purposes nearly obsolete. However, at MSKCC, we perform 3 to 4 myelograms a week for treatment-planning purposes in patients about to undergo radiosurgery. Myelography for diagnostic purposes is performed only in patients who cannot undergo an MRI, in whom the diagnosis of epidural spinal cord compression is suspected and when a complete spine CT is not definitive. A complete myelogram is essential for two reasons: (1) to define the length of the block, and (2) to determine whether additional lesions exist above the block encountered by the lumbar myelogram, which should be done first. Approximately 10% of patients with epidural lesions have more than one lesion.

In a patient suspected of having a block, 3 mL of myelographic contrast material is inserted in the lumbar area under fluoroscopic observation to screen the subarachnoid space. If a complete block is encountered, contrast material is inserted via a lateral C1 to C2 puncture into the cervical spinal canal, and the upper end of the block is discerned, completing the myelogram. CT is often combined with myelography to improve identification of epidural lesions.

The major disadvantage of myelography is that the lumbar puncture can alter cerebrospinal fluid (CSF) dynamics in such a way that patients with a complete block may worsen clinically (see Chapter 14). One series reported a 15% incidence of increased clinical symptoms following lumbar myelograms in patients with spinal cord tumors.[116] Our experience suggests that this is a rare problem. Nonetheless, clinical worsening is an occasional risk. Some investigators believe that lateral cervical puncture alters the CSF dynamics less and is thus safer. Empirical evidence that supports this view is unavailable, although one report[117] describes patients who developed spinal symptoms after undergoing ventricular decompression for hydrocephalus. If a complete block is suspected, corticosteroids should be given immediately before the myelogram. The rapid onset of steroid action may help prevent complications.

Differential Diagnosis

Many causes of myelopathy that may be confused clinically with epidural spinal cord compression (Table 6–8) are excluded by MRI. The MRI of intramedullary or extramedullary intradural tumors, leptomeningeal metastasis, radiation myelopathy, arteriovenous malformations, and subacute transverse myelopathy, all of which cause spinal cord dysfunction, are not easily confused with those of epidural spinal cord compression. Even epidural lipomatosis (see p 104) is distinguished from tumor by its density on CT or MRI. Those illnesses that can be confused, both clinically and radiographically, with epidural spinal cord compression from systemic tumor include herniated disks, epidural hematoma or abscess, and rarely, extradural hematopoiesis.

EPIDURAL HEMORRHAGE

On a few occasions, epidural hematomas are mistaken for spinal cord compression from a tumor and so treated. Epidural spinal hematomas associated with cancer usually occur in thrombocytopenic patients, are usually acute,

without preceding pain, and are generally (although not invariably) painful at onset. The signs and symptoms usually evolve rapidly over minutes to hours rather than days to weeks. On scan, there is no evidence of vertebral involvement by tumor, and the epidural block usually covers several segments rather than the one or two segments characteristic of epidural spinal cord compression from tumor. Furthermore, the density characteristics of hemorrhage on the MRI are different from those of tumor, except for a hemorrhagic tumor.[15] When these clinical and radiographic factors are considered, epidural hematomas can usually be distinguished from epidural metastases, although at times the differential diagnosis cannot be made without biopsy. Unfortunately, most patients with epidural hematomas are profoundly thrombocytopenic and cannot undergo operative removal of the hematoma, which would confirm the diagnosis and treat the illness. Epidural hemorrhages from systemic vasculitis, as in polyarteritis nodosa, also occasionally cause spinal cord compression.[118]

A radiation-induced cavernous malformation of the spinal cord (see Chapter 13) could be mistaken for an intramedullary metastasis. These can usually be distinguished clinically by the course and knowledge of the preceding history of radiation. Epidural cavernous malformations are rare causes of spinal cord dysfunction.[119] They have not been reported as a complication of RT.

EPIDURAL ABSCESS

Epidural abscesses are rare in patients with cancer, but they can represent a problem in the differential diagnosis. In cancer patients, an epidural abscess can arise from an epidural catheter inserted for pain control, osteomyelitis of a vertebral body, resulting from hematogenous spread of bacteremia, or from direct extension of an open wound of the head, neck, or back. The clinical picture of epidural abscess resembles tumor spinal cord compression. Pain, probably from vertebral body involvement, is the first symptom and is usually present for days or weeks before other neurologic signs appear, evolving either slowly or rapidly. Radiographs of the spine may be normal at onset but eventually almost always become abnormal; characteristically, two vertebral bodies across a disk space are destroyed.

Table 6–8. **Differential Diagnosis of Epidural Spinal Cord Compression**

Diagnosis	Example(s)	Diagnostic Test
Intramedullary tumor	Glioma	MR with gadolinium
	Metastasis	
Extramedullary–intradural tumor	Meningioma	MR with gadolinium
	Neurofibroma	
	Drop metastasis from primary brain tumor	
Leptomeningeal tumor	Primary lymphoma	MR with gadolinium, CSF cytology
Radiation myelopathy	Previous RT to spine	MR with gadolinium
Arteriovenous malformation		MR with gadolinium, myelogram, arteriogram
Transverse myelopathy	Postinfectious myelopathy	MR with gadolinium
Epidural hematoma	Thrombocytopenia (history of LP)	MR, CT
Epidural abscess	Sepsis	Culture/MR
Herniated disk		CT or MR
Osteoporosis	Vertebral collapse	MR/biopsy

MR, magnetic resonance scan; CSF, cerebrospinal fluid; RT, radiation therapy; LP, lumbar puncture; CT, computed tomography.

This is a hallmark of infection because it is rare for metastatic tumor to cross the disk space and involve two contiguous vertebral bodies, although two adjacent vertebral bodies may be involved independently with an intact disk between. The inflammatory response in the epidural space characteristically initiates a CSF pleocytosis; the CSF culture is negative. Most patients with epidural abscess, but not all, are febrile. To further complicate the diagnosis, epidural abscess may form at the site of a metastatic epidural tumor, but this is rare. If the diagnosis is in doubt, needle biopsy of the involved vertebral body will establish the diagnosis.[120] Often, the patient can be treated by needle drainage of an abscess.

HERNIATED DISK

Herniation of lumbar or cervical disks (rarely thoracic disks) can mimic epidural spinal cord or cauda equina compression, leading to inappropriate treatment.[121] The signs are local and radicular pain, sometimes accompanied by dermatomal sensory and motor loss; spinal cord involvement is rare. Characteristically, affected patients complain of pain when sitting or walking but usually have relief when lying down; conversely, spinal cord compression is usually associated with more pain in the lying position than when sitting or standing. MRI should establish the diagnosis and also identify those instances in which disk herniation is caused by a vertebral body tumor.[122]

OTHER EPIDURAL MASSES

Epidural spinal cord compression is occasionally caused by non-neoplastic masses, some of which are uncommon or rare and are listed in Table 6–9.

Extramedullary hematopoiesis usually occurs in patients with severe hematologic abnormalities including a myelodysplastic syndrome,[123] but also occasionally in patients with cancer who have widespread marrow destruction with compensatory hematopoiesis in the remaining vertebral marrow.[124] MRI helps establish the diagnosis.[125] Vertebral *hemangiomas* can also cause cord compression at single or multiple levels.[126,127] The MRI differs from that of vertebral metastases in that both T1- and T2-weighted images are hyperintense.[128] RT, embolization, and decompressive surgery are all effective treatments when appropriate.[126]

In patients on chronic steroid therapy, *lipomatosis* often develops, sometimes causing mediastinal widening or epidural cord compression.[129] Characteristically, the patient has

Table 6–9. **Some Unusual Causes of Spinal Cord Symptoms**

Idiopathic vertebral hyperostosis
Hyperparathyroidism
Gout
Pseudogout
Rheumatoid nodule(s)
Epidural lipomatosis
Sarcoidosis
Ossification of spinal ligaments
Calcification of arachnoid
Extramedullary hematopoiesis
Paget disease (compression or vascular
   cause uncertain)
Osteomalacia
Amyloid
Nodular fasciitis
Vertebral hemangioma
Spinal cord malformations

Table 6–10. **Treatment Options for Epidural Spinal Cord Compression from Metastatic Tumor**

Treat edema
   Corticosteroids
   Possibly mannitol
   Possibly other agents
Radiation therapy
   External beam high dose per fraction
   Stereotactic radiosurgery
Chemotherapy
   Hormonal inhibitors
   Antineoplastic agents
Surgical decompression
   Laminectomy
   Posterior–lateral approach
   Vertebral body resection

been on steroids for a prolonged period and usually has other features of steroid excess; back pain is not prominent. The fat proliferation is often located posteriorly to the spinal cord rather than anteriorly as in most epidural tumors; MRI signal of the epidural lesion is that of fat rather than tumor. Metastatic involvement of the vertebral body is absent.

Benign tumors of the spinal cord can also occur in patients with cancer. Meningiomas of the spinal canal are more common in women with breast carcinoma than in women without such cancer. MRI can usually distinguish a meningioma from a metastasis, but an exception occurs when tumor metastasizes to a meningioma.[130] Pachymeningeal *amyloidosis* from the polyneuropathy, organomegaly, endocrinopathy, monoclonal gammopathy, and skin changes (POEMS) syndrome (see Chapter 15) also can cause spinal cord compression,[131] as can *sarcoidosis,*[132] *nodular fasciitis,*[133] *rheumatoid nodules,*[134] and *tophaceous gout.*[135]

Treatment

Before considering treatment, it is important to have a method of evaluating its response. Accordingly, the American Spinal Injury Association (ASIA) has developed an impairment scale that allows one to track results of treatment (Table 1–13).

Several treatments aid in controlling symptoms of epidural spinal cord compression (Table 6–10).

For most patients, definitive treatment consists of either RT or surgical decompression followed by RT. No definitive, controlled trial is available to guide the clinician toward a scientifically based decision for optimal treatment, although several studies compare the outcome of surgery with RT and other treatment modalities[27,72,136]; which one to choose in a specific instance is based on judgment and experience.

CORTICOSTEROIDS

The role of corticosteroids in treating epidural spinal cord compression is discussed in Chapter 3, with a few additional comments here.[79,137,138] In some patients, corticosteroids alone have salutary effects on symptoms and signs, particularly pain.[79,136,138] A randomized trial of high-dose steroids compared 96 mg dexamethasone given intravenously followed by 24 mg 4 times a day for 3 days and then tapered over 10 days with a placebo. All of the patients received radiation. More patients in the steroid group remained ambulatory (81% vs. 63%) at

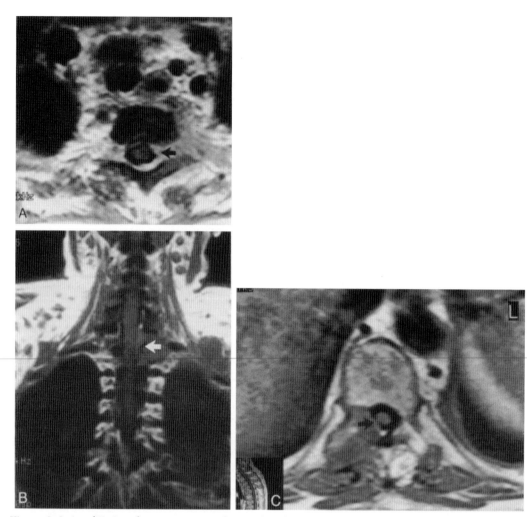

Figure 6–8. A and **B**: Lymphoma compressing the spinal cord. The tumor arose in the paravertebral space and grew through the intervertebral foramen to compress the spinal cord laterally (*black arrow*). The bones are entirely normal. The tumor is distinguished from normal fat, as seen on the opposite side, by the difference in the signal and by the fact that the spinal cord is compressed (*white arrow*). **C**: Metastases growing through the intervertebral foramen from a thyroid carcinoma. The spinal cord is compressed (*arrow*). Both anterior and posterior elements are partially destroyed by the primary growth, but the major growth is through the intervertebral foramen. The patient's clinical symptoms responded to radioactive iodine.

the end of therapy and continued ambulation at 6 months (59% vs. 33%).[139]

Steroids have two mechanisms of action: (1) they may have an oncolytic effect on the tumor (Figs. 6–8 and 6–9), and (2) they ameliorate spinal cord edema. Oncolytic effects occur primarily in patients with lymphoma or breast carcinoma. In lymphoma, the effect is so rapid and dramatic that, if lymphoma is a suspected but not established diagnosis, the clinician should withhold corticosteroids until the diagnosis can be confirmed by biopsy.

The length of time a clinician should wait to begin other therapy after corticosteroids are administered is not established. MSKCC's policy supports beginning definitive therapy, usually RT, immediately after the diagnosis is made by MRI or myelography. One might argue that definitive therapy should be deferred, as it is with brain metastasis, for 12 to 48 hours to allow corticosteroids to maximally shrink edema. Clinical data to refute or support such a position are not available. Unlike the situation in metastatic brain tumors, both

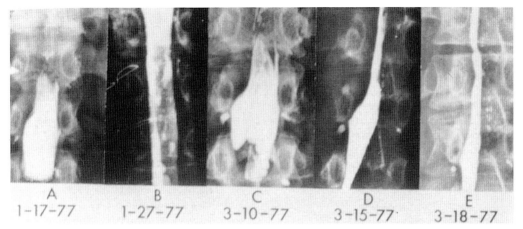

A
1-17-77

B
1-27-77

C
3-10-77

D
3-15-77

E
3-18-77

Figure 6–9. An oncolytic response to corticosteroids. This patient suffered epidural spinal cord compression from a metastatic malignant thymoma. **A**: A lumbar myelogram on January 17, 1977 reveals complete block of the contrast column at T11. The major defect is on the right. No bone involvement is seen. **B**: On January 27, after 10 days of steroid therapy, a refluoroscopic myelogram reveals no extradural tumor. **C**: On March 10, 9 days after steroids were discontinued, a recurrent complete block is at T11 again with the major extradural defect on the right. **D**: On March 15, after 5 additional days of steroid therapy, the previous complete block again resolves. A small extradural defect is seen on the right. **E**: On March 18, a right decubitus view shows nearly complete resolution of the extradural tumor. (From Ref. 149 with permission.)

experimental[52] and clinical evidence[140–143] suggest that large fractions of RT for epidural spinal cord compression do not cause immediate clinical worsening. Corticosteroids were not required in one phase II trial of 20 consecutive patients with spinal cord compression who were said to have no neurologic deficits except for radiculopathy, although four of the patients required assistance in walking.[144] The patients received 30 Gy in 10 fractions without neurologic deterioration; the four who required assistance recovered fully. Thus, corticosteroids are used not to protect against potential side effects of acute RT, as they are in patients with metastatic brain tumors, but to relieve pain and produce any other potential benefits that they may afford while awaiting the salutary effects of RT. Nevertheless, the randomized trial described earlier indicated a better outcome in those patients who received high-dose corticosteroids. Accordingly, we give corticosteroids to all patients who are to receive RT to the spinal cord, but we do not delay the radiation to allow full effect of the steroids, as we do in patients with brain tumors.

The optimal dose of corticosteroids is not established. The original decision to use the 100 mg loading dose[79] was based on animal experimentation, indicating a dose–response curve in experimental spinal cord compression.[51,145] The high-dose regimen has

been reported to cause significant steroid complications in three patients (11%) in one series[139] (two steroid psychoses and one gastric ulcer requiring surgery) and in four patients in another series.[146] In that series, there were eight complications in 28 patients on high doses (two GI perforations, two GI hemorrhages, two wound infections, one pneumonia, and one hyperglycemia). There were three complications in 38 patients on conventional treatment (two hyperglycemia, one wound infection). Our experience has been that if the high-dose corticosteroids are tapered rapidly, serious side effects are not common and those that occur can be treated effectively if diagnosed early.

A study of 37 patients randomized to receive an initial bolus of either 10 or 100 mg dexamethasone intravenously followed by 16 mg daily orally reported no difference in the two groups.[138] In that series, there was a significant decrease in pain at 3 hours in patients who received high-dose dexamethasone. A recent attempt to perform a randomized study comparing high-dose versus low-dose dexamethasone failed because of inadequate recruitment.[147]

Our current recommendation is to use high-dose corticosteroids in patients with substantial motor weakness or rapid deterioration in motor function. We give a bolus of 100 mg dexamethasone followed by 100 mg orally in divided doses tapering, by half every 2 to 3 days, depending on the patient's clinical situation. Patients who do not have substantial weakness can probably be treated with 16 mg orally daily. Some patients probably do not require steroids at all.[144]

HYPEROSMOLAR AGENTS

Mannitol and hyperosmolar glucose[148] draw water from the normal cord where the blood–spinal cord barrier is intact. Little or no empirical evidence supports their clinical use in spinal cord compression.

HORMONAL AND CHEMOTHERAPEUTIC AGENTS

Only a few studies have been published regarding the effects of hormonal and chemotherapeutic agents used to treat vertebral metastases with spinal cord compression. Because no blood–spinal cord barrier exists in epidural tumors, even water-soluble agents reach the spinal metastasis. In one animal study,[52] tumors causing cord compression were treated more effectively by cyclophosphamide than by either decompressive laminectomy or RT. Steroids, as oncolytic agents, successfully treated some lymphomas compressing the cord, as well as some other tumors[149] (Fig. 6–9). Occasional patients whose thyroid cancer metastasizes and compresses the spinal cord respond to thyroid suppression with thyroxin[150] and, conversely, develop spinal cord signs when thyroid suppression is withdrawn.[150,151] Androgen suppression with orchiectomy or ketoconazole in patients with metastatic spinal cord compression from previously untreated prostate cancer may relieve symptoms effectively.[152]

Several reports indicate that treating breast metastases with hormonal therapy,[153] or treating prostate metastases with estrogens or anti-androgens,[154] not only ameliorates spinal cord symptoms, particularly pain, but also serious motor weakness.[154] Tamoxifen may be useful even in estrogen receptor–negative primary breast tumors because these tumors can occasionally spawn estrogen receptor–positive spinal metastases. Carter[155] stresses the differences of methodology in reporting the outcome of spinal metastases from breast cancer treated with hormonal and chemotherapeutic agents. Usually, little change is noted in radiographs of the spine, but pain and survival are improved by chemotherapy. Few data report the salutary effect on other neurologic symptoms. However, one report[23] indicates that, in breast cancer with spinal cord compression, responses to systemic therapy equaled responses to RT. Positive responses to chemotherapy have also been reported in Hodgkin disease,[156,157] non-Hodgkin lymphoma,[18,158,159] osteogenic sarcoma (either as a preoperative treatment[160] or as sole therapy),[161] germ-cell tumors,[162,163] Ewing sarcoma,[164] and neuroblastoma.[164]

Specific hormonal and chemotherapeutic agents should be considered for treating spinal metastases and spinal cord compression in three settings: (1) An asymptomatic or mildly symptomatic patient with a vertebral metastasis with or without a small epidural mass might be given chemotherapy without other treatment; (2) A patient with spinal cord compression previously irradiated who is not a candidate for further radiation or surgical therapy should be considered for chemotherapy; (3) In the acute treatment of symptomatic epidural spinal cord compression, chemotherapy probably should be used only in conjunction with either surgery or RT. An exception to the above statement occurs in patients with lymphoma. In such patients, even with rapidly developing paralysis, the combination of corticosteroids and other cytotoxic agents often leads to resolution of the spinal cord compression without either surgery or RT.[165]

RADIATION THERAPY

Principles

RT is the mainstay of treatment even for patients who undergo surgical decompression (Fig. 6–10). Those cases in which the surgeon is generally unable to remove all or even most of the tumor require RT following surgery. All patients who do not undergo surgery require RT as definitive treatment except for those few in whom chemotherapy is effective. An optimal dose and fractionation schedule has not been established.[166] Most schedules yield equivalent results, although the more protracted schedules, for example 2 Gy in 20 treatments, resulted in fewer infield recurrences than the shorter schedules, for example 8 Gy in 1 day.[166] Doses beyond 30 Gy in 10 fractions do not add anything to smaller doses.[167]

Certain principles apply to most or all patients: The smallest radiation port consistent

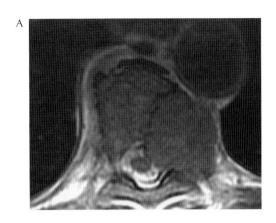

A

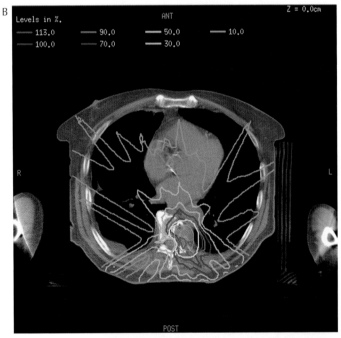

B

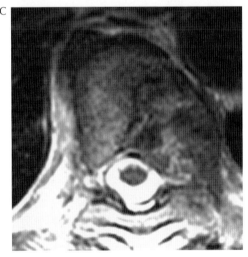

C

Figure 6–10. Response of tumor to radiation therapy. A 71-year-old man with hepatocellular carcinoma presented with severe back pain due to a T8 epidural mass. He was successfully treated with radiosurgery (2100 cGy in a single fraction). The pain rapidly resolved. **A**: Pretreatment magnetic resonance image demonstrating a large vertebral and pedicular mass compressing the spinal cord on its lateral aspect. **B**: Isodose curves showing that the tumor receives at least 92% of the dose (yellow line) whereas that portion of the normal spinal cord adjacent to the tumor receives less than 47% (aqua line). **C**: Four months after treatment, the spinal cord was no longer compressed and the patient was pain free.

219

Table 6–11. **Motor Improvement after RT in the 302 Patients with Motor Deficits Longer than 14 Days before Beginning Treatment Therapy in Relation to Tumor Type**

Subgroup	Breast Cancer (n = 98)		Prostate Cancer (n = 64)		Myeloma/ Lymphoma (n = 40)		Lung Cancer (n = 98)		Other Tumors (n = 73)		Entire Subgroup (n = 302)	
	Number	Percentage	Number	Percentage	Number	Percentage	Number	Percentage	Number	Percentage	Number	Percentage
Improvement of motor function	43	44	27	42	32	80	9	33	31	42	142	47
No progression of motor deficits[a]	98	100	64	100	40	100	25	93	73	100	300	99
Ambulatory patients after RT	97	99	60	94	40	100	24	89	70	96	291	96
Patients regaining walking ability	6/7	86	9/13	69	5/5	100	3/5	60	8/11	73	31/41	76
Patients maintaining walking ability	91	100	51	100	35	100	21	95	62	100	261	99.6
1-year local control		98		89		100		96		89		94
2-year local control		87		82		100		96		89		88
1-year survival		89		81		94		63		70		81
2-year survival		73		69		94		46		47		67

From Ref. 171 with permission.
[a] Improvement or no further progression.
RT, radiotherapy.

with effective treatment should be drawn. RT to the vertebral bodies destroys bone marrow. The port is generally approximately 8 cm wide centered at the midline of the spine. If a paravertebral mass exists, the port must be modified to include that mass. When an image defines the epidural tumor, the site should be marked by the radiologist and carefully checked by the radiation oncologist and neurologist. Because many patients have widespread systemic cancer requiring marrow-suppressing chemotherapy, the less marrow irradiated the better. In general, with conventional RT, for epidural masses that involve one vertebral body, the port is centered on that vertebral body and encompasses one to two vertebral bodies above and below. Second lesions sometimes later compress the cord near the site of the first lesion, perhaps representing regrowth at the margin of the RT port, and arguing for a more generous initial port.[168,169]

RT should begin as soon as possible after the diagnosis in patients with neurologic signs or symptoms other than pain. The sooner treatment is started, the sooner it will work. Because neurologic signs and symptoms may evolve rapidly, early treatment is essential.

As indicated in the preceding text, most fractionation schedules are equivalent at least for short-term outcome. A retrospective review of over 1000 patients with metastatic spinal cord compression comparing schedules of 8 Gy in 1 fraction (1 day), 4 Gy in 5 fractions (1 week), 3 Gy in 10 fractions (2 weeks), 2.5 Gy in 15 fractions (3 weeks), and 2 Gy in 20 fractions (4 weeks) yielded similar results, save for the fact that the more protracted schedules were associated with fewer infield recurrences.[166] Most radiation oncologists now choose to treat with 3 Gy in 10 fractions. Those with a very short-term prognosis from their systemic disease can probably be treated with shorter courses with the goal of relieving pain, whereas those with a good long-term prognosis should undergo a more protracted course.

Some radiotherapists fear that radiation may lead to spinal cord swelling and exacerbate neurologic symptoms. Rubin's study[141] in experimental animals with epidural tumor indicates that the spinal cord did not swell when it was irradiated. Even high doses of RT failed to worsen the animals. MSKCC studies[52] support Rubin's finding. Although animals given 10 Gy in a single dose were worse the day following treatment, the worsening was less than in animals treated with lower doses or not treated at all, and by day 2 following treatment, the animals were substantially better. Furthermore, animals given corticosteroids and then 10 Gy were better the day following treatment, although the long-term outcome resembled RT alone.

Worsening neurologic symptoms after apparently successful RT are not necessarily due to tumor recurrence. By eradicating tumor, RT may lead to further vertebral collapse and kyphosis, causing spinal cord compression that requires surgical decompression.[170]

Prognosis

Patients with spinal cord compression who are ambulatory at the onset of RT are likely to maintain ambulation.[171] In a series of 511 patients who had involvement of three or more vertebral bodies and did not have other bone or visceral metastases, motor function remained stable in 54%, improved in 40%, and deteriorated in only 7%.[171] Ninety-four percent of the ambulatory patients remained so and 58% of nonambulatory patients became ambulatory. Local control at 1, 2, and 3 years was 92%, 88%, and 78% respectively, better in those patients who received more protracted courses of radiation. As Table 6–11 indicates, relatively radiosensitive tumors such as breast, myeloma, and lymphoma do better than less radiosensitive tumors. Age does not seem to be a factor; functional outcome is about the same in the elderly (>75 years old) and in younger patients.[172]

In most series, patients who have only pain do not become paraplegic if treated with RT.[171,173] Depending on the tumor, about one-third to one-half of patients who are nonambulatory regain ambulation.[20,171] An occasional paraplegic patient regains ambulation,[174] but most do not recover. Recovery may occur weeks to several months after treatment. One of our patients with thoracic spinal cord compression from metastatic prostate cancer left the hospital unable to stand and was transferred to a nursing home. Several months later, with only minimal physical therapy, he walked out of the nursing home.

The rate at which the paralysis develops is more important in determining the eventual outcome of treatment than the duration of paralysis; a slow (>14 days) rather than fast rate predicts a better recovery.[171] We have also seen delayed responses to RT, but have not encountered recovery from total paralysis and sensory loss. Our policy is to treat all patients with spinal cord compression, whether paraplegic or not, recognizing that an occasional patient may have a dramatic response despite the statistics.

If the patient has normal bladder and bowel function before treatment, the vast majority usually retain those functions. If the patient requires a catheter before therapy, the likelihood of recovery varies from 20% to 40% in different series.[64,175]

Factors that influence local control and survival include the histologic type of tumor (breast, prostate, lymphoma/myeloma—favorable), the presence or absence of visceral metastases, the presence or absence of other bone metastases, ambulatory status, interval between tumor diagnosis and cord compression (greater than 15 months—favorable), and the time over which weakness develops (greater than 14 days—favorable). A more protracted RT course was probably also favorable.[176] Ambulatory patients and those who regain ambulation after treatment have a longer survival than those who do not.[177] Rades et al.[178] have developed a scoring system that predicts survival based on the six prognostic factors listed in this paragraph.

At MSKCC, we treat epidural spinal cord compression from relatively radiosensitive tumors (e.g., lymphoma, myeloma, breast cancer) with RT, 30 Gy in 10 treatments. Radioresistant tumors (e.g., melanoma, renal cancer) are treated surgically, as described in the paragraphs on surgery that follow. Residual tumor, no longer in contact with the spinal cord, is treated with image-guided intensity-modulated radiation therapy (IG-IMRT), as indicated in the following paragraphs.

RADIOSURGERY

In part because the tolerance of the spinal cord to radiation is limited and, in part, because many of the tumors compressing the spinal cord are resistant to standard doses of radiation, recent efforts have been directed toward delivering high doses of radiation, often in a single fraction, to an exquisitely focused target encompassing the tumor but not the contiguous spinal cord.[179,180] To deal with this problem, a number of techniques including IG-IMRT (e.g., Trilogy),[181,182] CyberKnife,[183]

and tomotherapy[184] have been devised. These techniques can be used both as primary treatment for a newly discovered vertebral metastasis or postoperatively after surgical treatment of spinal cord compression.[185] They allow delivery of radiation beams to within less than 2 mm accuracy.[186,187]

A recent study describes 500 patients so treated.[183] The treatment occurred in a variety of settings and included patients who had failed previous standard external beam radiation, as a boost after standard radiation, and as "consolidation" treatment after surgery. Only patients with overt spinal instability and a neurologic defect caused by bone compressing neural structures were excluded. Pain was the primary indication in 336 patients, but 35 had progressive neurological defects. Thirty of 35 (85%) patients with progressive neurologic deficits experienced improvement. Long-term pain improvement occurred in 86% of patients, including those with metastases from renal cell, breast and lung carcinomas, and melanoma. Long-term radiographic control was achieved in 88% overall, with breast and lung cancers achieving 100% control, renal cell 87%, and melanoma 75%.

Despite the tight conformal beam, some of the spinal cord will receive significant doses of radiation with these techniques.[186] In one series of 230 lesions in 177 patients treated with radiosurgery, the investigators calculated the partial volume tolerance was at least 10 Gy to 10% of the spinal cord volume.[186] In 86 of their patients who survived longer than 1 year, only one patient developed radiation-induced myelopathy, and in that patient the dose was lower than average.

At MSKCC, patients with metastatic spinal lesions not causing high-grade spinal cord or cauda equina compression are treated with high-dose, single fraction IG-IMRT using a medium dose of 24 Gy in a single fraction. The actuarial local control was 90%.[188] Acute toxicity (skin reactions, esophagitis) was mild; no radiation myelopathy was encountered.

RE-IRRADIATION

A significant number of patients who are successfully irradiated for spinal cord compression develop recurrence at the same site.[19,79,189] Radiation oncologists are often reluctant to re-irradiate, fearing radiation myelopathy.

Recent studies have addressed this issue.[190–192] In one study of 62 patients, 25 (40%) improved their motor function and six previously nonambulatory patients regained the ability to walk following re-irradiation.[191] Radiation myelopathy was not observed. In another report of 26 patients who had failed radiation (median dose 30 Gy/10 fractions) and had undergone re-irradiation with image-guided techniques, 70% of patients experienced durable palliation of symptoms and 54% had durable radiographic local control.[193] Nieder et al. concluded that patients whose initial biological dose was less than 102 Gy bioequivalent dose (BED) (see Chapter 13), and whose interval between treatments was greater than 6 months, could receive up to 135 Gy BED total dose without risk of myelopathy.[192] They devised a scoring system that included (1) the cumulative biologically effective dose, (2) the interval between courses, and (3) the biologically effective dose of each course. Patients were divided into a low-risk group (24 patients had no myelopathy), an intermediate-risk group (2/6 patients developed myelopathy), and a high-risk group (9/10 cases of myelopathy). As indicated earlier, radiosurgery after failure of standard radiation reduced the risk even further. Therefore, because untreated cord compression will lead to paralysis and paralysis from RT myelopathy may or may not occur, patients who develop recurrent spinal cord compression, and for whom no other therapeutic options exist, probably deserve re-irradiation despite the risk. At MSKCC, recurrence in the vertebral body or the paravertebral space is treated using IG-IMRT.[193] The usual dose is 30 Gy in 5 fractions. The spinal cord receives about 990 cGy. If the recurrence compresses the spinal cord, surgery is performed first, followed by IG-IMRT.

BACK BRACES, BED REST AND KYPHOSIS

Back braces are prescribed for many patients with spinal cord compression, whether the spine is unstable or not. Many other patients are placed at bed rest when the initial diagnosis is made and are often kept there during the course of therapy. Neither bed rest nor back braces are indicated for most patients with epidural spinal cord compression. In the thoracic area, the ribs assist in stabilizing the spine, and back braces add little additional stability. Little or no evidence is available concerning acute damage to the spine with movement in patients with thoracic metastases, and the risks of bed rest in these patients who can be hypercoagulable (see Chapter 9) seem greater than any potential damage to the spine by ambulation. To the extent that it controls the pain when the patient is up and about, bracing the back can be tried, but the braces are cumbersome and patients rarely use them for any extended period. In the lumbar spine, bracing can also be tried for pain relief, but neither bed rest nor bracing appears necessary unless spinal instability is evident. For the cervical spine, a soft cervical collar, especially when the patient is riding in an automobile, is sufficient to relieve pain on motion and stabilize the spine adequately. Other more restrictive collars, such as Philadelphia collars, are prescribed, although patients find them uncomfortable and are reluctant to wear them. Bilsky recommends that patients with atlantooccipital cervical spine subluxation less than 5 mm can be irradiated in a hard collar and maintained in that collar for 6 weeks after the completion of radiation.[72] If major instability in the cervical spine occurs, cervical stabilization should be considered.

In lieu of an external orthosis, many patients with axial load pain resulting from thoracic or lumbar burst fractures may benefit from vertebral bone cement augmentation via procedures such as kypho- or vertebroplasty. Unfortunately, patients with high-grade spinal cord compression or overt spinal instability are excluded from consideration. These procedures require percutaneous cannulation of the vertebral body via the pedicle under local or general anesthesia.[194,195] A recent review of 64 pathologic fractures treated in 31 patients showed a significant improvement in disability and pain.[194] Eight patients had radiographic cement extravasation from the anterior vertebral body, but this was not clinically significant in any patient. As experience grows with this technique and long-term follow-up is accrued, this procedure will probably become more common for pathologic fractures.

SURGERY

In recent years it has become increasingly clear that those patients who are candidates for surgical therapy of epidural spinal cord compression

Table 6–12. **Surgical Approach to Spinal Neoplasms**

Level	Anterior	Posterior
Upper cervical	Transoral	Midline
	Extraoral	Extreme lateral
Lower cervical	Southwick–Robinson	Midline
Cervicothoracic/ upper thoracic	Sternal splitting	Midline
	Low anterior costotransversectomy ± endoscopy	Costotransversectomy
Thoracic	Thoracotomy	Midline
	Costotransversectomy ± endoscopy	Costotransversectomy
		Transpedicular
Thoracolumbar	11th rib extrapleural–retroperitoneal	Midline
		Posterolateral
Lumbar	Retroperitoneal	Midline
	Transabdominal	Transpedicular
		Posterolateral

From Ref. 8 with permission.

should receive that therapy. Exceptions are those with highly radiosensitive tumors such as lymphoma and myeloma. A recent meta-analysis comparing 999 patients in 24 surgical studies and 543 patients in four radiation studies indicates that surgical patients were 1.3 times more likely to be ambulatory after treatment and twice as likely to regain ambulatory function as those treated by radiation alone. The overall success rate for surgery was 85% as compared to 64% for radiation.[27] Surgery has several advantages over conventional radiation: (1) It establishes the diagnosis in patients who present with cord compression but were not known to have cancer. A stereotactic needle biopsy might also establish the diagnosis, but in surgical candidates one operation is better than two. (2) It rapidly decompresses the compressed spinal cord, relieving symptoms. (3) It restores stability to the unstable spine. (4) It removes all or most of the metastasis, increasing the likelihood that further therapy will prevent recurrence.

One randomized controlled trial comparing standard external beam RT with modern surgery followed by radiation was stopped after the interim analysis, when it was discovered that patients treated surgically had an ambulatory rate of 84%, whereas those treated with radiation had an ambulatory rate of 57%.[196] Three of

10 patients who failed to respond to radiation regained ambulation after salvage surgery. This study strongly validates the important role of surgery, but patients were highly selected and only patients with epidural tumor compressing the spinal cord, but not the cauda equina, were enrolled in the study, limiting the applicability of these results to all patients with epidural spinal cord compression.

The exact surgical approach will depend in part on the location of the tumor, both in the cephalic–caudal plane (Table 6–12) and the ventral–dorsal plane (Fig. 6–1). The standard laminectomy proved no better than external beam RT[20,197] and, except in rare instances such as zone IA tumor (Fig. 6–1), is no longer part of the surgical approach to epidural spinal cord compression. The various surgical approaches are indicated in Table 6–12. After removal of as much tumor as possible, the spine is stabilized using a variety of hardware including posterior fixation with screw–rod systems and anterior fixation using allograft bone or cages, anterior plates, or screw–rod constructs.[72] Whenever possible, one should choose stabilizing materials that do not distort postoperative imaging,[72] such as polyetheretherketone (PEEK) carbon fiber or titanium.

In a series of patients with thoracic spinal cord compression who underwent single-stage posterolateral transpedicular surgery, pain was substantially relieved in over 90% and one-half of the patients with motor weakness improved[72] (Table 6–13).

However, surgery has complications. In 140 patients with resection of epidural metastases, the 30-day mortality was 4.3% and morbidity was 14.3% including wound infection/dehiscence, pneumonia, and pulmonary embolism.[198] A national survey of over 26,000 hospital admissions for the surgical treatment of spinal metastasis between 1993 and 2002 indicated an in-hospital mortality rate of 5.6% and a total complication rate of 21.9%, with complications occurring in 17.5% patients.[199] Complications were more likely in older patients and in patients with comorbidities. Some evidence suggests that mortality was decreasing with time. The most common complications were pulmonary and hemorrhagic. Less than 1% of patients were thought to suffer neurologic complications. Despite the significant complication rate, even in the best of hands, surgery should be the first consideration when clinically appropriate. Surgery is even more complicated and has a higher morbidity in patients who have received prior radiation.

Indications for Surgery

Our indications for surgery include the following.

Unknown primary tumor: If a patient with cord compression is not known to have cancer, a diagnosis from tissue samples is essential. An image-assisted percutaneous needle biopsy of the vertebral body is reported to give an appropriate diagnosis in 90% of instances, with a complication rate of less than 1%.[200,201] Needle biopsy of blastic lesions has a lower yield. Alternatively, the patient may undergo a more definitive surgical procedure, allowing both diagnosis and treatment.

Relapse after RT: If the patient cannot tolerate further RT and if the patient's general condition warrants, surgery is the best approach to relieve symptoms. In many patients, however, re-treatment with RT often preserves ambulation, with low risk of radiation myelopathy (see preceding text).

Progression while on RT: Perhaps the most difficult decision for the physician is treating the patient who deteriorates during RT for epidural spinal cord compression. Increasing the corticosteroid dose occasionally helps, but most physicians seek surgical consultation. We often do the same, although our experience has been disappointing; surgery usually fails to halt the inexorable progression of paraplegia. Persevering with RT may eventually give a satisfactory response.[174]

Spinal instability: In some patients with vertebral metastases, symptoms are primarily caused by spinal instability,[67,72] often with subluxation as a consequence of the tumor and progressive kyphosis with extrusion of bone and disk into the spinal canal. Kyphosis can result not only from destruction of the vertebral body by tumor, but also from successful treatment of the tumor by RT. Spinal instability, which accounts for pain in approximately 10% of patients with vertebral metastases,[67] is likely to be present if (1) the odontoid process of the atlas has been destroyed, leading to significant atlantoaxial subluxation more than 5 mm or more than 3.5 mm subluxation and 11 degrees of angulation of C1 and C2; (2) the tumor causes a burst or compression fracture in the subaxial cervical or thoracic spine with lytic involvement of at least a single facet joint at the involved level; (3) lumbar mechanical radiculopathy exists, which is caused by a burst or compression fracture of the lumbar spine leading to radiculopathy on axial loading, such as sitting or standing; and (4) any combination of the aforementioned occurs.

Spinal instability is characterized clinically by severe pain at the site of the lesion on attempted movement. This pain does not respond to corticosteroids and will not improve with RT; surgery is required. With odontoid fractures and the atlantooccipital dislocations, patients hold their head and neck stiffly, sometimes in a slightly awkward position, and refuse either to move it actively or to allow themselves to be moved passively. Occasionally, numbness is felt in the tongue (neck–tongue syndrome), a result of compression of afferent fibers from the lingual nerve traveling via the hypoglossal nerve to the second cervical root.[93] Pain on neck rotation alone does not denote instability. The subluxed vertebral column may compress the cord, causing quadriparesis and respiratory embarrassment. In spinal instability at lower spinal levels, patients generally complain of severe pain when turning over in bed or attempting to get up. Of note, thoracic instability may cause severe pain when the patient lies

Table 6–13. Results of Vertebral Body Resection (*n* = 140)

Post-Operative Score

NEUROLOGIC FUNCTION

Pre-Operative Score	E	D	C	B	A
E	53 (39%)	2 (1%)			
D	36 (26%)	32 (23%)	1 (1%)		
C		7 (5%)	3 (2%)	1 (1%)	1 (1%)
B	1 (1%)				
A					

(Worse / Stable / Improved diagonal indicators)

PAIN

Pre-Operative Score	0	1	2	3
0	1 (1%)			
1	6 (4%)	3 (2%)		
2	10 (7%)	12 (9%)	1 (1%)	
3	23 (17%)	69 (51%)	10 (7%)	

(Worse / Stable / Improved diagonal indicators)

PERFORMANCE ECOG

Pre-Operative Score	0	1	2	3	4	5
0	7 (5%)					
1	18 (13%)	28 (21%)				
2	3 (2%)	21 (16%)	6 (4%)			
3	3 (2%)	15 (11%)	12 (9%)	5 (4%)	1 (1%)	
4		3 (2%)	5 (4%)	4 (3%)	3 (2%)	
5						

(Worse / Stable / Improved diagonal indicators)

E = Normal
D = Needs support
C = Not ambulatory
B = Paraplegic with sensation
A = Paraplegia
Data from Ref. 70

0 = None
1 = Mild
2 = Moderate
3 = Severe

0 = Full function
1 = Mild restriction
2 = Self care
3 = Limited activity
4 = Disabled
5 = Dead

flat, resulting from the straightening of an unstable kyphosis. Usually the patient is unwilling to move the affected part and exhibits tenderness to palpation or percussion over the area. The diagnosis of subluxation is made easily by plain radiographs of the spine. Spinal instability without major subluxation may be more difficult to diagnose; flexion and extension X-rays of the affected spinal segment may be required to establish the diagnosis. The patient should never be forced to move further than comfort allows. For atlantoaxial and subaxial cervical stable pathologic fractures, patients can be irradiated wearing a collar for the period of radiation and 6 weeks following[202]; a review of 25 patients so treated reported that 23 had pain resolution within 2 weeks.

Decision-Making in Epidural Spinal Cord Compression

Bilsky has proposed a scheme in which four factors are considered: The patient's neurological status (N), the oncological status (O), the mechanical state of the spine (M), and systemic and medical comorbidities (S), thus "NOMS."[72] The neurologic and oncologic assessments are made together. As Figure 6–11 illustrates, patients with neurologic symptomatology (N) are judged on the presence of myelopathy, radiculopathy, and the degree of epidural spinal or cauda equina compression. Oncologic concerns include the radio- and chemosensitivity of the tumor. Patients with radiosensitive tumors (O), such as hematologic malignancies (e.g., lymphoma, multiple myeloma), are considered for standard fraction radiation even in the presence of high-grade spinal cord compression or myelopathy. Patients with solid tumors with only bone involvement or minor epidural impingement (N) should receive RT as initial treatment. As more data accrue, radioresistant tumors without significant epidural disease are being considered for stereotactic radiosurgery. Radioresistant tumors with low alpha/beta ratios (see Chapter 13) respond better to high dose per fraction. Increasing numbers of patients will be treated with radiosurgery as the technology becomes more widely available. Patients with radioresistant tumors (O) and high-grade spinal cord compression (N) should be considered for initial decompression followed by radiosurgery. On the basis of Patchell's data,[196] this gives the patient the best chance for recovery of ambulation and good neurologic outcomes.

Mechanical instability is a separate assessment from the neurologic and oncologic considerations. Gross spinal instability (M) is an indication for initial surgery regardless of the degree of spinal cord compression (N) or tumor histology (O). For instance, a patient with lymphoma and no spinal cord compression who is judged to be unstable should have a stabilizing procedure. A number of patients have axial back pain, but no overt instability. These patients may be candidates for vertebral body bone cement augmentation with procedures such as kyphoplasty or vertebroplasty.

Finally, patients with extensive systemic disease with a short predicted lifespan may be better treated by radiation or radiosurgery, with extensive surgery reserved for patients with more limited disease. However, these and other guidelines in this chapter (Fig. 6–7) are just that, guidelines. The physician must take all factors into consideration and make a choice for each individual patient.

Specific Tumors

The approaches described in the preceding text are generally applicable to all primary tumors causing spinal cord compression. However, there are some differences among tumors that dictate both evaluation and treatment. Some of these are considered in the following paragraphs.

PROSTATE CANCER

Because the tumor is so common and often affects the spine, there is a rather large literature on the treatment of spinal metastases from prostate cancer.[152,203–205] In one series, 119/634 patients with prostate cancer (18.8%) had "significant" spinal metastases.[204] The majority were lower thoracic, lumbar, and sacral metastases. At least one-half were osteoblastic and one-quarter were combined osteoblastic and osteolytic. In addition to the usual radiation and surgery, hormonal therapy and chemotherapy seemed to improve symptoms in some patients. In another series[152] of nine patients with spinal cord compression treated

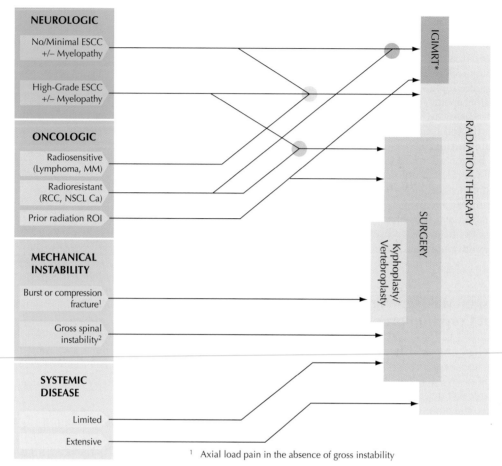

Figure 6–11. NOMS decision-making framework. ESCC, epidural spinal cord compression; MM, multiple myeloma; NOMS, neurologic, oncologic, mechanical stability, and systemic disease and comorbidities; NSCLC, non–small cell lung carcinoma; RCC, renal cell cancer; ROI, region of interest; IGRT, image-guided radiation therapy. (From Ref. 72 with permission.)

with hormone therapy, two became ambulatory and three improved. Chemotherapy alone was not efficacious. In one series, only 3.1% of the 147 patients with metastatic prostate cancer became paraplegic or quadriplegic. Thus, it would appear that in newly diagnosed patients who are not chemotherapy or hormone refractory, and have little or no neurologic disability, a trial of these agents might be indicated. One report describes two patients with symptomatic spinal cord compression who improved after hormone therapy, obviating the need for surgical intervention.[154] In the vast majority of

patients, one should follow a recently published algorithm for the treatment of spinal cord compression from prostate cancer,[203] which recommends radiation for patients whose compression is not caused by bony destruction and whose signs are stable, and surgery for the others.

In one study of 281 patients who received RT for epidural spinal cord compression from prostate cancer, the overall response rate was 86% (37% improvement of motor function, 53% no further progression); 33% of nonambulatory patients regained the ability to walk.[206] In one series of 150 patients screened by spine

MRI, because they were believed to be at high risk for developing spinal cord compression, 41 were found to have asymptomatic spinal cord compression. Treatment of these and other patients with vertebral metastases, with or without back pain, effectively prevented the development of spinal cord compression at the radiated site in 80% to 90% of these high-risk patients.[207]

One must be aware that androgen-suppressive therapy may cause an initial flare with exacerbation of bone pain in the first weeks following treatment. In one series of 46 men with untreated advanced prostate cancer, two developed symptoms of epidural spinal cord compression within 1 week of the initiation of hormonal therapy.[208]

BREAST CANCER

Most spinal metastases from breast cancer are relatively radiosensitive and some respond to hormonal therapy or chemotherapy alone.[153] For the majority of patients, RT will control symptoms. For most patients, RT is the treatment of choice unless the spine is unstable or the patient deteriorates after radiation. In one series of 335 irradiated patients, 31% improved motor function and 39% nonambulatory patients regained the ability to walk.[209] Patients receiving short-course radiation (1 × 8 Gy or 5 × 4 Gy) were more likely to recur than those receiving longer courses (10 × 3 Gy or 20 × 2.5 Gy). Thus, the authors recommended a shorter course for patients with poor expected survival, but a longer course for those with a good prognosis.

A recent review of 87 patients with breast cancer who underwent surgery for spinal metastases identified a median survival of 21 months after the first spinal surgery; 62% of patients were alive at 1 year, 44% at 2 years, and 24% at 5 years.[210] Tumor recurrence was common (23%), both at the operated site (35%) or a new site (50%). Interestingly, a cervical spine metastasis was a negative prognostic variable, although visceral or other bony metastases did not affect survival.

LYMPHOMA/MYELOMA

These tumors are radiosensitive. RT should be the first consideration. A report of 25 patients with non–Hodgkin lymphoma presenting with spinal cord compression demonstrated effectiveness of a combination of chemotherapy and RT. Despite the fact that most patients were nonambulatory and suffered sphincter dysfunction, the majority regained function and the 5-year survival was 59%.[158] Spinal cord compression is somewhat less common in Hodgkin than in non–Hodgkin lymphoma, but it also responds to chemotherapy usually followed by RT.[211] Myeloma is also radiosensitive. A review of 172 myeloma patients with spinal cord compression who received RT described motor improvement in 52%. Of the nonambulatory patients, 47% regained the ability to walk. Longer courses of radiation resulted in better function than shorter courses.[212]

LUNG CANCER

Small cell lung cancer is fairly radiosensitive, but non–small cell lung cancer is not. A review of 252 patients with non–small cell lung cancer, who developed spinal cord compression and were treated with RT, described improvement in only 14% with deterioration in 32%.[213]

OTHER CANCERS

A recent report of 32 patients with bladder cancer who received radiotherapy for cord compression described improvement in only two patients (6%), with stabilization in 25 patients (78%)[214]; median survival was only 4 months, but 16% of patients lived a year. These data suggest surgery should be the first consideration. In the absence of a surgical option, a short course of radiation seemed to give the same functional result as a longer course.[214]

Colorectal cancer is relatively radioresistant. A recent series of 81 patients with cord compression described improvement of motor function in only 14% but stabilization in 68%. No significant differences were found whether the patient received a short or prolonged course of radiation.[215]

Ovarian cancer rarely causes cord compression; brain and leptomeningeal metastases are more common. One report of seven patients treated with RT described improvement of motor function in three and recovery of ambulation in two of five patients who were nonambulatory.[216] Both short and long courses were used, but the numbers were too small for adequate comparison.

Renal cell carcinoma is relatively radioresistant. If surgery is an option it should be performed. Because these tumors are often highly vascular (the same is true of thyroid metastases), preoperative embolization should be considered.[217] If surgery is not feasible, one should consider radiosurgery. In a series of 87 patients treated with radiation, 29% showed improvement in motor function; 25% of non-ambulatory patients became ambulatory. Short-course radiation appeared as efficacious as the long course.[218] The authors suggested that 8 Gy in one treatment might be best.

Germ-cell tumors respond to chemotherapy as well as radiation. One report describes three patients who had complete neurological recovery after treatment with cisplatin-based chemotherapy.[219]

Sometimes the patient presents with spinal cord compression without a primary cancer being identified (cancer of unknown primary). In one series, 143 such patients were radiated for cord compression. Motor function improved in 10% with stabilization in 53%.[220] No differences were seen between short and long course radiation. These patients should undergo surgery if at all possible.

INTRADURAL AND INTRAMEDULLARY METASTASES

Intradural metastases from systemic cancer are uncommon.[4,221] A literature survey encompassing the years 1950 to 1964 identified only 59 such cases.[221] Most intradural extramedullary tumors are either primary spine tumors or drop metastases from primary brain tumors.[7] Most intradural metastases from systemic tumor are also drop metastases, but from metastatic brain tumors, particularly posterior fossa metastases that have been treated surgically.[222,223] In one series, 8 of 10 patients with intradural metastases had antecedent intracranial metastases.[221] Because most are drop metastases, they tend to seed the lower thoracic and lumbar spine[221] and many are associated with more widespread leptomeningeal metastases (see Chapter 7). However, the intradural tumor may present as a single mass lesion compressing the spinal cord or cauda equina and thus be amenable to surgical therapy.[4,221] Occasionally, intradural metastases arise when the tumor cells reach the subarachnoid space either by hematogenous dissemination, by perineural growth along nerve roots through the intervertebral foramen, or by invasion of the dura from an epidural mass.

By definition intradural extramedullary spinal metastases cause symptoms by forming a mass and compressing the spinal cord. Perrin et al.[4] reported 10 such patients with clinical features essentially identical to those of patients with epidural spinal cord compression. The primary tumors included breast (4), lung (3), melanoma (2), and uterus (1). The lesion occurred at T5 in one patient and was found entangled in the nerve roots about the thoracolumbar junction in the remaining patients. Of the 10 tumors, nine appeared to have spread from a brain metastasis via the CSF.

Chow and McCutcheon[221] operated on 10 patients with intradural metastases from systemic tumor, among 266 spinal operations for neoplastic disease performed at their hospital between 1980 and 1994. As in other series, pain and weakness were the predominant symptoms and the lower thoracic and lumbar spine were the predominant sites. They reported improved symptoms in four patients, no change in four patients, and worsening in two patients. The intradural location of the tumor was identified by gadolinium-enhanced MRI. Because the tumors are often entangled in the roots within the subarachnoid space, they are rarely completely resectable; therefore, the treatment of choice is often RT, perhaps with the addition of intrathecal chemotherapy (see Chapter 7).

Intramedullary metastases arise either from growth of subarachnoid tumor along nerve roots directly into the spinal cord or by direct hematogenous spread to the parenchyma of the cord. They are rare, affecting only 0.1% to 0.4% of all cancer patients and comprising only 1% to 3% of all intramedullary spinal cord neoplasms.[224] In 1972, Edelson et al.[3] found only 70 such cases reported in the English language literature and added nine cases of their own (Table 6–14). Costigan and Winkelman[2] found 13 patients with intramedullary spinal cord lesions among 627 patients with invasive carcinoma whose spines were examined at autopsy. Of these 13 patients, nine had hematogenous metastases to spinal cord substance; of those nine, only four were symptomatic, but in three, myelopathy was the presenting symptom of systemic cancer. In the other four patients,

Table 6–14. **Intramedullary Spinal Cord Metastases by Primary Tumor Site (%)**

Primary Type	Percentage
Bronchogenic	41–85
Breast	11–13
Melanoma	8–9
Lymphoma	4–5
Renal	4–9
Colorectal	3
Thyroid	2
Ovarian	1–2
Unknown	3–6

From Ref. 225 with permission.

leptomeningeal tumor invaded the spinal cord; two were symptomatic. In only 2 of 13 were the spinal lesions solitary central nervous system lesions; the others had brain metastases as well. A review of the English language literature up to 2004 identified 284 patients with intramedullary metastases[224]; 32 were treated surgically.

Although any systemic cancer can metastasize to the spinal cord, lung cancer is the most common primary tumor (Table 6–14), representing over 50% of the cases[226] equally divided between small cell[227] and non–small cell tumors. Small cell lung cancer is particularly likely to cause intramedullary spinal metastasis or leptomeningeal metastasis after prophylactic brain RT.[228] Breast cancer accounts for approximately 15% of the metastases, and lymphoma (both Hodgkin[229] and non–Hodgkin lymphoma) for another 10%. In an occasional patient, a single intramedullary metastasis is the only site of metastatic disease, and myelopathy may be the first evidence that the patient has cancer.[230,231] However, over one-half of the patients have brain metastases and one-quarter have leptomeningeal disease.[225,226]

Neither the patient's symptoms nor neurologic signs unequivocally distinguish intramedullary metastasis from epidural spinal cord compression[3,225,232] (Table 6–15). However, patients with intramedullary metastases generally do not have a long period of neck or back pain without other symptoms, and once neurologic symptoms develop they usually progress rapidly. Those signs said to be more suggestive

of intramedullary than extramedullary lesions included somewhat less severe pain,[226,233] sparing of sensation in the sacral area (sacral sparing), early onset of autonomic dysfunction, and especially Brown–Sequard syndrome.[226] They are, in our experience, not helpful in distinguishing different sites of metastases.

The tumor can involve any area of the spinal cord, with some series showing tumors about equally divided among cervical, thoracic, and lumbar areas.[234] Most metastases are single, but approximately 15% are multiple.[234] The tumor is often located in the ventral posterior horn and the medial lateral column, corresponding to the terminal supply of the central artery.[235] It may be associated with hemorrhage, pencil-shaped softening, or syrinx formation.[236]

Diagnosis

A spinal MRI establishes the diagnosis[237,238] (Fig. 6–12). One-half of the lesions do not distort the anatomy of the spinal cord, the reason why myelography fails to identify so many lesions. The lesions are isointense on T1 and hyperintense on T2. They always contrast enhance, usually in a homogeneous or nodular pattern, but sometimes with ring enhancement.[237] The enhancing area is often surrounded by edema. These lesions cannot be unequivocally distinguished from primary spinal tumors (or even an infection or demyelinating lesion), but in the appropriate clinical setting, the MRI is pathognomonic. On the other hand, myelography demonstrates the lesion in only approximately 50% of symptomatic patients.[3]

Once epidural lesions have been excluded, the differential diagnosis includes primary intramedullary tumor such as glioma or ependymoma, radiation myelopathy in previously irradiated patients, spinal cord infarction from embolus or vascular occlusion, and paraneoplastic necrotic myelopathy. Cavernous angiomas of the spinal cord with acute bleeding, sometimes related to prior RT (Fig. 13–10), can also mimic spinal cord tumors. If bleeding is not evident, the characteristic MRI narrows the differential diagnosis to primary and metastatic tumor. In a patient known to have cancer, particularly with widespread metastases, the diagnosis is obvious. In patients not known to have cancer, biopsy may be required to definitively establish the diagnosis.

Table 6–15. **Symptoms at Presentation and Signs on Physical Examination in Patients Diagnosed with Intramedullary Spinal Cord Metastases**

	Percentage
Presenting Complaint	
Sensory loss	42.5–79
Pain	30–38
Weakness	30–90
Gait abnormality	5
Incontinence	2.5
Physical Examination	
Weakness	93
Sensory loss	78–87
Pain	52–72
Brown–Sequard syndrome	23
Bowel/bladder dysfunction	60–62
Systemic symptoms	37

From Ref. 225 with permission.

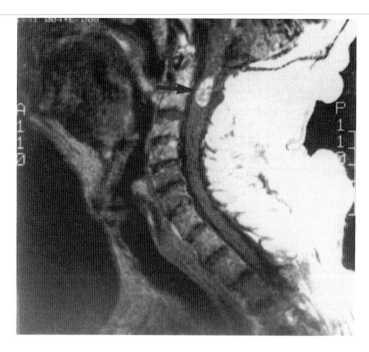

Figure 6–12. An intramedullary metastasis from lung carcinoma and an irregularly contrast-enhancing mass in the cord (*arrow*), consistent with the findings of a myelopathy at the upper cervical level.

Treatment

Most patients with intramedullary metastases develop the disorder late in the course of widespread cancer and live briefly, although they may respond temporarily to RT with or without chemotherapy.[234] In those few patients in whom the metastasis is isolated and in whom the spread has been hematogenous rather than via the subarachnoid space, the physician should consider surgical removal, particularly if the diagnosis is doubtful and surgical exploration can both establish the diagnosis and help the patient.[224,239,240]

REFERENCES

1. Harrington KD. Metastatic disease of the spine. In: Harrington KD, ed. *Orthopaedic Management of Metastatic Bone Disease.* St. Louis: The C.V. Mosby Company, 1988. pp. 309–383.
2. Costigan DA, Winkelman MD. Intramedullary spinal cord metastasis. A clinicopathological study of 13 cases. *J Neurosurg* 1985;62:227–233.
3. Edelson RN, Deck MD, Posner JB. Intramedullary spinal cord metastases. Clinical and radiographic findings in nine cases. *Neurology* 1972;22:1222–1231.
4. Perrin RG, Livingston KE, Aarabi B. Intradural extramedullary spinal metastasis. A report of 10 cases. *J Neurosurg* 1982;56:835–837.
5. Post MJD, Quencer RM, Green BA, et al. Intramedullary spinal cord metastases, mainly of nonneurogenic origin. *Am J Roentgenol* 1987; 148:1015–1022.
6. Lee SS, Kim MK, Sym SJ, et al. Intramedullary spinal cord metastases: a single-institution experience. *J Neuro-Oncol* 2007;84:85–89.
7. Traul DE, Shaffrey ME, Schiff D. Part I: Spinal-cord neoplasms-intradural neoplasms. *Lancet Oncol* 2007;8:35–45.
8. Anderson ME, McLain RF. Tumors of the spine. In: Herkowitz HN, Garfin SR, Eismont FJ, et al., eds. *The Spine.* 5th ed. New York: Saunders, Elsevier, 2006. pp. 1235–1264.
9. Barron KD, Hirano A, Araski S, et al. Experiences with metastatic neoplasms involving the spinal cord. *Neurology* 1959;9:91–106.
10. Ecker RD, Endo T, Wetjen NM, et al. Diagnosis and treatment of vertebral column metastases. *Mayo Clin Proc* 2005;80:1177–1186.
11. Ortiz Gomez JA. The incidence of vertebral body metastases. *Int Orthop* 1995;19:309–311.
12. Bubendorf L, Schöpfer A, Wagner U, et al. Metastatic patterns of prostate cancer: an autopsy study of 1,589 patients. *Hum Pathol* 2000;31:578–583.
13. Loblaw DA, Laperriere NJ, Mackillop WJ. A population-based study of malignant spinal cord compression in Ontario. *Clin Oncol (R Coll Radiol)* 2003;15:211–217.
14. Hildebrand J, Leenaerts L, Nubourgh Y, et al. Epidural spinal cord compression in acute myelogenous leukemia. *Arch Neurol* 1980;37:319.
15. Wong MC, Krol G, Rosenblum MK. Occult epidural chloroma complicated by acute paraplegia following lumbar puncture. *Ann Neurol* 1992;31:110–112.
16. Abdel-Dayem HM, Oh YS, Sil R. Treated stage IIB Hodgkin's disease complicated by late paraplegia. *Am J Roentgenol* 1979;132:265–266.
17. Oviatt DL, Kirshner HS, Stein RS. Successful chemotherapeutic treatment of epidural compression in non-Hodgkin's lymphoma. *Cancer* 1982;49:2446–2448.
18. Perry JR, Deodhare SS, Bilbao JM, et al. The significance of spinal cord compression as the initial manifestation of lymphoma. *Neurosurgery* 1993;32:157–162.
19. Gilbert RW, Kim J-H, Posner JB. Epidural spinal cord compression from metastatic tumor: diagnosis and treatment. *Ann Neurol* 1978;3:40–51.
20. Mut M, Schiff D, Shaffrey ME. Metastasis to nervous system: spinal epidural and intramedullary metastases. *J Neuro-Oncol* 2005;75:43–56.
21. Sciubba DM, Gokaslan ZL. Diagnosis and management of metastatic spine disease. *Surg Oncol* 2006;16(10):1659–1667.
22. Boogerd W, van der Sande JJ, Kroger R. Early diagnosis and treatment of spinal epidural metastasis in breast cancer: a prospective study. *J Neurol Neurosurg Psychiatry* 1992;55:1188–1193.
23. Findlay GF. Adverse effects of the management of malignant spinal cord compression. *J Neurol Neurosurg Psychiatry* 1984;47:761–768.
24. Husband DJ. Malignant spinal cord compression: prospective study of delays in referral and treatment. *Br Med J* 1998;317:18–21.
25. Kwok Y, Tibbs PA, Patchell RA. Clinical approach to metastatic epidural spinal cord compression. *Hematol Oncol Clin North Am* 2006;20:1297–1305.
26. Klimo P, Jr, Thompson CJ, Kestle JR, et al. A meta-analysis of surgery versus conventional radiotherapy for the treatment of metastatic spinal epidural disease. *Neuro Oncol* 2005;7:64–76.
27. Loblaw DA, Perry J, Chambers A, et al. Systematic review of the diagnosis and management of malignant extradural spinal cord compression: the cancer care Ontario practice guidelines initiative's neuro-oncology disease site group. *J Clin Oncol* 2005;23(9):2028–2037.
28. Prasad D. Malignant spinal-cord compression. *Lancet Oncol* 2005;6:15–24.
29. Abraham JL, Banffy MB, Harris MB. Spinal cord compression in patients with advanced metastatic cancer. *JAMA* 2008;299:937–946.
30. Fornasier VL, Horne JG. Metastases to the vertebral column. *Cancer* 1975;36:590–594.
31. Wong DA, Fornasier VL, MacNab I. Spinal metastases: the obvious, the occult, and the imposters. *Spine* 1990;15:1–4.
32. Galasko CSB. The anatomy and pathways of skeletal metastases. In: Weiss L, Gilbert HA, eds. *Bone Metastasis.* Boston: G.K. Hall, 1981. pp. 49–63.
33. Byrne TN, Waxman SG. *Spinal Cord Compression. Diagnosis and Principles of Management.* Philadelphia: F.A. Davis, 1990.
34. Lipton A. Pathophysiology of bone metastases: how this knowledge may lead to therapeutic intervention. *J Support Oncol* 2004;2:205–213.
35. Clouston PD, DeAngelis LM, Posner JB. The spectrum of neurologic disease in patients with systemic cancer. *Ann Neurol* 1992;31:268–273.
36. Byrne TN. Spinal cord compression from epidural metastases. *N Engl J Med* 1992;327:614–619.
37. Schiff D, O'Neill BP, Suman VJ. Spinal epidural metastasis as the initial manifestation of malignancy: clinical features and diagnostic approach. *Neurology* 1997;49:452–456.
38. Stark RJ, Henson RA, Evans SJ. Spinal metastases. A retrospective survey from a general hospital. *Brain* 1982;105:189–213.
39. Costans JP, de Divitiis E, Donzelli R, et al. Spinal metastases with neurological manifestations. Review of 600 cases. *J Neurosurg* 1983;59:111–118.

40. Loberg RD, Bradley DA, Tomlins SA, et al. The lethal phenotype of cancer: the molecular basis of death due to malignancy. *CA Cancer J Clin* 2007;57:225–241.

41. Veri A, D'Andrea MR, Bonginelli P, Gasparini G. Clinical usefulness of bisphosphonates in oncology: treatment of bone metastases, antitumoral activity and effect on bone resorption markers. *Int J Biol Markers* 2007;22:24–33.

42. Layman R, Olson K, Van PC. Bisphosphonates for breast cancer: questions answered, questions remaining. *Hematol Oncol Clin North Am* 2007;21:341–367.

43. Weinstein JN. Surgical approach to spine tumors. *Orthopedics* 1989;12:897–905.

44. Lecouvet FE, Geukens D, Stainier A, et al. Magnetic resonance imaging of the axial skeleton for detecting bone metastases in patients with high-risk prostate cancer: diagnostic and cost-effectiveness and comparison with current detection strategies. *J Clin Oncol* 2007;25:3281–3287.

45. Khaw FM, Worthy SA, Gibson MJ, et al. The appearance on MRI of vertebrae in acute compression of the spinal cord due to metastases. *J Bone Joint Surg [Br]* 1999;81B:830–834.

46. Kato A, Ushio Y, Hayakawa T, et al. Circulatory disturbance of the spinal cord with epidural neoplasm in rats. *J Neurosurg* 1985;63:260–265.

47. Feldenzer JA, McKeever PE, Schaberg DR, et al. The pathogenesis of spinal epidural abscess: microangiographic studies in an experimental model. *J Neurosurg* 1988;69:110–114.

48. Doppman JL. The mechanism of ischemia in anteroposterior compression of the spinal cord. *Invest Radiol* 1975;10:543–551.

49. Beggs JL, Waggener JD. Transendothelial vesicular transport of protein following compression injury to the spinal cord. *Lab Invest* 1976;34:428–439.

50. Ushio Y, Posner R, Posner JB, et al. Experimental spinal cord compression by epidural neoplasm. *Neurology* 1977;27:422–429.

51. Delattre J-Y, Arbit E, Thaler HT, et al. A dose-response study of dexamethasone in a model of spinal cord compression caused by epidural tumor. *J Neurosurg* 1989;70:920–925.

52. Ushio Y, Posner R, Kim J-H, et al. Treatment of experimental spinal cord compression by extradural neoplasms. *J Neurosurg* 1977;47:380–390.

53. Siegal T, Shohami E, Shapira Y, et al. Indomethacin and dexamethasone treatment in experimental neoplastic spinal cord compression. Part 2: Effect on edema and prostaglandin synthesis. *Neurosurgery* 1988;22:334–339.

54. Siegal T. Serotonergic manipulations in experimental neoplastic spinal cord compression. *J Neurosurg* 1993;78:929–937.

55. Siegal T. Current considerations in the management of neoplastic spinal cord compression. *Spine* 1989;14:223–228.

56. Siegal T, Lossos F. Experimental neoplastic spinal cord compression: effect of anti-inflammatory agents and glutamate receptor antagonists on vascular permeability. *Neurosurgery* 1990;26:967–970.

57. Tarlov IM, Klinger H. Spinal cord compression studies. II. Time limits for recovery after acute compression in dogs. *Arch Neurol Psychiatry* 1954;71:271–290.

58. Gledhill RF, Harrison BM, McDonald WI. Demyelination and remyelination after acute spinal cord compression. *Exp Neurol* 1973;38:472–487.

59. Smith R. An evaluation of surgical treatment for spinal cord compression due to metastatic carcinoma. *J Neurol Neurosurg Psychiatry* 1965;28:152–158.

60. McAlhany HJ, Netsky MG. Compression of the spinal cord by extramedullary neoplasms. A clinical and pathologic study. *J Neuropathol Exp Neurol* 1955;14:276–287.

61. Kakulas BA, Harper CG, Shibasaki K, et al. Vertebral metastases and spinal cord compression. *Clin Exp Neurol* 1978;15:98–113.

62. Hashizume Y, Iljima S, Kishimoto H, et al. Pencil-shaped softening of the spinal cord. Pathologic study in 12 autopsy cases. *Acta Neuropathol (Berl)* 1983;61:219–224.

63. Hormigo A, Lobo-Antunes JL, Bravo-Marques JM, et al. Syringomyelia secondary to compression of the cervical spinal cord by an extramedullary lymphoma. *Neurosurgery* 1990;27:834–836.

64. Bach F, Larsen BH, Rohde K, et al. Metastatic spinal cord compression. Occurrence, symptoms, clinical presentations and prognosis in 398 patients with spinal cord compression. *Acta Neurochir* 1990;107:37–43.

65. Helweg-Larsen S, Sorenson PS. Symptoms and signs in metastatic spinal cord compression: a study of progression from first symptom until diagnosis in 153 patients. *Eur J Cancer* 1994;30A:396–398.

66. Galasko CS, Sylvester BS. Back pain in patients treated for malignant tumours. *Clin Oncol* 1978;4:273–283.

67. Galasko CS. Spinal instability secondary to metastatic cancer. *J Bone Joint Surg [Br]* 1991;73:104–108.

68. Helweg-Larsen S, Sorensen PS, Kreiner S. Prognostic factors in metastatic spinal cord compression: a prospective study using multivariate analysis of variables influencing survival and gait function in 153 patients. *Int J Radiat Oncol Biol Phys* 2000;46:1163–1169.

69. O'Rielly DD, Loomis CW. Spinal prostaglandins facilitate exaggerated A- and C-fiber-mediated reflex responses and are critical to the development of allodynia early after L5-L6 spinal nerve ligation. *Anesthesiology* 2007;106:795–805.

70. McMahon SB, Cafferty WB, Marchand F. Immune and glial cell factors as pain mediators and modulators. *Exp Neurol* 2005;192:444–462.

71. Bernat JL, Greenberg ER, Barrett J. Suspected epidural compression of the spinal cord and cauda equina by metastatic carcinoma. Clinical diagnosis and survival. *Cancer* 1983;51:1953–1957.

72. Bilsky MH. New therapeutics in spine metastases. *Expert Rev Neurother* 2005;5:831–840.

73. Dodge HW, Svien HJ, Camp JD, et al. Tumors of the spinal cord without neurologic manifestations, producing low back and sciatic pain. *Proc Staff Meet Mayo Clin* 1951;26:88–96.

74. Love JG, Rivers MH. Spinal cord tumors simulating protruded intervertebral disks. *JAMA* 1962;179:878–881.

75. O'Rourke T, George CB, Redmond J, III, et al. Spinal computed tomography and computed tomographic metrizamide myelography in the early diagnosis of metastatic disease. *J Clin Oncol* 1986;4:576–583.

76. Christy WC, Powell DL. Knee pain exacerbated by recumbency: an unusual manifestation of spinal cord involvement by diffuse histiocytic lymphoma. *Arthritis Rheum* 1984;27(3):341–343.

77. Scott M. Lower extremity pain simulating sciatica. Tumors of the high thoracic and cervical cord as causes. *JAMA* 1956;160:528–534.

78. Fagius J, Westerberg C-E. Pseudoclaudication syndrome caused by a tumour of the cauda equina. *J Neurol Neurosurg Psychiatry* 1979;42:187–189.

79. Greenberg HS, Kim J-H, Posner JB. Epidural spinal cord compression from metastatic tumor: results with a new treatment protocol. *Ann Neurol* 1980;8:361–366.

80. Koehler PJ, Endtz LJ. The Brown-Sequard syndrome. True or false? *Arch Neurol* 1986;43:921–924.

81. King RB, Stoops WL. Cervical myelopathy with fasciculations in the lower extremities. *J Neurosurg* 1963;20:948–952.

82. Aisen ML, Brown W, Rubin M. Electrophysiologic changes in lumbar spinal cord after cervical cord injury. *Neurology* 1992;42(3 Pt 1):623–626.

83. Hirano H, Suzuki H, Sakakibara T, et al. Foramen magnum and upper cervical cord tumors. Diagnostic problems. *Clin Orthop* 1983;176:171–177.

84. Broager B. Lhermitte's sign in thoracic spinal tumour. *Acta Neurochir (Wein)* 1978;41:127–135.

85. Ventafridda V, Caraceni A, Martini C, et al. On the significance of Lhermitte's sign in oncology. *J Neuro-Oncol* 1991;10:133–137.

86. Newton HB, Rea GL. Lhermitte's sign as a presenting symptom of primary spinal cord tumor. *J Neuro-Oncol* 1996;29:183–188.

87. Lee YY, Tsay WL, Lou MF, Dai YT. The effectiveness of implementing a bladder ultrasound programme in neurosurgical units. *J Adv Nurs* 2007;57:192–200.

88. Gudesblatt M, Cohen JA, Gerber O, et al. Truncal ataxia presumably due to malignant spinal cord compression. Letter to the editor. *Ann Neurol* 1987;21:511–512.

89. Hainline B, Tuzynski MH, Posner JB. Ataxia in epidural spinal cord compression. *Neurology* 1992;42:2193–2195.

90. Mittal MM, Gupta NC, Sharma ML. Spinal epidural meningioma associated with increased intracranial pressure. *Neurology* 1970;20:818–820.

91. Nicola GC, Nizzoli V. Increased intracranial pressure and papilloedema associated with spinal tumors. *Neurochirurgia* 1969;12:138–144.

92. Koehler PJ, Wijngaard PR. Brown-Sequard syndrome due to spinal cord infarction after subclavian vein catheterisation. Letter to the editor. *Lancet* 1986;2:914–915.

93. Lance JW, Anthony M. Neck-tongue syndrome on sudden turning of the head. *J Neurol Neurosurg Psychiatry* 1980;43:97–101.

94. Denny-Brown D, Yanagisawa N. The function of the descending root of the fifth nerve. *Brain* 1973;96:783–814.

95. Petren-Mallmin M. Clinical and experimental imaging of breast cancer metastases in the spine. *Acta Radiol* 1994;391:1–23.

96. Guillevin R, Vallee JN, Lafitte F, et al. Spine metastasis imaging: review of the literature. *J Neuroradiol* 2007;35:311–321.

97. Venkitaraman R, Sohaib SA, Barbachano Y, et al. Detection of occult spinal cord compression with magnetic resonance imaging of the spine. *Clin Oncol (R Coll Radiol)* 2007;19:528–531.

98. Portenoy RK, Galer BS, Salamon O, et al. Identification of epidural neoplasm: radiography and bone scintigraphy in the symptomatic and asymptomatic spine. *Cancer* 1989;64:2207–2213.

99. van der Sande JJ, Kroger R, Boogerd W. Multiple spinal epidural metastases: an unexpectedly frequent finding. *J Neurol Neurosurg Psychiatry* 1990;53:1001–1003.

100. Portenoy RK, Lipton RB, Foley KM. Back pain in the cancer patient: an algorithm for evaluation and management. *Neurology* 1987;37:134–138.

101. Kamholtz R, Sze G. Current imaging in spinal metastatic disease. *Semin Oncol* 1991;18:158–169.

102. Rodichok LD, Ruckdeschel JC, Harper GR. Early detection and treatment of spinal epidural metastases: the role of myelography. *Ann Neurol* 1986;20:696–702.

103. Santangelo JR, Thomson JD. Childhood leukemia presenting with back pain and vertebral compression fractures. *Am J Orthop* 1999;28:257–260.

104. Ribeiro RC, Pui CH, Schell MJ. Vertebral compression fracture as a presenting feature of acute lymphoblastic leukemia in children. *Cancer* 1988;61:589–592.

105. Taoka T, Mayr NA, Lee HJ, et al. Factors influencing visualization of vertebral metastases on MR imaging versus bone scintigraphy. *Am J Roentgenol* 2001;176:1525–1530.

106. Roland J, van den Weygaert D, Krug B, et al. Metastases seen on SPECT imaging despite a normal planar bone scan. *Clin Nucl Med* 1995;20:1052–1054.

107. Buyukdereli G, Ermin T, Kara O, et al. Tc-99m MIBI uptake in traumatic vertebral fractures and metastatic vertebral lesions: comparison with Tc-99m MDP. *Adv Ther* 2006;23:33–38.

108. Moore KR. Radiology of metastatic spine cancer. *Neurosurg Clin N Am* 2004;15:381–389.

109. Weissman DE, Gilbert M, Wang H, et al. The use of computed tomography of the spine to identify patients at high risk for epidural metastases. *J Clin Oncol* 1985;3:1541–1544.

110. Redmond J, III, Spring DB, Munderloh SH, et al. Spinal computed tomography scanning in the evaluation of metastatic disease. *Cancer* 1984;54:253–258.

111. Bohdiewicz PJ, Wong CY, Kondas D, et al. High predictive value of F-18 FDG PET patterns of the spine for metastases or benign lesions with good agreement between readers. *Clin Nucl Med* 2003;28:966–970.

112. Metser U, Lerman H, Blank A, et al. Malignant involvement of the spine: assessment by ^{18}F-FDG PET/CT. *J Nucl Med* 2004;45:279–284.

113. Andreula C, Murrone M. Metastatic disease of the spine. *Eur Radiol* 2005;15:627–632.

114. Cuenod CA, Laredo JD, Chevret S, et al. Acute vertebral collapse due to osteoporosis or malignancy: appearance on unenhanced and gadolinium-enhanced MR images. *Radiology* 1996;199:541–549.

115. Ruckdeschel JC. Rapid, cost-effective diagnosis of spinal cord compression due to cancer. *Cancer Control* 1995;2:320–323.

116. Hollis PH, Malis LI, Zappulla RA. Neurological deterioration after lumbar puncture below complete spinal subarachoid block. *J Neurosurg* 1986;64:253–256.

117. Jooma R, Hayward RD. Upward spinal coning: impaction of occult spinal tumours following relief of hydrocephalus. *J Neurol Neurosurg Psychiatry* 1984;47:386–390.

118. Rodgers H, Veale D, Smith P, et al. Spinal cord compression in polyarteritis nodosa. *J Royal Soc Med* 1992;85:707–708.

119. Zevgaridis D, Büttner A, Weis S, et al. Spinal epidural cavernous hemangiomas—report of three cases and review of the literature. *J Neurosurg* 1998;88:903–908.

120. Kang M, Gupta S, Khandelwal N, et al. CT-guided fine-needle aspiration biopsy of spinal lesions. *Acta Radiol* 1999;40:474–478.

121. Goodkin R, Carr BI, Perrin RG. Herniated lumbar disc disease in patients with malignancy. *J Clin Oncol* 1987;5:667–671.

122. Schramm J, Umbach W. Simultaneous occurrence of spinal tumor and lumbar disk herniation. *Neurochirgica* 1977;20:22–28.

123. Dibbern DA, Jr, Loevner LA, Lieberman AP, et al. MR of thoracic cord compression caused by epidural extramedullary hematopoiesis in myelodysplastic syndrome. *Am J Neuroradiol* 1997;18:363–366.

124. Barton JC, Conrad ME, Poon M-C. Pseudochloroma: extramedullary hematopoietic nodules in chronic myelogenous leukemia. *Ann Intern Med* 1979;91:735–738.

125. Ma SK, Chan JC, Wong KF. Diagnosis of spinal extramedullary hemopoiesis by magnetic resonance imaging. *Am J Med* 1993;95:111–112.

126. Fox MW, Onofrio BM. The natural history and management of symptomatic and asymptomatic vertebral hemangiomas. *J Neurosurg* 1993;78:36–45.

127. Zito G, Kadis GN. Multiple vertebral hemangiomas resembling metastases with spinal cord compression. *Arch Neurol* 1980;37:247–248.

128. Ross JS, Masaryk TJ, Modic MT, et al. Vertebral hemangiomas: MR imaging. *Radiology* 1987;165:165–169.

129. Fogel GR, Cunningham PY, III, Esses SI. Spinal epidural lipomatosis: case reports, literature review and meta-analysis. *Spine J* 2005;5:202–211.

130. Aghi M, Kiehl TR, Brisman JL. Breast adenocarcinoma metastatic to epidural cervical spine meningioma: case report and review of the literature. *J Neuro-Oncol* 2005;75:149–155.

131. Toyokuni S, Ebina Y, Okada S, et al. Report of a patient with POEMS Takatsuki/Crow-Fukase syndrome associated with focal spinal pachymeningeal amyloidosis. *Cancer* 1992;70:882–886.

132. Weissman MN, Lange R, Kelley C, et al. Intraspinal epidural sarcoidosis: case report. *Neurosurgery* 1996;39:179–181.

133. Sengupta RP, So SC, Perry RH. Nodular fasciitis: an unusual cause of extradural spinal cord compression. *Br J Surg* 1975;62:573–575.

134. Sasaki S, Nakamura K, Oda H, et al. Thoracic myelopathy due to intraspinal rheumatoid nodules. *Scand J Rheumatol* 1997;26:227–228.

135. Dhote R, Roux FX, Bachmeyer C, et al. Extradural spinal tophaceous gout: evolution with medical treatment. *Clin Exp Rheumatol* 1997;15:421–423.

136. Byrne TN, Borges LF, Loeffler JS. Metastatic epidural spinal cord compression: update on management. *Semin Oncol* 2006;33:307–311.

137. Cantu RC. Corticosteroids for spinal metastases. Letter to the editor. *Lancet* 1968;2:912.

138. Vecht CJ, Haaxma-Reiche H, van Putten WLJ, et al. Initial bolus of conventional versus high-dose dexamethasone in metastatic spinal cord compression. *Neurology* 1989;39:1255–1257.

139. Sorensen PS, Helweg-Larsen S, Mouridsen H, et al. Effect of high-dose dexamethasone in carcinomatous metastatic spinal cord compression treated with radiotherapy: a randomised trial. *Eur J Cancer* 1994;30A:22–27.

140. Millburn L, Hibbs GG, Hendrickson FR. Treatment of spinal cord compression from metastatic carcinoma. Review of literature and presentation of a new method of treatment. *Cancer* 1968;21:447–452.

141. Rubin P. Extradural spinal cord compression by tumor. I. Experimental production and treatment trials. *Radiology* 1969;93:1243–1248.

142. Tefft M, Mitus A, Schulz MD. Initial high dose irradiation for metastases causing spinal cord compression in children. *Am J Roentgenol* 1969;106:385–393.

143. Turner S, Marosszeky B, Timms I, et al. Malignant spinal cord compression: a prospective evaluation. *Int J Radiat Oncol Biol Phys* 1993;26:141–146.

144. Maranzano E, Latini P, Beneventi S, et al. Radiotherapy without steroids in selected metastatic spinal cord compression patients. A phase II trial. *Am J Clin Oncol* 1996;19:179–183.

145. Delattre JY, Arbit E, Rosenblum MK, et al. High dose versus low dose dexamethasone in experimental epidural spinal cord compression. *Neurosurgery* 1988;22:1005–1007.

146. Heimdal K, Hirschberg H, Slettebo H, et al. High incidence of serious side effects of high-dose dexamethasone treatment in patients with epidural spinal cord compression. *J Neuro-Oncol* 1992;12:141–144.

147. Graham PH, Capp A, Delaney G, et al. A pilot randomised comparison of dexamethasone 96 mg vs 16 mg per day for malignant spinal-cord compression treated by radiotherapy: TROG 01.05 Superdex study. *Clin Oncol (R Coll Radiol)* 2006;18:70–76.

148. Deutsch AD, Levin BE, Nathanson DC, et al. Temporary reversal of cord compression with hyperosmolar glucose. Letter. *Neurology* 1992;42:2220.

149. Posner JB, Howieson J, Cvitkovic E. "Disappearing" spinal cord compression: oncolytic effect of glucocorticosteroids (and other chemotherapeutic agents) on epidural metastases. *Ann Neurol* 1977;2:409–413.

150. Goldberg LD, Ditchek NT. Thyroid carcinoma with spinal cord compression. *JAMA* 1981;245:953–954.

151. Shortliffe EH, Crapo LM. Thyroid carcinoma with spinal cord compression. Letter to the editor. *JAMA* 1982;247:1565–1566.

152. Nagata M, Ueda T, Komiya A, et al. Treatment and prognosis of patients with paraplegia or quadriplegia because of metastatic spinal cord compression in prostate cancer. *Prostate Cancer Prostatic Dis* 2003;6:169–173.

153. Boogerd W, van der Sande JJ, Kroger R, et al. Effective systemic therapy for spinal epidural metastases

from breast carcinoma. *Eur J Cancer Clin Oncol* 1989;25:149–153.

154. Sasagawa I, Gotoh H, Miyabayashi H, et al. Hormonal treatment of symptomatic spinal cord compression in advanced prostatic cancer. *Int Urol Nephrol* 1991;23:351–356.

155. Carter SK. Methodology of data reporting in advanced breast cancer trials. *Cancer Chemother Pharmacol* 1979;3:1–5.

156. Burch PA, Grossman SA. Treatment of epidural cord compressions from Hodgkin's disease with chemotherapy. A report of two cases and a review of the literature. *Am J Med* 1988;84:555–558.

157. Higgins SA, Peschel RE. Hodgkin's disease with spinal cord compression: a case report and a review of the literature. *Cancer* 1995;75:94–98.

158. McDonald AC, Nicoll JAR, Rampling RP. Non-Hodgkin's lymphoma presenting with spinal cord compression: a clinicopathological review of 25 cases. *Eur J Cancer [A]* 2000;36:207–213.

159. Wong ET, Portlock CS, O'Brien JP, DeAngelis LM. Chemosensitive epidural spinal cord disease in non-Hodgkins lymphoma. *Neurology* 1996;46:1543–1547.

160. Sundaresan N, Rosen G, Fortner JG, et al. Preoperative chemotherapy and surgical resection in the management of posterior paraspinal tumors. Report of three cases. *J Neurosurg* 1983;58:446–450.

161. Ogihara Y, Sekiguchi K, Tsuruta T. Osteogenic sarcoma of the fourth thoracic vertebra. Long-term survival by chemotherapy only. *Cancer* 1984;53:2615–2618.

162. Cooper K, Bajorin D, Shapiro W, et al. Decompression of epidural metastases from germ cell tumors with chemotherapy. *J Neuro-Oncol* 1990;8:275–280.

163. Gale GB, O'Connor DM, Chu JY, et al. Successful chemotherapeutic decompression of epidural malignant germ cell tumor. *Med Pediatr Oncol* 1986;14:97–99.

164. Hayes FA, Thompson EI, Hvizdala E, et al. Chemotherapy as an alternative to laminectomy and radiation in the management of epidural tumor. *J Pediatr* 1984;104:221–224.

165. Matsubara H, Watanabe K, Sakai H, et al. Rapid improvement of paraplegia caused by epidural involvements of Burkitt's lymphoma with chemotherapy. *Spine* 2004;29:E4–E6.

166. Rades D, Stalpers LJ, Veninga T, et al. Evaluation of five radiation schedules and prognostic factors for metastatic spinal cord compression. *J Clin Oncol* 2005;23:3366–3375.

167. Rades D, Karstens JH, Hoskin PJ, et al. Escalation of radiation dose beyond 30 Gy in 10 fractions for metastatic spinal cord compression. *Int J Radiat Oncol Biol Phys* 2007;67:525–531.

168. Bates T. A review of local radiotherapy in the treatment of bone metastases and cord compression. *Int J Radiat Oncol Biol Phys* 1992;23:217–221.

169. Kaminski HJ, Diwan VG, Ruff RL. Second occurrence of spinal epidural metastases. *Neurology* 1991;41:744–746.

170. Lord CF, Herndon JH. Spinal cord compression secondary to kyphosis associated with radiation therapy for metastatic disease. *Clin Orthop* 1986;210:120–127.

171. Rades D, Veninga T, Stalpers LJ, et al. Outcome after radiotherapy alone for metastatic spinal cord compression in patients with oligometastases. *J Clin Oncol* 2007;25:50–56.

172. Rades D, Hoskin PJ, Karstens JH, et al. Radiotherapy of metastatic spinal cord compression in very elderly patients. *Int J Radiat Oncol Biol Phys* 2007;67:256–263.

173. Ratanatharathorn V, Powers WE. Epidural spinal cord compression from metastatic tumor: diagnosis and guidelines for management. *Cancer Treat Rev* 1991;18:55–71.

174. Helweg-Larsen S, Rasmusson B, Sorensen PS. Recovery of gait after radiotherapy in paralytic patients with metastatic epidural spinal cord compression. *Neurology* 1990;40:1234–1236.

175. Maranzano E, Latini P, Beneventi S, et al. Comparison of two different radiotherapy schedules for spinal cord compression in prostate cancer. *Tumori* 1998;84:472–477.

176. Rades D, Fehlauer F, Schulte R, et al. Prognostic factors for local control and survival after radiotherapy of metastatic spinal cord compression. *J Clin Oncol* 2006;24:3388–3393.

177. Rades D, Veninga T, Stalpers LJ, et al. Improved posttreatment functional outcome is associated with better survival in patients irradiated for metastatic spinal cord compression. *Int J Radiat Oncol Biol Phys* 2007;67:1506–1509.

178. Rades D, Dunst J, Schild SE. The first score predicting overall survival in patients with metastatic spinal cord compression. *Cancer* 2008;112:157–161.

179. Rock JP, Ryu S, Yin FF, et al. The evolving role of stereotactic radiosurgery and stereotactic radiation therapy for patients with spine tumors. *J Neuro-Oncol* 2004;69:319–334.

180. Niranjan A, Maitz AH, Lunsford A, et al. Radiosurgery techniques and current devices. *Prog Neurol Surg* 2007;20:50–67.

181. Klish MD, Watson GA, Shrieve DC. Radiation and intensity-modulated radiotherapy for metastatic spine tumors. *Neurosurg Clin N Am* 2004;15:481–490.

182. Jin JY, Chen Q, Jin R, et al. Technical and clinical experience with spine radiosurgery: a new technology for management of localized spine metastases. *Technol Cancer Res Treat* 2007;6:127–133.

183. Gerszten PC, Welch WC. Cyberknife radiosurgery for metastatic spine tumors. *Neurosurg Clin N Am* 2004;15:491–501.

184. Rock JP, Ryu S, Yin FF. Novalis radiosurgery for metastatic spine tumors. *Neurosurg Clin N Am* 2004;15:503–509.

185. Rock JP, Ryu S, Shukairy MS, et al. Postoperative radiosurgery for malignant spinal tumors. *Neurosurgery* 2006;58:891–898.

186. Ryu S, Jin JY, Jin R, et al. Partial volume tolerance of the spinal cord and complications of single-dose radiosurgery. *Cancer* 2007;109:628–636.

187. Bilsky MH, Yamada Y, Yenice KM, et al. Intensity-modulated stereotactic radiotherapy of paraspinal tumors: a preliminary report. *Neurosurgery* 2004;54:823–830.

188. Yamada Y, Bilsky MH, Lovelock DM, et al. High-dose, single-fraction image-guided intensity-modulated radiotherapy for metastatic spinal lesions. *Int J Radiat Oncol Biol Phys*, in press.

189. Maranzano E, Trippa F, Pacchiarini D, et al. Reirradiation of brain metastases and metastatic spinal cord compression: clinical practice suggestions. *Tumori* 2005;91:325–330.

190. Nieder C, Grosu AL, Andratschke NH, et al. Update of human spinal cord reirradiation tolerance based on additional data from 38 patients. *Int J Radiat Oncol Biol Phys* 2006;66:1446–1449.

191. Rades D, Stalpers LJ, Veninga T, et al. Spinal reirradiation after short-course RT for metastatic spinal cord compression. *Int J Radiat Oncol Biol Phys* 2005;63:872–875.

192. Nieder C, Grosu AL, Andratschke NH, et al. Proposal of human spinal cord reirradiation dose based on collection of data from 40 patients. *Int J Radiat Oncol Biol Phys* 2005;61:851–855.

193. Wright JL, Lovelock DM, Bilsky MH, et al. Clinical outcomes after reirradiation of paraspinal tumors. *Am J Clin Oncol* 2006;29:495–502.

194. Pflugmacher R. Balloon kyphoplasty for the treatment of pathological fractures in the thoracic and lumbar spine caused by metastasis: one-year follow-up. *Acta Radiologica* 2007;48:89–95.

195. Hentschel SJ. Percutaneous vertebroplasty and kyphoplasty performed at a cancer center: refuting proposed contraindications. *J Neurosurg Spine* 2005;2:436–440.

196. Patchell RA, Tibbs PA, Regine WF, et al. Direct decompressive surgical resection in the treatment of spinal cord compression caused by metastatic cancer: a randomised trial. *Lancet* 2005;366:643–648.

197. Young RF, Post EM, King GA. Treatment of spinal epidural metastases. Randomized prospective comparison of laminectomy and radiotherapy. *J Neurosurg* 1980;53:741–748.

198. Wang JC, Boland P, Mitra N, et al. Single-stage posterolateral transpedicular approach for resection of epidural metastatic spine tumors involving the vertebral body with circumferential reconstruction: results in 140 patients. Invited submission from the Joint Section Meeting on Disorders of the Spine and Peripheral Nerves. *J Neurosurg Spine* 2004;1:287–298.

199. Patil CG, Lad SP, Santarelli J, et al. National inpatient complications and outcomes after surgery for spinal metastasis from 1993–2002. *Cancer* 2007;110:625–630.

200. Bender CE, Berquist TH, Wold LE. Imaging-assisted percutaneous biopsy of the thoracic spine. *Mayo Clin Proc* 1986;61:942–950.

201. Akhtar I, Flowers R, Siddiqi A, et al. Fine needle aspiration biopsy of vertebral and paravertebral lesions: retrospective study of 124 cases. *Acta Cytol* 2006;50:364–371.

202. Bilsky MH, Shannon FJ, Sheppard S, et al. Diagnosis and management of a metastatic tumor in the atlantoaxial spine. *Spine* 2002;27:1062–1069.

203. Chen TC. Prostate cancer and spinal cord compression. *Oncology (Williston Park)* 2001;15:841–855.

204. Cereceda LE, Flechon A, Droz JP. Management of vertebral metastases in prostate cancer: a retrospective analysis in 119 patients. *Clin Prostate Cancer* 2003;2:34–40.

205. Tazi H, Manunta A, Rodriguez A, et al. Spinal cord compression in metastatic prostate cancer. *Eur Urol* 2003;44:527–532.

206. Rades D, Stalpers LJ, Veninga T, et al. Evaluation of functional outcome and local control after radiotherapy for metastatic spinal cord compression in patients with prostate cancer. *J Urol* 2006;175:552–556.

207. Venkitaraman R, Barbachano Y, Dearnaley DP, et al. Outcome of early detection and radiotherapy for occult spinal cord compression. *Radiother Oncol* 2007;85:469–472.

208. Ahmann FR, Citrin DL, deHaan HA, et al. Zoladex: a sustained-release, monthly luteinizing hormone-releasing hormone analogue for the treatment of advanced prostate cancer. *J Clin Oncol* 1987;5:912–917.

209. Rades D, Veninga T, Stalpers LJ, et al. Prognostic factors predicting functional outcomes, recurrence-free survival, and overall survival after radiotherapy for metastatic spinal cord compression in breast cancer patients. *Int J Radiat Oncol Biol Phys* 2006;64:182–188.

210. Sciubba DM, Gokaslan ZL, Suk I, et al. Positive and negative prognostic variables for patients undergoing spine surgery for metastatic breast disease. *Eur Spine J* 2007;16(10):1659–1667.

211. Higgins SA, Peschel RE. Hodgkin's disease with spinal cord compression. A case report and a review of the literature. *Cancer* 1995;75:94–98.

212. Rades D, Hoskin PJ, Stalpers LJ, et al. Short-course radiotherapy is not optimal for spinal cord compression due to myeloma. *Int J Radiat Oncol Biol Phys* 2006;64:1452–1457.

213. Rades D, Stalpers LJ, Schulte R, et al. Defining the appropriate radiotherapy regimen for metastatic spinal cord compression in non-small cell lung cancer patients. *Eur J Cancer* 2006;42:1052–1056.

214. Rades D, Walz J, Schild SE, et al. Do bladder cancer patients with metastatic spinal cord compression benefit from radiotherapy alone? *Urology* 2007;69:1081–1085.

215. Rades D, Dahm-Daphi J, Rudat V, et al. Is short-course radiotherapy with high doses per fraction the appropriate regimen for metastatic spinal cord compression in colorectal cancer patients? *Strahlenther Onkol* 2006;182:708–712.

216. Rades D, Schild SE, Dunst J. Radiotherapy is effective for metastatic spinal cord compression in patients with epithelial ovarian cancer. *Int J Gynecol Cancer* 2007;17:263–265.

217. Gottfried ON, Schloesser PE, Schmidt MH, et al. Embolization of metastatic spinal tumors. *Neurosurg Clin N Am* 2004;15:391–399.

218. Rades D, Walz J, Stalpers LJ, et al. Short-course radiotherapy (RT) for metastatic spinal cord compression (MSCC) due to renal cell carcinoma: results of a retrospective multi-center study. *Eur Urol* 2006;49:846–852.

219. Cooper K, Bajorin D, Shapiro W, et al. Decompression of epidural metastases from germ cell tumors with chemotherapy. *J Neuro-Oncol* 1990;8:275–280.

220. Rades D, Fehlauer F, Veninga T, et al. Functional outcome and survival after radiotherapy of

metastatic spinal cord compression in patients with cancer of unknown primary. *Int J Radiat Oncol Biol Phys* 2007;67:532–537.

221. Chow TSF, McCutcheon IE. The surgical treatment of metastatic spinal tumors within the intradural extramedullary compartment. *J Neurosurg* 1996;85:225–230.

222. Mirimanoff RO, Choi NC. Intradural spinal metastases in patients with posterior fossa brain metastases from various primary cancers. *Oncology* 1987;44:232–236.

223. Mirimanoff RO, Choi NC. The risk of intradural spinal metastases in patients with brain metastases from bronchogenic carcinomas. *Int J Radiat Oncol Biol Phys* 1986;12:2131–2136.

224. Kalayci M, Cagavi F, Gul S, et al. Intramedullary spinal cord metastases: diagnosis and treatment—an illustrated review. *Acta Neurochir (Wien)* 2004;146:1347–1354.

225. Chi JH, Parsa AT. Intramedullary spinal cord metastasis: clinical management and surgical considerations. *Neurosurg Clin N Am* 2006;17:45–50.

226. Schiff D, O'Neill BP. Intramedullary spinal cord metastases: clinical features and treatment outcome. *Neurology* 1996;47:906–912.

227. Murphy KC, Feld R, Evans WK, et al. Intramedullary spinal cord metastases from small cell carcinoma of the lung. *J Clin Oncol* 1983;1:99–106.

228. Holoye P, Libnoch J, Cox J, et al. Spinal cord metastasis in small cell lung carcinoma of the lung. *Int J Radiat Oncol Biol Phys* 1984;10:349–356.

229. Lyding JM, Tseng A, Newman A, et al. Intramedullary spinal cord metastasis in Hodgkin's disease. Rapid diagnosis and treatment resulting in neurologic recovery. *Cancer* 1987;60:1741–1744.

230. Grasso G, Meli F, Patti R, et al. Intramedullary spinal cord tumor presenting as the initial manifestation of metastatic colon cancer: case report and review of the literature. *Spinal Cord* 2007;45:793–796.

231. Donovan DJ, Freeman JH. Solitary intramedullary spinal cord tumor presenting as the initial manifestation of metastatic renal cell carcinoma: case report. *Spine* 2006;31:E460–E463.

232. Winkelman MD, Adelstein DJ, Karlins NL. Intramedullary spinal cord metastasis. Diagnostic and therapeutic considerations. *Arch Neurol* 1987;44:526–531.

233. Grem JL, Burgess J, Trump DL. Clinical features and natural history of intramedullary spinal cord metastasis. *Cancer* 1985;56:2305–2314.

234. Connolly ES, Jr, Winfree CJ, McCormick PC, et al. Intramedullary spinal cord metastasis: report of three cases and review of the literature. *Surg Neurol* 1996;46:329–337.

235. Hashizume Y, Hirano A. Intramedullary spinal cord metastasis. Pathologic findings in five autopsy cases. *Acta Neuropathol (Berl)* 1983;61:214–218.

236. Foster O, Crockard HA, Powell MP. Syrinx associated with intramedullary metastasis. *J Neurol Neurosurg Psychiatry* 1987;50:1067–1070.

237. Crasto S, Duca S, Davini O, et al. MRI diagnosis of intramedullary metastases from extra-CNS tumors. *Eur Radiol* 1997;7:732–736.

238. Watanabe M, Nomura T, Toh E, et al. Intramedullary spinal cord metastasis: a clinical and imaging study of seven patients. *J Spinal Disord Tech* 2006;19:43–47.

239. Sutter B, Arthur A, Laurent J, et al. Treatment options and time course for intramedullary spinal cord metastasis. Report of three cases and review of the literature. *Neurosurg Focus* 1998;4:e3.

240. Findlay JM, Bernstein M, Vanderlinden RG, et al. Microsurgical resection of solitary intramedullary spinal cord metastases. *Neurosurgery* 1987;21:911–915.

Chapter 7

Leptomeningeal Metastases

INTRODUCTION

Cancer cells that metastasize to any part of the nervous system in contact with the cerebrospinal fluid (CSF) may be shed into the CSF and float along CSF pathways to other areas of the nervous system where they may settle and grow (Fig. 7–1). These cells may seed the meninges focally, multifocally, or diffusely. The resulting leptomeningeal tumor may be visible grossly or only on microscopic examination. The tumor can remain within the meninges or invade the parenchyma of the brain, spinal cord, or cranial or peripheral nerves. The terms applied to this disorder have varied. When the primary disease is leukemia, the term "meningeal leukemia" is generally used.[1,2] When the primary

tumor is not leukemia, the general term "meningeal carcinomatosis"[3] has been applied, even if the tumor is a lymphoma or sarcoma. The terms "carcinomatous meningitis"[4] or "neoplastic meningitis"[5,6] are frequently used misnomers that suggest a leptomeningeal inflammatory response, which may or may not accompany tumor in the CSF.[7] The terms "leptomeningeal seeding" or "leptomeningeal metastases"[8,9] seem more appropriate.

First described by Eberth[10] in 1870 and named "meningitis carcinomatosa" by Siefert[11] in 1902, leptomeningeal metastases were once thought to be rare, a subject of only individual case reports[12] and a diagnosis made only at autopsy. In 1974, two large clinical series of leptomeningeal metastases from solid tumors[7,13]

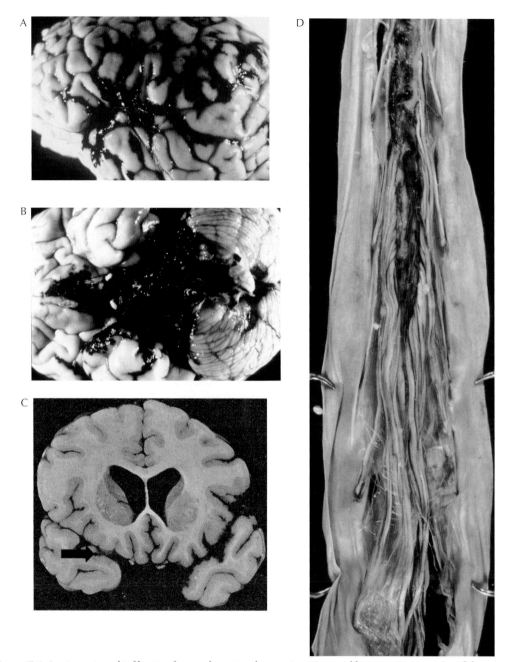

Figure 7–1. Leptomeningeal infiltration from malignant melanoma in a 23-year-old woman. Pigmentation of the tumor provides a marker demonstrating the predilection of tumor to concentrate in particular anatomic areas. **A**: Lateral view of cerebral hemisphere with tumor concentrated within the Sylvian fissure and its connecting sulci over the surface. Tumor is relatively absent near the dorsal surface. **B**: Ventral surface of the brain showing tumor concentrated within the basal cisterns, extending laterally at the optic chiasm to fill the cistern connecting with the Sylvian fissure. **C**: Coronal section of the hemispheres through the anterior portion of the Sylvian fissure shows the same pattern. A single, small, discrete cortical nodule is in the temporal lobe (*arrow*). Note the hydrocephalus that is due to obstruction of spinal fluid absorption pathways in the subarachnoid space. **D**: The spinal cord showing tumor over the cord and nodular swellings along nerve roots. (From Ref. 7 with permission.)

could cite fewer than 125 previous reports.[13] By that time, however, meningeal leukemia was well recognized clinically. Since the 1970s, leptomeningeal metastases from solid tumors have assumed increasing importance in neuro-oncology because of their apparently increasing frequency and the severe neurologic disability associated with the disorder.[5,6,14–16]

FREQUENCY

The exact incidence of leptomeningeal metastases is difficult to determine, but several autopsy studies (Table 7–1) have found an overall incidence varying from 0.8% to 8%; the higher figure is probably the more accurate. Part of the difficulty in determining the incidence of leptomeningeal metastases is that tumor may be inapparent to gross inspection at autopsy. Furthermore, tumor may seed the leptomeninges in a multifocal rather than diffuse fashion, so it may not be identified on microscopic examination unless the pathologist examines multiple areas likely to be invaded by disease. For example, in some patients, only the spinal meninges may be involved[17,18] (Fig. 7–2), and unless the spinal cord is examined, the diagnosis is missed. The clinical estimates of incidence may also be inaccurate because even easily identifiable clinical signs may be attributed to peripheral metastases or nonmetastatic neurologic disorders. In one series,[18] only 7 of 16 patients with autopsy-proved spinal leptomeningeal metastases were diagnosed antemortem.[18]

Leptomeningeal metastases from some primary neoplasms are increasing while those from other primary tumors are decreasing. The incidence of meningeal metastases from acute lymphoblastic leukemia and some lymphomas has decreased from a high of 66% of patients to approximately 2% to 10%,[22,23] because of effective prophylaxis.[24] The opposite is true of small cell lung cancer (SCLC)[25,26] and breast cancer,[27–29] in which clinical evidence suggests an increasing incidence.

One reason for the rising incidence of leptomeningeal metastases is that the subarachnoid space, like the brain, may serve as a sanctuary for tumor cells in patients whose systemic cancer has been successfully treated with chemotherapy that does not cross the blood–CSF barrier (see Chapter 5). This has been observed in a variety of settings, including an apparent increase in leptomeningeal metastases in patients with breast cancer treated with trastuzumab[30,31] or paclitaxel, where isolated leptomeningeal metastases were observed in 12%,[32] and patients with non–small cell lung cancer (NSCLC) who respond to gefitinib.[33] A second reason is that better detection methods, particularly magnetic resonance imaging (MRI), allow early identification of mildly symptomatic individuals. In addition, the increasing use of surgery to treat brain metastases, particularly those in the posterior fossa, may also play a role.[34,35]

Whether or not the incidence is actually increasing, reports of antemortem diagnosis of leptomeningeal metastasis causing clinical complications are more numerous than before. Aisner et al.[36] reported a 10% incidence and Rosen et al.[25] an 11% incidence in patients with SCLC. The actuarial incidence of leptomeningeal metastasis rises from 0.5% at presentation

Table 7–1. Incidence of Leptomeningeal Metastasis at Autopsy in Patients with Cancer

Reference	Dates	No. of Autopsies	No. (%) of Patients with Leptomeningeal Tumor	No. (%) of Patients with Leptomeningeal Tumor without Other Intracranial Masses
Posner and Chernik[19]	1970–1976	2374	184 (8)	63 (3)
Gonzalez-Vitale and Garcia-Bunuel[20]	Before 1976	2227	–	18 (0.8)
Takakura et al.[21]	1950–1970	3359	118 (3.5)	–

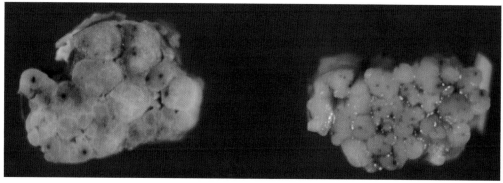

Figure 7–2. Cauda equina of a person who died without neurologic disease (*right*), and a patient who died with leptomeningeal metastasis from Hodgkin disease (*left*). The spinal roots involved by tumor are enlarged and matted when compared with the smaller, discrete roots of the nondiseased person.

to 25% if the patient survives for 3 years. Increased awareness of the problem led to an increase in antemortem diagnosis, from 39% before 1977 to 88% after that year. Aroney et al.[37] reported that 42% of patients with SCLC who relapsed after initial treatment did so in the meninges; in 27%, the meninges was the sole site of relapse. Prophylactic radiation therapy (RT) to the brain (see Chapter 5), which effectively prevents brain metastases, does not prevent leptomeningeal metastasis in patients with SCLC.[38] A more recent study by Seute et al. described a 2% prevalence of leptomeningeal metastases from SCLC and a 2-year cumulative incidence of 10%.[26]

The incidence of leptomeningeal metastasis from breast cancer at autopsy varies from 3% to 40%, depending on the series.[39,40] A study of leptomeningeal metastasis from infiltrating lobular breast carcinoma reports a clinical incidence of 2.7%, which contrasts with a 5.2% incidence of brain metastases.[41] As in other series, leptomeningeal metastasis was usually a late consequence of the tumor. In a Memorial Sloan-Kettering Cancer Center (MSKCC) study, we found the leptomeninges involved as the sole site of intracranial disease in 3% of patients with breast cancer (Table 7–2). Leptomeningeal metastasis with or without other intracranial lesions was present at autopsy[19] in 8% of all patients with cancer, and was usually accompanied by other metastatic central nervous system (CNS) tumors, primarily dural or parenchymal metastases. Leptomeningeal metastasis was the sole intracranial lesion in only 40 of 2088 patients who had cancer other than leukemia, an overall frequency of 1.9%.[19]

Any systemic cancer can seed the leptomeninges. Carcinoma of the breast[27,29] and lung[42] (particularly SCLC[25,26]), melanoma,[35,43,44] leukemia, and lymphoma are the common offenders. Several reports summarize the clinical incidence of leptomeningeal metastasis from particular tumors, including acute nonlymphocytic leukemia (5%–15%),[45,46] non–Hodgkin lymphoma (6%),[47,48] head and neck cancer (1%),[49] breast cancer (5%),[28] and SCLC (10%–15%).[25,26] Gastric carcinoma, once believed to be a major cause of leptomeningeal tumor,[50] is now rarely encountered in the United States. A report from Korea describes 19 patients with confirmed leptomeningeal metastases from gastric cancer identified over an 11-year period; in two patients, leptomeningeal symptoms preceded identification of the primary cancer that in almost all cases had poorly differentiated or signet ring histology. The median survival was 4 weeks; intrathecal chemotherapy appeared to be helpful.[51]

Other less common tumors that can metastasize to the leptomeninges include mycosis fungoides,[52–54] multiple myeloma,[55,56] squamous cell carcinoma,[57] thyroid cancer,[58] rectal cancer,[59] carcinoid,[60] rhabdomyosarcoma,[61] chronic lymphocytic leukemia,[62] Hodgkin disease,[63] and neuroblastoma.[64]

Not all leptomeningeal tumors are metastatic. Primary lymphomas,[65] melanomas,[66–68] and rhabdomyosarcomas[69] of the leptomeninges are less frequent than metastatic lesions, but can arise as neoplasms confined to the leptomeninges. Most primary CNS lymphomas are parenchymal, often with leptomeningeal spread;[70] most metastatic lymphomas, on the

Table 7–2. **Frequency of Leptomeningeal Metastases at Autopsy as the Sole Intracranial Lesion for Specific Primary Tumors**

Primary Tumor	Total No. of Autopsies	No. (%) of Patients with Leptomeningeal Metastases
Leukemia	287	28 (10)
Acute lymphocytic leukemia	87	21 (24)
Acute myelogenous leukemia	103	5 (5)
Lymphoma	309	15 (4)
Hodgkin	119	2 (2)
Non–Hodgkin	190	13 (7)
Breast	324	11 (3)
Melanoma	125	6 (5)
Lung	297	4 (1)
Gastrointestinal	311	3 (1)
Sarcoma	126	1 (1)

Data from Ref. 19.

Table 7–3. **Primary Tumors in Clinical Cases of Leptomeningeal Metastases from MSKCC* (N = 140)**

Primary Tumor	No. of Patients MSKCC	Percentage Solid Tumors From Ref. 19
Breast	81 (58)	12%–34%
Lung	24 (17)	10%–26%
Melanoma	17 (12)	17%–25%
Lymphoma	14 (10)	
Unknown primary	5 (4)	1%–7%
Renal	3 (2)	
Prostate	2 (1)	
Pancreas	2 (1)	
Sarcoma	2 (1)	
Nasopharynx	1 (0.7)	
Esthesioneuroblastoma	1 (0.7)	

other hand, are meningeal,[71] but exceptions are not rare.[65,72,73]

Leptomeningeal metastases, although occasionally a first sign of initial or recurrent cancer, are usually a late manifestation of systemic cancer and often accompany relapse elsewhere in the body.[7,25,28] Some patients develop symptoms before the primary tumor has been discovered, whereas others, particularly those with breast carcinoma, may become symptomatic many years after the initial diagnosis and treatment of the primary cancer. Similar late

occurrences have been reported with leukemia.[74] Of 50 patients with solid tumors, 11 had no systemic metastases at the time neurologic symptoms occurred; the cancer appeared to have been cured in five patients.[7]

Table 7–3 lists the primary tumors in patients with leptomeningeal metastases encountered at MSKCC. The number of lymphomas is apparently small because the largest of our studies was devoted to solid tumors. The percentage of patients with various solid tumors culled from the literature is indicated on the right side of

Table 7–4. How Tumor Cells Reach the Leptomeninges

Direct extension from preexisting CNS tumors, primary or metastatic
  Spontaneous
  Post-surgery
Hematogenous—arterial
  From choroid plexus
  From arachnoid
Hematogenous—venous
  From vertebral metastasis
  Via Batson plexus
Direct extension along peripheral nerves

Table 7–5. Pathophysiology of Neurologic Symptoms and Signs

Hydrocephalus
Increased intracranial pressure
Invasion of CNS parenchyma
Ischemia and infarction
Metabolic competition
Compression
Immune response
Inflammation
Blood–CSF barrier disruption

CNS, central nervous system; CSF, cerebrospinal fluid.

the table. It is obvious that the vast majority of patients who do not have leukemia or lymphoma have either breast or lung cancer or melanoma.

As discussed in Chapter 2, tumor cells may reach the leptomeninges by one of several routes (Table 7–4) (see also Fig. 2–8).

PATHOPHYSIOLOGY OF SIGNS AND SYMPTOMS

Hydrocephalus

Leptomeningeal metastases can cause CNS dysfunction in several ways (Table 7–5).

One of the most common symptoms is hydrocephalus (see Fig. 7–1C). The tumor invades the base of the brain and occludes the CSF outflow foramina of the fourth ventricle. It also frequently infiltrates the Sylvian fissure and, at times, the arachnoid villi. The meningeal tumor and its accompanying inflammatory response, if any, also increase resistance to CSF absorption, leading to hydrocephalus. Intracranial pressure is usually elevated, but slowly developing hydrocephalus may cause ventricular dilatation without elevation of CSF pressure, so-called normal-pressure hydrocephalus. Conversely, rapid occlusion of the subarachnoid spaces by tumor, particularly when localized near the sagittal sinus, may elevate intracranial pressure but cause slight or no dilatation of the ventricles.[76] Both phenomena occur with leptomeningeal metastasis

and may lead to diagnostic confusion, particularly when either hydrocephalus or isolated increased intracranial pressure is the first or only symptom. Leptomeningeal leukemia may raise intracranial pressure sufficiently to cause tentorial and cerebellar herniation, sometimes leading to death.[77] Elevated intracranial pressure is often missed, because the symptoms are attributed to direct effects of the tumor in the CSF. The opening pressure is not always measured when a lumbar puncture is performed, and elevated pressure cannot always be appreciated simply from the rate of CSF draining from the spinal needle. We have seen patients undergo an extensive gastrointestinal (GI) workup before it was appreciated that their nausea and vomiting were a consequence of elevated intracranial pressure.

Even in the absence of hydrocephalus or increased intracranial pressure, substantial abnormalities of CSF flow dynamics often occur.[78–80] Studying CSF bulk flow with a radioisotope instilled in the cerebral ventricle, Grossman et al.[81] found abnormalities in 70% of their patients with leptomeningeal metastasis. The abnormalities were of three types. The radionuclide tracer failed to (1) leave the ventricular system at a normal rate, (2) reach the lumbar sac at a normal rate, or (3) ascend over the cortical convexities within 24 hours. The results imply that chemotherapeutic agents injected into the ventricle or lumbar space of such patients may not distribute uniformly throughout the subarachnoid space. Furthermore, if ventricular outflow is completely obstructed, chemotherapeutic agents administered into the cerebral ventricles become trapped in the ventricular system and diffuse across the ependyma into the brain substance, increasing the likelihood of neurotoxicity.[81] Chamberlain and Corey-Bloom[82] reported finding unsuspected physiologic blocks to radionuclide flow in the spinal canal of patients with leptomeningeal metastasis. CSF flow studies in patients with leukemia or lymphoma are normal, unless there are leptomeningeal metastases.[83]

Parenchymal Invasion

Leptomeningeal tumor can grow along Virchow–Robin spaces into the brain to cause neurologic dysfunction, such as focal or generalized seizures[84] (Fig. 7–3). Growth of such tumors may remain restricted to the perivascular spaces without actually infiltrating the brain itself. Conversely, the tumor may directly invade brain, spinal cord, and cranial and peripheral nerve roots to cause neurologic symptoms (Fig. 7–4).

Ischemia

Tumor invasion into Virchow–Robin spaces down to the arteriolar level probably interferes with the blood supply and/or consumes oxygen meant for neurons. Microscopic examination of the brain and spinal cord of patients with leptomeningeal tumor has identified ischemic changes or frank infarction in cortical areas that tumor has invaded (Fig. 7–3).[18,85,86] In some patients with leptomeningeal metastases, angiography reveals narrowing or obliteration of cerebral blood vessels.[86,87] Transient ischemic attacks[86] and stroke[85] are clinical consequences of these vascular changes. Cerebral blood flow is decreased in some patients with leptomeningeal tumor; the magnitude of this decrease correlates roughly with cognitive changes caused by the leptomeningeal disease.[88]

Metabolic Competition

Symptoms may be caused by competition between the tumor and neurons for essential metabolites such as glucose.[89] Leptomeningeal tumor invasion causes hypoglycorrhachia by decreasing glucose transport, which has the effect of restricting access to this critical nutrient. In experimental animals, investigators found decreased cerebral glucose utilization in areas of brain underlying leptomeningeal tumor and also at anatomically remote areas that are functionally related to structures subjacent to the tumor.[89] The syndrome of *hypothalamic leukemia*[90,91] can be explained by metabolic competition between neurons and infiltrating tumor cells. In this disorder, now no longer seen, leukemic patients in systemic remission inexplicably gained weight. CSF examination revealed meningeal leukemia, and when CNS leukemia was treated, the weight returned to normal. In the few pathologically reported cases, the hypothalamus was infiltrated by tumor,[92] presumably entering the brain from the third ventricle. Because other hypothalamic functions appeared to be normal and the situation was reversible, the best hypothesis is that neurons in the hypothalamus sensitive to the glucose concentration perceive a reduced glucose level due to competition from adjacent tumor cells. An 18-year-old girl at MSKCC, in remission after treatment for acute lymphoblastic leukemia, gained 20 kg during her first college semester despite being athletically active and obtaining high grades. Only when she discovered foot weakness did she return to her oncologist, who found 1000 malignant cells and a glucose concentration of 7 mg/dL in her CSF. RT to the brain and intrathecal methotrexate (MTX) induced a weight loss of exactly 20 kg, returning her to her original weight.

Immune Responses

Several reports suggest that leptomeningeal metastases trigger an immune response in the CNS.[93–95] Weller et al.[96] studied 47 paired serum and spinal fluid samples from patients with leptomeningeal metastasis from a variety of carcinomas and identified elevated immunoglobulin G (IgG) and immunoglobulin M (IgM) indices, implying CSF production of antibody, oligoclonal bands, and increased CSF interleukin 6. CSF tumor necrosis factor α was elevated in patients with malignant melanoma. The findings were not helpful diagnostically, and it is not clear whether these proteins play a role in clinical symptomatology.[96–98]

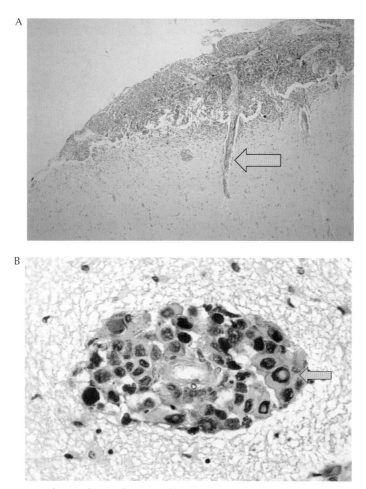

Figure 7–3. Leptomeningeal tumor from malignant melanoma. **A**: The tumor involves the leptomeninges over the brain's surface and grows down the Virchow–Robin spaces (*arrow*), but the brain parenchyma is not invaded. The area immediately underlying the tumor mass is pale, probably from ischemia. **B**: Examined at a higher power, the pigmented malignant cells can be seen localized to a Virchow–Robin space.

Figure 7–4. Infiltration of a cauda equina spinal root by leptomeningeal tumor. The tumor can be seen both on the surface of the root and within its depths. The area of infiltration is marked by pallor, indicating demyelination.

Inflammation

It is common for the CSF of patients with leptomeningeal metastases to contain many more inflammatory cells, usually T-lymphocytes, than tumor cells. In a few patients, symptoms of leptomeningeal metastases may mimic those of infectious meningitis. Headache, nuchal rigidity, and Kernig and Brudzinski signs are not rare, but fever does not occur. In one of our patients, nuchal rigidity causing opisthotonus was the presenting symptom.

Blood–Cerebrospinal Fluid Barrier Disruption

The relationship between leptomeningeal metastases and the blood–CSF barrier[99,100] is diagnostically and therapeutically important (see Chapter 3). Because leptomeningeal leukemia was not treated effectively by parenteral chemotherapy, the concept arose that the blood–CSF barrier remains intact when leptomeningeal metastases develop. However, clinical and experimental evidence indicate the contrary. Bleyer and Poplack[101] demonstrated that, after systemic injection, higher levels of MTX are achieved in the CSF of patients with meningeal leukemia than in those without the disease. This finding indicates that the barrier to the passage of water-soluble MTX is disrupted when the leptomeninges are infiltrated by tumor. Furthermore, leptomeningeal metastases can often be identified on MRI by contrast enhancement of the meninges in the brain or spinal cord, indicating disruption of the blood–CSF barrier.[102–104] Ushio et al.[105] have shown, by entry of horseradish peroxidase into leptomeningeal tumor in an experimental rat model, that the blood–CSF barrier is disrupted through vascularization of the tumor by fenestrated capillaries. They also demonstrated that, although parenterally administered MTX is ineffective in prolonging survival in animals with leptomeningeal tumor when given before the fifth day after injection of tumor cells into the CSF, survival is prolonged if the drug is administered after the fifth day. These data suggest that leptomeningeal tumor does not substantially alter the blood–CSF barrier until the tumor's growth has been sufficient to stimulate its own vasculature. At that point,

the barrier is at least partially breached, and chemotherapeutic agents can enter to a greater degree. Diagnostically, only thick or nodular growth of leptomeningeal metastases can be visualized on contrast imaging. Therapeutically, clinical evidence indicates that the blood–CSF barrier disruption is incomplete, and parenteral chemotherapy with water-soluble agents is not always effective in eradicating and certainly not in preventing leptomeningeal metastases. However, drug penetration is not the sole determinant of response of leptomeningeal metastasis to systemic chemotherapy. Some investigators have demonstrated good responses to systemic chemotherapy in some patients with leptomeningeal metastasis.[105–108] One report describes a good response of leptomeningeal prostate metastasis to hormonal therapy.[109]

CLINICAL FINDINGS

Leptomeningeal metastases affect the CNS at more than one site, thereby causing multifocal signs and symptoms. Physicians should suspect leptomeningeal metastasis in patients with signs and symptoms indicative of nervous system dysfunction at more than one level. For that purpose, it is useful to divide neurologic signs and symptoms into those originating from three separate anatomic areas—the brain, the cranial nerves, and the spinal roots[7,75,85,110,111] (Table 7–6). A fourth set of symptoms arises from irritation of the leptomeninges induced by the tumor or tumor-induced inflammation.

Two important clues should lead the physician to suspect strongly the clinical diagnosis of leptomeningeal metastasis: (1) a careful history often identifies symptoms suggesting involvement of more than one anatomic area, and (2) even if the patient's symptoms suggest involvement of a single anatomic area, careful neurologic examination may reveal multifocal abnormalities, indicating that other areas of the nervous system are involved asymptomatically. For example, a patient may complain only of focal seizures (brain), but when an absent ankle or knee jerk (spinal root) is discovered on examination, meningeal seeding becomes a likely diagnosis. Signs and symptoms indicative of meningeal irritation, such as nuchal

Table 7–6. **Symptoms and Signs in 140 Patients with Leptomeningeal Metastases from Solid Tumors at MSKCC**

Symptom	Initial (%)	At Any Time (%)	Sign	Initial (%)	At Any Time (%)
Cerebral					
Headache	38	40	Papilledema	12	12
Mental change	25	30	Abnormal mental state	50	50
Nausea and vomiting	12	20	Seizures	14	15
Gait difficulty	46	68	Extensor plantar(s)	50	66
Cranial Nerves					
Visual loss	8	12	Ocular muscle paresis	30	38
Diplopia	8	20	Trigeminal neuropathy	12	14
Hearing loss	6	9	Facial weakness	25	26
Dysphagia	2	4	Hearing loss	20	20
Spinal					
Pain	25	40	Nuchal rigidity	16	17
Back	18	50	Reflex absence/decrease	60	76
Radicular	12	25	Dermatomal sensory loss	50	50
Paresthesias	10	42	Lower motor neuron weakness	78	78
Weakness	22	50			

rigidity in the afebrile patient, also suggest the diagnosis.

Not all patients have symptoms or signs involving the three anatomic areas. In one series,[85] 50% of patients complained of symptoms in only one area at diagnosis, but only 25% had signs and symptoms limited to that area. Another 50% of patients had signs or symptoms suggesting the involvement of two anatomic areas, and approximately 25% had involvement of all three anatomic regions. The earlier the diagnosis is considered, the greater the likelihood of unifocal abnormalities. The most common site is the spinal area, involved in more than 75% of patients. A little more than one-half have cranial nerve involvement, and one-half also have cerebral signs and symptoms. Leptomeningeal signs, such as stiff neck and pain on straight-leg raising, are present in only 15% of patients.[85]

The following paragraphs and Table 7–6 summarize the symptoms and signs of leptomeningeal metastasis. The data are culled from several series of patients, as well as some individual case reports.

Cerebral (Brain) Symptoms and Signs

HEADACHE

Headache affects almost 25% of the patients and is usually among the first symptoms, but the type of headache is nonspecific. In some patients it is bifrontal, while in others it is diffuse or located at the base of the skull, radiating into the neck. Headache is associated with nausea, vomiting, or lightheadedness in some patients. Rarely, the headache mimics migraine or cluster headache.[112] Severe episodic headache indicates plateau waves associated with increased intracranial pressure (see Chapter 3).

GAIT DIFFICULTY

Many patients appear to have mild gait apraxia[113,114] characterized by a broad-based stance and difficulty lifting their feet from the floor when attempting to walk (magnetic gait). Others are mildly ataxic, unsteady on turning or when attempting tandem walking, probably because of cerebellar pathway dysfunction. Some patients have leg weakness (see following text), further compromising gait. In most instances, the gait difficulty is hard to define on examination and probably has multifactorial causes.

COGNITIVE DYSFUNCTION

Approximately 25% of patients present with this complaint, and over 33% complain of cognitive dysfunction at some time during the course of the illness. In approximately 50% of patients with leptomeningeal metastases, a careful mental status examination reveals difficulty with recent memory and sometimes with concentration.

EPISODIC LOSS OF CONSCIOUSNESS

This complaint usually results from seizures,[110] but plateau waves are also an occasional cause. Increased intracranial pressure with resulting plateau waves may occur in the absence of identifiable hydrocephalus and may be relieved by ventriculoperitoneal shunting. Generalized or focal seizures are common. Nonconvulsive status epilepticus[84,115] is easily mistaken for a confusional state or psychosis has also been reported (see Chapter 4).

OTHER CEREBRAL COMPLAINTS

Other complaints include dizziness or lightheadedness and speech difficulty, either dysphasia or dysarthria. Diabetes insipidus (DI), a common complication of breast carcinoma,[116] usually results from metastases to the sella turcica with secondary involvement of the posterior pituitary. Less commonly, leptomeningeal metastasis leads to tumor invasion of the pituitary stalk as it passes through the subarachnoid space.[117,118] In one series, 20% of all patients with DI seen in a mixed general and cancer hospital had metastatic invasion of the posterior pituitary as the cause. DI was found in almost 1% of patients with breast cancer.[119]

CEREBRAL SIGNS

Many patients have hyperreflexia and extensor plantar responses, but signs of major parenchymal dysfunction, such as aphasia, hemiparesis, hemisensory loss, or visual field defects, are uncommon. Visual field defects may occasionally arise from invasion of the optic chiasm or optic tract (see following text), but focal hemispheric signs usually suggest the presence of parenchymal metastasis.

Invasion of the cortex by leptomeningeal tumor can cause cortical brain dysfunction. Madow and Alpers[120] described an unusual and rare manifestation of leptomeningeal tumor with cortical invasion, the encephalitic form of metastatic carcinoma. In this disorder, neurologic signs result from tumor invading the underlying cortex.[121] The invasion is widespread, and, because it is microscopic, imaging studies may be negative, but malignant cells are usually found in the CSF. Patients generally present with confusion and disorientation, as well as frequent seizures and focal neurologic signs such as hemiparesis or a hemisensory defect. One report[122] describes the triphasic waves typical of metabolic brain disease on the EEG (electroencephalogram) of such a patient (see Chapter 11).

A unique patient with leptomeningeal tumor from adenocarcinoma of the lung suffered from cerebral salt wasting leading to hyponatremia.[123] Another patient with leptomeningeal tumor presented with central neurogenic hyperventilation,[124] but these manifestations are rare.

Cranial Nerve Symptoms and Signs

Cranial nerve abnormalities are not usually major presenting complaints of patients with leptomeningeal tumor, although on careful examination, mild abnormalities of ocular movement, facial weakness, or decreased facial sensation and hearing loss are often found. Both signs and symptoms of cranial nerve dysfunction become more prominent as the disease progresses (Table 7–6). The most common complaint is diplopia; patients also complain of hearing loss, visual difficulties, and less commonly, decreased taste, swallowing difficulty, and hoarseness.

Blindness has been described as the first or most prominent symptom in a number of patients when tumor directly invades the optic nerves or optic chiasm.[125,126] The phenomenon is most common in leukemias, lymphomas, and breast carcinoma. Because blindness can be a complication of chemotherapy (see Chapter 12), the diagnosis may be confusing if the visual loss occurs during treatment.[127] Sudden bilateral hearing loss also has been reported.[128] Vertigo and hearing loss resembling Meniere syndrome can occur as a complication of subarachnoid tumor that has spread to the labyrinth and the cochlea.[129,130] In fact, Meniere's original patient may have had leukemic involvement of the labyrinth.[131] Isolated oculomotor weakness can be a presenting sign,[132] but a more common finding is multiple unilateral oculomotor pareses, involving the oculomotor, trochlear, and abducens cranial nerves and usually associated with meningeal invasion of the cavernous sinus (see Chapter 8). Unilateral facial palsy mimicking Bell palsy is also a relatively common presenting symptom.[133]

A clue that cranial involvement is secondary to leptomeningeal tumor rather than to a metastasis at the skull base (see Chapter 8) is that multiple cranial nerves are usually affected.[134] If the involvement is bilateral, it is even more likely that the pathology lies in the subarachnoid space rather than at the base of the brain. The differential diagnosis can be difficult if only one nerve is involved.

Spinal Symptoms and Signs

Spinal symptoms affect more than 50% of patients with leptomeningeal metastasis (Table 7–6). Leptomeningeal tumor frequently invades the nerve roots of the cauda equina, sometimes forming masses that can be seen on MRI[135] (see following text). Occasionally, leptomeningeal tumor is restricted to the spinal meninges.[17] Symptoms can be divided into two broad categories: (1) those caused by invasion of spinal roots (neural dysfunction) (see Fig. 7–4) and (2) those caused by invasion of the leptomeninges alone (meningeal irritation). Meningeal signs include pain in the neck or back, sometimes associated with nuchal rigidity that can mimic acute meningitis in severity. When the lumbar puncture reveals increased pressure with a large number of white blood cells (WBCs) but few or no tumor cells, the differential diagnosis may be difficult, but the absence of fever or organisms and the presence of malignant cells by cytologic examination

usually establish the diagnosis, as discussed later in this chapter.

Direct invasion of nerve roots frequently causes radicular pain, sometimes mimicking the pain from a herniated disk. In addition to pain, patients complain of weakness, paresthesias, and bladder or bowel dysfunction. The most characteristic sign of bladder dysfunction from leptomeningeal tumor is asymptomatic bladder enlargement. The patient is unaware of the sensation of bladder or bowel fullness and often does not report a change in urination.

Spinal root signs, primarily the absence of one or more deep tendon reflexes, are found in approximately 70% of patients on initial examination. A cauda equina syndrome with bilateral leg weakness, foot numbness, absent ankle reflexes, and diminished rectal tone is relatively common.

LABORATORY TESTS

Those laboratory tests that help establish the diagnosis of leptomeningeal metastasis are examination of the CSF and MRI of brain and spine. Myelography, cerebral arteriography, and other laboratory tests are now rarely indicated but are described in the paragraphs that follow (Table 7–7).

Magnetic Resonance Imaging

An MRI of the neuraxis (brain and spinal cord) should be the first test performed in patients suspected of suffering leptomeningeal metastases[135,137–139] (Table 7–8).

The MRI is best performed before lumbar puncture because a lumbar puncture can cause intracranial hypotension, leading to enhancement of the pachymeninges of the brain and spinal cord. Pachymeningeal enhancement, which is thick, linear, or sometimes nodular along the undersurface of the calvarium, falx, and tentorium,[140] can usually be differentiated from leptomeningeal enhancement, which follows the convolutions of the gyri and is often present around the basal cisterns and within the cerebellar folia. However, pachymeningeal enhancement is often misinterpreted as leptomeningeal metastases. Moreover, intracranial hypotension may also be associated with an

Table 7–7. **Laboratory and Imaging Findings in Patients with Leptomeningeal Metastases**

Diagnostic Test	Abnormal Finding	Percentage of Patients with Abnormal Finding
Lumbar puncture	Elevated CSF protein	75
	Elevated CSF pressure	50
	Low CSF glucose	40
	Positive CSF cytology on initial lumbar puncture	50
	Positive CSF cytology after three lumbar punctures	85
Neuro-imaging studies	Meningeal enhancement	50
	Enlarged ventricles	<25
	Spinal subarachnoid masses	<25

From Ref. 136 with permission.
CSF, cerebrospinal fluid.

increased cell count and elevated protein that can suggest leptomeningeal tumor.[141]

Our attention to the syndrome of intracranial hypotension was required when two patients with severe headache and spinal fluid pleocytosis were admitted to our hospital for evaluation of leptomeningeal metastases of unknown primary; both had idiopathic intracranial hypotension, but no cancer.[141]

Several findings help identify leptomeningeal metastases by MRI (Table 7–8). Contrast enhancement identifies areas of disrupted blood–CSF barrier, both cerebral (Fig. 7–5) and spinal (Fig. 7–6).[104,143] Common locations are cortical sulci, cerebellar folia, and cauda equina. Lesions may appear simply as linear enhancement or as nodules on nerve roots, particularly in the cauda equina; the former may be difficult to distinguish from normal vascular markings. On occasion, FLAIR (fluid-attenuated inversion recovery) images will identify hyperintensity in subarachnoid spaces when enhancement is minimal or absent (Fig. 7–5). Clumping of nerve roots of the cauda equina also suggests the diagnosis. These changes are not pathognomonic because similar images can be caused by acute or chronic inflammatory lesions of the leptomeninges. When the clinical suspicion of leptomeningeal metastasis is strong, however, an MRI showing contrast enhancement in the basal cisterns, cauda equina, or ependyma provides sufficient confirmation for treatment even in the absence of identifiable malignant cells in

Table 7–8. **Magnetic Resonance Imaging Findings with Leptomeningeal Metastases**

Definite
Intradural enhancing nodules in spinal canal
Enhancement and enlargement of cranial nerves
Superficial linear sulcal, cisternal, or dural enhancement
Irregular tentorial enhancement
Irregular ependymal enhancement
Cisternal or sulcal obliteration
Subarachnoid-enhancing nodules
Intraventricular-enhancing nodules

Suggestive
Hydrocephalus
Multiple small superficial brain nodules
Spinal linear enhancement
Spinal cord enlargement
Asymmetry of the roots

From Ref. 75 with permission.

the CSF. Definitive findings for leptomeningeal metastasis, such as nodules on the cauda equina or enhancement of CSF in a patient with known cancer, is sufficient to establish the diagnosis and CSF analysis is not essential unless the patient has symptoms of increased intracranial pressure and measurement of the pressure is necessary. Frequent false-negative imaging in cytologically

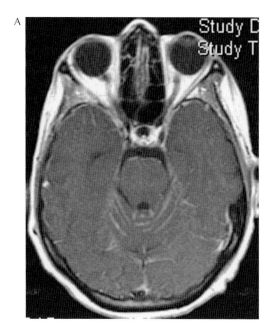

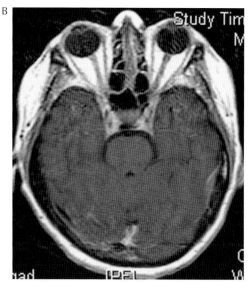

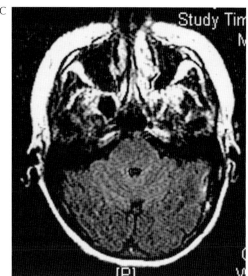

Figure 7–5. Leptomeningeal metastases from breast cancer. **A**: Enhancing tumor can be seen within the cerebellar folia and along the ventral surface of the pons. **B**: Although no tumor can be seen in the cerebellar folia of this patient on unenhanced T1-weighted image, the folia appear indistinct. **C**: A fluid-attenuation inversion recovery (FLAIR) image of the same patient showing hyperintensity in the cerebellar folia; leptomeningeal tumor sometimes can be seen only on FLAIR images.

proved leptomeningeal metastases[144] may be decreased by using higher doses of gadolinium. However, care must be taken when interpreting magnetic resonance scans enhanced with higher doses of contrast material. Mild enhancement of otherwise normal-appearing spinal roots may not be pathologic.

In the absence of contrast enhancement of the basal meninges, communicating hydrocephalus suggesting obstruction of CSF absorptive pathways may suggest leptomeningeal tumor. This change is nonspecific because hydrocephalus can be caused by various lesions, but it may support a diagnosis of leptomeningeal metastasis in a patient strongly suspected of having this disorder.

Enlarged ventricles do not establish the diagnosis of hydrocephalus because cerebral atrophy, sometimes induced by RT or chemotherapy, also enlarges the ventricles. However,

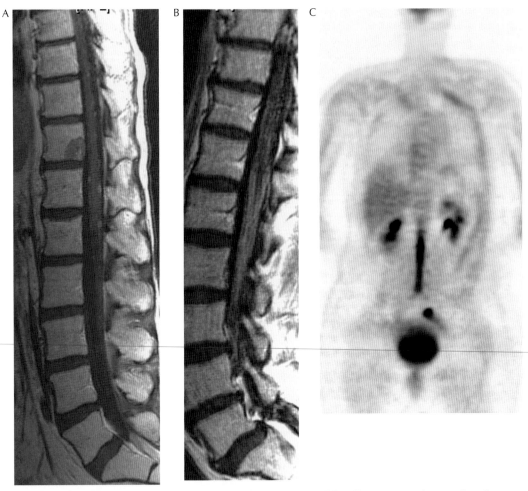

Figure 7–6. A: An enhanced magnetic resonance image demonstrating nodules of leptomeningeal tumor along the nerve roots of the cauda equina. **B**: In another patient, nodules are not identified, but streaks of linear enhancement representing leptomeningeal metastases are prominent. If these had been slightly less evident, they might have been mistaken for normal vasculature. **C**: An FDG/PET in a patient with leptomeningeal lymphoma. Both the leptomeninges and the exiting sacral nerve root are hypermetabolic.

an MRI may help differentiate these entities, particularly if white matter hyperintensity is present at the ventricular frontal horns, suggesting transependymal absorption of CSF because of obstructed ventricles. White matter hyperintensity can also be caused by RT and chemotherapy, but the changes tend to involve the white matter more diffusely, rather than just the periventricular area.

In patients with solid tumors, MRI is highly sensitive and specific.[145] In hematologic malignancies, although specificity is high, sensitivity is less than 50%. In one series the sensitivity of MRI for identifying solid tumors was 100%, with a specificity of 92%.[145] In other series, sensitivity has varied from 76% to 87% with specificity being about the same. This implies that patients with negative imaging who are strongly suspected for leptomeningeal metastasis need a comprehensive CSF analysis (see following text). In addition, if leptomeningeal metastasis is diagnosed on an initial test, either by imaging or CSF analysis, the patient should have complete neuraxis imaging performed to identify areas of bulky disease and to perform an initial assessment of CSF pathway integrity. CT scans are a poor substitute for MRI and should be performed only if MRI is contraindicated.

Lumbar Puncture

Although MRI should be the first test performed to establish a diagnosis of leptomeningeal metastasis, CSF examination is the most definitive diagnostic test.[110,146,147] The test should be deferred if the patient is also suspected of harboring an intracranial or spinal mass lesion(s) because of the possibility of causing herniation (see Chapter 14). In this situation, an MRI should be done first. If a mass is evident, lumbar puncture, if still clinically indicated, can wait for the few days it takes to control intracranial pressure with steroids.

At the time the lumbar puncture is performed, care should be taken to procure enough CSF for a full diagnostic evaluation, as discussed in Chapter 10. This includes a minimum of 4 cc for cytologic examination. (Table 10–5 details the volume of fluid required for other diagnostic tests.) Routine examination of the CSF in patients suspected of harboring leptomeningeal metastasis should also include careful measurement of pressure, counting of red blood cells (RBCs) and WBCs, and analysis of protein and glucose concentration. Culture for organisms is unnecessary unless one suspects infection. A full diagnostic evaluation is important because the classical findings of high opening pressure, low glucose concentration, high protein concentration, lymphocytic pleocytosis, and malignant cells on cytologic examination are present in only a minority of patients. It is rare for the spinal fluid to be entirely normal in patients with solid tumors, but normal CSF is seen with some reliability in patients with leukemia[85] (Table 7–9).

CEREBROSPINAL FLUID PRESSURE

Because leptomeningeal metastases can obstruct CSF absorptive pathways, CSF pressure is elevated in at least 50% of patients. Pressure elevations frequently occur without hydrocephalus on imaging. However, the physician should be careful to exclude other causes of increased intracranial pressure common in patients with cancer. For instance, because of its effect on cerebral blood flow and volume, elevated PCO_2 levels substantially raise intracranial pressure. An increase in PCO_2 of approximately 4 mm Hg doubles the intracranial pressure. Thus, patients with respiratory failure from lung and other cancers may have increased intracranial pressure. Also, intracranial pressure reflects cerebral venous pressure which in turn reflects systemic venous pressure. Compression of the superior vena cava, congestive heart failure, and jugular vein compression raise intracranial pressure. These phenomena must be excluded for the physician to attribute an elevation of intracranial pressure, as measured by lumbar puncture, to intracranial disease. Accurate pressure measurement requires that the patient's head be at the same level as the middle of the back where the needle is inserted. One of our patients was reported to have an elevated pressure of 440 mm of CSF despite a normal neurologic examination, normal imaging, and other CSF values being normal. The neurology consultant was puzzled, until the intern reported that he had done the

Table 7–9. **Leptomeningeal Metastases from Solid Tumors in 90 Patients: Cerebrospinal Fluid Findings**

	Initial	Percentage	Subsequent	Percentage
Pressure > 160 mm CSF	45	50	64	71
Cells > 5/mm³	51	57	65	72
Protein > 50 mg/dL	73	81	80	89
Glucose < 40 mg/dL	28	31	37	41
Positive cytology	49	54	82	91
Normal	3	3	1	1

From Ref. 85 with permission.
CSF, cerebrospinal fluid.

puncture and measured the pressure in the sitting position.

The range of normal intracranial pressure, measured by lumbar puncture in the lateral recumbent position with the patient relaxed, varies between 90 and 250 mm H_2O of CSF.[148] In most patients, however, particularly the middle-aged or debilitated, the pressure is less than 160 mm CSF; higher levels should be viewed with suspicion. Levels greater than 180 mm CSF should enhance the clinician's suspicion that intracranial pressure is abnormal. The pressure should be measured as quickly as possible after the patient is relaxed; if one waits for several minutes, leakage of spinal fluid around the needle can give a falsely low reading of a truly elevated intracranial pressure.[149]

The only accurate indication of increased intracranial pressure is direct measurement. Some investigators report that the identification of venous pulsations on ocular funduscopic examination indicates normal intracranial pressure but, in our experience, that sign is not reliable.[150] The presence of venous pulsations usually is well correlated with normal intracranial pressure, but their absence does not correlate well with elevated pressure. Venous pulsations persist in a few patients with substantially increased intracranial pressure. Direct measurement of CSF pressure is only accurate when measured by lumbar puncture. In patients with an Ommaya reservoir, assessment of increased intracranial pressure via a tap of the reservoir usually underestimates the pressure. This is usually a consequence of the technique necessary to access the CSF through a reservoir with a small gauge butterfly needle attached to a long, thin tube before the CSF reaches the manometer. We have seen many patients with symptoms of increased intracranial pressure who had a normal pressure measured from a reservoir, but had elevated intracranial pressure on a lumbar puncture.

CELL COUNT

The WBC count is increased in more than one-half of patients with leptomeningeal metastases. The WBCs are usually lymphocytes, although occasional polymorphonuclear leukocytes are found. The cell count may vary from just a few cells to many hundreds, raising the suspicion of an infectious meningitis. Eosinophils in the CSF, usually suggestive of a parasitic infection, have been reported in a number of patients with leptomeningeal metastasis from Hodgkin disease,[151] lymphoma[152] and, in one case, an epithelial tumor.[153] The eosinophils may occur in the absence of malignant cells in the CSF and can be a strong clue to the presence of lymphomatous infiltration of the meninges in clinically appropriate circumstances. A patient with both eosinophilic and basophilic meningitis associated with leptomeningeal leukemia has been reported.[154] A note of caution is required: CSF eosinophilia can complicate therapy with ibuprofen,[155] a drug frequently administered for pain management to patients with cancer.

RBCs associated with xanthochromia result from bleeding of leptomeningeal tumor. Bleeding can occur with any tumor invading the leptomeninges but frequently complicates leptomeningeal melanoma. The yellow color of the CSF is due to bilirubin and should not be mistaken for melanin, which, when present in the CSF, is black. Frank subarachnoid hemorrhage is rare.[156]

PROTEIN CONCENTRATION

The CSF protein concentration is frequently elevated, probably a combined result of breakdown of the blood–CSF barrier with passage of serum proteins into the CSF and of tumor and WBC breakdown directly related to leptomeningeal infiltration. CSF protein concentrations of greater than 2 g/dL occur occasionally in patients with leptomeningeal metastasis and are at times sufficiently high for the fluid to actually clot.

The physician should assess CSF protein levels from Ommaya reservoirs with caution. The normal ventricular protein level of approximately 10 mg/dL is much lower than that in the cistern or the lumbar sac. A level of 30 or 40 mg/dL is thus a substantial elevation in a ventricular sample. In addition, a dead space is present in the reservoir. Protein appears to accumulate in this space if the reservoir is not tapped frequently. One of our patients, a 17-year-old boy with leukemia, underwent a puncture of an Ommaya reservoir that had not been disturbed for a year. The fluid in the 2.5 mL dead space of the reservoir was not discarded but was instead assayed and found to have a protein concentration of 366 mg/dL. A second procedure performed the following day revealed a normal ventricular protein concentration of 6 mg/dL.

In addition to the generalized increase in protein, increases may occur in specific proteins. Elevated CSF IgM in patients with leptomeningeal myeloma[157] in the absence of an elevated albumin is diagnostic because IgM does not cross a relatively intact blood–CSF barrier and thus must be produced by cells within the subarachnoid space. Schipper et al.[93] have reported that 8 of 22 patients with leptomeningeal metastases from various tumors had elevated IgG in the CSF with an elevated IgG index or oligoclonal bands suggesting local production. Fractionation of CSF protein by protein electrophoresis or immunoelectrophoresis often gives additional clues in the analysis of patients with suspected leptomeningeal tumor. Elevated myelin basic protein concentration in the CSF is present in many patients with leptomeningeal metastases,[158] usually clearing after successful treatment, but it may follow treatment of meningeal leukemia, where it is observed in patients with treatment-related leukoencephalopathy.[159]

GLUCOSE CONCENTRATION

Hypoglycorrhachia, although not common, is highly specific. A glucose level less than 40 mg/dL is found in 25% to 30% of patients with leptomeningeal metastases.[85,160] In one patient, hypoglycorrhachia was an isolated finding of leptomeningeal metastases.[161] In the absence of a clear-cut infection, hypoglycorrhachia strongly indicates leptomeningeal tumor. Some caveats, however, are necessary. A persistently elevated or depressed serum glucose level can alter the CSF glucose level, which is generally 50% to 60% of the serum level. Simultaneous measurement of blood and CSF glucose is not useful if the patient's serum glucose levels are changing rapidly, because it takes more than an hour for the CSF level to equilibrate with the serum concentration. We use an absolute level of less than 40 mg/dL as indicative of hypoglycorrhachia, recognizing that some higher levels may indicate a depressed CSF glucose in diabetic patients. Under normal circumstances, the CSF glucose concentration, unlike the protein concentration, does not differ, depending upon the level of the neuraxis from which the CSF was sampled.

The cause of hypoglycorrhachia is unclear. Two possibilities exist: (1) diminished carrier-mediated transport of glucose across the blood–CSF or blood–brain barrier,[162,163] and (2) glucose metabolism by malignant cells. Support for the former hypothesis comes from the study of a patient in whom IV infusion of glucose did not increase the CSF glucose concentration, nor did it increase the high basal CSF lactate concentration (see following text), suggesting an impaired transport mechanism.[163] The latter hypothesis is supported by the close inverse correlation between the levels of glucose and lactate,[164] suggesting that the glucose is being metabolized intrathecally. Glucose is metabolized not only by malignant cells, which favor anaerobic glycolysis, but also by the reactive WBCs and proliferating pial cells induced by the malignant infiltration. The degree of hypoglycorrhachia does not correlate with the density of cells found in the CSF, just as the number of malignant cells measured in a CSF sample does not correlate with the number of cells infiltrating the leptomeninges.

CYTOLOGIC EXAMINATION

Cytologic analysis of CSF is the definitive diagnostic test and virtually the *sine qua non* for establishing leptomeningeal metastasis[145,165–167] (Fig. 7–7). A minimum of 4 cc and preferably 10 cc of CSF[165] should be procured and either hand carried to the laboratory or placed in fixative and then sent to the laboratory, where it is centrifuged, fixed on a slide with albumin, and stained.

Table 7–9 indicates the results of CSF examination from 90 patients with leptomeningeal metastases from solid tumors confirmed by either clinical course or microscopic analysis at autopsy. Malignant cells were encountered in the CSF of approximately 50% of the patients on the initial lumbar puncture; the other 50% presumably had "false-negative" test results. However, the CSF was virtually always abnormal in some respect. Sometimes repeated lumbar punctures are helpful; in over 90% of the patients studied, malignant cells were eventually identified. Glantz et al. compiled reports of 532 patients and identified positive cytologies in 71, 86, 90, and 98%, respectively after 1, 2, 3, or more samples per patient.[165] Despite these figures, some patients have persistently negative cytologic examinations despite clear evidence of leptomeningeal metastases on imaging[104] (see following text).

A

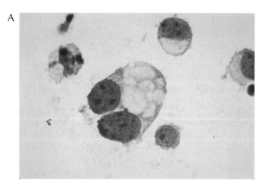

B

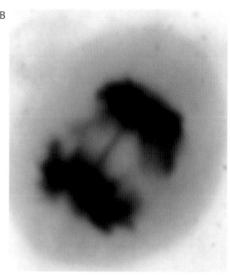

Figure 7–7. Malignant cells in the cerebrospinal fluid. **A**: Cytologic preparation showing multiple malignant cells. **B**: A single malignant cell in mitosis.

These false-negative cytologic tests are more likely when the leptomeninges are infiltrated sparsely with tumor, but sometimes a false-negative cytology occurs even with extensive infiltration of the leptomeninges of the cauda equina.[146] Apparently, in these instances, the tumor adheres tightly and does not exfoliate easily into the CSF. Our data indicate that a false-negative cytology is encountered more commonly with carcinomas than with acute lymphoblastic or other leukemias, although Kaplan et al.[110] found no difference. Hodgkin disease infrequently metastasizes to the subarachnoid space, and Reed–Sternberg cells, the diagnostic malignant cell in Hodgkin disease, are rarely found in CSF even when the meninges are heavily infiltrated.

Striking differences may be evident in both biochemistry and cytology between the lumbar and ventricular CSF.[168] In one series of 60 patients in whom both lumbar and ventricular CSF were examined, findings were discordant in 19 (32%).[160] The lumbar fluid was more likely to be positive and ventricular fluid negative if the patient had mainly spinal signs, whereas the findings were opposite if the patient had mainly cerebral signs. Thus, one can decrease the number of false-negative findings by sampling fluid near the site of major symptomatology, for example, lumbar fluid for spinal signs, cisternal fluid[169,170] for cranial nerve signs, and ventricular fluid for brain signs.[165] Cisternal

fluid can be sampled by a lateral cervical puncture at C2 under fluoroscopic visualization (Table 7–10).

In addition to cytology, the chemical composition of the CSF can vary at different levels. Protein concentration, normal in the ventricle, may be grossly elevated in the lumbar space, and hypoglycorrhachia may be present in the lumbar space and absent in the ventricle. Thus, both for the initial diagnosis and for monitoring the effects of therapy, lumbar CSF samples are vital.[169,171]

A false-positive cytology (cells that are apparently malignant in patients who do not have leptomeningeal tumor) occurs in three settings other than technical errors where slides are mislabeled:

1. The literature reports a 20% to 40% incidence of positive cytology in patients with metastasis in the brain but without leptomeningeal tumor.[172] This has not been our experience. When positive CSF cytology was correlated with pathologic examination of the nervous system at autopsy, every patient with a positive cytology had at least one detectable focus of leptomeningeal tumor.[146] At times, however, a patient with a brain metastasis that focally infiltrates the leptomeninges may develop a positive cytology. In our experience at MSKCC, the incidence of such focal infiltration–yielding malignant

Table 7–10. **Effect of Site of Cerebrospinal Fluid Sampling**

	Location of Metastases			
	Cranial (24 Paired Samples)	Spinal (31 Paired Samples)	Both (4 Paired Samples)	Total (59 Paired Samples)
Lumbar Cytology				
Positive	14	26	4	44
Negative	10	5	0	15
Ventricular Cytology				
Positive	19	20	4	43
Negative	5	11	0	16

From Ref. 168 with permission.

cells on CSF examination is very low but must be considered. It might be wise to treat such patients because it is possible that, if untreated, diffuse leptomeningeal seeding will develop.

2. Malignant cells are sometimes found in the CSF of asymptomatic patients, especially with breast or SCLC. These are not truly false-positive results, but represent early development of leptomeningeal tumor and are analogous to identifying malignant cells discovered on a routine lumbar puncture in patients with leukemia and lymphoma; in one series, 10% of lymphoma patients who were asymptomatic[71] had malignant cells in the CSF. These patients require CNS treatment.

3. Sometimes reactive lymphocytes are misinterpreted as malignant, or vice versa, particularly in patients with lymphoproliferative disorders who suffer CNS viral infections (e.g., Herpes zoster).[71,146,173] In general, these errors can be avoided if the patient's physician reviews the cytologic slides with the pathologist, giving him or her the benefit of the full history and clinical situation. Furthermore, reactive lymphocytes in CSF are almost always T-cells and most lymphomas are clonal B-cell tumors. The presence of a large number of B-cells, particularly if they are clonal, is diagnostic of meningeal lymphoma.

IMMUNOCYTOCHEMISTRY

Pathologic examination of the CSF using monoclonal antibodies has been reported to enhance routine cytologic examination.[145,174,175] In one series, combined use of routine cytology and immunocytochemistry led to a 9% increase in sensitivity for detecting malignant cells. Thus, the technique is useful only in a limited number of patients. Monoclonal antibodies against surface markers on lymphocytes can be useful to diagnose CNS lymphoma. They are particularly helpful if cytologic evaluation fails to distinguish between reactive and malignant lymphocytes. Demonstration by surface markers of a monoclonal population of lymphocytes establishes the diagnosis of tumor. Unfortunately, polyclonality does not exclude tumor because a combination of malignant lymphocytes and lymphocytes reactive to the malignancy may lead to polyclonal proliferation.[176–178] Immunocytochemical analysis has proved extremely sensitive in hematologic malignancies, ranging from 90 to 95%, and almost 100% specific even when MRI (see following text) was negative.[145] In solid tumors, immunocytochemical analysis was much less sensitive, with 46% sensitivity and 79% specificity.

FLOW CYTOMETRY

Flow cytometry is a sensitive method for detecting malignant cells in the CSF of patients with hematologic cancers. It is widely available and has proven valuable in the early detection of leptomeningeal infiltration even in asymptomatic patients. It doubles the detection rate over cytopathology alone. Patients with malignancy detectable in the CSF by flow cytometry alone usually have lower CSF cell counts, protein concentration, and negative imaging, suggesting the disease is at an earlier stage. This is confirmed by their better prognosis compared to patients with a positive CSF cytology.

Flow cytometry occasionally detects abnormal cells when the results of routine cytologic analysis are negative and when biochemical markers are absent.[179,180] The presence of aneuploid cells and, in particular, hyperdiploid cells in CSF strongly suggests leptomeningeal tumor. In addition, CSF cells can be examined by flow cytometry for the presence of carcinoembryonic antigen (CEA) on cell surfaces, which occasionally proves to be diagnostically useful. Flow cytometry is particularly useful in the diagnosis of hematologic malignancies.[181]

BIOCHEMICAL/MOLECULAR MARKERS

The markers listed in Table 7–11 have assumed more importance in the diagnosis of systemic cancer in recent years and have been analyzed in the CSF as a means of identifying leptomeningeal metastases. CEA, alpha fetoprotein (AFP), beta human chorionic gonadotropin (βhCG), and other biochemical markers have proved useful not only in the diagnosis of

Table 7–11. Biochemical/Molecular Markers in Cerebrospinal Fluid

Marker	Associated Tumor	Positive
Routine Biochemical Markers		
CEA	GI, lung	>1% of serum
AFP	Germ cell	>1% of serum
βHCG	Germ cell	>1% of serum
Melanin	Melanoma	Presence in the CSF
CA 125	Ovary	>1% of serum
CA15-3	Breast	>1% of serum
5-HIAA	Carcinoid	High elevation/usual CSF
PSA	Prostate	>1% of serum
Beta-2 microglobulin	Lymphoma, infection	
CA19-9	Adenocarcinoma, biliary disease	>1% of serum
Biomolecular Techniques		
Flow cytometry		Aneuploid cells
		Hyperdiploid cells
DNA single cell cytometry	Hematopoietic, GI, lung tumors	Presence of CEA on cell surface
FISH (interphase cytogenetics)	Solid tumors	Numerical or structural aberrations in interphase nucleus
RT-PCR markers MAGE, MART-1, tyrosinase	Melanoma	
Immunocytochemical Analysis		
Protein S-100	Melanoma	
HMB45	Melanoma	
TTF 1	If primary cancer unknown: suggests lung or thyroid carcinoma	
VEGF index	Elevated in LM	
tPA index	Depressed in LM	

Modified from Ref. 75 with permission.

malignant disease but also in following the course of treatment.

Biochemical markers can be divided into specific and nonspecific categories. When found in CSF, specific markers, such as those listed in the Table 7–11, are virtually diagnostic of leptomeningeal metastasis from a specific tumor except when a very high serum concentration of the marker has led to seepage into the CSF across a normal blood–brain barrier or a blood–brain barrier partially disrupted by a metastatic brain tumor. To distinguish between these possibilities, the serum and CSF concentrations should be compared. For practical purposes, a rapid calculation of the expected CSF–serum ratio (1% with CEA) will determine whether the CSF is disproportionately elevated. Metastases in the brain do not seem to have a substantial effect on the CSF concentration of these markers. Gross elevation of substances normally found at low levels in the CSF, such as IgM in patients with multiple myeloma or 5-hydroxyindoleacetic acid (5-HIAA) in patients with carcinoid,[60] are also diagnostic.

A recent report describes the combination of an elevated VEGF index and a depressed tPA index as being 100% sensitive for leptomeningeal tumor and 73% specific.[182] We have no personal experience with this measurement.

Nonspecific markers, such as β-glucuronidase,[183] isozyme V of lactic acid dehydrogenase,[184] β2-microglobulin,[185] and myelin basic protein,[158,159] although elevated in many patients with leptomeningeal tumors (the β2-microglobulin particularly in leukemias and lymphomas), may also be elevated in patients with inflammatory disorders of the nervous system.[186,187] If a physician can confidently exclude inflammatory meningitis, striking elevations of these biochemical markers indicate the presence of leptomeningeal tumor. In general, all markers are lower in concentration in ventricular than in lumbar fluid.[85]

Other Diagnostic Tests

It is a rare patient in whom a diagnosis of leptomeningeal metastasis cannot be made by MRI or CSF examination. The tests listed herein are rarely used but are described for those instances in which they may be of some use.

MYELOGRAPHY

Myelography can identify leptomeningeal tumor when an MRI cannot be done.[188] In 49 consecutive patients[85] with leptomeningeal metastasis, the myelogram was diagnostic in 13. A myelographic finding of small nodules along nerve roots suggests leptomeningeal metastasis. Although this finding can be present in patients with primary meningeal tumors such as neurofibromas and the radiation-induced lower motor neuron syndrome (see Chapter 13), multiple nodules along roots, particularly in a patient with systemic cancer, strongly suggest leptomeningeal metastasis.

CEREBRAL ARTERIOGRAPHY

Cerebral arteriograms can help support a diagnosis of leptomeningeal metastasis.[85–87] Multiple constrictions of meningeal blood vessels suggest invasion or compression by leptomeningeal disease. These findings are also encountered in some patients with tuberculosis and other granulomatous infections of the nervous system, including vasculitis.

A report of CT angiography described five patients whose scan showed "blurred margin of blood vessels in the interhemispheric fissure, an increase in the number of vessels and a pathologic configuration of the vessels in the interhemispheric fissure." All five patients were subsequently discovered to have leptomeningeal metastases.[189]

ELECTROENCEPHALOGRAPHY

The EEG is nonspecific. Results may be normal or show focal or generalized slow or sharp wave activity. Triphasic waves, generally believed to indicate metabolic encephalopathy, have been reported in leptomeningeal tumor.[122]

ELECTROMYOGRAPHY

Electromyography assists in diagnosing leptomeningeal metastasis in some instances. Prolonged F-wave latencies (i.e., waves that mark conduction of the most proximal portions of the motor roots[190,191]) have been reported of value in diagnosing leptomeningeal tumor in a number of patients. We have rarely found the test necessary.

MENINGEAL BIOPSY

A biopsy of the leptomeninges is occasionally diagnostic when all other tests fail,[192] although we have encountered at least one negative biopsy in a patient who was later found to have extensive leptomeningeal tumor on autopsy examination. In one study of 37 patients with chronic meningitis, meningeal biopsy established the diagnosis in only 16 of 41 biopsies (39%), but in those 15 patients with enhancement on MRI, a final diagnosis was made in 12 (80%). Leptomeningeal metastasis was the final diagnosis in seven patients; one required two biopsies to establish that diagnosis.[192]

An intriguing report of direct visualization of spinal roots through a lumbar puncture needle by fiberoptic endoscopy suggests the possibility of direct visualization to localize the biopsy site of leptomeningeal tumor[193] in those rare patients in whom a diagnosis cannot be made by other means.

POSITRON EMISSION TOMOGRAPHY

Positron emission tomography (PET) with 2-fluorodeoxy-D-glucose (FDG) or C11-methionine has been reported to identify increased uptake in the leptomeninges of patients with cancer, strongly suggestive of leptomeningeal metastases.[194,195] We have encountered occasional such patients, usually in the course of doing body PET scanning to identify metastatic lesions (Fig. 7–6C). We have also encountered several patients with unequivocal leptomeningeal metastases in whom the scans were negative.

CEREBROSPINAL FLUID FLOW STUDIES

If one chooses to treat leptomeningeal metastases with intrathecal chemotherapy (see following text), a radionuclide flow study will establish the distribution of the chemotherapeutic agent(s) that is to be injected.[78,79] CSF flow is abnormal in a substantial percentage of patients with leptomeningeal tumor. The isotope can fail to leave the ventricular system because of obstruction to flow at the outlet foramina of the fourth ventricle, or it can be obstructed anywhere along the spinal canal or within the subarachnoid cisterns or subarachnoid space over the cerebral convexities.[79] When flow is obstructed, intrathecal drugs will not be distributed evenly. If injected into the ventricle when the outlet foramina of the fourth ventricle is obstructed, the agent may diffuse into the brain, causing severe neurotoxicity. Sometimes focal treatment, such as radiation of the site of the block, reestablishes flow.[78,79]

DIFFERENTIAL DIAGNOSIS

Some of the illnesses that may mimic leptomeningeal metastasis are listed in Table 7–12. An algorithm for differentiating among the various causes of *chronic meningitis* has been published.[196]

Table 7–12. **Differential Diagnosis of Leptomeningeal Metastases**

Primary Tumor
  Meningeal seeding from primary brain tumor
    Medulloblastoma
    Glioma
    Germ cell tumor
    Lymphoma
  Primary leptomeningeal tumors
    Melanoma
    Lymphoma
    Sarcoma

Opportunistic Infections (see Chapter 10)
  Infectious meningitis

Autoimmune Disorders
  Vasculitis
  Sarcoidosis
  Wegener granulomatosis
  Langerhan cell histiocytosis
  Bell palsy

Others
  Guillain-Barré
  Radiotherapy-induced lower motor neuron syndrome (see Chapter 13)
  Intracranial hypotension
  Post–lumbar puncture
  Post–spinal surgery
  Idiopathic
  Enhancing meningeal blood vessels

The clinical suspicion of leptomeningeal metastasis rests on finding either multifocal neurologic symptoms or signs, signs suggestive of meningeal irritation, or both. The first finding can be mimicked by multiple brain or epidural metastases and the second by CNS infection. Differentiation of leptomeningeal metastases from multiple parenchymal and/or epidural metastases requires imaging of the brain and spine to look for mass lesions and CSF analysis to search for malignant cells. If malignant cells are present, the patient has leptomeningeal tumor. Differentiating leptomeningeal metastasis from CNS infection, particularly in patients with lymphomas who are susceptible to both illnesses, is more difficult.[85,197] The diagnosis can often be established only by careful and repeated examination of the nervous system.

In general, patients with leptomeningeal metastasis develop identifiable neurologic signs, particularly cranial or spinal nerve dysfunction, early, when meningeal signs and even changes in the CSF such as pleocytosis and high protein are mild. Patients with CNS infections, on the other hand, tend to develop cranial and spinal nerve abnormalities late in the course, if at all. Thus, signs of meningeal irritation accompanied by fever and abnormal CSF with a normal segmental neurologic examination suggest CNS infection, whereas cranial and spinal nerve dysfunction without meningeal signs or fever, and with only modest CSF changes, suggests tumor.

Diagnostic Approach

The clinical diagnosis of leptomeningeal metastasis usually suggests itself in one of four settings:

1. As indicated previously, when a patient presents with symptoms or signs of neurologic dysfunction at several levels of the neuraxis.
2. When a patient has neurologic disease that may initially suggest a single CNS lesion, but imaging of the area in question does not reveal a metastatic mass. Microscopic tumor involving the leptomeninges and the brain or spinal cord must then be considered.
3. When a patient presents with headache, nuchal rigidity, and, sometimes, radicular pain suggesting an inflammatory meningitis. Such a patient is usually afebrile, and a diagnosis of leptomeningeal involvement by tumor must be considered as well.
4. When an image of the brain or spine, taken to rule out a mass lesion, shows evidence of either leptomeningeal enhancement or obstruction of CSF pathways (e.g., hydrocephalus). As indicated in the preceding text, care must be taken to differentiate the linear enhancement along the surface of the spinal cord and the cauda equina from normal blood vessels. Furthermore, pathologic enhancement of the roots of the cauda equina can occur with inflammatory diseases such as infection or the Guillain–Barré syndrome.

Whatever the setting, the first step (Fig. 7–8) is a gadolinium-enhanced MRI of the symptomatic site. If an alternative diagnosis is excluded or leptomeningeal metastasis is identified on the initial MRI, the entire neuraxis should be scanned promptly. If the scan does not establish the diagnosis of leptomeningeal tumor, lumbar puncture is the next diagnostic test. If malignant cells are found in the CSF, the diagnosis of leptomeningeal metastasis is established and treatment should be started. Always send CSF for the tumor markers relevant to the patient's primary tumor, if known (Table 7–11). If the results are negative, the lumbar puncture should be repeated up to three times within the next several days to look for malignant cells.

Other laboratory tests that mandate treatment in the absence of malignant cells, but in the presence of appropriate clinical findings include (1) enhancing nodules on cranial nerve on MRI, if one can be certain that the patient does not suffer from multiple neurofibromas, (2) diffuse enhancement of the leptomeninges on MRI, if infection can be excluded, and (3) tumor-specific CSF markers (Table 7–11) higher in CSF than in serum, or much higher than can be attributed to diffusion across the blood–brain barrier.

Hydrocephalus alone on CT or MRI does not establish the diagnosis of leptomeningeal metastasis, and that finding does not mandate specific treatment; meningeal tumor is an uncommon cause of normal-pressure hydrocephalus.

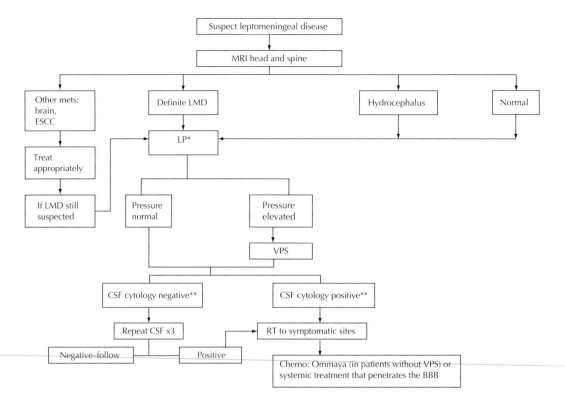

* Must measure pressure (see Figure 1–5)
** Malignant cells not required to initiate therapy in patients with
definitive imaging findings

Figure 7–8. Approach to the patient with meningeal symptoms. LP, lumbar puncture; RT, radiation therapy; ESCC, epidural spinal cord compression; VSP, ventriculoperitoneal shunt; BBB, blood-brain barrier.

In a few patients, even after a thorough workup, the diagnosis remains uncertain; malignant cells are not identified, images are not diagnostic, or clinical symptoms could be due to nonmetastatic disease. In those instances, it is wise to wait at least 2 weeks, repeat the lumbar puncture, consider a lateral cervical puncture to sample cisternal fluid, and repeat the scans using an increased dose of contrast. Virtually all of the problems will clarify themselves within a few weeks.

Occasionally a lumbar puncture or scan performed for other reasons suggests the presence of leptomeningeal metastasis. Even in the absence of symptoms, if the diagnosis can be established by laboratory tests, treatment should be undertaken. We usually defer RT for asymptomatic patients and treat with intrathecal drugs alone.

In rare instances, a leptomeningeal biopsy may be necessary to establish the diagnosis, particularly in patients who are not known to have an underlying neoplasm and whose clinical symptoms or laboratory results arouse suspicion of leptomeningeal tumor. Because the disorder may be multifocal rather than diffuse, the biopsy sample should be chosen on the basis of clinical symptoms, sites of enhancement on MRI,[192] and the knowledge of the loci of leptomeningeal spread (Fig. 7–1), which is best detected at the cisterns at the base of the brain and the leptomeninges of the cauda equina.

TREATMENT

The major therapeutic modalities (Table 7–13) are RT and chemotherapy.[5,6,14,15,75] Surgery has a minor role, as indicated in the following text. The major problem in treating leptomeningeal metastasis is that the entire neuraxis must be treated. If only symptomatic areas are treated, that area will rapidly be re-seeded from tumor elsewhere in the meninges.

Table 7–13. Treatment of Leptomeningeal Metastases

Radiation
   Whole neuraxis
   Symptomatic sites

Chemotherapy/immunotherapy
   Intrathecal
      Lumbar
      Ventricular
   Systemic

Surgery
   Biopsy for diagnosis
   Insertion of ventricular cannula and reservoir for therapy
   Insertion of ventriculoperitoneal shunt for increased intracranial pressure

Radiation Therapy

External-beam RT can be delivered either to the entire neuraxis or to symptomatic sites (Table 7–14), leaving the remainder of the neuraxis to be treated by chemotherapy.[16,198] Intrathecal radiation with radioactive nuclides such as gold[199–201] or radiolabeled monoclonal antibodies[202–204] is another method of delivering RT, but it is subject to the same limitations as intrathecal chemotherapy (see following text). RT is the most effective method of relieving symptoms.[205] Ideally, the entire neuraxis should be radiated, but delivering the necessary dose to the entire neuraxis in patients suffering from carcinoma often causes significant morbidity (acute GI toxicity, mucositis) and suppresses the bone marrow, compromising completion of the RT and subsequent chemotherapy for the systemic tumor.[205]

Our policy has been to irradiate symptomatic areas, using chemotherapy to treat the rest of the neuraxis.[5,75,85]

Asymptomatic areas that show leptomeningeal mass lesions unlikely to be penetrated by intrathecal drugs should also be irradiated, unless one plans to use systemic rather than intrathecal chemotherapy. Because RT and chemotherapy with methotrexate (MTX) appear to be synergistic in causing CNS toxicity (see Chapter 12). We prefer not to irradiate the brain unless symptoms or signs require it. In those patients with cranial nerve palsies as

Table 7–14. Indications for Radiotherapy

Focal
Symptoms (even in the absence of abnormal imaging)
Bulky disease
Cerebrospinal fluid flow block
Encasement of spinal roots with tumor
Diffuse involvement of cranial meninges
Radiographic involvement of cranial nerves and nerve roots
Cauda equina syndrome
Cranial nerve dysfunction referable to tumor encasement
Central nervous system prophylaxis with whole-brain irradiation for leukemia and lymphoblastic lymphomas (rarely used)

Radicular pain from involvement of spinal roots
Refractory nausea and vomiting caused by tumor deposits in 4th ventricle

Craniospinal
Neoplastic meningitis refractory to chemotherapy in patients with good performance status
Relapsed central nervous system leukemia
Primary brain tumors
   Recurrent intracranial germinoma
   Medulloblastoma
   Disseminated ependymomas
Germ cell tumors
   Embryonal carcinoma
   Choriocarcinoma
   Yolk sac tumors
Primitive neuroectodermal tumors
   Pineoblastoma
   Ependymoblastoma
   Medulloepithelioma
Choroid plexus carcinoma
Malignant rhabdoid tumors

Modified from Ref. 205 with permission.

the major symptom, RT can be delivered to the base of the skull, sparing most of the brain from RT side effects. However, one should choose a fractionation scheme that uses smaller fractions, for example, 2.5 Gy in 10 fractions to the base of the skull.[205] This will allow the brain to be irradiated subsequently, if it becomes necessary. In patients presenting with weak lower extremities or bladder and bowel dysfunction, the lumbosacral spine (i.e., the cauda equina) is irradiated. Patients with hydrocephalus or focal seizures receive whole-brain RT, usually 30 Gy in 10 fractions, provided intracranial pressure is not elevated. Whenever possible, one should use conformal radiation, both to preserve bone marrow and to decrease CNS toxicity.[205]

As Table 7–14 indicates, craniospinal irradiation is generally reserved for patients with primary brain tumors that seed the leptomeninges. Exceptions include leptomeningeal tumors known to be refractory to chemotherapy in those patients who have a good performance status and, rarely, leukemias or lymphomas.

Chemotherapy

ROUTE OF ADMINISTRATION

Chemotherapy can be given either systemically or directly into the CSF.[14,75,206] Systemic chemotherapy is sometimes helpful when the blood–brain barrier has been extensively disrupted by the tumor, when higher doses of an agent (e.g., high-dose MTX) yield therapeutic CSF levels,[207] or if the tumor is sensitive to a lipid-soluble chemotherapeutic agent.[208] For theoretical reasons, it would appear that chemotherapeutic agents injected into the CNS are superior. By definition, all leptomeningeal tumor is in contact with CSF and, thus, injection into CSF should achieve high concentrations of the chemotherapeutic agent at the tumor site without excessively exposing the bone marrow or the rest of the body to toxicity. In addition, because most leptomeningeal metastases are only a few cells thick and do not form mass lesions, all of the cells in contact with the subarachnoid space should be bathed in the chemotherapeutic drug. Other clinicians have objected to this theoretic postulate on the grounds that tumor cells are often buried in Virchow–Robin spaces deep in the sulci, where circulating chemotherapeutic agents might not reach. Studies to prove or refute this point are unavailable. One study concerning the distribution of radioactive MTX and cytosine arabinoside (ara-C) after intrathecal injection suggests rapid and extensive penetration into spinal cord parenchyma.[209]

The distribution of radiolabeled monoclonal antibody was also revealed to be widespread in a postmortem study done on one patient.[202]

A randomized trial of 35 patients with leptomeningeal metastases from breast cancer compared intrathecal chemotherapy with systemic chemotherapy.[210] The authors found no difference in survival or neurological response but did find an increase in toxicity with the intrathecal treatment. Despite these findings, the issue of the best route of administration is still unsettled, in part because individual clinical situations often limit treatment options, and in part because treatment by either route has limited effectiveness.

Systemic chemotherapy is being used with increasing frequency. It can penetrate into areas of bulky disease, which intrathecal drugs are not always able to do. It can be used in patients with poor CSF circulation without undue toxicity, and even drugs that do not normally penetrate the blood–brain barrier have been reported to be effective in some patients. There are some data to suggest that administration of systemic chemotherapy prolongs survival in patients with leptomeningeal metastasis. It would appear that patients with a positive CSF cytology, but minimal or no abnormalities on MRI, may be best served by intrathecal drug administration, whereas those with bulky disease on imaging and poor CSF flow require systemic treatment.

If the clinician chooses to inject chemotherapeutic agents into the CSF, a route must be determined. The agents can be injected directly into the lumbar CSF or into the lateral ventricle of the brain by the use of a ventricular cannula with a subcutaneous reservoir (Ommaya device) placed in the scalp (Fig. 7–9A). The Ommaya reservoir is the preferred approach.

Chemotherapeutic agents injected by lumbar puncture can be injected inadvertently into epidural or subdural spaces. In one series, this occurred 10% of the time[211] even though the lumbar puncture and intrathecal injection appeared successful to the physician. One is more likely to achieve a successful injection on the first lumbar puncture and to fail on subsequent efforts in part because a subdural or epidural CSF collection may have developed as a result of the previous puncture. Such a collection collapses the normal subarachnoid space but yields CSF through the needle, giving the physician a false impression of having the needle in the subarachnoid space. An incorrect injection is particularly likely in a patient who is suffering from post-lumbar puncture

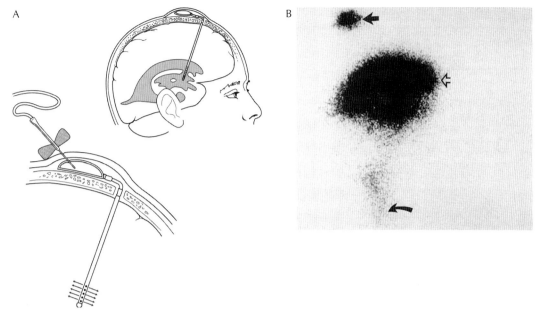

Figure 7–9. A: Ommaya device. A 23-gauge scalp vein needle is used to enter the reservoir implanted over the bone but beneath the scalp (*bottom*), and the drug is injected into the anterior horn of the lateral ventricle. The catheter enters the brain through a burr hole, anterior to the reservoir site, so that the needle cannot penetrate the brain. **B**: Radionuclide ventriculography minutes after injection of ^{131}I albumin into an Ommaya reservoir. Residual radioactivity is present in the reservoir (*closed straight arrow*), and the ventricular system is uniformly filled with radionuclide (*open arrow*). Areas at the base of the brain and down the spinal cord (*closed curved arrow*) are beginning to accumulate the radionuclide.

headache from a prior procedure, indicating a CSF leak. To avoid subdural or epidural injection, some clinicians have placed lumbar catheters attached to subcutaneous reservoirs, but this has no advantage over the intraventricular Ommaya reservoir.[212]

The presence of a subdural collection of fluid, occurring either spontaneously from a tear in the dura or after a lumbar puncture, can be confusing diagnostically as well. In patients suffering from postural headache, an attempt at lumbar puncture may sample only the subdural collection, yielding a fluid that may contain WBCs but not tumor cells and an elevated protein concentration.[141] Such fluid may not truly reflect findings in the subarachnoid space and may mislead the observer into believing that the patient has leptomeningeal infiltration with tumor or infection when, in fact, the CSF is normal.[141]

Even if the chemotherapy is placed correctly into the subarachnoid space via lumbar puncture, the entire neuraxis, particularly the ventricular system, may not achieve a high concentration of the agent. Lumbar injection does not guarantee therapeutic levels in the ventricle[213] even when CSF flow is normal.

Intraventricular injection reliably produces therapeutic drug levels in both ventricle and lumbar sac when CSF flow is patent.

Once an Ommaya reservoir is placed, ventricular fluid can be sampled and the chemotherapeutic agent injected with little or no discomfort to the patient. Repeated lumbar punctures are often uncomfortable and meet with considerable patient resistance. Occasional lumbar puncture is necessary, however, to follow the progress of the disease.

Empiric evidence supports the superiority of ventricular over lumbar therapy.[101,214] An Ommaya device can be used for diagnostic purposes even when the platelet count is too low to perform a lumbar puncture safely (see Chapter 9). However, when platelets are below 50,000/mm³, the sample should not be aspirated, but allowed to drip into the collection tubes. Placing the patient in the Trendelenburg position facilitates this.

CHEMOTHERAPEUTIC AGENTS

Intrathecal Agents

MTX and cytarabine are the most widely used agents; other agents are also listed in Table 7–15.

Table 7–15. **Drugs for Intrathecal (Ventricular) Instillation**

Drug	Dose	Schedule	Reference
Conventional Drugs			
Methotrexate	12 mg	2× a week × 3 weeks, then weekly for 4 weeks, then monthly	14
Thiotepa	10–15 mg	2× a week for 3 weeks, weekly for 4 weeks, then monthly	14
Cytarabine	30–60 mg	2× a week for 3 weeks, weekly for 4 weeks, then monthly	14
DepoCyt	50 mg	Every 2 weeks	215
Experimental Drugs (Human and Animal)			
Topotecan	0.4 mg	2× a week	216
Mafosfamide	20 mg	1–2× a week	217
Diaziquone	1 mg	2× a week	218
Busulfan	13 mg	2× a week ×2	219
Rituximab	10–25 mg	q 2 weeks	220
Trastuzumab	6 mg/kg	q 3 weeks	221
IL-2	12×10^6 IV	Weekly	222
Interferon α	1×10^6 IV	3× a week for 4 weeks	222

The dose of MTX for adults is generally 12 mg twice a week, although Strother et al.[223] have demonstrated that a schedule consisting of an initial dose of 6 mg followed by supplemental doses of 6, 4, or 2 mg at 24 and 48 hours, according to serial measurements of MTX levels in the ventricle, will maintain the therapeutic level for 72 hours. The volume of CSF reaches the adult level by the age of 4 years, so these recommendations apply to most children as well.[224] Bleyer and Poplack[101] have shown that a dose of 12 mg of MTX intrathecally in children older than 2 years of age, given prophylactically, was more effective in preventing CNS leukemia than the previously used schedule of 12 mg/m². When 12 mg is used as a fixed dose, MTX levels reliably exceed and remain greater than therapeutic levels for at least 48 hours.[224] Higher doses are not necessary and lower doses are inadequate. Initially, MTX is administered twice a week, so that therapeutic CSF levels are maintained almost continuously, but after six injections over 3 weeks, the frequency of MTX administration is decreased; eventually, intrathecal MTX may be administered monthly for maintenance. Intrathecal therapy should be continued for at least 3 to 6 months and perhaps indefinitely; the most effective duration of treatment has not been established. Some investigators combine drugs and add hydrocortisone to decrease acute toxicity (aseptic meningitis) of the MTX[225] (see Chapter 12), but data do not demonstrate increased efficacy of chemotherapeutic combinations over single agents. Furthermore, intrathecal hydrocortisone can cause a chemical meningitis, resulting in increased toxicity. MTX is not metabolized by the nervous system and is reabsorbed into the bloodstream by bulk flow and active transport by choroid plexus[225a] of the 4th ventricle.

In patients being treated with intrathecal MTX, oral leucovorin is given twice a day (5–10 mg bid), beginning the day of treatment and for the following 3 days, to prevent systemic toxicity (usually mucositis) from MTX being reabsorbed into the systemic circulation. Leucovorin does not cross the blood–brain barrier in amounts sufficient to interfere with the effect of MTX on the CNS tumor.[226]

A recent report of 10 patients with metastatic leptomeningeal metastases (6 breast) describes continuous infusion of methotrexate into the lateral ventricle. Ten milligrams of methotrexate along with 50 mg of prednisolone was injected over a period of 120 hours biweekly for 2 to 7 cycles. Toxicity was mild, and the responses were observed.[227]

Cytarabine (ara-C) is the only widely used alternative to MTX. The drug is not metabolized by the CNS and is given in a dose of approximately 50 mg twice a week. Some physicians who treat leukemia and lymphoma alternate MTX and ara-C to decrease toxicity. The drug is probably less toxic but is also less effective than MTX in most instances. Cytarabine can also be administered in a liposomal preparation, which releases the drug slowly,[228] so that the drug can be detected in CSF up to 14 days after injection.[229] Thus, the drug can be administered only every 2 weeks, which is more convenient. However, it is associated with a high incidence of chemical meningitis and is also very costly. Consequently, the standard preparation is usually preferred.

Thiotepa is administered at 10 mg twice a week. Its rapid absorption from the CSF (leaving the CSF within an hour) raises questions about its efficacy. However, we have seen patients respond to this agent, especially those with breast cancer. Two monoclonal antibodies, 3F8 and 8H9, attached to radioactive iodine have proved effective in treating some patients with leptomeningeal neuroblastoma and some other tumors expressing the antigens recognized by those antibodies.[204,230]

Systemically Administered Agents

Systemic chemotherapy offers a wider variety of options and the potential to select a drug that may be effective against the patient's underlying primary. MTX and cytarabine were administered intrathecally because they cross the blood–CSF barrier poorly. However, high doses of methotrexate (3–8 g/m^2) or cytarabine (3 g/m^2) allow sufficient entry to produce therapeutic levels within the CSF. Thiotepa crosses the blood–brain barrier easily to produce therapeutic levels within the CSF. Capecitabine crosses the blood–brain barrier well and has been used to treat leptomeningeal tumor from breast cancer successfully.[231] Temozolomide crosses the blood–CSF barrier and has been reported to be effective when combined with other agents that do not cross the blood–CSF barrier.[232] Small molecules such as gefitinib have been reported to be effective. A case report describes a woman with NSCLC and leptomeningeal metastases who had failed therapy with erlotinib in combination with carboplatin and paclitaxel, but had a dramatic clinical and MRI response to gefitinib.[233] The response lasted 8 months, until she died of aspiration pneumonia.

It would seem that where the systemic agent can cross the blood–CSF barrier and particularly if there is bulky disease in the leptomeninges, one should try systemic agents either alone or in combination with intrathecal drugs.

Surgery

The primary purposes of surgery are (1) to treat intractable elevation of intracranial pressure with or without hydrocephalus and (2) to achieve access for intraventricular chemotherapy. A recent report describing navigation-guided reservoir placement indicates that it is a less invasive technique and is at least as good as the standard placement techniques.[234] It may be associated with fewer complications (1.2%).

On rare occasions, surgery may be indicated to treat a mass lesion within the subarachnoid space that is causing major symptomatology or to biopsy the meninges to make a diagnosis.

Patients with symptomatic increased intracranial pressure caused by leptomeningeal metastases are candidates for a CSF-systemic shunt. A ventriculoperitoneal shunt is the procedure of choice.[235] Patients selected for shunting include those with severe and intractable headache, severe papilledema with potential visual loss, stupor or obtundation, or repetitive plateau waves. Occasionally, steroids will lower intracranial pressure enough so that shunting is not required, allowing time for other therapy to treat the CSF blockage.

One report describes 37 patients who required a ventriculoperitoneal shunt for management of intracranial hypertension.[235] These patients represented 6% of patients evaluated and treated over an 8-year period at MSKCC. Symptoms improved in 77% of patients, especially headache, nausea/vomiting, and alertness; complications were few. Three patients experienced shunt malfunction and one developed a subdural hematoma that required drainage. There were no infections or tumor spread to the peritoneal cavity. However, median survival was only 2 months, although one patient survived more than 3 years. Programmable valves allowed the surgeon to adjust the intracranial pressure to an appropriate level for the patient. Treatment of the patient with intrathecal chemotherapy after shunting is a dilemma. Some physicians place "on–off" valves in the shunting system, allowing the shunt to be turned off for several hours after drugs are injected into the ventricular system. The need for a shunt implies obstruction of CSF pathways. Therefore, drugs injected into the ventricular system will not exit during the time the valve is closed off and thus will not circulate throughout the subarachnoid space, leaving most tumor untreated. In addition, prolonged exposure of the ependymal surface to high concentrations of drug leads to transependymal absorption of the chemotherapeutic agent that results in neurotoxicity. Furthermore, some patients cannot tolerate valve closure for even a few hours.

Our approach has been to perform the shunt when necessary and give intrathecal drugs by lumbar puncture after the shunting is complete. The chemotherapeutic agent then encounters the spinal leptomeninges and basal cisterns before entering the ventricular system and exiting through the shunt. Alternatively, one can give systemic chemotherapy.

Some physicians are concerned about the possibility of seeding the peritoneal cavity from the shunt. This presents a theoretical but not a practical problem (see Chapter 14). One unique patient with ovarian carcinoma in the peritoneal cavity seeded the ependyma of the lateral ventricles through a ventriculoperitoneal shunt that had been placed years before for another problem.[236]

COMPLICATIONS

At MSKCC, complications of the placement and use of Ommaya reservoirs and shunts are uncommon.[235] However, some reports described complications in up to 17% of patients[237–239] (Table 7–16).

Hemorrhage at the time of placement occurs in fewer than 1% of patients. In addition, a small number of reservoirs fail after placement. They may have been misplaced, such as through the floor of the third ventricle into the basal cisterns, into brain substance, or, as on one occasion, into the Sylvian fissure. If inappropriately placed, they may become occluded with detritus. To ensure proper placement, a CT scan should be done after Ommaya insertion but before chemotherapy is administered through the device. If the reservoir is obstructed, or if the ventricular pressure exceeds tissue pressure, the injected material backs up along

the catheter and can form a fluid mass in the brain (see Chapter 14).

Infection complicates approximately 5% (2%–16%)[248] of reservoirs at some time during therapy. Common organisms are *Staphylococcus aureus*, coagulase negative *Staphylococcus*, and *Propionibacterium acnes*. Other organisms are occasional offenders. Most reservoir infections are asymptomatic, but a few patients develop symptoms of meningitis (headache, stiff neck, and fever). We have seen occasional *P. acnes* infections that appeared to clear without treatment, but most of the infections require therapy. Instillation of antibiotics into the reservoir often sterilizes the system effectively.[249] At times it is necessary to remove the reservoir, but a new one can be inserted and treatment continued.[250] Rarely, infections can become evident many years after the last puncture, likely because of slow growth of these indolent organisms.[251] Table 7–17 lists the neurologic complications of treatment for leptomeningeal metastasis. Further information can be found in Chapter 12.

PROGNOSIS

Evidence from retrospective analyses indicates that untreated patients, once they have become symptomatic, generally progress rapidly and succumb to their leptomeningeal tumor within 6 weeks to 2 months from the initial diagnosis.[7,85] The patient usually succumbs to progressive neurologic disability. Treatment is not very effective in reversing fixed neurologic deficits. In a rare patient, the disease is more indolent. Two MSKCC patients with breast carcinoma have had extraordinarily long courses. One presented with hydrocephalus that responded to shunting; 2 years later, she developed weakness in her legs. At that time, malignant cells were found in the CSF, and treatment was undertaken. The second patient complained of weakness and numbness in the lower extremities for approximately 18 months before clinical evaluation revealed malignant cells. Because of the patient's reluctance, another 6 months elapsed before treatment was undertaken, with only very slow progression of the illness. A third patient also had a 2-year course of leg weakness before CSF examination revealed malignant adenocarcinoma cells. The primary tumor site was never found.

Table 7–16. **Complications Associated with Placement and Use of the Ommaya Device**

Study	Patients (n)	Catheter Misplacement (%)	Hemorrhage (%)	Late Reservoir Malfunction (%)	Bacterial Meningitis or Ventriculitis (%)	Other Complications
Ongerboer-de Visser et al.[240]	19		16	10		
Obbens et al.[239]	387	7		5		
Hitchins et al.[241]	42				14	28
Dinmindorf and Bleyer[242]	32				16	
Pfeffer et al.[243]	98	7	5	10		
Siegal et al.[244]	66				12	
Lishner et al.[245]	106	10	3	5	9	1
Boogerd et al.[27]	44		11		2	
Siegal et al.[105]	31	10			4	
Grant et al.[246]	36	13		9		9
Chamberlain et al.[247]	120	2	5		8	2
Sandberg et al.[238]	107	5	3		2	

Modified from Ref. 248 with permission.

Table 7-17. **Neurologic Toxicities and Complications of Treatment for Leptomeningeal Metastases**

Nature	Training	Agents	Clinical and Radiological Findings	Pathological Findings	Treatment and Course
Aseptic meningitis	Several hours after injection	Any IT agent, especially Depocyt	Mimics bacterial meningitis; CSF: pleocytosis, ↑ protein	Culture negative	Oral anti-pyretics, anti-emetics and steroids; reversible within 1-3 days
Acute encephalopathy	Within 24-48 hr after treatment	IT MTX or Ara C, IV HD MTX	Seizures, confusion, disorientation, and lethargy		Further treatment possible; usually totally reversible
Myelopathy	Within 48 hr to months	IT MTX, Ara C, Thiotepa, ? any agent	Myelopathy; CSF: ↑ protein; MR: spinal cord swelling, ↑ signal on T_2WI	Demyelination	Poor prognosis with persistent paraparesis (60%)
Subacute encephalopathy	5-6 days after treatment	HD IV MTX	Stroke-like syndrome; CSF and MR: normal		Reversible within 48-72 hr; further treatment possible
Acute cerebellar syndrome	2-5 days after treatment	HD IV Ara C (<3 g/m²)	Encephalopathy immediately followed by cerebellar syndrome	Diffuse loss of Purkinje cells ± WM demyelination	Recovery after treatment discontinuation; but may be permanent
Other: seizures, encephalopathy, myelopathy			MR: cerebellar atrophy, reversible and diffuse leukoencephalopathy		
Delayed leukoencephalopathy	Months to years after treatment	Typically combined RT + IV/IT CT	Subcortical-frontal syndrome; CSF: ↑ protein; MR: cerebral atrophy, diffuse WM ↑ signal on FLAIR and T_2WI, ventricular dilatation	Disseminated foci of demyelination	Permanent and progressive axonal loss

From Ref. 75 with permission.

CSF, cerebrospinal fluid; CT, chemotherapy; ↑, elevated; HD, high-dose; IT, intrathecal; IV, intravenous; MR, magnetic resonance; MTX, methotrexate; RT, radiation therapy; T_2WI, T_2-weighted images; WM, white matter; Ara C, cytarabine.

Several reports detail the outcome of different treatment protocols. In general, the results with leukemia are excellent. Because these tumors are sensitive to MTX and ara-C, leukemia[71,243,252] can often be eradicated from the nervous system, although relapse does occur.[243] Patients can even be cured following relapse. Patients with lymphoma may also respond, but they do not do as well as those with CNS leukemia. In studies with solid tumors, the results are not as good. In approximately 40% of patients, neurologic symptoms improve[8] more often in response to RT than chemotherapy, regardless of the route of administration. In many of these patients, treatment results in several months of good-quality life. Approximately 25% of patients grow progressively worse despite treatment. Those patients who are relatively neurologically normal at the time treatment is undertaken do better than those patients with fixed neurologic deficits. Exceptions do occur, making patients with advanced neurologic symptomatology worthy of treatment. For example, marked improvement in visual or hearing loss is occasionally reported. In one of our patients with adenocarcinoma of the lung metastatic to the leptomeninges, vision improved from less than 20/200 to 20/20 and remained stable for several months. Of solid tumors, breast carcinomas and SCLCs respond best; melanoma responds worst.

Along with the clinical response, CSF findings often improve. In some patients, malignant cells disappear and protein and glucose concentrations return to normal. In others, cytology is persistently positive despite clinical improvement and improvement in other CSF variables. Some believe that no treatment helps and that apparent benefit derives from patient selection, but this view appears unduly pessimistic.

Despite initial clinical improvement, most patients have a short survival, usually only a few months. Approximately 15% of patients with breast cancer survive more than a year, but such survival is rare with lung cancer or melanoma. In one recent series,[8] the median survival of patients with leptomeningeal metastasis from solid tumors was 3.3 months, but from lymphoreticular tumors, it was 11.4 months. Nevertheless, 29% of patients with solid primaries were alive at 12 months and 17% at 24 months. Among solid tumor primaries, patients with breast cancer did best with

a median survival of 11.3 months, 48% alive at 12 months and 34% alive at 24 months. The longest breast cancer survival in that series was 74 months. Patients with lung cancer had a median survival of 1.8 months, with only 18% alive at 1 and 2 years; one patient lived 27 months. The median survival for melanoma was 4.7 months, with 26% alive at 1 year but none at 2 years. Occasional patients do have long and useful survival after treatment of established meningeal metastases from solid tumors.[253]

Prognostic factors play an important role. Herrlinger et al.[8] divided patients into three risk categories based on age (over 60 years), elevated CSF albumin levels, and elevated CSF lactate. Patients with none or one of these risk factors (low risk) survived 14 months; patients with two risk factors (medium risk) survived 5.9 months, and patients with three risk factors (high risk) survived 1.6 months. Other prognostic factors include degree of neurologic disability, tumor histology, and status of systemic disease.

Despite a rather extensive literature, often consisting of reports with small numbers of patients, a general agreement does not exist on how or even whether to treat leptomeningeal metastases from solid tumors other than lymphomas. Ongerboer de Visser et al.[240] reported good responses of breast cancer to intensive intraventricular MTX therapy, with a 25% 1-year survival. Grossman and Moynihan[254] reported no significant neurologic improvement in 59 adults with nonleukemic malignancies, 28 of whom had breast cancer. Pfeffer et al.[243] reported nine complete responses and six partial responses in 27 patients with breast cancer following a protocol using intrathecal MTX and RT; unfortunately, treatment complications occurred in 30% of patients, with four deaths. Boogerd et al.[27] studied 58 patients with breast cancer and leptomeningeal metastases, and a multivariate analysis of pretreatment characteristics showed a poor prognosis for patients older than age 55 with lung metastases, cranial nerve involvement, hypoglycorrhachia, and elevated protein concentrations. Patients with good risk factors had a better response to treatment and a longer survival rate. Siegal et al.[106] reported sustained responses to systemic therapy in 31 of 137 (23%) patients with leptomeningeal metastases. As indicated in the preceding text, Herrlinger et al. reported clinical improvement in 41% of patients.[8]

It seems obvious that leukemia and lymphomas involving the leptomeninges should be treated vigorously with intrathecal or systemic drugs and systemic RT. A significant percentage of patients with breast cancer respond with sufficient improvement in clinical symptoms and long enough survival to warrant their treatment as well. The same appears true of SCLC. The prognosis for NSCLC, melanoma, and other adenocarcinomas affecting the leptomeninges is so poor that one might question the usefulness of vigorous treatment. However, the occasional dramatic responses observed by us and reported by others deserve serious consideration for aggressive treatment.

REFERENCES

1. Glass J. Neurologic complications of lymphoma and leukemia. *Semin Oncol* 2006;33:342–347.
2. Nolan CP, Abrey LE. Leptomeningeal metastases from leukemias and lymphomas. *Cancer Treat Res* 2005;125:53–69.
3. Gasecki AP, Bashir RM, Foley J. Leptomeningeal carcinomatosis: a report of 3 cases and review of the literature. *Eur Neurol* 1992;32:74–78.
4. Jayson GC, Howell A. Carcinomatous meningitis in solid tumours. *Ann Oncol* 1996;7:773–786.
5. Jaeckle KA. Neoplastic meningitis from systemic malignancies: diagnosis, prognosis and treatment. *Semin Oncol* 2006;33:312–323.
6. Gleissner B, Chamberlain MC. Neoplastic meningitis. *Lancet Neurol* 2006;5:443–452.
7. Olson ME, Chernik NL, Posner JB. Infiltration of the leptomeninges by systemic cancer. A clinical and pathologic study. *Arch Neurol* 1974;30:122–137.
8. Herrlinger U, Forschler H, Kuker W, et al. Leptomeningeal metastasis: survival and prognostic factors in 155 patients. *J Neurol Sci* 2004;223:167–178.
9. Berg SL, Chamberlain MC. Current treatment of leptomeningeal metastases: systemic chemotherapy, intrathecal chemotherapy and symptom management. *Cancer Treat Res* 2005;125:121–146.
10. Eberth CJ. Zur Entwickelung des Epithelioms (Cholesteatoms) der Pia und der Lunge. *Virchows Archiv A Pathol Anat* 1870;49:51–63.
11. Siefert E. Uber die multiple Karzinomatose des Zentralnervensystems. *Munch Med Wochenschr* 1902;49:826–828.
12. Grain GO, Karr JP. Diffuse leptomeningeal carcinomatosis. Clinical and pathologic characteristics. *Neurology* 1955;5:706–722.
13. Little JR, Dale AJ, Okazaki H. Meningeal carcinomatosis. Clinical manifestations. *Arch Neurol* 1974;30:138–143.
14. Drappatz J, Batchelor TT. Leptomeningeal neoplasms. *Curr Treat Options Neurol* 2007;9:283–293.
15. DeAngelis LM, Boutros D. Leptomeningeal metastasis. *Cancer Invest* 2005;23:145–154.
16. O'Meara WP, Borkar SA, Stambuk HE, et al. Leptomeningeal metastasis. *Curr Probl Cancer* 2007;31:367–424
17. Parsons M. The spinal form of carcinomatous meningitis. *Q J Med N Ser* 1972;41:509–519.
18. Kizawa M, Mori N, Hashizume Y, et al. Pathological examination of spinal lesions in meningeal carcinomatosis. *Neuropathology*, in press.
19. Posner JB, Chernik NL. Intracranial metastases from systemic cancer. *Adv Neurol* 1978;19:575–587.
20. Gonzalez-Vitale JC, Garcia-Bunuel R. Meningeal carcinomatosis. *Cancer* 1976;37:2906–2911.
21. Takakura K, Sano K, Hojo S, et al. *Metastatic Tumors of the Central Nervous System.* Tokyo: Igaku-Shoin, 1982.
22. Pui CH. Central nervous system disease in acute lymphoblastic leukemia: prophylaxis and treatment. *Hematology Am Soc Hematol Educ Progr* 2006:142–146.
23. Littman P, Coccia P, Bleyer WA, et al. Central nervous system (CNS) prophylaxis in children with low risk acute lymphoblastic leukemia (ALL). *Int J Radiat Oncol Biol Phys* 1987;13:1443–1449.
24. Hill QA, Owen RG. CNS prophylaxis in lymphoma: who to target and what therapy to use. *Blood Rev* 2006;20:319–332.
25. Rosen ST, Aisner J, Makuch RW, et al. Carcinomatous leptomeningitis in small cell lung cancer: a clinicopathologic review of the National Cancer Institute experience. *Medicine* 1982;61:45–53.
26. Seute T, Leffers P, Ten Velde GP, et al. Leptomeningeal metastases from small cell lung carcinoma. *Cancer* 2005;104:1700–1705.
27. Boogerd W, Hart AA, van der Sande JJ, et al. Meningeal carcinomatosis in breast cancer. Prognostic factors and influence of treatment. *Cancer* 1991;67:1685–1695.
28. Yap H-Y, Yap B-S, Tashima CK, et al. Meningeal carcinomatosis in breast cancer. *Cancer* 1978;42:283–286.
29. Boogerd W. Central nervous system metastasis in breast cancer. *Radiother Oncol* 1996;40:5–22.
30. Bendell JC, Domchek SM, Burstein HJ, et al. Central nervous system metastases in women who receive trastuzumab-based therapy for metastatic breast carcinoma. *Cancer* 2003;97:2972–2977.
31. Lai R, Dang CT, Malkin MG, et al. The risk of central nervous system metastases after trastuzumab therapy in patients with breast carcinoma. *Cancer* 2004;101:810–816.
32. Kosmas C, Malamos NA, Tsavaris NB, et al. Isolated leptomeningeal carcinomatosis (carcinomatous meningitis) after taxane-induced major remission in patients with advanced breast cancer. *Oncology (Basel)* 2002;63:6–15.
33. Omuro AM, Kris MG, Miller VA, et al. High incidence of disease recurrence in the brain and leptomeninges in patients with nonsmall cell lung carcinoma after response to gefitinib. *Cancer* 2005;103:2344–2348.
34. Mirimanoff RO, Choi NC. Intradural spinal metastases in patients with posterior fossa brain metastases from various primary cancers. *Oncology* 1987;44:232–236.
35. Mirimanoff RO, Choi NC. The risk of intradural spinal metastases in patients with brain metastases from

bronchogenic carcinomas. *Int J Radiat Oncol Biol Phys* 1986;12:2131–2136.

36. Aisner J, Aisner SC, Ostrow S, et al. Meningeal carcinomatosis from small cell carcinoma of the lung. Consequence of improved survival. *Acta Cytol* 1979;23:292–296.

37. Aroney RS, Dalley DN, Chan WK, et al. Meningeal carcinomatosis in small cell carcinoma of the lung. *Am J Med* 1981;71:26–32.

38. Balducci L, Little DD, Khansur T, et al. Carcinomatous meningitis in small cell lung cancer. *Am J Med Sci* 1984;287:31–33.

39. Lee Y-T. Breast carcinoma: pattern of metastasis at autopsy. *J Surg Oncol* 1983;23:175–180.

40. Tsukada Y, Fouad A, Pickren JW, et al. Central nervous system metastasis from breast carcinoma. Autopsy study. *Cancer* 1983;52:2349–2354.

41. Smith DB, Howell A, Harris M, et al. Carcinomatous meningitis associated with infiltrating lobular carcinoma of the breast. *Eur J Surg Oncol* 1985;11:33–36.

42. Theodore WH, Gendelman S. Meningeal carcinomatosis. *Arch Neurol* 1981;38:696–699.

43. Inci S, Bozkurt G, Gulsen S, et al. Rare cause of subarachnoid hemorrhage: spinal meningeal carcinomatosis. Case report. *J Neurosurg Spine* 2005;2:79–82.

44. Raizer JJ, Hwu WJ, Panageas KS, et al. Brain and leptomeningeal metastases from cutaneous melanoma: survival outcomes based on clinical features. *Neuro Oncol*, in press.

45. Dekker AW, Elderson A, Punt K, et al. Meningeal involvement in patients with acute nonlymphocytic leukemia. Incidence, management, and predictive factors. *Cancer* 1985;56:2078–2082.

46. Peterson BA, Brunning RD, Bloomfield CD, et al. Central nervous system involvement in acute nonlymphocytic leukemia. A prospective study of adults in remission. *Am J Med* 1987;83:464–470.

47. Ersboll J, Schultz HB, Thomsen BL, et al. Meningeal involvement in non-Hodgkin's lymphoma: symptoms, incidence, risk factors and treatment. *Scand J Haematol* 1985;35:487–496.

48. Hoerni-Simon G, Suchaud JP, Eghbali H, et al. Secondary involvement of the central nervous system in malignant non-Hodgkin's lymphoma. A study of 30 cases in a series of 498 patients. *Oncology* 1987;44:98–101.

49. Redman BG, Tapazoglou E, Al-Sarraf M. Meningeal carcinomatosis in head and neck cancer. Report of six cases and review of the literature. *Cancer* 1986;58:2656–2661.

50. Moberg A, Reis GV. Carcinosis meningum. *Acta Med Scand* 1961;170:747–755.

51. Lee JL, Kang YK, Kim TW, et al. Leptomeningeal carcinomatosis in gastric cancer. *J Neuro-Oncol* 2004;66:167–174.

52. Hauch TW, Shelbourne JD, Cohen HJ, et al. Meningeal mycosis fungoides: clinical and cellular characteristics. *Ann Intern Med* 1975;82:499–505.

53. Lundberg WB, Cadman EC, Skeel RT. Leptomeningeal mycosis fungoides. *Cancer* 1976;38:2149–2153.

54. Wabula A, Imitola J, Satagata S, et al. Mycosis fungoides with leptomeningeal involvement. *J Clin Oncol* 2007;25:5658–5661.

55. Patriarca F, Zaja F, Silvestri F, et al. Meningeal and cerebral involvement in multiple myeloma patients. *Ann Hematol* 2001;80:758–762.

56. Maldonado JE, Kyle RA, Ludwig J, et al. Meningeal myeloma. *Arch Intern Med* 1970;126:660–663.

57. Weed JC, Jr, Creasman WT. Meningeal carcinomatosis secondary to advanced squamous cell carcinoma of the cervix: a case report. Meningeal metastasis of advanced cervical cancer. *Gynecol Oncol* 1975;3:201–204.

58. Barnard RO, Parsons M. Carcinoma of the thyroid with leptomeningeal dissemination following the treatment of a toxic goitre with 131-I and methyl thiouracil. Case with a co-existing intracranial dermoid. *J Neurol Sci* 1969;8:299–306.

59. Bresalier RS, Karlin DA. Meningeal metastasis from rectal carcinoma with elevated cerebrospinal fluid carcinoembryonic antigen. *Dis Colon Rectum* 1979;22:216–217.

60. Nagourney RA, Hedaya R, Linnoila M, et al. Carcinoid carcinomatous meningitis. *Ann Intern Med* 1985;102:779–782.

61. Berry MP, Jenkin RD. Parameningeal rhabdomyosarcoma in the young. *Cancer* 1981;48:281–288.

62. Cash J, Fehir KM, Pollack SM. Meningeal involvement in early stage chronic lymphocytic leukemia. *Cancer* 1987;59:798–800.

63. Bender BL, Mayernik DG. Hodgkin's disease presenting with isolated craniospinal involvement. *Cancer* 1986;58:1745–1748.

64. Matthay KK, Brisse H, Couanet D, et al. Central nervous system metastases in neuroblastoma: radiologic, clinical, and biologic features in 23 patients. *Cancer* 2003;98:155–165.

65. Lachance DH, O'Neill BP, Macdonald DR, et al. Primary leptomeningeal lymphoma: report of 9 cases, diagnosis with immunocytochemical analysis and review of the literature. *Neurology* 1991;41:95–100.

66. Aichner F, Schuler G. Primary leptomeningeal melanoma. Diagnosis by ultrastructural cytology of cerebrospinal fluid and cranial computed tomography. *Cancer* 1982;50:1751–1756.

67. Savitz MH, Anderson PJ. Primary melanoma of the leptomeninges: a review. *Mt Sinai J Med* 1974;41:774–791.

68. Silbert SW, Smith KR, Jr, Horenstein S. Primary leptomeningeal melanoma. An ultrastructural study. *Cancer* 1978;41:519–527.

69. Smith MT, Armbrustmacher VM, Violett TW. Diffuse meningeal rhabdomyosarcoma. *Cancer* 1981;47:2081–2086.

70. DeAngelis LM. Primary central nervous system lymphoma: a new clinical challenge. *Neurology* 1991;41:619–621.

71. Recht L, Straus DJ, Cirrincione C, et al. Central nervous system metastases from non-Hodgkin's lymphoma: treatment and prophylaxis. *Am J Med* 1988; 84:425–435.

72. Etzioni A, Levy J, Lichtig C, et al. Brain mass as a manifestation of very late relapse in nonendemic Burkitt's lymphoma. *Cancer* 1985;55:861–863.

73. Jones GR, Mason WH, Fishman LS, et al. Primary central nervous system lymphoma without intracranial mass in a child. Diagnosis by documentation of monoclonality. *Cancer* 1985;56:2804–2808.

74. Schweinle JE, Alperin JB. Central nervous system recurrence ten years after remission of acute lymphoblastic leukemia. *Cancer* 1980;45:16–18.

75. Taillibert S, Laigle-Donadey F, Chodkiewicz C, et al. Leptomeningeal metastases from solid malignancy: a review. *J Neuro-Oncol* 2005;75:85–99.

76. Hansen K, Gjerris F, Sorensen PS. Absence of hydrocephalus in spite of impaired cerebrospinal fluid absorption and severe intracranial hypertension. *Acta Neurochir (Wien)* 1987;86:93–97.

77. Sinniah D, Looi LM, Ortega JA, et al. Cerebellar coning and uncal herniation in childhood acute leukaemia. *Lancet* 1982;2:702–704.

78. Chamberlain MC. Radioisotope CSF flow studies in leptomeningeal metastases. *J Neuro-Oncol* 1998;38:135–140.

79. Mason WP, Yeh SDJ, DeAngelis LM. [111]Indium-diethylenetriamine pentaacetic acid cerebrospinal fluid flow studies predict distribution of intrathecally administered chemotherapy and outcome in patients with leptomeningeal metastases. *Neurology* 1998;50:438–444.

80. Glantz MJ, Hall WA, Cole BF, et al. Diagnosis, management, and survival of patients with leptomeningeal cancer based on cerebrospinal fluid-flow status. *Cancer* 1995;75:2919–2931.

81. Grossman SA, Trump DL, Chen DC, et al. Cerebrospinal fluid flow abnormalities in patients with neoplastic meningitis. An evaluation using [111]Indium-DTPA ventriculography. *Am J Med* 1982;73:641–647.

82. Chamberlain MC, Corey-Bloom J. Leptomeningeal metastasis: [111]indium-DTPA CSF flow studies. *Neurology* 1991;41:1765–1769.

83. Haaxma-Reiche H, Piers A, Beekhuis H. Normal cerebrospinal fluid dynamics. A study with intraventricular injection of ^{111}In-DTPA in leukemia and lymphoma without meningeal involvement. *Arch Neurol* 1989;46:997–999.

84. Broderick JP, Cascino TL. Nonconvulsive status epilepticus in a patient with leptomeningeal cancer. *Mayo Clin Proc* 1987;62:835–837.

85. Klein P, Haley EC, Wooten GF, et al. Focal cerebral infarctions associated with perivascular tumor infiltrates in carcinomatous leptomeningeal metastases. *Arch Neurol* 1989;46:1149–1152.

86. Wasserstrom W, Glass JP, Posner JB. Diagnosis and treatment of leptomeningeal metastases from solid tumors: experience with 90 patients. *Cancer* 1982;49:759–772.

87. Latchaw RE, Gabrielsen TO, Seeger JF. Cerebral angiography in meningeal sarcomatosis and carcinomatosis. *Neuroradiology* 1974;8:131–139.

88. Siegal T, Mildworf B, Stein D, et al. Leptomeningeal metastases: reduction of regional cerebral blood flow and cognitive impairment. *Ann Neurol* 1985;17:100–102.

89. Hiesiger EM, Picco-Del Bo A, Lipschutz LE, et al. Experimental meningeal carcinomatosis selectively depresses local cerebral glucose utilization in rat brain. *Neurology* 1989;39:90–95.

90. Greydanus DE, Burgert EO, Jr, Gilchrist GS. Hypothalamic syndrome in children with acute lymphocytic leukemia. *Mayo Clin Proc* 1978;53:217–220.

91. Pochedly C. Neurological manifestations in acute leukemia. II. Involvement of cranial nerves and hypothalamus. *NY State J Med* 1975;75:715–721.

92. Hadfield MG, Vennart GP, Rosenblum WI. Hypoglycemia: Invasion of the hypothalamus by lymphosarcoma. Metastasis to blood glucose regulating centers. *Arch Pathol* 1972;94:317–321.

93. Schipper HI, Bardosi A, Jacobi C, et al. Meningeal carcinomatosis: origin of local IgG production in the CSF. *Neurology* 1988;38:413–416.

94. Weller M, Stevens A, Sommer N, et al. Tumor cell dissemination triggers an intrathecal immune response in neoplastic meningitis. *Cancer* 1992;69:1475–1480.

95. Weller M, Stevens A, Sommer N, et al. Intrathecal IgM response in disseminated cerebrospinal metastasis of malignant melanoma. *J Neuro-Oncol* 1993;16:55–59.

96. Weller M, Stevens A, Sommer N, et al. Tumour necrosis factor-alpha in malignant melanomatous meningitis. Letter to the editor. *J Neurol Neurosurg Psychiatry* 1992;55:74.

97. Weller M, Sommer N, Stevens A, et al. Increased intrathecal synthesis of fibronectin in bacterial and carcinomatous meningitis. *Acta Neurol Scand* 1990;82:138–142.

98. Wiederkehr F, Bueler MR, Vonderschmitt DJ. Analysis of circulating immune complexes isolated from plasma, cerebrospinal fluid and urine. *Electrophoresis* 1991;12:478–486.

99. Alcolado R, Weller RO, Parrish EP, et al. The cranial arachnoid and pia mater in man: anatomical and ultrastructural observations. *Neuropath Appl Neurobiol* 1988;14:1–17.

100. Nicholas DS, Weller RO. The fine anatomy of the human spinal meninges. A light and scanning electron microscopy study. *J Neurosurg* 1988;69:276–282.

101. Bleyer WA, Poplack DG. Prophylaxis and treatment of leukemia in the central nervous system and other sanctuaries. *Semin Oncol* 1985;12:131–148.

102. Smirniotopoulos JG, Murphy FM, Rushing EJ, et al. Patterns of contrast enhancement in the brain and meninges. *Radiographics* 2007;27:525–551.

103. Frank JA, Girton M, Dwyer AJ, et al. Meningeal carcinomatosis in the VX2 rabbit tumor model: detection with Gd-DTPA-enhanced MR imaging. *Radiology* 1988;167:825–829.

104. Freilich RJ, Krol G, DeAngelis LM. Neuroimaging and cerebrospinal fluid cytology in the diagnosis of leptomeningeal metastasis. *Ann Neurol* 1995;38:51–57.

105. Ushio Y, Shimizu K, Aragaki Y, et al. Alteration of blood-CSF barrier by tumor invasion into the meninges. *J Neurosurg* 1981;55:445–449.

106. Siegal T, Lossos A, Pfeffer MR. Leptomeningeal metastases: analysis of 31 patients with sustained off-therapy response following combined-modality therapy. *Neurology* 1994;44:1463–1469.

107. Glantz MJ, Cole BF, Recht L, et al. High-dose intravenous methotrexate for patients with nonleukemic leptomeningeal cancer: is intrathecal chemotherapy necessary? *J Clin Oncol* 1998;16:1561–1567.

108. Bokstein F, Lossos A, Siegal T. Leptomeningeal metastases from solid tumors—a comparison of two prospective series treated with and without intra-cerebrospinal fluid chemotherapy. *Cancer* 1998;82:1756–1763.

109. Mencel PJ, DeAngelis LM, Motzer RJ. Hormonal ablation as effective therapy for carcinomatous meningitis from prostatic carcinoma. *Cancer* 1994;73:1892–1894.

110. Kaplan JG, DeSouza TG, Farkash A, et al. Leptomeningeal metastases: comparison of clinical features and laboratory data of solid tumors,

lymphomas and leukemias. *J Neuro-Oncol* 1990;9:225–229.

111. Balm M, Hammack J. Leptomeningeal carcinomatosis. Presenting features and prognostic factors. *Arch Neurol* 1996;53:626–632.

112. DeAngelis LM, Payne R. Lymphomatous meningitis presenting as atypical cluster headache. *Pain* 1987;30:211–216.

113. Fisher CM. Hydrocephalus as a cause of disturbances of gait in the elderly. *Neurology* 1982;32:1358–1363.

114. Sudarsky L, Ronthal M. Gait disorders among elderly patients. A survey of 50 patients. *Arch Neurol* 1983;40:740–743.

115. Dexter DD, Jr, Westmoreland BF, Cascino TL. Complex partial status epilepticus in a patient with leptomeningeal carcinomatosis. *Neurology* 1990;40:858–859.

116. Yap H-Y, Tashima CK, Blumenschein GR, et al. Diabetes insipidus and breast cancer. *Arch Intern Med* 1979;139:1009–1011.

117. Ra'anani P, Shpilberg O, Berezin M, et al. Acute leukemia relapse presenting as central diabetes insipidus. *Cancer* 1994;73:2312–2316.

118. Tham LC, Millward MJ, Lind MJ, et al. Metastatic breast cancer presenting with diabetes insipidus. Letter to the editor. *Acta Oncol* 1992;31(6):679–683.

119. Houck WA, Olson KB, Horton J. Clinical features of tumor metastasis to the pituitary. *Cancer* 1970;26:656–659.

120. Madow L, Alpers BJ. Encephalitic form of metastatic carcinoma. *Arch Neurol Psychiatry* 1951;65:161–173.

121. Floeter MK, So YT, Ross DA, et al. Miliary metastasis to the brain: clinical and radiologic features. *Neurology* 1987;37:1817–1818.

122. Miller JW, Klass DW, Mokri B, et al. Triphasic waves in cerebral carcinomatosis. Another nonmetabolic cause. *Arch Neurol* 1986;43:1191–1193.

123. Oster JR, Perez GO, Larios O, et al. Cerebral salt wasting in a man with carcinomatous meningitis. *Arch Intern Med* 1983;143:2187–2188.

124. Karp G, Nahum K. Hyperventilation as the initial manifestation of lymphomatous meningitis. *J Neuro-Oncol* 1992;13:173–175.

125. Altrocchi PH, Eckman PB. Meningeal carcinomatosis and blindness. *J Neurol Neurosurg Psychiatry* 1973;36:206–210.

126. Kattah JC, Suski ET, Killen JY, et al. Optic neuritis and systemic lymphoma. *Am J Ophthalmol* 1980;89:431–436.

127. Boogerd W, Moffie D, Smets LA. Early blindness and coma during intrathecal chemotherapy for meningeal carcinomatosis. *Cancer* 1990;65:452–457.

128. Houck JR, Murphy K. Sudden bilateral profound hearing loss resulting from meningeal carcinomatosis. *Otolaryngol Head Neck Surg* 1992;106:92–97.

129. LaVenuta F, Moore JA. Involvement of the inner ear in acute stem cell leukemia. Report of two cases. *Ann Otol Rhinol Laryngol* 1972;81:132–137.

130. Oshiro H, Perlman HB. Subarachnoid spread of tumor to the labyrinth. *Arch Otolaryngol* 1965;81:328–334.

131. Atkinson M. Meniere's famous autopsy and its interpretation. *Arch Otolaryngol* 1945;42:186–187.

132. Wilkins DE, Samhouri AM. Isolated bilateral oculomotor paresis due to lymphoma. *Neurology* 1979;29:1425–1428.

133. Van Rossum J, Zwaan FE, Bots GT. Facial palsy as the initial symptom of lymphoreticular malignancy. Case report. *Eur Neurol* 1979;18:212–216.

134. Ingram LC, Fairclough DL, Furman WL, et al. Cranial nerve palsy in childhood acute lymphoblastic leukemia and non-Hodgkin's lymphoma. *Cancer* 1991;67:2262–2268.

135. Singh SK, Agris JM, Leeds NE, et al. Intracranial leptomeningeal metastases: Comparison of depiction at FLAIR and contrast-enhanced MR imaging. *Radiology* 2000;217:50–53.

136. Grossman SA, Krabak MJ. Leptomeningeal carcinomatosis. *Cancer Treat Rev* 1999;25:103–119.

137. Collie DA, Brush JP, Lammie GA, et al. Imaging features of leptomeningeal metastases. *Clin Radiol* 1999;54:765–771.

138. Singh SK, Leeds NE, Ginsberg LE. MR imaging of leptomeningeal metastases: comparison of three sequences. *AJNR Am J Neuroradiol* 2002; 23:817–821.

139. Kremer S, Abu EM, Bierry G, et al. Accuracy of delayed post-contrast FLAIR MR imaging for the diagnosis of leptomeningeal infectious or tumoral diseases. *J Neuroradiol* 2006;33:285–291.

140. Kioumehr F, Dadsetan MR, Feldman N, et al. Postcontrast MRI of cranial meninges: Leptomeningitis versus pachymeningitis. *J Comput Assist Tomogr* 1995;19:713–720.

141. Panullo SC, Reich JB, Krol G, et al. MRI changes in intracranial hypotension. *Neurology* 1993;43:919–926.

142. River Y, Schwartz A, Gomori JM, et al. Clinical significance of diffuse dural enhancement detected by magnetic resonance imaging. *J Neurosurgery* 1996;85:777–783.

143. Straathof CS, de Bruin HG, Dippel DW, et al. The diagnostic accuracy of magnetic resonance imaging and cerebrospinal fluid cytology in leptomeningeal metastasis. *J Neurol* 1999;246:810–814.

144. Yousem DM, Patrone PM, Grossman RI. Leptomeningeal metastases: MR evaluation. *J Comput Assist Tomogr* 1990;14:255–261.

145. Zeiser R, Burger JA, Bley TA, et al. Clinical follow-up indicates differential accuracy of magnetic resonance imaging and immunocytology of the cerebral spinal fluid for the diagnosis of neoplastic meningitis—a single centre experience. *Br J Haematol* 2004;124:762–768.

146. Glass JP, Melamed M, Chernik NL, et al. Malignant cells in cerebrospinal fluid (CSF): the meaning of a positive CSF cytology. *Neurology* 1979;29:1369–1375.

147. Mahmoud HH, Rivera GK, Hancock ML, et al. Low leukocyte counts with blast cells in cerebrospinal fluid of children with newly diagnosed acute lymphoblastic leukemia. *N Engl J Med* 1993;329:312–319.

148. Corbett JJ, Mehta MP. Cerebrospinal fluid pressure in normal obese subjects and patients with pseudotumor cerebri. *Neurology* 1983;33:1386–1388.

149. Lundberg N, West KA. Leakage as a source of error in measurement of the cerebrospinal fluid pressure by lumbar puncture. *Acta Neurol Scand (Suppl)* 1965;41:115–121.

150. Van Uitert RL, Eisenstadt ML. Venous pulsations not always indicative of normal intracranial pressure. Letter to the editor. *Arch Neurol* 1978;35:550.

151. Mulligan MJ, Vasu R, Grossi CE, et al. Case report: neoplastic meningitis with eosinophilic pleocytosis in Hodgkin's disease: a case with cerebellar dysfunction and a review of the literature. *Am J Med Sci* 1988;296(5):322–326.

152. King DK, Loh KK, Ayala AG, et al. Eosinophilic meningitis and lymphomatous meningitis. Letter to the editor. *Ann Intern Med* 1975;82:228.

153. Conrad KA, Gross JL, Trojanowski JQ. Leptomeningeal carcinomatosis presenting as eosinophilic meningitis. *Acta Cytologica* 1986;30:29–31.

154. Budka H, Guseo A, Jellinger K, et al. Intermittent meningitic reaction with severe basophilia and eosinophilia in CNS leukaemia. *J Neurol Sci* 1976;28:459–468.

155. Quinn JP, Weinstein RA, Caplan LR. Eosinophilic meningitis and ibuprofen therapy. *Neurology* 1984;34:108–109.

156. Lossos A, Siegal T. Spinal subarachnoid hemorrhage associated with leptomeningeal metastases. *J Neuro-Oncol* 1992;12:167–171.

157. Siegal T, Shorr J, Lubetzki-Korn I, et al. Myeloma protein synthesis within the CNS by plasma cell tumors. *Ann Neurol* 1981;10:271–273.

158. Nakagawa H, Yamada M, Kanayama T, et al. Myelin basic protein in the cerebrospinal fluid of patients with brain tumors. *Neurosurgery* 1994;34:825–833.

159. Mahoney DH, Jr, Fernbach DJ, Glaze DG, et al. Elevated myelin basic protein levels in the cerebrospinal fluid of children with acute lymphoblastic leukemia. *J Clin Oncol* 1984;2:58–61.

160. DeVita VT, Canellos GP. Hypoglycorrhachia in meningeal carcinomatosis. *Cancer* 1966;19:691–694.

161. Kim P, Ashton D, Pollard JD. Isolated hypoglycorrhachia: leptomeningeal carcinomatosis causing subacute confusion. *J Clin Neurosci* 2005;12:841–843.

162. Fishman RA. Studies of the transport of sugars between blood and cerebrospinal fluid in normal states and in meningeal carcinomatosis. *Trans Am Neurol Assoc* 1963;88:114–118.

163. Jann S, Comini A, Pellegrini G. Hypoglycorrhachia in leptomeningeal carcinomatosis. A pathophysiological study. *Ital J Neurol Sci* 1988;9:83–88.

164. Schold SC, Wasserstrom WR, Fleisher M, et al. Cerebrospinal fluid biochemical markers of central nervous system metastases. *Ann Neurol* 1980;8:597–604.

165. Glantz MJ, Cole BF, Glantz LK, et al. Cerebrospinal fluid cytology in patients with cancer: minimizing false-negative results. *Cancer* 1998;82:733–739.

166. Grossman SA, Krabak MJ. Leptomeningeal carcinomatosis. *Cancer Treat Rev* 1999;25:103–119.

167. Kolmel HW. Cytology of neoplastic meningiosis. *J Neuro-Oncol* 1998;38:121–125.

168. Chamberlain MC, Kormanik PA, Glantz MJ. A comparison between ventricular and lumbar cerebrospinal fluid cytology in adult patients with leptomeningeal metastases. *Neuro-Oncol* 2001;3:42–45.

169. Murray JJ, Greco FA, Wolff SN, et al. Neoplastic meningitis. Marked variations of cerebrospinal fluid composition in the absence of extradural block. *Am J Med* 1983;75:289–294.

170. Rogers LR, Duchesneau PM, Nunez C, et al. Comparison of cisternal and lumbar CSF examination in leptomeningeal metastasis. *Neurology* 1992;42:1239–1241.

171. Twijnstra A, Ongerboer DE, Visser BW, et al. Serial lumbar and ventricular cerebrospinal fluid biochemical marker measurements in patients with leptomeningeal metastases from solid and hematological tumors. *J Neuro-Oncol* 1989;7:57–63.

172. Balhuizen JC, Bots GT, Schaberg A, et al. Value of cerebrospinal fluid cytology for the diagnosis of malignancies in the central nervous system. *J Neurosurg* 1978;48:747–753.

173. Kappel TJ, Manivel JC, Goswitz JJ. Atypical lymphocytes in spinal fluid resembling posttransplant lymphoma in a cardiac transplant recipient: a case report. *Acta Cytol* 1994;38:470–474.

174. Coakham HB, Garson JA, Brownell B, et al. Use of monoclonal antibody panel to identify malignant cells in cerebrospinal fluid. *Lancet* 1984;1:1095–1098.

175. Garson JA, Coakham HB, Kemshead JT, et al. The role of monoclonal antibodies in brain tumour diagnosis and cerebrospinal fluid (CSF) cytology. *J Neuro-Oncol* 1985;3:165–171.

176. Ezrin-Waters C, Klein M, Deck J, et al. Diagnostic importance of immunological markers in lymphoma involving the central nervous system. *Ann Neurol* 1984;16:668–672.

177. Kranz BR, Thierfelder S, Gerl A, et al. Cerebrospinal fluid immunocytology in primary central nervous system lymphoma. Letter to editor. *Lancet* 1992;340:727.

178. Li CY, Witzig TE, Phyliky RL, et al. Diagnosis of B-cell non-Hodgkin's lymphoma of the central nervous system by immunocytochemical analysis of cerebrospinal fluid lymphocytes. *Cancer* 1986;57:737–744.

179. Cibas ES, Malkin MG, Posner JB, et al. Detection of DNA abnormalities by flow cytometry in cells from cerebrospinal fluid. *Am J Clin Pathol* 1987;88:570–577.

180. Dux R, Kindler-Rohrborn A, Annas M, et al. A standardized protocol for flow cytometric analysis of cells isolated from cerebrospinal fluid. *J Neurol Sci* 1994;121:74–78.

181. Bromberg JE, Breems DA, Kraan J, et al. CSF flow cytometry greatly improves diagnostic accuracy in CNS hematologic malignancies. *Neurology* 2007;68:1674–1679.

182. van de Langerijt B, Gijtenbeek JM, de Reus HP, et al. CSF levels of growth factors and plasminogen activators in leptomeningeal metastases. *Neurology* 2006;67:114–119.

183. Tallman RD, Kimbrough SM, O'Brien JF, et al. Assay for b-glucuronidase in cerebrospinal fluid: usefulness for the detection of neoplastic meningitis. *Mayo Clin Proc* 1985;60:293–298.

184. Wasserstrom WR, Schwartz MK, Fleisher M, et al. Cerebrospinal fluid biochemical markers in central nervous system tumors: a review. *Ann Clin Lab Sci* 1981;11:239–251.

185. Hansen PB, Kjeldsen L, Dalhoff K, et al. Cerebrospinal fluid beta-2-microglobulin in adult patients with acute leukemia or lymphoma: a useful marker in early diagnosis and monitoring

of CNS-involvement. *Acta Neurol Scand* 1992; 85:224–227.

186. Mavligit GM, Stuckey SE, Cabanillas FF, et al. Diagnosis of leukemia or lymphoma in the central nervous system by beta-2-microglobulin determination. *N Engl J Med* 1980;303:718–722.

187. Peterslund NA, Black FT, Geil JP, et al. Beta-2-microglobulin in the cerebrospinal fluid of patients with infections of the central nervous system. *Acta Neurol Scand* 1989;80:579–583.

188. Krol G, Sze G, Malkin M, et al. MR of cranial and spinal meningeal carcinomatosis: Comparison with CT and myelography. *Am J Roentgenol* 1988;151:583–588.

189. Ertl-Wagner BB, Hoffmann RT, Bruening R, Herrmann K, Dichgans M, Reiser MF. Blurring of the vessels of the interhemispheric fissure in multislice CT angiography: a sign of meningeal carcinomatosis. *Eur Radiol* 2004;14:673–678.

190. Argov Z, Siegal T. Leptomeningeal metastases: peripheral nerve and root involvement—clinical and electrophysiological study. *Ann Neurol* 1985;17:593–596.

191. Kaplan JG, Portenoy RK, Pack DR, et al. Polyradiculopathy in leptomeningeal metastasis: the role of EMG and late response studies. *J Neuro-Oncol* 1990;9:219–224.

192. Cheng TM, O'Neill BP, Scheithauer BW, Piepgras DG. Chronic meningitis: the role of meningeal or cortical biopsy. *Neurosurgery* 1994;34:590–596.

193. Olinger CP, Ohlhaber RL. Eighteen-gauge microscopic-telescopic needle endoscope with electrode channel: potential clinical and research application. *Surg Neurol* 1974;2:151–160.

194. Padma MV, Jacobs M, Kraus G, et al. ^{11}C-methionine PET imaging of leptomeningeal metastases from primary breast cancer—a case report. *J Neuro-Oncol* 2001;55:39–44.

195. Komori T, Delbeke D. Leptomeningeal carcinomatosis and intramedullary spinal cord metastases from lung cancer: detection with FDG positron emission tomography. *Clin Nucl Med* 2001; 26:905–907.

196. Hildebrand J, Aoun M. Chronic meningitis: still a diagnostic challenge. *J Neurol* 2003;250:653–660.

197. Scully RE, Mark EJ, McNeely WF, et al., eds. Case records of the Massachusetts General Hospital. Case 14–1988. *N Engl J Med* 1988;318:903–915.

198. Mehta M, Bradley K. Radiation therapy for leptomeningeal cancer. *Cancer Treat Res* 2005; 125:147–158.

199. D'Angio GJ, French LA, Stadlan EM, et al. Intrathecal radioisotopes for the treatment of brain tumors. *Clin Neurosurg* 1968;15:288–299.

200. Doge H, Hliscs R. Intrathecal therapy with 198Au-colloid for meningosis prophylaxis. *Eur J Nucl Med* 1984;9:125–128.

201. Metz O, Stoll W, Plenert W. Meningiosis prophylaxis with intrathecal 198Au-colloid and methotrexate in childhood acute lymphocytic leukemia. *Cancer* 1982;49:224–228.

202. Benjamin JC, Moss T, Moseley RP, et al. Cerebral distribution of immunoconjugate after treatment for neoplastic meningitis using an intrathecal radiolabeled monoclonal antibody. *Neurosurgery* 1989;25:253–258.

203. Moseley RP, Davies AG, Richardson RB, et al. Intrathecal administration of ^{131}I radiolabelled monoclonal antibody as a treatment for neoplastic meningitis. *Br J Cancer* 1990;62:637–642.

204. Kramer K, Cheung NK, Humm JL, et al. Targeted radioimmunotherapy for leptomeningeal cancer using (131)I-3F8. *Med Pediatr Oncol* 2000; 35:716–718.

205. Chang EL, Maor MH. Standard and novel radiotherapeutic approaches to neoplastic meningitis. *Curr Oncol Rep* 2003;5:24–28.

206. Berg SL, Chamberlain MC. Systemic chemotherapy, intrathecal chemotherapy, and symptom management in the treatment of leptomeningeal metastasis. *Curr Oncol Rep* 2003;5:29–40.

207. Lassman AB, Abrey LE, Shah GD, et al. Systemic high-dose intravenous methotrexate for central nervous system metastases. *J Neuro-Oncol* 2006;78:255–260.

208. Ekenel M, Hormigo AM, Peak S, et al. Capecitabine therapy of central nervous system metastases from breast cancer. *J Neuro-Oncol* 2007;85(2):223–227.

209. Burch PA, Grossman SA, Reinhard CS. Spinal cord penetration of intrathecally administered cytarabine and methotrexate: a quantitative autoradiographic study. *J Natl Cancer Inst* 1988;80:1211–1216.

210. Boogerd W, van den Bent MJ, Koehler PJ, et al. The relevance of intraventricular chemotherapy for leptomeningeal metastasis in breast cancer: a randomised study. *Eur J Cancer [A]* 2004;40:2726–2733.

211. Larson SM, Schall GL, Di Chiro G. The influence of previous lumbar puncture and pneumoencephalography on the incidence of unsuccessful radioisotope cisternography. *J Nucl Med* 1971; 12:555–557.

212. Dyck P. Lumbar reservoir for intrathecal chemotherapy. *Cancer* 1985;55:2771–2773.

213. Shapiro WR, Young DF, Mehta BM. Methotrexate: distribution in cerebrospinal fluid after intravenous, ventricular and lumbar injections. *N Engl J Med* 1975;293:161–166.

214. Shapiro WR, Posner JB, Ushio Y, et al. Treatment of meningeal neoplasms. *Cancer Treat Rep* 1977;61:733–743.

215. Glantz MJ, Jaeckle KA, Chamberlain MC, et al. A randomized controlled trial comparing intrathecal sustained-release cytarabine (DepoCyt) to intrathecal methotrexate in patients with neoplastic meningitis from solid tumors. *Clin Cancer Res* 1999;5:3394–3402.

216. Gammon DC, Bhatt MS, Tran L, et al. Intrathecal topotecan in adult patients with neoplastic meningitis. *Am J Health Syst Pharm* 2006;63:2083–2086.

217. Slavc I, Schuller E, Czech T, et al. Intrathecal mafosfamide therapy for pediatric brain tumors with meningeal dissemination. *J Neuro-Oncol* 1998;38:213–218.

218. Berg SL, Balis FM, Zimm S, et al. Phase I/II trial and pharmacokinetics of intrathecal diaziquone in refractory meningeal malignancies. *J Clin Oncol* 1992;10:143–148.

219. Gururangan S, Petros WP, Poussaint TY, et al. Phase I trial of intrathecal spartaject busulfan in children with neoplastic meningitis: a Pediatric Brain Tumor

Consortium Study (PBTC-004). *Clin Cancer Res* 2006;12:1540–1546.

220. Rubenstein JL, Fridlyand J, Abrey L, et al. Phase I study of intraventricular administration of rituximab in patients with recurrent CNS and intraocular lymphoma. *J Clin Oncol* 2007;25:1350–1356.

221. Stemmler HJ, Schmitt M, Harbeck N, et al. Application of intrathecal trastuzumab for treatment of meningeal carcinomatosis in HER2-overexpressing metastatic breast cancer. *Oncol Rep* 2006;15:1373–1377.

222. Herrlinger U, Weller M, Schabet M. New aspects of immunotherapy of leptomeningeal metastasis. *J Neuro-Oncol* 1998;38:233–239.

223. Strother DR, Glynn-Barnhart A, Kovnar E, et al. Variability in the disposition of intraventricular methotrexate: a proposal for rational dosing. *J Clin Oncol* 1989;7:1741–1747.

224. Bleyer WA. Clinical pharmacology of intrathecal methotrexate. II. An improved dosage regimen derived from age-related pharmacokinetics. *Cancer Treat Rep* 1977;61:1419–1425.

225. Sullivan MP, Moon TE, Trueworthy R, et al. Combination intrathecal therapy for meningeal leukemia: two versus three drugs. *Blood* 1977;50:471–479.

225a. Bode U, Magrath IT, Bleyer WA, Poplack DG, Glaubiger DL. Active transport of methotrexate from cerebrospinal fluid in humans. *Cancer Res* 1980;40:2184–2187.

226. Mehta BM, Glass JP, Shapiro WR. Serum and cerebrospinal fluid distribution of 5-methyltetrahydrofolate after intravenous calcium leucovorin and intra-Ommaya methotrexate administration in patients with meningeal carcinomatosis. *Cancer Res* 1983;43:435–438.

227. Shinoura N, Tabei Y, Yamada R, et al. Continuous intrathecal treatment with methotrexate via subcutaneous port: implications for leptomeningeal dissemination of malignant tumors. *J Neuro-Oncol*, in press

228. Benesch M, Urban C. Liposomal cytarabine for leukemic and lymphomatous meningitis: recent developments. *Exp Opin Pharmacother* 2008;9:302–309.

229. Phuphanich S, Maria B, Braeckman R, et al. A pharmacokinetic study of intra-CSF administered encapsulated cytarabine (DepoCyt) for the treatment of neoplastic meningitis in patients with leukemia, lymphoma, or solid tumors as part of a phase III study. *J Neuro-Oncol* 2007;81:201–208.

230. Kramer K, Modak S, Kushner BH, et al. Metastatic neuroblastoma (NB) to the central nervous system (CNS): improved outcome with combined modality including 131-I-8H9 or 131-I-F8 radioimmunotherapy (RIG delivered through cerebral spinal fluid (CSF). *J Clin Oncol* 2007:abstract 2022.

231. Tham YL, Hinckley L, Teh BS, et al. Long-term clinical response in leptomeningeal metastases from breast cancer treated with capecitabine monotherapy: a case report. *Clin Breast Cancer* 2006;7:164–166.

232. Ku GY, Krol G, Ilson DH. Successful treatment of leptomeningeal disease in colorectal cancer with a regimen of bevacizumab, temozolomide, and irinotecan. *J Clin Oncol* 2007;25:e14–16.

233. Choong NW, Dietrich S, Seiwert TY, et al. Gefitinib response of erlotinib-refractory lung cancer involving meninges—role of EGFR mutation. *Nat Clin Pract Oncol* 2006;3:50–57.

234. Takahashi M, Yamada R, Tabei Y, et al. Navigation-guided Ommaya reservoir placement: implications for the treatment of leptomeningeal metastases. *Minim Invasive Neurosurg* 2007;50:340–345.

235. Omuro AM, Lallana EC, Bilsky MH, et al. Ventriculoperitoneal shunt in patients with leptomeningeal metastasis. *Neurology* 2005;64:1625–1627.

236. Eralp Y, Saip P, Aydin Z, et al. Leptomeningeal dissemination of ovarian carcinoma through a ventriculoperitoneal shunt. *Gynecol Oncol* 2008;108:248–250.

237. Machado M, Sacman M, Kaplan RS, et al. Expanded role of the cerebrospinal fluid reservoir in neurooncology: indications, causes of revision, and complications. *Neurosurgery* 1985;17:600–603.

238. Sandberg DI, Bilsky MH, Souweidane MM, et al. Ommaya reservoirs for the treatment of leptomeningeal metastases. *Neurosurgery* 2000;47:49–54.

239. Obbens EA, Leavens ME, Beal JW, et al. Ommaya reservoirs in 387 cancer patients: a 15-year experience. *Neurology* 1985;35:1274–1278.

240. Ongerboer de Visser BW, Somers R, Nooyen WH, et al. Intraventricular methotrexate therapy of leptomeningeal metastasis from breast carcinoma. *Neurology* 1983;33:1565–1572.

241. Hitchins RN, Bell DR, Woods RL, et al. A prospective randomized trial of single-agent versus combination chemotherapy in meningeal carcinomatosis. *J Clin Oncol* 1987;5:1655–1662.

242. Dinndorf PA, Bleyer WA. Management of infectious complications of intraventricular reservoirs in cancer patients: low incidence and successful treatment without reservoir removal. *Cancer Drug Deliv* 1987;4:105–117.

243. Pfeffer MR, Wygoda M, Siegal T. Leptomeningeal metastases—treatment results in 98 consecutive patients. *Isr J Med Sci* 1988;24:611–618.

244. Siegal T, Pfeffer MR, Steiner I. Antibiotic therapy for infected Ommaya reservoir systems. *Neurosurgery* 1988;22:97–100.

245. Lishner M, Perrin RG, Feld R, et al. Complications associated with Ommaya reservoirs in patients with cancer. The Princess Margaret Hospital experience and a review of the literature. *Arch Intern Med* 1990;150:173–176.

246. Grant R, Naylor B, Junck L, et al. Clinical outcome in aggressively treated meningeal carcinomatosis. *Neurology* 1992;42:252–254.

247. Chamberlain MC, Kormanik PA, Barba D. Complications associated with intraventricular chemotherapy in patients with leptomeningeal metastases. *J Neurosurg* 1997;87:694–699.

248. Siegal T. Toxicity of treatment for neoplastic meningitis. *Curr Oncol Rep* 2003;5:41–49.

249. Sutherland GE, Palitang EG, Marr JJ, et al. Sterilization of Ommaya reservoir by instillation of vancomycin. *Am J Med* 1981;71:1068–1070.

250. Trump DL, Grossman SA, Thompson G, et al. CSF infections complicating the management of neoplastic meningitis. Clinical features and results of therapy. *Arch Intern Med* 1982;142:583–586.

251. Park DM, DeAngelis LM. Delayed infection of the Ommaya reservoir. *Neurology* 2002; 59:956–957.

252. Gleissner B, Chamberlain M. Treatment of CNS dissemination in systemic lymphoma. *J Neuro-Oncol* 2007;84:107–117.

253. Kopelson G, Parkinson D, Rudders RA. Long-term survivors with leptomeningeal tumor involvement. Letter to the editor. *Int J Radiat Oncol Biol Phys* 1983;9:119–120.

254. Grossman SA, Moynihan TJ. Neoplastic meningitis. *Neurol Clin* 1991;9:843–856.

Chapter 8

Cancer Involving Cranial and Peripheral Nerves and Muscles

INTRODUCTION

Both metastatic and nonmetastatic complications of cancer can affect peripheral nerves. Metastases are the common cause of cranial nerve involvement in cancer patients; peripheral nerve involvement is more commonly nonmetastatic (see Chapter 12). Although peripheral nerve lesions generally lack the potential for neurologic devastation that occurs with brain, leptomeningeal, and spinal cord metastases, they often cause severe pain and, depending on the nerve involved, can cause substantial neurologic disability.

Tumors can involve nerves either by direct extension, as when the brachial plexus is compressed by a superior sulcus tumor of the lung, or by hematogenous metastases, as when leukemia metastasizes to the brachial plexus or when cranial nerves are invaded by a tumor that metastasizes to the base of the skull.[1] Tumors that directly extend to involve nerves and those that metastasize to nerves can cause damage either by compressing nerve fibers or by invading them (Fig. 8–1). Compressive lesions are fairly easy to identify radiologically, but such identification is often difficult with invasive lesions.

The peripheral nervous system and muscles may be involved by metastases either focally, multifocally, or rarely diffusely. This chapter emphasizes those cranial and peripheral nerve lesions that occur after nerves have exited from the subarachnoid space, thus distinguishing them from leptomeningeal metastases.[2,3] Table 8–1 classifies the lesions by anatomic site and contrasts them with nonmetastatic lesions of peripheral nerves discussed in other chapters.

Muscle metastases are uncommon, but can cause significant symptoms. They are also considered in this chapter.

FREQUENCY

The frequency of peripheral nerve involvement in patients with cancer is probably underdiagnosed for several reasons: (1) Peripheral nerve

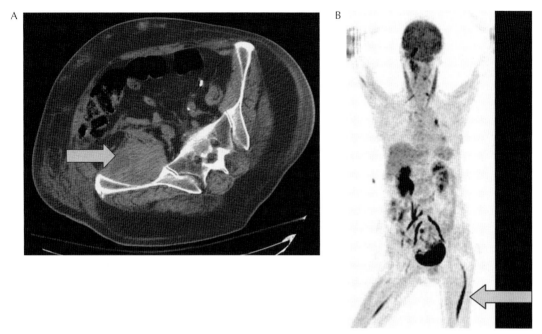

Figure 8–1. Metastases to peripheral nerves: compression versus invasion. **A**: A 69-year-old man with renal carcinoma developed severe pain in the right groin followed by progressive weakness of the leg. Examination revealed weakness of the iliopsoas and quadriceps and absent patellar reflex. A magnetic resonance image identified a large mass in the iliacus muscle compressing the femoral nerve (*arrow*). **B**: A 63-year-old woman with diffuse large cell lymphoma and a good response to chemotherapy later developed progressive weakness of her left leg. A positron emission tomography scan revealed hypermetabolism in the femoral nerve (*arrow*) as well as several other nerves throughout the body.

involvement is rarely included in the diagnostic coding of patients with widespread cancer. (2) The most frequent anatomic sites of nerve metastases, at the skull base and in the brachial plexus, are usually not carefully dissected during autopsy examinations. (3) Pain, the most common symptom of peripheral nerve involvement, is often believed to result from local metastases affecting nerve endings rather than from direct involvement of a nerve trunk or plexus.

A few studies address the incidence of peripheral and cranial nerve dysfunction in particular tumors. For example, facial nerve paralysis occurs in approximately 20% of patients with malignant parotid neoplasms.[4] In approximately 3% of patients with lung cancer, the cancer develops in the superior pulmonary sulcus (Pancoast tumor) and involves the brachial plexus.[5] Even in the absence of incidence data, it is clear that cranial and peripheral nerve dysfunction is a common and important clinical problem. Some tumors, particularly those involving the face, where there is a high

density of nerve endings, are neurotropic, often tracking back along cranial nerves to reach the subarachnoid space or even the brainstem[6–9] (Fig. 8–2). Neural invasion also occurs in several other cancers, including prostate,[10] breast,[11] lung,[12] pancreas,[13] and other gastrointestinal (GI) cancers[14–16]; most studies indicate that perineural invasion portends a worse prognosis.

CRANIAL NERVES

Each of the 12 cranial nerves can be involved selectively by tumor. Table 8–2 lists some common cranial neuropathies caused by invasion of cancer.

Tumors that involve cranial nerves usually either originate near the base of the skull (e.g., chordoma) or metastasize there from a distant primary site (e.g., prostate cancer). They can then either invade the cranial nerves or compress them. Tumors also may grow into or through neural foramina of the skull. Neural

Table 8–1. Cranial and Peripheral Neuropathies Caused by Cancer

	Tumor	Paraneoplastic (Chapter 15)	Treatment (Chapters 12–14)	Other (Chapters 9–10)
Cranial nerves	Leptomeningeal metastases, skull-base metastases, perineural spread	Brainstem encephalitis	Radiotherapy, chemotherapy, surgery	Herpes zoster
Sensory ganglia	Metastases	Sensory neuronopathy	Chemotherapy	Herpes zoster
Roots	Leptomeningeal metastases, compression from vertebral metastases		Radiotherapy	Herpes zoster
Plexuses				
Cervical	Head and neck tumor		Radiotherapy	
Brachial	Breast or lung cancer (mainly lower plexus)	Inflammatory plexopathy	Radiotherapy (mainly upper plexus)	
Lumbosacral	Prostate, gynecological, colon cancer	Inflammatory plexopathy	Radiotherapy (mainly upper plexus)	
Peripheral nerves				
Mononeuropathy	Metastases, compression, or invasion		Radiotherapy	Femoral nerve, iliopsoas hemorrhage, peroneal nerve palsy (weight loss)
Polyneuropathy	Neurolymphomatosis	Paraneoplastic neuropathy, vasculitis	Chemotherapy	Cachexia, metabolic
Neuromyotonia		Neuromyotonia with VGKC-Ab and thymoma	Chemotherapy (transient neuromyotonia), radiotherapy (focal neuromyotonia)	
Lower motor neuron		Paraneoplastic ALS	Radiotherapy	
Neuromuscular junction		Lambert–Eaton myasthenic syndrome (SCLC)	Myasthenia (interferon)?	Aminoglycoside antibiotics (Chapter 4)
Muscle	Hematogenous or direct invasion	Dermato (poly)myositis	Myositis	Steroids

Modified from Ref. 2.
VGKC-Ab, voltage-gated potassium channel antibodies; ALS, amyotrophic lateral sclerosis; SCLC, small cell lung cancer.

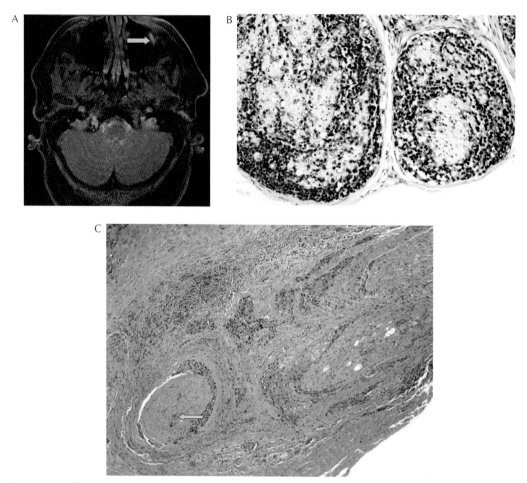

Figure 8–2. Infiltration of peripheral nerves by tumor. **A**: An MRI showing a thickened, enhancing inferior alveolar nerve (*arrow*). Several months after removal of a basal cell carcinoma of the face, this patient developed numbness and tingling over the maxillary sinus that progressed over several months. He then developed facial nerve weakness. Biopsy of the infraorbital nerve revealed infiltration with squamous cell carcinoma. **B**: Cross-section of a nerve invaded by lymphoma. The brown staining cells are lymphoma cells. The lighter staining structures represent axons and their myelin sheaths. Most of the invasion is in the perineurium. The nerve itself can be damaged by demyelination, direct axonal damage, or ischemia from invasion of vascular structures. **C**: An H&E section of a cranial nerve invaded by squamous cell carcinoma. The arrow points to tumor cells within the nerve fascicle. Note that a schematic cross-section of a peripheral nerve infiltrated by tumor is shown in Figure 2–9.

foramen invasion is a common complication of nasopharyngeal carcinoma; approximately 15% to 40% of patients first present with cranial nerve palsies,[17–19] a poor prognostic sign. A metastasis from a distant primary site (e.g., breast, lung, or prostate cancer) to the skull base can also invade or compress cranial nerves.[20–22] Cranial nerve dysfunction is sometimes the first manifestation of cancer.[23–25] Metastases to soft tissue or vascular structures at the skull base or in the upper neck can lead to cranial nerve damage. A common site is the cavernous sinus.[18]

Clinical Findings

Cranial nerve palsies may either be the first sign of cancer[23–25] or occur late in the course of the disease. In a study of 150 patients with nasopharyngeal cancer, 23 presented with cranial nerve dysfunction as the only symptom. Symptoms included diplopia, trigeminal neuralgia, dysarthria, dysphagia, anosmia, ptosis, and hoarseness. Neurologic symptoms eventually developed in 74 additional patients.[26] A cavernous sinus syndrome with visual loss

Table 8–2. **Metastatic Lesions Causing Cranial Neuropathies**

Lesion Site	Findings	Comments
Nasal cavity, anterior cranial fossa (olfactory nerves)	Hyposmia, anosmia, dysosmia	Esthesioneuroblastoma, nasopharyngeal cancers
Eye: globe, optic nerve	Decreased visual acuity; retinal detachment	Choroidal lesions are more common than retinal; pain, proptosis, and diplopia are rare; breast and lung cancer are common causes
Eye: orbital apex, superior orbital fissure, ocular motor nerves	Pain, proptosis, diplopia (single ocular nerve or muscle), sensory loss V_1; decreased visual acuity in one-third of cases, usually late	As common as choroidal metastases; breast, prostate cancer, and neuroblastoma are common causes
Parasellar plus trigeminal	Unilateral frontal headache, oculomotor palsies (III, IV, VI), sensory loss V_1	Vision is rarely affected, no proptosis
Sella turcica	Diabetes insipidus	Anterior pituitary insufficiency, and visual loss are uncommon; when present, they suggest a primary pituitary tumor
Middle cranial fossa	Facial numbness ($V_{2,3}$); abducens palsy in some (VI)	Lightning-like facial pain (trigeminal neuralgia) is rare in patients with neoplastic compression
Jugular foramen	Hoarseness, dysphagia, pain in pharynx (IX, X), sternocleidomastoid weakness (XI), occasionally tongue weakness (XII)	Papilledema may occur if a dominant jugular vein is compressed; glossopharyngeal neuralgia is uncommon
Occipital condyle	Unilateral occipital pain and neck stiffness, unilateral tongue weakness (XII)	Pain may radiate to frontal or temporal areas and may be the major complaint
Mandible	Unilateral numb chin and gum, "mental neuropathy"	Numbness also results from meningeal or base-of-skull metastases; breast cancer and lymphoma most common
Carotid sinus or glossopharyngeal nerve	Syncope, pharynx or neck pain on swallowing	Cardioinhibitory, vasodepressor, or both; head and neck cancer; indicates recurrent tumor; may be life threatening

as an early symptom may be the presenting sign of systemic lymphoma.[27,28]

Cranial neuropathies can also signal recurrence of a successfully treated cancer. This phenomenon is fairly common with neurotropic head and neck tumors that may cause their first neurologic symptom years after their initial surgical removal.[29] One study[30] estimates that cranial nerve dysfunction occurs in 13% of patients with breast carcinoma. The most frequently affected cranial nerves were trigeminal and facial; cranial nerve palsies are also relatively common in advanced prostate cancer.[22,31]

Radiation therapy (RT) is often effective in the treatment of cranial nerves infiltrated by cancer. In one series of 32 patients with prostate cancer and cranial nerve dysfunction, one-half improved with treatment.[32] Unfortunately, as in this situation, most patients have far advanced disease and the median survival was only 3 months. Clinical improvement is similar in patients with breast cancer but survival is longer.[30] The prognosis for recovery of cranial nerve function is better in patients with nasopharyngeal carcinoma,[33,34] likely because nerve compression occurs earlier in the course of the illness.

OLFACTORY NERVE (CRANIAL NERVE I)

The olfactory nerve is rarely tested in a routine oncologic or neurologic examination. This is unfortunate because abnormalities of taste and smell are common in patients with cancer (Chapter 4) and bedside testing of olfactory perception is fairly easy.[35] The alcohol sniff test can discriminate anosmia from hyposmia and from normal function by measuring the distance an alcohol pad must be held from the nose for the patient to perceive it; alcohol held directly under the nose can stimulate trigeminal receptors, causing the patient to report something even in the absence of adequate olfactory function. For more sophisticated testing, cards permeated with specific odors can be used.[36] Probably because olfaction is so seldom tested, only a few reports describe hyposmia or anosmia as an important finding in either primary or metastatic cancer. One report identifies anosmia in 3% of patients with nasopharyngeal cancer.[18] In another report of 366 patients with nasopharyngeal carcinoma, anosmia was not mentioned, although almost 20% presented with cranial nerve palsies.[37] As indicated in the following text, oculomotor and trigeminal dysfunction are the most frequently reported cranial nerve palsies in this disorder.[33,37] Esthesioneuroblastoma (olfactory neuroblastoma) presents with anosmia in 11% of patients.[38] Only in olfactory meningiomas is anosmia reported as a major symptom (58%).[39]

The olfactory system, which is an extension of the brain rather than a true cranial nerve, has a long intracranial course outside the cerebrum proper, along the base of the skull and through the cribriform plate to the olfactory epithelium (Fig. 8–3). Metastases to the base of the anterior fossa would be expected to compress the olfactory nerve, either unilaterally or bilaterally, causing loss of odor perception. The optic nerves lie nearby and are often involved as well. The olfactory groove is occasionally compressed and olfactory function lost as a false-localizing sign of a brain tumor or hydrocephalus.[40]

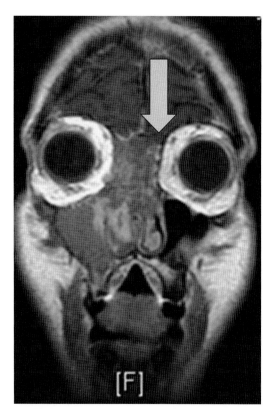

Figure 8–3. Olfactory nerve involvement by an esthesioneuroblastoma. A 53-year-old man presented to medical attention with nasal stuffiness and anosmia. A magnetic resonance image showed a large contrast-enhancing mass invading the maxillary sinus and the nasal cavity, destroying the cribriform plate (*arrow*). Initial treatment led to resolution of the tumor, but he subsequently succumbed to leptomeningeal spread.

Unlike most of the central and peripheral neurons, olfactory receptors (and taste buds) have a relatively high cellular turnover rate. Because these cells are rapidly reproducing, they can be damaged by radiation and chemotherapy (see Chapters 12 and 13), as well as by the cancer itself, often leading to abnormalities of odor and taste perception that can depress appetite, cause weight loss, and diminish quality of life.[41] However, their regenerative capacity often leads to restoration of function weeks to months after compression is relieved or the toxic insult has ended.

OPTIC NERVE (CRANIAL NERVE II)

Like the olfactory nerve, the optic nerve is part of the brain rather than a true cranial nerve.

Isolated hematogenous invasion of the optic nerve itself is uncommon,[42–44] but visual loss is a relatively common symptom of nervous system metastases[45] (Fig. 8–4). Acute blindness in cancer patients can arise from a variety of causes. One report describes 10 patients with acute blindness encountered over 20 months in a cancer center.[46] Three were due to stroke (see Chapter 9), two due to posterior reversible encephalopathy syndrome (see Chapter 12), two due to herniation syndromes (see Chapter 3), and only three due to direct involvement of the optic nerves, all from leptomeningeal metastases (see Chapter 7).

Intraocular metastases affect as many as 12% of patients who die of cancer and perhaps a third of those who die of breast cancer.[47] In one series of 716 eyes examined at autopsy from patients who died of malignant neoplasms, the overall incidence of ocular metastases was 9%.[48] The highly vascular choroid is the most common site of intraocular metastases, but visual dysfunction can be caused by metastases to the iris, ciliary body, vitreous, conjunctiva,

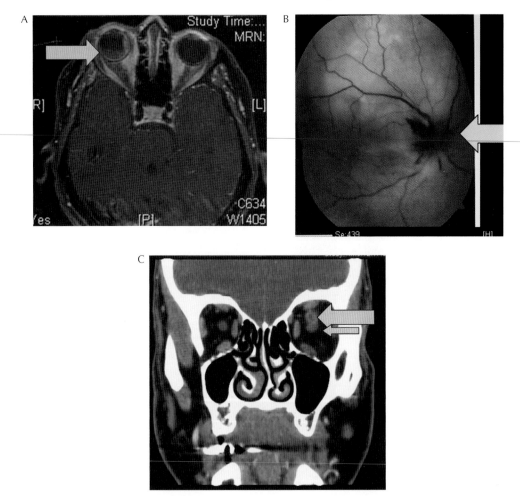

Figure 8–4. Metastasis to the globe and orbit. **A**: A woman with breast cancer complained of failing vision in the right eye. A magnetic resonance image (MRI) revealed a choroidal metastasis extending into the vitreous (*arrow*). **B**: A retinal photograph of a man with renal cell carcinoma complaining of failing vision. The upper part of the figure shows the retina to be elevated from a choroidal metastasis. The optic disk is hemorrhagic (*arrow*), representing either papilledema or more likely metastasis to the nerve. **C**: A 51-year-old man with known metastatic appendiceal carcinoma developed painless visual loss in the left eye. On direct questioning, he noticed discomfort, not actually pain, when he moved the eye. An MRI revealed metastatic tumor involving the optic nerve. The tumor (*large arrow*) sat just above the optic nerve (*small arrow*).

retina, eyelids, lacrimal glands, and even the avascular cornea.[44,49,50] Furthermore, visual loss may result from metastases to the skull base, orbit (orbital metastases cause diplopia but rarely visual loss [see following text]), optic chiasm, or the leptomeninges surrounding the optic nerves (see Chapter 7). Metastatic brain tumors may also cause visual loss. For example, cortical blindness can result from bilateral occipital metastases or may be a secondary effect of metastasis-induced increased intracranial pressure either compressing the optic nerves themselves or the posterior cerebral arteries that supply the visual cortex. Hydrocephalus can produce visual changes when the enlarging third ventricle compresses the optic chiasm.[51]

Ocular metastases usually occur in patients with known cancer, but may be the presenting complaint. Even when they are not the presenting complaint, ocular metastases may occur when the patient's primary tumor has been "successfully" treated and the patient is believed to be cured. Among the solid tumors, breast cancer is the most common,[45] whereas lung cancer, melanoma,[52] and other solid tumors are less common.

Ocular abnormalities are also common in the leukemias[53] and affect adults more often than children. Ocular changes are more common in those with myeloid leukemia and myelodysplastic syndromes[54] than in those with lymphoid leukemia. The majority of changes affect the retinal vasculature and are asymptomatic; however, in one series 10% of patients with leukemia had ophthalmologic symptoms at diagnosis.[53] Leukemic retinal infiltrates[55] or optic nerve invasion[56] can cause visual loss as an early symptom of leukemia or as the first symptom of recurrence because the optic nerve, like the rest of the CNS, can be a sanctuary for leukemic cells.

The diagnosis is usually suggested by the rapid onset of visual loss, often bilateral. The leukemic infiltrates can be identified on routine funduscopic examination, although the raised hemorrhagic disk may be confused with papilledema. Distinguishing papilledema caused by increased intracranial pressure from infiltration of the optic disk by tumor can be difficult, but papilledema rarely causes loss of visual acuity, except for transient visual obscurations that are common but not dangerous. Infiltration of the optic disk, on the other hand, usually causes severe visual loss and should be treated immediately. Additionally, optic nerve infiltration is typically unilateral, but is surprisingly often bilateral, whereas papilledema is almost always bilateral (the Foster–Kennedy syndrome is caused by compression of one optic nerve by tumor leading to atrophy. If the tumor raises intracranial pressure, papilledema occurs only in the contralateral eye). Other sites of leukemic sanctuary that may cause visual loss include the anterior chamber of the eye.[57–59] The presence of cells in the anterior chamber or the vitreous (uveitis) of patients with a history of leukemia requires vitreous biopsy to exclude recurrent leukemia.

Leukemia often affects the retina, as found in 8 of 60 children at autopsy.[60] Choroid infiltrates were found in 26, all asymptomatic during life, and 4 had optic nerve involvement associated with leptomeningeal leukemia.

Hodgkin and non–Hodgkin lymphomas and solid tumors, including carcinomas of the breast and lung, can also cause isolated hematogenous metastases to the optic nerve. When such tumors metastasize to the optic nerve head, they can be seen by funduscopic examination, but if the metastasis is retrobulbar, the disk may appear normal or atrophic despite severe visual loss.

Patients with choroid metastases usually complain of painless blurred vision, visual loss, or both, depending on the choroidal lesion site (Fig. 8–4). The lesion can often be seen on routine funduscopic examination as a mass elevating the retina, often associated with retinal detachment. However, indirect ophthalmoscopy through dilated pupils or ocular ultrasound may be required to identify peripheral lesions. If untreated, such lesions cause secondary glaucoma with intractable pain. The metastases usually respond to treatment with either radiation or chemotherapy.[47,61–63]

Orbital tumors (see following text) cause pain, proptosis, and diplopia from ocular muscle palsies, although vision may be lost if the optic nerve is also invaded or compressed. One unusual sign reported with primary optic nerve tumors is transient unilateral visual loss that occurs when the patient deviates his or her eyes in a particular direction. The sign is believed to be caused by arterial compression from the tumor.[64]

Metastases to the iris or ciliary body are uncommon and usually asymptomatic. Occasionally they bleed, causing hyphema as an initial sign of the cancer.[65] Ciliary body metastases may simulate uveitis but, as indicated in the preceding text, cytologic examination of the aqueous humor yields tumor rather than inflammatory cells. Breast cancer is the common offender.

Tumor need not be in the orbit or optic nerve to cause visual loss or ophthalmoplegia. Metastasis to, or primary carcinoma of, the sphenoid sinus often presents with visual loss when the tumor invades the anterior–superior portion of the sinus to involve the medial aspect of the optic canal. Diplopia results from involvement of the superior orbital fissure.[66]

OCULAR MOTOR NERVES (CRANIAL NERVES III, IV, AND VI)

Three nerves supply the six muscles (extraocular muscles) that move the globe within the orbit: oculomotor (cranial nerve III), trochlear (cranial nerve IV), and abducens (cranial nerve VI). Dysfunction of these nerves or their muscles, either alone or in combination, causes diplopia. Individual cranial nerves or the muscles they supply are more likely to be involved where they are anatomically further apart, as in the lateral orbit or the brainstem; they are likely to be involved in combination where they are close together, as in the orbital apex or cavernous sinus. An orbital metastasis can cause diplopia without either direct nerve or muscle damage simply by shifting the globe. When this occurs, neither the patient nor the ophthalmologist can move the globe through a full range of motion (forced ductions). With nerve or muscle involvement, the patient cannot move the globe but the ophthalmologist can. One of our patients with breast cancer and widespread bony metastases developed horizontal diplopia. She was believed to have a sixth nerve palsy from base of skull metastasis until it was demonstrated that the globe could not be moved laterally on forced duction testing. Careful imaging identified metastases in the lateral orbit.

The intraocular muscles include the pupillodilator fibers (sympathetic fibers) and the pupilloconstrictor and ciliary fibers that constrict the pupil and move the lens to accommodate close or distant vision. These enter the orbit as part of the oculomotor nerve, so impairment of oculomotor function may also be accompanied by pupillary or accommodation abnormalities.

Patients with involvement of extraocular muscles or nerves first complain of slightly blurred vision, like "a ghost on a television set," which clears when either eye is covered. (A simple bedside test for diplopia is asking the patient to cover the eyes, one at a time. If covering each eye clears up the blurring, the patient is suffering from diplopia. If the blurred vision persists, the problem is not diplopia from extraocular nerve dysfunction. Double vision with the eye covered, monocular diplopia, can occur with cerebral lesions, refractive abnormalities such as cataract or retinal detachment.[67] Detailed bedside testing of diplopia is described in a recent review.[68]) As the dysfunction progresses, the images become double. Involvement of the abducens nerve or the lateral rectus muscle causes horizontal diplopia. Involvement of the oculomotor nerve causes vertical diplopia with ptosis. Trochlear nerve paralysis causes oblique diplopia and may cause a head tilt away from the side of the lesion. The symptoms are usually progressive, although rare spontaneous remissions have been reported.[69] Paradoxically, the more severe the ocular motor paralysis, the less it bothers the patient. Images close together are much more disturbing than large deviations, in which one image can be relegated to the visual periphery. Because the binocular visual apparatus is so finely tuned, minimal dysfunction of ocular motor nerves causes symptoms. A patient often complains of diplopia undetected by the physician during a bedside examination of the ocular muscles. Conversely, diplopia found by the examiner at extremes of the patient's lateral gaze, but not otherwise noticed by the patient, is probably irrelevant.

Diplopia may be caused by lesions anywhere along the ocular motor apparatus, from the orbit to the brainstem, or by lesions that infiltrate ocular muscles.[70,71] Orbital metastases are common causes,[72,73] particularly those that emanate from breast[74] and prostate[75] cancers and from neuroblastomas.[76] However, metastatic tumors represent only approximately 10% of all orbital masses even in the older adult population.[77] Orbital metastasis may be the first evidence of cancer, 19% in one series,[72] or a late sign of widely metastatic disease. Affected patients usually present with diplopia with or without pain. Proptosis (exophthalmos), usually unilateral, is also an early symptom (Fig. 8–5). Occasional patients with scirrhous breast carcinoma develop diplopia associated with enophthalmos (retraction of the globe)

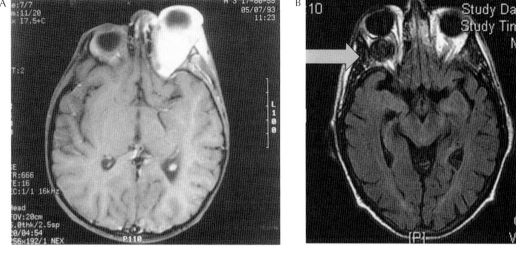

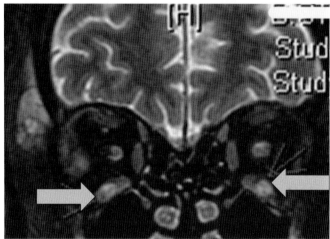

Figure 8–5. Orbital metastases. **A**: A child with a neuroblastoma presented with visual loss and severe proptosis. He was too young to have complained of diplopia, but that must have been the initial symptom. **B**: This patient with melanoma did complain of diplopia and was noted to have proptosis. The mass displaces the globe, which could be responsible for the diplopia but also probably damages the abducens nerve, producing a lateral rectus palsy. **C**: Bilateral metastases to the inferior rectus muscles (*arrows*) causing diplopia with neither visual loss nor proptosis.

caused by the metastatic fibrotic reaction.[78,79] Highly vascular metastases, such as renal cell carcinoma, may cause pulsating exophthalmos that can be mistaken for a vascular malformation or an arteriovenous fistula.[80] Orbital metastases are easily identified by computed tomography (CT) or magnetic resonance scan. If an orbital lesion is suspected, the physician must specifically request orbital views, preferably with fat-suppressed, post-gadolinium T1 sequences, because routine scanning may not detect small orbital lesions. In rare patients, metastases to the orbit that originate from colon

cancer or melanoma can calcify, confusing the diagnosis.[81]

Anteriorly placed orbital lesions usually affect only one of the three ocular nerves. Also, because they are distant from the optic nerve, these lesions usually do not cause visual loss until late. More posteriorly placed lesions, such as those at the superior orbital fissure or orbital apex, involve all three ocular nerves and, often, the first division of the trigeminal nerve as well, causing numbness of the forehead.[82] Orbital apex lesions involve the optic nerve and may cause early visual loss, whereas a superior orbital fissure lesion is slightly more anteriorly placed and affects vision later, if at all.

Base-of-skull tumors are a common cause of diplopia from involvement of ocular motor nerves and often provide the first clue that a patient has cancer or that a known primary cancer has metastasized. Two common sites of metastasis are the cavernous sinus and the parasellar area; tumors at both of these areas are likely to involve more than one ocular motor nerve and the trigeminal nerve as well (see following text). The clivus is also a common site of metastasis and can present as an isolated VI nerve palsy.

Although ocular motor palsies are common symptoms of metastatic tumor, tumor is not the most frequent cause. One series of 49 patients, aged 15 to 50, with isolated sixth nerve palsies revealed only two patients with metastatic tumors, one from breast carcinoma and one from a cylindroma.[83] Even when the paralysis lasted more than 3 months, as in another series of 38 patients, only three instances of paralysis were due to metastasis.[84] In 1000 unselected patients with paralysis of one or more ocular motor nerves, the paralysis was caused by neoplasms in 143 (14%). Of these, 69 were metastatic and another 6 were due to nasopharyngeal carcinomas. Of the 75 patients with metastatic or nasopharyngeal tumors that caused ocular motor paralysis, cranial nerve VI alone was affected in 35 patients, III alone in 12, and IV alone in 4. Combinations included III and IV in three patients, III and VI in seven patients, and all three nerves in 14 patients.[85]

Ocular Motor Syndromes

The orbital,[82] parasellar,[86] and cavernous sinus syndromes[87,88] are caused by direct extension of cancer from the skull base or by metastases to the area. Each has characteristic signs and symptoms.

The orbital syndrome (Fig. 8–5) occurs with carcinomas of the breast, lung, and prostate, and with non–Hodgkin lymphoma and neuroblastoma.[76,89] Tumor metastasizes to the skull and grows into the orbit or, rarely, metastasizes to the soft tissues of the orbit itself. Whichever the cause, the syndrome is usually characterized initially by a dull, continuous supraorbital ache. The first neurologic symptom other than pain is generally blurred binocular vision, soon followed by frank diplopia. Initially, the examination shows proptosis of the involved eye, accompanied by a degree of external ophthalmoplegia. Many patients also have decreased sensation in the ophthalmic division of the trigeminal nerve. Rarely, enophthalmos is present.[78] When the orbit is palpated, the tumor sometimes can be felt either as a discrete mass or as increased resistance to orbital displacement. At other times, a striking tenderness is noted when bone surrounding the orbit is palpated. Because the tumor generally stays outside of the sheath surrounding the optic nerve, decreased vision, visual field deficits, and papilledema rarely occur until very late in the course, when failing vision and optic atrophy appear. One of our patients with Ewing sarcoma developed an orbital metastasis characterized by diplopia and, later, optic atrophy. The lesion responded to radiation. A year later, he presented with headache and contralateral papilledema, or Foster Kennedy syndrome, and was found to have a skull metastasis compressing the sagittal sinus (Chapter 5). The atrophic optic nerve could not develop papilledema.

Orbital metastases are an unusual complication of systemic cancer. Ferry and Font[90] reported only 28 patients with orbital metastases from a series of 235 patients with metastatic disease of the eye; the remainder involved the choroid. Nevertheless, in 17 of these 28 patients, orbital metastases were the initial sign of malignancy. In a series of 1264 patients with orbital tumors, 91 (7%) were distant metastases of solid tumors, 130 (10%) lymphomas or leukemias, and 142 (11%) direct extension of tumor into the orbit.[89]

Parasellar metastases arise either in the sella turcica and extend into the cavernous space[86,91] or metastasize directly to the cavernous space[88,92] (Fig. 8–6). The parasellar syndrome is characterized by unilateral frontal headaches and ocular paresis, usually without proptosis. Because the veins draining from the orbit empty into the cavernous sinus, they may be compressed, leading to striking periorbital edema. Thus, edema is usually more prominent in the parasellar syndrome than in the orbital syndrome, but proptosis, if present, is less prominent. The unilateral headache is localized to the supraorbital frontal areas and is generally characterized as a dull, aching

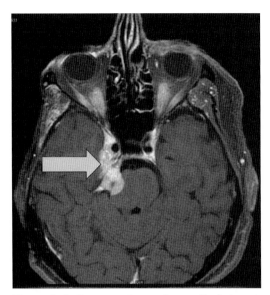

Figure 8–6. Cavernous sinus and parasellar lesions. A 65-year-old man suffered multiple squamous cell carcinomas of the face requiring many operations. He subsequently developed diplopia, with paralysis of all three extraocular muscles and a mid-position pupil fixed to light (both sympathetic and parasympathetic involvement). Tumor had tracked from the face into the cavernous sinus (*arrow*) and ultimately into the brainstem.

pain; some patients complain of episodic, sharp shooting pains that are not, however, typical of trigeminal neuralgia. Diplopia results from involvement of the ocular motor nerves as they traverse the cavernous sinus. Among our seven patients, four had oculomotor nerve paresis, one had trochlear paresis, and five had abducens paresis.[93] Some patients complained of paresthesias over the forehead, and sensory loss was detectable in one. Two patients had papilledema; none complained of visual loss. When the oculomotor nerve is involved, the pupil becomes dilated and fixed (8–9 mm). However, when the sympathetic fibers traversing the cavernous sinus are also involved, the fixed pupil may be mid-position (4–6 mm).

The metastatic parasellar syndrome is relatively common.[86] Roessmann et al.,[94] in an autopsy study of 60 consecutive cases of carcinoma, reported 16 parasellar bony lesions, of which nine were due to breast cancer. Conversely, in 102 patients whose parasellar syndrome was diagnosed clinically, 23 were caused by metastases, usually from cancers of the breast, prostate, or lung.[95] Most investigators agree that parasellar lesions usually begin

with metastasis to the bone, either the petrous apex or the sella turcica, rather than direct metastasis to the cavernous space. Cavernous sinus involvement is usually unilateral, but may be bilateral.[96–98] Magnetic resonance imaging (MRI) is the best diagnostic test.[91,99] Occasionally, a biopsy is necessary to establish the diagnosis.[100] Standard radiotherapy or radiosurgery is the best treatment.[101]

Carcinomas arising in the face, usually squamous but occasionally basal cell, may cause the cavernous sinus syndrome by growing microscopically along the first or second division of the trigeminal nerve to the cavernous sinus[102] (Fig. 8–6). In such instances, facial pain or sensory loss is the first symptom; diplopia occurs later. A mass lesion may not be identifiable on MRI, and diagnosis may require biopsy. The specimen should be taken from nerve endings at the site of the original lesion. In two recent patients, both with facial pain and numbness developing years after resection of basal cell carcinomas of the face, the diagnosis was delayed because images were all reported as normal. In both, careful inspection of the so-called normal MRIs by a neuroradiologist, after being told what to look for by the clinical neurologist, identified abnormal nerves, a superior orbital nerve in one and an infraorbital nerve in the other. In both instances, nerve biopsy established the diagnosis. However, implantation of normal squamous cells into nerve may occur at the original surgery for the cutaneous neoplasm and, at re-excision, be mistaken by the pathologist for perineural invasion by cancer.[103]

TRIGEMINAL NERVE (CRANIAL NERVE V)

Facial numbness (Fig. 8–7), with or without pain, is a common symptom of tumor involving the trigeminal nerve. Facial pain without numbness can also be a symptom of a nonmetastatic lung tumor[104–106] (see p 298). When facial pain or numbness is caused by a metastasis, the site can be anywhere along the course of the trigeminal nerve, from the skin of the face to the entry into the brainstem. The nerve can be involved either by direct infiltration[107] or by compression.[108] Characteristic syndromes related to metastatic cancer are the gasserian ganglion syndrome and the numb chin syndrome.

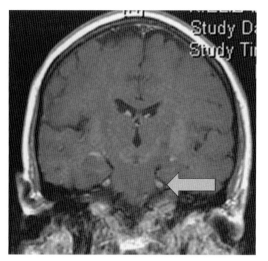

Figure 8–7. Trigeminal neuropathy. This woman with carcinoma of the breast complained of numbness without pain on both sides of the face. A magnetic resonance scan revealed bilateral enhancement in the trigeminal nerves (*arrow*). Lumbar puncture revealed leptomeningeal metastases.

Gasserian Ganglion Syndrome

Numbness, paresthesias, and usually pain referred to the trigeminal distribution are the typical presenting complaints.[93,109] The symptoms may be present for 2 weeks to 1 year (median = 3 weeks) before diagnosis. Painless numbness or facial paresthesias are common. The numbness begins close to the midline on the upper lip or chin and progresses laterally to the anterior part of the ear. Although uncommon, other sensory symptoms, for example, a gritty feeling in the eye, may occur. Pain consists of either a dull ache in the cheek, jaw, or forehead or lightning-like pain similar or identical to trigeminal neuralgia but unaccompanied by trigger points. Although lightning-like pain may be the only symptom, numbness and sensory loss appear rapidly, differentiating this entity from idiopathic trigeminal neuralgia, which is not associated with sensory loss. Headache is uncommon, in contrast to the parasellar syndrome, in which most patients suffer headache as an early and severe symptom.[93]

By the time of diagnosis, examination reveals sensory loss in the distribution of one or more trigeminal nerve roots in almost all patients, and there is often evidence of dysfunction of the motor root as well. Obvious weakness of the pterygoid or masseter muscles may be absent, with no deviation of the jaw visible on jaw opening, but the physician may see or feel slight atrophy of the temporalis and masseter muscles and note a delay in contraction and less palpable bulk of the ipsilateral masseter when the patient bites down. Even in the absence of clearly defined weakness or atrophy, electromyography often reveals denervation. Motor loss distinguishes involvement of the gasserian ganglion and trigeminal nerve by tumor from trigeminal sensory neuropathy (a non-neoplastic syndrome[110–112]). In addition, because the tumor often spreads beyond the area of the ganglion, other motor nerves, particularly the abducens, may become involved.

In patients with hematogenous metastases directly to the ganglion, radiographic evaluation may be normal, although MRI sometimes reveals enlargement of the gasserian ganglion and the tumor may enhance. A fine-needle biopsy can establish the diagnosis.[113] If the ganglion is secondarily involved from tumor metastasis to the petrous bone, bone destruction can also be identified on MRI, CT or radionuclide bone scan. The gasserian ganglion syndrome is so characteristic in patients known to be suffering from an underlying neoplasm, in particular breast or lung carcinoma, that focal radiation is indicated even in the absence of radiographic findings. A lumbar puncture should always be performed to evaluate for leptomeningeal metastases (see Chapter 7).

Numb Chin Syndrome

The numb chin syndrome usually results from dysfunction of the inferior alveolar nerve compressed by metastases to the mandible.[108] It may also be caused by metastases to the skull base or leptomeninges.[114] In one series, one-half of the patients had mandibular lesions; the others had either base-of-skull or leptomeningeal lesions. In a few, no lesion could be found.[115] The numb chin is often the presenting complaint of cancer in otherwise asymptomatic patients; in others, it may herald recurrence in patients believed to be cured of cancer.[114] Patients complain of numbness, almost always painless, over the chin and lower lip. The symptoms are usually unilateral; occasionally they occur on both sides simultaneously and, in rare cases, alternately. Signs consist of sensory loss over the chin and a contiguous area on the inside of the lip and gum. The sensory loss may be mild or severe, and the symptoms may persist or may disappear spontaneously after a few weeks or months. Breast cancer and lymphoma are common causes.[108,115] Panoramic radiographs of the mandible identify bone metastases in fewer than 50% of the patients, but an MRI is more often positive.[116] The CSF may contain malignant cells, suggesting leptomeningeal tumor. The diagnosis is usually made clinically. The disorder often responds to treatment of the metastasis: RT is directed to the mandible after leptomeningeal metastases are excluded. Chemotherapy can be given either systemically or intrathecally, depending on the site and nature of the lesion. The importance of the numb chin syndrome is that it may be the first sign of cancer or metastasis. In a patient not known to have cancer, it should prompt a search for an occult neoplasm, particularly a lymphoma,[117] breast,[25] or prostate[118] cancer. It is unclear why the symptoms sometimes remit spontaneously. The disorder has also been reported as a complication of sickle cell anemia,[119] presumably because of infarction of the nerve. In the elderly, mandibular atrophy may be the cause.[120]

Other Sites

Paresthesias and pain can accompany ophthalmoplegia when the ophthalmic (first) division of the trigeminal nerve is involved in the orbit (see preceding text). Both the ophthalmic and maxillary (second) divisions can be involved if the lesion is in the cavernous sinus. Numbness and pain in the distribution of the maxillary division of the trigeminal nerve can result from a tumor in the maxillary sinus involving the infraorbital branch of the trigeminal nerve. Zosteriform metastases of squamous cell carcinoma in the distribution of the maxillary division has been reported.[121] Trismus, presumably from involvement of the motor branch of the trigeminal nerve, has been reported as an occasional presenting complaint in patients with nasopharyngeal carcinoma.[122]

FACIAL NERVE (CRANIAL NERVE VII)

Like the trigeminal nerve, the facial nerve may be involved by metastasis anywhere along its course. Parotid tumors, either primary[123] or metastatic,[124] can present with facial weakness, usually painful, as can temporal bone metastases.[125] Facial paralysis is also a common manifestation of leptomeningeal metastases (see Chapter 7). Facial paralysis is particularly likely to be an isolated finding when neurotropic carcinomas of the skin and salivary glands invade branches of the nerve and grow perineurally.[29,126–128] Although the facial nerve is predominantly a motor nerve, facial pain may be an early symptom, followed by partial or complete facial paralysis. The onset is usually gradual and progressive, differentiating it from the more common, acute Bell palsy. However, sudden onset, even with remission, may be the initial symptom of tumor invasion of the facial nerve.[126] To compound the problem, CT and MRI may be normal early in the course of metastatic tumors involving the facial nerve[126]; conversely, the facial nerve may contrast enhance acutely in patients with Bell palsy, suggesting tumor infiltration. The facial nerve may be damaged during surgery for malignant parotid tumors,[123] but a postoperative injury is readily diagnosed. Depending on where along its course the nerve is damaged, other facial nerve branches, such as those subserving taste, hearing acuity, and tearing, may be involved. Loss of taste is unilateral, unlike the bilateral distortion of taste that is fairly common in patients with cancer. Bilateral taste loss reflects systemic rather than direct nerve involvement.[41,129]

Facial nerve paralysis accompanied by pain and otorrhea is usually infectious in origin, but metastatic tumor occasionally causes symptoms suggesting otitis.[130] Conversely, malignant external otitis, a potentially lethal but curable infection of elderly diabetics, usually due to *Pseudomonas* organisms, may be mistaken for metastatic tumor.[131]

In patients in whom the progressive development of facial paralysis suggests tumor even when imaging is negative, nerve biopsy will often establish the diagnosis.[126] In those instances, stereotactic radiosurgery may be useful.[132]

ACOUSTIC AND VESTIBULAR NERVES (CRANIAL NERVE VIII)

Loss of hearing (acoustic nerve) and vertigo (vestibular nerve) can result from leptomeningeal tumor[133] (see Chapter 7) or from metastases to the temporal bone,[134] usually easily identifiable by an MRI. Cerebellopontine angle metastases can mimic a vestibular schwannoma, causing hearing loss, vertigo, and tinnitus as well as facial and trigeminal involvement. Isolated metastases to the temporal bone or to the soft tissues of the inner ear occur by hematogenous spread from leukemias, lymphomas, and solid tumors, by direct extension of carcinomas of the head and neck region, or by invasion of nasopharyngeal tumors.[135] Symptoms are usually unilateral, but can be bilateral; they usually evolve over days or weeks. Sudden bilateral hearing loss has been reported with bilateral temporal bone metastases[136] and with leptomeningeal metastases.[133,137] The labyrinth and cochlea can serve as a sanctuary for leukemic cells.[138,139]

GLOSSOPHARYNGEAL AND VAGUS NERVES (CRANIAL NERVES IX AND X)

Several clinical disorders can be associated with metastatic tumor involving these lower cranial nerves. Metastases in the subarachnoid space or at the skull base usually cause dysfunction of other cranial nerves (e.g., hypoglossal) as well as the glossopharyngeal and vagus nerves (see section on Jugular Foramen Syndrome). More distal metastases may damage the two nerves either together or individually. Metastatic involvement of the glossopharyngeal nerve in isolation usually does not cause sensory or motor symptoms, but can stimulate baroreceptor reflexes, causing hypotension and syncope (see following text), or conversely, damage baroreceptor reflexes, causing hypertension.[140] Motor fibers of the vagus nerve innervate palatal muscles (on the ipsilateral side) and larynx muscles; their dysfunction leads to dysphagia, aspiration, and hoarseness from vocal cord paralysis. Involvement of afferent vagal fibers may also cause cardiovascular dysfunction and syncope (see following text). Unilateral facial pain can be an unusual symptom of involvement of the vagus nerve in the chest by lung cancer (see following text).

Patients in whom the motor portion of the vagus nerve is affected by tumor usually complain of dysphagia and hoarseness. If the involvement is proximal, examination usually reveals flaccidity of the ipsilateral palate and shift of the uvula to the contralateral side when the gag reflex is elicited or when the patient vocalizes. The gag reflex is absent on the ipsilateral side. More distal involvement of the vagus nerve can selectively involve only laryngeal muscles. Hoarseness and a weak cough follow. Even with involvement of the larynx alone, dysphagia with repeated aspiration of liquids is

often present. Surgical procedures can relieve the symptoms.[141]

Dysphagia is such a common problem when tumor metastasizes to lower cranial nerves that the physician sometimes fails to recognize dysphagia resulting from esophageal or cervical lesions rather than from neural involvement. Neural dysphagia presents with more difficulty in swallowing liquids, whereas esophageal or pharyngeal compression or obstruction make solids more difficult to swallow.[142] Non-neural dysphagia may be a presenting symptom of metastatic breast cancer.[143]

Jugular Foramen Syndrome

Patients with tumors involving the jugular foramen present with either hoarseness or dysphagia.[93] A dull, unilateral, aching pain localized behind the ear on the involved side is usually a prominent feature. Rarely, the metastasis causes glossopharyngeal neuralgia with very brief throat pain followed by syncope. Signs include weakness of the palate, vocal cord paralysis, weakness, and atrophy of the ipsilateral sternocleidomastoid muscle and the upper part of the trapezius, and sometimes Horner syndrome. Compression by the tumor of the transverse sinus or the jugular vein within the jugular foramen can cause papilledema.[144] Some patients have ipsilateral weakness and atrophy of the tongue, indicating extension of the tumor to the hypoglossal nerve. The diagnosis is usually easily established by MRI,[145] but slow flow in the jugular may at times be mistaken for tumor.[146]

Syncope

With or without pain, syncope may be the only symptom of metastatic involvement of the glossopharyngeal or vagus nerves[147–150] (Fig. 8–8). This condition commonly accompanies head and neck tumors and is particularly likely when the tumor recurs after initial treatment, especially after radical neck dissection. The tumor affects the glossopharyngeal nerve somewhere along its course from the carotid sinus to the skull base. Affected patients may complain of severe paroxysms of pain lasting from a few minutes to 30 minutes, or may be pain free. The pain, if present, may be in the neck, the ear, or the side of the head and is accompanied by syncope, the result of sudden hypotension. The hypotension is sometimes, but not always,

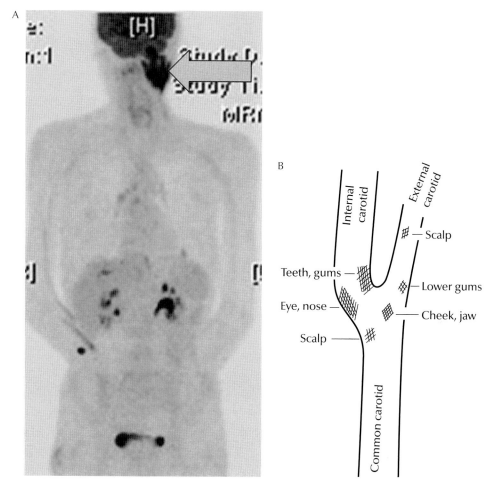

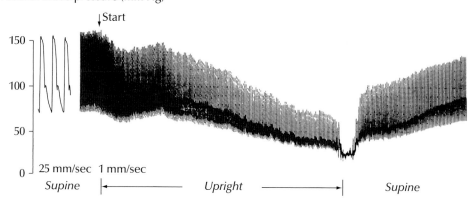

Figure 8–8. A: A 63-year-old man with carcinoma of the tonsil presented to neurologic attention with episodes of loss of consciousness. He reported identical episodes in which he developed pain in the face and ear followed by syncope. A positron emission tomography (PET) scan revealed metastasis involving the carotid artery (*arrow*). These episodes resolved spontaneously. **B**: The distribution of pain engendered by carotid sinus stimulation. The distribution is similar to that reported by the patient whose PET scan is shown here. **C**: This 72-year-old woman with breast cancer developed syncope preceded by severe neck pain. Despite a pacemaker, symptoms continued, indicating carotid sinus syncope. (B: From Ref. 151 with permission. C: From Ref. 252 with permission.)

accompanied by bradycardia and occasionally by cardiac arrest. The disorder may occur once every few weeks to several times a day. The patient usually recovers from each episode promptly, although the cardiac arrest may be life threatening. This disorder is probably a result of aberrant discharge of the damaged nerve, which stimulates brainstem nuclei to inhibit sympathetic vasoconstrictor tone.[148] Sometimes the condition resolves spontaneously as the tumor progresses to completely destroy the nerve.

Treatment is difficult. If the underlying tumor can be treated successfully, the syndrome may resolve.[153] Bradycardia or cardiac arrest is rarely the primary cause of the syncope so that a pacemaker is unlikely to correct the problem; the hypotensive episodes continue.[149] Several drugs, including anticonvulsants and sympathomimetic drugs,[147,148] have been used to treat the disorder but none have been entirely successful. Intracranial sectioning of the glossopharyngeal nerve usually cures the disorder but may cause transient hypertension, headache, or dysphagia.[149]

Another cause of syncope related to metastatic tumor is *swallow syncope*.[154,155] This rare syndrome can also result from involvement of glossopharyngeal or vagal afferents or from metastatic tumor in the esophagus. The patient suddenly loses consciousness during a hard swallow, usually because of intense bradycardia caused by stimulation of baroreceptor nerves. This disorder can often be blocked by atropine or similar drugs.

Facial Pain

Unilateral aching facial pain in and around the jaw, ear, and temporal area may be a presenting symptom of lung cancer (Fig. 8–9).[156,157] The pain is usually a constant aching pain, but may be intermittent, resembling a cluster headache.[105] Routine chest radiographs may be normal; a CT usually gives the diagnosis.[104] Unfortunately, unless the syndrome is recognized, the chest CT may not be done. The pain had been present as long as 4 years in some patients. The pain appears to result from stimulation of the vagus nerve in the chest by the lung cancer. The pain may disappear spontaneously as the vagus nerve is destroyed by the growing cancer.[106] The pain is usually on the right side because of the close anatomic relationship of the vagus nerve to mediastinal lymph nodes. Treatment of the lung tumor relieves the pain.[105]

Facial pain without syncope (see Syncope earlier) can be caused when tumor invades the carotid artery. The pain can involve the scalp, teeth, gums, cheek, jaw, eyes, or nose (Fig. 8–8B). It is often constant and severe, but may be episodic and throbbing, resembling migraine (carotidynia).[151] The carotid artery is often quite tender to palpation.

SPINAL ACCESSORY NERVE (CRANIAL NERVE XI)

The spinal accessory nerve innervates the trapezius and sternocleidomastoid muscles on the ipsilateral side. It arises from upper cervical segments, enters the skull, and then exits through the jugular foramen. It is rarely involved alone by tumor, although it is often damaged during radical neck dissection[159] (Chapter 14). Involvement of the nerve leads to drooping of the shoulder, aching pain, shoulder and arm discomfort, and some weakness on turning the head to the opposite side. On examination, weakness is rarely prominent; instead, the physician finds atrophy of the superior portion of the trapezius muscle and the sternocleidomastoid, particularly when the patient tries to contract those muscles. Electromyography settles the issue of whether the muscles are denervated.

HYPOGLOSSAL NERVE (CRANIAL NERVE XII)

The hypoglossal nerve leaves the skull through the hypoglossal foramen of the occipital bone and travels in the neck to as low as the second or third cervical vertebral body before curving back to innervate the tongue. Thus, ipsilateral tongue paralysis can occur when the nerve is involved by tumor in the subarachnoid space, at the hypoglossal foramen, or in the upper neck. Because the 12th nerves are close together at the lower end of the clivus before leaving the skull, metastases to that bone can cause bilateral hypoglossal palsies.[160] Bilateral hypoglossal paralysis can also complicate RT to the upper neck[161] (see Chapter 13) and intubation for surgical procedures (see Chapter 14).[162] Interestingly, MRI abnormalities develop acutely or subacutely in patients with hypoglossal palsy. The base of the ipsilateral tongue appears

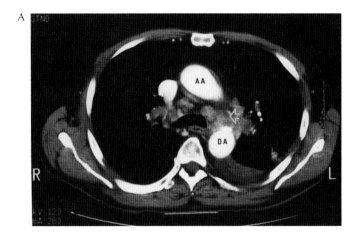

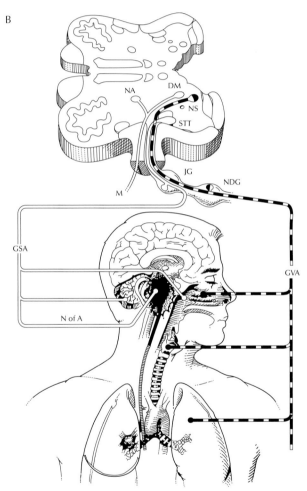

Figure 8–9. Face pain and headache from a lung lesion. **A**: A 67-year-old man developed severe pain in the left cheek, mandible, and sometimes the forehead. The pain was steady, lasting for hours, sometimes awakening him from sleep. A month later a chest film revealed lung cancer (*open arrow*). The pain improved with therapy. AA, aortic arch; DA, descending aorta. **B**: The anatomical connections that provide an explanation for how a lung mass can cause referred pain to the ipsilateral ear and face. GVA, general visceral afferents; GSA, general somatic afferents; JG, jugular ganglia; NDG, nodose ganglion; NA, nucleus ambiguus; DM, dorsal motor nucleus; NS, nucleus solitarius; ST-T, spinal trigeminal tract and nucleus; M, motor fibers; N of A, nerve of Arnold. (From Ref. 158 with permission.)

expanded and shows increased signal intensity on T2-weighted images.[163] At times the abnormality can be mistaken for an infiltrative mass in the tongue. The diagnosis is most challenging when the hypoglossal weakness is caused by early leptomeningeal metastases before there is radiographic evidence of tumor in the subarachnoid space.

Occipital Condyle Syndrome

A syndrome caused by metastatic tumor involving the occipital condyle is more common than the jugular foramen syndrome and slightly more difficult to diagnose.[20,93,164] The clinical picture is quite uniform. Patients complain of continuous, severe, localized, unilateral occipital pain made worse by neck flexion and/or head turning, often associated with a stiff neck. Sometimes the pain radiates anteriorly toward the ipsilateral temporal area or eye. About one-half of the patients complain of dysarthria, dysphagia, or both, specifically related to difficulty in moving the tongue. Symptoms usually appear weeks to months before a diagnosis is made. On neurologic examination, patients hold their neck stiffly and sometimes report tenderness to palpation over the occipital area on the involved side. The ipsilateral tongue is atrophic, and sometimes fasciculations are evident. The tongue deviates toward the weak side when it is protruded but cannot be deviated toward that side when in the mouth. When the tongue is resting in the mouth, the ipsilateral side is higher, giving the appearance of deviation to the other side.[165] The pain may precede tongue weakness by weeks or months.[164]

Diagnosis and Treatment

When tumors invade the base of the brain, MRI is the diagnostic test of choice.[1] It is particularly useful in evaluating tumor encasement of the arteries, invasion of the cavernous sinus, and the relationship of tumor to cranial nerves such as the optic nerve and optic chiasm. Bone destruction, however, is occasionally more easily observed on CT[166] or on radionuclide bone scans.[167] Negative imaging does not exclude early involvement. At times, diagnosis requires biopsy.

The treatment depends on the nature of the underlying tumor. RT and chemotherapy are usually the treatments of choice.[1,101] Stereotactic radiosurgery may be useful in particular situations such as prior RT to the involved region.

SPINAL ROOTS

Metastatic tumor can involve spinal roots in two ways. The less common way is by direct hematogenous metastasis to the nerve root or dorsal root ganglion[168–171]; more common is invasion or compression of roots by tumor in the paravertebral space. In the first instance, one or two roots are usually involved. In the second, tumor may grow longitudinally in the paravertebral space to involve multiple roots. A metastasis to a single dorsal root ganglion or spinal root may be asymptomatic because the sensory distribution of dorsal roots usually overlaps sufficiently so that that damage to a single root does not cause either sensory or motor loss.[168] On the other hand, hematogenous metastases to dorsal root ganglia can cause pain and paresthesias in the distribution of the dermatome(s) supplied by that root.[171]

Patients usually suffer symptoms when nerve roots or their sympathetic fibers are compressed or invaded by paravertebral tumors. The first symptom is chronic and aching pain in the sensory distribution of the root. The pain sometimes has the lightning-like characteristics of trigeminal neuralgia or tabes dorsalis. Pain may be the only neurologic symptom for a variable time, but eventually other signs of either hyperfunction or loss of function appear. Paresthesias usually tingle but may occasionally burn. Numbness and loss of sensation also eventually occur in the involved dermatomes. Loss of autonomic function leads to diminished or absent sweating and increased redness and warmth of the skin.

When upper thoracic segments and the stellate ganglion are involved, as occurs with Pancoast tumors, Horner syndrome develops. It is characterized by miosis, ptosis, and diminished sweating over the ipsilateral half of the face and (depending on the sympathetic roots involved) the upper extremity. When the arm is affected, the skin of the fingers fails to wrinkle after immersion in warm water because the sympathetic fibers that cause wrinkling are nonfunctioning.[172] Less common are signs of autonomic hyperfunction with focal hyperhidrosis,[173] piloerection, and sometimes coolness and pallor in the involved dermatome. Intermittent or continuous mydriasis may be evident, as well as gustatory sweating[174] (i.e., hemifacial ipsilateral sweating when eating spicy foods).

Motor signs of nerve root involvement by tumor include weakness and atrophy in the distribution of the involved root. Occasionally, fasciculations may be observed. Muscle cramps are often an early sign of nerve root or peripheral nerve damage in patients with cancer,[175] and the muscle innervated by the involved root is sometimes paradoxically enlarged.[176]

When nerve roots in the cervical or lumbosacral areas are involved, the diagnosis is usually obvious because weakness and loss of reflex function or Horner syndrome are associated with typical radicular pain in the upper or lower extremity. When thoracic and abdominal roots are involved, the diagnosis is more difficult. Involvement of a few dermatomes may cause only pain and paresthesias. Autonomic dysfunction, weakness, and sensory loss may be absent or inapparent. Sometimes one superficial abdominal reflex is absent, or an observant clinician may note failure of the abdominal muscles to contract symmetrically. A careful sensory examination of the trunk and inspection for autonomic dysfunction are mandatory. In patients complaining of shoulder or upper extremity pain, the presence of even a mild Horner syndrome may be helpful in establishing the diagnosis. In one series[177] of 216 patients with Horner syndrome, the symptoms were caused by malignant neoplasms involving sympathetic fibers in 58. Involvement of sympathetic fibers reaching the lower extremities is characterized by redness, increased temperature, and dryness of the leg. In one report, 2 of 15 patients with tumors involving the lumbosacral plexus and associated sympathetic fibers complained of a hot, dry foot, with or without pain as the first symptom of a lumbar plexopathy.[178] Conversely, patients with paravertebral tumor involving nerve roots may present with signs of sympathetic overactivity.[173]

Diagnosing nerve root involvement may be difficult. If microscopic metastasis has occurred by hematogenous spread to a root, the diagnosis is virtually impossible except at autopsy or during the course of ganglionectomies for pain treatment. Slightly larger lesions can sometimes be identified on MRI by noting enlargement of the root or the ganglion, by contrast enhancement, or by hypermetabolism on positron emission tomography (PET) scan. If a tumor in the paravertebral space compresses the root, resulting changes are often apparent radiographically. The best test is the MRI, which identifies a paravertebral mass and also may reveal that the tumor has grown through the intervertebral foramen to compress the spinal cord. Radiation is often effective in relieving symptoms, particularly pain.[179]

NERVE PLEXUSES

Four nerve plexuses are interposed between nerve roots and peripheral nerves: cervical, brachial, lumbar, and sacral[180] (Table 8–3). The latter two are often affected together by tumor, causing a disorder known as *lumbosacral plexus dysfunction*. Brachial plexus involvement by tumor is more common than the others and is also more serious because it affects the hand.

Table 8–3. **Neoplastic Plexopathy: Common Neoplasms by Location of Plexopathy**

Cervical		Brachial		Lumbosacral	
Tumor	Percentage	Tumor	Percentage	Tumor	Percentage
Lymphoma	Unknown	Lung	37	Colorectal	20
Head and neck		Breast	32	Sarcoma	16
Lung		Lymphoma	8	Breast	11
Breast		Sarcoma	5	Lymphoma	9
		Others	18	Cervix	7
				Others	37

Modified from Ref. 181 with permission.

In addition, pain and neurologic dysfunction of the brachial plexus are often more severe and difficult to treat.

Cervical Plexus

The cervical plexus arises from the upper four cervical roots. Its sensory fibers innervate skin behind the ear (greater auricular nerve), the occipital scalp (lesser occipital nerve), the skin of the supraclavicular fossa, as well as the anterior chest as far as the nipple and the posterior chest for a variable distribution around the scapula (see Fig. 14–2). The only motor innervation likely to be symptomatic is the phrenic nerve. Sympathetic fibers from the superior cervical ganglion also travel with it, so that Horner syndrome is a common accompaniment of cervical plexopathy. The major neoplasms involving the cervical plexus are head and neck cancers with cervical lymphadenopathy, metastatic lung and breast cancers, and lymphomas involving cervical lymph nodes. Pain is the most prominent symptom. It is generally severe, constant, and may be localized in the neck or referred to the throat, shoulder, or even face and head (see section on Facial Pain). The diagnosis is suggested by palpation of the neck mass and confirmed by MRI or PET, followed by biopsy, when appropriate. Treatment of the tumor may ameliorate the pain. A more common cause of cervical plexopathy is surgical damage (see Chapter 14) with pain, which may be unremitting.

Brachial Plexus

CLINICAL FINDINGS

The brachial plexus is interposed between cervical roots C5–T1 and the nerves that innervate the arm and hand (axillary, musculocutaneous, radial, median, and ulnar) (Fig. 8–10). The brachial plexus is commonly compressed or invaded by tumors arising at the lung apex (Pancoast tumor)[182–184] (Fig. 8–10) or by lymph node metastases arising from breast carcinoma.[185] In both instances, because of the anatomy of the brachial plexus, tumor tends to compress or invade the plexus from below, involving those fibers that begin as the C8 and T1 root and end as the ulnar nerve. The intercostobrachial nerve may be the first nerve affected, causing pain and numbness in the axilla, anterior chest, and dorsal medial upper arm.[186] Wherever the plexus is involved along its course, the primary symptom is pain. The pain usually begins as posterior or lateral shoulder pain of a dull, aching quality and rapidly expands to include the medial aspects of the upper arm, the elbow, and frequently the forearm. In many patients, the pain localizes to either the posterior shoulder or around the elbow; the local pain may so overshadow the other areas of the nerve's distribution that the physician is tempted to investigate the bony or soft tissue structures around the shoulder or elbow rather than the more proximal nerve plexus. Even when nerve compression is suspected, the diagnosis of cervical intervertebral disk disease is often made and treated. Involvement of the C8 and T1 roots by cervical disks is uncommon,[187] but it is the characteristic symptom of tumor involving the brachial plexus.

Paresthesias and numbness, and less often pain, are commonly perceived in the ring and little fingers. Significant sensory signs are rarely present unless the patient also complains of sensory symptoms; even then, they are less striking. Pre-existing sensory signs in the medial upper arm may have been unrecognized by the patient, however, and their discovery may confuse the diagnosis. For example, axillary node dissection may sacrifice the intercostobrachial nerve, causing sensory loss in the posterior and medial aspect of the arm.[188–190] Patients may experience pain or other abnormal sensations post operatively,[189] but are often unaware of the sensory loss that persists after the pain resolves. In this situation, the physician discovering the sensory loss on examination may be misled into thinking that a new brachial plexus lesion is present. Phantom sensations in the breast after mastectomy are frequent, and phantom pain is common[189] (see Chapter 14).

Accompanying the loss of sensation and the paresthesias of brachial plexus tumors is weakness, which usually begins in the small muscles of the hand, making it difficult to hold small objects such as pencils, or to oppose the thumb and fifth finger. Later, the weakness extends to involve finger (grip) flexors, wrist flexors and extensors, and elbow extensors (triceps muscle). Biceps, brachioradial, and deltoid muscles are often spared until later. Initially normal, the triceps reflex diminishes and finally disappears as the disease progresses. The biceps and

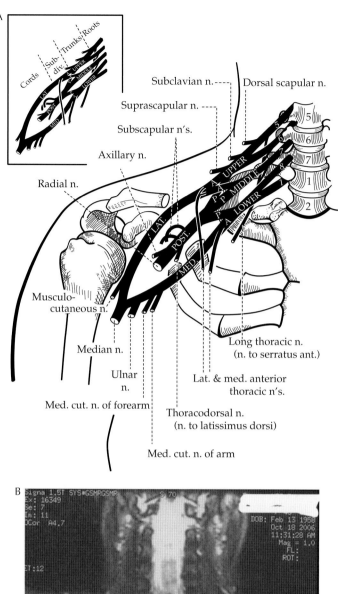

A

Cords | Sub-div. | Trunks/Roots

LAT.
POST.
MED.

5
6
7
1
2

UPPER
MIDDLE
LOWER

A UPPER
P
P MIDDLE
P
A LOWER

LAT.
POST.
MED.

Subclavian n. - - - -   Dorsal scapular n.

Suprascapular n. - - - -

Subscapular n's.

Axillary n.

Radial n.

Musculo-
cutaneous n.

Median n.

Ulnar
n.

Med. cut. n. of forearm

Thoracodorsal n.
(n. to latissimus dorsi)

Lat. & med. anterior
thoracic n's.

Long thoracic n.
(n. to serratus ant.)

Med. cut. n. of arm

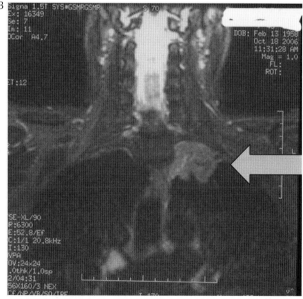

Figure 8–10. Tumor involving the brachial plexus. **A**: A schematic of the brachial plexus. **B**: A 48-year-old woman presented to medical attention with a several-month history of pain in the left chest and arm followed by weakness in her hand and Horner syndrome. A Pancoast tumor was revealed (*arrow*) only after neurologic consultation was secured and a magnetic resonance image of the brachial plexus was performed.

303

Table 8–4. Clinical Features of Brachial Plexus Syndromes in Patients with Breast Cancer

	Tumor Infiltration	Radiation Fibrosis	Acute Reversible Radiation Injury	Acute Ischemic Brachial Plexopathy
Frequency of pain	89%	18%	40%	Painless
Location of pain	Shoulder, upper arm, elbow, medial forearm	Shoulder, wrist, hand	Hand, forearm	None
Nature of pain	Dull aching in shoulder; lancinating pain in elbow and ulnar aspect of hand; occasional dysesthesias, burning, or freezing sensations	Aching pain in shoulder; paresthesias in C5-C6 distribution in hand	Pain in shoulder; paresthesias in hand and forearm	Paresthesias in hand and forearm
Severity of pain	Severe in 98% of patients	Usually absent or mild (severe in a few patients)	Mild	None
Course	Progressive neurologic dysfunction, atrophy, weakness C7-T1 distribution; pain persistent; Horner syndrome	Progressive weakness in C5-C6 distribution; pain stabilizes with appearance of weakness	Transient weakness and atrophy in C6-C7, T1; complete resolution of motor findings	Acute nonprogressive weakness and sensory changes
CT	Circumscribed mass with diffuse infiltration of tissue planes	Obliteration of tissue planes; scarring	Normal	Normal; angiography shows subclavian artery segmental obstruction
MRI	High signal intensity in circumscribed mass on T2-weighted images; may enhance with gadolinium	Diffuse low signal intensity on T2-weighted images; no change with gadolinium	No available data	Normal
FDG/PET	Hypermetabolic	Hypometabolic	Not known	Hypometabolic
Electromyographic findings	Segmental slowing, no myokymia	Myokymia	Segmental slowing or conduction block; no myokymia	Segmental slowing, no myokymia
Lymphedema	May be stable or increase	Usually increases	None	None
Treatment	RT, chemotherapy	? Neurolysis (see Chapter 13), PT	None; ? steroids	Vascular reconstruction

CT, computed tomography; MRI, magnetic resonance imaging; FDG/PET, 2-fluorodeoxy-D-glucose/positron emission tomography; RT, radiation therapy; PT, physical therapy.

brachioradialis reflexes are usually spared until later. An early and useful sign of lower plexus involvement is ipsilateral absence of the finger-stretch reflex, if the reflex is present on the contralateral side. Absent finger-stretch reflexes bilaterally in a patient with otherwise brisk reflexes are often an early sign of T1 vertebral body and root involvement with cord compression. If the tumor is paravertebral, the sympathetic trunk may be involved, and the observant patient may note the ptosis and hypohidrosis of Horner syndrome.

In typical patients, the neurologic examination may be normal early in the development of the illness, when pain and paresthesias are the only symptoms. Inspection of the chest may reveal fullness of the supraclavicular or infraclavicular fossa on the involved side, so that the sharp lines of the clavicle are effaced. Sometimes, an increased venous pattern over the anterior chest represents collateral circulation from an axillary or subclavian vein compressed by tumor. Palpation of the supraclavicular fossa and axilla may reveal adenopathy or diffuse thickening of tissues, suggesting tumor. Firm palpation or percussion, particularly in the supraclavicular area, may cause pain or paresthesias in the arm, suggesting demyelination of the brachial plexus in that area. Tumor can decrease the size of the thoracic outlet, leading to changes in pulse and blood pressure in the involved arm. Furthermore, as the involved plexus is stretched, pain is likely to be exacerbated and paresthesias appear. Thus, the shoulder should be examined through a full range of motion to determine in which position pain is increased, whether a bruit develops over the subclavian artery, or whether the blood pressure and pulse in the arm diminish. Except that pain is uncommon with radiation fibrosis, these findings of brachial plexus involvement do not distinguish tumor from radiation fibrosis (see Chapter 13), however, or even from a benign thoracic outlet syndrome.

LABORATORY TESTS

The first diagnostic laboratory test in a patient not known to have cancer is a chest radiograph. The film should be examined to be sure that the apices of the lungs are well seen and that one can determine if tumor is present in the superior portion of the lung or in the supraclavicular area. An AP (anteroposterior)

radiograph of the cervical spine may show the apex of the lung better than a plain chest X-ray.[182] An MRI of the brachial plexus is the best test;[191] however, CT is also excellent in patients unable to have an MRI. A PET scan might also be helpful,[192] particularly when the lesion infiltrates rather than compresses the plexus or the patient has received prior RT; PET can help distinguish tumor recurrence from radiation injury. Electromyography can identify the distribution of denervation and suggest the site of plexus involvement. Myokymia on electromyography suggests radiation fibrosis rather than tumor (see Chapter 13).[193]

In some patients, laboratory tests yield evidence of brachial plexus dysfunction, but fail to reveal the cause; exploration of the plexus with biopsy of surrounding tissue may be necessary.[194] Tender and Kline[195] propose a posterior exploration to examine the entire plexus. If one finds fibrous bands compressing the plexus instead of tumor, these can be removed. The widest possible exploration is necessary to be certain of the presence or absence of tumor. Even with exploration, the diagnosis may not be possible if the tumor does not cause a discernible mass that can be biopsied but instead invades the nerves, yielding no grossly visible evidence of its presence.

COURSE

Untreated, pain will grow progressively more severe. Similarly, neurologic dysfunction will spread from the lower to the upper plexus and eventually will paralyze the entire upper extremity. Lymphedema and an increased venous pattern are often late accompaniments. The pain is usually completely disabling and overshadows the functional disability. Often the pain develops into a complex regional pain syndrome[196] (reflex sympathetic dystrophy or sympathetically maintained pain) with trophic changes in the hand and nails.

DIFFERENTIAL DIAGNOSIS

The major problem in differential diagnosis occurs in patients previously irradiated to the upper thorax or axilla (e.g., after mastectomy for breast cancer). The development of a brachial plexopathy usually indicates tumor recurrence or radiation fibrosis; several clinical clues help distinguish the conditions (Table 8–4). Pain is a more common presenting complaint and is usually much more severe[3,181,185] in patients with tumor that invades the plexus than in patients with radiation fibrosis. Increasing lymphedema suggests radiation injury[185] rather than tumor. Radiation plexopathy often involves muscles of the shoulder girdle before it involves the hand, is painless, and is associated with lymphedema, whereas tumor involving the plexus often begins in the C8 and T1 roots, affects the hand before the shoulder, is painful, and is not associated with lymphedema. Horner syndrome is more common with tumor recurrence than with radiation plexopathy.[185] Other investigators have not found the motor distribution helpful in distinguishing between these two entities, but all agree that myokymia on electromyography is virtually pathognomonic of radiation fibrosis.[180,193]

An acute reversible brachial neuritis syndrome has been reported in a few patients shortly following RT for breast carcinoma.[197] Symptoms generally occur approximately four months after RT and are characterized by mild shoulder pain and arm paresthesias. Weakness may be severe but is reversible. The pathogenesis of the syndrome is unknown, but could be an immune-mediated acute brachial neuritis (see following text).

Acute ischemic brachial neuropathy results from occlusion of the subclavian artery, a late complication of RT. The disorder is predominantly motor, sudden in onset, and painless.[198,199] Paresthesias may be felt in the forearm and hand, and pulses are reduced or lost in the affected arm compared to the contralateral side.

The thoracic outlet syndrome[200] rarely causes diagnostic difficulties because the pain it generates is usually intermittent rather than continuous, is related to the position of the arm, and is rarely associated with severe motor or sensory dysfunction other than paresthesias in the extremities.

Primary nerve sheath tumors are distinguished by their history and the diagnosis is made with biopsy.[201,202] However, radiation-induced nerve sheath tumors (Chapter 13) may be confused initially with recurrence of the primary tumor or a late radiation plexopathy.

In patients not known to have cancer, the most common error in differential diagnosis is to assume that the symptoms are caused by cervical osteoarthritis or cervical disk disease.

This common error can usually be avoided by remembering that most cervical osteoarthritis involves the C5, C6, or C7 roots, causing pain in the outer aspect of the arm and lateral aspect of the forearm, and paresthesias in the thumb, index, and middle fingers rather than in the fifth finger. If weakness is present, it usually involves the triceps, deltoid, biceps, or brachioradialis; hand muscles are spared. Reflex loss affects triceps, biceps, and brachioradialis; finger-stretch reflexes are usually spared. It is unusual for brachial plexus tumor to involve these structures first. Although radiation plexopathy may involve more proximal muscles early, it can be distinguished from cervical disk disease by the absence of pain; cervical disk disease is rarely painless.

Another problem in differential diagnosis is distinguishing acute brachial neuritis[203] from tumor involvement of the plexus. Acute brachial neuritis is characterized by the sudden onset of severe pain referred to the tip of the shoulder, followed in several days by marked weakness, particularly of muscles around the shoulder girdle but sometimes of arm muscles as well. The deltoid and serratus anterior muscles are involved most commonly. Biceps, brachioradialis, and supraspinatus can also be involved. Motor changes are usually much more striking than sensory changes. After one to several weeks, the pain tends to clear spontaneously, and strength usually returns to normal without treatment after weeks to months. The acute onset, rapid progression, and relative absence of sensory changes, as well as the distribution of the muscle weakness, exclude the diagnosis of brachial plexus tumor and suggest brachial neuritis. Acute brachial neuritis is more common in patients with Hodgkin disease, as is Guillain-Barré syndrome, than it is in the general population. In patients with Hodgkin disease, brachial neuritis may occur early during RT,[204,205] after the first or second treatment, or it may be unassociated with radiation.

TREATMENT

Two general approaches are used to treat brachial plexus tumor.[206] The first is to try to diminish or eradicate the tumor by surgery, RT, or chemotherapy. Debulking a tumor mass is sometimes helpful in relieving pain, but may cause increased neurologic disability.[207]

Treatment of the tumor often can relieve the pain and diminish the tumor size but is usually less effective in restoring neurologic function, particularly if severe weakness or sensory loss is already present. The second approach is to treat the pain (see Chapter 4).[194] Pain from brachial plexus involvement is often intense and intractable.

Lumbosacral Plexus

Although anatomically separate, at least in part, the lumbar and sacral plexus are often considered together as the lumbosacral plexus.[180] The lumbar plexus (Fig. 8–11) arises from the nerve roots of L1–3 and part of L4; sometimes there is a branch from T12. It terminates as the

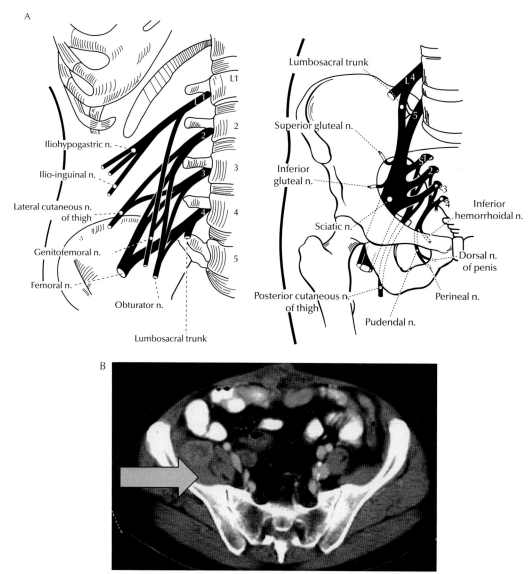

Figure 8–11. Tumor involving the lumbosacral plexus. **A**: A schematic of the lumbosacral plexus. Lumbar plexus is on the left image and the sacral plexus is on the right. **B**: A 56-year-old woman with colon cancer presented to neurologic attention with an 11-day history of numbness on the inside of the medial thigh and calf and weakness climbing stairs. She also had abdominal pain, but none in the leg. Examination revealed weakness of the iliopsoas and complete paralysis of the quadriceps and absent right knee jerk. The magnetic resonance image showed metastatic colon cancer involving the iliopsoas muscle and the lumbar plexus.

iliohypogastric, ilioinguinal, and genitofemoral nerves that supply the lower abdomen and the obturator, femoral, and lateral femoral cutaneous nerves that supply the leg. A portion of the L4 root fuses with the L5 root to form the lumbosacral trunk that, along with the sacral roots, forms the sacral plexus (Fig. 8–11). That plexus terminates to supply the gluteal nerves of the pelvis and the sciatic and posterior cutaneous nerve of the thigh.

Cancers may affect the plexus in one of several ways: (1) They may directly extend from an abdominopelvic primary tumor such as prostate or colon cancer or a chordoma. (2) Metastases may affect nearby soft or bony tissues, secondarily compressing the plexus. (3) Tumor such as breast cancer may metastasize directly to a plexus. (4) Tumor in lymph nodes may compress the plexus as, for example, in lymphomas. (5) Metastases to muscle, particularly iliopsoas (see following text), may compress the plexus. (6) Tumor extension to the plexus along nerves from the primary tumor (perineural spread) has been reported in prostate cancer.[208] In one series of 85 tumors causing lumbosacral plexopathy, colorectal cancer was the most common cause (17); sarcoma (14), breast cancer (9), lymphoma (8), and cervical cancer (6) were less common.[209]

LUMBAR PLEXUS

Clinical Findings

Tumors that frequently affect the lumbar plexus include colorectal carcinomas, retroperitoneal sarcomas, or metastatic tumors from breast, lymphoma, cervix, bladder, or prostate. The plexus may be involved by local extension of the tumor along lymphatic channels into the vertebral bodies or paravertebral space. Isolated lumbar plexopathy (Fig. 8–11) is less common than either isolated sacral plexopathy (Fig. 8–12) or lumbosacral plexopathy. The tumor generally is bulky and compresses the plexus, but it may invade neural structures without causing a mass. The severity and type of pain vary with the site of plexus involvement but is generally of two types (Table 8–5): (1) If the lumbar plexus is involved, dull, aching back pain occurs near the costovertebral angle. This local pain may be absent or completely overshadowed by a radicular distribution of pain involving the groin and anteromedial aspect of the thigh. (2) If the lower plexus is involved, the back pain is usually

slightly lower, often near the iliac crest, radiating into the buttocks and down the posterior aspect of the thigh and leg toward the ankle. Sometimes the entire plexus is involved. The aching pain extends anteriorly and posteriorly in the thigh and leg. Usually exacerbated either by sudden movement or, at times, by lying in the supine position, the pain is sometimes relieved by standing or walking, a phenomenon that helps distinguish plexus lesions from bony disease of the pelvis or hip. This pain may be present without other neurologic signs or symptoms. When neurologic symptoms other than pain occur, they usually appear first as motor weakness with mild sensory complaints. If the upper lumbar plexus is involved, the patient complains of difficulty climbing stairs and walks with his or her leg held in rigid extension to avoid knee buckling. The patient may also note atrophy of the anterior thigh and muscle tenderness in the quadriceps. If the lower lumbar plexus is involved, the patient may complain of foot drop.

On examination, focal findings are usually minimal; sometimes the tumor mass can be felt on rectal examination. The hip can generally be moved through a full range of motion without substantial pain, although extreme external rotation may cause pain, perhaps from irritation of a branch of the sciatic nerve to the capsule of the femoral head. Likewise, palpation of the pelvic bones usually has negative findings. If the lower lumbar plexus is involved, straight-leg raising may reproduce the patient's pain. In patients with upper plexus lesions, pain may be absent on straight-leg raising, but severe pain often occurs when the leg rapidly descends after it has been raised either voluntarily or passively, probably because the iliopsoas muscle is also affected by tumor, making sudden stretching painful. If the upper plexus is involved, the physician finds weakness of the iliopsoas and sometimes of the quadriceps muscle as well. The patellar reflex may be diminished or absent, but sensory changes are usually few and mild. If the lower lumbar or upper sacral plexus is involved, the weakness is generally most marked in the hamstring muscles and the dorsiflexors and plantar flexors of the foot; the Achilles reflex is absent. The proximal muscles that are supplied by the lower lumbar plexus, including adductor magnus, gluteus maximus and medius, and tensor fasciae latae also are weak. The focal weakness of proximal muscles

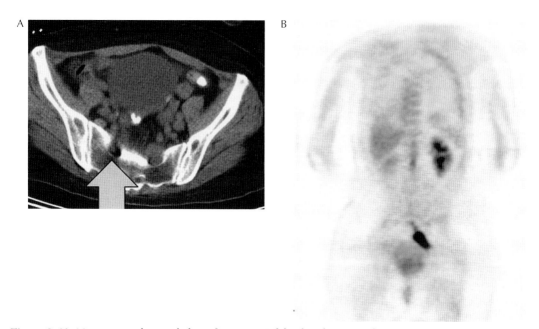

Figure 8–12. Metastasis to the sacral plexus from non–Hodgkin lymphoma. **A**: The magnetic resonance image shows a mass at the foramen of S1. The arrow points to the normal S1 foramen and root. On the opposite side, there is a mass infiltrating and enlarging the root. **B**: The coronal positron emission tomography image shows hypermetabolism at the foramen extending into the plexus. The patient complained of pain in the buttock and back of the leg. There was mild weakness of plantar flexion of the foot and an absent ankle jerk.

Table 8–5. **Symptoms and Signs of Tumor Involving Lumbosacral Plexus**

Clinical Level	Upper	Lower	Pan-Plexus
Number of patients	12	16	6
Most common tumor	Colorectal	Sarcoma	Genitourinary
Pain distribution			
Local	Lower abdomen	Buttock, perineum	Lumbosacral
Radicular	Anterolateral thigh	Posterolateral thigh, leg	Variable
Referred	Flank, iliac crest	Hip and ankle	Variable
Numbness/paresthesias	Anterior thigh	Perineum, thigh, sole	Anterior thigh, leg, foot
Motor and reflex changes	L2–4	L5-S1	L2-S2
Sensory loss	Anterolateral thigh	Posterior thigh, sole	Especially anterior thigh, leg
Tenderness	Lumbar	Sciatic notch, sacrum	Lumbosacral
Positive SLRT			
Direct	6/12	8/16	5/6
Reverse	2/12	8/16	5/6
Leg edema	5/12	6/16	5/6
Rectal mass	3/12	7/16	1/6
Sphincter weakness	0/12	8/16	0/6

From Ref. 209 with permission.
SLRT, straight-leg raising test.

suggests that the plexus is involved rather than more peripheral structures in the thigh or leg, such as the sciatic or peroneal nerve.[210,211]

Laboratory Tests

The laboratory approach indicated for brachial plexus lesions can be applied here. MRI is the best diagnostic test for identifying tumor involvement of the lumbar plexus. PET/CT imaging may be superior when the plexus is invaded rather than compressed, particularly with lymphoma.[212] On occasion, particularly with very difficult diagnoses, surgical exploration of the plexus is necessary.

Differential Diagnosis (Table 8–6)

Metastatic disease of the hip is identified on examination by severe pain on hip movement and evidence of tumor on a bone scan or MRI. Compression by tumor of the femoral, obturator, or sciatic nerve in the pelvis can be distinguished from plexus involvement by the distribution of muscles involved and also by location of the radiographic mass identified. However, referred pain may cause local tenderness in muscles and bone, often leading the clinician to misinterpret the tender area as being the lesion site.

Another source of possible misinterpretation is osteitis pubis (Fig. 8–13), an illness that is sometimes misdiagnosed as peripheral nerve or plexus tumor. Infection of the symphysis pubis and pubic bone may follow genitourinary surgery.[213] Symptoms do not appear for several weeks after the operation but then begin as severe pain down the medial aspects of both thighs, often with surprisingly little pubic pain. The pain is frequently so severe that the patient cannot walk. Neurologists are sometimes asked to consult in such instances because of suspected lumbar plexus or nerve root involvement. We recently saw such a patient who had not walked for 2 weeks after the onset of pain, leading to mild proximal leg muscle atrophy, which further confused the diagnosis. The neurologic examination is normal insofar as can be determined; patients frequently are unwilling to move their legs. The symphysis pubis, when palpated, is very tender; palpation produces pain locally and reproduces the spontaneous pain, if referred. CT of the pubic bone may reveal inflammatory

Table 8–6. Diagnosis of Lumbosacral Plexopathy in Cancer Patients

	Metastatic Plexopathy	Radiation Plexopathy	Lumbosacral Neuritis
Pain	Early, frequent, and severe	Mild; develops after numbness or paresthesias; can be severe in a few patients	Severe, short lived
Distribution of signs and symptoms	Unilateral > bilateral	Often bilateral	Unilateral
Distinctive features	Positive straight-leg raising test; rectal mass; warm and dry foot	No rectal mass, no sphincter involvement; absent autonomic dysfunction; lymphedema	Weakness, little sensory change
MRI	May show mass; often low T1 and high T2 signal, enhancement	No mass; often low T1 and T2 signal in plexus	May be normal
Nerve conduction velocity/EMG	Axonal loss or demyelination, no myokymia	Myokymia	Demyelination
Treatment	Underlying malignancy	Supportive measures; mild effects of neurolysis and omental flap (on pain), or anticoagulation	Corticosteroids

Modified from Ref. 3.
MRI, magnetic resonance imaging; EMG, electromyography.

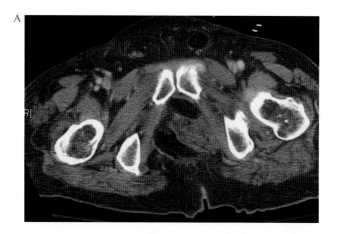

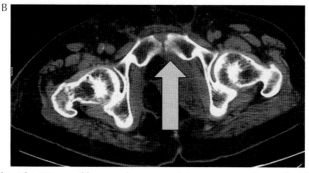

Figure 8–13. Osteitis pubis. This 65-year-old man underwent a pelvic exenteration for poorly differentiated carcinoma. A few weeks after the operation, he began to complain of bilateral groin pain radiating down the medial aspect of his thighs. He also complained of weakness that prevented walking. Neurological consultation demonstrated only apparent weakness caused by pain, but with normal reflexes. There was marked tenderness over the synthesis pubis. A preoperative computed tomography (CT) scan (**A**) showed normal bones of the pelvis. A postoperative CT scan (**B**) demonstrated osseous changes in the pubis symphysis, recognized by the radiologist only after the neurologist had made a suggestion of osteitis pubis.

changes, or early results may be normal. MRI[214] and PET are more helpful. The diagnosis is made with biopsy specimen and bone culture. Symptoms are relieved completely by appropriate antibiotic treatment. The organism is usually *Staphylococcus* or *Pseudomonas*.

Other bone and joint disorders complicating cancer and sometimes confused with nerve or plexus lesions are avascular necrosis (see Chapter 4) and diffuse bone and joint pain accompanying acute leukemia[215]; leukemic lesions are often misdiagnosed as osteomyelitis, delaying appropriate treatment.[216,217] Osteoporotic fractures of the pelvis may simulate metastatic disease.[218]

RT delivered to tumors in the lumbosacral area can cause radiation plexopathy, which generally begins a year or more after the original radiation was begun (see Chapter 13). Radiation damage is usually characterized by bilateral, indolent leg weakness, often accompanied by lymphedema, whereas patients with tumors involving the lumbosacral plexus usually have severe unilateral pain and weakness without lymphedema.

Treatment

The principles applied to treating brachial plexus involvement also apply to the lumbar and sacral plexuses. RT is delivered to the area and analgesic agents are administered for the pain. If this regimen is ineffective, epidural analgesics should be considered.

SACRAL PLEXUS

Clinical Findings

The sacral plexus is usually involved by tumor from the colon, prostate, bladder, or uterus, which extends directly to the sacrum (Fig. 8–12).

The sacral plexus is closer to the midline than is the lumbar plexus, and because the involvement of the sacral plexus is usually secondary to bony invasion of the sacrum, bilateral (albeit asymmetric) changes are common. The symptoms usually begin as dull, aching, midline pain, often exacerbated by lying in a supine position or sitting, and sometimes relieved by lying in a prone position or by standing or walking. The pain sometimes radiates into the buttocks but not into the posterior aspects of the thigh or leg. The pain is usually followed by numbness in the perianal region, beginning unilaterally but rapidly becoming bilateral. Numbness and paresthesias may involve the entire buttocks and often extend to involve the posterior aspect of the thigh. Associated with the numbness is loss of sensation of evacuating bladder and bowel; constipation is followed by urinary retention or incontinence. On examination, pain may occur if the sacrum is palpated. At times the tumor can be felt, or pain is stimulated, by rectal or pelvic examination with the finger palpating posteriorly. The rectal sphincter is often lax, and the bulbocavernosus and anal wink reflexes are absent. Sensory loss to pinprick and touch also is found over the penis, portions of the scrotum or labia, and the perianal area.

Laboratory Tests

Laboratory tests to identify sacral plexus involved by tumor are the same as those described previously for brachial and lumbar plexus involvement.

Differential Diagnosis

The major difficulty in differential diagnosis is distinguishing leptomeningeal metastases from sacral plexus involvement. Leptomeningeal metastases are usually more bilaterally symmetric from onset and characterized by less pain compared to sacral plexus involvement, in which pain is an early and prominent symptom and the findings are usually asymmetric until late. In patients suspected of harboring sacral plexus metastases, however, enhanced MR should be performed to examine the distal subarachnoid space. CSF should also be examined.

Treatment

The principles of treatment resemble those for other plexus involvement. RT is the treatment of choice. In many instances, patients have already received the maximum external RT and are not candidates for additional treatment. If the pain is severe and does not respond to RT, analgesics should be optimized; if drugs fail, other approaches may be considered, such as spinal opioids, chemical neurolysis, or cordotomy.

PERIPHERAL NERVES

Peripheral nerves can be involved by tumor either focally or diffusely. Peripheral neuropathy that is due to tumor is much less common than that which is due to chemotherapy (see Chapter 12). Diffuse involvement by tumor, causing a sensorimotor polyneuropathy, is rare and is usually caused by leukemia, lymphoma, or myeloma.[219–221] Also rare are tumors that are metastatic to individual peripheral nerves, also usually caused by leukemia or lymphoma, but sometimes by solid tumors.[222,223]

Mononeuropathies

CLINICAL FINDINGS

The most common cause of peripheral nerve involvement is tumor that invades a contiguous bone and compresses a nerve, usually where the nerve either passes directly over a bone (such as the radial nerve at the humerus) or through a bony canal (such as the obturator nerve through the obturator canal).[224] Other nerves sometimes affected include the ulnar nerve, affected by metastatic disease around the elbow or nodal metastases within the axilla; intercostal nerves, by rib metastases; the sciatic nerve, by involvement in the pelvis at several sites; and the peroneal nerve, as it passes behind the head of the fibula.

The signs and symptoms are initially pain and then sensory and motor loss in the distribution of the involved nerve. Careful neurologic evaluation with attention to the particular muscles and the distribution of the sensory loss easily distinguishes involvement of the peripheral nerve from that of more proximal structures such as plexuses and roots. In addition, radiating paresthesias can occur when the nerve is percussed near the site of tumor involvement (Tinel sign). The mass involving or compressing the nerve may be palpable,

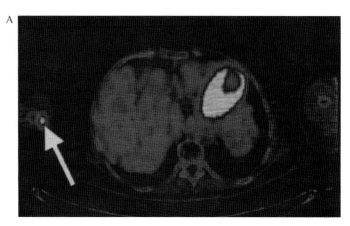

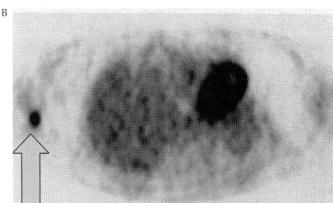

Figure 8–14. Metastasis to the radial nerve in a patient with acute myelogenous leukemia. The patient complained of pain in the arm and weakness of dorsiflexion of the wrist. The metastasis was identified both on magnetic resonance imaging (**A**, *arrow*) and as hypermetabolism on the positron emission tomography scan (**B**).

and palpating adjacent bones often reveals tenderness in those structures. Once the nerve involvement has been identified clinically, MRI[225–227] or PET[228] may visualize the lesion (Fig. 8–14). Occasionally, no mass lesion can be defined clearly; instead, a metastasis is present within the nerve itself. We have encountered this problem in breast carcinoma and melanoma. The diagnosis can be difficult. Usually, in addition to sensory and motor loss, the physician can identify severe tenderness at a single point along the distribution of the nerve, which sometimes indicates an area invaded by tumor.

Mononeuritis multiplex, the simultaneous involvement of several individual nerves but not of all nerves as occurs in polyneuropathy, suggests a vasculitis but may result from multifocal invasion by tumor.[229]

DIFFERENTIAL DIAGNOSIS

Considered in the differential diagnosis are entrapment or compression neuropathies.[230]

Sometimes an entrapment neuropathy is caused by cancer, as when lymphedema causes brachial plexus entrapment or carpal tunnel syndrome. Compression neuropathies, such as peroneal nerve palsy (see following text), are associated with weight loss because the nerve, no longer protected by soft tissue, is compressed against bony prominences. The nerves involved by entrapment and compression are frequently the same nerves as those involved by tumor. Also to be considered in the differential diagnosis are secondary effects of chemotherapy (see Chapter 12) and paraneoplastic vasculitis leading to mononeuritis multiplex (see Chapter 15).

A problem in differential diagnosis is distinguishing a peroneal nerve palsy from more proximal involvement of the sciatic nerve or the L5 root. The sciatic nerve and the L5 root control foot inversion; the peroneal nerve does not, so foot inversion should be spared with a peroneal nerve palsy. Furthermore, the hamstring muscles, which are supplied by the sciatic nerve and L5 and S1 roots, should not be weak when the peroneal nerve is the culprit. If the lesion is more proximal, an electromyogram should show weakness and denervation of the hamstrings. It is also difficult sometimes to distinguish femoral nerve involvement from lumbar plexus dysfunction. The distribution of the lumbar plexus includes the adductor thigh muscles as well as the iliopsoas and quadriceps; with femoral involvement, only the iliopsoas and the quadriceps are involved.

Painless peroneal nerve palsy of sudden onset in a patient who has recently lost weight and habitually sits with his or her legs crossed (the leg with the nerve palsy on top) suggests that the patient suffers a compression neuropathy. Rarely, a similar compression neuropathy occurs as the radial nerve passes behind the humerus (i.e., the Saturday night palsy), leading to painless wrist drop with little sensory change. Compression neuropathies of several other nerves can occur at the time of surgery for cancer (see Chapter 14).

Polyneuropathy

Direct invasion of multiple nerves by a neurotropic tumor occasionally causes acute or subacute sensorimotor polyneuropathies, often painful. Although rare, the disorder has been reported in patients with leukemias and lymphomas (neurolymphomatosis)[219,220,231] and may appear before the onset of other symptoms or when the patient is otherwise in remission. On some occasions, successful treatment of leptomeningeal leukemia with intrathecal drugs may be followed by a painful polyneuropathy that appears to be the result of tumor infiltrating nerve roots or peripheral nerves after they exit the leptomeninges. Whether it is the tumor in the leptomeninges that extends along nerve roots outside the central nervous system (CNS) is not certain.

It may be almost impossible to distinguish polyneuropathy caused by direct neoplastic involvement from the more common paraneoplastic polyneuropathies (see Chapter 15) or from idiopathic polyneuropathy. Even the finding of steroid responsiveness does not help, because lymphoma and leukemia sometimes are sensitive to steroids and the drug may cause a remitting and relapsing course. The only certain diagnostic test is biopsy. If the underlying problem is leukemia or lymphoma, lymphocyte markers must be used to distinguish an inflammatory neuropathy from a neoplastic one, and the patient must be off corticosteroids to avoid a false-negative biopsy specimen.

Accurate diagnosis of the cause of peripheral nerve dysfunction in the patient with cancer is often difficult. These patients are susceptible not only to peripheral nerve lesions associated with their cancer, but also to the far more common causes of peripheral neuropathy such as diabetes and alcoholism. Because smoking and alcohol intake are risk factors for head and neck cancer, and because diabetes is such a common disorder, these factors should always be considered in patients presenting with peripheral neuropathy. In patients with known cancer under active therapy, chemotherapeutic complications are more likely to be the cause than cancer invasion. In patients not known to have cancer, particularly those without weight loss, paraneoplastic polyneuropathy (see Chapter 15) requires careful consideration.

Because the disorder is unusual and other causes of polyneuropathy are much more common, the diagnosis of tumor invasion by metastases is made infrequently. Patients who have or have had leptomeningeal invasion, particularly if successfully treated, who then develop a painful polyneuropathy, should be suspects. In patients such as those with rapidly developing, painful polyneuropathy in whom a cause is not immediately clear (especially if they have not received neurotoxic chemotherapeutic agents), nerve biopsy is usually the best diagnostic approach.[232] Because invasion of peripheral nerves may not involve the entire nerve, a blind biopsy may be negative.[233] MRI neurography[226] and/or PET/CT[228] may identify a good biopsy site. Because peripheral nerves (but not the dorsal root ganglia) are protected by the blood–nerve barrier, standard doses of water-soluble agent such as methotrexate may not be effective, but higher doses may be.[234]

MUSCLES

Although muscle is often invaded by metastases to surrounding structures such as bone or by direct extension from a primary tumor such as breast, hematogenous metastases to skeletal muscle are generally believed to be rare[235] (Fig. 8–15). However, a disparity appears to exist between older autopsy studies in which muscle

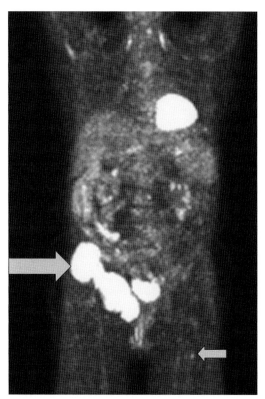

Figure 8–15. Two types of muscle metastases. This patient with thyroid cancer presented with hip pain. In addition to the mass in the thyroid, positron emission tomography scan revealed a large mass involving the iliac bone and invading the iliacus muscle. In addition, there was a small metastasis within a muscle in the contralateral thigh (*arrow*).

metastases were common and clinical studies in which they are rare. In one autopsy study of 110 individuals, 930 muscles were examined. In the 38 patients with cancer,[236] hematogenous metastases were found in 14 muscles from six individuals, an incidence of 16%. In another study,[237] hematogenous metastases to skeletal muscle were found in 34 of 194 autopsies of patients with cancer (18%). Diaphragm and iliopsoas were the most common muscles involved; carcinomas, particularly adenocarcinoma, were more common than leukemias and lymphomas. A study of 198 autopsies that excluded the diaphragm identified hematogenous metastases in only four patients. In 14 others, muscle was involved by direct spread from either the primary tumor or metastasis. An autopsy study of 82 patients with leukemia or lymphoma revealed involvement of one or more skeletal muscles in 43 patients (52%).[238]

Clinically, hematogenous metastases to the muscle are quite rare, in part because of the fact that they are rarely symptomatic. However, they appear to be rare even when the whole body is imaged. In one series of over 2000 magnetic resonance images performed on a large orthopedic oncology service, only 28 patients had skeletal muscle metastases.[239] In another series, at an institution that had examined over 54,000 new cancer patients, skeletal muscle metastases were found in only 15 patients.[240] Interestingly, 7 of the 15 patients were not known to have cancer at the time the metastases were identified. In some patients, a skeletal muscle metastasis was the first sign of cancer[241]; in others known to have cancer, the metastasis appeared many years (19 years in one case) after the primary tumor had been apparently cured.[242] The two largest clinical series include 21 patients who were referred to *neuromuscular oncologists*[243] and 28 patients from an orthopedic oncology center.[239] In the latter series of 28 patients with biopsy-proved skeletal muscle metastases, the tumor developed in 8 patients at the site of previously documented trauma.[239]

It is unclear why skeletal muscle metastases are so rare, but both mechanical and metabolic hypotheses have been advanced. The infrequency of metastases to skeletal muscle has been demonstrated experimentally by injection of tumor cells into the iliac artery.[244] Furthermore, conditioned media from muscle cell culture has been demonstrated to inhibit tumor growth.[245] Skeletal muscle comprises approximately 50% of body mass and during exercise receives a large portion of the cardiac output. When injected into the ventricle, cancer cells arrest in the microvasculature of muscle, but survive only in denervated muscle,[246] leading to the hypothesis that mechanical interaction within the microvasculature, perhaps because of differences in tumor cell adhesion, leads to rapid death of the cancer cells.[247] If one electrically stimulates muscle, causing more contraction, even fewer cancer cells survive than in normally innervated muscle. Some of the inhibition may be mediated through an adenosine receptor.[245] Lactic acid produced by muscle has also been implicated as a potential reason for muscle resistance to cancer.[248]

When skeletal muscle metastases present clinically, they usually do so as a painful mass.

They may also present occasionally as diffuse enlargement of one or more muscles,[249] or simply as focal weakness.[250] Occasionally, diffuse infiltration of proximal muscles by tumor can mimic a paraneoplastic myopathy.[251] Microscopic infiltration of muscle has been reported to elicit an inflammatory response leading to a focal myositis,[252] resembling a paraneoplastic syndrome. The presence of a skeletal muscle metastasis is usually obvious on either MRI or PET/CT[253] (Fig. 8–15). The scans do not distinguish primary from metastatic tumor, although the presence of surrounding edema in muscle suggests metastasis.[254] However, the situation is more difficult in patients who have undergone bone marrow or stem cell transplant. Here the differential diagnosis includes myositis due to graft-versus-host disease, polymyositis as an autoimmune disorder, or recurrent tumor restricted to skeletal muscle.[255–258]

Treatment depends on the situation. Single painful metastases can be excised.[254] Radiation and chemotherapy can also be effective.

REFERENCES

1. Laigle-Donadey F, Taillibert S, Martin-Duverneuil N, et al. Skull-base metastases. *J Neuro-Oncol* 2005;75:63–69.
2. Antoine JC, Camdessanche JP. Peripheral nervous system involvement in patients with cancer. *Lancet Neurol* 2007;6:75–86.
3. Ramchandren S, Dalmau J. Metastases to the peripheral nervous system. *J Neuro-Oncol* 2005;75:101–110.
4. Terhaard C, Lubsen H, Tan B, et al. Facial nerve function in carcinoma of the parotid gland. *Eur J Cancer* 2006;42:2744–2750.
5. Kraut MJ, Vallieres E, Thomas CR, Jr. Pancoast (superior sulcus) neoplasms. *Curr Prob Cancer* 2003;27:81–104.
6. Clouston PD, Sharpe DM, Corbett AJ, et al. Perineural spread of cutaneous head and neck cancer. *Arch Neurol* 1990;47:73–77.
7. Catalano PJ, Sen C, Biller HF. Cranial neuropathy secondary to perineural spread of cutaneous malignancies. *Am J Otol* 1995;16:772–777.
8. Hughes TA, McQueen IN, Anstey A, et al. Neurotropic malignant melanoma presenting as a trigeminal sensory neuropathy. *J Neurol Neurosurg Psychiatry* 1995;58:381–382.
9. Ojiri H. Perineural spread in head and neck malignancies. *Radiat Med* 2006;24:1–8.
10. Harnden P, Shelley MD, Clements H, et al. The prognostic significance of perineural invasion in prostatic cancer biopsies: a systematic review. *Cancer* 2007;109:13–24.
11. Duraker N, Caynak ZC, Turkoz K. Perineural invasion has no prognostic value in patients with invasive breast carcinoma. *Breast* 2006;15:629–634.
12. Sayar A, Turna A, Solak O, et al. Nonanatomic prognostic factors in resected nonsmall cell lung carcinoma: the importance of perineural invasion as a new prognostic marker. *Ann Thorac Surg* 2004;77:421–425.
13. Kayahara M, Nakagawara H, Kitagawa H, et al. The nature of neural invasion by pancreatic cancer. *Pancreas* 2007;35:218–223.
14. Duraker N, Sisman S, Can G. The significance of perineural invasion as a prognostic factor in patients with gastric carcinoma. *Surg Today* 2003;33:95–100.
15. Compton CC. Pathology report in colon cancer: what is prognostically important? *Dig Dis* 1999;17:67–79.
16. Fujita S, Nakanisi Y, Taniguchi H, et al. Cancer invasion to Auerbach's plexus is an important prognostic factor in patients with pT3-pT4 colorectal cancer. *Dis Colon Rectum* 2007;50:1860–1866.
17. Turgut M, Erturk O, Saygi S, et al. Importance of cranial nerve involvement in nasopharyngeal carcinoma. A clinical study comprising 124 cases with special reference to clinical presentation and prognosis. *Neurosurg Rev* 1998;21:243–248.
18. Leung SF, Tsao SY, Teo P, et al. Cranial nerve involvement by nasopharyngeal carcinoma: response to treatment and clinical significance. *Clin Oncol (R Coll Radiol)* 1990;2:138–141.
19. Thomas JE, Waltz AG. Neurological manifestations of nasopharyngeal malignant tumors. *JAMA* 1965;192:103–106.
20. Capobianco DJ, Brazis PW, Rubino FA, et al. Occipital condyle syndrome 2. *Headache* 2002;42:142–146.
21. Kocak Z, Celik Y, Uzal MC, et al. Isolated bilateral sixth nerve palsy secondary to metastatic carcinoma: a case report with a review of the literature. *Clin Neurol Neurosurg* 2003;106:51–54.
22. McDermott RS, Anderson PR, Greenberg RE, et al. Cranial nerve deficits in patients with metastatic prostate carcinoma—clinical features and treatment outcomes. *Cancer* 2004;101:1639–1643.
23. Shields JA, Shields CL, Ehya H. Lung cancer diagnosed by fine needle biopsy of the optic disk. *Retina* 2001;21:665–666.
24. Miro C, Orecchia R, Canas de Paz F. Skull metastasis from breast cancer presenting as monocular blindness. *Lancet Neurol* 2003;2:635–636.
25. Moris G, Perez-Pena M, Miranda E, et al. Trigeminal mononeuropathy: first clinical manifestation of breast cancer. *Eur Neurol* 2005;54:212–213.
26. Turgman J, Braham J, Modan B, et al. Neurological complications in patients with malignant tumors of the nasopharynx. *Eur Neurol* 1978;17:149–154.
27. Holte H, Saeter G, Dahl IM, et al. Progressive loss of vision in patients with high-grade non-Hodgkin's lymphoma. *Cancer* 1987;60:2521–2523.
28. Julien J, Ferrer X, Drouillard J, et al. Cavernous sinus syndrome due to lymphoma. *J Neurol Neurosurg Psychiatry* 1984;47:558–560.
29. Batsakis JG. Nerves and neurotropic carcinomas. *Ann Otol Rhinol Laryngol* 1985;94:426–427.
30. Hall SM, Buzdar AU, Blumenschein GR. Cranial nerve palsies in metastatic breast cancer due to osseous metastasis without intracranial involvement. *Cancer* 1983;52:180–184.
31. Ransom DT, DiNapoli RP, Richardson RL. Cranial nerve lesions due to base of the skull metastases in prostate carcinoma. *Cancer* 1990;65:586–589.
32. O'Sullivan JM, Norman AR, McNair H, et al. Cranial nerve palsies in metastatic prostate cancer—results

of base of skull radiotherapy. *Radiother Oncol* 2004;70:87–90.

33. Li JC, Mayr NA, Yuh WT, et al. Cranial nerve involvement in nasopharyngeal carcinoma: response to radiotherapy and its clinical impact. *Ann Otol Rhinol Laryngol* 2006;115:340–345.

34. Ozyar E, Atahan IL, Akyol FH, et al. Cranial nerve involvement in nasopharyngeal carcinoma: its prognostic role and response to radiotherapy. *Radiat Med* 1994;12:65–68.

35. Davidson TM, Murphy C. Rapid clinical evaluation of anosmia. The alcohol sniff test. *Arch Otolaryngol Head Neck Surg* 1997;123:591–594.

36. Doty RL, Marcus A, Lee WW. Development of the 12-item Cross-Cultural Smell Identification Test (CC-SIT). *Laryngoscope* 1996;106:353–356.

37. Chong VFH, Tsang SY. *Nasophyangeal Carcinoma.* 1st ed. Singapore: Armour, 1997.

38. Levine PA, Gallagher R, Cantrell RW. Esthesioneuroblastoma: reflections of a 21-year experience. *Laryngoscope* 1999;109:1539–1543.

39. Spektor S, Valarezo J, Fliss DM, et al. Olfactory groove meningiomas from neurosurgical and ear, nose, and throat perspectives: approaches, techniques, and outcomes. *Neurosurgery* 2005;57:268–280.

40. Cushing H. Anosmia and sellar distension as misleading signs in the localization of a cerebral tumor. *J Nerv Ment Dis* 1916;44:415–419.

41. Hutton JL, Baracos VE, Wismer WV. Chemosensory dysfunction is a primary factor in the evolution of declining nutritional status and quality of life in patients with advanced cancer. *J Pain Symptom Manage* 2007;33:156–165.

42. Allaire GS, Corriveau C, Arbour JD. Metastasis to the optic nerve: clinicopathological correlations. *Can J Ophthalmol* 1995;30:306–311.

43. Brown GC, Shields JA. Tumors of the optic nerve head. *Surv Ophthalmol* 1985;29:239–264.

44. De Potter PP. Ocular manifestations of cancer. *Curr Opin Ophthalmol* 1998;9:100–104.

45. Tang RA, Kellaway J, Young SE. Ophthalmic manifestations of systemic cancer. *Oncology (Williston Park)* 1991;5:59–71.

46. Chamberlain MC, Chalmers L. Acute binocular blindness. *Cancer* 2007;109:1851–1854.

47. Small W, Jr. Management of ocular metastasis. *Cancer Control* 1998;5:326–332.

48. Nelson CC, Hertzberg BS, Klintworth GK. A histopathologic study of 716 unselected eyes in patients with cancer at the time of death. *Am J Ophthalmol* 1983;95:788–793.

49. Shields CL, Shields JA, Gross NE, et al. Survey of 520 eyes with uveal metastases. *Ophthalmology* 1997;104:1265–1276.

50. Solomon SD, Smith JH, O'Brien J. Ocular manifestations of systemic malignancies. *Curr Opin Ophthalmol* 1999;10:447–451.

51. Bogdanovic MD, Plant GT. Chiasmal compression due to obstructive hydrocephalus. *J Neuroophthalmol* 2000;20:266–267.

52. Fahmy P, Heegaard S, Jensen OA, et al. Metastases in the ophthalmic region in Denmark 1969–98. A histopathological study. *Acta Ophthalmol Scand* 2003;81:47–50.

53. Reddy SC, Jackson N, Menon BS. Ocular involvement in leukemia—a study of 288 cases. *Ophthalmologica* 2003;217:441–445.

54. Kezuka T, Usui N, Suzuki E, et al. Ocular complications in myelodysplastic syndromes as preleukemic disorders. *Jpn J Ophthalmol* 2005;49:377–383.

55. Rudolph G, Haritoglou C, Schmid I, et al. Visual loss as a first sign of adult-type chronic myeloid leukemia in a child. *Am J Ophthalmol* 2005;140:750–751.

56. Schocket LS, Massaro-Giordano M, Volpe NJ, et al. Bilateral optic nerve infiltration in central nervous system leukemia. *Am J Ophthalmol* 2003;135:94–96.

57. Harnett AN, Plowman PN. The eye in acute leukaemia. 2. The management of solitary anterior chamber relapse. *Radiother Oncol* 1987;10:203–207.

58. Hurtado-Sarrio M, Duch-Samper A, Taboada-Esteve J, et al. Anterior chamber infiltration in a patient with Ph+ acute lymphoblastic leukemia in remission with imatinib. *Am J Ophthalmol* 2005;139:723–724.

59. MacLean H, Clarke MP, Strong NP, et al. Primary ocular relapse in acute lymphoblastic leukemia. *Eye* 1996;10 (Pt 6):719–722.

60. Robb RM, Ervin LD, Sallan SE. An autopsy study of eye involvement in acute leukemia of childhood. *Med Pediatr Oncol* 1979;6:171–177.

61. Manquez ME, Brown MM, Shields CL, et al. Management of choroidal metastases from breast carcinomas using aromatase inhibitors. *Curr Opin Ophthalmol* 2006;17:251–256.

62. Tsina EK, Lane AM, Zacks DN, et al. Treatment of metastatic tumors of the choroid with proton beam irradiation. *Ophthalmology* 2005;112:337–343.

63. Amichetti M, Caffo O, Minatel E, et al. Ocular metastases from breast carcinoma: a multicentric retrospective study. *Oncol Rep* 2000;7:761–765.

64. Bradbury PG, Levy IS, McDonald WI. Transient uniocular visual loss on deviation of the eye in association with intraorbital tumours. *J Neurol Neurosurg Psychiatry* 1987;50:615–619.

65. Sabbagh R, Shields CL, Shields JA, et al. Spontaneous hyphema. Initial manifestation of lung carcinoma. *JAMA* 1991;266:3194.

66. Harbison JW, Lessell S, Selhorst JB. Neuro-ophthalmology of sphenoid sinus carcinoma. *Brain* 1984;107:855–870.

67. Comer RM, Dawson E, Plant G, et al. Causes and outcomes for patients presenting with diplopia to an eye casualty department. *Eye* 2007;21:413–418.

68. Rucker JC, Tomsak RL. Binocular diplopia. A practical approach. *Neurologist* 2005;11:98–110.

69. Galetta SL, Sergott RC, Wells GB, et al. Spontaneous remission of a third-nerve palsy in meningeal lymphoma. *Ann Neurol* 1992;32:100–102.

70. Lacey B, Chang W, Rootman J. Nonthyroid causes of extraocular muscle disease. *Surv Ophthalmol* 1999;44:187–213.

71. Peckham EL, Giblen G, Kim AK, et al. Bilateral extraocular muscle metastasis from primary breast cancer. *Neurology* 2005;12:65:74.

72. Shields JA, Shields CL, Brotman HK, et al. Cancer metastatic to the orbit: the 2000 Robert M. Curts Lecture. *Ophthal Plast Reconstr Surg* 2001;17:346–354.

73. Holland D, Maune S, Kovacs G, et al. Metastatic tumors of the orbit: a retrospective study. *Orbit* 2003;22:15–24.

74. Dieing A, Schulz CO, Schmid P, et al. Orbital metastases in breast cancer: report of two cases and review of the literature. *J Cancer Res Clin Oncol* 2004;130:745–748.

75. Baltogiannis D, Kalogeropoulos C, Ioachim E, et al. Orbital metastasis from prostatic carcinoma. *Urol Int* 2003;70:219–222.

76. Ahmed S, Goel S, Khandwala M, et al. Neuroblastoma with orbital metastasis: ophthalmic presentation and role of ophthalmologists. *Eye* 2006;20:466–470.

77. Demirci H, Shields CL, Shields JA, et al. Orbital tumors in the older adult population. *Ophthalmology* 2002;109:243–248.

78. Goncalves AC, Moura FC, Monteiro ML. Bilateral progressive enophthalmos as the presenting sign of metastatic breast carcinoma. *Ophthal Plast Reconstr Surg* 2005;21:311–313.

79. Chang BY, Cunniffe G, Hutchinson C. Enophthalmos associated with primary breast carcinoma. *Orbit* 2002;21:307–310.

80. Howard GM, Jakobiec FA, Trokel SL, et al. Pulsating metastatic tumor of the orbit. *Am J Ophthalmol* 1978;85:767–771.

81. Froula PD, Bartley GB, Garrity JA, et al. The differential diagnosis of orbital calcification as detected on computed tomographic scans. *Mayo Clin Proc* 1993;68:256–261.

82. Yeh S, Foroozan R. Orbital apex syndrome. *Curr Opin Ophthalmol* 2004;15:490–498.

83. Moster ML, Savino PJ, Sergott RC, et al. Isolated sixth-nerve palsies in young adults. *Arch Ophthalmol* 1984;102:1328–1330.

84. Savino PJ, Hilliker JK, Casell GH, et al. Chronic sixth nerve palsies. Are they really harbingers of serious intracranial disease? *Arch Ophthalmol* 1982;100:1442–1444.

85. Rush JA, Younge BR. Paralysis of cranial nerves III, IV, and VI. Cause and prognosis in 1,000 cases. *Arch Ophthalmol* 1981;99:76–79.

86. Johnston JL. Parasellar syndromes. *Curr Neurol Neurosci Rep* 2002;2:423–431.

87. Kline LB, Hoyt WF. The Tolosa-Hunt syndrome. *J Neurol Neurosurg Psychiatry* 2001;71:577–582.

88. Post MJD, Mendez DR, Kline LB, et al. Metastatic disease to the cavernous sinus: clinical syndrome and CT diagnosis. *J Comput Assist Tomogr* 1985;9:115–120.

89. Shields JA, Shields CL, Scartozzi R. Survey of 1264 patients with orbital tumors and simulating lesions: the 2002 Montgomery Lecture, part 1. *Ophthalmology* 2004;111:997–1008.

90. Ferry AP, Font RL. Carcinoma metastatic to the eye and orbit. I. A clinicopathologic study of 227 cases. *Arch Ophthalmol* 1974;92:276–286.

91. Smith JK. Parasellar tumors: suprasellar and cavernous sinuses. *Top Magn Reson Imaging* 2005;16:307–315.

92. Keane JR. Cavernous sinus syndrome. Analysis of 151 cases. *Arch Neurol* 1996;53:967–971.

93. Greenberg HS, Deck MD, Vikram B, et al. Metastasis to the base of the skull: clinical findings in 43 patients. *Neurology* 1981;31:530–537.

94. Roessmann U, Kaufman B, Friede RL. Metastatic lesions in the sella turcica and pituitary gland. *Cancer* 1970;25:478–480.

95. Thomas JE, Yoss RE. The parasellar syndrome: problems in determining etiology. *Mayo Clin Proc* 1970;45:617–623.

96. Mills RP, Insalaco SJ, Joseph A. Bilateral cavernous sinus metastasis and ophthalmoplegia. Case report. *J Neurosurg* 1981;55:463–466.

97. Keane JR. Bilateral ocular paralysis: analysis of 31 inpatients. *Arch Neurol* 2007;64:178–180.

98. McAvoy CE, Kamalarajab S, Best R, et al. Bilateral third and unilateral sixth nerve palsies as early presenting signs of metastatic prostatic carcinoma. *Eye* 2002;16:749–753.

99. Rennert J, Doerfler A. Imaging of sellar and parasellar lesions. *Clin Neurol Neurosurg* 2007;109:111–124.

100. Kingdom TT, Delgaudio JM. Endoscopic approach to lesions of the sphenoid sinus, orbital apex, and clivus. *Am J Otolaryngol* 2003;24:317–322.

101. Iwai Y, Yamanaka K, Ishiguro T. Gamma knife radiosurgery for the treatment of cavernous sinus meningiomas. *Neurosurgery* 2003;52:517–524.

102. Zhu JJ, Padillo O, Duff J, et al. Cavernous sinus and leptomeningeal metastases arising from a squamous cell carcinoma of the face: case report. *Neurosurgery* 2004;54:492–498.

103. Beer TW. Reexcision perineural invasion: a mimic of malignancy. *Am J Dermatopathol* 2006;28:423–425.

104. Abraham PJ, Capobianco DJ, Cheshire WP. Facial pain as the presenting symptom of lung carcinoma with normal chest radiograph. *Headache* 2003;43:499–504.

105. Sarlani E, Schwartz AH, Greenspan JD, et al. Facial pain as first manifestation of lung cancer: a case of lung cancer-related cluster headache and a review of the literature. *J Orofac Pain* 2003;17:262–267.

106. Palmieri A. Lung cancer presenting with unilateral facial pain: remission after laryngeal nerve palsy. *Headache* 2006;46:813–815.

107. ten Hove MW, Glaser JS, Schatz NJ. Occult perineural tumor infiltration of the trigeminal nerve. Diagnostic considerations. *J Neuroophthalmol* 1997;17:170–177.

108. Laurencet FM, Anchisi S, Tullen E, et al. Mental neuropathy: report of five cases and review of the literature. *Crit Rev Oncol Hematol* 2000;34:71–79.

109. Nakano I, Iwasaki K, Kondo A. Solitary metastatic breast carcinoma in a trigeminal nerve mimicking a trigeminal neurinoma—case report. *J Neurosurg* 1996;85:677–680.

110. Dumas M, Perusse R. Trigeminal sensory neuropathy: a study of 35 cases. *Oral Surg Oral Med Oral Pathol Oral Radiol Endod* 1999;87:577–582.

111. Penarrocha M, Alfaro A, Bagan JV, et al. Idiopathic trigeminal sensory neuropathy. *J Oral Maxillofac Surg* 1992;50:472–476.

112. Hagen NA, Stevens CJ, Michet C. Trigeminal sensory neuropathy associated with connective tissue diseases. *Neurology* 1990;40:891–896.

113. Dresel SHJ, Mackey JK, Lufkin RB, et al. Meckel cave lesions: Percutaneous fine-needle-aspiration biopsy cytology. *Radiology* 1991;179:579–582.

114. Maillefert JF, Gazet-Maillefert MP, Tavernier C, et al. Numb chin syndrome. *Joint Bone Spine* 2000;67:86–93.

115. Lossos A, Siegal T. Numb chin syndrome in cancer patients: etiology, response to treatment, and prognostic significance. *Neurology* 1992;42:1181–1184.

116. Go JL, Kim PE, Zee CS. The trigeminal nerve. *Semin Ultrasound CT MR* 2001;22:502–520.

117. Baskaran RK, Krishnamoorthy, Smith M. Numb Chin Syndrome—a reflection of systemic malignancy. *World J Surg Oncol* 2006;4:52.

118. Halachmi S, Madeb R, Madjar S, et al. Numb chin syndrome as the presenting symptom of metastatic prostate carcinoma. *Urology* 2000;55:286.

119. Konotey-Ahulu FI. Mental-nerve neuropathy: a complication of sickle-cell crisis. Letter to the editor. *Lancet* 1972;2:388.

120. Furukawa T. Numb chin syndrome in the elderly. Letter to editor. *J Neurol Neurosurg Psychiatry* 1990;53:173.

121. Cohen JL, Barankin B, Zloty DM, et al. Metastatic zosteriform squamous cell carcinoma in an immunocompetent patient. *J Cutan Med Surg* 2004;8:438–441.

122. Ozyar E, Cengiz M, Gurkaynak M, et al. Trismus as a presenting symptom in nasopharyngeal carcinoma. *Radiother Oncol* 2005;77:73–76.

123. Huang CC, Tseng FY, Chen ZC, et al. Malignant parotid tumor and facial palsy. *Otolaryngol Head Neck Surg* 2007;136:778–782.

124. Park YW, Hlivko TJ. Parotid gland metastasis from renal cell carcinoma. *Laryngoscope* 2002; 112:453–456.

125. Streitmann MJ, Sismanis A. Metastatic carcinoma of the temporal bone. *Am J Otol* 1996;17:780–783.

126. Boahene DO, Olsen KD, Driscoll C, et al. Facial nerve paralysis secondary to occult malignant neoplasms. *Otolaryngol Head Neck Surg* 2004;130:459–465.

127. Bourne RG. The Costello Memorial Lecture. The spread of squamous carcinoma of the skin via the cranial nerves. *Australas Radiol* 1980;24:106–114.

128. Geopfert H, Dichtel WJ, Medina JE, et al. Perineural invasion in squamous cell skin carcinoma of the head and neck. *Am J Surg* 1984;148:542–547.

129. Dewys WD, Walters K. Abnormalities of taste sensation in cancer patients. *Cancer* 1975;36:1888–1896.

130. Martin DS, Benecke J, Maas C. Metastatic tumor presenting as chronic otitis and facial paralysis. *Ann Otol Rhinol Laryngol* 1992;101:280–281.

131. Mani N, Sudhoff H, Rajagopal S, et al. Cranial nerve involvement in malignant external otitis: implications for clinical outcome. *Laryngoscope* 2007;117:907–910.

132. Yao M, Nguyen T, Hansen MR, et al. Optically guided stereotactic radiotherapy for facial nerve paralysis secondary to occult malignant neoplasms. *Otolaryngol Head Neck Surg* 2006;135:657–659.

133. Uppal HS, Ayshford CA, Wilson F. Sudden onset bilateral sensorineural hearing loss: a manifestation of occult breast carcinoma. *J Laryngol Otol* 2001;115:907–910.

134. Gloria-Cruz TI, Schachern PA, Paparella MM, et al. Metastases to temporal bones from primary nonsystemic malignant neoplasms. *Arch Otolaryngol Head Neck Surg* 2000;126:209–214.

135. Berlinger NT, Koutroupas S, Adams G, et al. Patterns of involvement of the temporal bone in metastatic and systemic malignancy. *Laryngoscope* 1980;90:619–627.

136. Igarashi M, Card GG, Johnson PE, et al. Bilateral sudden hearing loss and metastatic pancreatic adenocarcinoma. *Arch Otolaryngol* 1979;105:196–199.

137. Wagemakers M, Verhagen W, Borne B, et al. Bilateral profound hearing loss due to meningeal carcinomatosis. *J Clin Neurosci* 2005;12:315–318.

138. Paparella MM, Berlinger NT, Oda M, et al. Otological manifestations of leukemia. *Laryngoscope* 1973;83:1510–1526.

139. Okura SI, Kaga K. Temporal bone pathology of leukemia and malignant lymphoma with middle ear effusion. *Auris Nasus Larynx* 1994;21:1–7.

140. Guasti L, Simoni C, Scamoni C, et al. Mixed cranial nerve neuroma revealing itself as baroreflex failure. *Auton Neurosci* 2006; (30)130:57–60.

141. Zeitels SM, Casiano RR, Gardner GM, et al. Management of common voice problems: committee report. *Otolaryngol Head Neck Surg* 2002; 126:333–348.

142. Javle M, Ailawadhi S, Yang GY, et al. Palliation of malignant dysphagia in esophageal cancer: a literature-based review. *J Support Oncol* 2006;4: 365–373, 379.

143. Nazareno J, Taves D, Preiksaitis HG. Metastatic breast cancer to the gastrointestinal tract: a case series and review of the literature. *World J Gastroenterol* 2006;12:6219–6224.

144. Graus F, Slatkin NE. Papilledema in the metastatic jugular foramen syndrome. *Arch Neurol* 1983; 40:816–818.

145. Lowenheim H, Koerbel A, Ebner FH, et al. Differentiating imaging findings in primary and secondary tumors of the jugular foramen. *Neurosurg Rev* 2006;29:1–11.

146. Widick MH, Haynes DS, Jackson CG, et al. Slow-flow phenomena in magnetic resonance imaging of the jugular bulb masquerading as skull base neoplasms. *Am J Otol* 1996;17:648–652.

147. Macdonald DR, Strong E, Nielsen S, et al. Syncope from head and neck cancer. *J Neuro-Oncol* 1983;1:257–268.

148. Onrot J, Wiley RG, Fogo A, et al. Neck tumour with syncope due to paroxysmal sympathetic withdrawal. *J Neurol Neurosurg Psychiatry* 1987; 50:1063–1066.

149. Cicogna R, Bonomi FG, Curnis A, et al. Parapharyngeal space lesions syncope-syndrome. A newly proposed reflexogenic cardiovascular syndrome. *Eur Heart J* 1993;14:1476–1483.

150. Chen-Scarabelli C, Kaza AR, Scarabelli T. Syncope due to nasopharyngeal carcinoma. *Lancet Oncol* 2005;6:347–349.

151. Raskin NH, Prusiner S. Carotidynia. *Neurology* 1977;27:43–46.

152. Osswald S, Troutan TG. Neurocardiogenic (vasodepressor) syncope. *N Engl J Med* 1993;329:30.

153. Choi YM, Mafee MF, Feldman LE. Successful treatment of syncope in head and neck cancer with induction chemotherapy. *J Clin Oncol* 2006;24:5332–5333.

154. Levin B, Posner JB. Swallow syncope: report of a case and review of the literature. *Neurology* 1972;22:1086–1093.

155. Omi W, Murata Y, Yaegashi T, et al. Swallow syncope, a case report and review of the literature. *Cardiology* 2006;105:75–79.

156. Bongers KM, Willigers HMM, Koehler PJ. Referred facial pain from lung carcinoma. *Neurology* 1992;42:1841–1842.

157. Capobianco DJ. Facial pain as a symptom of nonmetastatic lung cancer. *Headache* 1995;35:581–585.

158. Eross EJ, Dodick DW, Swanson JW, et al. A review of intractable facial pain secondary to underlying lung neoplasms. *Cephalalgia* 2003;23:2–5.

159. Swift TR. Involvement of peripheral nerves in radical neck dissection. *Am J Surg* 1970;119:694–698.

160. Rotta FT, Romano JG. Skull base metastases causing acute bilateral hypoglossal nerve palsy. *J Neurol Sci* 1997;148:127–129.

161. Johnston EF, Hammond AJ, Cairncross JG. Bilateral hypoglossal palsies: a late complication of curative radiotherapy. *Can J Neurol Sci* 1989;16:198–199.

162. Cinar SO, Seven H, Cinar U, et al. Isolated bilateral paralysis of the hypoglossal and recurrent laryngeal nerves (bilateral Tapia's syndrome) after transoral intubation for general anesthesia. *Acta Anaesthesiol Scand* 2005;49:98–99.

163. Batchelor TT, DeAngelis LM, Krol GS. Neuroimaging abnormalities with hypoglossal nerve palsies. *J Neuroimag* 1997;7:86–88.

164. Moris G, Roig C, Misiego M, et al. The distinctive headache of the occipital condyle syndrome: a report of four cases. *Headache* 1998;38:308–311.

165. Riggs JE. Distinguishing between extrinsic and intrinsic tongue muscle weakness in unilateral hypoglossal palsy. *Neurology* 1984;34:1367–1368.

166. Paling MR, Black WC, Levine PA, et al. Tumor invasion of the anterior skull base: a comparison of MR and CT studies. *J Comput Assist Tomogr* 1987;11:824–830.

167. Brillman J, Valeriano J, Adatepe MH. The diagnosis of skull base metastases by radionuclide bone scan. *Cancer* 1987;59:1887–1891.

168. Johnson PC. Hematogenous metastases of carcinoma to dorsal root ganglia. *Acta Neuropathol (Berl)* 1977;38:171–172.

169. Chason JL, Walker FB, Landers JW. Metastatic carcinoma in the central nervous system and dorsal root ganglia. A prospective autopsy study. *Cancer* 1963;16:781–787.

170. Dickenman RC, Chason JL. Alterations in the dorsal root ganglia and adjacent nerves in the leukemias, the lymphomas and multiple myeloma. *Am J Pathol* 1958;34:349–357.

171. Wigfield CC, Hilton DA, Coleman MG, et al. Metastatic adenocarcinoma masquerading as a solitary nerve sheath tumour. *Br J Neurosurg* 2003;17:459–461.

172. Braham J, Sadeh M, Sarova-Pinhas I. Skin wrinkling on immersion of hands. *Arch Neurol* 1979;36:113–114.

173. Walsh JC, Low PA, Allsop JL. Localized sympathetic overactivity: an uncommon complication of lung cancer. *J Neurol Neurosurg Psychiatry* 1976;39:93–95.

174. Friedman JH. Hemifacial gustatory sweating due to Pancoast's tumor. *Am J Med* 1987;82:1269–1272.

175. Steiner I, Siegal T. Muscle cramps in cancer patients. *Cancer* 1989;63:574–577.

176. Cooper WH, Ringel SP, Treihaft MM, et al. Calf enlargement from S-1 radiculopathy. Report of two cases. *J Neurosurg* 1985;62:442–444.

177. Giles CL, Henderson JW. Horner's syndrome: analysis of 216 cases. *Am J Ophthalmol* 1958; 46:289–296.

178. Dalmau J, Graus F, Marco M. "Hot and dry foot" as initial manifestation of neoplastic lumbosacral plexopathy. *Neurology* 1989;39:871–872.

179. Son YH. Effectiveness of irradiation therapy in peripheral neuropathy caused by malignant disease. *Cancer* 1967;20:1447–1451.

180. Wilbourn AJ. Plexopathies. *Neurol Clin* 2007; 25:139–171.

181. Villas C, Collia A, Aquerreta JD, et al. Cervicobrachialgia and pancoast tumor: value of standard anteroposterior cervical radiographs in early diagnosis. *Orthopedics* 2004;27:1092–1095.

182. Jaeckle KA. Neurological manifestations of neoplastic and radiation-induced plexopathies. *Semin Neurol* 2004;24:385–393.

183. Arcasoy SM, Jett JR. Superior pulmonary sulcus tumors and Pancoast's syndrome. *N Engl J Med* 1997 6;337:1370–1376.

184. Detterbeck FC. Pancoast (superior sulcus) tumors. *Ann Thorac Surg* 1997;63:1810–1818.

185. Kori SH, Foley KM, Posner JB. Brachial plexus lesions in patients with cancer: 100 cases. *Neurology* 1981;31:45–50.

186. Marangoni C, Lacerenza M, Formaglio F, et al. Sensory disorder of the chest as presenting symptom of lung cancer. *J Neurol Neurosurg Psychiatry* 1993;56:1033–1034.

187. Yoss RE, Corbin KB, MacCarty CS, et al. Significance of symptoms and signs in localization of involved root in cervical disk protrusion. *Neurology* 1957;7:673–683.

188. Loukas M, Hullett J, Louis RG, Jr, et al. The gross anatomy of the extrathoracic course of the intercostobrachial nerve. *Clin Anat* 2006;19:106–111.

189. Baron RH, Fey JV, Borgen PI, et al. Eighteen sensations after breast cancer surgery: a two-year comparison of sentinel lymph node biopsy and axillary lymph node dissection. *Oncol Nurs Forum* 2004;31:691–698.

190. Torresan RZ, Cabello C, Conde DM, et al. Impact of the preservation of the intercostobrachial nerve in axillary lymphadenectomy due to breast cancer. *Breast J* 2003;9:389–392.

191. Kichari JR, Hussain SM, Den Hollander JC, et al. MR imaging of the brachial plexus: current imaging sequences, normal findings, and findings in a spectrum of focal lesions with MR-pathologic correlation. *Curr Probl Diagn Radiol* 2003; 32:88–101.

192. Luthra K, Shah S, Purandare N, et al. F-18 FDG PET-CT appearance of metastatic brachial plexopathy in a case of carcinoma of the breast. *Clin Nucl Med* 2006;31:432–434.

193. Harper CM, Jr, Thomas JE, Cascino TL, et al. Distinction between neoplastic and radiation-induced brachial plexopathy, with emphasis on the role of EMG. *Neurology* 1989;39:502–506.

194. Bai LY, Chiu CF, Liao YM, et al. Recurrent lymphoma presenting as brachial plexus neuropathy. *J Clin Oncol* 2007;25:726–728.

195. Tender GC, Kline DG. Posterior subscapular approach to the brachial plexus. *Neurosurgery* 2005;57:377–381.

196. Eisenberg E, Geller R, Brill S. Pharmacotherapy options for complex regional pain syndrome. *Expert Rev Neurother* 2007;7:521–531.

197. Salner AL, Botnick LE, Herzog AG, et al. Reversible brachial plexopathy following primary radiation therapy for breast cancer. *Cancer Treat Rep* 1981;65:797–802.

198. Gerard JM, Franck N, Moussa Z, et al. Acute ischemic brachial plexus neuropathy following radiation therapy. *Neurology* 1989;39:450–451.

199. Rubin DI, Schomberg PJ, Shepherd RF, et al. Arteritis and brachial plexus neuropathy as delayed complications of radiation therapy. *Mayo Clin Proc* 2001;76:849–852.

200. Huang JH, Zager EL. Thoracic outlet syndrome. *Neurosurgery* 2004;55:897–902.

201. Kim DH, Murovic JA, Tiel RL, et al. A series of 146 peripheral non-neural sheath nerve tumors: 30-year experience at Louisiana State University Health Sciences Center. *J Neurosurg* 2005;102:256–266.

202. Kim DH, Murovic JA, Tiel RL, et al. A series of 397 peripheral neural sheath tumors: 30-year experience at Louisiana State University Health Sciences Center. *J Neurosurg* 2005;102:246–255.

203. Miller JD, Pruitt S, McDonald TJ. Acute brachial plexus neuritis: an uncommon cause of shoulder pain. *Am Fam Physician* 2000;62:2067–2072.

204. Malow BA, Dawson DM. Neuralgic amyotrophy in association with radiation therapy for Hodgkin's disease. *Neurology* 1991;41:440–441.

205. Churn M, Clough V, Slater A. Early onset of bilateral brachial plexopathy during mantle radiotherapy for Hodgkin's disease. *Clin Oncol (R Coll Radiol)* 2000;12:289–291.

206. Narayan S, Thomas CR, Jr. Multimodality therapy for Pancoast tumor. *Nat Clin Pract Oncol* 2006;3:484–491.

207. Gachiani J, Kin DH, Nelson A, et al. Management of metastatic tumors invading the peripheral nervous system. *Neurosurg Focus* 2007;22:E14.

208. Ladha SS, Spinner RJ, Suarez GA, et al. Neoplastic lumbosacral radiculoplexopathy in prostate cancer by direct perineural spread: an unusual entity. *Muscle Nerve* 2006;34:659–665.

209. Jaeckle KA, Young DF, Foley KM. The natural history of lumbosacral plexopathy in cancer. *Neurology* 1985;35:8–15.

210. Saphner T, Gallion HH, van Nagell JR, et al. Neurologic complications of cervical cancer. A review of 2261 cases. *Cancer* 1989;64:1147–1151.

211. Anderson TS, Regine WF, Kryscio R, et al. Neurologic complications of bladder carcinoma—a review of 359 cases. *Cancer* 2003;97:2267–2272.

212. Kanter P, Zeidman A, Streifler J, et al. PET-CT imaging of combined brachial and lumbosacral neurolymphomatosis. *Eur J Haematol* 2005;74:66–69.

213. Ross JJ, Hu LT. Septic arthritis of the pubic symphysis: review of 100 cases. *Medicine (Baltimore)* 2003;82:340–345.

214. Kunduracioglu B, Yilmaz C, Yorubulut M, et al. Magnetic resonance findings of osteitis pubis. *J Magn Reson Imaging* 2007;25:535–539.

215. Marsh WL, Jr, Bylund DJ, Heath VC, et al. Osteoarticular and pulmonary manifestations of acute leukemia. Case report and review of the literature. *Cancer* 1986;57:385–390.

216. Kai T, Ishii E, Matsuzaki A, et al. Clinical and prognostic implications of bone lesions in childhood leukemia at diagnosis. *Leuk Lymphoma* 1996;23:119–123.

217. Gallagher DJ, Phillips DJ, Heinrich SD. Orthopedic manifestations of acute pediatric leukemia. *Orthop Clin North Am* 1996;27:635–644.

218. Hauge MD, Cooper KL, Litin SC. Insufficiency fractures of the pelvis that simulate metastatic disease. *Mayo Clin Proc* 1988;63:807–812.

219. Diaz-Arrastia R, Younger DS, Hair L, et al. Neurolymphomatosis: A clinicopathological syndrome re-emerges. *Neurology* 1992;42:1136–1141.

220. Baehring JM, Damek D, Martin EC, et al. Neurolymphomatosis. *Neuro-Oncol* 2003;5:104–115.

221. Kelly JJ, Karcher DS. Lymphoma and peripheral neuropathy: a clinical review. *Muscle Nerve* 2005;31:301–313.

222. Varin S, Faure A, Bouc P, et al. Endoneural metastasis of the sciatic nerve disclosing the relapse of a renal carcinoma, four years after its surgical treatment. *Joint Bone Spine* 2006;73:760–762.

223. Di Tommaso L, Magrini E, Consales A, et al. Malignant granular cell tumor of the lateral femoral cutaneous nerve: report of a case with cytogenetic analysis. *Hum Pathol* 2002;33:1237–1240.

224. Rogers LR, Borkowski GP, Albers JW, et al. Obturator mononeuropathy caused by pelvic cancer: six cases. *Neurology* 1993;43:1489–1492.

225. Filler AG, Haynes J, Jordan SE, et al. Sciatica of non-disc origin and piriformis syndrome: diagnosis by magnetic resonance neurography and interventional magnetic resonance imaging with outcome study of resulting treatment. *J Neurosurg Spine* 2005; 2:99–115.

226. Filler AG, Maravilla KR, Tsuruda JS. MR neurography and muscle MR imaging for image diagnosis of disorders affecting the peripheral nerves and musculature. *Neurol Clin* 2004;22:643.

227. Kuntz C, Blake L, Britz G, et al. Magnetic resonance neurography of peripheral nerve lesions in the lower extremity. *Neurosurgery* 1996;39:750–756.

228. Bokstein F, Goor O, Shihman B, et al. Assessment of neurolymphomatosis by brachial plexus biopsy and PET/CT. Report of a case. *J Neuro-Oncol* 2005;72:163–167.

229. Lekos A, Katirji MB, Cohen ML, et al. Mononeuritis multiplex. A harbinger of acute leukemia in relapse. *Arch Neurol* 1994;51:618–622.

230. Ganel A, Engel J, Sela M, et al. Nerve entrapments associated with postmastectomy lymphedema. *Cancer* 1979;44:2254–2259.

231. Odabasi Z, Parrott JH, Reddy VV, et al. Neurolymphomatosis associated with muscle and cerebral involvement caused by natural killer cell lymphoma: a case report and review of literature. *J Peripher Nerv Syst* 2001;6:197–203.

232. Krendel DA, Albright RE, Graham DG. Infiltrative polyneuropathy due to acute monoblastic leukemia in hematologic remission. *Neurology* 1987; 37:474–477.

233. van den Bent MJ, de Bruin HG, Bos GM, et al. Negative sural nerve biopsy in neurolymphomatosis. *J Neurol* 1999;246:1159–1163.

234. Ghobrial IM, Buadi F, Spinner RJ, et al. High-dose intravenous methotrexate followed by autologous stem cell transplantation as a potentially effective therapy for neurolymphomatosis. *Cancer* 2004;100:2403–2407.

235. Nabi G, Gupta NP, Gandhi D. Skeletal muscle metastasis from transitional cell carcinoma of the urinary bladder: clinicoradiological features. *Clin Radiol* 2003;58:883–885.

236. Pearson CM. Incidence and type of pathologic alterations observed in muscle in a routine autopsy survey. *Neurology* 1959;9:757–766.

237. Acinas GO, Fernandez FA, Satue EG, et al. Metastasis of malignant neoplasms to skeletal muscle. *Rev Esp Oncol* 1984;31:57–67.

238. Buerger LF, Monteleone PN. Leukemic-lymphomatous infiltration of skeletal muscle. Systematic study of 82 autopsy cases. *Cancer* 1966; 19:1416–1422.

239. Magee T, Rosenthal H. Skeletal muscle metastases at sites of documented trauma. *Am J Roentgenol* 2002;178:985–988.

240. Herring CL, Jr, Harrelson JM, Scully SP. Metastatic carcinoma to skeletal muscle. A report of 15 patients. *Clin Orthop* 1998;355:272–281.

241. Glockner JF, White LM, Sundaram M, et al. Unsuspected metastases presenting as solitary soft tissue lesions: a fourteen-year review. *Skeletal Radiol* 2000;29:270–274.

242. Hur J, Yoon CS, Jung WH. Multiple skeletal muscle metastases from renal cell carcinoma 19 years after radical nephrectomy. *Acta Radiol* 2007;48:238–241.

243. Damron TA, Heiner J. Distant soft tissue metastases: a series of 30 new patients and 91 cases from the literature. *Ann Surg Oncol* 2000;7:526–534.

244. Luo C, Jiang Y, Liu Y, et al. Experimental study on mechanism and rarity of metastases in skeletal muscle. *Chin Med J (Engl)* 2002;115:1645–1649.

245. Bar-Yehuda S, Barer F, Volfsson L, et al. Resistance of muscle to tumor metastases: a role for a3 adenosine receptor agonists. *Neoplasia* 2001;3:125–131.

246. Weiss L. Biomechanical destruction of cancer cells in skeletal muscle: a rate-regulator for hematogenous metastasis. *Clin Exp Metastasis* 1989;7:483–491.

247. Schluter K, Gassmann P, Enns A, et al. Organ-specific metastatic tumor cell adhesion and extravasation of colon carcinoma cells with different metastatic potential. *Am J Pathol* 2006;169:1064–1073.

248. Seely S. Possible reasons for the high resistance of muscle to cancer. *Med Hypotheses* 1980;6:133–137.

249. Grem JL, Neville AJ, Smith SC, et al. Massive skeletal muscle invasion by lymphoma. *Arch Intern Med* 1985;145:1818–1820.

250. Kandel RA, Bedard YC, Pritzker KP, et al. Lymphoma: presenting as an intramuscular small cell malignant tumor. *Cancer* 1984;53:1586–1589.

251. Doshi R, Fowler T. Proximal myopathy due to discrete carcinomatous metastases in muscle. *J Neurol Neurosurg Psychiatry* 1983;46:358–360.

252. Uppal SS, Salopal TK, Singh H. Left gluteal focal myositis in a patient with signet ring adenocarcinoma of the stomach: not a paraneoplastic phenomenon. *Rheumatol Int* 2004;24:365–367.

253. Heffernan E, Fennelly D, Collins CD. Multiple metastases to skeletal muscle from carcinoma of the esophagus detected by FDG PET-CT imaging. *Clin Nucl Med* 2006;31:810–811.

254. Tuoheti Y, Okada K, Osanai T, et al. Skeletal muscle metastases of carcinoma: a clinicopathological study of 12 cases. *Jpn J Clin Oncol* 2004;34:210–214.

255. Sato N, Okamoto S, Mori T, et al. Recurrent acute myositis after allogeneic bone marrow transplantation for myelodysplasia. *Hematology* 2002;7:109–112.

256. Chim CS, Au WY, Poon C, et al. Primary natural killer cell lymphoma of skeletal muscle. *Histopathology* 2002;41:371–374.

257. Min HS, Hyun CL, Paik JH, et al. An autopsy case of aggressive CD30+ extra-nodal NK/T-cell lymphoma initially manifested with granulomatous myositis. *Leuk Lymphoma* 2006;47:347–352.

258. Fritz J, Vogel W, Claussen CD, et al. Generalized intramuscular granulocytic sarcoma mimicking polymyositis. *Skeletal Radiol* 2007;36(10):985–989.

PART 3

NONMETASTATIC COMPLICATIONS OF CANCER

Chapter 9

Vascular Disorders

INTRODUCTION

Vascular lesions complicating cancer are important to the clinician for three reasons:

First, a cerebrovascular event may precede identification of the cancer and be the first evidence that the patient suffers from disease of any kind.[1–3] In this situation, the physician must not only find the cause of the cerebrovascular event, but also consider an occult and potentially curable neoplasm. Although thrombotic or embolic cerebral infarction as a herald of an underlying occult neoplasm is too rare

to warrant a search for cancer in all patients who suffer a stroke, investigation of peripheral vascular events is more rewarding.[4] Some studies suggest that an unexplained pulmonary embolus[5] or deep vein thrombosis (DVT)[6] reflects a substantially increased risk of cancer appearing within the next 2 years, particularly in patients older than age 65. One study found a 10% incidence of cancer in patients with DVT; some tumors were quite small. In patients with *idiopathic* DVT, a malignancy was found in 23%.[7] In another series, 16% of patients with idiopathic DVT were found to have

325

cancer within 2 years of the vascular event.[8] Especially at risk are patients with recurrent thromboembolism. The incidence of underlying cancer is even higher in patients with DVT of the upper limbs,[9] with migratory superficial thrombophlebitis (Trousseau syndrome)[10–13] or nonbacterial thrombotic endocarditis (NBTE). Although uncommon, a recent report[4] indicates that ischemic heart disease may also be associated with an occult neoplasm.

Second, venous thromboembolism may also complicate the course of patients with cancer treated with chemotherapy. In one series, 7% of patients developed venous thromboembolism during a 3-month period after beginning chemotherapy. The annual incidence was 10%. Fluorouracil and leucovorin appeared to be particular offenders[14] (see Chapter 12).

Third, cerebral neoplasms may be mistaken for vascular events either because their symptoms develop suddenly, such as hemorrhage into a brain metastasis, or because computed tomography (CT) or magnetic resonance images initially resemble an infarct or a hemorrhage rather than a neoplasm.[15]

Central nervous system (CNS) vascular disease can be a major cause of disability in an otherwise functioning patient with cancer. Accurate diagnosis allows appropriate treatment, such as anticoagulation, or at least prevents inappropriate treatment, such as radiation therapy (RT) when a cerebral infarct is mistaken for a brain metastasis. Peripheral vascular lesions, such as DVT, also cause significant disability in some patients with cancer[16] and can be difficult to treat.[17]

FREQUENCY

CNS vascular disease is common in patients with cancer.[18–23] Autopsy data indicate that cerebrovascular lesions are found in approximately 15% of patients who die from cancer, 50% of whom had symptoms of the vascular disease while alive.[24] Hemorrhagic and ischemic lesions are equally frequent (Fig. 9–1). Arteriosclerosis and stroke are reported to be more common in patients who subsequently develop cancer. A study of over 100,000

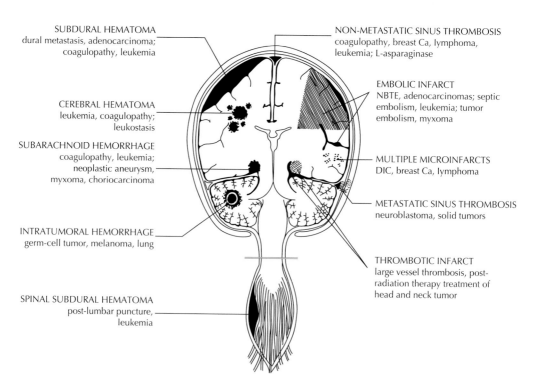

Figure 9–1. Cerebrovascular disorders encountered in patients with cancer. The left side of the figure illustrates hemorrhagic lesions and the right illustrates ischemic lesions. From Ref. 24 with permission.

stroke patients, followed after discharge from hospital, identified a highly significant increase in newly diagnosed cancers; 10% of patients had either a history of, or a concurrent malignancy.[21] Most of the increased risk occurred within a year of the diagnosis of stroke. Even more striking was the finding that women with arteriosclerotic diseases had more than a threefold relative risk of developing lung cancer.[20] Interestingly, lung cancer was not significantly increased in men.

Although atherosclerotic cerebral infarcts appear to be associated with cancer, autopsy findings suggest a lower incidence of severe arteriosclerosis in patients with cancer[25–28] (Table 9–1). One hypothesis to explain the disparity is that cancer may cause regression of preexisting atherosclerosis.[29] However, in an autopsy study, conventional atherosclerotic stroke was the most common cause of cerebral ischemia in adult patients with cancer.[24] Our clinical findings are slightly different, suggesting that an embolic stroke is more common.[22]

PATHOPHYSIOLOGY OF CEREBROVASCULAR DISEASE

Patients with cancer experience two types of hemostatic abnormalities: deficient coagulation (hypocoagulopathy), usually related to thrombocytopenia, which can cause cerebral hemorrhage,[30] or excessive coagulation (hypercoagulopathy), which can cause either CNS hemorrhage or infarction, or both[13,31–33] (Table 9–2). Even an occult cancer can precipitate a coagulopathy that causes cerebrovascular disease.[1,34] Over 50% of patients with cancer, and 90% of patients with metastatic cancer, develop laboratory evidence of a coagulation abnormality during the course of the cancer or its treatment.[32,35]

Hypocoagulation

The incidence of hemorrhage caused by deficient clotting mechanisms[24,36] is high in patients with cancer. Approximately 15%

Table 9–1. **Central Nervous System Vascular Disorders Found at Autopsy in Patients with Cancer**

Disorders	No. of Patients	No. of Patients with Symptoms
Cerebral Hemorrhage	244	138
Intracerebral hematoma		
Intratumoral	60	47
Secondary to coagulopathy	88	57
Hypertensive	9	8
Subdural hematoma	63	16
Subarachnoid hemorrhage	24	10
Cerebral Infarction	256	117
Athero sclerosis	73	17
Intravascular coagulation	39	28
Nonbacterial thrombotic endocarditis	42	32
Septic occlusion	33	22
Tumor embolus	12	4
Venous occlusion	33	6
Miscellaneous	24	8
Total	500	255

Adapted from Ref. 24.

Table 9–2. Coagulation Abnormalities that Cause Bleeding Disorders in Patients with Cancer

Platelet Abnormalities
Thrombocytopenia
Thrombocytopathia
Antiplatelet aggregation drugs

Vascular Defects
Infiltration by tumor
Infiltration by fungus
Hyperviscosity/leukostasis
Extramedullary hematopoiesis

Coagulation Factor Abnormalities
Liver dysfunction
Drugs
Fibrinolysis
Von Willebrand disease (acquired)
Disseminated intravascular coagulation

to 25% of patients with myeloproliferative syndromes experience bleeding, usually associated with platelet dysfunction.[36] In a review of 718 patients with solid tumors, all of whom received myelosuppressive agents, 10% experienced one or more hemorrhagic episodes.[37] The most common cause, thrombocytopenia, occurs in three settings: when bone marrow is replaced by the cancer, when chemotherapy or RT damages the bone marrow, or when disseminated intravascular coagulation (DIC) develops.[37] However, in another series of 1274 patients with solid tumors, 11% of whom developed severe thrombocytopenia (20,000/mm^3), only 15% of the thrombocytopenia group developed serious hemorrhage and none of these had a CNS bleed.[38] Interestingly, most of the bleeding occurred at platelet counts between 20/mm^3 and 50,000/mm^3.

In general, spontaneous CNS hemorrhage occurs only when the platelet count falls below 10,000 platelets/mm^3, and even then it is quite rare. Patients with such low platelet counts are also susceptible to intracranial hemorrhage following minor trauma or to spinal hemorrhage after lumbar puncture.[39]

A second cause of CNS bleeding in patients with cancer is direct involvement of blood vessels by the tumor. Certain primary cancers

metastatic to brain (e.g., malignant melanoma, germ cell tumors, renal cell carcinoma, leukemia, and lymphoma) may invade and destroy small vessels in the tumor mass. More rarely, a tumor embolus causes an aneurysm.[40] The damaged vessels then cause potentially fatal cerebral hemorrhages. Similar cerebral bleeding from fungal infections of the brain is discussed in Chapter 10. Leukostasis from hyperleukocytic leukemias can also lead to hemorrhage[41] (see following text).

Other causes of intracranial hemorrhage[36] include DIC, particularly from acute promyelocytic leukemia (APML). APML cells express both tissue factor (TF) resulting in factor VII activation and cancer procoagulants resulting in activation of factor X, causing thrombin generation independent of factor VII; the resulting DIC consumes both platelets and clotting factors faster than they can be replaced.[42] The risk of hemorrhagic death remains high, usually occurring within the first weeks of tumor development.[42] Acquired hemophilia, resulting from production of inhibitors against factor VIII, can cause hemorrhage in patients with both leukemias and solid tumors. Sometimes, immunosuppression reverses the hemophilia; at other times, it responds only to treatment of the tumor.[43]

Hypercoagulation

Several causes of excessive coagulation lead to CNS infarction (Table 9–3). Laboratory evidence for a hypercoagulable state can be found in many patients with cancer and is most marked in those with widespread cancer.[32,33,44,45] The hypercoagulable state may result from the cancer's production of coagulation promoters, from destruction of vital organs, or from cancer therapy.

Injection of Walker 256 carcinoma cells into the vein of an experimental animal causes hypercoagulability, with small-vessel occlusion, consumption of platelets and fibrinogen, and fibrin-split products in the serum.[46] When the liver is involved with cancer, production of coagulation cascade proteins is decreased, leading in turn to either excessive coagulation or deficient fibrinolysis. Liver dysfunction is a common cause of coagulopathy.[47,48]

Malnutrition engendered by widespread cancer and its treatment has also been

Table 9–3. **Some Causes of Thrombosis in Patients with Cancer**

Platelet Aggregation
Tumor-induced thrombin formation
Tumor ADP production
Arachidonate activation

Tumor Cell Procoagulants
TF production
Factor X activators

Monocyte Procoagulants

ADP, adenosine diphosphate; TF, tissue factor.

associated with coagulopathic abnormalities, perhaps through secondary involvement of the liver. Therapy may also may cause abnormal coagulation.[49,50] A striking example was superior sagittal sinus occlusion associated with treating acute leukemia using L-asparaginase[51]: the mechanism of the hypercoagulable state was decreased production of anti-thrombin III. The venous occlusion sometimes resulted in hemorrhage, infarction, or both. Similar abnormal coagulation sometimes follows head injury and may develop or worsen after intracranial surgery.[52] We rarely see this problem now.

Table 9–4 summarizes the usual pathophysiologic events leading to cerebrovascular disorders in patients with cancer.

Table 9–4. **Pathophysiology of Cerebrovascular Disease in Patients with Cancer**

Mechanism	Pathology	Typical Tumors
Direct Effect of Tumor		
Tumor embolism	Embolic infarction	Myxoma, lung cancer
	Neoplastic aneurysm	Myxoma, choriocarcinoma
Dural metastasis	Sagittal sinus thrombosis	Adenocarcinoma, neuroblastoma
	Subdural hematoma	Breast, prostate
Neoplastic infiltration of a cerebral vessel	Intratumoral hemorrhage	Melanoma, germ-cell tumors, lung
Leukostasis	Cerebral hemorrhage	Leukemia
Related to Sepsis		
Septic embolism	Embolic infarction	Leukemia
Vasculitis	Thrombotic microinfarction	Leukemia, lymphoma
Related to Coagulation Disorders		
Sinus thrombosis	Hemorrhagic infarction	Lymphoma, breast cancer
DIC	Thrombotic infarction	Leukemia, breast, lung cancer
NBTE	Embolic infarction	Any tumor
Thrombocytopenia	Cerebral hemorrhage	Leukemia, lymphoma
Hyperviscosity	Thrombotic infarction	Myeloma, lymphoma
Related to Treatment or Diagnostic Procedures		
Lumbar puncture	Subdural hematoma (spinal)	Leukemia
Radiation therapy	Thrombotic infarction	Head and neck, brain
Chemotherapy		
L-asparaginase	Sinus thrombosis	Leukemia
Mitomycin, bleomycin, platinum	Cerebral infarcts	Solid tumor

From Ref. 24 with permission.
DIC, disseminated intravascular coagulation; NBTE, nonbacterial thrombotic endocarditis.

CENTRAL NERVOUS SYSTEM HEMORRHAGE

Hemorrhages can occur in any part of the central or peripheral nervous system and can be classified by anatomic distribution as well as pathophysiologic cause (Table 9–1). In most patients, the site of CNS hemorrhage is the brain parenchyma. Epidural,[53] subdural,[54,55] and subarachnoid hemorrhages[56] are less common. Spinal hemorrhages are even rarer; they may be epidural (usually into tumor) as well as subdural, subarachnoid, or intraparenchymal (see following text). The three most common causes of CNS hemorrhage are thrombocytopenia, metastatic tumor, and leukostasis. Hypertensive hemorrhages are uncommon and can be distinguished easily from those related to cancer because they occur in the basal ganglia rather than in subcortical white matter.[24] Most subdural hematomas are related to a coagulopathy, including anticoagulant therapy, sometimes exacerbated by very minor trauma; others can develop within subdural metastases. Patients with cancer can also experience typical traumatic subdural hematomas because of their increased tendency to fall.[54] Spinal subdural hemorrhages usually occur in patients with thrombocytopenia or a coagulopathy who undergo lumbar puncture.[57] Leptomeningeal metastases, particularly from melanoma, can cause spinal subarachnoid hemorrhages.[58] Leptomeningeal melanoma metastases can also cause an asymptomatic hemorrhage that is detected only by lumbar puncture.

Spontaneous subarachnoid hemorrhages occasionally complicate hypocoagulable states, particularly thrombocytopenia, but spontaneous epidural hematomas are rare in either the brain or spinal canal. The low incidence of spinal epidural hematomas in patients with cancer is surprising because this lesion is more common in the noncancer population, complicating lumbar puncture in anticoagulated patients and appearing spontaneously in patients with coagulation abnormalities resulting from liver disease.

Cerebral hemorrhages are symptomatic in almost all patients, including those with cancer. Because many patients who suffer nonhypertensive cerebral hemorrhages do not succumb (and most who do are not examined postmortem), the actual number of cerebral hemorrhages encountered clinically is greater than the number indicated by autopsy series.

Hemorrhage into Brain Metastases

The most common cause of intracranial hemorrhage in patients with solid tumors is hemorrhage into a metastatic lesion [24,59] that lies either within the brain substance or in the subdural space.[54] Any brain metastasis has the potential for hemorrhage. The metastatic tumor most often causing intracerebral hemorrhage is lung cancer, simply because so many brain metastases are from lung cancer[59] (Fig. 9–2). Metastatic malignant melanoma and germ-cell tumors are more likely to bleed than lung cancer, but their fewer overall numbers make them less common causes of hemorrhage.[24] In a series from Taiwan, 12 of 129 (9.3%) metastatic brain tumors hemorrhaged; thyroid and hepatocellular carcinomas were the common causes. The authors noted that the hemorrhages tended to be at the borders of the tumor.[60] In a series of 45 intracranial metastases from hepatocellular carcinoma, also from Taiwan, intracranial hemorrhage was the initial manifestation in 18 (40%).[61] Other cancers that commonly cause hemorrhage within the substance of the brain are listed in Table 9–5.

Subdural hematomas occur in patients with breast and prostate cancer because these tumors commonly metastasize to the subdural space or calvarium.[55,62] Bleeding probably results from tumor-induced neovascularization. The tumor infiltration occurs within the inner dura mater layers, so most of the hemorrhage is actually intradural.[63] As in metastatic brain tumors, the dural metastasis may be small, so that virtually all of its symptoms are caused by the hemorrhage.

CLINICAL FINDINGS

When hemorrhage occurs in a brain metastasis, symptoms almost always occur suddenly, beginning with severe headache; focal neurologic signs such as hemiparesis follow within minutes. Many such patients were previously well neurologically, although usually, but not always, known to be suffering from cancer. Sometimes, the history reveals subtle neurologic symptoms

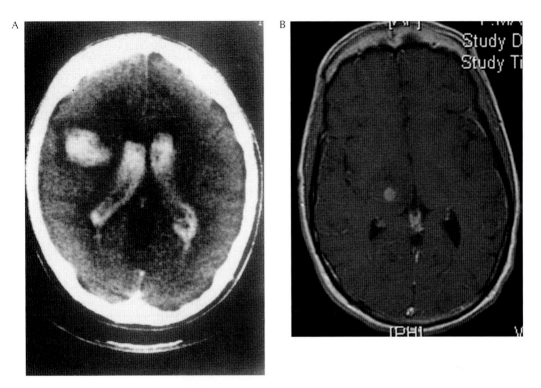

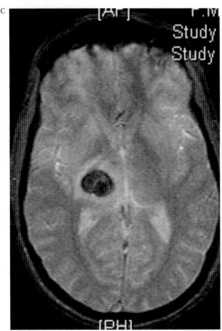

Figure 9–2. A hemorrhagic metastasis. **A:** An unenhanced CT scan illustrating acute cerebral hemorrhage in a brain metastasis from lung cancer. The patient was neurologically asymptomatic until the silent metastasis bled massively and ruptured into the ventricle. The patient died; small remnants of the underlying tumor were encountered at autopsy. **B:** A patient with known metastatic melanoma developed paresthesias of the left side of his body. The unenhanced T1 MRI image reveals hyperintensity within an otherwise hypointense metastatic lesion, suggesting recent hemorrhage. **C:** A gradient echo sequence (GRE) of the same patient, showing hemosiderin, indicating the diagnosis. Note that the lesion on GRE is larger than on the T1 MRI, indicating prior bleeding that was asymptomatic. Some melanomas are hyperintense on T1 and hypointense on gradient echo images, a result of the melanin in the tumor. This can confuse the diagnosis.

Table 9–5. **Onset of Clinical Symptoms in 60 Patients with Hemorrhagic Metastasis**

Type of Tumor	No. of Patients	Asymptomatic	Symptoms	
			Gradual Onset	Abrupt Onset (Stroke-Lsike)
Melanoma	22	2	14	6
Germ-cell	19°	4	6	9
Other[†]	19	7	7	5
Total	60	13	27	20

From Ref. 24 with permission.
° Choriocarcinoma component in 12 of 19 (63%) patients.
[†] Sarcoma, 4 (3 with sudden onset); lung, 8 (1 with sudden onset); kidney, 2; breast, 1; lymphoma, 2; unknown, 2 (1 with sudden onset).

preceding the hemorrhage by days or weeks, suggesting that the sudden ictus is hemorrhage into a pre-existing lesion. The focal signs progress over minutes to hours. With massive hemorrhage, signs of increased intracranial pressure and cerebral herniation may follow. Although death may occur within several hours, usually symptoms stabilize and the patient presents to the hospital with a headache and a focal neurologic deficit. The hemorrhages are usually hemispheral and in the white matter, leading to sensory or visual abnormalities or hemiplegia with or without aphasia. Hemorrhages are less common in the cerebellum or brainstem,[64] but the latter may be fatal.

DIAGNOSIS

The diagnosis of intracranial hemorrhage is suspected from the clinical history and findings and confirmed by CT or magnetic resonance imaging (MRI).[65] With acute hemorrhage (Fig. 9–2), the unenhanced CT shows the hemorrhage as a hyperdense abnormality. Gradient echo MRI images are sensitive to both macro and microhemorrhages; signal heterogeneity on MRI identifies nonhemorrhagic tissue within the hemorrhage and suggests an underlying tumor. The presence of significant edema around an acute hemorrhage suggests that a tumor caused the hemorrhage because, with uncomplicated hemorrhage, edema does not develop for days. Some colon cancers and some melanomas may be hyperdense on CT or hyperintense on MRI without hemorrhage. The increased signal results from either tight packing of the cells or from the presence of melanin in the case of melanoma. The density of a hemorrhage on CT is usually greater than that seen in nonhemorrhagic metastases, but less than in heavily calcified tumors. Interestingly, in patients with melanoma and multiple brain metastases, many of the lesions appear to contain hemorrhage of the same age, as if multiple lesions bled at the same time. Often only one is symptomatic (typically a seizure), but hemorrhage is identified in several lesions by imaging. The pathophysiology of this is unknown.

Lack of enhancement indicates either the absence of a pre-existing metastasis or obliteration of the tumor by the bleed. Contrast enhancement contiguous with the hemorrhage suggests that the bleed has occurred within a metastasis. Areas of enhancement elsewhere in the brain suggest that the patient has multiple metastases, only one of which has bled.

Enhancement around the hemorrhage seen acutely on the CT or MRI suggests an underlying tumor because hypertensive cerebral hemorrhages do not enhance for days to weeks after the ictus. Enhancement of a hemorrhagic metastasis is usually irregular and thick and can be distinguished easily from the serpiginous enhancement associated with an underlying vascular malformation.

Sometimes images do not assist in the diagnosis, but instead, the diagnosis is made by histological examination of the clot removed at surgery.[66] In a series of 31 patients in whom an intracranial blood clot was removed because the etiology of the hemorrhage had not been established, two metastatic tumors were found;

both were from non–small cell lung cancer.[66] In those instances where resection is not indicated, stereotactic needle biopsy may establish the diagnosis.[67]

DIFFERENTIAL DIAGNOSIS

The differential diagnosis includes nonhemorrhagic tumor with apoplectic onset mimicking a vascular event (see Chapter 5), hemorrhagic infarction from vascular occlusion, and brain hemorrhage not associated with metastatic tumor. Multiple cryptic venous angiomas of the brain have been mistaken for hemorrhagic metastases.[68,69]

Cerebral infarction, like cerebral hemorrhage, may be sudden in onset, but signs of increased intracranial pressure are uncommon because of less mass effect. The CT may be normal initially, but the MRI diffusion-weighted images become positive within minutes, revealing restricted diffusion probably as a result of cytotoxic edema; contrast enhancement of the involved gyri develops several days later. A hemorrhagic infarct may be dense on the precontrast CT, but not as dense as a cerebral hemorrhage; the increased density is absent at the symptom onset but develops slowly over several days as the infarct turns from an ischemic to a hemorrhagic lesion. Hemorrhagic infarcts can sometimes be distinguished from cerebral hemorrhages by combining the clinical and radiographic findings; in patients with either bland or hemorrhagic infarction, neurologic signs are usually more pronounced than suspected from the images, whereas with cerebral hemorrhage it is the opposite, that is, clinical findings often match or are less significant than those expected from the image. Exceptions occur, however, when edema or shifts of tissue cause symptoms at a distance from the infarcted site.

TREATMENT

The initial medical treatment of hemorrhage into a brain metastasis is directed at the mass. If the patient has evidence of increased intracranial pressure or cerebral herniation, the treatment is identical to that of patients suffering from brain metastases without hemorrhage (see Chapter 3). Lumbar puncture should be avoided because of the increased risk of herniation. If a coagulopathy is present, it should be treated (see following text). When the cerebral hemorrhage is unrelated to metastatic brain disease, corticosteroids are ineffective,[70] but in patients with metastatic hemorrhages, corticosteroids are useful in treating the underlying edema caused by the metastasis. Hyperosmolar agents, particularly mannitol, can increase cerebral blood volume,[71] thus possibly exacerbating the hemorrhage; however, it does not seem to increase regional cerebral blood flow as evaluated by single photon emission computed tomography (SPECT) scanning.[72] In patients with a marked increase in intracranial pressure resulting in cerebral herniation, however, the effectiveness of mannitol or hypertonic saline in decreasing intracranial pressure probably outweighs the theoretical risk of increasing the hemorrhage.[72,73] Hyperosmolar agents should be administered only to patients with progressive signs of increased intracranial pressure and cerebral herniation. Low-dose mannitol (100 mL of 20% mannitol every 4 hours $\times$ 5 days) is ineffective.[74]

The treatment of a hemorrhagic cerebral metastasis is directed at the metastasis itself. The criteria used to choose either surgery or RT as the primary treatment modality for metastatic brain tumors apply here (see Chapter 5). RT should be delayed until the patient stabilizes as increased edema associated with RT may worsen the clinical situation. In a stable patient, RT is usually begun after 2 to 3 days of corticosteroid treatment. Surgery is indicated when the hemorrhage is large and potentially life threatening or fails to respond to conservative therapy. The lesions are often relatively superficial, and many symptoms are caused by shifts of brain structure rather than tissue destruction; thus, the hemorrhage and its surrounding tumor can often be removed successfully, and the patient can make a good functional recovery.

A more difficult therapeutic decision exists in patients with cancer who suffer a cerebral hemorrhage but in whom the imaging or the clinical findings do not establish the presence of a metastasis. In this instance, the patient is usually followed carefully, and the MRI repeated every few weeks. If the hemorrhage resolves without further evidence of cerebral metastasis, the patient is not treated. If clear evidence of a cerebral metastasis appears as the hemorrhage resolves, treatment of the metastasis should be initiated.

The prognosis of a cerebral hemorrhage from tumor is similar to that of the metastatic tumor itself, that is, radiosensitive and chemosensitive tumors such as germ-cell tumors have a good prognosis, whereas resistant tumors such as melanoma have a poor prognosis unless they can be surgically removed or successfully treated by radiosurgery (see Chapter 5). If the tumor can be successfully treated and the hemorrhage has not caused severe neurologic disability, good recovery and long-term survival are possible. Because brain metastases usually bleed from veins or small arterioles and the site is often fairly superficial, the prognosis for immediate survival and functional recovery is better for metastasis-associated hemorrhage than for hypertensive basal ganglia hemorrhages.

Subdural Hemorrhage

CLINICAL FINDINGS

Subdural hematomas in patients with cancer can result either from dural metastases or from bleeding due to trauma, cancer-induced coagulopathy, or anticoagulant therapy. Occasionally, lumbar puncture, by lowering intracranial pressure, can lead to subdural effusions. Openshaw et al. identified 17 patients, over a 10 year period, who developed subdural hygromas or hematomas after stem cell transplant; most had had preceding lumbar punctures.[75] In one series, 26 of 38 patients with subdural hematomas complicating solid tumors had a history of either head trauma or anticoagulant therapy, suggesting that metastasis was the cause of the subdural hematoma in only a minority. Conversely, only 4 of 32 patients with hematologic neoplasms gave a history of either trauma or anticoagulant therapy, suggesting that either metastasis or cancer-induced coagulopathy played a major role.[54] Some tumors that metastasize to the subdural space, especially those from prostate cancer, reach it via Batson plexus (see Chapter 2). Other mechanisms include extension of skull metastases or hematogenous spread via the arterial circulation. The symptoms are caused either by the tumor mass or by subdural hematomas or hygromas, which arise from leaking tumor neovessels or from rupture of friable vessels by minor trauma.[24,54] An alternative hypothesis is that mechanical obstruction could lead to dilatation and eventual rupture of the capillaries in the intradural layer.[76] Coagulation abnormalities and increased capillary pressure caused by tumor-induced venous occlusion may exacerbate the bleeding.

Neurologic symptoms from subdural lesions usually develop more slowly than do those caused by hemorrhage into a brain metastasis. The first symptom is headache, generalized or localized over the hematoma, sometimes associated with tenderness to percussion over that site. The patient then develops one of several progressive clinical pictures that mark the growth of the subdural hematoma[77]:

1. Progressive neurologic dysfunction characterized by hemiparesis and hemisensory loss may develop over days to weeks, mimicking a brain metastasis.

2. Delirium or stupor with absent or only minor focal signs may suggest a metabolic disorder (see Chapter 11), often with unilateral asterixis (contralateral to the lesion) as the only focal sign.

3. Acute bleeding into the subdural space may cause false-localizing signs (see Chapter 5). Headache and acute encephalopathy may be accompanied by hemiparesis or hemiplegia ipsilateral to the subdural hematoma. The ipsilateral hemiplegia results from shift of the brainstem away from the side of the hematoma, compressing the cerebral peduncle (which carries motor fibers from brain to spinal cord) against the contralateral tentorium cerebelli (Kernohan notch) (see Figs. 1–6 and 3–6C). Compression of the ipsilateral third nerve causes first a dilated pupil and, subsequently, a complete third-nerve palsy on the side ipsilateral to the subdural hematoma, establishing the lesion side, even in the presence of an ipsilateral hemiparesis.

4. Rarely, subdural hematomas may cause unusual and confusing clinical symptoms, mimicking transient ischemic attacks,[78] Parkinson disease,[79] chorea,[80] dystonia,[81] or internuclear ophthalmoplegia.[82]

Focal signs, when present, can reflect direct compression of the brain by the overlying hematoma, compression of vascular structures, or brain edema. Edema is greater in the hemisphere underlying a subdural hematoma than in the opposite, less-compressed hemisphere.[83] The pathogenesis of the metabolic encephalopathy–like picture is not clear.

DIAGNOSIS

The diagnosis of subdural hematoma or hygroma is made by CT or MRI. Fresh blood is dense on CT, but chronic hematomas or hygromas becomes isodense with brain. However, effacement of the normal gyral pattern is usually sufficient to suggest the diagnosis. MRI clearly delineates subdural hematomas and hygromas of any age, particularly on T2-weighted images. Contrast injection may reveal tumor in the subdural space.

TREATMENT

A subdural hematoma that is due to tumor usually requires surgical evacuation. Often it can be evacuated through a burr hole, making a craniotomy unnecessary. Occasionally, placing a temporary drain or an Ommaya device (see Chapter 7) into the subdural space allows drainage of re-accumulated fluid; permanent shunts are rarely required. The drained fluid should be

examined cytologically or the subdural membrane should be biopsied to determine whether tumor is causing the subdural blood; cytologic examination may be negative even when tumor is present.[84] If the subdural hematoma has resulted from subdural tumor, surgical evacuation of the hematoma should be followed by radiation to treat the neoplasm and thus prevent re-accumulation of the subdural fluid.

Coagulopathic Hemorrhage

INTRACRANIAL HEMORRHAGE

Cerebral hemorrhage because of coagulopathy occurs more commonly in patients with hematologic malignancies, especially myelocytic leukemias, than in patients with solid tumors (Table 9–6). Patients with myeloproliferative syndromes such as polycythemia and plasma-cell dyscrasias also bleed occasionally,[85] although symptoms from hyperviscosity (see

Table 9–6. Factors Leading to Intracranial Hemorrhage in Patients with Leukemia

	Hemorrhage without CNS Leukemic Infiltration	Hemorrhage with CNS Leukemic Infiltration	
		Parenchymal Infiltrates with Leukostasis	Arachnoidal Infiltrates without Leukostasis
No. of patients	50	13	6
No. of symptomatic patients	38 (76%)	8 (61.5%)	3 (50%)
Histologic type*			
ALL	5 (2)	3 (2)	1 (1)
AML	19 (16)	2 (1)	3 (1)
APML	9 (7)	–	1 (1)
CML	5 (5)	3 (2)	1
Other	12 (8)	5 (3)	–
Hemorrhage at time of diagnosis of leukemia	7	5	–
Fever	68.4%	37.5%	100%
WBC count (per mm³)	8000 (100–104,000)	260,000 (70,000–730,000)	36,000 (1000–97,000)
Platelet count (per mm³)	13,500 (2000–52,000)	36,000 (10,000–50,000)	32,000 (3000–65,000)
Multiple hematomas	12%	62.5%	16.6%

From Ref. 24 with permission.
* Total number and (number of symptomatic patients).
CNS, central nervous system; ALL, acute lymphoblastic leukemia; AML, acute myelogenous leukemia; APML, acute promyelocytic leukemia; CML, chronic myelogenous leukemia; WBC, white blood cell.

the following section on hyperviscosity) are more common than from hemorrhage in these patients. In patients with APML, DIC with intracranial hemorrhage is a common cause of death.[42] DIC is also a common complication of sepsis.[86] The hemorrhagic form of DIC, also called consumption coagulopathy, is caused by low levels of platelets and procoagulant factors associated with massive coagulation activation. Cerebral hemorrhage is a common complication. Although frequently treated with platelet transfusion and fresh frozen plasma, the effectiveness of this therapy is questionable. In a study of 555 patients with APML, 29 (5%) died of bleeding within the presentation period; one-half of the deaths occurred within the first week of treatment despite intensive blood product support.[42]

DIC is an acquired syndrome characterized by the intravascular activation of coagulation factors. The disorder can cause damage to the microvasculature that is sufficiently severe to produce widespread organ failure.[87,88] DIC is a syndrome with many causes.[87] In patients with cancer, the disorder can be caused by the malignancy itself, or by sepsis associated with the cancer. The widespread intravascular coagulation can damage any organ, including the brain. Encephalopathy, with or without fleeting focal neurologic signs, is a hallmark of the disorder.[89] The diagnosis is based on clinical suspicion, that is, does the patient have an underlying disorder known to be associated with DIC and are coagulation tests abnormal? The coagulation tests are highly sensitive, but not very specific. However, algorithms scoring the importance of individual coagulation tests have been developed and can assist in the diagnosis (see paragraphs on DIC that follow). The tests generally available that yield positive findings include low and dropping platelet counts, elevated fibrin degradation products or D-dimer, prolonged prothrombin time (PT), and low fibrinogen level.[87] In patients with widespread organ failure due to DIC, anticoagulation with heparin may be helpful. Recently, the suggestion has been made that restoration of coagulation inhibitors such as anti-thrombin III and activated protein C might be beneficial.[87]

Intracranial hemorrhage is less common in other myelocytic leukemias than in APML and clearly defined DIC is encountered rarely.[90] Instead, the hematologic defect is usually thrombocytopenia, with platelet counts of fewer than 10,000/mm³. However, thrombocytopenia

alone is not the entire explanation for the bleeding. Intracranial hemorrhage much more commonly accompanies acute myelogenous leukemia with thrombocytopenia than it does lymphocytic leukemia or solid tumors that cause equal thrombocytopenia. Nor does leukostasis explain the difference in the frequency of intracranial hemorrhage in these groups.

The most common pathologic finding in leukemia is a large hemorrhage, usually found in the centrum semiovale. A single ruptured vessel is usually not identified, and the rather slow onset of clinical symptoms suggests that veins or small arterioles have bled. At times, leukemic infiltrates surround blood vessels within or adjacent to the hemorrhage, but most often the pathologist finds only a hematoma without an excessive number of white blood cells (WBCs) and often with little or no inflammatory reaction. A careful search at a distance from the massive hemorrhage site often discloses microscopic perivascular hemorrhages.

Clinical Findings

The signs and symptoms depend partially upon the compartment into which the bleeding has occurred (i.e., subdural space or brain parenchyma) and partially on the rapidity of bleeding. Headache and neurologic signs usually develop more gradually than with hemorrhage into a metastatic tumor, but otherwise the signs are essentially identical. The usual setting is an APML patient who is undergoing treatment and has profound thrombocytopenia but no neurologic symptoms until headache and neurologic signs prompt hospitalization. A CT establishes the diagnosis.

Treatment

Medical treatment aims to reduce brain edema and incipient herniation (see Chapter 3) and is also directed at the coagulopathy. In patients with documented DIC (a minority), heparin has been used to reverse the consumption of coagulation factors. We have no experience with heparin to treat intracerebral hemorrhage, but reports[91] suggest that heparin can stop GI and other bleeding abnormalities caused by DIC. Also, intracerebral hemorrhages appear to be fewer in patients with DIC who are treated with IV heparin.[92] Tranexamic acid may also control hemorrhage in patients with APML.[93]

All-trans-retinoic acid usually rapidly corrects APML's coagulopathy. Immediate treatment of

APML with retinoic acid prevents the coagulopathy and associated brain hemorrhage. Although it occasionally occurs (see Fig. 3–6), intracerebral hemorrhage is now an uncommon cause of death or disability in newly diagnosed patients. Platelet transfusion for thrombocytopenia and replacement of coagulation factors with fresh frozen plasma and cryoprecipitate to maintain fibrinogen levels of greater than 100 mg/dL are helpful.

If the patient deteriorates neurologically while receiving optimal medical treatment, the physician should consider surgical evacuation of the lesion. Despite the threat of continued bleeding because of the coagulopathy, subdural hematomas can be successfully evacuated in some thrombocytopenic patients. The platelet count must be restored to greater than 100,000/mm³ before and after the operation by the transfusion of platelets and other coagulation factors. Continued bleeding may be a problem in these patients despite platelet transfusions, however; the physician should approach surgical extirpation only as a last resort when medical treatment has failed.

SPINAL HEMORRHAGE

Spontaneous spinal hemorrhages are rare in patients with cancer, but spinal subdural hematomas can follow lumbar puncture in thrombocytopenic patients[57,94] (Fig. 9–3). The hematoma probably results from damage by the needle to small radicular vessels as they pass through the subdural space to supply the cauda equina.[94,95] Unlike intracerebral hemorrhage, spinal subdural hemorrhage is as common in patients with lymphoblastic leukemia as it is in patients with myelogenous leukemia, suggesting that thrombocytopenia and the lumbar puncture are sufficient cause for the hemorrhage and that additional coagulation deficits are not required.

Characteristically, patients with spinal hemorrhage either have a stable platelet count of fewer than 10,000/mm³ or a rapidly dropping count rarely as high as 50,000/mm³. Rarely, patients with adequate platelet counts (>100,000/mm³) develop spinal subdural hematomas after lumbar puncture when platelet function is poor, particularly in the setting of renal dysfunction. Several hours following a lumbar puncture, the patient develops back pain radiating down the legs, followed by leg weakness and sensory changes indicating either

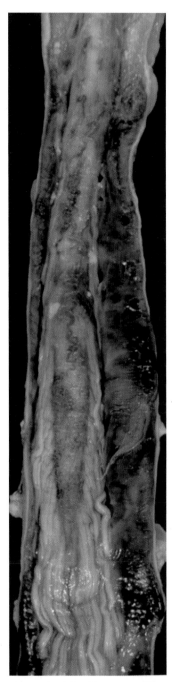

Figure 9–3. Spinal subdural hematoma following lumbar puncture. A patient with thrombocytopenia from acute lymphoblastic leukemia underwent lumbar puncture to instill methotrexate. The patient complained of pain and weakness in the legs within hours following the lumbar puncture. Over the next 48 hours, the patient became paraplegic, with a sensory level at T4. She remained paraplegic until her death from leukemia a year later. The organized hematoma compressed both the cauda equina roots and the spinal cord.

cauda equina or spinal cord dysfunction. If paraplegia occurs, it usually appears within 24 hours of the lumbar puncture, but sometimes is delayed for several days. The bleeding may dissect upward in the subdural space (Fig. 9–3) to affect the spinal cord as high as the upper thoracic level. Some patients with paraplegia recover, particularly when the cauda equina rather than the spinal cord has been involved. Subdural bleeding should be suspected whenever severe back pain follows a lumbar puncture. Unfortunately, our experience is that the patient's complaint of pain is often ignored until weakness develops. The diagnosis is established by CT or MRI.

The best treatment is prevention. Patients with rapidly dropping platelet counts, or those with fewer than 20,000/mm[3], should be transfused before and during lumbar puncture. However, in one series of children, lumbar puncture performed 199 times in patients with platelet counts under 20,000/mm[3] yielded no adverse effects.[96] The lower the platelet count the more likely to get red cells (traumatic tap) even without serious clinical effects.[97] The puncture should be performed with a No. 20 or smaller needle by the most skilled physician available (see Chapter 14). Since we have adopted this policy, the frequency of clinically significant subdural hematomas has dropped. We have found small subdural hematomas at autopsy in a number of thrombocytopenic patients who have undergone lumbar puncture shortly before death. Some of these small hematomas probably correlate with the clinical symptom of mild-to-moderate back pain unassociated with neurologic dysfunction. Patients with other hemostatic abnormalities seldom have similar problems, but special care should be taken when performing the lumbar puncture.

When patients develop post-lumbar puncture weakness, we give them platelet transfusions, but we have not tried to evacuate the hematoma because of their severe thrombocytopenia. If medical treatment fails, a case might be made for evacuation of the hematoma[98]; needle aspiration might also be considered.[99] The decision must be made rapidly, however. Once patients become paraplegic, they are unlikely to recover function even if the hematoma is successfully treated.

We have encountered only a few spontaneous spinal hemorrhages in thrombocytopenic patients.[100,101] The patient complains of sudden onset of severe back pain; neurologic signs develop over minutes to hours, sometimes leading to paraplegia. In contradistinction to cord compression from metastatic tumor (see Chapter 6), spinal radiographs are usually unrevealing. The MRI typically indicates an epidural mass usually more extensive than that encountered with metastatic spinal cord compression. The signal intensity on CT or MRI is that of blood rather than tumor. Bleeding into a spinal epidural tumor may be impossible to differentiate from a spontaneous hematoma.

Intraparenchymal spinal hemorrhage may complicate intramedullary metastases[102,103] or be a late result of RT (see Chapter 13).

Leukostasis

Leukostasis refers to leukemic cells accumulating in small vessels of the brain[104] and in the perivascular spaces surrounding them.[41] Leukostasis occurs in patients with either lymphocytic or nonlymphocytic leukemia[105] who have a high circulating WBC count.[106] Hyperleukocytic leukemias are defined as those with a white cell count above 100,000/mm[3]. This phenomenon has a 20% to 40% risk of early mortality.[41] Mortality results from either pulmonary failure or intracerebral hemorhage. Leukocytic counts greater than 300,000/mm[3] create a 60% risk of intracerebral hemorrhage. Leukostasis is more likely to occur at lower counts with granulocytic blast cells than with lymphocytic cells; the granulocytic blast increases blood viscosity more than the lymphoblast, is more rigid, and occludes small vessels more readily. A review of children with acute lymphoblastic leukemia (ALL) and leukocyte counts greater than 200,000/mm[3] found that only 4 patients of 178 (2%), all with initial leukocyte counts greater than 400,000/mm[3], suffered a CNS hemorrhage.[107] Leukostasis also leads to vessel wall damage. Although the mechanism is unclear, it is believed that endothelial damage results from soluble cytokines released during the interaction between leukemic cells and vascular endothelial cells.[41] Hypotheses regarding possible mechanisms include ischemia caused by high oxygen consumption by leukemic cells in the cerebral microcirculation, mechanical disruption of the vessel wall by invasive tumor, or damage by thrombosis. Whatever the mechanism, when the vessel wall is sufficiently damaged, either microscopic or gross hemorrhage occurs. In some patients, the hemorrhages are large enough to cause death.[41,108]

Clinically, leukostatic hemorrhages are indistinguishable from hemorrhages caused by coagulation defects. Patients often complain of headache, dizziness, blurred vision, and ataxia. They are usually found to be confused and may become stuporous or comatose; there may be focal or generalized seizures. Papilledema, retinal hemorrhages, and retinal vein distention are important clinical findings.[41] However, the diagnosis can be established only by the presence of leukostatic changes at autopsy. Leukostasis should be suspected when a patient with intracranial hemorrhage associated with nonlymphocytic leukemia, but without severe thrombocytopenia or other coagulation abnormalities, has or has had a WBC count greater than 200,000/mm[3]. The presence of concomitant respiratory symptoms from leukostasis of pulmonary vessels supports the clinical diagnosis. Unfortunately, some patients with leukostatic lesions also have underlying abnormalities of coagulation, including thrombocytopenia, and they may not have high circulating WBC counts, making diagnosis difficult. In such patients, hyperviscosity, suggested clinically by retinal hemorrhages and encephalopathy, may be the cause (see section on Hyperviscosity).[109]

Patients with high WBC counts may develop multiple small hemorrhages that are sometimes too small to be identified on CT, but may be seen on gradient echo MRI. Such lesions cause delirium and focal or generalized convulsions resembling the syndrome of intravascular coagulation.

Although not all patients with hyperleukocytic acute myeloid leukemia will develop leukostasis, it is difficult to predict which ones will, and therefore all should be treated. Aggressive treatment includes cytoreduction with leukapheresis and hydroxyurea and immediate application of induction chemotherapy. The treatment of hemorrhages due to leukostasis is similar to that of other intracerebral hemorrhages except that whole-brain radiation (12–24 Gy) is often given if the patient survives the acute episode. RT is given to eliminate the abnormal white cells from the brain in order to prevent future hemorrhages. Some recommend whole-brain radiation (e.g., 6 Gy in one dose) be given prophylactically, particularly in patients with minor neurological symptoms.

Hyperviscosity

Blood hyperviscosity occurs in leukemia,[110] paraproteinemias[111] including multiple myeloma,[112–114] Waldenstrom macroglobulinemia, essential thrombocytosis, and polycythemia vera. The hyperviscosity syndrome can occur in leukemia when the WBC count exceeds 200,000/mm[3]. Measurable hyperviscosity is found in approximately 5% of patients with immunoglobulin G (IgG) myeloma and in 22% of patients with immunoglobulin M (IgM) values greater than 5 g/dL. Clinical symptoms usually occur only when the serum relative viscosity exceeds 4.0 centipoise. Patients may complain of headache, lethargy, dizziness, vertigo, and visual disturbances. As the viscosity rises, they may develop hemiparesis, seizures, and acute confusional states that progress to coma, so-called *coma paraproteinemia*.

Although the neurologic examination is usually not specific, clues to hyperviscosity or venous thrombosis include retinopathy (Fig. 9–4), characterized by tortuous and congested veins that probably represent compensatory dilatation in response to high viscosity, sometimes papilledema, and multiple retinal hemorrhages. The patient's history may also relate easy bruising, mucosal bleeding, or bleeding from the nose or uterus. The disorder is usually treatable with therapeutic apheresis,[110] phlebotomy for polycythemia vera, or initiation of chemotherapy for the underlying disorder. Blood or red cell transfusion is contraindicated until the hyperviscosity is controlled. The red cells increase viscosity and can lead to the production of clinical symptoms.

Hypertension

Hypertensive intracranial hemorrhages sometimes affect patients with cancer. The patient may suffer hypertension either as a coincidental disease, a result of renal dysfunction associated with the cancer, or as a consequence of antitumor agents that target the vascular endothelial growth factors (VEGF). A common setting is a patient with cervical carcinoma who develops a pelvic recurrence leading to ureteral obstruction, hydronephrosis, renal damage, and hypertension. The neurologic findings in such patients are those of hypertensive hemorrhages and hypertensive encephalopathy, the same as with severe essential hypertension.

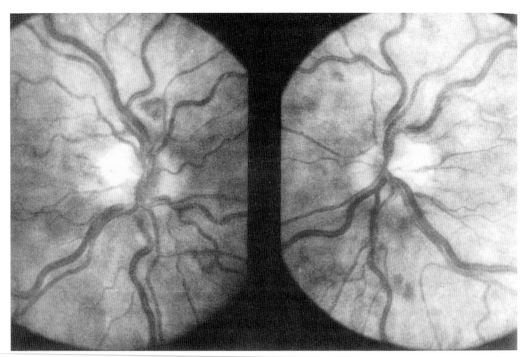

Figure 9–4. Abnormal retinal blood vessels in a patient with a hyperviscosity syndrome. This patient with chronic myelogenous leukemia developed hemorrhages in both eye grounds. Grossly dilated retinal veins with arterioles of normal size and blurred disk margins are noted.

The symptoms begin suddenly with headache and focal neurologic signs. A hypertensive hemorrhage is usually more acute in onset, rapid in progression, and serious in outcome than other intracranial hemorrhages. The hemorrhages typically affect the basal ganglia, obviating surgery as a therapeutic option. Patients are treated conservatively, and the prognosis is poor.

Hypertension can also cause posterior reversible leukoencephalopathy (PRES), described in Chapter 12, and has also been reported to complicate induction chemotherapy of ALL.[115]

CENTRAL NERVOUS SYSTEM INFARCTION

CNS infarcts, whether bland or hemorrhagic, may result from arterial or venous occlusion. Arteries may be occluded either by thrombus in situ or by embolization from a distant site. Venous occlusions are almost always due to in situ thrombus formation. Table 9–1 lists the causes of cerebral infarction, and Table 9–4 classifies the pathophysiologic mechanisms of CNS infarction in patients with cancer; the following paragraphs detail these causes.

Atherosclerosis

Cerebral atherosclerosis is usually less severe in patients dying from cancer than in those dying from other causes.[25–28] The reason is not entirely clear. It may be an artifact of the phenomenon that the two most common causes of death are cancer and heart disease. If all patients dying in a general hospital are studied, more atherosclerosis will likely be found in the noncancer population because patients dying from heart disease usually have relatively severe cerebral atherosclerosis. A second possibility is that the genetic or biochemical defect underlying the cancer also reduces atherosclerosis. A third possibility is that, when a patient develops cancer, tumor-generated substances or malnutrition associated with cancer may reverse previously established atherosclerosis so that the lesion is less severe by the time the patient dies from the cancer.[29,116] Whatever the cause, at all ages, patients with lung and breast carcinomas, malignant melanoma, and hematologic malignancies have significantly less atherosclerosis in the circle of Willis than do patients dying from noncancerous causes. Between ages 50 and 70, this reduced incidence of atherosclerosis in the circle of Willis is also true for patients suffering from colon and head and neck carcinomas.[25] After the age of 70, the incidence of cerebral atherosclerosis with cancers of the head and neck and colon exceeds that of the control population, whereas the incidence remains low in patients with other cancers.[25]

Despite the autopsy findings suggesting an inverse relationship between atherosclerosis and cancer, some clinical findings suggest the contrary. An epidemiological study of over 100,000 stroke patients found 5151 cases of cancer during a mean follow-up time of 2.4 years.[21] There was a statistically significant, but only slightly increased relative risk (RR = 1.12) of cancer occurring in the stroke patients. Most of the increased risk occurred in the first year after the stroke. Some of this may have been due to more careful medical surveillance in these already ill patients. Interestingly, 10% of all patients discharged with the diagnosis of stroke had a previous, or a concurrent, cancer. These patients were excluded from analysis, but if included, would have increased the relationship between cancer and stroke.

However, another study of over 69,000 patients with atherosclerosis (defined as *cerebral or peripheral vascular disease*) found no increased incidence of non–smoking-related cancers in these patients.[117] This study did note that women with arteriosclerotic disease had a significantly elevated risk of lung cancer compared to women without atherosclerotic disease, unrelated to tobacco use (RR = 3.26).[20] The authors suggested that oxidative stress due to episodes of ischemia/reperfusion increased the risk of lung cancer, perhaps because of gender-specific susceptibility to oxidative DNA damage.

As discussed in detail in Chapter 13, RT accelerates the development of atherosclerosis. This is particularly important in patients with Hodgkin disease who are young when they get the disease and have many years of life after the illness is cured.[118] Patients with head and neck cancer were also at increased risk.[119]

Some scientific evidence supports a relationship between atherosclerosis and cancer. Both diseases are characterized by uncontrolled regulation of cellular growth and differentiation and share many common genomic targets during the course of growth dysregulation.[120] Thus, some suggest that atherosclerotic plaques can be viewed as neoplasms of blood vessels.

Autopsy findings suggest that, when compared with the noncancer population, atherosclerotic occlusion is a less common cause of symptomatic stroke in patients with cancer. As Table 9–1 shows, of 256 patients with cerebral infarction, only 73 were atherosclerotic, and only 17 of 117 with symptomatic infarcts had atherosclerotic occlusions. When atherosclerotic occlusions are responsible for cerebral ischemia or infarction, accelerated atherosclerosis caused by RT delivered to a blood vessel is sometimes the culprit (Fig. 9–5) (see Chapter 13).[121] Accelerated atherosclerosis and infarction have been reported in patients with head and neck tumors, nasopharyngeal carcinomas, and lymphomas.[121] It sometimes occurs in children whose brains have been irradiated and results in moyamoya disease (the term is from the Japanese for "puff of smoke," describing the appearance of the brain's collateral circulation after occlusion of major blood vessels).[122] Cerebral infarcts that are due to large vessel atherosclerosis do occur in patients with cancer and in one retrospective series accounted for 33% of ischemic events in such patients. Hypercoagulable states accounted for 30% and cardioembolism for 21%. In 15%, the cause was uncertain.[123]

A study of 96 patients with radiographically confirmed stroke encountered at Memorial Sloan-Kettering Cancer Center (MSKCC) between 1997 and 2000 found that the incidence of stroke in patients with cancer was substantially lower than that in the general population, at least as indicated by hospital admission rate. Moreover, the pathogenesis differed in that the strokes were embolic in 54% and non-embolic (probably atherosclerotic) in only 46%.[22] If the pathogenesis of stroke in this population were identical to that in the general population, one would expect an age-adjusted distribution of primary neoplasms to reflect the prevalence of the most common malignancies (i.e., lung, breast, prostate). However, this and other large series show a much wider variety of tumor types including, interestingly enough, brain tumors (Table 9–7).

Disseminated Intravascular Coagulation

DIC, a common hematologic abnormality in patients with cancer,[35,124] is more common in the leukemias, particularly APML.[125] However, the disorder may complicate any cancer at any

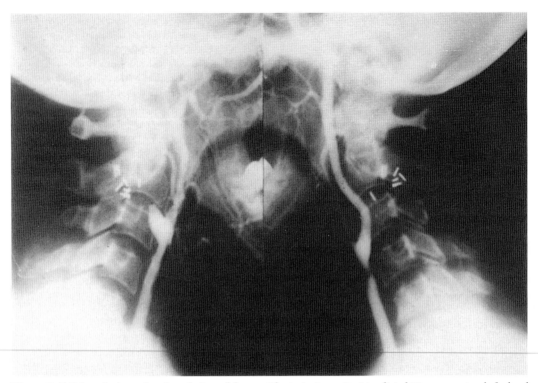

Figure 9–5. Bilateral atherosclerotic occlusions of the carotid arteries in a patient irradiated 15 years previously for head and neck cancer. The patient suffered transient ischemic attacks. The right carotid artery is occluded and the left shows major stenosis. Note that the occlusion on the right is well below the bifurcation of the common carotid artery into internal and external carotids; most arteriosclerotic occlusions occur at the bifurcation.

Table 9–7. **Primary of Cancer in Patients with Comorbid Stroke**

Overall Prevalence, ACS* (2005); n = 1,372,910 Tumor (%)	Lindvig et al.[21] (1990); n = 5151 Tumor (%)	Chaturvedi et al.[123] (1994); n = 33 Tumor (%)	Cestari et al.[22] (2004); n = 96 Tumor (%)
Prostate (17)	Gastrointestinal (27.7)	Gynecologic (20.6)	Lung (30)
Breast (15)	Lung (13.4)	Renal/genitourinary (10.5)	Brain (9)
Lung (13)	Skin (12.4)	Gastrointestinal (10.5)	Prostate (9)
Colon and rectum (10)	Genital (10.2)	Lymphoma (8)	Leukemia (6)
Urinary/bladder (5)	Urinary (8.6)	Prostate (7.5)	Lymphoma (6)
Melanoma (5)	Brain (6.7)	Lung (5.1)	Gynecologic (6)
NHL† (4)	Breast (6.6)	Breast (1.9)	Bladder (6)
Ovarian and uterine (4)	Hematologic (4.9)	Other (35.9)	Gastroesophageal (6)
Pancreas (2)	Other (4.5)		Breast (4)
Other (25)			Other (20)

° American Cancer Society 2005 data.
† Non–Hodgkin lymphoma.

time, sometimes appearing when the underlying disease is occult,[124] or later as a response to disseminated[126] or terminal cancer. Up to 75% of patients with disseminated cancer develop chronic DIC,[127] although many patients have no clinical symptoms.[88] DIC may be caused by nonmetastatic complications of cancer, including sepsis[56] (particularly with Gram-negative organisms), RT or chemotherapy, and liver disease.[88]

DIC results when a variety of insults activate the coagulation system, causing an increase in the amount of circulating thrombin and plasmin.[88] Marked increases in thrombin lead to acute DIC and smaller increases to chronic, fluctuating, and often asymptomatic DIC. The hypothrombinemic state can be activated by release of TF, endothelial damage, or a variety of cytokines. Once initiated, the pathophysiology of DIC is similar, whatever the cause, resulting in intravascular clotting. If the clotting factors are exhausted, a hypocoagulable state causing hemorrhage can develop.[88,128]

A strict definition of DIC requires a coagulation profile showing active coagulation and fibrinolysis. Characteristically, in acute DIC, thrombocytopenia, hypofibrinogenemia, an elevated PT, and fibrinogen degradation products (FDP) are noted. DIC can be chronic, existing even when the platelet count, fibrinogen level, and PT are normal. The diagnosis rests on the demonstration of FDP and, more specifically, D-dimer, which is formed only when intravascular coagulation occurs. Severely ill, hospitalized patients may suffer vitamin K deficiency that can be mistaken for or may complicate DIC.[129,130] DIC as defined hematologically may be entirely asymptomatic and may wax and wane, at times ceasing spontaneously.

The exact pathogenesis of DIC is unknown. The thrombotic form of DIC, also called purpura fulminans, is caused by widespread microvascular thrombosis.[86] It is probably more common than the hemorrhagic form except in patients with APML. DIC develops when there is widespread systemic activation of coagulation, resulting in fibrin deposition in small- and medium-sized vessels. In DIC, coagulation proceeds via the (extrinsic pathway) TF/factor VIIa route and there is simultaneously occurring depression of inhibitory mechanisms such as anti-thrombin III and protein C system.[131] Inflammatory cytokines can activate the coagulation system, and conversely, activated proteases and protease inhibitors may modulate inflammation by interacting with specific cell receptors. Strategies aimed at the inhibition of coagulation activation may be beneficial, including inhibition of TF-mediated activation of coagulation or restoration of physiologic anticoagulant pathways, for example by administration of activated protein C.[131] Some case reports suggest that treatment with drotrecogin alfa (activated) or substitution of protein C may successfully treat sepsis-related purpura fulminans.[86]

PATHOLOGY

Two pathologic abnormalities occur in patients with DIC. The first, as the name implies, consists of small, fibrin-platelet thrombi that occlude arterioles and/or venules. Larger fibrin accumulations may at times cause local thrombus formation in veins, including those as large as the superior sagittal sinus (see following text), or vegetations on heart valves (i.e., NBTE, a discussion of which follows). DIC does not cause arterial thrombi, but small-vessel arteriolar occlusion can lead to small infarcts. The infarcts can appear in any organ, but the most common site of both vascular occlusion and infarction is the brain. In pathology series[132,133] from general hospitals, the brain is involved in approximately 70% of patients, followed in order by the heart, kidneys, and spleen. The disorder may be unsuspected during life.

The second pathologic change is hemorrhage. When intravascular coagulation is severe, consumption of coagulation factors leads to spontaneous bleeding. The bleeding occurs commonly in the gastrointestinal (GI) tract, associated with local erosion,[124] but hematuria, subcutaneous bleeding, and bleeding into other organs can also occur. Gross bleeding into the brain is rare except with APML, in which massive cerebral hemorrhage can be a rare cause of death (see preceding text). Microscopic bleeding in the brain in the form of hemorrhagic infarction or petechial hemorrhages surrounding occluded vessels is relatively common.[89]

Pathologic changes in the brain are interesting and unusual.[89] Grossly, the brain may appear entirely normal or may show small, scattered petechial hemorrhages in either gray or white matter. Characteristically, the petechiae occur in white matter, often in the corpus callosum. In some patients with the same clinical syndrome, small hemorrhages may select gray matter rather than white matter. The causes of this selection are uncertain. Occasionally, bland infarcts can be seen grossly, but usually the infarcts and occluded blood vessels can be identified only microscopically.

Microscopically, the disease is characterized by fibrin-platelet occlusions of small arterial and venous vessels of less than 50 μ. The venous occlusions permit the pathologist to conclude that the disorder is thrombotic and not embolic. Occlusions, surrounded by bland or hemorrhagic infarction, are scattered throughout both hemispheres and are not concentrated in the distribution of a single large cerebral vessel. In some instances, the infarcts are all of the same age, but in others, the disease process appears to have waxed and waned, causing some old and some recent infarcts. The vessel walls appear unaffected.

Because the occluded vessels are small and the size of the resulting infarctions are usually only millimeters, no symptoms appear in most organs. Small infarcts at strategic sites in the brain, however, can be responsible for clinical symptoms. When many infarcts are scattered throughout the brain, the symptoms reflect diffuse rather than focal brain disease.

Vascular occlusions and infarcts begin when ample concentrations of clotting factors are still present in the blood. Since these substances have considerable reserves, neurologic symptoms often precede thrombocytopenia and hypofibrinogenemia. The first abnormalities are the presence of FDP and D-dimer, indicating that widespread clotting has occurred and that fibrinolysis is active. These tests are particularly helpful when platelet count, PT, and partial thromboplastin time are normal[88,134] (see laboratory evaluation under section on Laboratory Tests).

INCIDENCE

DIC affecting the brain is not rare. Pathologic changes may be found in 1% to 2% of autopsies in a general hospital,[132] and the clinical syndrome is encountered commonly, particularly in patients suffering from severe and widespread disease such as sepsis or cancer.[126,135] In a cancer hospital, DIC usually complicates the course of leukemias and lymphomas rather than solid tumors, and it is the single most common cause of thrombotic infarction in the brain. In our pathologic study,[24] nearly 25% (28 of 117) of brain infarcts in symptomatic patients occurred as a result of DIC. This figure excludes other complications of intravascular coagulation, such as NBTE or cerebral venous sinus occlusion.

CLINICAL FINDINGS

A patient suffering from advanced cancer, usually leukemia or lymphoma, often complicated by sepsis, suddenly becomes confused.[89] All of our patients (Table 9–8) were confused and disoriented, the confusional state divided almost equally between agitated and lethargic delirium. The confusion, sometimes associated with asterixis or multifocal myoclonus, is complicated by generalized seizures in about one-third of patients. The confusion may be episodic but, if untreated, usually progresses to stupor or coma, although sometimes it clears spontaneously.

In more than one-half of our patients, evidence of diffuse brain disease was accompanied by additional evidence of focal brain disease (Table 9–8), characteristically mild and fleeting. Usually, each focal neurologic abnormality lasted several hours or a day or two but then cleared, only to be replaced by evidence of neurologic disease elsewhere in the brain. The episodes are usually longer than transient ischemic attacks[136] but rarely as prolonged or severe as a large-vessel occlusion.

Table 9–8. Neurologic Signs and Symptoms of Intravascular Coagulation in Patients with Cancer

Signs and Symptoms	No. of Patients
Generalized Brain Disease	12
Confusion or disorientation	12
Agitated delirium	4
Lethargy or stupor	4
Coma	4
Asterixis or multifocal myoclonus	2
Generalized seizure	4
Focal Brain Disease	7
Hemiparesis	2
Cortical blindness	2
Aphasia	1
Focal seizures	4
Epilepsia partialis continua	2
Ataxia	2
Cranial nerve abnormalities	1

From Ref. 89 with permission.

Because sepsis is a common inciting factor in intravascular coagulation, patients may be febrile. Petechial hemorrhages in the skin are rare. Because small hemorrhages are occasionally encountered in the eye grounds, a careful funduscopic examination through dilated pupils may give a clue to the cause of a mysterious delirium. Otherwise, physical examination is of little help. Other systemic findings sometimes encountered include purpura, hemorrhagic bullae, acral cyanosis, and oozing from a venipuncture, surgical wound, or trauma.[88]

LABORATORY TESTS

Patients are often anemic and slightly thrombocytopenic, but these changes are usually attributable to previous chemotherapy or the effects of the tumor itself. Coagulation profiles may initially be entirely normal but become abnormal during the course of the neurologic illness. The most reliable laboratory tests are measurement of profragment 1+2, D-dimer, and anti-thrombin III.[88] Even these are not entirely specific or sensitive. Yu et al.[137] and Levi[87] have devised algorithms for the diagnosis that include an appropriate clinical setting, evidence of stromal hemorrhagic events, and abnormal laboratory findings as indicated in the preceding text (Table 9–9). Blood cultures may be helpful because many patients suffer from sepsis from either Gram-negative organisms or fungi. At autopsy, 2 of our 12 patients had fungal infections in the brain.

Other laboratory tests are rarely valuable. Hemorrhagic infarcts more than 5 mm in diameter can be identified on CT or MRI. Such large lesions, however, suggest NBTE rather than DIC. One report[138] suggests that in a similar (but not cancer-related) disorder, thrombotic thrombocytopenic purpura (TTP) (see section on Thrombotic Microangiopathy), an abnormal CT scan portends a poor outcome whereas a normal scan, even in the presence of severe neurologic signs, portends a good outcome. In TTP, biopsy of skin or gums has helped to establish the diagnosis,[139] but the utility of systemic biopsy has not been examined in DIC.

DIFFERENTIAL DIAGNOSIS

Metabolic brain disease is usually the major consideration. If the patient has fleeting,

Table 9–9. Diagnostic Algorithm for the Diagnosis of Overt Disseminated Intravascular Coagulation

Risk assessment: Does the patient have an underlying disorder known to be associated with overt DIC?

If yes, proceed. If no, do not use this algorithm.

Order global coagulation tests (platelet count, PT, fibrinogen, soluble fibrin monomers, or fibrin degradation products).

Score global coagulation test results:

Platelet count ($>100 \times 10^9$/L = 0, $<100 \times 10^9$/L = 1, $<50 \times 10^9$/L = 2)

Elevated fibrin-related marker (e.g., soluble fibrin monomers/fibrin degradation products) (no increase = 0, moderate increase = 2, strong increase = 3)

Prolonged INR (<3 sec = 0, but <6 sec = 1, >6 sec= 2)

Fibrinogen level (>10 g/l = 0, <10 g/l = 1)

Calculate score.

If ≥ 5: compatible with overt DIC; repeat scoring daily

If < 5: suggestive (not affirmative) for non-overt DIC; repeat next 1–2 d

From Ref. 87 with permission.

episodic focal signs, the clinician should suspect DIC rather than metabolic brain disease, particularly if no underlying metabolic defect can explain the patient's encephalopathy. Opportunistic CNS infections can resemble intravascular coagulation, but headache is more common with infection, and fleeting focal signs are more common with DIC. Sometimes the two complications of cancer accompany each other because infection may cause DIC. The clinical picture of DIC differs from that of metastasis and an MRI establishes the presence of a metastasis. In NBTE, which is part of the DIC picture, larger cerebral arteries are occluded, usually leading to a stroke rather than an encephalopathy.

TREATMENT

The best treatment is to identify and treat the underlying disorder. Because the disorder results from excessive coagulation, anticoagulation with heparin[11,140] or low molecular weight heparin[141] may be useful, particularly in chronic DIC. However, many are reluctant to use these agents because of the possibility of increased bleeding. Other recommended agents include anti-thrombin concentrates, protein C, TF pathway inhibitor, anti-fibrinolytic agents such as tranexamic acid or epsilon aminocaproic acid, thrombomodulin, activated factor VII, Gabexate mesylate and hirudin.[128]

Unfortunately, except in APML, the clinical diagnosis is often established too late or the pat-ient's underlying disease is so severe that treatment is not useful. We have treated only a few patients with heparin; the treatment appeared to ameliorate the neurologic symptoms.[134,142] Warfarin is probably ineffective.

Arterial Occlusion by Extrinsic Tumor

In rare patients, an extrinsic tumor occludes a large artery by either compressing or invading the vessel wall, causing cerebral infarction associated with thrombus formation.[143,144] A review of the literature describing 40 patients with arterial compression by tumor notes that only four suffered from metastatic tumor.[145] Most patients with arteries compressed by tumors have meningiomas growing in the cavernous sinus, compressing the carotid artery. Arteries embedded in tumors lose smooth muscle and eventually become amuscular tubes. In addition, such arteries are often deformed into aneurysmal or crumpled shapes. These phenomena may make arteries incapable of normal blood flow regulation even if they are not occluded by the tumor.[146] Tumors do not often invade the walls of large arteries, so that arterial rupture or occlusion from tumor growing into the vessel is uncommon. Arterial rupture sometimes occurs in the carotid blow-out syndrome,[147] but this is usually a result of surgery and radiation rather than tumor growing directly in the wall.

Veins are much more commonly occluded by extrinsic tumor compression; the superior vena caval syndrome is a frequent presenting complaint of mediastinal tumor, and the cerebral sagittal or lateral sinuses can be occluded by skull metastases (see Chapter 5).

CLINICAL FINDINGS

When tumor involves the carotid artery, particularly with head and neck tumors and metastatic breast cancer, the patient may complain of severe headache or pain in the eye, nose, scalp, teeth, or gums (Fig. 8–8B); the syndrome resembles the migraine variant called carotidynia.[148] Similar phenomena have been reported after internal carotid artery dissection and carotid endarterectomy. In some patients, syncope (usually but not always associated with headache) occurs as the carotid sinus is invaded (see Chapter 8). Characteristically, arterial compression is otherwise asymptomatic unless the patient suffers an acute vessel occlusion, leading to cerebral infarction. A cerebral infarct can be identified on MRI and, in some instances, a contrast-enhancing mass can be identified at the base of the brain surrounding the carotid or vertebral artery. A CT or magnetic resonance angiogram is more likely to yield useful information, demonstrating not only the occluded vessel but the site of extrinsic compression from the tumor. The treatment should be directed at the tumor.

Cerebral Emboli

Cerebral emboli are the most common cause of infarction in patients with cancer.[22] Emboli to cerebral vessels can cause bland or hemorrhagic infarction. The emboli are usually fibrin/platelet clots; less common causes include tumor emboli, emboli of infectious material, and more exotic sources[149] (Table 9–10).

Table 9–10. Causes of Embolic Cerebral Infarction in Patients with Cancer

Atheromatous plaques
Infected material
    Bacterial or fungal endocarditis
    Fungal lung lesions
Nonbacterial thrombotic endocarditis
Tumor
Mucin
Fat
Bone marrow
Calcified valves
Silicone particles
Lymphangiography
Talc
Air

NONBACTERIAL THROMBOTIC ENDOCARDITIS

NBTE is characterized by uninfected fibrin vegetations on one or more heart valves.[150] The underlying heart valves are usually free of other pathologic abnormalities. The disorder is common in patients with cancer, but also occurs in patients who die from other systemic disorders and, rarely, in those with no systemic illness (Tables 9–11 and 9–12). NBTE can happen at any time in the course of the cancer and may be the first evidence of malignancy. More commonly, NBTE occurs late, when patients have disseminated cancer. Its overall incidence at autopsy in general hospitals is approximately 1% (Table 9–11).

In most patients, emboli are detected in more than one organ, such as spleen, kidney, and heart; approximately 50% have emboli in the brain with or without infarction[151,152] (Fig. 9–6). Approximately 7% of patients with adenocarcinoma of the lung have identifiable vegetations on their valves at autopsy.[153] Other cancers, including carcinoma of the pancreas, have a lesser incidence of NBTE. In one series of 311 patients with pancreatic adenocarcinoma, only two patients had arterial embolization and only one had identifiable NBTE.[154] Contrary to former belief, mucin-secreting adenocarcinomas do not usually cause NBTE or cerebral embolization,[48,155] although mucin itself may embolize to cerebral vessels. Kearsley and Tattersall[156] reported eight patients with acute

Table 9–11. Prevalence of Nonbacterial Thrombotic Endocarditis and Comorbid Malignancy

Study	Unselected Autopsies, n	NBTE, n (%)	NBTE + Cancer, n (%)
Chomette et al. (1980)	6000	130 (2.2)	108 (83)
Pomerance (1975)	4720	119 (2.5)	n/a
Loire et al. (1981)	5100	40 (0.8)	24 (60)
Biller et al. (1982)	13,913	99 (0.7)	42 (42)
Kim et al. (1977)	4783	36 (0.8)	18 (50)
Total	34,516	424 (1.2)	192 (63)*

From Ref. 19 with permission.
* Denominator excludes Pomerance data.
NBTE, nonbacterial thrombotic endocarditis; n/a, not available.

Table 9–12. Prevalence of Nonbacterial Thrombotic Endocarditis and Strokes in Patients with Cancer

Study	Selected Autopsies; N (%)	NBTE + Cancer; N (%)	Pathological Strokes; N (%)	Clinical Strokes; N (%)
Kearsley and Tattersall (1982)	233	5 (2.1)	n/a	n/a
Chomette et al. (1980)	2287	108 (4.7)	n/a	n/a
Graus et al. (1985)	3426	115 (3.4)	42 (49)*	32 (76)*
Tsanaclis and Robert (1976)	2208	37 (1.7)	24 (65)	16 (67)
Kooiker et al. (1976)	n/a	18	9 (50)	8 (89)
Ojeda et al. (1985)	n/a	16	7 (44)	7 (100)
Gonzalez et al. (1991)	n/a	8	4 (50)	4 (100)

From Ref. 19 with permission.
* Eighty-six of 115 patients with NBTE had their brain examined at autopsy.
NBTE, nonbacterial thrombotic endocarditis; n/a, not available.

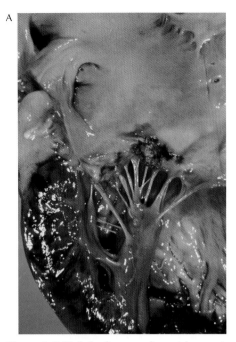

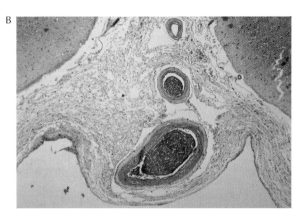

Figure 9–6. Embolus from a nonbacterial vegetation. This patient with known colon cancer suffered the sudden onset of a right hemiplegia and aphasia. She died within a matter of days and was found to have a massive infarct in her left hemisphere. **A**: At autopsy, vegetations were found on the mitral valve. **B**: Leptomeningeal and parenchymal vessels including the middle cerebral artery were filled with fibrin clots.

cerebral emboli among 3000 patients with cancer. Of the 233 patients examined at autopsy, five had NBTE. In another study,[157] valvular vegetations were found by echocardiogram in 5 of 30 patients with myeloproliferative disorders (i.e., polycythemia and thrombocythemia). In a study of 200 unselected ambulatory patients with tumors that included lymphoma (26%), GI tumors (20%), and lung carcinoma (16%), cardiac valvular vegetations were found by echocardiography in 38 patients (19%).[158] The common offenders were pancreatic carcinoma (3/6), lung cancer (9/32), and lymphoma (10/52). Twenty-two patients suffered thromboembolism, but only two of these were cerebral infarcts. D-dimer levels were elevated in 90% of patients with thromboembolism, but also in 32% without thromboembolism. In a group of 100 consecutive patients without overt heart disease who were referred to echocardiography for detection of an occult embolic source, only 2 were found to have cardiac valvular vegetations.

Most patients suffering from cancer-associated NBTE have identifiable coagulation abnormalities. Only 6 of 42 patients whom we studied with cerebral infarction from NBTE

had an entirely normal coagulation profile.[24,48] The coagulation abnormalities may be relatively mild and not reflect the extent of the clotting present in the patient. Many patients with NBTE also have other evidence of hypercoagulability, including DVT or fulminant DIC. Not all patients with NBTE, however, are symptomatic. Because the heart valve disease is usually asymptomatic, the abnormality is recognized clinically only when emboli occlude vessels supplying important organs. Arteries of any size may be occluded. Medium-sized vessels, such as the middle cerebral artery and its branches, are common occlusion sites. In one of our patients, the internal carotid artery was occluded by an embolus. The brain is a common site of cerebral embolization and infarction from NBTE, but of 86 such patients examined at autopsy, only 47 had identifiable cerebral infarcts, and in only 42 of these could the infarct be attributed with certainty to embolization from the heart valve.[48]

Clinical Findings

Although children are occasionally affected,[159] patients are usually adults who suffer from solid tumors, most commonly non–small cell lung cancer. The clinical syndrome usually occurs late in the course of cancer. Of our 42 patients, 27 suffered from progressive cancer and were being treated actively; 6 had widespread disease but were clinically stable, and 7 were in the early stages of cancer. In two patients, neurologic symptoms preceded the diagnosis of malignancy,[48] a phenomenon probably much more prominent in a general hospital than in a cancer hospital. By the time our patients died, the tumor was widely disseminated in most of them, but six had only localized tumor; three had carcinoma restricted to the lung, and three had carcinoma restricted to the cervix.

Neurologic findings can be divided into the following three groups (Table 9–13): (1) those suggesting focal brain disease, probably resulting from a single large embolus; (2) those suggesting diffuse brain disease, probably resulting from multiple small emboli scattered throughout the brain; and (3) those with a combination of focal and diffuse abnormalities.

Of our 32 patients with clinical symptoms, 18 suffered from focal brain disease, suggesting a stroke. The onset was usually sudden and caused fixed neurologic symptoms, usually hemiparesis or hemiplegia (Fig. 9–7). In four patients, the disease was clearly multifocal, suggesting occlusion of vessels in more than one major arterial distribution. In a minority of patients, the symptoms were transient, similar to transient ischemic attacks. Some patients with focal signs at onset subsequently developed evidence of more generalized brain dysfunction.

Table 9–13. **Initial Neurologic Signs in 32 Patients with Nonbacterial Thrombotic Endocarditis and Cerebral Infarction**

Focal Only	18°
Aphasia	13
With right hemiparesis	6
With right hemiparesis and sensory loss	2
With right hemiparesis and hemianopia	1
With hemianopia	1
Without other signs	3
Focal seizures	3
Hemiparesis, hemihypoesthesia	2
Herniation	1
Monocular blindness	1
Diplopia, vertigo	1
Cortical blindness	1
Diffuse Only	9
Confusion	5
Coma	3
Lethargy	1
Focal and Diffuse	5

° Some patients had more than one focal sign.

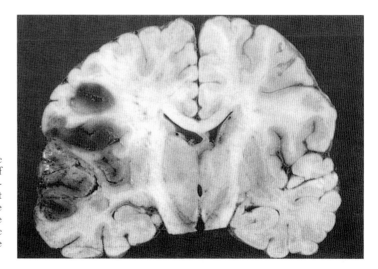

Figure 9–7. Multiple hemorrhagic infarcts in the right hemisphere of a patient with nonbacterial thrombotic endocarditis and malignant melanoma. Most of the infarcts are in the distribution of the middle cerebral artery, causing hemorrhagic infarction of cortex as well as white matter.

Nine patients showed evidence of diffuse brain dysfunction, including confusion, stupor, or coma without focal abnormalities; the disorder resembled either metabolic brain disease or DIC. Five patients suffered combined diffuse brain dysfunction plus the sudden onset of focal symptoms, suggesting major vascular occlusions, clinical findings that are virtually pathognomonic.

Findings outside the nervous system sometimes suggest NBTE, including occlusive disease of vessels elsewhere in the body, such as DVT with pulmonary embolization, arterial embolization to the extremities, acute myocardial infarction, central retinal artery occlusion, and bleeding abnormalities such as GI bleeding, hematuria, and bleeding from venipuncture sites. Unlike infective endocarditis, NBTE does not cause a cardiac murmur.

Laboratory Tests

Transthoracic echocardiography sometimes establishes the diagnosis, but is often negative[48,160–162]; transesophageal echocardiography is more rewarding (Fig. 9–8).[160,163–165] In one series of 51 consecutive patients with cancer and evidence of cerebral ischemia referred for transesophageal echocardiography, valvular vegetations were found in nine (18%).[166] CT or MRI identifies cerebral infarction and, if the infarcts are multiple and in the distribution of more than one cerebral vessel, the diagnosis is obvious. Diffusion-weighted images were abnormal in all but two of the 35 patients with thrombotic endocarditis who underwent the imaging tests. Of the nine patients with NBTE, all demonstrated lesions in multiple vascular territories. Of 27 patients with infective endocarditis, all had either a single lesion or lesions in a single vascular distribution.[167] Even in the absence of infarction, cerebral angiograms may reveal evidence of multiple embolic occlusions in more than one vascular distribution, showing that the emboli must have come from the heart. In the appropriate clinical setting, this result is virtually pathognomonic of NBTE. CT or MR angiography may identify embolic occlusion noninvasively.

Pathology

The pathologic examination usually reveals vegetations on the valves.[48] In some patients, however, all of the vegetative material has embolized and the valves appear normal. In one of our patients, who died with multiple embolic infarctions in the brain, a vegetation seen at the time the heart was opened at autopsy was accidentally dislodged by the prosector. The results of subsequent gross and microscopic examination of the valves were entirely normal. Some clinical evidence indicates that the disease may either wax and wane or be self-limited. Thus, the patient may have suffered a stroke in the distant past but have no evidence of disease on the heart valves at autopsy. Microscopically, the heart valves may be entirely normal or may show some evidence of microscopic vegetations in place of previously larger ones.

The brain usually shows evidence of multiple, grossly visible infarcts that are bland in

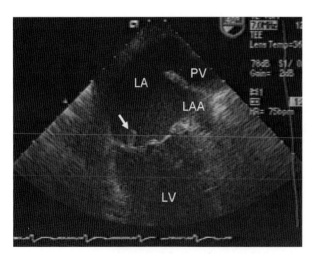

Figure 9–8. A transesophageal echocardiogram showing vegetation on a valve in a patient with nonbacterial thrombotic endocarditis. The patient suffered from leukemia and presented with multiple seizures. Bilateral posterior cerebral infarcts were found on the MR scan and large vegetations on the aortic valve (*arrows*).

approximately one-half of patients and hemorrhagic in the others.[48] In occasional patients, only a single infarct is discovered. Careful search of medium-sized blood vessels reveals platelet-fibrin thrombi in almost all patients. At times, vessels are occluded in areas without infarction; conversely, an infarct may be present without a vessel occlusion, the embolus having broken up and dispersed. The disease often occurs in conjunction with other vascular complications of cancer, including DIC. We found pathologic evidence of that disorder in 11 of the 42 brains examined in our NBTE series.

Treatment

Although no controlled studies validate the treatment of NBTE, most observers consider anticoagulation to be beneficial. In two patients, transient neurologic symptoms resolved with anticoagulation; in one, neurologic symptoms recurred on two occasions when heparin doses were subtherapeutic, and the symptoms promptly disappeared when appropriate anticoagulation was achieved.[48] Anecdotal evidence similarly suggests that anticoagulation with warfarin is ineffective in NBTE and other coagulopathies associated with cancer.[10] Previously, we have treated patients with IV heparin for 10 days to 2 weeks and then attempted subcutaneous heparin maintenance on an outpatient basis. Currently, we start with low molecular weight heparin and use unfractionated heparin only if the patient continues to embolize through therapeutic doses of low molecular weight heparin.

The possibility that anticoagulation increases hemorrhagic infarction seems to be more theoretical than real. We compared brain hemorrhages in patients treated with anticoagulants to those not so treated; no difference was found in either the number of patients with hemorrhagic infarctions or the degree of hemorrhage.[48] Interestingly, fewer cases of DIC were noted in the treated patients than in the nontreated group.

SEPTIC ARTERIAL OCCLUSIONS

Arterial or arteriolar occlusions caused by infectious organisms, sometimes bacteria but especially fungi, with resulting bland or hemorrhagic infarction of the brain, are encountered relatively frequently. Such infarcts are the third most common symptomatic cause of cerebral infarction in the cancer population, after DIC and NBTE (see Chapter 10). The embolus may arise from infective endocarditis, pulmonary infection, or an infected venous catheter.[160,168] In one study of 654 consecutive cancer patients suspected of suffering from endocarditis, a positive echocardiogram was identified in 45 (7%). Positive blood cultures were found in 26 (58%) of the 45; the remainder were culture negative and probably suffered from NBTE. The diagnoses were supported by the fact that peripheral septic emboli were more common in those who were culture positive than in those who were culture negative. Cerebrovascular events were identified in three (12%) of the culture-positive patients, but in 7 of the 19 (37%) culture-negative patients, suggesting that NBTE is more likely to cause cerebral infarcts, perhaps because the lesions are more easily dislodged from the valves.

Clinical Findings

The clinical signs and symptoms depend on the number and size of vessels occluded. In most instances, small vessels of arteriolar size are occluded, and the vascular occlusions are scattered throughout both hemispheres. When this occurs, symptoms are usually those of a diffuse encephalopathy, sometimes accompanied by fleeting and mild focal signs resembling the clinical picture of DIC. Frequently, the patients are febrile, but the fever is often attributed to systemic infection, for which most patients are already being treated.

Because, in some patients, larger vessels are occluded and the area of infarction is large, approximately 33% of patients develop acute onset of relatively severe and fixed focal neurologic deficits with or without delirium, similar to that described for NBTE. One of our patients suffered a septic occlusion of the anterior spinal artery, causing paraplegia; in another, occlusion of the basilar artery caused brainstem infarction.

The general physical evaluation may help with the diagnosis. Septic occlusion of small vessels may lead to hemorrhagic skin lesions, particularly in *Pseudomonas* infections. A biopsy of these lesions may yield the organism. Blood cultures are occasionally positive, particularly when the offending organism is a bacterium. The chest radiograph may show bilateral lesions, particularly with *Aspergillus* fungi; aspiration of the pulmonary lesion may yield the organism. Lumbar puncture seldom helps. Occasionally, pleocytosis is present, and sometimes a few red blood cells (RBCs). The CSF protein is usually increased, and the CSF glucose is usually normal. Organisms are rarely identified. CT or MRI usually reveals multiple infarcts, often hemorrhagic.

Pathology

The pathologic changes result in part from the size of the vessel occluded and in part from the duration of the occlusion. The organisms promote coagulation locally and thus generally become enmeshed in a fibrin clot. They then invade the vessel wall, usually causing an inflammatory vasculitis and sometimes causing a bland infarct to become hemorrhagic. After the vessel wall is invaded, the organism may grow freely within the substance of the brain, producing abscesses of varying size. Disseminated microabscesses are the most common pathologic feature of *Phycomycetes* and *Candida* infection, as well as of *Pseudomonas* infection. *Aspergillus* organisms often form one or only a few large infarcts; thus, the *Aspergillus* organism is more likely to cause severe focal neurologic disability (see Chapter 10).

The pathogenesis of the lesions is not fully established. The vascular occlusions could be caused by either thrombosis or embolization. If thrombosis is the mechanism, the circulating organisms probably lodge in small arterioles and trigger a focal coagulation process. The organisms that cause vascular occlusions in the brain are also those that have a predilection to be associated with DIC. In favor of in situ thrombosis is the absence of vegetation on valves or in heart chambers that might lead to embolization.

An embolic pathogenesis is suggested by (1) the frequent presence of pulmonary lesions that may serve as a source for embolization, (2) infectious vasculitis that is much more prominent on the arterial than the venous side of the circulation, and (3) the occasional occlusion of large blood vessels (e.g., carotid siphon) by what appear to be fungal emboli. Both thrombosis and embolism are probable mechanisms in some patients.

Treatment

The best treatment is prevention. If the offending organism can be treated while the infection is limited to systemic organs, CNS infection and vascular occlusion may be prevented. In an MSKCC series collected from 1970 to 1973, 18% of the patients suffering from CNS vascular disease had septic occlusions of cerebral vessels. Between 1977 and 1979, the same number of patients with CNS vascular disease was encountered as in the earlier era, but only 5% suffered from septic occlusions, presumably because the earlier use of antibiotics controls the primary infection.[24] The incidence of septic embolization as a percentage of autopsies showed a similar decrease. The incidence is even lower now.

TUMOR EMBOLI

Embolization of tumor from a primary or metastatic site outside the nervous system to the cerebral vasculature is a rare cause of cerebral vascular occlusion but a more common cause of pulmonary emboli. Mucin-secreting tumors may embolize to brain. Large or small arteries are occluded by mucin or a mucin and fibrin clot with or without tumor cells.[169,170] Tumor emboli are not necessarily restricted to the brain; major peripheral arterial occlusions can also be caused by malignant tumor embolization.[171,172] Atrial myxomas are a rare but well-recognized cause of cerebral or spinal infarction.[173] O'Neill et al.[174] found seven adults with clinical symptoms and pathologically proved non-myxomatous tumor emboli to the brain at the Mayo Clinic between 1951 and 1984. Four of the tumors were from the lung, one each from colon and hypopharynx and one from an unknown primary source. Three presented with cerebral infarction as the initial manifestation of cancer.

Tumors that embolize to the brain usually come from the lungs (Fig. 9–9). The primary pulmonary tumor or metastasis invades a pulmonary vein, where a portion of the tumor breaks off and travels through the heart to enter the cerebral circulation. In one patient, an embolus appeared to come from a metastasis on cardiac valves. How non–lung tumor emboli reach the arterial circulation in other patients is not clear. Paradoxical embolization is one possible route. One patient suffered a

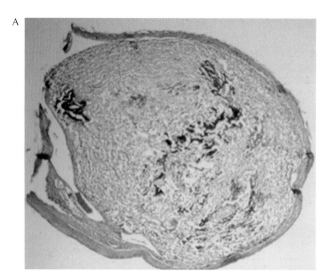

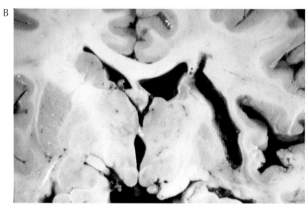

Figure 9–9. Tumor embolus. An 18-year-old girl underwent resection of a pulmonary metastasis from osteosarcoma. She awakened from surgery paralyzed on the right side. She made a partial recovery, but died of widespread tumor several months later. **A**: Middle cerebral artery was occluded with osteogenic sarcoma tissue that presumably broke off from the pulmonary metastasis during surgery. **B**: A healed infarct was found within the distribution of the left middle cerebral artery.

middle cerebral artery occlusion from testicular carcinoma not metastatic to the lungs. The occlusion was found at autopsy along with a tumor thrombus sitting astride a patent foramen ovale, indicating that the embolus had reached the cerebral circulation via that route.[172] Paradoxical embolization is a relatively common cause of stroke in the general population. If a DVT first causes a pulmonary embolus, the sudden rise in right atrial pressure can open a potentially patent foramen ovale, allowing a second embolus to cross from the right to the left atrium. Because cerebral blood flow is 15% to 20% of total blood flow in the resting state, a high likelihood exists that an embolus will reach the CNS once it has entered the arterial circulation. Furthermore, the probability is high that an embolus of any size to the cerebral circulation will cause clinical symptoms,

whereas a small embolus in a systemic organ is often asymptomatic. In patients with cancer, paradoxical emboli can result from extremity or pelvic thrombosis, as in a patient whose cerebral embolus was the initial symptom of a stage IC ovarian cancer associated with venous thrombi.[175]

Tumor emboli may be large enough to occlude large vessels, including the middle cerebral, basilar,[174] or angular artery.[176] The result is a bland or hemorrhagic infarction. A fibrin reaction may be set up within the vessel, but usually the tumor does not appear to invade the vessel wall. On occasion, cerebral embolization occurs during a thoracotomy for removal of a pulmonary tumor.[174] Pulmonary procedures can also cause arterial air embolus. Transthoracic pulmonary needle biopsy and intrathoracic anticancer drug administration have been reported as causes (Fig. 9–9).[177]

Clinically, the patient suffers the sudden onset of a focal neurologic defect, which may remain unchanged until death. In a few instances, repetitive showers of small emboli cause either transient focal neurologic defects or a diffuse encephalopathy much like DIC.

Although the diagnosis may be suspected from the clinical history and clinical findings closely resembling NBTE, the correct diagnosis is rarely considered initially. The absence of coagulation abnormalities, and the presence of a large primary tumor or metastases in the lung, may lead one to suspect the diagnosis. CT or MR angiography will reveal the arterial occlusion but not its source; however, there is no treatment. Occasionally, the diagnosis is only established later when a brain metastasis appears months after the patient had an ischemic infarct that was likely due to a cerebral tumor embolus (Chapter 5).

Thrombotic Microangiopathy

Several other causes of cerebral infarction (Table 9–14) are occasionally encountered in the cancer population.[24] Among these is thrombotic microangiopathy, which can occur as a complication of cancer, especially gastric, breast, lung, lymphoma,[178,179] Hodgkin disease, and myeloma[180] or cancer therapy[181] (see Chapter 12) (Table 9–15) including transplant.[182] Occasionally, it may develop in patients apparently cured of cancer.[178]

The clinical symptoms resemble DIC and thrombotic thrombocytopenic purpura (TTP)[183] and are characterized by encephalopathy and fleeting multifocal neurologic deficits. In one patient, progressive brainstem dysfunction culminated in coma and respiratory arrest.[184]

Thrombotic microangiopathy is believed to result from the formation of circulating immune complexes or autoantibodies. The disorder is similar to microangiopathic hemolytic anemia, TTP, and the hemolytic uremic syndrome not associated with cancer.[183] In one series of 351 patients who presented with symptoms suggesting TTP, 10 were subsequently discovered to be suffering from previously undiagnosed disseminated cancer.[183] Pathologic examination reveals widespread multi vascular thrombosis with platelets and fibrin associated with micro infarction. The heart, kidneys, and brain are the most severely affected organs. The diagnosis can be suspected from blood evaluation, which shows a hemolytic anemia; the peripheral smear shows the characteristic RBC changes of a microangiopathic process with schistocytes, burr cells, and helmet cells. The platelet count decreases, and platelet survival is shortened. Evidence of DIC may coexist,[185]

Table 9–14. **Other Causes of Infarction in Patients with Cancer**

Disorder	Association
Thrombotic microangiopathy	Malignancy (gastric, breast, lung), Chemotherapy (mitomycin, cisplatin)
Vasculitis	Hairy cell leukemia, Hodgkin disease
Intravascular neoplasia	Lymphoma
Thrombocytosis	Myeloproliferative disorders
Miscellaneous (owing to decreased blood flow)	
Hypotension	Anesthesia, sepsis, chemotherapy
Carotid artery rupture	Head and neck cancer, radiation therapy
Carotid cavernous fistula	Surgical complication
Arteriovenous shunt	Thyroid adenoma (not malignant)
Other	
Bone marrow embolus	
Air embolus	Venous catheter placement
Fat embolus	Orthopedic procedures

Table 9–15. **Thrombotic Microangiopathy in the Patient with Cancer**

Distinguishing Features	Carcinoma-Associated MAHA	Chemotherapy-Related HUS
Three most common tumors	Gastric, breast, lung	Gastric, colon, breast
Tumor status	Usually widely metastatic	Clinical remission common
Male:female ratio	1:2.1	1:1.9
Principal site of microvascular lesions	Lung	Kidney
Renal failure	Unusual	Characteristic
Severe hypertension	Uncommon	Common
Pulmonary edema	Unusual	Common
Adverse reactions to RBC transfusion	Not described	Common
Leukoerythroblastic picture	Common	Unusual
Laboratory evidence of intravascular coagulation	Common	Frequently absent
Proposed initiating factor(s)	Tumor emboli-derived factors	Chemotherapy toxicity + circulating immune complexes
Recommended treatment	Specific antitumor therapy ± heparin	Immunoperfusion or plasma exchange ± antiplatelet agents

From Ref. 185 with permission.
MAHA, microangiopathic hemolytic anemia; HUS, hemolytic-uremic syndrome; RBC, red blood cell.

but often the coagulation profile is normal. The MRI T2-weighted images may show multiple, small, hyperintense lesions corresponding to the areas of microangiopathy. Sometimes MRI changes include posterior leukoencephalopathy (p 451); more rarely, ischemic infarction is identified.[186,187] Unlike TTP, which is believed to result from deficiency of von Willebrand factor–cleaving protease and which responds to plasma exchange, no such deficiency is found in the neoplastic illness, and plasma exchange is ineffective.[183,188] Instead, the underlying malignancy must be treated, but even then the prognosis is poor.

Cerebral Vasculitis

Vasculitis, either systemic or restricted to the nervous system, may complicate cancers,[189] particularly hairy cell leukemia,[190] Hodgkin disease,[191,192] and other lymphoproliferative disorders.[193] The pathology of the vasculitis is variable. In some patients, it is associated with cryoglobulinemia; in others it is characteristic of granulomatous angiitis, polyarteritis nodosa, or leukoclastic vasculitis.[190,193] In a report of 47 patients with hairy cell leukemia and vasculitis, 17 had a picture of polyarteritis nodosa, 21

leukoblastic vasculitis, and 9 had direct invasion of the vessel wall by hairy cells.[190] Most patients do not have neurologic symptomatology.

Granulomatous angiitis restricted to the nervous system may complicate the course of Hodgkin disease or be the first symptom[191]; the disorder may be caused by Herpes zoster (see Chapter 10). In a series of 271 consecutive patients with giant-cell arteritis, 20 had current malignancies, representing over 7% of all cases; 27 additional patients were identified in published reports.[189] In these 47 patients, hematologic malignancy accounted for 45% of those with cancer; the others were a variety of solid tumors. Temporal artery biopsy of patients with giant-cell arteritis and chronic lymphocytic leukemia revealed only T-cells in the inflammatory infiltrates.[194]

When vasculitis involves the nervous system either along with systemic involvement or exclusively, the disorder is characterized by remitting and relapsing focal neurologic signs involving either brain or spinal cord. Patients often have headache, fever, and delirium. The MRI reveals evidence of multiple small infarcts, particularly in the white matter of the brain; angiography may show beading or occlusion of small arterial vessels.[195] White blood cells may be found in the CSF, although frequently the diagnosis can be made only by biopsy. Corticosteroids are said to be helpful in some instances, but most patients progress and then die from the neurologic disorder. The combination of cyclophosphamide and corticosteroids may be more effective than corticosteroids alone.[196]

Intravascular Lymphoma

Intravascular lymphoma, once called neoplastic angioendotheliomatosis or angiotropic lymphoma, is an extranodal diffuse large B-cell lymphoma characterized by the presence of neoplastic lymphocytes only in the lumen of small vessels, particularly capillaries (Fig. 9–10).[197,198] This systemic disorder involving the CNS and peripheral nervous system is believed to cause symptoms when tumor cells occlude small blood vessels in the brain, spinal cord, and peripheral nerves. However, in some patients, symptoms are relieved by plasma exchange,[199] suggesting antibody-mediated clotting factors. In one patient, inhibitory antibodies to von Willebrand factor–cleaving protease were found.[200] Structural alterations of several chromosomes have been found in the tumor cells and one investigator has postulated that an area on chromosome 6 (6q 21–24) may be the site of a tumor suppressor gene.[201]

Symptoms may include a subacutely developing dementia with or without focal neurologic signs. When signs are focal they may suggest brain, spinal cord,[202] muscle, or peripheral nerve involvement. In one series of eight patients, seven had encephalopathy or dementia, five had epileptic seizures, and two had myelopathy.[203] The MRI may show multiple hyperintense lesions, usually nonenhancing, in the subcortical white matter and cerebellar hemisphere.[204] Skin lesions including petechiae, purpura plaques, and discoloration, as well as evidence of other organ involvement, should raise the suspicion of this disorder.[205] A biopsy of skin lesions, muscle, or cerebral vasculature may show small vessels occluded by lymphoma cells. As with other widespread small-vessel disease, even when the disease is disseminated, signs and symptoms are often restricted to the nervous system. Plasmapheresis[199] should be considered. Anthracycline-based chemotherapy has been reported to have a 60% response rate and a 3-year overall survival rate higher than 30%.[197] One patient treated with rituximab developed an acute intracerebral hemorrhage.[206]

Thrombocytosis

Thrombocytosis may occur either as a reactive process or as a specific myelodysplastic syndrome (essential thrombocythemia).[207] At least 50% of patients with essential thrombocythemia have a gain of function mutation involving the JAK2 tyrosine kinase gene.[208] Recently proposed criteria for the diagnosis of essential thrombocythemia include all of the following: (1) sustained platelet count >450,000/mm³; (2) bone marrow biopsy showing proliferation of megakaryocyte lineage

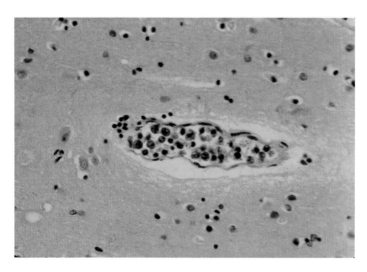

Figure 9–10. Intravascular lymphomatosis. A 65-year-old man developed transient neurologic symptoms including episodic confusion, extremity paresthesias, and incontinence. When he eventually succumbed, examination of the brain and other organs revealed occlusion of small vessels by malignant lymphocytes. All of the malignant cells remained within the vessel. None had infiltrated the brain.

with increased numbers of enlarged mature megakaryocytes and no significant increase or left shift of neutrophil granulopoiesis or erythropoiesis; (3) not meeting the criteria for other myelodysplastic syndromes; (4) demonstration of the gene mutation or no evidence of reactive thrombocytosis.[209] Platelet counts of more than 1,000,000/mm^3 sometimes occur in patients with severe myeloproliferative disorders, including chronic myelogenous leukemia, myelofibrosis, polycythemia vera, and primary thrombocythemia.[210] The disorder produces both arterial and microvascular thrombotic events and occasional hemorrhage.

Neurologic symptoms include headache, paresthesias, transient ischemic attacks in anterior or posterior circulation or both, and occasionally seizures. The clinical symptoms are usually fleeting but may be associated with small, hyperintense lesions on a T2-weighted MRI. A single patient has been reported with a lateral sinus thrombus related to thrombocytosis.[211] Peripheral microvascular occlusions can produce erythromelalgia (see Chapter 12). With counts above 1 million/mm^3, bleeding, usually in the skin or mucous membranes, may occur.[207]

Symptomatic patients can be treated with hydroxyurea, alpha-interferon, or anagrelide.[212,213] Inhibitors of the JAK2 pathway are currently under investiation. Asymptomatic patients are often placed on aspirin to prevent platelet aggregation and thrombosis.

VENOUS OCCLUSIONS

Occlusion of the large venous sinuses can result from compression or invasion by metastatic tumor (e.g., tumor in the skull overlying the sagittal sinus [see Chapter 5]) or from a coagulopathy.[51,214–216] Sudden occlusion of a large venous sinus, particularly the posterior sagittal sinus, can cause hemorrhagic infarcts (Fig. 9–11A). The most commonly involved sinus is the superior sagittal sinus (Figs. 5–9 and 9–11B), although other large cerebral veins may be affected.[217]

Compressive Venous Sinus Occlusion

Compressive occlusion of the venous sinuses can be caused by any tumor that metastasizes to the skull vault.[218] Neuroblastoma and breast or prostate cancers are common causes. The tumor may occlude or compress the sinus anywhere along its course, but symptoms are unlikely unless the posterior portion of the sagittal sinus or the larger of the two lateral sinuses is compressed or occluded. Accordingly, in symptomatic patients, the lesion is usually in the occipital bone overlying the confluence of venous sinuses (torcular Herophili) (Fig. 5–9).[219,220] In one patient, compression of the internal jugular vein by a neck metastasis led to progressive thrombosis of the transverse and sagittal sinus that was ultimately lethal.[221]

CLINICAL FINDINGS

The typical clinical picture can be called *pseudo-pseudotumor*. Patients with known cancer and bony metastases present to the physician either complaining of headache or sometimes with florid papilledema (Fig. 5–9C) without complaints of either headache or visual alterations.

Cerebral venous sinus occlusion from tumor compression seldom causes seizures or infarction because it develops slowly, allowing ample time for collateral circulation to develop and prevent infarction but not enough to maintain normal intracranial pressure. The diagnosis is suspected by the clinical picture and is usually established by an MRI. The sinus need not be totally occluded. Substantial, but incomplete, compression can raise intracranial pressure sufficiently to cause papilledema. If lumbar puncture is performed, the pressure is markedly elevated, but the fluid is otherwise normal.

DIFFERENTIAL DIAGNOSIS

The differential diagnosis includes brain metastases and leptomeningeal metastases, but florid papilledema in the absence of other neurologic signs effectively rules these out. Rarely, sinus occlusion accompanies leptomeningeal cancer.[222] Idiopathic true *pseudotumor cerebri* must be considered if the patient is an obese, middle-aged woman. In a very young or very old woman or in any man, the diagnosis of idiopathic pseudotumor is unlikely. The MRI changes of skull metastasis and sinus compression are conclusive.

TREATMENT

Treatment is RT to the site of the lesion. If the tumor is radiosensitive, it shrinks and decompresses the venous sinus. Even if the tumor does not shrink substantially, collateral circulation eventually decreases intracranial pressure, and the papilledema resolves. We have not encountered any patients with substantial visual loss from the papilledema.

Venous Sinus Thrombosis

Spontaneous thrombosis of a venous sinus is often a more serious illness than external compression. This complication of hypercoagulability can occur in patients with carcinoma of the breast and lung and other solid tumors, but is more common in those with hematologic malignancies. In one series of 624 adult patients with cerebral vein and dural sinus thrombosis, 46 (7.4%) were associated with malignancy. In this series, 18 tumors were hematologic, 20 were solid tumors arising outside the brain, and 14 were brain tumors (the disparity in numbers was present in the original report).[223] The disorder can occur when the patient's cancer is relatively quiescent and need not represent terminal disease.[216] One clinical setting was in patients with newly diagnosed leukemia receiving chemotherapy with L-asparaginase and prednisone.[224] In a series of 288 patients with ALL, 17 (5.9%) developed cerebral venous thrombosis, of whom 3 died.[225] The problem is more common when the patient has prothrombotic risk factors, including the MTHFR TT 67 genotype, the prothrombin G20210A variant, and others.[226] When dexamethasone was used instead of prednisone, the incidence of venous thrombosis decreased.[227] The disorder has also been reported following allogeneic bone marrow transplantation.[228] Several thrombophilic genetic variants may also play a role.[229]

CLINICAL FINDINGS

The clinical picture depends upon the size of the vessel occluded, the potential for rapidly developing collateral circulation, and whether the initial occlusion extends from its original site. The disease occurs in the following three forms:

1. A sudden onset of headache is associated with seizures and focal neurologic signs, often more striking in the legs than in the arms, followed by rapid progression to stupor, coma, cerebral herniation, and death (see Fig. 9–11).
2. In its less florid form, a sudden onset of headache is sometimes associated with focal seizures and mild focal neurologic signs that rapidly resolve.
3. A sudden onset of headache occurs without other neurologic signs, followed by progressive neurologic symptoms that begin suddenly or gradually. These indicate extension of the thrombus from its original site. In one of our patients with breast cancer, pain behind the ear suggested a lateral sinus thrombosis; in the succeeding 3 weeks, the clot progressively occluded the rest of her venous sinus system, leading to coma and death from large, bilateral cerebral infarcts.

The diagnosis can be suspected by the clinical setting and findings confirmed by MRI or MR venography.[228] If sinus occlusion is suspected, a gradient echo magnetic resonance study should be performed to evaluate hemorrhage in the brain.

The CSF occasionally has a few RBCs or WBCs but usually is normal. The lumbar spinal fluid pressure typically is increased, but in some patients it remains normal, even with papilledema and unequivocal sinus occlusion. In these patients, there has probably been enough molding of the temporal lobes through the tentorium, and the tonsils through the foramen magnum, to isolate the intracranial compartment from the spinal compartment and give an artificially low pressure in the lumbar sac. Such a phenomenon could lead to cerebral herniation, although we have not encountered it. It is probably wise not to perform lumbar punctures in these patients if the diagnosis can be made by other means and if meningeal tumor or infection is not suspected.

TREATMENT

The best treatment for the patient with cancer is not established. Usually, the disease takes a benign course, whether or not it is treated. Patients who suffer the acute onset of severe neurologic disability are probably destined to herniate and die no matter what the treatment. The only ones likely to benefit from treatment

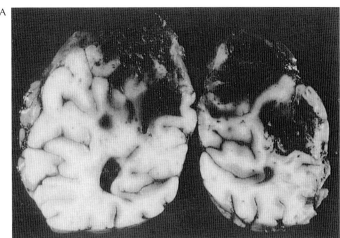

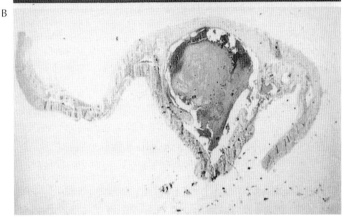

Figure 9–11. Sagittal sinus occlusion. **A**: Multiple hemorrhagic infarcts from sagittal sinus occlusion. This patient had just completed the induction phase of treatment for acute myelogenous leukemia and was about to be discharged when she suffered sudden headache and collapse. She was found unconscious with bilateral cortical spinal tract signs most marked in the lower extremities. She died from cerebral herniation. A fresh occlusion of the sagittal sinus was found, with bilateral (largely parasagittal) hemorrhagic infarcts. **B**: The occluded sinus.

are those who have a mild onset and then progress. Such patients cannot be identified in advance, raising the question of whether treatment with its attendant risk should be delivered to all patients.

The treatment has two aspects. The first is dealing with the increased intracranial pressure. We have not used steroids in most patients because the pressure signs are usually not disabling. When infarction has occurred and threatens to cause cerebral herniation, we have used hyperosmolar agents and steroids in the same manner as for metastatic tumors, but these agents have not been very helpful. One patient with intractable headache required steroids for several weeks to keep the headache under control; in this situation, they appeared to be beneficial.

The second aspect of treatment is lysis of the clot or prevention of the clot from propagating. Anticoagulant therapy with heparin is the appropriate treatment for most patients with cerebral venous thrombosis.[230–232] Endovascular thrombolysis for symptomatic cerebral venous thrombosis is sometimes useful.[233] Patients at MSKCC who have been heparinized have not experienced severe cerebral hemorrhage or a change of a bland infarction to a hemorrhagic one. On the other hand, whether these patients would have recovered under any circumstance is not clear. The use of IV heparin might very well prevent the clot from propagating, but it is doubtful that it can effectively deal with a clot already present. To complicate this issue, heparin-induced thrombocytopenia has been reported to cause a sagittal sinus occlusion.[215] Clot-lysing agents, such as urokinase[233] or streptokinase, and thrombectomy[234] have been reported effective in some patients, but have not been tried by us.

PROGNOSIS AND SEQUELAE

The prognosis is generally good. Most patients recover fully after the acute episode and have no continuing neurologic disability. In one instance, an arteriogram repeated 3 weeks later revealed a patent sinus. One patient had severe headaches for a number of weeks, and another had early-morning headaches for several weeks following her occlusion. One patient developed pulmonary emboli, possibly from the sagittal sinus occlusion, and required late heparinization. With these exceptions, the course has been benign. We have followed two patients for more than 10 years who have had no further neurologic disability.

Tumor Emboli

Tumor emboli could theoretically reach the cerebral venous system either via Batson plexus or by retrograde movement up a jugular vein invaded by head and neck cancer. We have encountered only one such case.

OTHER DISORDERS

Systemic Thrombophlebitis

Peripheral thrombophlebitis with or without pulmonary embolization is a common manifestation of both occult and established cancer (Chapter 4).[235–237] The disorder is particularly common following surgery for primary[238] or metastatic brain tumors.[239]

Episodic Neurologic Dysfunction in Patients with Hodgkin Disease

Although it is not certain that this disorder has a vascular pathogenesis, it is discussed here because its symptoms mimic those of transient ischemic attacks.[240] Some patients, mostly women, believed to be cured of Hodgkin disease, suffer transient episodes of neurologic dysfunction lasting 15 seconds to 45 minutes. Most had been in remission for months to years before developing the neurologic disorder. The unilateral or bilateral attacks are usually visual, with scintillating scotomata or transient visual loss. Other symptoms include focal weakness, sensory loss, or aphasia. Some patients develop weakness or sensory changes in their extremities. The episodes recur at intervals of days to weeks; they may be identical but more often they vary from episode to episode. In most patients, the attacks clear after months or years, either spontaneously or in response to taking low-dose aspirin.

The pathophysiology of these transient episodes is unknown. It is not even certain that they are related to the underlying Hodgkin disease. Levy[241] has reported similar transient neurologic episodes in healthy people without a history of cancer. No epidemiologic study has been done to determine the frequency of such episodes in the general population or in the Hodgkin disease population.

Whether or not these episodes are vascular is also uncertain. Migraine variants, as described by Fisher,[242] are another possible pathophysiologic mechanism. Although the episodes are usually benign, careful examination of the arterial system to look for a source of emboli is probably indicated when a patient treated for Hodgkin disease presents with such symptoms.

Systemic Hypotension

Because many patients with cancer are desperately ill, they can suffer from systemic hypotension. Of 256 infarcts encountered during autopsy examinations, four were located in watershed areas, presumably owing to systemic hypotension from the patient's underlying illness (see Fig. 2–4).[24] Carotid artery rupture in patients with head and neck cancer led to four additional watershed infarcts. One patient endured a hemorrhagic infarct after a surgically caused carotid cavernous fistula (Fig. 9–12); another unusual patient suffered cerebral infarction from a bone marrow embolus. Highly vascular tumors can compete with the nervous system for blood in much the same way as an arteriovenous malformation does. One unusual patient with an asymptomatic carotid artery occlusion[243] had transient ischemic attacks when a highly vascular thyroid adenoma apparently took enough cardiac output to decrease blood flow to the brain. Cerebral infarcts related to chemotherapy are discussed in Chapter 12.

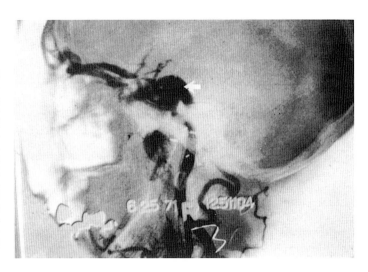

Figure 9–12. Surgically induced carotid cavernous fistula. An angiogram showing a carotid cavernous fistula in a patient who had recently undergone surgery for carcinoma of the ear. During the course of the operation, the temporal bone was fractured. A bone spicule entered the cavernous sinus and carotid artery. A carotid cavernous fistula was created (*arrow*). The entire carotid blood flow emptied into the cavernous sinus, depriving the cerebral hemisphere of blood. Pressure in the sinus caused proptosis and ocular muscle paralysis. The large cerebral infarct caused the patient's death.

Table 9–16. Cerebrovascular Disease in Patients with Cancer

CNS Lesion	Primary Cancer	Mechanism
At Diagnosis or Initial Therapy		
Cerebral hemorrhage	Acute promyelocytic leukemia	Disseminated intravascular coagulation
	Acute nonlymphoblastic leukemia	Leukostasis
Sinus thrombosis	Acute lymphoblastic leukemia	L-asparaginase
At Relapse/Progressive Disease		
Cerebral hemorrhage	Acute nonlymphocytic leukemia	Coagulopathy
Thrombotic microinfarcts	Lymphoma/leukemia	Intravascular coagulation
Embolic infarct	Lymphoma/leukemia	Septic (fungal) embolism
Sinus thrombosis	Lymphoma	Coagulopathy
At Diagnosis/During Remission		
Cerebral hemorrhage	Lung, melanoma, germ-cell tumor	Intratumoral bleeding
Embolic infarct	Adenocarcinoma	NBTE
	Myxoma	Tumor embolism
Subdural hematoma	Prostate, breast	Dural metastasis
Thrombotic infarct	Head and neck carcinoma	Atherosclerosis
At Relapse/Active Disseminated Disease		
Cerebral hemorrhage	Melanoma, germ-cell tumor, lung	Intratumoral bleeding
Subdural hematoma	Breast, prostate	Dural metastasis
Thrombotic microinfarcts	Breast carcinoma	Intravascular coagulation
Sinus thrombosis	Breast, lung	Coagulopathy
Embolic infarct	Lung carcinoma	NBTE

The table details the likely cerebrovascular lesions in patients with cancer. The lesions differ depending on whether they appear early or late in the course of the cancer.
CNS, central nervous system; NBTE, nonbacterial thrombotic endocarditis.

Air Embolism

Once believed to be almost exclusively a problem encountered at craniotomy done in the sitting position, *cerebral air embolism* is probably now more common as a complication of non-surgical procedures such as venous catheterization.[244,245] The phenomenon may occur during insertion of the catheter, its removal or accidental disconnection,[246] or after needle biopsy of the lung or mediastinal biopsy. In one review of 11,583 central venous catheters, air embolism was reported in 15 patients.[245] The phenomenon may be asymptomatic, cause pulmonary symptomatology or cerebral symptoms.[244] The emboli reach the brain either via a pulmonary shunt or through a patent foramen ovale. We had one patient who developed an air embolism during placement of a central venous catheter who had asymptomatic and undiagnosed multiple pulmonary emboli causing pulmonary hypertension, which likely opened a right-to-left cardiac shunt. Neurologic symptoms can be divided into two groups: In the first a diffuse encephalopathy sometimes leads to coma or death.[246] In the second, focal signs, particularly left hemiparesis are found. The air reaches the right side of the brain because the air flows more easily into the right brachiocephalic artery, the first branch of the aortic arch. The diagnosis is suspected by the clinical situation in which a patient develops neurologic symptoms when catheters are inserted or removed. A CT or MRI will identify air in cerebral vessels, although the air may be gone within 24 hours. Treatment consists of placing the patient in the head down position, supplying oxygen and perhaps using hyperbaric oxygenation.[244,246]

Cerebral fat embolism is much rarer[247] in the cancer patient. It may follow surgical procedures on bone. The symptoms are similar to air embolism with the addition that cutaneous petechiae can often be identified. Brain images identify fat in blood vessels, sometimes surrounded by an area of infarction.[248]

APPROACH TO THE PATIENT

Tables 9–16 details the most likely cerebrovascular lesions to be found in patients with hematologic neoplasms. The lesions differ depending on whether the vascular disorder appears at diagnosis and initial treatment of the neoplasm or during progressive disease or relapse.

REFERENCES

1. Cornuz J, Bogousslavsky J, Schapira M, et al. Ischemic stroke as the presenting manifestation of localized systemic cancer. *Schweiz Arch Neurol Psychiatr* 1988;139(2):5–11.
2. Navi BB, DeAngelis LM, Segal AZ. Multifocal strokes as the presentation of occult lung cancer. *J Neuro-Oncol* 2007;85:307–309.
3. Di MP, Coppola L, Diadema MR, et al. Internal jugular vein thrombosis as first sign of metastatic lung cancer. *Tumori* 2003;89:448–451.
4. Naschitz JE, Yeshurun D, Lev LM. Thromboembolism in cancer. Changing trends. *Cancer* 1993; 71:1384–1390.
5. Gore JM, Appelbaum JS, Greene HL, et al. Occult cancer in patients with acute pulmonary embolism. *Ann Intern Med* 1982;96:556–560.
6. Aderka D, Brown A, Zelikovski A, et al. Idiopathic deep vein thrombosis in an apparently healthy patient as premonitory sign of occult cancer. *Cancer* 1986;57:1846–1849.
7. Monreal M, Lafoz E, Casals A, et al. Occult cancer in patients with deep venous thrombosis. A systematic approach. *Cancer* 1991;67:541–545.
8. Prandoni P, Lensing AW, Buller HR, et al. Deep-vein thrombosis and the incidence of subsequent symptomatic cancer. *N Engl J Med* 1992;327:1128–1133.
9. Girolami A, Prandoni P, Zanon E, et al. Venous thromboses of upper limbs are more frequently associated with occult cancer as compared with those of lower limbs. *Blood Coagul Fibrinolysis* 1999;10:455–457.
10. Bell WR, Starksen NF, Tong S, et al. Trousseau's syndrome. Devastating coagulopathy in the absence of heparin. *Am J Med* 1985;79:423–430.
11. Sack GH, Jr, Levin J, Bell WR. Trousseau's syndrome and other manifestations of chronic disseminated coagulopathy in patients with neoplasms: clinical, pathophysiologic, and therapeutic features. *Medicine* 1977;56:1–37.
12. Trousseau A. Phlegmasia alba dolens. *Clin Med Hotel Dieu de Paris* 1865;3:94.
13. Nijziel MR, van OR, Hillen HF, et al. From Trousseau to angiogenesis: the link between the haemostatic system and cancer. *Neth J Med* 2006;64:403–410.
14. Otten HM, Mathijssen J, ten Cate H, et al. Symptomatic venous thromboembolism in cancer patients treated with chemotherapy—an underestimated phenomenon. *Arch Intern Med* 2004;164:190–194.
15. Weisberg LA, Nice CN. Intracranial tumors simulating the presentation of cerebrovascular syndromes. Early detection with cerebral computed tomography (CCT). *Am J Med* 1977;63:517–524.
16. Schiff D, DeAngelis LM. Therapy of venous thromboembolism in patients with brain metastases. *Cancer* 1994;73:493–498.
17. Molgaard CP, Yucel EK, Geller SC, et al. Access-site thrombosis after placement of inferior vena cava filters with 12-14-F delivery sheaths. *Radiology* 1992;185:257–261.

18. Rogers LR. Cerebrovascular complications in cancer patients. *Oncology* 1994;8:23–30.
19. Nguyen T, DeAngelis LM. Stroke in cancer patients. *Curr Neurol Neurosci Rep* 2006;6:187–192.
20. Dreyer L, Prescott E, Gyntelberg F. Association between atherosclerosis and female lung cancer—a Danish cohort study. *Lung Cancer* 2003;42:247–254.
21. Lindvig K, Moller H, Mosbech J, et al. The pattern of cancer in a large cohort of stroke patients. *Int J Epidemiol* 1990;19:498–504.
22. Cestari DM, Weine DM, Panageas KS, et al. Stroke in patients with cancer—incidence and etiology. *Neurology* 2004;62:2025–2030.
23. Katz JM, Segal AZ. Incidence and etiology of cerebrovascular disease in patients with malignancy. *Curr Atheroscler Rep* 2005;7:280–288.
24. Graus F, Rogers LR, Posner JB. Cerebrovascular complications in patients with cancer. *Medicine* 1985;64:16–35.
25. Chernik NL, Loewenson RB, Posner JB, et al. Cerebral atherosclerosis and stroke in cancer patients. Abstract. *Neurology* 1978;28:350.
26. Klassen AC, Loewenson RB, Resch JA. Cerebral atherosclerosis in selected chronic disease states. *Atherosclerosis* 1973;18:321–336.
27. Sternby NH. Atherosclerosis and malignant tumours. *Bull World Health Organ* 1976;53:555–561.
28. Sternby NH, Berge T. Atherosclerosis and malignant tumors. *APMIS Sect A Pathol* 1973;236:34–44.
29. Malinow MR, Senner JW. Arterial pathology in cancer patients suggests atherosclerosis regression. *Med Hypoth* 1983;11:353–357.
30. Johnson MJ. Bleeding, clotting and cancer. *Clin Oncol (R Coll Radiol)* 1997;9:294–301.
31. Eidelberg D, Sotrel A, Horoupian S, et al. Thrombotic cerebral vasculopathy associated with herpes zoster. *Ann Neurol* 1986;19:7–14.
32. De CM. The prothrombotic state in cancer: pathogenic mechanisms. *Crit Rev Oncol Hematol* 2004;50:187–196.
33. Zwicker JI, Furie BC, Furie B. Cancer-associated thrombosis. *Crit Rev Oncol Hematol* 2007;62:126–136.
34. Gutmann DH, Cantor CR, Piacente GJ, et al. Cerebral vasculopathy and infarction in a woman with carcinomatous meningitis. Letter to the editor. *J Neuro-Oncol* 1990;9:183–185.
35. Bick RL. Coagulation abnormalities in malignancy: a review. *Semin Thromb Hemost* 1992;18:353–372.
36. Ey FS, Goodnight SH. Bleeding disorders in cancer. *Semin Oncol* 1990;17:187–197.
37. Belt RJ, Leite C, Haas CD, et al. Incidence of hemorrhagic complications in patients with cancer. *JAMA* 1978;239:2571–2574.
38. Dutcher JP, Schiffer CA, Aisner J, et al. Incidence of thrombocytopenia and serious hemorrhage among patients with solid tumors. *Cancer* 1984;53:557–562.
39. Wirtz PW, Bloem BR, van der Meer FJ, et al. Paraparesis after lumbar puncture in a male with leukemia. *Pediatr Neurol* 2000;23:67–68.
40. Murata J-I, Sawamura Y, Takahashi A, et al. Intracerebral hemorrhage caused by a neoplastic aneurysm from small-cell lung carcinoma: Case report. *Neurosurgery* 1993;32:124–126.
41. Porcu P, Cripe LD, Ng EW, et al. Hyperleukocytic leukemias and leukostasis: a review of pathophysiology, clinical presentation and management. *Leuk Lymphoma* 2000;39:1–18.
42. Arbuthnot C, Wilde JT. Haemostatic problems in acute promyelocytic leukaemia. *Blood Rev* 2006;20:289–297.
43. Delgado J, Jimenez-Yuste V, Hernandez-Navarro F, et al. Acquired haemophilia: review and meta-analysis focused on therapy and prognostic factors. *Br J Haematol* 2003;121:21–35.
44. Franchini M, Montagnana M, Targher G, et al. Pathogenesis, clinical and laboratory aspects of thrombosis in cancer. *J Thromb Thrombolysis* 2007;24:29–38.
45. Kwaan HC, Vicuna B. Incidence and pathogenesis of thrombosis in hematologic malignancies. *Semin Thromb Hemost* 2007;33:303–312.
46. Hilgard P, Gordon-Smith EC. Microangiopathic haemolytic anaemia and experimental tumour-cell emboli. *Br J Haematol* 1974;26:651–659.
47. Nand S, Fisher SG, Salgia R, et al. Hemostatic abnormalities in untreated cancer: incidence and correlation with thrombotic and hemorrhagic complications. *J Clin Oncol* 1987;5:1998–2003.
48. Rogers LR, Cho E-S, Kempin S, et al. Cerebral infarction from non-bacterial thrombotic endocarditis. Clinical and pathological study including the effects of anticoagulation. *Am J Med* 1987;83:746–756.
49. Levine MN, Gent M, Hirsh J, et al. The thrombogeneic effect of anticancer drug therapy in women with stage II breast cancer. *N Engl J Med* 1988;318:404–407.
50. Ruiz MA, Marugan I, Estelles A, et al. The influence of chemotherapy on plasma coagulation and fibrinolytic systems in lung cancer patients. *Cancer* 1989;63:643–648.
51. Feinberg WM, Swenson MR. Cerebrovascular complications of l-asparaginase therapy. *Neurology* 1988;38:127–133.
52. van der Sande JJ, Veltkamp JJ, Bouwhuis-Hoogerwerf ML. Hemostasis and intracranial surgery. *J Neurosurg* 1983;58:693–698.
53. Simmons NE, Elias WJ, Henson SL, et al. Small cell lung carcinoma causing epidural hematoma: case report. *Surg Neurol* 1999;51:56–59.
54. Minette SE, Kimmel DW. Subdural hematoma in patients with systemic cancer. *Mayo Clin Proc* 1989;64:637–642.
55. Laigle-Donadey F, Taillibert S, Mokhtari K, et al. Dural metastases. *J Neuro-Oncol* 2005;75:57–61.
56. Clifford JR, Kirgis HD, Connolly ES. Metastatic melanoma of the brain presenting as subarachnoid hemorrhage. *South Med J* 1975;68:206–208.
57. Edelson RN, Chernik NL, Posner JB. Spinal subdural hematomas complicating lumbar puncture. *Arch Neurol* 1974;31:134–137.
58. Inci S, Bozkurt G, Gulsen S, et al. Rare cause of subarachnoid hemorrhage: spinal meningeal carcinomatosis. Case report. *J Neurosurg Spine* 2005;2:79–82.
59. Mandybur TI. Intracranial hemorrhage caused by metastatic tumors. *Neurology* 1977;27:650–655.
60. Lieu AS, Hwang SL, Howng SL, et al. Brain tumors with hemorrhage. *J Formos Med Assoc* 1999;98:365–367.
61. Chang L, Chen YL, Kao MC. Intracranial metastasis of hepatocellular carcinoma: review of 45 cases. *Surg Neurol* 2004;62:172–177.
62. Posner JB, Chernik NL. Intracranial metastases from systemic cancer. *Adv Neurol* 1978;19:575–587.

63. Haines DE, Harkey HL, Al-Mefty O. The "sub-dural" space: a new look at an outdated concept. *Neurosurgery* 1993;32:111–120.

64. O'Laoire SA, Crockard HA, Thomas DG, et al. Brain-stem hematoma: a report of six surgically treated cases. *J Neurosurg* 1982;56:222–227.

65. Kim D, Liebeskind DS. Neuroimaging advances and the transformation of acute stroke care. *Semin Neurol* 2005;25:345–361.

66. Abrahams NA, Prayson RA. The role of histopathologic examination of intracranial blood clots removed for hemorrhage of unknown etiology: a clinical pathologic analysis of 31 cases. *Ann Diagn Pathol* 2000;4:361–366.

67. Niizuma H, Nakasato N, Yonemitsu T, et al. Intracerebral hemorrhage from a metastatic brain tumor. Importance of differential diagnosis preceding stereotaxic hema-toma aspiration. *Surg Neurol* 1988;29:232–236.

68. Requena I, Arias M, Lopez-Ibor L, et al. Cavernomas of the central nervous system: clinical and neuroimag-ing manifestations in 47 patients. *J Neurol Neurosurg Psychiatry* 1991;54:590–594.

69. Wolf PA, Rosman NP, New PFJ. Multiple small cryp-tic venous angiomas of brain mimicking cerebral metastases. *Neurology* 1967;17:491–501.

70. Poungvarin N, Bhoopat W, Viriyavejakul A, et al. Effects of dexamethasone in primary supratento-rial intracerebral hemorrhage. *New Engl J Med* 1987;316:1229–1233.

71. Ravussin P, Abou-Madi M, Archer D, et al. Changes in CSF pressure after mannitol in patients with and without elevated CSF pressure. *J Neurosurg* 1988;69:869–876.

72. Kalita J, Misra UK, Ranjan P, et al. Effect of mannitol on regional cerebral blood flow in patients with intra-cerebral hemorrhage. *J Neurol Sci* 2004;224:19–22.

73. Misra UK, Kalita J, Vajpayee A, et al. Effect of single mannitol bolus in intracerebral hemorrhage. *Eur J Neurol* 2007;14:1118–1123.

74. Misra UK, Kalita J, Ranjan P, et al. Mannitol in intra-cerebral hemorrhage: a randomized controlled study. *J Neurol Sci* 2005;234:41–45.

75. Openshaw H, Ressler JA, Snyder DS. Lumbar puncture and subdural hygroma and hematomas in hemopoietic cell transplant patients. *Bone Marrow Transplant*, in press.

76. Caputi F, Lamaida E, Gazzeri R. Acute subdural hematoma and pachymeningitis carcinomatosa: case report. *Rev Neurol (Paris)* 1999;155:383–385.

77. Cameron MM. Chronic subdural haematoma: a review of 114 cases. *J Neurol Neurosurg Psychiatry* 1978;41:834–839.

78. Arseni C, Stanciu M. Particular clinical aspects of chronic subdural haematoma in adults. *Eur Neurol* 1969;2:109–122.

79. Trosch RM, Ransom BR. Levodopa-responsive par-kinsonism following central herniation due to bilateral subdural hematomas. *Neurology* 1990;40:376–377.

80. Kotagal S, Shuter E, Horenstein S. Chorea as a mani-festation of bilateral subdural hematoma in an elderly man. Clinical note. *Arch Neurol* 1981;38:195.

81. Nobbe FA, Krauss JK. Subdural hematoma as a cause of contralateral dystonia. *Clin Neurol Neurosurg* 1997;99:37–39.

82. Devereaux MW, Brust JC, Keane JR. Internuclear ophthalmoplegia caused by subdural hematoma. *Neurology* 1979;29:251–255.

83. Browder J, Rabiner AM. Regional swelling of the brain in subdural hematoma. *Ann Surg* 1970;134:369–375.

84. Tseng SH, Liao CC, Lin SM, et al. Dural metastasis in patients with malignant neoplasm and chronic sub-dural hematoma. *Acta Neurol Scand* 2003;108:43–46.

85. Newton LK. Neurologic complications of poly-cythemia and their impact on therapy. *Oncology (Williston Park)* 1990;4:59–66.

86. Dempfle CE. Coagulopathy of sepsis. *Thromb Haemost* 2004;91:213–224.

87. Levi M. Current understanding of dissemi-nated intravascular coagulation. *Br J Haematol* 2004;124:567–576.

88. Bick RL. Disseminated intravascular coagulation—current concepts of etiology, pathophysiology, diag-nosis, and treatment. *Hematol Oncol Clin North Am* 2003;17:149–176.

89. Collins RC, Al-Mondhiry H, Chernik NL, et al. Neurologic manifestations of intravascular coagu-lation in patients with cancer. A clinicopathologic analysis of 12 cases. *Neurology* 1975;25:795–806.

90. Dixit A, Chatterjee T, Mishra P, et al. Disseminated intravascular coagulation in acute leukemia at pre-sentation and during induction therapy. *Clin Appl Thromb Hemost* 2007;13:292–298.

91. Gralnick HR, Bagley J, Abrell E. Heparin treatment for the hemorrhagic diathesis of acute promyelocytic leukemia. *Am J Med* 1972;52:167–174.

92. Imaoka S, Ueda T, Shibata H, et al. Fibrinolysis in patients with acute promyelocytic leukemia and dis-seminated intravascular coagulation during heparin therapy. *Cancer* 1986;58:1736–1738.

93. Avvisati G, ten Cate JW, Buller HR, et al. Tranexamic acid for control of haemorrhage in acute promyelo-cytic leukaemia. *Lancet* 1989;2:122–124.

94. Breuer AC, Tyler HR, Marzewski DJ, et al. Radicular vessels are the most probable source of needle-induced blood in lumbar puncture: significance for the thrombocytopenic cancer patient. *Cancer* 1982;49:2168–2172.

95. Masdeu JC, Breuer AC, Schoene WC. Spinal subarach-noid hematomas: clue to a source of bleeding in trau-matic lumbar puncture. *Neurology* 1979;29:872–876.

96. Howard SC, Gajjar A, Ribeiro RC, et al. Safety of lumbar puncture for children with acute lympho-blastic leukemia and thrombocytopenia. *JAMA* 2000;284:2222–2224.

97. Howard SC, Gajjar AJ, Cheng C, et al. Risk fac-tors for traumatic and bloody lumbar puncture in children with acute lymphoblastic leukemia. *JAMA* 2002;288:2001–2007.

98. Blade J, Gaston F, Montserrat E, et al. Spinal suba-rachnoid hematoma after lumbar puncture causing reversible paraplegia in acute leukemia. Case report. *J Neurosurg* 1983;54:438–439.

99. Solymosi L, Wappenschmidt J. A new neuroradio-logic method for therapy of spinal epidural hemato-mas. *Neuroradiology* 1985;27:67–69.

100. Gustafsson H, Rutberg H, Bengtsson M. Spinal hae-matoma following epidural analgesia. Report of a patient with ankylosing spondylitis and a bleeding diathesis. *Anaesthesia* 1988;43:220–222.

101. Mattle H, Sieb JP, Rohner M, et al. Nontraumatic spinal epidural and subdural hematomas. *Neurology* 1987;37:1351–1356.

102. Li Y, Takayasu M, Takagi T, et al. [Intramedullary spinal cord metastasis associated with hemorrhage: a case report]. *No Shinkei Geka* 2000;28:453–457.

103. Hashizume Y, Hirano A. Intramedullary spinal cord metastasis. Pathologic findings in five autopsy cases. *Acta Neuropathol (Berl)* 1983;61:214–218.

104. Azzarelli B, Itani A-L, Catanzaro PT. Cerebral phlebothrombosis. A complication of lymphoma. *Arch Neurol* 1980;37:126–127.

105. Hug V, Keating M, McCredie K, et al. Clinical course and response to treatment of patients with acute myelogenous leukemia presenting with a high leukocyte count. *Cancer* 1983;52:773–779.

106. McKee LC, Jr, Collins RD. Intravascular leukocyte thrombi and aggregates as a cause of morbidity and mortality in leukemia. *Medicine* 1974;53:463–478.

107. Lowe EJ, Pui CH, Hancock ML, et al. Early complications in children with acute lymphoblastic leukemia presenting with hyperleukocytosis. *Pediatr Blood Cancer* 2005;45:10–15.

108. Creutzig U, Ritter J, Budde M, et al. Early deaths due to hemorrhage and leukostasis in childhood acute myelogenous leukemia. Associations with hyperleukocytosis and acute monocytic leukemia. *Cancer* 1987;60:3071–3079.

109. Baer MR, Stein RS, Dessypris EN. Chronic lymphocytic leukemia with hyperleukocytosis. The hyperviscosity syndrome. *Cancer* 1985;56:2865–2869.

110. Zarkovic M, Kwaan HC. Correction of hyperviscosity by apheresis. *Semin Thromb Hemost* 2003;29:535–542.

111. Mehta J, Singhal S. Hyperviscosity syndrome in plasma cell dyscrasias. *Semin Thromb Hemost* 2003;29:467–471.

112. Mueller J, Hotson JR, Lanston JW. Hyperviscosity-induced dementia. *Neurology* 1983;33:101–103.

113. Pruzanski W, Watt JG. Serum viscosity and hyperviscosity syndrome in IgG multiple myeloma. *Ann Intern Med* 1972;77:853–860.

114. Somer T. Hyperviscosity syndrome in plasma cell dyscrasias. *Adv Microcirc* 1975;6:1–55.

115. Pihko H, Tyni T, Virkola K, et al. Transient ischemic cerebral lesions during induction chemotherapy for acute lymphoblastic leukemia. *J Pediatr* 1993;123:718–724.

116. Hennerici M, Rautenberg W, Trockel U, et al. Spontaneous progression and regression of small carotid atheroma. *Lancet* 1985;1:1415–1419.

117. Dreyer L, Olsen JH. Risk for non-smoking-related cancer in atherosclerotic patients. *Cancer Epidemiol Biomarkers Prev* 1999;8:915–918.

118. Bowers DC, McNeil DE, Liu Y, et al. Stroke as a late treatment effect of Hodgkin's disease: a report from the Childhood Cancer Survivor Study. *J Clin Oncol* 2005;23:6508–6515.

119. Dorresteijn LD, Kappelle AC, Boogerd W, et al. Increased risk of ischemic stroke after radiotherapy on the neck in patients younger than 60 years. *J Clin Oncol* 2002;20:282–288.

120. Ramos KS, Partridge CR. Atherosclerosis and cancer: flip sides of the neoplastic response in mammalian cells? *Cardiovasc Toxicol* 2005;5:245–255.

121. Elerding SC, Fernandez RN, Grotta JC, et al. Carotid artery disease following external cervical irradiation. *Ann Surg* 1981;194:609–615.

122. Barcos M, Lane W, Gomez GA, et al. An autopsy study of 1206 acute and chronic leukemias (1958 to 1982). *Cancer* 1987;60:827–837.

123. Chaturvedi S, Ansell J, Recht L. Should cerebral ischemic events in cancer patients be considered a manifestation of hypercoagulability? *Stroke* 1994;25:1215–1218.

124. Mant MJ, Fisk RL, Amy RW. Case report: chronic disseminated intravascular coagulation due to occult carcinoma. *Am J Med Sci* 1977;274:69–74.

125. Kwaan HC, Wang J, Boggio LN. Abnormalities in hemostasis in acute promyelocytic leukemia. *Hematol Oncol* 2002;20:33–41.

126. Kim H-S, Suzuki M, Lie JT, et al. Nonbacterial thrombotic endocarditis (NBTE) and disseminated intravascular coagulation (DIC): autopsy study of 36 patients. *Arch Pathol Lab Med* 1977;101:65–68.

127. Baker WF, Jr. Clinical aspects of disseminated intravascular coagulation: a clinician's point of view. *Semin Thromb Hemost* 1989;15:1–57.

128. Saba HI, Morelli GA. The pathogenesis and management of disseminated intravascular coagulation. *Clin Adv Hematol Oncol* 2006;4:919–926.

129. Green D. Management of bleeding complications of hematologic malignancies. *Semin Thromb Hemost* 2007;33:427–434.

130. Alperin JB. Coagulopathy caused by vitamin K deficiency in critically ill, hospitalized patients. *JAMA* 1987;258:1916–1919.

131. Levi M, De Jonge E, van der Poll T. New treatment strategies for disseminated intravascular coagulation based on current understanding of the pathophysiology. *Ann Med* 2004;36:41–49.

132. Kim H-S, Suzuki M, Lie JT, et al. Clinically unsuspected disseminated intravascular coagulation (DIC). *Am J Clin Pathol* 1976;66:31–39.

133. Shimamura K, Oka K, Nakazawa M, et al. Distribution patterns of microthrombi in disseminated intravascular coagulation. *Arch Pathol Lab Med* 1983;107:543–547.

134. Clark J, Rubin RN. A practical approach to managing disseminated intravascular coagulation. *J Crit Illness* 1994;9:265–280.

135. Seifter EJ, Parker RI, Gralnick HR, et al. Abnormal coagulation results in patients with Hodgkin's disease. *Am J Med* 1985;78:942–950.

136. Ois A, Gomis M, Rodriguez-Campello A, et al. Factors associated with a high risk of recurrence in patients with transient ischemic attack or minor stroke. *Stroke*, in press.

137. Yu M, Nardella A, Pechet L. Screening tests of disseminated intravascular coagulation: guidelines for rapid and specific laboratory diagnosis. *Crit Care Med* 2000;28:1777–1780.

138. Kay AC, Solberg LA, Jr, Nichols DA, et al. Prognostic significance of computed tomography of the brain in thrombotic thrombocytopenic purpura. *Mayo Clin Proc* 1991;66:602–607.

139. Goodman A, Ramos R, Petrelli M, et al. Gingival biopsy in thrombotic thrombocytopenic purpura. *Ann Intern Med* 1978;89:501–504.

140. Miyahara S, Yasu T, Yamada Y, et al. Subcutaneous injection of heparin calcium controls chronic disseminated intravascular coagulation associated with inoperable dissecting aortic aneurysm in an outpatient clinic. *Intern Med* 2007;46:727–732.

141. Feinstein DI. Diagnosis and management of disseminated intravascular coagulation: the role of heparin therapy. *Blood* 1982;60:284–287.
142. Wheeler A, Rubenstein EB. Current management of disseminated intravascular coagulation. *Oncology (Huntingt)* 1994;8:69–79.
143. Mori K, Takeuchi J, Ishikawa M, et al. Occlusive arteriopathy and brain tumor. *J Neurosurg* 1978;49:22–35.
144. Sacher M, Som PM, Lanzieri CF, et al. Total internal carotid artery occlusion by a benign carotid body tumor: a rare occurrence. *J Comput Tomogr* 1985;9:213–217.
145. Launay M, Fredy D, Merland JJ, et al. Narrowing and occlusion of arteries by intracranial tumors. Review of the literature and report of 25 cases. *Neuroradiology* 1977;14:117–126.
146. Yaegashi H, Takahashi T. Encasement and other deformations of tumor-embedded host arteries due to loss of medial smooth muscles. Morphometric and three-dimensional reconstruction studies on some human carcinomas. *Cancer* 1990;65:1097–1103.
147. Upile T, Triaridis S, Kirkland P, et al. The management of carotid artery rupture. *Eur Arch Otorhinolaryngol* 2005;262:555–560.
148. Raskin NH, Prusiner S. Carotidynia. *Neurology* 1977;27:43–46.
149. Ghatak NR. Pathology of cerebral embolization caused by nonthrombotic agents. *Hum Pathol* 1975;6:599–610.
150. el-Shami K, Griffiths E, Streiff M. Nonbacterial thrombotic endocarditis in cancer patients: Pathogenesis, diagnosis, and treatment. *Oncologist* 2007;12:518–523.
151. Biller J, Challa VR, Toole JF, et al. Nonbacterial thrombotic endocarditis: a neurologic perspective of cliniciopathologic correlations of 99 patients. *Arch Neurol* 1982;39:95–98.
152. Deppisch LM, Fayemi AO. Non-bacterial thrombotic endocarditis: clinicopathologic correlations. *Am Heart J* 1976;92:723–729.
153. Rosen P, Armstrong D. Nonbacterial thrombotic endocarditis in patients with malignant neoplastic diseases. *Am J Med* 1973;54:23–29.
154. Schattner A, Klepfish A, Huszar M, et al. Two patients with arterial thromboembolism among 311 patients with adenocarcinoma of the pancreas. *Am J Med Sci* 2002;324:335–338.
155. Bryan CS. Nonbacterial thrombotic endocarditis with malignant tumors. *Am J Med* 1969;46:787–793.
156. Kearsley JH, Tattersall MH. Cerebral embolism in cancer patients. *Q J Med* 1982;51:279–291.
157. Reisner SA, Rinkevich D, Markiewicz W, et al. Cardiac involvement in patients with myeloproliferative disorders. *Am J Med* 1992;93:498–504.
158. Edoute Y, Haim N, Rinkevich D, et al. Cardiac valvular vegetations in cancer patients: A prospective echocardiographic study of 200 patients. *Am J Med* 1997;102:252–258.
159. Young RSK, Zalneraitis EL. Marantic endocarditis in children and young adults: clinical and pathological findings. *Stroke* 1981;12:635–639.
160. Yusuf SW, Ali SS, Swafford J, et al. Culture-positive and culture-negative endocarditis in patients with cancer: a retrospective observational study, 1994–2004. *Medicine (Baltimore)* 2006;85:86–94.
161. Estevez CM, Corya BC. Serial echocardiographic abnormalities in nonbacterial thrombotic endocarditis of the mitral valve. *Chest* 1976;69:801–804.
162. Hofmann T, Kasper W, Meinertz T, et al. Echocardiographic evaluation of patients with clinically suspected arterial emboli. *Lancet* 1990;336:1421–1424.
163. Khandheria BK, Seward JB, Tajik AJ. Transesophageal echocardiography. *Mayo Clin Proc* 1994;69:856–863.
164. Lee RJ, Bartzokis T, Yeoh T-K, et al. Enhanced detection of intracardiac sources of cerebral emboli by transesophageal echocardiography. *Stroke* 1991;22:734–739.
165. Reeder GS, Khandheria BK, Seward JB, et al. Transesophageal echocardiography and cardiac masses. *Mayo Clin Proc* 1991;66:1101–1109.
166. Dutta T, Karas MG, Segal AZ, et al. Yield of transesophageal echocardiography for nonbacterial thrombotic endocarditis and other cardiac sources of embolism in cancer patients with cerebral ischemia. *Am J Cardiol* 2006;97:894–898.
167. Singhal AB, Topcuoglu MA, Buonanno FS. Acute ischemic stroke patterns in infective and nonbacterial thrombotic endocarditis: a diffusion-weighted magnetic resonance imaging study. *Stroke* 2002;33:1267–1273.
168. Ghanem GA, Boktour M, Warneke C, et al. Catheter-related *Staphylococcus aureus* bacteremia in cancer patients: high rate of complications with therapeutic implications. *Medicine (Baltimore)* 2007;86:54–60.
169. Amico L, Caplan LR, Thomas C. Cerebrovascular complications of mucinous cancers. *Neurology* 1989;39:522–526.
170. Deck JH, Lee MA. Mucin embolism to cerebral arteries: a fatal complication of carcinoma of the breast. *Can J Neurol Sci* 1978;5:327–330.
171. Prioleau PG, Katzenstein AL. Major peripheral arterial occlusion due to malignant tumor embolism: histologic recognition and surgical management. *Cancer* 1978;42:2009–2014.
172. Thompson T, Evans W. Paradoxical embolism. *Q J Med* 1930;23:135–150.
173. Hirose G, Kosoegawa H, Takado M, et al. Spinal cord ischemia and left atrial myxoma. *Arch Neurol* 1979;36:439.
174. O'Neill BP, DiNapoli RP, Okazaki H. Cerebral infarction as a result of tumor emboli. *Cancer* 1987;60:90–95.
175. Wada Y, Takahashi R, Yanagihara C, et al. Paradoxical cerebral embolism as the initial symptom in a patient with ovarian cancer. *J Stroke Cerebrovasc Dis* 2007;16:88–90.
176. Nakagawa Y, Tashiro K, Isu T, et al. Occlusion of cerebral artery due to metastasis of chorioepithelioma. Case report. *J Neurosurg* 1979;51:247–250.
177. Yamashita Y, Mukaida H, Hirabayashi N, et al. Cerebral air embolism after intrathoracic anticancer drug administration. *Ann Thorac Surg* 2006;82:1121–1123.
178. Carey RW, Harris N. Thrombotic microangiopathy in three patients with cured lymphoma. *Cancer* 1989;63:1393–1397.
179. Nordström B, Strang P. Microangiopathic hemolytic anemias (MAHA) in cancer. A case report and review. *Anticancer Res* 1993;13:1845–1850.

180. Dawson TM, Lavi E, Raps EC, et al. Thrombotic microangiopathy isolated to the central nervous system. *Ann Neurol* 1991;30:843–846.

181. Zakarija A, Bennett C. Drug-induced thrombotic microangiopathy. *Semin Thromb Hemost* 2005;31:681–690.

182. Qu L, Kiss JE. Thrombotic microangiopathy in transplantation and malignancy. *Semin Thromb Hemost* 2005;31:691–699.

183. Francis KK, Kalyanam N, Terrell DR, et al. Disseminated malignancy misdiagnosed as thrombotic thrombocytopenic purpura: a report of 10 patients and a systematic review of published cases. *Oncologist* 2007;12:11–19.

184. Walker RW, Rosenblum MK, Kempin SJ, et al. Carboplatin-associated thrombotic microangiopathic hemolytic anemia. *Cancer* 1989;64:1017–1020.

185. Murgo AJ. Thrombotic microangiopathy in the cancer patient including those induced by chemotherapeutic agents. *Semin Hematol* 1987;24:161–177.

186. Garewal M, Yahya S, Ward C, et al. MRI changes in thrombotic microangiopathy secondary to malignant hypertension. *J Neuroimaging* 2007; 17:178–180.

187. Hawley JS, Ney JP, Swanberg MM. Thrombotic thrombocytopenic purpura-induced posterior leukoencephalopathy in a patient without significant renal or hypertensive complications. *J Postgrad Med* 2004;50:197–199.

188. Werner TL, Agarwal N, Carney HM, et al. Management of cancer-associated thrombotic microangiopathy: what is the right approach? *Am J Hematol* 2007;82:295–298.

189. Liozon E, Loustaud V, Fauchais AL, et al. Concurrent temporal (giant cell) arteritis and malignancy: report of 20 patients with review of the literature. *J Rheumatol* 2006;33:1606–1614.

190. Hasler P, Kistler H, Gerber H. Vasculitides in hairy cell leukemia. *Semin Arthritis Rheum* 1995;25:134–142.

191. Rosen CL, DePalma L, Morita A. Primary angiitis of the central nervous system as a first presentation in Hodgkin's disease: a case report and review of the literature. *Neurosurgery* 2000;46:1504–1508.

192. Delobel P, Brassat D, Danjoux M, et al. Granulomatous angiitis of the central nervous system revealing Hodgkin's disease. *J Neurol* 2004; 251:611–612.

193. Wooten MD, Jasin HE. Vasculitis and lymphoproliferative diseases. *Semin Arthritis Rheum* 1996;26:564–574.

194. Martinez-Taboada V, Brack A, Hunder GG, et al. The inflammatory infiltrate in giant cell arteritis selects against B lymphocytes. *J Rheumatol* 1996; 23:1011–1014.

195. Wynne PJ, Younger DS, Khandji A, et al. Radiographic features of central nervous system vasculitis. *Neurol Clin* 1997;15:779–804.

196. Abu-Shakra M, Khraishi M, Grosman H, et al. Primary angiitis of the CNS diagnosed by angiography. *Q J Med* 1994;87:351–358.

197. Ponzoni M, Ferreri AJ, Campo E, et al. Definition, diagnosis, and management of intravascular large B-cell lymphoma: proposals and perspectives from an international consensus meeting. *J Clin Oncol* 2007;25:3168–3173.

198. Ponzoni M, Ferreri AJ. Intravascular lymphoma: a neoplasm of "homeless" lymphocytes? *Hematol Oncol* 2006;24:105–112.

199. Harris CP, Sigman JD, Jaeckle KA. Intravascular malignant lymphomatosis: amelioration of neurological symptoms with plasmapheresis. *Ann Neurol* 1994;35:357–359.

200. Kawahara M, Kanno M, Matsumoto M, et al. Diffuse neurodeficits in intravascular lymphomatosis with ADAMTS13 inhibitor. *Neurology* 2004;63:1731–1733.

201. Mandal AK, Savvidou L, Slater RM, et al. Angiotropic lymphoma: associated chromosomal abnormalities. *Eur J Intern Med* 2007;18:432–434.

202. Dubas F, Saint-Andre JP, Pouplard-Barthelaix A, et al. Intravascular malignant lymphomatosis (so-called malignant angioendotheliomatosis): a case confined to the lumbosacral spinal cord and nerve roots. *Clin Neuropathol* 1990;9:115–120.

203. Beristain X, Azzarelli B. The neurological masquerade of intravascular lymphomatosis. *Arch Neurol* 2002;59:439–443.

204. Iijima M, Fujita A, Uchigata M, et al. Change of brain MRI findings in a patient with intravascular malignant lymphomatosis. *Eur J Neurol* 2007;14:e4–e5.

205. Lui PC, Wong GK, Poon WS, et al. Intravascular lymphomatosis. *J Clin Pathol* 2003;56:468–470.

206. Ganguly S. Acute intracerebral hemorrhage in intravascular lymphoma: a serious infusion related adverse event of rituximab. *Am J Clin Oncol* 2007;30:211–212.

207. Sanchez S, Ewton A. Essential thrombocythemia: a review of diagnostic and pathologic features. *Arch Pathol Lab Med* 2006;130:1144–1150.

208. Tefferi A. Classification, diagnosis and management of myeloproliferative disorders in the JAK2V617F. *Hematol Am Soc Hematol Educ Program* 2006;240–245.

209. Tefferi A, Thiele J, Orazi A, et al. Proposals and rationale for revision of the World Health Organization diagnostic criteria for polycythemia vera, essential thrombocythemia, and primary myelofibrosis: recommendations from an ad hoc international expert panel. *Blood* 2007;110:1092–1097.

210. Buss DH, Stuart JJ, Lipscomb GE. The incidence of thrombotic and hemorrhagic disorders in association with exreme thrombocytosis: an analysis of 129 cases. *Am J Hematol* 1985;20:365–372.

211. Mitchell D, Fisher J, Irving D, et al. Lateral sinus thrombosis and intracranial hypertension in essential thrombocythaemia. Letter to the editor. *J Neurol Neurosurg Psychiatry* 1986;49:218–219.

212. Gisslinger H. Update on diagnosis and management of essential thrombocythemia. *Semin Thromb Hemost* 2006;32:430–436.

213. Penninga EI, Bjerrum OW. Polycythaemia vera and essential thrombocythaemia: current treatment strategies. *Drugs* 2006;66:2173–2187.

214. Hickey WF, Garnick MB, Henderson IC, et al. Primary cerebral venous thrombosis in patients with cancer—a rarely diagnosed paraneoplastic syndrome. Report of three cases and review of the literature. *Am J Med* 1982;73:740–750.

215. Kyritsis AP, Williams EC, Schutta HS. Cerebral venous thrombosis due to heparin-induced thrombocytopenia. *Stroke* 1990;21:1503–1505.

216. Sigsbee B, Deck MD, Posner JB. Nonmetastatic superior sagittal sinus thrombosis complicating systemic cancer. *Neurology* 1979;29:139–146.

217. Hagner G, Iglesias-Rozas JR, Kolmel HW, et al. Hemorrhagic infarction of the basal ganglia. An unusual complication of acute leukemia. *Oncology* 1983;40:387–391.

218. Plant GT, Donald JJ, Jackowski A, et al. Partial, non-thrombotic, superior sagittal sinus occlusion due to occipital skull tumours. *J Neurol Neurosurg Psychiatry* 1991;54:520–523.

219. Kim AW, Trobe JD. Syndrome simulating pseudotumor cerebri caused by partial transverse venous sinus obstruction in metastatic prostate cancer. *Am J Ophthalmol* 2000;129:254–256.

220. Goldsmith P, Burn DJ, Coulthard A, et al. Extrinsic cerebral venous sinus obstruction resulting in intracranial hypertension. *Postgrad Med J* 1999;75:550–551.

221. Lopez-Pelaez MF, Millan JM, de Vergas J. Fatal cerebral venous sinus thrombosis as major complication of metastatic cervical mass: computed tomography and magnetic resonance findings. *J Laryngol Otol* 2000;114:798–801.

222. Brown MT, Freidman HS, Oakes WJ, et al. Saggittal sinus thrombosis and leptomeningeal medulloblastoma: resolution without anticoagulation. *Neurology* 1991;41:455–456.

223. Ferro JM, Canhao P, Stam J, et al. Prognosis of cerebral vein and dural sinus thrombosis: results of the International Study on Cerebral Vein and Dural Sinus Thrombosis (ISCVT). *Stroke* 2004;35:664–670.

224. Nowak-Gottl U, Heinecke A, von Kries R, et al. Thrombotic events revisited in children with acute lymphoblastic leukemia: impact of concomitant *Escherichia coli* asparaginase/prednisone administration. *Thromb Res* 2001;103:165–172.

225. Wermes C, Fleischhack G, Junker R, et al. Cerebral venous sinus thrombosis in children with acute lymphoblastic leukemia carrying the MTHFR TT677 genotype and further prothrombotic risk factors. *Klin Padiatr* 1999;211:211–214.

226. Nowak-Gottl U, Wermes C, Junker R, et al. Prospective evaluation of the thrombotic risk in children with acute lymphoblastic leukemia carrying the MTHFR TT 677 genotype, the prothrombin G20210A variant, and further prothrombotic risk factors. *Blood* 1999;93:1595–1599.

227. Nowak-Gottl U, Ahlke E, Fleischhack G, et al. Thromboembolic events in children with acute lymphoblastic leukemia (BFM protocols): prednisone versus dexamethasone administration. *Blood* 2003;101:2529–2533.

228. Harvey CJ, Peniket AJ, Miszkiel K, et al. MR angiographic diagnosis of cerebral venous sinus thrombosis following allogeneic bone marrow transplantation. *Bone Marrow Transplant* 2000; 25:791–795.

229. Reuner KH, Jenetzky E, Aleu A, et al. Factor XII C46T gene polymorphisms and the risk of cerebral venous thrombosis. *Neurology* 2008;70:129–132.

230. Stam J, de Bruijn S, deVeber G. Anticoagulation for cerebral sinus thrombosis. *Stroke* 2003;34:1054–1055.

231. Stam J, de Bruijn SF, deVeber G. Anticoagulation for cerebral sinus thrombosis. *Cochrane Database Syst Rev* 2002:CD002005.

232. Einhaupl K, Bousser MG, de Bruijn SF, et al. EFNS guideline on the treatment of cerebral venous and sinus thrombosis. *Eur J Neurol* 2006;13:553–559.

233. Philips MF, Bagley LJ, Sinson GP, et al. Endovascular thrombolysis for symptomatic cerebral venous thrombosis. *J Neurosurg* 1999;90:65–71.

234. Ferro JM, Canhao P. Acute treatment of cerebral venous and dural sinus thrombosis. *Curr Treat Options Neurol* 2008;10:126–137.

235. Piccioli A, Falanga A, Baccaglini U, et al. Cancer and venous thromboembolism. *Semin Thromb Hemost* 2006;32:694–699.

236. Prandoni P, Falanga A, Piccioli A. Cancer and venous thromboembolism. *Lancet Oncol* 2005;6:401–410.

237. Sallah S, Wan JY, Nguyen NP. Venous thrombosis in patients with solid tumors: determination of frequency and characteristics. *Thromb Haemost* 2002;87:575–579.

238. Semrad TJ, O'Donnell R, Wun T, et al. Epidemiology of venous thromboembolism in 9489 patients with malignant glioma. *J Neurosurg* 2007;106:601–608.

239. Sawaya R, Zuccarello M, Elkalliny M, et al. Postoperative venous thromboembolism and brain tumors. Part I. Clinical profile. *J Neuro-Oncol* 1992;14:119–125.

240. Feldmann E, Posner JB. Episodic neurologic dysfunction in patients with Hodgkins disease. *Arch Neurol* 1986;43:1227–1233.

241. Levy DE. Transient CNS deficits: a common, benign syndrome in young adults. *Neurology* 1988;38:831–836.

242. Fisher CM. Late-life migraine accompaniments as a cause of unexplained transient ischemic attacks. *Can J Neurol Sci* 1980;7:9–17.

243. Sugarbaker EV, Chretien PB. A tumor shunt syndrome. Transient cerebral ischemia induced by a large thyroid adenoma. *Arch Surg* 1972;104:213–215.

244. Mirski MA. Diagnosis and treatment of vascular air embolism. *Anesthesiology* 2007;106:164–177.

245. Vesely TM. Air embolism during insertion of central venous catheters. *J Vasc Interv Radiol* 2001;12:1291–1295.

246. Heckmann JG. Neurologic manifestations of cerebral air embolism as a complication of central venous catheterization. *Crit Care Med* 2000;28:1621–1625.

247. Glover P. Fat embolism. *Crit Care Resuscitation* 1999;1:276–284.

248. Simon AD. Contrast-enhanced MR imaging of cerebral fat embolism: case report and review of the literature. *Am J Neuroradiol* 2003;24:97–101.

Chapter 10

Central Nervous System Infections

INTRODUCTION

Central nervous system (CNS) infections, now relatively uncommon complications of cancer, have been decreasing in frequency over the years as a result of earlier and more vigorous antibiotic use in patients with or at risk for (prophylactic antibiotics are routine in many settings) systemic infection[1] and the use of colony-stimulating factors to treat neutropenia. From 1962 to 1981, we encountered 29 patients with neutropenic meningitis, but from 1993 to 2004, we encountered only 10. However, as the incidence of meningitis due to immunosuppression has decreased at Memorial Sloan-Kettering Cancer Center (MSKCC), the incidence of meningitis resulting from neurosurgical procedures has increased.

Between 1955 and 1971, 30 (30%) of 104 patients with meningitis developed the infection as a result of neurosurgical manipulation.[2] Between 1993 and 2004, 58 of 77 infections occurred in patients with Ommaya reservoirs, ventriculoperitoneal shunts, or hardware placed during surgical procedures. These statistics are the result of an increase in the neurosurgical procedures, not in the proportion of post-operative infections. Not all of these infections were a result of the procedure; many were caused by subsequent use of the Ommaya reservoir.

Despite their decrease, CNS infections are more common in patients with cancer than in the general population and are more challenging to treat. In 1970, 0.2% of all admissions to MSKCC were for CNS infections (i.e.,

369

positive cerebrospinal fluid [CSF] cultures).[2] Beginning in 1971, the number of infections decreased, so that by 1974 the incidence was approximately 0.05%.[3] Despite these improvements, CNS infections can be a problem. At the Massachusetts General Hospital, between 1985 and 1990, one of every six patients with a CNS infection had an underlying malignancy.[4]

Patients with specific types of cancer are more likely than others to develop a CNS infection. Approximately 85% of intracranial infections occur in the 12% of patients hospitalized with lymphoma, acute leukemia, bone marrow or stem cell transplants, or a surgically induced communication between the subarachnoid space and the surface of the body. CNS infections other than herpes zoster strike 2.7% of patients with Hodgkin disease, 0.5% of patients with non–Hodgkin lymphoma, and 2.5% of patients with chronic lymphocytic leukemia (CLL),[5] although, as indicated earlier, these are old data and may not be valid anymore.

The diagnosis of a CNS infection in a patient with cancer can be difficult. The usual florid symptoms apparent in immunocompetent patients with CNS infection are often absent in those with cancer. Instead of severe headache, high fever, nuchal rigidity, and focal neurologic signs, patients may be afebrile or have only a slight increase in body temperature. Increasing confusion or lethargy may be the only additional sign to suggest a CNS infection.[5,6] In one series[7] of 13 patients with cryptococcal meningitis, only 8 had headache, 2 had nuchal rigidity, and 2 had fever. The diagnosis of brain abscess may be even more difficult because headache and fever are characteristically absent and focal neurologic signs may be absent as well. Even the MRI may not distinguish a brain abscess from a tumor[8,9] (see section on Brain Abscess).

The organisms causing CNS infection in the cancer patient are common in the environment, but are uncommon causes of infection in the normal host.[10,11] Cryptococcal and listerial organisms are the major causes of meningitis in patients with cancer, except those who have had a neurosurgical procedure such as a ventriculoperitoneal shunt or ventricular catheters where staphylococcal and *Propionibacterium* infections are common. This change has occurred in our institution as the incidence of non-neurosurgical CNS infections has decreased and the incidence of neurosurgical

procedures has increased. *Streptococcus pneumoniae*, the number one cause of meningitis in normal hosts, is very rare in our institution. *Toxoplasma* and *Aspergillus* organisms are the major causes of brain abscess. *Haemophilus influenzae*, *Neisseria meningitidis*, and surprisingly, *Mycobacterium tuberculosis* rarely cause CNS infections; systemic tuberculosis is increased in foreign-born patients, especially those with hematologic malignancies. With the exception of head and neck cancers, American-born patients with solid tumors did not have an increased incidence of systemic tuberculosis.[12]

Because most CNS infections in patients with cancer result from an abnormality of host defense mechanisms, and because the particular abnormality is often characteristic of the underlying cancer or its treatment, the clinician can often predict with reasonable confidence the likely invading organism(s).[2,6,11,13,14] For example, acute meningitis in a patient with Hodgkin disease with a normal white blood cell (WBC) count is likely to result from a *Listeria* or *Cryptococcus* infection. If that patient has undergone a splenectomy in the past, *Streptococcus pneumoniae* is also a likely cause of meningitis, even in vaccinated patients.[15,16] If, however, that same patient has been under intensive chemotherapy and the absolute neutrophil count is fewer than 1000/mm[3], Gram-negative organisms such as *Pseudomonas aeruginosa* or *Escherichia coli* are likely offenders.[2]

Once the organism is known, vigorous antibiotic treatment often eradicates the infection. However, the same host factors that led to the patient becoming infected may make treatment difficult; relapse and superinfection are frequent. In one study,[5] in 19 (39%) of 49 patients with CNS infections associated with systemic infections, the CNS was infected by different organisms from those infecting the rest of the body, or multiple organisms infected the CNS, either simultaneously or sequentially.

On the following pages, we discuss the pathophysiology, diagnostic approach, and common causes of CNS infection in patients with cancer. Many considerations that apply to cancer-associated infections also apply to infections complicating other immune disorders such as organ and bone marrow transplantation, autoimmune diseases being treated with immunosuppressive drugs, and the immunosuppression caused by AIDS. Several reviews address CNS infections in patients with cancer[5,11] and CNS infections in general.[17]

PATHOPHYSIOLOGY OF CENTRAL NERVOUS SYSTEM INFECTION

Host Defenses

PHYSICAL BARRIERS

For an offending organism to reach the nervous system, it must penetrate two barriers, one between the body's surface and its interior and the second between the body's interior and the CNS (Figs. 10–1 and 10–2). The first barrier includes skin, mucous membranes, and epithelium of the gastrointestinal (GI) tract, and respiratory and genitourinary systems. Several factors affecting patients with cancer make it easier for organisms to penetrate the first barrier.

Tumors, wounds, or mucosal erosion (e.g., from chemotherapy) may disrupt the integrity of these barriers. Even in the absence of disruption of mucosa or epithelium, bacteria can reach the bloodstream.[19] For example, pneumococci, bacteria that reside in the nasopharynx, bind a polymeric immunoglobulin that translocates them across the nasopharyngeal mucosa.[20] Previous antimicrobial therapy may have altered normal flora, enabling organisms to penetrate the barrier better. The cell's normal ciliary action may be damaged by chemotherapy, preventing mechanical clearing of the organisms.

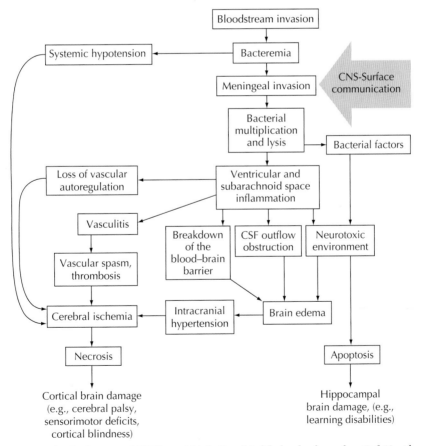

Figure 10–1. Steps in the development of CNS bacterial infection. (Modified and redrawn from Ref. 18 with permission.)

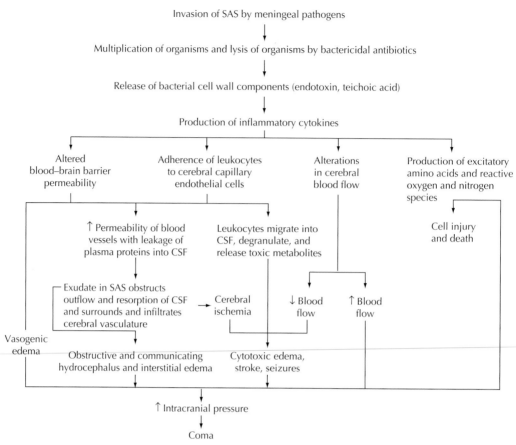

Figure 10–2. The pathophysiology of neurologic complications of bacterial meningitis. This flow diaphragm, courtesy of Dr. Karen Roos from *Harrison's Principles of Internal Medicine* (with permission), illustrates how, once organisms reach the SAS, the organisms themselves and the immunological response of the host lead to a variety of biochemical and structural abnormalities that can culminate in coma. Abbreviations: SAS, subarachnoid space; CSF, cerebrospinal fluid.

Diagnostic or therapeutic procedures such as indwelling venous, epidural (for pain), or urethral catheters provide a conduit for organisms to bypass normal barriers.

The CNS–surface barrier may be penetrated directly by surgery on the head or spine that creates a direct communication between the surface and the CSF, but more often the invading organisms reach the CNS via the arterial circulation. Bacteria reach the CNS from the blood in different ways[18,19] (Fig. 10–3). For example, pneumococci bind a receptor for platelet-activating factor on brain capillary endothelium and are transcytosed into the CNS, dependent on the presence of pneumococcal choline-binding protein A.[21] *Escherichia coli* proteins interact with endothelial receptors present on brain microvasculature but not on systemic vascular endothelial cells.[22] Because organisms that

penetrate surface barriers enter the venous circulation before the arterial circulation, they are likely to infect liver or lungs before the brain. Thus, brain abscesses caused by *Nocardia* or *Aspergillus* organisms usually develop after similar lesions have formed in the lung and can be identified on chest imaging; these lesions are hypermetabolic on the PET/CT (positron emission tomography/computed tomography), similar to malignant tumors. Blood cultures are commonly positive in patients with cancer suffering from bacterial meningitis, but not meningitis caused by fungal organisms. Nevertheless, as already mentioned, in approximately 40% of patients, the organisms infecting the CNS differ from those causing concomitant systemic infection[5]; sometimes the organism reaches the CNS via the bloodstream without evidence of infection elsewhere in the body.

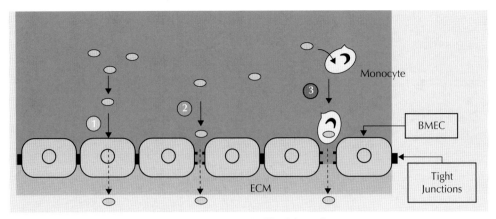

Figure 10–3. Cellular pathways for microbial invasion of the blood–brain barrier. Microorganisms may penetrate the blood–brain barrier and enter the central nervous system by at least one of the following three pathways: (1) transcellular invasion of brain microvascular endothelial cells (BMEC); (2) paracellular entry after disruption of the tight junctions of the blood–brain barrier; (3) transport within circulating phagocytic cells such as monocytes (Trojan horse mechanism). Abbreviation: ECM, extracellular matrix. (From Ref. 44 with permission.)

IMMUNOLOGIC DEFENSES

Even when epithelial barriers are penetrated, normal host defenses usually eradicate the organisms. In patients with cancer, the following abnormalities of host defense mechanisms predispose the patient to CNS infection[2,11,13] (Table 10–1).

Impaired cellular immunity from T-cell and mononuclear phagocyte abnormalities may occur as a result of the cancer or its treatment. This defense mechanism is particularly impaired in patients with lymphoma, especially those with Hodgkin disease, and in patients receiving corticosteroids or other cytotoxic drugs.

Impaired neutrophil function (absolute neutropenia of <1000 cells/mm³) is common in patients with acute leukemia and during the chemotherapy of many cancers.[23] Qualitative abnormalities of neutrophil activity can also follow chemotherapy even when neutrophils are not quantitatively depressed[24–26]; examples include steroid therapy and several specific diseases of neutrophil function.[27] The absence of neutrophils may prevent the CNS from mounting a cellular response to the infection, causing a falsely reassuring acellular CSF.[6] Appropriate use of colony-stimulating factors considerably ameliorates this problem.[28]

Low or dysfunctional immunoglobulin production occurs in patients with multiple myeloma, CLL, or congenital hypogammaglobulinemia. Such patients cannot mount an immunoglobulin response to an antigenic challenge.[29] A similar phenomenon often occurs following combined radiation therapy (RT) and multidrug chemotherapy or following stem cell transplantation. Opsonizing and bactericidal antibodies decline and susceptibility to infection with encapsulated bacteria increases.

Splenectomy was once part of the treatment of some patients with Hodgkin disease.[29,30] The result was low immunoglobulin production in response to an antigenic challenge, inability to produce immunoglobulin M (IgM)-opsonizing antibodies, and loss of the spleen as a filter of organisms.[31] The combined effects heightened vulnerability to encapsulated bacterial organisms, particularly *Pneumococcus*.

An anatomic communication between the CNS and an epithelial surface occurs in patients with tumors of the skull or spine, either when the tumor erodes into the CNS or a surgical procedure creates a fistula between the CNS and the surface. Deliberate disruption of the barrier is more common, as when a ventricular cannula is placed for the injection of chemotherapeutic agents or a ventriculoperitoneal shunt is used to divert obstructed CSF (see Chapter 14).

MULTIPLE DEFECTS

Many patients with cancer suffer a multiplicity of immune system and/or anatomic abnormalities. For example, a splenectomized patient

Table 10–1. Host Defense Abnormalities in Patients with Cancer

Major Host Defect	Usual Cancers	Other Risk Factors
T-lymphocyte, mononuclear phagocyte defects	Hodgkin disease Non–Hodgkin lymphoma Chronic lymphocytic leukemia	Steroids Alkylating agents Antimetabolites Antitumor antibiotics
Neutrophil defects (granulocytopenia)	Acute leukemias	Chemotherapy Solid tumors receiving chemotherapy
Abnormal immunoglobulins	Chronic lymphocytic leukemia	Steroids
Splenectomy	Multiple myeloma; chronic myelogenous leukemia	Hodgkin disease; hairy cell leukemia
Communication between CSF and surface	Spinal column tumors Skull tumors Head and neck cancers	Head and spine surgery Cerebroventricular reservoirs, shunts

with Hodgkin disease who is being treated with chemotherapeutic agents may have abnormalities of both T- and B-cell function as well as neutropenia. Many agents used for cancer treatment have multiple effects on the immune system. Corticosteroids are probably the most frequent and serious offenders encountered in neuro-oncology. They suppress antibody production, decrease both the acute and chronic inflammatory response, interfere with T cell–mediated immunity, reduce interferon production, impair wound healing, and interfere with the clearance of foreign materials. They also attenuate the inflammation of meningitis[32] and appear to interfere with the entry of antibiotics into the CSF.[33] However, corticosteroids, probably because they suppress inflammation, prevent some of the long-term adverse effects of CNS infection and constitute appropriate treatment for patients with bacterial meningitis.[34] Some cytotoxic drugs also have such multiple effects.

Infection Sites within the Central Nervous System

Compartmentalization of infection within the nervous system is common (Table 10–2).

Some organisms, such as *Listeria* and *Cryptococcus*, usually attack only the leptomeninges, causing either acute or chronic meningitis, but can occasionally cause encephalitis, cerebritis, or brain abscesses.[35–41] Organisms, such as *Nocardia* and *Toxoplasma*, settle focally within the CNS parenchyma, causing focal encephalitis or brain abscesses. Organisms such as varicella-zoster virus invade the brain diffusely; they cause diffuse encephalitis with or without an accompanying meningitis. Finally, some organisms with a predilection to thrombose blood vessels of the brain (e.g., *Aspergillus*) can cause either cerebral infarction or hemorrhage. Varicella-zoster can also cause a vasculitis without other evidence of CNS infection.

Secondary involvement of the cerebral vasculature in infection is also common. Even without an inflammatory response, necrosis and occlusion of small vessels within the brain sometimes occur with acute bacterial meningitis[6] (Fig. 10–4). Vasculitis may narrow or occlude major arteries at the base of the brain, small vessels within the brain parenchyma, sulcal veins, and venous sinuses. Patients with meningitis who show such abnormalities on an angiogram have an unfavorable prognosis.[42]

MENINGITIS

In the absence of a direct communication between the CSF and the surface, bacteria must reach the leptomeninges and cause meningitis by crossing the blood–brain/CSF barriers, either at the choroid plexus (e.g., *H. influenzae*) or across brain endothelium (e.g., *S. pneumoniae*).[43] Pathogens may cross

Table 10–2. **Infection Sites within the Central Nervous System and Resultant Complications**

Primary Site	Infections	Causal Organisms	Complications
Leptomeninges	Meningitis	*Listeria*	Hydrocephalus (± intracranial hypertension), arachnoiditis
		Cryptococcus	Cranial nerve palsies; vasculitis (infarcts)
Brain parenchyma	Focal cerebritis	*Nocardia*	Cerebral edema, herniation, dementia
	Abscess	*Toxoplasma*	
Cerebral cortex ± meninges	Encephalitis, meningoencephalitis	Varicella-zoster, herpes simplex	Seizures
Cerebral blood vessels	Vasculitis	Varicella-zoster	Cerebral infarction or hemorrhage
		Aspergillus *Mucorales*	

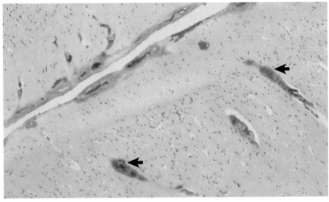

Figure 10–4. Bacterial meningitis causing cerebral vasculitis. Photomicrograph of a brain section from a neutropenic patient who died of Gram-negative meningitis. The cortical section shows a bacterial vasculopathy. The cells surrounding the blood vessels are all bacteria; white cells were absent in the field. The patient died despite antibiotic treatment.

between endothelial or ependymal cells by disrupting tight junctions, cross through the cell by active or passive transcytosis, or traverse the barrier together with or inside of leukocytes (Trojan horse mechanism).[43] Current evidence suggests that the major CNS pathogens in cancer patients, including *Listeria, S. pneumoniae,* and *Cryptococcus* use the transcellular mechanism[44] (Fig. 10–3).

Once in the CNS, bacteria can multiply rapidly because many host defense mechanisms found elsewhere in the body, such as complement and immunoglobulins, are present only at very low levels in CSF. Bacteria release inflammatory and toxic compounds by autolysis and secretion.[19] Pneumococci mediate brain damage both by eliciting an inflammatory host response (less vigorous in the immunocompromised host) and by release of substances that cause neuronal and microglial apoptosis.[45]

Bacteria reaching the subarachnoid space evoke hyperemia in meningeal vessels, leading to the migration of polymorphonuclear leukocytes (if the patient has any) into the subarachnoid space.[43] The WBCs and the engulfed invading organisms form an exudate that surrounds the blood vessels in the Virchow–Robin spaces, but the organisms rarely enter into the brain parenchyma except in areas where a concomitant vasculitis has led to infarction. Nevertheless, the brain is not spared in meningitis. The exudate at the base of the brain may impede CSF absorption, causing hydrocephalus with increased intracranial pressure. Vasculitis produced by the organism may cause cerebral infarction with secondary abscess formation.[42] Stenosis of cerebral vessels may be identified by magnetic resonance angiography or transcranial Doppler, although the two techniques are not always in agreement.[46,47] Toxic substances secreted by the organisms and the WBCs responding to them may cause cerebral edema[48] or other forms of CNS damage.[49,50] The combined insults occasionally cause cerebral herniation and death.[51,52]

The combined bacterial invasion and inflammatory response, particularly the production of cytokines and chemokines, have several deleterious consequences. As a result, even those patients who survive are at high risk (35%) for suffering neurologic sequelae such as hearing loss, cognitive abnormalities, or focal brain deficits.[53] Both reactive oxygen (e.g., superoxide, hydrogen peroxide) and reactive nitrogen species (nitric oxide) are important in mediating the meningitis-associated brain damage.[54] Some recent evidence indicates that lipid peroxidation and poly(ADP-ribose) polymerase (PARP), an enzyme released by damaged DNA, may play an important role in brain damage.[53] The blood–brain barrier breaks down, probably mediated by matrix metalloproteinases,[43] as both tight junctions open and vesicular transport increases, leading to vasogenic brain edema.[55] However, cytotoxic edema, probably mediated by upregulation of aquaporin-4 (see Chapter 3), may play an even more significant role.[56] An initial increase in cerebral blood flow is followed by a decrease.[42] The combination of brain edema and increased blood flow causes intracranial hypertension and can lead to cerebral herniation. Autoregulation of cerebral blood flow ceases,[57,58] so that the blood flow rate varies passively with the systemic blood pressure. Thus, hypertension increases blood flow, causing cerebral edema, whereas hypotension may lead to focal or generalized cerebral ischemia and sometimes infarction. An increase in resistance to CSF absorption across the arachnoid granulations contributes to the increased intracranial pressure and perhaps causes the late development of hydrocephalus.[59,60] Many of these changes are believed to be cytokine or chemokine related and can be prevented or ameliorated by corticosteroids. Thus, although the clinician may be reluctant to use corticosteroids in an infected and immunosuppressed patient, judicious use can prevent permanent brain damage associated with bacterial meningitis.[34]

CEREBRITIS AND BRAIN ABSCESS

Brain abscess[61,62] and cerebritis (a focal area of inflammation that is not walled off as an abscess)[62,63] differ from meningitis because of the focal parenchymal involvement. Like meningitis, most of the organisms causing them are blood-borne and lodge in small distal vessels at the gray matter–white matter interface, similar to blood-borne tumor emboli (see Chapter 2). The organisms and accompanying WBCs breach the vascular wall and cause an area of local infection or cerebritis. The brain responds with an intense vascular proliferation; the vigor of this response depends on the type of organism and the patient's immune status. The neovascularization at the edge of the cerebritis does not possess a normal blood–brain barrier, so a ring of enhancement can be seen on a magnetic resonance imaging (MRI) or CT scan. The ring enhancement does not necessarily mean that a capsule has formed. With time, a collagenous capsule does form, effectively walling off the abscess from the normal brain. The capsule is characteristically thicker near the cortex, where oxygenation is better, and somewhat thinner near the ventricular surface.[62] Thus, abscess rupture, when it occurs, is usually into the ventricle rather than into the subarachnoid space. Spinal epidural abscess can result either from contiguous infection of vertebrae or from blood-borne organisms.[64]

MENINGOENCEPHALITIS

In encephalitis or meningoencephalitis, the invading organism may affect the entire brain and the leptomeninges as well.[65] The pathologic picture depends on the nature of the organism and the patient's immune response. The inflammatory response of the leptomeninges and the cortex may be widespread, with microabscess formation, but if the organisms invade without a response by the patient's immune system, little inflammation may occur.

CLINICAL FINDINGS

Signs and Symptoms

The signs and symptoms of CNS infection in patients with cancer (Table 10–3) are determined by the lesion's anatomic site (leptomeninges, brain parenchyma, or both), the virulence of the infecting organism, and the host's reaction to the infection. In general, the signs and symptoms are less fulminant, less severe, and often evolve more slowly than in the immunocompetent patient. The infections are also more likely to be lethal because the immune system of the patient with cancer is usually compromised. The most common symptoms are as follows:

Headache occurs in some but not all patients with meningitis, encephalitis, and brain abscesses.[2,6] Because the headache may be mild, patients suffering other pain from their cancer may not notice or volunteer the complaint. To complicate the issue, many patients are febrile from the cancer or systemic infection, and the fever itself can cause mild-to-moderate headache.

A body temperature greater than 38.5°C (101.5°F) is the most common sign of CNS infection.[6] However, *fever* may be absent, particularly with brain abscess. Many patients with meningitis and meningoencephalitis are already febrile from systemic infection. An increase in an already present fever, when accompanied by headache, suggests CNS infection.

Personality change is often the only clue to CNS infection.[5,6] Behavioral changes may be mild, characterized only by lethargy and irritability, or severe, characterized by delirium, stupor or even coma. Because such patients are often severely ill systemically, mild changes in behavior may pass unnoticed.

Seizures are common in some series of patients but uncommon in others.[66] They were present in only 7 of 55 infections in immunocompromised hosts[5] and in none of 15 patients with cryptococcal meningitis in an MSKCC series.[7] In another series,[6] 40% of neutropenic patients with meningitis had seizures. When seizures do occur, they are as likely to be a manifestation of meningitis or meningoencephalitis as of brain abscess.

Nuchal rigidity is surprisingly uncommon in patients with CNS infection, even those who have meningitis. In one series[5] of immunocompromised patients, only 6 of 10 patients with *Listeria* meningitis, and only 4 of 11 with meningitis caused by more conventional bacteria, had nuchal rigidity.

Focal neurologic signs, such as hemiparesis or aphasia, are more common in cerebritis and brain abscess than in meningitis but can occur in the latter as a result of cerebral infarction from vasculitis.[42,67] Rarely, vasculitis may lead to aneurysm formation with subarachnoid hemorrhage.[68] The particular focal neurologic signs in a given patient depend on the locus of the lesion. In patients with meningoencephalitis, in which the infection is diffuse, focal signs are often mild and fluctuating; they more closely resemble the neurologic findings in disseminated intravascular coagulation (see Chapter 9) than in metastasis.

The evolution of symptoms and signs is usually subacute or chronic rather than acute. For example, in 6 of 10 patients with *Listeria* meningitis, symptoms evolved over 2 to 10 days and, in two patients, over more than 10 days.[5] Only two patients developed their symptoms in less than 48 hours. Slow evolution is particularly common in cryptococcal or toxoplasma infection. Unless seizures or cerebral infarction causes acute symptoms, the course may be indolent, even in patients infected with organisms that cause acute symptoms in the normal host.

Meningitis

The clinical picture of meningitis in patients with cancer varies from an acute, fulminant

---

Table 10–3. Signs and Symptoms of Central Nervous System Infections in Patients with Cancer

Frequent
Headache: typically mild to moderate
Fever: new or increased in an already febrile
    patient
Personality changes, delirium
Seizures

Uncommon
Nuchal rigidity
Focal neurologic deficit

infection with headache, fever, stiff neck, and stupor to an indolent infection causing only malaise at the time of discovery by CSF examination.[6] The diagnosis is easiest in a patient with lymphoma who is in remission and not under active treatment but presents with sudden headache, fever, stiff neck, and delirium. If the patient has not had a splenectomy, the most likely cause is *Listeria* organisms; if the patient has had a splenectomy, the cause is either *Listeria* or *S. pneumoniae*. Even the pneumococcal vaccine does not always prevent pneumococcal infection.[69] Generally speaking, pneumococcal infections are usually more fulminant, evolving over just a few hours. *Listeria* infections tend to evolve over days, although by the time the patient reaches the hospital, both infections may be equally severe. A less likely diagnosis is cryptococcal meningitis; although this is a common cause of meningitis in the lymphoma population, the symptoms tend to evolve more slowly. A very rare cause of acute meningitis in lymphoma patients is toxoplasmosis.

The diagnosis is more difficult in patients under active treatment for cancer. Such patients are frequently symptomatic from the underlying disease or its treatment when CNS infection develops, so that early symptoms are often attributed to the underlying malignancy rather than to a new and superimposed illness. The major manifestations of meningitis, however, are usually the same and only partially masked by the systemic cancer. Most patients who are alert enough to give an adequate history complain of headache. Fever is almost always present. However, corticosteroids given as part of the therapy for cancer or cancer symptoms may prevent fever from occurring. The onset of meningitis is usually heralded by a fever spike of 0.5°C to 1.0°C (1–2°F) higher than the patient's baseline temperature. Most patients are delirious, and many become stuporous or comatose within the first 24 hours of symptoms. Nuchal rigidity is frequently absent, particularly in patients with low peripheral WBC counts, possibly because the meninges do not mount an adequate inflammatory response to the invading organisms.

GENERAL PHYSICAL EXAMINATION

Evidence of infection elsewhere in the body may suggest not only the source but also the nature of a CNS infection. For instance, most patients with *Nocardia*[70] or *Aspergillus meningitis*[71] have evident pulmonary disease when symptoms of meningitis develop. Patients with *Candida meningitis*, an extremely rare problem at our institution, usually have evidence of candidiasis elsewhere, such as a nodular rash. Gram-negative organisms generally cause septicemia before meningitis develops, unless a fistula connects the CSF to the surface.

NEUROLOGIC EXAMINATION

The neurologic examination rarely helps with the diagnosis. Sometimes the optic fundi reveal a suggestion of infection in the form of Roth spots.[72] However, these retinal hemorrhages with a white center are nonspecific and are found in many embolic and thrombotic disorders including infection.[65,73] Cranial nerve palsies, lower motor neuron weakness, and absent reflexes are uncommon. When they are present, the clinician should suspect that the meninges have been invaded by tumor rather than by organisms (see Chapter 7).

DIAGNOSIS

Lumbar Puncture

The diagnosis of meningitis is established by examining the CSF.[74,75] The quandary is when to perform a lumbar puncture in a patient with cancer suspected of CNS infection (Table 10–4).

CSF examination is essential for appropriate diagnosis and treatment, but if the neurologic symptoms result from a mass lesion, a lumbar puncture may do more harm than good. Rarely, a lumbar puncture performed in a patient without a mass lesion but with meningitis,[51] or even pseudotumor cerebri, causes death from herniation of a swollen brain. The decision to perform a lumbar puncture should be based on the acuteness and nature of the patient's symptoms. The problem usually arises in a patient with the acute onset of headache, delirium, and fever, in whom the suspicion of acute meningitis is strong. If the patient is suffering from symptoms of acute meningitis, the time between onset of symptoms and death may be only a few hours if treatment is not begun immediately, making it dangerous to wait for a brain CT before starting treatment. Some suggest drawing blood for culture, starting antibiotics believed appropriate for the clinical situation, procuring a CT, and then performing a lumbar puncture. Although this technique

Table 10–4. **Considerations for Performing a Lumbar Puncture in Patients with Cancer**

Indications for Test
Absolute
  Suspicion of meningitis
Relative
  Suspicion of nervous system disease, especially encephalitis or meningoencephalitis
  Intrathecal therapy for meningeal leukemia or fungal meningitis
  Symptomatic treatment of severe headache from subarachnoid hemorrhage or leptomeningeal
    metastases
  Before taking anticoagulant therapy for cerebrovascular disease

Contraindications
Absolute
  Tissue infection in region of puncture site
Relative
  Brain tumor or abscess
  Spinal cord tumor
  Thrombocytopenia or coagulopathy
  Increased intracranial pressure

*Complications of Test**
Common
  Headache
  Backache
Less common
  Lumbar radicular pain
  Intracranial hypotension[76]
  Hearing abnormalities (hearing or sensation of echoing in the head)
  Diplopia (sixth-nerve palsy)
  Enhanced meninges on magnetic resonance scan
Rare
  Transtentorial or foramen magnum herniation
  Worsening of spinal tumor symptoms
  Spinal hematoma (patients with bleeding tendency)
  Herniated or infected disk
  Reaction to anesthetic agent
  Meningitis (contaminated needle, children with sepsis)
  Subdural effusion

* See also Chapter 14.

may prevent identification of organisms by culture from CSF, other techniques such as polymerase chain reaction (PCR) will continue to be positive. However, sometimes organisms growing in the CSF are different from those in the blood in immunosuppressed patients. Furthermore, the CT scan and papilledema are not the best predictors of herniation after lumbar puncture[51]; clinical signs of impending herniation[77] (see Chapter 3) are. In those cases, appropriate treatment for herniation should be started before the lumbar puncture.

In most instances, this situation is not that acute and when lumbar puncture is indicated (suspicion of CNS infection, unexplained delirium[78]), it should be done promptly. To avoid complications, this should be done by the most skilled physician available. Ideally, atraumatic needles should be used and the stylet should be inserted before needle removal.[74] When a standard needle is used, the bevel should be pointed in a longitudinal (cephalad to caudad) direction.[79] CSF should be removed slowly in quantities sufficient to establish a definitive diagnosis (Table 10–5). Most complications of lumbar puncture occur because of a further leak of fluid through the dural tear made by the needle, not because of the volume of CSF removed at the time of the lumbar puncture. If the CSF pressure is very high, the patient should be placed under constant observation after the lumbar puncture. Corticosteroid treatment should be considered. A hyperosmolar agent such as mannitol should be available at the bedside to treat herniation if warning symptoms develop (see Chapter 3).

If the CSF from a patient suspected of acute meningitis is cloudy, therapy should be started immediately, even before the spinal fluid is fully evaluated (Table 10–6).

Cerebrospinal Fluid Evaluation

The CSF of most patients with bacterial or fungal meningitis resembles CSF from patients without cancer: the pressure is elevated, the fluid is cloudy with both neutrophils and lymphocytes, the protein concentration is elevated, and the glucose concentration is depressed.[74,75,80] In many patients with severe leukopenia, and even in occasional otherwise healthy individuals, the expected CSF pleocytosis is not found, at least initially. Of 26 patients with a peripheral WBC count of less than 1000 cells/mm^3, 12 had 5 or fewer WBCs/mm^3 in the CSF at clinical onset[6] of the infectious meningitis (Fig. 10–5A). Organisms were cultured from all 12. An additional three patients with normal or elevated peripheral WBC counts also had fewer than 5 WBCs/mm^3 in the CSF. Meningeal infections with normal CSF WBC counts occurred in patients with meningitis from *Pseudomonas, Listeria, S. pneumoniae, E. coli,* and *Proteus.* Of the 15 patients with normal CSF WBC counts, 3 had hypoglycorrhachia.

Why hypoglycorrhachia sometimes occurs in the absence of a CSF pleocytosis is unclear. Carrier-mediated glucose transport, utilizing the Glut1 carrier transporter,[81] could be diminished by the infection, but Fishman[82] has shown that diffusion of glucose into the CSF is increased by acute meningitis because the blood–brain barrier is disrupted. Glucose metabolism probably is increased partly by organisms and partly by WBCs, but Petersdorf and colleagues[83] showed that an extraordinarily large number of organisms are necessary to metabolize glucose to hypoglycorrhachic levels. Activation and proliferation of pial and arachnoid cells responding to the inflammation may also promote increased glucose metabolism.[84] As the glucose concentration diminishes, the lactate concentration increases. An increase in the lactate concentration is common in bacterial meningitis and the degree of increase is said to correlate with the severity of infection.[85,86]

Sufficient CSF must be sent to the laboratory to ensure growth of even a few organisms in CSF. If meningitis is suspected and the initial fluid is not purulent, send 10 mL of CSF to the laboratory for smear, and culture for aerobic and anaerobic bacteria and fungi. Cell count and differential requires 1 mL (see Table 10–5). The remainder of the specimen should be centrifuged to concentrate the organisms. The supernatant fluid can then be used

Table 10–5. Cerebrospinal Fluid Volume for Common Diagnostic Tests

Test	Volume of Cerebrospinal Fluid Required (mL)
Culture	5–10°
Cytology (malignant cells)	4
Cell count and differential	0.5–1.5
Glucose and protein	0.5†
Serologic test for syphilis	0.5†
Cryptococcal antigen	0.5†
Oligoclonal bands	2†
Viral culture/polymerase chain reaction	2

° Larger volumes are better.
† These tests can be done on supernatants from culture.

Table 10–6. **Empiric Treatment of Central Nervous System Infection in Patients with Cancer**

Primary Immune Defect	Blood Leukocyte Count	CSF Findings	Recommended Therapy	Alternate Therapy
T-lymphocyte, mononuclear phagocyte defect (e.g., Hodgkin)	Normal	Pleocytosis, no organisms, negative cryptococcal antigen	Ampicillin for listerial or pneumococcal organisms; add ceftriaxone and vancomycin if worried about pneumococcus	Trimethoprim–sulfamethoxazole (cotrimoxazole) or chloramphenicol and erythromycin
	Normal	No organisms ± pleocytosis, cryptococcal antigen positive	Obtain more CSF; consider amphotericin B + 5-FC for cryptococcus	Fluconazole or itraconazole
	Low	No organisms ± pleocytosis	Ampicillin + ceftazidime; + gentamicin IV + IT for Gram-negative meningitis	Ampicillin + imipenem
Neutrophil defect (e.g., chemotherapy, myelogenous leukemia)	Low	No organisms ± pleocytosis	Cefepime plus gentamicin IV ± IT	Imipenem
	Normal	No organisms ± pleocytosis	Cefepime plus gentamicin IV ± IT	Imipenem
Splenectomy	Normal	No organisms, pleocytosis	Ceftriaxone ± vancomycin	Ampicillin or cotrimoxazole or both
Surgery (brain or spine)	Normal	No organisms, pleocytosis	Cefepime+ gentamicin IV ± IT and oxacillin	Vancomycin can be substituted for oxacillin

CSF, cerebrospinal fluid; IT, intrathecal.

for chemical and serologic assays and the sediment placed in culture. An additional 4 mL of CSF should be analyzed cytologically for tumor cells because the major alternative diagnosis to acute or subacute meningitis is leptomeningeal infiltration. Additional fluid should be saved for molecular analysis (PCR).[80,87]

The treating physician should examine the CSF. By holding a tube of CSF to natural light and shaking it, as few as 25 to 50 WBCs/mm^3 can be identified by the naked eye.[88] Identification of this small number of cells in the CSF depends on the Tyndall phenomenon, that is, light rays refracted by their encounter with the cells. Opalescent or turbid spinal fluid suggests the presence of several hundred or a few thousand cells. A few hundred red blood cells (RBCs) may lend turbidity to the spinal fluid without altering its color.

A cell count should be performed in a counting chamber and a spun specimen stained with both Gram and Wright stain. A careful search for bacterial and fungal organisms, as well as malignant cells or parasites, should be made on the Gram and Wright stained slides. Organisms that can be identified on Gram stain include most of the bacteria, some parasites such as *Trichomonas* or *Amoeba*, and cryptococci; the latter are also identified by an India ink preparation (Fig. 10–6).

Serologic tests of CSF should include cryptococcal antigen, which is virtually always positive

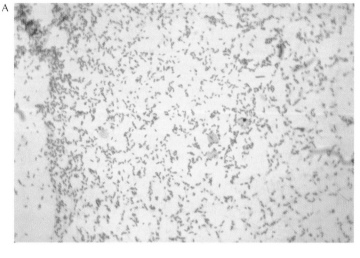

A

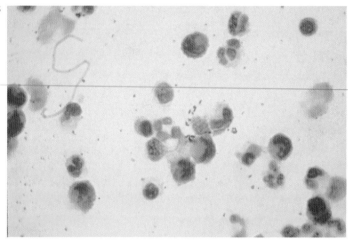

B

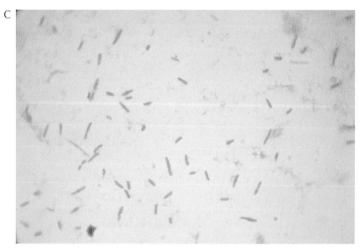

C

Figure 10–5. (*cont.*)

D

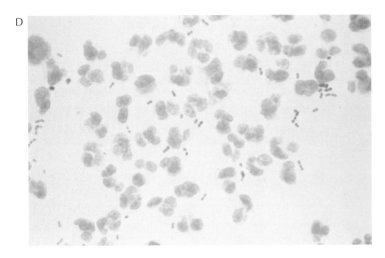

E

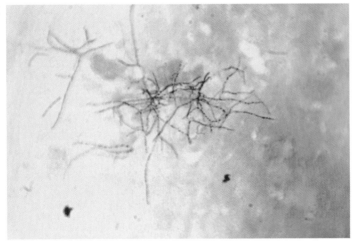

Figure 10–5. A montage of organisms responsible for central nervous system (CNS) infection in patients with cancer. **A**: A photomicrograph of an uncentrifuged specimen from the CNS showing Gram-negative rods (*Escherichia coli*) in the cerebrospinal fluid of a neutropenic patient with acute lymphocytic leukemia. No white blood cells are seen in the field. The patient had an elevated protein and a depressed glucose concentration in the cerebrospinal fluid. He died despite antibiotic therapy. A photomicrograph of his brain is shown in Figure 10–4. **B**: Cerebrospinal fluid (CSF) Gram stain of the patient with *Listeria* meningitis; note the Gram-positive rods. At times the organisms destain excessively and are mistaken for Gram-negative organisms. This patient suffered from Waldenstrom macroglobulinemia and presented with a subacute onset of headache and confusion. He eventually required a shunt for hydrocephalus. **C**: *Klebsiella ventriculitis* (Gram-negative rods) in a neutropenic patient being treated for leptomeningeal leukemia through an Ommaya reservoir (Fig. 7–9A). **D**: Gram-positive cocci in the CSF of a patient with pneumococcal meningitis. The patient had a splenectomy for Hodgkin disease several years before. **E**: Nocardial organisms in a patient with chronic lymphocytic leukemia. The patient had multiple abscesses in the brain. The delicately beaded organisms are best seen on modified acid fast stains.

in patients with cryptococcal meningitis and is diagnostic whether or not the organism grows in culture or is seen on Gram stain. When bacterial meningitis is suspected, antigens for pneumococcus, meningococcus, and haemophilus can be measured in CSF. IgM toxoplasma antibodies are occasionally positive in infection of the nervous system from *Toxoplasma* organisms

even when the IgG serology is negative. Certain enzymes in the CSF (e.g., β-glucuronidase) are increased in infection, but they do not help to differentiate the kind of infection or even to distinguish between infection and leptomeningeal metastases (see Chapter 7).

PCR testing is available for a number of viruses. In the immunocompromised patient,

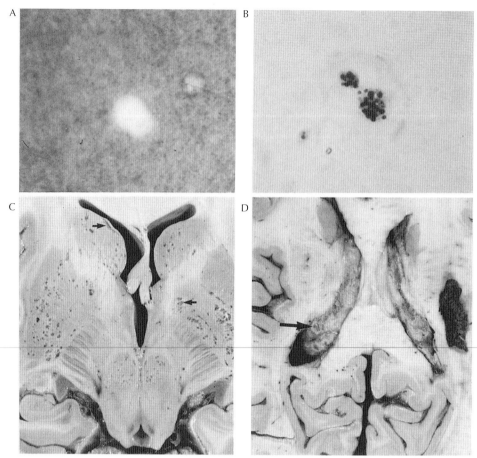

Figure 10–6. Cryptococcal organisms in the cerebrospinal fluid. **A**: A budding organism stained with India ink. **B**: A Gram stain. The patient had lymphoma and cryptococcal meningitis. **C**, **D**: Two pathologic examples of cryptococcosis. **C**: The typical soap-bubble appearance of *Cryptococcus* involving the basal ganglia (*arrows*). **D**: The fibromucoid appearance of the choroid plexus involved by *Cryptococcus* (*arrow*).

PCR is especially useful for identification of herpes viruses and the JC virus (progressive multifocal leukoencephalopathy [PML] [see following text]). PCR testing is also available for many bacteria, including mycobacteria tuberculosis and for some parasites including toxoplasmosis.[87]

Imaging
In patients with meningitis, the CT or MRI may show multifocal enhancement of the leptomeninges.[89,90] This finding is indistinguishable from that of leptomeningeal metastases and hard to distinguish from meningeal enhancement following lumbar puncture (see Chapter 14). Features suggesting meningitis include loss of definition of cerebral sulci and evidence of hydrocephalus.

Meningoencephalitis

The clinical pictures of encephalitis and meningoencephalitis vary with the pathology and offending organism. In general, when the organisms cause meningitis with secondary encephalitis, the signs and symptoms (fever and headache) resemble those of meningitis alone, but delirium is more prominent and focal signs, particularly focal and generalized convulsions, are more frequent. Seizures and focal signs that occur early with meningitis are probably caused by direct invasion of the brain by the organism. When such signs occur late, particularly after treatment is well under way, vasculitis with thrombosis and infarction is a more likely cause. Such vascular changes may lead to permanent neurologic impairment despite successful treatment of the underlying infection.

Contrast enhancement of the basal cisterns and cortical sulci on CT or MRI may attest to the intensity of the inflammation.[89,90] Contrast enhancement of cortical gyri suggests that infection and inflammation involve the brain as well as the meninges. FDG (2-fluorodeoxy-d-glucose)-PET may show areas of hypermetabolism.[91]

When the process is primarily encephalitic (i.e., without important meningeal involvement), the usual signs of meningitis may be absent. Instead, the illness begins with focal or generalized convulsions, or both, accompanied or followed by delirium often associated with asterixis and multifocal myoclonus (see Chapter 11). Fixed focal signs, such as hemiparesis or visual field deficits, may be present as well. The differential diagnosis includes metabolic encephalopathy (see Chapter 11) and focal brain lesions such as metastases. A severe and sustained delirium preceding seizures, especially when associated with intense multifocal myoclonus, can occur with either CNS infection or metabolic encephalopathy, but metabolic encephalopathy usually does not cause fever, and prominent focal signs are uncommon. When focal signs predominate in a delirious patient, they suggest encephalitis rather than a metabolic process. Cerebral infarcts or metastases can cause focal signs but they cause delirium infrequently; even spinal infarcts are seen occasionally (Fig. 10–7).

Because most encephalitides are caused by viral and nonbacterial infection, CSF is less helpful diagnostically. An increased cell count is usually but not always present; it may be predominantly lymphocytic. The protein and glucose concentrations are frequently normal; cultures are negative because the infection is usually viral. Serologic examination of the spinal fluid may help. Polymerase chain reaction (PCR) is the most specific diagnostic test. The diagnostic approach, differential diagnosis, and treatment are the same as with meningitis.

Brain Abscess

The symptoms and signs of brain abscess depend on the causal agent. *Toxoplasma* abscesses have

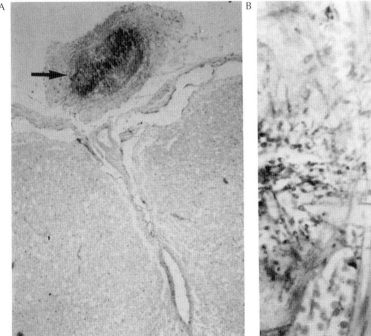
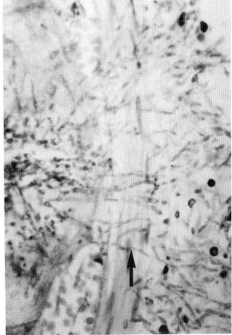

Figure 10–7. Anterior spinal artery occluded by aspergillosis in a neutropenic patient with acute lymphocytic leukemia. The patient suffered sudden onset of paraplegia days before death. **A**: At autopsy, the anterior spinal artery was occluded (*arrow*). **B**: The fungi can be seen growing through the vessel wall (*arrow*).

a predilection for the basal ganglia and may lodge there as a single mass. The subacute development of neurologic signs, especially hemiparesis, is indistinguishable from that of brain metastasis. *Nocardia* abscesses (Fig. 10–8) are often multifocal and cause either slowly progressive multifocal signs, similar to those of multiple metastases, or focal or generalized seizures. Although the diagnosis of brain abscess may be suspected based on the underlying cancer, differentiation from a brain metastasis may be possible only with biopsy.

CT and MRI help identify brain abscesses.[89,90] Typically, scans reveal a hypointense area before contrast, with a ring of enhancement after contrast. Some small abscesses appear as a solid mass, and occasional abscesses are hyperintense before contrast. An abscess can sometimes be differentiated from tumor by the nature of the ring enhancement. The enhancing ring of an abscess is generally thinner and more uniform than the ring of a tumor. Diffusion-weighted MRI may also be helpful. In general, a brain abscess demonstrates restricted diffusion, whereas tumors show increased diffusion; a hypointense rim on the T2-weighted image also suggests abscess.[92] However, these findings are not entirely diagnostic.[93]

With suspected *Toxoplasma* abscesses, the diagnosis may be inferred by demonstration of serum anti-toxoplasmosis antibodies and then established by the results of a therapeutic trial (see p 400). With other suspected abscesses, stereotactically directed needle biopsy performed early in the diagnostic workup reveals the organism(s), so that appropriate antibiotic therapy may be instituted.

Vascular Lesions

Vascular lesions caused by infection produce sudden focal neurologic signs, usually in a patient known to be suffering from systemic infection[94] or meningitis[42,67] (Fig. 10–7). The clinical picture and imaging of a cerebral infarct from bacterial endocarditis resemble nonbacterial thrombotic endocarditis (NBTE) (Chapter 9) or other causes of cerebral infarction (see Chapter 9). CSF examination often fails to help, although pleocytosis may be present and organisms very rarely can be grown, especially *Aspergillus* and *Zygomycetes*, and even less often *Candida*. The diagnosis of a vascular lesion due to CNS infection is usually easy because vascular disorders typically occur late in patients with overwhelming bacterial or fungal infection; many are being treated for the infection at the time the stroke occurs. Varicella-zoster vasculitis may cause a stroke after other signs of infection have resolved (see following text). The diagnosis of a vascular lesion due to endocarditis is discussed in Chapter 9.

APPROACH TO THE PATIENT

The physician must consider both the primary cancer and the impaired host defense mechanisms that are likely to lead to CNS infection, as shown in Table 10–1. Infections are most likely to occur in patients who are least likely to have other forms of neurologic dysfunction from cancer. For example, infection in Hodgkin disease is common but CNS metastases are rare; by contrast, CNS infections are rare in patients with solid tumors but metastases are common. The major factors predisposing to infection are as follows: (1) Hodgkin disease or non–Hodgkin lymphoma (Fig. 10–5D); (2) leukopenia from either disease or chemotherapy; (3) long-term immunosuppressive drugs, particularly corticosteroids; (4) a surgical or tumor-induced communication between the CNS and an epithelial surface.

A scrupulous search for a systemic nidus of infection is essential. Skin lesions are sometimes found with *Nocardia* infections[95,96] and also may complicate *Pseudomonas* sepsis. Chest radiographs or CT may reveal *Aspergillus*, *Zygomycetes*, or *Nocardia* abscesses that can be diagnosed with lung biopsy. Examination of blood, urine, or other bodily fluids may either identify the organism on culture or identify antibody titers that strongly suggest an active infection. In patients suffering pain, a bone scan may reveal an increased area of isotope uptake in the skull or spine, the first visual change of osteomyelitis. A gallium scan or PET may detect a deep-seated systemic abscess, or a CT may identify an abscess in the paravertebral region potentially communicating with the subarachnoid space.

The physician should search for communication between the CNS and the outside environment.[97] CSF otorrhea or rhinorrhea in patients suffering from a cancer involving the skull may

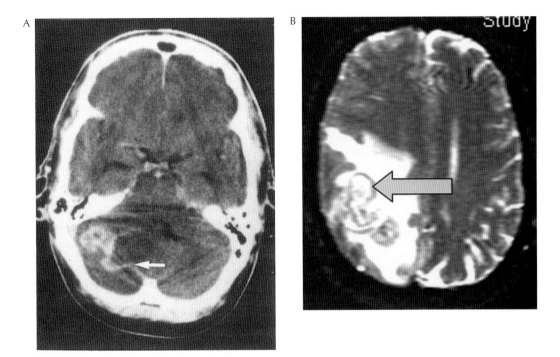

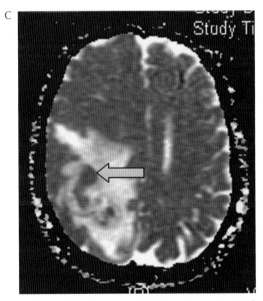

Figure 10–8. A: Computed tomography scan showing a nocardial abscess in a patient with chronic lymphocytic leukemia. The patient presented with headache and acute cerebellar signs but no fever. The vascular structure draining the abscess (*arrow*) led to an initial diagnosis of a highly vascular tumor. The abscess was discovered on resection. The patient made a complete recovery. **B**: A patient different from the one described in A: the same patient as shown in Figure 1–4. This patient without cancer presented with nocardial lesions in the lung and brain. Restricted diffusion suggested to the neuroradiologist an abscess rather than a brain metastasis from lung cancer. The central portion of the lesion was hyperintense on the diffusion-weighted image (*arrow*). **C**: To assure this was not "T2 shine through," an apparent diffusion coefficient demonstrated hypointensity in the central portion (*arrow*) establishing restricted diffusion in **B**. Although occasional metastases demonstrate restricted diffusion, the lesion is more likely to be an infection than a metastasis.

presage CNS infection, and CSF may sometimes leak from poorly healing back wounds in patients who have undergone laminectomy. In one of our patients, a careful examination of an exenterated eye socket detected an infected CSF fistula that explained the patient's delirium.

The neurologic examination usually does not help either to establish a diagnosis of CNS infection or to determine its specific cause. Instead, the physician must rely on laboratory test results. When clinical evidence indicates acute meningitis, the CSF must be examined. MRI with and without contrast should be performed to look for meningitis, encephalitis, or abscess. When clinically indicated, a CT with *bone windows* will identify erosion of the skull base or the orbits that may form a communication between the CNS and the surface. If the diagnosis cannot be made in any other way, a brain biopsy may be necessary.

Table 10–6 outlines an empirical approach to treating patients suffering from an acute CNS infection before the organism is identified. The physician first attempts to identify the underlying immune deficit, usually determined by the patient's primary illness and blood neutrophil count, and then chooses antibiotics based on the likely organism causing CNS infection in that setting.[11,75] Antibiotics that cross the blood–brain barrier easily are more effective than those that do not.

In patients with T-lymphocyte or mononuclear phagocyte defects caused by lymphoma but with normal blood neutrophil counts, *Listeria* is the most likely infecting organism, particularly if cryptococcal antigen is absent from CSF. Because so many patients are acutely ill, antibacterial antibiotics (we use ampicillin, ceftriaxone and vancomycin to cover the possibility of pneumococcal infection) should be started immediately after the spinal fluid is sampled, even before the outcome of the Gram stain is known. Although *Cryptococcus* is a common offending organism, antifungal agents can be deferred safely for a few hours until the cryptococcal antigen determination or the Gram or India ink stain implicates a cryptococcal infection. Once this cause of meningitis is identified, amphotericin B lipid complex and 5-flucytosine should be given[98] pending final culture results. In patients suffering from T-cell defects with low peripheral neutrophil counts, cefepime and gentamicin should also be administered. Because gentamicin crosses the blood–brain barrier relatively poorly, the physician should consider intrathecal or intraventricular gentamicin in a dose of 5 to 8 mg/day for those with proven or highly probable Gram-negative meningitis.

If the underlying abnormality is a defect of neutrophil function, cefepime and gentamicin are the drugs of choice as initial therapy, whether the blood neutrophil count is normal or low. Other third-generation cephalosporins and imipenem (or meropenem) are successful against enteric Gram-negative meningitis and appear to be reasonable alternatives.

In patients suffering from immunoglobulin abnormalities or who have been splenectomized, ceftriaxone should be started immediately, along with penicillin, pending the outcome of the cultures. Penicillin resistance is so common among the pneumococci that we use vancomycin until susceptibility studies are available. In those with a suspected CSF–surface communication, a combination of drugs, including oxacillin (for staphylococcal infection), ceftazidime, and gentamicin, is the initial treatment of choice. If methicillin-resistant infections or coagulase-negative staphylococcal infections are likely, vancomycin should be used instead of oxacillin. The role of quinolones, such as ciprofloxacin, has not been established.

The antibiotic agents mentioned are designed to treat the common bacterial and fungal infections that begin acutely or subacutely in patients with cancer. Modifications of the regimen are necessary when the physician suspects other organisms, when a different offending organism such as *Toxoplasma* or a viral agent is identified, or when careful monitoring of the patient indicates that the selected treatment is not working. More detailed guidelines for the treatment of specific CNS infections are presented in the following paragraphs (see also Ref. 11).

SPECIFIC ORGANISMS CAUSING CENTRAL NERVOUS SYSTEM INFECTION IN PATIENTS WITH CANCER

T-Lymphocyte and Mononuclear Phagocyte Defects

Table 10–7 lists the organisms likely to cause CNS infections in patients with impaired

Table 10–7. **Central Nervous System Infection with T-Lymphocyte and Mononuclear Phagocyte Defect**

Infection	Bacteria	Fungi	Parasites	Viruses
Meningitis or meningoencephalitis	*Listeria*	*Cryptococcus* *Coccidioides* *Histoplasma*	*Toxoplasma* *Strongyloides*	Varicella-zoster Papovavirus (PML) Cytomegalovirus
Brain abscess or encephalitis	*Nocardia* *Listeria*	*Cryptococcus* *Histoplasma*	*Toxoplasma*	Cytomegalovirus Herpes simplex

PML, progressive multifocal leukoencephalopathy.

cellular immunity. These patients either have underlying disorders, such as lymphoma or chronic leukemia, or are taking steroids or other immunosuppressive agents such as azathioprine. Neutrophil function may be normal, and patients generally have a normal (or even elevated) WBC count.

BACTERIA

Listeria

This organism is the most common cause of bacterial meningitis in patients with cancer, exclusive of those who have had neurosurgery; virtually all patients infected by *Listeria* are taking steroids or have recently done so. An aerobic Gram-positive rod, *Listeria* is widespread in the environment, being found in tap water, sewage, milk products, and stools of healthy people. *Listeria* has been implicated in food-borne outbreaks involving milk products and processed meats and it affects patients with poor cellular immunity, including patients with cancer, the elderly, neonates, patients with cirrhosis, pregnant women,[99] and occasionally people who are otherwise healthy. A review in Denmark of 299 cases of *Listeria* between 1994 and 2003, involving either the bloodstream or the CNS, indicated that 16% were associated with cancer (13% hematologic cancers, 3% solid tumors), 14% had received immunosuppressive treatment, and 47% had no identifiable underlying disease.[99] A review of 820 patients with CNS *Listeria* infections found that 97% suffered from meningitis or meningoencephalitis, the rest having rhomboencephalitis,[100]

cerebritis,[101] or brain abscess. Of 455 patients in whom an underlying illness was reported, 14% had a hematologic malignancy and a 7% solid tumor.[102] Organ or stem cell transplant is also a known predisposing condition.[103]

Listeria meningitis is usually associated with lymphoma or chronic leukemia. The peripheral WBC count may be normal or elevated. The usual clinical signs are of acute or subacute meningitis,[102] slightly less fulminant than pneumococcal meningitis but more than cryptococcal meningitis. In some instances, the illness may begin indolently, with mild headache and personality change, but generally severe headache, nausea, vomiting, fever, and a stiff neck dominate the picture; patients look quite ill. In a report of 367 episodes of *Listeria* meningitis/meningoencephalitis, fever was present in 92%, an altered sensorium in 65%, headache in 46%, focal neurologic signs in 18%, seizures in 5%, and photophobia in 3%.[102]

The organism induces an intense inflammatory reaction in the leptomeninges, which may occlude CSF absorptive pathways and lead to hydrocephalus. *Listeria* also tends to invade brain parenchyma, causing a focal cerebritis, diffuse encephalitis, or an abscess (Fig. 10–9). Any portion of the brain or spinal cord[41] can be involved, but the organism has a peculiar tendency to invade the brainstem, causing focal brainstem signs.[100] The clinical course of the brainstem infection is usually biphasic; headache, vomiting, fever, and leukocytosis precede the brainstem signs by 4 to 10 days.

The diagnosis of *Listeria* meningitis is suspected from the clinical findings and is

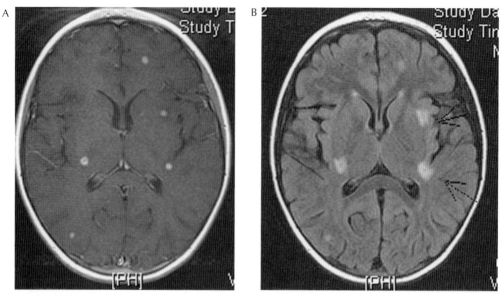

Figure 10–9. Multiple *Listeria* abscesses in a child with biphenotypic leukemia. **A:** A T1-enhanced magnetic resonance image shows multiple enhancing lesions scattered throughout the cerebral hemispheres but not the brainstem. There was no enhancement of the leptomeninges. **B:** The FLAIR sequences showed some edema surrounding the lesion. Despite the lack of leptomeningeal enhancement the cerebrospinal fluid was positive for *Listeria* and the patient responded to antibiotic therapy.

confirmed by CSF examination. The CSF usually contains several hundred WBCs, with a neutrophilic predominance despite the organism's name.[102] The glucose concentration is often low, and the protein concentration is elevated. Organisms are identified by Gram stain of CSF sediment in approximately 30% of cases.[102] Even when organisms are found, destaining may cause them to appear Gram-negative rather than Gram-positive rods, leading to an incorrect diagnosis. They may also mimic Gram-positive cocci. Because the illness is potentially lethal, a patient with a T-lymphocyte or mononuclear phagocyte defect who has a normal or elevated peripheral WBC count and symptoms of meningitis should be considered to have *Listeria* meningitis and be treated empirically with ampicillin. Isolation from culture may take 2 to 3 days, so it is not helpful in acute treatment decisions. Acute or subacute development of focal brain or brainstem dysfunction may result from cerebral infarction, bacteria-induced vasculitis, or direct invasion of brain by the organisms. If the organisms reach the brain hematogenously, the meninges may be spared and the CSF can be entirely normal. In such patients, blood

cultures are usually positive for *Listeria*. All patients with *Listeria* infections, including septicemia from *Listeria* without neurologic signs, require both CSF examination and an MRI.

The drugs of choice are ampicillin, 2 g every 4 hours, or penicillin, 2×10^6 U every 4 hours, for at least 3 to 4 weeks. Patients treated for a shorter period often relapse, particularly if they have cerebritis. For severely ill patients, gentamicin may be added because of synergism with penicillin-like drugs. It is given both systemically and intrathecally. Trimethoprim–sulfamethoxazole, vancomycin (a drug with poor CSF penetration that may allow relapse), erythromycin, chloramphenicol, or tetracycline have been used successfully in patients allergic to penicillin.[104]

If *Listeria* meningitis is diagnosed early, it can usually be treated effectively. Those who succumb are typically either terminal from their cancer or have suffered a delay in the diagnosis of meningitis. Most respond rapidly to appropriate antibiotics, but an occasional patient, particularly if stuporous or comatose on admission, may respond more slowly (e.g., after 72 hours). Slow recovery suggests direct invasion of the brain by the bacteria or, more likely, meningitis-induced hydrocephalus. If the patient's mental state has not returned to normal within a week, and if the ventricles are obviously enlarged, shunting should be considered before the hydrocephalus causes irreversible brain damage.

Two findings in experimental animals are of some interest. One study suggests that morphine increases the susceptibility of experimental animals to *Listeria*. Animals receiving morphine all died when exposed to the bacterium, whereas untreated animals lived.[105] The second finding is that recombinant *Listeria* co-expressing viral proteins and tumor antigens can induce immunity against metastatic brain tumors expressing the tumor antigens,[106] suggesting the bacterium may be useful as a vaccine.

Nocardia

This aerobic, filamentous-branching, Gram-positive rod, which may be weakly acid-fast, belongs to the same family as the actinomycetes and mycobacteria. In patients with impaired cellular immunity, infection spreads from the lung[107] to cause either brain abscesses[108] (usually multiple[109]) or, less commonly, meningitis.[110] Up to 50% of patients with *Nocardia* pulmonary infection develop CNS lesions.[111] The abscesses are usually indolent, presenting with the clinical symptoms of single or multiple mass lesions resembling metastases rather than infection (Fig. 10–8). Chorioretinitis also occurs.[112] Organisms are infrequently identified in the spinal fluid. Even when present, they grow slowly; the microbiology laboratory should be alerted to the possibility of nocardial infection so that they will keep the culture for several days.

Nocardia infection in the CNS is suspected when lymphoma or chronic lymphatic leukemia patients, or those receiving steroid therapy,[108] develop focal neurologic signs associated with one or more contrast-enhancing lesions in the brain. The patient usually also has lung or, less commonly, skin lesions from which cultures can often be positive. Infection with multiple organisms can occur, however, so if the brain lesions do not respond clinically and radiographically within 7 to 10 days, the physician should consider stereotactic needle biopsy of a brain lesion. Imipenem with ampicillin is first-line treatment at MSKCC, followed by trimethoprim–sulfamethoxazole. When the patient is stable, treatment with sulfadiazine at 120 mg/kg per day for 1 or 2 weeks and 60 mg/kg per day for several months, or trimethoprim–sulfamethoxazole, has also been used with success. No comparative clinical trials have been reported. It is very important to identify the sensitivity of the organism from at least one site because as many as 20% of patients may be resistant to the trimethoprim–sulfamethoxazole combination. Cefuroxime has been effective for pulmonary disease; linezolid may also be useful.[113]

FUNGI

Fungal infections are serious and increasing problems in patients with cancer[114–116] and those undergoing transplantation. The common pathogen in patients with T-cell defects is *Cryptococcus*.[117] In some geographic areas, histoplasmosis[118] and coccidioidomycosis[118] are also culprits. Cryptococcal infection usually causes a subacute or chronic meningitis,[117] but may cause local abscess formation[119,120] or ventriculitis without meningitis. Between 1955 and 1974, 40 patients with cryptococcal meningitis were diagnosed at MSKCC before death (two per year),[3] whereas between 1993 and 2004, only six additional patients were identified (0.5 per year); three had hematologic tumors and three had solid tumors.[121]

The cryptococcal organism is ubiquitous in soil and generally spreads to the CNS via the lung. The disorder may affect seemingly healthy individuals as well as the obviously immunosuppressed. In a study of 37 HIV negative patients with cryptococcosis, only 6 had cancer; 15 were receiving immunosuppressive drugs, usually corticosteroids.[122] In another series of 94 non-HIV patients, only 42 had a clearly identifiable underlying disease, usually diabetes mellitus.[123] In patients with cancer, the clinical course is usually subacute, with headache and fever. In occasional patients, the disorder develops insidiously and may last months to years before the diagnosis is made. Nuchal rigidity is uncommon. The onset is usually slower than with pneumococcal or *Listeria* meningitis but more rapid than in patients suffering from leptomeningeal metastases. Usually patients do not have neurologic signs other than mild encephalopathy, although cranial nerve palsies (especially facial and acoustic) and papilledema occur occasionally. However, cranial nerve palsies, especially when accompanied by absent deep tendon reflexes, should lead the physician to suspect leptomeningeal metastasis instead of cryptococcal meningitis.

The CT or MRI may be helpful. Meningeal enhancement that may be nodular is often present.[123,124] Symmetric nonenhancing gelatinous pseudocysts in the basal ganglia and thalami, or solid ring-enhancing masses located in the choroid plexus, have also been described.[119] Magnetic resonance spectroscopy (MRS) may show a "taller" lactate plus lipid peak as well as low myoinositol and NAA peaks.[124]

The diagnosis of cryptococcal meningitis is made by CSF examination that reveals elevated CSF pressure, an increased number of lymphocytes, a low glucose, and a high protein concentration. The India ink preparation identifies an organism with a clear halo due to the polysaccharide capsule surrounding the yeast, but the inexperienced observer may mistake RBCs, lymphocytes, and occasional inanimate debris in spinal fluid for fungi and should consider an India ink preparation positive only if budding yeast forms are clearly identified (Fig. 10–6). Gram stain reveals an irregularly stippled Gram-positive organism with a faintly Gram-negative capsule. Organisms are visible in the CSF in only about one-half of cases, and growth in culture may not be apparent for several days. Testing the CSF for cryptococcal antigen is highly reliable. A titer of greater than 1:4 establishes the diagnosis; a titer of greater than 1:256 carries a poor prognosis. Cryptococcal antigen tests may take 2 to 6 weeks to become positive after the infection begins, and test results may be negative if the patient is infected with unusual, small capsular forms of the organism.

If the patient has a cryptococcal abscess rather than meningitis, the CSF may be normal and the diagnosis can be established only by stereotactic needle biopsy. In an appropriate clinical setting, high titers of serum cryptococcal antigen or positive blood cultures may prompt the physician to treat the brain lesions empirically.

Guidelines for the management of cryptococcal disease have been developed by the Infectious Diseases Society of America.[125] They recommend combined induction therapy with amphotericin B with or without flucytosine, followed by chronic treatment with fluconazole. Lipid formulations of amphotericin are substituted in patients with renal disease.

Itraconazole may be substituted in patients who are unable to tolerate fluconazole.

In severe or fulminant cases, intrathecal amphotericin B (administered intraventricularly through an Ommaya reservoir) may be lifesaving.[126] In one series,[126] six of seven recent patients with cryptococcal meningitis so treated responded to therapy, whereas five of six patients who were less severely ill and did not receive intrathecal drug died of the infection. Intrathecal amphotericin B has also proved helpful in coccidioidal meningitis.[127] Intrathecal treatment should begin with 0.01 mg of amphotericin B and escalate at 12- to 24-hour intervals up to 0.5 mg/day as long as the CSF antigen remains positive. Administration by intraventricular reservoir rather than by lumbar puncture is preferred for the reasons given in Chapter 7, and also because intrathecal administration of amphotericin B by lumbar puncture can cause severe leg and back pain. Fluconazole, 200 to 400 mg/24 hours, should be considered for maintenance in patients with sustained immunosuppression[128]; the role of other antifungal agents is less certain.

The sequelae of cryptococcal meningitis include cranial nerve palsies (particularly deafness), occasionally visual loss[122,123] and hydrocephalus that may require shunting. With early and aggressive treatment, most patients, even those suffering from lymphoma, do well if their underlying disease remains in remission.

VIRUSES

The herpesviruses (varicella-zoster, herpes simplex, or cytomegalovirus [CMV]) may invade the CNS of patients with T-cell deficiencies.[129,130] The latter two are particular problems after transplantation[131] as is Herpesvirus 6.[132]

Varicella-Zoster

A major cause of nervous system dysfunction in patients with cancer, varicella-zoster virus also occurs in the apparently immunocompetent. The disorder may affect over 50% of those with Hodgkin disease.[133] Concomitant bacterial infection is common.[134] Table 10–8 shows the range of neurologic complications of varicella-zoster infection.[135]

Varicella: Before varicella vaccination was widely available, chickenpox could be a severe infection, more so in children with cancer than in healthy children. In one study[136]

Table 10–8. **Neurologic Complications of Varicella-Zoster Virus Infections**

Associated with Varicella
Meningoencephalitic syndromes
  Primary viral encephalitis
  Postinfectious encephalomyelitis
  Acute cerebellar ataxia
Intracerebral hemorrhage secondary to
  coagulation disorders
Myelitis
Aseptic meningitis and suppurative meningitis
Guillain-Barré syndrome
Reye syndrome
Congenital varicella syndrome

Associated with Cutaneous Zoster
Postherpetic neuralgia
Segmental motor weakness
Encephalomyelitis
Aseptic meningitis
Myelitis alone
Cranial nerve palsies
Guillain–Barré syndrome
Cerebral vasculitis
Multifocal leukoencephalitis syndrome
Myositis
Urinary retention

of 77 children with cancer who developed varicella, visceral dissemination occurred in 19 and fatal encephalitis in 2. The most common neurologic complication of chickenpox is an acute ataxic meningoencephalitis, more likely to occur in the immunosuppressed than in the general population. It is characterized by sudden ataxia, sometimes with dysarthria and nystagmus. Neurologic symptoms may precede or follow the rash by days or, rarely, weeks. More widespread evidence of encephalitis may include lethargy, seizures, and other focal signs. Most patients recover fully. The pathogenesis of the disorder is not known, but is hypothesized to be either direct invasion of the CNS by the virus or an immune encephalomyelitic response to the viral infection. Varicella-zoster virus can reach the nervous system by either neural or hematogenous spread.[137]

Reye Syndrome: Once a feared and sometimes fatal metabolic disorder following a viral infection, usually varicella or influenza,[138,139] Reye syndrome is now very rare or perhaps nonexistent.[138] The disorder was characterized by the sudden onset of severe, recurrent vomiting and the rapid development of irritability, confusion, and coma. Intracranial pressure was increased, and death resulted from cerebral herniation. At autopsy, acute fatty changes were found in the liver, and the brain was edematous. The disorder was much more common in children but had also been reported in adults.[140] It does not appear to be more common in the immunosuppressed than in the immunocompetent population; one report described three children who developed the syndrome during chemotherapy for leukemia or lymphoma.[141] Hypoglycemia from the liver disorder was a common complication. Ingestion of aspirin during the acute viral infection was implicated as a risk factor for the development of Reye syndrome, but that finding is controversial[142]; the exact cause of the metabolic disorder has not been established.

Cutaneous Zoster (Shingles): Herpes zoster[143,144] is much more common in the immunosuppressed patient. The frequency is approximately 2% in patients with CLL receiving imatinib, 10% to 15% in patients with CLL receiving fludarabine or alemtuzumab, 25% in patients with Hodgkin disease, 25% in patients receiving autologous stem cell transplant, and 45% to 60% among allogeneic stem cell transplant recipients.[130] Acyclovir prophylaxis may decrease the incidence.[131] Whether these figures will decrease as more patients contracting these cancers have been immunized in childhood against chickenpox is unclear. We recommend zoster vaccine only for patients with cancer in remission who have had no immunosuppressive or chemotherapeutic agents for at least 6 months. What effect the vaccine will have is unclear. One of our pediatric oncologists has the impression that zoster is less common in children with leukemia or lymphoma than heretofore.

The attack rate in patients with cancer is higher in children than adults, whereas in the immunocompetent population, the elderly are much more likely to be affected.[133] The peak incidence occurs within 6 months following RT and chemotherapy. In some instances, the zoster infection erupts at the site of a developing neoplasm or an RT portal. Herpes zoster is somewhat more likely to disseminate in patients with cancer than in the immunocompetent population, but in both groups the disorder usually runs a benign course. Cutaneous herpes zoster typically begins with pain, itching, or paresthesias in the dermatomal distribution in which the virus, latent in dorsal root ganglion neurons, reactivates. A rash follows in 3 to 7 days. The dermatome involved occasionally indicates metastatic disease of that dorsal root, but this phenomenon is so uncommon that it does not warrant extensive evaluation such as MRI unless other symptoms indicate neoplastic involvement of that site. The rash usually begins as patchy erythema in a dermatome, but after approximately 12 to 24 hours, vesicles appear that become pustules in a couple of days and then dry up within a week. Usually, the rash is mild and clears within a few weeks, but in patients with cancer the local rash may be severe, leading to ulceration and sloughing of the skin, which may require grafting. Many patients with herpes zoster have a few vesicles outside the involved dermatome, but widespread dissemination of the rash is more likely in cancer patients. The pain is usually maximal before and when the rash is present; it diminishes as the rash clears, leaving areas of hyperpigmentation or depigmentation with variable degrees of sensory loss in the dermatomal area. Occasionally, metastases to the skin may develop in a zosteriform distribution.[145,146] More than one-half of such patients were initially diagnosed as having herpes zoster and treated with antiviral agents. In one patient, cutaneous melanoma developed in the distribution of a previous infection with zoster.[147]

Complications. The virus may also spread proximally along the dorsal root into the spinal cord to cause a myelitis, may disseminate into the subarachnoid space to cause meningitis, or may seed the bloodstream with secondary spread to the nervous system causing an encephalitis.[135,148,149] Rarely, particularly when the ophthalmic division of the trigeminal ganglion is involved, the virus may cause a local vasculitis with carotid occlusion and cerebral infarction,[150,151] carotid dissection[152] or rarely a mycotic aneurysm.[153]

Several distinct neurologic syndromes are associated with herpes zoster:

Postherpetic neuralgia: In as many as 4% of patients with cancer and herpes zoster, severe pain outlasts the resolution of the rash. Postherpetic neuralgia is more common in those with cancer but is not directly related to the severity of the initial rash. The pain has two components: (1) lancinating or lightning-like pains in the dermatome, which usually clear spontaneously, and (2) a constant, burning pain, which may be incapacitating. Carbamazepine, pregabalin,[154] and other anticonvulsants relieve the lancinating component but usually have little effect on the burning pain. Early treatment with corticosteroids (see following text) may prevent the neuralgia. Other agents including topical lidocaine, antidepressants, and capsaicin may have some effect.[155] One report, suggesting that a low-grade viral infection in the dorsal root ganglion contributes to the pain, treated patients with postherpetic neuralgia with intravenous acyclovir followed by oral valacyclovir. About half of patients reported improvement in the pain.[156]

Paralysis: A few patients develop motor paresis involving neighboring spinal or cranial nerves. When weakness occurs, it usually appears a few days to a few weeks after the rash, although on occasion it precedes the rash. The weakness usually begins suddenly, reaching its maximum within a day or so. The muscles paralyzed are those supplied by the spinal root affected by the herpes, although motor weakness may be more widespread than the cutaneous abnormality and occasionally is distant from the cutaneous change. Most patients recover fully, but a few are left with permanent paralysis. Muscle paralysis, which affects approximately 0.5% of herpes zoster patients, appears to be no more frequent or severe in those with cancer.[157] The facial nerve and geniculate ganglion are commonly involved in Ramsay Hunt syndrome. Auditory and vestibular symptoms may accompany this syndrome. Paralysis of ocular or jaw muscles may complicate herpes involving the ophthalmic division of the trigeminal nerve.

In rare instances, either the pain of herpes zoster or motor loss associated with a rising herpes zoster titer has been reported without a cutaneous rash (zoster sine herpete).[158–161]

Myelitis: The virus may directly invade the spinal cord, causing a local inflammatory response. Symptoms usually appear several days (median =12 days) after the rash; rarely, the rash may be absent.[162,163] Myelitis begins with ipsilateral weakness and progresses over several days. The course may be remitting and exacerbating.[164] Neurologic dysfunction varies in severity from minor long-tract signs to a complete transverse myelopathy[165] resulting from spinal cord necrosis. The major pathologic alterations are in the dorsal root ganglia and adjacent posterior horn. Abnormalities include demyelination and necrosis, the latter from vasculitis. Viral particles are found in oligodendroglia, some astrocytes, and ependymal cells. The virus may spread vertically in the cord, causing symptoms and signs rostral to the dermatome with the cutaneous rash.[137] Detection of persistent viral DNA in CSF may aid in the diagnosis.[164]

Urinary retention: Rarely, zoster can affect the sacral roots, with or without rash. Patients may experience perineal pain and acute urinary retention. When the rash is absent, diagnosis can be difficult. The diagnosis is established by changing serum titers or detecting zoster DNA in the CSF by PCR.

Encephalitis: Encephalitis[166,167] is characterized by fever, headache, and delirium, sometimes with nuchal rigidity and pleocytosis. Pleocytosis is found in as many as 50% of uncomplicated cutaneous zoster infections, indicating that minor, clinically inapparent meningitis, myelitis, or encephalitis may be frequent. Antibodies have been found in CSF,[168,169] suggesting intrathecal synthesis. Three distinct pathological patterns have been identified[167]: (1) bland or hemorrhagic infarction resulting from large vessel occlusions; (2) deep white matter, ovoid mixed necrotic and demyelinated lesions resulting from small vessel vasculopathy as well as direct oligodendrocyte infection by the virus; (3) ependymal and periventricular necrosis from vasculopathy of subependymal vessels and secondary infection of ependymal and other glial cells in the periventricular region.[167]

One immunosuppressed patient developed a clinical picture suggesting limbic encephalitis with symmetrical medial temporal lesions on MRI (see Chapter 15). CSF PCR revealed varicella-zoster DNA; the patient responded to treatment.[170]

Cerebral vasculitis (Fig. 10–10): Two overlapping syndromes result from cerebral

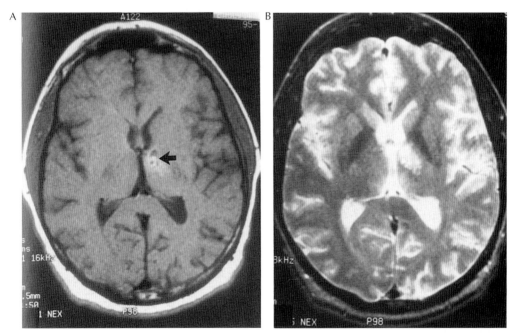

Figure 10–10. Cerebral infarct from herpes zoster vasculitis. This 38-year-old man with non-Hodgkin lymphoma suffered a sudden hemiparesis while recovering from cervical (C-2,C-3) herpes zoster. **A**: A T1-weighted MR scan demonstrated a basal ganglia infarct (*arrow*). **B**: A T2-weighted image confirmed findings suggesting that the infarct had resulted from a vasculitis involving penetrating vessels of the middle cerebral artery.

vasculitis.[171] One is characterized by head-ache and confusion progressing to stupor or coma, sometimes with multifocal or gener-alized convulsions. Focal neurologic signs, including hemiplegia, are sometimes present; pleocytosis is characteristic. Pathologically, vasculitis involves small vessels in the hemi-sphere ipsilateral to a trigeminal herpes, or it may be more diffuse in patients whose her-petic rashes have been elsewhere in the body and in whom the virus presumably reached the cerebrovascular structures by hemato-genous spread. Pontine infarction may follow cervical zoster[172] or thalamic infarction may follow lingual zoster[173] (Fig. 10–10). Viral particles have been identified in the vascular structures.[150,174] The infection may be associ-ated with either an intense inflammatory reac-tion[174] (i.e., granulomatous arteritis) or little or no inflammation.[150] The relationship, if any, between zoster vasculitis and the granuloma-tous angiitis associated with Hodgkin disease is unclear (see Chapter 9).

A second clinical syndrome, which occurs when large- or medium-sized arteries are infected, is characterized by the sudden onset of headache and contralateral hemiplegia fol-lowing ophthalmic-division trigeminal her-pes.[175] Confusion, stupor, and coma are less common in this disorder than in the more dif-fuse form. An occlusion of the carotid artery is visible on MR arteriography. A similar angiitis has been reported following primary varicella infection.[176]

Multifocal encephalopathy: Clinically similar to PML (a discussion of which fol-lows), this disorder affects patients with lymphoma or Hodgkin disease who have suffered from herpes zoster in the remote past. The diagnosis can be established only with brain biopsy and identification of the virus at the lesion site.[177] On MRI, it may resemble a bacterial brain abscess with ring enhancement.

Diagnosis: Diagnosing herpes zoster infection is usually easy. In the first few days the pain is dermatomal without rash. The cli-nician might suspect tumor in a spinal root, but the sudden onset without preceding ver-tebral pain suggests herpes zoster. The diag-nosis may be more difficult if motor changes precede the rash or if the rash develops in

an inconspicuous area. The sudden onset of Bell palsy, with or without eighth-nerve dysfunction in a lymphoma patient also sug-gests leptomeningeal metastasis and a modest pleocytosis seems to support that diagnosis. Nevertheless, a careful examination of the auditory canals, palate, and the scalp under the hair may yield evidence of sensory loss or a few herpetic vesicles. Serum titers for vari-cella-zoster virus should be drawn in the acute state and repeated a couple of weeks later. A fourfold rise suggests herpes zoster infec-tion. Occasionally, the rash of herpes simplex may also appear in a dermatomal distribution and be mistaken for herpes zoster. If doubt persists, the virus can usually be recovered from cutaneous vesicles.

The diagnosis of herpes zoster complica-tions is usually not difficult if the patient has had a recent cutaneous viral infection. Cerebrovascular involvement or delayed encep-halitis are more difficult diagnoses. In the appropriate clinical setting, cerebrovascular involvement can sometimes be established by arteriography demonstrating vasculitis and MRI showing multiple discrete subcortical nonenhancing lesions that coalesce and become enhancing and later involve gray matter sug-gesting varicella-zoster encephalitis.[178–180] Both PCR and antibody measurements in CSF are useful. In zoster vasculopathy, CSF antibodies may be positive when PCR is negative.[181] The virus can be found in the saliva of patients with active disease.[182] However, establishing the diagnosis of herpes zoster encephalitis often requires biopsy.

Ideally, herpes zoster is better prevented than treated, but the role of herpes zoster vac-cine is unclear. It should not be given to amelio-rate active infection or postherpetic neuralgia. Some have suggested that high-risk patients (e.g., stem cell transplant patients) be given prophylactic antiviral agents[183]; postexposure prophylaxis has also been recommended.[184] Valacyclovir 1 g po 3 times a day on postexpo-sure days 3 to 22 is the MSKCC recommen-dation.[184] All immunosuppressed patients with herpes zoster infections should be treated with acyclovir, famciclovir, or valacyclovir. These drugs may decrease late complications, including postherpetic neuralgia.[185,186] Corti-costeroids also reduce the incidence of pos-therpetic neuralgia if given in the acute stage of the rash, but they probably should not be given to immunosuppressed patients suffering from the disorder. The treatment of postherpetic neuralgia is often frustrating for both the physician and patient. Various approaches are discussed in detail elsewhere.[155]

Herpes Simplex

Systemic herpes simplex infections are common in the immunosuppressed host.[130] Patients with CLL treated with fludarabine have a 15% incidence; patients receiving stem cell transplants or those with acute leukemia, a 90% incidence.[130] Herpes simplex infection is usually characterized by ulcerative mucositis.[187,188] Herpes simplex encephalitis appears to be increased in the immunosuppressed population (Fig. 10–11). We have encountered several patients with lymphoma[189,190] or primary[190] or metastatic brain tumors treated with corticosteroids who developed herpes simplex encephalitis, either type 1 or type 2.[191] The disorder has also been reported to follow craniotomy for either benign or malignant brain tumors.[192,193] Case reports describe the illness during radiotherapy for a pontine glioma,[194] during adjuvant chemotherapy for breast cancer,[195] and during chemotherapy for a pineal tumor (in that case associated with herpetic necrosis of the retina).[196] One case report describes a herpes type-2 brainstem encephalitis associated with B-cell lymphoma.[197] Characteristically, an otherwise stable patient suddenly exhibits altered behavior and consciousness and possibly headache or language disturbances, sometimes associated with focal or generalized seizures but usually without fever. In most, the disease course appears no different from that of herpes simplex encephalitis in the immunocompetent population.[198] Abnormalities in the medial temporal lobe on MR scan resembled those in the immunocompetent patient (Fig. 10–11). A single case report[199] describes a patient with Hodgkin disease and impaired immunity who had an atypical slowly progressive clinical course leading to death. The pathologic features consisted of widespread neuronal destruction, astrocytic proliferation, and inclusion bodies without the typical inflammatory changes or hemorrhagic necrosis generally found in the medial temporal lobes.

The diagnostic problem is particularly challenging in a patient with pre-existing brain disease such as metastases. Clinical suspicion of viral encephalitis should lead to immediate antiviral treatment.[198] The CSF may contain RBCs and WBCs. Cultures are negative, but PCR, which identifies herpes simplex virus DNA in the CSF, allows an early diagnosis in most instances.[87] The EEG may demonstrate partial complex seizures that may suggest the diagnosis. Depending on the clinical situation, a brain biopsy may be necessary to establish a definitive diagnosis. Histologic changes at autopsy are usually typical of those found in the immunocompetent host, with inflammation and necrosis primarily in the medial temporal lobes. Both meningitis and myelitis may accompany herpes simplex type-2 infections.[200]

Progressive Multifocal Leukoencephalopathy

An infection of oligodendrogliocytes, PML is caused by the JC virus, a papovavirus (Fig. 10–12).[201] The virus is acquired by most people in childhood and is not associated with a significant illness. The virus remains latent, and with impaired cellular immunity, particularly with lymphoma or CLL, it may reactivate and cause a rapidly progressive demyelinating illness characterized by multifocal signs (usually hemispheral but sometimes cerebellar) that progress over 3 to 6 months to bilateral paralysis and stupor or coma. The illness has also been reported in lymphoma patients treated with rituximab[202] or fludarabine.[203] The diagnosis should be suspected in patients with depressed cellular immunity from disease or prolonged chemotherapy who develop progressive neurologic dysfunction, particularly when the signs are multifocal and unaccompanied by headache, seizures, or changes in the CSF. The MRI[124] may be normal very early in the course of the illness but eventually reveals multifocal (occasionally single), punched-out lesions of the white matter that usually do not contrast enhance and rarely cause mass effect (Fig. 10–12). MRS reveals a lactate/lipid peak with myoinositol preserved or elevated. The choline creatinine ratio may be increased.[124] Occasionally, the scan of nonenhancing lymphomas may be similar. Occasional patients

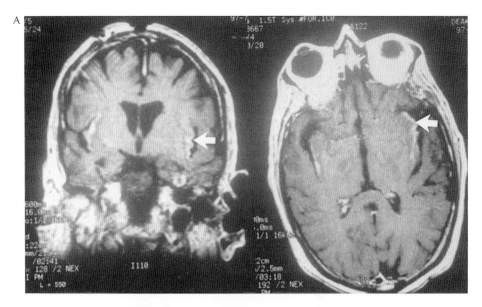

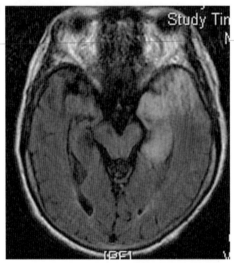

Figure 10–11. Herpes simplex encephalitis. **A**: A contrast-enhanced magnetic resonance (MR) scan of a man with stage IV non–Hodgkin lymphoma. He was being treated for epidural spinal cord compression and suddenly became confused and disoriented. Lumbar puncture revealed white cells in the spinal fluid, and the contrast-enhanced MR scan showed enhancement with bilateral involvement of the medial temporal cortex (*arrow*). An initial diagnosis of leptomeningeal tumor was made. At autopsy, the patient was discovered to have herpes simplex encephalitis. **B**: This patient presented with acute confusion while on chemotherapy for her cancer. The fluid attenuated inversion recovery (FLAIR) MR image showed unilateral hyperintensity in the medial temporal lobe. Cerebrospinal fluid (CSF) polymerase chain reaction (PCR) was positive for herpes simplex. With antiviral therapy, she made a good recovery.

with partially preserved immunity may show contrast enhancement, suggesting a better prognosis. The diagnosis is made by CSF PCR for JC virus[203]; there is no effective treatment.[204]

In a rare patient, especially one who can mount an inflammatory response, the disease remits spontaneously or stabilizes for long periods.[205] Likewise, if one can restore immunity, the disease may remit. One patient with lymphoma developed PML while being treated with interferon immunotherapy. After the drug was discontinued, the condition improved spontaneously. The patient had a ring-enhancing lesion.[206]

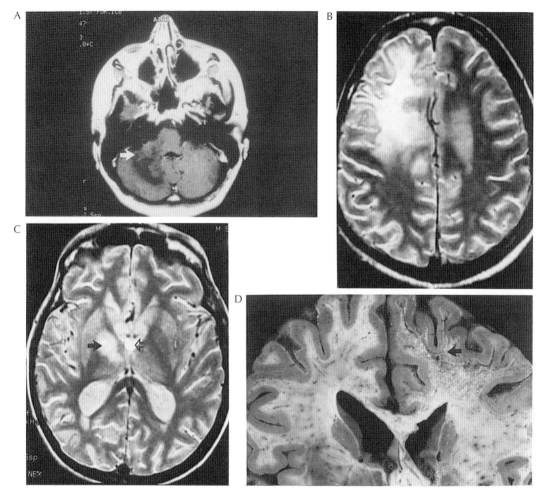

Figure 10–12. Progressive multifocal leukoencephalopathy (PML). **A:** A patient with chronic lymphocytic leukemia presented with a single cerebellar lesion (*arrow*). The lesion was hypointense on T1-weighted images and did not enhance. The spinal fluid examination was normal. The patient's only symptom was progressive cerebellar dysfunction, which was originally thought to be paraneoplastic. A needle biopsy specimen revealed PML. **B:** Another case of PML with the typical hyperintense T2-weighted magnetic resonance image restricted to white matter, sparing the cortex. **C:** In this patient, however, the virus also involved basal ganglia (*solid arrow*) and brainstem. An incidental colloid cyst (*open arrow*) caused mild hydrocephalus. **D:** The autopsy specimen shows the typical white matter abnormality (*arrow*) that spares the cortex.

Cytomegalovirus

CMV infects approximately 50% of persons by the time they are 50 years of age, but more than 90% of these are asymptomatic.[207] The virus in the nervous system may reactivate in patients with depressed cell-mediated immunity, particularly after transplantation and during chemotherapy with T-cell depleting agents.[130,188] The incidence of infection is from 5% to 75% in patients with hematologic malignancies.[130] The disorder can involve the brain, causing symptoms of subacute encephalitis with headache, confusion, and lethargy. One of our patients had focal brainstem signs only, accompanied by a contrast-enhancing lesion on MRI. The virus has a predilection for ependymal cells of the lateral ventricles[207,208] and may be suspected when MRI shows contrast-enhancing lesions as streaks surrounding the lateral ventricles (owl eyes)[209] in a patient with clinical encephalitis. CMV also causes a retinitis characterized by large yellow or white exudates as well as perivascular hemorrhage.[210]

Seroconversion, or an increase in the already positive CMV antibody titer, supports the diagnosis. The virus can sometimes be cultured from the urine, saliva, or buffy coat, although rarely from CSF or brain tissue. A plasma PCR test for CMV DNA may detect clinically relevant evidence of CMV viremia. Brain biopsy specimens reveal periventricular necrosis and giant cells, fused macrophages (cytomegaly) for which the virus, found in intracellular inclusions, is named. CSF PCR is diagnostic.

Prophylaxis for at-risk patients (those with lymphoma or CLL receiving alemtuzumab) with oral valganciclovir may prevent infection.[211] Treatment with ganciclovir, or oral valgangciclovir with or without foscarnet has been effective in the treatment of CMV retinitis and encephalitis.[211]

The virus has also been associated with a Guillain-Barre–like polyneuritis in both immunocompetent and immunosuppressed hosts.[212] The polyneuropathy sometimes resembles a cauda equina syndrome with urinary retention and lower extremity sensory loss. The CSF contains WBCs, often with a polymorphonuclear predominance, increased protein concentration and hypoglycorrhachia. The virus can often be isolated from the CSF and PCR may identify viral DNA in CSF.[212] Pathologic changes include a vasculitis of nerve roots and CMV inclusions.[213]

Other Viruses

Although their incidence is unknown, several other viruses have been reported to infect the CNS in patients suffering from cancer.[130] They include measles,[214,215] adenoviruses,[216] and others. Human herpesvirus 6 has been reported to cause a meningoencephalitis in patients after allogeneic stem cell transplantation.[132] The disorder occurred in 11 of 1148 patients (0.96%) in one series.[132] Patients developed fever and a rash followed by confusion, sometimes with seizures and/or short-term memory loss. The MRI was usually abnormal with hyperintensity in the medial temporal lobes and hypothalamus. CSF PCR was positive. Treatment with ganciclovir and foscarnet may be effective.

West Nile virus can affect immunocompromised patients, causing encephalitis,[217] myelitis with flaccid paralysis resembling poliomyelitis,[218] meningoencephalitis,[219] chorioretinitis, and optic neuropathy.[220] The illness resembles that seen in the immunocompetent host, with fever and myalgia followed by neurologic symptomatology. In the patient with encephalitis, the presence of chorioretinitis or optic neuropathy suggests the diagnosis, as does a relative lymphopenia and EEG showing more prominent slowing in anterior than posterior regions.[217] The MRI may show hyperintensities in the basal ganglia and thalami.[221] The CSF has a pleocytosis and viral RNA on PCR.[222] There is no established treatment. An investigational immunoglobulin product is being studied.[222]

PARASITES

Toxoplasma

Toxoplasmosis gondii is the most common parasitic CNS infection in patients with cancer and impaired cellular immunity. Before the AIDS epidemic, Hodgkin disease was the primary offender, with leukemias and other lymphomas being somewhat less common.[223,224] Once extremely common in patients with AIDS, the incidence is now decreasing[225,226] as patients receive anti-retroviral therapy. In patients with Hodgkin disease, *Toxoplasma* brain abscesses are more common than brain metastases, but the reverse is true in non–Hodgkin lymphomas. Toxoplasmosis also sometimes affects those with other cancers who are treated with steroids or chemotherapy[224] and patients who receive allogeneic stem cell transplants.[227] The pathologic changes of CNS toxoplasmosis may consist of (1) a meningoencephalitis with multiple small necrotic abscesses with little inflammatory response, scattered throughout the cortex and subcortex, that are sometimes not visible on MRI, or (2) one or more large mass lesions with a predilection for the basal ganglia.[228-230] Clinically, the disorder is characterized either by focal neurologic signs, usually in the absence of headache and fever, or by a diffuse encephalopathy accompanied by seizures. When results are positive, the MRI reveals one or more ring-enhancing lesions with a predilection for basal ganglia and cerebral cortex (i.e., gray matter rather than white matter). Occasionally, the lesions are hyperintense without contrast enhancement or are homogeneously enhancing, the latter resembling CNS lymphoma. A major problem is differentiating toxoplasmosis from metastatic or primary CNS lymphoma. Unfortunately, imaging does not differentiate the two reliably.[124] Both can show restricted diffusion[93]; SPECT imaging is sometimes, but not always, helpful. The lymphoma, unless necrotic, demonstrates an increased nuclide uptake, whereas the toxoplasmosis does not.[231] MRS usually reveals a large lipid signal. A lipid signal is also present in lymphoma but is generally smaller.[124] The presence of a deep-seated mass lesion in the brain of a Hodgkin disease patient is virtually pathognomonic of *Toxoplasma* abscess, and the patient can be treated on that basis.

Acquired *Toxoplasma* retinitis that causes acute visual loss has been reported[232] in several patients with depressed immunity. The organism may also cause an acute or subacute meningitis. If the patient has meningitis, the CSF has a lymphocytic pleocytosis with high protein and sometimes low glucose concentrations.[233] The organism is usually not isolated from the spinal fluid, but PCR is virtually always positive when the nervous system is involved.[227,234]

Treatment with sulfadiazine and pyrimethamine usually gives a rapid response and is quite safe; therefore, the therapy often establishes the diagnosis without biopsy. Biopsy should be considered only if this treatment does not lead to clinical and radiographic improvement within a week to 10 days. PCR analysis of stored peripheral blood samples may predict patients at risk who are receiving allogeneic stem cell transplant.[227]

Whether the brain infection is a re-exacerbation of an old asymptomatic infection or a new one is not clear. Usually, IgM and IgG antibodies are positive and rise during the infection; some exceptions occur, particularly in patients with severe hypogammaglobulinemia or with AIDS. Antibody titers should be measured in both blood and CSF; they may be positive in CSF when absent from blood. PCR for *Toxoplasma* should be sent from CSF samples. Treatment includes sulfadiazine, 1 g po qid, and pyrimethamine, 100 mg/day for 1 day and then 25 to 50 mg each day thereafter. When sulfadiazine is unavailable, treatment with cotrimoxazole may be used. The optimal duration of treatment is unknown. Folinic acid, 10 mg/24 hours, prevents pancytopenia.

Strongyloides Stercoralis

A worm that resides in the upper intestine of individuals in the Southeastern and South Central United States,[235,236] *Strongyloides stercoralis* causes CNS infection in patients with impaired cellular immunity, particularly those who have been on high-dose corticosteroids[237,238] or after stem cell transplant.[238] The reason corticosteroids are so strongly implicated in the pathophysiology of this syndrome is that they suppress eosinophils that are probably required to keep the infection under control. In infected, but usually asymptomatic patients, the larvae escape from the bowel and disseminate hematogenously to the CNS, causing a meningoencephalitis, sometimes with a concomitant bacterial meningitis and pneumonitis from enteric bacteria carried by the worms. The clinician should suspect this disorder when a patient on corticosteroids develops upper abdominal cramping pain associated with a headache and stiff neck. The patient may become encephalopathic and lapse into coma. Cerebral blood flow studies may suggest vasculitis, probably because the larvae occlude small vessels in the brain.[239] If several Gram-negative organisms are found in CSF or blood, the diagnosis is more strongly suspected. The worm itself has been found in the CSF[240–242] in a few patients, but the diagnosis is usually established by finding larvae in the stool, sputum, or duodenal contents. The patient is treated with thiabendazole, 25 mg/kg bid, for 7 days. Prevention is far better than treatment.[242] Examination of stool or duodenal aspirates should be performed in patients from endemic areas who are about to be given immunosuppressive drugs. If the results of one of these tests are positive, the patient should be treated prophylactically.

Eosinophilic Meningitis

Eosinophilic meningitis (>10% eosinophils in CSF) is usually caused by parasitic CNS infections or coccidiomycosis.[243] However, in some patients with Hodgkin disease,[244,245] lymphoma,[246] or leukemia[247,248] and rarely leptomeningeal carcinoma,[249] eosinophils appear to be a reaction to meningeal infiltration by neoplasm.[249] One report[247] describes basophilic meningitis in a similar situation. Eosinophilic meningitis may also complicate foreign bodies such as shunt tubing in the CSF.[250,251] Other causes of eosinophilic meningitis[243] include sarcoidosis,[252] nonsteroidal anti-inflammatory drugs,[253] and the hypereosinophilic syndrome.[254] Eosinophils may themselves cause CNS dysfunction,[255,256] presumably by cytokine release.

Neutrophil Defects

Table 10–9 lists the organisms that commonly invade the CNS of patients with granulocytopenia or other defects of neutrophil function.

Table 10–9. **Central Nervous System Infections in Patients with Neutrophil Defects (Granulocytopenia)**

Infection	Bacteria	Fungi
Meningitis	Enteric bacilli	*Candida* sp.
	Pseudomonas aeruginosa	
	Escherichia coli	
	Klebsiella pneumoniae	
	Listeria monocytogenes	
	Streptococcus pneumoniae	
	Staphylococcus aureus	
Meningoencephalitis	Same	*Aspergillus* sp.
Abscess	Same	*Mucoraceae*
		Candida sp.

These defects account for approximately 25% of CNS infections in patients with cancer. The most common cause of neutrophil defects in patients with cancer is neutropenia due to cytotoxic chemotherapy or bone marrow replacement by tumor. Corticosteroids also impair neutrophil function and, when used to treat such patients, may enhance the likelihood of CNS infection. Although certain organisms, such as *Listeria,* do affect patients with neutrophil defects as well as those with T-cell deficits, by and large different organisms affect the two groups. The major offenders in those with neutrophil defects are enteric bacilli and several fungal organisms that are usually nonpathogenic. Parasitic and viral infections are not a specific complication of neutrophil defects.

Neutropenic CNS bacterial infections develop when the absolute neutrophil count falls lower than 1000/mm³.[6] *Pseudomonas* and *E. coli* are the major offenders.[6] Approximately 50% of neutropenic patients who develop CNS infections have also been receiving steroid therapy, and a few have ventricular catheters in place. The underlying disorder is usually acute leukemia or lymphoma. Patients present with either a new fever or an increasing body temperature if already febrile. Most develop acute or subacute changes in mental state, with confusion, lethargy, and occasionally coma. Some have headache, and approximately 40% have seizures, but few have nuchal rigidity.

The diagnosis is difficult because most patients are already seriously ill systemically. Even when the clinician suspects infection because of increasing confusion and fever, examining the CSF for cells may not be useful because many patients lack pleocytosis, but the results of Gram stain and culture are usually positive. In the original series,[6] 14 of 31 patients had fewer than five white cells in the CSF; in the most recent series, three of six patients had an entirely normal spinal fluid.[121]

ENTERIC BACTERIA

Neutropenic patients or those with abnormal communication between the CSF and an epithelial surface, particularly the bowel, are susceptible to CNS infection caused by enteric Gram-negative bacilli that reside in the patient's own GI tract.[257] On occasion, Gram-negative meningitis complicates the course of patients with cell-mediated immune disorders.[257] At MSKCC, *P. aeruginosa* was the most common offending organism in the original series, followed by *E. coli* and *Klebsiella, Enterobacter,* and *Proteus* species. Salmonella was a rare offender.[257] The organisms may cause either meningitis or a brain abscess which may contain more than one organism. *P. aeruginosa* often causes a vasculitis that may, in the septic patient, lead to multiple small cerebral infarcts.

CNS infection from enteric Gram-negative bacilli is usually an acute meningitis. The CSF is sometimes cloudy, but may be devoid of WBCs, particularly when the patient is severely neutropenic. Despite the absence of WBCs, the glucose concentration is usually low and the protein concentration high. Organisms are readily apparent in the Gram stain specimen.

Treatment is outlined in Table 10–10. In some patients whose initial response is unsatisfactory, treatment may be improved by the placement of an Ommaya reservoir and direct intraventricular injection of the antimicrobial agents. Nevertheless, the prognosis in our original series was poor. Survival rates at MSKCC varied between 20% and 40%. Many patients, however, are desperately ill from cancer, side effects of its therapy such as thrombocytopenia and anemia and systemic infection. In those patients who are in relatively good condition at the time the Gram-negative meningitis develops, vigorous treatment usually yields a satisfactory outcome without significant sequelae.

In hospitals whose patients are infected with organisms that have extended spectrum beta-lactamase inhibitors (ESBLs) use of all beta-lactams (including penicillins and their derivatives and cephalosporins), regardless of

Table 10–10. Treatment of Common Central Nervous System Pathogens in Patients with Cancer

Organism	Recommended Therapy	Alternate Therapy*
Listeria monocytogenes	Ampicillin IV, 2–3 g q4h + gentamicin 1.25 mg/kg q6h	Cotrimoxazole IV 5 mg/kg q6h
Streptococcus pneumoniae	Penicillin IV, 2 × 10⁶ U q4h; ceftriaxone and vancomycin	Ceftriaxone 1–2 g q 12h or cefotaxime 2 g q4h; Vancomycin 1 g q12h
Cryptococcus neoformans	Amphotericin B IV 0.7–1.0 mg/kg per day plus flucytosine po 75–100 mg/kg per day in 4 divided doses ± intraventricular amphotericin B	Fluconazole IV 800 mg qd†
Escherichia coli	Ceftazidime [cefepime] 2 g q8h + gentamicin 1.25 mg/kg IV q6h and gentamicin IT 0.1 mg/kg q12–24h	Cotrimoxazole IV 5 mg/kg q6h trimethoprim + gentamicin IV and IT; ciprofloxacin IV 400 mg q12h
Klebsiella pneumoniae	A third- or fourth-generation cephalosporin + gentamicin IV 1.25 mg/kg q6h + gentamicin IT 0.1 mg/kg q12–24h	Ciprofloxacin IV 400 mg q12h; cotrimoxazole/trimethoprim IV 5 mg/kg q6h + gentamicin IV and IT
Pseudomonas aeruginosa	Ceftazidime 2 g q8h + tobramycin IV 1.25 mg/kg q6h + tobramycin IT 0.1 mg/kg	Imipenem 1 g q6h + tobramycin IV and IT
Nocardia asteroides	Sulfadiazine IV 15 mg/kg q6h or 1.5 g orally	Cotrimoxazole as above; imipenem IV 500 mg q6h + amikacin 7.5 mg/kg q12h
Toxoplasma gondii	Sulfadiazine IV 15 mg/kg + pyrimethamine 100 mg po × 1, then 25–50 mg qd	Clindamycin 600–900 mg q6–8h + pyrimethamine
Strongyloides stercoralis	Thiabendazole po 25 mg/kg q12h x 10 days	Ivermectin 200 mg/kg per day; Albendazole po 400 mg qd × 10 days
Staphylococcus aureus	Oxacillin IV 2 g q4h‡	Vancomycin IV 1 g q12h + rifampin 300 mg q12h
Staphylococcus epidermidis	Vancomycin IV 1 g q12h; rifampin 300 mg q12h	
Corynebacterium sp.	Penicillin IV 2 × 10⁶ U q4h	Vancomycin IV 1 g q12h

*Other semisynthetic penicillins, such as nafcillin, may be substitutes for oxacillin; aminoglycosides, such as amikacin or tobramycin, may be substituted for gentamicin; cotrimoxazole is the same preparation as trimethoprim-sulfamethoxazole.
† Use only in patients unable to tolerate amphotericin B.
‡ If oxacillin resistance is common, vancomycin should be used initially; change to oxacillin if the organism is sensitive.
IT, intrathecal; IV, intravenous; po, oral.

the results reported on the susceptibility panel should be avoided. Rather, the carbapenam class of antibiotics (imipenem and meropenem) should be used. In addition, empiric use of carbepenams should be considered if dictated by local hospital epidemiology.

Treating Gram-negative brain abscesses is more difficult. In the cancer population, these abscesses usually develop when the patient is terminally ill, so it is difficult to judge the results of antimicrobial treatment. In most instances, the offending organism is not known but can be suspected from the clinical setting and the presence of sepsis with positive blood cultures. The initial treatment is outlined in Table 10–6.

Brain abscesses can be treated successfully with antibiotics alone, even in immunosuppressed patients. Therefore, when a brain abscess caused by Gram-negative organisms is suspected, treatment may begin with antibiotic therapy and the patient followed up by MRI. Stereotactic needle aspiration of abscess contents can both establish a diagnosis and relieve symptoms. Only when antibiotic therapy and stereotactic aspiration have failed is therapeutic excision considered, provided that the patient is in sufficiently good systemic condition. The decision for surgery is based on the patient's clinical response to treatment, not only on the scan; the contrast-enhancing abnormality may persist long after the patient has become asymptomatic and antibiotic therapy has been discontinued. In a patient with a good clinical response, we discontinue antibiotic treatment after 4 weeks and follow the MRI. If the lesion is either shrinking or stable, no further therapy is given. If the lesion begins to enlarge or the patient relapses, surgical excision is considered.

FUNGI

In patients with neutrophil deficits, fungal infections are typically nosocomial superinfections, and the particular fungus varies with the hospital environment. Current evidence[258] suggests an increasing incidence of nosocomial systemic fungal infections.[258] Furthermore, there appears to be an increase in emerging uncommon fungal pathogens.[259]

The fungi usually invade the brain parenchyma, causing abscesses. When meningitis occurs, it is usually caused by *Candida*.[258] Some of the organisms invade blood vessels and thus can cause cerebral infarction from thrombosis or cerebral hemorrhage from disruption of the blood vessel wall[260] (Fig. 10–7). In most instances, the organisms spread from lung to brain, although in some patients, they may spread from paranasal sinuses and cause rhinocerebral mucormycosis (see the following).[260] In general, a diagnosis can be made only by biopsy of the involved area. Fungi should be considered a potential cause of CNS infections in any neutropenic patient who fails to respond to appropriate antibiotics. Recent reviews address problems of fungal infections and their treatment in the immunocompromised host.[260,259]

Aspergillus

This fungus causes brain abscess, cerebral infarction, cerebral hemorrhage, or mycotic aneurysms.[259,261] A primary infection is usually present in the lung, and approximately 10% of pulmonary infections disseminate to the brain. The disease is rarely isolated to the CNS, usually entering via the respiratory tract and paranasal sinuses. It typically appears in an ill patient who is neutropenic, septic, and treated with multiple antibiotics. Brain involvement is suspected only when the patient becomes delirious or develops seizures or focal signs.[262] The MRI may show single or multiple ring-enhancing lesions, or lesions suggesting infarction (either bland or hemorrhagic) in more than one vascular territory; MR angiography may give evidence of multiple vascular occlusions. The CSF is usually normal except for a slightly elevated protein concentration, and no organisms are isolated. However, CSF PCR, enzyme-linked immunosorbent assay, and latex agglutination tests are positive and establish the diagnosis.[263] Serum and CSF galactomannan assays may also be useful both for establishing the diagnosis and following treatment.[264]

The differential diagnosis usually includes cerebral vascular disease that is due to embolization or cerebral abscess caused by other organisms. Because most patients are terminally ill when the neurologic symptoms develop, an antemortem diagnosis is rarely made and clinically diagnosed patients are much less common than those found at autopsy.[258] Treatment should be started empirically in the appropriate clinical situation. Amphotericin is the standard treatment, but recent studies suggest that voriconazole may be more effective.[265–267] Early diagnosis and treatment of systemic or sinus infection may eradicate the disease before it spreads to the CNS.

Zygomycetes

Zygomycetes (also called zygomycosis) is usually a fatal fungal infection caused by fungi from the *Mucorales* order. It occurs in patients with leukemia and lymphoma[258] with neutropenia, as well as in patients with organ transplants, diabetic ketoacidosis, or metabolic acidosis that is due to renal failure or diarrhea. The classic form of CNS *Zygomycetes* is the syndrome of rhinocerebral mucormycosis.[268–270] This infection begins in the nasopharynx or pharynx and spreads to involve the paranasal sinuses, the orbit, and then the frontal lobe of the brain. The clinical findings first suggest sinusitis followed by unilateral proptosis, ptosis, and ophthalmoplegia. Decreased vision occurs late. Once the brain is involved, the patient develops seizures, focal neurologic signs, or delirium. Careful examination of the nose, nasopharynx, or palate may reveal a black, necrotic ulcer that, when scraped, will yield the organism. *Aspergillus* hyphae may cause the same clinical symptoms, but the two organisms are easily differentiated: the *Mucorales* forms are irregular, wide, ribbon-like hyphae that branch at right or nearly right angles, whereas the *Aspergillus*, with uniform caliber and cross septations, branch at acute angles.

Successful treatment of rhinocerebral mucormycosis requires surgical debridement of infectious tissue,[268] treatment of the underlying condition, and parenteral amphotericin B, but in the cancer population, the disorder is usually fatal. Posaconazole is a recently approved oral azole that is active in vitro and approved for this indication.[271]

Mucormycosis also causes brain abscesses by hematogenous dissemination from pulmonary lesions. The single or multiple necrotic mass lesions resemble those caused by aspergillosis. Occasionally, they occlude cerebral vessels leading to cerebral infarction (Fig. 10–13).[272] The diagnosis of this particular form of mucormycosis is rarely made antemortem and

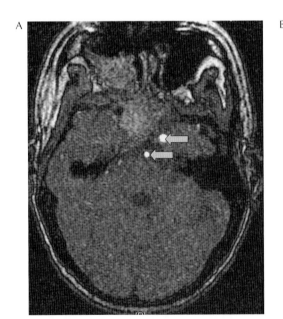 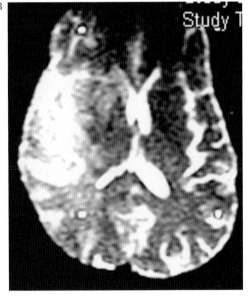

Figure 10–13. Rhinocerebral zygomycosis. A 52-year-old man, previously in good health, complained of headache. A computed tomography scan revealed sinusitis but because of abnormal blood counts a bone marrow was performed, revealing Burkitt lymphoma. While under treatment for the lymphoma, his headache intensified; he became hemiplegic on the left side and eventually became comatose and died. Autopsy revealed fungal infection in the sinuses with occlusion of the carotid artery. **A**: Magnetic resonance angiography showed hyperintensity in the sinuses and epidural space with a normal left carotid and basilar artery (*arrows*) but an absent right carotid artery. **B**: A diffusion-weighted image showing hyperintensity in the right hemisphere, indicating a large acute stroke.

can only be established by brain biopsy. As with aspergillosis, the CSF is usually normal. In patients suffering from acute cerebral disease, the presence of the organism in skin or lung is sufficient to warrant treatment.

Candida

The most frequent systemic fungal infections, *Candida albicans* and *C. tropicalis*, in cancer patients[258,265] cause meningitis or, more commonly, meningoencephalitis.[273,274] Multiple small abscesses are scattered throughout the cerebral cortex. Vertebral osteomyelitis has also been reported.[275] *Candida* may also cause vascular occlusions with infarction. The organisms usually disseminate hematogenously from the GI tract, from infected IV catheters, or from other sources of *Candida* sepsis. Unlike aspergillosis and mucormycosis, CSF pleocytosis, elevated protein, and depressed glucose concentrations are often present. CSF cultures may not test positive unless large samples of CSF (20–30 mL) are concentrated and the sediment cultured. If repeated lumbar punctures fail to reveal an organism, cisternal puncture may be positive. In some instances, the organism is never identified in the CSF even when the patient suffers from clinical meningitis[273]; CSF PCR should be diagnostic.[276] If PCR is negative, in selected cases, biopsy of the meninges may be helpful. The biopsy specimen should be taken from a symptomatic area,[277] usually at the base of the brain where the infected material tends to concentrate or where enhancement is present on MRI.[277] In one patient, a hemispheral biopsy failed to reveal the organism even though the patient died of candidal meningitis a few weeks later. The diagnosis of a CNS infection may be helped by the fact that disseminated candidiasis sometimes causes fever, rash, and myalgias from direct infection of muscle.[278] The treatment is amphotericin B and flucytosine.[265]

Splenectomy or B-Cell Abnormalities

Although splenectomy is no longer a part of the treatment of Hodgkin disease, splenectomized individuals still are encountered. Table 10–11 lists the common infections in Hodgkin disease patients who have undergone a splenectomy and in patients with immunoglobulin deficits (i.e., B-cell abnormalities). The most common infections are bacterial meningitis (most often acute pneumococcal meningitis), usually caused by the encapsulated organisms that cause meningitis in the general population.

BACTERIA

Streptococcus Pneumoniae

In the past, pneumococcal meningitis was relatively common in patients with Hodgkin disease who had undergone splenectomy as well as intensive RT and chemotherapy. Systemic infection is now more common in patients undergoing stem cell transplant though meningitis is quite rare.[15] *Pneumococcus* may also attack a patient with a CSF fistula following surgery. The splenectomized patient may be in complete remission or cured of the cancer when the meningitis begins, usually with the

Table 10–11. **Meningitis in Patients with Splenectomy or B-Cell Abnormalities (Hypogammaglobulinemia)**

	Bacteria	Fungi	Parasites	Viruses
Splenectomy	*Staphylococcus pneumoniae*	—	—	—
	Haemophilus influenzae			Rare
	Neisseria meningitidis			Rare
Hypogammaglobulinemia	Same	—	—	Echovirus Coxsackie B

typically fulminant onset found in the general population. The CSF is purulent, with many hundreds or thousands of neutrophils, but in some patients, WBCs may be absent from the CSF early in the illness, even when the peripheral WBC count is normal. The glucose concentration is typically low, and the protein concentration is high. Organisms are usually visible on Gram stain; CSF PCR when positive is diagnostic.[279] An immunochromatographic test is also useful.[280] The organism tends to cause a vasculitis that often leads to brain infarction with attendant focal neurologic signs.

Ceftriaxone, cefotaxime, or vancomycin are used for penicillin-resistant organisms. Ceftriaxone and cefotaxime-resistant organisms are increasingly common, so vancomycin should be included in any initial regimen, pending susceptibility results. If the diagnosis is made early and effective treatment begun, the patient usually recovers. Permanent sequelae are uncommon. A polyvalent vaccine against pneumococcal polysaccharide is approximately 80% effective in preventing pneumonia in healthy adults, but almost 40% of bacteremic pneumococcal infections found at MSKCC were caused by non-vaccine serotypes.[10] Because the vaccine is safe, however, it should be given to patients who are at high risk for developing pneumococcal infections, particularly patients who have undergone splenectomy or who have CLL. As indicated in the preceding text, corticosteroids are indicated for the treatment of bacterial meningitis.[34]

Although *H. influenzae* type B and *N. meningitides* should be common CNS infections in the same population vulnerable to pneumococcal infections, these two organisms rarely cause CNS infections in the cancer population. Only one instance of either infection in the CNS has been reported at MSKCC.

Cerebrospinal Fluid–Surface Communication

Table 10–12 lists the common infections in cancer patients with a CSF–surface communication caused by either intraventricular shunts or reservoirs or from surgery around the head or spine. As indicated previously, the incidence of meningitis due to immunosuppression has decreased at MSKCC, but the incidence of meningitis resulting from neurosurgical procedures has increased. Between 1955 and 1971, 30 (30%) of 104 patients with meningitis developed the infection as a result of neurosurgical manipulation.[2] Between 1993 and 2004, 58 of 77 (75%) infections occurred in patients with Ommaya reservoirs, ventriculoperitoneal shunts, or hardware placed during surgical procedures.[121] These data reflect an increase in the number of neurosurgical procedures being done. Not all of these infections occurred shortly following the procedure; many were the result of manipulation of the Ommaya reservoir.

BACTERIA

Staphylococci are the common offenders when a CSF leak develops following a neurosurgical procedure. In the patient with cancer and a CSF–surface anatomic connection,

Table 10–12. **Central Nervous System Infections with Cerebrospinal Fluid–Surface Communication**

	Bacteria	**Fungi**	**Parasites**	**Viruses**
Meningitis	*Staphylococcus aureus*	*Candida*	Trichomonadida	—
	Coagulase-negative staphylococcus			
	Propionibacterium acnes			
	Enteric bacilli			
Meningoencephalitis	Same	Same	Same	—
Abscess	Same	Same	—	—

Staphylococcus aureus usually causes an acute or subacute infection with headache, fever, stiff neck, and abundant WBCs and organisms in the CSF. The diagnosis is suggested on Gram stain, and the infection can be treated with oxacillin. For methicillin-resistant strains, vancomycin and rifampin are the drugs of choice. Repairing the leak is essential.

A more common clinical problem is infection of a ventricular shunt or an Ommaya device[281–283] (see Chapter 7). These infections are usually indolent, and the patient may be entirely asymptomatic or suffer only from malaise. In a few instances, the infection was discovered years after the reservoir had last been accessed.[284,285] The CSF usually, but not always, contains an increased number of WBCs, predominantly mononuclear. The organism may or may not be seen on Gram stain but grows on culture. In patients with Ommaya devices, pleocytosis and organisms may be identified from ventricular but not lumbar CSF or vice versa. If the clinician suspects CNS infection in this setting, both ventricular and lumbar CSF must be cultured. Because the infection is so indolent and because staphylococci commonly reside on the skin, the positive culture may represent contamination and not a true CNS infection. Thus, unless other clear evidence of CNS infection such as fever, headache, or pleocytosis is noticed, treatment should not be undertaken until cultures have been positive on more than one occasion. These infections are rarely an emergency, and treatment, based on the sensitivity of the organism, should be delivered both parenterally and directly into the shunt or reservoir; it should be continued for 4 to 6 weeks. In some instances, direct treatment of the indwelling device may preserve it,[282,286] but often it cannot be sterilized. If it must be removed, a new device may be placed at the time the old one is removed. Parenteral antibiotics prevent the new device from being contaminated by the organism.

Propionibacterium acnes, an anaerobic Gram-positive rod, is present on the skin of healthy individuals and is a common cause of Ommaya device infections. The organism causes an indolent infection that may be characterized by only a few cells in the CSF, a feeling of generalized malaise, and a positive culture, but it can also cause a clinically significant meningitis[287] or brainstem encephalitis.[288]

Like coagulase-negative *staphylococcus*, *propionibacterium* may contaminate cultures, so that repeatedly positive cultures are necessary to be sure that the CSF is actually infected. Specific treatment depends on sensitivity testing, and as with other infected shunts or reservoirs, drugs should be delivered both parenterally and directly into the reservoir. Intrathecal drugs successful in treating Ommaya devices have included oxacillin and vancomycin. Daily treatment begins with 25 mg of oxacillin and increases to 50 to 100 mg/24 hours. Vancomycin has also been reported to be effective at 75 mg/24 hours. Serious reactions to an oxacillin injection are rare. Throbbing headaches may occur as a complication of intraventricular injection of vancomycin. Other drugs such as chloramphenicol have been instilled into the ventricle to treat infections, but we have no experience with these agents. An unusual complication of infected ventriculo–systemic shunts is immune complex–mediated nephritis (*shunt nephritis*) caused by either coagulase-negative *staphylococcus* or *P. acnes*.[289]

Other anaerobic organisms that occasionally cause meningitis or abscesses in the cancer population include *Fusobacterium*, *Bacteroides*, *Clostridium*, and *Actinomyces* species. The infections can occur either with a CSF–surface communication or are seen in severely neutropenic patients. Both CSF and CNS tissues (when available at biopsy) should be cultured for anaerobic as well as aerobic organisms. No special handling of CSF by the clinician is necessary for anaerobic cultures. The treatment depends on the sensitivity of the organisms. Metronidazole, 500 mg qid, may be an effective treatment of many anaerobic organisms causing either meningitis or brain abscess.

PARASITES

Parasites may invade the nervous system via a CSF fistula in patients with cancer. We have encountered one patient with *Trichomonas meningitis*[290] whose nervous system was invaded via a GI tract–CNS fistula.

Other Infections

Two infections, infectious endocarditis and *malignant* external otitis, also begin outside the brain and secondarily affect the CNS but do not easily fit the classification used in this chapter.

INFECTIOUS ENDOCARDITIS

Bacterial endocarditis in patients with cancer is less common than NBTE, but the increasing use of indwelling venous catheters, with their high rate of infection, suggests that this disorder will be encountered more commonly in the future. Neurologic complications of infective endocarditis have been reviewed extensively.[291–293] These include cerebral emboli causing either symptoms of a diffuse encephalopathy or focal neurologic signs, the same as those in NBTE (see Chapter 9). Because the emboli are infected, they may weaken the blood vessel walls in which they lodge and cause mycotic aneurysms with rupture and subarachnoid hemorrhage,[294] as well as meningitis, meningoencephalitis, or brain abscess. Seizures are common. The usual infecting organisms are streptococci or staphylococci (if catheter-related), but many other bacteria and fungi also can be responsible.[94] The diagnostic clues to the presence of infective endocarditis include conjunctival petechiae, Roth spots, other retinal hemorrhages, and Osler nodes. *Splinter* hemorrhages under the nails are nonspecific, as are Roth spots and retinal hemorrhages, but an Osler node may establish the diagnosis. All of these changes may also be seen in patients with NBTE.[295]

MALIGNANT EXTERNAL OTITIS

Malignant (also *invasive* or *necrotizing*) external otitis is an unusual infection that occurs primarily in elderly diabetic patients and also occasionally in immunosuppressed individuals.[296,297] The usual infecting organism is *P. aeruginosa* in more than 90% of patients,[296] but other bacteria (e.g., *S. aureus*) or fungi (an *Aspergillus* species) also cause the disorder. Severe ear pain is often referred to the temporomandibular joint, which impinges on the external ear canal, and sometimes, to the temporal and occipital areas of the head. The pain, often nocturnal, is usually associated with granulation tissue or pus and edema in the external auditory canal. The tympanic membrane is intact. After days to weeks of local discomfort, the patient develops facial nerve paralysis caused by inflammation at the stylomastoid foramen. As the infection invades more centrally, other cranial nerves that exit from the skull base, including glossopharyngeal, vagus, and spinal accessory nerves (jugular foramen syndrome), and the hypoglossal nerve may be affected.[297] Less commonly, mycotic aneurysms, cerebral venous sinus thrombosis, or meningitis complicate the course. Because the disorder is so indolent, it is often mistaken for tumor invasion rather than infection, and the CT scan showing bone destruction may support the incorrect diagnosis. An ear culture confirms the diagnosis, and appropriate antibiotics are the effective treatment.

REFERENCES

1. Bucaneve G, Micozzi A, Menichetti F, et al. Levofloxacin to prevent bacterial infection in patients with cancer and neutropenia. *N Engl J Med* 2005; 353:977–987.
2. Chernik NL, Armstrong D, Posner JB. Central nervous system infections in patients with cancer. *Medicine* 1973;52:563–581.
3. Chernik NL, Armstrong D, Posner JB. Central nervous system infections in patients with cancer. Changing patterns. *Cancer* 1977;40:268–274.
4. Pruitt AA. Nervous system infections in patients with cancer. *Neurol Clin* 2003;21:193–219.
5. Hooper DC, Pruitt AA, Rubin RH. Central nervous system infection in the chronically immunosuppressed. *Medicine* 1982;61:166–188.
6. Lukes SA, Posner JB, Nielsen S, et al. Bacterial infections of the CNS in neutropenic patients. *Neurology* 1984;34:269–275.
7. White M, Cirrincione C, Blevens A, et al. Cryptococcal meningitis: Outcome in patients with AIDS and patients with neoplastic disease. *J Infect Dis* 1992; 165:960–963.
8. Mishra AM, Gupta RK, Jaggi RS, et al. Role of diffusion-weighted imaging and in vivo proton magnetic resonance spectroscopy in the differential diagnosis of ring-enhancing intracranial cystic mass lesions. *J Comput Assist Tomogr* 2004;28:540–547.
9. Al-Okaili RN, Krejza J, Woo JH, et al. Intraaxial brain masses: MR imaging-based diagnostic strategy— initial experience. *Radiology* 2007;243:539–550.
10. Armstrong D, Polsky B. Central nervous system infections in the compromised host. In: Rubin RH, Young LS, eds. *Clinical Approach to Infection in the Compromised Host.* New York: Plenum, 1987. pp. 163–194.
11. Pruitt AA. Central nervous system infections in cancer patients. *Semin Neurol* 2004;24:435–452.
12. Kamboj M, Sepkowitz KA. The risk of tuberculosis in patients with cancer. *Clin Infect Dis* 2006; 42:1592–1595.
13. Cunha BA. Central nervous system infections in the compromised host: a diagnostic approach. *Infect Dis Clin North Am* 2001;15:567–590.

14. Polsky B, Armstrong D. Infectious complications of neoplastic disease. *Am J Infect Control* 1985;13:199–209.

15. Youssef S, Rodriguez G, Rolston KV, et al. *Streptococcus pneumoniae* infections in 47 hematopoietic stem cell transplantation recipients: clinical characteristics of infections and vaccine-breakthrough infections, 1989–2005. *Medicine (Baltimore)* 2007;86:69–77.

16. Omlin AG, Muhlemann K, Fey MF, et al. Pneumococcal vaccination in splenectomised cancer patients. *Eur J Cancer* 2005;41:1731–1734.

17. Scheld WM, Whitley RJ, Marra CM, eds. *Infections of the Central Nervous System*. 3rd ed. Philadelphia: Lippincott, 2004.

18. Leib SL, Tauber MG. Pathogenesis and pathophysiology of bacterial infections. In: Scheld WM, Whitley RJ, Marra CM, eds. *Infections of the CNS*. 3rd ed. Philadelphia: Lippincott, 2004.pp. 331–346.

19. Nau R, Brück W. Neuronal injury in bacterial meningitis: mechanisms and implications for therapy. *Trends Neurosci* 2002;25:38–45.

20. Zhang JR, Mostov KE, Lamm ME, et al. The polymeric immunoglobulin receptor translocates pneumococci across human nasopharyngeal epithelial cells. *Cell* 2000;102:827–837.

21. Ring A, Weiser JN, Tuomanen EI. Pneumococcal trafficking across the blood-brain barrier. Molecular analysis of a novel bidirectional pathway. *J Clin Invest* 1998;102:347–360.

22. Kim KS. Escherichia coli translocation at the blood-brain barrier. *Infect Immunol* 2001;69:5217–5222.

23. Safdar A, Armstrong D. Infectious morbidity in critically ill patients with cancer. *Crit Care Clin* 2001;17:531–570.

24. Kuijpers TW, Weening RS, Roos D. Clinical and laboratory work-up of patients with neutrophil shortage or dysfunction. *J Immunol Methods* 1999;2:211–229.

25. Hubel K, Hegener K, Schnell R, et al. Suppressed neutrophil function as a risk factor for severe infection after cytotoxic chemotherapy in patients with acute nonlymphocytic leukemia. *Ann Hematol* 1999;78:73–77.

26. Hersh EM, Freireich EJ. Host defense mechanisms and their modification by cancer chemotherapy. *Meth Cancer Res* 1968;4:355–435.

27. Lehrer RI, Ganz T, Selsted ME, et al. Neutrophils and host defense [clinical conference]. *Ann Intern Med* 1988;109:127–142.

28. Clark OA, Lyman GH, Castro AA, et al. Colony-stimulating factors for chemotherapy-induced febrile neutropenia: a meta-analysis of randomized controlled trials. *J Clin Oncol* 2005;23:4198–4214.

29. Chou M-Y, Brown AE, Blevins A, et al. Severe pneumococcal infection in patients with neoplastic disease. *Cancer* 1983;51:1546–1550.

30. Walzer PD, Armstrong D, Weisman P, et al. Serum immunoglobulin levels in childhood Hodgkin's disease. Effect of splenectomy and long-term follow-up. *Cancer* 1980;45:2084–2089.

31. Hollis N, Marsh RH, Marshall RD, et al. Overwhelming pneumococcal sepsis in healthy adults years after splenectomy. Letter to the editor. *Lancet* 1987;1:110–111.

32. Koedel U, Pfister H-W, Tomasz A. Methylprednisolone attenuates inflammation, increase of brain water content and intracranial pressure, but does not influence cerebral blood flow changes in experimental pneumococcal meningitis. *Brain Res* 1994;644:25–31.

33. Gaillard J-L, Abadie V, Cheron G, et al. Concentrations of ceftriaxone in cerebrospinal fluid of children with meningitis receiving dexamethasone therapy. *Antimicrob Agents Chemother* 1994;38:1209–1210.

34. van de Beek D, de Gans J. Dexamethasone in adults with community-acquired bacterial meningitis. *Drugs* 66:415–27.

35. Rambeloarisoa J, Batisse D, Thiebaut JB, et al. Intramedullary abscess resulting from disseminated cryptococcosis despite immune restoration in a patient with AIDS. *J Infect* 2002;44:185–188.

36. Vakis A, Koutentakis D, Karabetsos D, et al. Intracerebral CSF collection mimicking cerebral abscess in a patient suffering from cryptococcal meningitis. *J Infect* 2005;51:e233–e235.

37. Armstrong RW, Fung PC. Brainstem encephalitis (rhombencephalitis) due to Listeria monocytogenes: case report and review. *Clin Infect Dis* 1993;16:689–702.

38. Bach MC, Davis KM. Listeria rhombencephalitis mimicking tuberculous meningitis. *Rev Infect Dis* 1987;9:130–133.

39. Dee RR, Lorber B. Brain abscess due to Listeria monocytogenes: case report and literature review. *Rev Infect Dis* 1986;8:968–977.

40. Fujita NK, Reynard M, Sapico FL, et al. Cryptococcal intracerebral mass lesions: the role of computed tomography and nonsurgical management. *Ann Intern Med* 1981;94:382–388.

41. Morrison RE, Brown J, Gooding RS. Spinal cord abscess caused by Listeria monocytogenes. *Arch Neurol* 1980;37:243–244.

42. Pfister HW, Borasio GD, Dirnagl U, et al. Cerebrovascular complications of bacterial meningitis in adults. *Neurology* 1992;42:1497–1504.

43. Koedel U, Scheld WM, Pfister HW. Pathogenesis and pathophysiology of pneumococcal meningitis. *Lancet Infect Dis* 2002;2:721–736.

44. Huang SH, Jong AY. Cellular mechanisms of microbial proteins contributing to invasion of the blood-brain barrier. *Cell Microbiol* 2001;3:277–287.

45. Braun JS, Sublett JE, Freyer D, et al. Pneumococcal pneumolysin and H_2O_2 mediate brain cell apoptosis during meningitis. *J Clin Invest* 2002;109:19–27.

46. Chang WN, Lu CH, Chang HW, et al. Time course of cerebral hemodynamics in cryptococcal meningitis in HIV-negative adults. *Eur J Neurol* 2007;14:770–776.

47. Lu CH, Chang HW, Lui CC, et al. Cerebral haemodynamics in acute bacterial meningitis in adults. *Q J Med* 2006;99:863–869.

48. O'Brien SN, Blijlevens NM, Mahfouz TH, Anaissie EJ. Infections in patients with hematological cancer: recent developments. *Hematology Am Soc Hematol Educ Prog* 2003:438–472.

49. Scheld WM, Koedel U, Nathan B, et al. Pathophysiology of bacterial meningitis: mechanism(s) of neuronal injury. *J Infect Dis* 2002;186 (Suppl 2):S225–S233.

50. Weber JR, Tuomanen EI. Cellular damage in bacterial meningitis: an interplay of bacterial and host driven toxicity. *J Neuroimmunol* 2007;184:45–52.

51. Joffe AR. Lumbar puncture and brain herniation in acute bacterial meningitis: a review. *J Intensive Care Med* 2007;22:194–207.

52. Horwitz SJ, Boxerbaum B, O'Bell J. Cerebral herniation in bacterial meningitis in childhood. *Ann Neurol* 1980;7:524–528.

53. Koedel U, Winkler F, Angele B, et al. Meningitis-associated central nervous system complications are mediated by the activation of poly(ADP-ribose) polymerase. *J Cereb Blood Flow Metab* 2002;22:39–49.

54. Klein M, Koedel U, Pfister HW. Oxidative stress in pneumococcal meningitis: a future target for adjunctive therapy? *Prog Neurobiol* 2006;80:269–280.

55. Quagliarello VJ, Ma A, Stukenbrok H, et al. Ultrastructural localization of albumin transport across the cerebral microvasculature during experimental meningitis in the rat. *J Exp Med* 1991; 174:657–672.

56. Papadopoulos MC, Verkman AS. Aquaporin-4 gene disruption in mice reduces brain swelling and mortality in pneumococcal meningitis. *J Biol Chem* 2005;280:13906–13912.

57. Pedersen M, Brandt CT, Knudsen GM, et al. Cerebral blood flow autoregulation in early experimental *S. pneumoniae* meningitis. *J Appl Physiol* 2007; 102:72–78.

58. Pedersen M, Brandt CT, Knudsen GM, et al. The effect of *S. pneumoniae* bacteremia on cerebral blood flow autoregulation in rats. *J Cereb Blood Flow Metab*, in press.

59. Dubey AK, Rao KL. Pathology of postmeningitic hydrocephalus. *Indian J Pediatr* 1997;64:30–33.

60. Kramer AH, Bleck TP. Neurocritical care of patients with central nervous system infections. *Curr Infect Dis Rep* 2007;9:308–314.

61. Ariza J, Casanova A, Fernandez, et al. Etiological agent and primary source of infection in 42 cases of focal intracranial suppuration. *J Clin Microbiol* 1986;24:899–902.

62. Kastenbauer SWB, Pfister H-W, Scheld WM. Brain abscess. In: Scheld WM, Whitley RJ, Durack DT, eds. *Infections of the Central Nervous System.* 3rd ed. Pliladelphia: Lippincott, 2004. pp. 479–508.

63. Salata RA, King RE, Gose F, et al. Listeria monocytogenes cerebritis, bacteremia, and cutaneous lesions complicating hairy cell leukemia. *Am J Med* 1986;81:1068–1072.

64. Darouiche RO, Hamill RJ, Greenberg SB, et al. Bacterial spinal epidural abscess. Review of 43 cases and literature survey. *Medicine* 1992;71:369–385.

65. Grosu I, Ghekiere O, Layios N, et al. Toxoplasma encephalitis after autologous stem cell transplantation. *Leuk Lymphoma* 2007;48:201–203.

66. Singh G, Rees JH, Sander JW. Seizures and epilepsy in oncological practice: causes, course, mechanisms and treatment. *J Neurol Neurosurg Psychiatry* 2007;78:342–349.

67. Weststrate W, Hijdra A, de Gans J. Brain infarcts in adults with bacterial meningitis. *Lancet* 1996;347:399.

68. Perry JR, Bilbao JM, Gray T. Fatal basilar vasculopathy complicating bacterial meningitis. *Stroke* 1992;23:1175–1178.

69. Simberkoff MS, Cross AP, Al-Ibrahim M, et al. Efficacy of pneumococcal vaccine in high-risk patients. Results of a Veterans Administration Cooperative Study. *N Engl J Med* 1986;315:1318–1327.

70. Al Soub H. Primary Nocardia meningitis in a patient without a predisposing condition: case report and review of the literature. *Scand J Infect Dis* 2007;39:737–741.

71. Saitoh T. Successful treatment with voriconazole of Aspergillus meningitis in a patient with acute myeloid leukemia. *Ann Hematol* 2007;86:697–698.

72. Jarrett WH, Christy JH. Retinal hole formation from septic embolization in acute bacterial endocarditis. *Am J Ophthalmol* 1967;64:472–474.

73. Rodriguez-Adrian LJ, King RT, Tamayo-Derat LG, et al. Retinal lesions as clues to disseminated bacterial and candidal infections: frequency, natural history, and etiology. *Medicine (Baltimore)* 2003;82:187–202.

74. Straus SE, Thorpe KE, Holroyd-Leduc J. How do I perform a lumbar puncture and analyze the results to diagnose bacterial meningitis? *JAMA* 2006;296:2012–2022.

75. Fitch MT, van de Beek D. Emergency diagnosis and treatment of adult meningitis. *Lancet Infect Dis* 2007;7:191–200.

76. Pannullo SC, Reich JB, Krol G, et al. MRI changes in intracranial hypotension. *Neurology* 1993;43:919–926.

77. Posner JB, Saper CB, Schiff ND, et al. *Plum and Posner's Diagnosis of Stupor and Coma.* 4th ed. New York: Oxford University Press, 2007.

78. Shah K, Richard K, Edlow JA. Utility of lumbar puncture in the afebrile vs. febrile elderly patient with altered mental status: a pilot study. *J Emerg Med* 2007;32:15–18.

79. Richman JM. Bevel direction and postdural puncture headache: a meta-analysis. *Neurologist* 2006; 12:224–228.

80. Greenlee JE, Carroll KC. Cerebrospinal fluid in central nervous system infections. In: Scheld WM, Whitley RJ, Marra CM, eds. *Infections of the Central Nervous System.* Philadelphia: Lippincott, 2004. pp. 5–30.

81. Klepper J. Impaired glucose transport into the brain: the expanding spectrum of glucose transporter type 1 deficiency syndrome. *Curr Opin Neurol* 2004;17:193–196.

82. Fishman RA. Carrier transport and the concentration of glucose in cerebrospinal fluid in meningeal diseases. Editorial. *Ann Intern Med* 1965;63:153–155.

83. Petersdorf RG, Swarner DR, Garcia M. Studies on the pathogenesis of meningitis during pneumococcal bacteremia. *J Clin Invest* 1962;41:320–327.

84. Nelson E, Blinzinger K, Hager H. An electron-microscopic study of bacterial meningitis. *Arch Neurol* 1962;6:390–403.

85. Holub M, Beran O, Dzupova O, et al. Cortisol levels in cerebrospinal fluid correlate with severity and bacterial origin of meningitis. *Crit Care* 2007;11:R41.

86. Brivet FG, Ducuing S, Jacobs F, et al. Accuracy of clinical presentation for differentiating bacterial from viral meningitis in adults: a multivariate approach. *Intensive Care Med* 2005;31:1654–1660.

87. Lawrence RH. The role of lumbar puncture as a diagnostic tool in 2005. *Crit Care Resusc* 2005;7:213–220.

88. Simon RP, Abele JS. Spinal-fluid pleocytosis estimated by the Tyndall effect. *Ann Intern Med* 1978;89:75–76.

89. Smirniotopoulos JG, Murphy FM, Rushing EJ, et al. Patterns of contrast enhancement in the brain and meninges. *Radiographics* 2007;27:525–551.

90. Ferreira NP, Otta GM, do Amaral LL, et al. Imaging aspects of pyogenic infections of the central nervous system. *Top Magn Reson Imaging* 2005;16:145–154.

91. Lee BY, Newberg AB, Liebeskind DS, et al. FDG-PET findings in patients with suspected encephalitis. *Clin Nucl Med* 2004;29:620–625.

92. Fertikh D, Krejza J, Cunqueiro A, et al. Discrimination of capsular stage brain abscesses from necrotic or cystic neoplasms using diffusion-weighted magnetic resonance imaging. *J Neurosurg* 2007;106:76–81.

93. Schroeder PC, Post MJ, Oschatz E, et al. Analysis of the utility of diffusion-weighted MRI and apparent diffusion coefficient values in distinguishing central nervous system toxoplasmosis from lymphoma. *Neuroradiology* 2006;48:715–720.

94. Rosen P, Armstrong D. Infective endocarditis in patients treated for malignant neoplastic diseases: a postmortem study. *Am J Clin Pathol* 1973;60:241–250.

95. Wilson PJ, Turner HR, Kirchner KA, et al. Nocardial infections in renal transplant recipients. *Medicine* 1989;68:38–57.

96. Young LS, Armstrong D, Blevins A, et al. Nocardia asteroides infection complicating neoplastic disease. *Am J Med* 1971;50:356–367.

97. Levin S, Nelson KE, Spies HW, et al. Pneumococcal meningitis: the problem of the unseen cerebrospinal fluid leak. *Am J Med Sci* 1972;264:319–327.

98. Johnson MD, Perfect JR. Combination antifungal therapy: what can and should we expect? *Bone Marrow Transplant* 2007;40:297–306.

99. Gerner-Smidt P, Ethelberg S, Schiellerup P, et al. Invasive listeriosis in Denmark 1994-2003: a review of 299 cases with special emphasis on risk factors for mortality. *Clin Microbiol Infect* 2005;11:618–624.

100. Reynaud L, Graf M, Gentile I, et al. A rare case of brainstem encephalitis by Listeria monocytogenes with isolated mesencephalic localization. Case report and review. *Diagn Microbiol Infect Dis* 2007;58:121–123.

101. Cunha BA, Fatehpuria R, Eisenstein LE. Listeria monocytogenes encephalitis mimicking Herpes Simplex virus encephalitis: the differential diagnostic importance of cerebrospinal fluid lactic acid levels. *Heart Lung* 2007;36:226–231.

102. Mylonakis E, Hohmann EL, Caderwood SB. Central nervous system infection with *Listeria monocytogenes*—33 years' experience at a general hospital and review of 776 episodes from the literature. *Medicine* 1998;77:313–336.

103. Radice C, Munoz V, Castellares C, et al. Listeria monocytogenes meningitis in two allogeneic hematopoietic stem cell transplant recipients. *Leuk Lymphoma* 2006;47:1701–1703.

104. Wacker P, Ozsahin H, Groll AH, et al. Trimethoprim-sulfamethoxazole salvage for refractory listeriosis during maintenance chemotherapy for acute lymphoblastic leukemia. *J Pediatr Hematol Oncol* 2000;22:340–343.

105. Asakura H, Kawamoto K, Igimi S, et al. Enhancement of mice susceptibility to infection with Listeria monocytogenes by the treatment of morphine. *Microbiol Immunol* 2006;50:543–547.

106. Prins RM, Bruhn KW, Craft N, et al. Central nervous system tumor immunity generated by a recombinant listeria monocytogenes vaccine targeting tyrosinase related protein-2 and real-time imaging of intracranial tumor burden. *Neurosurgery* 2006;58:169–178.

107. Agterof MJ, van der Bruggen T, Tersmette M, et al. Nocardiosis: a case series and a mini review of clinical and microbiological features. *Neth J Med* 2007;65:199–202.

108. Borges AA, Krasnow SH, Wadleigh RG, et al. Nocardiosis after corticosteroid therapy for malignant thymoma. *Cancer* 1993;71:1746–1750.

109. Durmaz R, Atasoy MA, Durmaz G, et al. Multiple nocardial abscesses of cerebrum, cerebellum and spinal cord, causing quadriplegia. *Clin Neurol Neurosurg* 2001;103:59–62.

110. Al SH, Almaslamani M, Al KJ, et al. Primary Nocardia meningitis in a patient without a predisposing condition: case report and review of the literature. *Scand J Infect Dis* 2007;39:737–741.

111. Barnicoat MJ, Wierzbicki AS, Norman PM. Cerebral nocardiosis in immunosuppressed patients: five cases. *Q J Med* 1989;72:689–698.

112. Meyer SL, Font RL, Shaver RP. Intraocular nocardiosis: report of three cases. *Arch Ophthalmol* 1970;83:536–541.

113. Ntziora F, Falagas ME. Linezolid for the treatment of patients with central nervous system infection. *Ann Pharmacother* 2007;41:296–308.

114. Mattiuzzi G, Giles FJ. Management of intracranial fungal infections in patients with haematological malignancies. *Br J Haematol* 2005;131:287–300.

115. Pagano L, Caira M, Candoni A, et al. The epidemiology of fungal infections in patients with hematologic malignancies: the SEIFEM-2004 study. *Haematologica* 2006;91:1068–1075.

116. Walsh TJ, Groll AH. Emerging fungal pathogens: evolving challenges to immunocompromised patients for the twenty-first century. *Transpl Infect Dis* 1999;1:247–261.

117. Chayakulkeeree M, Perfect JR. Cryptococcosis. *Infect Dis Clin North Am* 2006;20:507–544.

118. Davis LE. Fungal infections of the central nervous system. *Neurol Clin* 1999;17:761–781.

119. Garcia CA, Weisberg LA, Lacorte WS. Cryptococcal intracerebral mass lesions: CT-pathologic considerations. *Neurology* 1985;35:731–734.

120. Shen CC, Cheng WY, Yang MY. Isolated intramedullary cryptococcal granuloma of the conus medullaris: case report and review of the literature. *Scand J Infect Dis* 2006;38:562–565.

121. Safdieh J, Mead P, Sepkowitz K, et al. Bacterial and fungal meningitis in cancer patients: a retrospective analysis. Neurology 2008;70:943–947.

122. Kiertiburanakul S, Wirojtananugoon S, Pracharktam R, et al. Cryptococcosis in human immunodeficiency virus-negative patients. *Int J Infect Dis* 2006;10:72–78.

123. Shih CC, Chen YC, Chang SC, et al. Cryptococcal meningitis in non-HIV-infected patients. *Q J Med* 2000;93:245–251.

124. Kingsley PB, Shah TC, Woldenberg R. Identification of diffuse and focal brain lesions by clinical magnetic resonance spectroscopy. *NMR Biomed* 2006;19:435–462.

125. Saag MS, Graybill RJ, Larsen RA, et al. Practice guidelines for the management of cryptococcal disease. Infectious Diseases Society of America. *Clin Infect Dis* 2000;30:710–718.

126. Polsky B, Depman MR, Gold JW. Intraventricular therapy of cryptococcal meningitis via a subcutaneous reservoir. *Am J Med* 1986;81:25–28.

127. Labadie EL, Hamilton RH. Survival improvement in coccidioidal meningitis by high-dose intrathecal amphotericin B. *Arch Intern Med* 1986; 146:2013–2018.

128. Bozzette SA, Larsen RA, Chiu J, et al. A placebo-controlled trial of maintenance therapy with fluconazole after treatment of cryptococcal meningitis in the acquired immunodeficiency syndrome. California Collaborative Treatment Group. *N Engl J Med* 1991;324:580–584.

129. Straus SE, Ostrove JM, Inchauspe G, et al. NIH Conference. Varicella-zoster virus infections. Biology, natural history, treatment, and prevention. *Ann Intern Med* 1988;108:221–237.

130. Wade JC. Viral infections in patients with hematological malignancies. *Hematology Am Soc Hematology Educ Program* 2006;368–374.

131. Kim DH, Messner H, Minden M, et al. Factors influencing varicella zoster virus infection after allogeneic peripheral blood stem cell transplantation: low-dose acyclovir prophylaxis and pre-transplant diagnosis of lymphoproliferative disorders. *Transpl Infect Dis* 2008;10:90–98.

132. Fujimaki K, Mori T, Kida A, et al. Human herpesvirus 6 meningoencephalitis in allogeneic hematopoietic stem cell transplant recipients. *Int J Hematol* 2006;84:432–437.

133. Guinee VF, Guido JJ, Pfalzgraf KA, et al. The incidence of herpes zoster in patients with Hodgkin's disease. An analysis of prognostic factors. *Cancer* 1985;56:642–648.

134. Maiche AG, Kajanti MJ, Pyrhonen S. Simultaneous disseminated herpes zoster and bacterial infection in cancer patients. Letter to the editor. *Acta Oncol* 1992;31:681–683.

135. Nagel MA, Gilden DH. The protean neurologic manifestations of varicella-zoster virus infection. *Cleve Clin J Med* 2007;74:489.

136. Feldman S, Hughes WT, Daniel CB. Varicella in children with cancer: seventy-seven cases. *Pediatrics* 1975;56:388–397.

137. Schmidbauer M, Budka H, Pilz P, et al. Presence, distribution and spread of productive varicella zoster virus infection in nervous tissues. *Brain* 1992;115:383–398.

138. Mizuguchi M, Yamanouchi H, Ichiyama T, et al. Acute encephalopathy associated with influenza and other viral infections. *Acta Neurol Scand* 2007;115:45–56.

139. Lichtenstein PK, Heubi JE, Daugherty CC, et al. Grade I Reye's syndrome. A frequent cause of vomiting and liver dysfunction after varicella and upper-respiratory-tract infection. *N Engl J Med* 1983;309:133–139.

140. Van Coster RN, De Vivo DC, Blake D, et al. Adult Reye's syndrome: A review with new evidence for a generalized defect in intramitochondrial enzyme processing. *Neurology* 1991;41:1815–1821.

141. Rubie H, Guillot S, Netter JC, et al. Acute hepatopathy compatible with Reye's syndrome in 3 children treated by chemotherapy. *Arch Pediatr* 1994;1:573–577.

142. Schror K. Aspirin and Reye syndrome: a review of the evidence. *Paediatr Drugs* 2007;9:195–204.

143. Heininger U, Seward JF. Varicella. *Lancet* 2006;368:1365–1376.

144. Wareham DW, Breuer J. Herpes zoster. *BMJ* 2007;334:1211–1215.

145. Niiyama S, Satoh K, Kaneko S, et al. Zosteriform skin involvement of nodal T-cell lymphoma: a review of the published work of cutaneous malignancies mimicking herpes zoster. *J Dermatol* 2007;34:68–73.

146. Kikuchi Y, Matsuyama A, Nomura K. Zosteriform metastatic skin cancer: report of three cases and review of the literature. *Dermatology* 2001;202:336–338.

147. Zalaudek I, Leinweber B, Richtig E, et al. Cutaneous zosteriform melanoma metastases arising after herpes zoster infection: a case report and review of the literature. *Melanoma Res* 2003;13:635–639.

148. Hughes BA, Kimmel DW, Aksamit AJ. Herpes zoster-associated meningoencephalitis in patients with systemic cancer. *Mayo Clin Proc* 1993;68:652–655.

149. Kleinschmidt-DeMasters BK, Gilden DH. Varicella-zoster virus infections of the nervous system: clinical and pathologic correlates. *Arch Pathol Lab Med* 2001;125:770–780.

150. Eidelberg D, Sotrel A, Horoupian S, et al. Thrombotic cerebral vasculopathy associated with herpes zoster. *Ann Neurol* 1986;19:7–14.

151. Womack LW, Liesegang TJ. Complications of herpes zoster ophthalmicus. *Arch Ophthalmol* 1983;101:42–45.

152. Constantinescu CS. Association of varicella-zoster virus with cervical artery dissection in 2 cases. *Arch Neurol* 2000;57:427.

153. O'Donohue JM, Enzmann DR. Mycotic aneurysm in angiitis associated with herpes zoster ophthalmicus. *Am J Neuroradiol* 1987;8:615–619.

154. Blommel ML, Blommel AL. Pregabalin: an antiepileptic agent useful for neuropathic pain. *Am J Health Syst Pharm* 2007;64:1475–1482.

155. Christo PJ, Hobelmann G, Maine DN. Post-herpetic neuralgia in older adults: evidence-based approaches to clinical management. *Drugs Aging* 2007;24:1–19.

156. Quan D, Hammack BN, Kittelson J, et al. Improvement of postherpetic neuralgia after treatment with intravenous acyclovir followed by oral valacyclovir. *Arch Neurol* 2006; 63:940–942.

157. Thomas JE, Howard FM, Jr. Segmental zoster paresis—a disease profile. *Neurology* 1972;22:459–466.

158. Dueland AN, Devlin M, Martin JR, et al. Fatal varicella-zoster virus meningoradiculitis without skin involvement. *Ann Neurol* 1991;29:569–572.

159. Easton HG. Zoster sine herpete causing acute trigeminal neuralgia. *Lancet* 1970;2:1065–1066.

160. Gilden DH, Wright RR, Schneck SA, et al. Zoster sine herpete: a clinical variant. *Ann Neurol* 1994;35:530–533.

161. Mayo DR, Booss J. Varicella zoster-associated neurologic disease without skin lesions. *Arch Neurol* 1989;46:313–315.

162. Heller HM, Carnevale NT, Steigbigel RT. Varicella zoster virus transverse myelitis without cutaneous rash. Case report. *Am J Med* 1990;88:550–551.

163. Gilden D. Varicella zoster virus and central nervous system syndromes. *Herpes* 2004;11 (Suppl 2): 89A–94A.

164. Gilden DH, Beinlich BR, Rubinstein EM, et al. Varicella-zoster virus myelitis: an expanding spectrum. *Neurology* 1994;44:1818–1823.

165. Devinsky O, Cho E-S, Petito CK, et al. Herpes zoster myelitis. *Brain* 1991;114 (Pt 3):1181–1196.

166. Jemsek J, Greenberg SB, Taber L, et al. Herpes zoster-associated encephalitis: Clinicopathologic report of 12 cases and review of the literature. *Medicine* 1983;62:81–97.

167. Kleinschmidt-DeMasters BK, Amlie-Leford C, Gilden DH. The patterns of varicella zoster virus encephalitis. *Hum Pathol* 1996;27:927–938.

168. Andiman WA, White-Greenwald M, Tinghitella T. Zoster encephalitis. Isolation of virus and measurement of varicella-zoster-specific antibodies in cerebrospinal fluid. *Am J Med* 1982;73:769–772.

169. Beuche W, Thomas RS, Felgenhauer K. Demonstration of zoster virus antibodies in cerebrospinal fluid cells. *J Neurol* 1989;236:26–28.

170. Tattevin P, Schortgen F, de Broucker T, et al. Varicella-zoster virus limbic encephalitis in an immunocompromised patient. *Scand J Infect Dis* 2001;33:786–788.

171. Nagel MA, Cohrs RJ, Mahalingam R, et al. Varicella-zoster virus vasculopathies: Clinical, CSF, imaging and virologic features. Neurology 2008;70: 853–860.

172. Ross MH, Abend WK, Schwartz RB, et al. A case of C2 herpes zoster with delayed bilateral pontine infarction. *Neurology* 1991;41:1685–1686.

173. Geny C, Yulis J, Azoulay A, et al. Thalamic infarction following lingual herpes zoster. Case report. *Neurology* 1991;41:1846.

174. Linnemann CC, Jr, Alvira MM. Pathogenesis of varicella-zoster angiitis in the CNS. *Arch Neurol* 1980;37:239–240.

175. Hilt DC, Buchholz D, Krumholz A, et al. Herpes zoster ophthalmicus and delayed contralateral hemiparesis caused by cerebral angiitis: diagnosis and management approaches. *Ann Neurol* 1983;14:543–553.

176. Schwid S, Ketonen L, Betts R, et al. Cerebrovascular complications after primary varicella-zoster infection. Letter to the editor. *Lancet* 1992;340:669.

177. Horten B, Price RW, Jimenez D. Multifocal varicella-zoster virus leukoencephalitis temporally remote from herpes-zoster. *Ann Neurol* 1981;9:251–266.

178. Weaver S, Rosenblum MK, DeAngelis LM. Herpes varicella-zoster encephalitis in immunocompromised patients. *Neurology* 1999;52:193–195.

179. Lentz D, Jordan JE, Pike GB, et al. MRI in varicella-zoster virus leukoencephalitis in the immunocompromised host. *J Comput Assist Tomogr* 1993;17:313–316.

180. Blumenthal DT, Salzman KL, Baringer JR, et al. MRI abnormalities in chronic active varicella zoster infection. *Neurology* 2004;63:1538–1539.

181. Nagel MA, Forghani B, Mahalingam R, et al. The value of detecting anti-VZV IgG antibody in CSF to diagnose VZV vasculopathy. Neurology 2007; 68:1069–1073.

182. Mehta SK, Tyring SK, Gilden DH, et al. Varicella-zoster virus in the saliva of patients with herpes zoster. J Infect Dis 2008;197:654–657.

183. Boeckh M. Prevention of VZV infection in immunosuppressed patients using antiviral agents. *Herpes* 2006;13:60–65.

184. Weinstock DM, Boeckh M, Sepkowitz KA. Postexposure prophylaxis against varicella zoster virus infection among hematopoietic stem cell transplant recipients. *Biol Blood Marrow Transplant* 2006;12:1096–1097.

185. Alper BS. Does treatment of acute herpes zoster prevent or shorten postherpetic neuralgia? *J Fam Pract* 2000;49:255–264.

186. Jackson JL. The effect of treating herpes zoster with oral acyclovir in preventing postherpetic neuralgia. A meta-analysis. *Arch Intern Med* 1997; 157:909–912.

187. Bustamante CI, Wade JC. Herpes simplex virus infection in the immunocompromised cancer patient. *J Clin Oncol* 1991;9:1903–1915.

188. Wingard JR. Management of infectious complications of bone marrow transplantation. *Oncology (Williston Park)* 1990;4:69–82.

189. Rothman AL, Cheeseman SH, Lehrman SN, et al. Herpes simplex encephalitis in a patient with lymphoma: relapse following acyclovir therapy. *JAMA* 1988;259:1056–1057.

190. Schiff D, Rosenblum MK. Herpes simplex encephalitis (HSE) and the immunocompromised patient: a clinical and autopsy study of HSE in the settings of cancer and human immunodeficiency virus-type 1 infection. *Hum Pathol* 1998;29:215–222.

191. Manz HJ, Phillips TM, McCullough DC. Herpes simplex type 2 encephalitis concurrent with known cerebral metastases. *Acta Neuropathol [Berl]* 1979;47:237–240.

192. Spuler A, Blaszyk H, Parisi JE, et al. Herpes simplex encephalitis after brain surgery: case report and review of the literature. *J Neurol Neurosurg Psychiatry* 1999;67:239–242.

193. Ploner M, Turowski B, Wobker G. Herpes encephalitis after meningioma resection. *Neurology* 2005;65:1674–1675.

194. Riel-Romero RM, Baumann RJ. Herpes simplex encephalitis and radiotherapy. *Pediatr Neurol* 2003;29:69–71.

195. Cathomas R, Pelosi E, Smart J, et al. Herpes simplex encephalitis as a complication of adjuvant chemotherapy treatment for breast cancer. *Clin Oncol (R Coll Radiol)* 2005;17:292–293.

196. Nolan RC, Van GH, Byrne M. An unusual complication of chemotherapy: herpes simplex meningoencephalitis and bilateral acute retinal necrosis. *Clin Oncol (R Coll Radiol)* 2004;16:81–82.

197. Harrison NA, MacDonald B, Scott G, et al. Atypical herpes type 2 encephalitis associated with normal MRI imaging. *J Neurol Neurosurg Psychiatry* 2003;74:974–976.

198. Whitley RJ. Viral encephalitis. *N Engl J Med* 1990;323:242–250.

199. Price R, Chernik NL, Horta-Barbosa L, et al. Herpes simplex encephalitis in an anergic patient. *Am J Med* 1973;54:222–228.

200. Folpe A, Lapham LW, Smith HC. Herpes simplex myelitis as a cause of acute necrotizing myelitis syndrome. *Neurology* 1994;44:1955–1957.

201. Sabath BF, Major EO. Traffic of JC virus from sites of initial infection to the brain: the path to progressive multifocal leukoencephalopathy. *J Infect Dis* 2002;186 (Suppl 2):S180–S186.

202. Freim Wahl SG, Folvik MR, Torp SH. Progressive multifocal leukoencephalopathy in a lymphoma patient with complete remission after treatment with cytostatics and rituximab: case report and review of the literature. *Clin Neuropathol* 2007;26:68–73.

203. Kiewe P, Seyfert S, Korper S, et al. Progressive multifocal leukoencephalopathy with detection of JC virus in a patient with chronic lymphocytic leukemia parallel to onset of fludarabine therapy. *Leuk Lymphoma* 2003;44:1815–1818.

204. Hou J, Major E. Management of infections by the human polyomavirus JC: past, present and future. *Expert Rev Anti Infect Ther* 2005;3:629–640.

205. Schlitt M, Morawetz RB, Bonnin J, et al. Progressive multifocal leukoencephalopathy: three patients diagnosed by brain biopsy, with prolonged survival in two. *Neurosurgery* 1986;18:407–414.

206. Re D, Bamborschke S, Feiden W, et al. Progressive multifocal leukoencephalopathy after autologous bone marrow transplantation and alpha-interferon immunotherapy. *Bone Marrow Transplant* 1999; 23:295–298.

207. Bale JF, Jr. Human cytomegalovirus infection and disorders of the nervous system. *Arch Neurol* 1984;41:310–320.

208. Suzumiya J, Marutsuka K, Ueda S, et al. An autopsy case of necrotizing ventriculo-encephalitis caused by cytomegalovirus in Hodgkin's disease. *Acta Pathol Jpn* 1991;41:291–298.

209. Rubin DI. NeuroImages. "Owl's eyes" of CMV ventriculitis. *Neurology* 2000;54:2217.

210. Pollard RB, Egbert PR, Gallagher JG, et al. Cytomegalovirus retinitis in immunosuppressed hosts. I. Natural history and effects of treatment with adenine arabinoside. *Ann Intern Med* 1980;93:655–664.

211. O'Brien SM, Keating MJ, Mocarski ES. Updated guidelines on the management of cytomegalovirus reactivation in patients with chronic lymphocytic leukemia treated with alemtuzumab. *Clin Lymphoma Myeloma* 2006;7:125–130.

212. Gilden DH, Mahalingam R, Cohrs RJ, et al. Herpesvirus infections of the nervous system. *Nat Clin Pract Neurol* 2007;3:82–94.

213. Wolf DG, Spector SA. Diagnosis of human cytomegalovirus central nervous system disease in AIDS patients by DNA amplification from cerebrospinal fluid. *J Infect Dis* 1992;166:1412–1415.

214. Freeman AF, Jacobsohn DA, Shulman ST, et al. A new complication of stem cell transplantation: measles inclusion body encephalitis. *Pediatrics* 2004;114:e657–e660.

215. Wolinsky JS, Swoveland P, Johnson KP, et al. Subacute measles encephalitis complicating Hodgkin's disease in an adult. *Ann Neurol* 1977;1:452–457.

216. Davids D, Henslee PJ, Markesbery WR. Fatal adenovirus meningoencephalitis in a bone marrow transplant patient. *Ann Neurol* 1988;23:385–389.

217. Cunha BA. West Nile virus encephalitis: clinical, diagnostic and prognostic indicators in compromised hosts. *Clin Infect Dis* 2006;43:117.

218. Sejvar JJ, Haddad MB, Tierney BC, et al. Neurologic manifestations and outcome of West Nile virus infection. *JAMA* 2003;290:511–515.

219. Kleinschmidt-DeMasters BK, Marder BA, Levi ME, et al. Naturally acquired West Nile virus encephalomyelitis in transplant recipients: clinical, laboratory, diagnostic, and neuropathological features. *Arch Neurol* 2004;61:1210–1220.

220. Anninger WV, Lomeo MD, Dingle J, et al. West Nile virus-associated optic neuritis and chorioretinitis. *Am J Ophthalmol* 2003;136:1183–1185.

221. Rosas H, Wippold FJ. West Nile virus: case report with MR imaging findings. *AJNR Am J Neuroradiol* 2003;24:1376–1378.

222. Penn RG, Guarner J, Sejvar JJ, et al. Persistent neuroinvasive West Nile virus infection in an immunocompromised patient. *Clin Infect Dis* 2006;42:680–683.

223. Carey RM, Kimball AC, Armstrong D, et al. Toxoplasmosis. Clinical experiences in a cancer hospital. *Am J Med* 1973;54:30–38.

224. Ruskin J, Remington JS. Toxoplasmosis in the compromised host. *Ann Intern Med* 1976;84:193–199.

225. Silva MT, Araujo A. Highly active antiretroviral therapy access and neurological complications of human immunodeficiency virus infection: impact versus resources in Brazil. *J Neurovirol* 2005;11 (Suppl 3):11–15.

226. Subsai K, Kanoksri S, Siwaporn C, et al. Neurological complications in AIDS patients receiving HAART: a 2-year retrospective study. *Eur J Neurol* 2006; 13:233–239.

227. Matsuo Y, Takeishi S, Miyamoto T, et al. Toxoplasmosis encephalitis following severe graft-vs.-host disease after allogeneic hematopoietic stem cell transplantation: 17 yr experience in Fukuoka BMT group. *Eur J Haematol* 2007;79:317–321.

228. Best T, Finlayson M. Two forms of encephalitis in opportunistic toxoplasmosis. *Arch Pathol Lab Med* 1979;103:693–696.

229. Ghatak NR, Sawyer DR. A morphologic study of opportunistic cerebral toxoplasmosis. *Acta Neuropath (Berl)* 1978;42:217–221.

230. Powell HC, Gibbs CJ, Jr, Lorenzo AM, et al. Toxoplasmosis of the central nervous system in the adult. Electron microscopic observations. *Acta Neuropath (Berl)* 1978;41:211–216.

231. Ketonen L, De la Pena RC, Villanueva-Meyer J. MR and TL-210 spect imaging with pathologic correlation for the assessment of CNS lymphoma vs. toxoplasmosis in AIDS patients. *J NeuroAIDS* 1998; 2:21–42.

232. Martin WG, Brown GC, Parrish RK, et al. Ocular toxoplasmosis and visual field defects. *Am J Ophthalmol* 1980;90:25–29.

233. Grines C, Plouffe JF, Baird IM, et al. Toxoplasma meningoencephalitis with hypoglycorrhachia. *Arch Intern Med* 1981;141:935.

234. Vidal JE, Colombo FA, de Oliveira AC, et al. PCR assay using cerebrospinal fluid for diagnosis of cerebral toxoplasmosis in Brazilian AIDS patients. *J Clin Microbiol* 2004;42:4765–4768.

235. Genta RM. Global prevalence of strongyloidiasis: critical review with epidemiologic insights into the prevention of disseminated disease. *Rev Infect Dis* 1989;11:755–767.

236. Genta RM, Miles P, Fields K. Opportunistic *Strongyloides stercoralis* infection in lymphoma patients. Report of a case and review of the literature. *Cancer* 1989;63:1407–1411.

237. Cruz T, Rebouças G, Rocha H. Fatal strongyloidiasis in patients receiving corticosteroids. *New Engl J Med* 1966;275:1093–1096.

238. Orlent H, Crawley C, Cwynarski K, et al. Strongyloidiasis pre and post autologous peripheral blood stem cell transplantation. *Bone Marrow Transplant* 2003;32:115–117.

239. Wachter RM, Burke AM, MacGregor RR. Strongyloides stercoralis hyperinfection masquerading as cerebral vasculitis. *Arch Neurol* 1984; 41:1213–1216.

240. Dutcher JP, Marcus SL, Tanowitz HB, et al. Disseminated strongyloidiasis with central nervous system involvement diagnosed antemortem in a patient with acquired immunodeficiency syndrome and Burkitts lymphoma. *Cancer* 1990; 66:2417–2420.

241. Meltzer RS, Singer C, Armstrong D, et al. Case report: antemortem diagnosis of central nervous system strongyloidiasis. *Am J Med Sci* 1979; 277:91–98.

242. Takayanagui OM, Lofrano MM, Araugo MB, et al. Detection of *Strongyloides stercoralis* in the cerebrospinal fluid of a patient with acquired immunodeficiency syndrome. *Neurology* 1995; 45:193–194.

243. Lo RV, III, Gluckman SJ. Eosinophilic meningitis. *Am J Med* 2003;114:217–223.

244. Patchell R, Perry MC. Eosinophilic meningitis in Hodgkin disease. *Neurology* 1981;31:887–888.

245. Mulligan MJ, Vasu R, Grossi CE, et al. Case report: neoplastic meningitis with eosinophilic pleocytosis in Hodgkin's disease: a case with cerebellar dysfunction and a review of the literature. *Am J Med Sci* 1988;296(5):322–326.

246. King DK, Loh KK, Ayala AG, et al. Eosinophilic meningitis and lymphomatous meningitis. Letter to the editor. *Ann Intern Med* 1975;82:228.

247. Budka H, Guseo A, Jellinger K, et al. Intermittent meningitic reaction with severe basophilia and eosinophilia in CNS leukaemia. *J Neurol Sci* 1976;28:459–468.

248. Behrendt H, de Korte D, Van Gennip AH, et al. Eosinophilic meningitis preceding meningeal relapse in a child with acute lymphoblastic leukemia: abnormal nucleotide content of eosinophils. *Pediatr Hematol Oncol* 1987;4:261–267.

249. Conrad KA, Gross JL, Trojanowski JQ. Leptomeningeal carcinomatosis presenting as eosinophilic meningitis. *Acta Cytol* 1986;30:29–31.

250. Traynelis VC, Powell RG, Koss W, et al. Cerebrospinal fluid eosinophilia and sterile shunt malfunction. *Neurosurgery* 1988;23:645–649.

251. Tung H, Raffel C, McComb JG. Ventricular cerebrospinal fluid eosinophilia in children with ventriculoperitoneal shunts. *J Neurosurg* 1991; 75:541–544.

252. Scott TF. A new cause of cerebrospinal fluid eosinophilia: neurosarcoidosis. Letter to the editor. *Am J Med* 1988;84:973–974.

253. Quinn JP, Weinstein RA, Caplan LR. Eosinophilic meningitis and ibuprofen therapy. *Neurology* 1984;34:108–109.

254. Weingarten JS, O'Sheal SF, Margolis WS. Eosinophilic meningitis and the hypereosinophilic syndrome. Case report and review of the literature. *Am J Med* 1985;78:674–676.

255. Weaver DF, Heffernan LP, Purdy RA, et al. Eosinophil-induced neurotoxicity: axonal neuropathy, cerebral infarction, and dementia. *Neurology* 1988;38:144–146.

256. Weller PF. The immunobiology of eosinophils. *N Engl J Med* 1991;324:1110–1118.

257. Bolivar R, Bodey GP, Velasquez WS. Recurrent Salmonella meningitis in a compromised host. *Cancer* 1982;50:2034–2036.

258. Pagano L, Caira M, Falcucci P, et al. Fungal CNS infections in patients with hematologic malignancy. *Expert Rev Anti Infect Ther* 2005;3:775–785.

259. Walsh TJ, Groll A, Hiemenz J, et al. Infections due to emerging and uncommon medically important fungal pathogens. *Clin Microbiol Infect* 2004;10 (Suppl 1):48–66.

260. Lowe JT, Jr, Hudson WR. Rhinocerebral phycomycosis and internal carotid artery thrombosis. *Arch Otolaryngol* 1975;101:100–103.

261. Pagano L, Caira M, Picardi M, et al. Invasive Aspergillosis in patients with acute leukemia: update on morbidity and mortality—SEIFEM-C report. *Clin Infect Dis* 2007;44:1524–1525.

262. Sparano JA, Gucalp R, Llena JF, et al. Cerebral infection complicating systemic aspergillosis in acute leukemia: clinical and radiographic presentation. *J Neuro-Oncol* 1992;13:91–100.

263. Kami M, Ogawa S, Kanda Y, et al. Early diagnosis of central nervous system aspergillosis using polymerase chain reaction, latex agglutination test, and enzyme-linked immunosorbent assay. *Br J Haematol* 1999;106:536–537.

264. Viscoli C, Machetti M, Gazzola P, et al. Aspergillus galactomannan antigen in the cerebrospinal fluid of bone marrow transplant recipients with probable cerebral aspergillosis. *J Clin Microbiol* 2002; 40:1496–1499.

265. Black KE, Baden LR. Fungal infections of the CNS: treatment strategies for the immunocompromised patient. *CNS Drugs* 2007;21:293–318.

266. Mattei D, Mordini N, Lo NC, et al. Voriconazole in the management of invasive aspergillosis in two patients with acute myeloid leukemia undergoing stem cell transplantation. *Bone Marrow Transplant* 2002;30:967–970.

267. Camarata PJ, Dunn DL, Farney AC, et al. Continual intracavitary administration of amphotericin B as an adjunct in the treatment of aspergillus brain abscess: case report and review of the literature. *Neurosurgery* 1992;31:575–579.

268. Munir N, Jones NS. Rhinocerebral mucormycosis with orbital and intracranial extension: a case report and review of optimum management. *J Laryngol Otol* 2007;121:192–195.

269. Nenoff P, Kellermann S, Schober R, et al. Rhinocerebral zygomycosis following bone marrow transplantation in chronic myelogenous leukaemia. Report of a case and review of the literature. *Mycoses* 1998;41:365–372.

270. O'Neill BM, Alessi AS, George EB, et al. Disseminated rhinocerebral mucormycosis: a case report and review of the literature. *J Oral Maxillofac Surg* 2006;64:326–333.

271. Sable CA, Strohmaier KM, Chodakewitz JA. Advances in antifungal therapy. *Annu Rev Med*, 2008;59:361–369.

272. Galetta SL, Wulc AE, Goldberg HI, et al. Rhinocerebral mucormycosis: Management and survival after carotid occlusion. *Ann Neurol* 1990;28:103–107.

273. Edelson RN, McNatt EN, Porro RS. Candida meningitis with cerebral arteritis. *NYS J Med* 1975;75:900–904.

274. Gorell JM, Palutke WA, Chason JL. Candida pachymeningitis with multiple cranial nerve pareses. *Arch Neurol* 1979;36:719–720.

275. Shaikh BS, Appelbaum PC, Aber RC. Vertebral disk space infection and osteomyelitis due to Candida albicans in a patient with acute myelomonocytic leukemia. *Cancer* 1980;45:1025–1028.

276. Schabereiter-Gurtner C, Selitsch B, Rotter ML, et al. Development of novel real-time PCR assays for detection and differentiation of eleven medically important Aspergillus and Candida species in clinical specimens. *J Clin Microbiol* 2007;45:906–914.

277. Cheng TM, O'Neill BP, Scheithauer BW, et al. Chronic meningitis: the role of meningeal or cortical biopsy. *Neurosurgery* 1994;34:590–596.

278. Arena FP, Perlin M, Brahman H, et al. Fever, rash, and myalgias of disseminated candidiasis during antifungal therapy. Case report. *Arch Intern Med* 1981;141:1233.

279. Welinder-Olsson C, Dotevall L, Hogevik H, et al. Comparison of broad-range bacterial PCR and culture of cerebrospinal fluid for diagnosis of community-acquired bacterial meningitis. *Clin Microbiol Infect* 2007;13:879–886.

280. Saha SK, Darmstadt GL, Yamanaka N, et al. Rapid diagnosis of pneumococcal meningitis: implications for treatment and measuring disease burden. *Pediatr Infect Dis J* 2005;24:1093–1098.

281. Obbens EA, Leavens ME, Beal JW, et al. Ommaya reservoirs in 387 cancer patients: a 15-year experience. *Neurology* 1985;35:1274–1278.

282. Dinndorf PA, Bleyer WA. Management of infectious complications of intraventricular reservoirs in cancer patients: low incidence and successful treatment without reservoir removal. *Cancer Drug Deliv* 1987;4:105–117.

283. Lishner M, Perrin RG, Feld R, et al. Complications associated with Ommaya reservoirs in patients with cancer. The Princess Margaret Hospital experience and a review of the literature. *Arch Intern Med* 1990;150:173–176.

284. Park DM, DeAngelis LM. Delayed infection of the Ommaya reservoir. *Neurology* 2002;59:956–957.

285. Mechleb B, Khater F, Eid A, et al. Late onset Ommaya reservoir infection due to *Staphylococcus aureus*: case report and review of Ommaya infections. *J Infect* 2003;46:196–198.

286. Lishner M, Scheinbaum R, Messner HA. Intrathecal vancomycin in the treatment of Ommaya reservoir infection by staphylococcus epidermidis. *Scand J Infect Dis* 1991;23:101–104.

287. Ueunten D, Tobias J, Sochat M, et al. An unusual cause of bacterial meningitis in the elderly. Propionibacterium acnes. *Arch Neurol* 1983;40:388–389.

288. Camarata PJ, McGeachie RE, Haines SJ. Dorsal midbrain encephalitis caused by propionibacterium acnes. Report of two cases. *J Neurosurg* 1990;72:654–659.

289. Frank JA, Jr, Friedman HS, Davidson DM, et al. Propionibacterium shunt nephritis in two adolescents with medulloblastoma. *Cancer* 1983;52:330–333.

290. Masur H, Hood E, III, Armstrong D. A trichomonas species in a mixed microbial meningitis. *JAMA* 1976;236:1978–1979.

291. Lester SJ, Wilansky S. Endocarditis and associated complications. *Crit Care Med* 2007;35:S384–S391.

292. Paterick TE, Paterick TJ, Nishimura RA, et al. Complexity and subtlety of infective endocarditis. *Mayo Clin Proc* 2007;82:615–621.

293. Singhal AB, Topcuoglu MA, Buonanno FS. Acute ischemic stroke patterns in infective and non-bacterial thrombotic endocarditis: a diffusion-weighted magnetic resonance imaging study. *Stroke* 2002;33:1267–1273.

294. Venger BH, Aldama EA. Mycotic vasculitis with repeated intracranial aneurysmal hemorrhage. Case report. *J Neurosurg* 1988;69:775–779.

295. Rogers LR, Cho E-S, Kempin S, et al. Cerebral infarction from non-bacterial thrombotic endocarditis. Clinical and pathological study including the effects of anticoagulation. *Am J Med* 1987;83:746–756.

296. Rubin GJ, Branstetter BF, Yu VL. The changing face of malignant (necrotising) external otitis: clinical, radiological, and anatomic correlations. *Lancet Infect Dis* 2004;4:34–39.

297. Mani N, Sudhoff H, Rajagopal S, et al. Cranial nerve involvement in malignant external otitis: implications for clinical outcome. *Laryngoscope* 2007;117:907–910.

Chapter 11

Delirium and Metabolic and Nutritional Complications of Cancer

INTRODUCTION

Metabolic and nutritional disorders are common in patients with cancer. *Metabolic disorders* occur under the following circumstances: (1) the cancer invades a vital organ such as the liver or kidney, whose normal function is detoxification. The organ damage leads to the accumulation of toxic products that cause neurologic dysfunction[1]; (2) the patient may be taking a number of drugs for the treatment of the cancer or its symptoms that interfere with or alter the body's metabolism;[2,3] or

(3) the tumor itself may produce substances that impair the metabolic function of normal organs or tissues.[4]

Malnutrition can be caused by the cancer itself or by its treatment.[4–6] Taste buds can be affected, leading to a dysgeusia[7–9] that sometimes alters total food intake, or specific nutritional disorders may arise even when the patient continues to eat.[10] Several malignancies directly cause weight loss and malnutrition (i.e., cancer cachexia) even in individuals who appear to be eating an adequate and well-balanced diet.

418

Metabolic and nutritional disorders can affect any portion of the nervous system (Table 11–1), most frequently the central nervous system (CNS). Interference with brain metabolism induces a metabolic encephalopathy with confusion, disorientation, lethargy, and sometimes stupor or coma, even when the brain is structurally normal. Specific nutritional disturbances may affect particular areas of brain function, such as when a thiamine deficiency causes brainstem encephalopathy.[10,11]

Peripheral nerves are less commonly affected than the brain by metabolic or nutritional disorders, but both diffuse and focal neuropathies can occur. For example, polyneuropathy can be caused by uremia, vitamin-B deficiency, or by critical illness itself.[12–14] Cachexia can cause focal peroneal nerve palsy by decushioning the nerve so that it is vulnerable to compression behind the head of the fibula.[15] Critical-illness polyneuropathy may render peripheral nerves susceptible to pressure neuropathies related to immobility.[16]

Muscles also may be affected by metabolic or nutritional disorders. Most patients with cancer who lose weight do not develop specific abnormalities of muscle function. They may complain of "weakness," but they usually mean fatigability and are not weak.[17] However, Erlington et al.[18] found one or more measures of weakness in 44% of 150 patients with small cell lung cancer. Twenty-one percent were unable to rise from a squatting position. A few patients develop specific nutritional myopathies that lead to weakness of large proximal muscle groups.[19,20] There are reports that the neuromuscular junction may be adversely affected in malnourished patients, but the evidence is not compelling[20] (see also Chapter 15).

DELIRIUM ASSOCIATED WITH CANCER

The second most common reason for neurologic consultation at Memorial Sloan-Kettering Cancer Center (MSKCC) is an altered mental state (Fig. 11–1), second only to pain.[21] In some patients, the altered mental state results from focal abnormalities of cognitive function such as aphasia, apraxia, or agnosia, which are usually caused by brain or leptomeningeal metastases. In most, however, the disorder is more diffuse; these patients are delirious.

Definition

Delirium is defined by the *Diagnostic and Statistical Manual of Mental Disorders* (4th edition) as a disturbance of consciousness with reduced ability to focus, sustain or shift attention, a change in cognition or the development of perceptual disturbances that develop over a short period of time and tend to fluctuate

Table 11–1. **Nervous System Complications of Cancer Caused by Metabolic and Nutritional Disorders**

Brain
Diffuse encephalopathy
    Lethargic
    Agitated
Focal encephalopathy

Spinal Cord
Nutritional myelopathy

Peripheral Nerves
Polyneuropathy
Focal neuropathy(ies)

Neuromuscular Junction and Muscle
Nutritional myopathy

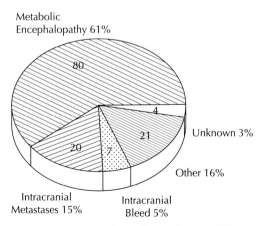

Figure 11–1. Causes of altered mental state in 132 inpatients with cancer in whom neurologic consultation was requested. (From Ref. 21 with permission.)

during the course of the day. Cognitive altera-tions can be found in attention and alertness, orientation and grasp, cognition, memory (particularly short-term), affect and comport-ment and perception (including illusions and hallucinations).[1] Most observers believe that failure of attention is the key phenomenon.

Incidence

In an unpublished survey of 100 consecutive admissions to the medical service at MSKCC, careful neurologic evaluation revealed that 15 patients suffered from encephalopathy, pre-sumably of metabolic origin. These figures are surprisingly close to those reported by Engel and Romano[22] in a similar study conducted on the medical wards of a general hospital. Furthermore, the primary physician and nurs-ing staff were aware that only one-third of those patients were delirious. In a study of terminal cancer patients with delirium (defined as a score of less than 24 or a drop of greater than 30% on the Mini-Mental Status Examination [see Table 1–13]), the cognitive failure was not detected by physicians in 23% of patients and by nurses in 20%.[23] A review of the incidence of delirium in elderly and advanced cancer patients indicates that approximately 20% of elderly general hospital patients are delirious; the percentages are higher in advanced cancer patients, ranging in various series from 20% to 86%.[24]

Pathophysiology

At least two factors conspire to make delirium common in patients with cancer. The first is that most patients with cancer are elderly and their brains are less able to compensate for the rav-ages of systemic illness. The second is that the cancer itself and the immune system response to the cancer produce cytokines and chemokines that affect brain function (Fig. 4–1).[4,25] Exactly how these factors cause brain dysfunction that manifests itself as delirium is not known. One hypothesis suggests that the final common pathway is a result of acetylcholine deficiency and dopamine excess.[26] Alterations of amino acid concentrations in the serum that might affect neurotransmitter levels have been iden-tified in a small number of studies.[27,28] Others

have suggested that endogenous anticholinergic substances may play a role.[29] As indicated in the preceding text, cytokines and chemokines secreted by the tumor or the immune system may also play a role, although this has been dis-puted.[30] The role of neurotransmitters in meta-bolic encephalopathy is discussed elsewhere.[1]

Clinical Findings

A core abnormality in the delirious patient is a disorder of attention and alertness. Three dif-ferent disorders of attention occur in delirious patients: (1) distractibility (the patient shifts attention from the examiner to noises in the hallway or other extraneous stimuli); (2) per-severation (patients answer a new question or respond to a new stimulus with the same response they gave to a previous stimulus); and (3) poor focusing (after being distracted by another stimulus, the patient will forget to return to the activity in which he or she was engaged before distraction).

Delirious patients are often disoriented; the first abnormality is usually disorientation in time. The patient may not know the date, the day of the week, the time of day or even the month and year. Disorientation for place may occur (if the patient is in the hospital, he or she may not know which hospital or where the hospital is located). Cognitive functions such as errors in arithmetic and memory (e.g., remembering three unrelated words for 5 min-utes) are easy to test at the bedside and can be re-examined repeatedly during the course of the hospitalization to determine the patient's clinical course (Table 1–13). Affect is usually abnormal. Some patients are withdrawn and apathetic; these patients are often believed to be depressed by relatives and other observ-ers. Other patients are disinhibited, often making inappropriate or embarrassing com-ments. Perceptual abnormalities often occur. Interpretation of ongoing stimuli, often with a paranoid flavor and hallucinations (particularly visual) are common. Often these are not men-tioned by the patient and must be elicited by the physician.

Cognitive and behavioral changes charac-terize all patients with delirium. The subtypes of delirium are hyperactive, hypoactive and mixed.[31] In the beginning, the changes are often so subtle that they escape recognition by the patient's family or physicians. Such patients may have mild difficulty concentrating and be unable to sustain attention. They may appear disinterested or have difficulty understanding explanations of impending medical procedures. Some may appear apathetic, preferring to lie in bed sleeping or staring at the wall rather than reading, watching television, or being active. If the patients read, they may find themselves spending an unusual amount of time on one paragraph or one page. Often the paragraph must be reread several times before the contents can be related to a questioner. As the illness progresses, behavioral abnormalities become more apparent. Apathy may worsen; patients may become lethargic, drowsy, and unable to attend to external stimuli for an appreciable time. On the other hand, some become restless, develop insomnia, or have vivid nightmares; they may be so distractible that they cannot concentrate or effectively communicate to others. Some patients appear silly or frivolous during the physician's explanation of their illness; conversations in the hall or other noises in the room distract them from the matter at hand. Some become fearful, anxious, and depressed. Patients who retain insight into the problem may express fear of going crazy.

Table 11–2 categorizes the clinical findings in an MSKCC study of 140 patients with cancer and altered mental state. Nearly one-half of the patients were alert or hyperalert and the remainder were lethargic. Cognitive abnormalities such as memory impairment, inattention, and dyscalculia were present in more than 90%, but language impairment was identified in only 35%. Over 80% were disoriented and about one-half displayed motor agitation. Delusions or hallucinations were identified in about one-third of patients; asterixis was present in a similar number. Seizures occurred in approximately 10%.

Other studies give similar results.[32] A study of 325 elderly patients admitted to a hospital for acute medical problems revealed *DSM-III-R* delirium in 38%. Hyperactive delirium (including hypervigilance, restlessness, fast or loud speech, irritability, and distractibility) occurred in 15%, whereas hypoactive delirium (including decreased alertness, slow speech, lethargy, psychomotor slowness, and apathy) was present in 19%. A mixed picture occurred in 52%, and 14% were neither excessively active nor excessively quiet.[33] Single signs of delirium,

Table 11–2. **Clinical Findings in 140 Patients with Cancer and Altered Mental Status**

Specific Findings	Percentage of Patients
Consciousness	
Alert	46
Lethargic	52
Comatose	2
Cognition	
Memory poor	91
Inattentive	92
Disoriented	83
Dyscalculia	92
Language impaired	35
Behavior	
Agitated	44
Delusions and/or hallucinations	28
Motor Behavior	
Asterixis	36
Seizures	9
Focal signs	40

From Ref. 35 with permission.

such as disorientation or lethargy, are often present in elderly patients who do not meet all of the *DSM-III-R* criteria for full-blown delirium.[34]

Drug withdrawal, corticosteroids, fever, and postoperative delirium are more likely causes of hyperactive delirium and, because of their florid behavior, are usually easily recognized. The diagnosis of delirium in a patient with a hypoactive variety may easily be missed. Whatever variety, diagnosis is important because treatment promotes survival and increases quality of life.

Diagnosis

The simplest and best-validated test for abnormalities of cognitive function at the bedside is the Mini-Mental State Examination (Table 1–13). The confusion assessment method (Table 1–14) can quickly evaluate patients in intensive care units as well as those on the general medical and surgical units.

Table 11–3. **Some Diffuse, Multifocal, or Metabolic Causes of Delirium, Stupor, and Coma**

A. Deprivation of Oxygen, Substrate, or Metabolic Cofactors

1. Hypoxia (interference with oxygen supply to the entire brain with normal CBF)°

   a. Decreased blood pO_2 and O_2 content: pulmonary disease; alveolar hypoventilation; decreased atmospheric oxygen tension

   b. Decreased blood O_2 content, pO_2 normal: "anemic anoxia"; carbon monoxide poisoning; methemoglobinemia

2. Ischemia (diffuse or widespread multifocal interference with blood supply to brain)°

   a. Decreased CBF resulting from decreased cardiac output: Stokes-Adams attack; cardiac arrest; cardiac arrhythmias; myocardial infarction; congestive heart failure; aortic stenosis; pulmonary embolism

   b. Decreased CBF resulting from decreased peripheral resistance in systemic circulation: syncope (see Table 5–8); carotid sinus hypersensitivity; low blood volume

   c. Decreased CBF associated with generalized or multifocal increased vascular resistance: hyperventilation syndrome; hyperviscosity (polycythemia, cryoglobulinemia, or macroglobulinemia, sickle cell anemia); subarachnoid hemorrhage; bacterial meningitis; hypertensive encephalopathy

   d. Decreased CBF owing to widespread small-vessel occlusions: disseminated intravascular coagulation; systemic lupus erythematosis; subacute bacterial endocarditis; fat embolism; cerebral malaria; cardiopulmonary bypass

3. Hypoglycemia resulting from exogenous insulin: spontaneous (endogenous insulin, liver disease, etc.)°

4. Cofactor deficiency

   a. Thiamine (Wernicke encephalopathy)

   b. Niacin

   c. Pyridoxine

   d. Cyanocobalamin

   e. Folic acid

B. Toxicity of Endogenous Products

1. Due to organ failure

   a. Liver (hepatic coma)

   b. Kidney (uremic coma)

   c. Lung (CO_2 narcosis)

   d. Pancreas (exocrine pancreatic encephalopathy)

2. Due to hyper- and/or hypofunction of endocrine organs: pituitary; thyroid (myxedema-thyrotoxicosis); parathyroid (hypo- and hyperparathyroidism); adrenal (Addison disease, Cushing disease, pheochromocytoma); pancreas (diabetes, hypoglycemia)

3. Due to other systemic diseases: diabetes, cancer, sepsis

C. Toxicity of Exogenous Poisons

1. Sedative drugs: hypnotics, tranquilizers, ethanol, opiates°

2. Acid poisons or poisons with acidic breakdown products: paraldehyde; methyl alcohol; ethylene glycol; ammonium chloride

3. Psychotropic drugs: tricyclic antidepressants and anticholinergic drugs; amphetamines; lithium; phencyclidine; phenothiazines; LSD and mescaline; monoamine oxidase inhibitors

4. Others: penicillin; anticonvulsants; steroids; cardiac glycosides; trace metals; organic phosphates; cyanide; salicylate

D. Abnormalities of Ionic or Acid–Base Environment of CNS

1. Water and sodium (hyper- and hyponatremia)

2. Acidosis (metabolic and respiratory)

3. Alkalosis (metabolic and respiratory)

Continued on following page

Table 11–3.—*continued*

---

    4. Magnesium (hyper- and hypomagnesemia)

    5. Calcium (hyper- and hypocalcemia)

    6. Phosphorus (hypophosphatemia)

E. Disordered Temperature Regulation

    1. Hypothermia

    2. Heat stroke, fever

F. Infections or Inflammation of CNS

    1. Leptomeningitis

    2. Encephalitis

    3. Acute "toxic" encephalopathy

    4. Parainfectious encephalomyelitis

    5. Cerebral vasculitis/vasculopathy

    6. Progressive multifocal leukoencephalopathy

    7. Subarachnoid hemorrhage

G. Primary Neuronal or Glial Disorders

    1. Creutzfeldt-Jakob disease

    2. Marchiafava-Bignami disease

    3. Adrenoleukodystrophy

    4. Gliomatosis, lymphomatosis cerebri

H. Miscellaneous Disorders of Unknown Cause

    1. Seizures and postictal states

    2. Concussion

    3. Acute delirious states: sedative drugs and withdrawal; "postoperative" delirium; intensive care unit delirium; drug intoxications°

From Ref. 1 with permission.

° Alone or in combination, the most common causes of delirium seen on medical or surgical wards.

CBF, cerebral blood flow; CNS, central nervous system.

Causes of Delirium

Delirium can be caused by a bewildering variety of structural and metabolic disorders (Table 11–3). In an MSKCC study, a definite single cause of delirium was identified in only about one-third of patients[35] (Table 11–4); the other two-thirds suffered multifactorial delirium (Table 11–5). When a single cause was identified, structural disease, in the form of multiple brain metastases, bacterial meningitis, or cerebral infarcts accounted for the delirium in one-half the patients, and metabolic brain disease accounted for the other half (Table 11–4).[35]

Other series are similar. In one series of 229 elderly patients in a general hospital, 22% were delirious. A single definite cause was identified in only 36%, and a single cause was probable in another 20%. The remaining patients had an average of 2.8 causative factors.[36] Delirious patients had a longer hospital stay and were more likely to die or be institutionalized.[36] Among the 67% in whom the cause of the altered mental state was multifactorial, the major contributing factors were more often metabolic than structural; psychoactive drugs and single- or multiple-organ failure were the most common contributing factors (Table 11–5). Structural diseases of the brain leading to delirium are discussed in Chapters 5, 7, 9, 10, and 12. Metabolic encephalopathy is discussed in the paragraphs that follow.

Table 11–4. Single Causes of Delirium in 140 Patients with Cancer

Poor oxygen supply	11
Hypoxia	2
Hypoperfusion (shock)	6
Disseminated intravascular coagulation	3
Organ failure	3
Liver	2
Kidney	1
Hyperosmolality	2
Brain metastases	19
Other focal lesions (bacterial meningitis, cerebral infarction)	2
Drugs (steroids, opioids)	6
Total	43

Table 11–5. Major Factors Contributing to Delirium in 140 Patients with Cancer

Factors	Percentage of Patients
Drugs	59
Organ failure	51
Fluid or electrolyte imbalance	45
Infection	45
Hypoxia	35
Brain lesions	21
Environment	21

Data from Ref. 35.

METABOLIC BRAIN DISEASE IN PATIENTS WITH CANCER

Metabolic brain disease or metabolic encephalopathy is defined as an abnormality of brain function resulting from interference with brain metabolism by extracerebral factors. The clinical result is delirium or an acute confusional state as described in the preceding text. Occasionally, focal signs such as ataxia may predominate, confusing the diagnosis.[37] Extracerebral disorders that lead to metabolic brain disease (Table 11–6) include deprivation of essential substrates, primarily glucose or oxygen; failure of vital organs;

Table 11–6. Causes of Metabolic and Nutritional Encephalopathy in Patients with Cancer

Drugs°
Opioids
Sedatives
Sepsis°
Oxygen deprivation°
Hypoxia
Ischemia
Fluid and electrolyte imbalance
Calcium°
Phosphorus
Magnesium
Osmolar imbalance
Hypo-osmolality (hyponatremia)°
Hyperosmolality (hyperglycemia)
Hypoglycemia
Uremia
Hepatic failure
Vitamin deficiencies
Thiamine (Wernicke encephalopathy)
Niacin (pellagra)
Hypercarbia
Endocrine disorders
Adrenal
Thyroid

° Particularly common in patients with cancer.

fluid, electrolyte, and acid–base abnormalities; and ingestion of toxic drugs. Because the brain is so sensitive to perturbations of its internal milieu, many systemic illnesses can cause neurologic dysfunction (Table 11–3).

Metabolic encephalopathy is an important neurologic complication of cancer for two reasons: (1) It is common. As indicated previously, delirium is a common symptom in both the cancer and general hospital populations, and in more than 50% of delirious patients, the encephalopathy is metabolic in origin. (2) It can often be treated. If the cause is treated, a patient suffering from a toxic or metabolic disorder can recover normal neurologic function.

Table 11–3 classifies the metabolic encephalopathies; those encountered most frequently in patients with cancer are indicated by an asterisk. The physician must recognize, however, that metabolic brain disease often has no single cause but results from the concurrence of multiple systemic metabolic abnormalities, none of which alone causes brain dysfunction. A positive consequence of the multiplicity of mild systemic metabolic abnormalities is that, in many instances, if one of the abnormalities can be identified and reversed, the patient often recovers.

Incidence

The exact incidence of metabolic encephalopathy in patients with cancer is unknown. Metabolic disorders account for about one-half of patients with delirium; one might expect that the figure also applies to cancer patients in hospital with delirium.

A study[35] defined the following risk factors that predispose patients with cancer to metabolic encephalopathy: The cancer is usually disseminated and the patients are older, with a mean age of 64 years. Although old age is a risk factor for the development of delirium, the occurrence of delirium in a young person with cancer is a poor prognostic sign for survival, presumably because a more severe systemic insult is required to cause encephalopathy in a younger brain. In addition to younger age, hypoxia and kidney or liver failure are major risk factors for death. Approximately 33% of delirious patients had an altered mental status at the time of hospital admission. Delirium was associated with a 25% mortality rate within 30 days and a 44% mortality rate within 6 months. Nevertheless, two-thirds of the patients were improved by the time of discharge.

An earlier MSKCC study attempted to identify the frequency and causes of metabolic encephalopathy in patients referred for neurologic consultation. In that study of 721 consecutive neurologic consultations during 5 months, 132 patients showed signs of diffuse encephalopathy.[21] Of these, a final diagnosis of metabolic encephalopathy was made in more than 60% (Fig. 11–1); the other 40% suffered from metastatic, vascular, or infectious diseases. These figures undoubtedly underestimate the true frequency of metabolic encephalopathy in hospitalized patients because minor cognitive changes were probably overlooked. In addition, when the cause of the encephalopathy was obvious, such as drug overdose or hypoxia,

the patients were often managed without neurologic consultation. These figures, therefore, can be considered the minimal frequency and probably skew the list of causes toward those most difficult to diagnose. Narcotic drug overdose, hypercalcemia, and hepatic encephalopathy are probably more common causes of encephalopathy than we have indicated.

Clinical Signs

The cognitive and behavioral changes in metabolic encephalopathy are the same as those of delirium and are detailed in the preceding paragraphs. Physical signs that suggest metabolic encephalopathy other than the cognitive changes include asterixis, abnormalities of respiration (hyperventilation causing metabolic acidosis or hypoventilation suggesting respiratory acidosis), small reactive pupils except for those in an agitated state where the pupils may be enlarged, and spontaneously roving eye movements with active doll's head eyes[38] as the patient becomes stuporous.[1]

Laboratory Tests

Two types of laboratory tests are useful in the diagnosis of metabolic encephalopathy. The first, which includes magnetic resonance imaging (MRI) and lumbar puncture, identifies structural lesions of the brain and distinguishes them from metabolic causes of encephalopathy. The second, which includes biochemical tests of blood and cerebrospinal fluid (CSF), identifies specific metabolic abnormalities causing delirium.

The MRI is particularly useful in identifying metastatic brain tumors, abscesses, or subdural effusions. These disorders can affect the brain diffusely and may cause clinical signs and symptoms that mimic those of metabolic brain disease. A lumbar puncture can identify leptomeningeal metastases and infection. Some causes of metabolic brain illness, such as uremia or hypothyroidism, may be associated with an elevated CSF protein concentration; rarely, uremic encephalopathy has been reported to be associated with pleocytosis. The electroencephalogram (EEG) can help to distinguish delirious patients from non-delirious patients (alpha frequency) and those with dementia (theta and delta activity); the EEG also can identify patients with nonconvulsive status epilepticus whose mental state may mimic delirium.[39,40]

The EEG is usually abnormal, even in mildly encephalopathic patients, often before the clinical features become apparent.[40] Abnormalities include excessive theta activity, slowing of background with prominent delta activity, and occasionally, triphasic waves. The EEG shows suppression or burst-suppression activity when the encephalopathy becomes severe enough.

Some tests of blood and CSF that may be helpful in the diagnosis of metabolic brain disease are listed in Table 11–7. Unless the cause of the encephalopathy is immediately apparent, the physician should err on the side of ordering too many rather than too few laboratory tests because unexpected diagnostic results may be revealed.

Table 11–7. Laboratory Evaluation of the Delirious Patient

MR scan (gadolinium enhanced)
    Metastases
    Infection
Lumbar puncture
    Leptomeningeal tumor
    Infection
EEG
Blood cultures
    Sepsis
    Septic emboli
Complete blood count
Blood gases: pO_2, pCO_2, pH
Electrolytes: Na, K, Ca, Mg, PO_4
BUN and creatinine
Lactate
Liver function tests including ammonia
Coagulation profile: PT, PTT, FSP, D-dimer
Vitamin B_{12}, folic acid, thiamine
Endocrine tests: T4, cortisol
Glucose
Drug levels: theophylline, digoxin, anticonvulsants, and others

MR, magnetic resonance; EEG, electroencephalogram; BUN, blood urea nitrogen; FSP, fibrin split products; PT, prothrombin time; PTT, partial thromboplastin time.

Differential Diagnosis

The differential diagnosis of metabolic brain disease has two aspects. (1) The physician must first determine whether the patient suffers from a metabolic or structural cause of delirium. (2) Having made that determination, the specific metabolic abnormality causing the neurologic dysfunction must be found. Making the first judgment depends on both the clinical and laboratory evaluation. In general, patients should be suspected of suffering from metabolic brain disease if behavioral and cognitive changes represent the earliest or the only signs, if motor signs such as tremor and asterixis are bilateral and symmetric, and if pupillary reactions are preserved even if the patient is comatose. The patient should be suspected of suffering from structural brain disease, either alone or in combination with metabolic brain disease, if focal motor abnormalities (including focal seizures) occur; if specific changes in cognitive function, such as aphasia, acalculia, or agnosia, appear out of proportion to a general overall decrease in mental state or if the patient is at particular risk for contracting one of the neurologic complications of cancer that may mimic metabolic brain disease, particularly disseminated intravascular coagulation or meningitis. Although clinical signs are often helpful, too much overlap exists to confirm a diagnosis on clinical examination alone. Some patients with hepatic encephalopathy or hypoglycemia may develop hemiparesis or visual field abnormalities strongly suggestive of structural brain disease, whereas other patients with multiple brain metastases may show nothing other than a global alteration of cognitive function. Thus, the laboratory tests are essential to exclude structural disease of the nervous system and to ensure that the problem is definitely metabolic in origin.

Treatment

The treatment of delirium depends on identifying the specific cause(s) and treating them when possible. Sometimes just correcting one of multiple causes will reverse the delirium. However, certain general principles apply to most patients with delirium (Table 11–8). More details on this approach can be found in Chapter 7 of Ref. 1.

Table 11–8. **Principles of Management of Delirious Patients**

1. Control agitation
2. Evaluate for seizures with EEG
3. Control glucose
4. Administer thiamine
5. Consider specific antidotes (e.g., naloxone, flumazenil, etc.)
6. Treat infections
7. Adjust body temperature
8. Ensure oxygenation
9. Maintain circulation
10. Lower intracranial pressure
11. Restore acid-base balance and electrolyte balance

For agitated patients, sedative drugs should be avoided if possible; however, in some patients, small doses of lorazepam 0.5 to 1.0 mg orally may be helpful. Keeping the patient in a lighted room with a relative or staff member sitting at the bedside and talking reassuringly may help reorient and quiet the patient. If these measures fail, haloperidol or other antipsychotics[40a] in gradually escalating doses are often effective. Restraints should be avoided whenever possible. When they are required, care should be taken so that they do not interfere with breathing or blood flow, or compress peripheral nerves.

Because nonconvulsive status epilepticus is an occasional cause of delirium, one should examine patients carefully looking for evidence of minor convulsive movements and consider an electroencephalogram. If that disorder is suspected anticonvulsants are indicated.

As indicated elsewhere in this chapter, both hypo-and hyperglycemia deleteriously affect brain function. Blood glucose should be carefully monitored and kept within the normal range. In the malnourished patient, thiamine should be administered before glucose.

Because many patients have been treated with opioids, one should consider opioid intoxication; naloxone is appropriate only if the patient has depressed respirations and has not been receiving opioids chronically. In patients receiving chronic opioids, naloxone may precipitate acute opioid withdrawal. To avoid this, the dose should be given in the manner described in the following paragraph on drugs. Some specific antidotes are also available for other drugs.

Sepsis is an important cause of delirium and should be investigated and treated. Severely hyperthermic patients should be cooled.

To avoid permanent brain damage, careful monitoring is required to insure adequate blood flow, oxygenation, and nutrition to the brain.

SPECIFIC METABOLIC CAUSES OF DELIRIUM

Postoperative Delirium

Postoperative delirium[41] can be one of the most florid and frightening complications to confront the physician. The clinical picture varies from mild cognitive impairment, often unrecognized, to an acute hyperactive delirium that can cause physical damage. The pathogenesis is unknown. Abnormal melatonin secretion (either increased or decreased) accompanies postoperative delirium in elderly patients.[42] Markedly increased levels are associated with postoperative complications such as pneumonia, shock, or heart failure. Low levels are associated with uncomplicated delirium. Whether the melatonin abnormalities are a cause or effect is unknown. One case report describes a typical postoperative delirium apparently caused by thiamine deficiency in a patient with cirrhosis and hepatocellular carcinoma.[43] The patient had been eating well just before surgery, but the liver dysfunction may have decreased the ability of that organ to store thiamine.

Drugs

Certain drugs associated with delirium in patients with cancer are listed in Table 11–9.

Both overdose and withdrawal of the drugs from the tolerant patient can cause delirium. In general, drug overdose causes a passive delirium characterized by drowsiness, inattention, disorientation, and excessive sleeping, whereas the delirium of drug withdrawal is more likely to be characterized by tremulousness, hallucinations, and sometimes seizures.[44] These distinctions are not absolute. For example, high doses of opioids, particularly meperidine, can cause multifocal myoclonus and seizures, usually phenomena of drug withdrawal.[45] Antibiotics may cause seizures, particularly in patients

Table 11–9. **Drugs Associated with Delirium in 140 Patients with Cancer**

Classes of Drugs	No. (%) of Patients Taking At Least One Drug
Opioids	75 (53.6)
Benzodiazepines	33 (23.6)
Corticosteroids	30 (21.5)
H$_2$ blockers	27 (19.3)
Anticholinergics	9 (6.4)
Anticonvulsants	8 (5.7)
Antihistamines	6 (4.3)
Others	12 (8.6)

Data from Ref. 35.

Table 11–10. **Clinical Features of Opioid Intoxication and Withdrawal**

Intoxication	Withdrawal
Stupor or coma	Anxiety, restlessness
Symmetric, pinpoint, reactive pupils	Insomnia
	Chills, hot flashes
Hypothermia	Myalgias, arthralgias
Bradycardia	Nausea, anorexia
Hypotension	Abdominal cramping
Skin cool, moist	Vomiting, diarrhea
Hypoventilation (respiratory slowing, irregular breathing, apnea)	Yawning
	Dilated pupils
	Tachycardia, hypertension (mild)
Pulmonary edema	Hyperthermia (mild), diaphoresis, lacrimation, rhinorrhea
Seizures (meperidine, propoxyphene, morphine)	
Reversal with naloxone	Piloerection
	Spontaneous ejaculation

with renal failure. Penicillin and its analogs are the major culprits, perhaps through activation of N-methyl-d-aspartate (NMDA) receptors.[46] Fluoroquinolones may exert their convulsant effect by increasing hippocampal glutamate levels.[47] Patients with drug withdrawal can become lethargic and withdrawn rather than hyperactive. Furthermore, in many patients, especially those who are elderly, a standard dose, even when the measured blood level is within the "nontoxic" range of a potentially toxic drug, may cause encephalopathy, particularly when other potential causes of metabolic brain disease are also present.[48]

In any patient taking opioids, the drug should be considered as either the sole cause or a contributor to a quiet, diffuse encephalopathy. Long-acting opioid agents, such as methadone and levorphanol, can accumulate gradually over days, particularly as the dose increases. After days with little pain relief, the patient may suddenly become confused and disoriented, lapsing into stupor or coma. Other characteristic signs are indicated in Table 11–10. This disorder is particularly likely when the pain has finally been relieved. Many patients who are wide awake while receiving large doses of opioids that fail to relieve pain may become confused and lethargic on the same dose if pain is alleviated either by the drugs or by other means.

A stuporous or comatose patient receiving opioids should be given an opioid antagonist such as naloxone. One ampule (0.4 mg) should be diluted in 10 mL of saline and injected gradually over several minutes while monitoring the patient's respiratory rate, pupils, and state of consciousness. No more drug than necessary to restore respiration and slightly arouse the patient should be given; in a tolerant patient, larger doses can cause acute withdrawal syndromes, triggering an agitated delirium and exacerbating pain. Naloxone is a short-acting drug, and many patients who are aroused from stupor relapse into that state if left alone. The relapse can be prevented either by a slow naloxone drip or by careful observation, repeating the naloxone as required.

Opioid withdrawal can also cause delirium, usually in tolerant patients whose pain is successfully relieved by another mechanism. Because these patients are not psychologically dependent on opioids, they stop requesting them once the pain is relieved. The physician must ensure that the drug is tapered rather than abruptly discontinued. The patients, being unaware of the effects of opiate withdrawal, fail to recognize its premonitory symptoms, such as runny nose, myalgias, yawning, and diarrhea. They frequently tell the doctor that they believe they are getting the flu. If the doctor is as unaware as the patient, florid signs of withdrawal (see Table 11–10) may ensue before the cause is recognized and the opioids are restarted at lower doses.

Repetitive doses of meperidine, which is metabolized to normeperidine, can cause delirium, multifocal myoclonus, and seizures, particularly in patients with renal failure.[45]

Consequently, any patient receiving meperidine who becomes tremulous should be observed carefully for seizures and switched to another agent. Because of its potential toxicity, meperidine should be avoided in the management of acute or chronic cancer pain. Other opioids only occasionally cause myoclonus.[49]

Some drugs, such as benzodiazepines, are rarely the sole cause of delirium in patients with cancer but frequently contribute to it, particularly in patients receiving other medications. In critically ill patients, even short-acting drugs may accumulate because the patient's ability to detoxify them is impaired. Measurement of blood levels of the drug may assist the physician in determining if a given drug is contributing to the patient's encephalopathy.[50]

Substrate Deprivation (Hypoxia–Ischemia–Hypoglycemia)

The average adult brain weighs approximately 1400 g, which is approximately 2% of body weight, but consumes approximately 20% of the body's oxygen and 65% of glucose production. Cortical gray matter, only 20% of total brain mass, uses approximately 75% of total brain oxygen. The brain uses no exogenous substrate other than glucose, and 85% of all glucose utilization is aerobic. Consequently, the brain is exceedingly vulnerable to diminution of its supply of either oxygen or glucose. In partial compensation for this vulnerability, the brain can regulate its own blood supply by the process called *autoregulation*. If systemic blood pressure falls, cerebral vessels dilate and the blood flow is maintained, at least until the mean pressure falls lower than 50 mm Hg. The brain also protects itself against hypertension by vasoconstricting as blood pressure rises, thus preventing increases in intracranial pressure. In individuals with normal pre-illness blood pressure, cerebral autoregulation maintains normal cerebral blood flow until mean blood pressure either falls below 50 mm Hg or rises higher than 160 mm Hg. These figures are higher for chronically hypertensive persons. In the resting state, the brain and the heart also extract more oxygen from the blood than do other tissues. As its blood supply falls, the brain can extract even more oxygen.

Hypoxemia and hypoperfusion are common metabolic disorders in patients with cancer. Thus, it is not surprising that lack of oxygen or glucose should be important causes of delirium, given the vulnerability of the nervous system to hypoxemia and hypoperfusion (Table 11–4). These two disorders are also common as associated factors in patients with other metabolic abnormalities.

HYPOXIA

The term *hypoxia* implies deprivation of oxygen to the brain when blood flow is maintained. Hypoxia results either from a low partial pressure of oxygen (PO_2) delivered to the brain, as in pneumonia or other lung disease, or from a low oxygen-carrying capacity of blood, as in severe anemia. The healthy brain can tolerate a PO_2 down to approximately 40 mm Hg. Levels lower than 40 mm Hg can result in cerebral dysfunction, as registered by the EEG and other sensitive tests of brain function, but clinical symptoms are not apparent until the PO_2 falls lower than 30 mm Hg. Nevertheless, even in healthy adults at high altitudes, a steady-state[51] PO_2 of 30 mm Hg at rest can cause persistent brain dysfunction characterized by abnormalities of motor, intellect, and psychological function. Oxygen deprivation at higher levels of PO_2 may cause neurologic dysfunction in patients who are already compromised by pre-existing brain disease. Hypoxia in the postoperative period is a significant risk factor for the development of delirium.[52]

The clinical symptoms of diffuse hypoxia resemble those of other metabolic encephalopathies. Patients usually become inattentive to their surroundings and are somewhat drowsy.[53] Hyperactivity and hallucinations are uncommon. Judgment is impaired early. Early-morning headache is a frequent sign of both hypoxia and hypercarbia; tremor, asterixis, and multifocal myoclonus are generally late signs.

ISCHEMIA

Ischemia is the term applied when the brain is deprived of blood, either focally or diffusely.

Generalized or global ischemia occurs with cardiac arrest. Focal or multifocal ischemia occurs with occlusion of large vessels supplying the brain, as may occur in nonbacterial thrombotic endocarditis (NBTE), or of multiple small vessels in the brain, as in disseminated intravascular coagulation (DIC). In patients with ischemia, the brain is deprived not only of oxygen but also of glucose. Focal ischemia causes focal signs that reflect the part of the brain deprived of substrate. When the focal ischemia involves the right parietal lobe, an agitated delirium without motor or sensory signs is characteristic.[54] Patients with elevated cortisol levels associated with focal ischemia are more likely to be delirious.[55]

Delirium can follow cardiac surgery, as well as cardiac arrest. The overall prevalence of postoperative delirium after cardiac surgery is 8.4%.[56] Patients likely to develop postoperative delirium include those with vascular disease, diabetes (insipidus or mellitus), cardiac dysfunction, or prolonged surgery.

Severe global ischemia, as in cardiac arrest, causes coma. In its most severe form, the patient is unconscious and does not respond in any way to painful stimuli. Brainstem reflexes and other motor responses are absent. In a less severe form, brainstem reflexes, including pupillary light responses, may be present, and the patient may develop extensor posturing either spontaneously or in response to noxious stimuli. The prognosis for recovery after cardiorespiratory arrest with global brain ischemia and coma is poor. In the patient with cancer, whose vital organs may be compromised already, the prognosis is worse.

Multifocal ischemia usually results from either multiple small emboli from NBTE or from DIC (see Chapter 9). Occasionally, it results from a vasculitis.

HYPOGLYCEMIA

Approximately two-thirds of the body's glucose consumption in the resting state occurs in the brain. During fasting, plasma glucose levels are maintained by the breakdown of liver glycogen or by glucose synthesis. Hypoglycemia without ischemia is an uncommon cause of metabolic encephalopathy in the patient with cancer. Fasting hypoglycemia can occur either when glucose synthesis is reduced or when utilization is increased. In patients with cancer, a tumor in the liver, either primary hepatoma or metastatic cancer, can decrease glucose synthesis. Metastases to the adrenal glands or the pituitary gland may cause adrenal or growth hormone insufficiency, which may interfere with stimulation of hepatic glucose production. Increases in glucose utilization occur either when some retroperitoneal tumors metabolize large quantities of glucose or when tumors produce insulin-like growth factors that may mimic the biologic actions of insulin.[57–59] Typically, large retroperitoneal tumors induce repeated episodes of hypoglycemia, causing hunger, confusion, disorientation, and sometimes stupor or coma. Plasma glucose levels during these episodes are low, but respond to infusion of glucose. Insulin-like growth factors, insulin, or insulin antibodies can be measured to determine the underlying cause. The disorder is best addressed by treating the tumor, but glucagon has also been reported to be successful except in those with poor hepatic glycogen reserve or liver failure.[58] We also successfully treated such a patient, who presented with intermittent confusion, with dexamethasone, which prevented episodes of hypoglycemia.

Insulinomas, tumors of pancreatic beta cells, produce insulin and cause hypoglycemia. Hypoglycemia may also be caused by anti-insulin antibodies spontaneously produced by myeloma cells,[60] or it may arise in patients with diabetes and cancer who are not eating well; their usual dose of insulin may lead to the disorder when caloric intake falls. Finally, hypoglycemia may follow the use of drugs such as pentamidine, used in patients with cancer for prophylaxis of *Pneumocystis carinii* pneumonia.[61,62]

The clinical picture of hypoglycemia can range from a quiet delirium to coma, depending on the level of the glucose and the rapidity with which it has fallen.[63] In patients with diabetes, hypoglycemia can be associated with symptoms of autonomic failure.[64] No clinically distinct characteristics allow the clinician to determine that an encephalopathy is due to hypoglycemia. In a few patients, hypoglycemia causes focal signs, especially hemiplegia, without clinical evidence of diffuse encephalopathy. Such findings can markedly delay the diagnosis, resulting in permanent neurologic disability. Prompt administration of glucose will reverse the focal signs and prevent permanent damage. However, in malnourished patients, an acute glucose load may precipitate Wernicke encephalopathy[65] (see also p 439). Therefore, care should be taken not to overcorrect hypoglycemia.

One of our patients suffering from a pelvic tumor that had been heavily irradiated developed a vascular occlusion with gangrene of one leg. Although her appetite was poor, she ate and drank. One morning, she was found aphasic with a right hemiplegia. A fingerstick serum glucose level was less than 20 mg/dL; her hemiplegia and aphasia resolved immediately after she received 50 mL of 50% dextrose intravenously; she was then given an infusion of 5% dextrose in water. Later that day, while being observed by the nursing staff, she suddenly became diaphoretic and lethargic, appearing confused and refusing to speak. Right-sided weakness was not present. Again, her glucose level was low, and she responded to immediate IV glucose. Hypoglycemic episodes continued despite glucose infusions but ceased immediately after the gangrenous leg was amputated. The exact cause of the hypoglycemia remained unknown. In this patient, two episodes of hypoglycemia on the same day caused first a focal and then a diffuse encephalopathy. The patient showed no evidence, either before or after the hypoglycemia, of structural disease of the left hemisphere or any other portion of the brain.

Hyperglycemia

Hyperglycemia is common in patients with cancer who have other neurologic complications. Some hyperglycemic patients are known to be diabetic, but most are not until they begin receiving corticosteroids for a CNS complication of cancer. In a few patients, phenytoin also appears to precipitate hyperglycemia.[66,67] As with hypoglycemia, two clinical syndromes are identified. The first is a diffuse encephalopathy generally referred to as *nonketotic hyperosmolar coma*.[68] Most of the patients are not known diabetics, but develop diabetes when placed on corticosteroids. In symptomatic patients blood sugars may range from 800 to 1200 g/dL with a serum osmolality in excess of 350.[69] A rare complication is acute nontraumatic rhabdomyolysis that may lead to renal failure.[70] The pathogenesis of the encephalopathy is believed to be cellular dehydration of the brain caused by the very high glucose level and often a moderately elevated adjusted serum sodium level.[71] Patients appear dehydrated and become lethargic and ultimately comatose, usually without focal signs. They awaken when the hyperglycemia is corrected. Most patients with nonketotic hyperosmolar coma have serum glucose levels greater than 800 mg/dL.

The second syndrome is a focal encephalopathy characterized by recurrent or continuous focal seizures without a change in consciousness.[72] The seizures do not respond to anticonvulsant medications and generally cease promptly when the hyperglycemia is corrected with insulin and rehydration. Serum glucose levels are usually in the range of 500 to 800 mg/dL, somewhat lower than in patients with nonketotic hyperosmolar coma.

Increasing evidence suggests that hyperglycemia may be damaging to the brain.[73,74] In a study of adults with either type I or type 2 diabetes, blood glucose levels greater than 270 mg/dL were associated with impaired cognitive performance tests, affecting approximately 50% of the 100 subjects investigated.[75] In intensive care unit patients[74] and patients undergoing cardiac surgery,[76] hyperglycemia is an independent risk factor for poor outcome. Fingerstick measurement of blood glucose may be inaccurate in the critically ill patient. Whole blood measurements are probably more accurate.[77]

Sepsis

The cause of septic encephalopathy is unknown. Possible pathogenetic mechanisms are shown in Figure 11–2.[78,79]

Blood–brain barrier derangement, permitting entry of otherwise excluded neurotoxic substances, has been implicated as a potential mechanism[80]; other postulated causes include brain inflammation, apoptosis, and endothelial activation.[78] Regardless of the cause, many patients who become septic, with or without cancer, develop severe delirium.[78,81,82] Septic encephalopathy is a relatively frequent, sole cause of delirium in patients with cancer and an extremely common cofactor in patients with multifactorial delirium. The symptoms of delirium may precede the onset of fever and be present throughout the course of an afebrile septic episode. Characteristically, affected patients become confused and are often agitated, exhibiting tremor, asterixis, and myoclonus[83]; many patients develop generalized seizures.[84] Encephalopathy is more common with Gramnegative infections but can occur with any form of sepsis. Appropriate antibiotic treatment usually leads to its resolution. Long-term treatment

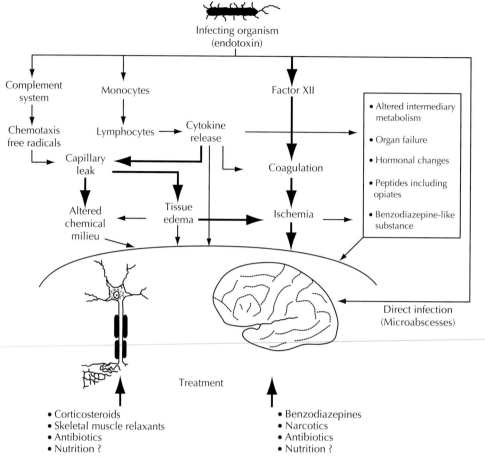

Figure 11–2. The multifactorial pathogenesis of septic encephalopathy. The infecting organism, the immune system, and drugs the patient is receiving all play a role. Mechanisms for septic encephalopathy and critical-illness polyneuropathy are suggested. Arrows pointing to the curved line indicate mechanisms that may apply to both the central and the peripheral nervous system. The lower arrows designate treatments that may affect these systems independently. The heavy arrows highlight the most likely mechanisms. These hypotheses are complex but involve the infecting organism's inducing chemical, microvascular and metabolic effects that may act independently or in concert. The release of cytokines from macrophages and thence from T-lymphocytes may directly affect the brain or act indirectly on the blood–brain barrier and microcirculation. Such vascular effects are abetted by activation of the complement system and factor XII. The encephalopathy may also be due to the failure of other organs or to direct infection of the brain, with the formation of microabscesses. Critical-illness polyneuropathy may be due to disturbances of the microcirculation of peripheral nerves through vascular effects similar to those affecting the brain. Various treatments used in the critical care unit may play an additive role for both the encephalopathy and polyneuropathy. (From Ref. 81 with permission.)

with anticonvulsants is unnecessary even in patients who have had seizures as part of the encephalopathy.

CRITICAL-ILLNESS WEAKNESS

Because sepsis and multiple-organ failure play such a large role in weakness encountered in critically ill patients, this problem is considered along with sepsis. The disorder is not restricted to patients with cancer, although cancer is one of the risk factors for developing the syndrome, if such patients become critically ill with sepsis or multiple-organ failure. The disorder, generally referred to as *critical-illness neuromyopathy*,[85–87] causes weakness and delays in weaning from a respirator. Critical-illness neuromyopathy is characterized by weakness caused by an axonal polyneuropathy or myopathy, often both. The patient who may or may not have been weak prior to admission to the intensive care unit is discovered to have diffuse weakness of muscles, including respiratory muscles, often with disappearance of deep tendon reflexes and sometimes distal sensory loss. As few as 18 hours of complete diaphragmatic inactivity in a patient on a respirator results in marked atrophy of diaphragm myofibers.[88] Once the reason for admission to the intensive care unit has resolved, the weakness becomes apparent. The patient usually recovers over days to weeks, although, in some, the weakness persists.

The pathophysiology of critical-illness neuromyopathy is not fully understood. Cytokines and chemokines released during sepsis are probably important factors. Oxidative stress, increased ubiquitin proteasomes, calpain activities, and glutamate release from muscles may all play a role.[86] Some of these same pathophysiologic factors are also believed to play a role in cancer cachexia. Risk factors for the development of the syndrome include the presence of a serious underlying disease (cancer), sepsis, hyperglycemia, corticosteroid medication, neuromuscular blocking agents, aminoglycosides, poor nutrition, and immobility.[87]

The diagnosis can be made clinically in patients who are not too delirious to be tested adequately. Diffuse weakness is present in both the neuropathy and the myopathy, but the absence of deep tendon reflexes suggests that neuropathy is predominant. If the patient cannot be evaluated clinically or serious questions arise concerning neuropathy versus myopathy, electrophysiologic tests are indicated. Patients with polyneuropathy show a decline in compound motor action potentials, usually with preservation of conduction velocities, indicating that the neuropathy is axonal and not demyelinating. Sensory nerve action potentials may be lost. Needle electromyography (EMG) reveals positive sharp waves and fibrillation potentials. In patients with myopathy, sensory and motor potentials are preserved. The needle EMG also reveals short duration low amplitude motor unit potentials with normal recruitment.[85] Creatine kinase is usually normal even with myopathy. A few patients, particularly those who received aminoglycosides, have abnormalities of neuromuscular transmission. This will induce clinical weakness similar to polyneuropathy. Neuromuscular transmission studies will reveal the defect.

Electrolyte Imbalance

HYPERCALCEMIA

Hypercalcemia, a common complication of cancer,[89–91] occurs in 10% to 20% of patients with solid tumors, particularly multiple myeloma and breast cancer. In one series from Okinawa, T-cell leukemia was the most common cause.[92] The incidence has fallen substantially in recent years because of the prophylactic use of bisphosphonates. Hypercalcemia is generally caused either by accelerated bone resorption in patients with widespread bony metastases or by tumor production of a parathormone-related protein. Less common causes include decreased renal calcium excretion, increased GI calcium absorption, and production by the tumor of vitamin D or cytokines.[90,93] Hypercalcemia is an occasional cause of delirium in patients with cancer. Characteristically, such patients become withdrawn and confused and complain of thirst or develop a markedly increased fluid intake. Constipation is common. As the hypercalcemia worsens, confusion and disorientation occur. Proximal muscle weakness develops and reflexes diminish. An occasional patient has an agitated or paranoid delirium, and seizures are a rare complication. The encephalopathy generally resolves when a normal calcium concentration is restored by any one of several treatments.[93] Bisphosphonates effectively treat hypercalcemia. Other agents such as gallium nitrate, calcitonin, and corticosteroids are rarely required.[90] In dehydrated patients, isotonic saline followed by diuretics once the patient is hydrated sometimes helps.

HYPOCALCEMIA

Hypocalcemia is not a major problem in patients with cancer. The disorder may occur as part of the tumor lysis syndrome (see following text),[94] after head and neck surgery,[95] or as a side effect of biphosphonate therapy.[96] Much like hypercalcemia, hypocalcemia can cause encephalopathy and muscle weakness. However, the major symptom is tetany. One of our patients, a young woman who was cured of osteosarcoma with cisplatin chemotherapy, presented years later with episodes of numbness and tingling of both hands and arms spreading into the face and followed by spasms of her arms that lasted several hours. A diagnosis of panic attacks was made and after sedation her symptomsing improved. Recurrent attacks prompted neurologic consultation, where brief hyperventilation at the bedside reproduced the tetany and Trousseau sign was present. Serum calcium levels were low.

HYPONATREMIA

The syndrome of inappropriate antidiuretic hormone (SIADH) secretion causing hyponatremia is extremely common in patients with cancer, as it is in the general population of hospitalized patients (~1%).[97–99] In one prospective series, 4% of hospitalized cancer patients were hyponatremic.[100] SIADH was the most common cause (30%), but volume depletion, diuretic use, hypervolemia, renal failure, and infusion of hypotonic fluids[101] can also cause hyponatremia. In neurosurgical patients, cerebral salt wasting can cause hyponatremia.[102] Hyponatremia was not the sole cause of encephalopathy in any of 140 patients studied at MSKCC,[35] but it was a contributing cause in a high percentage of those with multifactorial encephalopathy. Hyponatremia, with its associated hypo-osmolality, causes water to shift from the body to the brain, leading to brain edema.[103] Severe or acute hyponatremia (Fig. 11–3) may cause sufficient brain swelling to lead to cerebral herniation and death. The symptoms of less severe hyponatremia depend on the age of the patient and are said to be more pronounced at higher levels of serum sodium in women than in men.[97,104] Acute lowering of the serum sodium level to less than 130 mEq/L will cause headaches, nausea, vomiting, and encephalopathy. The rapid lowering of the serum sodium level to 120 mEq/L can cause seizures, coma, and sometimes respiratory arrest. Because the decrease of the serum sodium level is not acute in most patients with cancer, patients with striking hypo-osmolality may be only mildly encephalopathic. We have encountered occasional patients with serum sodium levels as low as 100 mEq/L who had only mild symptoms and recovered without sequelae as osmolality was restored. In patients with concomitant structural brain disease or metabolic derangements, however, relatively mild hyponatremia may contribute significantly to delirium.

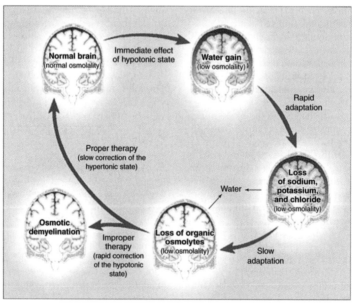

Figure 11–3. Effects of hyponatremia on the brain and adaptive responses. Within minutes after the development of hypotonicity, water gain causes swelling of the brain and a decrease in osmolality of the brain. Partial restoration of brain volume occurs within a few hours as a result of cellular loss of electrolytes (rapid adaptation). The normalization of brain volume is completed within several days through loss of organic osmolytes from brain cells (slow adaptation). Low osmolality in the brain persists despite the normalization of brain volume. Proper correction of hypotonicity re-establishes normal osmolality without risking damage to the brain. Overly aggressive correction of hyponatremia can lead to irreversible brain damage. (From Ref. 103 with permission.)

Patients who have been severely hyponatremic are susceptible to the development of demyelination in the brain,[105,106] particularly in the central pons (Fig. 11–4). The pathogenesis of central pontine myelinolysis is not completely understood. Most investigators believe it results from rapid correction of hyponatremia and advise that severe hyponatremia should be corrected slowly.[105]

Treating hyponatremia depends on its cause and severity. Fluid restriction will often suffice for people with mild, inappropriate antidiuretic hormone secretion. Hypertonic saline may be necessary for people with severe symptoms of hyponatremia. A new agent, conivaptan, a vasopressin receptor antagonist, can help correct hyponatremia from SIADH.[107] It is useful for patients with persistent hyponatremia, but is only available in an IV formulation. It is probably safe to correct hyponatremia with sodium, 1 to 2 mmol/hr, until symptoms are relieved and then to restrict water. Several formulas have been published for deciding the amount of sodium chloride necessary to restore normal natremia.[98]

HYPERNATREMIA

Hypernatremia occasionally causes delirium in the cancer population and leads to a hyperosmolar state and decrease in brain volume[108] (Fig. 11–5). Hypernatremia can stem from impaired thirst mechanisms from brain disease, from the inability to ingest adequate amounts of fluids, and from excessive fluid losses that are not replaced.[109–111] Diabetes insipidus, from the destruction of the pituitary gland by metastases, can cause hypernatremia in patients with breast

cancer. Fever causing excessive fluid loss also is a common cause. Patients characteristically have dry mucous membranes, poor skin turgor, and a quiet encephalopathy. Seizures, asterixis, and multifocal myoclonus are uncommon. Rapid correction of hypernatremia may cause symptoms of hyponatremia even at serum sodium levels considered normal. Hypernatremia has been reported to cause a myopathy associated with an elevated serum creatine kinase concentration and electromyographic abnormalities suggesting a muscle disorder. The syndrome reversed when the serum sodium was corrected.[112]

HYPOMAGNESEMIA

Hypomagnesemia typically develops in patients with cancer who are being treated with cisplatin or cetuximab (see Chapter 12). However, symptomatic hypomagnesemia is extremely rare; encephalopathy characterized by irritability, somnolence, confusion, and temporal spatial disorientation has been reported in a patient who became profoundly hypomagnesemic after cetuximab for metastatic colorectal cancer.[113] Seizures associated with cisplatin are more likely a result of hypo-osmolality than of hypomagnesemia. *Hypermagnesemia* is very rare in patients with cancer.

HYPOPHOSPHATEMIA

Mild hypophosphatemia, a common abnormality in hospitalized patients with cancer, can result from decreased dietary intake, decreased intestinal absorption, or shifts of phosphorus into cells such as occur with respiratory

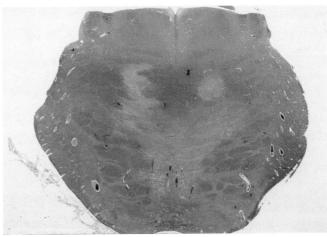

Figure 11–4. Central pontine myelinolysis: the effect of too rapid correction of hyponatremia. After rapid correction of his hyponatremia, this patient developed spastic weakness of all four extremities and diminished consciousness. He subsequently died. This myelin-stained photomicrograph of the pons shows the normal blue myelin surrounding an area of pink demyelination in the central pons.

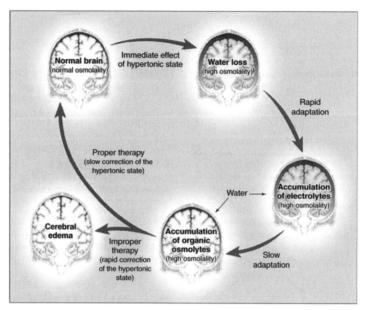

Figure 11–5. Effects of hypernatremia on the brain and adaptive responses. Within minutes after the development of hypertonicity, loss of water from brain cells causes shrinkage of the brain and an increase in osmolality. Partial restitution of brain volume occurs within a few hours as electrolytes enter the brain cells (rapid adaptation). The normalization of brain volume is completed within several days as a result of the intracellular accumulation of organic osmolytes (slow adaptation). The high osmolality persists despite the normalization of brain volume. Slow correction of the hypertonic state re-establishes normal brain osmolality without inducing cerebral edema, as the dissipation of accumulated electrolytes and organic osmolytes keeps pace with water repletion. In contrast, rapid correction may result in cerebral edema as water uptake by brain cells outpaces the dissipation of accumulated electrolytes and organic osmolytes. Such overly aggressive therapy carries the risk of serious neurologic impairment due to cerebral edema. (From Ref. 108 with permission.)

alkalosis[114] and some chemotherapeutic agents such as imatinib.[115,116] Rapidly proliferating malignant cells, particularly in leukemia and lymphoma, can absorb phosphorus, causing acute hypophosphatemia. Oncogenic osteomalacia, a paraneoplastic syndrome, can also cause hypophosphatemia.[117,118]

Although rare, severe acute hypophosphatemia can cause encephalopathy as well as a peripheral neuropathy. Patients develop irritability, apprehension, muscle weakness, numbness, paresthesias, dysarthria, confusion, obtundation, and convulsive seizures that can progress to coma and death. They sometimes have visual hallucinations[114,119,120] with or without other aspects of an agitated delirium. Respiratory weakness leading to hypercarbia, profound muscle weakness, sensory loss, and diminution of deep tendon reflexes may also occur, occasionally mimicking the Guillain–Barré syndrome. Peripheral neuropathy has also been reported.[121]

Endocrine and Other Organ Dysfunctions

ADRENAL DYSFUNCTION

Adrenal Failure

Adrenal failure can be caused by metastases to the adrenal gland or hypothalamus,[122,123] by bilateral adrenal lymphoma,[124,125] or may follow withdrawal from corticosteroid therapy[126,127] (see Chapter 4). In patients with brain tumors, the symptoms of adrenal failure may be mistaken for tumor progression; in others, the symptoms may be attributed to the underlying disease rather than adrenal failure. Drugs such as mitotane, aminoglutethimide, and megestrol, given for appetite enhancement, may cause adrenal failure.[128] Occasionally, replacement hydrocortisone is inadvertently discontinued in patients with adrenal insufficiency who are begun on dexamethasone, causing florid mineralocorticoid insufficiency. This occurred in the past in patients with breast cancer who had undergone adrenalectomy. These patients were maintained on replacement hydrocortisone, which was changed to dexamethasone when the patient developed CNS metastases. Because dexamethasone has no mineralocorticoid effect, if the cortisol was withdrawn (or rarely even if it was continued), patients developed acute adrenal crisis from inadequate mineralocorticoid. The phenomenon was accompanied by cardiovascular collapse but no encephalopathy.

Adrenal failure that deprives patients of both glucocorticoids and mineralocorticoids causes symptoms of a quiet delirium followed by cardiovascular collapse from fluid and electrolyte losses.[129] If one thinks of the possibility of adrenal failure, the diagnosis is easily made by electrolyte and endocrine measurements.[129]

Hyperadrenalcorticoism

Adrenal excess can occur either as a paraneoplastic effect due to increased secretion of adrenocorticotropic hormone-like peptides, usually from lung cancer,[130,131] as a result of chronic dexamethasone therapy, or from hormone-secreting tumors of the pituitary or adrenal gland. In these disorders, patients develop metabolic alkalosis. In the paraneoplastic syndrome, compensatory respiratory insufficiency can lead to hypoxia. Respiratory compensation for the metabolic alkalosis of exogenous steroid therapy is unusual; however, as indicated in Chapter 4, steroids can cause weakness of respiratory muscles, mimicking the compensatory decrease in respiration. In addition, depression is a common psychological symptom with Cushing syndrome, whether paraneoplastic or not, whereas mania is more common with exogenous steroid excess (see Chapter 4).

HYPER/HYPOTHYROIDISM

Hypothyroidism follows radiation to the neck or head (see Chapter 13) and occasionally metastases to the thyroid gland.[132,133] Some chemotherapeutic agents (e.g., sunitinib[134]) can also cause hypothyroidism (see Chapter 12). One report describes a high prevalence of hypothyroidism in men with cutaneous melanoma.[135]

Transient hyperthyroidism sometimes follows radiation therapy (RT) to the thyroid gland. It is a rare complication of cancer, unlike hypothyroidism, which is a frequent consequence of RT (see Chapter 13).

Organ Failure

Patients with cancer are susceptible to respiratory failure causing hypercarbia, hepatic failure causing hepatic encephalopathy,[136,137] and renal failure causing uremia.[138]

Hepatic encephalopathy and uremia are occasionally the sole causes of delirium in patients with cancer; more commonly, the conditions are part of a multifactorial delirium. Clinical manifestations resemble those caused by organ failure from other causes.[1,139,140]

Hyperammonemia is occasionally the sole cause of encephalopathy or even coma.[90] The clinical symptoms are similar to hepatic coma. The disorder can occur in the tumor lysis syndrome (see following text), after high-dose chemotherapy[141,142] or bone marrow transplant,[143] and rarely as a manifestation of myeloma.[144] The anticonvulsant valproic acid can cause hyperammonemic encephalopathy.[145]

Tumor Lysis Syndrome

The lysis of tumors can lead to several metabolic disorders that can secondarily affect the brain.[90,146] Gross elevations of uric acid, phosphorus, potassium, urea, and ammonia have been reported as part of the syndrome. Hypocalcemia and precipitation of calcium phosphate in soft tissues caused by hyperphosphatemia can produce the symptoms described under those headings in the previous paragraphs. Uric acid probably has no direct effect on the brain, but it can cause acute uremia as a result of its effect on the kidneys.

Lactic Acidosis

The exact cause of lactic acidosis is unknown, but it may result from tissue hypoxia or hypoperfusion (type A) or it may complicate hematologic malignancy or solid tumor, even when tissue oxygenation is adequate (type B).[147,148] Patients with these disorders become confused, but systemic symptoms usually outweigh the neurologic ones.

NUTRITIONAL DISORDERS ASSOCIATED WITH CANCER

Malnutrition and cachexia are frequent complications of advanced cancer.[5,149,150] Two general but overlapping categories of malnutrition affect patients with cancer. The first is generalized malnutrition with weight loss, which rarely affects the peripheral nervous system and the muscles. The second category involves deprivation of specific nutrients. These disorders often affect the nervous system, most often the CNS.

Malnutrition and Cachexia

Many studies implicate malnutrition as a cause of neuromuscular disorders in critically ill patients with or without cancer.[151,152] Hawley et al.,[153] who studied 71 patients with small cell lung cancer, reported that mononeuropathies from nerve compression occurred in 13%. All patients eventually experienced polyneuropathy, but its severity was less than in patients with chronic alcoholism who had lost a similar amount of weight. These investigators concluded that peripheral neuropathy associated with small cell lung cancer was a result of weight loss and malnutrition. Hildebrand and Coers[20] examined biopsy specimens from a number of patients with neuromuscular disorders associated with cancer and found abnormalities at the neuromuscular junction in many. The degree of abnormality was associated with the amount of weight loss. They concluded that the abnormalities were related to cachexia. Histologic changes included axonal degeneration of intramuscular nerve fibers, atrophy of type I and especially type II muscle fibers, and the sprouting of nerve endings.[154] Elrington et al. examined 150 patients with small cell lung cancer, 53 of whom were anorexic and 51 of whom had lost weight.[18] They found that 44% of subjects had one or more measures of weakness; 21 of the subjects were unable to rise from a squatting position. None of these patients had typical paraneoplastic syndromes (see Chapter 15).

Similar alterations are found in otherwise healthy patients with anorexia nervosa. Of 51 patients with anorexia nervosa, 4 had electrodiagnostic evidence of sensorimotor peripheral neuropathy, compared to none in a matched control group. Three others had an isolated peroneal nerve palsy, presumably from nerve compression with the associated loss of subcutaneous tissue.[155] In a study of individuals participating in a long hunger strike, there were only modest changes in compound motor action potentials, and EMG findings were normal.[152]

Cancer cachexia is not simply malnutrition from loss of appetite and not eating. Although its pathophysiology is not fully understood, it appears to result in breakdown of muscle, different from simple starvation in which muscles are spared as fat is metabolized. The catabolic mediators in patients with the cancer cachexia syndrome include several substances secreted by the tumor as well as systemic humeral responses to the tumor.[149] Among the tumor-derived catabolic mediators are proteolysis-inducing factor; among the systemic factors are interleukin 6 and interferon gamma as well as tumor necrosis factor.[149] The result is loss of muscle mass and substantial weight loss, often associated with hypermetabolism, even in patients who are eating well; most are not substantially weak.

The strength of most cachectic patients with cancer is surprisingly normal. This finding is true despite the fact that a systematic review of symptom prevalence in patients with cancer found that weakness was the fourth most common complaint in 60% of patients.[156] This complaint may have been the patient's interpretation of fatigue, which was the most common complaint (74% of patients). Nonetheless, occasionally the clinician encounters patients with one or both of two clinical syndromes that affect muscle or the neuromuscular junction. The first is probably a nutritional myopathy or perhaps a disorder of neuromuscular transmission in which patients complain of proximal weakness. Like polymyositis or steroid myopathy, the disorder is characterized by difficulty in rising from a low chair or the toilet seat and in climbing stairs. Raising heavy objects over the head or even combing one's hair may also be challenging. Deep tendon reflexes may be preserved. Wasting and weakness of muscles, particularly proximal muscles, also occurs. Myoedema is often apparent when the muscle is struck.[157] The creatine kinase level is characteristically normal. Electromyograms show little or no abnormality. Patients whose cancer is successfully treated and who gain weight generally recover muscle function. The pathogenesis of this uncommon disorder is not clear. It must be distinguished from paraneoplastic polymyositis and the much more common steroid myopathy (see Chapter 4).

The second disorder is a sensorimotor peripheral neuropathy. Patients complain of numbness and tingling in the toes. Examination frequently reveals distal weakness with absent deep tendon reflexes, particularly ankle jerks and, later, knee jerks.

When the two disorders previously described are combined, they match the diagnostic criteria of Croft and Wilkinson[158] for "neuromyopathy" associated with cancer. Malnutrition may be the major cause of such *paraneoplastic neuromyopathy* (see Chapter 15).

Disorders of Nutrient Deprivation Associated with Cancer

A number of disorders that are due to deprivation of single nutrients have been reported in patients with cancer, including the Wernicke–Korsakoff syndrome from vitamin B_1 and vitamin B_{12} deficiency and pellagra associated with carcinoid tumors.

WERNICKE–KORSAKOFF SYNDROME

This disorder, also called *thiamine deficiency encephalopathy*, is characterized by an acute delirium associated with nystagmus, ocular palsies, and ataxia. Some patients also develop autonomic insufficiency with severe postural hypotension. Unless treated rapidly, patients develop an irreversible amnestic syndrome known as Korsakoff psychosis. Pathologically, the disorder is characterized by focal, sometimes hemorrhagic, abnormalities in mamillary bodies, thalamus, and brainstem. The disorder can be entirely prevented by thiamine replacement, but is sometimes precipitated acutely by the injudicious use of glucose in a thiamine-deprived patient. In one autopsy study of 24 patients with leukemia and lymphoma, 8 had the neuropathologic features of Wernicke encephalopathy (Fig. 11–6), although clinical symptoms were not recognized in any of the patients.[159]

The patient may become confused and disoriented, but more characteristically develops nystagmus on lateral gaze, followed by lateral gaze paresis. The ocular motor signs may be accompanied by gait ataxia. This disorder generally begins acutely and, if treated early with thiamine, resolves entirely. Nystagmus may improve within a few minutes after an injection of thiamine. Nevertheless, if treatment is not begun early, the patient may develop a chronic amnestic dementia with severe memory loss but otherwise normal cognitive functions, mimicking the dementia of limbic encephalitis (see Chapter 15). Ataxia may also be permanent, but, unlike the ataxia of paraneoplastic cerebellar degeneration, the patient walks with a stiff-legged, wide-based gait and has little or no ataxia of the upper extremities, no nystagmus, and no dysarthria. This clinical difference corresponds to the histopathology of the two syndromes.[160,161] In nutritional cerebellar degeneration, destruction of Purkinje cells is limited to the vermis and posterior lobe of the cerebellum. In paraneoplastic cerebellar degeneration, Purkinje cells are lost diffusely. Polyneuropathy or cognitive impairment may be the only signs[162] of Wernicke encephalopathy, but some patients present comatose.[163] Typical MRI findings are symmetrical, high signal lesions surrounding the third ventricle and Sylvian aqueduct on FLAIR image, and sometimes enhancement in the mammillary bodies and periaqueductal gray matter.[164]

Any malnourished patient with an encephalopathy and nystagmus should be treated with thiamine. Thiamine deficiency should be considered in delirious cancer patients, especially those with lymphoma,[165] even when they appear to be adequately nourished.[10] Interestingly, healthy patients who participated in a long hunger strike developed encephalopathy as their major neurologic manifestation.[152] Severely malnourished patients who are going to be given a glucose infusion should also be treated with parenteral thiamine. Only a few milligrams a day are necessary to treat Wernicke encephalopathy, but larger doses do no harm and ensure replenishment of thiamine stores. We have encountered the disorder occasionally in hospitalized patients receiving intravenous fluids that did not include vitamins.[164]

VITAMIN B_{12}/FOLATE DEFICIENCY

Vitamin B_{12} deficiency is uncommon in patients with cancer,[166,167] although some patients with normal levels of B_{12} have low levels of holotranscobalamin II,[166] the latter finding associated with low vitamin D levels for unclear reasons.[166] Symptomatic vitamin B_{12} deficiency is rare. In many patients with cancer, vitamin B_{12} levels are increased, a poor prognostic sign.[168]

Vitamin B_{12} deficiency was found in 10 of 41 patients who underwent radiotherapy for bladder carcinoma[169]; serum folate levels were normal in all of the patients. Neurologic

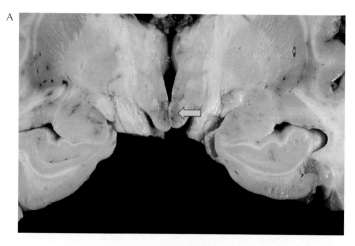

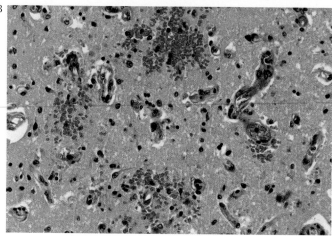

Figure 11–6. Wernicke encephalopathy. This woman with a diffuse large B-cell lymphoma was admitted to hospital after a syncopal episode. She was not eating and was nourished with IV fluids not containing vitamins. After several days, she became confused and developed nystagmus and ataxia that were attributed to metastases. When she died, there were no metastases. Instead, the brain showed hemorrhages in the mammillary bodies (*arrow*) that could be seen both grossly (**A**) and microscopically (**B**).

disturbances were not obvious on examination, but after B_{12} was given, the patients reported that previous weariness and muscular weakness had disappeared. Four patients observed an improvement of numbness and paresthesias in their fingers and toes. The disorder probably represents malabsorption of B_{12} related to radiation of the terminal ilium. Surprisingly, despite the malabsorption, most patients did not have diarrhea or other significant evidence of malabsorption. Vitamin E was not measured in this study, nor was there a careful evaluation of brain function.

Vitamin B_{12} deficiency can also cause a diffuse encephalopathy, spinal cord signs, and a peripheral neuropathy, sometimes without the well-known hematologic manifestations. One of our patients, a 35-year-old man who had undergone allogeneic bone marrow transplant for aplastic anemia, developed dysgeusia, orthostasis, psychological depression, paresthesias, and ataxia. An MRI (Fig. 11–7) demonstrated demyelination in the spinal cord. Symptoms resolved after vitamin B_{12} therapy.

CARCINOID SYNDROME

Carcinoids are interesting neuroendocrine tumors that can secrete several substances that cause systemic and neurologic symptoms.[170,171] The most frequently encountered syndrome is that of flushing and diarrhea. Serotonin

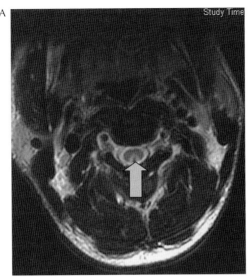

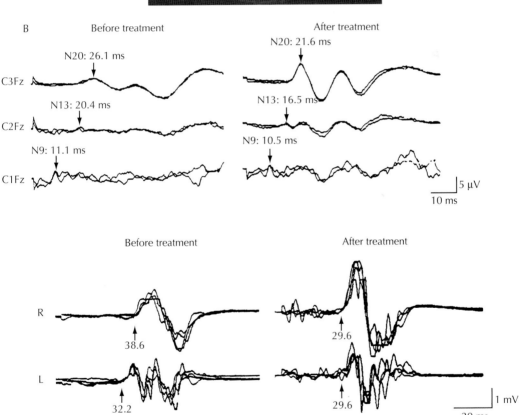

Figure 11–7. The MRI of a man with a vitamin B_{12} deficiency. **A**: This axial T2-weighted image shows demyelination in the posterior columns (*arrow*). A 35-year-old man developed paresthesias in his lower extremities and an ataxic gait after an allogeneic bone marrow transplant. His vitamin B_{12} level was 2 pg/mL (normal, 199–732 pg/mL). He responded well to treatment. **B**: Electrodiagnostic tests showing evoked potentials before and after treatment. Note the increase in the size of the central nervous system evoked–potential after successful treatment (*arrows*).

appears to play a major role, but histamine, bradykinin, and other vasoactive peptides may also contribute to the syndrome.[172] Serotonin, which requires niacin for its synthesis from tryptophan, also appears to play a role in carcinoid myopathy (see Chapter 15). Pellagra, a disease of niacin deficiency, occurs rarely in patients with a carcinoid syndrome; the disorder has also been reported in patients with anorexia nervosa.[173] The disorder appears related to high levels of serotonin produced by the tumor. Under normal circumstances, only 1% of tryptophan is metabolized to serotonin, but in the carcinoid syndrome, as much as 60% of tryptophan may be directed into this pathway. Diversion of tryptophan from its other metabolic pathways impairs both protein synthesis and niacin production and can lead to pellagra.[172]

Pellagra is clinically characterized by a triad of dementia, diarrhea, and dermatitis. The cutaneous eruption appears as an erythematous rash in sun-exposed areas and progresses to produce sharply marginated, symmetrically hyperpigmented and hyperkeratotic lesions, particularly over the dorsum of the hands.[174] Because malnutrition produces so many skin disorders, careful examination of the skin in cancer patients who appear malnourished may help in identifying the nature of the neurologic disorder.[175] The mental symptoms may be those of an agitated delirium; if untreated, the disorder progresses to a stuporous state and eventually to coma and death. Patients respond to replacement of niacin. An alternate explanation for some of the symptomatology may be a tryptophan deficiency in the brain, causing low levels of brain serotonin despite the production of serotonin by the tumor.[176] A review of the mental state of 22 patients with carcinoid syndrome revealed depression in 50%, anxiety in 35%, and altered levels of consciousness in 35%.[176]

An increased production of serotonin and other vasoactive peptides can also cause vasospasm; one case led to ischemic neuropathy of the lower extremities.[177]

REFERENCES

1. Posner JB, Saper CB, Schiff ND, et al. *Plum and Posner's Diagnosis of Stupor and Coma.* 4th ed. New York: Oxford University Press, 2007.

2. Gaudreau JD, Gagnon P, Roy MA, et al. Opioid medications and longitudinal risk of delirium in hospitalized cancer patients. *Cancer* 2007;109:2365–2373.

3. Gaudreau JD, Gagnon P, Harel F, et al. Psychoactive medications and risk of delirium in hospitalized cancer patients. *J Clin Oncol* 2005;23:6712–6718.

4. Cleeland CS, Bennett GJ, Dantzer R, et al. Are the symptoms of cancer and cancer treatment due to a shared biologic mechanism? A cytokine-immunologic model of cancer symptoms. *Cancer* 2003;97:2919–2925.

5. Retornaz F, Seux V, Sourial N, et al. Comparison of the health and functional status between older inpatients with and without cancer admitted to a geriatric/internal medicine unit. *J Gerontol A Biol Sci Med Sci* 2007;62:917–922.

6. Bernstein IL. Etiology of anorexia in cancer. *Cancer* 1986;58 (Suppl 8):1881–1886.

7. Deems DA, Doty RL, Settle RG, et al. Smell and taste disorders, a study of 750 patients from the University of Pennsylvania Smell and Taste Center. *Arch Otolaryngol Head Neck Surg* 1991;117:519–528.

8. Ovesen L, Sorensen M, Hannibal J, et al. Electrical taste detection thresholds and chemical smell detection thresholds in patients with cancer. *Cancer* 1991;68:2260–2265.

9. Dewys WD, Walters K. Abnormalities of taste sensation in cancer patients. *Cancer* 1975;36:1888–1896.

10. Yae S, Okuno S, Onishi H, Kawanishi C. Development of Wernicke encephalopathy in a terminally ill cancer patient consuming an adequate diet: a case report and review of the literature. *Palliat Support Care* 2005;3:333–335.

11. O'Keefe ST, Tormey WP, Glasgow R, et al. Thiamine deficiency in hospitalized elderly patients. *Gerontology* 1994;40:18–24.

12. Wijdicks EFM, Litchy WJ, Harrison BA, et al. The clinical spectrum of critical illness polyneuropathy. *Mayo Clin Proc* 1994;69:955–959.

13. Zochodne DW, Bolton CF, Wells GA, et al. Critical illness polyneuropathy. A complication of sepsis and multiple organ failure. *Brain* 1987;110:819–842.

14. Stevens RD, Dowdy DW, Michaels RK, et al. Neuromuscular dysfunction acquired in critical illness: a systematic review. *Intensive Care Med* 2007;33:1876–1891.

15. Sotaniemi KA. Slimmer's paralysis—peroneal neuropathy during weight reduction. *J Neurol Neurosurg Psychiatry* 1984;47:564–566.

16. Tan MJ, Kandler R, Baxter PS. Focal neuropathy in children with critical illness. *Neuropediatrics* 2003;34:149–151.

17. Bruera E, Brenneis C, Michaud M, et al. Muscle electrophysiology in patients with advanced breast cancer. *J Natl Cancer Inst* 1988;80:282–285.

18. Elrington GM, Murray NM, Spiro SG, et al. Neurological paraneoplastic syndromes in patients with small cell lung cancer. A prospective survey of 150 patients. *J Neurol Neurosurg Psychiatry* 1991;54:764–767.

19. Gomm SA, Thatcher N, Barber PV, et al. A clinicopathological study of the paraneoplastic neuromuscular syndromes associated with lung cancer. *Q J Med* 1990;278:577–595.

20. Hildebrand J, Coers C. The neuromuscular function in patients with malignant tumours. Electromyographic and histological study. *Brain* 1967;90:67–82.

21. Clouston PD, DeAngelis LM, Posner JB. The spectrum of neurologic disease in patients with systemic cancer. *Ann Neurol* 1992;31:268–273.

22. Engel GL, Romano J. Delirium, a sydrome of cerebral insufficiency. *J Chron Dis* 1959;9:260–277.

23. Bruera E, Miller L, McCallion J, et al. Cognitive failure in patients with terminal cancer: a prospective study. *J Pain Symptom Manage* 1992;7:192–195.

24. Centeno C, Sanz A, Bruera E. Delirium in advanced cancer patients. *Palliat Med* 2004;18:184–194.

25. Dantzer R, Kelley KW. Twenty years of research on cytokine-induced sickness behavior. *Brain Behav Immun* 2007;21:153–160.

26. Trzepacz PT. Is there a final common neural pathway in delirium? Focus on acetylcholine and dopamine. *Semin Clin Neuropsychiatry* 2000;5:132–148.

27. Flacker JM, Lipsitz LA. Large neutral amino acid changes and delirium in febrile elderly medical patients. *J Gerontol [A]* 2000;55:B249–B252.

28. Van der Mast RC, Fekkes D. Serotonin and amino acids: partners in delirium pathophysiology? *Semin Clin Neuropsychiatry* 2000;5:125–131.

29. Flacker JM, Wei JY. Endogenous anticholinergic substances may exist during acute illness in elderly medical patients. *J Gerontol A Biol Sci Med Sci* 2001;56:M353–M355.

30. van den Biggelaar AH, Gussekloo J, de Craen AJ, et al. Inflammation and interleukin-1 signaling network contribute to depressive symptoms but not cognitive decline in old age. *Exp Gerontol* 2007; 42:693–701.

31. Meagher DJ, O'Hanlon D, O'Mahony E, et al. Relationship between symptoms and motoric subtype of delirium. *J Neuropsychiatry Clin Neurosci* 2000;12:51–56.

32. Levkoff SE, Evans DA, Liptzin B, et al. Delirium. The occurrence and persistence of symptoms among elderly hospitalized patients. *Arch Intern Med* 1992; 152:334–340.

33. Liptzin B, Levkoff SE. An empirical study of delirium subtypes. *Br J Psychiatry* 1992;161:843–845.

34. Levkoff S, Cleary P, Liptzin B, et al. Epidemiology of delirium: an overview of research issues and findings. *Int Psychogeriatr* 1991;3:149–167.

35. Tuma R, DeAngelis LM. Altered mental status in patients with cancer. *Arch Neurol* 2000; 57:1727–1731.

36. Francis J, Martin D, Kapoor WN. A prospective study of delirium in hospitalized elderly. *JAMA* 1990;263:1097–1101.

37. Kelsey SM, Williams AC, Corbin D. Hyponatraemia as a cause of reversible ataxia. Short report. *BMJ* 1986;293:1346.

38. Roberts TA, Jenkyn LR, Reeves AG. On the notion of doll's eyes. *Arch Neurol* 1984;41:1242–1243.

39. Jacobson SA, Leuchter AF, Walter DO. Conventional and quantitative EEG in the diagnosis of delirium among the elderly. *J Neurol Neurosurg Psychiatry* 1993;56:153–158.

40. Young GB, Bolton CF, Archibald YM, et al. The electroencephalogram in sepsis-associated encephalopathy. *J Clin Neurophysiol* 1992;9:145–152.

40a. Lonergan E, Britton AM, Luxenberg J, Wyller T. Antipsychotics for delirium. *Cochrane Database Syst Rev* 2007;18:CD005594.

41. Winawer N. Postoperative delirium. *Med Clin North Am* 2001;85:1229–1239.

42. Shigeta H, Yasui A, Nimura Y, et al. Postoperative delirium and melatonin levels in elderly patients. *Am J Surg* 2001;182:449–454.

43. Onishi H, Sugimasa Y, Kawanishi C, et al. Wernicke encephalopathy presented in the form of postoperative delirium in a patient with hepatocellular carcinoma and liver cirrhosis: a case report and review of the literature. *Palliat Support Care* 2005; 3:337–340.

44. Wijdicks EF, Sharbrough FW. New-onset seizures in critically ill patients. *Neurology* 1993;43:1042–1044.

45. Kaiko RF, Foley KM, Grabinski PY, et al. Central nervous system excitatory effects of meperidine in cancer patients. *Ann Neurol* 1983;13:180–185.

46. Koppel BS, Hauser WA, Politis C, et al. Seizures in the critically ill: the role of imipenem. *Epilepsia* 2001;42:1590–1593.

47. Smolders I, Gousseau C, Marchand S, et al. Convulsant and subconvulsant doses of norfloxacin in the presence and absence of biphenylacetic acid alter extracellular hippocampal glutamate but not gamma-aminobutyric acid levels in conscious rats. *Antimicrob Agents Chemother* 2002;46:471–477.

48. Brown AS, Rosen J. Lithium-induced delirium with therapeutic serum lithium levels: a case report. *J Geriatr Psychiatry Neurol* 1992;5:53–55.

49. Vella-Brincat J. Adverse effects of opioids on the central nervous systems of palliative care patients. *J Pain Palliat Care Pharmacother* 2007;21:15–25.

50. Marcantonio ER, Juarez G, Goldman L, et al. The relationship of postoperative delirium with psychoactive medications. *JAMA* 1994;272:1518–1522.

51. Hornbein TF, Townes BD, Schoene RB, et al. The cost to the central nervous system of climbing to extremely high altitude. *N Engl J Med* 1989;321:1714–1719.

52. Aakerlund LP, Rosenberg J. Postoperative delirium: treatment with supplementary oxygen. *Br J Anaesth* 1994;72:286–290.

53. Flenly DC. Clinical hypoxia: causes, consequences and correction. *Lancet* 1978;1:542–546.

54. Mori E, Yamadori A. Acute confusional state and acute agitated delirium. Occurrence after infarction in the right middle cerebral artery territory. *Arch Neurol* 1987;44:1139–1143.

55. Gustafson Y, Olsson T, Asplund K, et al. Acute confusional state (delirium) soon after stroke is associated with hypercortisolism. *Cerebrovasc Dis* 1993; 3:33–38.

56. Bucerius J, Gummert JF, Borger MA, et al. Predictors of delirium after cardiac surgery delirium: effect of beating-heart (off-pump) surgery. *J Thorac Cardiovasc Surg* 2004;127:57–64.

57. Daughaday WH, Emanuele MA, Brooks MH, et al. Synthesis and secretion of insulin-like growth factor II by a leiomyosarcoma with associated hypoglycemia. *N Engl J Med* 1988;319:1434–1440.

58. Hoff AO, Vassilopoulou-Sellin R. The role of glucagon administration in the diagnosis and treatment of patients with tumor hypoglycemia. *Cancer* 1998;82:1585–1592.

59. Davda R, Seddon BM. Mechanisms and management of non-islet cell tumour hypoglycaemia in gastrointestinal stromal tumour: case report and a review of published studies. *Clin Oncol (R Coll Radiol)* 2007;19:265–268.

60. Redmon B, Pyzdrowski KL, Elson MK, et al. Hypoglycemia due to an insulin-binding monoclonal antibody in multiple myeloma. *N Engl J Med* 1992;326:994–998.

61. Chan JC. Drug-induced disorders of glucose metabolism. Mechanisms and management. *Drug Safety* 1996;15:135–157.

62. Scheen AJ. Drug interactions of clinical importance with antihyperglycaemic agents: an update. *Drug Safety* 2005;28:601–631.

63. Cryer PE. Symptoms of hypoglycemia, thresholds for their occurrence, and hypoglycemia unawareness. *Endocrinol Metab Clin North Am* 1999;28:495–500.

64. Cryer PE. Mechanisms of sympathoadrenal failure and hypoglycemia in diabetes. *J Clin Invest* 2006;116:1470–1473.

65. Nolli M. Wernicke's encephalopathy in a malnourished surgical patient: clinical features and magnetic resonance imaging. *Acta Anaesthesiol Scand* 2005;49:1566–1570.

66. Malherbe C, Burrill K, Levin SR, et al. Effect of diphenylhydantoin on insulin secretion in man. *N Engl J Med* 1972;286:339–342.

67. al-Rubeaan K, Ryan EA. Phenytoin-induced insulin insensitivity. *Diabet Med* 1991;8:968–970.

68. Arieff AI, Carroll JH. Cerebral edema and depression of sensorium in nonketotic hyperosmolar coma. *Diabetes* 1974;23:525–531.

69. Nugent BW. Hyperosmolar hyperglycemic state. *Emerg Med Clin North Am* 2005;23:629–648.

70. Ka T, Takahashi S, Tsutsumi Z, et al. Hyperosmolar nonketotic diabetic syndrome associated with rhabdomyolysis and acute renal failure: a case report and review of literature. *Diabetes Nutr Metab* 2003;16:317–322.

71. Daugirdas JT, Kronfol NO, Tzamaloukas AH, et al. Hyperosmolar coma: cellular dehydration and the serum sodium concentration. *Ann Intern Med* 1989;110:855–857.

72. Hennis A, Corbin D, Fraser H. Focal seizures and non-ketotic hyperglycaemia. *J Neurol Neurosurg Psychiatry* 1992;55:195–197.

73. Baird TA, Parsons MW, Phanh T, et al. Persistent poststroke hyperglycemia is independently associated with infarct expansion and worse clinical outcome. *Stroke* 2003;34:2208–2214.

74. Rady MY, Johnson DJ, Patel BM, et al. Influence of individual characteristics on outcome of glycemic control in intensive care unit patients with or without diabetes mellitus. *Mayo Clin Proc* 2005;80:1558–1567.

75. Cox DJ, Kovatchev BP, Gonder-Frederick LA, et al. Relationships between hyperglycemia and cognitive performance among adults with type 1 and type 2 diabetes. *Diabetes Care* 2005;28:71–77.

76. Gandhi GY. Intraoperative hyperglycemia and perioperative outcomes in cardiac surgery patients. *Mayo Clin Proc* 2005;80:862–866.

77. Desachy A, Vuagnat AC, Ghazali Ad, et al. Accuracy of bedside glucocometry in critically ill patients: influence of clinical characteristics and perfusion index. *Mayo Clin Proc* 2008;83:400–405.

78. Ebersoldt M. Sepsis-associated delirium. *Intensive Care Med* 2007;33:941–950.

79. Hotchkiss RS, Karl IE. The pathophysiology and treatment of sepsis 1. *N Engl J Med* 2003;348:138–150.

80. Jeppsson B, Freund HR, Gimmon Z, et al. Blood-brain barrier derangement in sepsis: cause of septic encephalopathy? *Am J Surg* 1981;141:136–142.

81. Bolton CF, Young GB, Zochodne DW. The neurological complications of sepsis. *Ann Neurol* 1993;33:94–100.

82. Young GB, Bolton CF, Austin TW, et al. The encephalopathy associated with septic illness. *Clin Invest Med* 1990;13:297–304.

83. Harris RL, Musher DM, Bloom K, et al. Manifestations of sepsis. *Arch Intern Med* 1987;147:1895–1906.

84. Guess HA, Resseguie LJ, Melton LJ, III, et al. Factors predictive of seizures among intensive care unit patients with Gram-negative infections. *Epilepsia* 1990;31:567–573.

85. Bird SJ. Diagnosis and management of critical illness polyneuropathy and critical illness myopathy. *Curr Treat Options Neurol* 2007;9:85–92.

86. de Jonghe JB, Lacherade JC, Durand MC, et al. Critical illness neuromuscular syndromes. *Crit Care Clin* 2007;23:55–69.

87. Johnson KL. Neuromuscular complications in the intensive care unit: critical illness polyneuromyopathy. *AACN Adv Crit Care* 2007;18:167–180.

88. Levine S, Nguyen T, Taylor N, et al. Rapid disuse atrophy of diaphragm fibers in mechanically ventilated humans. *N Engl J Med* 2008;358:1392–1394.

89. Theriault RL. Hypercalcemia of malignancy: pathophysiology and implications for treatment. *Oncology (Williston Park)* 1993;7:47–55.

90. Spinazze S, Schrijvers D. Metabolic emergencies. *Crit Rev Oncol Hematol* 2006;58:79–89.

91. Body JJ. Hypercalcemia of malignancy. *Semin Nephrol* 2004;24:48–54.

92. Tokuda Y, Maezato K, Stein GH. The causes of hypercalcemia in Okinawan patients: an international comparison. *Intern Med* 2007;46:23–28.

93. Warrell RP, Jr. Etiology and current management of cancer-related hypercalcemia. *Oncology (Williston Park)* 1992;6:37–43.

94. Pumo VV. Tumor lysis syndrome in elderly. *Crit Rev Oncol Hematol* 2007;64:31–42.

95. Hsu AK. Recurrent laryngeal cancer presenting as delayed hypoparathyroidism. *Head Neck* 2007;29:889–891.

96. Tanvetyanon T. Management of the adverse effects associated with intravenous bisphosphonates. *Ann Oncol* 2006;17:897–907.

97. Arieff AI, Ayus JC. Pathogenesis of hyponatremic encephalopathy: current concepts. *Chest* 1993;103:607–610.

98. Ellison DH. Clinical practice. The syndrome of inappropriate antidiuresis. *N Engl J Med* 2007;356:2064–2072.

99. Upadhyay A. Incidence and prevalence of hyponatremia. *Am J Med* 2006;119 (Suppl):s30–s35.

100. Berghmans T. Paesmans M, Body JJ. A prospective study on hyponatraemia in medical cancer patients: epidemiology, aetiology and differential diagnosis. *Supportive Care Cancer* 2000;8:192–197.

101. Moritz ML, Ayus JC. The pathophysiology and treatment of hyponatraemic encephalopathy: an update. *Nephrol Dial Transplant* 2003;18:2486–2491.

102. Cole CD, Gottfried ON, Liu JK, et al. Hyponatremia in the neurosurgical patient: diagnosis and management. *Neurosurg Focus* 2004;16:E9.

103. Adrogue HJ, Madias NE. Hyponatremia. *N Engl J Med* 2000;342:1581–1589.

104. Arieff AI. Influence of hypoxia and sex on hyponatremic encephalopathy. *Am J Med* 2006; 119(suppl1):s59–s64.

105. Fraser CL, Arieff AI. Epidemiology, pathophysiology, and management of hyponatremic encephalopathy. *Am J Med* 1997;102:67–77.

106. Lampl C, Yazdi K. Central pontine myelinolysis. *Eur Neurol* 2002;47:3–10.

107. Metzger BL, Devita MV, Michelis MF. Observations regarding the use of the aquaretic agent conivaptan for treatment of hyponatremia. *Int Urol Nephrol*, 2008:March 25 [Epubahead of print].

108. Adrogue HJ, Madias NE. Hypernatremia. *N Engl J Med* 2000;342:1493–1499.

109. Snyder NA, Fiegal DW, Arieff AI. Hypernatremia in elderly patients. A heterogeneous, morbid, and iatrogenic entity. *Ann Intern Med* 1987;107:309–319.

110. Berk L. Hypovolemia and dehydration in the oncology patient. *J Support Oncol* 2006;4:447–454.

111. Fall PJ. Hyponatremia and hypernatremia. A systematic approach to causes and their correction. *Postgrad Med* 2000;107:75–82.

112. Hiromatsu K, Kobayashi T, Fujii N, etal. Hypernatremic myopathy. *J Neurol Sci* 1994;122:144–147.

113. Perrin C, Fabre C, Raoul JL, et al. Behavioral disorders secondary to profound hypomagnesemia in a patient given cetuximab for metastatic colorectal cancer hypomagnesemia due to cetuximab treatment. *Acta Oncol* 2006;45:1135–1136.

114. Miller DW, Slovis CM. Hypophosphatemia in the emergency department therapeutics. *Am J Emerg Med* 2000;18:457–461.

115. Osorio S, Noblejas AG, Duran A, et al. Imatinib mesylate induces hypophosphatemia in patients with chronic myeloid leukemia in late chronic phase, and this effect is associated with response. *Am J Hematol* 2007;82:394–395.

116. Gollob JA, Rathmell WK, Richmond TM, et al. Phase II trial of sorafenib plus interferon alfa-2b as first- or second-line therapy in patients with metastatic renal cell cancer. *J Clin Oncol* 2007;25:3288–3295.

117. Folpe AL, Fanburg-Smith JC, Billings SD, etal. Most osteomalacia-associated mesenchymal tumors are a single histopathologic entity—an analysis of 32 cases and a comprehensive review of the literature. *Am J Surg Pathol* 2004; 28:1–30.

118. Kumar R. Tumor-induced osteomalacia and the regulation of phosphate homeostasis. *Bone* 2000;27:333–338.

119. Barbe B, Lejoyeux M, Bouleau JH, et al. Visual hallucination related to severe hypophosphataemia. Letter to the Editor. *Lancet* 1991;338:1083.

120. Berner YN, Shike M. Consequences of phosphate imbalance. *Annu Rev Nutr* 1988;8:121–148.

121. Silvis SE, Paragas PD, Jr. Paresthesias, weakness, seizures, and hypophosphatemia in patients receiving hyperalimentation. *Gastroenterology* 1972;62:513–520.

122. Modhi G, Bauman W, Nicolis G. Adrenal failure associated with hypothalamic and adrenal metastases: a case report and review of the literature. *Cancer* 1981;47:2098–2101.

123. Sheeler LR, Myers JH, Eversman JJ, et al. Adrenal insufficiency secondary to carcinoma metastatic to the adrenal gland. *Cancer* 1983;52:1312–1316.

124. Park CK, Miller C, Lawrence G. Addison's disease from non-Hodgkin's lymphoma with normal-size adrenal glands. *J Clin Oncol* 2007;25:2322–2324.

125. Mantzios G, Tsirigotis P, Veliou F, et al. Primary adrenal lymphoma presenting as Addison's disease: case report and review of the literature. *Ann Hematol* 2004;83:460–463.

126. Einaudi S, Bertorello N, Masera N, et al. Adrenal axis function after high-dose steroid therapy for childhood acute lymphoblastic leukemia. *Pediatr Blood Cancer* 2008;50;537–541.

127. Da Silva AN, Schiff D. Adrenal insufficiency secondary to glucocorticoid withdrawal in patients with brain tumor. *Surg Neurol* 2007;67:508–510.

128. Dev R, Del FE, Bruera E. Association between megestrol acetate treatment and symptomatic adrenal insufficiency with hypogonadism in male patients with cancer. *Cancer* 2007;110:1173–1177.

129. Torrey SP. Recognition and management of adrenal emergencies. *Emerg Med Clin North Am* 2005;23:687–702.

130. Terzolo M, Reimondo G, Ali A, et al. Ectopic ACTH syndrome: molecular bases and clinical heterogeneity. *Ann Oncol* 2001;12 (Suppl 2):S83–S87.

131. Ilias I, Torpy DJ, Pacak K, et al. Cushing's syndrome due to ectopic corticotropin secretion: twenty years' experience at the National Institutes of Health. *J Clin Endocrinol Metab* 2005;90:4955–4962.

132. Wang SA, Rahemtullah A, Faquin WC, et al. Hodgkin's lymphoma of the thyroid: a clinicopathologic study of five cases and review of the literature. *Mod Pathol* 2005;18:1577–1584.

133. Youn JC, Rhee Y, Park SY, et al. Severe hypothyroidism induced by thyroid metastasis of colon adenocarcinoma: a case report and review of the literature. *Endocr J* 2006;53:339–343.

134. Wong E, Rosen LS, Mulay M, et al. Sunitinib induces hypothyroidism in advanced cancer patients and may inhibit thyroid peroxidase activity. *Thyroid* 2007;17:351–355.

135. Shah M, Orengo IF, Rosen T. High prevalence of hypothyroidism in male patients with cutaneous melanoma. *Dermatol Online J* 2006;12:1.

136. Alexopoulou A, Koskinas J, Deutsch M, et al. Acute liver failure as the initial manifestation of hepatic infiltration by a solid tumor: report of 5 cases and review of the literature. *Tumori* 2006;92:354–357.

137. Wang WH, Wang HL. Fulminant adenovirus hepatitis following bone marrow transplantation. A case report and brief review of the literature. *Arch Pathol Lab Med* 2003;127:e246–e248.

138. Bolton CF, Young GB. *Neurological Complications of Renal Disease*. Boston: Butterworths, 1990.

139. Arieff AI, Griggs RCE. *Metabolic Brain Dysfunction in Systemic Disorders*. Boston: Little, Brown, 1992.

140. Marsano L, McClain C. How to manage both acute and chronic hepatic encephalopathy. *J Crit Illness* 1993;8:579–600.

141. Mitchell RB, Wagner JE, Karp JE, et al. Syndrome of idiopathic hyperammonemia after high-dose chemotherapy: review of nine cases. *Am J Med* 1988;85:662–667.

142. del Rosario M, Werlin SL, Lauer SJ. Hyperammonemic encephalopathy after chemotherapy. Survival after treatment with sodium benzoate and sodium phenylacetate. *J Clin Gastroenterol* 1997;25:682–684.

143. Tse N, Cederbaum S, Glaspy JA. Hyperammonemia following allogeneic bone marrow transplantation. *Am J Hematol* 1991;38:140–141.

144. Caminal L, Castellanos E, Mateos V, et al. Hyperammonaemic encephalopathy as the presenting feature of IgD multiple myeloma. *J Intern Med* 1993;233:277–279.

145. Wadzinski J, Franks R, Roane D, et al. Valproate-associated hyperammonemic encephalopathy. *J Am Board Fam Med* 2007;20:499–502.

146. Hande KR, Garrow GC. Acute tumor lysis syndrome in patients with high-grade non-Hodgkin's lymphoma. *Am J Med* 1993;94:133–139.

147. Sillos EM. Lactic acidosis: a metabolic complication of hematologic malignancies: case report and review of the literature. *Cancer* 2001;92:2237–2246.

148. Flombaum CD. Metabolic emergencies in the cancer patient. *Semin Oncol* 2000;27:322–334.

149. Argiles JM, Busquets S, Moore-Carrasco R, et al. Targets in clinical oncology: the metabolic environment of the patient. *Front Biosci* 2007;12:3024–3051.

150. Sarhill N, Mahmoud FA, Christie R, et al. Assessment of nutritional status and fluid deficits in advanced cancer. *Am J Hosp Palliat Care* 2003;20:465–473.

151. Russell DM, Prendergast PJ, Darby PL, et al. A comparison between muscle function and body composition in anorexia nervosa: the effect of refeeding. *Am J Clin Nutr* 1983;38:229–237.

152. Basoglu M, Yetimalar Y, Gurgor N, et al. Neurological complications of prolonged hunger strike. *Eur J Neurol* 2006;13:1089–1097.

153. Hawley RJ, Cohen MH, Saini N, et al. The carcinomatous neuromyopathy of oat cell lung cancer. *Ann Neurol* 1980;7:65–72.

154. Wokke JHJ, Jennekens FG, van den Oord CJ, et al. Histological investigations of muscle atrophy and end plates in two critically ill patients with generalized weakness. *J Neurol Sci* 1988;88:95–106.

155. MacKenzie JR, LaBan MM, Sackeyfio AH. The prevalence of peripheral neuropathy in patients with anorexia nervosa. *Arch Phys Med Rehab* 1989;70:827–830.

156. Teunissen SC, Wesker W, Kruitwagen C, et al. Symptom prevalence in patients with incurable cancer: a systematic review. *J Pain Symptom Manage* 2007;34:94–104.

157. Salick AI, Pearson CM. Electrical silence of myoedema. *Neurology* 1967;17:899–901.

158. Croft PB, Wilkinson M. The incidence of carcinomatous neuromyopathy in patients with various types of carcinomas. *Brain* 1965;88:427–434.

159. De Reuck J, Sieben G, De Coster W, et al. Prospective neuropathologic study on the occurrence of Wernicke's encephalopathy in patients with tumors of the lymphoid-hemopoietic systems. *Acta Neuropathol (Berl) Suppl* 1981;7:356–358.

160. Victor M. The effects of nutritional deficiency on the nervous system. A comparison with the effects of carcinoma. In: Brain L, Norris F, Jr, eds. *Contemporary Neurology Symposia, Vol. 1: The Remote Effects of Cancer on the Nervous System.* New York: Grune & Stratton, 1965. pp. 134–161.

161. Victor M, Adams RD, Collins GH. The Wernicke-Korsakoff syndrome. A clinical and pathological study of 245 patients, 82 with post-mortem examinations. *Contemp Neurol* 1971;7:1–206.

162. Lossos A, Siegal T. Thiamine (B1) deficiency in cancer out-patients: possible causes and neurological manifestations. *J Neuro-Oncol* 1994;21:73.

163. Pearce SHS, Rees CJ. Coma in Wernicke's encephalopathy. *Postgrad Med J* 1994;70:597.

164. Lacasse L. Wernicke encephalopathy in a patient with T-cell leukemia and severe malnutrition. *Can J Neurol Sci* 2004;31:97–98.

165. Boniol S, Boyd M, Koreth R, et al. Wernicke encephalopathy complicating lymphoma therapy: case report and literature review. *South Med J* 2007;100:717–719.

166. Vu T, Amin J, Ramos M, et al. New assay for the rapid determination of plasma holotranscobalamin II levels: preliminary evaluation in cancer patients. *Am J Hematol* 1993;42:202–211.

167. Plant AS, Tisman G. Frequency of combined deficiencies of vitamin D and holotranscobalamin in cancer patients. *Nutr Cancer* 2006;56:143–148.

168. Geissbuhler P, Mermillod B, Rapin CH. Elevated serum vitamin B12 levels associated with CRP as a predictive factor of mortality in palliative care cancer patients: a prospective study over five years. *J Pain Symptom Manage* 2000;20:93–103.

169. Kinn A-C, Lantz B. Vitamin B_{12} deficiency after irradiation for bladder carcinoma. *J Urol* 1984;131:888–890.

170. Modlin IM, I. A 5-decade analysis of 13,715 carcinoid tumors. *Cancer* 2003;97:934–959.

171. Robertson RG, Geiger WJ, Davis NB. Carcinoid tumors. *Am Fam Physician* 2006;74:429–434.

172. van der Horst-Schrivers AN, Wymenga AN, Links TP, et al. Complications of midgut carcinoid tumors and carcinoid syndrome. *Neuroendocrinology* 2004;80 (Suppl 1):28–32.

173. Prousky JE. Pellagra may be a rare secondary complication of anorexia nervosa: a systematic review of the literature. *Altern Med Rev* 2003;8:180–185.

174. Hegyi J. Pellagra: dermatitis, dementia, and diarrhea. *Int J Dermatol* 2004;43:1–5.

175. Strumia R. Dermatologic signs in patients with eating disorders. *Am J Clin Dermatol* 2005;6:165–173.

176. Major LF, Brown GL, Wilson WP. Carcinoid and psychiatric symptoms. *South Med J* 1973; 66:787–790.

177. Kucuk O, Noskin G, Petersen K, et al. Lower extremity vasospasm associated with ischemic neuropathy, dermal fibrosis, and digital gangrene in a patient with carcinoid syndrome. *Cancer* 1988;62:1026–1029.

Chapter 12

Side Effects of Chemotherapy

INTRODUCTION

Patients with cancer often take several different types of drugs (Table 12–1). Some drugs, including antineoplastic, hormonal, or biologic agents, treat the cancer. Others treat symptoms related to the cancer or the side effects of antineoplastics. In addition, because many patients with cancer are elderly, they may also be taking drugs for comorbid illnesses such as hypertension or cardiac disease that are unrelated to the underlying cancer. Finally, a significant number of patients take unconventional agents,[1] often not informing the physician that they are doing so. Some drugs from each category, acting alone or with other agents, are neurotoxic. Experimental antineoplastic agents, the side effects of which are not completely known, may also be neurotoxic. Thus, neurotoxicity caused by medication must be high on the list of potential causes of otherwise unexplained neurologic symptoms. A careful history to elicit all of the agents that a patient with neurologic symptoms is taking

is essential. For example, one of our patients, being treated for colon cancer, developed a peripheral neuropathy that appeared to be unrelated to his chemotherapy. An alert resident ordered a heavy-metal screen; a high arsenic level explained the neuropathy. An assay of the Chinese herbal medicine he had been taking, unknown to his oncologist, revealed arsenic. This chapter considers the neurotoxicity of agents used to treat cancer and its complications. Chapter 4 has already considered supportive care medications.

That neurotoxicity is a major side effect of antineoplastic agents is surprising. Most antineoplastic agents affect rapidly dividing tumor cells, so the clinician expects to see side effects directed at relatively rapidly dividing normal cells such as those in bone marrow and the gastrointestinal (GI) tract. The nervous system consists of cells that either do not divide, such as most neurons (olfactory neurons and taste receptors that do reproduce are susceptible to chemotherapeutic agents,[2,3]) or divide slowly,

447

Table 12–1. Agents Used to Treat Patients with Cancer

Antineoplastic agents
    Antineoplastic chemicals (chemotherapeutics)
    Hormones and hormone antagonists
Biologics
    Therapeutic growth factors
Supportive care agents (see Chapter 4)
    Analgesics
    Corticosteroids
    Anticonvulsants
    Anti-emetics
    Antibiotics
    Anticoagulants
Agents unrelated to cancer therapy
    Antihypertensives
    Cardiac medications
    Hypoglycemics
    Others

such as glia in the central nervous system (CNS) and Schwann cells in the peripheral nervous system. Furthermore, save for a few areas such as the circumventricular organs of the brain and the dorsal root ganglia, the nervous system is protected against easy entry of water-soluble agents (see Chapter 3). As a result, most chemotherapeutic agents given parenterally achieve much lower concentrations in the nervous system than elsewhere in the body. Despite these theoretical considerations, the nervous system is frequently the site of symptomatic toxicity of antineoplastic agents.

Neurotoxicity can cause significant disability, sometimes first appearing after the agent has been discontinued and persisting after a patient has been treated effectively or even cured of the cancer. Because regeneration of nervous system structures is poor, once an agent has caused severe damage to nerve cells or supporting structures, recovery is unlikely. Even when regeneration is possible under optimal conditions, as it is in some peripheral neuropathies, the poor nutritional state of many patients may preclude useful recovery. In addition, even minor toxicity to the nervous system causes symptoms. Toxicity of the liver, lung, or kidney, unless severe, may cause abnormal laboratory tests but usually no symptoms, whereas similarly mild toxicity to the CNS is almost certain to cause at least discernible encephalopathy, and to the peripheral nervous system, distressing paresthesias. Nervous system toxicity often either limits the dose of useful chemotherapeutic agents (e.g., cisplatin, fludarabine) or prevents a new chemotherapeutic agent from reaching the market.

CLINICAL FINDINGS AND DIAGNOSIS

Diagnosing the nature of neurologic abnormalities in patients receiving treatment for cancer is often difficult. A given set of symptoms may result from metastatic or nonmetastatic (e.g., infectious or vascular) complications of the cancer, a paraneoplastic syndrome, or a therapeutic agent. Thus, the diagnosis of a treatment-induced neurologic complication can be established only by clinical inference (Table 12–2). The physician is forced to make the diagnosis on the basis of the proximity of the neurologic complication to the treatment and on the knowledge that previous patients have suffered similar complications from these or similar treatments. Recently published reviews[4–11] help identify drugs that cause neurotoxic side effects, but new agents, new combinations of agents, and new dosage schedules are often associated with new and different neurologic symptomatology. To complicate matters, nervous system dysfunction may be caused by either antineoplastic drugs or supportive care agents such as anticonvulsants, corticosteroids, anti-emetics, opioids, and immunosuppressive agents. For example, the P450 isozyme CYP3A4 metabolizes Ifosfamide, etoposide, cyclophosphamide, tamoxifen, docetaxel, teniposide, paclitaxel, vinca alkaloids, and retinoic acid. The enzyme is induced by glucocorticoids and anticonvulsants,[12] thus increasing the metabolism of these drugs and decreasing their therapeutic effectiveness at standard doses. On the other hand, anti-fungal agents such as ketoconazole and itraconazole[13] inhibit the enzyme, sometimes leading to toxicity even when generally safe doses of the chemotherapeutic agent are administered. Furthermore, drugs may act synergistically either with other drugs or with other anticancer treatments, such as radiation therapy (RT), and may cause more serious neurotoxicity in patients who are receiving or have received chemotherapy. Vincristine and corticosteroids together may cause more weakness from combined vincristine neuropathy

Table 12–2. **Neurologic Signs Caused by Agents Commonly Used in Patients with Cancer**

Acute encephalopathy (delirium)
  Corticosteroids
  Methotrexate (high-dose IV, IT)
  Cisplatin
  Vincristine
  Asparaginase
  Procarbazine
  5-Fluorouracil (± levamisole)
  Cytarabine (high-dose)
  Nitrosoureas (high-dose or arterial)
  Cyclosporin
  Interleukin 2
  Ifosfamide
  Interferons
  Tamoxifen
  Etoposide (high-dose)

Chronic encephalopathy (dementia)
  Methotrexate
  Carmustine
  Cytarabine
  Carmofur
  Fludarabine

Visual loss
  Tamoxifen
  Gallium nitrate
  Nitrosoureas (intra-arterial)
  Cisplatin
  Fludarabine

Cerebellar dysfunction/ataxia
  5-Fluorouracil (± levamisole)
  Cytarabine
  Phenytoin
  Procarbazine
  Hexamethylmelamine
  Vincristine
  Cyclosporin

Aseptic meningitis
  Trimethoprim-sulfamethoxazole (Co-trimoxazole)
  IVIg
  NSAIDs
  Levamisole

  Monoclonal antibodies
  Metrizamide
  OKT3
  Cytarabine
  Carbamazepine
  Methotrexate (IT)

Headaches without meningitis
  Retinoic acid
  Trimethoprim-sulfamethoxazole
  Cimetidine
  Corticosteroid withdrawal
  Tamoxifen

Seizures
  Methotrexate
  Etoposide (high-dose)
  Cisplatin
  Vincristine
  Asparaginase
  Nitrogen mustard
  Carmustine
  Dacarbazine (intra-arterial or high-dose)
  Busulfan (high-dose)
  Cyclosporin
  Beta-lactam antibiotics
  Iodinated contrast material (IV or IT)

Myelopathy (intrathecal drugs)
  Methotrexate
  Cytarabine
  Thiotepa

Peripheral neuropathy
  Vinca alkaloids
  Cisplatin
  Hexamethylmelamine
  Procarbazine
  5-Azacytidine
  Etoposide
  Teniposide
  Cytarabine
  Taxanes
  Suramin
  Mitotane
  Bortezomib

IT, intrathecal; IV, intravenous; IVIg, intravenous gamma globulin; NSAIDs, nonsteroidal anti-inflammatory drugs; OKT3, orthoclone.

and corticosteroid myopathy than either alone (see Fig 4–2).

Even when the physician believes that a complication is therapy related, alternate diagnoses must be considered. In addition, other factors may exacerbate the neurotoxicity of antineoplastic drugs, so that treatment-induced neurotoxicity can be considered the sole cause only after other possible causes are excluded. For example, a seizure associated with chemotherapy in a patient with cancer requires a brain image to identify structural disease such as a metastasis, and possibly a cerebrospinal fluid (CSF) evaluation for an infectious cause. If a metastatic or infectious disease is identified, it may not be the sole cause of the neurologic complication. Ample evidence confirms that neurotoxicity from antineoplastic agents is more likely to occur and more likely to be severe in patients with preexisting nervous system disease, as exemplified by the devastating effect of vinca alkaloids on patients with hereditary neuropathies.[14]

This chapter first considers the wide variety of neurologic abnormalities that chemotherapeutic agents can cause (Table 12–2) and then describes the neurotoxic side effects of individual agents. Because the peripheral nervous system is the most common site of neurotoxicity, it is discussed first.

Peripheral Nervous System

The peripheral nervous system can be damaged either diffusely, usually causing a symmetrical distal neuropathy, or focally, involving individual nerves.

DIFFUSE NEUROPATHY

Diffuse peripheral nervous system neurotoxicity[15] falls into three broad categories: (1) An acute or subacutely developing sensorimotor peripheral neuropathy, often predominantly motor, which resembles the Guillain–Barre syndrome. The drugs probably cause demyelination, and the disorder is often reversible when the drug is discontinued. Suramin is the prototypic culprit.[16–19] (2) A distal sensorimotor neuropathy primarily affecting axons. This disorder begins with paresthesias, but motor weakness soon becomes apparent. Exemplified by vincristine neuropathy,[20,21] this disorder

may reverse with time; even deep tendon reflexes, which usually disappear early, may eventually return. (3) A pure sensory neuropathy, often painful, involving either large fibers or both large and small fibers, probably originating from damage to dorsal root ganglion cells. The patient first complains of numbness and tingling in fingers and toes, but sometimes only fingers.[22] In platinum neuropathies,[23,24] only large fibers are involved, causing loss of proprioception; pin and temperature sensation and motor power are preserved. In taxane neuropathy,[25,26] both large and small fibers are lost to an approximately equal degree. Depending on the degree of damage to the dorsal root ganglion cells, the disease may reverse itself if the patient survives in otherwise good health for a year or more after discontinuation of the chemotherapeutic agent. A recent review analyzes the efficacy of the various neuroprotective agents that have been proposed for preventing or treating chemotherapy-induced neuropathy.[11]

FOCAL NEUROPATHY

Neuropathies that affect only one or a few nerves are uncommon but can complicate both vincristine and platinum therapy. Focal neuropathy may occur at the site of nerve compression, as in the peroneal nerve palsy from crossed legs, or may be independent of any other focal injury, as in recurrent laryngeal or phrenic nerve paralysis.[27]

Specific findings in patients with chemotherapy-induced peripheral neuropathy suggest the offending agent. For example, Lhermitte sign (electric shock–like paresthesias in the arms or legs precipitated by neck flexion and indicative of demyelination in the posterior columns of the spinal cord) is associated with platinum therapy[28] but can also be caused by taxanes.[29] Lhermitte sign may also follow therapeutic irradiation to the spinal cord (see Chapter 13) and rarely occurs in patients who have not received platinum or radiation but who have had a bone marrow transplant.[30] Muscle cramps also appear to be more common in platinum neuropathy[31] but also occur with vincristine therapy.[32] Cramps may persist for more than a year after chemotherapy is discontinued. Paresthesias limited to the fingertips are a more common early sign of vincristine neuropathy. Severe, acute muscle aching, possibly neurogenic, occurs after vincristine injection, steroid withdrawal, and taxane infusion, but is not a common manifestation of other chemotherapeutic involvement.

Central Nervous System

ACUTE ENCEPHALOPATHY

Acute encephalopathy is characterized by confusion, disorientation, and altered behavior. The patient with a quiet delirium may appear simply depressed and withdrawn; others are agitated and fearful. Most patients develop insomnia. A few suffer florid hallucinations, but they may not report them unless asked. Multifocal myoclonus and generalized seizures often accompany the encephalopathy and give a physical clue to the presence of this condition. Seizures may be the only manifestation of an acute encephalopathy. Encephalopathy induced by chemotherapeutic agents cannot be distinguished clinically from that caused by metabolic disorders (see Chapter 11).

The posterior reversible encephalopathy syndrome (PRES),[33] also called posterior leukoencephalopathy syndrome, is characterized by the rapid onset of headache, confusion, and often seizures and cortical blindness (blindness associated with normal pupillary responses; sometimes the patient does not recognize that he is blind: Anton syndrome[34]). The magnetic resonance image (MRI) is characterized by diffuse hyperintensity on fluid-attenuated inversion recovery (FLAIR) images. The abnormalities selectively involve parieto-occipital white matter, but can be much more widespread. The disorder occurs in severely hypertensive patients (hypertensive encephalopathy) and in patients with eclampsia, renal failure, and a variety of agents used in the treatment of cancer. Culprits include cytotoxic drugs, including cytarabine, cisplatin, cyclophosphamide, vincristine, ifosfamide, etoposide, and gemcitabine[35] and the tumor lysis syndrome.[36] Other agents in the treatment of cancer that cause the syndrome include bevacizumab,[37] erythropoietin, cyclosporin, tacrolimus, interferon alpha, and high-dose steroids. The disorder is probably caused by damage to blood vessels with disruption of normal autoregulation. The posterior hemispheres are more susceptible, probably because they have a lesser sympathetic nerve supply.[38] The disorder is almost always fully reversible. Hypertension, if present, should be treated.

Thrombotic microangiopathy is another cause of acute encephalopathy. Several chemotherapeutic agents, especially mitomycin C, but also gemcitabine, interferon alpha, cyclosporin, and tacrolimus can cause the disorder, probably by direct toxicity to endothelial cells. Neurologically, the disorder is characterized by confusion and fleeting neurologic signs, usually with renal failure (hemolytic/uremic syndrome).[39] A more complete description can be found in Chapter 9.

CHRONIC ENCEPHALOPATHY

Chronic encephalopathy is characterized by dementia and sometimes seizures. A subacutely developing encephalopathy is often reversible, whereas an encephalopathy that develops more slowly is usually irreversible. Patients are characteristically quiet and withdrawn, often appearing apathetic. They may sleep more than usual or sleep during the day and stay awake at night. They often sit motionless for long periods and do not spontaneously initiate conversation. They suffer moderate-to-severe memory loss and disorientation, particularly with respect to time. When pressed, however, many patients can (if given sufficient time) answer questions better than the examiner would suspect on the basis of their overall demeanor. These characteristics, often called *subcortical dementia*, are caused by several chemotherapeutic agents, especially methotrexate (MTX).

Although severe dementia is rare in adults who have not received RT to the brain, lesser abnormalities of cognitive function may be relatively common.[40] Many patients under treatment for cancer complain of memory loss (recent as opposed to remote memory) and difficulty concentrating; patients with breast cancer have given this a name—*chemobrain*.[41,42] However, many patients with breast cancer show cognitive abnormalities on formal neuropsychologic testing before chemotherapy and although there are changes after treatment, they fail to reach statistical significance.[42] Furthermore, there appears to be a poor correlation between patients' complaints of cognitive dysfunction and results of neuropsychologic testing; instead, patients' complaints correlate better with the presence of anxiety and depression.[43] Nevertheless, a subset of patients undergoing chemotherapy experience cognitive decline significant enough to affect quality-of-life.[44] Areas affected include attention and concentration, verbal memory, visual memory, and speed of information processing.[45] Risk factors and the number of patients affected by various treatment modalities remain to be defined[46] but probably include premorbid IQ and education, psychological factors, genetic factors, and gender.[40,47] Mechanisms are as yet undefined.[41,48] One study describes small but statistically significant decreases in cerebral glucose metabolism in patients undergoing systemic chemotherapy with high-dose MTX, CHOP (cyclophosphamide, doxorubicin, vincristine, prednisone), or BACOD-VP16 (bleomycin, doxorubicin, vincristine, cyclophosphamide, dexamethasone, etoposide) at conventional doses.[49] It is known that, at least in patients with brain tumors, corticosteroids decrease overall brain glucose metabolism.[50]

White-matter hyperintensity seen on T2-weighted images on MR scans often progresses for several years.[51] Transient changes in white matter with hyperintensity on MR images are often found in young children who undergo chemotherapy for acute lymphoblastic leukemia. Most of these children also suffer neuropsychological deficits.[52]

Whether chemotherapy alone, in the absence of radiation, affects long-term cognitive function in children is not fully established. Many survivors of acute lymphoblastic leukemia and small cell lung cancer have received combined chemotherapy and RT either prophylactically or therapeutically. These patients suffer long-term cognitive defects, particularly with memory and attention.[53,54]

Preirradiation MTX appears to be less toxic to the brain than postirradiation MTX, particularly in young girls. Many investigators believe, however, that patients who undergo intensive systemic chemotherapy, particularly young children and the elderly, are at risk for long-term neuropsychological deficits involving memory and attention whether or not the brain has been irradiated. These deficits are not as prominent or severe as those that occur with RT plus chemotherapy, or even with RT alone; they generally do not incapacitate the patient. Animal models may elucidate some of these issues.[55]

OTHER NEUROLOGIC DISORDERS

Focal disorders include acute cerebellar syndromes, acute or subacute myelopathy, visual loss, hearing loss, taste and smell dysfunction, headache, and aseptic meningitis. The cerebellar syndrome caused by 5-fluorouracil (5-FU) and high-dose cytarabine is characterized by truncal and appendicular ataxia. In its mildest form, only gait ataxia is found. Sometimes they are reversible, although permanent neurologic deficits characterized pathologically by diffuse loss of Purkinje cells have been reported. A transverse myelopathy following intrathecal therapy can be caused by MTX, cytarabine or, rarely, thiotepa. Several types of visual loss can complicate chemotherapy: (1) tamoxifen therapy causes reversible retinopathy, (2) gallium nitrate and several other drugs cause an optic neuropathy, and (3) high-dose fludarabine and cisplatin can cause cortical blindness. Isolated headache, seizures, and aseptic meningitis can be caused by several medications (Table 12–2). Hearing loss is a common complication of cisplatin therapy. Several chemotherapeutic agents alter perception of taste and/or smell.[3,56,56a] One or both modalities may be lost or distorted, leading to abnormal perception. Some drugs (e.g., cisplatin) cause a sensation of a metallic taste during infusion, lasting hours to days afterward. Patients may also complain that they smell the drug being infused. Loss of these modalities is not surprising because their receptors turn over fairly rapidly. Taste receptors are said to have a half-life of approximately 10 days, and olfactory receptors approximately 30 days.[2] Almost as distressing as the loss of these sensations is their perversion. Food once enjoyed by the patient may taste as if it were spoiled. Among the drugs believed to alter taste and smell are the platins, cyclophosphamide, anthracyclines, fluorouracil, and MTX. One recent study describes reduction of olfactory function occurring in only 1 of 21 patients receiving cisplatin.[57] Patients receiving high-dose chemotherapy with stem cell rescue appear to be at increased risk.[2]

Non-Neurologic Toxicity

Several non-neurologic chemotherapy toxicities are briefly described here. These include the cutaneous abnormalities[58,59] of hyperpigmentation and alopecia, nail changes including banding, discoloration or, at times, onycholysis. A variety of rashes as well as photosensitivity can affect the skin. In some instances, particularly with epidermal growth factor receptor agents, the rash presages a response to therapy.

The hand-foot syndrome (palmar–plantar erythrodysesthesia) is sometimes mistaken for a peripheral neuropathy.[60,61] The syndrome typically starts with paresthesias of the hands and feet beginning a few days after the administration of chemotherapy, particularly capecitabine, fluorouracil, and liposomal doxorubicin. After 3 to 4 days, patients develop symmetrical edema of the palms and soles that can be so painful that the patient is unable to walk. Erythematous, edematous plaques appear that may lead to skin breakdown. At least one of our patients had severe pain and dysesthesias with only mild cutaneous changes, leading his physicians to believe he was suffering from a peripheral neuropathy.

Radiation recall[58] is a phenomenon in which chemotherapy (e.g., with anthracyclines) causes a cutaneous reaction in an area previously radiated. One of our patients with breast cancer, who had been irradiated for vertebral metastasis a year before, developed a myelopathy during chemotherapy with doxorubicin. Cutaneous hyperemia marked the RT port.

Graft-versus-host disease after allogeneic transplant is usually characterized by cutaneous changes. Isolated reports describe polymyositis[62] or a rash reminiscent of dermatomyositis[63] in a few patients. An inflammatory lesion of the brain (encephalitis) has also been described.[64,65]

SPECIFIC AGENTS

Cytotoxic Agents

Drugs used to treat cancer include chemotherapeutic agents, hormones, hormone antagonists, biologicals, small molecules, immunotherapeutic agents, and chemicals that act as adjuvants to RT or chemotherapy (see Table 12–1). Antineoplastic chemotherapeutic agents, generally the most neurotoxic, can be subdivided by their mechanism of action and chemical structure into alkylating agents, antibiotics, antimetabolites, plant alkaloids, camptothecins and a miscellaneous group (Table 12–3).

Table 12–3. Classification of Antineoplastic Drugs

Class	Example(s)
Alkylating agents	Cisplatin
	Carmustine (BCNU)
	Cyclophosphamide, ifosfamide
	Procarbazine
Antibiotics	
Anthracyclines	Doxorubicin
Others	Bleomycin
Antimetabolites	Methotrexate
	Cytarabine
	5-Fluorouracil, capecitabine
	Fludarabine
Plant alkaloids	Vincristine
	Etoposide
	Paclitaxel (Taxol)
Miscellaneous	Asparaginase
	Suramin
	Irinotecan (CPT-11)

ALKYLATING AGENTS

Alkylating agents work by covalent bonding of alkyl groups to DNA to form reactive intermediates that attack nucleophilic sites.[66] *Classic* alkylating agents usually contain a chloromethyl group that bifunctionally alkylates macromolecules. *Nonclassic* alkylating agents (e.g., cyclophosphamide) contain an N-methyl group and generally must be metabolically activated, usually in the liver, to assume the active form. Individual alkylating agents differ in site and degree of toxicity and in antitumor activity. Drugs such as cyclophosphamide and busulfan cause little or no neurotoxicity, whereas some of the platinum drugs such as cisplatin cause dose-limiting neurotoxicity. The alkylating agents also differ by site of action. The nitrosoureas, such as 1,3-bis (2-chloroethyl)-1-nitrosourea (BCNU), affect the O^6 position of guanine. Cells that contain methyl guanine methyl transferase (MGMT), as many glial cells do, may be resistant to the nitrosoureas. Failure of a tumor to be sensitive to one alkylating agent does not predict resistance to other alkylating agents. Table 12–4

Table 12–4. **Alkylating Agents**

Platins
  Cisplatin
  Carboplatin
  Oxaliplatin

Nitrosoureas
  Carmustine (BCNU)
  Lomustine (CCNU)
  Streptozotocin
  Chlorozotocin
  Semustine (methyl CCNU)
  ACNU
  HECNU

Mustards
  Ifosfamide (with mesna)
  Cyclophosphamide
  Melphalan
  Chlorambucil
  Mechlorethamine (nitrogen mustard)

Aziridines
  Thiotepa
  Diaziquone (AZQ)
  Mitomycin C

Aklylsulfonates

Table 12–5. **Neurotoxicity of Cisplatin**

Common
  Peripheral neuropathy (large fiber, sensory)
  Lhermitte sign
  Hearing loss (high frequency)
  Tinnitus

Uncommon
  Encephalopathy
  Visual loss
    Retinal toxicity
    Optic neuropathy
    Cortical blindness
  Seizures
  Cerebral herniation (hydration related)
  Electrolyte imbalance (Ca^{++}, Mg^{++}, Na^+, SIADH)
  Vestibular toxicity
  Autonomic neuropathy

SIADH, syndrome of inappropriate secretion of anti-diuretic hormone.

lists the commonly used conventional alkylating agents and some experimental ones.

Platinum-Based Agents

Platinum-based chemotherapeutic agents covalently bind to DNA bases and thus disrupt DNA function. These drugs are generally administered intravenously, but the intra-arterial and intracavitary (peritoneal cavity) routes have also been used. Intracarotid platinums have been used to treat tumors of the head, neck, and brain. The platinum drugs are effective against various tumors, including ovarian and testicular cancer.

Cisplatin: This drug binds to plasma proteins (95%) and crosses the blood–brain barrier poorly. The CSF-plasma ratio in experimental animals is less than 0.04.[67] In patients with brain tumors, however, the CSF peak concentration may be as high as 40% of

non-protein–bound platinum[68] and is reached approximately 60 minutes after an IV bolus. Intrathecal injection of more than 10 nmol to rats is toxic.[69] Although the drug does not easily enter the normal brain or spinal cord, it does enter and accumulate in dorsal root ganglia[70] and peripheral nerves, where concentrations are four to five times the brain concentration,[70] perhaps because gene expression of the multidrug resistant transporter MRP2 is much lower in dorsal root ganglion than in brain or spinal cord.[71] The platinum binding to DNA in dorsal root ganglion neurons leads to their apoptosis. Nerve growth factor does not prevent the binding but does prevent apoptosis.[72] In experimental animals, cisplatin also binds the endothelial cells of the vasa nervorum of peripheral nerves, decreasing blood flow and perhaps causing ischemic neuropathy. Vascular endothelial growth factor (VEGF) attenuates endothelial cell apoptosis, inhibits destruction of the nerve vasculature and preserves nerve function.[73]

As with other heavy-metal compounds, peripheral neuropathy is the most important neurotoxic effect. Cisplatin also causes ototoxicity, CNS toxicity, and probably CNS vascular toxicity (Table 12–5).

Peripheral nerves: Peripheral neuropathy occasionally follows cumulative doses of as little as 200 mg/m^2 and is usual after 400 mg/m^2. The degree of peripheral neuropathy correlates with the total cumulative dose, but not with dose intensity.[74] Retreatment of patients after prior treatment of 400 to 450 mg/m^2 is generally safe up to a dose of 800 to 900 mg/m^2.[75] Focal lumbosacral plexopathies or mononeuropathies have been reported following infusion of internal or external iliac arteries[76] Optic neuropathy has followed carotid infusion (see following text).[77]

The peripheral neuropathy begins with tingling that is occasionally painful in the toes and later the fingers. It then spreads proximally to affect both legs and arms. The deep tendon reflexes disappear (ankle jerks first), and proprioceptive loss (vibration sense first) is sometimes so severe that patients are unable to feed themselves or walk (sensory ataxia). Pseudoathetoid movements of the outstretched hands are common. Pin and temperature sensation are spared, and motor power is normal. Autonomic dysfunction is uncommon,[78–80] but the sympathetic skin response is often diminished or absent[81]; in experimental animals, neurogenic bronchial constriction is attenuated, a result of failure of sensory neurons to release neuropeptides.[82] Electrical studies reveal decreased conduction velocities in sensory nerves and diminished amplitude of sensory nerve potentials compatible with a sensory axonopathy.[78,83] An early decrease in the sural nerve action potential of greater than 50% of baseline value predicts a more severe neuropathy.[84] In some patients, proprioceptive loss is so severe that patients appear weak when attempting to sustain a muscle contraction not under direct vision. Instructing the patient to look at the limb being tested usually increases strength to normal.

The first neuropathic symptoms may not appear until the cisplatin treatment is completed and then may progress for several months before stabilizing.[31,85,86] If the patient survives the cancer, the neuropathy usually improves and may clear entirely after months or years.[86] However, some patients do not recover.[87]

Cisplatin neuropathy is often confused with paraneoplastic sensory neuronopathy (see Chapter 15) but differs from that disorder in that paraneoplastic sensory neuronopathy usually affects all sensory modalities equally, rarely improves (but often stabilizes) and is often associated with an antineuronal autoantibody in the patient's serum. Other immune disorders (Sjögren syndrome, Miller Fisher syndrome) and other toxic neuropathies (pyridoxine overdose) may also mimic cisplatin neuropathy.[22]

Pathologic examination of nerve roots reveals axonal loss with secondary demyelination.[88,89] Axonal loss is also found in dorsal but not ventral roots and there is secondary degeneration of posterior columns.[90] However, the primary site of pathology is loss of the large dorsal root ganglion neurons,[70,91] resulting in loss of large sensory axons distally.[92,93] Several agents have been proposed to prevent or treat cisplatin neuropathy: Amifostine, a radiation protector (see Chapter 13) that decreases the incidence of cisplatin-induced renal toxicity, does not appear to either prevent or treat peripheral neuropathy.[94] Although relatively safe, amifostine has been associated with hypotension leading to syncope.[95,96] Vitamin E appears to be protective. Two randomized trials of either 300 mg[97] or 600 mg[98] given concomitantly with cisplatin therapy substantially reduced the incidence and severity of neurotoxicity. In experimental animals, the drug did not interfere with the effectiveness of the chemotherapy.[97] Acetyl-L-carnitine prevents the development of cisplatin-induced neuropathy in experimental animals without affecting antitumor activity.[99,100] Preliminary data in humans suggest that the drug may both prevent and treat cisplatin-induced neuropathy without significant toxicity.[102] Protection against the development of cisplatin neuropathy had been reported with the drug ORG 2766, an adrenocorticotropic hormone 4–9 analog,[103] but a randomized placebo-controlled trial of 196 women with ovarian cancer failed to show efficacy.[104] A recent paper reviews agents that may be neuroprotective during cancer chemotherapy[105]; however, more clinical studies are necessary.

Several agents have shown efficacy in experimental animals. These include erythropoietin[106,107] and several neurotrophic factors,[108] but they have shown little or no clinical efficacy,[109] and one neurotrophic factor, BDNF, may inhibit cisplatin-induced cytotoxicity.[110]

Because at the present state of our knowledge we can neither prevent nor reverse cisplatin-induced neuropathy, a number of agents have been tried to reduce the pain and discomfort associated with that neuropathy. The most widely used are gabapentin and pregabalin. Gabapentin failed to show efficacy in a phase 3 trial compared to placebo.[111] However, in many patients, these drugs effectively reduce symptoms and improve quality of life.[112]

Cranial nerve neuropathy: Cranial nerve neuropathy develops in approximately 6% of patients after intra-arterial infusion of the common carotid artery with cisplatin for head and neck cancer. Any or all of the cranial nerves[113,114] can be affected. In one of our patients, sensory loss in the distribution of the ipsilateral superior orbital nerve followed intracarotid infusion. She recovered completely.

Lhermitte sign: Lhermitte sign, during or shortly after treatment with cisplatin,[28,115,116] suggests a transient demyelinating lesion in the posterior columns of the spinal cord. It also can be caused by spinal cord compression and cervical radiation.[116,117] Lhermitte sign is considered here, along with peripheral nerves, because the posterior columns are an extension of sensory neurons in the dorsal root ganglion, a part of the peripheral nervous system. Sensory conduction may be slowed in the spinal cord.[115] Lesions are not visible on MRI; positron emission tomography may show hypermetabolic activity in the cervical cord.[117] In some instances, patients may experience distal paresthesias when they abduct their arms, suggesting that they are stretching a demyelinated brachial plexus.[28]

Muscles: Muscle cramps unrelated to electrolyte imbalance (see following text) are also common; they are probably a result of hyperactivity of peripheral nerves. Like Lhermitte sign, muscle cramps usually resolve spontaneously.[118]

Hearing/vestibular deficits: Cisplatin can cause ototoxicity resulting in hearing loss[119,120] and perhaps vestibular dysfunction.[121,122] The damage is predominantly to hair cells in the organ of Corti, but the stria vascularis and spiral ganglion may be damaged as well.[123] Cisplatin suppresses endogenous levels of glutathione, leading to the production of reactive oxygen species causing apoptosis of hair cells.[120,124] Polymorphism in glutathione S-transferases affect sensitivity to hearing loss and other neurotoxicity.[124a] The hearing loss is often subclinical and affects primarily the high-frequency range (>4000 Hz). In most patients, hearing thresholds decrease at doses above 60 mg/m². Tinnitus may precede hearing loss; tinnitus is believed to result from either damage to the inferior colliculus or disinhibition of that structure from the cochlear damage.[125] Rarely, high-dose cisplatin causes acute deafness. Risk factors for ototoxicity include previous cranial irradiation, young age, and presence of a brain tumor.[126] If cisplatin is given before radiation, ototoxicity is not increased.[127] Aminoglycoside antibiotics, diuretics, such as furosemide, and carboplatin may also exacerbate cisplatin ototoxicity.

In experimental animals, a number of agents (generally the same ones used to prevent neuropathy [see preceding text]) have appeared to protect against ototoxicity; none has yet proved effective in the clinic although vitamin E appears promising.[128,129]

Although experimental animals demonstrate clinical and pathological abnormalities in vestibular function including loss of hair cells of the cristae ampullares and maculae, these changes are much less severe than those of the cochlea.[130,131] Some doubt that cisplatin causes clinical or laboratory evidence of vestibular loss.[121] However, some reports describe vestibular toxicity, characterized by vertigo, oscillopsia, ataxia, and benign paroxysmal positional vertigo[132] associated with cisplatin treatment. Vestibular dysfunction may occur either with or without other symptoms of ototoxicity and may be exacerbated by the previous use of aminoglycoside antibiotics.[133,134]

Visual/ocular toxicity [135,136]: Rarely, ocular toxicity may include retinopathy (usually after intracarotid infusion[137,138] but at times after intravenous infusion[139]), papilledema, or retrobulbar neuritis. Color perception may be disturbed, probably because of retinal cone dysfunction.[140] Cortical visual loss (homonymous hemianopia and cortical blindness) is sometimes part of the encephalopathy occasionally caused by cisplatin (see following text).[141,142] Nonvisual ocular toxicity includes lacrimal duct stenosis leading to tearing (epiphora), conjunctival injection, corneal edema, and corneal opacities.[136] Most of these symptoms clear after chemotherapy is completed.

Encephalopathy: Characterized by seizures and focal brain dysfunction, particularly cortical blindness, encephalopathy is rare following IV infusion[143–145] but more common following intra-arterial infusion.[146] The symptoms are usually reversible, but may be fatal.[147] The MRI may show widespread gray (basal ganglia)[147] and white matter damage, particularly in the posterior portions of the cerebral hemispheres (reversible posterior leukoencephalopathy syndrome[148]). Both gray and white matter changes may be found at autopsy.[145]

Encephalopathy and seizures caused by cisplatin itself must be differentiated from those caused by the hydration preceding cisplatin therapy (water intoxication with herniation[149]) or by hyponatremia (syndrome of inappropriate antidiuretic hormone [SIADH])[150,151] or salt-wasting nephropathy.[152] Cisplatin nephropathy can also cause hypocalcemia and hypomagnesemia that may rarely cause tetany, encephalopathy, or seizures (see Chapter 11).[153] Hypozincemia[154] has also been reported.

Cisplatin has been implicated in the vascular toxicity (cardiac,[155] cerebral infarction,[156] and Raynaud phenomenon[157]) that sometimes follows multiagent chemotherapy (Fig. 12–1). Some researchers believe that the late toxicity is caused by the cisplatin, either related to hypomagnesemia or as a direct effect of the drug on endothelial cells[158,159]; others believe that bleomycin is more likely to be the major culprit.[157] Other rare complications

of cisplatin include irreversible myelopathy,[160] taste disturbances,[161,162] and a myasthenic syndrome.[163]

Carboplatin: This drug possesses many of the salutary effects of cisplatin but little neurotoxicity. The drug reaches higher concentrations in CSF than does cisplatin but still crosses the blood–brain barrier poorly.[164] It has been given to patients and animals intra-arterially with little neurotoxicity other than retinopathy.[165] Hearing loss has been reported in very young children receiving the drug for treatment of retinoblastoma.[166] Very high doses may cause a peripheral neuropathy.[167] Retinal toxicity may follow intra-arterial injection.[165] One case of carboplatin-induced thrombotic microangiopathy resulted in multiple small cortical infarcts leading to coma and death,[168] and two patients have developed cortical blindness.[169] Hypersensitivity reactions with dyspnea, hypotension, anginal pain, and rash are not rare.[170,171]

Oxaliplatin: This platinum-containing agent is useful in the treatment of advanced colorectal cancer.[112] The drug can cause both an acute and chronic sensory neuropathy. Similar to cisplatin, the dose-limiting side effect is a chronic sensory peripheral neuropathy. The disorder develops in 10% to 15% of patients who receive a cumulative dose of 780 to 850 mg/m^2.[172] The sensory neuropathy is clinically similar to that of cisplatin except that it is usually less severe and more likely to reverse over time, although in many patients, particularly those who receive more than 1 g/m^2, the neuropathy is persistent.[173] In addition to the sensory changes in the extremities, Lhermitte sign and rare autonomic dysfunction (urinary retention) may occur.[174] Recurrent priapism[175] occurring after each infusion and lasting several hours has been reported in a single case. Pathologic changes in the dorsal root ganglia consist of atrophy of large sensory neurons,[176] without cell loss.

Unlike cisplatin, the drug does not cause nephropathy. Also unlike cisplatin, the drug causes acute sensory and motor changes that appear either during or shortly following the infusion, are exacerbated by environmental cold, and are transient.[177,178] Patients develop cold-induced paresthesias predominately in the hands and throat, and sometimes in the extremities. In addition, patients may experience pain in the jaw, eyes, and extremities.

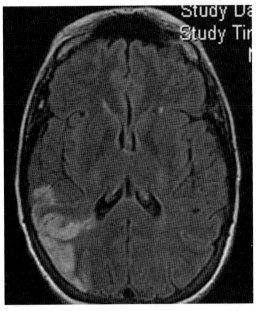

Figure 12–1. A FLAIR MRI demonstrating a cerebral infarct in a patient receiving cisplatin-based chemotherapy.

Muscle cramps and delayed relaxation, resembling neuromyotonia, are also present.[177] The cramps may persist for days after treatment. All of the symptoms are usually worse after the first treatment. Pregabalin may ameliorate the symptoms.[177a]

The acute sensorimotor changes are believed to be caused by a decrease in the amplitude of the voltage-gated sodium current, resulting in nerve hyperexcitability,[179] probably because one of the drug's metabolites, oxalate, binds the channel.[180] However, the symptoms do not respond to carbamazepine.

Infusion of calcium gluconate and magnesium sulfate appears to decrease both the acute and chronic neuropathic symptoms.[181] Glutathione infusions were also reported to be effective.[182] A recent study suggests that xaliproden, a non-peptide neurotrophic agent, is effective in reducing severe neuropathy, but the overall incidence of neuropathy remains the same.[183] Oxcarbazepine (600 mg twice a day) strikingly decreased the incidence of a subtle platinum-induced neuropathy in an open label controlled trial.[184] In experimental animals, erythropoietin, both protected against the development of cisplatin-induced neuropathy and promoted recovery.[185] A risk factor for both acute and chronic neurotoxicity is a minor haplotype of the enzyme AGXT.[180]

Nitrosoureas

The clinically used nitrosoureas include lomustine (CCNU), carmustine (BCNU), streptozotocin, and ACNU, the last widely used in Japan. These drugs are highly lipid soluble and cross the blood–brain barrier easily. The nitrosoureas are used to treat primary brain tumors as well as melanoma and lymphoma. ACNU has been given intrathecally to experimental animals without toxicity.[186] The nitrosoureas alkylate the O^6 position of guanine in DNA. As indicated many tumors contain MGMT, which produces resistance to the nitrosoureas. The enzyme can be inhibited by O^6-benzyl guanine, a drug that has been used in conjunction with BCNU to overcome resistance.[187] Neither this drug nor the nitrosoureas in conventional doses causes neurologic toxicity. Both cause bone marrow suppression,[187] the dose-limiting toxicity. Other uncommon side effects of the nitrosoureas include pulmonary fibrosis[188,189] (one of our patients with an

anaplastic astrocytoma developed a severe pulmonary reaction after the first dose. He recovered; the chemotherapy was discontinued. He is still well, without recurrence, 15 years later), renal failure, hepatotoxicity, myelofibrosis,[190] and myocardial ischemia.[191]

Patients with CNS tumors who have received previous RT, especially those treated with high-dose IV or intra-arterial BCNU, may develop ocular toxicity[192] and encephalopathy (Fig. 12–2).[193] Sudden blindness due to optic neuropathy is a rare complication of oral CCNU when combined with brain irradiation.[194] After intracarotid infusion with BCNU, the first problem is usually visual loss with both retinal and optic nerve damage. The brain disorder is sometimes heralded by seizures and generally characterized by slowly progressive neurologic dysfunction, the exact signs depending on the area infused by the intra-arterial injection. A problem with intra-arterial drug infusion is that laminar flow often prevents uniform perfusion of the entire area supplied by the artery, so that the tumor sometimes receives a lower dose than normal areas of brain in the same arterial distribution. Rapid or retrograde injection of the drug, by increasing turbulent flow, sometimes prevents this problem. White matter hypodensity can be seen on CT (hyperintensity on T2-weighted MRI), sometimes at a site distant from the tumor (Fig. 12–2) but in the same arterial territory. The affected white matter may calcify, and ipsilateral gyral enhancement has been reported.[195]

The pathology of nitrosourea leukoencephalopathy is necrotizing encephalopathy, similar to radiation or MTX leukoencephalopathy. If the drug has been given intra-arterially, the necrosis is confined to the vascular territories perfused by the BCNU.[190,196,197] Both vascular and direct neural damage appear to participate in the pathogenesis of the lesion. Experimental animals receiving intracarotid BCNU develop increased blood–brain barrier permeability.[5,198]

BCNU is also administered directly to brain tumors by implanting biodegradable polymer wafers (Gliadel wafers) impregnated with the drug.[199,200] In one study,[201] 8 of 17 patients so treated developed complications. These included wound infection (4) with sepsis in three patients, CSF leak in two patients,

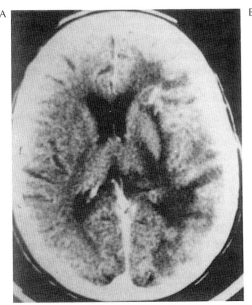
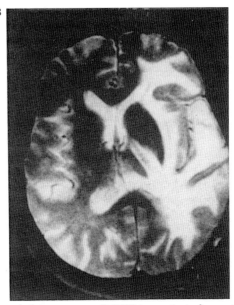

Figure 12–2. Leukoencephalopathy caused by 1,3-bis(2-chloroethyl)-1-nitrosourea (BCNU) in a patient who received radiation therapy followed by intracarotid administration of BCNU. Hypointensity on the CT scan (**A**) and hyperintensity on the T2-weighted image (**B**) of the MRI demonstrates leukoencephalopathy with relative sparing of the gray matter. The entire hemisphere perfused by the BCNU was involved, but the opposite hemisphere, which had received only radiation therapy, was less affected. Much of the white matter damage was at a site in the hemisphere remote from the tumor. Intra-arterial chemotherapy is now rarely used to treat brain tumors.

malignant cerebral edema in one patient,[202] and seizures in two patients. Gliadel is currently under study for the treatment of brain metastases.[203]

Mustards

This group of highly useful alkylating agents is not, except for ifosfamide, particularly neurotoxic; some are prodrugs that cannot become active without first being metabolized by the liver.

Ifosfamide: Ifosfamide requires hydroxylation by microsomal liver enzymes to produce biologically active metabolites. The drug and some of its metabolites cross the blood–brain barrier. It is excreted in the urine and requires the use of a uroprotective agent (mesna) to prevent severe bladder toxicity; mesna itself is probably not neurotoxic.

Depending on dose and other factors, between 5% and 30% of patients develop an encephalopathy,[204–206] usually reversible and characterized by cerebellar dysfunction, extrapyramidal signs, hallucinations, seizures (including nonconvulsive status epilepticus[207]),

quiet or agitated delirium, and sometimes coma. Severe asterixis as an isolated sign associated with triphasic waves on an electroencephalograph (EEG) and a normal MRI has been reported.[208] The encephalopathy usually begins within 24 hours of the drug's infusion but may be delayed for 4 to 6 days.[209] The encephalopathy usually clears in 3 to 4 days (range, 1–12 days) but persistent symptoms or even fatality may occur.[209] Encephalopathy usually contraindicates further use of the drug at that dose, although some patients have tolerated retreatment without recurrence of the encephalopathy.[210] Methylene blue (orally or IV) may reverse the encephalopathy,[211] as can thiamine (100 mg IV q4 hr until improved).[212] Peripheral neuropathy has also been reported.[213]

The risk factors for ifosfamide encephalopathy include high dosage, renal impairment, hepatic impairment, low serum albumin level[214] hypocalcemia, underlying brain disease,[215] phenobarbital,[216] and previous cisplatin therapy.[217] The EEG during the encephalopathy slows diffusely to 1 to 3 Hz or may show epileptic activity.[207] The mechanism of the neurotoxicity is unknown;[218,219] chloroacetalde-hyde, a metabolic product with effects similar to those of acetaldehyde, may be the culprit. If so, drugs such as phenobarbital, which can hasten the metabolic breakdown of ifosfamide, may increase neurotoxicity. Ifosfamide itself may cause the neurotoxicity, perhaps by inter-fering with thiamine metabolism[212]; rapid infu-sion appears to be more neurotoxic than slow infusion.

Cyclophosphamide: Cyclophosphamide (Cytoxan) also requires activation by the liver for antitumor activity. Its metabolite, 4-hydroxyperoxycyclophosphamide, has been used experimentally by intrathecal injection to treat leptomeningeal metastases; it appears to be effective and non-neurotoxic except at high doses, which can cause lethargy, seizures, or both.[220,221] Vasculitis of superficial arteries has also been observed. Cyclophosphamide intra-venously or by mouth has little or no neuro-toxicity, although reversible visual blurring,[222] dizziness, and confusion[223] have been reported in a few patients receiving high-dose therapy. Hyponatremia[224] has also been reported. A rare, peculiar rheumatic disorder with myalgias and arthralgias has been reported to begin 1 to 3 months after completion of adjuvant chemo-therapy for breast cancer; cyclophosphamide and 5-FU were given to all of these patients.[225]

Melphalan: This non-neurotoxic drug crosses the blood–brain barrier via the neutral amino acid transporter.[226] As a result, although not much drug enters the brain under normal circumstances, a diet lowering amino acids may promote entry.[227] At high concentration, melphalan may also open the blood–brain barrier, promoting entry of other molecules. A severe encephalopathy complicated high-dose melphalan therapy in two women who also had renal failure; both were comatose, and one died.[228] Intrathecal injection is safe and effec-tive in rats.[229]

Chlorambucil: Chlorambucil causes ence-phalopathy and seizures when given at high doses[230] or in accidental overdose.[231,232] Rarely, at standard doses, the drug may induce seizures,[230] multifocal myoclonus,[233] or ocular toxicity including keratitis, retinal edema, and hemorrhage.[234]

Other Alkylating Agents

Busulfan: This bifunctional alkylating agent is used at high doses to prepare for a stem cell transplant. This drug easily crosses the blood–brain barrier; the CSF concentration equals that in the blood in children receiving high doses.[235] With high-dose therapy[236,237] or with accidental overdosage,[238] seizures are common, usually occurring 2 to 4 hours after the dose. The seizures can sometimes[237] be prevented with prophylactic anticonvulsants and do not occur with conventional doses of busulfan. Electroencephalographic abnormali-ties occur in the absence of seizures or may persist for many days after a seizure.[239] Isolated case reports indicate that myasthenia gravis,[240] adrenal failure, Sjogren syndrome,[241] and per-manent alopecia[242] may complicate chronic busulfan treatment.

Dacarbazine/Temozolomide: These simi-lar prodrugs, the former used to treat malig-nant melanoma and the latter brain tumors, rarely produce neurotoxic side effects, although seizures (in the absence of underlying brain disease), encephalopathy, and occasionally dementia have been reported[243] with dacarba-zine (DTIC). With temozolomide, constipation is a major problem. Headache, confusion, visual blurring, and seizures have been described[244] but are more likely due to the disease (brain tumor) than the drug. Like the nitrosoureas, temozolomide alkylates the O[6] position of guanine and is less effective in tumors contain-ing MGMT.

Nitrogen Mustard (Mechlorethamine): This drug (the M in MOPP) is rarely neurotoxic, but brain necrosis and necrotizing uveitis following intracarotid infusion of nitrogen mustard have been reported.[245] Both encephalopathy[246,247] and hearing loss[248] occur rarely after conventional IV doses.

Procarbazine: This weak monoamine oxidase (MAO) inhibitor rapidly crosses the blood–brain barrier. Because it is an MAO inhibitor, patients are advised to avoid foods containing tyramine, although the importance of this precaution is doubtful. Procarbazine's exact mechanism of action is not known, although it is believed to interfere with DNA synthesis. The drug is given orally because experience with IV administration has revealed unacceptable neurotoxicity. Even with the oral form, however, encephalopathy may rarely occur, ranging from mild drowsiness to stu-por.[249] Occasionally, patients develop confu-sion, agitation, or even psychosis. A reversible peripheral neuropathy with distal paresthesias, decreased deep tendon reflexes, and myalgias[250] has been reported in 10% to 20% of patients. Hypersensitivity reactions, including skin rash, interstitial pneumonitis, and hepatitis may be seen, particularly in patients with glioma.[251]

Mitotane: This drug is an adrenal cytotoxic agent that acts against adrenal carcinoma. It crosses the blood–brain barrier. It causes a reversible encephalopathy with lethargy and somnolence, dizziness, and vertigo as well as neuropsychological impairments in visual spatial tasks, language and memory.[252] In some patients, cerebellar ataxia and extrapyramidal signs develop. The EEG may show delta and beta activity but imaging is normal.[252] Low-dose, monitored treatment can achieve therapeutic blood levels with manageable side effects.[253]

Altretamine (hexamethylmelamine): This drug, whose mechanism of action is unknown, has some effectiveness against ovarian cancer. The drug itself has poor CNS penetration. The CSF concentration is less than 6% of the plasma concentration. Its demethylated metabolites occur in high concentrations and reach CSF-to-plasma ratios close to 1. These metabolites may be concentrated in the brain. Toxicity consists of mood changes, including lethargy, depression, agitation, hallucinations, and sometimes coma. Peripheral neuropathy occurs rarely. Dose-limiting vomiting and anorexia may be an effect of the drug on the CNS.[254,255]

Thiotepa: This drug and its major metabolite, TEPA, cross the blood–brain barrier easily. Systemic administration of thiotepa provides prolonged CSF exposure to TEPA. When the drug is administered intrathecally, as is sometimes done to treat leptomeningeal tumor, the rapid clearance of thiotepa results in an uneven neuraxial distribution.[256] The drug causes marrow suppression. At high doses used for stem cell transplant, the drug can cause a severe encephalopathy, sometimes fatal.[257,258] Myelopathy and radiculopathy have been reported[259–261] in a few patients after intrathecal injections.

ANTINEOPLASTIC ANTIBIOTICS

The antineoplastic antibiotics (Table 12–6) consist of the anthracyclines, such as doxorubicin (Adriamycin), and other agents such as mitomycin and bleomycin. All have little or no neurotoxicity in humans.

Table 12–6. Antineoplastic Antibiotics

Anthracyclines
  Doxorubicin (Adriamycin)
  Daunorubicin
  Idarubicin
  Epirubicin
  Mitoxantrone

Other antibiotics
  Bleomycin
  Mitomycin C
  Dactinomycin

Anthracyclines

The anthracyclines are active against Hodgkin disease, non–Hodgkin lymphoma, acute myelogenous leukemia, and some epithelial tumors including breast cancer. The drugs are topoisomerase II inhibitors and also intercalate into double-stranded DNA to cause structural changes that interfere with synthesis. Except for idarubicin, which has high lipid solubility and the liposomal anthracyclines, these drugs do not cross the blood–brain barrier easily.[262,263] Tumors develop resistance to anthracyclines by several mechanisms: P-glycoprotein, the protein product of the multidrug-resistance gene, is present in cell membranes and at the blood–brain barrier. It promotes efflux of several chemotherapeutic agents, including anthracyclines, vinca alkaloids, epipodophyllotoxins, and others from the cell and possibly from the brain.[264] Other mechanisms of anthracycline resistance include ATP-dependent efflux proteins, altered topoisomerase activity, altered free radical biochemistry, and overexpression of anti-apoptosis genes.

Doxorubicin: This agent causes severe and often fatal myelopathy and encephalopathy after accidental intrathecal injection.[265,266] Necrosis and hemorrhagic infarcts in the brain sometimes follow intracarotid injection associated with osmotic blood–brain barrier opening.[267] After systemic administration, cardiac thrombi associated with doxorubicin-induced cardiac toxicity[268] may cause transient cerebral ischemia or infarction. One report[269] suggests that the combination of doxorubicin and cyclosporin may increase doxorubicin concentration in the brain and lead to fatal encephalopathy. Surprisingly, although the drug damages dorsal root ganglia in experimental animals,[270] it does not cause a peripheral neuropathy in humans. The drug is a vesicant and may cause severe skin and subcutaneous reactions if it extravasates during intravenous infusion. Liposomal doxorubicin may cause the hand-foot syndrome (see p 453).[60] The liposomal preparation may also cause acute infusion reactions characterized by back pain, chest tightness, and flushing during the first 5 minutes of the first infusion, possibly related to capillary plugging by neutrophils.[60]

Idarubicin, Epirubicin, Daunorubicin: Although these anthracycline antibiotics do not appear to be neurotoxic in humans, epirubicin causes neuronal damage to mice after IV injection.[262] Harmless levels of idarubicin have been detected in the CSF of children after systemic therapy.[271]

Mitoxantrone: An analog of the anthracyclines with a similar mechanism of action, mitoxantrone has a narrow spectrum of antitumor activity confined to breast cancer, leukemias, and lymphomas. The drug causes less cardiac toxicity than the anthracyclines, but cardiac toxicity is a concern when the drug is used to treat multiple sclerosis[272] and when given in high doses.[273] It is not known to be neurotoxic when given intravenously, but when given intrathecally, the drug caused nerve root damage and myelopathy[274] so that intrathecal therapy is no longer used. The drug also causes a bluish discoloration of the sclera, fingernails, and urine.

Other Antibiotics

Bleomycin: This drug is a mixture of peptides containing bleomycinic acid, derived from the fungus *Streptomyces verticillus*. The peptides cleave DNA strands via free radicals. Bleomycin acts against Hodgkin disease, non–Hodgkin lymphoma, testicular cancer, and head and neck cancers. The primary toxicity is directed at lung and skin. Subacute or chronic interstitial pneumonitis leading to pulmonary fibrosis can be fatal. Raynaud phenomenon affects 5% to 7% of patients and usually occurs when bleomycin is used in combination with vinca alkaloids, with or without cisplatin.[275] In most instances, the disorder is mild and not disabling.

More disturbing are delayed effects of bleomycin on the vasculature of the brain and heart, usually when used as part of cisplatin-based chemotherapy. Cerebral and myocardial infarcts have occurred in a very small number of patients who were apparently cured of testicular or other cancers by the use of bleomycin, cisplatin, and other drugs.[157] Which of the drugs is at fault is unclear, although bleomycin has been suspected as the main culprit. Strokes and myocardial infarcts are rare but are historically clearly related to the underlying chemotherapy. Hypomagnesemia has been implicated as a risk factor. Recent reviews summarize the vascular toxicity.[157,276]

Mitomycin C: Originating from *Streptomyces caespitosus*, mitomycin C is useful against several solid cancers, usually without neurotoxicity. When the drug extravasates after intravenous administration, there may be severe ulceration. Veno-occlusive disease has been reported after high doses of the agent.[277] Ocular toxicity is an occasional complication of topical application of the drug,[278,279] as are airway complications.[280] The drug may cause thrombotic microangiopathy with encephalopathy.[39]

Dactinomycin (actinomycin D): An inhibitor of RNA and protein synthesis, dactinomycin is active against several tumors, including choriocarcinoma, Wilm tumor, and neuroblastoma. Neurotoxicity after IV therapy in humans has not been reported. Intrathecal injection in animals is fatal,[281] causing a status spongiosus. The drug is a vesicant, can cause veno-occlusive disease, and may induce the radiation recall phenomenon (see p 453). One report describes the posterior encephalopathy syndrome (see p 451) induced in children by a combination of high-dose MTX, cyclophosphamide, and dactinomycin.[282]

ANTIMETABOLITES

The antimetabolites (Table 12–7) can be subdivided into antifolates, cytidine analogs, fluorinated pyrimidines, and purine antimetabolites. Antimetabolites are widely used in cancer therapy and most have some neurotoxicity.

Antifolates

The antifolates include MTX and pemetrexed. Antifolates inhibit dihydrofolate reductase,

Table 12–7. **Antimetabolites**

Antifolates
   Methotrexate
   Pemetrexed

Cytidine analogs
   Cytarabine
   5-Azacytidine
   Gemcitabine

Fluorinated Pyrimidines
   Fluorouracil
   Capecitabine (Xeloda)
   Floxuridine
   Fotofur
   Carmofur
   5-deoxy-5-fluorouridine

Purine Analogs
   Fludarabine
   Cladribine (2Cda)
   Pentostatin
   6-Mercaptopurine
   6-Thioguanine

Others
   Hydroxyurea

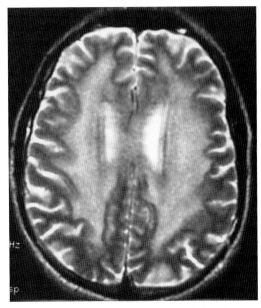

Figure 12–3. Methotrexate leukoencephalopathy in a patient with breast cancer heavily treated with IV MTX. The patient did not receive brain irradiation or intrathecal drugs.

partially depleting reduced folates and causing inhibition of purine and thymidine biosynthesis.

Methotrexate: The most widely used antimetabolite in cancer chemotherapy, MTX has been given orally, intravenously, intraarterially, and intrathecally. The drug is S-phase, cell cycle–specific and acts primarily against leukemias and lymphomas. MTX crosses the blood–brain barrier poorly; after a continuous IV infusion, the brain-to-blood partition coefficient is 0.1.[283] MTX levels in the brain can be increased with very high doses intravenously followed by a reversal of the potentially fatal systemic toxicity with folinic acid,[284] by coated nanoparticles,[285] or by blood–brain barrier disruption.[286,287] Folinic acid crosses the blood–brain barrier poorly and thus does not interfere with the CNS effects of MTX, even when it completely prevents systemic toxicity.[284] Intrathecal injection, either by intraventricular injection through an Ommaya device (see Chapter 7) or by lumbar puncture, is used for prophylaxis or treatment of leptomeningeal tumor. Orally administered folinic acid can prevent toxicity from systemic absorption of intrathecal MTX. When injected intrathecally, MTX penetrates the brain variably. In a rabbit study, some drug had reached 40% of the brain 1 hour after intraventricular injection of MTX.[288] In the larger human brain, penetration is less, so that MTX effectively treats only tumor in the leptomeninges, not in the brain.

Acute or chronic neurotoxicity after oral, IV (Fig. 12–3), or intrathecal injection of MTX is a well-recognized complication of therapy (Table 12–8). Risk factors include dose, route of administration, and other concurrent treatments such as RT. Although MTX itself can damage neural tissue, most clinically encountered neurotoxicity results from additive or synergistic effects of cranial irradiation.[289] The choroid plexus[290,291] actively transports MTX out of the nervous system, but cytocidal concentrations in CSF are easily achieved after intraventricular injection.[292,293]

Aseptic meningitis: Aseptic meningitis is the most common form of acute MTX neurotoxicity, complicating intrathecal administration.[294,295] Aseptic meningitis occurs in 10%

Table 12–8. **Methotrexate Neurotoxicity**

Route of Administration	Dose	Toxic Effect
Oral or intravenous	Conventional	Leukoencephalopathy (if prior brain irradiation)
Intravenous	High-dose	Acute: transient encephalopathy Chronic: leukoencephalopathy
Intra-arterial	Conventional	Hemorrhagic cerebral infarction
Intrathecal	Conventional	Acute: aseptic meningitis, paraplegia, seizures Chronic: leukoencephalopathy, cerebral atrophy, calcification

to 60%[295] of patients who receive intrathecal MTX and is more frequent after lumbar than intraventricular injection. The clinical syndrome is marked by the abrupt onset of headache, stiff neck, nausea, vomiting, lethargy, and fever, usually occurring 2 to 4 hours after MTX instillation and typically lasting for 12 to 72 hours. Examination of the CSF usually reveals a pleocytosis. The disorder may mimic bacterial meningitis, but it occurs too soon after drug instillation to be due to bacterial growth from injection of contaminated material; CSF cultures are negative. The syndrome resolves spontaneously and does not appear to have any long-term sequelae, although one report[296] suggests that such reactions may predispose to later leukoencephalopathy. The reaction may follow either the first or any subsequent injection of MTX, although patients usually do not experience difficulty with continued injections. Risk factors may include higher dose, more frequent instillation, and the presence of leptomeningeal tumor. The pathogenesis is unknown. Some investigators have instilled hydrocortisone with MTX in an attempt to prevent the reaction[297]; others use oral dexamethasone (4 mg twice a day), a technique that decreased the incidence of meningitis in one series from 60% to 12%.[295] The inflammatory reaction may cause arachnoiditis leading to cauda equina dysfunction.[298] One report describes a patient who developed aseptic meningitis following intramuscular MTX.[299] Other toxic effects of MTX may cause a transverse myelopathy or an encephalopathy as indicated in the paragraphs that follow.

Encephalomyelopathy: An acute fatal encephalomyelopathy with seizures, paralysis, and coma follows massive overdoses of intrathecal MTX, for example, 650 mg compared to the normal dose of 12 mg.[300] Toxicity from lesser overdoses has been ameliorated by treatment with systemic[301] or intrathecal[302,303] folinic acid or carboxypeptidase G2,[304] or by CSF exchange.[305] More perplexing are the occasional reports of severe reactions, sometimes fatal, to conventional doses of intrathecal MTX.[306–310] One report describes reversible posterior leukoencephalopathy following an intrathecal injection of MTX and cytarabine.[311] A transient encephalopathy with confusion, disorientation, seizures, and a focal finding resembling stroke occasionally occurred several days following an intrathecal injection of MTX.[312,313] The syndrome is accompanied by focal changes in diffusion-weighted images that also are transient. A similar syndrome following high-dose intravenous MTX is more common and is described in the following text. Rarely, conventional doses of intrathecal MTX cause acute noncardiogenic pulmonary edema,[314] pneumonitis,[315] or acute tumor lysis syndrome.[316]

Transverse myelopathy: Transverse myelopathy, a rare complication of intrathecal MTX, usually occurs after several treatments, although it may follow the first, and generally presents within 12 to 48 hours after injection.[298] However, the onset may be delayed for weeks.[317,318] The patient complains of back pain with or without radiation into the legs. The pain is followed by sensory loss, paraplegia, and bowel and bladder dysfunction.[317,318] Recovery varies; some severely affected patients recover fully, others not at all. The exact pathogenesis of MTX myelopathy is unknown but it is believed to be an idiosyncratic drug reaction. One investigator has reported increased levels of homocysteine in the spinal fluid in patients suffering neurotoxicity after either intravenous or intrathecal MTX.[319] Some patients appeared to respond to oral dextromethorphan. The authors postulated that the direct vascular toxicity of homocysteine and the excitation of NMDA receptors by its metabolites caused the toxicity that could be reversed by dextromethorphan.

Pathologic examination has revealed vacuolar demyelination and necrosis of the spinal cord, sometimes accompanied by brainstem and cerebral hemisphere abnormalities without striking inflammatory or vascular changes.[320] The major site of damage is to areas of the brain and spinal cord closest to the CSF, such as root entry zones, the base of the brain, midbrain, and colliculi. Treatment is usually ineffective, and the clinician cannot predict which patients will be affected, although the presence of active CNS leukemia or prior spinal irradiation may be predisposing factors.

MRI often reveals clumped, contrast-enhancing nerves of the cauda equina[298,317] and may show patchy enhancement of the spinal cord,[318] suggesting tumor infiltration. In one of our patients, biopsy did not reveal tumor. Somatosensory-evoked potentials may reveal slowing in the spinal cord even in asymptomatic patients who have received intrathecal MTX.[321] Other symptoms that may rarely follow intrathecal MTX include cranial nerve paresis[322,323] and transient or permanent hemiparesis.[323]

Acute/subacute encephalopathy: This disorder sometimes follows the instillation of MTX into the cerebral ventricles, particularly if exodus from the CSF is obstructed (Fig. 12–4).[292] A radionuclide flow study that demonstrates normal flow from ventricles into the subarachnoid space will help prevent this complication (see Chapter 7). Encephalopathy has also been reported following injection of MTX into the cerebral white matter through a misplaced ventricular cannula.[324–326]

High-dose MTX given intravenously can also cause an acute encephalopathy characterized by confusion, disorientation, somnolence, and sometimes seizures (Table 12–9). The onset is within 24 to 48 hours of the treatment and generally resolves without neurologic disability. Prior brain irradiation may be a risk factor.

In some patients, a subacute encephalopathy follows the administration of weekly IV high-dose MTX (HD MTX). A stroke-like

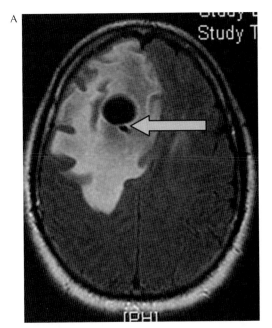
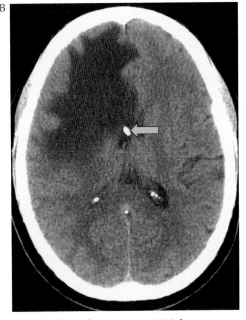

Figure 12–4. Methotrexate leukoencephalopathy after intraventricular chemotherapy. **A**: An MRI showing a necrotic cyst surrounded by edema; the ventricular catheter can be seen below the cyst (*arrow*). **B**: A CT demonstrating the tip of the Ommaya reservoir (hyperdense) in the appropriate place at the foramen of Monro (*arrow*). In a few patients, the methotrexate injected into the ventricle tracks back along the ventricular catheter to cause focal damage to the brain. Biopsy of such patients reveals typical methotrexate necrosis.

Table 12–9. **High-Dose Methotrexate Neurotoxicity**

Characteristics	Acute Encephalopathy	Subacute Encephalopathy	Chronic Leukoencephalopathy
Onset	<48 hours	3–10 days	≥3 months
Symptoms	Confusion, lethargy, headaches, seizures	Multifocal deficits	Spasticity, dementia, seizures
Abnormal MRI	No	DWI	White matter hyperintensity, calcifications
Clinical outcome	Full recovery	Usually full recovery	Persistent neurologic deficits, death

DWI, diffusion weighted image.

syndrome in either adults or children[327–330] typically follows the second or third treatment by 5 or 6 days. Patients present with an altered mental status ranging from inappropriate laughter to stupor, usually accompanied by hemiparesis that may fluctuate from one side to the other. Other focal findings include aphasia, coma, and sometimes seizures. Patients generally recover spontaneously within 48 to 72 hours without sequelae and can usually be retreated with HD MTX without recurrence, although a subsequent episode does occur rarely. Serum MTX levels are nontoxic at the time of the syndrome's onset. The EEG may show diffuse slowing without epileptiform discharges. The MRI often shows focal areas of restricted diffusion suggesting ischemia.[328] Brain glucose metabolism decreases following IV infusion with HD MTX, suggesting that derangements in regional glucose metabolism play a role in the clinical presentation.[331] Homocysteine elevations have been reported, suggesting that the pathogenesis is related to damage to blood vessels.[329] One report describes impurities in the MTX administered.[330] However, it is highly likely that the syndrome results from a direct toxic effect, albeit transient, of MTX although the exact pathogenesis is unknown. As indicated in the preceding text, a similar syndrome can follow intrathecal MTX given in standard doses.[331a]

HD MTX can cause *dry eyes* with burning and itching,[332] pleuritis,[333] or osteoporosis with back pain.[334] Retinal damage can follow intra-arterial MTX after the blood–brain barrier is opened with hyperosmolar agents.[335] Dizziness, headache, and "fuzzy-headedness", reversible after ceasing treatment, have been reported to affect as many as 25% of patients receiving weekly low-dose systemic MTX,[336]

but a more recent report suggests a lower incidence.[337]

Chronic encephalopathy: Diffuse leukoencephalopathy is the most devastating form of MTX neurotoxicity. The disorder generally follows repeated doses of IV HD MTX[338] or intrathecal MTX, but it also may occur after standard-dose IV MTX.[339,340] Leukoencephalopathy may appear months to years following therapy, beginning either insidiously or abruptly, with personality changes and learning disability[341] and sometimes behavioral changes.[342,343] Most of the series involved children treated prophylactically to prevent CNS leukemia and young age appears to be a major risk factor. However, the disorder also occurs in adults (particularly the elderly) and is particularly apparent in patients treated for primary CNS lymphoma.[344,345] Children cured of leukemia who have received only IV HD MTX or intrathecal MTX with cranial irradiation show a decrease in IQ after therapy, often by more than 15 points.[341,346] Seizures can occur but usually only late in the course. The clinical course varies. Patients may recover slowly over weeks or months, their symptoms may stabilize with a mild-to-moderate dementia, or progress and may be relentless with spastic hemiparesis or quadriparesis, severe dementia, and coma, ending in death.

The myelin basic protein concentration may be elevated in CSF, presumably because of myelin breakdown.[347] The CT reveals cerebral atrophy, bilateral and diffuse white matter hypodensities, ventricular dilatation, and sometimes cortical calcifications, findings that are occasionally seen in asymptomatic patients with leukemia.[348] Focal enhancement may be present in the early stages[349] and may signify a poor prognosis.[350] Abnormalities are even more apparent on MRI.[350–352] Scans may be normal even in patients with diminished intellect, and vice versa,[352] but usually changes in white matter on the MRI precede neurologic signs.[353] Such changes should warn the physician that the patient is at risk for leukoencephalopathy. Some of the white matter changes resolve over time.[354] Quantitative MR imaging assessment may be more sensitive than standard testing.[338,354]

Several different pathologic abnormalities have been reported. The most common is disseminated foci of white matter degeneration characterized by demyelination, axonal swelling, and dystrophic mineralization of axonal debris. These necrotizing changes may occasionally be accompanied by fibrinoid necrosis of small blood vessels.[343] Histologically, leukoencephalopathy from RT (see chapter 13) cannot easily be distinguished from that of MTX, although axonal swelling is much more characteristic of MTX-induced leukoencephalopathy than of radiation effect.

An unusual pathologic and sometimes radiologic change (now rarely encountered), virtually restricted to children, is mineralizing microangiopathy characterized by non-inflammatory fibrosis and calcification of arterial capillaries and venules, particularly in the basal ganglia.[355]

Pathophysiology: The pathophysiology of MTX neurotoxicity is not completely understood.[356] It is possible that each of the neurologic symptoms described in the preceding text has a different pathophysiology. The depletion of reduced folates in the brain, inhibition of cerebral protein or glucose metabolism, injury to cerebral vascular endothelium resulting in increased blood–brain barrier permeability, and inhibition of catecholamine neurotransmitter synthesis have all been implicated. Elevation of homocysteine may be a major risk factor leading to vascular damage. Decreased methionine synthesis may also play a role.[357] Other abnormalities include changes in excitatory amino acids, nitric oxide,[358] adenosine deaminase,[358] and monoamines.[359] Isolated reports describe effective treatment with aminophyllin and high-dose folinic acid[360] and caffeic acid phenethyl ester.[358]

Pemetrexed: This recently approved antifol, similar to MTX in that it crosses the blood–brain barrier poorly,[361] is effective against mesothelioma, non–small cell lung cancer and probably other solid tumors.[362] Many of its toxic side effects can be prevented by use of prophylactic dexamethasone (to decrease rash), vitamin B_{12}, and folic acid.[362]

Significant neurotoxicity has not been described.

Cytidine Analogs

Cytosine Arabinoside (cytarabine, ara-C): This pyrimidine analog is an S-phase cell cycle–specific drug that inhibits DNA polymerase, is incorporated into DNA, and terminates DNA chain elongation. Although now produced synthetically, arabinose nucleotides were originally isolated from a sponge. The drug treats leukemia and lymphoma. It is not useful as a single agent against most solid tumors, probably because it is poorly activated in solid tumors and has its best effect against rapidly dividing cells. However, the drug is used intrathecally to treat solid tumors.[295,363] Cytarabine can be given systemically and intrathecally[364,365] in conventional doses and sometimes intravenously in a high-dose regimen.[366] A liposomal preparation (DepoCyt) is available for intrathecal injection.[367] At conventional doses, the drug crosses the blood–brain barrier poorly[368,369] with blood-to-CSF transport occurring by facilitated diffusion across the choroid plexus. High-dose cytarabine yields higher CNS levels; the concentration of cytarabine in CSF reaches 6% to 22% of plasma values.[370] IV cytarabine does not cause neurotoxicity at conventional doses, but can be neurotoxic at high[371] or intermediate doses.[372] The clinical findings of cytarabine toxicity depend on the route of administration, as well as on patient age, renal function, drug dosage, and the frequency of administration.

Both intrathecal cytarabine and the liposomal preparation can be neurotoxic.[367] One hour after intrathecal injection of cytarabine, more than 67% of the rabbit spinal cord is exposed to the drug[365] but exposure is certainly much less in humans. Aseptic meningitis and, rarely, myelopathy, both clinically similar to MTX toxicity,[373,374] can follow intrathecal cytarabine. Meningeal irritation causing headache (more common with the liposomal preparation probably because of the longer exposure of the meninges to the drug), stiff neck, and pleocytosis have been encountered in approximately 10% of patients given the drug, although no direct relationship between this syndrome and the individual or cumulative dose has been established. The myelopathy begins with back or radicular leg pain. Weakness, sensory alterations, and bowel or bladder dysfunction occur any time from a few days to months following treatment. The clinical picture usually evolves rapidly, sometimes rendering the patient paraplegic. Signs typically persist, but some patients recover.[375] Examination of the CSF usually reveals an elevated protein level and a modest pleocytosis; an elevated myelin basic protein level has also been reported.[320,376] Pathologically, portions of the spinal cord reveal demyelination with associated white matter vacuolization, histologically indistinguishable from MTX-induced myelopathy.[320,377] Forty-eight hours after a single dose of 100 mg of intrathecal cytarabine in conjunction with IV cytarabine, cisplatin, and doxorubicin, one patient developed a *locked-in syndrome*, in which consciousness was preserved but paralysis prevented communication except by eye movement. At autopsy, extensive brainstem necrosis was discovered.[378] One patient with leukemia who received intrathecal therapy with MTX, cytarabine and corticosteroids (triple intrathecal chemotherapy) developed acute cerebellitis, leading to herniation, presumably because of cerebellar toxicity of cytarabine.[379]

Rarely, intrathecal administration of cytarabine is associated with seizures, an acute or subacute encephalopathy, or both.[374,380] Higher doses and greater frequency of administration increase the risk of intrathecal cytarabine neurotoxicity.[381] The neurotoxic effects of prior radiation or intrathecal MTX treatment may be synergistic with those of intrathecal cytarabine,[382] although patients who have suffered MTX neurotoxicity have been successfully treated with intrathecal cytarabine without adverse effects. Liposomal cytarabine, when combined with high-dose systemic chemotherapy, was particularly toxic when used as prophylaxis for leukemia. One death from encephalopathy occurred.[382a]

IV HD cytarabine[371] may cause several neurologic disorders.[383] Cerebellar dysfunction occurs more frequently in older patients and usually at a cumulative dose of at least 36 g/m^2, but it has been reported after only 3 g/m^2.[384,385] Patients present with dysarthria, nystagmus, and appendicular and gait ataxia.[386,387] They may also develop confusion, lethargy, and somnolence. With cessation of the drug, complete resolution of signs and symptoms generally occurs within 2 weeks. Predisposing factors include abnormal renal function, elevated alkaline phosphatase, prior neurologic disorders, and age beyond 60 years.[371] The incidence of neurotoxicity in patients with renal insufficiency may be reduced by lowering the dose from 3 to 2 g/m^2, modifying the dose based on daily creatinine values and administering the dose once rather than twice daily.[366] The EEG may show modest slowing of brain waves. The MRI is usually normal, although cerebellar atrophy and reversible white matter abnormalities may be seen.[388,389] Single photon emission computed tomography (SPECT) may show diffuse heterogeneous brain hypoperfusion in both cerebrum and cerebellum.[390] The CSF is normal.

Although the cerebellar disorder is not fatal, when patients die from their malignancy, neuropathologic changes include widespread loss of Purkinje cells, most pronounced in the deeper portion of the primary and secondary cerebellar sulci. The remainder of the CNS appears to be largely unaffected, although white matter demyelination and filamentous degeneration of neurons of brainstem and spinal cord[391] have also been reported.[385,392]

Peripheral neuropathy, either axonal, demyelinating, or both,[393,394] is a rare complication of cytarabine, although one report describes a 1% incidence of demyelinating polyneuropathy after high-dose cytarabine.[395] In most patients, high-dose cytarabine was given along with other potentially neurotoxic agents, including fludarabine.[396] Also reported occasionally are seizures[380]; reversible ocular toxicity,[397] including blurred vision, photophobia, and burning eye pain; blindness[398]; bulbar and pseudobulbar palsy[399]; Horner syndrome[400]; the "painful leg, moving toes syndrome"[401]; brachial plexus neuropathy[402]; reversible bilateral lateral rectus palsies[402]; acute aseptic meningitis (after IV injection)[404]; and anosmia.[405] One patient who received a cumulative dose of 72 g developed a parkinsonian syndrome that resolved within 12 weeks[406]; reversible posterior leukoencephalopathy has been reported in one patient after receiving intermediate-dose cytarabine (500 mg/m^2).[372]

The mechanism of cytarabine toxicity is unknown. Cytarabine kills neurons by apoptosis[407] (especially immature neurons) in culture. Postulated mechanisms include oxidative stress,[408] blocking nerve growth factor activity,[409] interfering with cyclin-dependent kinases,[410] and activation of astrocytes, increasing susceptibility to excitotoxicity.[411]

There is no generally accepted treatment; one report describes a patient with cerebellar and peripheral nerve toxicity responding to methylprednisolone.[412]

5-Azacytidine/Decitabine: Cytidine analogs that incorporate into DNA and RNA and prevent DNA methylation treat myelodysplastic disorders.[413] The drugs are not usually neurotoxic, but 5-azacytidine was reported to be hepatotoxic. Fatal hepatic coma occurred in four patients with extensive hepatic metastases.[414] A neuromuscular syndrome, characterized by muscle tenderness and weakness, may develop after the second or third dose of 200 mg/m^2 per day, as well as encephalopathy, lethargy, and confusion; in one patient, coma was reported.[415] Dizziness and fatigue are also side effects.[413]

Gemcitabine: Gemcitabine, a prodrug that is activated intracellularly, is active against several solid tumors, especially pancreatic cancer. Gemcitabine crosses the blood–brain barrier poorly with a CSF:plasma ratio of 6.7%.[416] The drug can cause pulmonary toxicity (pneumonitis),[417] vasculitis[418] with retinopathy,[419] a capillary leak syndrome,[420] and thrombotic microangiopathy,[421] but is generally believed not to be neurotoxic even when combined with oxaliplatin.[422] However, paresthesias occur in approximately 6% of patients, and an autonomic neuropathy has been reported.[423] Severe neurotoxicity characterized by confusion and myoclonic seizures has also been reported.[424] The drug can also cause radiation recall syndrome (see p 453), characterized by dermatitis and myositis within the irradiated portal.[425,426] One of our patients developed severe anterior abdominal wall pain while being treated with gemcitabine following RT for pancreatic carcinoma (Fig. 12–5). There was marked tenderness of the abdominal musculature. An MRI demonstrated increased signal and abnormal enhancement of the rectus abdominis muscle.[426] Symptoms rapidly resolved with corticosteroids and did not recur when steroids were discontinued.

Fluorinated Pyrimidines

Fluorinated pyrimidines (primarily 5-FU and capecitabine) take advantage of neoplastic cells absorbing uracil more avidly than nonmalignant cells. These drugs inhibit thymidylate synthase and are incorporated into DNA and RNA, interfering with its function.[427,428] Tumors expressing low levels of thymidine synthase or thymidine phosphorylase are more sensitive to these drugs, as are tumors expressing low levels of dihydropyrimidine dehydrogenase, an enzyme necessary for the catabolism of fluorouracil. Because the drug causes apoptosis via the p53 pathway, wild type p53 enhances cytotoxicity.[429]

5-Fluorouracil: A pyrimidine analog that binds thymidylate synthase, thereby interfering with DNA synthesis,[427,428] 5-FU is useful in GI tumors and breast cancer. In monkeys[430] it crosses the blood–brain barrier by simple diffusion to achieve a CSF concentration of 11% to 50% of plasma. In humans, CNS penetration is probably minimal during conventional infusion therapy.[431,432] The drug is found in higher concentrations in the cerebellum, especially in granular and Purkinje cell layers, than in the forebrain.[433,434] One experimental study indicated that 5-FU "opened" the blood–brain barrier by promoting pinocytotic vesicular transport[434]; another study failed to demonstrate that finding.[435]

Neurotoxicity is rare with conventional doses of the drug, except for patients with dihydropyrimidine dehydrogenase deficiency, the initial rate-limiting enzyme in pyrimidine catabolism. These patients, whose deficiency can result from either a mutated gene or a methylated promoter,[436] can develop severe ataxia at low doses.[437–439] The deficiency is more prevalent in black women (12.3%) and black men (4.0%) than in white women (3.5%) or men (1.9%).[440] A uracil breath test is both sensitive and specific for the deficit.[441] The primary neurotoxicity with higher doses of 5-FU is also a cerebellar syndrome, clinically indistinguishable from paraneoplastic or cytarabine-induced cerebellar disorders, consisting of truncal and limb ataxia, dysmetria, nystagmus, and dysarthria.[431,442] The mechanism of this or other neurotoxicity is unknown. The signs usually reverse within a week after the drug is discontinued but may recur with reintroduction of 5-FU. In experimental animals, 5-FU damages Purkinje cells, granule cells, and neurons of the inferior olive and vestibular nuclei.[442] 5-FU and some of its derivatives also damage myelin.[443] We have had the impression that patients on standard doses of fluorouracil, particularly the elderly, appear to have more difficulty walking heel-to-toe

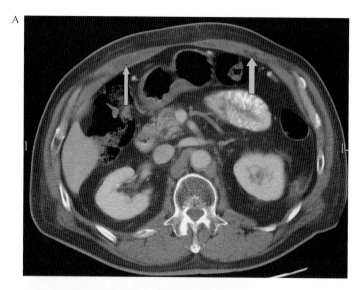

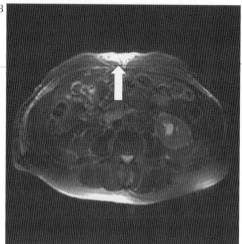

Figure 12–5. Radiation recall induced myositis. A 62-year-old man with pancreatic carcinoma was treated with radiation and gemcitabine. Five months after completion of radiation therapy, while still on gemcitabine, he developed severe right upper quadrant abdominal wall pain that was present only upon movement and spread from the right to the left side. On examination, the anterior abdominal wall muscles were extremely tender to touch although the overlying skin was non-tender. A CT scan (**A**) revealed hypointensity in the rectus abdominis muscles bilaterally (*arrows*). An MRI (**B**) showed a focal area of hyperintensity within the subcutaneous fat of the anterior abdominal wall and the underlying muscle (*arrow*). He was treated with prednisone and made a complete recovery.

along a chalk line (tandem gait) than they did before therapy was begun.

A less common complication is a diffuse encephalopathy with or without cerebellar signs.[444,445] The encephalopathy has an acute onset during chemotherapy treatment and usually resolves spontaneously. The encephalopathy may be severe enough to cause coma and/or seizures.[446] The disorder has been reported to be associated with hyperammonemia and lactic acidosis,[447,448] particularly with high-dose fluorouracil. In some patients with encephalopathy, diffusion-weighted MRI suggests restricted diffusion in white matter that resolves as symptoms improve.[449,450] Focal encephalopathy, with symptoms resembling Parkinson disease, has also been described.[451] Ocular toxicity[452] including blepharitis, conjunctivitis, lacrimal duct stenosis, and excessive lacrimation,[453,454] as well as optic neuropathy,[455] and ocular muscle abnormalities,[452,454,456] have all been described. Cerebral infarcts have been reported to complicate fluorouracil chemotherapy, although all of the patients reported had also received other chemotherapeutic agents.[457] Peripheral neuropathy, probably demyelinating, is also a rare complication of fluorouracil therapy.[458–460] One patient with a chronic peripheral neuropathy developed an exacerbation when treated with topical fluorouracil.[461] Palmar-plantar erythrodysesthesia (see hand–foot syndrome p 453) is a rare complication, as is the radiation recall syndrome (see p 453).

Several modulators have been used to try to enhance the effectiveness of 5-FU or reduce its toxicity in normal tissues. These include folinic acid (leucovorin), levamisole, allopurinol, thymidine, phosphonoacetyl-L-aspartate (PALA), and eniluracil, an inhibitor of dihydropyrimidine dehydrogenase. Combining 5-FU with allopurinol, however, can cause acute and subacute cerebellar dysfunction, visual disturbance, dizziness, and, rarely, seizures.[462] The neurotoxicity may be related to reduced 5-FU clearance in the presence of allopurinol. PALA inhibits de novo pyrimidine biosynthesis and enhances 5-FU cytotoxicity by increasing its incorporation into RNA. PALA itself may cause ataxia, encephalopathy, and seizures. PALA combined with 5-FU frequently causes cerebellar dysfunction, tremor, memory loss, hallucinations, and seizures.[463] Thymidine may be neurotoxic when administered in high doses and also appears to enhance the neurotoxicity of 5-FU. Patients develop confusion, disorientation, and lethargy, and a few have experienced nystagmus and dysmetria. When PALA, thymidine, and 5-FU are combined, neurotoxicity is dose limiting, with dizziness, confusion, ataxia, agitated delirium, and aphasia.[464] When levamisole is combined with 5-FU, a few patients develop an inflammatory multifocal leukoencephalopathy (Fig. 12–6).[465] The disorder usually presents with confusion and focal signs,

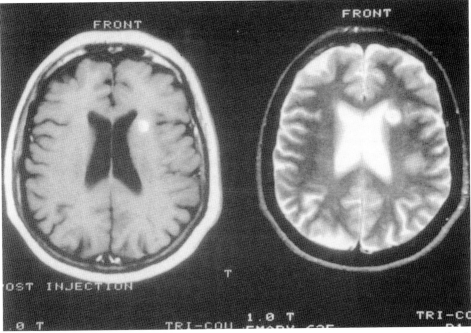

Figure 12–6. Encephalopathy caused by 5-FU with levamisole. This patient with Duke B colon cancer developed a mild hemiparesis and underwent an MRI (*left*) that showed a contrast-enhancing lesion in the left hemisphere. Several other lesions did not contrast-enhance but were apparent as hyperintensities on the T2-weighted image (*right*). After a diagnosis of metastatic tumor, the patient received 3000 cGy to the whole brain. She became demented and bedridden. A needle biopsy specimen of the lesion following the RT revealed the typical demyelination reported as a side effect of 5-FU-levamisole therapy. Metastatic disease was not present. Levamisole is rarely used now, having given way to the less toxic leucovorin.

including ataxia or hemiparesis associated with multiple contrast-enhancing lesions on MR scan which may be confused with metastases. Levamisole alone is rarely neurotoxic.[466]

The mechanism by which 5-FU and its analogs cause neurotoxicity is not entirely known. In addition to the acidosis and hyperammonemia alluded to earlier, a report describes Wernicke encephalopathy as a neurotoxic complication.[467] Thiamine should probably be given to all patients exhibiting cerebellar signs.

Capecitabine (Xeloda): An oral analog of 5-FU,[468] the drug has a similar toxicity profile. Like fluorouracil, neurotoxicity is uncommon at standard doses. However, acute encephalopathy with[469] or without cerebellar signs, including coma[470] and seizures,[471] have been reported. The MRI may show reversible white matter abnormalities.[471,472] Peripheral neuropathy has also been reported.[473] The hand-foot syndrome is relatively common.[468] Arthralgias and myalgias occasionally occur; one patient presented with trismus along with more diffuse CNS and peripheral symptoms.[474] Both 5-FU and capecitabine can potentiate the effects of warfarin and lead to bleeding[475]; the mechanism is believed to be inhibition of cytochrome P450 activity.

Other analogs: Floxuridine (5-fluoro-2'-deoxyuridine [FUDR]), Fotofur, Carmofur, and 5'-deoxy-5-fluorouridine (DFURD) are analogs of 5-FU rarely used in this country. Their toxicity is similar to that of 5-FU.

Purines and Other Antimetabolites

Adenosine Analogs: Three analogs of adenosine, fludarabine, cladribine, and pentostatin, are active against indolent lymphoid malignancies.[476] The drugs have similar neurotoxicity, that can result from direct effects on nerve tissue,[477] secondarily from nervous system infections resulting from the profound immunosuppression that the drugs cause,[478] or from metabolic abnormalities resulting from tumor lysis.[479]

Fludarabine: Fludarabine is an inhibitor of DNA polymerase and ribonucleotide reductase. In lower doses, the drug causes little neurotoxicity except for transient episodes of somnolence and fatigue,[480] or peripheral neuropathy with either paresthesias or weakness.[480] A few patients have reversible focal CNS signs; T2-weighted MR images show foci of white matter hyperintensity.[481] At doses greater than 40 mg/m²/day, a delayed progressive encephalopathy, characterized by cortical blindness, dementia, coma, and death, can develop.[482,483] The drug can also cause myelopathy (Fig. 12–7). Autopsy shows a diffuse necrotizing leukoencephalopathy most severe in the occipital lobes, medullary pyramids, and posterior columns.[477,482,484] This toxicity may increase when fludarabine is combined with cytarabine.[396] The combination of the immunosuppression caused by lymphoid neoplasms and that caused by fludarabine make opportunistic infections a serious problem[478,485] (see Chapter 10). Infections that can affect the nervous system can be caused by bacteria

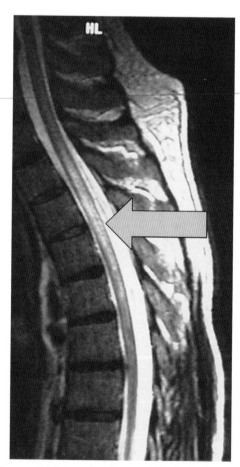

Figure 12–7. A myelopathy caused by fludarabine in a patient with leukemia. The patient developed weakness of the lower extremities during the course of treatment and made a good recovery over several months. An MRI while he was weak demonstrated hyperintensity within the thoracic spinal cord (*arrow*). His symptoms and the MRI eventually reverted to normal.

(Listeria), fungi (Candida, Aspergillus, cryptococcus), viruses (zoster, CMV, progressive multifocal leukoencephalopathy,)[486] or Pneumocystis carinii.

Cladribine (2Cda): Cladribine (2Cda) also causes little neurotoxicity at usual doses. However, at higher doses, it can cause paralysis, either paraplegia or quadriplegia, from damage to the spinal cord.[477] An acute polyneuropathy (Guillain–Barré syndrome) has also been described.[487] Cortical blindness can also occur. Infectious complications are similar to those from fludarabine, but bacterial infections (streptococci, Gram-negative rods, Neisseria meningitides) have also been reported.[485]

Pentostatin(deoxycoformycin): Pentostatin (deoxycoformycin) is an adenosine deaminase inhibitor that causes less myelosuppression than the other two drugs because it is relatively specific for toxicity of lymphoid cells.[488] The drug crosses the blood–brain barrier to achieve CSF levels 10% of plasma.[489] In high doses, it causes an encephalopathy characterized by somnolence or agitation and sometimes coma,[489] but at usual doses, severe encephalopathy is uncommon.[477,478] The symptoms reverse when the drug is discontinued. Conjunctivitis and complaints of muscle and joint pain have also been reported.

Guanine analogs: The guanine analogs 6-mercaptopurine, 6-thioguanine, and azathioprine, can cause myelosuppression and rarely a mild encephalopathy resulting from hepatotoxicity. They are not generally considered to be neurotoxic.

Hydroxyurea: An inhibitor of ribonucleotide reductase, an enzyme that catalyzes the rate-limiting step in the de novo synthesis of nucleotide triphosphates required for DNA synthesis, hydroxyurea easily enters CSF[490] to achieve a level of 10% to 25% that of plasma. The drug enhances the effect of ionizing radiation and has been used to treat head and neck cancers,[491] meningiomas,[492] and glioblastomas.[493] Neurologic toxicity is rare, but when it occurs it includes headache, drowsiness, confusion, and dizziness. One report[494] described data from the spontaneous reporting of adverse drug reactions to the Medicines Control Agency of the United Kingdom from August 1967 to January 1998. Of 162 such adverse effects, only 11 were neurologic. The 11 adverse reactions included fatal coma, seizures, trigeminal neuralgia, tremor, headache, peripheral neuropathy, and delirium. How many of these were, in fact, actually due to hydroxyurea is unclear.

PLANT ALKALOIDS

The plant alkaloids all bind to tubulin, the microtubular protein, but the antineoplastic mechanism is different for each group of drugs (Table 12–10). All can be neurotoxic, primarily affecting the peripheral nervous system; some also have CNS effects.

Vinca Alkaloids

The clinically important vinca alkaloids include vincristine, vinblastine, vindesine, and vinorelbine. These agents bind to tubulin and block its polymerization, preventing microtubule formation, thereby arresting cells in metaphase.[495,496] The tubulin binding interferes with axonal transport, probably the mechanism of the neuropathy. The major site of damage appears to be in the sensory neurons of the dorsal root ganglia,[497] a structure that does not have a blood–nerve barrier. The drugs cross the blood–brain barrier poorly.[498] Inadvertent intrathecal injection is usually lethal.[499–500] However, one patient treated with 6 days of CSF irrigation plus intrathecal fresh frozen plasma and systemic pyridoxine, folic acid and glutamine recovered, albeit with a paraparesis below T9.[501]

Table 12–10. **Plant Alkaloids**

Vincas
  Vincristine
  Vindesine
  Vinblastine
  Vinorelbine

Podophyllins
  Etoposide (VP-16)
  Teniposide (VM-26)

Taxanes
  Paclitaxel (Taxol)
  Docetaxel (Taxotere)

Others
  Estramustine
  Epothilones

Vincristine: The most neurotoxic vinca, vincristine, primarily affects the peripheral nerves, although it can also be toxic to the CNS, cranial nerves, and autonomic nervous system (Table 12–11). A dose-limiting sensorimotor neuropathy appears in most patients.[21] The earliest complaint is tingling and paresthesias of the fingertips and, later, of the toes. Fine movements are often impaired.[502] Loss of ankle jerks is the earliest sign. With continued drug administration, other reflexes also disappear. Muscle cramps, usually diurnal, affect arms and legs and may be the first symptom of neurotoxicity and the last to recover.[32]

Objective sensory loss is uncommon, but weakness, especially of the dorsiflexors of the feet and the wrist extensors, is typical. Foot-drop is either unilateral or bilateral. Unilateral foot-drop is especially common in the superior leg of patients who have lost weight and habitually sit with crossed legs. The weakness seen with vincristine is usually tolerable, but rare patients become bed-bound or quadriparetic,[503] particularly if they have a preexisting neuropathy.[504] The sensory symptoms, weakness, and lost reflexes are reversible, although recovery may require several months after the medication is stopped.[505] In some instances, symptoms may progress for several weeks after discontinuation of the drug.[21]

The neuropathy may be painful and the pain may sometimes be disabling. In addition, some patients suffer muscle cramps[32] or acute muscle pain following injection; others complain of face pain.[506] Whether the pain is neurogenic or myogenic is not clear.

Vincristine neurotoxicity can be enhanced either by interaction with other drugs[507] or by comorbid neurologic disease.[508,509] Drugs include the azole anti-fungal agents (e.g., itraconazole), cyclosporin and isoniazid, all of which inhibit cytochrome P450, decreasing the metabolism of vincristine and prolonging its half-life. Nifedipine, often used to treat steroid-induced hypertension in children with leukemia, inhibits P-glycoprotein and may promote the retention of vincristine in cells, thus increasing toxicity.[507] One report describes severe vincristine neuropathy when the drug was associated with administration of hemopoietic colony stimulating factors.[510] Preexisting peripheral nerve disease, even if asympomatic,[504] can also enhance neurotoxicity.[509] A particular problem affects patients with lymphoma and preexisting Charcot-Marie-Tooth disease (hereditary motor and sensory neuropathy type-1).[509] One of our patients, a young woman with Hodgkin disease, developed quadriparesis and became bedridden after one dose of vincristine. It was 4 months before she recovered neurological

Table 12–11. **Spectrum of Vincristine Neurotoxicity**

Toxic Effect	Subacute (1 day to 2 weeks)	Intermediate (1 to 4 weeks)	Chronic (>4 weeks)
Peripheral neuropathy	Depressed Achilles reflex, paresthesias	Other tendon reflexes depressed, paresthesias	Sensory loss, weakness, foot-drop
Myopathy	Muscle pain, tenderness, (especially quadriceps); jaw pain		
Autonomic neuropathy	Ileus with abdominal cramping pain	Constipation, urinary hesitancy, impotence, orthostatic hypotension	
Cranial neuropathy (uncommon)			Optic atrophy; ptosis; sixth, seventh and eighth cranial nerve dysfunction; hoarseness; dysphagia
"Central" toxicity		Seizures, SIADH	

SIADH, syndrome of inappropriate secretion of antidiuretic hormone.

function. A careful family history and neurologic examination with attention to the peripheral nervous system may warn the physician of the possibility of unusual sensitivity to vincristine.[508] A single report describes the uneventful administration of four courses of 100 mg/m[2] of cisplatin in a patient with X-linked Charcot-Marie-Tooth disease with a subjective increase in paresthesias, but no objective change in neurologic function.[511] There are no reports of paclitaxel, another neurotoxic chemotherapeutic agent, exacerbating preexisting neurologic disease.

A previous history of paralytic poliomyelitis also may enhance vincristine neurotoxicity. Giving two substantially neurotoxic drugs may also enhance neurotoxicity. Neurotoxicity has been reported with a combination of vinorelbine and paclitaxel, whereas vinorelbine as a single agent rarely causes significant neuropathy.[512,513]

Risk factors include age (adults are more severely affected than children), dose, nutritional state, abnormal liver enzymes,[514] and possibly lymphoma.[515] Other therapeutic agents such as etoposide,[516] teniposide,[517] cyclosporin,[518] and limb radiation probably also enhance toxicity.[519]

A number of agents have been reported to help prevent or relieve vincristine neurotoxicity. Most are ineffective. Such agents include gangliosides,[520,521] isaxonine,[522] prednisone,[514] pyridoxine,[523] nerve growth factor,[524] and Org 2766.[525] More promising are folinic acid[526] and glutamate.[498] A controlled trial of glutamate reduced paresthesias and preserved ankle jerks.[498] Transiently occluding the arterial supply to the dominant hand during vincristine infusion is also said to be helpful.[527]

Vincristine occasionally causes unilateral or bilateral focal neuropathies of peripheral or cranial nerves.[27,528] Findings include ptosis,[529] ophthalmoplegia,[529] and vocal cord paralysis.[506,530,531] The facial nerve,[529] acoustic nerve,[532] and optic nerve are also occasionally affected.[533-535] Night blindness from retinal damage has also been reported.[536]

Autonomic neuropathy, characterized by colicky abdominal pain and constipation, occurs in as many as 33% of patients. Paralytic ileus is especially frequent in children but is rarely fatal. Amelioration of vincristine-induced ileus with metoclopramide has been reported[537] but preventing constipation is essential; all patients receiving vincristine should follow a prophylactic bowel regimen of stool softeners and laxatives. This precaution is particularly important when patients are also receiving corticosteroids, because constipation is a major risk factor for steroid-induced bowel perforation (see Chapter 4). Other, infrequent manifestations of autonomic dysfunction include bladder atony, impotence, postural hypotension,[538] heart rate abnormalities,[539] and dysphagia.[540]

Vincristine crosses the blood–brain barrier poorly. After a bolus injection of 1.5 mg/m[2], no drug was detected in CSF.[541] The absence of a blood–peripheral nerve barrier in the dorsal root ganglion probably explains peripheral toxicity. CNS toxicity of vincristine includes hyponatremia from SIADH.[542,543] Encephalopathy[544] and focal or generalized seizures not due to SIADH[545,546] have also been reported. Cortical blindness (probably a result of posterior reversible leukoencephalopathy [see p 451])[547,548] and other focal cerebral signs such as athetosis, ataxia,[549] and parkinsonian symptoms[550] usually reverse after treatment is discontinued.

Diagnosing vincristine neurotoxicity is usually easy, even though the motor weakness typically develops a few weeks following therapy and progresses for several additional weeks. Neurophysiologic studies are usually unnecessary, but when undertaken, they show the features of an axonal neuropathy. These findings have been confirmed by nerve biopsy.[551]

In experimental animals,[552] vincristine causes focal axonal swelling, primarily in the proximal portion of the peripheral nerves. The swelling results from accumulated misaligned neurofilaments with secondary perinodal demyelination. Other changes include a decreased number of microtubules.[553] In one patient with optic neuropathy, selective loss of beta-tubulin in the optic nerve was found.[554] What appears to be a painful neuropathy, characterized at onset by cold allodynia, serves as an experimental model, in rodents, for vincristine-induced peripheral neuropathy.[555]

Vindesine, Vinblastine, and Vinorelbine: These drugs are less neurotoxic than vincristine,[556] with vindesine being relatively more neurotoxic than vinblastine. Vinorelbine is the least toxic, perhaps because it binds axonal tubulin less well.[557] However, the same peripheral and CNS side effects as with vincristine have been reported with vindesine,[558,559] vinblastine,[560] and vinorelbine.[561] The relative cytotoxicity of these vinca alkaloids in cell culture parallels their clinical neurotoxicity.[556]

Podophyllins

Epipodophyllotoxin is an antimitotic agent extracted from the May apple or mandrake plant. The agent binds to tubulin at a different site than the vinca alkaloids. Two synthetic derivatives, etoposide (VP-16) and teniposide (VM-26), are used against many human malignancies, including lymphomas, germ-cell tumors, and small cell lung cancer. These drugs are believed to damage DNA by forming a complex involving the drug, DNA, and topoisomerase type II. Thus, they do not exert their major cytotoxicity through tubulin binding, although this is the mechanism of their peripheral nerve toxicity. Although neurotoxicity is rare, the drug can cause an axonal sensory neuropathy or neuronopathy that usually, but not always, reverses after therapy is discontinued.[562] Podophyllotoxins found in herbal medications, particularly those from China, can also cause peripheral neuropathy, as does podophyllin, used topically to treat condyloma acuminatum. Physicians evaluating a patient with cancer for neurologic symptoms of either peripheral nervous system or CNS origin[563] must always consider the possibility that the patient is receiving, but not reporting, herbal medicines containing podophyllins or other neurotoxins from an "alternative medicine" practitioner.

Etoposide (VP-16): This drug penetrates CSF poorly, yielding CSF concentrations that are less than 5% of simultaneously measured plasma levels.[564] It has been given intrathecally to animals without toxicity,[565] but has been reported to occasionally cause a mild and reversible peripheral neuropathy in patients. Intra-arterial injection of etoposide opens the blood–brain barrier.[566,567] When combined with other water-soluble agents, it may allow more of those agents to enter the CNS than would enter otherwise. When used in high doses in patients with bone marrow transplantation, etoposide may cause seizures, confusion, and somnolence.[568]

Teniposide (VM-26): This drug crosses the blood–brain barrier less well than etoposide.[565] Unlike etoposide, teniposide rarely causes peripheral neuropathy,[569] but has been reported to enhance the neurotoxicity of vincristine.[517]

Taxanes

Paclitaxel and Docetaxel: (Taxol[570] and Taxotere) are agents derived from the Pacific yew tree. Both drugs bind to tubulin. But instead of inhibiting polymerization, as do vinca alkaloids, they stabilize and promote microtubular assembly. How this interaction with microtubules causes cytotoxicity is not clear. Both drugs have been useful in treating ovarian, breast, and other cancers. Their interaction with microtubules prevents anterograde axonal transport, leading to their major toxicity, peripheral neuropathy.[571,572] The incidence of peripheral neuropathy depends on the treatment schedule, dose per cycle, duration of infusion, cumulative dose, treatment with other cytotoxic chemotherapeutic agents,[573] and comorbidities including diabetes and peripheral neuropathies of other causes.[574] In some schedules, as many as 30% of patients develop grade 3 or grade 4 peripheral neuropathy (see Table 1–17).[574] The neuropathy is largely sensory, beginning with paresthesias first in the toes and then in the fingers. The paresthesias may be accompanied by numbness and pain. More severely involved patients demonstrate stocking-glove loss of all sensory modalities and of deep tendon reflexes. In a few patients, symptoms may progress even after the drug is discontinued,[575] but in most, the symptoms do not progress and may even resolve despite continued therapy.[576] Itching can be a prominent manifestation of the neuropathy. Arthralgias and myalgias often appear 2 to 5 days after the drug is given, may curtail activity, and may last for several days[577]; one report describes relief using antihistamines.[578] The sensation of light flashing across the visual field (photopsia) has been reported in some patients during paclitaxel infusion.[579]

Autonomic dysfunction,[580] Lhermitte sign,[581] encephalopathy,[582,583] seizures,[584] cerebral infarction,[584] phantom limb pain,[585] and motor weakness are uncommon complications. The motor weakness may be predominantly proximal,[586] but is probably neuropathic rather than myopathic. A sensory or a sensorimotor neuropathy is usually also present. The weakness may progress or resolve with repeated treatment trials.

A number of neuroprotective agents have been—or are being—tried to ameliorate the effects of the taxanes on the nervous system.[105,574] Amifostine[587] and corticosteroids[588] are probably ineffective. Some reports described lessening of peripheral neuropathic symptoms by vitamin E.[589] Acetyl-L-carnitine[590] and glutamine[591] may have some effect, but do not appear sufficiently effective to warrant widespread use until more controlled clinical trials have been completed. Tricyclic antidepressants and antiepileptic agents such as gabapentin may help control pain.[574]

Several chemotherapeutic agents, particularly docetaxel,[592] cause damage to nails and nail beds including onycholysis,[593] separation of the nail from its bed. This often painful disorder appears to have a neurogenic cause as denervation of the extremity may prevent it from occurring.[592] Cooling the hand with a glove also appears effective.[594]

Other Tubulin Binding Agents

Estramustine: Estramustine, originally believed to be an alkylating agent, probably exerts its cytotoxicity via binding to tubulin. Peripheral neuropathy is usually quite mild[595]; thromboembolism is a major complication.[596] A rapid intravenous infusion can cause severe perineal burning sensations similar to those produced by rapid infusions of dexamethasone (see Chapter 4).[597] Slowing the infusion generally prevents the symptoms.

Epothilones: Epothilones[598] are microtubule inhibitors with a distinct mode of binding to tubulin that makes them effective against neoplasms resistant to taxanes and other microtubule inhibitors.[599] Epothilones now in clinical trials include ixabepilone[600] and patuplione and cause toxicity similar to that of other tubulin-binding agents, including the taxanes.[574] Fatigue and myalgia are relatively common.

MISCELLANEOUS CYTOTOXIC AGENTS

L-Asparaginase: This agent catalyzes the hydrolysis of L-asparagine to aspartic acid and ammonia, depleting asparagine and thus decreasing protein and glycoprotein synthesis. It does not cross the blood–brain barrier but has been given intrathecally.[601] At high doses, encephalopathy (probably related to liver toxicity) was a problem, but current regimens using lower doses are generally not plagued by this complication.[602,603]

L-Asparaginase interferes with coagulation by inhibiting antithrombin III, possibly leading to thrombosis or hemorrhage. The onset of seizures, headache, or focal neurologic signs should alert the physician to the possibility of cerebral venous or venous sinus thrombosis with secondary cerebral ischemia or infarction.[604,605] In one series, three of four adults with acute lymphoblastic leukemia undergoing induction therapy with asparaginase developed cerebral infarction, two due to venous sinus thrombosis.[606] Typically, the disorder appears after a few weeks of therapy but sometimes not until therapy is completed. A definitive diagnosis is made by magnetic resonance venography (MRV). Treatment is controversial; some investigators recommend anticoagulation, but others favor administration of fresh-frozen plasma to replace antithrombin III and depleted fibrinogen. Prophylactic antithrombin may prevent the disorder.[606] The use of prednisone may be a risk factor, but dexamethasone is not.[607] Steroids alleviate the headache; low doses may be necessary for several days to weeks.

Amsacrine (m-ASMA):   This drug,[608] a topoisomerase II inhibitor, can cause headache, dizziness myalgia, cardiac arrhythmias (usually associated with hypokalemia), and (rarely) neurotoxicity that includes transient peripheral neuropathy.[608–610]

Suramin: A polysulfonated urea, and an antihelminthic agent that is useful to treat prostate and some other cancers,[611] this drug inhibits reverse transcriptase and the binding of several growth factors to their specific receptors.[612,613] Suramin serves as a radiation enhancer as well as a cytotoxic agent. Because this drug inhibits cortisol secretion, patients are given hydrocortisone replacement.

Peripheral neuropathy is the major neurotoxicity.[17,614] The neuropathy takes two forms. It can present as a Guillain–Barre–like syndrome, with the rapid onset of a predominantly motor neuropathy with four-extremity paralysis[17,615] and sometimes bulbar and respiratory paralysis. Paresthesias in the face and limbs are common before the weakness, which often begins proximally. The disorder probably results when demyelination causes a conduction block in the peripheral nerves that can be identified by electromyographic study.[616] The disorder reverses itself after the drug is discontinued. The second form is a mild axonal, length-dependent, sensorimotor polyneuropathy.[17] Some studies suggest that the neuropathy is dose dependent, becoming almost invariable when peak concentrations reach 350 mcg/ml.[17] Others have not reported dose dependency.[617] Nevertheless, careful monitoring of blood levels is recommended by most. Visual symptoms, either corneal abnormalities or refractive errors causing visual blurring, are common and easily correctable.[618] Suramin-induced neurite growth inhibition can be overcome by nerve growth factor.[619] Whether this is of clinical importance is unclear.

Camptothecins

Irinotecan (CPT-11): The camptothecins include topotecan and irinotecan (CPT-11). They are topoisomerase 1 inhibitors. Irinotecan is useful in the treatment of colon cancer and has little neurotoxicity. The drug, but not its active metabolite, crosses the blood–brain barrier.[620] A few patients develop CNS toxicity characterized by dysarthria, sometimes with ataxia, during infusion of the drug.[621] In one patient, the dysarthria was followed by transient aphasia.[622] When they occur, symptoms clear rapidly after the infusion, but may recur when the drug is infused again. Irinotecan is an inhibitor of acetylcholinesterase. In some patients, an acute cholinergic syndrome, characterized by abdominal cramping, diarrhea, salivation, dizziness, and visual blurring may occur during or shortly following the infusion.[623,624] Atropine prevents the symptoms.

Angiogenesis Inhibitors

Thalidomide: Thalidomide is an angiogenesis inhibitor used to treat myeloma and some lymphomas. The drug causes a sensorimotor peripheral neuropathy that is predominantly sensory, with pain, dysesthesias, and tremor.[625,626] The development of an axonal neuropathy appears more dependent on the duration the drug is administered than on other risk factors, including dose.[627] Serial monitoring and electrodiagnostic studies do not appear useful.[628] The drug, first developed as a sedative, not surprisingly causes somnolence and some patients complain of dizziness. However, encephalopathy and seizures are quite rare.[629] *Lenalidomide,*[630] (Revlimid) a thalidomide analog, is an immunomodulatory agent in the treatment of myelodysplastic syndrome and multiple myeloma. The drug appears to have less neurotoxicity than thalidomide although both drugs promote thromboembolism[631] perhaps exacerbated by corticosteroids.[631a] (See also section on Biologic Agents for more on angiogenesis inhibitors.)

COMBINATION CHEMOTHERAPY

Most patients with cancer who are treated with chemotherapeutic agents receive more than one agent because single-agent therapy is rarely curative. Combining chemotherapeutic agents may cause more and sometimes different neurotoxicity than single-agent therapy. For example, a regimen for treating acute lymphoblastic leukemia included vincristine, L-asparaginase, daunorubicin, and prednisone followed by consolidation with oral MTX and weekly intrathecal injections of MTX, cytarabine, and hydrocortisone and ending the consolidation with IV VP-16 and cytarabine. This regimen led to seizures or episodes of transient neurologic dysfunction 10 days after the administration of IV cytarabine, VP-16, and triple intrathecal therapy. Discontinuing the intrathecal cytarabine did not prevent the neurotoxicity, but giving oral leucovorin 24 to 36 hours after the MTX did.[632] In another study of multiagent chemotherapy and hormonal therapy for breast cancer, over 1% of patients suffered arterial thromboses, most of which caused cerebral infarcts; four of the nine cerebral infarcts were fatal.[633] The events occurred while patients received chemotherapy, suggesting that the multiagent chemotherapy was causal. One patient with a urethral cancer developed status epilepticus while being treated with a combination of MTX, vincristine, epirubicin, and cisplatin.[634] Combination chemotherapy is generally more neurotoxic when administered with cranial RT than it is alone. Complications of RT are discussed in Chapter 13.

Antineoplastic Hormones

Both hormone agonists and antagonists have selective antineoplastic activity (Table 12–12). Most are used against breast and prostate carcinomas. The two most widely used are tamoxifen, and anastrozole for breast cancer, and adrenocorticosteroids (such as dexamethasone and prednisone) for lymphomas and breast cancer; both can be neurotoxic. The other

Table 12–12. **Antineoplastic Hormones**

Type	Example
Anti-estrogens	
Selective estrogen receptor modulator	Tamoxifen, raloxifene (Evista)
Aromatase inhibitors	Anastrozole (Arimidex)
Progestin analogs	Medroxyprogesterone (Megace)
Anti-androgens	
Non-steroidal	Bicalutamide (Casodex)
LHRH analog or antagonist	Goserelin (Zoladex)
	Leuprolide (Lupron)
Somatostatin analogs	Octreotide

LHRH, luteinizing hormone-releasing hormone.

hormonal agents have expected endocrine side effects, but neurotoxicity is not a major problem. Corticosteroid side effects are discussed in Chapter 4.

ANTI-ESTROGENS

Tamoxifen appears to cross the blood–brain barrier to concentrate both in the brain and brain metastases,[635] although levels in CSF are low. Although quite uncommon, the most common neurologic side effect is reversible, decreased visual acuity with macular edema.[636,637] Other uncommon side effects include leg cramps, priapism,[638] acute reversible encephalopathy characterized by delusions and hallucinations,[639] depression, irritability, insomnia, headache, poor concentration,[640] cerebellar dysfunction,[641] and radiation recall syndrome.[642] The drug increases the incidence of thromboembolic disease, including stroke.[643,644] Other anti-estrogens include raloxifene[645] and toremifene.[646] Their side effects are similar although the incidence is perhaps less. Aromatase inhibitors are rapidly replacing selective estrogen receptor modulators because of their lesser toxicity and greater efficacy.[647] Although most side effects are less frequent with these drugs, arthralgias[648,649] and paresthesias, including carpal tunnel syndrome, are more frequent.[647]

Medroxyprogesterone acetate (Megace), a progestin analog, has been used to treat advanced breast[650] and endometrial[651] cancer, but now has its primary role in increasing appetite in patients with cancer cachexia (see Chapter 4). The drug has no significant neurotoxicity. It can cause paresthesias, tremor, and muscle cramps and is associated with an increased incidence of thrombophlebitis.[650] The drug lowers bone density.[652] Galactorrhea has been reported[653] as has Addison disease associated with clinical features of Cushing syndrome.[654] The Cushingoid appearance, noted in several patients with adrenal insufficiency, was probably a direct result of the Megace.[654]

ANTI-ANDROGENS

Non-steroidal anti-androgens are used to treat prostate cancer.[655] Bicalutamide (Casodex) is the most widely used. Anti-androgens, including flutamide and nilutamide, cause gynecomastia, chest pain, and hot flashes. These symptoms are generally mild and the drugs help retain libido and lean muscle mass. Nilutamide uniquely causes delayed visual dark adaptation.[655]

Luteinizing hormone-releasing hormone analogs such as goserelin (Zoladex) and leuprolide (Lupron) suppress estrogen and testosterone; they are mostly used to treat prostate cancer.[656,657] Hot flashes, loss of libido, gynecomastia, and decreased bone marrow density[658] are the major side effects.[656] Because there is initial increase in hormone output before suppression, a tumor flare is possible[659] and most serious in that it may increase symptoms of spinal cord compression. Rarer complications include depression[660] and, in one patient, a pain syndrome associated with muscle weakness.[661]

Octreotide, a somatostatin analog, has been used to treat neuroendocrine tumors,[662] including those causing acromegaly.[663] It has also been effective in treating diarrhea induced by chemotherapy.[664] Its toxicity is generally mild and includes fatigue, headache, myalgias, and feelings of weakness.[664,665] Psychological changes including manic episodes have been described in a few patients.[666]

Biologic Agents

Several biologic approaches may be used to manipulate the immune system to destroy neoplastic cells (Table 12–13). Antibodies, and either peripheral-blood or tumor-infiltrating lymphocytes (TIL) from the patient enhanced in vitro (e.g., lymphokine-activated killer [LAK] cells, dendritic cells), directly attack the tumor. Tumor vaccines, cytokines, or nonspecific immune-enhancing agents such as levamisole strengthen the patient's intrinsic immunity. Conversely, immunosuppressive drugs may be used to prevent the graft-versus-host reaction in patients undergoing bone marrow transplantation. Also available are agents that wage a biologic attack on the tumor by causing differentiation of tumor cells or by inhibiting tumor angiogenesis.

A few of these agents, to be discussed shortly, cause notable nervous system toxicity (Table 12–14).

Table 12–13. **Antineoplastic Biologics**

Type	Example(s)
Antibodies	Bevacizumab (Avastin)
	Rituximab
Adaptive cellular therapy	Dendritic cells
Vaccines	
Cytokines	Interleukin 1,2
	Interferons
	Tumor necrosis factor
Immune enhancers	Levamisole
Growth factors	Erythropoietin

ANTIBODIES

Table 12–14 lists some monoclonal antibodies used to treat various cancers. Almost all have been described to cause occasional headache, fatigue, and paresthesias, but most have no significant neurotoxicity. Most of them rarely cause serious allergic infusion reactions, including a systemic inflammatory response syndrome[667]; those directed against lymphocytes can lead to recrudescence of infections, including herpes zoster and cytomegalovirus.[668] Because they do not cross the blood–brain barrier, they are not effective against tumors within the CNS that have an intact blood–brain barrier. In fact, trastuzumab may increase the risk of isolated CNS metastases from breast cancer (see Chapter 5).[669,670]

Bevacizumab (Avastin), a monoclonal antibody against VEGF, is proving useful in a number of cancers,[671] including brain tumors. The drug does not appear to have direct neurotoxicity, but does cause hypertension and has been reported to cause posterior reversible leukoencephalopathy.[672,673] The drug may also promote hemorrhage,[674] but brain hemorrhage does not appear to be a major problem in those patients being treated for brain tumors.

Rituximab (Rituxan) is used to treat lymphoid malignancies and peripheral neuropathy associated with monoclonal gammopathy,[675] rheumatoid arthritis,[676] and a variety of other immune-mediated disorders. The drug appears

Table 12–14. **Some Monoclonal Antibodies**

Antibody	Trade Name	Target Antigen	Antibody Type	References
Rituximab	Rituxan	CD20	Chimeric	812
[131]I-tositumomab	Bexxar	CD20	Murine radiolabeled	813
Alemtuzumab	Campath	CD52	Humanized	666, 814
Gemtuzumab ozogamicin	Mylotarg	CD33	Humanized with toxin	815
Bevacizumab	Avastin	VEGF	Humanized	673
Cetuximab	Erbitux	EGFR	Humanized	816
Panitumumab		EGFR	Humanized	681
Trastuzumab	Herceptin	HER2	Humanized	682
Pertuzumab		HER2	Humanized	684
3F8		GD2	Murine	681
Muronomab	Orthoclone (OKT3)	CD3	Murine	687
81C6		Tenascin	Murine	692

quite safe although its long-term effects are not yet established. Gemtuzumab can cause a veno-occlusive syndrome with liver failure,[677] in addition to bleeding, including fatal intracranial hemorrhage[678] and infection. Cetuximab (Erbitux), in addition to hypersensitivity reactions and rash, can cause occasional ocular toxicity characterized by blepharitis and conjunctivitis[679]; hypomagnesemia has also been reported probably because there is strong expression of epidermal growth factor receptor in the loop of Henle.[680] Panitumumab, the first fully human monoclonal antibody, is similar to cetuximab in that it blocks epidermal growth factor receptor[681]; the toxicity, including hypomagnesemia, is also similar. Trastuzumab (Herceptin) can be cardiotoxic[682]; the drug has been given intrathecally for leptomeningeal metastases from breast cancer with safety but no established efficacy.[683] Pertuzumab inhibits the dimerization of HER2 and, like trastuzumab, might be effective against breast and ovarian cancer.[684] The drug does not appear to be neurotoxic, but causes fatigue and rash as well as peripheral edema and sometimes abdominal pain.[684]

An antiganglioside monoclonal antibody raised against neuroblastoma, 3F8, reacts with dorsal root ganglion cells to cause severe pain, often necessitating a continuous IV infusion of morphine. No permanent neurotoxicity has been identified.[685] The antibody has been given intrathecally.[686]

OTK3, an anti–T-cell monoclonal antibody, is sometimes used to suppress the graft-versus-host reaction.[687] This antibody occasionally causes aseptic meningitis,[688] cerebritis with seizures,[689] akinetic mutism,[690] and rarely, at high doses, cerebral infarction.[691]

An I-131 murine antibody against tenascin has been injected into the surgical resection cavity of patients with primary and metastatic brain tumors.[692] Neurotoxicity, characterized by headache, aphasia, and motor weakness was usually reversible.[692]

CELLULAR THERAPY

Cellular therapy, including the injection of natural killer cells, LAK cells, TIL, and dendritic cells, either alone or genetically modified, has little or no neurotoxicity. These cells do occasionally cause headache and fatigue, probably related to cytokine release. However, interleukin 2 (IL-2) used to activate the lymphocytes is neurotoxic (see following text). Cryopreserved stem cells used for bone marrow rescue have been reported to cause encephalopathy, usually reversible, and cerebral infarction.[693,694] The MRI may show hyperintense lesions in cerebral cortex and white matter sometimes with contrast enhancement.[693] The toxicity is probably due to the DMSO preservative.

VACCINES

Vaccines,[695] including those that act against glioma cells,[696] have been injected into patients for many years. At one time, physicians feared that contaminating myelin might cause an immune-mediated encephalomyelitis. The vaccines are not neurotoxic[696] but, unfortunately, are also not very effective for most tumors.

Table 12–15 lists the neurotoxic effects of some immunotherapeutic agents.

CYTOKINES

Cytokines used clinically include IL-2, IL-1, tumor necrosis factor, and the interferons.[697,698] Colony-stimulating factors such as granulocyte colony-stimulating factor (G-CSF) and granulocyte macrophage colony-stimulating factor (GM-CSF) counteract the marrow-suppressive effects of anticancer drugs. Most cross the blood–brain barrier poorly and have little neurotoxicity.

Interleukins

Although IL-2 can cross the blood–brain barrier[699,700] and can be administered safely intrathecally, it causes capillaries, including those in the brain, to leak.[701] After giving patients IL-2, increases in brain water content were inferred from MRI.[702] IL-2, particularly at high doses,[702] can also cause a severe encephalopathy,[702,704] which is sometimes fatal,[705] but the patients often are so systemically ill that the encephalopathy is not noticed. It is usually reversible if the patient recovers from the acute systemic toxicity.[706] Transient focal neurologic signs, such as visual loss, ataxia, hemiparesis, and seizures have also been reported,[706] as have occasional transient hypothyroidism,[707] myalgias, and headache.[708] Acute brachial plexopathy is a rare effect.[709] A patient given intraventricular IL-2 for leptomeningeal disease developed progressive and cognitive dysfunction beginning 3 months after the treatment irreversible[710]; white matter abnormalities were seen on MRI. The neurotoxicity of IL-2 can be

Table 12–15. **Neurotoxicity of Some Immunotherapeutic Agents**

	Acute (<1 week)	Early-Delayed (1 to 4 weeks)	Late-Delayed (>1 month)
OTK3 antibody	Aseptic meningitis Seizures Cerebral infarction		
Interleukin 2	Cerebral edema Encephalopathy (frequent) Leukoencephalopathy (rare) Nerve compression (during vascular syndrome) Brachial plexopathy	Transient focal deficits	
Interferon α	Acute paresthesias Loss of taste/smell Encephalopathy Leukoencephalopathy (intrathecal)	Brachial plexopathy Encephalopathy Headache	Encephalopathy/dementia Parkinsonism Sensory and sensorimotor polyneuropathy Third-nerve palsy Optic neuropathy
Levamisole	Aseptic meningitis		Leukoencephalopathy "MS-like" with 5-FU
Cyclosporin	Seizures Burning feet and hands during infusion	Seizures Tremor Cortical blindness Myelopathy PRES	Myopathy Sensorimotor polyneuropathy

5-FU, 5-fluorouracil; MS, multiple sclerosis PRES, posterior reversible encephalopathy syndrome.

minimized by scrupulous attention to the supportive care.[703] In experimental animals, histamine protects against the pulmonary vascular leak syndrome.[711] A low-toxicity IL-2 devoid of vaso-permeability activity has been developed.[712] High-dose IL-2 therapy in patients with brain metastases causes no increased neurotoxicity when compared to those without brain lesions.[713]

IL-1 can cause encephalopathy and, infrequently, seizures.[714] IL-4 causes fatigue, anorexia, headache, and myalgia in some patients,[715] and occasionally, hyponatremia. The other interleukins (interleukin 11, interleukin 12, tumor necrosis factor) can cause headache and fatigue; some cause vascular leak but to a lesser degree than interleukin 2. Neurotoxicity is not a major problem.

Interferons

The acute toxicity of interferons includes headaches, myalgias, and fatigability.[716,717] Chronic toxicity includes encephalopathy, which may be irreversible.[718,719] Milder changes include nightmares, headache, myalgias, fatigue, and tremor.[716] Bilateral brachial plexopathy[720] oculomotor palsies,[721] hearing loss,[722] and endocrine disorders including transient thyrotoxicosis followed by hypothyroidism have been reported.[717] Meningitis and blood–brain barrier disruption have been observed in rats.[723] One of our patients with metastatic melanoma, in remission on interferon, presented for neurologic consultation complaining of burning dysesthesias in both hands.[724] Examination suggested bilateral compression of the median nerves in the carpal tunnel. Thyroid-stimulating hormone (TSH) was grossly elevated.[725] His symptoms responded to thyroid replacement.

IMMUNE ENHANCERS

Nonspecific agents that enhance natural immunity are sometimes used as adjuvants to chemotherapy. The most common of these is levamisole, an antihelminthic agent that enhances immunity.[726] Neurotoxicity is dose limiting.[726] Levamisole can cause aseptic meningitis with headaches, stiff neck, fever, and CSF lymphocytic pleocytosis, as well as vertigo[466] and seizures. It sometimes alters taste and causes a hyperalert state with anxiety and irritability. When combined with 5-FU, or even when given alone,[727] it can cause demyelination of the brain[728] characterized by multifocal neurologic symptoms associated with multiple demyelinating lesions of the white matter resembling multiple sclerosis.[729,730] The lesions may contrast-enhance on computed tomography (CT) or MRI, but are usually cold on thallium SPECT scanning[731] (Fig. 12–6). The lesions may be mistaken for brain metastases, leading to incorrect and probably deleterious treatment with RT.[732]

IMMUNOSUPPRESSIVE AGENTS

Some antineoplastic agents, such as cyclophosphamide and corticosteroids, are also immuno-suppressive. Other agents, including cyclosporin, tacrolimus, sirolimus, and IV gamma globulin, are sometimes prescribed to prevent graft rejection after transplantation. Cyclosporine and tacrolimus, calcineurin inhibitors, can be neurotoxic. Both cause reversible posterior leukoencephalopathy[733] (see p 451). Ence-phalopathy,[734] neuropathic pain,[735] cerebral hemorrhage,[736] and a reversible demyelinating polyneuropathy have also been reported.[737] The drugs are also associated with thrombotic microangiopathy.[738] This disorder is discussed in Chapter 9. Sirolimus (Rapamycin),[739] not a calcineurin inhibitor, and mycophenolate mofetil (CellCept)[740] do not produce these complications after transplantation. Rapamycin has not only immunosuppressive properties, but also inhibits a protein kinase called *mammalian target of rapamycin* (mTOR), interfering with cell cycle progression, cell growth, and proliferation in malignant cells from a variety of solid tumors. Two derivatives of rapamycin, temsirolimus, and everolimus,[741,742] are being tested against leukemia and solid tumors. Unlike tacrolimus, the drugs do not appear to be neurotoxic, although they have been reported to produce a rash, hyperglycemia, and hypophosphatemia.

Intravenous immunoglobulin

Intravenous immunoglobulin (IVIg) causes minor symptoms such as headache and malaise in a significant number of patients,[743] but is generally non-neurotoxic. Rarely, IVIg can cause aseptic meningitis[744] and, more rarely, cerebral infarction[745] or cranial pachymeningitis.[746] The pathogenesis of cerebral infarction is unknown, but may be related to hyperviscosity caused by the high protein load.

GROWTH FACTORS

Table 12–16 lists some hematopoietic growth factors.

Erythropoietin

Human recombinant erythropoietin is used to treat the anemia of cancer and its chemotherapy as well as that of chronic renal failure.[747–749] When used with recommended doses and time schedules, it has no significant neurotoxicity. However, both hypertension leading to the posterior leukoencephalopathy syndrome[750] and venous thromboses including cerebral venous thromboses[751] occasionally have been reported with the drug, especially when using unconventional doses or treatment schedules. Minor symptoms sometimes include fatigue, dizziness, and paresthesias. Pain at the site of injection is common, but mild; bone pain is less common.[748] One case report describes both acute gout and a vasculitis-like syndrome responsive to steroids.[752] The drug appears to reduce axonal and myelin degeneration as well as pain-related behavior in peripheral nerve injury in experimental animals; the mechanism appears to be downregulation of TNF-α by erythropoietin.[753]

Granulocyte-Stimulating Factors

G-CSF and GM-CSF are used to prevent or reverse chemotherapy-induced neutropenia[754] or to mobilize peripheral blood cells for use in stem cell transplant. Significant neurotoxicity is uncommon but headache and bone, muscle, and joint pain are relatively common.[754] The hand-foot syndrome has been described in a rare patient who developed a reversible sensory neuropathy.[754] Encephalopathy[755] and the posterior reversible leukoencephalopathy syndrome occur rarely.[756] Whether the drugs cause genetic alterations in stem cell donors is controversial.[757,758] A case of cerebral arterial thrombosis in a patient receiving G-CSF along with multiagent chemotherapy has been reported.[759] One patient developed a fatal

Table 12–16. **Hematopoietic Growth Factors**

Cytokine	Generic Names	Brand Names	Endogenous Sources
Granulocyte colony-stimulating factor	Filgrastim Lenograstim	Neupogen (Amgen) Granocyte (Chugai)	Monocytes/macrophages, fibroblasts, endothelial cells, keratinocytes
Granulocyte-macrophage colony-stimulating factor	Sargramostim Molgramostim Regramostim	Leukine (Immunex) Leucomax (Schering)	T-lymphocytes, monocytes/macrophages, fibroblasts, endothelial cells, osteoblasts, epithelial cells
Erythropoietin	Epoetin α Epoetin α Epoetin β Darbepoeitin α	Epogen (Amgen) Procrit/Eprex (Ortho) NeoRecormon (Roche) Aranesp (Amgen)	Renal cells, hepatocytes
Interleukin 11	Oprelvekin	Neumega (Genetics Institute)	Stromal fibroblasts, trophoblasts
Stem cell factor, steel factor, mast cell growth factor, c-kit ligand	Ancestim	Stemgen (Amgen)	Endothelial cells, fibroblasts circulating mononuclear cells, bone marrow stromal cells
Thrombopoietin, megakaryocyte growth and development factor			Liver, kidney

leukoencephalopathy after receiving G-CSF and IVIg.[760]

Interleukin 11

This drug is used to increase platelets in patients with chemotherapy-induced thrombocytopenia. Its toxicity includes headache, arthralgias, myalgias, and fatigue.[761,762] Papilledema that resolved after discontinuation of the drug has been reported.[762,763]

Keratinocyte Growth Factor

A number of agents are being used—or are under investigation—to prevent and treat oral and GI mucositis.[764] Palifermin, a recombinant keratinocyte growth factor, has proved effective in the treatment of oral mucositis[765] and in experimental animals has decreased chemotherapy-induced diarrhea.[766] Toxicity is minor and includes rash and erythema, as well as paresthesias, mouth and tongue discomfort, and alterations of taste.[765]

SMALL MOLECULES

Table 12–17 lists some small molecules used to treat cancer.

Several small molecule inhibitors interfere with signal transduction pathways and control the growth of various cancers.

Imatinib (Gleevec): This drug targets BCR-ABL kinases and c-kit. It treats chronic myelogenous leukemia and gastrointestinal stromal tumors (GIST) and is being tested against a variety of other cancers.[767] Rash,[768] hypophosphatemia,[769] peripheral anterior orbital edema, headache, fatigue, myalgias, and muscle cramps are common but usually mild.[767] Muscle cramps can be controlled with benzodiazepines.[770] Cerebral infarction and hemorrhage are probably the result of the underlying illness rather than the drug.[771]

Erlotinib (Tarceva): This drug is an inhibitor of epidermal growth factor receptor tyrosine kinase, and targets lung cancer as well as other solid tumors expressing specific mutations in the receptor. Diarrhea and rash are the common complications.[772,773] The drug causes eyelashes to grow that may lead to corneal irritation. Another epidermal growth factor receptor inhibitor, gefitinib *(Iressa)* has similar toxicity.[774]

Sunitinib (Sutent): This drug inhibits several receptor tyrosine kinases and has both anti-angiogenic and antitumor

Table 12–17. **Small Molecules**

Agent	Trade Name	Target	Indication	Reference
Gefitinib	Iressa	EGFR	NSCLC	772
Imatinib	Gleevec	BCR-ABL, cKIT	CML, GIST	767
Erlotinib	Tarceva	EGFR	NSCLC	817
Sunitinib	Sutent	Multi-kinase	GI, Renal	776
Sorafenib	Neaxavar	Multi-kinase	Renal	778
Lapatinib	Tykerb	HER2	Breast	781
Tipifarnib	Zarnestra	Farnesyltransferase	Several	782
Bortezomib	Velcade	Proteasome	Several	785

properties.[775,776] Toxicity includes the hand-foot syndrome, glossodynia, fatigue and anorexia.[776] Hypothyroidism, often subclinical, but rarely causing coma,[776a,776b] is a relatively frequent complication.[777]

Sorafenib (Nexavar): The drug is a multi-kinase inhibitor that targets renal cell carcinoma.[778,779] Its toxicity includes rash, hypertension, fatigue, and the hand-foot syndrome, perhaps with an added, nonprogressive sensory neuropathy.[778]

Lapatinib: This drug is a tyrosine kinase inhibitor of HER-2/neu and is used in the treatment of breast cancer.[780] It does not appear to have significant neurotoxicity.[780,781]

Tipifarnib (Zarnestra): This drug is an inhibitor of the enzyme farnesyltransferase, an enzyme necessary for the activation of the *Ras* oncogene. It is used to treat patients with both leukemia and solid tumors.[782,783] Toxicity includes nausea, vomiting, rash, and fatigue; a few patients developed a reversible sensory neuropathy.[782,783]

Bortezomib (Velcade): The proteasome is responsible for the breakdown of intracellular proteins, including those that regulate the cell cycle and prevent apoptosis. Accordingly, the proteasome might also control the growth of some cancers.[784] Bortezomib, a proteasome inhibitor, has been used to treat myeloma,[785] solid tumors,[786] and some lymphomas.[787] Peripheral and CNS as well as muscle toxicity have been reported. Peripheral neuropathy, primarily sensory, appears to be relatively common.[787,788,788a] CNS toxicity is less common, but includes dizziness, confusion, and aphasia.[788] Hyponatremia has also been reported; this could cause CNS symptoms.[788] Pain, not well characterized but perhaps neuropathic or myopathic, is also relatively common.[785]

Rhabdomyolysis has been described in one patient.[789] Because the drug is relatively new, neurotoxicity has not been well characterized.

Differentiation Agents

Retinoic acid, hexamethylene bisacetamide, phenylacetate, and other agents promote the differentiation of tumor cells into their more normal counterparts.

All-*trans*-retinoic acid (ATRA) is the drug of choice for acute promyelocytic leukemia. The most serious side effect is the retinoic acid syndrome characterized by fever, pulmonary edema, pleural and pericardial effusion, bone pain, headache, and renal failure.[790] The disorder is not restricted to the use of this drug; a similar syndrome can be caused by arsenic trioxide used to treat relapsed acute promyelocytic leukemia.[791] The neurotoxicity of ATRA is similar to that of vitamin A neurotoxicity, both causing the syndrome of pseudotumor cerebri,[792] characterized by headache, elevated CSF pressure, and papilledema. A few case, reports describe abducens nerve palsies without elevated intracranial pressure.[793,794] One patient also developed other mononeuropathies.[794] In addition, a few cases of myositis with painful swollen muscles and, in one case, cardiac myositis, have been reported.[795] Arsenic trioxide causes peripheral neuropathy in a significant percentage of patients[791]; one report describes Wernicke encephalopathy associated with the drug.[796]

Other differentiating agents, including histone deacetylase inhibitors such as vorinostat,[797,798] appear to lack significant neurotoxicity. The toxicity of valproic acid, also a histone deacetylase inhibitor, is discussed in Chapter 4.

Hematopoietic Stem Cell Transplantation (HSCT)

Most drugs that effectively treat cancer cause dose-limiting bone marrow suppression. Replacing the bone marrow or stem cells either with the patient's own bone marrow/stem cells that have been taken and stored before high-dose chemotherapy (autologous), or with marrow/stem cells from an HLA identical or closely matched donor (allogeneic), effectively treats some leukemias and lymphomas. The procedure has four phases[799]: (1) Cells are harvested from either the patient or the donor. (2) The patient's marrow is destroyed by high-dose chemotherapy with or without total body irradiation. (3) The period before engraftment is marked by severe pancytopenia. (4) The marrow or stem cells reconstitute the patient's bone marrow.

During each of these phases neurotoxicity may occur.[739,799–802] The first phase is generally without neurologic complications. When the procedure involved marrow harvested from donors via iliac crest biopsy, usually under general anesthesia, one of our rather obese donors awakened and developed severe headache and vomiting on getting out of bed. She recovered spontaneously, but undoubtedly suffered a CSF leak from inadvertent puncture of the subarachnoid space.[803] When colony-stimulating factors are used to increase stem cell yield, underlying autoimmune diseases can be exacerbated,[804] but this is a rare phenomenon that can probably be prevented by using cyclophosphamide along with a colony-stimulating factor.[804] In the second and third phases, neurotoxicity can be caused by side effects of chemotherapeutic agents discussed elsewhere in this chapter, by opportunistic infections discussed in Chapter 10, by vascular complications discussed in Chapter 9, and by metabolic abnormalities discussed in Chapter 11. Sometimes it is impossible to identify one particular cause of neurologic signs; more commonly, the neurotoxicity is multifactorial. Metabolic encephalopathy and seizures are common acutely after stem cell transplant. The seizure is usually a single generalized convulsion and, although most patients are thrombocytopenic at the time, is almost never related to CT-detectable bleeding in the brain. The seizure usually does not recur even if the patient does not receive anticonvulsant drugs, and it does not appear to create a long-term risk for the development of epilepsy. The pathogenesis of the seizure, or the metabolic encephalopathy often accompanying it, is not generally identifiable, but the symptoms are usually reversible. The MRI is usually normal, but focal abnormalities may result from hemorrhage, thrombosis, or posterior leukoencephalopathy. Some of these changes are not reversible.

After the marrow reconstitutes, patients may remain on immunosuppressive therapy and suffer opportunistic infections (see Chapter 10). Graft-versus-host or graft-versus-tumor disease occurring after allogeneic bone marrow transplantation is generally not associated with neurotoxicity, but a polymyositis,[805] demyelinating myelopathy (Fig 12–8),[806] cerebral angiitis,[807] subacute panencephalitis,[64] and microangiopathy[808] have all been reported in a few patients.

Adjuvant Agents

Some patients with cancer receive drugs that do not have antineoplastic effects.[94] Instead, they (1) enhance the effects of antineoplastic agents, (2) protect against toxicity,[11] or (3) treat symptoms associated with the cancer or its treatment; in other words, they provide supportive care (see Chapter 4) (Table 12–18).

A sulfhydryl chemo-radioprotector, *amifostine*, penetrates the blood–brain barrier poorly, but does not appear to protect brain or spinal cord, as it protects other normal tissues, from the damaging effects of RT. It also shields most normal tissues against the cytotoxicity of cisplatin, cyclophosphamide, and some mustards.[94,809] The drug generally causes mild somnolence, possible hypotension during infusion, and sometimes hypocalcemia that is not usually symptomatic. The pathophysiology of the somnolence is not understood.

Allopurinol has both enhancing and protecting aspects. It prolongs the half-life of 6-mercaptopurine by decreasing the rate of metabolic elimination, thus increasing the cytotoxicity of that drug. It also prevents formation of toxic 5-FU metabolites, however, decreasing the toxicity of that drug. It does not have significant neurotoxicity. Other non-neurotoxic protectors include *mesna*,[94] *leucovorin*, and *uridine*. In experimental animals, CpG oligodeoxynucleotides enhance chemotherapeutic responses.[810] The drug is apparently not neurotoxic.

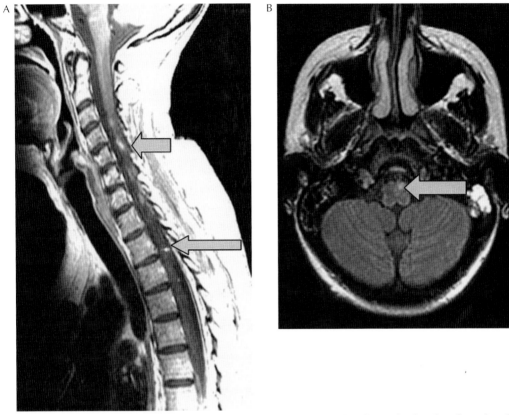

Figure 12–8. A rare complication of bone marrow transplant. A patient with leukemia developed weakness largely in the lower extremities and some paresthesias following the transplant. **A**: An MRI of the spine showed multiple areas of enhancement suggesting acute demyelination (*arrows*). **B**: An MRI of the brain was essentially normal save for one area of hyperintensity in the brainstem (*arrow*), probably also demyelination that was asymptomatic. The patient made a gradual but complete recovery.

Table 12–18. **Protectors and Rescue Agents**

Amifostine
Allopurinol
Mesna
Leucovorin
Uridine
Bisphosphonates

Zoledronic acid (Zometa) and other bis-phosphonates were used to treat hypercalcemia and bone metastases. These drugs are generally not neurotoxic, but can cause an acute systemic inflammatory reaction, scleritis, hypocalcemia, hypophosphatemia, hypomagnesemia, and occasionally hypermagnesemia.[811] An uncommon complication,

osteonecrosis of the jaw, presents with pain and sometimes trismus that might be mistaken for trigeminal neuropathy.

REFERENCES

1. Ernst E, Cassileth BR. The prevalence of complementary/alternative medicine in cancer: a systematic review. *Cancer* 1998;83:777–782.
2. Epstein JB, Phillips N, Parry J, et al. Quality of life, taste, olfactory and oral function following high-dose chemotherapy and allogeneic hematopoietic cell transplantation. *Bone Marrow Transplant* 2002;30:785–792.
3. Comeau TB, Epstein JB, Migas C. Taste and smell dysfunction in patients receiving chemotherapy: a review of current knowledge. *Support Care Cancer* 2001;9:575–580.
4. Sul JK, DeAngelis LM. Neurologic complications of cancer chemotherapy. *Semin Oncol* 2006;33:324–332.
5. Perry A, Schmidt RE. Cancer therapy-associated CNS neuropathology: an update and review of the literature. *Acta Neuropathol (Berl)* 2006;111:197–212.

6. Dropcho EJ. Neurotoxicity of cancer chemotherapy. *Semin Neurol* 2004;24:419–426.

7. Cavaliere R, Schiff D. Neurologic toxicities of cancer therapies. *Curr Neurol Neurosci Rep* 2006;6:218–226.

8. Cavaletti G, Marmiroli P. Chemotherapy-induced peripheral neurotoxicity. *Expert Opin Drug Saf* 2004;3:535–546.

9. Ocean AJ, Vahdat LT. Chemotherapy-induced peripheral neuropathy: pathogenesis and emerging therapies. *Support Care Cancer* 2004;12:619–625.

10. Schiff D, Wen P. Central nervous system toxicity from cancer therapies. *Hematol Oncol Clin North Am* 2006;20:1377–1398.

11. Walker M, Ni O. Neuroprotection during chemotherapy: A systematic review. *Am J Clin Oncol* 2007; 30:82–92.

12. Vecht CJ, Wagner GL, Wilms EB. Interactions between antiepileptic drugs and chemotherapeutic agents. *Lancet Neurol* 2003;2:404–409.

13. Kamataki T, Yokoi T, Fujita K, et al. Preclinical approach for identifying drug interactions. *Cancer Chemother Pharmacol* 1998;42 (Suppl):S50–S53.

14. McGuire SA, Gospe SM, Jr, Dahl G. Acute vincristine neurotoxicity in the presence of hereditary motor and sensory neuropathy type I. *Med Pediatr Oncol* 1989;17:520–523.

15. Quasthoff S, Hartung HP. Chemotherapy-induced peripheral neuropathy. *J Neurol* 2002;249:9–17.

16. Bachmann T, Koetter KP, Muhler J, et al. Guillain-Barre syndrome after simultaneous therapy with suramin and interferon-alpha. *Eur J Neurol* 2003;10:599.

17. Chaudhry V, Eisenberger MA, Sinibaldi VJ, et al. A prospective study of suramin-induced peripheral neuropathy. *Brain* 1996;119:2039–2052.

18. Russell JW, Gill JS, Sorenson EJ, et al. Suramin-induced neuropathy in an animal model. *J Neurol Sci* 2001;192:71–80.

19. Gill JS, Windebank AJ. Suramin induced ceramide accumulation leads to apoptotic cell death in dorsal root ganglion neurons. *Cell Death Differ* 1998;5:876–883.

20. Forman A. Peripheral neuropathy in cancer patients: clinical types, etiology, and presentation. Part 2. *Oncology* 1990;4:85–89.

21. Verstappen CCP, Koeppen S, Heimans JJ, et al. Dose-related vincristine-induced peripheral neuropathy with unexpected off-therapy worsening. *Neurology* 2005;64(6):1076–1077.

22. Kuntzer T, Antoine JC, Steck AJ. Clinical features and pathophysiological basis of sensory neuronopathies (ganglionopathies). *Muscle Nerve* 2004;30:255–268.

23. Pasetto LM, D'Andrea MR, Rossi E, et al. Oxaliplatin-related neurotoxicity: how and why? *Crit Rev Oncol Hematol* 2006;59:159–168.

24. Alberts DS, Noel JK. Cisplatin-associated neurotoxicity: can it be prevented? *Anticancer Drugs* 1995;6:369–383.

25. Mielke S, Sparreboom A, Mross K. Peripheral neuropathy: a persisting challenge in paclitaxel-based regimes. *Eur J Cancer* 2006;42:24–30.

26. Markman M. Management of toxicities associated with the administration of taxanes. *Expert Opin Drug Saf* 2003;2:141–146.

27. Levitt LP, Prager D. Mononeuropathy due to vincristine toxicity. *Neurology* 1975;25:894–895.

28. Lossos A, Siegal T. Electric shock-like sensations in 42 cancer patients: clinical characteristics and distinct etiologies. *J Neuro-Oncol* 1996;29:175–181.

29. Kuroi K. Neurotoxicity of taxanes: symptoms and quality of life assessment. *Breast Cancer* 2004;11:92–99.

30. Wen PY, Blanchard KL, Block CC, et al. Development of Lhermitte's sign after bone marrow transplantation. *Cancer* 1992;69:2262–2266.

31. Siegal T, Haim N. Cisplatin-induced peripheral neuropathy. Frequent off-therapy deterioration, demyelinating syndromes, and muscle cramps. *Cancer* 1990;66:1117–1123.

32. Haim N, Barron SA, Robinson E. Muscle cramps associated with vincristine therapy. *Acta Oncol* 1991;30:707–711.

33. Garg RK. Posterior leukoencephalopathy syndrome. *Postgrad Med J* 2001;77:24–28.

34. Kahana A, Rowley HA, Weinstein JM. Cortical blindness: clinical and radiologic findings in reversible posterior leukoencephalopathy syndrome: case report and review of the literature. *Ophthalmology* 2005;112:e7–e11.

35. Tam CS, Galanos J, Seymour JF, et al. Reversible posterior leukoencephalopathy syndrome complicating cytotoxic chemotherapy for hematologic malignancies. *Am J Hematol* 2004;77:72–76.

36. Kaito E, Terae S, Kobayashi R, et al. The role of tumor lysis in reversible posterior leukoencephalopathy syndrome. *Pediatr Radiol* 2005;35:722–727.

37. Ozcan C, Wong SJ, Hari P. Reversible posterior leukoencephalopathy syndrome and bevacizumab. *N Engl J Med* 2006;354:980–982.

38. Vaughan CJ, Delanty N. Hypertensive emergencies. *Lancet* 2000;356:411–417.

39. Zakarija A, Bennett C. Drug-induced thrombotic microangiopathy. *Semin Thromb Hemost* 2005; 31:681–690.

40. Minisini A, Atalay G, Bottomley A, et al. What is the effect of systemic anticancer treatment on cognitive function? *Lancet Oncol* 2004;5:273–282.

41. Nelson CJ. Chemotherapy and cognitive deficits: mechanisms, findings, and potential interventions. *Palliat Support Care* 2007;5:273–280.

42. Wefel JS, Lenzi R, Theriault R, et al. "Chemobrain" in breast carcinoma? *Cancer* 2004;101:466–475.

43. Poppelreuter M, Weis J, Kulz AK, et al. Cognitive dysfunction and subjective complaints of cancer patients. A cross-sectional study in a cancer rehabilitation centre. *Eur J Cancer* 2004;40:43–49.

44. Kayl AE, Meyers CA. Side-effects of chemotherapy and quality of life in ovarian and breast cancer patients. *Curr Opin Obstet Gynecol* 2006;18:24–28.

45. Tannock IF, Ahles TA, Ganz PA, et al. Cognitive impairment associated with chemotherapy for cancer: report of a workshop. *J Clin Oncol* 2004;22:2233–2239.

46. Kayl AE, Wefel JS, Meyers CA. Chemotherapy and cognition: effects, potential mechanisms, and management. *Am J Ther* 2006;13:362–369.

47. Ahles TA, Saykin A. Cognitive effects of standard-dose chemotherapy in patients with cancer. *Cancer Invest* 2001;19:812–820.

48. Saykin AJ, Ahles TA, McDonald BC. Mechanisms of chemotherapy-induced cognitive disorders: neuropsychological, pathophysiological, and neuroimaging perspectives. *Semin Clin Neuropsychiatry* 2003;8:201–216.

49. Okada J, Yoshikawa K, Imazeki K, et al. Change of cerebral glucose metabolism by antineoplastic drug. *Am J Physiol Imaging* 1991;6:162–166.

50. Fulham MJ, Brunetti A, Aloj L, et al. Decreased cerebral glucose metabolism in patients with brain tumors: an effect of corticosteroids. *J Neurosurg* 1995;83:657–664.

51. Johnson BE, Patronas N, Hayes W, et al. Neurologic, computed cranial tomographic, and magnetic resonance imaging abnormalities in patients with small-cell lung cancer: further follow-up of 6- to 13-year survivors. *J Clin Oncol* 1990;8:48–56.

52. Wilson DA, Nitschke R, Bowman ME, et al. Transient white matter changes on MR images in children undergoing chemotherapy for acute lymphocytic leukemia: correlation with neuropsychologic deficiencies. *Radiology* 1991;180:205–209.

53. Rubenstein CL, Varni JW, Katz ER. Cognitive functioning in long-term survivors of childhood leukemia: a prospective analysis. *J Dev Behav Pediatr* 1990;11:301–305.

54. Waber DP, Tarbell NJ, Kahn CM, et al. The relationship of sex and treatment modality to neuropsychologic outcome in childhood acute lymphoblastic leukemia. *J Clin Oncol* 1992;10:810–817.

55. Yanovski JA, Packer RJ, Levine JD, et al. An animal model to detect the neuropsychological toxicity of anticancer agents. *Med Pediatr Oncol* 1989; 17:216–221.

56. Doty RL, Bromley SM. Effects of drugs on olfaction and taste. *Otolaryngol Clin North Am* 2004;37:1229–1254.

56a. Scully C, Bagan JV. Adverse drug reactions in the orofacial region. *Crit Rev Oral Biol Med* 2004; 15;221–239.

57. Yakirevitch A, Talmi YP, Baram Y, et al. Effects of cisplatin on olfactory function in cancer patients. *Br J Cancer* 2005;92:1611–1613.

58. Wyatt AJ, Leonard GD, Sachs DL. Cutaneous reactions to chemotherapy and their management. *Am J Clin Dermatol* 2006;7:45–63.

59. Guillot B, Bessis D, Dereure O. Mucocutaneous side effects of antineoplastic chemotherapy. *Expert Opin Drug Saf* 2004;3:579–587.

60. Alberts DS, Muggia FM, Carmichael J, et al. Efficacy and safety of liposomal anthracyclines in phase I/II clinical trials. *Semin Oncol* 2004;31:53–90.

61. Scheithauer W, Blum J. Coming to grips with hand-foot syndrome. Insights from clinical trials evaluating capecitabine. *Oncology (Williston Park)* 2004;18:1161–1168, 1173.

62. Lin PC, Hsiao LT, Chen PM. Acute polymyositis after donor lymphocyte infusion. *Eur J Haematol* 2005;74:166–168.

63. Arin MJ, Scheid C, Hubel K, et al. Chronic graft-versus-host disease with skin signs suggestive of dermatomyositis. *Clin Exp Dermatol* 2006; 31:141–143.

64. Iwasaki Y, Sako K, Ohara Y, et al. Subacute panencephalitis associated with chronic graft-versus-host disease. *Acta Neuropathol* 1993;85:566–572.

65. Marosi C, Budka H, Grimm G, et al. Fatal encephalitis in a patient with chronic graft-versus-host disease. *Bone Marrow Transplant* 1990;6:53–57.

66. Chabner BA, Longo DLE. *Cancer and Chemotherapy and Biotherapy: Principles and Practice.* 4th ed. Philadelphia: Lippincott, 2006.

67. Gormley PE, Gangji D, Wood JH, et al. Pharmacokinetic study of cerebrospinal fluid penetration of cis-diamminedichloroplatinum (II). *Cancer Chemother Pharmacol* 1981;5:257–260.

68. DeGregorio M, Wilbur B, King O, et al. Peak cerebrospinal fluid platinum levels in a patient with ependymoma: evaluation of two different methods of cisplatin administration. *Cancer Treat Rep* 1986;70:1437–1438.

69. Olivi A, Gilbert M, Duncan KL, et al. Direct delivery of platinum-based antineoplastics to the central nervous system: a toxicity and ultrastructural study. *Cancer Chemother Pharmacol* 1993;31:449–454.

70. Gregg RW, Molepo JM, Monpetit VJ, et al. Cisplatin neurotoxicity: the relationship between dosage, time, and platinum concentration in neurologic tissues, and morphologic evidence of toxicity. *J Clin Oncol* 1992;10:795–803.

71. Balayssac D, Cayre A, Authier N, et al. Involvement of the multidrug resistance transporters in cisplatin-induced neuropathy in rats. Comparison with the chronic constriction injury model and monoarthritic rats. *Eur J Pharmacol* 2006;544:49–57.

72. McDonald ES, Randon KR, Knight A, et al. Cisplatin preferentially binds to DNA in dorsal root ganglion neurons in vitro and in vivo: a potential mechanism for neurotoxicity. *Neurobiol Dis* 2005;18:305–313.

73. Kirchmair R, Walter DH, Ii M, et al. Antiangiogenesis mediates cisplatin-induced peripheral neuropathy: attenuation or reversal by local vascular endothelial growth factor gene therapy without augmenting tumor growth. *Circulation* 2005;111:2662–2670.

74. Hilkens PH, van der Burg ME, Moll JW, et al. Neurotoxicity is not enhanced by increased dose intensities of cisplatin administration. *Eur J Cancer* 1995;31A:678–681.

75. van den Bent MJ, van Putten WLJ, Hilkens PHE, et al. Retreatment with dose-dense weekly cisplatin after previous cisplatin chemotherapy is not complicated by significant neuro-toxicity. *Eur J Cancer [A]* 2002;38:387–391.

76. Castellanos AM, Glass JP, Yung WKA. Regional nerve injury after intra-arterial chemotherapy. *Neurology* 1987;37:834–837.

77. Shimamura Y, Chikama M, Tanimoto T, et al. Optic nerve degeneration caused by supraophthalmic carotid artery infusion with cisplatin and ACNU. Case report. *J Neurosurg* 1990;72:285–288.

78. Boogerd W, ten Bokkel Huinink WW, Dalesio O, et al. Cisplatin induced neuropathy: central, peripheral and autonomic nerve involvement. *J Neuro-oncol* 1990;9:255–263.

79. Cohen SC, Mollman JE. Cisplatin-induced gastric paresis. *J Neuro-oncol* 1987;5:237–240.

80. Rosenfeld CS, Broder LE. Cisplatin-induced autonomic neuropathy. *Cancer Treat Rep* 1984;68:659–660.

81. Argyriou AA, Koutras A, Polychronopoulos P, et al. The impact of paclitaxel or cisplatin-based chemotherapy on sympathetic skin response: a prospective study. *Eur J Neurol* 2005;12:858–861.

82. Horvath P, Szilvassy Z, Peitl B, et al. Changes in tracheo-bronchial sensory neuropeptide receptor gene expression pattern in rats with cisplatin-induced sensory neuropathy. *Neuropeptides* 2006;40:77–83.

83. Riggs JE, Ashraf M, Snyder RD, et al. Prospective nerve conduction studies in cisplatin therapy. *Ann Neurol* 1988;23:92–94.

84. Argyriou AA, Polychronopoulos P, Koutras A, et al. Peripheral neuropathy induced by administration of cisplatin- and paclitaxel-based chemotherapy. Could it be predicted? *Support Care Cancer* 2005;13:647–651.

85. Grunberg SM, Sonka S, Stevenson LL, et al. Progressive paresthesias after cessation of therapy with very high-dose cisplatin. *Cancer Chemother Pharmacol* 1989;25:62–64.

86. Mollman JE, Hogan WM, Glover DJ, et al. Unusual presentation of cis-platinum neuropathy. *Neurology* 1988;38:488–490.

87. van der Hoop RG, van der Burg MEL, ten Bokkel Huinink WW, et al. Incidence of neuropathy in 395 patients with ovarian cancer treated with or without cisplatin. *Cancer* 1990;66:1697–1702.

88. Ongerboer De Visser BW, Tiessens G. Polyneuropathy induced by cisplatin. *Prog Exp Tumor Res* 1985;29:190–196.

89. Thompson SW, Davis LE, Kornfeld M, et al. Cisplatin neuropathy. Clinical, electrophysiologic, morphologic, and toxicologic studies. *Cancer* 1984; 54:1269–1275.

90. Walsh TJ, Clark AW, Parhad IM, et al. Neurotoxic effects of cisplatin therapy. *Arch Neurol* 1982; 39:719–720.

91. Krarup-Hansen A, Rietz B, Krarup C, et al. Histology and platinum content of sensory ganglia and sural nerves in patients treated with cisplatin and carboplatin: an autopsy study. *Neuropathol Appl Neurobiol* 1999;25:29–40.

92. Krarup-Hansen A, Fugleholm K, Helweg-Larsen S, et al. Examination of distal involvement in cisplatin-induced neuropathy in man. An electrophysiological and histological study with particular reference to touch receptor function. *Brain* 1993;116 (Pt 5): 1017–1041.

93. Yamamoto M, Kachi T, Yamada T, et al. Sensory conduction study of cisplatin neuropathy: preservation of small myelinated fibers. *Intern Med* 1997;36:829–833.

94. Schuchter LM, Hensley ML, Meropol NJ, et al. 2002 update of recommendations for the use of chemotherapy and radiotherapy protectants: clinical practice guidelines of the American Society of Clinical Oncology. *J Clin Oncol* 2002;20:2895–2903.

95. Thorstad WL, Chao KS, Haughey B. Toxicity and compliance of subcutaneous amifostine in patients undergoing postoperative intensity-modulated radiation therapy for head and neck cancer. *Semin Oncol* 2004;31:8–12.

96. Rades D, Fehlauer F, Bajrovic A, et al. Serious adverse effects of amifostine during radiotherapy in head and neck cancer patients. *Radiother Oncol* 2004;70:261–264.

97. Pace A, Savarese A, Picardo M, et al. Neuroprotective effect of vitamin E supplementation in patients treated with cisplatin chemotherapy. *J Clin Oncol* 2003;21:927–931.

98. Argyriou AA, Chroni E, Koutras A, et al. Vitamin E for prophylaxis against chemotherapy-induced neuropathy: a randomized controlled trial. *Neurology* 20051;64:26–31.

99. Pisano C, Pratesi G, Laccabue D, et al. Paclitaxel and Cisplatin-induced neurotoxicity: a protective role of acetyl-l-carnitine. *Clin Cancer Res* 2003;9:5756–5767.

100. Ghirardi O, Vertechy M, Vesci L, et al. Chemotherapy-induced allodynia: neuroprotective effect of acetyl-l-carnitine. *In Vivo* 2005;19:631–637.

101. Bianchi G, Vitali G, Caraceni A, et al. Symptomatic and neurophysiological responses of paclitaxel- or cisplatin-induced neuropathy to oral acetyl-l-carnitine. *Eur J Cancer* 2005;41:1746–1750.

102. Maestri A, De Pasquale CA, Cundari S, et al. A pilot study on the effect of acetyl-l-carnitine in paclitaxel- and cisplatin-induced peripheral neuropathy. *Tumori* 2005;91:135–138.

103. van der Hoop RG, Vecht CJ, van der Burg ME, et al. Prevention of cisplatin neurotoxicity with an ACTH (4–9) analogue in patients with ovarian cancer. *N Engl J Med* 1990;322:89–94.

104. Roberts JA, Jenison EL, Kim K, et al. A randomized, multicenter, double-blind, placebo-controlled, dose-finding study of ORG 2766 in the prevention or delay of cisplatin-induced neuropathies in women with ovarian cancer. *Gynecol Oncol* 1997;67:172–177.

105. Walker M. Neuroprotection during chemotherapy: a systematic review. *Am J Clin Oncol* 2007;30:82–92.

106. Bianchi R, Brines M, Lauria G, et al. Protective effect of erythropoietin and its carbamylated derivative in experimental Cisplatin peripheral neurotoxicity. *Clin Cancer Res* 2006;12:2607–2612.

107. Orhan B, Yalcin S, Nurlu G, et al. Erythropoietin against cisplatin-induced peripheral neurotoxicity in rats. *Med Oncol* 2004;21:197–203.

108. Apfel SC, Kessler JA. Neurotrophic factors in the treatment of peripheral neuropathy. *Ciba Found Symp* 1996;196:98–108.

109. Apfel SC. Is the therapeutic application of neurotrophic factors dead? *Ann Neurol* 2002;51:8–11.

110. Jaboin J, Hong A, Kim CJ, et al. Cisplatin-induced cytotoxicity is blocked by brain-derived neurotrophic factor activation of TrkB signal transduction path in neuroblastoma. *Cancer Lett* 2003;193:109–114.

111. Rao RD, Michalak JC, Sloan JA, et al. Efficacy of gabapentin in the management of chemotherapy-induced peripheral neuropathy: A phase 3 randomized, double-blind, placebo-controlled, crossover trial (N00C3). *Cancer* 2007;110:2110–2118.

112. Grothey A. Clinical management of oxaliplatin-associated neurotoxicity. *Clin Colorectal Cancer* 2005;5 (Suppl 1):S38–S46.

113. Frustaci S, Barzan L, Comoretto R, et al. Local neurotoxicity after intra-arterial cisplatin in head and neck cancer. *Cancer Treat Rep* 1987;71:257–259.

114. Alderson LM, Noonan PT, Choi IS, et al. Regional subacute cranial neuropathies following internal carotid cisplatin infusion. *Neurology* 1996;47:1088–1090.

115. Walther PJ, Rossitch E, Jr, Bullard DE. The development of Lhermitte's sign during cisplatin chemotherapy. Possible drug-induced toxicity causing spinal cord demyelination. *Cancer* 1987;60:2170–2172.

116. Ventafridda V, Caraceni A, Martini C, et al. On the significance of Lhermitte's sign in oncology. *J Neurooncol* 1991;10:133–137.

117. Esik O, Csere T, Stefanits K, et al. Increased metabolic activity in the spinal cord of patients with long-standing Lhermitte's sign. *Strahlenther Onkol* 200;179:690–693.

118. Steiner I, Siegal T. Muscle cramps in cancer patients. *Cancer* 1989;63:574–577.

119. Rademaker-Lakhai JM, Crul M, Zuur L, et al. Relationship between cisplatin administration and the development of ototoxicity. *J Clin Oncol* 2006; 24:918–924.

120. Shulman A. The cochleovestibular system/ototoxicity/clinical issues. *Ann N Y Acad Sci* 1999; 884:433–436.

121. Myers SF, Blakley BW, Schwan S. Is cis-platinum vestibulotoxic? *Otolaryngol Head Neck Surg* 1993;108:322–328.

122. Prim MP, de Diego JI, de Sarria MJ, et al. [Vestibular and oculomotor changes in subjects treated with cisplatin]. *Acta Otorrinolaringol Esp* 2001;52:367–370.

123. van Ruijven MW, de Groot JC, Klis SF, et al. The cochlear targets of cisplatin: an electrophysiological and morphological time-sequence study. *Hear Res* 2005;205:241–248.

124. Rybak LP, Whitworth CA. Ototoxicity: therapeutic opportunities. *Drug Discov Today* 2005;10:1313–1321.

124a. Oldenburg J, Kraggerud SM, Brydoy M, et al. Association between long-term neurotoxicities in testicular cancer and polymorphisms in glutathione-s-transferase-P1 and M1, a cross sectional study. *J Transl Med* 2007;27:70.

125. Rybak LP. Neurochemistry of the peripheral and central auditory system after ototoxic drug exposure: implications for tinnitus. *Int Tinnitus J* 2005;11:23–30.

126. Schell MJ, McHaney VA, Green AA, et al. Hearing loss in children and young adults receiving cisplatin with or without prior cranial irradiation. *J Clin Oncol* 1989;7:754–760.

127. Kretschmar CS, Warren MP, Lavally BL, et al. Ototoxicity of preradiation cisplatin for children with central nervous system tumors. *J Clin Oncol* 1990;8:1191–1198.

128. Kalkanis JG, Whitworth C, Rybak LP. Vitamin E reduces cisplatin ototoxicity. *Laryngoscope* 2004; 114:538–542.

129. Weijl NI, Elsendoorn TJ, Lentjes EG, et al. Supplementation with antioxidant micronutrients and chemotherapy-induced toxicity in cancer patients treated with cisplatin-based chemotherapy: a randomised, double-blind, placebo-controlled study. *Eur J Cancer* 2004;40:1713–1723.

130. Sergi B, Ferraresi A, Troiani D, et al. Cisplatin ototoxicity in the guinea pig: vestibular and cochlear damage. *Hear Res* 2003;182:56–64.

131. Nakayama M, Riggs LC, Matz GJ. Quantitative study of vestibulotoxicity induced by gentamicin or cisplatin in the guinea pig. *Laryngoscope* 1996;106:162–167.

132. Black FO, Pesznecker SC, Homer L, et al. Benign paroxysmal positional nystagmus in hospitalized subjects receiving ototoxic medications. *Otol Neurotol* 2004;25:353–358.

133. Black FO, Myers EN, Schramm VL, et al. Cisplatin vestibular ototoxicity: preliminary report. *Laryngoscope* 1982;92:1363–1368.

134. Moroso MJ, Blair RL. A review of cis-platinum ototoxicity. *J Otolaryngol* 1983;12:365–369.

135. Al-Tweigeri T, Nabholtz JM, Mackey JR. Ocular toxicity and cancer chemotherapy—a review. *Cancer* 1996;78:1359–1373.

136. Schmid KE, Kornek GV, Scheithauer W, et al. Update on ocular complications of systemic cancer chemotherapy. *Surv Ophthalmol* 2006;51:19–40.

137. Kupersmith MJ, Frohman LP, Choi IS, et al. Visual system toxicity following intra-arterial chemotherapy. *Neurology* 1988;38:284–289.

138. Wu HM, Lee AG, Lehane DE, et al. Ocular and orbital complications of intraarterial cisplatin. A case report. *J Neuroophthalmol* 1997; 17:195–198.

139. Katz BJ, Ward JH, Digre KB, et al. Persistent severe visual and electroretinographic abnormalities after intravenous Cisplatin therapy. *J Neuroophthalmol* 2003;23:132–135.

140. Wilding G, Caruso R, Lawrence TS, et al. Retinal toxicity after high-dose cisplatin therapy. *J Clin Oncol* 1985;3:1683–1689.

141. Berman IJ, Mann MP. Seizures and transient cortical blindness associated with cis-platinum (II) Diamminedichloride (PDD) therapy in a thirty-year-old man. *Cancer* 1980;45:764–766.

142. Cohen RJ, Cuneo RA, Cruciger MP, et al. Transient left homonymous hemianopsia and encephalopathy following treatment of testicular carcinoma with cisplatinum, vinblastine, and bleomycin. *J Clin Oncol* 1983;1:392–393.

143. Cattaneo MT, Filipazzi V, Piazza E, et al. Transient blindness and seizure associated with cisplatin therapy. *J Cancer Res Clin Oncol* 1988;114:528–530.

144. Verschraegen C, Conrad CA, Hong WK. Subacute encephalopathic toxicity of cisplatin. *Lung Cancer* 1995;13:305–309.

145. Steeghs N, de Jongh FE, Sillevis Smitt PA, et al. Cisplatin-induced encephalopathy and seizures. *Anticancer Drugs* 2003;14:443–446.

146. Newton HB, Page MA, Junck L, et al. Intra-arterial cisplatin for the treatment of malignant gliomas. *J Neuro-oncol* 1989;7:39–45.

147. Cossaart N, SantaCruz KS, Preston D, et al. Fatal chemotherapy-induced encephalopathy following high-dose therapy for metastatic breast cancer: a case report and review of the literature. *Bone Marrow Transplant* 2003;31:57–60.

148. Ito Y, Arahata Y, Goto Y, et al. Cisplatin neurotoxicity presenting as reversible posterior leukoencephalopathy syndrome. *Am J Neuroradiol* 1998; 19:415–417.

149. Walker RW, Cairncross JG, Posner JB. Cerebral herniation in patients receiving cisplatin. *J Neuro-oncol* 1988;6:61–65.

150. Kusuki M, Iguchi H, Nakamura A, et al. The syndr-ome of inappropriate antidiuretic hormone secretion associated with chemotherapy for hypopharyngeal cancer. *Acta Otolaryngol Suppl* 2004; 74–77.

151. Ishii K, Aoki Y, Sasaki M, et al. Syndrome of inappropriate secretion of antidiuretic hormone induced by intraarterial cisplatin chemotherapy. *Gynecol Oncol* 2002;87:150–151.

152. Cao L, Joshi P, Sumoza D. Renal salt-wasting syndrome in a patient with cisplatin-induced hyponatremia: case report. *Am J Clin Oncol* 2002; 25:344–346.

153. Bellin SL, Selim M. Cisplatin-induced hypomagnesemia with seizures: a case report and review of the literature. *Gynecol Oncol* 1988;30:104–113.

154. Sweeney JD, Ziegler P, Pruet C, et al. Hyperzincuria and hypozincemia in patients treated with cisplatin. *Cancer* 1989;63:2093–2095.

155. Pai VB, Nahata MC. Cardiotoxicity of chemotherapeutic agents: incidence, treatment and prevention. *Drug Saf* 2000;22:263–302.

156. Li SH, Chen WH, Tang Y, et al. Incidence of ischemic stroke post-chemotherapy: a retrospective review of 10,963 patients. *Clin Neurol Neurosurg* 2006;108:150–156.

157. Chaudhary UB, Haldas JR. Long-term complications of chemotherapy for germ cell tumours. *Drugs* 2003;63:1565–1577.

158. Gerl A, Clemm C, Wilmanns W. Acute and late vascular complications following chemotherapy for germ cell tumors. *Onkologie* 1993;16:88–92.

159. Icli F, Karaoguz H, Dincol D, et al. Severe vascular toxicity associated with cisplatin-based chemotherapy. *Cancer* 1993;72:587–593.

160. List AF, Kummet TD. Spinal cord toxicity complicating treatment with cisplatin and etoposide. *Am J Clin Oncol* 1990;13:256–258.

161. Just T, Pau HW, Bombor I, et al. Confocal microscopy of the peripheral gustatory system: comparison between healthy subjects and patients suffering from taste disorders during radiochemotherapy. *Laryngoscope* 2005;115:2178–2182.

162. Minakata Y, Yamagata T, Nakanishi H, et al. Severe gustatory disorder caused by cisplatin and etoposide. *Int J Clin Oncol* 2002;7:124–127.

163. Wright DE, Drouin P. Cisplatin-induced myasthenic syndrome. *Clin Pharm* 1982;1:76–78.

164. Riccardi R, Riccardi A, Di Rocco C, et al. Cerebrospinal fluid pharmacokinetics of carboplatin in children with brain tumors. *Cancer Chemother Pharmacol* 1992;30:21–24.

165. Stewart DJ, Belanger G, Grahovac Z, et al. Phase I study of intracarotid administration of carboplatin. *Neurosurgery* 1992;30:512–517.

166. Smits C, Swen SJ, Theo GS, et al. Assessment of hearing in very young children receiving carboplatin for retinoblastoma. *Eur J Cancer* 2006; 42:492–500.

167. Heinzlef O, Lotz JP, Roullet E. Severe neuropathy after high dose carboplatin in three patients receiving multidrug chemotherapy. *J Neurol Neurosurg Psychiatry* 1998;64:667–669.

168. Walker RW, Rosenblum MK, Kempin SJ, et al. Carboplatin-associated thrombotic microangiopathic hemolytic anemia. *Cancer* 1989;64:1017–1020.

169. O'Brien MER, Tonge K, Blake P, et al. Blindness associated with high-dose carboplatin. Letter to the editor. *Lancet* 1992;339:558.

170. Navo M, Kunthur A, Badell ML, et al. Evaluation of the incidence of carboplatin hypersensitivity reactions in cancer patients. *Gynecol Oncol* 2006; 103(2):608–613.

171. Weidmann B, Mulleneisen N, Bojko P, et al. Hypersensitivity reactions to carboplatin. *Cancer* 1994;73:2218–2222.

172. Gamelin E, Gamelin L, Bossi L, et al. Clinical aspects and molecular basis of oxaliplatin neurotoxicity: current management and development of preventive measures. *Semin Oncol* 2002;29:21–33.

173. Pietrangeli A, Leandri M, Terzoli E, et al. Persistence of high-dose oxaliplatin-induced neuropathy at long-term follow-up. *Eur Neurol* 2006;56:13–16.

174. Taieb S, Trillet-Lenoir V, Rambaud L, et al. Lhermitte sign and urinary retention. *Cancer* 2002;94:2434–2440.

175. Adenis A, Mailliez A, Rigot J-M, et al. Recurrent priapism related to oxaliplatin infusion. *J Clin Oncol* 2008;26:1016–1017.

176. Jamieson SM, Liu J, Connor B, et al. Oxaliplatin causes selective atrophy of a subpopulation of dorsal root ganglion neurons without inducing cell loss. *Cancer Chemother Pharmacol* 2005;56:391–399.

177. Wilson RH, Lehky T, Thomas RR, et al. Acute oxaliplatin-induced peripheral nerve hyperexcitability. *J Clin Oncol* 2002;20:1767–1774.

177a. Saif MW, Hashmi S. Successful amelioration of oxaliplatin-induced hyperexcitability syndrome with the antiepileptic pregabalin in a patient with pancreatic cancer. *Cancer Chemother Pharmacol* 2008;61:349–54.

178. Lehky RJ, Leonard GD, Wilson RH, et al. Oxaliplatin-induced neurotoxicity: acute hyperexcitability and chronic neuropathy. *Muscle Nerve* 2004; 29:387–392.

179. Grolleau F, Gamelin L, Boisdron-Celle M, et al. A possible explanation for a neurotoxic effect of the anticancer agent oxaliplatin on neuronal voltage-gated sodium channels. *J Neurophysiol* 2001; 85:2293–2297.

180. Gamelin L. Predictive factors of oxaliplatin neurotoxicity: the involvement of the oxalate outcome pathway. *Clin Cancer Res* 2007;13:6359–6368.

181. Gamelin L, Boisdron-Celle M, Delva R, et al. Prevention of oxaliplatin-related neurotoxicity by calcium and magnesium infusions: a retrospective study of 161 patients receiving oxaliplatin combined with 5-Fluorouracil and leucovorin for advanced colorectal cancer. *Clin Cancer Res* 2004; 10:4055–4061.

182. Cascinu S, Catalano V, Cordella L, et al. Neuroprotective effect of reduced glutathione on oxaliplatin-based chemotherapy in advanced colorectal cancer: a randomized, double-blind, placebo-controlled trial. *J Clin Oncol* 2002;20:3478–3483.

183. Cassidy J, Bjarnason GA, Hickish T. Randomized double-blind placebo controlled phase III study assessing the efficacy of xaliproden in reducing the cumulative peripheral sensory neuropathy induced by oxaliplatin and 5-FU/LV combination. *J Clin Oncol* 2006;24. Abstract.

184. Argyriou AA, Chroni E, Polychronopoulos P, et al. Efficacy of oxcarbazepine for prophylaxis against cumulative oxaliplatin-induced neuropathy. *Neurology* 2006;67:2253–2255.

185. Bianchi R, Gilardini A, Rodriguez-Menendez V, et al. Cisplatin-induced peripheral neuropathy: neuroprotection by erythropoietin without affecting tumour growth. *Eur J Cancer* 2007;43:710–717.

186. Kochi M, Kuratsu J, Mihara Y, et al. Neurotoxicity and pharmacokinetics of intrathecal perfusion of ACNU in dogs. *Cancer Res* 1990;50:3119–3123.

187. Batts ED. O6-benzylguanine and BCNU in multiple myeloma: a phase II trial. *Cancer Chemother Pharmacol* 2007;60:415–421.

188. Lehne G, Lote K. Pulmonary toxicity of cytotoxic and immunosuppressive agents. A review. *Acta Oncol* 1990;29:113–124.

189. Lieberman A, Ruoff M, Estey E, et al. Irreversible pulmonary toxicity after single course of BCNU. *Am J Med Sci* 1980;279:53–56.

190. McKenney SA, Fehir KM. Myelofibrosis following treatment with a nitrosourea for malignant glioma. *Cancer* 1987;58:1426–1427.

191. Kanj SS, Sharara AI, Shpall EJ, et al. Myocardial ischemia associated with high-dose carmustine infusion. *Cancer* 1991;68:1910–1912.

192. Shingleton BJ, Bienfang DC, Albert DM, et al. Ocular toxicity associated with high-dose carmustine. *Arch Ophthalmol* 1982;100:1766–1772.

193. Rosenblum MK, Delattre J-Y, Walker RW, et al. Fatal necrotizing encephalopathy complicating treatment of malignant gliomas with intra-arterial BCNU and irradiation: a pathological study. *J Neuro-oncol* 1989;7:269–281.

194. Wilson WB, Perez GM, Kleinschmidt-DeMasters BK. Sudden onset of blindness in patients treated with oral CCNU and low-dose cranial irradiation. *Cancer* 1987;59:901–907.

195. Mahaley MS, Jr, Whaley RA, Blue M, et al. Central neurotoxicity following intracarotid BCNU chemotherapy for malignant gliomas. *J Neuro-oncol* 1986;3:297–314.

196. Burger PC, Kamenar E, Schold SC, et al. Encephalomyelopathy following high-dose BCNU therapy. *Cancer* 1981;48:1318–1327.

197. Kleinschmidt-DeMasters BK, Geier JM. Pathology of high-dose intra-arterial BCNU. *Surg Neurol* 1989;31:435–443.

198. Nagahiro S, Yamamoto YL, Diksic M, et al. Neurotoxicity after intracarotid 1,3-bis(2-chloroethyl)-1-nitrosourea administration in the rat: hemodynamic changes studied by double-tracer autoradiography. *Neurosurgery* 1991;29:19–25.

199. Westphal M, Ram Z, Riddle V, et al. Gliadel wafer in initial surgery for malignant glioma: long-term follow-up of a multicenter controlled trial. *Acta Neurochir (Wien)* 2006;148:269–275.

200. Brem H, Piantadosi S, Burger PC, et al. Placebo-controlled trial of safety and efficacy of intraoperative controlled delivery by biodegradable polymers of chemotherapy for recurrent gliomas. The Polymer-Brain Tumor Treatment Group. *Lancet* 1995; 345:1008–1012.

201. Subach BR, Witham TF, Kondziolka D, et al. Morbidity and survival after 1,3-bis(2-chloroethyl)-1 nitrosourea wafer implantation for recurrent glioblastoma: a retrospective case matched cohort series. *Neurosurgery* 1999;45:17–22.

202. Weber EL, Goebel EA. Cerebral edema associated with Gliadel wafers: two case studies. *Neuro-oncology* 2005;7:84–89.

203. Lin SH, Kleinberg LR. Carmustine wafers: Localized delivery of chemotherapeutic agents in CNS malignancies. *Exp Rev Anticancer Ther* 2008; 8:343–359.

204. Nicolao P, Giometto B. Neurological toxicity of ifosfamide. *Oncology (Basel)* 2003;65:11–16.

205. David KA, Picus J. Evaluating risk factors for the development of ifosfamide encephalopathy. *Am J Clin Oncol* 2005;28:277–280.

206. Rieger C, Fiegl M, Tischer J, et al. Incidence and severity of ifosfamide-induced encephalopathy. *Anticancer Drugs* 2004;15:347–350.

207. Kilickap S, Cakar M, Onal IK, et al. Nonconvulsive status epilepticus due to ifosfamide. *Ann Pharmacother* 2006;40:332–335.

208. Meyer T, Ludolph AC, Münch C. Ifosfamide encephalopathy presenting with asterixis. *J Neurol Sci* 2002;199:85–88.

209. Watkin SW, Husband DJ, Green JA, et al. Ifosfamide encephalopathy: a reappraisal. *Eur J Cancer Clin Oncol* 1989;25:1303–1310.

210. Pratt CB, Green AA, Horowitz ME, et al. Central nervous system toxicity following the treatment of pediatric patients with ifosfamide/mesna. *J Clin Oncol* 1986;4:1253–1261.

211. Patel PN. Methylene blue for management of Ifosfamide-induced encephalopathy. *Ann Pharmacother* 2006;40:299–303.

212. Buesa JM, Garcia-Teijido P, Losa R, et al. Treatment of ifosfamide encephalopathy with intravenous thiamin. *Clin Cancer Res* 2003;9:4636–4637.

213. Patel SR, Forman AD, Benjamin RS. High-dose ifosfamide-induced exacerbation of peripheral neuropathy. *J Natl Cancer Inst* 1994;86:305–306.

214. Brunello A, Basso U, Rossi E, et al. Ifosfamide-related encephalopathy in elderly patients: report of five cases and review of the literature. *Drugs Aging* 2007;24:967–73.

215. Chastagner P, Sommelet-Olive D, Kalifa C, et al. Phase II study of ifosfamide in childhood brain tumors: a report by the French Society of Pediatric Oncology (SFOP). *Med Pediatr Oncol* 1993; 21:49–53.

216. Ghosn M, Carde P, Leclerq B, et al. Ifosfamide/ mesna related encephalopathy: a case report with a possible role of phenobarbital in enhancing neurotoxicity. *Bull Cancer* 1988;75:391–392.

217. Pratt CB, Goren MP, Meyer WH, et al. Ifosfamide neurotoxicity is related to previous cisplatin treatment for pediatric solid tumors. *J Clin Oncol* 1990;8:1399–1401.

218. Ajithkumar T. Ifosfamide encephalopathy. *Clin Oncol* 2007;19:108–114.

219. Lokiec F. Ifosfamide: pharmacokinetic properties for central nervous system metastasis prevention. *Ann Oncol* 2006;17 (Suppl 4):iv33–iv36.

220. Phillips PC, Than TT, Cork LC, et al. Intrathecal 4-hydroperoxycyclophosphamide: neurotoxicity, cerebrospinal fluid pharmacokinetics, and antitumor activity in a rabbit model of VX2 leptomeningeal carcinomatosis. *Cancer Res* 1992;52:6168–6174.

221. Nagarajan R, Peters C, Orchard P, et al. Report of severe neurotoxicity with cyclophosphamide. *J Pediatr Hematol Oncol* 2000;22:544–546.

222. Kende G, Sirkin SR, Thomas PR, et al. Blurring of vision: a previously undescribed complication of cyclophosphamide therapy. *Cancer* 1979;44:69–71.

223. Tashima CK. Immediate cerebral symptoms during rapid intravenous administration of cyclophosphamide (NSC-26271). *Cancer Chemo Rep* 1975;59:441–442.

224. DeFronzo RA, Braine H, Colvin OM, et al. Water intoxication in man after cyclophosphamide therapy. Time course and relation to drug activation. *Ann Intern Med* 1973;78:861–869.

225. Loprinzi CL, Duffy J, Ingle JN. Postchemotherapy rheumatism. *J Clin Oncol* 1993;11:768–770.

226. Cornford EM, Young D, Paxton JW, et al. Melphalan penetration of the blood-brain barrier via the neutral amino acid transporter in tumor-bearing brain. *Cancer Res* 1992;52:138–143.

227. Groothuis DR, Lippitz BE, Fekete I, et al. The effect of an amino acid-lowering diet on the rate of melphalan entry into brain and xenotransplanted glioma. *Cancer Res* 1992;52:5590–5596.

228. Schuh A, Dandridge J, Haydon P, et al. Encephalopathy complicating high-dose melphalan. *Bone Marrow Transplant* 1999;24:1141–1143.

229. Friedman HS, Archer GE, McLendon RE, et al. Intrathecal melphalan therapy of human neoplastic meningitis in athymic nude rats. *Cancer Res* 1994;54:4710–14.

230. Salloum E, Khah KK, Cooper DL, et al. Chlorambucil-induced seizures. *Cancer* 1997;79:1009–1013.

231. Blumenreich MS, Woodcock TM, Sherrill EJ, et al. A phase I trial of chlorambucil administered in short pulses in patients with advanced malignancies. *Cancer Invest* 1988;6:371–375.

232. Vandenberg SA, Kulig K, Spoerke DG, et al. Chlorambucil overdose: accidental ingestion of an antineoplastic drug. *J Emerg Med* 1988;6:495–498.

233. Wyllie AR, Bayliff CD, Kovacs MJ. Myoclonus due to chlorambucil in two adults with lymphoma. *Ann Pharmacother* 1997;31:171–174.

234. Burns LJ. Ocular toxicities of chemotherapy. *Semin Oncol* 1992;19:492–500.

235. Pichini S, Altieri I, Bacosi A, et al. High-performance liquid chromatographic-mass spectrometic assay of busulfan in serum and cerebrospinal fluid. *J Chromatogr* 1992;581:143–146.

236. Murphy CP, Harden EA, Thompson JM. Generalized seizures secondary to high-dose busulfan therapy. *Ann Pharmacother* 1992;26:30–31.

237. De La CR, Tomas JF, Figuera A, et al. High dose busulfan and seizures. *Bone Marrow Transplant* 1991;7:363–364.

238. Jenke A, Freiberg-Richter J, Wilhelm S, et al. Accidental busulfan overdose during conditioning for stem cell transplantation. *Bone Marrow Transplant* 2005;35:125–128.

239. La MC, Mondini S, Guarino M, et al. Busulfan neurotoxicity and EEG abnormalities: a case report. *Neurol Sci* 2004;25:95–97.

240. Djaldetti M, Pinkhas J, Vries Ad, et al. Myasthenia gravis in a patient with chronic myeloid leukemia treated by busulfan. *Blood* 1968;32:336–340.

241. Sidi Y, Douer D, Pinkhas J. Sicca syndrome in a patient with toxic reaction to busulfan. Case report. *JAMA* 1977;238:1951.

242. Tosti A, Piraccini BM, Vincenzi C, et al. Permanent alopecia after busulfan chemotherapy. *Br J Dermatol* 2005;152:1056–1058.

243. Paterson AHG, McPherson TA. A possible neurologic complication of DTIC. Letter to the editor. *Cancer Treat Rep* 1978;61:105–106.

244. Cohen MH, Johnson JR, Pazdur R. Food and Drug Administration Drug approval summary: temozolomide plus radiation therapy for the treatment of newly diagnosed glioblastoma multiforme. *Clin Cancer Res* 2005;11:6767–6771.

245. Ariel IM. Intra-arterial chemotherapy for metastatic cancer to the brain. *Am J Surg* 1961;102:647–650.

246. Bethlenfalvay NC, Bergin JJ. Severe cerebral toxicity after intravenous nitrogen mustard therapy. *Cancer* 1972;29:366–369.

247. Zaniboni A, Simoncini E, Marpicati P, et al. Severe delayed neurotoxicity after accidental high-dose nitrogen mustard. *Am J Hematol* 1988;27:305.

248. Segal GM, Duckert LG. Reversible mechlorethamine-associated hearing loss in a patient with Hodgkin's disease. *Cancer* 1986;57:1089–1091.

249. Postma TJ, Van Groeningen CJ, Witjes RJ, et al. Neurotoxicity of combination chemotherapy with procarbazine, CCNU and vincristine (PCV) for recurrent glioma. *J Neuro-oncol* 1998; 38:69–75.

250. Spivack SD. Drugs five years later: Procarbazine. *Ann Intern Med* 1974;81:795–800.

251. Coyle T, Bushunow P, Winfield J, et al. Hypersensitivity reactions to procarbazine with mechlorethamine, vincristine, and procarbazine chemotherapy in the treatment of glioma. *Cancer* 1992;69:2532–2540.

252. Lanser JB, van Seters AP, Moolenaar AJ, et al. Neuropsychologic and neurologic side effects of mitotane and reversibility of symptoms. Letter to the editor. *J Clin Oncol* 1992;10:1504.

253. Terzolo M, Pia A, Berruti A, et al. Low-dose monitored mitotane treatment achieves the therapeutic range with manageable side effects in patients with adrenocortical cancer. *J Clin Endocrinol Metab* 2000;85:2234–2238.

254. Lee CR, Faulds D. Altretamine. A review of its pharmacodynamic and pharmacokinetic properties, and therapeutic potential in cancer chemotherapy. *Drugs* 1995;49:932–953.

255. Keldsen N, Havsteen H, Vergote I, et al. Altretamine (hexamethylmelamine) in the treatment of platinum-resistant ovarian cancer: a phase II study. *Gynecol Oncol* 2003;88:118–122.

256. Heideman RL, Cole DE, Balis F, et al. Phase I and pharmacokinetic evaluation of thiotepa in the cerebrospinal fluid and plasma of pediatric patients: evidence for dose-dependent plasma clearance of thiotepa. *Cancer Res* 1989;49:736–741.

257. Papadopoulos KP, Garvin JH, Fetell M, et al. High-dose thiotepa and etoposide-based regimens with autologous hematopoietic support for high-risk or recurrent CNS tumors in children and adults. *Bone Marrow Transplant* 1998;22:661–667.

258. Cairncross G, Swinnen L, Bayer R, et al. Myeloablative chemotherapy for recurrent aggressive oligodendroglioma. *Neuro-oncology* 2000;2:114–119.

259. Gutin PH, Weiss HD, Wiernik PH, et al. Intrathecal N,N',N"-triethylenethiophosphoramide [thio-TEPA (NSC 6396)] in the treatment of malignant meningeal disease: phase I-II study. *Cancer* 1976; 38:1471–1475.

260. Gutin PH, Levi JA, Wiernik PH, et al. Treatment of malignant meningeal disease with intrathecal thioTEPA: a phase II study. *Cancer Treat Rep* 1977;61:885–887.

261. Martin AS, Henriquez I, Rebollo J, et al. Severe polyneuropathy and motor loss after intrathecal thiotepa combination chemotherapy: description of two cases. *Anticancer Drugs* 1990;1:33–35.

262. Bigotte L, Olsson Y. Distribution and toxic effects of intravenously injected epirubicin on the central nervous system of the mouse. *Brain* 1989;112 (Pt 2):457–469.

263. Vail DM, Amantea MA, Colbern GT, et al. Pegylated liposomal doxorubicin: proof of principle using pre-clinical animal models and pharmacokinetic studies. *Semin Oncol* 2004;31:16–35.

264. Mizutani T, Hattori A. New horizon of MDR1 (P-glycoprotein) study. *Drug Metab Rev* 2005; 37:489–510.

265. Arico M, Nespoli L, Porta F, et al. Severe acute encephalopathy following inadvertent intrathecal doxorubicin administration. *Med Pediatr Oncol* 1990;18:261–263.

266. Mortensen ME, Cecalupo AJ, Egorin MJ, et al. Inadvertent intrathecal injection of daunoru-bicin with fatal outcome. *Med Pediatr Oncol* 1992;20:249–253.

267. Neuwelt EA, Pagel M, Barnett P, et al. Pharmacology and toxicity of intracarotid Adriamycin administra-tion following osmotic blood-brain barrier modifica-tion. *Cancer Res* 1981;41:4466–4470.

268. Schachter S, Freeman R. Transient ischemic attack and Adriamycin cardiomyopathy. *Neurology* 1982;32:1380–1381.

269. Beck WT, Kuttesch JF. Neurological symptoms associated with cyclosporin plus doxorubicin. Letter to the editor. *Lancet* 1992;340:496.

270. Jortner BS, Cho E-S. Neurotoxicity of Adriamycin in rats: a low-dose effect. *Cancer Treat Rep* 1980;64:257–261.

271. Reid JM, Pendergrass TW, Krailo MD, et al. Plasma pharmacokinetics and cerebrospinal fluid concen-trations of idarubicin and idarubicinol in pediatric leukemia patients: a Children's Cancer Study Group report. *Cancer Res* 1990;50:6525–6528.

272. Fox EJ. Management of worsening multiple scle-rosis with mitoxantrone: a review. *Clin Ther* 2006;28:461–474.

273. Seiter K, Liu D, Feldman E, et al. Long-term follow-up of high-dose mitoxantrone-based induc-tion therapy for patients with newly-diagnosed acute myelogenous leukemia. Twelve year results from a single institution. *Leuk Lymphoma* 2006; 47:425–432.

274. Hall C, Dougherty WG, Lebish IJ, et al. Warning against use of intrathecal mitoxantrone. Letter to the editor. *Lancet* 1989;1:734.

275. Berger CC, Bokemeyer C, Schneider M, et al. Secondary Raynaud's phenomenon and other late vascular complications following chemo-therapy for testicular cancer. *Eur J Cancer [A]* 1995;31A:2229–2238.

276. Shahab N. Vascular toxicity of antineoplastic agents. *Sem Oncol* 2006;33:121–138.

277. Lazarus HM, Gottfried MR, Herzig RH, et al. Veno-occlusive disease of the liver after high-dose mitomycin C therapy and autologous bone marrow transplantation. *Cancer* 1982;49:1789–1795.

278. Hardten DR, Samuelson TW. Ocular toxicity of mitomycin-C. *Int Ophthalmol Clin* 1999;39:79–90.

279. Mietz H. The toxicology of mitomycin C on the cili-ary body. *Curr Opin Ophthalmol* 1996;7:72–79.

280. Hueman EM, Simpson CB. Airway complications from topical mitomycin C. *Otolaryngol Head Neck Surg* 2005;133:831–835.

281. Taira M, Kojima K, Takeuchi H. A comparative study of the action of actinomycin D and actinomycinic

ACID on the central nervous system when injected into the cerebrospinal fluid of higher animals. *Epilepsia* 1972;13:649–662.

282. Sánchez-Carpintero R, Narbona J, De Mesa RL, et al. Transient posterior encephalopathy induced by chemotherapy in children. *Pediatr Neurol* 2001;24:145–148.

283. Phillips PC, Thaler HT, Berger CA, et al. Acute high-dose methotrexate neurotoxicity in the rat. *Ann Neurol* 1986;20:583–589.

284. Cohen IJ. Defining the appropriate dosage of foli-nic acid after high-dose methotrexate for childhood acute lymphatic leukemia that will prevent neu-rotoxicity without rescuing malignant cells in the central nervous system. *J Pediatr Hematol Oncol* 2004;26:156–163.

285. Gao K, Jiang X. Influence of particle size on trans-port of methotrexate across blood brain barrier by polysorbate 80-coated polybutylcyanoacrylate nano-particles. *Int J Pharm* 2006;310:213–219.

286. Fortin D, Desjardins A, Benko A, et al. Enhanced chemotherapy delivery by intraarterial infusion and blood-brain barrier disruption in malignant brain tumors: the Sherbrooke experience. *Cancer* 2005;103:2606–2615.

287. Erdlenbruch B, Alipour M, Fricker G, et al. Alkylglycerol opening of the blood-brain barrier to small and large fluorescence markers in normal and C6 glioma-bearing rats and isolated rat brain capil-laries. *Br J Pharmacol* 2003;140:1201–1210.

288. Grossman SA, Reinhard CS, Loats HL. The intra-cerebral penetration of intraventricularly adminis-tered methotrexate: a quantitative autoradiographic study. *J Neuro-Oncol* 1989;7:319–328.

289. Spiegler BJ, Kennedy K, Maze R, et al. Comparison of long-term neurocognitive outcomes in young children with acute lymphoblastic leukemia treated with cranial radiation or high-dose or very high-dose intravenous methotrexate. *J Clin Oncol* 2006;24:3858–3864.

290. Bode U, Magrath IT, Bleyer WA, et al. Active trans-port of methotrexate from cerebrospinal fluid in humans. *Cancer Res* 1980;40:2184–2187.

291. Breen CM, Sykes DB, Baehr C, et al. Fluorescein-methotrexate transport in rat choroid plexus ana-lyzed using confocal microscopy. *Am J Physiol Renal Physiol* 2004;287:F562–F569.

292. Shapiro WR, Chernik NL, Posner JB. Necrotizing encephalopathy following intraventricular instilla-tion of methotrexate. *Arch Neurol* 1973;28:96–102.

293. Morikawa N, Mori T, Kawashima H, et al. Pharmacokinetics of anticancer drugs in cerebrospi-nal fluid. *Ann Pharmacother* 1998;32:1008–1012.

294. Mott MG, Stevenson P, Wood CB. Methotrexate meningitis. Letter to the editor. *Lancet* 1972; 2:656.

295. Glantz MJ, Jaeckle KA, Chamberlain MC, et al. A randomized controlled trial comparing intrath-ecal sustained-release cytarabine (DepoCyt) to intrathecal methotrexate in patients with neoplas-tic meningitis from solid tumors. *Clin Cancer Res* 1999;5:3394–3402.

296. Boogerd W, vd Sande JJ, Moffie D. Acute fever and delayed leukoencephalopathy following low dose intraventricular methotrexate. *J Neurol Neurosurg Psychiatry* 1988;51:1277–1283.

297. Pullen J, Boyett J, Shuster J, et al. Extended triple intrathecal chemotherapy trial for prevention of CNS relapse in good-risk and poor-risk patients with B-progenitor acute lymphoblastic leukemia: a Pediatric Oncology Group study. *J Clin Oncol* 1993;11:839–849.

298. Bay A, Oner AF, Etlik O, et al. Myelopathy due to intrathecal chemotherapy: report of six cases. *J Pediatr Hematol Oncol* 2005;27:270–272.

299. Hawboldt J. Intramuscular methotrexate-induced aseptic meningitis. *Ann Pharmacother* 2007; 41:1906–1911.

300. Ettinger LJ. Pharmacokinetics and biochemical effects of a fatal intrathecal methotrexate overdose. *Cancer* 1982;50:444–450.

301. Ettinger LJ, Freeman AI, Creaven PJ. Intrathecal methotrexate overdose without neurotoxicity: case report and literature review. *Cancer* 1978; 41:1270–1273.

302. Finkelstein Y, Zevin S, Heyd J, et al. Emergency treatment of life-threatening intrathecal methotrexate overdose. *Neurotoxicology* 2004;25:407–410.

303. Jardine LF, Ingram LC, Bleyer WA. Intrathecal leucovorin after intrathecal methotrexate overdose. *J Pediatr Hematol Oncol* 1996;18:302–304.

304. Adamson PC, Balis FM, McCully CL, et al. Rescue of experimental intrathecal methotrexate overdose with carboxypeptidase-G2. *J Clin Oncol* 1991;9:670–674.

305. Jakobson AM, Kreuger A, Mortimer O, et al. Cerebrospinal fluid exchange after intrathecal methotrexate overdose. A report of two cases. *Acta Paediatr* 1992;81:359–361.

306. Ten Hoeve RF, Twijnstra A. A lethal neurotoxic reaction after intraventricular methotrexate administration. *Cancer* 1988;62:2111–2113.

307. Shore T, Barnett MJ, Phillips GL. Sudden neurologic death after intrathecal methotrexate. *Med Pediatr Oncol* 1990;18:159–161.

308. Szawarski P, Chapman CS. A woman who couldn't speak: report of methotrexate neurotoxicity. *Postgrad Med J* 2005;81(953):194–195.

309. Brock S, Jennings HR. Fatal acute encephalomyelitis after a single dose of intrathecal methotrexate. *Pharmacotherapy* 2004;24:673–676.

310. Boran P, Tokuc G, Boran BO, et al. Intracerebral hematoma as a complication of intrathecal methotrexate administration. *Pediatr Blood Cancer* 2008;50:152–154.

311. Henderson RD, Rajah T, Nicol AJ, Read SJ. Posterior leukoencephalopathy following intrathecal chemotherapy with MRA-documented vasospasm. *Neurology* 2003;60:326–328.

312. Kuker W, Bader P, Herrlinger U, et al. Transient encephalopathy after intrathekal methotrexate chemotherapy: diffusion-weighted MRI. *J Neuro-oncol* 2005;73:47–49.

313. Rollins N, Winick N, Bash R, et al. Acute methotrexate neurotoxicity: findings on diffusion-weighted imaging and correlation with clinical outcome. *Am J Neuroradiol* 2004;25:1688–1695.

314. Bernstein ML, Sobel DB, Wimmer RS. Noncardiogenic pulmonary edema following injection of methotrexate into the cerebrospinal fluid. *Cancer* 1982;50:866–868.

315. Gutin PH, Green MR, Bleyer WA, et al. Methotrexate pneumonitis induced by intrathecal methotrexate

therapy: a case report with pharmacokinetic data. *Cancer* 1976;38:1529–1534.

316. Simmons ED, Somberg KA. Acute tumor lysis syndrome after intrathecal methotrexate administration. *Cancer* 1991;67:2062–2065.

317. Anderson SC, Baquis GD, Jackson A, et al. Ventral polyradiculopathy with pediatric acute lymphocytic leukemia. *Muscle Nerve* 2002;25:106–110.

318. McLean DR, Clink HM, Enst P, et al. Myelopathy after intrathecal chemotherapy. A case report with unique magnetic resonance imaging changes. *Cancer* 1994;73:3037–3040.

319. Drachtman RA, Cole PD, Golden CB, et al. Dextromethorphan is effective in the treatment of subacute methotrexate neurotoxicity. *Pediatr Hematol Oncol* 2002;19:319–327.

320. Clark AW, Cohen SR, Nissenblatt MJ, et al. Paraplegia following intrathecal chemotherapy: neuropathologic findings and elevation of myelin basic protein. *Cancer* 1982;50:42–47.

321. Vainionpaa L, Kovala T, Tolonen U, et al. Chemotherapy for acute lymphoblastic leukemia may cause subtle changes of the spinal cord detectable by somatosensory evoked potentials. *Med Pediatr Oncol* 1997;28:41–47.

322. Boogerd W, Moffie D, Smets LA. Early blindness and coma during intrathecal chemotherapy for meningeal carcinomatosis. *Cancer* 1990;65:452–457.

323. Yim YS, Mahoney DH, Jr, Oshman DG. Hemiparesis and ischemic changes of the white matter after intrathecal therapy for children with acute lymphocytic leukemia. *Cancer* 1991;67:2058–2061.

324. Lemann W, Wiley RG, Posner JB. Leukoencephalopathy complicating intraventricular catheters: clinical, radiographic and pathologic study of 10 cases. *J Neurooncol* 1988;6:67–74.

325. Packer RJ, Zimmerman RA, Rosenstock J, et al. Focal encephalopathy following methotrexate therapy. Administration via a misplaced intraventricular catheter. *Arch Neurol* 1981;38:450–452.

326. Page KA, Vogel H, Horoupian DS. Intracerebral (parenchymal) infusion of methotrexate: report of a case. *J Neuro-oncol* 1992;12:181–186.

327. Walker RW, Allen JC, Rosen G, et al. Transient cerebral dysfunction secondary to high-dose methotrexate. *J Clin Oncol* 1986;4:1845–1850.

328. Haykin ME, Gorman M, van HJ, et al. Diffusion-weighted MRI correlates of subacute methotrexate-related neurotoxicity. *J Neuro-oncol* 2006; 76:153–157.

329. Valik D, Sterba J, Bajciova V, et al. Severe encephalopathy induced by the first but not the second course of high-dose methotrexate mirrored by plasma homocysteine elevations and preceded by extreme differences in pretreatment plasma folate. *Oncology* 2005;69:269–272.

330. Lee AC, Li CH, Wong YC. Transient encephalopathy following high-dose methotrexate. *Med Pediatr Oncol* 2003;41:101.

331. Phillips PC, Dhawan V, Strother SC, et al. Reduced cerebral glucose metabolism and increased brain capillary permeability following high-dose methotrexate chemotherapy: a positron emission tomographic study. *Ann Neurol* 1987;21:59–63.

331a. Baehring JM, Fulbright RK. Delayed leukoencephalopathy with stroke-like presentation in chemotherapy recipients. *J Neurol Neurosurg Psychiatry* 2008;79:535–9.

332. Doroshow JH, Locker GY, Gaasterland DE, et al. Ocular irritation from high-dose methotrexate therapy: pharmacokinetics of drug in the tear film. *Cancer* 1981;48:2158–2162.

333. Urban C, Nirenberg A, Caparros B, et al. Chemical pleuritis as the cause of acute chest pain following high-dose methotrexate treatment. *Cancer* 1983; 51:34–37.

334. Ragab AH, Frech RS, Vietti TJ. Osteoporotic fractures secondary to methotrexate therapy of acute leukemia in remission. *Cancer* 1970;25:580–585.

335. Millay RH, Klein ML, Shults WT, et al. Maculopathy associated with combination chemotherapy and osmotic opening of the blood-brain barrier. *Am J Ophthalmol* 1986;102:626–632.

336. Wernick R, Smith DL. Central nervous system toxicity associated with weekly low-dose methotrexate treatment. *Arthritis Rheum* 1989; 32:770–775.

337. van Jaarsveld CH, Jahangier ZN, Jacobs JW, et al. Toxicity of anti-rheumatic drugs in a randomized clinical trial of early rheumatoid arthritis. *Rheumatology (Oxford)* 2000;39:1374–1382.

338. Glass JO, Reddick WE, Li CS, et al. Computer-aided detection of therapy-induced leukoencephalopathy in pediatric acute lymphoblastic leukemia patients treated with intravenous high-dose methotrexate. *Magn Reson Imaging* 2006;24:785–791.

339. Price RA, Jamieson PA. The central nervous system in childhood leukemia. II. Subacute leukoencephalopathy. *Cancer* 1975;35:306–318.

340. Rubinstein LJ, Herman MM, Long TG, et al. Disseminated necrotizing leukoencephalopathy: a complication of treated central nervous system leukemia and lymphoma. *Cancer* 1975;35:291–305.

341. Ochs J, Mulhern R, Fairclough D, et al. Comparison of neuropsychologic functioning and clinical indicators of neurotoxicity in long-term survivors of childhood leukemia given cranial radiation or parenteral methotrexate: a prospective study. *J Clin Oncol* 1991;9:145–151.

342. Antunes NL, Souweidane MM, Rosenblum MK, et al. Methotrexate leukoencephalopathy presenting as Kluver-Bucy syndrome and uncinate seizures. *Pediatr Neurol* 2002;26:305–308.

343. Robb J, Chalmers L, Rojiani A, et al. Multifocal necrotizing leukoencephalopathy: an unusual complication of acute leukemia. *Arch Neurol* 2006;63:1028–1029.

344. Lai R, Abrey LE, Rosenblum MK, et al. Treatment-induced leukoencephalopathy in primary CNS lymphoma: a clinical and autopsy study. *Neurology* 2004;62:451–456.

345. Gavrilovic IT, Hormigo A, Yahalom J, et al. Long-term follow-up of high-dose methotrexate-based therapy with and without whole brain irradiation for newly diagnosed primary CNS lymphoma. *J Clin Oncol* 2006;24:4570–4574.

346. Montour-Proulx I, Kuehn SM, Keene DL, et al. Cognitive changes in children treated for acute lymphoblastic leukemia with chemotherapy only according to the Pediatric Oncology Group 9605 protocol. *J Child Neurol* 2005;20(2):129–133.

347. Gangji D, Reaman GH, Cohen SR, et al. Leukoencephalopathy and elevated levels of myelin basic protein in the cerebrospinal fluid of patients with acute lymphoblastic leukemia. *N Engl J Med* 1980;303:19–21.

348. Peylan-Ramu N, Poplack DG, Pizzo PA, et al. Abnormal CT scans of the brain in asymptomatic children with acute lymphocytic leukemia after prophylactic treatment of the central nervous system with radiation and intrathecal chemotherapy. *N Engl J Med* 1978;298:815–818.

349. Uldry PA, Teta D, Regli L. Focal cerebral necrosis caused by intraventricular chemotherapy with methotrexate. *Neurochirurgie* 1991;37:72–74.

350. Oka M, Terae S, Kobayashi R, et al. MRI in methotrexate-related leukoencephalopathy: disseminated necrotising leukoencephalopathy in comparison with mild leukoencephalopathy. *Neuroradiology* 2003;45:493–497.

351. Reddick WE, Glass JO, Helton KJ, et al. Prevalence of leukoencephalopathy in children treated for acute lymphoblastic leukemia with high-dose methotrexate. *Am J Neuroradiol* 2005;26:1263–1269.

352. Paakko E, Harila-Saari A, Vanionpaa L, et al. White matter changes on MRI during treatment in children with acute lymphoblastic leukemia: correlation with neuropsychological findings. *Med Pediatr Oncol* 2000;35:456–461.

353. Lien HH, Blomlie V, Saeter G, et al. Osteogenic sarcoma: MR signal abnormalities of the brain in asymptomatic patients treated with high-dose methotrexate. *Radiology* 1991;179:547–550.

354. Reddick WE, Glass JO, Helton KJ, et al. A quantitative MR imaging assessment of leukoencephalopathy in children treated for acute lymphoblastic leukemia without irradiation. *Am J Neuroradiol* 2005;26:2371–2377.

355. Price RA, Birdwell DA. The central nervous system in childhood leukemia. III. Mineralizing microangiopathy and dystrophic calcification. *Cancer* 1978; 42:717–728.

356. Vezmar S, Becker A, Bode U, et al. Biochemical and clinical aspects of methotrexate neurotoxicity. *Chemotherapy* 2003;49:92–104.

357. Linnebank M, Pels H, Kleczar N, et al. MTX-induced white matter changes are associated with polymorphisms of methionine metabolism. *Neurology* 2005; 64(5):912–913.

358. Uzar E, Sahin O, Koyuncuoglu HR, et al. The activity of adenosine deaminase and the level of nitric oxide in spinal cord of methotrexate administered rats: protective effect of caffeic acid phenethyl ester. *Toxicology* 2006;218:125–133.

359. Madhyastha S, Somayaji SN, Rao MS, et al. Effect of intracerebroventricular methotrexate on brain amines. *Indian J Physiol Pharmacol* 2005;49:427–435.

360. Jaksic W, Veljkovic D, Pozza C, et al. Methotrexate-induced leukoencephalopathy reversed by aminophylline and high-dose folinic acid. *Acta Haematol* 2004;111:230–232.

361. Stapleton SL, Reid JM, Thompson PA, et al. Plasma and cerebrospinal fluid pharmacokinetics of pemetrexed after intravenous administration in non-human primates. *Cancer Chemother Pharmacol* 2007;59(4):461–466.

362. Villela LR, Stanford BL, Shah SR. Pemetrexed, a novel antifolate therapeutic alternative for cancer chemotherapy. *Pharmacotherapy* 2006;26:641–654.

363. Cole BF, Glantz MJ, Jaeckle KA, et al. Quality-of-life-adjusted survival comparison of sustained-release cytosine arabinoside versus intrathecal methotrexate for treatment of solid tumor neoplastic meningitis. *Cancer* 2003;97:3053–3060.

364. Band PR, Holland JF, Bernard J, et al. Treatment of central nervous system leukemia with intrathecal cytosine arabinoside. *Cancer* 1973;32:744–748.

365. Burch PA, Grossman SA, Reinhard CS. Spinal cord penetration of intrathecally administered cytarabine and methotrexate: a quantitative autoradiographic study. *J Natl Cancer Inst* 1988; 80:1211–1216.

366. Smith GA, Damon LE, Rugo HS, et al. High-dose cytarabine dose modification reduces the incidence of neurotoxicity in patients with renal insufficiency. *J Clin Oncol* 1997;15:833–839.

367. Glantz MJ, LaFollette S, Jaeckle KA, et al. Randomized trial of a slow-release versus a standard formulation of cytarabine for the intrathecal treatment of lymphomatous meningitis. *J Clin Oncol* 1999;17:3110–3116.

368. Damon LE, Plunkett W, Linker CA. Plasma and cerebrospinal fluid pharmacokinetics of 1-β-D-arabinofuranosylcytosine and 1-β-D-arabinofuranosyluracil following the repeated intravenous administration of high- and intermediate-dose 1-β-D-arabinofuranosylcytosine. *Cancer Res* 1991; 51:4141–4145.

369. Scott-Moncrieff JCR, Chan TCK, Samuels ML, et al. Plasma and cerebrospinal fluid pharmacokinetics of cytosine arabinoside in dogs. *Cancer Chemother Pharmacol* 1991;29:13–18.

370. Slevin ML, Piall EM, Aherne GW, et al. Effect of dose and schedule on pharmacokinetics of high-dose cytosine arabinoside in plasma and cerebrospinal fluid. *J Clin Oncol* 1983;1:546–551.

371. Baker WJ, Royer GL, Jr, Weiss RB. Cytarabine and neurologic toxicity. *J Clin Oncol* 1991;9:679–693.

372. Saito B, Nakamaki T, Nakashima H, et al. Reversible posterior leukoencephalopathy syndrome after repeat intermediate-dose cytarabine chemotherapy in a patient with acute myeloid leukemia. *Am J Hematol* 2007;82(4):304–306.

373. Dunton SF, Nitschke R, Spruce WE, et al. Progressive ascending paralysis following administration of intrathecal and intravenous cytosine arabinoside. A Pediatric Oncology Group study. *Cancer* 1986;57:1083–1088.

374. Resar LM, Phillips PC, Kastan MB, et al. Acute neurotoxicity after intrathecal cytosine arabinoside in two adolescents with acute lymphoblastic leukemia of B-cell type. *Cancer* 1993;71:117–123.

375. Saleh MN, Christian ES, Diamond BR. Intrathecal cytosine arabinoside-induced acute, rapidly reversible paralysis. *Am J Med* 1989;86:729–730.

376. Bates S, Raphaelson MI, Price RA, et al. Ascending myelopathy after chemotherapy for central nervous system acute lymphoblastic leukemia: correlation with cerebrospinal fluid myelin basic protein. *Med Pediatr Oncol* 1985;13:4–8.

377. Breuer AC, Pitman SW, Dawson DM, et al. Paraparesis following intrathecal cytosine arabinoside: a case report with neuropathologic findings. *Cancer* 1977;40:2817–2822.

378. Kleinschmidt-DeMasters BK, Yeh M. "Locked-in syndrome" after intrathecal cytosine arabinoside therapy for malignant immunoblastic lymphoma. *Cancer* 1992;70:2504–2507.

379. Legrand F, Dorgeret S, Saizou C, et al. Cerebellar herniation after intrathecal chemotherapy including cytosine arabinoside in a boy with T acute lymphoblastic leukemia. *Leukemia* 2002; 16:2454–2455.

380. Eden OB, Goldie W, Wood T, et al. Seizures following intrathecal cytosine arabinoside in young children with acute lymphoblastic leukemias. *Cancer* 1978;42:53–58.

381. Wolff L, Zighelboim J, Gale RP. Paraplegia following intrathecal cytosine arabinoside. *Cancer* 1979;43:83–85.

382. Gay CT, Bodensteiner JB, Nitschke R, et al. Reversible treatment-related leukoencephalopathy. *J Child Neurol* 1989;4:208–213.

382a. Jabbour E, O'Brien S, Kantarjian H, et al. Neurologic complications associated with intrathecal liposomal cytarabine given prophylactically in combination with high-dose methotrexate and cytarabine to patients with acute lymphocytic leukemia. *Blood* 2007;109:3214–18.

383. Barnett MJ, Richards MA, Ganesan TS, et al. Central nervous system toxicity of high-dose cytosine arabinoside. *Semin Oncol* 1985;12 (2 Suppl 3): 227–232.

384. Benger A, Browman GP, Walker IR, et al. Clinical evidence of a cumulative effect of high-dose cytarabine on the cerebellum in patients with acute leukemia. A leukemia intergroup report. *Cancer Treat Rep* 1985;69:240–241.

385. Hwang TL, Yung WK, Lee Y-Y, et al. High dose Ara-C related leukoencephalopathy. *J Neuro-oncol* 1986;3:335–339.

386. Dworkin LA, Goldman RD, Zivin LS, et al. Cerebellar toxicity following high-dose cytosine arabinoside. *J Clin Oncol* 1985;3:613–616.

387. Herzig RH, Hines JD, Herzig GP, et al. Cerebellar toxicity with high-dose cytosine arabinoside. *J Clin Oncol* 1987;5:927–932.

388. Vaughn DJ, Jarvik JG, Hackney D, et al. High-dose cytarabine neurotoxicity: MR findings during the acute phase. *Am J Neuroradiol* 1993;14:1014–1016.

389. Miller L, Link MP, Bologna S, et al. Cerebellar atrophy caused by high-dose cytosine arabinoside: CT and MR findings. *Am J Roentgenol* 1989; 152:343–344.

390. Vera P, Rohrlich P, Stievenart JL, et al. Contribution of single-photon emission computed tomography in the diagnosis and follow-up of CNS toxicity of a cytarabine-containing regimen in pediatric leukemia. *J Clin Oncol* 1999;17:2804–2810.

391. Vogel H, Horoupian DS. Filamentous degeneration of neurons. A possible feature of cytosine arabinoside neurotoxicity. *Cancer* 1993;71:1303–1308.

392. Winkelman MD, Hines JD. Cerebellar degeneration caused by high-dose cytosine arabinoside: a clinicopathological study. *Ann Neurol* 1983;14:520–527.

393. Borgeat A, DeMuralt B, Stalder M. Peripheral neuropathy associated with high-dose Ara-C therapy. *Cancer* 1986;58:852–854.

394. Powell BL, Capizzi RL, Lyerly ES, et al. Peripheral neuropathy after high-dose cytosine arabinoside,

daunorubicin, and asparaginase consolidation for acute nonlymphocytic leukemia. *J Clin Oncol* 1986;4:95–97.

395. Openshaw H, Slatkin NE, Stein AS, et al. Acute polyneuropathy after high dose cytosine arabinoside in patients with leukemia. *Cancer* 1996;78:1899–1905.

396. Kornblau SM, Cortes-Franco J, Estey E. Neurotoxicity associated with fludarabine and cytosine arabinoside chemotherapy for acute leukemia and myelodysplasia. *Leukemia* 1993;7:378–383.

397. Ritch PS, Hansen RM, Heuer DK. Ocular toxicity from high-dose cytosine arabinoside. *Cancer* 1983;51:430–432.

398. Margileth DA, Poplack DG, Pizzo PA, et al. Blindness during remission in two patients with acute lymphoblastic leukemia: a possible complication of multimodality therapy. *Cancer* 1977;39:58–61.

399. Shaw PJ, Procopis PG, Menser MA, et al. Bulbar and pseudobulbar palsy complicating therapy with high- dose cytosine arabinoside in children with leukemia. *Med Pediatr Oncol* 1991;19:122–125.

400. Nevill TJ, Benstead TJ, McCormick CW, et al. Horner's syndrome and demyelinating peripheral neuropathy caused by high-dose cytosine arabinoside. *Am J Hematol* 1989;32:314–315.

401. Malapert D, Degos JD. Painful legs and moving toes. Neuropathy caused by cytarabine. *Rev Neurol (Paris)* 1989;145:869–871.

402. Scherokman B, Filling-Katz MR, Tell D. Brachial plexus neuropathy following high-dose cytarabine in acute monoblastic leukemia. *Cancer Treat Rep* 1985;69:1005–1006.

403. Ventura GJ, Keating MJ, Castellanos AM, et al. Reversible bilateral lateral rectus muscle palsy associated with high-dose cytosine arabinoside and mitoxantrone therapy. *Cancer* 1986;58:1633–1635.

404. Thordarson H, Talstad I. Acute meningitis and cerebellar dysfunction complicating high-dose cytosine arabinoside therapy. *Acta Med Scand* 1986;220:493–495.

405. Hoffman DL, Howard JR, Jr, Sarma R, et al. Encephalopathy, myelopathy, optic neuropathy, and anosmia associated with intravenous cystosine arabinoside. *Clin Neuropharmacol* 1993;16:258–262.

406. Luque FA, Selhorst JB, Petruska P. Parkinsonism induced by high-dose cytosine arabinoside. *Mov Disord* 1987;2:219–222.

407. Besirli CG, Deckwerth TL, Crowder RJ, et al. Cytosine arabinoside rapidly activates Bax-dependent apoptosis and a delayed Bax-independent death pathway in sympathetic neurons. *Cell Death Differ* 2003;10:1045–1058.

408. Geller HM, Cheng KY, Goldsmith NK, et al. Oxidative stress mediates neuronal DNA damage and apoptosis in response to cytosine arabinoside. *J Neurochem* 2001;78:265–275.

409. Chang JY, Brown S. Cytosine arabinoside differentially alters survival and neurite outgrowth of neuronal PC12 cells. *Biochem Biophys Res Commun* 1996;218:753–758.

410. Courtney MJ, Coffey ET. The mechanism of Ara-C-induced apoptosis of differentiating cerebellar granule neurons. *Eur J Neurosci* 1999;11:1073–1084.

411. Ahlemeyer B, Kolker S, Zhu Y, et al. Cytosine arabinofuranoside-induced activation of astrocytes increases the susceptibility of neurons to glutamate due to the release of soluble factors. *Neurochem Int* 2003;42:567–581.

412. Malhotra P, Mahi S, Lal V, et al. Cytarabine-induced neurotoxicity responding to methyl prednisolone. *Am J Hematol* 2004;77:416.

413. Kuykendall JR. 5-azacytidine and decitabine monotherapies of myelodysplastic disorders. *Ann Pharmacother* 2005;39:1700–1709.

414. Bellet RE, Mastrangelo MJ, Engstrom PF, et al. Hepatotoxicity of 5-azacytidine (NSC-102816) (a clinical and pathologic study). *Neoplasma* 1973;20:303–309.

415. Levi J, Wiernik P. A comparative clinical trial of 5-azacytidine and guanazole in previously treated adults with acute nonlymphocytic leukemia. *Cancer* 1976;38:36–41.

416. Kerr JZ, Berg SL, Dauser R, et al. Plasma and cerebrospinal fluid pharmacokinetics of gemcitabine after intravenous administration in nonhuman primates. *Cancer Chemother Pharmacol* 2001;47:411–414.

417. Vander Els NJ, Miller V. Successful treatment of gemcitabine toxicity with a brief course of oral corticosteroid therapy. *Chest* 1998;114:1779–1781.

418. Voorburg AM, van Beek FT, Slee PH, et al. Vasculitis due to gemcitabine. *Lung Cancer* 2002 May;36:203–205.

419. Banach MJ, Williams GA. Purtscher retinopathy and necrotizing vasculitis with gemcitabine therapy. *Arch Ophthalmol* 2000;118:726–727.

420. De PT, Curigliano G, Franceschelli L, et al. Gemcitabine-induced systemic capillary leak syndrome. *Ann Oncol* 2001;12:1651–1652.

421. Ruiz I, Del VJ, Gomez A. Gemcitabine and haemolytic-uraemic syndrome. *Ann Oncol* 2004;15:1575–1576.

422. Gan HK, Mitchell PL, Galettis P, et al. A phase 1 and pharmacokinetic study of gemcitabine and oxaliplatin in patients with solid tumors. *Cancer Chemother Pharmacol* 2006;58:157–164.

423. Dormann AJ, Grunewald T, Wigginghaus B, et al. Gemcitabine-associated autonomic neuropathy. *Lancet* 1998;351:644.

424. Larsen FO, Hansen SW. Severe neurotoxicity caused by gemcitabine treatment. *Acta Oncol* 2004;43:590–591.

425. Ganem G, Solal-Celigny P, Joffroy A, et al. Radiation myositis: the possible role of gemcitabine. *Ann Oncol* 2000;11:1615–1616.

426. Friedlander PA, Bansal R, Schwartz L, et al. Gemcitabine-related radiation recall preferentially involves internal tissue and organs. *Cancer* 2004;100:1793–1799.

427. Chu E, Callender MA, Farrell MP, et al. Thymidylate synthase inhibitors as anticancer agents: from bench to bedside. *Cancer Chemother Pharmacol* 2003;52 (Suppl 1):S80–S89.

428. El-Khoueiry AB, Lenz HJ. Should continuous infusion 5-fluorouracil become the standard of care in the USA as it is in Europe? *Cancer Invest* 2006;24:50–55.

429. Soong R, Diasio RB. Advances and challenges in fluoropyrimidine pharmacogenomics and pharmacogenetics. *Pharmacogenomics* 2005;6:835–847.

430. Kerr IG, Zimm S, Collins JM, et al. Effect of intravenous dose and schedule on cerebrospinal fluid pharmacokinetics of 5-fluorouracil in the monkey. *Cancer Res* 1984;44:4929–4932.

431. Horton J, Olson KB, Sullivan J, et al. 5-Flourouracil in cancer: an improved regimen. *Ann Intern Med* 1970;73:897–900.

432. Liss RH, Chadwick M. Correlation of 5-fluorouracil (NSC-19893) distribution in rodents with toxicity and chemotherapy in man. *Cancer Chemother Rep* 1974;58:777–786.

433. Chadwick M, Rogers WI. The physiological disposition of 5-fluorouracil in mice bearing solid L1210 lymphocytic leukemia. *Cancer Res* 1972;32:1045–1056.

434. MacDonell LA, Potter PE, Leslie RA. Localized changes in blood-brain barrier permeability following the administration of antineoplastic drugs. *Cancer Res* 1978;38:2930–2934.

435. Neuwelt EA, Barnett P, Barranger J, et al. Inability of dimethyl sulfoxide and 5-fluorouracil to open the blood-brain barrier. *Neurosurgery* 1983;12:29–34.

436. Ezzeldin HH, Lee AM, Mattison LK, et al. Methylation of the DPYD promoter: an alternative mechanism for dihydropyrimidine dehydrogenase deficiency in cancer patients. *Clin Cancer Res* 2005;11:8699–8705.

437. Diasio RB. Clinical implications of dihydropyrimidine dehydrogenase on 5-FU pharmacology. *Oncology (Williston Park)* 2001;15:21–26.

438. Johnson MR, Hageboutros A, Wang K, et al. Life-threatening toxicity in a dihydropyrimidine dehydrogenase-deficient patient after treatment with topical 5-fluorouracil. *Clin Cancer Res* 1999;5:2006–2011.

439. Takimoto CH, Lu Z-H, Zhang R, et al. Severe neurotoxicity following 5-fluorouracil-based chemotherapy in a patient with dihydropyrimidine dehydrogenase deficiency. *Clin Cancer Res* 1996;2:477–481.

440. Mattison LK, Fourie J, Desmond RA, et al. Increased prevalence of dihydropyrimidine dehydrogenase deficiency in African-Americans compared with Caucasians. *Clin Cancer Res* 2006;12:5491–5495.

441. Mattison LK, Fourie J, Hirao Y, et al. The uracil breath test in the assessment of dihydropyrimidine dehydrogenase activity: pharmacokinetic relationship between expired $^{13}CO_2$ and plasma [2–13C] dihydrouracil. *Clin Cancer Res* 2006;12:549–555.

442. Riehl J-L, Brown WJ. Acute cerebellar syndrome secondary to 5-fluorouracil therapy. *Neurology* 1965;14:961–967.

443. Akiba T, Okeda R, Tajima T. Metabolites of 5-fluorouracil, a-fluoro-b-alanine and fluoroacetic acid, directly injure myelinated fibers in tissue culture. *Acta Neuropathol (Berl)* 1996;92:8–13.

444. Langer CJ, Hageboutros A, Kloth DD, et al. Acute encephalopathy attributed to 5-FU. *Pharmacotherapy* 1996;16:311–313.

445. Pirzada NA, Ali II, Dafer RM. Fluorouracil-induced neurotoxicity. *Ann Pharmacother* 2000;34:35–38.

446. Elkiran ET, Altundag K, Beyazit Y, et al. Fluorouracil-induced neurotoxicity presenting with generalized tonic-clonic seizure. *Ann Pharmacother* 2004;38:2171.

447. Liaw CC, Wang HM, Wang CH, et al. Risk of transient hyperammonemic encephalopathy in cancer patients who received continuous infusion of 5-fluorouracil with the complication of dehydration and infection. *Anticancer Drugs* 1999;10:275–281.

448. Yeh KH, Cheng AL. High-dose 5-fluorouracil infusional therapy is associated with hyperammonaemia, lactic acidosis and encephalopathy. *Br J Cancer* 1997;75:464–465.

449. Tha KK, Terae S, Sugiura M, et al. Diffusion-weighted magnetic resonance imaging in early stage of 5-fluorouracil-induced leukoencephalopathy. *Acta Neurol Scand* 2002;106:379–386.

450. Valik D. Encephalopathy, lactic acidosis, hyperammonaemia and 5-fluorouracil toxicity. *Br J Cancer* 1998;77:1710–1712.

451. Nichols M, Bergevin PR, Vyas AC, et al. Neurotoxicity from 5-fluorouracil (NSC-19893) administration reproduced by mitomycin C (NSC-26980). *Cancer Treat Rep* 1976;60:293–294.

452. Forbes JE, Brazier DJ, Spittle M. 5-Fluorouracil and ocular toxicity. *Br J Ophthalmol* 1993;77:465–466.

453. Fraunfelder FT, Meyer SM. Ocular toxicity of antineoplastic agents. *Ophthalmology* 1983;90:1–3.

454. Stevens A, Spooner D. Lacrimal duct stenosis and other ocular toxicity associated with adjuvant cyclophosphamide, methotrexate and 5-fluorouracil combination chemotherapy for early stage breast cancer. *Clin Oncol (R Coll Radiol)* 2001;13:438–440.

455. Delval L, Klastersky J. Optic neuropathy in cancer patients. Report of a case possibly related to 5 fluorouracil toxicity and review of the literature. *J Neuro-Oncol* 2002;60:165–169.

456. Jansman FG, Sleijfer DT, de Graaf JC, et al. Management of chemotherapy-induced adverse effects in the treatment of colorectal cancer. *Drug Saf* 2001;24:353–367.

457. El Amrani M, Heinzlef O, Debroucker T, et al. Brain infarction following 5-fluorouracil and cisplatin therapy. *Neurology* 1998;51:899–901.

458. van Laarhoven HW, Verstappen CC, Beex LV, et al. 5-FU-induced peripheral neuropathy: a rare complication of a well-known drug. *Anticancer Res* 2003;23:647–648.

459. Saif MW, Wilson RH, Harold N, et al. Peripheral neuropathy associated with weekly oral 5-fluorouracil, leucovorin and eniluracil. *Anticancer Drugs* 2001;12:525–531.

460. Stein ME, Drumea K, Yarnitsky D, et al. A rare event of 5-fluorouracil-associated peripheral neuropathy: a report of two patients. *Am J Clin Oncol* 1998;21:248–249.

461. Saif MW, Hashmi S, Mattison L, et al. Peripheral neuropathy exacerbation associated with topical 5-fluorouracil. *Anticancer Drugs* 2006;17:1095–1098.

462. Howell SB, Pfeifle CE, Wung WE. Effect of allopurinol on the toxicity of high-dose 5-fluorouracil administered by intermittent bolus injection. *Cancer* 1983;51:220–225.

463. Muggia FM, Camacho FJ, Kaplan BH, et al. Weekly 5-fluorouracil combined with PALA: toxic and therapeutic effects in colorectal cancer. *Cancer Treat Rep* 1987;71:253–256.

464. Wiley RG, Gralla RJ, Casper ES, et al. Neurotoxicity of the pyrimidine synthesis inhibitor N-phosphonoacetyl-l-aspartate. *Ann Neurol* 1982;12:175–183.

465. Kimmel DW, Schutt AJ. Multifocal leukoenceph-alopathy: occurrence during 5-fluorouracil and levamisole therapy and resolution after discontinuation of chemotherapy. *Mayo Clin Proc* 1993; 68:363–365.

466. Parkinson DR, Cano PO, Jerry LM, et al. Complications of cancer immunotherapy with Levamisole. *Lancet* 1977;1:1129–1132.

467. Kondo K, Fujiwara M, Murase M, et al. Severe acute metabolic acidosis and Wernicke's encephalopathy following chemotherapy with 5-fluorouracil and cisplatin: case report and review of the literature. *Jpn J Clin Oncol* 1996;26:234–236.

468. Walko CM, Lindley C. Capecitabine: a review. *Clin Ther* 2005;27:23–44.

469. Renouf D, Gill S. Capecitabine-induced cerebellar toxicity. *Clin Colorectal Cancer* 2006;6:70–71.

470. Formica V, Leary A, Cunningham D, et al. 5-Fluorouracil can cross brain-blood barrier and cause encephalopathy: should we expect the same from capecitabine? A case report on capecitabine-induced central neurotoxicity progressing to coma. *Cancer Chemother Pharmacol* 2006;58:276–278.

471. Niemann B, Rochlitz C, Herrmann R, et al. Toxic encephalopathy induced by capecitabine. *Oncology (Basel)* 2004;66:331–335.

472. Videnovic A, Semenov I, Chua-Adajar R, et al. Capecitabine-induced multifocal leukoen-cephalopathy: a report of five cases. *Neurology* 2005;65:1792–1794.

473. Saif MW, Wood TE, McGee PJ, et al. Peripheral neuropathy associated with capecitabine. *Anticancer Drugs* 2004;15:767–771.

474. Couch LS, Groteluschen DL, Stewart JA, et al. Capecitabine-related neurotoxicity presenting as trismus. *Clin Colorectal Cancer* 2003;3:121–123.

475. Buyck HC, Buckley N, Leslie MD, et al. Capecitabine-induced potentiation of warfarin. *Clin Oncol (R Coll Radiol)* 2003;15:297.

476. Ho AD, Hensel M. Pentostatin and purine analogs for indolent lymphoid malignancies. *Future Oncol* 2006;2:169–183.

477. Cheson BD, Vena DA, Foss FM, et al. Neurotoxicity of purine analogs: a review. *J Clin Oncol* 1994;12:2216–2228.

478. Cheson BD. Infectious and immunosuppressive complications of purine analog therapy. *J Clin Oncol* 1995;13:2431–2448.

479. Cheson BD, Frame JN, Vena D, et al. Tumor lysis syndrome: an uncommon complication of fludarabine therapy of chronic lymphocytic leukemia. *J Clin Oncol* 1998;16:2313–2320.

480. Grever MR, Kopecky KJ, Coltman CA, et al. Fludarabine monophosphate: a potentially useful agent in chronic lymphocytic leukemia. *Nouv Rev Fr Hematol* 1988;30:457–459.

481. Cohen RB, Abdallah JM, Gray JR, et al. Reversible neurologic toxicity in patients treated with standard-dose fludarabine phosphate for mycosis fungoides and chronic lymphocytic leukemia. *Ann Intern Med* 1993;118:114–116.

482. Chun HG, Leyland-Jones B, Caryk SM, et al. Central nervous system toxicity of fludarabine phosphate. *Cancer Treat Rep* 1986;70:1225–1228.

483. Warrell RP, Jr, Berman E. Phase I and II study of fludarabine phosphate in leukemia: therapeutic efficacy with delayed central nervous system toxicity. *J Clin Oncol* 1986;4:74–79.

484. Spriggs DR, Stopa E, Mayer RJ, et al. Fludarabine phosphate (NSC 312878) infusions for the treatment of acute leukemia: phase I and neuropathological study. *Cancer Res* 1986;46:5953–5958.

485. Samonis G, Kontoyiannis DP. Infectious complications of purine analog therapy. *Curr Opin Infect Dis* 2001;14:409–413.

486. Saumoy M, Castells G, Escoda L, et al. Progressive multifocal leukoencephalopathy in chronic lympho-cytic leukemia after treatment with fludarabine. *Leuk Lymphoma* 2002;43:433–436.

487. Sarmiento MA, Neme D, Fornari MC, et al. Guillain-Barre syndrome following 2-chlorodeoxy-adenosine treatment for Hairy Cell Leukemia. *Leuk Lymphoma* 2000;39:657–659.

488. Ho AD, Hensel M. Pentostatin for the treatment of indolent lymphoproliferative disorders. *Semin Hematol* 2006;43:S2–S10.

489. Major PP, Agarwal RP, Kufe DW. Deoxycoformycin: neurological toxicity. *Cancer Chemother Pharmacol* 1981;5:193–196.

490. Beckloff GL, Lerner HJ, Frost D, et al. Hydroxyurea (NSC-32065) in biologic fluids: dose-concentration relationship. *Cancer Chemother Rep* 1963;48:57–58.

491. Beitler JJ, Smith RV, Owen RP, et al. Phase II clinical trial of parenteral hydroxyurea and hyperfractionated, accelerated external beam radiation therapy in patients with advanced squamous cell carcinoma of the head and neck: toxicity and efficacy with continuous ribonu-cleoside reductase inhibition. *Head Neck* 2006; 29(1):18–25.

492. Ragel BT, Gillespie DL, Kushnir V, et al. Calcium channel antagonists augment hydroxyu-rea- and ru486-induced inhibition of menin-gioma growth in vivo and in vitro. *Neurosurgery* 2006;59:1109–1120.

493. Reardon DA, Egorin MJ, Quinn JA, et al. Phase II study of imatinib mesylate plus hydroxyurea in adults with recurrent glioblastoma multiforme. *J Clin Oncol* 2005;23:9359–9368.

494. Barry M, Clarke S, Mulcahy F, et al. Hydroxyurea-induced neurotoxicity in HIV disease. *AIDS* 1999;13:1592–1594.

495. Gidding CE, Kellie SJ, Kamps WA, et al. Vincristine revisited. *Crit Rev Oncol Hematol* 1999;29:267–287.

496. Owellen RJ, Hartke CA, Dickerson RM, et al. Inhibition of tubulin-microtubule polymerization by drugs of the Vinca alkaloid class. *Cancer Res* 1976;36:1499–1502.

497. Topp KS, Tanner KD, Levine JD. Damage to the cytoskeleton of large diameter sensory neurons and myelinated axons in vincristine-induced painful peripheral neuropathy in the rat. *J Comp Neurol* 2000;424:563–576.

498. Jackson DV, Wells HB, Atkins JN, et al. Amelioration of vincristine neurotoxicity by glutamic acid. *Am J Med* 1988;84:1016–1022.

499. Alcaraz A, Rey C, Concha A, et al. Intrathecal vin-cristine: fatal myeloencephalopathy despite cere-bral spinal fluid perfusion. *J Toxicol Clin Toxicol* 2002;40:557–561.

500. Gaidys WG, Dickerman JD, Walters CL, et al. Intrathecal vincristine. Report of a fatal case despite CNS washout. *Cancer* 1983;52:799–801.

501. Williams ME, Walker AN, Bracikowski JP, et al. Ascending myeloencephalopathy due to intrathecal vincristine sulfate. A fatal chemotherapeutic error. *Cancer* 1983;51:2041–2047.

502. DeAngelis LM, Gnecco C, Taylor L, et al. Evolution of neuropathy and myopathy during intensive vincristine/corticosteroid chemotherapy for non-Hodgkin's lymphoma. *Cancer* 1991;67:2241–2246.

503. Moudgil SS, Riggs JE. Fulminant peripheral neuropathy with severe quadriparesis associated with vincristine therapy. *Ann Pharmacother* 2000;34:1136–1138.

504. Chauvenet AR, Shashi V, Selsky C, et al. Vincristine-induced neuropathy as the initial presentation of charcot-marie-tooth disease in acute lymphoblastic leukemia: a Pediatric Oncology Group study. *J Pediatr Hematol Oncol* 2003;25:316–320.

505. Postma TJ, Benard BA, Huijgens PC, et al. Long term effects of vincristine on the peripheral nervous system. *J Neuro-Oncol* 1993;15:23–27.

506. McCarthy GM, Skillings JR. A prospective cohort study of the orofacial effects of vincristine neurotoxicity. *J Oral Pathol Med* 1991;20:345–349.

507. Sathiapalan RK, El Solh H. Enhanced vincristine neurotoxicity from drug interactions: case report and review of literature. *Pediatr Hematol Oncol* 2001;18:543–546.

508. Naumann R, Mohm J, Reuner U, et al. Early recognition of hereditary motor and sensory neuropathy type 1 can avoid life-threatening vincristine neurotoxicity. *Br J Haematol* 2001;115:323–325.

509. Hildebrandt G, Holler E, Woenkhaus M, et al. Acute deterioration of Charcot-Marie-Tooth disease IA (CMT IA) following 2 mg of vincristine chemotherapy. *Ann Oncol* 2000;11:743–747.

510. Weintraub M, Adde MA, Venzon DJ, et al. Severe atypical neuropathy associated with administration of hematopoietic colony-stimulating factors and vincristine. *J Clin Oncol* 1996;14:935–940.

511. Cowie F, Barrett A. Uneventful administration of cisplatin to a man with X-linked Charcot-Marie-Tooth disease (CMT). *Ann Oncol* 2001;12:422.

512. Parimoo D, Jeffers S, Muggia FM. Severe neurotoxicity from vinorelbine-paclitaxel combinations. *J Natl Cancer Inst* 1996;88:1079–1080.

513. Fazeny B, Zifko U, Meryn S, et al. Vinorelbine induced neurotoxicity in patients with advanced breast cancer pretreated with paclitaxel—a phase II study. *Cancer Chemother Pharmacol* 1996;39:150–156.

514. Perrault DJ, Turley J, Quirt I, et al. A prospective study of vincristine neuropathy. Abstract. *Proc Am Soc Clin Oncol* 1988;7:A1115.

515. Watkins SM, Griffin JP. High incidence of vincristine-induced neuropathy in lymphomas. *Br Med J* 1978;1:610–612.

516. Thant M, Hawley RJ, Smith MT, et al. Possible enhancement of vincristine neuropathy by VP-16. *Cancer* 1982;49:859–864.

517. Griffiths JD, Stark RJ, Ding JC, et al. Vincristine neurotoxicity enhanced in combination chemotherapy including both teniposide and vincristine. *Cancer Treat Rep* 1986;70:519–521.

518. Weber DM, Dimopoulos MA, Alexanian R. Increased neurotoxicity with VAD-cyclosporin in multiple myeloma. Letter to the editor. *Lancet* 1993; 341:558–559.

519. Cassady JR, Tonnesen GL, Wolfe LC, et al. Augmentation of vincristine neurotoxicity by irradiation of peripheral nerves. *Cancer Treat Rep* 1980; 64:963–965.

520. Di Gregorio F, Favaro G, Panozzo C, et al. Efficacy of ganglioside treatment in reducing functional alterations induced by vincristine in rabbit peripheral nerves. *Cancer Chemother Pharmacol* 1990;26:31–36.

521. Houi K, Mochio S, Kobayashi T. Gangliosides attenuate vincristine neurotoxicity on dorsal root ganglion cells. *Muscle Nerve* 1993;16:11–14.

522. Le Quesne PM, Fowler CJ, Harding AE. A study of the effects of isaxonine on vincristine-induced peripheral neuropathy in man and regeneration following peripheral nerve crush in the rat. *J Neurol Neurosurg Psychiatry* 1985;48:933–935.

523. Jackson DV, Jr, Pope EK, McMahan RA, et al. Clinical trial of pyridoxine to reduce vincristine neurotoxicity. *J Neuro-oncol* 1986;4:37–41.

524. Hayakawa K, Itoh T, Niwa H, et al. NGF prevention of neurotoxicity induced by cisplatin, vincristine and taxol depends on toxicity of each drug and NGF treatment schedule: in vitro study of adult rat sympathetic ganglion explants. *Brain Res* 1998;794:313–319.

525. Koeppen S, Verstappen CC, Korte R, et al. Lack of neuroprotection by an ACTH (4–9) analogue. A randomized trial in patients treated with vincristine for Hodgkin's or non-Hodgkin's lymphoma. *J Cancer Res Clin Oncol* 2004;130:153–160.

526. Grush OC, Morgan SK. Folinic acid rescue for vincristine toxicity. *Clin Toxicol* 1979;14:71–78.

527. Paladine W, Belle N, Weaver JW. Reduction of vincristine (VCR) paresthesias of the hand. Abstract. *Proc Am Soc Clin Oncol* 1989; 8:A1290.

528. Verstappen CC, Heimans JJ, Hoekman K, et al. Neurotoxic complications of chemotherapy in patients with cancer: clinical signs and optimal management. *Drugs* 2003;63:1549–1563.

529. Albert DM, Wong VG, Henderson ES. Ocular complications of vincristine therapy. *Arch Ophthalmol* 1967;78:709–713.

530. Ryan SP, DelPrete SA, Weinstein PW, et al. Low-dose vincristine-associated bilateral vocal cord paralysis. *Conn Med* 1999;63:583–584.

531. Delaney P. Vincristine-induced laryngeal nerve paralysis. *Neurology* 1982;32:1285–1288.

532. Mahajan SL, Ikeda Y, Myers TJ, et al. Acute acoustic nerve palsy associated with vincristine therapy. *Cancer* 1981;47:2404–2406.

533. Shurin SB, Rekate HL, Annable W. Optic atrophy induced by vincristine. *Pediatrics* 1982;70:288–291.

534. Sanderson PA, Kuwabara T, Cogan DG. Optic neuropathy presumably caused by vincristine therapy. *Am J Ophthalmol* 1976;81:146–150.

535. Teichmann KD, Dabbagh N. Severe visual loss after a single dose of vincristine in a patient with spinal cord astrocytoma. *J Ocul Pharmacol* 1988;4:117–121.

536. Ripps H, Carr RE, Siegel IM, et al. Functional abnormalities in vincristine-induced night blindness. *Invest Ophthalmol Vis Sci* 1984;25:787–794.

537. Garewal HS, Dalton WS. Metaclopramide in vincristine-induced ileus. *Cancer Treat Rep* 1985;69:1309–1311.

538. Carmichael SM, Eagleton L, Ayers CR, et al. Orthostatic hypotension during vincristine therapy. *Arch Intern Med* 1970;126:290–293.

539. Roca E, Bruera E, Politi PM, et al. Vinca alkaloid-induced cardiovascular autonomic neuropathy. *Cancer Treat Rep* 1985;69:149–151.

540. Wang WS, Chiou TJ, Liu JH, et al. Vincristine-induced dysphagia suggesting esophageal motor dysfunction: a case report. *Jpn J Clin Oncol* 2000;30:515–518.

541. Kellie SJ, Barbaric D, Koopmans P, et al. Cerebrospinal fluid concentrations of vincristine after bolus intravenous dosing—a surrogate marker of brain penetration. *Cancer* 2002;94:1815–1820.

542. O'Callaghan MJ, Ekert H. Vincristine toxicity unrelated to dose. *Arch Dis Childhood* 1976; 51:289–292.

543. Robertson GL, Bhoopalam N, Zelkowitz LJ. Vincristine neurotoxicity and abnormal secretion of antidiuretic hormone. *Arch Intern Med* 1973;132:717–720.

544. Whittaker JA, Parry DH, Bunch C, et al. Coma associated with vincristine therapy. *Br Med J* 1973;4:335–337.

545. Dallera F, Gamoletti R, Costa P. Unilateral seizures following vincristine intravenous injection. *Tumori* 1984;70:243–244.

546. Hurwitz RL, Mahoney DH, Jr, Armstrong DL, et al. Reversible encephalopathy and seizures as a result of conventional vincristine administration. *Med Pediatr Oncol* 1988;16:216–219.

547. Byrd RL, Rohrbaugh TM, Raney RB, Jr, et al. Transient cortical blindness secondary to vincristine therapy in childhood malignancies. *Cancer* 1981;47:37–40.

548. Scheithauer W, Ludwig H, Maida E. Acute encephalopathy associated with continuous vincristine sulfate combination therapy: case report. *Invest New Drugs* 1985;3:315–318.

549. Carpentieri U, Lockhart LH. Ataxia and athetosis as side effects of chemotherapy with vincristine in non-Hodgkin's lymphoma. *Cancer Treat Rep* 1978;62:561–562.

550. Boranic M, Raci F. A Parkinson-like syndrome as side effect of chemotherapy with vincristine and adriamycin in a child with acute leukaemia. *Biomedicine* 1979;31:124–125.

551. McLeod JG, Penny R. Vincristine neuropathy: an electrophysiological and histological study. *J Neurol Neurosurg Psychiatry* 1969;32:297–304.

552. Cho E-S, Lowndes HE, Goldstein BD. Neurotoxicology of vincristine in the cat. Morphological study. *Arch Otoxicol* 1983;52:83–90.

553. Sahenk Z, Brady ST, Mendell JR. Studies on the pathogenesis of vincristine-induced neuropathy. *Muscle Nerve* 1987;10:80–84.

554. Munier F, Perentes E, Herbort CP, et al. Selective loss of optic nerve β-tubulin in vincristine-induced blindness. Letter to the editor. *Am J Med* 1992;93:232–234.

555. Lynch JJ, III, Wade CL, Zhong CM, et al. Attenuation of mechanical allodynia by clinically utilized drugs in a rat chemotherapy-induced neuropathic pain model. *Pain* 2004;110:56–63.

556. King KL, Boder GB. Correlation of the clinical neurotoxicity of the vinca alkaloids vincristine, vinblastine, and vindesine with their effects on cultured rat midbrain cells. *Cancer Chemother Pharmacol* 1979;2:239–242.

557. Lobert S, Vulevic B, Correia JJ. Interaction of vinca alkaloids with tubulin: a comparison of vinblastine, vincristine, and vinorelbine. *Biochemistry* 1996;35:6806–6814.

558. Focan C, Olivier R, Le Hung S, et al. Neurological toxicity of vindesine used in combination chemotherapy of 51 human solid tumors. *Cancer Chemother Pharmacol* 1981;6:175–181.

559. Heran F, Defer G, Brugieres P, et al. Cortical blindness during chemotherapy: clinical, CT, and MR correlations. *J Comput Assist Tomogr* 1990;14:262–266.

560. Conter V, Rabbone ML, Jankovic M, et al. Overdose of vinblastine in a child with Langerhans' cell histiocytosis: toxicity and salvage therapy. *Pediatr Hematol Oncol* 1991;8:165–169.

561. Pace A, Bove L, Nistico C, et al. Vinorelbine neurotoxicity: clinical and neurophysiological findings in 23 patients. *J Neurol Neurosurg Psychiatry* 1996;61:409–411.

562. Chang M-H, Liao K-K, Wu Z-A, et al. Reversible myeloneuropathy resulting from podophyllin intoxication: an electrophysiological follow up. Letter to the editor. *J Neurol Neurosurg Psychiatry* 1992;55:235–236.

563. Ng THK, Chan YW, Yu YL, et al. Encephalopathy and neuropathy following ingestion of a Chinese herbal broth containing podophyllin. *J Neurol Sci* 1991;101:107–113.

564. Slevin ML. The clinical pharmacology of etoposide. *Cancer* 1991;67:319–329.

565. Savaraj N, Feun LG, Lu K, et al. Pharmacology of intrathecal VP-16–213 in dogs. *J Neuro-Oncol* 1992;13:211–215.

566. Ogasawara H, Kiya K, Kurisu K, et al. Effect of intracarotid infusion of etoposide with angiotensin II-induced hypertension on the blood-brain barrier and the brain tissue. *J Neuro-oncol* 1992;13:111–117.

567. Spigelman MK, Zappulla RA, Strauchen JA, et al. Etoposide induced blood-brain barrier disruption in rats: duration of opening and histological sequelae. *Cancer Res* 1986;46:1453–1457.

568. Leff RS, Thompson JM, Daly MB, et al. Acute neurologic dysfunction after high-dose etoposide therapy for malignant glioma. *Cancer* 1988; 62:32–35.

569. O'Dwyer PJ, Alonso MT, Leyland-Jones B, et al. Teniposide: a review of 12 years experience. *Cancer Treat Rep* 1984;68:1455–1466.

570. Einzig AI, Wiernik PH, Sasloff J, et al. Phase II study and long-term follow-up of patients treated with taxol for advanced ovarian adenocarcinoma. *J Clin Oncol* 1992;10:1748–1753.

571. Rowinsky EK, Chaudhry V, Cornblath DR, et al. Neurotoxicity of taxol. *J Natl Cancer Inst Monogr* 1993;15:107–115.

572. Theiss C, Meller K. Taxol impairs anterograde axonal transport of microinjected horseradish peroxidase in dorsal root ganglia neurons in vitro. *Cell Tissue Res* 2000;299:213–224.

573. Chaudhry V, Rowinsky EK, Sartorius SE, et al. Peripheral neuropathy from Taxol and cisplatin combination chemotherapy: clinical and electrophysiological studies. *Ann Neurol* 1994;35:304–311.

574. Lee JJ, Swain SM. Peripheral neuropathy induced by microtubule-stabilizing agents. *J Clin Oncol* 2006;24:1633–1642.

575. Hilkens PH, Verweij J, Vecht CJ, et al. Clinical characteristics of severe peripheral neuropathy induced by docetaxel (Taxotere). *Ann Oncol* 1997;8:187–190.

576. Forsyth PA, Balmaceda C, Peterson K, et al. Prospective study of paclitaxel-induced peripheral neuropathy with quantitative sensory testing. *J Neuro-Oncol* 1997;35:47–53.

577. Holmes FA, Walters RS, Theriault RL, et al. Phase II trial of taxol, an active drug in the treatment of metastatic breast cancer. *J Natl Cancer Inst* 1991;83:1797–1805.

578. Martoni A, Zamagni C, Gheka A, et al. Antihistamines in the treatment of taxol-induced paroxystic pain syndrome. *J Natl Cancer Inst* 1993;85:676.

579. Seidman AD, Barrett S. Photopsia during 3-hour paclitaxel administration at doses > 250 mg/m². *J Clin Oncol* 1994;12:1741–1742.

580. Jerian SM, Sarosy GA, Link CJ, Jr, et al. Incapacitating autonomic neuropathy precipitated by taxol. *Gynecol Oncol* 1993;51:277–280.

581. van den Bent MJ, Hilkens PHE, Smitt PAES, et al. Lhermitte's sign following chemotherapy with docetaxel. *Neurology* 1998;50:563–564.

582. Nieto Y, Cagnoni PJ, Bearman SI, et al. Acute encephalopathy: a new toxicity associated with high-dose paclitaxel. *Clin Cancer Res* 1999;5:501–506.

583. Perry JR, Warner E. Transient encephalopathy after paclitaxel (Taxol) infusion. *Neurology* 1996;46:1596–1599.

584. Chan AT, Yeo W, Leung WT, et al. Thromboembolic events with paclitaxel. *Clin Oncol (R Coll Radiol)* 1996;8:133.

585. Khattab J, Terebelo HR, Dabas B. Phantom limb pain as a manifestation of paclitaxel neurotoxicity. *Mayo Clin Proc* 2000;75:740–742.

586. Freilich RJ, Balmaceda C, Seidman AD, et al. Motor neuopathy due to docetaxel and paclitaxel. *Neurology* 1996;47:115–118.

587. Gelmon K, Eisenhauer E, Bryce C, et al. Randomized phase ii study of high-dose paclitaxel with or without amifostine in patients with metastatic breast cancer. *J Clin Oncol* 1999;17:3038–3047.

588. Pronk LC, Hilkens PH, van den Bent MJ, et al. Corticosteroid co-medication does not reduce the incidence and severity of neurotoxicity induced by docetaxel. *Anticancer Drugs* 1998;9:759–764.

589. Argyriou AA, Chroni E, Koutras A, et al. Preventing paclitaxel-induced peripheral neuropathy: a phase II trial of vitamin E supplementation. *J Pain Symptom Manage* 2006;32:237–244.

590. Flatters SJ, Xiao WH, Bennett GJ. Acetyl-l-carnitine prevents and reduces paclitaxel-induced painful peripheral neuropathy. *Neurosci Lett* 2006;397:219–223.

591. Stubblefield MD, Vahdat LT, Balmaceda CM, et al. Glutamine as a neuroprotective agent in high-dose paclitaxel-induced peripheral neuropathy: a clinical and electrophysiologic study. *Clin Oncol (R Coll Radiol)* 2005;17:271–276.

592. Wasner G, Hilpert F, Schattschneider J, et al. Docetaxel-induced nail changes—a neurogenic mechanism: a case report. *J Neuro-oncol* 2002; 58:167–174.

593. Hussain S, Anderson DN, Salvatti ME, et al. Onycholysis as a complication of systemic chemotherapy: report of five cases associated with prolonged weekly paclitaxel therapy and review of the literature. *Cancer* 2000;88:2367–2371.

594. Scotte F, Tourani JM, Banu E, et al. Multicenter study of a frozen glove to prevent docetaxel-induced onycholysis and cutaneous toxicity of the hand. *J Clin Oncol* 2005;23:4424–4429.

595. Athanasiadis A, Tsavdaridis D, Rigatos SK, et al. Hormone refractory advanced prostate cancer treated with estramustine and paclitaxel combination. *Anticancer Res* 200;23:3085–3088.

596. Ferrari AC, Chachoua A, Singh H, et al. A phase I/II study of weekly paclitaxel and 3 days of high dose oral estramustine in patients with hormone-refractory prostate carcinoma. *Cancer* 2001;91:2039–2045.

597. Hudes G, Haas N, Yeslow G, et al. Phase I clinical and pharmacologic trial of intravenous estramustine phosphate. *J Clin Oncol* 2002;20:1115–1127.

598. Sordet O, Goldman A, Pommier Y. Topoisomerase II and tubulin inhibitors both induce the formation of apoptotic topoisomerase I cleavage complexes. *Mol Cancer Ther* 2006;5:3139–3144.

599. Vahdat L. Ixabepilone: a novel antineoplastic agent with low susceptibility to multiple tumor resistance mechanisms. *Oncologist* 2008;13:214–221.

600. Lee JJ, Low JA, Croarkin E, et al. Changes in neurologic function tests may predict neurotoxicity caused by ixabepilone. *J Clin Oncol* 2006;24:2084–2091.

601. Riccardi R, Holcenberg JS, Glaubiger DL, et al. l-Asparaginase pharmacokinetics and asparagine levels in cerebrospinal fluid of rheusus monkeys and humans. *Cancer Res* 1981;41:4554–4558.

602. Leonard JV, Kay JD. Acute encephalopathy and hyperammonaemia complicating treatment of acute lymphoblastic leukemia with asparaginase. *Lancet* 1986;1:162–163.

603. Steiner M, Attarbaschi A, Kastner U, et al. Distinct fluctuations of ammonia levels during asparaginase therapy for childhood acute leukemia. *Pediatr Blood Cancer* 2006;49(5):640–642.

604. Cairo MS, Lazarus K, Gilmore RL, et al. Intracranial hemorrhage and focal seizures secondary to use of l-asparaginase during induction of therapy of acute lymphocytic leukemia. *J Pediatr* 1980;97:829–833.

605. Feinberg WM, Swenson MR. Cerebrovascular complications of l-asparaginase therapy. *Neurology* 1988;38:127–133.

606. Elliott MA, Wolf RC, Hook CC, et al. Thromboembolism in adults with acute lymphoblastic leukemia during induction with l-asparaginase-containing multi-agent regimens: incidence, risk factors, and possible role of antithrombin. *Leuk Lymphoma* 2004;45:1545–1549.

607. Nowak-Gottl U, Ahlke E, Fleischhack G, et al. Thromboembolic events in children with acute lymphoblastic leukemia (BFM protocols): prednisone versus dexamethasone administration. *Blood* 2003;101:2529–2533.

608. Van Echo DA, Chiuten DF, Gormley PE, et al. Phase 1 clinical and pharmacological study of

4'-(9-acridinylamino)-methanesulfon-m-anisidide using an intermittent biweekly schedule. *Cancer Res* 1979;39:3881–3884.

609. Cabanillas F, Legha SS, Bodey GP, et al. Initial experience with AMSA as single agent treatment against malignant lymphoproliferative disorders. *Blood* 1981;57:614–616.

610. Sung WJ, Kim DH, Sohn SK, et al. Phase II trial of amsacrine plus intermediate-dose Ara-C (IDAC) with or without etoposide as salvage therapy for refractory or relapsed acute leukemia. *Jpn J Clin Oncol* 2005;35:612–616.

611. Myers C, Cooper M, Stein C, et al. Suramin: a novel growth factor antagonist with activity in hormone-refractory metastatic prostate cancer. *J Clin Oncol* 1992;10:881–889.

612. Minniti CP, Maggi M, Helman LJ. Suramin inhibits the growth of human rhabdomyosarcoma by interrupting the insulin-like growth factor II autocrine growth loop. *Cancer Res* 1992;52:1830–1835.

613. Taylor CW, Lui R, Fanta P, et al. Effects of suramin on in vitro growth of fresh human tumors. *J Natl Cancer Inst* 1992;84:489–494.

614. Soliven B, Dhand UK, Kobayashi K, et al. Evaluation of neuropathy in patients on suramin treatment. *Muscle Nerve* 1997;20:83–91.

615. La Rocca RV, Meer J, Gilliatt RW, et al. Suramin-induced polyneuropathy. *Neurology* 1990; 40:954–960.

616. Di Paolo A, Danesi R, Innocenti F, et al. Chronic administration of suramin induces neurotoxicity in rats. *J Neurol Sci* 1997;152:125–131.

617. Hussain M, Fisher EI, Petrylak DP, et al. Androgen deprivation and four courses of fixed-schedule suramin treatment in patients with newly diagnosed metastatic prostate cancer: a Southwest Oncology Group Study. *J Clin Oncol* 2000;18:1043–1049.

618. Hemady RK, Sinibaldi VJ, Eisenberger MA. Ocular symptoms and signs associated with suramin sodium treatment for metastatic cancer of the prostate. *Am J Ophthalmol* 1996;121:291–296.

619. Russell JW, Windebank AJ, Podratz JL. Role of nerve growth factor in suramin neurotoxicity studies in vitro. *Ann Neurol* 1994;36:221–228.

620. Blaney SM, Takimoto C, Murry DJ, et al. Plasma and cerebrospinal fluid pharmacokinetics of 9-aminocamptothecin (9-AC), irinotecan (CPT-11), and SN-38 in nonhuman primates. *Cancer Chemother Pharmacol* 1998;41:464–468.

621. Hamberg P, Donders RC, ten Bokkel HD. Central nervous system toxicity induced by irinotecan. *J Natl Cancer Inst* 2006;98:219.

622. De MS, Squilloni E, Vigna L, et al. Irinotecan chemotherapy associated with transient dysarthria and aphasia. *Ann Oncol* 2004;15:1147–1148.

623. Gandia D, Abigerges D, Armand JP, et al. CPT-11-induced cholinergic effects in cancer patients. *J Clin Oncol* 1993;11:196–197.

624. Dodds HM, Bishop JF, Rivory LP. More about: irinotecan-related cholinergic syndrome induced by coadministration of oxaliplatin. *J Natl Cancer Inst* 1999;91:91–92.

625. Isoardo G, Bergui M, Durelli L, et al. Thalidomide neuropathy: clinical, electrophysiological and neuroradiological features. *Acta Neurol Scand* 2004; 109:188–193.

626. Chaudhry V, Cornblath DR, Corse A, et al. Thalidomide-induced neuropathy. *Neurology* 2002;59:1872–1875.

627. Tosi P, Zamagni E, Cellini C, et al. Neurological toxicity of long-term (>1 yr) thalidomide therapy in patients with multiple myeloma. *Eur J Haematol* 2005;74:212–216.

628. Mileshkin L, Stark R, Day B, et al. Development of neuropathy in patients with myeloma treated with thalidomide: patterns of occurrence and the role of electrophysiologic monitoring. *J Clin Oncol* 2006;24:4507–4514.

629. Sohlbach K, Heinze S, Shiratori K, et al. Encephalopathy in a patient after long-term treatment with thalidomide. *J Clin Oncol* 2006; 24:4942–4944.

630. Sharma RA, Steward WP, Daines CA, et al. Toxicity profile of the immunomodulatory thalidomide analogue, lenalidomide: phase I clinical trial of three dosing schedules in patients with solid malignancies. *Eur J Cancer* 2006;42:2318–2325.

631. Bennett CL, Angelotta C, Yarnold PR, et al. Thalidomide- and lenalidomide-associated thromboembolism among patients with cancer. *JAMA* 2006;296:2558–2560.

631a. Menon SP, Rajkumar SV, Lacy M, et al. Thromboembolic events with lenalidomide-based therapy for multiple myeloma. *Cancer* 2008; 112:1522–28.

632. Winick NJ, Bowman WP, Kamen BA, et al. Unexpected acute neurological toxicity in the treatment of children with acute lymphoblastic leukemia. *J Natl Cancer Inst* 1992;84:252–256.

633. Wall JG, Weiss RB, Norton L, et al. Arterial thrombosis associated with adjuvant chemotherapy for breast carcinoma: a Cancer and Leukemia Group B study. *Am J Med* 1989;87:501–504.

634. Meessen S, Riedel RR, Bruhl P. Status epilepticus in MVEC chemotherapy of urothelial cancer. *Urologe [A]* 1990;29:348–349.

635. Lien EA, Wester K, Lonning PE, et al. Distribution of tamoxifen and metabolites into brain tissue and brain metastases in breast cancer patients. *Br J Cancer* 1991;63:641–645.

636. Nayfield SG, Gorin MB. Tamoxifen-associated eye disease. A review. *J Clin Oncol* 1996;14:1018–1026.

637. Gianni L, Panzini I, Li S, et al. Ocular toxicity during adjuvant chemoendocrine therapy for early breast cancer: results from International Breast Cancer Study Group trials. *Cancer* 2006;106:505–513.

638. Fernando IN, Tobias JS. Priapism in patient on tamoxifen. Letter to the editor. *Lancet* 1989;1:436.

639. Ron IG, Inbar MJ, Barak Y, et al. Organic delusional syndrome associated with tamoxifen treatment. *Cancer* 1992;69:1415–1417.

640. Love RR. Tamoxifen therapy in primary breast cancer: biology, efficacy, and side effects. *J Clin Oncol* 1989;7:803–815.

641. Pluss JL, DiBella NJ. Reversible central nervous system dysfunction due to tamoxifen in a patient with breast cancer. *Ann Intern Med* 1984;101:652.

642. Parry BR. Radiation recall induced by tamoxifen. Letter to the editor. *Lancet* 1992;340:49.

643. Saphner T, Tormey DC, Gray R. Venous and arterial thrombosis in patients who received adjuvant therapy for breast cancer. *J Clin Oncol* 1991; 9:286–294.

644. Gradishar WJ. Safety considerations of adjuvant therapy in early breast cancer in postmenopausal women. *Oncology* 2005;69:1–9.

645. Barrett-Connor E, Mosca L, Collins P, et al. Effects of raloxifene on cardiovascular events and breast cancer in postmenopausal women. *N Engl J Med* 2006;355:125–137.

646. Harvey HA, Kimura M, Hajba A. Toremifene: an evaluation of its safety profile. *Breast* 2006;15:142–157.

647. Buzdar A, Howell A, Cuzick J, et al. Comprehensive side-effect profile of anastrozole and tamoxifen as adjuvant treatment for early-stage breast cancer: long-term safety analysis of the ATAC trial. *Lancet Oncol* 2006;7:633–643.

648. Pandya N, Morris GJ. Toxicity of aromatase inhibitors. *Semin Oncol* 2006;33:688–695.

649. Ohsako T, Inoue K, Nagamoto N, et al. Joint symptoms: a practical problem of anastrozole. *Breast Cancer* 2006;13:284–288.

650. Pannuti F, Martoni A, Di Marco AR, et al. Prospective, randomized clinical trial of two different high dosages of medroxyprogesterone acetate (MAP) in the treatment of metastatic breast cancer. *Eur J Cancer* 1979;15:593–601.

651. Carey MS, Gawlik C, Fung-Kee-Fung M, et al. Systematic review of systemic therapy for advanced or recurrent endometrial cancer. *Gynecol Oncol* 2006;101:158–167.

652. Albertazzi P, Bottazzi M, Steel SA. Bone mineral density and depot medroxyprogesterone acetate. *Contraception* 2006;73:577–583.

653. Omar HA, Zakharia RM, Kanungo S, et al. Incidence of galactorrhea in young women using depot-medroxyprogesterone acetate. *Scientific World J* 2006;6:538–541.

654. Dux S, Bishara J, Marom D, et al. Medroxyprogesterone acetate-induced secondary adrenal insufficiency. *Ann Pharmacother* 1998;32:134.

655. Gillatt D. Antiandrogen treatments in locally advanced prostate cancer: are they all the same? *J Cancer Res Clin Oncol* 2006;132 (Suppl 13):17–26.

656. Gommersall LM, Hayne D, Shergill IS, et al. Luteinising hormone releasing hormone analogues in the treatment of prostate cancer. *Expert Opin Pharmacother* 2002;3:1685–1692.

657. Cheer SM, Plosker GL, Simpson D, et al. Goserelin: a review of its use in the treatment of early breast cancer in premenopausal and perimenopausal women. *Drugs* 2005;65:2639–2655.

658. Smith MR, Boyce SP, Moyneur E, et al. Risk of clinical fractures after gonadotropin-releasing hormone agonist therapy for prostate cancer. *J Urol* 2006;175:136–139.

659. Massoud W, Paparel P, Lopez JG, et al. Discovery of a pituitary adenoma following treatment with a gonadotropin-releasing hormone agonist in a patient with prostate cancer. *Int J Urol* 2006;13:87–88.

660. Kohen I, Koppel J. Goserelin-induced new-onset depressive disorder. *Psychosomatics* 2006;47:360–361.

661. Ernst G, Gericke A, Berg P. Central pain and complex motoric symptoms after gosarelin therapy of prostate cancer. *Scientific World J* 2004; 4:969–973.

662. Artale S, Giannetta L, Cerea G, et al. Treatment of metastatic neuroendocrine carcinomas based on WHO classification. *Anticancer Res* 2005;25:4463–4469.

663. Grottoli S, Celleno R, Gasco V, et al. Efficacy and safety of 48 weeks of treatment with octreotide LAR in newly diagnosed acromegalic patients with macroadenomas: an open-label, multicenter, non-comparative study. *J Endocrinol Invest* 2005; 28:978–983.

664. Rosenoff SH, Gabrail NY, Conklin R, et al. A multicenter, randomized trial of long-acting octreotide for the optimum prevention of chemotherapy-induced diarrhea: results of the STOP trial. *J Support Oncol* 2006;4:289–294.

665. Matulonis UA, Seiden MV, Roche M, et al. Long-acting octreotide for the treatment and symptomatic relief of bowel obstruction in advanced ovarian cancer. *J Pain Symptom Manage* 2005;30:563–569.

666. Fernandez-Real JM, Recasens M, Ricart W. Octreotide-induced manic episodes in a patient with acromegaly. *Ann Intern Med* 2006;2:144:704.

667. Seifert G, Reindl T, Lobitz S, et al. Fatal course after administration of rituximab in a boy with relapsed all: a case report and review of literature. *Haematologica* 2006;91:ECR23.

668. Osterborg A, Karlsson C, Lundin J, et al. Strategies in the management of alemtuzumab-related side effects. *Semin Oncol* 2006;33:S29–S35.

669. Burstein HJ, Lieberman G, Slamon DJ, et al. Isolated central nervous system metastases in patients with HER2-overexpressing advanced breast cancer treated with first-line trastuzumab-based therapy. *Ann Oncol* 2005;16:1772–1777.

670. Abrey LE. Does trastuzumab increase the risk of isolated CNS metastases in patients with breast cancer? *Nat Clin Pract Neurol* 2006;2:302–303.

671. Sandler A, Gray R, Perry MC, et al. Paclitaxel-carboplatin alone or with bevacizumab for non-small-cell lung cancer. *N Engl J Med* 2006;355:2542–2550.

672. Glusker P, Recht L, Lane B. Reversible posterior leukoencephalopathy syndrome and bevacizumab. *N Engl J Med* 2006 2;354:980–982.

673. Allen JA, Adlakha A, Bergethon PR. Reversible posterior leukoencephalopathy syndrome after bevacizumab/FOLFIRI regimen for metastatic colon cancer. *Arch Neurol* 2006;63:1475–1478.

674. Gordon MS, Cunningham D. Managing patients treated with bevacizumab combination therapy. *Oncology* 2005;69 (Suppl 3):25–33.

675. Zivkovic SA. Rituximab in the treatment of peripheral neuropathy associated with monoclonal gammopathy. *Expert Rev Neurother* 2006; 6:1267–1274.

676. Kavanaugh AF. B-cell targeted therapies: safety considerations. *J Rheumatol Suppl* 2006;77:18–23.

677. Lannoy D, Decaudin B, Grozieux de LA, et al. Gemtuzumab ozogamicin-induced sinusoidal obstructive syndrome treated with defibrotide: a case report. *J Clin Pharm Ther* 2006;31:389–392.

678. Sievers EL, Larson RA, Stadtmauer EA, et al. Efficacy and safety of gemtuzumab ozogamicin in patients with CD33-positive acute myeloid leukemia in first relapse. *J Clin Oncol* 2001;19:3244–3254.

679. Dranko S, Kinney C, Ramanathan RK. Ocular toxicity related to cetuximab monotherapy in patients with colorectal cancer. *Clin Colorectal Cancer* 2006;6:224–225.

680. Fakih MG, Wilding G, Lombardo J. Cetuximab-induced hypomagnesemia in patients with colorectal cancer. *Clin Colorectal Cancer* 2006; 6:152–156.

681. Cohenuram M, Saif MW. Panitumumab the first fully human monoclonal antibody: from the bench to the clinic. *Anticancer Drugs* 2007;18:7–15.

682. Yeon CH, Pegram MD. Anti-erbB-2 antibody trastuzumab in the treatment of HER2-amplified breast cancer. *Invest New Drugs* 2005;23:391–409.

683. Platini C, Long J, Walter S. Meningeal carcinomatosis from breast cancer treated with intrathecal trastuzumab. *Lancet Oncol* 2006;7:778–780.

684. Gordon MS, Matei D, Aghajanian C, et al. Clinical activity of pertuzumab (rhuMAb 2C4), a HER dimerization inhibitor, in advanced ovarian cancer: potential predictive relationship with tumor HER2 activation status. *J Clin Oncol* 2006;24:4324–4332.

685. Kushner BH, Kramer K, LaQuaglia MP, et al. Neuroblastoma in adolescents and adults: the Memorial Sloan-Kettering experience. *Med Pediatr Oncol* 2003;41:508–515.

686. Kramer K, Humm JL, Souweidane MM, et al. Phase I study of targeted radioimmunotherapy for leptomeningeal cancers using intra-Ommaya 131-I-3F8. *J Clin Oncol* 2007;25:5465–5470.

687. Benekli M, Hahn T, Williams BT, et al. Muromonab-CD3 (Orthoclone OKT3), methylprednisolone and cyclosporine for acute graft-versus-host disease prophylaxis in allogeneic bone marrow transplantation. *Bone Marrow Transplant* 2006;38:365–370.

688. Walker RW, Brochstein JA. Neurologic complications of immunosuppressive agents. *Neurol Clin* 1988;6:261–278.

689. Capone PM, Cohen ME. Seizures and cerebritis associated with administration of OKT3. *Pediatr Neurol* 1991;7:299–301.

690. Pittock SJ, Rabinstein AA, Edwards BS, et al. OKT3 neurotoxicity presenting as akinetic mutism. *Transplantation* 2003;75:1058–1060.

691. Raasveld MHM, Surachno S, Hack CE, et al. Thromboembolic complications and dose of monoclonal OKT3 antibody. Letter to the editor. *Lancet* 1992;339:1363–1364.

692. Reardon DA, Akabani G, Coleman RE, et al. Salvage radioimmunotherapy with murine iodine-131-labeled antitenascin monoclonal antibody 81C6 for patients with recurrent primary and metastatic malignant brain tumors: phase II study results. *J Clin Oncol* 2006;24:115–122.

693. Bauwens D, Hantson P, Laterre PF, et al. Recurrent seizure and sustained encephalopathy associated with dimethylsulfoxide-preserved stem cell infusion. *Leuk Lymphoma* 2005;46:1671–1674.

694. Hoyt R, Szer J, Grigg A. Neurological events associated with the infusion of cryopreserved bone marrow and/or peripheral blood progenitor cells. *Bone Marrow Transplant* 2000;25:1285–1287.

695. Saito H, Frleta D, Dubsky P, et al. Dendritic cell-based vaccination against cancer. *Hematol Oncol Clin North Am* 2006;20:689–710.

696. Wheeler CJ, Black KL. Dendritic cell vaccines and immunity in glioma patients. *Front Biosci* 2005;10:2861–2881.

697. Dezfouli S, Hatzinisiriou I, Ralph SJ. Use of cytokines in cancer vaccines/immunotherapy: recent developments improve survival rates for patients with metastatic malignancy. *Curr Pharm Des* 2005;11:3511–3530.

698. Parton M, Gore M, Eisen T. Role of cytokine therapy in 2006 and beyond for metastatic renal cell cancer. *J Clin Oncol* 2006;24:5584–5592.

699. Martin R, Schwulera U, Menke G, et al. Interleukin-2 and blood brain barrier in cats: pharmacokinetics and tolerance following intrathecal and intravenous administration. *Eur Cytokine Netw* 1992;3:399–406.

700. Alexander JT, Saris SC, Oldfield EH. The effect of interleukin-2 on the blood-brain barrier in the 9L gliosarcoma rat model. *J Neurosurg* 1989;70:92–96.

701. Saris SC, Patronas NJ, Rosenberg SA, et al. The effect of intravenous interleukin-2 on brain water content. *J Neurosurg* 1989;71:169–174.

702. Siegel JP, Puri RK. Interleukin-2 toxicity. *J Clin Oncol* 1991;9:694–704.

703. Schwartz RN, Stover L, Dutcher J. Managing toxicities of high-dose interleukin-2. *Oncology (Williston Park)* 2002;16:11–20.

704. Merrill JE. Interleukin-2 effects in the central nervous system. *Ann N Y Acad Sci* 1990;594:188–199.

705. Vecht CJ, Keohane C, Menon RS, et al. Acute fatal leukoencephalopathy after interleukin-2 therapy. *N Engl J Med* 1990;323:1146–1147.

706. Meyers CA, Valentine AD, Wong FC, et al. Reversible neurotoxicity of interleukin-2 and tumor necrosis factor: correlation of SPECT with neuropsychological testing. *J Neuropsychiatry Clin Neurosci* 1994;6:285–288.

707. Weijl NI, Van Der Harst D, Brand A, et al. Hypothyroidism during immunotherapy with interleukin-2 is associated with antithyroid antibodies and response to treatment. *J Clin Oncol* 1993; 11(7):1376–1383.

708. Urba WJ, Steis RG, Longo DL, et al. Immunomodulatory properties and toxicity of interleukin 2 in patients with cancer. *Cancer Res* 1990;50:185–192.

709. Loh FL, Herskovitz S, Berger AR, et al. Brachial plexopathy associated with interleukin-2 therapy. *Neurology* 1992;42:462–463.

710. Meyers CA, Yung WK. Delayed neurotoxicity of intraventricular interleukin-2: a case report. *J Neuro-oncol* 1993;15:265–267.

711. Hornyak SC, Orentas DM, Karavodin LM, et al. Histamine improves survival and protects against interleukin-2-induced pulmonary vascular leak syndrome in mice. *Vascul Pharmacol* 2005; 42:187–193.

712. Hu P, Mizokami M, Ruoff G, et al. Generation of low-toxicity interleukin-2 fusion proteins devoid of vasopermeability activity. *Blood* 2003; 101:4853–4861.

713. Guirguis LM, Yang JC, White DE, et al. Safety and efficacy of high-dose interleukin-2 therapy in patients with brain metastases. *J Immunother* 2002;25:82–87.

714. Neta R, Oppenheim JJ. IL-1: can we exploit Jekyll and subjugate Hyde? *Biol Ther Cancer Updates* 1992;2:1–11.

715. Atkins MB, Vachino G, Tilg HJ, et al. Phase I evaluation of thrice-daily intravenous bolus interleukin-4 in patients with refractory malignancy. *J Clin Oncol* 1992;10:1802–1809.

716. Caraceni A, Gangeri L, Martini C, et al. Neurotoxicity of interferon-α in melanoma therapy: results from a randomized controlled trial. *Cancer* 1998;83:482–489.

717. Jonasch E, Haluska FG. Interferon in oncological practice: review of interferon biology, clinical applications, and toxicities. *Oncologist* 2001;6:34–55.

718. Merimsky O, Reider-Groswasser IR, Inbar M, et al. Interferon-related mental deterioration and behavioral changes in patients with renal cell carcinoma. *Eur J Cancer* 1990;26:596–600.

719. Meyers CA, Scheibel RS, Forman AD. Persistent neurotoxicity of systemically administered interferon-alpha. *Neurology* 1991;41:672–676.

720. Bernsen PL, Wong Chung RE, Vingerhoets HM, et al. Bilateral neurologic amyotrophy induced by interferon treatment. *Arch Neurol* 1988;45:449–451.

721. Bauherz G, Soeur M, Lustman F. Oculomotor nerve paralysis induced by alpha II-interferon. *Acta Neurol Belg* 1990;90:111–114.

722. Kanda Y, Shigeno K, Kinoshita N, et al. Sudden hearing loss associated with interferon. *Lancet* 1994;343:1134–1135.

723. Pavlovsky L, Seiffert E, Heinemann U, et al. Persistent BBB disruption may underlie alpha interferon induced seizures. *J Neurol* 2005;252:42–46.

724. Duyff RF, Van den BJ, Laman DM. Neuromuscular findings in thyroid dysfunction: a prospective clinical and electrodiagnostic study. *J Neurol Neurosurg Psychiatry* 2000;68:750–755.

725. Carella C, Mazziotti G, Morisco F, et al. Long-term outcome of interferon-alpha-induced thyroid autoimmunity and prognostic influence of thyroid autoantibody pattern at the end of treatment. *J Clin Endocrinol Metab* 2001;86:1925–1929.

726. O'Connell MJ, Sargent DJ, Windschitl HE, et al. Randomized clinical trial of high-dose levamisole combined with 5-fluorouracil and leucovorin as surgical adjuvant therapy for high-risk colon cancer. *Clin Colorectal Cancer* 2006;6:133–139.

727. Kimmel DW, Wijdicks EFM, Rodriguez M. Multifocal inflammatory leukoencephalopathy associated with levamisole therapy. *Neurology* 1995;45:374–376.

728. Hook CC, Kimmel DW, Kvols LK, et al. Multifocal inflammatory leukoencephalopathy with 5-fluorouracil and levamisole. *Ann Neurol* 1992;31:262–267.

729. Liu HM, Hsieh WJ, Yang CC, et al. Leukoencephalopathy induced by levamisole alone for the treatment of recurrent aphthous ulcers. *Neurology* 2006;67:1065–1067.

730. Wu VC, Huang JW, Lien HC, et al. Levamisole-induced multifocal inflammatory leukoencephalopathy: clinical characteristics, outcome, and impact of treatment in 31 patients. *Medicine (Baltimore)* 2006;85:203–213.

731. Savarese DM, Gordon J, Smith TW, et al. Cerebral demyelination syndrome in a patient treated with 5-fluorouracil and levamisole. The use of Thallium SPECT imaging to assist in noninvasive diagnosis—a case report. *Cancer* 1996;77:387–394.

732. Peterson K, Rosenblum MK, Powers JM, et al. Effect of brain irradiation on demyelinating lesions. *Neurology* 1993;43:2105–2112.

733. Wong R, Beguelin GZ, de Lima M., et al. Tacrolimus-associated posterior reversible encephalopathy syndrome after allogeneic haematopoietic stem cell transplantation. *Br J Haematol* 2003;122:128–134.

734. Woo M, Przepiorka D, Ippoliti C, et al. Toxicities of tacrolimus and cyclosporin A after allogeneic blood stem cell transplantation. *Bone Marrow Transplant* 1997;20:1095–1098.

735. Fujii N, Ikeda K, Koyama M, et al. Calcineurin inhibitor-induced irreversible neuropathic pain after allogeneic hematopoietic stem cell transplantation. *Int J Hematol* 2006;83:459–461.

736. Tsutsumi Y, Kanamori H, Mashiko S, et al. Leukoencephalopathy with cerebral hemorrhage following acute pancreatitis due to tacrolimus in a case of allogeneic peripheral blood stem cell transplantation. *Leuk Lymphoma* 2006;47:943–947.

737. Liedtke W, Quabeck K, Beelen DW, et al. Recurrent acute inflammatory demyelinating polyradiculitis after allogeneic bone marrow transplantation. *J Neurol Sci* 1994;125:110–111.

738. Qu L, Kiss JE. Thrombotic microangiopathy in transplantation and malignancy. *Semin Thromb Hemost* 2005;31:691–699.

739. Vivarelli M, Vetrone G, Zanello M, et al. Sirolimus as the main immunosuppressant in the early postoperative period following liver transplantation: a report of six cases and review of the literature. *Transpl Int* 2006;19:1022–1025.

740. Froud T, Baidal DA, Ponte G, et al. Resolution of neurotoxicity and beta-cell toxicity in an islet transplant recipient following substitution of tacrolimus with MMF. *Cell Transplant* 2006; 15:613–620.

741. Patel PH, Chadalavada RS, Chaganti RS, et al. Targeting von Hippel-Lindau pathway in renal cell carcinoma. *Clin Cancer Res* 2006;12:7215–7220.

742. Tedesco-Silva H, Jr, Vitko S, Pascual J, et al. 12-Month safety and efficacy of everolimus with reduced exposure cyclosporine in de novo renal transplant recipients. *Transpl Int* 2007;20:27–36.

743. Orange JS, Hossny EM, Weiler CR, et al. Use of intravenous immunoglobulin in human disease: a review of evidence by members of the Primary Immunodeficiency Committee of the American Academy of Allergy, Asthma and Immunology. *J Allergy Clin Immunol* 2006;117:S525–S553.

744. Watson JD, Gibson J, Joshua DE, et al. Aseptic meningitis associated with high dose intravenous immunoglobulin therapy. *J Neurol Neurosurg Psychiatry* 1991;54:275–276.

745. Silbert PL, Knezevic WV, Bridge DT. Cerebral infarction complicating intravenous immunoglobulin therapy for polyneuritis cranialis. *Neurology* 1992;42:257–259.

746. Marie I, Herve F, Lahaxe L, et al. Intravenous immunoglobulin-associated cranial pachymeningitis. *J Intern Med* 2006;260:164–167.

747. Cheer SM, Wagstaff AJ. Epoetin beta: a review of its clinical use in the treatment of anaemia in patients with cancer. *Drugs* 2004;64:323–346.

748. Perez-Oliva JF, Casanova-Gonzalez M, Garcia-Garcia I, et al. Comparison of two recombinant erythropoietin formulations in patients with anemia due to end-stage renal disease on hemodialysis: a parallel, randomized, double blind study. *BMC Nephrol* 2005;6:5.

749. Ross SD, Allen IE, Henry DH, et al. Clinical benefits and risks associated with epoetin and darbepoetin in patients with chemotherapy-induced anemia: a systematic review of the literature. *Clin Ther* 2006;28:801–831.

750. Delanty N, Vaughan C, Frucht S, et al. Erythropoietin-associated hypertensive posterior leukoencephalopathy. *Neurology* 1997;49:686–689.

751. Finelli PF, Carley MD. Cerebral venous thrombosis associated with epoetin alfa therapy. *Arch Neurol* 2000;57:260–262.

752. Buchbinder A, Adler H, Ballard H. An unusual and unreported toxicity to erythropoietin. *Am J Hematol* 1993;42:412–413.

753. Campana WM, Li X, Shubayev VI, et al. Erythropoietin reduces Schwann cell TNF-α, Wallerian degeneration and pain-related behaviors after peripheral nerve injury. *Eur J Neurosci* 2006;23:617–626.

754. Burstein HJ, Parker LM, Keshaviah A, et al. Efficacy of pegfilgrastim and darbepoetin alfa as hematopoietic support for dose-dense every-2-week adjuvant breast cancer chemotherapy. *J Clin Oncol* 2005;23:8340–8347.

755. Kastrup O, Diener HC. Granulocyte-stimulating factor filgrastim and molgramostim induced recurring encephalopathy and focal status epilepticus. *J Neurol* 1997;244:274–275.

756. Leniger T, Kastrup O, Diener HC. Reversible posterior leukencephalopathy syndrome induced by granulocyte stimulating factor filgrastim. *J Neurol Neurosurg Psychiatry* 2000;69:280–281.

757. Nagler A, Korenstein-Ilan A, Amiel A, et al. Granulocyte colony-stimulating factor generates epigenetic and genetic alterations in lymphocytes of normal volunteer donors of stem cells. *Exp Hematol* 2004;32:122–130.

758. Pamphilon D, Mackinnon S, Nacheva E, et al. The use of granulocyte colony-stimulating factor in volunteer blood and marrow registry donors. *Bone Marrow Transplant* 2006;38:699–700.

759. Hotton KM. A phase Ib/II trial of granulocyte-macrophage-colony stimulating factor and interleukin-2 for renal cell carcinoma patients with pulmonary metastases: a case of fatal central nervous system thrombosis. *Cancer* 2000;88:1892–1901.

760. Okeda R. An autopsy case of drug-induced diffuse cerebral axonopathic leukoencephalopathy: the pathogenesis in relation to reversible posterior leukoencephalopathy syndrome. *Neuropathology* 2007;27:364–370.

761. Herrlinger KR, Witthoeft T, Raedler A, et al. Randomized, double blind controlled trial of subcutaneous recombinant human interleukin-11 versus prednisolone in active Crohn's disease. *Am J Gastroenterol* 2006;101:793–797.

762. Smith JW. Tolerability and side-effect profile of rhIL-11. *Oncology (Williston Park)* 2000;14:41–47.

763. Bussel JB, Mukherjee R, Stone AJ. A pilot study of rhIL-11 treatment of refractory ITP. *Am J Hematol* 2001;66:172–177.

764. von BI, Brennan MT, Spijkervet FK, et al. Growth factors and cytokines in the prevention and treatment of oral and gastrointestinal mucositis. *Support Care Cancer* 2006;14:519–527.

765. Spielberger R, Stiff P, Bensinger W, et al. Palifermin for oral mucositis after intensive therapy for hematologic cancers. *N Engl J Med* 2004;351:2590–2598.

766. Gibson RJ, Bowen JM, Keefe DM. Palifermin reduces diarrhea and increases survival following irinotecan treatment in tumor-bearing DA rats. *Int J Cancer* 2005;116:464–470.

767. Guilhot F. Indications for imatinib mesylate therapy and clinical management. *Oncologist* 2004;9:271–281.

768. Robert C, Soria JC, Spatz A, et al. Cutaneous side-effects of kinase inhibitors and blocking antibodies. *Lancet Oncol* 2005;6:491–500.

769. Berman E, Nicolaides M, Maki RG, et al. Altered bone and mineral metabolism in patients receiving imatinib mesylate. *N Engl J Med* 2006;354:2006–2013.

770. Medeiros BC, Lipton JH. Chlordiazepoxide for imatinib-induced muscular cramps. *Eur J Haematol* 2006;77:538.

771. Kantarjian H, Sawyers C, Hochhaus A, et al. Hematologic and cytogenetic responses to imatinib mesylate in chronic myelogenous leukemia. *N Engl J Med* 2002;346:645–652.

772. Shah NT, Kris MG, Pao W, et al. Practical management of patients with non-small-cell lung cancer treated with gefitinib. *J Clin Oncol* 2005; 23:165–174.

773. Sandler AB. Nondermatologic adverse events associated with anti-EGFR therapy. *Oncology (Williston Park)* 2006;20:35–40.

774. Bareschino MA, Schettino C, Troiani T, et al. Erlotinib in cancer treatment. *Ann Oncol* 2007;18 (Suppl 6):vi35–vi41.

775. Motzer RJ, Hutson TE, Tomczak P, et al. Sunitinib versus interferon alfa in metastatic renal-cell carcinoma. *N Engl J Med* 2007;356:115–124.

776. Demetri GD, van Oosterom AT, Garrett CR, et al. Efficacy and safety of sunitinib in patients with advanced gastrointestinal stromal tumour after failure of imatinib: a randomised controlled trial. *Lancet* 2006;368:1329–1338.

776a. Mannavola D, Coco P, Vannucchi G, et al. A novel tyrosine-kinase selective inhibitor, sunitinib, induces transient hypothyroidism by blocking iodine uptake. *J Clin Endocrinol Metabol* 2007;92:3531–34.

776b. Arnaud L, Schartz NE, Kerob D, et al. Transient sunitinib-induced coma in a patient with fibromyxoid sarcoma. *J Clin Oncol* 2008;26:1569–71.

777. Desai J, Yassa L, Marqusee E, et al. Hypothyroidism after sunitinib treatment for patients with gastrointestinal stromal tumors. *Ann Intern Med* 2006;145:660–664.

778. Kane RC, Farrell AT, Saber H, et al. Sorafenib for the treatment of advanced renal cell carcinoma. *Clin Cancer Res* 2006;12:7271–7278.

779. Escudier B, Eisen T, Stadler WM, et al. Sorafenib in advanced clear-cell renal-cell carcinoma. *N Engl J Med* 2007;356:125–134.

780. Moy B, Goss PE. Lapatinib: current status and future directions in breast cancer. *Oncologist* 2006;11:1047–1057.

781. Geyer CE, Forster J, Lindquist D, et al. Lapatinib plus capecitabine for HER2-positive advanced breast cancer. *N Engl J Med* 2006;355:2733–2743.

782. Lancet JE, Gojo I, Gotlib J, et al. A phase II study of the farnesyltransferase inhibitor tipifarnib in poor-risk and elderly patients with previously untreated acute myelogenous leukemia. *Blood* 2006;109(4):1387–1394.

783. Siegel-Lakhai WS, Crul M, De PP, et al. Clinical and pharmacologic study of the farnesyltransferase

inhibitor tipifarnib in cancer patients with normal or mildly or moderately impaired hepatic function. *J Clin Oncol* 2006;24:4558–4564.

784. Ludwig H, Khayat D, Giaccone G, et al. Proteasome inhibition and its clinical prospects in the treatment of hematologic and solid malignancies. *Cancer* 2005;104:1794–1807.

785. Richardson PG, Barlogie B, Berenson J, et al. A phase 2 study of bortezomib in relapsed, refractory myeloma. *N Engl J Med* 2003;348:2609–2617.

786. Fanucchi MP, Fossella FV, Belt R, et al. Randomized phase II study of bortezomib alone and bortezomib in combination with docetaxel in previously treated advanced non-small-cell lung cancer. *J Clin Oncol* 2006;24:5025–5033.

787. Belch A, Kouroukis CT, Crump M, et al. A phase II study of bortezomib in mantle cell lymphoma: the National Cancer Institute of Canada Clinical Trials Group trial IND.150. *Ann Oncol* 2006;18(1):116–121.

788. Faderl S, Rai K, Gribben J, et al. Phase II study of single-agent bortezomib for the treatment of patients with fludarabine-refractory B-cell chronic lymphocytic leukemia. *Cancer* 2006;107:916–924.

788a. Richardson PG, Briemberg H, Jagannath S, et al. Frequency, characteristics, and reversibility of peripheral neuropathy during treatment of advanced multiple myeloma with bortezomib. *J Clin Oncol* 2006;24:3113–20.

789. Cibeira MT, Mercadal S, Arenillas L, et al. Bortezomib-induced rhabdomyolysis in multiple myeloma. *Acta Haematol* 2006;116:203–206.

790. Tallman MS, Andersen JW, Schiffer CA, et al. Clinical description of 44 patients with acute promyelocytic leukemia who developed the retinoic acid syndrome. *Blood* 2000;95:90–95.

791. Soignet SL, Frankel SR, Douer D, et al. United States multicenter study of arsenic trioxide in relapsed acute promyelocytic leukemia. *J Clin Oncol* 2001;19:3852–3860.

792. Schroeter T, Lanvers C, Herding H, et al. Pseudotumor cerebri induced by all-trans-retinoic acid in a child treated for acute promyelocytic leukemia. *Med Pediatr Oncol* 2000;34:284–286.

793. Alemdar M, Iseri P, Selekler HM, et al. Isolated abducens nerve palsy associated with retinoic acid therapy: a case report. *Strabismus* 2005;1 3:129–132.

794. Yamaji S, Kanamori H, Mishima A, et al. All-trans retinoic acid-induced multiple mononeuropathies. *Am J Hematol* 1999;60:311.

795. Fabbiano F, Magrin S, Cangialosi C, et al. All-trans retinoic acid induced cardiac and skeletal myositis in induction therapy of acute promyelocytic leukaemia. *Br J Haematol* 2005;129:444–445.

796. Yip SF, Yeung YM, Tsui EYK. Severe neurotoxicity following arsenic therapy for acute promyelocytic leukemia: potentiation by thiamine deficiency. *Blood* 2002;99:3481–3482.

797. O'connor OA. Clinical experience with the novel histone deacetylase inhibitor vorinostat (suberoylanilide hydroxamic acid) in patients with relapsed lymphoma. *Br J Cancer* 2006;95 (Suppl 1):S7–S12.

798. Fouladi M. Histone deacetylase inhibitors in cancer therapy. *Cancer Invest* 2006;24:521–527.

799. Rosenfeld MR, Pruitt A. Neurologic complications of bone marrow, stem cell, and organ transplantation in patients with cancer. *Semin Oncol* 2006;33:352–361.

800. Faraci M, Lanino E, Dini G, et al. Severe neurologic complications after hematopoietic stem cell transplantation in children. *Neurology* 2002;59:1895–1904.

801. Sostak P, Padovan CS, Yousry TA, et al. Prospective evaluation of neurological complications after allogeneic bone marrow transplantation. *Neurology* 2003;60:842–848.

802. Mathew RM, Rosenfeld MR. Neurologic complications of bone marrow and stem cell transplantation in patients with cancer. *Curr Treat Options Neurol* 2007;9:308–314.

803. Leiberman F, Gulati S. Headaches after inadvertent lumbar puncture during bone marrow harvest. *Neurology* 1996;46:268–269.

804. Burt RK, Fassas A, Snowden J, et al. Collection of hematopoietic stem cells from patients with autoimmune diseases. *Bone Marrow Transplant* 2001;28:1–12.

805. Parker P, Chao NJ, Ben-Ezra J, et al. Polymyositis as a manifestation of chronic graft-versus-host disease. *Medicine* 1996;75:279–285.

806. Openshaw H, Slatkin NE, Parker PM, et al. Immune-mediated myelopathy after allogeneic marrow transplantation. *Bone Marrow Transplant* 1995;15:633–636.

807. Ma M, Barnes G, Pulliam J, et al. CNS angiitis in graft vs host disease. *Neurology* 2002;59:1994–1997.

808. Martinez MT, Bucher C, Stussi G, et al. Transplant-associated microangiopathy (TAM) in recipients of allogeneic hematopoietic stem cell transplants. *Bone Marrow Transplant* 2005; 36:993–1000.

809. Brizel DM, Overgaard J. Does amifostine have a role in chemoradiation treatment? *Lancet Oncol* 2003;4:378–381.

810. Mason KA, Neal R, Hunter N, et al. CpG oligodeoxynucleotides are potent enhancers of radio- and chemoresponses of murine tumors. *Radiother Oncol* 2006;80:192–198.

811. Tanvetyanon T, Stiff PJ. Management of the adverse effects associated with intravenous bisphosphonates. *Ann Oncol* 2006;17:897–907.

812. Coiffier B. Rituximab in the treatment of diffuse large B-cell lymphomas. *Semin Oncol* 2002;29:30–35.

813. Horning SJ, Younes A, Jain V, et al. Efficacy and safety of tositumomab and iodine-131 tositumomab (Bexxar) in B-cell lymphoma, progressive after rituximab. *J Clin Oncol* 2005;23:712–719.

814. Cheson BD. Monoclonal antibody therapy for B-cell malignancies. *Semin Oncol* 2006;33:S2–S14.

815. van der Heiden PL, Jedema I, Willemze R, et al. Efficacy and toxicity of gemtuzumab ozogamicin in patients with acute myeloid leukemia. *Eur J Haematol* 2006;76:409–413.

816. Patel DD, Goldberg RM. Cetuximab-associated infusion reactions: pathology and management. *Oncology (Williston Park)* 2006;20:1373–1382.

817. Bareschino MA, Schettino C, Troiani T, et al. Erlotinib in cancer treatment. *Ann Oncol* 2007;18 (Suppl 6):vi35–vi41.

Chapter 13

Side Effects of Radiation Therapy

INTRODUCTION

Radiation therapy (RT) kills tumor cells. Given a sufficient dose of RT, all neoplasms can be sterilized and the cancer cured. Unfortunately, the curative potential of this therapy is often limited by potential damage to normal tissue within the radiation portal. Nowhere is potential damage to normal tissue more devastating than to the nervous system. As Table 13–1 suggests, any part of the nervous system can suffer direct damage from RT, and some areas endure indirect damage, such as cerebral infarcts caused by accelerated atherosclerosis to carotid arteries after radiation for head and neck cancers (see following text and also Chapter 9). Furthermore, damage to the nervous system can occur acutely (during the course of radiation), within several weeks following RT (early-delayed radiation toxicity), or many months to years after the completion of radiation (late-radiation toxicity). Differentiation of these time-related effects are important because acute and early-delayed toxicity are usually reversible whereas the late effects are usually not.

BIOLOGY OF RADIATION DAMAGE

Ionization

X-rays, gamma rays, and atomic particles such as electrons, neutrons, and protons, as well as heavier particles such as helium or carbon ions, cause ionization when they interact with tissue. The average number of ionizations per unit of path length is referred to as the linear energy transfer (LET). Types of radiation with higher LET (e.g., neutron, proton radiation) cause more biologic damage than low LET radiation (e.g., X-rays). Ionizing radiation causes cellular damage in several ways:

1. It breaks DNA. Single-stranded breaks usually repair themselves, but double-stranded breaks are often repaired either in a mutated form or not at all. The repair

511

Table 13–1. **Sites of Radiation Therapy–Induced Injury to the Nervous System**

Direct Injury	Indirect Injury
Brain	Blood vessels
Spinal cord	Endocrine organs
Cranial nerves	RT-induced tumors
Peripheral nerves	Fibrosis
Nerve roots	
Nerve plexus	
Individual nerves	
Muscles	

of sublethal damage usually occurs within 6 hours, although some cells in the nervous system may take longer.[1,2] O-2A cells, the precursor cells for oligodendrocytes and some astrocytes, do not repair sublethal damage, but are relatively radioresistant.[3] Failure to repair sublethal damage leads to either immediate cell death or delayed death at the first or a subsequent attempt of the damaged cell to reproduce. When RT has been directed at the peripheral nervous system or CNS, cell death may occur months or years later because glial and Schwann cells reproduce so slowly.

2. Ionizing radiation may attack RNA, lipids, and proteins, causing cellular damage and probably cell death, but most calculations of radiobiological effectiveness consider only single- and double-stranded DNA breaks.

3. Radiation stimulates apoptosis, particularly in endothelial cells and oligodendroglia.[4,5] Although the exact mechanisms are not fully understood, critical genes, including $p53$, ataxia-telangiectasia mutated (ATM), and Bax all play a role.[5,6]

Fractionation

Therapeutic external RT is usually given in a series of equal-sized fractions, either daily or more than once a day for several weeks. Each fraction kills a similar proportion of tumor cells, resulting in a logarithmic decline in the number of surviving cells as the number of fractions increases. As a result, at any given dose, smaller tumors are more susceptible to

complete eradication than larger tumors, and the differential effect on tumor tissue becomes greater with a larger number of fractions.

Several techniques attempt to decrease RT toxicity to normal tissue surrounding the tumor. These include (1) radiation sensitizers,[7–9] (2) protectors of normal tissue against radiation damage,[7–9] and (3) focusing the radiation beam to decrease the dose to surrounding normal tissue.[10,11] *Radiation sensitizers* increase the susceptibility of the tumor to lethal damage (e.g., hypoxic cells are radioresistant and because tumors are often hypoxic but the surrounding normal tissue is not, increasing oxygen supply may serve to sensitize tumor). *Radiation protectors* (the best example is amifostine[9]) target the normal tissue and scavenge free radicals that contribute to radiation injury. Several methods efficient in *focusing the radiation beam* include using high-energy particles such as protons; conformal radiation, particularly intensity-modulated radiation therapy (IMRT); and radiosurgery and brachytherapy (implantation of the tumor with radioactive sources). Neither radiation sensitizers, with the possible exception of motexafin gadolinium and efaproxiral as discussed in Chapter 5, nor protectors have yet proved effective in treating CNS tumors. A recent report describes protection of normal tissue using an activator of NK-κB that prevents apoptosis in normal cells[11a]. High-energy particles have proved particularly effective against pituitary and some base of skull tumors. IMRT uses multiple beams of nonuniform radiation intensity within each field. Dose distribution curves generated by this technique exquisitely conform to the shape of the target volume and sculpt around adjacent critical normal tissues such as the optic nerve and spinal cord.[12] When combined with image guidance (IG), IMRT (IG-IMRT)[12a] allows exquisite localization of tumor abutting the spinal cord or sensitive brain areas. However, IMRT may have a downside in that it may increase the number of second cancers because a bigger volume of normal tissue is exposed to low doses of radiation.[13]

Mechanisms of Nervous System Damage

Damage to the nervous system from RT depends on many exogenous and endogenous factors (Table 13–2).

Table 13–2. **Risk Factors for Radiation Damage to the Nervous System**

Radiation Factors	Host Factors
Doses per fraction	Age
Total dose	Sex
Total duration of therapy	Genetic predisposition
	Preexisting CNS damage
Volume of tissue Irradiated	Infection
Energy of radiation	Systemic diseases (e.g., hypertension, diabetes)
	Chemotherapy
	Life style choices (e.g., smoking)

CNS, central nervous system.

The mechanism(s) by which the nervous system is damaged is not fully understood. Three hypotheses, not necessarily mutually exclusive, have been proposed. The first, and probably the most important, is that ionizing radiation directly damages normal cells within the nervous system. Apoptosis of brain endothelial cells is probably responsible for the acute disruption of the blood–brain barrier caused by ionizing radiation.[4] Mature oligodendroglia appear to be the most vulnerable brain cells. Their peripheral nervous system counterpart, Schwann cells, are more radioresistant, but peripheral neuropathy is a major dose-limiting complication of intraoperative RT.[14] In experimental animals, a single dose of 10 or 20 Gy (Gy, Gray = unit of absorbed dose of radiation; $1 Gy = 100 cGy$) to the brain causes apoptotic cell death of oligodendrocytes within hours,[15] perhaps explaining the early-delayed radiation syndrome as well as late-delayed leukoencephalopathy. Mature neurons[16] and immature neuronal precursors[17] are both susceptible to double-stranded DNA breaks that are poorly repaired; the cells undergo apoptosis after exposure to RT. Radiation can also induce inflammation characterized by activated glial cells and entry of peripheral monocytes into the radiated area.[18,19]

Compounding the direct damage to brain cells is the damage to brain vasculature. Both the acute and late effects of radiation on the CNS are due in part to endothelial cell damage.[4] Endothelial cell damage disrupts the blood–brain barrier, leading to brain edema; this occurs within a few hours of standard doses of RT and is a result of endothelial apoptosis.[20,21] In experimental animals, a number of substances cause endothelial apoptosis including retinoids, tissue inhibitors of matrix metalloproteinases (TIMP1),[22] fibroblast growth factor, and sphingomyelinase deficiency.[20,23] Radiation-induced apoptosis appears to be p53 dependent.[5] Late effects of radiation on the blood–brain barrier are the result of permeability changes caused by increased amounts of vascular endothelial growth factor (VEGF).[21]

A third source of damage, proposed by Lampert et al.[24] and Lampert and Davis,[25] suggests that injured glial cells release antigens identified as foreign by the immune system. An immune response leads to the necrosis and vascular changes of a hypersensitivity reaction. Virtually no direct evidence exists for this hypothesis, although in one patient a focal vasculitis in an irradiated area of brain was associated with circulating immune complexes.[26]

The Linear-Quadratic Concept

The linear-quadratic concept was developed to assign a numerical score when comparing different fractionation schedules (Fig. 13–1); it posits a double mechanism for cell killing. The concept assumes that double-stranded DNA breaks are both necessary and sufficient for cell killing; killing is defined as the loss of the ability of the cell to divide successfully. If the double-stranded break is caused by a single electron or two closely spaced electrons, the result is the first, or linear component, also called

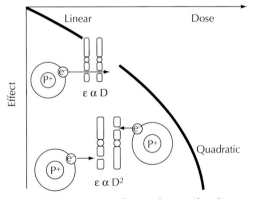

Figure 13–1. Linear-quadratic relation. This diagram shows how breaks in two chromosomes may be caused by a single electron track or by two separate electron tracks. The probability of these events will be proportional to the dose and the square of the dose, respectively. (From Ref. 27 with permission.)

nonrepairable or alpha (α). Tissues with a high α suffer early effects of RT.[27] The probability of an effect is proportional to the dose (D) of RT. The second, or quadratic component, also called repairable or beta (β), results from two non-simultaneous single-strand breaks. Tissues with a high β suffer late effects. The probability of an effect is proportional to the square of the dose (D^2). The overall effect is expressed as $E = \alpha D + \beta D^2$. The α/β ratio (Fig. 13–1) is the dose at which cell killing by the linear and quadratic components is equal; thus, the ratio measures fractionation sensitivity. For early reactions of normal tissue, such as skin desquamation, α/β is large, approximately 10 Gy, far in excess of all standard clinically administered RT fractions (but not radiosurgery). Thus, one rarely sees serious skin reactions during a course of RT. For tissue such as the nervous system, which suffers little early damage but much late damage, the α/β ratio is small (one estimate is 2 Gy for cervical and thoracic spinal cord and 4 Gy for lumbar cord[28]), implying that the size of the RT fractions can have an important influence for subsequent neurotoxicity.

The linear-quadratic formula allows the physician to explain the total effect (E) of radiation based on the number of fractions of RT administered (N), the dose per fraction (D), and the α and β factors (Table 13–3). Available evidence[29] indicates that it is reasonable to use α/β values of 8 to 10 Gy for tissue that suffers early damage, such as rapidly growing tumor, and of 1.5 to 4 Gy for tissue such as the brain, which suffers late damage. Based on these formulas, higher daily doses of radiation, 1.2 Gy bid, can be delivered to the brain with greater biologic effect on the tumor and a biologic

Table 13–3. Tolerance of Normal Tissues to Therapeutic Irradiation

Tissue	Injury	TD$_{5/5}$*, Gy	TD$_{50/5}$[†], Gy	Field Size
Brain	Infarction, necrosis	45	60	Whole
		60	75	1/3 organ
Brainstem	Necrosis	60	–	Whole
Spinal cord	Infarction, necrosis	47	–	20 cm
		50	70	5 or 10 cm
Brachial plexus	Paralysis	60	–	Whole
Eye				
Retina	Visual loss	45	65	Whole
Cornea	Visual loss	50	6	Whole
Lens	Cataract	10	18	Whole
Endocrine				
Thyroid	Hypothyroidism	45	150	Whole
Pituitary	Hypopituitarism	45	200	Whole
Peripheral nerves	Neuritis	60	100	
Optic nerve	Blindness	50		1/3
		80		Whole
Optic chiasm	Blindness	50		Whole
Ear				
Middle	Serous otitis	30	40	No volume
Vestibular	Ménière syndrome	60	70	effect
Muscle				
Child	Atrophy	20	40	Whole
Adult	Fibrosis	60	80	Whole
Large arteries and veins	Sclerosis	80	100	10 cm^2

Modified from Refs. 27 and 30.
* TD$_{5/5}$ = 5% of patients affected at 5 years.
[†] TD$_{50/5}$ = 50% of patients affected at 5 years.

effect on the normal brain similar to that obtained with conventional daily doses. Thus, the biological effective dose (BED) = (nd) (1 + d/α/β), where n = fraction number, d = fraction size, and α/β for brain = 2.

Table 13–3 lists the normal tissue tolerance of nervous system tissues.[27,30]

DIAGNOSIS OF RADIATION-INDUCED NEUROTOXICITY

The diagnosis of RT neurotoxicity may be difficult. In any patient with neurologic symptoms or signs who is receiving or has had RT previously, the physician should first determine if the clinical and imaging signs are appropriate to radiation toxicity. For example, radiation fibrosis of the brachial plexus is likely to be painless whereas recurrent tumor is usually painful. Radiation necrosis of a metastatic lesion is usually hypometabolic on positron emission tomography (PET) scan, whereas recurrent metastasis is usually hypermetabolic. Because there are exceptions to these statements, the physician should allow wide latitude in assessing the clinical and imaging situation. Having judged that radiation damage is a possible cause of the neurologic symptoms, the physician should determine whether the portal and dose (using one of the formulas for radiation toxicity found in Table 13–3), as well as the other factors delineated in Table 13–2, make radiation damage likely. The clinical effects of radiation neurotoxicity, both direct and indirect, are delineated in the following paragraphs.

Direct Radiation Damage to the Nervous System

BRAIN

Radiation injury can occur at almost any time, from seconds to years, after the therapy is delivered.[31] The side effects of nervous system RT can generally be divided into those that are acute, usually observed within the first few days; early-delayed, seen within 4 weeks to 4 months following RT; or late-delayed, appearing a few months to many years after RT is completed. Late-delayed RT-induced brain dysfunction can take several forms (Table 13–4).

Acute Encephalopathy

The disorder usually follows large fractions (>3 Gy) delivered to a large volume of brain in patients with increased intracranial pressure from primary or metastatic brain tumor.[32] Absence of corticosteroid treatment increases the risk. Immediately or within a few hours following the first treatment, susceptible patients develop headache, nausea, vomiting, somnolence, fever, and worsening of preexisting neurologic symptoms.[32–34] In rare instances, the disorder culminates in cerebral herniation (Fig. 13–2) and death. Young et al.[32] reported acute complications, usually after the first dose, in 41 of 83 patients given 15 Gy in 2 fractions over 3 days to treat brain metastases. Hindo and coworkers[34] reported four deaths within 48 hours of receiving 10 Gy in 1 fraction. However, 216 patients treated with 6 cGy of whole-brain radiation given in 3 fractions over 3 days had no complications.[35] It is not clear if these patients received corticosteroids. Acute encephalopathy usually follows the first radiation dose and becomes progressively less severe with each ensuing fraction. A mild form of the disorder is common and consists of headache and nausea immediately following radiation, reported in one series to occur in 40% of patients, a figure that seems very high to us.[36]

The pathogenesis of acute encephalopathy is probably disruption of the blood–brain barrier by endothelial apoptosis,[20,21] causing increased cerebral edema and a rise in intracranial pressure. Evidence indicates that a single dose of 3 Gy to the brain in an experimental animal causes substantial disruption of the blood–brain barrier if measured 2 hours after the radiation. After 24 hours, the barrier reconstitutes itself, but 30 Gy in 10 doses—a standard treatment regimen for brain metastases—leads to a progressive increase in blood–brain barrier permeability for up to 4 weeks after treatment.[37] Similar changes in vascular permeability occur in the rat lung and can be ameliorated by dexamethasone.[38] In humans, corticosteroids substantially diminish the disruption of the blood–brain barrier and prevent most clinical symptoms (see Chapter 3). A few investigators have noted an increase in intracranial pressure following a single, high dose of radiation,[32] but others have failed to document such an increase either in humans or in experimental animals, even when clinical symptoms develop.[33] Most imaging studies[39] in humans do not provide

Table 13–4. **Cerebral Radiation Injury**

Designation	Time after RT	Clinical Findings	Pathogenesis	Outcome
Acute	Immediate (minutes to hours)	Headache, vomiting Neurologic signs	Increased intracranial pressure (edema)	Recovery (usually)
Early-delayed	4 to 16 weeks	Somnolence Increased focal signs Worsening MR scan	Demyelination Possibly cerebral edema	Recovery
Late-delayed	Months to years			
Necrosis		Focal signs MR enhancement	Brain necrosis Possibly vascular	Responds to steroids Surgical removal
Atrophy		Memory loss Dementia Incontinence Gait ataxia MR atrophy, ventriculomegaly, leukoencephalopathy	Cellular loss Demyelination Possibly hydrocephalus	Modest response to shunting in some cases
Hemorrhage		Focal signs	Telangiectasia	Partial recovery
Infarction		Focal signs	Cerebral/carotid vasculopathy	Variable
Encephalopathy		Confusion, disorientation	Hypothyroidism	Recovery
Neoplasm		Focal signs	RT-induced neoplasm	Usually poor

RT, radiation therapy; MR, magnetic resonance.

evidence of increased cerebral edema, but such analyses may be insensitive to small, acute changes. A study of 14 patients receiving 4 Gy of whole-brain irradiation for 5 days demonstrated a 19% increase in capillary surface area permeability on the first day after radiation. Values returned to normal by the fifth day of radiation.[40] The increase in capillary permeability correlated with an increase in headache. These patients were receiving corticosteroids. A PET study measuring transport of rubidium (an analog of potassium) into brain and brain tumor revealed no increase in capillary permeability or plasma water volume 60 to 90 minutes after 2 to 6 Gy whole-brain RT.[41] These patients were also receiving corticosteroids.

Other experimental animal studies[42] indicate that even a single, low dose of radiation can affect brain function. A single dose of 0.50 to 1.5 Gy alters the late components of visual-evoked responses of the brain; the abnormalities peak at approximately 6 hours after the radiation. Low-dose total-body radiation causes a learning disability in rats[43]; the effect can be ameliorated by liposome-entrapped Cu/Zn superoxide dismutase.

Acute encephalopathy has also been reported in children after the initiation of low-dose cranial RT, often given in conjunction with intrathecal methotrexate (MTX), for prophylaxis of meningeal leukemia. In these patients, the acute disorder was characterized by headache, increased intracranial pressure, ataxia, and depressed state of consciousness, and was ameliorated by corticosteroids.[44–46]

These observations send two clinical messages. First, patients harboring large brain tumors, particularly tumors causing signs of increased intracranial pressure, should probably be treated with fractions no larger than

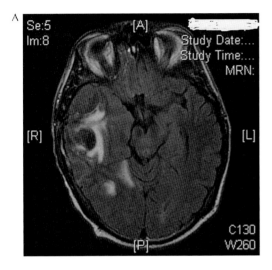
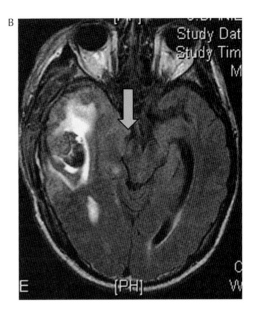

Figure 13–2. Herniation during radiation therapy. A 31-year-old man with non-seminomatous germ-cell tumor of the testis was found to have multiple brain metastases, the largest in the right temporal lobe. Whole-brain radiation was prescribed. **A:** An MRI the day before radiation revealed a large temporal lobe metastasis with a moderate amount of surrounding edema and mild compression of the midbrain. **B**: After the second radiation treatment (2.5 Gy × 2), his headaches increased substantially and he became confused and lethargic. The MRI revealed increased edema and uncal herniation (*arrow*), further compressing the midbrain. He recovered cognitive function after surgical resection of the metastasis.

2 Gy. Second, most patients undergoing brain radiation should be protected with corticosteroids such as dexamethasone, 8 mg or 16 mg/24 hr or more, if increased intracranial pressure is symptomatic. The drugs should probably be administered at least 48 to 72 hours before initiating RT.

Other acute effects following brain radiation include radiation-induced dermatitis and alopecia. Hair may regrow or the alopecia may be permanent, largely dependent on the dose of radiation.[47] Men who are partially bald have a tendency toward permanent alopecia, as do those receiving concomitant chemotherapy.[47] If the radiation portal to the brain extends low enough to encompass salivary glands, mucositis and xerostomia[48] may develop. Hyperamylasemia with or without parotitis may develop as well.[49] Early development of dermatitis of the scalp and ears in patients receiving antiepileptic drugs should suggest the possibility of Stevens–Johnson syndrome (see Chapter 4).

Although headache, nausea and vomiting are uncommon complications of cranial radiation when patients are protected by corticosteroids, fatigue is common,[50] particularly if the radiation course continues more than a few days. The fatigue seems to worsen with each day of radiation, only to ameliorate somewhat during the weekend, when no radiation is given; the fatigue returns when radiation is reinstituted on Monday. Fatigue and its treatment are addressed in Chapter 4.

Early-Delayed Encephalopathy
This disorder usually begins 2 weeks to 4 months after RT. The syndrome can take one of several forms, depending on the dose, the volume irradiated, and the presence or absence of underlying brain disease.

Neurologic Deterioration: If the patient suffers from a primary or metastatic brain tumor, the symptoms of early-delayed encephalopathy may simulate tumor progression. The patient develops headache, lethargy, and worsening or reappearance of the original neurologic symptoms. The disorder usually peaks 2 months after RT and resolves within 6 months.[51–53] Hoffman et al.[54] report such early-delayed symptoms in 7 of 51 patients radiated for glioma. The symptoms suggested tumor progression within 4 months of RT, but improved spontaneously over the ensuing 4 months. Magnetic resonance (MR) or computed tomography (CT) scans reveal an increase in the contrast-enhancing area and surrounding edema (pseudo-progression)[53]; this finding has been reported in about 50% of patients with malignant glioma who have developed increased enhancement on the first postRT MRI.[53,55] In some instances, new areas of contrast enhancement appear. In some asymptomatic patients, contrast-enhanced lesions are encountered on CT or magnetic resonance imaging (MRI) at routine follow-up examinations. Both clinical and radiologic changes resolve spontaneously, hastened by corticosteroids. After corticosteroids are discontinued, the patient with radiation encephalopathy remains improved, as verified by scans, whereas the patient with tumor progression deteriorates. The clinical and scan findings of early-delayed radiation encephalopathy can follow RT to the whole-brain or to the tumor alone by conventional external-beam RT or radiosurgery. The clinical implication is that failure of the patient or the scan to improve, or even worsen within 2 or 3 months of RT, does not mean the therapy has failed.[56] The patient should be supported with corticosteroids, close follow-up examinations, and scanning. A glucose PET scan that is hypometabolic suggests radiation damage, whereas hypermetabolism suggests tumor regrowth. Increased uptake of thallium in the abnormal area is a more reliable sign of recurrence than MRI.[57] Magnetic resonance spectroscopy (MRS) showing a decline of the choline peak in the presence of a lipid peak (the latter suggesting necrosis) may also be a useful test.[58,59] However, none of these tests is 100% reliable. Clinical judgment and careful follow-up are required, and if the situation does not clarify itself, a biopsy may be necessary.

Occasional patients develop disturbances of cognition with akinesia and tremor, resembling Parkinson disease, within a few months after RT for brain tumors.[60] The clinical symptoms are accompanied by brain atrophy characterized by ventriculomegaly, attenuation of the white matter, and increased sulcal size on CT. In those patients who died, the cortex appeared normal but the white matter was characterized by swelling, loss of myelin sheaths, and

reactive astrocytosis.[61] A more common finding is a transitory decline in cognitive function, specifically in retrieval of verbal information from long-term memory storage, 1 to 2 months after brain RT, followed by recovery 4 to 8 months later.[62,63] These early cognitive changes are not predictive of cognitive changes that occur later[64] (see p 523).

Radiation Somnolence Syndrome: Characterized by somnolence often associated with headache, nausea, vomiting, anorexia, occasional fever, and rarely, papilledema, the radiation somnolence syndrome[65–67] appears in patients without tumor, particularly children, and after whole-brain RT for prevention of meningeal leukemia. In 1929, Druckmann[68] noted marked somnolence in 3% of children treated with low-dose RT for scalp ringworm. The incidence is 40% to 60% in children administered 18 to 24 Gy for prophylaxis of leukemia.[69] It can also follow total-body radiation given before bone marrow transplantation.[70] Although more common in children, the syndrome also occurs in adults[69]; it has been reported in one adult following radiation to the pineal region.[71] The brain's electrical activity slows diffusely,[72] and focal neurologic signs are absent. The MRI may indicate demyelination.[73] The syndrome is ameliorated by corticosteroids and is prevented by giving adequate doses (4 mg/m^2 dexamethasone) of steroids during RT.[74] Without treatment, it usually resolves in 3 to 6 weeks. Somnolence does not predict late-delayed effects.[67]

Focal Encephalopathy: A third early-delayed syndrome is a focal encephalopathy that follows high-dose RT to extracranial tumors such as those in an ear, an eye, or the face when a portion of the brain, especially the brainstem, has been included in the treated field. The symptoms develop 8 to 11 weeks after RT and depend on the portion of the brain irradiated. Brainstem signs include ataxia, diplopia, dysarthria, and nystagmus. Transient white matter hyperintensity suggesting demyelination may be found on MR scan. Most patients recover spontaneously within 6 to 8 weeks. Transient metabolic changes that can be identified by MRS may assist in the diagnosis; these changes can be found even in the normal white matter of patients treated with radiation.[59,75,76] Symptoms can progress to stupor, coma, and death.[24,25,77–79]

Pathogenesis: The pathogenesis of early-delayed encephalopathy is believed to be demyelination resulting from transient damage to oligodendroglia with the subsequent breakdown of myelin sheaths. The best evidence supporting that hypothesis comes from pathologic studies in patients with early-delayed brainstem encephalopathy in whom confluent areas of demyelination with varying degrees of axonal loss were found in areas receiving the radiation.[24,25,78] Associated were loss of oligodendrocytes and abnormal, often multinucleated, giant astrocytes. Other pathologic changes included perivenous inflammatory infiltrates. Necrosis and vascular changes were uncommon.[58,75,76,79–81] Coincidental with the onset of symptoms in humans, a global decline in cerebral blood flow occurs, as measured by the xenon inhalation method. One case report describes a patient who developed neurologic symptoms 8 weeks after a pituitary adenoma was radiated and who died at 14 weeks; the brain was demyelinated, with loss of oligodendroglia and with abnormal giant astrocytes. Vascularity was not increased and no pathologic changes were found in the vessel wall, suggesting that demyelination was the cause of the neurologic symptoms.[77]

Clinical evidence suggests that, in patients with multiple sclerosis, brain irradiation may enhance the demyelinating process and result in permanent disability.[82]

Late-Delayed Encephalopathy

Late-delayed radiation damage, occurring 90 days or more following radiation to the brain, can be direct or indirect in nature (Table 13–4), depending on the dose, the volume of tissue irradiated, the radiation portal, and the disease treated.[31] In patients treated for a primary or metastatic brain tumor with focal or whole-brain RT, and in some patients treated for extracerebral tumors in whom part of the nervous system was included in the radiation portal, radiation necrosis is the characteristic late-delayed radiation complication. Radiosurgery, either alone, or in combination with whole-brain RT, can cause focal necrosis.[83,84] Evidence indicates that the risk of necrosis increases rapidly after an equivalent single dose of 12 or 13 Gy.[83] In one series radiation necrosis occurred in 12 (23%) of 53 patients within 6 months of treatment of brain metastases.[84]

Radiation Necrosis: Usually beginning 1 to 2 years after RT is completed, radiation necrosis can start as early as 3 months after treatment or be delayed for several years[85–88] (Fig. 13–3). Risk factors for radiation necrosis include total radiation dose, size of each fraction, the duration of treatment, irradiated volume, and the nature of the tumor being treated. Because these factors vary from series to series, the exact incidence is hard to determine. Furthermore, radiation necrosis can be defined either by imaging or by biopsy and can be either symptomatic or asymptomatic. Thus, it is not surprising that the incidence varies from as low as less than 5%[86] to as high as 24%.[89] Various calculations of radiation necrosis range from 13%[86] with a plateau at 3 years to 83% after 5 years.[90] Biologic variability from patient to patient and complicating factors such as chemotherapy and patient age make it impossible to calculate exactly the dose that will induce radiation necrosis (Table 13–3). Depending on the series, 60 Gy to the whole brain in 30 fractions will yield radiation necrosis in 5% to 15% of patients.[86,87,91] The dose–response curve unfortunately is steep; small increments beyond the generally tolerated dose rapidly increase the likelihood of radiation necrosis.[86]

In patients treated for primary or metastatic brain tumors, the symptoms, when present, generally recapitulate those of the brain tumor, leading the physician to suspect tumor recurrence. Rarely, new focal signs indicate a lesion distant from the original lesion.[92] The CT or MRI may show increased contrast enhancement at the original tumor site or, more rarely, at a remote distance from the original tumor.[93] Evidence of surrounding cerebral edema increases the clinical suspicion that tumor has recurred. However, neither the clinical symptoms nor imaging unequivocally distinguishes tumor recurrence from necrosis; imaging does help.[94] Fluorodeoxyglucose (FDG) PET scans may demonstrate hypometabolism in areas of necrosis and hypermetabolism in areas containing tumor.[95] However, areas of radiation necrosis may be infiltrated by a number of macrophages that can give a false-positive hypermetabolic FDG PET. Amino acid PET scans generally show increased uptake in tumor, but not in radiation necrosis.[96] A standard MRI usually shows enhancement, sometimes with

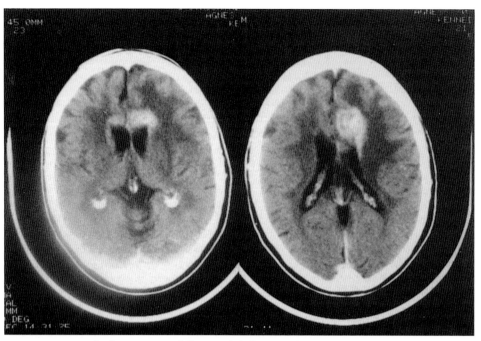

Figure 13–3. CT scans of radiation necrosis following whole-brain RT for metastatic lung cancer. Three years following surgical extirpation of a tumor in the left frontal lobe and whole-brain RT, the patient developed contrast-enhancing lesions in the corpus callosum and the left frontal lobe. A needle biopsy specimen revealed radiation necrosis without evidence of tumor. The patient deteriorated intellectually and eventually became demented and bedridden.

a *Swiss cheese* or *soap bubble* pattern, suggestive of necrosis.[89] Perfusion MRI is more likely to show increased blood volume in tumor than in necrosis.[97] Diffusion tensor MRI may show decreased directionality of water diffusion in radiation necrosis.[98] Magnetic resonance spectroscopy usually reveals a significantly higher choline/creatinine and choline/NAA (*N*-acetyl-aspartate) ratio in recurrent tumor than radiation injury.[81] In a thallium single-photon emission computed tomography (SPECT) scan, increased uptake of thallium suggests tumor recurrence.[99] Despite all of these tests, many of which are not readily available, only a biopsy can clearly distinguish radiation necrosis from recurrent tumor. Furthermore, in many patients, both processes are present in the same area. Radiation necrosis usually occurs near the tumor site, even when the patient has received whole-brain RT, suggesting that tissue damaged either by tumor or by surrounding cerebral edema is more susceptible to radiation damage.

The situation is easier when normal brain has been included in a radiation portal used to treat an extracerebral tumor.[88] Examples include RT of head and neck tumors[88] such as nasopharyngeal[100] and pituitary tumors.[101] Radiation necrosis is characterized by new focal neurologic signs, depending on the site radiated. Amnesia, indicating medial temporal lobe destruction, may occur after RT for nasopharyngeal or pituitary tumors. Hemiparesis or personality changes follow RT for eye or maxillary sinus tumors because of damage to frontal or temporal lobes. Symptoms of increased intracranial pressure with headache, lethargy, and papilledema may also be noted.[88] CT or MRI show contrast-enhancing mass lesions with surrounding edema (Fig. 13–4). An arteriogram may show an avascular mass containing vessels with beading suggesting a vasculopathy.[88]

Grossly, the typical radiation necrosis lesion consists of coagulative necrosis in the white matter with relative sparing of the overlying cortex and deep gray matter.[102] Microscopically, striking abnormalities are found in blood vessels, with hyalinized thickening and fibrinoid necrosis often associated with vascular thrombosis, hemorrhage, and accumulated perivascular fibrinoid material.[88,102] These pathologic changes cannot be distinguished from the

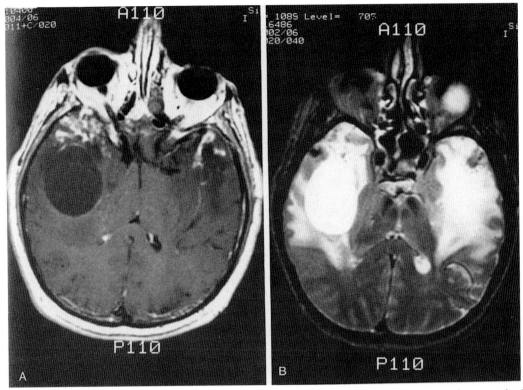

Figure 13–4. Radiation necrosis of the temporal lobes following RT for nasopharyngeal cancer. **A:** The T1-weighted MR scan shows a cystic mass in the right temporal lobe and a somewhat smaller mass in the left temporal lobe. Contrast enhancement is appreciated anterior to both cysts. **B:** The hyperintensity on the T2-weighted image demonstrates edema surrounding the cysts. The patient was demented and aphasic.

white matter damage caused by MTX (see Chapter 12). In other organs, standard vascular risk factors such as smoking and hypertension may increase the risk of radionecrosis[103]; it is unknown if these factors contribute to cerebral radiation damage. In less severely affected areas, demyelination is noted with a loss of oligodendrocytes, variable axonal loss, axonal swellings, dystrophic calcifications (that may be visible on MRI[104]), fibrillary gliosis, and scattered perivascular infiltrates and mononuclear cells. Telangiectatic vessels also form there and may be responsible for late hemorrhages, particularly in the spinal cord.

Radiation necrosis is usually treated by surgical resection. Most patients respond transiently to steroids but relapse when steroids are discontinued, although reports have described prolonged responses after steroid therapy without surgery.[85,105] One of our patients developed radiation necrosis of the temporal lobe after receiving RT for an ocular tumor. She became

aphasic and hemiparetic with a contrast-enhancing mass on MR scan. Both the clinical signs and the necrosis revealed on the scan resolved with corticosteroids. She has remained well 3 years after steroid treatment (Fig. 13–5). A similar patient is well 10 years after surgical resection of the anterior temporal lobe.[106]

Other suggested treatments presupposing a vascular mechanism have included aspirin, anticoagulation, and hyperbaric oxygenation,[85,107–109] but they are not, in our opinion, strikingly effective.[110] A patient of ours developed biopsy-confirmed RT necrosis, failed to respond to steroids, and had a dramatic resolution of enhancement on MRI with hyperbaric oxygen, but no clinical improvement. A recent report describes resolution of MRI abnormalities in patients with radiation necrosis treated with bevacizumab (see Chapter 12).[111] These findings imply that VEGF released in response to hypoxia may play a role in the pathogenesis of radiation necrosis.[86,87,91]

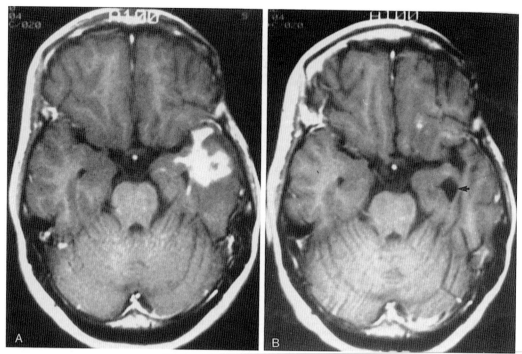

Figure 13–5. Radiation necrosis of the temporal lobe treated by corticosteroids. Three years following exenteration and radiation to the eye for a lacrimal gland tumor, the patient suddenly developed aphasia and a right hemiparesis. **A**: An MRI revealed a contrast-enhancing lesion with mass effect in the left temporal lobe. It was treated with corticosteroids, with amelioration of clinical symptoms. The steroids were then discontinued. The patient has been well for several years. **B**: The temporal lobe now reveals only a dilated ventricle (*arrow*) with no enhancement.

Cerebral Atrophy and Leukoencephalopathy: If substantial volumes of brain tissue are irradiated either for therapy or prophylaxis of leukemia or small cell lung cancer, most patients develop CT- or MRI-documented evidence of enlarged cerebral sulci and ventriculomegaly (Fig. 13–6).[112–115] MRI may reveal hyperintensity, most marked in cerebral white matter around the ventricles on the T2-weighted or fluid-attenuated inversion recovery (FLAIR) image; hypodensity is noted on the CT scan. These changes, which may worsen over time, almost invariably follow whole-brain RT with 30 Gy in 10 fractions to treat brain metastases or more than 50 Gy delivered in smaller fractions. The changes documented on the scan are more severe in the very young and the elderly and increase with both total dose and dose per fraction. Concomitant chemotherapy enhances the process.[116] The abnormalities are usually clear-cut by 1 year following RT, but occasionally they may become obvious within 2 or 3 months after RT is completed and persist or progress thereafter. One report comparing patients with low-grade glioma who were irradiated with those who were not revealed no difference in cerebral atrophy, at least in the few months after radiation.[117]

Long-term alterations in the electroencephalograph (EEG) and visual-evoked responses have been reported after low-dose (1.3 Gy) RT to the brains of children to treat tinea capitis,[118] suggesting that even very low doses of RT may cause mild brain damage in these patients. This suggestion is supported by a recent epidemiologic study of boys treated under the age of 18 months for cutaneous hemangiomas of the scalp. The proportion who attended high school decreased with increasing doses of radiation to both frontal and posterior parts of the brain.[119]

Sooner or later, almost all patients who receive more than 30 Gy of whole-brain or focal radiation develop white matter changes identifiable as T2 and FLAIR hyperintensity, either diffusely in the case of whole-brain radiation (Fig. 13–7) or focally with limited field radiation.[120] In children, quantitative T1 MRI identifies white matter changes in those

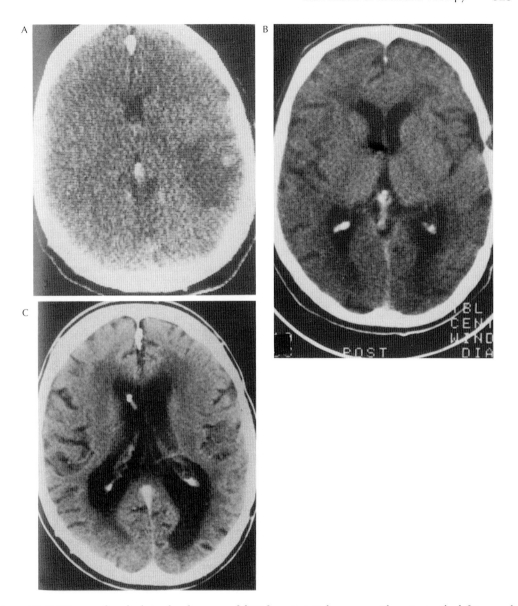

Figure 13–6. CT scans of cerebral atrophy after successful RT for metastatic lung cancer. The patient with a left temporal brain metastasis, whose ventricles were initially small (**A**), developed progressive enlargement of the ventricles several months (**B**) after treatment. Neurologic symptoms after 3 years included dementia, gait ataxia, and incontinence (**C**). The patient clinically responded transiently to ventriculoperitoneal shunting but eventually became demented and bedridden. The CT shows progressive atrophy and ventricular dilatation despite the shunt. The tip of the shunt can be seen in the right frontal horn.

who receive more than 20 Gy.[121,122] A recent report describes five young men treated with whole-brain RT and chemotherapy who developed severe leukoencephalopathy 9 months to 6 years after the RT.[122a]

Children whose brains are radiated with therapeutic doses of RT (e.g., for medulloblastoma) declined 2–4 IQ points each year, with the exception of verbal memory and receptive vocabulary. For young children, the decline begins early but levels off after several years.[123] The relationship between RT-induced CT or MRI abnormalities and clinical symptomatology is uncertain. Many patients with such changes have no symptoms[124] but in those who do, the severity of symptoms roughly

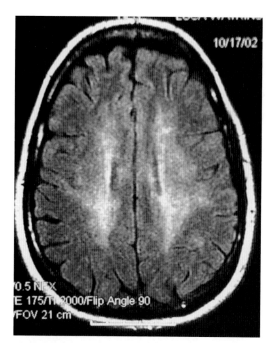

Figure 13–7. Leukoencephalopathy after radiation in infancy. A 50-year-old woman developed headaches, ultimately diagnosed as migraine. An MRI revealed white matter hyperintensity surrounding the ventricular system. Her past history indicated that she had received radiation therapy to her head for a hemangioma when she was an infant. A second lesion in the shoulder was also radiated. She subsequently developed a basal cell carcinoma within the shoulder radiation portal. Despite the changes on the MRI she was neurologically normal.

correlates with scan findings.[115,125,126] In our experience, virtually all patients in whom a substantial portion of the brain is irradiated to treat primary or metastatic brain tumors complain of short-term memory loss,[127,128] resembling that experienced by most older people (e.g., forgetting names, telephone numbers, appointments, and recent events but remembering remote events well). The memory loss may even prevent the individual from returning to gainful employment. An analysis of long-term survivors of RT for glioma indicates that approximately 60% were employed at jobs comparable to those they held before receiving RT.[124] In a minority of patients, perhaps only 10% to 20%, who have been treated for brain metastases with 30 Gy in 10 fractions, the memory loss progresses and affects other cognitive functions, leading to a more severe dementia.[92] Some of these patients also have gait abnormalities (gait apraxia) and

urinary urgency followed by incontinence, a triad suggestive of the syndrome of normal-pressure hydrocephalus.[129,130] If the scan suggests hydrocephalus, that is, if ventricular dilatation is greater than cortical atrophy, the patient may respond at least temporarily to ventricular shunting. These disturbances usually develop 12 to 14 months after RT but occasionally appear earlier or later.

The pathogenesis of the ventricular dilatation, cerebral atrophy, and white matter changes is uncertain. In some patients, true communicating hydrocephalus, perhaps from RT-induced arachnoiditis or obliteration of pacchionian granulations, appears to be causal. In others, a loss of cerebral substance seems to result from direct radiation damage. Direct damage to neurons and neuron precursors, as indicated in the previous paragraphs, may play a role. White matter changes may represent demyelination resulting from apoptosis of oligodendrocytes or chronic ischemia from vascular change. Diffusion tensor MRI demonstrates an early dose-dependent progressive demyelination accompanied by mild axonal degeneration. With the passage of time, the demyelination becomes independent of dose, suggesting that low doses of RT may be associated with structural changes of brain.[131] One report suggests that post-radiotherapy leukoencephalopathy, which occurs in all patients who receive more than 45 Gy, is analogous to vascular dementia.[132] Except in those patients who respond to shunting, cerebral atrophy is at present untreatable. However, prophylactic treatment with hyperbaric oxygenation has been reported to prevent white matter lesions in patients receiving stereotactic radiosurgery for brain metastases,[133] and lithium prophylaxis has been reported to prevent hippocampal damage in radiated experimental animals.[134]

The whole question of whether standard doses of radiation cause cognitive dysfunction, with or without changes on MRI, is still open. The problem is confounded by the fact that in most patients, brain tumors in and of themselves cause cognitive dysfunction. Studies of patients with low-grade gliomas indicate that cognitive difficulties are caused by the tumor and use of antiepileptic drugs rather than the radiation.[135] The same seems true of patients irradiated for brain metastases.[136] Prophylactic brain irradiation for leukemia in children and small cell lung cancer in adults seemed to offer a better opportunity to evaluate the cognitive effects of radiation. However, both groups also received chemotherapy, which might affect cognitive function, and much evidence suggests that the adults with small cell lung cancer may have had cognitive dysfunction before treatment, perhaps as a paraneoplastic syndrome[137] (see Chapter 15).

Radiation-Induced Hemorrhage: Formation of radiation-induced cavernous malformations can cause cerebral hemorrhages.[138] The lesions may appear many years after RT, the mean latency before detection was 8.9 years.[138] However, we have seen a few patients who had been treated for either metastases or primary brain tumors and who developed cavernous angiomas within 2 years after RT. The lesions are detected either incidentally on surveillance MRIs, where they may be mistaken for tumor recurrence, or when the patient presents with symptoms of a cerebral hemorrhage or a new-onset seizure. Some hemorrhagic cavernomas must be surgically removed.

Posterior Radiation "Migraine:" An enigmatic late response of brain irradiation is the so-called SMART syndrome—stroke-like migraine attacks after RT. This late complication affects both children[139] and adults.[140,141] The disorder is characterized by severe migraine-like headache with or without neurologic disability. The episodes are transient, but may last several days to weeks and may be associated with reversible gadolinium enhancement on MRI.[142] In one of our patients, the syndrome was associated with prolonged seizures. The pathogenesis of these symptoms is not known. A SPECT scan on one patient showed hyperperfusion in the left parieto-occipital region.[143]

Treatment

As indicated in the preceding text, there is no truly effective therapy for radiation injury.[144,145] Among those treatments that may have some efficacy is methylphenidate,[146–148] which appears to temporarily reduce some attentional and social deficits. The anticholinergic inhibitor, donepezil, has improved functioning in some trials[149] but not in others.[150] As indicated earlier, we have not found hyperbaric oxygenation, anticoagulation, or aspirin to be particularly useful.

SPINAL CORD

Table 13–5 classifies the varieties of radiation injury to the spinal cord. Clinical syndromes are described in the following paragraphs.

Acute Myelopathy

RT to the spinal cord does not cause acute clinical symptoms. Although some physicians are concerned that high doses of RT delivered to the spinal cord to treat epidural

Table 13–5. **Spinal Cord Radiation Injury**

Designation	Time after RT	Clinical Findings	Pathogenesis	Outcome
Acute	During treatment	None	–	–
Early-delayed	2 to 37 weeks	Lhermitte sign	Demyelination	Recovery
Late-delayed	Months to years			
Transverse myelopathy		Paraplegia-quadriplegia	Necrosis	Irreversible
		Brown–Séquard syndrome		
		Spastic paraparesis		
Motor neuron dysfunction		Leg weakness	Ventral roots	Irreversible
Hemorrhagic myelopathy		Acute paraparesis	Telangiectasia/cavernous angioma	Reversible

RT, radiation therapy.

spinal cord compression may worsen neuro-logic symptoms,[151] clinical and experimental evidence refutes this concern. Single fractions of 10 Gy delivered to the human spinal cord to treat spinal cord compression did not, unlike brain RT, cause acute toxicity.[152,153] Likewise, spinal cord dysfunction does not worsen when large fractions are delivered to experimental animals, whether or not the animal is treated with corticosteroids.[154,155] Patients with epi-dural spinal cord compression who suddenly deteriorate during RT are probably suffer-ing from the natural history of the tumor or from sudden hemorrhage into the tumor. Experimental evidence, however, does indicate that spinal cord irradiation acutely increases the blood-to-spinal cord transfer constant for water-soluble agents, that is, it disrupts the blood–spinal cord barrier.[156] Disruption of the blood–spinal cord barrier probably results from a combination of increased VEGF, the intercel-lular adhesion molecule ICAM-1, and endothe-lial cell apoptosis.[157] However, at least one study failed to find disruption of the barrier until 3 months after a single 25 Gy dose to the cervi-cal spinal cord. After 3 months, barrier disrup-tion appeared in white but not gray matter.[158]

Early-Delayed Myelopathy

This myelopathy is characterized by paresthe-sias or electric shock–like sensations called Lhermitte sign (Table 13–5), radiating down the spine or into the extremities when the patient flexes the neck. This symptom probably results from demyelination of the spinal pos-terior columns, leading to a spontaneous dis-charge of sensory axons when the spinal cord is stretched by neck flexion. The patient may not be aware initially that the experience requires neck flexion, but this can be demonstrated eas-ily at the bedside. A tingling paresthesia resem-bling a mild electric shock, Lhermitte sign is unpleasant but not painful. It usually begins 12 to 20 weeks after RT has been admin-istered and generally resolves during a few months to a year. In occasional patients, it may persist for several years. The MRI is usually normal, but a few studies have demonstrated increased glucose uptake on PET scanning.[159] Prolonged somatosensory-evoked responses (i.e., decreased conduction velocity in the spinal cord) have been reported by some but not all investigators.[160,161] The demyelination

hypothesis is not based on histologic evidence because none exists. One autopsy study of a patient who developed chronic radiation dam-age 5 months following an RT overdose to the spinal cord, and who partially recovered over 9 years, showed demyelination with only mini-mal thickening of capillary walls.[162]

The presence of Lhermitte sign does not pre-dict late-delayed radiation spinal cord injury. In experimental animals, myelin basic protein is transiently increased in the CSF shortly fol-lowing RT,[163] supporting the demyelination hypothesis for early-delayed myelopathy. The level rises again if a late-delayed myelopathy develops.[163]

Lhermitte sign is common after therapeutic radiation is delivered to portals that include[164] the cervical and sometimes the thoracic spinal cord,[159,165,166] especially after RT for Hodgkin disease or neck and mediastinal tumors. Word et al.[166] estimated its frequency at 15% follow-ing mantle RT (i.e., RT to cervical, supraclavic-ular, infraclavicular, axillary, mediastinal, and hilar lymph nodes in a single portal). Higher doses are more likely to cause the disorder.[166] A study of 1171 patients with nasopharyngeal cancer who received RT, with or without che-motherapy, identified transient radiation myel-opathy in 10%. The symptoms developed from 2 weeks to 72 months after completion of radi-ation (median 3 months) and resolved after 1 to 82 weeks (median 17 weeks). Patients older than 60 years were less likely to develop a prob-lem.[167] In another study of 1112 patients who received more than 30 Gy to at least 2 cm of the cervical spinal cord, the syndrome devel-oped in 40 (3.6%).[164] Higher doses and larger fractions were more likely to cause the syn-drome. No patient developed permanent spinal cord damage.[159–163]

The clinical history establishes the diag-nosis. The differential diagnosis includes cervical spinal cord compression, multiple scle-rosis, chemotherapy-induced Lhermitte sign, particularly from cisplatin or paclitaxel, and vitamin B_{12} deficiency. Because the sensation is not painful, reassurance is the only therapy that is necessary.

Late-Delayed Myelopathy

Late-delayed radiation myelopathy has three forms: a progressive or sometimes remitting myelopathy, a lower motor neuron syndrome, and hemorrhage in the spinal cord. The possible pathological lesions are listed in Table 13–6.

Progressive Myelopathy: The most common form is progressive myelopathy. The initial symptoms begin 12 to 50 months after RT and progress during weeks or months to paraparesis or quadriparesis.[168–170] Usually, the symptoms progress subacutely, but in some patients they advance slowly over several years, whereas in others they may stabilize, leaving the patient with only mild or moderate paraparesis. In some instances, the disorder may improve spontaneously, at least partially.[162] Late-delayed radiation myelopathy is dose dependent[171] and affects as many as 5% of patients who survive 18 months after receiving 50 Gy of mediastinal RT for lung tumors. Both total dose and fraction size are risk factors.[172] A safe dose is generally considered to be 45 Gy in 20 to 25 daily fractions, with the risk of myelopathy less than 1% at a 50 Gy dose, but 5% at 60 Gy administered in 1.8 to 2.0 Gy fractions.[56] We encountered one patient who developed radiation myelopathy as a radiation recall syndrome.[173] The patient had received RT for an epidural mass but was asymptomatic until doxorubicin was started many months later. She improved after the drug was stopped.

The first symptom is usually a Brown–Sequard syndrome with paresthesias and weakness in one leg and a decrease in temperature and pain sensation in the other. Although unusual, pain sometimes occurs first at the lesion site and then radiates into the arm, leg, or trunk. The symptoms begin distally and ascend to reach the irradiated level of the spinal cord. At worst, all motor, sensory, and autonomic functions can be lost below the level of the lesion. Some patients do not develop a Brown–Sequard syndrome, but instead exhibit a transverse myelopathy with both legs equally affected by weakness and sensory loss that rise to the level of the radiation portal. We have encountered a few patients with progressive weakness, hyperactive reflexes, and extensor plantar responses associated with sensory loss to position and vibration sense but with sparing of pain and temperature sensation. The progression of the lesion seemed much slower in these patients; several continued to be ambulatory many years after the onset of the syndrome.

Radiation myelopathy must be differentiated from epidural spinal cord compression or intramedullary metastases, which can be excluded by MRI, and from subacute necrotic myelopathy as a paraneoplastic syndrome (see Chapter 15). The latter is so rare, however, particularly in patients who are known to have cancer, that it is not a major consideration. The CSF is usually normal but may show an increased protein level. The MRI may initially be normal or, in the acute stages, show swelling[174] and contrast enhance-ment[175–179] (Fig. 13–8); however, the enhancement is usually patchy and rarely confused with an intramedullary metastasis. Identification of hyperintensity of the vertebral bodies on MRI often outlines the irradiated portal, which is helpful if RT records are unavailable. In later stages, the cord appears to be atrophic.[175,180] The motor conduction velocity in spinal cord pathways is reduced.[181]

Table 13–6. **Pathology of Radiation Myelopathy**

White Matter Lesions
Myelin breakdown of isolated nerve fibers
Myelin breakdown of groups of nerve fibers
 (spongiosis)
"Inactive" necrosis/malacia
  Spongiosis, spheroids
  Scar
"Active" necrosis/malacia
  Coagulative necrosis/malacia
  Liquefactive necrosis/malacia
    Amorphous
    Foam cell fields
    Cystic
Hemorrhagic necrosis/malacia

Vasculopathies
Morphologically intact blood vessel
Altered vascularity
Telangiectasia or vascular dilatation
Hyaline degeneration and thickening
Edema and fibrin exudation
Perivascular fibrosis and inflammation
Vasculitis
Fibrinoid degeneration/necrosis
Stagnation of erythrocytes or thrombosis
Hemorrhage

Modified from Ref. 182.

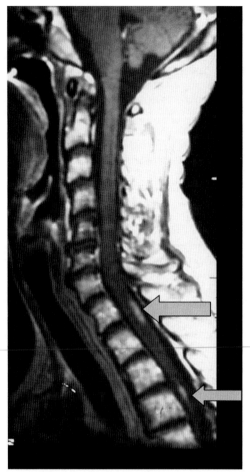

Figure 13–8. Radiation myelopathy. A 42-year-old woman developed supraclavicular adenopathy and received a diagnosis of non–Hodgkin lymphoma. She was treated with radiation therapy and chemotherapy and made a complete recovery. Two years later she noted weakness of her legs and a "painful pressure sensitivity" in her legs. Symptoms progressed to unsteadiness on her feet and urgency incontinence. An MRI revealed enhancing lesions within the radiation portal in the thoracic spinal cord (*arrows*). After several months the symptoms stabilized, the enhancement disappeared, and her neurologic symptoms improved, perhaps associated with hyperbaric oxygen treatment. Even without such treatment, symptoms either stabilize or resolve in some patients.

Disruption of the blood–spinal cord barrier, especially in the white matter, probably precedes other pathologic changes in the spinal cord.[156,157] The disruption of the blood–spinal cord barrier causes edema of the white matter, which in turn leads to the first clear pathologic change, that of demyelination in the posterior columns. The demyelination progresses to frank necrosis, first focally and then more diffusely.[156,182] Vascular changes, including fibrinoid necrosis of vessel walls, hyaline thickening, obliteration of the lumen, telangiectasia, and occasional inflammatory infiltrates[182] that resemble those in the brain are much less striking. Spinal cord gray matter appears strikingly resistant to the effects of radiation.[183]

The pathogenesis of radiation myelopathy is not fully understood, but the same processes that occur in the brain in response to RT, such as blood–spinal cord barrier disruption, and oligodendrocytes and endothelial apoptosis, are probably the major offenders. Pre-existing vascular risk factors may also contribute.[174] The pathogenesis of these changes has been considered earlier. Whether peroxidation of lipids plays any role in spinal cord damage is disputed. One study showed no difference in the degree of damage in animals deprived of or given excessive amounts of vitamin E,[184] whereas another study suggested that a combination of magnesium and vitamin E protected against spinal cord RT damage by lowering the amount of lipid peroxides.[185]

Steroids sometimes delay progression of the lesion.[170] One report describes sustained improvement following treatment with high doses of corticosteroids (methylprednisolone 1 g a day for 5 days followed by 80 mg a day tapered over 3 weeks).[186] Anticoagulation with heparin or warfarin has been reported to ameliorate symptoms,[187] but we have not been convinced that it is very effective. Hyperbaric oxygenation has been reported to stabilize or improve symptoms[188,189]; we have seen occasional patients who stabilized, but sometimes stability occurs spontaneously. Prevention is the only effective treatment. When radiation portals are designed to spare the spinal cord, radiation myelopathy does not occur; modern forms of RT, including stereotactic and intensity-modulated RT, should decrease the incidence of this devastating disorder (see Chapter 6).

Lower Motor Neuron Syndrome: This disorder, originally thought to result from anterior horn cell damage, is now recognized to result from damage to ventral roots within the spinal canal.[190,191] Characteristically, it follows pelvic RT for testicular tumors,[190,192–195] but has also occurred after lumbosacral RT for other tumors[190,196–198] or after craniospinal RT for medulloblastoma.[199] The disorder may occur 3 months to 23 years following RT and is characterized by the subacute onset of a flaccid weakness of the legs,[190,197] affecting both distal and proximal muscles with atrophy, fasciculations, and areflexia. It is usually bilateral and symmetric but may begin in or remain restricted to one leg. Sensory changes are absent, and bowel, bladder, and sexual functions are normal. The CSF may contain an increased protein concentration. The lumbar spine may be normal or show enhancing nodules on nerve roots that on biopsy are benign (Fig. 13–9).[200] Although electromyography reveals varying degrees of denervation, sensory and motor conduction velocities are normal. The deficit usually stabilizes after several months to a few years; often the patient can still walk. One of our patients can walk after 12 years of weakness from abdominal RT. Occasionally, patients become paraplegic.

One observer, noting that the syndrome often occurs at doses of radiation not usually associated with nervous system damage, suggested that a preexisting or concomitant viral infection may be required in order for the syndrome to develop.[190]

Impossible to differentiate from a pure motor polyneuropathy or isolated motor neuron loss, it also resembles the paraneoplastic syndrome, subacute motor neuronopathy (see Chapter 15). A single report describes a brachial motor neuron syndrome 3 years following RT that was associated with a cystic hypodense cavity affecting the spinal cord from C4 to C6.[201]

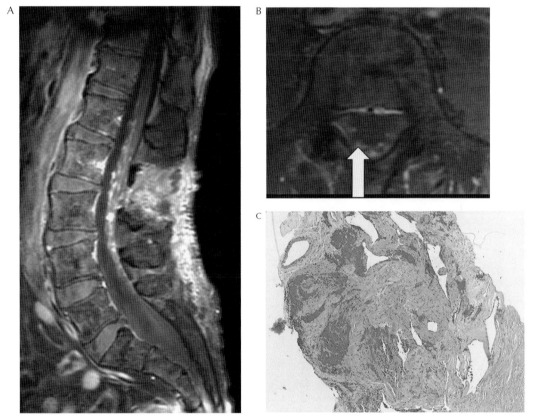

Figure 13–9. Radiation-induced lower motor neuron syndrome. A 53-year-old man presented to neurologic attention in 2000 complaining of weakness in both legs. He had received radiation therapy for testicular cancer in 1971. Examination revealed weakness and atrophy of both lower extremities without significant sensory symptoms or signs. An MRI revealed enhancing nodules on the roots of the cauda equina (**A**) and thickened roots on the axial scan (**B**) (*arrow*). Suspecting schwannomas, a biopsy was performed. However, the pathologic specimen revealed only fibrosis (**C**). The symptoms slowly progressed, although in 2007 he was still walking.

Pathologic changes include demyelination and axon loss, vascular changes,[202] and fibrosis in both sensory and motor roots, with focal areas of complete demyelination. The roots involved are primarily those of the cauda equina; some anterior horn cells (motor neurons) in the lumbar cord exhibited chromatolysis, suggesting secondary damage.[203]

Spinal Hemorrhage: The third form of late-delayed radiation myelopathy, hemorrhage in the spinal cord, develops many years after RT.[204,205] Characteristically, 8 to 30 years following RT to the spinal cord, a patient without prior neurologic symptoms suddenly develops back pain and leg weakness. In two of our patients, the syndrome began when the patient was taking a nonsteroidal anti-inflammatory agent (NSAID) for an unrelated disorder. The symptoms may evolve over a few hours to a few days. The MRI suggests acute or subacute hemorrhage in the spinal cord (Fig. 13–10). The cord may be slightly atrophic, but no other lesions are found. Characteristically, after several days, the patient begins to improve and the neurologic symptoms may resolve entirely. A few patients have had recurrent episodes of spinal cord hemorrhage. The pathogenesis is believed to be hemorrhage from cavernous

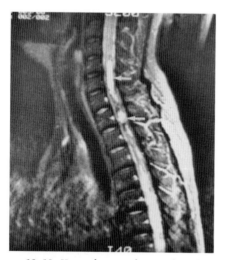

Figure 13–10. Hemorrhage in the spinal cord several years following successful mantle RT for Hodgkin disease. The patient was taking nonsteroidal anti-inflammatory drugs (NSAIDs) for unrelated pain when she developed sudden back pain associated with weak legs. An MRI revealed an acute hemorrhage in the spinal cord (*arrows*). She subsequently made a full recovery.

angiomas caused by RT. A biopsy sample of the spinal cord from one patient was said to show an arteriovenous malformation,[204] but it is more likely that these were cavernous malformations.[205] Affected patients should be admonished to avoid aspirin and NSAIDs, which might lead to subsequent hemorrhage, but no treatment helps once the hemorrhage has occurred.

PERIPHERAL NERVOUS SYSTEM

Table 13–7 classifies the various types of injuries that can affect the peripheral nervous system, including cranial nerves (except for the olfactory and optic nerves that are part of the brain) and peripheral spinal roots, nerve plexuses, and individual nerves. The clinical syndromes are described in the following paragraphs.

Cranial Nerves

RT directed at either the head or the neck can damage cranial nerves (Table 13–7).[206–208] Radiation-induced cranial palsies are uncommon and usually a late-delayed effect. Radiation damage to olfactory and optic nerves, which strictly speaking, are part of the brain and not truly nerves, and early-delayed conductive hearing loss, which may mimic damage to the acoustic nerve, are more common (see following text). Incidence data are sparse; one author estimates the incidence to range from 0.3% to 6% of irradiated head and neck tumors.[207] Another report describes 19 patients with cranial nerve palsies after radiation for nasopharyngeal carcinoma from a cohort of over 1200.[208] A randomized trial of altered fractionated radiotherapy for nasopharyngeal carcinoma identified cranial nerve palsies in 9 of 82 patients treated with conventional radiotherapy, and in 7 of 77 with hyperfractionated therapy.[206] Although the incidence of cranial nerve palsies was about the same in both groups, the total incidence of CNS complications was 23% in the first group and 49% in the second.

Cranial and other peripheral nerves are susceptible to direct damage. Axonal shrinkage and neurofilament alterations develop before Schwann cell or endothelial capillary damage and before fibrosis of surrounding connective tissue. Nevertheless, most late radiation

Table 13–7. **Cranial/Peripheral Nerve Radiation Injury**

Designation	Time after RT	Clinical Findings	Pathogenesis	Outcome
Acute	During treatment	Paresthesias	Direct stimulation	Benign
Early-delayed	2 to 30 weeks	Weakness, paresthesias	Possibly autoimmune	Benign
		Demyelination		
Late-delayed				
Focal	Months to years	Focal neuropathy	Fibrosis or plexopathy (painless)	Permanent
Diffuse	Years	Polyneuropathy/ myopathy	Hypothyroidism	Reversible
Neoplastic	Years	Weakness, pain/ tumor mass	RT-induced tumor	Permanent/ often fatal

RT, radiation therapy.

damage to the lower cranial nerves found in the neck probably results from fibrosis of connective tissue causing nerve damage by compression or ischemia.[209]

Olfactory Nerve: Olfactory neurons are continuously replenished, with a turnover rate in mice of approximately 3% per month.[210] As a result, these neurons might be expected to be sensitive to radiation. While undergoing head irradiation, some patients perceive an odor[211] probably due to stimulation of olfactory nerve endings by the radiation. Olfactory function, particularly odor discrimination,[212] is profoundly affected by RT for head and neck tumors.[213,214] Thresholds for perception rise markedly during the course of radiation, remain high at 1 month, but recover in some patients by 6 months,[214] although another report described continued deterioration in olfactory threshold scores at 12 months[215]; these deficits were not noticed by the patient. Because many patients confuse taste and smell, and because taste is so often affected by RT (see following text), the clinician must directly test odor perception.[216]

Optic Nerve: The visual system from the cornea to the occipital cortex may be affected by RT,[217–219] including the dry eye syndrome, glaucoma, cataracts,[220] retinopathy, or optic neuropathy. The only acute effects of RT on the visual system are those resulting from stimulation of retinal photoreceptors by the photons. Patients in a dark room or with closed eyes can perceive visual sensations

when the radiation beam is turned on,[217,221] a phenomenon not widely appreciated. One of our patients, a young man being irradiated for a pineal region germinoma, reported seeing a blue light every time "the beam was turned on." The technician suggested it was all "in his imagination."

Radiation retinopathy[217,222] is characterized by visual loss with microaneurysms, cotton-wool spots, capillary dilatations, telangiectasia, and capillary occlusion. The vascular changes may cause retinal edema, disk neovascularization, vitreous hemorrhages, and retinal detachment. Total dose and fraction size are important factors in the development of this disorder, as are chemotherapy and comorbid illnesses including diabetes and hypertension. Intravitreal bevacizumab has been reported to improve visual function.[223]

Late-delayed optic neuropathy has followed whole-brain RT for brain metastases, pituitary RT with photons or protons to treat adenomas, and RT for nasopharyngeal tumors.[218,224] The most important risk factor appears to be the total dose.[225,226]

Optic neuropathy begins months to years after RT and is characterized by the painless development of monocular or bilateral blindness. Visual acuity is reduced and is often associated with an altitudinal field defect. If the damage occurs anteriorly, the funduscopic examination will show papilledema, telangiectasia, and prepapillary as well as premacular hemorrhages. Cotton-wool spots, soft exudates, and retinal arterial narrowing mark damage by RT to the retina. If most of the damage occurs in the retrobulbar area, as with pituitary irradiation, the optic fundus may be normal, but MRI may reveal enhancing lesions of the optic nerve or chiasm that may precede visual loss[227] and may persist[228] or spontaneously resolve over time.[229]

The pathogenesis is thought to be ischemia from vascular occlusion,[230] producing either anterior or posterior ischemic optic neuropathy.[219] Apoptosis of oligodendrocytes may also play a role.[231] Histologic changes include axonal loss, demyelination, gliosis, and thickening and hyalinization of vessel walls as well as loss of endothelial cells.[232] Most affected patients do not recover vision.

A number of prophylactic[233,234] endeavors have been reported. The angiotensin-converting enzyme inhibitor, ramipril, has been reported to reduce radiation injury in experimental animals[233,234] when started 2 weeks after radiation. Treatments have included hyperbaric oxygenation,[235] anticoagulation,[236] bevacizumab,[237] optic nerve fenestration,[238] and intravitreal corticosteroid[239]; none have proved uniformly effective.

Ocular Motor Nerves: Radiation damage to the motor nerves that move the eye is rarely reported, in part because many affected patients suffer a concomitant optic neuropathy that precludes diplopia as a symptom, and in part because these nerves are more resistant to radiation than the optic nerve.[240] Nevertheless, ocular motor paralysis has been reported after radiation to cavernous sinus or pituitary tumors.[240–242] Recovery, either partial or complete, after RT suggests that demyelination, not radiation fibrosis, causes these disorders.

An unusual ocular complication of RT is *ocular neuromyotonia*[243–245] characterized by spontaneous spasm of eye muscles secondary to spontaneous discharges of ocular nerves. It occurs after RT for sella turcica or cavernous sinus tumors. Affected patients complain of intermittent diplopia, sometimes associated with a feeling of movement of the involved eye. The episodes are painless, can occur as frequently as several times an hour, and last only a few seconds. No electromyographic or pathologic studies are available to determine the pathogenesis of this disorder, although

the clinical symptoms are similar to the myokymia and neuromyotonia that occur after radiation of other cranial or peripheral nerves (see brachial or lumbosacral plexus in following text). Both disorders represent hyperexcitability of nerve. Myokymic discharges are brief bursts of single motor unit potentials firing at rates of 5 to 150 Hz. When clinically evident, they are seen as undulating muscle said to resemble tiny snakes wiggling under the skin.[246] Neuromyotonic discharges are similar, but the bursts are 150 to 300 Hz. When clinically evident, they produce either episodic spasms of muscle or sustained muscle contraction. Both disorders are relieved by anticonvulsants such as carbamazepine. A related syndrome, called *oculomotor nerve paresis with cyclic spasms*,[247] consists of episodic unilateral lid retraction with ipsilateral esotropia occurring about every 2 minutes and lasting 10 to 30 seconds. Between episodes there is ptosis and exotropia, indicating partial oculomotor paralysis. A single case report describes seesaw nystagmus following a combination of whole-brain RT and intrathecal MTX for lymphoma. Although no lesion was found on imaging, it is likely that this represents upper brainstem damage.[248]

Trigeminal Nerve: Trigeminal neuropathy has been reported rarely[249] following standard RT to the head, but has been reported following stereotactic radiosurgery for trigeminal lesions including schwannomas,[250] trigeminal neuralgia[251] and, in particular, intractable chronic cluster headache.[252] A more common cause of trigeminal neuropathy after head and neck tumor therapy is microscopic invasion of the trigeminal nerve by neurotropic tumors (see Chapter 8). Both myokymia and neuromyotonia of muscles supplied by the motor branch of the trigeminal nerve have been reported[253,254]; the disorder usually responds to carbamazepine.

The most serious trigeminal problem is radiation-induced trismus. This phenomenon is believed to result from fibrosis of masseter muscles and perhaps ligaments,[255,256] possibly related to damage to the trigeminal nerve from RT. We have seen one patient who developed radiation-induced trismus associated with typical neuromyotonia of the masseter muscle. Trismus can also originate from brainstem lesions.[257] Another patient of ours developed acute trismus immediately following surgery for a brainstem tumor. If carbamazepine is ineffective, botulinum toxin injection may be helpful.

Facial Nerve: The motor portion of the facial nerve is rarely, if ever, damaged by RT.[258] If late-delayed facial nerve paralysis occurs, it is most likely due to neurotropic tumor invasion until proved otherwise. Neuromyotonia has been reported,[253] in one case so severe that it resembled facial dystonia (Meige syndrome).[259]

Taste perception is a facial nerve function susceptible to radiation damage. Taste buds are a rapidly regenerating tissue, turning over about every 10 days. Many patients complain of ageusia (loss of taste) during the course of RT.[260] Taste usually returns within a year.[261] Confounding factors such as loss of smell and diminished salivary gland output, also from the RT, make it difficult to define the degree of taste loss. Patients with taste loss during RT have greater weight loss than those who do not suffer ageusia.[262] Many patients with cancer complain of abnormal taste even before RT[263]; the gustatory loss during RT can occur whether or not the taste buds and facial nerve are included in the radiation portal.

Acoustic/Vestibular Nerve: RT can affect any portion of the acoustic/vestibular (eighth cranial nerve) system.[264] Cutaneous changes within the radiation portal occurring during radiation can affect both the auricle and the external auditory canal.[265] These do not affect hearing or vestibular function except in the rare instances of canal stenosis.[266] In one series of 325 patients with a variety of head and neck tumors, mostly nasopharyngeal, external ear reactions occurred in 108 (33.2%). Ninety-three of these involved acute otitis externa occurring during radiation. The others occurred late, with acute and chronic otitis externa as well as atrophy and ulceration of the external auditory canal and canal stenosis.[264]

Middle ear involvement can cause conductive hearing loss (air conduction is decreased but bone conduction remains normal) usually associated with a middle ear effusion. This effusion can occur either during the course of radiation or later and can be either sterile or infected; the latter can lead to mastoiditis. It is a common problem following head and neck irradiation, affecting 98/325 (28.6%) of patients in a recent study.[264] It is

not a common complication of whole-brain RT for brain metastases but is common with the higher doses of RT for malignant brain tumors. Radiation-induced chronic otitis media is often accompanied by additional eighth-nerve problems. When compared with standard chronic otitis media, vestibular-evoked myogenic potentials suggest that RT caused additional retro labyrinthine or brainstem damage not found in nonirradiated patients.[264,267] Obstruction of the Eustachian tube also occurs over time.[268] With time, the tube becomes patulous, correcting the middle ear abnormality.[269] Patients who lose hearing from middle ear effusions can be treated by myringotomy if more conservative methods fail.[267,270] Radiation damage to the middle ear caused chorda tympani dysfunction in 23 of 325 patients (7.1%)[264]; the result was loss of taste on the ipsilateral tongue.[271]

Damage to the inner ear by RT often causes sensorineural hearing loss and less frequently vestibular dysfunction.[264,271] In the 325 patients referred to in the preceding text,[264] 49 (15.1%) developed sensorineural hearing loss that was persistent, 44 (13.5%) developed tinnitus, and 49 (15.1%) complained of vertigo or imbalance. The vestibular symptoms were usually mild. The symptoms result from direct radiation damage to the cochlea and labyrinth. Examination of the temporal bones after RT reveals thickened mucosa of the middle ear, fluid in the utricle and saccule, atrophy of stria vascularis, and decreased spiral ganglion and inner and outer hair cells.[272] Acute auditory and vestibular dysfunction can occur when radiosurgery is used to treat vestibular schwannomas.[273,274] In one instance, the dysfunction was caused by acute hemorrhage into the cochlea.[273] One of our patients, who had received two courses of whole-brain RT for brain metastases from breast cancer, developed sudden hearing loss in her left ear. Extensive workup for brainstem metastases and leptomeningeal tumor, as well as consideration of radiation-induced damage, preceded examination of the ear, which revealed the external canal filled with cerumen. Removal of the cerumen cured the hearing loss.

All of these complications are dose related. High fractions and concomitant chemotherapy also play a role, particularly with ototoxic drugs such as cisplatin.[264] Unexpectedly, cochlear implantation can restore hearing in patients who have RT-induced deafness.[275,276]

Lower Cranial Nerves: The lower cranial nerves (i.e., the glossopharyngeal, vagus, spinal accessory, and hypoglossal nerves) are partially located in the neck and are thus susceptible to radiation damage when RT is directed at head or neck tumors.[208] Nineteen such patients were identified from over 1200 patients with nasopharyngeal carcinoma treated at a single center. Hypoglossal paralysis (17) was the most common, followed by vagus (11), recurrent laryngeal nerve (6), and accessory nerves (2); hearing loss was not considered. Risk factors include total dose and dose per fraction.[206] The time of onset varies inversely with the radiation dose. The disorders occur many months to years after RT and are generally associated with obvious fibrosis of cervical soft tissues that has affected the nerves. The glossopharyngeal nerve has not been reported as a target of radiation damage but is difficult to assess clinically in isolation. Damage to the vagus nerve leads to unilateral paralysis of the palate and vocal cords, or the vocal cord alone (recurrent laryngeal nerve) when the neck or mediastinum are irradiated. Left-sided vocal cord paralysis may be a late effect of RT to the mediastinum, developing as late as 25 years post-RT.[277] Aspiration can be detected on physical examination when a patient coughs after swallowing water, indicating that aspiration is present even if the palate is not obviously paralyzed on examination.[278] Damage to the recurrent laryngeal nerve, a branch of the vagus nerve, paralyzes the vocal cords only, causing hoarseness, but also often causing aspiration. Examining the vocal cords identifies the abnormality.

Involvement of the spinal accessory nerve leads to shoulder droop and sometimes chronic arm pain. One of our patients received radiation for a tumor of the epiglottis. Shortly afterwards, she developed a Lhermitte sign that improved after several months but was still mildly present. Years after radiation, while still free of cancer, she began to note involuntary twitching of the right side of the neck. Examination revealed prominent myokymia, sometimes sufficient to move her head.

Horner syndrome may result from radiation damage to sympathetic fibers in the neck; it may be encountered in association with lower cranial nerve dysfunction or in isolation.

The hypoglossal nerve is probably the most commonly involved lower cranial nerve. One report of 234 patients with nasopharyngeal carcinoma identified 14 with hypoglossal palsy following radiation[279]; only one was bilateral. Unilateral paralysis is often asymptomatic; at most, affected patients complain of having difficulty with food caught between the teeth and the cheek; dysarthria is uncommon. Bilateral lingual paralysis, however, is extremely disabling and can cause severe dysarthria as well as swallowing difficulty.

The diagnosis is usually evident based on the history and the presence of obvious fibrosis in the neck. The MRI usually does not show any abnormality in radiation-induced cranial neuropathies, but does exclude recurrent tumor. Hypoglossal palsy is evident indirectly on MRI by signal abnormalities in the denervated tongue rather than an abnormality in the nerve itself.[280]

Peripheral Nerves

Although generally considered relatively radioresistant, any peripheral nerve can be damaged by a sufficient dose of RT. Most peripheral nerve damage results from doses greater than 60 Gy.[281] The α/β ratio of nerve is approximately 2 Gy.[281] Fraction size and chemotherapy also play a role. Most clinically encountered radiation-induced peripheral nerve damage is probably secondary to radiation-induced fibrosis. The major sites of damage encountered clinically are the brachial plexus after treatment for breast or lung cancer (and when RT was the norm for Hodgkin disease) and the lumbosacral plexus after treatment to the pelvis and lower abdomen for a number of cancers (Table 13–7). In experimental animals, radiation injury to peripheral nerves or plexuses is a dose-limiting factor[14,282] for intraoperative RT. It has not been a major clinical problem in patients treated for rectal or gynecological cancers, or retroperitoneal sarcomas.[283]

Peripheral nerves probably do not suffer acute damage from RT, although Haymaker and Lindgren[284] mention that patients "under the beam" sometimes experience acute paresthesias. Biochemical and pathological abnormalities, particularly disruption of the blood–nerve barrier, can be identified experimentally within 2 days of the radiation.[285] These alterations are dose dependent and probably reversible. Later changes, including demyelination and axonal damage, are secondary to fibrosis and vascular damage.

In most patients with peripheral nerve damage, weakness and atrophy are apparent, although in women with breast cancer, lymphedema may mask the atrophy. One patient developed hypertrophy in the trapezius muscle associated with radiation damage.[286]

Brachial Plexus: Brachial plexopathy is now fortunately much less common, with improvements in radiation technique; it probably affects fewer than 1% of radiated patients.[287] The disorder is characterized by hand and arm paresthesias; loss of sensation in the thumb and index finger; painless weakness of shoulder muscles, biceps, and brachioradialis,[288–291] and frequently, lymphedema and palpable induration in the supraclavicular fossa. In a few patients who are asymptomatic, the only finding is a decrease in the biceps and brachioradial reflexes. Myokymia, detected by an electromyogram, affects muscles innervated by the involved nerve trunks and is a useful criterion for differentiating radiation damage from tumor infiltration of the plexus.[291–294] The disorder often progresses to a panplexopathy that paralyzes the entire arm, usually without severe pain, although on very rare occasions, the disorder may be mild and reversible.[295] Symptoms often begin within months of the radiation and may represent an early-delayed process.) Sometimes the disorder begins in the lower plexus, affecting the hand muscles before the arm is involved. In one series[296] of 79 patients radiated for breast cancer, 28 had clinical evidence of brachial plexopathy, 15 of whom had significant disability, and most had panplexopathy. Surgery on the axilla and adjuvant chemotherapy may have been risk factors. In another series[297] of 161 patients irradiated for breast cancer, 5% had late-delayed disabling plexopathy and 9% had mild plexopathy. Cytotoxic therapy, young age, and large RT fractions are risk factors.

The important differential diagnosis in plexus lesions lies between radiation damage and recurrent tumor (Table 8–4). Pain is the single most important clinical differentiating feature. Severe pain is uncommon with radiation fibrosis, but is almost invariably present if tumor is the culprit. Proximal distribution of the weakness suggests radiation fibrosis but

may also be seen with recurrent tumor. The CT or MRI may identify a discrete tumor mass but often reveals only diffuse, nonspecific loss of tissue plane in radiation fibrosis. An FDG PET scan should be hypermetabolic in patients with tumor but not in those with radiation fibrosis. Diagnostic certainty may require surgical exploration.[298] Even this may not exclude tumor if the nerves are infiltrated microscopically without bulky disease. Recurrent tumor may be treated with RT or chemotherapy. No effective treatment exists for radiation plexopathy.[299] Because radiation-induced fibrosis is believed to be the major factor causing radiation-induced plexopathy, the combination of pentoxifylline and vitamin E, reported to be effective in superficial radiation fibrosis, has been tried as treatment.[109,300] In our experience, it has not been successful. A few reports suggest that surgical neurolysis may help.[301,302] One report describes a radiation-induced conduction block of the brachial plexus resolving after anticoagulation.[303] The patient did not have a typical brachial plexopathy; it is difficult to know what the actual diagnosis was in this patient. A randomized double-blind phase II study of hyperbaric oxygen in patients with radiation-induced brachial plexopathy did not give evidence of significant improvement.[304]

Acute radiation-induced brachial plexopathy is a clinical disorder that is no longer seen, probably because of changes in therapy. It affected some patients with Hodgkin disease who developed an acute brachial plexus palsy, usually painful but reversible, and clinically indistinguishable from acute idiopathic brachial neuritis. The disorder occurred abruptly a few days to weeks after RT began.[305,306] Although its pathogenesis is unknown (it is probably an immune-mediated neuropathy precipitated by the radiation),[307] the RT can be completed safely.

Early-delayed brachial plexopathy is reversible[308] and characterized by paresthesias in the hand and forearm, sometimes associated with pain and accompanied by weakness and atrophy in a C6 to T1 distribution. The disorder begins approximately 4 months after RT. Nerve conduction velocities reveal segmental slowing; recovery occurs within a few weeks or months. This disorder affected 14% of patients with breast carcinoma who were irradiated in one institution during a 10-year period. One series reports reversible brachial plexopathy occurring in 16 patients, usually younger women, irradiated for breast cancer. In these patients, the symptoms were mild, with numbness or paresthesias in the hand or arm, weakness, and decreased deep tendon reflexes.[295] Symptoms often began within months of the radiation and may have represented an early-delayed process. The disorder is no longer observed. The pathogenesis of the disorder is not clear, but it does not appear to predispose to late-delayed radiation damage.[308] A similar reversible disorder of lumbar plexopathy may begin 4 months after wide-field pelvic RT.[309,310]

Late-delayed radiation plexopathy can affect either the brachial or lumbosacral plexus, although the former is much more common. It usually occurs a year or more after radiation. Eleven of 12 patients radiated for breast cancer from 1963 to 1965, and followed for many years after 57 Gy to the brachial plexus, eventually developed arm paralysis.[311] It is not rare for patients to develop the syndrome decades after the radiation.[287–291,293–299,301–304]

Acute ischemic brachial plexopathy can occur years after RT to the breast and is caused by occlusion of the subclavian artery. It is a rare late-delayed complication.[312] The plexopathy is painless, acute in onset, and nonprogressive, although not reversible.

Lumbosacral Plexus: This disorder causes weakness of one or both legs.[303,309,313,314] As with brachial plexopathy, pain is usually absent and, if present, generally mild. The disorder often affects the foot first. Most patients also have sensory loss. The electromyogram frequently reveals myokymic discharges that help differentiate the RT-induced disorder from tumor plexopathy. RT-induced lumbosacral plexopathy often progresses slowly over many years, but it may stabilize while the patient is still functional or resolve spontaneously.[309] Diabetes seems to be a risk factor.[315] CT, MRI, or PET can usually exclude recurrent tumor.

Sacral Plexus: Radiation-associated erectile impotence is common following treatment for prostate cancer. In one study, 80% of previously potent patients complained of decreased erectile capacity. Arteriography in some patients revealed bilateral occlusive disease of internal pudendal and penile arteries, suggesting a vascular cause rather than direct damage to sacral nerves.[316] One patient with a sacral plexopathy related to osteonecrosis of the sacrum improved with hyperbaric oxygen therapy.[317]

Muscle

Skeletal muscle is relatively resistant to radiation. The α/β ratio is approximately 4 Gy. However, large doses (>63 Gy) given to large volumes can cause weakness and muscle swelling.[318] Loss of muscle fibers, vascular lesions, inflammation, and fibrosis all play a role in the pathogenesis.[318] In experimental animals, intraoperative irradiation can cause substantial muscle damage,[318] but with modern techniques, this does not appear to be a significant clinical problem.

Two troubling clinical syndromes are caused in part by muscle damage: the first is trismus and the second is the dropped head syndrome.

Trismus: Trismus is a term used to describe any restriction to mouth opening. After RT to portals involving muscles of mastication, the ability to open the mouth fully begins to decrease within a few months and progresses for approximately 4 years.[255] The degree of mouth opening may be sufficiently restricted to compromise eating and dental hygiene. It is believed to result from masseter fibrosis. However, as indicated in the discussion of trigeminal neuropathy, damage to the motor branch of the trigeminal nerve may also play a role. Physical therapy, pentoxifylline,[319] and botulinum toxin[320] have been used to treat the disorder; some patients benefit symptomatically.

Dropped Head Syndrome: The *dropped head syndrome* is a term used to describe patients with focal weakness of neck extensors, such that they are unable to keep their head upright.[321] Most patients who have received RT for the treatment of Hodgkin disease develop some atrophy of neck muscles. In many, strength is virtually normal and in some men the neck atrophy is only noticed by a decrease in collar size. Because extensors are weak out of proportion to flexors, patients often discover that their chin falls on their chest, particularly when the head is jostled, such as during a sudden stop in an automobile. At its worst, the head remains in the dropped position. The disorder probably results from damage both to the muscles and nerves supplying those muscles. There is no treatment; a cervical collar maintains the head upright.

Indirect Nervous System Damage

RADIOGENIC TUMORS

Ionizing radiation not only treats but also causes tumors.[322,323] The problem exists at all ages but is particularly important in children, who are substantially more susceptible and have longer survival after treatment of their initial neoplasm.

Cranial radiation even at low doses is carcinogenic in genetically susceptible individuals.[324] Of over 10,000 children treated for tinea capitis in the 1950s with a mean estimated radiation dose of 1.5 Gy, 67 developed benign meningiomas and 31 malignant intracranial tumors, a substantially greater incidence than nonirradiated individuals[325] (Fig. 13–11). Radiation-induced meningiomas are more likely to be atypical or anaplastic than non-radiation–induced tumors, and are more likely to develop in those who have a genetic susceptibility, although the exact gene (or genes) is unknown.[324]

Of 426 patients irradiated for pituitary adenomas, 11 (2.4%) developed an intracranial tumor within 20 years; six were meningiomas, and five malignant brain tumors.[326]

Radiation-induced intracranial tumors occurred in 22 of 1612 patients who received prophylactic brain irradiation for acute lymphoblastic leukemia[327]; 10 were low-grade gliomas, 11 meningiomas, and 1 high-grade glioma. New intracranial tumors also occur at increased frequency in children irradiated for brain tumors.[328] Of 14,361 5-year survivors of childhood cancer, 116 developed intracranial tumors, 40 developed gliomas, and 66 developed meningiomas.[329] Most of these children had been treated originally with cranial irradiation for leukemia or CNS tumors.[330]

Even radiosurgery can cause CNS neoplasms. In one case, malignant transformation of a vestibular schwannoma occurred 6 years after gamma knife radiosurgery.[331] The malignant tumor had a p53 mutation that was not present in the primary tumor. Radiosurgery for a meningioma was followed 7 years later by a glioblastoma within the radiation field.[332] Radiosurgery for an arteriovenous malformation was followed 6½ years later by a glioblastoma.[333]

Radiation carcinogenesis probably results from radiation damage to DNA with abnormal repair, although the pathogenesis is not fully established. Biologic variables affect the likelihood that a given patient will develop a tumor. For example, patients with neurofibromatosis are more likely to develop malignant peripheral nerve sheath tumors than are normal subjects receiving the same RT.[334] In experimental animals, genetic factors, sex, age, and hormonal influences appear to play a role.[335]

The signs and symptoms of tumors arising in the brain or spinal cord after previous RT do not differ from those of spontaneous tumors. RT-induced tumors can be either malignant or benign and must be examined by biopsy to establish whether the lesion has a different histology from the original tumor irradiated many years before. Rare late relapses of the original tumor are possible. We have encountered patients with pilocytic astrocytomas recurring as late as 30 years following initial resection and RT.

Malignant or atypical nerve sheath tumors can develop following RT of the brachial, cervical, and lumbar plexuses (Fig. 13–12). Of the 14,372 patients enrolled in the Childhood

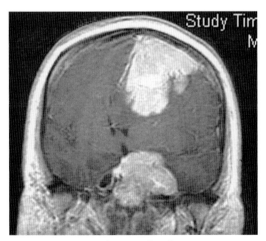

Figure 13–11. Radiation-induced meningiomas. A 54-year-old Russian-born woman presented to neurologic attention with electric shock–like pain involving her right upper gum. An MRI revealed two large meningiomas that on biopsy were benign, but showed some cytologic atypia. Past history revealed she had received radiation for tinea capitis in infancy (see text).

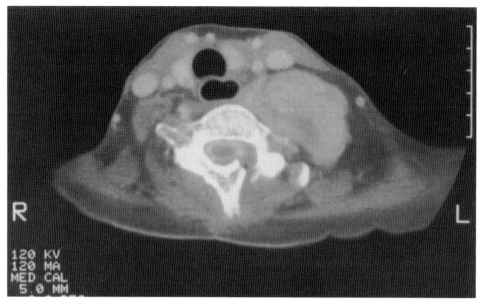

Figure 13–12. RT-induced malignant peripheral nerve sheath tumor of the brachial plexus. This patient, successfully treated with mantle radiation therapy for Hodgkin disease 12 years previously, developed painful weakness of the left upper extremity. A mass was palpated in the supraclavicular fossa that, when biopsied, proved to be a neurogenic sarcoma. The spinal cord was compressed. Continued growth caused the patient's death despite additional RT to the lesion.

Cancer Survivor Study, 108 developed secondary sarcomas, nine times higher than the general population[330]; 19 of these were malignant peripheral nerve sheath tumors. It should not be assumed that patients who have been irradiated previously for breast, abdominal, or pelvic tumors, and who many years later develop painful mass lesions of the brachial or lumbosacral plexus, are experiencing recurrence of their primary tumor. As mentioned previously, the new lesion should be examined by biopsy or resected before other treatment is initiated. The treatment of nerve sheath tumors is surgical excision where possible and, with malignant tumors, as much external-beam irradiation as the patient can tolerate. Radiosurgery, brachytherapy, and chemotherapy should also be considered if sufficiently large doses of external-beam RT cannot be administered because of the prior RT.

RADIATION-INDUCED VASCULOPATHY

RT can cause late-delayed effects on blood vessels of all sizes (Fig. 13–13) and can cause either hemorrhage[336,337] or infarction.[338,339] As indicated previously, small-vessel vascular occlusions play an important role in the development of both brain and spinal cord necrosis. RT-induced cavernous malformations can lead to delayed hemorrhage in the brain and spinal cord[204,205,340] (radiosurgery can also cause cavernous malformations[286]).

Mantle RT for Hodgkin disease[341] or cranial RT for leukemia or brain tumors[337,342] substantially increases the risk of stroke. The relative risk for those with Hodgkin disease was 4.32, those with leukemia was 6.4, and those with brain tumors 29.0. The increased risk occurred in those who received more than 30 Gy. We have encountered patients who, some years after RT for brain tumors (medulloblastomas, low-grade gliomas), developed small, asymptomatic contrast-enhancing lesions that disappeared without treatment. Whether these represented subacutely evolving infarcts or not is unclear. What is important is that they should not be confused with recurrent tumor.

Late-delayed effects on large vessels can have major secondary effects on CNS function. Lesions of large intracranial or extracranial blood vessels may be apparent 4 months to 20 years after RT (Fig. 13–14). The pathology of large-vessel damage is accelerated atherosclerosis. Radiation for parotid tumors leads to

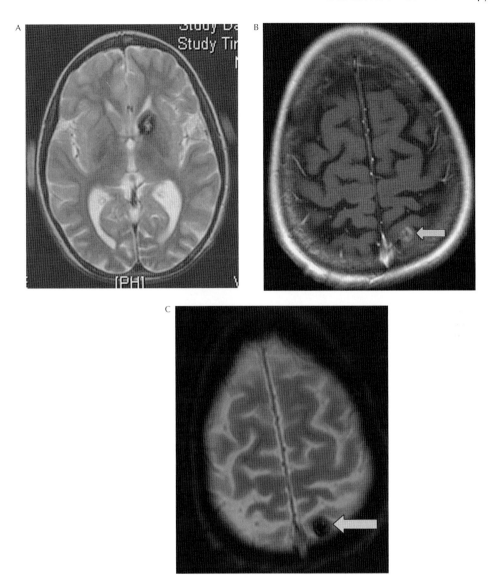

Figure 13–13. Radiation-induced cavernous malformations. **A:** A 19-year-old man received an MRI of the head before treatment with growth hormone. He had been treated with chemotherapy and 24 Gy of cranial radiotherapy 16 years previously for acute leukemia. The scan revealed an asymptomatic radiation-induced cavernous angioma. It has remained asymptomatic. **B:** A 60-year-old woman with metastatic breast cancer was found to have asymptomatic brain metastases and was treated with whole-brain radiation in addition to chemotherapy for her systemic metastases. Thirteen months later a new enhancing lesion was found in the left parietal-occipital lobe (B) (*arrow*). A gradient echo image (**C**) revealed the lesion to be hemorrhagic (*arrow*), suggesting a radiation-induced cavernous angioma rather than a new metastasis.

increased carotid wall thickness.[343] The most vulnerable vessel of neurologic interest is the carotid artery.[338,343,344] The incidence of carotid stenosis following head and neck irradiation varies from 30% to 50%.[338] RT-induced carotid stenosis or occlusion can cause transient ischemia or cerebral infarction.[338,339] Interestingly, supraclavicular radiation for women with breast cancer does not appear to increase their incidence of stroke,[345] or has only a modest effect on stroke risk.[346]

The diagnosis of RT-induced carotid occlusion is suggested when a patient's carotid arteries were included in the RT portal. MR or contrast angiography supports the diagnosis if the site of carotid stenosis or occlusion is within

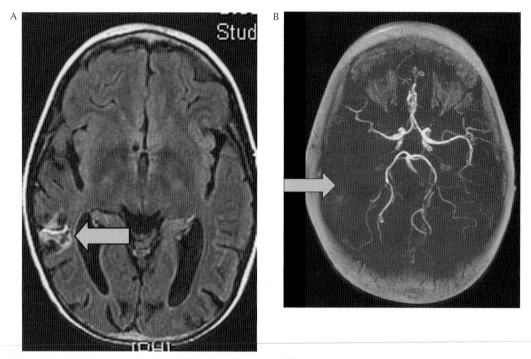

Figure 13–14. Radiation-induced cerebral infarct. A 30-year-old woman was treated for leukemia at age 4 with whole-brain radiation and chemotherapy. She presented at age 30 with transient episodes of numbness and tingling of the left side of her body. An MRI (**A**) revealed an infarct in the right temporoparietal area that had not been present on a routine scan 2 years before (*arrow*). An MRA (**B**) demonstrated a paucity of small vessels in that region (*arrow*).

the radiation portal and is different from its usual location at the bifurcation in nonirradiated patients. RT-induced occlusion can be intracranial or extracranial, depending on the RT portal.

Cranial radiotherapy, particularly in children and individuals with neurofibromatosis-1, can cause occlusion of large vessels of the circle of Willis and lead to the moyamoya syndrome.[347–349] (Moyamoya, Japanese for "cloud of smoke," describes collateral vessels that arise to supply the brain after large-vessel occlusions.) The disorder may be asymptomatic or can lead to stroke or hemorrhage.

The treatment of RT-induced atherosclerosis causing neurologic symptoms resembles that of typical atherosclerosis. Experimental evidence indicates that radiation doses insufficient to cause vascular damage in experimental animals may produce significant atheromatous lesions when combined with hyperlipidemia. Thus, lowering serum cholesterol may be helpful in patients. Antiplatelet agents should be administered. Carotid endarterectomy, when indicated, should be performed, but the surgery may be more difficult technically than usual

because periarterial fibrosis may make it difficult to separate the intima from the media.[350] Carotid stenting has been reported to be safe and effective.[351]

Cerebral aneurysms occasionally occur after RT.[352,353] The aneurysms differ from ordinary saccular aneurysms in their location and histologic features.

ENDOCRINE DYSFUNCTION

RT-induced endocrine dysfunction is relatively common. When severe, abnormal hormonal levels can secondarily affect the nervous system, causing cognitive changes, sexual dysfunction, and other neurologic disability in adults, as well as growth failure in children.[354] Such abnormalities occur after RT that includes the pituitary and hypothalamus, or after RT to the neck when the thyroid and parathyroid glands are within the portal.

RT-induced dysfunction of the hypothalamic–pituitary axis[355,356] is not rare after treatment for nasopharyngeal carcinoma and pituitary tumors; it also may follow whole-brain irradiation for treatment or prophylaxis.[357] The dose that causes radiation damage to the hypothalamic–pituitary axis is uncertain. Clinical evidence suggests that, in at least some patients, conventional doses and fractionation schedules can cause substantial damage when hormonal levels are measured several years after RT. The radiation damage site in the hypothalamic–pituitary axis differs from patient to patient, depending in part on the fractionation of RT, the radiation portal, and the patient's underlying disease. Patients who have had pituitary surgery are more likely to experience pituitary failure, whereas children with acute lymphoblastic leukemia (ALL) tend to suffer hypothalamic failure. Furthermore, each portion of the hypothalamic–pituitary axis has its own intrinsic susceptibility to RT. Growth hormone is most sensitive, thyroid-stimulating hormone the least, with luteinizing hormone/follicle-stimulating hormone and adrenocorticotropic hormone in between.

The most common clinical symptom is growth failure in children. In adult men treated for brain tumors, approximately 67% complain of sexual difficulties, usually decreased libido and impotence, within 2 years of RT. Serum testosterone levels may be reduced. The presumed basis is hypothalamic damage. Adrenal failure and hypothyroidism are rare, but many patients who have received RT to the brain complain of feeling cold chronically, even when thyroid function test results are normal.

Diagnosing hypothalamic–pituitary failure is often difficult. When the condition is florid, standard pituitary hormones are measurably low in the serum, except for the prolactin level, which is usually high because of a failure to secrete dopamine, the prolactin inhibitory factor. However, in a few patients who have symptoms suggesting endocrine dysfunction and low levels of hormone at rest, the results of stimulation testing may be normal, making the diagnosis and identification of the exact site of failure difficult.

Of 31 adults with nasopharyngeal carcinoma who received approximately 40 Gy to the hypothalamus and 60 Gy to the pituitary, more than 50% developed endocrine dysfunction within 5 years of cranial RT. Clinically apparent hypopituitarism was common following a mean latent interval of 3.8 years.[356]

Of 47 patients receiving RT for acromegaly, hypopituitarism was present in only 33% at baseline, but developed in 85% by 15 years.[358] Of 107 patients with base of skull tumors, hyperprolactinemia was present in 84%,

hypothyroidism in 63%, hypogonadism in 36%, and hypoadrenalism in 28% at 10 years.[359]

A recent report examining 75 randomly selected long-term survivors of childhood ALL who received cranial radiotherapy indicated that 60% had developed some components of the metabolic syndrome compared to 20% of those who had not received cranial radiation. The major abnormalities were growth hormone deficiency with resultant lower insulin-like growth factor 1, higher fasting insulin levels, abdominal obesity, and dyslipidemia.[357]

Also of neurologic interest is hypothyroidism following neck irradiation.[360,361] The problem arises particularly in patients who have been treated successfully for Hodgkin disease. The actuarial risk of developing thyroid disease 20 years after mantle RT is approximately 52%. Hypothyroidism is the most common complication, although Grave disease with ophthalmopathy can occur, as can thyroid cancer.[361] Neurologic symptoms from hypothyroidism can involve the peripheral nervous system or CNS. They include encephalopathy, ataxia, and peripheral neuropathy, which can develop at virtually any time following RT and may occur in the absence of other systemic symptoms of hypothyroidism. Hypothyroidism causes elevated spinal fluid protein, which may prompt an extensive and unnecessary neurologic evaluation before recognition that the neurologic disorder is caused by hypothyroidism. Thyroid function studies should be conducted in any neurologic patient who has received prior RT to the neck.

Hypercalcemic hyperparathyroidism has also been reported to follow RT.[362]

RADIOTHERAPY AND DRUGS IN COMBINATION

Drugs including cytotoxic agents, molecularly targeted drugs, and others are often used in conjunction with RT to augment its effects. The rationales are several (Table 13–8).[363] (1) Spatial cooperation: the drugs may cooperate spatially, for example when cytotoxic drugs are used to treat micrometastases in a patient who will be radiated to control local disease. (2) Cytotoxic enhancement: the drug may enhance the efficacy of RT by reducing radiation resistance. (3) Biological cooperation: the drugs may target tumor cell populations resistant to RT. (4) Temporal modulation: the drugs may reduce cell proliferation between fractions

Table 13–8. Examples of Clinical Trials of Drug–Radiation Combinations and Potential Mechanisms that Underline Therapeutic Benefit

Mechanism	Clinical Model	Drug Combined with RT	Outcome (level of evidence)	Comments
Spatial cooperation	Stage I–II breast cancer	Adjuvant chemotherapy, hormonal therapy, trastuzumab	RT-reduced locoregional recurrence risk by a factor of 3 (level 1)	No proven benefit of RT on systemic relapse rate
Cytotoxic enhancement	Glioblastoma	Temozolomide	2-year survival increased from 10% with RT alone to 26% (level 1)	Biological modifiers will be investigated in future trials
	Non-metastatic, inoperable, non–small cell lung cancer	Low-dose cisplatin	2-year survival improved from 13% with RT alone to 26% (level 1)	Novel cytotoxic agents to be tested concomitantly with RT
Biological cooperation	Stage III–IV, non-metastatic head and neck cancer	Tirapazamine	Higher response rate in poorly oxygenated tumors (level 3)	Independent validation in prospective study should be sought
Temporal modulation	Stage III–IV, non-metastatic head and neck cancer	Cetuximab	3-year overall survival improved from 44% with RT alone to 57% (level 1)	To be tested in combination with chemoradiation therapy
Normal tissue protection	Head and neck cancer, all stages	Amifostine	Moderate and severe late xerostomia reduced from 57% with RT alone to 34%. Locoregional control reduced from 63% to 58% (not significant; level 1)	Trials completed to date may not have sufficient statistical power to reveal a clinically relevant outcome

From Ref. 357 with permission.
RT, radiation therapy.

of RT. (5) Normal tissue protection: radiation protectors[9] may allow higher doses of radiation with fewer side effects.[363]

Table 13–9 lists some drugs that sensitize tissues to radiation as well as a few that serve to protect normal tissues against the deleterious effects of ionizing radiation.

Unfortunately, as indicated in the previous chapter, the neurotoxicity of chemotherapy and other agents such as small molecules, antibodies, and hormonal agents may be increased when used in combination with radiation. Table 13–10

lists some chemotherapeutic agents well known to enhance radiation neurotoxicity. Other agents undoubtedly are just as capable of increasing neurotoxicity as those listed.[364]

TREATMENT

Table 13–11 lists some of the many approaches that have been taken to prevent or treat radiation toxicity. Some of these are described in the paragraphs on specific types of injuries.

Table 13–9. **Some Radiation Sensitizers and Protectors**

Radiation Sensitizers[°]	
Platins (cisplatin, oxaliplatin)	Trastuzumab
Gemcitabine	Cetuximab
Irinotecan	Tryosine kinase inhibitors
Temozolomide	Gefitinib
Fluorouracil/capecitabine	Erlotinib
Taxanes	Bevacizumab
Tirapazamine	Bortezomib
Valproic acid[358a]	Cox-2 inhibitors
Efaproxiral[358b]	Motexafin gadolinium[358c]
Radiation Protectors	
Amifostine (and others)	Antioxidants[358d]
Erythropoetin	NF-κB[358e]

[°]Data from Refs. 9 and 358, unless specified.

Table 13–10. **Neurotoxic Interactions between Chemotherapy and RT**

Agent	Route of Administration	Neurotoxicity
Methotrexate	IV, IT	Leukoencephalopathy Mineralizing microangiopathy Cognitive impairment Myelopathy
Nitrosoureas	IA, high-dose IV	Leukoencephalopathy Cognitive impairment Possibly myelopathy
Cytosine arabinoside	High-dose IV, IT	Leukoencephalopathy Possibly cognitive impairment Myelopathy
Multidrug regimens	IV	Cognitive impairment Leukoencephalopathy Pontine leukoencephalopathy
Vincristine	IV	Myelopathy Peripheral neuropathy

Adapted from Refs. 365 and 366.
IA, intra-arterial; IT, intrathecal; IV, intravenous.

Table 13–11. Possible Preventive and Therapeutic Interventions for Radiation-Induced Nervous System Injury

Cytokines (growth factors)
   IGF1
   CNTF
   PDGF
   VEGF
   bFGF

Antioxidants/free radical scavengers
   NAC
   Cysteine
   Methionine
   Glutathione
   Sodium thiosulfate
   Melatonin
   Vitamins C and E
   WR2721 (Amifostine)
   MnSOD
   PUFAs

Other agents
   Donepezil
   Cox-1 and Cox-2 inhibitors
   Vasoactive drugs
   Angiotensin inhibitors
   DMFO
   Ginkgo biloba
   Ginseng
   Erythropoetin/Darbopoetin

Regenerative approaches
   Glial cell transplantation
   (O-2A progenitor cells, mature
   oligodendrocytes)
   Neural stem cell transplantation
   Combined agents/approaches

Hyperbaric oxygen

Corticosteroids

Anticoagulants

Bevacizumab

Botulinum toxin (myokymia)

Modified from Ref. 144.

The sad fact is that there is no proven method of either preventing or treating most radiation injuries.[143]

REFERENCES

1. Landuyt W, Fowler J, Ruifrok A, et al. Kinetics of repair in the spinal cord of the rat. *Radiother Oncol* 1997;45:55–62.
2. Lavey RS, Johnstone AK, Taylor JM, et al. The effect of hyperfractionation on spinal cord response to radiation. *Int J Radiat Oncol Biol Phys* 1992; 24:681–686.
3. Philippo H, Winter EA, van der Kogel AJ, et al. Recovery capacity of glial progenitors after in vivo fission-neutron or X irradiation: age dependence, fractionation and low-dose-rate irradiations. *Radiat Res* 2005;163:636–643.
4. Li YQ, Chen P, Haimovitz-Friedman A, et al. Endothelial apoptosis initiates acute blood-brain barrier disruption after ionizing radiation. *Cancer Res* 2003;63:5950–5956.
5. Chow BM, Li YQ, Wong CS. Radiation-induced apoptosis in the adult central nervous system is p53-dependent. *Cell Death Differ* 2000;7:712–720.
6. Chong MJ, Murray MR, Gosink EC, et al. Atm and Bax cooperate in ionizing radiation-induced apoptosis in the central nervous system. *Proc Natl Acad Sci USA* 2000;97:889–894.
7. Karagiannis TC, El-Osta A. The paradox of histone deacetylase inhibitor-mediated modulation of cellular responses to radiation. *Cell Cycle* 2006; 5:288–295.
8. Poggi MM, Coleman CN, Mitchell JB. Sensitizers and protectors of radiation and chemotherapy. *Curr Prob Cancer* 2001;25:334–411.
9. Spalding AC, Lawrence TS. New and emerging radiosensitizers and radioprotectors. *Cancer Invest* 2006;24:444–456.
10. Klish MD, Watson GA, Shrieve DC. Radiation and intensity-modulated radiotherapy for metastatic spine tumors. *Neurosurg Clin N Am* 2004; 15:481–490.
11. Teh BS, Mai WY, Grant WH, III, et al. Intensity modulated radiotherapy (IMRT) decreases treatment-related morbidity and potentially enhances tumor control. *Cancer Invest* 2002;20:437–451.
11a. Burdelya LG, Krivokrysendo VI, Tallant TC, et al. An agonist of toll-like receptor 5 has radioprotective activity in mouse and primate models. *Science* 2008;320:226–30.
12. ten Haken RK, Lawrence TS. The clinical application of intensity-modulated radiation therapy. *Semin Radiat Oncol* 2006;16:224–231.
12a. Gong Y, Wang J, Bai S, et al. Conventionally- fractionated image-guided intensity modulated radiotherapy (IG-IMRT): A safe and effective treatment for cancer spinal metastasis. *Radiat Oncol* 2008; 3:11.
13. Hall EJ. The inaugural Frank Ellis Lecture— Iatrogenic cancer: the impact of intensity-modulated radiotherapy. *Clin Oncol (R Coll Radiol)* 2006;18:277–282.
14. Vujaskovic Z, Gillette SM, Powers BE, et al. Effects of intraoperative irradiation and intraoperative

hyperthermia on canine sciatic nerve: neurologic and electrophysiologic study. *Int J Radiat Oncol Biol Phys* 1996;34:125–131.

15. Kurita H, Kawahara N, Asai A, et al. Radiation-induced apoptosis of oligodendrocytes in the adult rat brain. *Neurol Res* 2001;23:869–874.

16. Gobbel GT, Bellinzona M, Vogt AR, et al. Response of postmitotic neurons to X-irradiation: implications for the role of DNA damage in neuronal apoptosis. *J Neurosci* 1998;18:147–155.

17. Otsuka S, Coderre JA, Micca PL, et al. Depletion of neural precursor cells after local brain irradiation is due to radiation dose to the parenchyma, not the vasculature. *Radiat Res* 2006;165:582–591.

18. Rola R, Raber J, Rizk A, et al. Radiation-induced impairment of hippocampal neurogenesis is associated with cognitive deficits in young mice. *Exp Neurol* 2004;188:316–330.

19. Rola R, Sarkissian V, Obenaus A, et al. High-LET radiation induces inflammation and persistent changes in markers of hippocampal neurogenesis. *Radiat Res* 2005;164:556–560.

20. Pena LA, Fuks Z, Kolesnick RN. Radiation-induced apoptosis of endothelial cells in the murine central nervous system: protection by fibroblast growth factor and sphingomyelinase deficiency. *Cancer Res* 2000;60:321–327.

21. Nordal RA, Wong CS. Molecular targets in radiation-induced blood-brain barrier disruption. *Int J Radiat Oncol Biol Phys* 2005;62:279–287.

22. Vorotnikova E, Tries M, Braunhut S. Retinoids and TIMP1 prevent radiation-induced apoptosis of capillary endothelial cells. *Radiat Res* 2004;161:174–184.

23. Kolesnick R, Fuks Z. Radiation and ceramide-induced apoptosis. *Oncogene* 2003;22:5897–5906.

24. Lampert P, Tom MI, Rider WD. Disseminated demyelination of the brain following Co^{60} (gamma) radiation. *Arch Pathol (Chicago)* 1959;68:322–330.

25. Lampert PW, Davis RL. Delayed effects of radiation on the human central nervous system: "early" and "late" delayed reactions. *Neurology* 1964; 14:912–917.

26. Groothuis DR, Mikhael MA. Focal cerebral vasculitis associated with circulating immune complexes and brain irradiation. *Ann Neurol* 1986;19:590–592.

27. Hall EJ. *Radiobiology for the Radiologist.* 6th ed. Philadelphia: Lippincott, 2006.

28. Nieder C, Grosu AL, Andratschke NH, et al. Proposal of human spinal cord reirradiation dose based on collection of data from 40 patients. *Int J Radiat Oncol Biol Phys* 2005;61(3):851–855.

29. van den Bogaet W, Horiot J-C, van der Schueren E. Radiotherapy with multiple fractions per day. In: Steel GG, Adams GE, Horwich A, eds. *The Biological Basis of Radiotherapy.* 2nd ed. Amsterdam: Elsevier Science, 1989. pp. 209–222.

30. Emami B, Lyman J, Brown A, et al. Tolerance of normal tissue to therapeutic irradiation. *Int J Radiat Oncol Biol Phys* 1991;21:109–122.

31. Armstrong CL, Gyato K, Awadalla AW, et al. A critical review of the clinical effects of therapeutic irradiation damage to the brain: the roots of controversy. *Neuropsychol Rev* 2004;14:65–86.

32. Young DF, Posner JB, Chu F, et al. Rapid-course radiation therapy of cerebral metastases: results and complications. *Cancer* 1974;34:1069–1076.

33. Hakansson CH. Effect of irradiation of brain tumours on ventricular fluid pressure. *Acta Radiol Ther Phys Biol* 1967;6:22–32.

34. Hindo WA, DeTrana FA, III, Lee M-S, et al. Large dose increment irradiation in treatment of cerebral metastases. *Cancer* 1970;26:138–141.

35. Haie-Meder C, Pellae-Cosset B, Laplanche A, et al. Results of a randomized clinical trial comparing two radiation schedules in the palliative treatment of brain metastases. *Radiother Oncol* 1993;26:111–116.

36. Italian Group for Antiemetic Research in Radiotherapy. Radiation-induced emesis: a prospective observational multicenter Italian trial. *Int J Radiat Oncol Biol Phys* 1999;44:619–625.

37. Phillips PC, Delattre J-Y, Berger CA, et al. Early and progressive increases in regional brain capillary permeability following single- and fractionated-dose cranial radiation in the rat. Abstract. *Neurology* 1987;37 (Suppl 1):301.

38. Evans ML, Graham MM, Mahler PA, et al. Use of steroids to suppress vascular response to radiation. *Int J Radiat Oncol Biol Phys* 1987;13:563–567.

39. Deck MD. Imaging techniques in the diagnosis of radiation damage to the central nervous system. In: Gilbert HA, Kagan AR, eds. *Radiation Damage to the Nervous System.* New York: Raven, 1980. pp. 107–127.

40. Millar BA, Purdie TG, Yeung I, et al. Assessing perfusion changes during whole brain irradiation for patients with cerebral metastases. *J Neuro-Oncol* 2005;71:281–286.

41. Jarden JO, Dhawan V, Poltorak A, et al. Positron emission tomographic measurement of blood-to-brain and blood-to-tumor transport of 82Rb: the effect of dexamethasone and whole-brain radiation therapy. *Ann Neurol* 1985;18:636–646.

42. Minamisawa T, Tsuchiya T, Eto H. Changes in the averaged evoked potentials of the rabbit during and after fractionated x-irradiation. *Electroencephalogr Clin Neurophysiol* 1972;33:591–601.

43. Lamproglou I, Magdelenat H, Boisserie G, et al. An experimental model of acute encephalopathy after total body irradiation in the rat: effect of liposome-entrapped Cu/Zn superoxide dismutase. *Int J Radiat Oncol Biol Phys* 1998;42:179–184.

44. Casteels-van Daele M, Van de Casseye W. Acute encephalopathy after initiation of cranial irradiation for meningeal leukaemia. Letter to the editor. *Lancet* 1978;2:834–835.

45. Oliff A, Bleyer WA, Poplack DG. Acute encephalopathy after initiation of cranial irradiation for meningeal leukaemia. *Lancet* 1978;2:13–15.

46. Shalev O, Silverberg R. Dexamethasone for acute radiation encephalopathy. Letter to the editor. *Lancet* 1978;2:574–575.

47. Lawenda BD, Gagne HM, Gierga DP, et al. Permanent alopecia after cranial irradiation: dose-response relationship. *Int J Radiat Oncol Biol Phys* 2004;60:879–887.

48. Stone HB, Coleman CN, Anscher MS, et al. Effects of radiation on normal tissue: consequences and mechanisms. *Lancet Oncol* 2003;4:529–536.

49. Cairncross JG, Salmon J, Kim J-H, et al. Acute parotitis and hyperamylasemia following whole-brain radiation therapy. *Ann Neurol* 1980;7:385–387.

50. Jacobsen PB, Thors CL. Fatigue in the radiation therapy patient: current management and investigations. *Semin Radiat Oncol* 2003;13:372–380.

51. Boldrey E, Sheline G. Delayed transitory clinical manifestations after radiation treatment of intracranial tumors. *Acta Radiol (Ther)* 1966;5:5–10.

52. Graeb DA, Steinbok P, Robertson WD. Transient early computed tomographic changes mimicking tumor progression after brain tumor irradiation. *Radiology* 1982;144:813–817.

53. Watne K, Hager B, Heier M, et al. Reversible oedema and necrosis after irradiation of the brain. Diagnosis procedures and clinical manifestations. *Acta Oncol* 1990;29:891–895.

54. Hoffman WF, Levin VA, Wilson CB. Evaluation of malignant glioma patients during the post-irradiation period. *J Neurosurg* 1979;50:624–628.

55. Taal W, Brandsma D, de Bruin HG, et al. Incidence of early pseudo-progression in a cohort of malignant glioma patients treated with chemoirradiation with temozolomide. *Cancer* 2008. In press.

56. Schultheiss TE, Kun LE, Ang KK, et al. Radiation response of the central nervous system. *Int J Radiat Oncol Biol Phys* 1995;31:1093–1112.

57. Lorberboym M, Baram J, Feibel M, et al. A prospective evaluation of Thallium-201 single photon emission computerized tomography for brain tumor burden. *Int J Radiat Oncol Biol Phys* 1995;32:249–254.

58. Kimura T, Sako K, Tanaka K, et al. Evaluation of the response of metastatic brain tumors to stereotactic radiosurgery by proton magnetic resonance spectroscopy,[201] TICI single-photon emission computerized tomography, and gadolinium-enhanced magnetic resonance imaging. *J Neurosurg* 2004;100:835–841.

59. Esteve F, Rubin C, Grand S, et al. Transient metabolic changes observed with proton MR spectroscopy in normal human brain after radiation therapy. *Int J Radiat Oncol Biol Phys* 1998;40:279–286.

60. Voermans NC, Bloem BR, Janssens G, et al. Secondary parkinsonism in childhood: a rare complication after radiotherapy. *Pediatr Neurol* 2006;34:495–498.

61. Asai A, Matsutani M, Kohno T, et al. Subacute brain atrophy after radiation therapy for malignant brain tumor. *Cancer* 1989;63:1962–1974.

62. Armstrong C, Ruffer J, Corn B, et al. Biphasic patterns of memory deficits following moderate-dose partial-brain irradiation: neuropsychologic outcome and proposed mechanisms. *J Clin Oncol* 1995;13:2263–2271.

63. Armstrong CL, Corn BW, Ruffer JE, et al. Radiotherapeutic effects on brain function: double dissociation of memory systems. *Neuropsychiatry Neuropsychol Behav Neurol* 2000;13:101–111.

64. Cappuzzo F, Mazzoni F, Maestri A, et al. Medical treatment of brain metastases from solid tumours. *Forum (Genova)* 2000;10:137–148.

65. Freeman JE, Johnston PG, Voke JM. Somnolence after prophylactic cranial irradiation in children with acute lymphoblastic leukaemia. *BMJ* 1973;4:523–525.

66. Littman P, Rosenstock J, Gale G, et al. The somnolence syndrome in leukemic children following reduced daily dose fractions of cranial radiation. *Int J Radiat Oncol Biol Phys* 1984;10:1851–1853.

67. Mandell LR, Walker RW, Steinherz P, et al. Reduced incidence of the somnolence syndrome in leukemic children with steroid coverage during prophylactic cranial radiation therapy. Results of a pilot study. *Cancer* 1989;63:1975–1978.

68. Druckmann A. Schlafsucht als Folge der Rontegenbestrahlung. Beitrag zur Strahlenempfindlichkeit des Gehirns. *Strahlentherapie* 1929;33:382–384.

69. Faithfull S, Brada M. Somnolence syndrome in adults following cranial irradiation for primary brain tumours. *Clin Oncol (R Coll Radiol)* 1998;10:250–254.

70. Miyahara M, Azuma E, Hirayama M, et al. Somnolence syndrome in a child following 1200-cGy total body irradiation in an unrelated bone marrow transplantation. *Pediatr Hematol Oncol* 2000;17:489–495.

71. Kelsey CR, Marks LB. Somnolence syndrome after focal radiation therapy to the pineal region: case report and review of the literature. *J Neuro-Oncol* 2006;78:153–156.

72. Garwicz S, Aronson S, Elmqvist D, et al. Postirradiation syndrome and EEG findings in children with acute lymphoblastic leukaemia. *Acta Paediatr Scand* 1975;64:399–403.

73. Miyatake S-I, Kikuchi H, Oda Y, et al. A case of treatment-related leukoencephalopathy: sequential MRI, CT and PET findings. *J Neuro-Oncol* 1992;14:143–149.

74. Uzal D, Ozyar E, Hayran M, et al. Reduced incidence of the somnolence syndrome after prophylactic cranial irradiation in children with acute lymphoblastic leukemia. *Radiother Oncol* 1998;48:29–32.

75. Lee MC, Pirzkall A, McKnight TR, et al. 1H-MRSI of radiation effects in normal-appearing white matter: dose-dependence and impact on automated spectral classification. *J Magn Reson Imaging* 2004;19:379–388.

76. Zeng QS, Li CF, Liu H, et al. Distinction between recurrent glioma and radiation injury using magnetic resonance spectroscopy in combination with diffusion-weighted imaging. *Int J Radiat Oncol Biol Phys* 2007;68:151–158.

77. Monro P, Mair WG. Radiation effects on the human central nervous system 14 weeks after x-radiation. *Acta Neuropathol* 1968;11:267–274.

78. Rider WD. Radiation damage to the brain—a new syndrome. *J Can Assoc Radiol* 1963;14:67–69.

79. Kleinschmidt-DeMasters BK. Necrotizing brainstem leukoencephalopathy six weeks following radiotherapy. *Clin Neuropathol* 1995;14:63–68.

80. de Wit MC, de Bruin HG, Eijkenboom W, et al. Immediate post-radiotherapy changes in malignant glioma can mimic tumor progression. *Neurology* 2004;63:535–537.

81. Sundgren PC, Fan X, Weybright P, et al. Differentiation of recurrent brain tumor versus radiation injury using diffusion tensor imaging in patients with new contrast-enhancing lesions. *Magn Reson Imaging* 2006;24:1131–1142.

82. Peterson K, Rosenblum MK, Powers JM, et al. Effect of brain irradiation on demyelinating lesions. *Neurology* 1993;43:2105–2112.

83. Hopewell JW, Millar WT, Ang KK. Toward improving the therapeutic ratio in stereotactic radiosurgery: Selective modulation of the radiation responses of both normal tissues and tumor. *J Neurosurg* 2007;107:84–93.

84. Chang SD, Lee E, Sakamoto GT, et al. Stereotactic radiosurgery in patients with multiple brain metastases. *Neurosurg Focus* 2000;9:e3.

85. Giglio P, Gilbert MR. Cerebral radiation necrosis. *Neurology* 2003;9:180–188.

86. Ruben JD, Dally M, Bailey M, et al. Cerebral radiation necrosis: incidence, outcomes, and risk factors with emphasis on radiation parameters and chemotherapy. *Int J Radiat Oncol Biol Phys* 2006;65:499–508.

87. Marks JE, Wong J. The risk of cerebral radionecrosis in relation to dose, time and fractionation: a follow-up study. *Prog Exp Tumor Res* 1985;29:210–218.

88. Rottenberg DA, Chernik NL, Deck MD, et al. Cerebral necrosis following radiotherapy of extracranial neoplasms. *Ann Neurol* 1977;1:339–357.

89. Kumar AJ, Leeds NE, Fuller GN, et al. Malignant gliomas: MR imaging spectrum of radiation therapy- and chemotherapy-induced necrosis of the brain after treatment. *Radiology* 2000;217:377–384.

90. Nieder C, Leicht A, Motaref B, et al. Late radiation toxicity after whole brain radiotherapy: the influence of antiepileptic drugs. *Am J Clin Oncol* 1999; 22:573–579.

91. Sheline GE, Wara WM, Smith V. Therapeutic irradiation and brain injury. *Int J Radiat Oncol Biol Phys* 1980;6:1215–1228.

92. DeAngelis LM, Delattre J-Y, Posner JB. Radiation-induced dementia in patients cured of brain metastases. *Neurology* 1989;39:789–796.

93. Shewmon DA, Masdeu JC. Delayed radiation necrosis of the brain contralateral to original tumor. *Arch Neurol* 1980;37:592–594.

94. Evans ES, Hahn CA, Kocak Z, et al. The role of functional imaging in the diagnosis and management of late normal tissue injury. *Semin Radiat Oncol* 2007;17:72–80.

95. Chao ST, Suh JH, Raja S, et al. The sensitivity and specificity of FDG PET in distinguishing recurrent brain tumor from radionecrosis in patients treated with stereotactic radiosurgery. *Int J Cancer* 2001;96:191–197.

96. Hustinx R, Pourdehnad M, Kaschten B, et al. PET imaging for differentiating recurrent brain tumor from radiation necrosis. *Radiol Clin North Am* 2005;43:35–47.

97. Sugahara T, Korogi Y, Tomiguchi S, et al. Posttherapeutic intraaxial brain tumor: the value of perfusion-sensitive contrast-enhanced MR imaging for differentiating tumor recurrence from non-neoplastic contrast-enhancing tissue. *ANJR Am J Neuroradiol* 2000;21:901–909.

98. Kashimura H, Inoue T, Beppu T, et al. Diffusion tensor imaging for differentiation of recurrent brain tumor and radiation necrosis after radiotherapy—three case reports. *Clin Neurol Neurosurg* 2007;109:106–110.

99. Serizawa T, Saeki N, Higuchi Y, et al. Diagnostic value of thallium-201 chloride single-photon emission computerized tomography in differentiating tumor recurrence from radiation injury after gamma knife surgery for metastatic brain tumors. *J Neurosurg* 2005;102 (Suppl):266–271.

100. Lee AW, Kwong DL, Leung SF, et al. Factors affecting risk of symptomatic temporal lobe necrosis: significance of fractional dose and treatment time. *Int J Radiat Oncol Biol Phys* 2002; 53:75–85.

101. Grattan-Smith PJ, Morris JG, Langlands AO. Delayed radiation necrosis of the central nervous system in patients irradiated for pituitary tumours. *J Neurol Neurosurg Psychiatry* 1992;55:949–955.

102. Perry A, Schmidt RE. Cancer therapy-associated CNS neuropathology: an update and review of the literature. *Acta Neuropathol (Berl)* 2006; 111:197–212.

103. Shimizu N, Okamoto H, Fukuda T, et al. Late laryngeal radionecrosis in severe arteriosclerosis. *J Laryngol Otol* 2005;119:922–925.

104. Suzuki S, Nishio S, Takata K, et al. Radiation-induced brain calcification: paradoxical high signal intensity in T1-weighted MR images. *Acta Neurochir (Wien)* 2000;142:801–804.

105. Shaw PJ, Bates D. Conservative treatment of delayed cerebral radiation necrosis. *J Neurol Neurosurg Psychiatry* 1984;47:1338–1341.

106. Delattre J-Y, Fuks Z, Krol G, et al. Cerebral necrosis following neutron radiation of an extracranial tumor. *J Neuro-Oncol* 1988;6:113–117.

107. Hsu YC, Wang LF, Lee KW, et al. Cerebral radionecrosis in patients with nasopharyngeal carcinoma. *Kaohsiung J Med Sci* 2005;21:452–459.

108. Delanian S, Lefaix JL. Current management for late normal tissue injury: radiation-induced fibrosis and necrosis. *Semin Radiat Oncol* 2007; 17:99–107.

109. Chuba PJ, Aronin P, Bhambhani K, et al. Hyperbaric oxygen therapy for radiation-induced brain injury in children. *Cancer* 1997;80:2005–2012.

110. Bennett MH, Feldmeier J, Hampson N, et al. Hyperbaric oxygen therapy for late radiation tissue injury. *Cochrane Database Syst Rev* 2005;CD005005.

111. Gonzalez J, Kumar AJ, Conrad CA, et al. Effect of bevacizumab on radiation necrosis of the brain. *Int J Radiat Oncol Biol Phys* 2007;67:323–326.

112. Omuro AM, Ben-Porat LS, Panageas KS, et al. Delayed neurotoxicity in primary central nervous system lymphoma. *Arch Neurol* 2005;62: 1595–1600.

113. Bhansali A, Banerjee AK, Chanda A, et al. Radiation-induced brain disorders in patients with pituitary tumours. *Australas Radiol* 2004;48:339–346.

114. Laukkanen E, Klonoff H, Allan B, et al. The role of prophylactic brain irradiation in limited stage small cell lung cancer: clinical, neuropsychologic, and CT sequelae. *Int J Radiat Oncol Biol Phys* 1988;14:1109–1117.

115. Lee JS, Unsawasdi T, Lee Y-Y, et al. Neurotoxicity in long-term survivors of small cell lung cancer. *Int J Radiat Oncol Biol Phys* 1986;12:313–321.

116. Brecher ML, Berger P, Freeman AI, et al. Computerized tomography scan findings in children with acute lymphocytic leukemia treated with three different methods of central nervous system prophylaxis. *Cancer* 1985;56:2430–2433.

117. Armstrong CL, Hunter JV, Hackney D, et al. MRI changes due to early-delayed conformal radiotherapy and postsurgical effects in patients with brain tumors. *Int J Radiat Oncol Biol Phys* 2005;63:56–63.

118. Yaar I, Ron E, Modan B, et al. Long-lasting cerebral functional changes following moderate dose

x-radiation treatment to the scalp in childhood: an electroencephalographic power spectral study. *J Neurol Neurosurg Psychiatry* 1982;45:166–169.

119. Hall P, Adami HO, Trichopoulos D, et al. Effect of low doses of ionising radiation in infancy on cognitive function in adulthood: Swedish population based cohort study. *BMJ* 2004;328:19.

120. Fujii O, Tsujino K, Soejima T, et al. White matter changes on magnetic resonance imaging following whole-brain radiotherapy for brain metastases. *Radiat Med* 2006;24:345–350.

121. Steen RG, Koury M, Granja CI, et al. Effect of ionizing radiation on the human brain: white matter and gray matter T1 in pediatric brain tumor patients treated with conformal radiation therapy. *Int J Radiat Oncol Biol Phys* 2001 1;49:79–91.

122. Steen RG, Spence D, Wu SJ, et al. Effect of therapeutic ionizing radiation on the human brain. *Ann Neurol* 2001;50:787–795.

122a. Doyle DM, Einhorn LH. Delayed effects of whole brain radiotherapy in germ cell tumor patients with central nervous system metastases. *Int J Radiat Oncol Biol Phys* 2008;70:1361–1364.

123. Spiegler BJ, Bouffet E, Greenberg ML, et al. Change in neurocognitive functioning after treatment with cranial radiation in childhood. *J Clin Oncol* 2004;22:706–713.

124. Kleinberg L, Wallner K, Malkin MG. Good performance status of long-term disease-free survivors of intracranial gliomas. *Int J Radiat Oncol Biol Phys* 1993;26:129–133.

125. Pavlovsky S, Fismann N, Arizaga R, et al. Neuropsychological study in patients with ALL: two different CNS prevention therapies—cranial irradiation plus IT methotrexate vs. IT methotrexate alone. *Am J Pediatr Hematol/Oncol* 1983; 5:79–86.

126. Valk PE, Dillon WP. Radiation injury of the brain. *AJNR Am J Neuroradiol* 1991;12:45–62.

127. Crossen JR, Garwood D, Glatstein E, et al. Neurobehavioral sequelae of cranial irradiation in adults: a review of radiation-induced encephalopathy. *J Clin Oncol* 1994;12:627–642.

128. Mulhern RK, Ochs J, Fairclough D, et al. Intellectual and academic achievement status after CNS relapse: a retrospective analysis of 40 children treated for acute lymphoblastic leukemia. *J Clin Oncol* 1987; 5:933–940.

129. Thiessen B, DeAngelis LM. Hydrocephalus in radiation leukoencephalopathy—results of ventriculoperitoneal shunting. *Arch Neurol* 1998;55:705–710.

130. Perrini P, Scollato A, Cioffi F, et al. Radiation leukoencephalopathy associated with moderate hydrocephalus: intracranial pressure monitoring and results of ventriculoperitoneal shunting. *Neurol Sci* 2002;23:237–241.

131. Nagesh V, Tsien CI, Chenevert TL, et al. Radiation-induced changes in normal appearing white matter in patients with cerebral tumors: A diffusion tensor imaging study. *Int J Radiat Oncol Biol Phys* 2008;70:1002–1010.

132. Moretti R, Torre P, Antonello RM, et al. Neuropsychological evaluation of late-onset post-radiotherapy encephalopathy: a comparison with vascular dementia. *J Neurol Sci* 2005;229:195–200.

133. Ohguri T, Imada H, Kohshi K, et al. Effect of prophylactic hyperbaric oxygen treatment for radiation-induced brain injury after stereotactic radiosurgery of brain metastases. *Int J Radiat Oncol Biol Phys* 2007;67:248–255.

134. Yazlovitskaya EM, Edwards E, Thotala D, et al. Lithium treatment prevents neurocognitive deficit resulting from cranial irradiation. *Cancer Res* 2006;66:11179–11186.

135. Correa DD, DeAngelis LM, Shi W, et al. Cognitive functions in low-grade gliomas: disease and treatment effects. *J Neuro-Oncol* 2007;81:175–184.

136. Li J, Bentzen SM, Renschler M, et al. Regression after whole-brain radiation therapy for brain metastases correlates with survival and improved neurocognitive function. *J Clin Oncol* 2007;25: 1260–1266.

137. Komaki R, Meyers CA, Shin DM, et al. Evaluation of cognitive function in patients with limited small cell lung cancer prior to and shortly following prophylactic cranial irradiation. *Int J Radiat Oncol Biol Phys* 1995;33:179–182.

138. Nimjee SM, Powers CJ, Bulsara KR. Review of the literature on de novo formation of cavernous malformations of the central nervous system after radiation therapy. *Neurosurg Focus* 2006;21:E4.

139. Shuper A, Packer RJ, Vezina LG, et al. Complicated migraine-like episodes' in children following cranial irradiation and chemotherapy. *Neurology* 1995;45:1837–1840.

140. Bartleson JD, Krecke KN, O'Neill BP, et al. Reversible, strokelike migraine attacks in patients with previous radiation therapy. *Neuro-Oncol* 2003;5:121–127.

141. Black DF, Bartleson JD, Bell ML, et al. SMART: stroke-like migraine attacks after radiation therapy. *Cephalalgia* 2006;26:1137–1142.

142. Partap S, Walker M, Longstreth WT, Jr, et al. Prolonged but reversible migraine-like episodes long after cranial irradiation. *Neurology* 2006;66:1105–1107.

143. Cordato DJ, Brimage P, Masters LT, et al. Post-cranial irradiation syndrome with migraine-like headaches, prolonged and reversible neurological deficits and seizures. *J Clin Neurosci* 2006;13:586–590.

144. Shaw EG, Robbins ME. The management of radiation-induced brain injury. *Cancer Treat Res* 2006;128:7–22.

145. Butler JM, Rapp SR, Shaw EG. Managing the cognitive effects of brain tumor radiation therapy. *Curr Treat Options Oncol* 2006;7:517–523.

146. Conklin HM, Khan RB, Reddick WE, et al. Acute neurocognitive response to methylphenidate among survivors of childhood cancer: a randomized, double-blind, cross-over trial. *J Pediatr Psychol* 2007;32:1127–1139.

147. Mulhern RK, Khan RB, Kaplan S, et al. Short-term efficacy of methylphenidate: a randomized, double-blind, placebo-controlled trial among survivors of childhood cancer. *J Clin Oncol* 2004; 22:4795–4803.

148. Meyers CA, Weitzner MA, Valentine AD, et al. Methylphenidate therapy improves cognition, mood, and function of brain tumor patients. *J Clin Oncol* 1998;16:2522–2527.

149. Shaw EG, Rosdhal R, D'Agostino RB, Jr, et al. Phase II study of donepezil in irradiated brain tumor

patients: effect on cognitive function, mood, and quality of life. *J Clin Oncol* 2006;24:1415–1420.

150. Jatoi A, Kahanic SP, Frytak S, et al. Donepezil and vitamin E for preventing cognitive dysfunction in small cell lung cancer patients: preliminary results and suggestions for future study designs. *Support Care Cancer* 2005;13:66–69.

151. Goldwein JW. Radiation myelopathy: a review. *Med Pediatr Oncol* 1987;15:89–95.

152. Millburn L, Hibbs GG, Hendrickson FR. Treatment of spinal cord compression from metastatic carcinoma. Review of literature and presentation of a new method of treatment. *Cancer* 1968;21:447–452.

153. Tefft M, Mitus A, Schulz MD. Initial high dose irradiation for metastases causing spinal cord compression in children. *Am J Roentgenol* 1969;106:385–393.

154. Rubin P. Extradural spinal cord compression by tumor. I. Experimental production and treatment trials. *Radiology* 1969;93:1243–1248.

155. Ushio Y, Posner R, Posner JB, et al. Experimental spinal cord compression by epidural neoplasm. *Neurology* 1977;27:422–429.

156. Delattre J-Y, Rosenblum MK, Thaler HT, et al. A model of radiation myelopathy in the rat. Pathology, regional capillary permeability changes and treatment with dexamethasone. *Brain* 1988;111 (Pt 6): 1319–1336.

157. Nordal RA, Wong CS. Intercellular adhesion molecule-1 and blood-spinal cord barrier disruption in central nervous system radiation injury. *J Neuropathol Exp Neurol* 2004;63:474–483.

158. Stewart PA, Vinters HV, Wong CS. Blood-spinal cord barrier function and morphometry after single doses of X-rays in rat spinal cord. *Int J Radiat Oncol Biol Phys* 1995;32:703–711.

159. Esik O, Csere T, Stefanits K, et al. A review on radiogenic Lhermitte's sign. *Pathol Oncol Res* 2003;9:115–120.

160. Dorfman LJ, Donaldson SS, Gupta PR, et al. Electrophysiologic evidence of subclinical injury to the posterior columns of the human spinal cord after therapeutic radiation. *Cancer* 1982;50:2815–2819.

161. Lecky BR, Murray NM, Berry RJ. Transient radiation myelopathy: spinal somatosensory evoked responses following incidental cord exposure during radiotherapy. *J Neurol Neurosurg Psychiatry* 1980;43:747–750.

162. Lengyel Z, Reko G, Majtenyi K, et al. Autopsy verifies demyelination and lack of vascular damage in partially reversible radiation myelopathy. *Spinal Cord* 2003;41:577–585.

163. Chiang CS, Mason KA, Withers HR, et al. Alteration in myelin-associated proteins following spinal cord irradiation in guinea pigs. *Int J Radiat Oncol Biol Phys* 1992;24:929–937.

164. Fein DA, Marcus RB, Jr, Parsons JT, et al. Lhermitte's sign: incidence and treatment variables influencing risk after irradiation of the cervical spinal cord. *Int J Radiat Oncol Biol Phys* 1993;27:1029–1033.

165. Jones A. Transient radiation myelopathy. *Br J Radiol* 1964;37:727–744.

166. Word JA, Kalokhe UP, Aron BS, et al. Transient radiation myelopathy (Lhermitte's sign) in patients with Hodgkin's disease treated by mantle irradiation. *Int J Radiat Oncol Biol Phys* 1980;6:1731–1733.

167. Leung WM, Tsang NM, Chang FT, et al. Lhermitte's sign among nasopharyngeal cancer patients after radiotherapy. *Head Neck* 2005;27:187–194.

168. Behin A, Delattre JY. Complications of radiation therapy on the brain and spinal cord. *Semin Neurol* 2004;24:405–417.

169. Dische S, Warburton MF, Saunders MI. Radiation myelitis and survival in the radiotherapy of lung cancer. *Int J Radiat Oncol Biol Phys* 1988;15:75–81.

170. Godwin-Austen RB, Howell DA, Worthington B. Observations on radiation myelopathy. *Brain* 1975; 98:557–568.

171. Schultheiss TE, Stephens LC, Maor MH. Analysis of the histopathology of radiation myelopathy. *Int J Radiat Oncol Biol Phys* 1988;14:27–32.

172. McCunniff AJ, Liang MJ. Radiation tolerance of the cervical spinal cord. *Int J Radiat Oncol Biol Phys* 1989;16:675–678.

173. Putnik K, Stadler P, Schafer C, et al. Enhanced radiation sensitivity and radiation recall dermatitis (RRD) after hypericin therapy—case report and review of literature. *Radiat Oncol* 2006;1:32.

174. Koehler PJ, Verbiest H, Jager J, et al. Delayed radiation myelopathy: serial MR-imaging and pathology. *Clin Neurol Neurosurg* 1996;98(2):197–201.

175. Melki PS, Halimi P, Wibault P, et al. MRI in chronic progressive radiation myelopathy. *J Comput Assist Tomogr* 1994;18:1–6.

176. Michikawa M, Wada Y, Sano M, et al. Radiation myelopathy: significance of gadolinium-DPTA enhancement in the diagnosis. *Neuroradiology* 1991;33:286–289.

177. Rubin P, Whitaker JN, Ceckler TL, et al. Myelin basic protein and magnetic resonance imaging for diagnosing radiation myelopathy. *Int J Radiat Oncol Biol Phys* 1988;15:1371–1381.

178. Alfonso ER, De Gregorio MA, Mateo P, et al. Radiation myelopathy in over-irradiated patients: MR imaging findings. *Eur Radiol* 1997; 7(3):400–404.

179. Yasui T, Yagura H, Komiyama M, et al. Significance of gadolinium-enhanced magnetic resonance imaging in differentiating spinal cord radiation myelopathy from tumor. Case report. *J Neurosurg* 1992;77(4):628–631.

180. Wang PY, Shen WC, Jan JS. Serial MRI changes in radiation myelopathy. *Neuroradiology* 1995; 37(5):374–377.

181. Snooks SJ, Swash M. Motor conduction velocity in the human spinal cord: slowed conduction in multiple sclerosis and radiation myopathy. *J Neurol Neurosurg Psychiatry* 1985;48:1135–1139.

182. Okada S, Okeda R. Pathology of radiation myelopathy. *Neuropathology* 2001;21(4):247–265.

183. Bijl HP, Van LP, Coppes RP, et al. Regional differences in radiosensitivity across the rat cervical spinal cord. *Int J Radiat Oncol Biol Phys* 2005; 61(2):543–551.

184. Gutin PH, Levin KJ, McDermott MW, et al. Lipid peroxidation does not appear to be a factor in late radiation injury of the cervical spine cord of rats. *Int J Radiat Oncol Biol Phys* 1993;25:67–72.

185. Peker S, Abacioglu U, Sun I, et al. Prophylactic effects of magnesium and vitamin E in rat spinal cord

radiation damage: evaluation based on lipid peroxidation levels. *Life Sci* 2004;75(12):1523–1530.

186. Genc M, Genc E, Genc BO, et al. Significant response of radiation induced CNS toxicity to high dose steroid administration. *Br J Radiol* 2006;79(948):e196–e199.

187. Liu CY, Yim BT, Wozniak AJ. Anticoagulation therapy for radiation-induced myelopathy. *Ann Pharmacother* 2001;35:188–191.

188. Calabro F, Jinkins JR. MRI of radiation myelitis: a report of a case treated with hyperbaric oxygen. *Eur Radiol* 2000;10:1079–1084.

189. Angibaud G, Ducasse JL, Baille G, et al. [Potential value of hyperbaric oxygenation in the treatment of post-radiation myelopathies]. *Rev Neurol (Paris)* 1995;151:661–666.

190. Esik O, Vonoczky K, Lengyel Z, et al. Characteristics of radiogenic lower motor neurone disease, a possible link with a preceding viral infection. *Spinal Cord* 2004;42:99–105.

191. Feistner H, Weissenborn K, Munte TF, et al. Post-irradiation lesions of the caudal roots. *Acta Neurol Scand* 1989;80:277–281.

192. Fossa SD, Aass N, Kaalhus O. Long-term morbidity after infradiaphragmatic radiotherapy in young men with testicular cancer. *Cancer* 1969;64:404–408.

193. Grunewald RA, Chroni E, Panayiotopoulos CP, et al. Late onset radiation-induced motor neuron syndrome. Letter to the editor. *J Neurol Neurosurg Psychiatry* 1992;55:741–742.

194. Jackson M. Post radiation monomelic amyotrophy. Letter to the editor. *J Neurol Neurosurg Psychiatry* 1992;55:629.

195. Tallaksen CM, Jetne V, Fossa S. Postradiation lower motor neuron syndrome -- a case report and brief literature review. *Acta Oncol* 1997;36:345–347

196. DeGreve JLP, Bruyland M, DeKeyser J, et al. Lower motor neuron disease in a patient with Hodgkin's disease treated with radiotherapy. *Clin Neurol Neurosurg* 1984;86:43–46.

197. Kristensen O, Melgard B, Schiodt AV. Radiation myelopathy of the lumbo-sacral spinal cord. *Acta Neurol Scand* 1977;56:217–222.

198. Lamy C, Mas JL, Varet B, et al. Postradiation lower motor neuron syndrome presenting as monomelic amyotrophy. *J Neurol Neurosurg Psychiatry* 1991;64:648–649.

199. Sadowsky CH, Sachs E, Jr, Ochoa J. Postradiation motor neuron syndrome. *Arch Neurol* 1976;33:786–787.

200. Bowen J, Gregory R, Squier M, et al. The post-irradiation lower motor neuron syndrome—neuronopathy or radiculopathy? *Brain* 1996;119:1429–1439.

201. Malapert D, Brugieres P, Degos JD. Motor neuron syndrome in the arms after radiation treatment. *J Neurol Neurosurg Psychiatry* 1991;54:1123–1124.

202. Labauge P, Lefloch A, Chapon F, et al. Postirradiation spinal root cavernoma. *Eur Neurol* 2006;56:256–257.

203. Berlit P, Schwechheimer K. Neuropathological findings in radiation myelopathy of the lumbosacral cord. *Eur Neurol* 1987;27:29–34.

204. Allen JC, Miller DC, Budzilovich GN, et al. Brain and spinal cord hemorrhage in long-term survivors of malignant pediatric brain tumors: a possible late effect of therapy. *Neurology* 1991;41:148–150.

205. Jabbour P, Gault J, Murk SE, et al. Multiple spinal cavernous malformations with atypical phenotype after prior irradiation: case report. *Neurosurgery* 2004;55:1431.

206. Teo PM, Leung SF, Chan AT, et al. Final report of a randomized trial on altered-fractionated radiotherapy in nasopharyngeal carcinoma prematurely terminated by significant increase in neurologic complications. *Int J Radiat Oncol Biol Phys* 2000;48:1311–1322.

207. Chong VFH, Khoo JBK, Chan LL, Rumpel H. Neurological changes following radiation therapy for head and neck tumours. *Eur J Radiol* 2002;44:120–129.

208. Lin YS, Jen YM, Lin JC. Radiation-related cranial nerve palsy in patients with nasopharyngeal carcinoma. *Cancer* 2002;95:404–409.

209. Giese WL, Kinsella TJ. Radiation injury to peripheral and cranial nerves. In: Gutin PH, Leibel SA, Sheline GE, eds. *Radiation Injury to the Nervous System.* New York: Raven, 1991. pp. 383–403.

210. Mizrahi A, Lu J, Irving R, et al. In vivo imaging of juxtaglomerular neuron turnover in the mouse olfactory bulb. *Proc Natl Acad Sci USA* 2006;103:1912–1917.

211. Sagar SM, Thomas RJ, Loverock LT, et al. Olfactory sensations produced by high-energy photon irradiation of the olfactory receptor mucosa in humans. *Int J Radiat Oncol Biol Phys* 1991;20:771–776.

212. Holscher T, Seibt A, Appold S, et al. Effects of radiotherapy on olfactory function. *Radiother Oncol* 2005;77:157–163.

213. Hua MS, Chen ST, Tang LM, et al. Olfactory function in patients with nasopharyngeal carcinoma following radiotherapy. *Brain Injury* 1999;13:905–915.

214. Ophir D, Guterman A, Gross-Isseroff R. Changes in smell acuity induced by radiation exposure of the olfactory mucosa. *Arch Otolaryngol Head Neck Surg* 1988;114:853–855.

215. Ho WK, Kwong DL, Wei WI, et al. Change in olfaction after radiotherapy for nasopharyngeal cancer—a prospective study. *Am J Otolaryngol* 2002;23:209–214.

216. Beidler LM, Smith JC. Effects of radiation therapy and drugs on cell turnover and taste. In: Getchell TV, Doty RL, Bartoshuk LM, et al., eds. *Smell and Taste in Health and Disease.* New York: Raven, 1991. pp. 753–763.

217. Gupta A, Dhawahir-Scala F, Smith A, et al. Radiation retinopathy: case report and review. *BMC Ophthalmol* 2007;7:6.

218. Durkin SR, Roos D, Higgs B, et al. Ophthalmic and adnexal complications of radiotherapy. *Acta Ophthalmol Scand* 2007;85:240–250.

219. Parsons JT, Bova FJ, Mendenhall WM, et al. Response of the normal eye to high dose radiotherapy. *Oncology* 1996;10:837–847.

220. Ferrufino-Ponce ZK, Henderson BA. Radiotherapy and cataract formation. *Semin Ophthalmol* 2006;21:171–180.

221. Steidley KD, Eastman RM, Stabile RJ. Observations of visual sensations produced by cerenkov radiation from high-energy electrons. *Int J Rad Oncol Biol Phys* 1989;17:685–690.

222. Finger PT. Radiation therapy for choroidal melanoma. *Surv Ophthalmol* 1997;42:215–232.

223. Finger PT, Chin K. Anti-vascular endothelial growth factor bevacizumab (avastin) for radiation retinopathy. *Arch Ophthalmol* 2007;125:751–756.

224. Özyar E, Yildz F, Akyol FH, et al. Adjuvant high-dose-rate brachytherapy after external beam radiotherapy in nasopharyngeal carcinoma. *Int J Radiat Oncol Biol Phys* 2002;52:101–108.

225. Bhandare N, Monroe AT, Morris CG, et al. Does altered fractionation influence the risk of radiation-induced optic neuropathy? *Int J Radiat Oncol Biol Phys* 2005;62:1070–1077.

226. Monroe AT, Bhandare N, Morris CG, et al. Preventing radiation retinopathy with hyperfractionation. *Int J Radiat Oncol Biol Phys* 2005;61(3):856–864.

227. Lessell S. Magnetic resonance imaging signs may antedate visual loss in chiasmal radiation injury. *Arch Ophthalmol* 2003;121:287–288.

228. Piquemal R, Cottier JP, Arsene S, et al. Radiation-induced optic neuropathy 4 years after radiation: report of a case followed up with MRI. *Neuroradiology* 1998;40:439–441.

229. Guy J, Mancuso A, Beck R, et al. Radiation-induced optic neuropathy: a magnetic resonance imaging study. *J Neurosurg* 1991;74:426–432.

230. Singh R, Trobe JD, Hayman JA, et al. Ophthalmic artery occlusion secondary to radiation-induced vasculopathy. *J Neuroophthalmol* 2004;24:206–210.

231. Nagayama K, Kurita H, Nakamura M, et al. Radiation-induced apoptosis of oligodendrocytes in the adult rat optic chiasm. *Neurol Res* 2005; 27:346–350.

232. Levin LA, Gragoudas ES, Lessell S. Endothelial cell loss in irradiated optic nerves. *Ophthalmology* 2000;107:370–374.

233. Ryu S, Kolozsvary A, Jenrow KA, et al. Mitigation of radiation-induced optic neuropathy in rats by ACE inhibitor ramipril: importance of ramipril dose and treatment time. *J Neuro-Oncol* 2007; 82:119–124.

234. Kim JH, Brown SL, Kolozsvary A, et al. Modification of radiation injury by ramipril, inhibitor of angiotensin-converting enzyme, on optic neuropathy in the rat. *Radiat Res* 2004;161:137–142.

235. Borruat FX, Schatz NJ, Glaser JS, et al. Visual recovery from radiation-induced optic neuropathy. The role of hyperbaric oxygen therapy. *J Clin Neuroophthalmol* 1993;13:98–101.

236. Danesh-Meyer HV, Savino PJ, Sergott RC. Visual loss despite anticoagulation in radiation-induced optic neuropathy. *Clin Experiment Ophthalmol* 2004;32:333–335.

237. Finger PT. Anti-VEGF bevacizumab (Avastin) for radiation optic neuropathy. *Am J Ophthalmol* 2007;143:335–338.

238. Mohamed IG, Roa W, Fulton D, et al. Optic nerve sheath fenestration for a reversible optic neuropathy in radiation oncology. *Am J Clin Oncol* 2000;23:401–405.

239. Shields CL, Demirci H, Marr BP, et al. Intravitreal triamcinolone acetonide for acute radiation papillopathy. *Retina* 2006;26:537–544.

240. Leber KA, Berglöff J, Pendl G. Dose-response tolerance of the visual pathways and cranial nerves of the cavernous sinus to stereotactic radiosurgery. *J Neurosurg* 1998;88:43–50.

241. Morita A, Coffey RJ, Foote RL, et al. Risk of injury to cranial nerves after gamma knife radiosurgery for skull base meningiomas: experience in 88 patients. *J Neurosurg* 1999;90:42–49.

242. Kuo JS, Chen JCT, Yu C, et al. Gamma knife radiosurgery for benign cavernous sinus tumors: quantitative analysis of treatment outcomes. *Neurosurgery* 2004;54:1385–1393.

243. Lessell S, Lessell IM, Rizzo JF, III. Ocular neuromyotonia after radiation therapy. *Am J Ophthalmolmol* 1986;102:766–770.

244. Shults WT, Hoyt WF, Behrens M, et al. Ocular neuromyotonia. A clinical description of six patients. *Arch Ophthalmol* 1986;104:1028–1034.

245. Yee RD, Purvin VA. Ocular neuromyotonia: three case reports with eye movement recordings. *J Neuroophthalmol* 1998;18:1–8.

246. Gutmann L, Gutmann L. Myokymia and neuromyotonia 2004. *J Neurol* 2004;251:138–142.

247. Miller NR, Lee AG. Adult-onset acquired oculomotor nerve paresis with cyclic spasms: relationship to ocular neuromyotonia. *Am J Ophthalmol* 2004;137:70–76.

248. Epstein JA, Moster ML, Spiritos M. Seesaw nystagmus following whole brain irradiation and intrathecal methotrexate. *J Neuroophthalmol* 2001; 21:264–265.

249. Berger PS, Bataini JP. Radiation-induced cranial nerve palsy. *Cancer* 1977;40:152–155.

250. Pan L, Wang EM, Zhang N, et al. Long-term results of Leksell gamma knife surgery for trigeminal schwannomas. *J Neurosurg* 2005;102 (Suppl):220–224.

251. Gorgulho A, De Salles AA, McArthur D, et al. Brainstem and trigeminal nerve changes after radiosurgery for trigeminal pain. *Surg Neurol* 2006;66:127–135.

252. Donnet A, Tamura M, Valade D, et al. Trigeminal nerve radiosurgical treatment in intractable chronic cluster headache: unexpected high toxicity. *Neurosurgery* 2006;59:1252–1257.

253. Martí-Fàbregas J, Montero J, López-Villegas D, et al. Post-irradiation neuromyotonia in bilateral facial and trigeminal nerve distribution. *Neurology* 1997;48:1107–1109.

254. Diaz JM, Urban ES, Schiffman JS, et al. Post-irradiation neuromyotonia affecting trigeminal nerve distribution: an unusual presentation. *Neurology* 1992;42:1102–1104.

255. Wang CJ, Huang EY, Hsu HC, et al. The degree and time-course assessment of radiation-induced trismus occurring after radiotherapy for nasopharyngeal cancer. *Laryngoscope* 2005;115:1458–1460.

256. O'Sullivan B, Levin W. Late radiation-related fibrosis: pathogenesis, manifestations, and current management. *Semin Radiat Oncol* 2003;13:274–289.

257. Kellett MW, Humphrey PR, Tedman BM, et al. Hyperekplexia and trismus due to brainstem encephalopathy. *J Neurol Neurosurg Psychiatry* 1998;65:122–125.

258. Catterall M, Errington RD. The implications of improved treatment of malignant salivary gland tumors by fast neutron radiotherapy. *Int J Radiat Oncol Biol Phys* 1987;13:1313–1318.

259. Jankelowitz SK, Clouston PD. Radiation-induced "Meige syndrome". *Mov Disord* 2000;15:1287–1288.

260. Vissink A, Jansma J, Spijkervet FK, et al. Oral seque-lae of head and neck radiotherapy. *Crit Rev Oral Biol Med* 2003;14:199–212.

261. Ruo Redda MG, Allis S. Radiotherapy-induced taste impairment. *Cancer Treat Rev* 2006;32:541–547.

262. Bolze MS, Fosmire GJ, Stryker JA, et al. Taste acuity, plasma zinc levels, and weight loss during radiotherapy: a study of relationships. *Radiology* 1982;144:163–169.

263. Dewys WD, Walters K. Abnormalities of taste sensa-tion in cancer patients. *Cancer* 1975; 36:1888–1896.

264. Bhandare N, Antonelli PJ, Morris CG, et al. Ototoxicity after radiotherapy for head and neck tumors. *Int J Radiat Oncol Biol Phys* 2007;67:469–479.

265. Robinson AC. Management of radiation-induced otitis externa. *J Laryngol Otol* 1990;104:458–459.

266. Carls JL, Mendenhall WM, Morris CG, et al. External auditory canal stenosis after radiation ther-apy. *Laryngoscope* 2002;112:1975–1978.

267. Yang TL, Young YH. Radiation-induced otitis media—study of a new test, vestibular-evoked myogenic potential. *Int J Radiat Oncol Biol Phys* 2004;60:295–301.

268. Kew J, King AD, Leung SF, et al. Middle ear effu-sions after radiotherapy: correlation with pre-radiotherapy nasopharyngeal tumor patterns. *Am J Otol* 2000;21:782–785.

269. Young YH, Cheng PW, Ko JY. A 10-year longitudinal study of tubal function in patients with nasopharyn-geal carcinoma after irradiation. *Arch Otolaryngol Head Neck Surg* 1997;123:945–948.

270. Hwang CF, Chien CY, Lin HC, et al. Laser myrin-gotomy for otitis media with effusion in nasopha-ryngeal carcinoma patients. *Otolaryngol Head Neck Surg* 2005;132:924–927.

271. Johannesen TB, Rasmussen K, Winther FO. Late radiation effects on hearing, vestibular function, and taste in brain tumor patients. *Int J Radiat Oncol Biol Phys* 2002;53:86–90.

272. Hoistad DL, Ondrey FG, Mutlu C, et al. Histopath-ology of human temporal bone after cis-platinum, radiation, or both. *Otolaryngol Head Neck Surg* 1998;118:825–832.

273. Franco-Vidal V, Songu M, Blanchet H, et al. Intracochlear hemorrhage after gamma knife radio-surgery. *Otol Neurotol* 2007;28:240–244.

274. Pollack AG, Marymont MH, Kalapurakal JA. Acute neurological complications following gamma knife surgery for vestibular schwannoma. Case report. *J Neurosurg* 2005;103:546–551.

275. Adunka OF, Buchman CA. Cochlear implantation in the irradiated temporal bone. *J Laryngol Otol* 2007;121:83–86.

276. Low WK, Gopal K, Goh LK, et al. Cochlear implan-tation in postirradiated ears: outcomes and chal-lenges. *Laryngoscope* 2006;116:1258–1262.

277. Johansson S, Löfroth PO, Denekamp J. Left sided vocal cord paralysis: a newly recognized late compli-cation of mediastinal irradiation. *Radiother Oncol* 2001;58:287–294.

278. DePippo KL, Holas MA, Reding MJ. Validation of the 3-oz water swallow test for aspiration following stroke. *Arch Neurol* 1992;49:1259–1261.

279. King AD, Leung SF, Teo P, et al. Hypoglossal nerve palsy in nasopharyngeal carcinoma. *Head Neck* 1999;21:614–619.

280. King AD, Ahuja A, Leung SF, et al. MR features of the denervated tongue in radiation induced neuro-pathy. *Br J Radiol* 1999;72:349–353.

281. Gillette EL, Mahler PA, Powers BE, et al. Late radiation injury to muscle and peripheral nerves. *Int J Radiat Oncol Biol Phys* 1995;31:1309–1318.

282. Vujaskovic Z, Powers BE, Paardekoper G, et al. Effects of intraoperative irradiation (IORT) and intraoperative hyperthermia (IOHT) on canine sciatic nerve: histopathological and morphomet-ric studies. *Int J Radiat Oncol Biol Phys* 1999; 43:1103–1109.

283. Willett CG, Czito BG, Tyler DS. Intraoperative radi-ation therapy. *J Clin Oncol* 2007;25:971–977.

284. Haymaker W, Lindgren M. Nerve disturbances fol-lowing exposure to ionizing radiation. In: Vinken PJ, Bruyn GW, eds. *Handbook of Clinical Neurology. Part I. Diseases of Nerves, Vol. 7.* Amsterdam: North-Holland, 1970. pp. 388–401.

285. Mendes DG, Nawalkar RR, Eldar S. Post-irradiation femoral neuropathy. A case report. *J Bone Joint Surg Am* 1991;73:137–140.

286. Gorkhaly MP, Lo YL. Segmental neurogenic muscle hypertrophy associated with radiation injury. *Clin Neurol Neurosurg* 2002;105:32–34.

287. Galecki J, Hicer-Grzenkowicz J, Grudzien-Kowalska M, et al. Radiation-induced brachial plexopathy and hypofractionated regimens in adjuvant irradiation of patients with breast cancer—a review. *Acta Oncol* 2006;45:280–284.

288. Bagley FH, Walsh JW, Cady B, et al. Carcinomatous versus radiation-induced brachial plexus neuropathy in breast cancer. *Cancer* 1978;41:2154–2157.

289. Kori SH, Foley KM, Posner JB. Brachial plexus lesions in patients with cancer: 100 cases. *Neurology* 1981;31:45–50.

290. Lederman RJ, Wilbourn AJ. Brachial plexopathy: recurrent cancer or radiation? *Neurology* 1984; 34:1331–1335.

291. Thomas JE, Colby MY. Radiation-induced or metas-tatic brachial plexopathy? *JAMA* 1972; 222:1392–1395.

292. Wilbourn AJ. Plexopathies. *Neurol Clin 2007* 2007;25:139–171.

293. Ferrante MA, Wilbourn AJ. Electrodiagnostic approach to the patient with suspected brachial plexopathy. *Neurol Clin* 2002;20:423–450.

294. Roth G, Magistris MR, Le-Fort D, et al. Post-radiation brachial plexopathy. Persistent conduction block. Myokymic discharges and cramps. *Rev Neurol* 1988;144(3):173–180.

295. Pierce SM, Recht A, Lingos TI, et al. Long-term radiation complications following conservative sur-gery (CS) and radiation therapy (RT) in patients with early stage breast cancer. *Int J Radiat Oncol Biol Phys* 1992;23:915–923.

296. Olsen NK, Pfeiffer P, Mondrup K, et al. Radiation-induced brachial plexus neuropathy in breast cancer patients. *Acta Oncol* 1990;29:885–890.

297. Olsen NK, Pfeiffer P, Johannsen L, et al. Radiation-induced brachial plexopathy: neurological follow-up in 161 recurrence-free breast cancer patients. *Int J Radiat Oncol Biol Phys* 1993;26:43–49.

298. Tender GC, Kline DG. Posterior subscapular approach to the brachial plexus. *Neurosurgery* 2005;57:377–381.

299. Schierle C, Winograd JM. Radiation-induced brachial plexopathy: review. Complication without a cure. *J Reconstr Microsurg* 2004;20:149–152.

300. Delanian S, Porcher R, Balla-Mekias S, et al. Randomized, placebo-controlled trial of combined pentoxifylline and tocopherol for regression of superficial radiation-induced fibrosis. *J Clin Oncol* 2003;21:2545–2550.

301. LeQuang C. Postirradiation lesions of the brachial plexus. Results of surgical treatment. *Hand Clin* 1989;5(1):23–32.

302. Narakas AO. Operative treatment for radiation-induced and metastatic brachial plexopathy in 45 cases, 15 having an omentoplasty. *Bull Hosp Jt Dis Orthop Inst* 1984;44(2):354–375.

303. Jaeckle KA. Neurological manifestations of neoplastic and radiation-induced plexopathies. *Semin Neurol* 2004;24:385–393.

304. Pritchard J, Anand P, Broome J, et al. Double-blind randomized phase II study of hyperbaric oxygen in patients with radiation-induced brachial plexopathy. *Radiother Oncol* 2001;58:279–286.

305. Fulton DS. Brachial plexopathy in patients with breast cancer. *Dev Oncol* 1987;51:249–257.

306. Lachance DH, O'Neill BP, Harper CM, Jr, et al. Paraneoplastic brachial plexopathy in a patient with Hodgkin's disease. *Mayo Clin Proc* 1991;66:97–101.

307. Malow BA, Dawson DM. Neuralgic amyotrophy in association with radiation therapy for Hodgkin's disease. *Neurology* 1991;41:440–441.

308. Salner AL, Botnick LE, Herzog AG, et al. Reversible brachial plexopathy following primary radiation therapy for breast cancer. *Cancer Treat Rep* 1981;65:797–802.

309. Enevoldson TP, Scadding JW, Rustin GJ, et al. Spontaneous resolution of a postirradiation lumbosacral plexopathy. *Neurology* 1992;42:2224–2225.

310. Schiodt AV, Kristensen O. Neurologic complications after irradiation of malignant tumors of the testis. *Acta Radiol Oncol Radiat Phys Biol* 1978;17:369–378.

311. Johansson S, Svensson H, Larsson LG, et al. Brachial plexopathy after postoperative radiotherapy of breast cancer patients—a long-term follow-up. *Acta Oncologica* 2000;39:373–382.

312. Gerard JM, Franck N, Moussa Z, et al. Acute ischemic brachial plexus neuropathy following radiation therapy. *Neurology* 1989;39:450–451.

313. Georgiou A, Grigsby PW, Perez CA. Radiation induced lumbosacral plexopathy in gynecologic tumors: clinical findings and dosimetric analysis. *Int J Radiat Oncol Biol Phys* 1993;26:479–482.

314. Pettigrew LC, Glass JP, Maor M, et al. Diagnosis and treatment of lumbosacral plexopathies in patients with cancer. *Arch Neurol* 1984;41:1282–1285.

315. Dahele M, Davey P, Reingold S, et al. Radiation-induced lumbo-sacral plexopathy (RILSP): an important enigma. *Clin Oncol (R Coll Radiol)* 2006;18:427–428.

316. Goldstein I, Feldman MI, Deckers PJ, et al. Radiation-associated impotence. *JAMA* 1984;251:903–910.

317. Videtic GM, Venkatesan VM. Hyperbaric oxygen corrects sacral plexopathy due to osteoradionecrosis appearing 15 years after pelvic irradiation. *Clin Oncol (R Coll Radiol)* 1999;11:198–199.

318. Powers BE, Gillette EL, Gillette SL, et al. Muscle injury following experimental intraoperative irradiation. *Int J Radiat Oncol Biol Phys* 1991; 20:463–471.

319. Chua DT, Lo C, Yuen J, Foo YC. A pilot study of pentoxifylline in the treatment of radiation-induced trismus. *Am J Clin Oncol* 2001;24:366–369.

320. Stubblefield MD, Levine A, Custodio CM, et al. The role of botulinum toxin type A in the radiation fibrosis syndrome: A preliminary report. *Arch Phys Med Rehabil* 2008;89:417–421.

321. Rowin J, Cheng G, Lewis SL, et al. Late appearance of dropped head syndrome after radiotherapy for Hodgkin's disease. *Muscle Nerve* 2006;34:666–669.

322. Hall EJ. Intensity-modulated radiation therapy, protons, and the risk of second cancers. *Int J Radiat Oncol Biol Phys* 2006;65:1–7.

323. Kirova YM, Gambotti L, De RY, et al. Risk of second malignancies after adjuvant radiotherapy for breast cancer: a large-scale, single-institution review. *Int J Radiat Oncol Biol Phys* 2007; 68:359–363.

324. Flint-Richter P, Sadetzki S. Genetic predisposition for the development of radiation-associated meningioma: an epidemiological study. *Lancet Oncol* 2007;8:403–410.

325. Sadetzki S, Chetrit A, Freedman L, et al. Long-term follow-up for brain tumor development after childhood exposure to ionizing radiation for tinea capitis. *Radiat Res* 2005;163:424–432.

326. Minniti G, Traish D, Ashley S, et al. Risk of second brain tumor after conservative surgery and radiotherapy for pituitary adenoma: Update after an additional 10 years. *J Clin Endocrinol Metab* 2005;90(2):800–804.

327. Walter AW, Hancock ML, Pui CH, et al. Secondary brain tumors in children treated for acute lymphoblastic leukemia at St Jude Children's Research Hospital. *J Clin Oncol* 1998;16:3761–3767.

328. Duffner PK, Krischer JP, Horowitz ME, et al. Second malignancies in young children with primary brain tumors following treatment with prolonged postoperative chemotherapy and delayed irradiation: a Pediatric Oncology Group study. *Ann Neurol* 1998;44:313–316.

329. Neglia JP, Robison LL, Stovall M, et al. New primary neoplasms of the central nervous system in survivors of childhood cancer: a report from the Childhood Cancer Survivor Study. *J Natl Cancer Inst* 2006;98:1528–1537.

330. Henderson TO, Whitton J, Stovall M, et al. Secondary sarcomas in childhood cancer survivors: a report from the Childhood Cancer Survivor Study. *J Natl Cancer Inst* 2007;99:300–308.

331. Shin M, Ueki K, Kurita H, et al. Malignant transformation of a vestibular schwannoma after gamma knife radiosurgery. *Lancet* 2002;360:309–310.

332. Yu JS, Yong WH, Wilson D, et al. Glioblastoma induction after radiosurgery for meningioma. *Lancet* 2000;356:1576–1577.

333. Kaido T, Hoshida T, Uranishi R, et al. Radiosurgery-induced brain tumor. Case report. *J Neurosurg* 2001;95:710–713.

334. Foley KM, Woodruff JM, Ellis FT, et al. Radiation-induced malignant and atypical peripheral nerve sheath tumors. *Ann Neurol* 1980;7:311–318.

335. Bernstein M, Laperriere N. Radiation-induced tumors of the nervous system. In: Gutin PH, Leibel

SA, Sheline GE, eds. *Radiation Injury to the Nervous System.* New York: Raven, 1991. pp. 455–472.

336. Rogers LR. Cerebrovascular complications in cancer patients. *Neurol Clin* 2003;21:167–192.

337. Bowers DC, Liu Y, Leisenring W, et al. Late-occurring stroke among long-term survivors of childhood leukemia and brain tumors: a report from the Childhood Cancer Survivor Study. *J Clin Oncol* 2006;24:5277–5282.

338. Abayomi OK. Neck irradiation, carotid injury and its consequences. *Oral Oncol* 2004;40:872–878.

339. Dorresteijn LDA, Kappelle AC, Boogerd W, et al. Increased risk of ischemic stroke after radiotherapy on the neck in patients younger than 60 years. *J Clin Oncol* 2002;20:282–288.

340. Nimjee SM, Powers CJ, Bulsara KR. Review of the literature on de novo formation of cavernous malformations of the central nervous system after radiation therapy. *Neurosurg Focus* 2006;21:e4.

341. Bowers DC, McNeil DE, Liu Y, et al. Stroke as a late treatment effect of Hodgkin's Disease: a report from the Childhood Cancer Survivor Study. *J Clin Oncol* 2005;23:6508–6515.

342. Bowers DC, Mulne AF, Reisch JS, et al. Nonperioperative strokes in children with central nervous system tumors. *Cancer* 2002;94:1094–1101.

343. Dorresteijn LD, Kappelle AC, Scholz NM, et al. Increased carotid wall thickening after radiotherapy on the neck. *Eur J Cancer* 2005;41:1026–1030.

344. Cheng SW, Ting AC, Lam LK, et al. Carotid stenosis after radiotherapy for nasopharyngeal carcinoma. *Arch Otolaryngol Head Neck Surg* 2000;126:517–521.

345. Woodward WA, Giordano SH, Duan Z, et al. Supraclavicular radiation for breast cancer does not increase the 10-year risk of stroke. *Cancer* 2006;106:2556–2562.

346. Jagsi R, Griffith KA, Koelling T, et al. Stroke rates and risk factors in patients treated with radiation therapy for early-stage breast cancer. *J Clin Oncol* 2006;24:2779–2785.

347. Ishikawa N, Tajima G, Yofune N, et al. Moyamoya syndrome after cranial irradiation for bone marrow transplantation in a patient with acute leukemia. *Neuropediatrics* 2006;37:364–366.

348. Desai SS, Paulino AC, Mai WY, et al. Radiation-induced moyamoya syndrome. *Int J Radiat Oncol Biol Phys* 2006;65:1222–1227.

349. Ullrich NJ, Robertson R, Kinnamon DD, et al. Moyamoya following cranial irradiation for primary brain tumors in children. *Neurology* 2007;68:932–938.

350. Loftus CM, Biller J, Hart MN, et al. Management of radiation-induced accelerated carotid atherosclerosis. *Arch Neurol* 1987;44:711–714.

351. Ting AC, Cheng SW, Yeung KM, et al. Carotid stenting for radiation-induced extracranial carotid artery occlusive disease: efficacy and midterm outcomes. *J Endovasc Ther* 2004;11:53–59.

352. Yucesoy K, Feiz-Erfan I, Spetzler RF, et al. Anterior communicating artery aneurysm following radiation therapy for optic glioma: report of a case and review of the literature. *Skull Base* 2004;14:169–173.

353. Lau WY, Chow CK. Radiation-induced petrous internal carotid artery aneurysm. *Ann Otol Rhinol Laryngol* 2005;114:939–940.

354. Duffner PK. Long-term effects of radiation therapy on cognitive and endocrine function in children with leukemia and brain tumors. *Neurologist* 2004;10:293–310.

355. Constine LS, Wolff PD, Cann D, et al. Hypothalamic-pituitary dysfunction after radiation for brain tumors. *N Engl J Med* 1993;328:87–94.

356. Lam KS, Tse VK, Wang C, et al. Effects of cranial irradiation on hypothalamic-pituitary function—a 5-year longitudinal study in patients with nasopharyngeal carcinoma. *Q J Med* 1991;78:165–176.

357. Gurney JG, Ness KK, Sibley SD, et al. Metabolic syndrome and growth hormone deficiency in adult survivors of childhood acute lymphoblastic leukemia. *Cancer* 2006;107:1303–1312.

358. Minniti G, Jaffrain-Rea ML, Osti M, et al. The long-term efficacy of conventional radiotherapy in patients with GH-secreting pituitary adenomas. *Clin Endocrinol (Oxf)* 2005;62:210–216.

358a. Harikrishnan NK, Karagiannis TC, Chow MZ, et al. Effect of valproic acid on radiation-induced DNA damage in euchromatic and heterochromatic compartments. *Cell Cycle* 2008;7:468–476.

358b. Scott C, Suh J, Baldassarre S, et al. Improved survival, quality of life, and quality-adjusted survival in breast cancer patients treated with Efaproxiral (Efaproxyn) plus whole-brain radiation therapy for brain metastases. *Am J Clin Oncol* 2007;30:580–587.

358c. Forouzannia A, Richards GM, Khuntia D, et al. Motexafin gadolinium: a novel radiosensitizer for brain tumors. *Expert Rev Anticancer Ther* 2007;7:785–794.

358d. Okunieff P, Swarts S, Keng P, et al. Antioxidants reduce consequences of radiation exposure. *Adv Exp Med Biol* 2008;614:165–178.

358e. Burdelya LG, Krivokrysenko VI, Tallant TC, et al. An agonist of toll-like receptor 5 has radioprotective activity in mouse primate models. *Science* 2008;320:226–230.

359. Pai HH, Thornton A, Katznelson L, et al. Hypothalamic/pituitary function following high-dose conformal radiotherapy to the base of skull: demonstration of a dose-effect relationship using dose-volume histogram analysis. *Int J Radiat Oncol Biol Phys* 2001;49:1079–1092.

360. Alterio D, Jereczek-Fossa BA, Franchi B, et al. Thyroid disorders in patients treated with radiotherapy for head-and-neck cancer: a retrospective analysis of seventy-three patients. *Int J Radiat Oncol Biol Phys* 2007;67:144–150.

361. Jereczek-Fossa BA, Alterio D, Jassem J, et al. Radiotherapy-induced thyroid disorders. *Cancer Treat Rev* 2004;30:369–384.

362. Schneider AB, Gierlowski TC, Shore-Freedman E, et al. Dose-response relationships for radiation-induced hyperparathyroidism. *J Clin Endocrinol Metab* 1995;80:254–257.

363. Bentzen SM, Harari PM, Bernier J. Exploitable mechanisms for combining drugs with radiation: concepts, achievements and future directions. *Nat Clin Pract Oncol* 2007;4:172–180.

364. DeAngelis LM, Shapiro WR. Drug/radiation inter-actions and central nervous system injury. In: Gutin PH, Leibel SA, Sheline GE, eds. *Radiation Injury to the Nervous System*. New York: Raven, 1991. pp. 361–382.

365. Choy H, Kim DW. Chemotherapy and irradiation interaction. *Semin Oncol* 2003;30: (Suppl 9) 3–10.

366. Bleyer WA, Griffin TW. White matter necrosis, min-eralizing microangiopathy, and intellectual abilities in survivors of childhood leukemia: associations with central nervous system irradiation and methotrexate therapy. In: Gilbert HA, Kagan AR, eds. *Radiation Damage to the Nervous System. A Delayed Therapeutic Hazard*. New York: Raven, 1980. pp. 155–174.

Chapter 14

Neurotoxicity of Surgical and Diagnostic Procedures

INTRODUCTION

For most patients with cancer, iatrogenic neurotoxicity results from either chemotherapy (see Chapter 12) or ionizing radiation (see Chapter 13). In a few, anesthesia, tumor surgery, diagnostic tests, and other procedures related to patient care may result in significant neurologic dysfunction. Some of these nervous system abnormalities are described in this chapter.

DISORDERS RESULTING FROM ANESTHESIA

General Anesthesia

CENTRAL NERVOUS SYSTEM COMPLICATIONS

Neurologic disorders following anesthesia are uncommon.[1] Anesthesia results in death

in less than one in 10,000 operations, but may contribute to death in as many as one in 1700 operations.[1] One study of 1784 legal claims related to anesthetic care in the 1990s implicated nerve damage in 21%, brain damage in 9%, headache in 5%, and backache in 5%.[1] Neurologic disorders, ranging from postoperative delirium to severe and prolonged muscle paralysis in the intensive care unit (see also Chapter 11) (Table 14–1), can also complicate general anesthesia for patients undergoing treatment for non-neoplastic diseases.[2]

Because surgical procedures in patients with cancer are likely to be more complex, anesthesia is more prolonged, leading to more frequent complications than found in less complex surgical procedures. The type of neurologic disorder depends on the type of anesthesia, the patient's position during surgery, and often the patient's age and general preoperative

Table 14–1. **Neurotoxicity of Anesthesia**

General Anesthesia
  CNS toxicity
    Awakening during anesthesia
    Postoperative delirium
      Emergence delirium
      Delayed delirium
      Cognitive dysfunction
  Peripheral nerve toxicity
    Intensive care myopathy/neuropathy
    Compressive neuropathies

Regional Anesthesia (Epidural, Spinal, Nerve Block)
  CNS toxicity
    ? Cognitive dysfunction
    Intracranial hypotension
    Spinal epidural hematoma, infection
    Myelopathy, cauda equina syndrome
  Peripheral nerve toxicity
    Positional
    Toxic

CNS, central nervous system.

condition. Damage to the nervous system includes the following:

Postoperative Delirium/Cognitive Dysfunction

Three types of cognitive dysfunction can complicate general anesthesia[3]:

1. Emergence delirium occurs as the patient is awakening from anesthesia and mimics the excitation stage of ether anesthesia. The delirium usually resolves within minutes or hours and is of no long-term consequence.

2. Postoperative delirium occurs after a lucid interval, usually between 1 and 3 postoperative days. The incidence of postoperative delirium in patients with cancer is unknown, but it is common particularly in older adults.[3–5] Dasgupta and Drumbell reported the incidence after noncardiac surgery as ranging from 5.1% to 52.2%, with hip fracture and aortic surgery the major offenders.[6] However, clinically significant delirium that interferes with smooth postoperative care and recovery probably affects less than 1% of patients. When it occurs, it frightens patients, family, and sometimes the medical and

nursing staff, and considerably complicates recovery. Although a few patients are delirious immediately upon awakening from surgery, a lucid interval of 24 to 48 hours when the patient seems quite normal is more typical. This lucid period is followed by the sudden or gradual onset of confusion, disorientation, and clouding of consciousness. Most patients are agitated and suffer from delusions, often of persecution, or from visual and auditory hallucinations. The delusions and hallucinations can cause combative behavior; the patient may pull at dressings and tubing and cause extreme difficulty in nursing care. Tachycardia, diaphoresis, and tremor are frequently present even without fever. The disorder may last several days, but the patient virtually always returns to his or her preoperative state.[5,7] Persistent delirium may indicate unsuspected preexisting or concurrent structural brain disease and warrants neuroradiologic investigation with computed tomography (CT) or magnetic resonance imaging (MRI). Occult cerebral metastases, compensated hydrocephalus, and cerebrovascular disease are but a few conditions that may become apparent only in the postoperative period as a delirium.

The pathogenesis of most postoperative delirium is unknown. Risk factors[3,6] include brain dysfunction before surgery; old age; alcoholism; psychological factors, including extreme anxiety in the preoperative period and the sensory deprivation that can affect patients who are kept in windowless intensive care units[8,9] or whose eyes are patched after surgery. Postoperative drugs, particularly opioids and anticholinergics may be culprits.[10]

A perioperative cerebral infarct, especially if located in the territory of the nondominant (right) middle cerebral artery or the left posterior cerebral artery, may also cause a severely agitated delirium with minimal focal signs[11,12] and should always be considered as a possible cause of postoperative confusion.

The treatment is to restore cognition as soon as possible while protecting the patient from self-induced harm. Contributing organic factors should be corrected, and metabolic abnormalities should be reversed. The patient should be in a lighted room, preferably with a relative at the bedside to supply reassurance and assist with orientation. Sedative drugs should be withheld. If the patient's behavior has to be controlled to prevent self-injury, haloperidol or an atypical antipsychotic such as olanzapine[13] rather than benzodiazepines or opioids should be administered. When opioids are necessary for pain relief, they should be closely titrated to avoid excessive sedation. Because restraints increase agitation, they should be used only when absolutely necessary. Prophylactic low-dose haloperidol may reduce severity and duration of delirium in patients undergoing hip surgeries.[14] Low-dose prophylactic benzodiazepines may be appropriate for patients with known alcoholism who have been drinking up to the day of surgery. If the patient has stopped drinking at least 5 days before surgery, the risk of postoperative delirium tremens is minimal and does not warrant prophylaxis. While alcoholism may be present in any patient, in the cancer population, those with head and neck cancer have a high frequency of associated alcoholism.

3. A number of reports describe abnormalities of cognitive function that may affect 5% to 10% of patients and that may last for several months after anesthesia for noncardiac surgery.[3,5] The etiology of this dysfunction is unknown; anesthetic agents affect gene expression, neurotransmitters and their receptors and may, in some experimental situations, induce apoptosis.[15] However, almost all reports describe individuals with postoperative cognitive defects as returning to normal after a year.[3] Curiously, the dysfunction appears as common after regional anesthesia as general anesthesia.[3]

PERIPHERAL NERVOUS SYSTEM COMPLICATIONS

The patient's position during anesthesia may cause nerve injuries from compression.[16,17] The most common of these is a *brachial plexus* palsy that occurs when the arm is abducted above the head during thoracotomy or mastectomy. Abduction to more than 90 degrees causes the head of the humerus to descend into the axilla. The brachial plexus, which passes caudal to the head of the humerus, is stretched and compressed by that structure, especially if the patient remains in that position for several hours. Alternately, a shoulder brace with the patient in Trendelenburg position, may also exert pressure on the plexus. Usually, affected patients awaken to find the arm completely paralyzed and numb. Within hours to a few days, the lower portion of the plexus recovers; the patient is then weak in the deltoid and other shoulder muscles, generally with little numbness or sensory loss. In most instances, recovery is complete within a few weeks. Permanent loss of neurologic function is rare. In addition, acute brachial neuritis, an immune-mediated neuropathy, may follow any surgical procedure.[18]

Ulnar nerve paralysis can result from compression at the elbow by arm boards used to secure the IV line. The patient awakens from surgery with a numb fifth finger; the numbness usually also involves the ulnar aspect of the fourth finger, but the median aspect of the fourth finger, innervated by the median nerve, is spared. The small hand muscles are often weak. Recovery usually occurs in days to weeks, but numbness and weakness is occasionally permanent.

The *radial nerve* is vulnerable to compression as it winds around the humerus. The nerve can be compressed between the table edge and the humerus or, if the patient is in the lateral decubitus position, between the body and the humerus. Affected patients awaken with a wrist drop and inability to extend the fingers. The brachioradialis muscle is weak, but the triceps muscle is spared because its nerve supply exits the radial nerve above the humeral groove. The patient often complains of a weak grip, suggesting median nerve damage as well, but if the examiner passively extends the wrist, the grip is normal. Sensory loss or paresthesias are usually mild and may not be present at all. If present, they are evident over the dorsum of the median nerve side of the hand, the thumb, and index finger. Rarely, a cerebral infarct in its early stages may mimic a radial nerve palsy, but with an infarct both the brachioradialis and triceps are weak.

Compression of the *femoral nerve* can result from retraction during surgery.[19] The patient suffers weakness when extending the knee and a modest sensory loss in the anterior thigh and medial leg. The patellar reflex is absent. Other thigh and leg muscles are normal, permitting the patient to walk, although the knee may sometimes buckle. Recovery is usually complete after several weeks.

Sciatic neuropathy sometimes follows surgery, particularly if the patient was operated on in the sitting position, for example, to remove a posterior fossa metastasis. Weakness in the leg and foot is evident immediately after surgery, specifically affecting dorsiflexion and plantar flexion of the foot and flexion of the knee. Recovery is usually, but not always, complete.

Peroneal nerve paralysis may follow compression of that nerve at the fibular head from positioning of the patient or from intraoperative compression boots used to prevent thrombophlebitis.[20] In contradistinction to sciatic neuropathy, plantar flexion and inversion of the foot remain normal, whereas dorsiflexion and eversion are weak. Sensory loss, if any, is restricted to a small area between the great and second toes. Recovery is usually complete.

Other less common nerve injuries include those to median, long thoracic, musculocutaneous, axillary, tibial, saphenous, obturator, pudendal, and lateral femoral cutaneous nerves.[17]

Nerve damage may also result from delayed hemorrhage after surgery. Patients usually exhibit normal nerve function when they awaken from surgery, but within the first 24 to 48 hours, they begin to complain of pain and paresthesias in the distribution of affected nerves. The hemorrhage typically is found at such sites as the brachial plexus, where it causes arm paralysis, or at the lumbar plexus, in and around the iliopsoas muscle, where it paralyzes thigh flexion and knee extension. Hematomas in other areas are usually rapidly discovered before nerve compression occurs. CT or MRI can easily identify an acute hematoma of the brachial or lumbar plexus. When appropriate, draining the hematoma usually relieves neurologic symptoms, although recovery is typical even without drainage.

Prolonged use of propofol anesthesia in the intensive care unit can lead to the propofol infusion syndrome characterized by myocardial failure, metabolic acidosis, and rhabdomyolysis that may present as limb weakness.[21]

Regional Anesthesia

Complications of regional anesthesia are quite rare.[22] A recent review reports the incidence of permanent injury after epidural spinal anesthesia at well under 0.10%.[22] Complications include paraplegia and a cauda equina syndrome with urinary incontinence or retention. Transient weakness or sensory loss, including a cauda equina syndrome, is slightly more common.[22] Other complications may include cerebrospinal fluid (CSF) leaks with intracranial hypotension causing headache and in rare instances cranial subdural hematomas, spinal epidural hematomas, or infections,[23] and compressive injuries of peripheral nerves, as described in the section on General Anesthesia. One review suggests that postoperative delirium and cognitive changes can also occur after regional anesthesia.[3] Selective transforaminal block and epidural corticosteroids for the relief of back pain have caused spinal cord infarction when the corticosteroid is inadvertently injected into a radicular artery.[24]

Peripheral nerve blockade for anesthesia is rarely used at Memorial Sloan-Kettering Cancer Center (MSKCC), except for awake craniotomy (see following text). One report describes nerve block–related neuropathy as occurring in 0.22% of over 1000 patients.[25] In most instances, the pathogenesis is uncertain and possibly multifactorial. Pathogenetic factors include direct injury from the needle or from a hematoma, nerve compression or stretch, ischemia and direct toxicity of the anesthetic agent. Symptoms are usually transient.

Patients undergoing awake craniotomy receive local anesthetic blocks to occipital and supraorbital nerves as well as areas of the scalp. On rare occasions (about 1 in 500 awake craniotomies at MSKCC), injection in the preauricular area may temporarily paralyze the facial nerve as it exits from the skull at the stylomastoid foramen. Patients always fully recover.

Intubation

Tapia syndrome, bilateral paralysis of hypoglossal and recurrent laryngeal nerves after transoral intubation for general anesthesia, is a rare phenomenon.[26] Unilateral paralysis of one or both nerves also occurs. In one of our patients, an extensive search for base of skull metastasis was conducted after atrophy of the tongue was noted on routine examination several weeks after surgery. The patient reported that when he awoke from surgery for resection of an osteosarcoma of the leg, he noted dysarthria and some mild difficulty in eating, but did not report it because he thought it was a result of the operation. He was correct; no metastasis was found. Also possible is hypoglossal or vagal paralysis from carotid dissection following neck manipulation for intubation.

SURGERY

Neurologic complications can follow surgery on almost any area of the anatomy, but especially after craniotomy (Table 14–2).

Only some of these complications can be considered here and in the paragraphs that follow. In general, 5% to 10% of patients undergoing craniotomy for tumor suffer worsening of their neurologic symptoms; the rest either remain unchanged (many patients with brain tumors who were symptomatic on presentation became asymptomatic before surgery after receiving corticosteroids) or are improved neurologically (at least one-third of patients).[27,28]

Table 14–2. **Neurotoxicity of Craniotomy**

Vascular
    Arterial infarction
    Hemorrhage
    Venous occlusion/infarction
Infection
    Wound infection
    Meningitis
Seizures
Aseptic meningitis
Pneumocephalus
Optic neuropathy
Headache
Craniotomy defect—CSF leak
Cerebellar mutism
Electrolyte disorders

CSF, cerebrospinal fluid.

Craniotomy

Several complications, both neurologic and systemic, can follow craniotomy for either primary or metastatic brain tumors.[27] Table 14–3 details the variety of complications, both major and minor and their incidence.

VASCULAR LESIONS

Three different types of vascular complications can follow craniotomy: hemorrhage either into the tumor bed or at a distance[29]; cerebral infarcts from arterial occlusion or vasospasm; venous or venous sinus occlusion that may or may not result in brain infarction.[30,31]

Some blood in the surgical bed following craniotomy is inevitable. However, clinically significant hematomas are uncommon. In one series, only 8 of 400 craniotomies were followed by hematoma, and two of these were minor.[27] Even less common are intracerebral hemorrhages that are remote from the craniotomy site,[29,32,33] particularly posterior fossa hemorrhage after supratentorial surgery.[33] Coagulation abnormalities, patient positioning, and intraoperative hypertension all appear to play a role.[33]

Cerebral infarcts from arterial occlusion or vasospasm are an occasional complication of surgery, particularly surgery around the Sylvian fissure where there are branches of the middle cerebral artery. Symptomatic strokes are uncommon, the infarcts being identified only by diffusion-weighted images on postoperative MRI; however, most patients with a diffuse-weighted abnormality on the postoperative MRI have incomplete recovery of a surgically induced postoperative deficit.[34]

Dural venous sinus occlusion is a rare but recognized complication of suboccipital and translabyrinthine craniotomy.[35] In the neuro-oncology patient, hypercoagulability may also play a role. Occlusion can lead to cortical hemorrhagic infarcts. Diagnosis can be established by magnetic resonance venography. In appropriate circumstances, thrombolysis may be considered.[35]

INFECTIONS

With modern surgical techniques and prophylactic antibiotics, postoperative infections are

Table 14–3. **Postoperative Complications after Craniotomy in 327 Patients with Brain Tumors**

Complication Type°	No. of Complications		
	Minor (%)	Major (%)	Either (%)
Neurological	49 (12.25)	34 (8.50)	83 (20.75)
Motor or sensory deficit	37 (9.25)	30 (7.50)	67 (16.75)
Aphasia/dysphasia	16 (4.00)	2 (0.50)	18 (4.50)
Visual field deficit	0 (0.00)	2 (0.50)	2 (0.50)
Regional	16 (4.00)	12 (3.00)	28 (7.00)
Seizure	10 (2.50)	0 (0.00)	10 (2.50)
Hematoma	2 (0.50)	6 (1.50)	8 (2.00)
Hydrocephalus	0 (0.00)	1 (0.25)	1 (0.25)
Pneumocephalus	0 (0.00)	1 (0.25)	1 (0.25)
Wound infection	2 (0.50)	1 (0.25)	3 (0.75)
Cerebrospinal fluid leak	4 (1.00)	3 (0.75)	7 (1.75)
Meningitis	NA	4 (1.00)	4 (1.00)
Systemic	20 (5.00)	11 (2.75)	31 (7.75)
Deep vein thrombosis	1 (0.25)	4 (1.00)	5 (1.25)
Pulmonary infection	6 (1.50)	4 (1.00)	10 (2.50)
Urinary tract infection	10 (2.50)	0 (0.00)	10 (2.50)
Pulmonary embolism	NA	1 (0.25)	1 (0.25)
Other systemic	6 (1.50)	2 (0.50)	8 (2.00)
Any[†]	75 (18.75)	53 (13.25)	128 (32)

From Ref. 27 with permission.
° Number of patients having complications in each category.
[†] Number of patients having any of these complications.
NA, not applicable.

uncommon.[36] In one series of 400 craniotomies, there were three wound infections and four cases of postoperative meningitis, some of which may have been aseptic (see following text).[27] In another series of 208 patients operated for brain metastases, there were 6 instances of wound infection and one case of meningitis.[28] Wound infections are more likely to follow reoperation, particularly in patients who have received radiation therapy and prolonged corticosteroids. CSF leaks, occurring in 7 of 400 craniotomies,[27] are a significant risk factor.[36]

SEIZURES

Postoperative seizures occurred in 10 of 400 craniotomies performed for brain tumors in one series[27] and 2 of 208 patients operated for brain metastases.[28] Virtually all patients undergoing craniotomy for brain tumors are placed on prophylactic anticonvulsants.[37,38] As

indicated in Chapter 4, these drugs are effective in preventing early postoperative seizures (first 2 weeks), but not in preventing later seizures.[39] Patients who experience an early postoperative seizure, but who have not had preoperative seizures, probably do not require prolonged anticonvulsant therapy.

ASEPTIC MENINGITIS

Some patients develop a non-infective post-craniotomy meningitis, most often after posterior fossa surgery or when the surgeon violates the cerebral ventricular system. Occasionally, the syndrome occurs after spinal operations if the subarachnoid space has been entered. The disorder is caused by blood and blood products in the subarachnoid space (hemogenic meningitis).[40,41] In one series of 1146 operations for cerebellopontine angle tumors, symptomatic chemical meningitis occurred in 30 patients; bacterial meningitis occurred in 10.[42] Characteristically, several days after the operation, when the corticosteroid dose is being tapered, the patient complains of headache or backache and develops a low-grade fever. Nuchal rigidity may or may not be present. A few patients become frankly delirious. The CSF pressure is usually high, and the cell count varies from a few hundred to thousands of white blood cells (WBCs), initially dominated by neutrophils, which are then replaced by lymphocytes. The protein concentration is raised. The glucose concentration usually remains normal, but sometimes is reduced. Results of CSF cultures are negative but one report suggests that polymerase chain reaction may reveal bacteria that do not grow in culture.[43] The lumbar puncture may be therapeutic; if not, the aseptic meningitis is treated by increasing the corticosteroids dose and then tapering more slowly. The inflammatory reaction may be severe enough to cause hydrocephalus, although a shunt is rarely required. Occasionally, the aseptic meningitis persists for months.

Distinguished from bacterial meningitis by its more indolent course and the absence of bacteria in the CSF, this syndrome can also be distinguished from leptomeningeal metastasis, which also may follow posterior fossa craniotomy, by the absence of malignant cells in the CSF. Furthermore, leptomeningeal metastasis usually becomes evident months, not days, after surgery.

PNEUMOCEPHALUS

Craniotomies allow air to enter the intracranial cavity. The air usually resolves but may remain under pressure, causing tension pneumocephalus.[44] Efflux of CSF from the intracranial space during surgery creates a negative pressure that is filled by the entry of air, the *inverted pop bottle* phenomenon (Fig. 14–1).[44–46] Clinical signs usually appear 2 to 4 days postoperatively. Occasionally, the air appears months to years after the procedure.[47] Patients complain of headache, may be delirious, and may develop focal neurologic signs depending on the location of the air. One patient developed a *locked-in* syndrome (a state of paralysis of all 4 extremities and most cranial nerves leading to the appearance of unconsciousness in a fully awake patient) from tension pneumocephalus after posterior fossa surgery.[48] CT or MRI clearly identifies the offending agent as air rather than hematoma. Usually the air resolves by itself; at times, it must be aspirated. Occasionally, surgery may be required to repair a skull-base fracture or CSF leak.

DECOMPRESSION OPTIC NEUROPATHY

Blindness is a rare but devastating complication of craniotomy performed to remove a brain tumor.[49] Following apparently uncomplicated craniotomy, the disorder occurs in

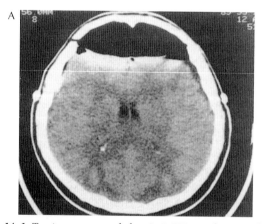
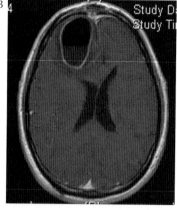

Figure 14–1. Tension pneumocephalus. **A**: A CT scan showing a large amount of air entrapped in the frontal region, displacing and compressing the frontal lobes. The frontal horns of the lateral ventricles are deformed and asymmetric. **B**: One month after an anterior craniofacial operation, the patient became lethargic and confused. An MRI revealed an air- and fluid-filled mass in the right frontal lobe. Because of the concern that the lesion was an abscess, an open craniotomy was performed and the air was evacuated. Because there was no cerebrospinal fluid rhinorrhea after surgery, the air was thought to have arisen from a lumbar puncture performed at surgery. (A: From Ref. 46 with permission.)

patients with markedly increased intracranial pressure and papilledema. Most patients do not show preoperative evidence of visual field abnormalities, yet the postoperative blindness does not reverse. The pathogenesis of this rare disorder is unknown, but the sudden lowering of intracranial pressure probably alters the vascular supply to the nerves. A similar disorder occasionally follows ventriculostomy for relief of increased intracranial pressure.

HEADACHE

Headache in the first few days following craniotomy is common and sometimes severe[50,51]; it is often under-treated for fear of side effects of opioid analgesics.[52] In most patients, the headache resolves in a few days. Chronic headache is much less common. However, a significant number of patients undergoing posterior fossa craniotomy, especially for vestibular schwannoma, develop chronic headaches that may be debilitating and are sometimes relieved by cranioplasty.[53–55] The exact pathogenesis is unclear. Although many patients undergoing supratentorial craniotomy complain of brief episodes of lightning-like pain at the site of the craniotomy persisting for months after surgery, these are rarely disabling and respond to simple reassurance. Other headaches after supratentorial craniotomy are quite rare.

CEREBELLAR MUTISM

This disorder, better named the *cerebellar cognitive affective syndrome*, occurs following posterior fossa surgery.[56–58] Much more common in children than in adults, this syndrome follows surgery affecting the vermis and posterior lobes of the cerebellum. Patients are awake but somnolent. Whatever their level of alertness, they do not speak and often behave abnormally, either by not responding to the examiner or by behaving inappropriately. Patients may refuse to swallow food although they are not dysphagic. The syndrome begins hours to days after awakening from anesthesia and is largely reversible although some behavioral abnormalities may persist. Its pathophysiology is unknown. There is no specific treatment.

CRANIECTOMY DEFECT/CSF LEAK

Unlike CSF leaks after spinal surgery (see following text), CSF leaks after craniotomy may lead to infection but do not cause neurologic symptoms. However, a large craniotomy defect without a CSF leak may lead to symptoms similar to those of spinal CSF leaks.[59] Characteristically, when an otherwise asymptomatic patient stands, neurologic symptoms including hemiparesis may develop. The craniotomy defect changes from convex to concave and the symptoms clear when the patient assumes the recumbent position. In other patients, symptoms are present in all positions and may be progressive. In all patients with the syndrome, the skin flap is at times concave, so much so that some have named the disorder the *sinking scalp flap syndrome*.[59] The disorder is corrected by cranioplasty. Even without florid symptoms, a large craniectomy defect impaired blood flow to the involved hemisphere both at rest and on assuming the upright posture. It also decreased cerebral metabolism in the involved hemisphere. These defects were corrected by cranioplasty.[60]

ELECTROLYTE AND OTHER DISORDERS

When patients do not do as well as expected after craniotomy, electrolyte disorders such as hyponatremia,[61,62] or even adrenal failure,[63] should be suspected. Other electrolyte or endocrine disorders may follow surgery involving the hypothalamus and pituitary area.

Base of Skull Surgery

Table 14–4 outlines some of the complications of surgery of the base of skull.

Anterior skull base surgery (craniofacial resection) done for tumors such as esthesioneuroblastoma has, as its most common complication, wound infection.[64–66] Central nervous system (CNS) complications include CSF leak with or without meningitis, olfactory nerve dysfunction, which is inevitable, and pneumocephalus. Mortality is about 5%.[64] Damage to the frontal lobes can lead to a postoperative delirium, as can tension pneumocephalus.

Transsphenoidal excision of pituitary tumors (primary or metastases) can cause diabetes insipidus as well as CSF leak. Damage to the optic chiasm may cause visual loss or visual field defect.[67] Extraocular nerves are sometimes damaged.

Table 14–4. **Neurotoxicity of Base of Skull Surgery**

Infection
  Wound
  Meningitis
  Osteomyelitis
CSF leak
Cranial nerve damage
  Olfactory
  Others
Pneumocephalus
Carotid damage
Hemorrhage

Table 14–5. **Neurotoxicity of Spinal Surgery**

Vascular disorder
  Infarction
  Hemorrhage
Wound dehiscence
  Wound infection
  Epidural infection
  Vertebral infection
Spinal instability
CSF leak
  Infection
  Intracranial hypotension
  Cerebral subdural hygromas
Myelopathy
Cerebral disorders
  Visual loss
  Pituitary apoplexy
  Pneumocephalus

CSF, cerebrospinal fluid.

Injury to the carotid artery can include occlusion, aneurysm formation or, as in one of our patients, a carotid-cavernous fistula (Fig. 9–12).

Cranial nerves are vulnerable as they exit the skull. In treating malignant tumors or olfactory groove meningiomas, there is often no way to preserve olfactory nerves; the facial and acoustic nerves are at risk during surgery for vestibular schwannomas; other cranial nerves can usually be spared.[66]

Spinal Surgery

Table 14–5 lists some of the complications that follow surgery of the spine or spinal cord in patients with cancer.

Hemorrhage during or after the surgery may result from tumor hypervascularity (e.g., renal tumors), dilated epidural veins, soft tissue paraspinal blood vessels, and sometimes uninvolved bone.[68] Preoperative embolization of the tumor can prevent some hemorrhages. Inadvertent surgical damage to radicular arteries can cause spinal cord infarction. Wound dehiscence is more likely to occur after upper thoracic operations where the underlying soft tissues are sparse. Risk factors include spinal implants, high-dose corticosteroids, malnutrition, and previous irradiation.[68] Postoperative infections can follow wound dehiscence and may involve the bone or epidural space. Meningitis is an uncommon but feared complication. If the dura has been violated during surgery, a postoperative CSF leak can lead to infection, postural headache, chronic subdural hygromas (intracranial hypotension syndrome; see page 569), and cortical vein thrombosis).[69,69a] Motor and sensory dysfunction may occur as a result of damage to the spinal cord, particularly during surgery on the cord itself (intramedullary metastasis) or subdural tumors.

Rarely, spinal surgery can result in cerebral dysfunction. As indicated in the preceding text, intracranial hypotension can lead to subdural hygromas. Rare instances of unilateral or bilateral visual loss from ischemic optic neuropathy[70] or pituitary apoplexy[71] have been reported.

Head and Neck Surgery

Surgery on the neck or head may damage branches of peripheral or cranial nerves. For instance, injury to the infraorbital branch of the trigeminal nerve during maxillary sinus surgery leads to a loss of facial sensation lateral to the nose. Another example is damage to the facial nerve during parotid surgery. Table 14–6 and Figure 14–2 illustrate the nerve damage caused by radical neck dissection. Modern neck dissection[72] obviates some but not all of the damage identified in the table.[73,74] The marginal mandibular branch of

Table 14–6. Involvement of Peripheral Nerves in 33 Radical Neck Dissections

Nerve	No. Involved in Radical Neck Dissection	No. Noted to Be Excised in Operative Report	Percentage Involved
Facial (excluding mandibular, platysmal and marginal branches)	22	3	67
Glossopharyngeal	3	2	10
Vagus	5	4	15
Spinal accessory	33	33	100
Hypoglossal	13	7	39
Dorsal scapular (rhomboids)	2	0	6
Medial and lateral thoracic nerves (pectorals)	2	0	6
Sympathetic	11	0	33
Cervical plexus and supraclavicular nerves	33	0	100
Suprascapular	2	0	6
Phrenic	3	0	10

Adapted from Ref. 73.

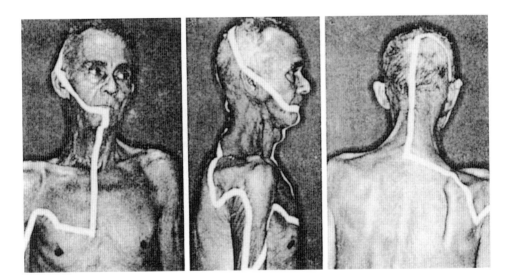

Figure 14–2. Neurologic findings after radical neck dissection. Note that the sensory loss extends to the cheek and chin and to areas generally thought to be supplied by the trigeminal nerve. Sensory loss also occurs to a variable degree in the anterior and posterior chest and the lateral arm. The spread of sensory loss varies from patient to patient and is always quite extensive. (From Ref. 73 with permission.)

the facial nerve may be damaged during neck dissection.[73] This thin twig descends below the mandible into the upper neck before re-ascending into the face to innervate the lower lip. The twig is so fine that even if the surgeon identifies the nerve and preserves it anatomically, it often suffers sufficient damage to impair function. Affected patients cannot fully depress the lower lip when grimacing; many women find it difficult to apply lipstick to that area. Aside from the mild cosmetic change, the injury has no physiologic consequence. With modern surgical techniques, the weakness is often transient.

The vagus nerve runs a long course through the neck and can easily be damaged during neck surgery. Symptomatic damage to branches of the vagus nerve causes palatal and vocal cord paralysis.

Fibers from the cervical sympathetic chain travel with the vagus, as does the glossopharyngeal nerve, which is also vulnerable during neck surgery. Sympathetic damage causes an ipsilateral Horner syndrome after some neck dissections. Damage to recurrent laryngeal nerve fibers causes unilateral vocal cord paralysis, a problem that sometimes also follows thyroid surgery.

Florid perspiration of the cheek, jaw, temple, and sometimes neck while eating, known as gustatory sweating, follows aberrant regeneration of surgically damaged autonomic fibers that travel with the glossopharyngeal and vagus nerves. The disorder can follow neck dissection.[75] It is more a nuisance than a disability.

The spinal accessory nerve travels through the neck to innervate the trapezius and sternocleidomastoid muscles. It was regularly damaged during radical neck dissection, but less extensive neck dissections may still injure the nerve even if it is not severed. Damage leads to drooping of the shoulders that can generate chronic arm pain.[76] Most patients benefit from postoperative physical therapy to the shoulders.

The hypoglossal nerve descends into the neck before innervating the ipsilateral half of the tongue; occasional injury during neck surgery causes slurred speech and difficulty eating, particularly clearing fluid between the teeth and cheek. Both are usually mild.

Sensory branches of supraclavicular and infraclavicular nerves and those of the second and third cervical roots are often sectioned during radical neck dissection.[73] Sensory loss may affect the anterior chest, sometimes extending to the nipples; the posterior chest, sometimes to the middle of the scapula; the lateral arm to the lower end of the deltoid; the lateral face anterior to the ear; and the lateral aspect of the jaw. The physician may mistake these last changes for trigeminal nerve involvement, although careful sensory evaluation reveals that the medial portion of the face supplied by the trigeminal nerve, particularly the upper and lower lip, is spared.

The most serious consequence of neck dissection is a rare but extremely painful disorder of the jaw and face, often chronic and disabling.

Characteristically, the patient is numb in the C2 and C3 distribution over the jaw and lateral face and, within weeks to a few months following surgery, begins to note sharp, lancinating pains, often precipitated by touching the scar. Patients will generally not allow an examination because touch precipitates lancinating pain. The sharp pains are soon replaced by a chronic dysesthetic pain in the same distribution. The sharp, lancinating pains are often relieved by carbamazepine or gabapentin, but the chronic dysesthetic pain, resembling postherpetic neuralgia, is not relieved by most medications. Exploration and excision of the scar may reveal neuromas, but too often this does not relieve the chronic pain, and many patients never obtain relief.[77,78]

Carotid sinus syncope due to tumor involving the carotid sinus (see p 296) usually affects patients who have had a prior neck dissection. The surgery opens the carotid sheath and exposes the sinus to tumor invasion if the tumor recurs.

Carotid damage either during or at some time following radical neck dissection, often after RT, may cause bleeding.[79] When carotid leakage occurs, the artery must be ligated. Many patients tolerate ligation without substantial neurologic injury, but a few may develop cerebral infarction with hemiparesis or hemiplegia. In rare instances, brain swelling, herniation, and death may follow the infarct. The symptoms usually begin immediately after the artery is occluded, but can begin at any time within the first several hours. The neurologic damage is usually maximal within a few hours of onset. In patients with cerebral swelling, the symptoms may progress for 24 to 48 hours and then begin to resolve.

Breast Surgery

About 50% of women undergoing a mastectomy experience chronic discomfort or pain after the procedure.[80] During a mastectomy, the intercostal–brachial nerve is often sacrificed or damaged. In one series, sensory loss was identified in 84% of patients in whom the nerve was sacrificed and 53% of patients in whom the nerve was preserved.[81] Patients with damage to the nerve develop sensory loss in the posterior upper arm, of which they may be unaware.[82] However, 15% to 20% of patients with intercostal–brachial nerve damage complain of a tight, burning sensation in that distribution. The pain may be present immediately after surgery or begin several weeks later[83] and can be associated with paresthesias in the fingers. The differential diagnosis is outlined in Table 14–7. In a small percentage of patients, the pain may be severe, intractable, and lead to chronic disability. In a few instances, the patient or doctor discovers the sensory loss caused by intercostobrachial nerve damage long after the surgery. Not being aware of the entity, this may lead to a fruitless search for recurrent tumor.

About 20% of patients undergoing mastectomy also experience a phantom breast that is painful in only 7% at 6 weeks and 1% after 2 years.[83–85] Other sensations following either a sentinel node biopsy or an axillary lymph node dissection include tenderness, soreness, and a pulling sensation that tends to persist but is rarely severe or distressing.[86] All sensory postmastectomy symptoms are more marked when the surgery includes a lymph node dissection or sentinel node biopsy.

Thoracotomy

Chronic pain after thoracotomy (post-thoracotomy pain syndrome) may affect more than 50% of patients undergoing that procedure.[80,87] In one series, the incidence of post-thoracotomy pain was 80% at 3 months and 60% at 1 year

Table 14–7. **Causes of Arm Pain Following Mastectomy**

Postmastectomy pain (intercostal-brachial neuralgia)
Brachial plexus neuropathy
    Tumor infiltration
    Radiation fibrosis
    Lymphedema entrapment
    Transient neuritis
Cervical radiculopathy
    Vertebral metastatic compression
    Degenerative spine disease
Carpal tunnel syndrome
    With lymphedema
    Without lymphedema
Pericapsulitis of shoulder joint

Adapted from Ref. 83.

after surgery; however, severe pain affected only 3% to 5%.[87] The incidence is substantially less after video-assisted thoracoscopic surgery.[80,88] Damage to intercostal nerves is an inevitable result of thoracotomy. The large overlap of sensory fields of the thoracic roots usually precludes identifiable sensory loss when only one or two nerves have been sectioned, but a few patients develop painful dysesthetic and lancinating sensations in the damaged area. This post-thoracotomy syndrome usually begins within a few weeks after surgery and may persist for many years. It must be distinguished from tumor recurrence invading the chest wall and compressing intercostal nerves or from vertebral body metastasis compressing thoracic roots. The time course and chronicity of the pain usually make the distinction fairly simple.

A rare complication of thoracotomy is the development of tension pneumocephalus resulting from an iatrogenic subarachnoid-pleural fistula.[89] The diagnosis should be considered in individuals who develop changes in mental status after thoracotomy. Neuroimaging easily identifies air in the cranium causing compression. CSF leak is a rare complication of thoracotomy. Patients usually present with postural headache resulting from intracranial hypotension (see p 569). When chest tubes are in place, the leak is particularly difficult to identify as the CSF is drained by the chest tube and the surgeon may be puzzled as to why the flow of fluid does not diminish as expected the first few days after surgery.

Abdominal/Pelvic Surgery

Damage to lower extremity nerves during abdominal or pelvic surgery is uncommon. Retractor injuries sometimes cause femoral nerve damage that is almost always reversible. Painful postoperative syndromes of the lower extremities are also rare. However, autonomic dysfunction after retroperitoneal lymph node dissection may occur on the side not operated. The contralateral leg is cold and sweaty, probably because of compensatory sympathetic hyperfunction[90] due to disruption of sympathetic fibers on the operated side. Painful and nonpainful phantom sensation may follow rectal amputation.[91] Osteitis pubis (see Chapter 8) can cause severe lower extremity pain, often mistaken for a neuropathy or plexopathy.

Extremity Surgery

When a limb is amputated because of bone or soft-tissue sarcomas, most patients are left with a phantom sensation.[92,93] The phantom may initially be painful; the incidence of pain tends to decrease over time but persists in a significant number of patients.[93] Luckily, with the use of limb-sparing surgery, this is less of a problem for patients with cancer than in the past. Evidence indicates that the patient who has experienced pain before the amputation is more likely to have postoperative pain, at least acutely. As the phantom sensations diminish, the more distal parts of the phantom seem to move proximally so that the patient may be left feeling a foreshortened limb with fingers or toes still discernible. A prosthesis is reported to hasten the resolution of the phantom sensation. Painful phantoms are characterized by chronic, unremitting pains often perceived as a cramped, abnormal position of the phantom or as a burning, dysesthetic sensation. The pain does not respond to analgesic agents. Similar phantom sensations rarely follow nerve injury without amputation, especially when the brachial plexus is affected. A clinical report describes two patients with prior amputations who developed phantom pain only after receiving therapy with the neurotoxic agent, paclitaxel.[94]

DIAGNOSTIC PROCEDURES CAUSING NEUROTOXICITY

Computed Tomography

Iodinated contrast material used for diagnostic procedures is generally well tolerated, particularly in patients without previous renal dysfunction. Various adverse reactions may affect systems other than the kidney, however.[95] The agents are epileptogenic, and if a brain lesion such as a metastasis causes blood–brain barrier disruption, a seizure may develop when the intravenously injected, iodinated contrast material leaks across the barrier. About 15% of patients with brain metastases who undergo a CT with contrast suffer a focal or generalized seizure.[96] If a patient about to undergo a CT is suspected of harboring a brain metastasis or other lesion that impairs the blood–brain barrier, administering an oral dose (5–10 mg) of

diazepam 30 minutes before the procedure considerably lowers the likelihood of an associated seizure. The risk of seizures does not apply to gadolinium-contrast enhancement used in MRI.

In addition to seizures, other nervous system complications include transient global amnesia, cortical blindness, visual blurring, and worsening of preexisting myasthenia gravis.[95]

Lumbar Puncture

Lumbar puncture, whether used diagnostically or therapeutically, can precipitate neurologic complications[97] (Table 14–8).

CEREBRAL HERNIATION

Continuing spinal fluid leakage sometimes occurs after lumbar puncture, lowering pressure in the spinal canal and causing a pressure differential between the intracranial and the spinal spaces in those patients in whom cerebral or spinal tumors obstruct free passage of CSF. In patients with large intracranial lesions

Table 14–8. **Complications of Lumbar Puncture**

Uncal or tonsillar herniation
Reversible tonsillar descent
Spinal coning in patients with rostral
    subarachnoid block
Post-dural puncture headache
Cranial neuropathies
Nerve root irritation, herniation, and transection
Low back pain
Implantation of epidermal tumors
Infections
Bleeding complications
    Intracranial bleeding
    Traumatic lumbar puncture
    Spinal hematomas
Other complications
    Vasovagal syncope
    Cardiac arrest
    Seizures
    Incorrect laboratory analysis of cerebrospinal
        fluid

From Ref. 97 with permission.

and increased intracranial pressure, the result may be cerebral herniation and sometimes death.[98,99] Characteristically, patients who suffer cerebral herniation induced by lumbar puncture are already partially herniated when the lumbar puncture is performed. Lumbar CSF pressure may appear normal because the downwardly displaced brain obstructs the foramen magnum, thereby separating the intracranial and spinal cavities. Immediately after the lumbar puncture, the patient appears to tolerate the procedure well, but some hours later develops headache, worsening neurologic signs, and then sudden respiratory arrest from tonsillar herniation and brainstem compression. The premonitory signs may evolve rapidly; occasionally, the patient may be found dead the following morning. The apparently normal CSF pressure at the time of the lumbar puncture can give the physician a false sense of security, but it is exactly the apparently normal pressure that increases the risk of herniation. Except for those patients with suspected bacterial meningitis, lumbar puncture should not be performed in patients suspected of having an intracranial mass until a CT or MRI of the intracranial contents excludes such lesions. Even in the absence of a mass lesion, lumbar puncture may rarely cause herniation if intracranial pressure is raised by meningitis[100] or leptomeningeal tumor.

The frequency of cerebral herniation in patients with increased intracranial pressure is unknown but certainly low. Most patients with brain tumors can safely undergo lumbar puncture. Patients whose large lesions are in the posterior fossa or the temporal lobe, or whose scan shows evidence of massive shift and early herniation, are at increased risk of herniation, and lumbar puncture should be avoided. Herniation cannot be prevented by removing a smaller quantity of CSF because the disorder arises mostly from the post-puncture CSF leak through the needle hole. The amount of fluid removed at the time of lumbar puncture probably is irrelevant. Thus, if the needle has already entered the subarachnoid space, the physician should remove as much fluid as is necessary for the appropriate examinations.

Any patient who has elevated intracranial pressure from a mass lesion and who requires lumbar puncture should be started on corticosteroids to decrease the intracranial pressure before conducting the lumbar puncture, unless the lesion is believed to be a lymphoma.

INTRACRANIAL HYPOTENSION

The CSF leak from the needle piercing the arachnoid may chronically lower the intracranial pressure and lead to persistent intracranial hypotension.[69,97] This leak causes headache in approximately 30% of patients who undergo lumbar puncture. The headache usually begins 24 hours (range, 12 hours to 6 days) after the procedure and persists for 3 or 4 days or at times may be chronic. If the patient suffers a block to free passage of CSF by an epidural tumor, the headache may appear later, after relief of the epidural block by radiation.[101]

Post-lumbar puncture headache is usually absent when the patient is recumbent but begins seconds or minutes after the patient assumes an upright position. The pain usually starts in the occipital region and the neck, radiating to the shoulders and then to the forehead. The pain can be mild or severe; severe pain can be associated with autonomic symptoms, including pallor, sweating, nausea, and vomiting. The headache usually disappears promptly when the patient returns to the recumbent position. A Valsalva maneuver, a cough, or a sneeze, all of which increase intracranial pressure, may precipitate the headache, even if the patient is in the recumbent position. Requiring the patient to be flat for several hours after a lumbar puncture does not prevent post-lumbar puncture headache.[101a]

The headache that occurs when the erect position is assumed is believed to result in part from traction on pain-sensitive structures as the brain descends, unbuoyed by the CSF, and in part from compensatory vasodilatation, which also results from the decreased volume of CSF. In a few patients, additional symptoms include tinnitus, diplopia from abducens nerve palsy, vertigo, hearing loss,[97] and visual alterations.[102]

Persistent intracranial hypotension can cause subdural effusions that may become large and require surgical evacuation (Fig. 14–3). Patients who suffer persistent intracranial hypotension develop contrast enhancement of the pachymeninges identified on MR and occasionally on CT (Fig. 14–4).[103,104] Unlike the contrast enhancement of leptomeningeal

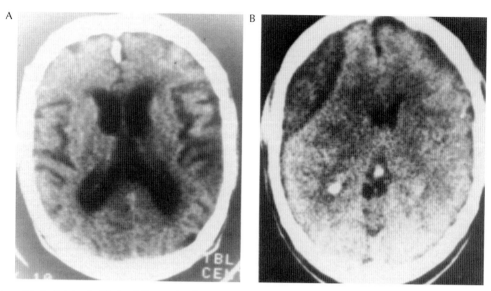

Figure 14–3. Subdural effusion following lumbar puncture. A patient with lung cancer developed slowly progressive dementia. Leptomeningeal metastases were suspected, although the CT (**A**) showed only cerebral atrophy and hydrocephalus ex vacuo. Results of a lumbar puncture were normal. The patient worsened several days after the lumbar puncture. A repeat CT (**B**) showed small ventricles compressed by large subdural effusions.

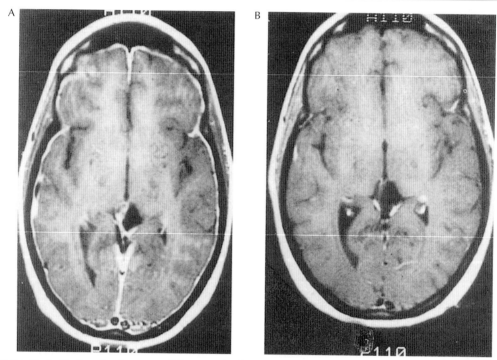

Figure 14–4. Meningeal enhancement following a lumbar puncture. MRIs demonstrating intense pachymeningeal enhancement following a lumbar puncture to evaluate possible leptomeningeal metastases in a patient with breast cancer. Persistent post-lumbar puncture headache led to an MRI that showed the intense diffuse pachymeningeal enhancement (**A**). The cerebrospinal fluid was otherwise normal. Several weeks later (**B**), after the headaches resolved spontaneously, the MRI showed no evidence of enhancement. Note that a dural reaction to diffuse calvarial metastases and dural invasion by metastatic tumor can give a similar picture. Note also that the enhancement is pachymeningeal, not leptomeningeal. (From Ref. 69 with permission.)

tumor, which is usually patchy and involves the leptomeninges (skull or dural metastases may lead to enhancement of the pachymeninges), the contrast enhancement following lumbar puncture is usually diffuse and does not involve the leptomeninges. If a patient with cancer undergoes a lumbar puncture for any reason and then develops headache leading to an MRI, meningeal enhancement may suggest the diagnosis of leptomeningeal tumor when the headache is only the result of the lumbar puncture–induced intracranial hypotension. Occasionally, meningeal enhancement can be observed after lumbar puncture in the absence of headache and other signs of intracranial hypotension, further obscuring the diagnosis. This diffuse pachymeningeal enhancement may also be present following ventriculoperitoneal shunting and can be evident on MRI for many years. Rarely, the disorder may lead to cortical vein thrombosis with hemorrhagic infarction.[69a]

For patients with prolonged postural headache, an epidural blood patch usually relieves the symptoms within a few hours.[104] At times, after immediate relief, the headache recurs; a second blood patch may be necessary.[105]

SPINAL HERNIATION

In occasional patients with asymptomatic epidural spinal cord compression, the lumbar puncture performed for diagnostic reasons or for spinal anesthesia may cause acute paraplegia, presumably from the sharp reduction of CSF pressure below the level of the epidural compression.[106–108] Typical is the patient with prostate carcinoma who may have vertebral metastases but shows no evidence of neurologic disability.[108] A lumbar puncture is performed for spinal anesthesia during prostate surgery. Postoperative numbness from the spinal anesthesia does not resolve, and the patient's condition progresses to paraplegia. A rare situation is paraplegia from upward spinal coning following ventricular shunting to relieve hydrocephalus.[109]

SPINAL HEMORRHAGE

Spinal hemorrhage,[110] probably resulting from the inadvertent puncture of vessels in the subdural space (Fig. 14–5), particularly affects patients who are thrombocytopenic,[111] with a platelet count lower than 20,000 or lower than

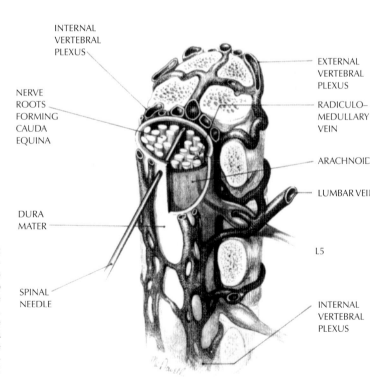

Figure 14–5. Spinal contents at the fourth and fifth lumbar vertebrae, showing the relationship of a lumbar puncture needle to the major vessels at this level. The major radiculomedullary vein, shown accompanying the L5 nerve root, is situated far laterally to a needle correctly positioned in the dural sac midline. Note the avascular subdural space. (From Ref. 110 with permission.)

INTERNAL VERTEBRAL PLEXUS

NERVE ROOTS FORMING CAUDA EQUINA

DURA MATER

SPINAL NEEDLE

EXTERNAL VERTEBRAL PLEXUS

RADICULO-MEDULLARY VEIN

ARACHNOID

LUMBAR VEIN

L5

INTERNAL VERTEBRAL PLEXUS

50,000 but dropping rapidly. Patients who are anticoagulated with heparin or warfarin and patients who have a dysfunctional clotting system, such as that caused by liver failure, or have dysfunction of platelets as seen in chronic renal failure, are also at risk. Most patients with such hemorrhages have suffered multiple attempts to enter the lumbar sac. The CSF may or may not be bloody. The patient characteristically tolerates the procedure well, but within a few hours complains of severe back pain that may radiate into the legs. The pain is followed by paresthesias, weakness of the legs, urinary retention, and constipation. In some instances, the subdural hematoma dissects upward, causing paraplegia with sensory and motor levels reaching as high as the upper thoracic area[111] (Fig. 9–3). Autopsy examination in such instances reveals a large subdural hematoma compressing the spinal cord and cauda equina, probably caused by laceration of radicular vessels by the lumbar puncture needle.[112] A similar problem occasionally follows lateral cervical puncture, causing quadriparesis and compromising respiratory muscles.[113,114]

A spinal MRI or CT visualizes the hematoma and establishes the diagnosis (Fig. 14–6). Generally, it is impossible to decompress the subdural hematoma surgically because the coagulopathy precludes surgery. When a lumbar puncture is performed in a patient at risk, the most experienced physician available should do it, using a needle no larger than 20 gauge. Although some patients remain paraplegic,

many recover neurologic function spontaneously following a hematoma.

MENINGITIS

Lumbar puncture almost never causes acute bacterial meningitis, at least in patients older than 1 year, even when performed in patients with known sepsis.[115,116] An exception is the patient harboring a skin or epidural abscess at the site of the needle puncture. Acute aseptic meningitis was rarely reported to follow lumbar puncture in the years that preceded the use of disposable spinal needles.

Arterial/Venous Catheters

Although infection (Chapter 10) is the most common complication of the placement of indwelling vascular catheters[117] (urokinase rinses reduce their incidence.[117a]) for the infusion of chemotherapy or other agents, neurologic complications also occur (Chapter 9). Procedures to place subclavian vein catheters may lead to bleeding and compression of the brachial plexus from hematomas. Intrathoracic hemorrhage following laceration of a vertebral artery requiring surgical repair has been reported. The local anesthesia for internal jugular vein catheterization may cause a transient brachial plexopathy.[118] A brachial plexopathy also has followed *Aspergillus* infection of a Hickman catheter.[119] Several

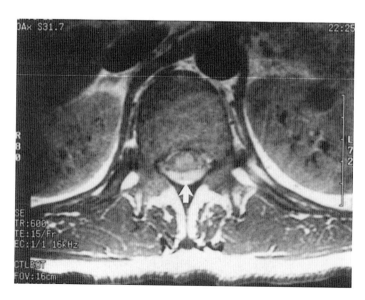

Figure 14–6. MR scan of a 35-year-old man with leukemia who underwent a lumbar puncture 3 days before this scan. Six hours after the lumbar puncture, he began to develop back pain that radiated toward the buttock. He did not develop weakness. The MR scan, however, reveals subdural blood posteriorly from T12 to L5 (*arrow*). The hyperintense area in the center is the normal conus medullaris.

cases of stroke temporally associated with subclavian or internal jugular vein catheterization have been reported in patients treated for cancer.[120–122] The stroke occurs in the anterior or posterior cerebral circulation and may be caused by paradoxic embolization from a clot on the end of the catheter or by inadvertent arterial damage with subsequent embolization during catheter placement. A Brown–Sequard syndrome due to spinal cord infarction has been reported after subclavian vein catheterization.[123] Similar complications can follow arterial catheterization for the performance of angiograms.

Air embolus from inadvertent disconnection of subclavian catheters has been reported. The negative pressure in the thorax allows air to enter the catheter when it is open to atmospheric pressure[121] (see Chapter 9).

Epidural/Intrathecal Catheters

Catheters placed in the epidural or subarachnoid space can be used for analgesia[124] or relief of spasticity.[125] Catheters may become displaced or fracture. Infections[126] and the development of granulomas[124,127] are uncommon but serious complications.

REFERENCES

1. Aitkenhead AR. Injuries associated with anaesthesia. A global perspective. *Br J Anaesth* 2005;95:95–109.
2. Hindman BJE. *Neurological and Psychological Complications of Surgery and Anesthesia*. Boston: Little, Brown, 1986.
3. Silverstein JH, Timberger M, Reich DL, et al. Central nervous system dysfunction after noncardiac surgery and anesthesia in the elderly. *Anesthesiology* 2007;106:622–628.
4. Winawer N. Postoperative delirium. *Med Clin North Am* 2001;85:1229–1239.
5. Bekker AY, Weeks EJ. Cognitive function after anaesthesia in the elderly. *Best Pract Res Clin Anaesthesiol* 2003;17:259–272.
6. Dasgupta M, Dumbrell AC. Preoperative risk assessment for delirium after noncardiac surgery: a systematic review. *J Am Geriatr Soc* 2006;54:1578–1589.
7. Mesulam M-M, Geschwind N. Disordered mental states in the postoperative period. *Urol Clin North Am* 1976;3:199–215.
8. Wilson LM. Intensive care delirium. The effect of outside deprivation in a windowless unit. *Arch Intern Med* 1972;130:225–226.
9. Dyson M. Intensive care unit psychosis, the therapeutic nurse-patient relationship and the influence of the intensive care setting: analyses of interrelating factors. *J Clin Nurs* 1999;8:284–290.
10. Brown DV, Heller F, Barkin R. Anticholinergic syndrome after anesthesia: a case report and review. *Am J Therapeut* 2004;11:144–153.
11. Devinsky O, Bear D, Volpe BT. Confusional states following posterior cerebral artery infarction. *Arch Neurol* 1988;45:160–163.
12. Mori E, Yamadori A. Acute confusional state and acute agitated delirium. Occurrence after infarction in the right middle cerebral artery territory. *Arch Neurol* 1987;44:1139–1143.
13. Lonergan E, Britton AM, Luxenberg J, et al. Antipsychotics for delirium. *Cochrane Database Syst Rev* 2007;2:CT 005594.
14. Siddiqi N, Stockdale R, Britton AM, et al. Interventions for preventing delirium in hospitalised patients. *Cochrane Database Syst Rev* 2007:CD005563.
15. Perouansky M. Liaisons dangereuses? General anaesthetics and long-term toxicity in the CNS. *Eur J Anaesthesiol* 2007;24:107–115.
16. Sawyer RJ, Richmond MN, Hickey JD, et al. Peripheral nerve injuries associated with anaesthesia. *Anaesthesia* 2000;55:980–991.
17. Winfree CJ, Kline DG. Intraoperative positioning nerve injuries. *Surg Neurol* 2005;63:5–18.
18. Malamut RI, Marques W, England JD, et al. Postsurgical idiopathic brachial neuritis. *Muscle Nerve* 1994;17:320–324.
19. Al-Hakim M, Katirji MB. Femoral mononeuropathy induced by the lithotomy position: a report of 5 cases with a review of the literature. *Muscle Nerve* 1993;16:891–895.
20. Lachmann EA, Rook JL, Tunkel R, et al. Complications associated with intermittent pneumatic compression. *Arch Phys Med Rehab* 1992;73:482–485.
21. Kam PC, Cardone D. Propofol infusion syndrome. *Anaesthesia* 2007;62:690–701.
22. Brull R, McCartney CJ, Chan VW, et al. Neurological complications after regional anesthesia: contemporary estimates of risk. *Anesth Analg* 2007;104:965–974.
23. Ruppen W, Derry S, McQuay H, et al. Incidence of epidural hematoma, infection, and neurologic injury in obstetric patients with epidural analgesia/anesthesia. *Anesthesiology* 2006;105:394–399.
24. Tiso RL, Thomas PS, Macadaeg K. Adverse central nervous system sequelae after selective transforaminal block: the role of corticosteroids. *Spine* 2004;4:468–474.
25. Watts SA, Sharma DJ. Long-term neurological complications associated with surgery and peripheral nerve blockade: outcomes after 1065 consecutive blocks. *Anaesth Intensive Care* 2007;35:24–31.
26. Cinar SO, Seven H, Cinar U, et al. Isolated bilateral paralysis of the hypoglossal and recurrent laryngeal nerves (bilateral Tapia's syndrome) after transoral intubation for general anesthesia. *Acta Anaesthesiol Scand* 2005;49:98–99.
27. Sawaya R, Hammoud M, Schoppa D, et al. Neurosurgical outcomes in a modern series of 400 craniotomies for treatment of parenchymal tumors. *Neurosurgery* 1998;42:1044–1055.
28. Paek SH, Audu PB, Sperling MR, et al. Reevaluation of surgery for the treatment of brain metastases: review of 208 patients with single or multiple brain metastases treated at one institution with modern neurosurgical techniques. *Neurosurgery* 2005;56:1021–1034.

29. Brisman MH, Bederson JB, Sen CN, et al. Intracerebral hemorrhage occurring remote from the craniotomy site. *Neurosurgery* 1996;39:1114–1121.

30. Keiper GL Jr, Sherman JD, Tomsick TA, et al. Dural sinus thrombosis and pseudotumor cerebri: Unexpected complications of suboccipital craniotomy and translabyrinthine craniectomy. J Neurosurg 1999; 91:192–197.

31. Lega BC, Yoshor D. Postoperative dural sinus thrombosis in a patient in a hypercoagulable state. Case report. *J Neurosurg* 2006;105:772–774.

32. Rapaná A, Lamaida E, Pizza V. Multiple postoperative intracerebral haematomas remote from the site of craniotomy. *Br J Neurosurg* 1998;12:364–368.

33. Koller M, Ortler M, Langmayr J, et al. Posterior-fossa haemorrhage after supratentorial surgery—report of three cases and review of the literature. *Acta Neurochir* (Wien) 1999;141:587–592.

34. Khan RB, Gutin PH, Rai SN, et al. Use of diffusion weighted magnetic resonance imaging in predicting early postoperative outcome of new neurological deficits after brain tumor resection. *Neurosurgery* 2006;59:60–66.

35. Keiper GL, Jr, Sherman JD, Tomsick TA, et al. Dural sinus thrombosis and pseudotumor cerebri: unexpected complications of suboccipital craniotomy and translabyrinthine craniectomy. *J Neurosurg* 1999;91:192–197.

36. Blomstedt GC. Craniotomy infections. *Neurosurg Clin N Am* 1992;3:375–385.

37. Temkin NR. Prophylactic anticonvulsants after neurosurgery. *Epilepsy Curr* 2002;2:105–107.

38. Temkin NR. Antiepileptogenesis and seizure prevention trials with antiepileptic drugs: meta-analysis of controlled trials. *Epilepsia* 2001;42:515–524.

39. Manaka S, Ishijima B, Mayanagi Y. Postoperative seizures: epidemiology, pathology, and prophylaxis. *Neurol Med Chir* 2003;43:589–600.

40. Carmel PW, Greif LK. The aseptic meningitis syndrome: a complication of posterior fossa surgery. *Pediatr Neurosurg* 1993;19(5):276–280.

41. Finlayson AI, Penfield W. Acute postoperative aseptic leptomeningitis: review of cases and discussion of pathogenesis. *Arch Neurol Psychiatry* 1941;46:250–276.

42. Sanchez GB, Kaylie DM, O'Malley MR, et al. Chemical meningitis following cerebellopontine angle tumor surgery. *Otolaryngol Head Neck Surg* 2008;138:368–373.

43. Druel B, Vandenesch F, Greenland T, et al. Aseptic meningitis after neurosurgery: a demonstration of bacterial involvement. *Clinical Microbiol Iinfect* 1996;1:230–234.

44. Sprague A, Poulgrain P. Tension pneumocephalus: a case report and literature review. *J Clin Neurosci* 1999;6:418–424.

45. Clevens RA, Marentette LJ, Esclamado RM, et al. Incidence and management of tension pneumocephalus after anterior craniofacial resection: case reports and review of the literature. *Otolaryngol Head Neck Surg* 1999;120:579–583.

46. Arbit E, Shah J, Bedford R, et al. Tension pneumocephalus: treatment with controlled decompression via a closed water-seal drainage system. Case report. *J Neurosurg* 1991;74:139–142.

47. Posner JB, Saper CS, Schiff ND, et al. Diagnosis of Stupor and Coma, 4th ed. Oxford University press, New York.

48. Biyani N, Silbiger A, Ben-Ari J, et al. Postoperative brain stem tension pneumocephalus causing transient locked-in syndrome. *Pediatr Neurosurg* 2007;43:414–417.

49. Beck RW, Greenberg HS. Post-decompression optic neuropathy. *J Neurosurg* 1985;63:196–199.

50. Talke PO, Gelb AW. Postcraniotomy pain remains a real headache! *Eur J Anaesthesiol* 2005;22:325–327.

51. de Gray LC, Matta BF. Acute and chronic pain following craniotomy: a review. *Anaesthesia* 2005;60:693–704.

52. Roberts GC. Post-craniotomy analgesia: current practices in British neurosurgical centres—a survey of post-craniotomy analgesic practices. *Eur J Anaesthesiol* 2005;22:328–332.

53. Schaller B, Baumann A. Headache after removal of vestibular schwannoma via the retrosigmoid approach: a long-term follow-up-study. *Otolaryngol Head Neck Surg* 2003;128:387–395.

54. Rimaaja T, Haanpaa M, Blomstedt G, et al. Headaches after acoustic neuroma surgery. *Cephalalgia* 2007; 27:1128–1135.

55. Soumekh B, Levine SC, Haines SJ, et al. Retrospective study of postcraniotomy headaches in suboccipital approach: Diagnosis and management. Am J Otol 1996;17:617–619.

56. Ozgur BM, Berberian J, Aryan HE, et al. The pathophysiologic mechanism of cerebellar mutism. *Surg Neurol* 2006; 66:18–25.

57. Schmahmann JD, Sherman JC. The cerebellar cognitive affective syndrome. *Brain* 1998;121:561–579.

58. Robertson PL, Muraszko KM, Holmes EJ, et al. Incidence and severity of postoperative cerebellar mutism syndrome in children with medulloblastoma: a prospective study by the Children's Oncology Group. *J Neurosurg* 2006;105:444–451.

59. Schiffer J, Gur R, Nisim U, et al. Symptomatic patients after craniectomy. *Surgical Neurol* 1997;47:231–237.

60. Winkler PA, Stummer W, Linke R, et al. The influence of cranioplasty on postural blood flow regulation, cerebrovascular reserve capacity, and cerebral glucose metabolism. *Neurosurg Focus* 2000;8:1–9.

61. Guerrero R, Pumar A, Soto A, et al. Early hyponatraemia after pituitary surgery: cerebral salt-wasting syndrome. *Eur J Endocrinol* 2007;156:611–616.

62. Hassan ZU, Kruer JJ, Fuhrman TM. Electrolyte changes during craniotomy caused by administration of hypertonic mannitol. *J Clin Anesth* 2007; 19:307–309.

63. Gutenberg A, Lange B, Gunawan B, et al. Spontaneous adrenal hemorrhage: a little-known complication of intracranial tumor surgery. Case report. *J Neurosurg* 2007;106:1086–1088.

64. Bentz BG, Bilsky MH, Shah JP, et al. Anterior skull base surgery for malignant tumors: a multivariate analysis of 27 years of experience. *Head Neck* 2003;25:515–520.

65. Deschler DG, Gutin PH, Mamelak AN, et al. Complications of anterior skull base surgery. *Skull Base Surg* 1996;6:113–118.

66. Ganly I, Gross ND, Patel SG, et al. Outcome of craniofacial resection in patients 70 years of age and older. *Head Neck* 2007;29:89–94.

67. Persky MS, Brunner E, Cooper PR, et al. Perioperative complications of transseptosphenoidal excision for pituitary adenomas. *Skull Base Surg* 1996; 6:231–235.

68. Bilsky MH, Fraser JF. Complication avoidance in vertebral column spine tumors. *Neurosurg Clin N Am* 2006;17:317–29, vii.

69. Panullo SC, Reich JB, Krol G, et al. MRI changes in intracranial hypotension. *Neurology* 1993;43:919–926.

69a. Cornips EM, Staals J, Stavast A, et al. Fatal cerebral and cerebellar hemorrhagic infarction after thoracoscopic microdiscectomy. Case report. *J Neurosurg Spine* 2007;6:276–9.

70. Halfon MJ, Bonardo P, Valiensi S, et al. Central retinal artery occlusion and ophthalmoplegia following spinal surgery. *Br J Ophthalmol* 2004;88:1350–1352.

71. Lennon M, Seigne P, Cunningham AJ. Pituitary apoplexy after spinal anaesthesia. *Br J Anaesth* 1998;81:616–618.

72. Rigual NR, Wiseman SM. Neck dissection: current concepts and future directions. *Surg Oncol Clin N Am* 2004;13:151–166.

73. Swift TR. Involvement of peripheral nerves in radical neck dissection. *Am J Surg* 1970;119:694–698.

74. Warpeha RL. Head and neck surgery. *Surg Clin North Am* 1977;57:1357–1363.

75. Myers EN, Conley J. Gustatory sweating after radical neck dissection. *Arch Otolaryngol* 1970;91:534–542.

76. Witt RL, Rejto L. Spinal accessory nerve monitoring in selective and modified neck dissection. *Laryngoscope* 2007;117:776–780.

77. Portenoy RK. Cancer pain: pathophysiology and syndromes. *Lancet* 1992;339:1026–1031.

78. Vecht CJ, Hoff AM, Kansen PJ, et al. Types and causes of pain in cancer of the head and neck. *Cancer* 1992;70:178–184.

79. Leikensohn J, Milko D, Cotton R. Carotid artery rupture. Management and prevention of delayed neurologic sequelae with low-dose heparin. *Arch Otolaryngol* 1978;104:307–310.

80. Perkins FM, Kehlet H. Chronic pain as an outcome of surgery. A review of predictive factors. *Anesthesiology* 2000;93:1123–1133.

81. Abdullah TI, Iddon J, Barr L, et al. Prospective randomized controlled trial of preservation of the intercostobrachial nerve during axillary node clearance for breast cancer. *Br J Surg* 1998;85:1443–1445.

82. Paredes JP, Puente JL, Potel J. Variations in sensitivity after sectioning the intercostobrachial nerve. *Am J Surg* 1990;160:525–528.

83. Vecht CJ, Van de Brand HJ, Wajer OJ. Post-axillary dissection pain in breast cancer due to a lesion of the intercostobrachial nerve. *Pain* 1989;38:171–176.

84. Dijkstra PU, Rietman JS, Geertzen JH. Phantom breast sensations and phantom breast pain: a 2-year prospective study and a methodological analysis of literature. *Eur J Pain* 2007;11:99–108.

85. Jung BF, Ahrendt GM, Oaklander AL, et al. Neuropathic pain following breast cancer surgery: proposed classification and research update. *Pain* 2003;104:1–13.

86. Baron RH, Fey JV, Borgen PI, et al. Eighteen sensations after breast cancer surgery: a two-year comparison of sentinel lymph node biopsy and axillary lymph node dissection. *Oncol Nurs Forum* 2004;31:691–698.

87. Perttunen K, Tasmuth T, Kalso E. Chronic pain after thoracic surgery: a follow-up study. *Acta Anaesthesiol Scand* 1999;43:563–567.

88. Landreneau RJ, Wiechmann RJ, Hazelrigg SR, et al. Effect of minimally invasive thoracic surgical approaches on acute and chronic postoperative pain. *Chest Surg Clin N Am* 1998;8:891–906.

89. Bilsky MH, Downey RJ, Kaplitt MG, et al. Tension pneumocephalus resulting from iatrogenic subarachnoid-pleural fistulae: report of three cases. *Ann Thoracic Surg* 2001;71:455–457.

90. Heier MS, Aass N, Ous S, et al. Asymmetrical autonomic dysfunction of the feet after retroperitoneal surgery in patients with testicular cancer: 2 case reports. *J Urol* 1992;147:470–471.

91. Ovesen P, Kroner K, Ornsholt J, et al. Phantom-related phenomena after rectal amputation: prevalence and clinical characteristics. *Pain* 1991; 44:289–291.

92. Marchettini P, Formaglio F, Lacerenza M. Iatrogenic painful neuropathic complications of surgery in cancer. *Acta Anaesthesiol Scand* 2001;45:1090–1094.

93. Mishra S, Bhatnagar S, Gupta D, et al. Incidence and management of phantom limb pain according to World Health Organization analgesic ladder in amputees of malignant origin. *Am J Hosp Palliat Care* 2007;24:455–462.

94. Khattab J, Terebelo HR, Dabas B. Phantom limb pain as a manifestation of paclitaxel neurotoxicity. *Mayo Clin Proc* 2000;75:740–742.

95. Bilazarian SD, Mittal S, Mills RM, Jr. Recognizing the extrarenal hazards of intravascular contrast agents. *J Crit Illness* 1991;6:859–869.

96. Pagani JJ, Hayman LA, Bigelow RH, et al. Diazepam prophylaxis of contrast media-induced seizures during computed tomography of patients with brain metastases. *AJNR Am J Neuroradiol* 1983;140:787–792.

97. Evans RW. Complications of lumbar puncture. *Neurol Clin* 1998;16:83–105.

98. Duffy GP. Lumbar puncture in the presence of raised intracranial pressure. *BMJ* 1969;1:407–409.

99. Van Crevel H, Hijdra A, De Gans J. Lumbar puncture and the risk of herniation: when should we first perform CT? *J Neurol* 2002;249:129–137.

100. Joffe AR. Lumbar puncture and brain herniation in acute bacterial meningitis: a review. *J Intensive Care Med* 2007;22:194–207.

101. Parlow JL, Einarson DW. An unusual cause of delayed postmyelogram headache. *Anesthesiology* 1991;75:145–146.

101a. Tejavanija S, Sithinamsuwan P, Sithinamsuwan N, et al. Comparison of prevalence of post-dural puncture headache between six hour-supine recumbence and early ambulation after lumbar puncture in Thai patients: A randomized controlled study. *J Med Assoc Thai* 2006;89:814–20.

102. Horton JC, Fishman RA. Neurovisual findings in the syndrome of spontaneous intracranial hypotension from dural cerebrospinal fluid leak. *Ophthalmology* 1994;101:244–251.

103. Mokri B. Low cerebrospinal fluid pressure syndromes. *Neurol Clin* 2004;22:55–74.

104. Ahmed SV, Jayawarn C, Jude E. Post lumbar puncture headache: diagnosis and management. *Postgrad Med J* 2006;82:713–716.

105. Eustace N, Hennessy A, Gardiner J. The management of dural puncture in obstetrics and the efficacy of epidural blood patches. *Ir Med J* 2004; 97:298–300.

106. Graham GP, Dent CM, Mathews P. Paraplegia following spinal anaesthesia in a patient with prostatic metastases. *Br J Urol* 1992;70:445.

107. Hollis PH, Malis LI, Zappulla RA. Neurological deterioration after lumbar puncture below complete spinal subarachoid block. *J Neurosurg* 1986;64:253–256.

108. Mutoh S, Aikou I, Ueda S. Spinal coning after lumbar puncture in prostate cancer with asymptomatic vertebral metastasis: a case report. *J Urol* 1991;145:834–835.

109. Jooma R, Hayward RD. Upward spinal coning: impaction of occult spinal tumours following relief of hydrocephalus. *J Neurol Neurosurg Psychiatry* 1984;47:386–390.

110. Edelson RN, Chernik NL, Posner JB. Spinal subdural hematomas complicating lumbar puncture. *Arch Neurol* 1974;31:134–137.

111. Wirtz PW, Bloem BR, van der Meer FJ, et al. Paraparesis after lumbar puncture in a male with leukemia. *Pediatr Neurol* 2000;23:67–68.

112. Masdeu JC, Breuer AC, Schoene WC. Spinal subarachnoid hematomas: clue to a source of bleeding in traumatic lumbar puncture. *Neurology* 1979;29:872–876.

113. Abla AA, Rothfus WE, Maroon JC, et al. Delayed spinal subarachnoid hematoma: a rare complication of C1-C2 cervical myelography. *AJNR Am J Neuroradiol* 1986;7:526–528.

114. Mapstone TB, Rekate HL, Shurin SB. Quadriplegia secondary to hematoma after lateral C-1, C-2 puncture in a leukemic child. *Neurosurgery* 1983;12:230–231.

115. Yaniv LG, Postasman I. Iatrogenic meningitis: Increasing role for resistant viridans streptococci? Case report and review of the last 20 years. *Scand J Infect Dis* 2000;32:693–696.

116. Baer ET. Post-dural puncture bacterial meningitis. *Anesthesiology* 2006;105:381–393.

117. Raad I, Hanna H, Maki D. Intravascular catheter-related infections: advances in diagnosis, prevention, and management. *Lancet Infect Dis* 2007;7:645–657.

117a. van Rooden CJ, Schippers EF, Guiot HF, et al. Prevention of coagulase-negative staphylococcal central venous catheter-related infection using urokinase rinses: a randomized double-blind controlled trial in patients with hematologic malignancies. *J Clin Oncol* 2008;26:428–33.

118. Sylvestre DL, Sandson TA, Nachmanoff DB. Transient brachial plexopathy as a complication of internal jugular vein cannulation. Letter to the editor. *Neurology* 1991;41:760.

119. Krol TC, O'Keefe P. Brachial plexus neuritis and fatal hemorrhage following Aspergillus infection of a Hickman catheter. *Cancer* 1982;50:1214–1217.

120. Hurwitz BJ, Posner JB. Cerebral infarction complicating subclavian vein catheterization. *Ann Neurol* 1977;1:253–254.

121. Ploner F, Saltuari L, Marosi MJ, et al. Cerebral air emboli with use of central venous catheter in mobile patient. Letter to the editor. *Lancet* 1991;338:1331.

122. Sloan MA, Mueller JD, Adelman LS, et al. Fatal brainstem stroke following internal jugular vein catherization. *Neurology* 1991;41:1092–1095.

123. Koehler PJ, Wijngaard PR. Brown-Sequard syndrome due to spinal cord infarction after subclavian vein catheterisation. Letter to the editor. *Lancet* 1986;2:914–915.

124. Cabbell KL, Taren JA, Sagher O. Spinal cord compression by catheter granulomas in high-dose intrathecal morphine therapy: case report. *Neurosurgery* 1998;42:1176–1180.

125. Murphy PM, Skouvaklis DE, Amadeo RJ, et al. Intrathecal catheter granuloma associated with isolated baclofen infusion. *Anesth Analg* 2006;102:848–852.

126. Okano K, Kondo H, Tsuchiya R, et al. Spinal epidural abscess associated with epidural catheterization: report of a case and a review of the literature. *Jpn J Clin Oncol* 1999;29:49–52.

127. Bejjani GK, Karim NO, Tzortzidis F. Intrathecal granuloma after implantation of a morphine pump: case report and review of the literature. *Surg Neurol* 1997;48:288–291.

Chapter 15

Paraneoplastic Syndromes

INTRODUCTION

In patients with cancer, all neurologic abnormalities not caused by the cancer's spread to the nervous system are, by definition, paraneoplastic. Most physicians, however, use this term to refer to a group of disorders, also called "remote effects of cancer on the nervous system,"[1–3] that are not caused by those non-metastatic complications of cancer considered in the preceding five chapters. The paraneoplastic disorders differ from other non-metastatic complications in several respects:

1. The neurologic symptoms of paraneoplastic syndromes usually precede the identification of the cancer, whereas neurologic symptoms of other non-metastatic disorders usually occur in patients with known cancer.
2. Even when the paraneoplastic-related cancer is identified, it is often small,

non-metastatic, and grows indolently, in contradistinction to the other non-metastatic complications of cancer, in which the tumors are often large and widely disseminated. Rarely, there are spontaneous regressions.[4–6,6a]

3. The neurologic disability caused by paraneoplastic syndromes is often profound in the absence of any other cancer symptoms. In the other non-metastatic disorders, systemic disability is usually prominent and often overshadows the neurologic dysfunction.
4. Paraneoplastic syndromes are often irreversible but, as in other non-metastatic disorders, treating the cancers often ameliorates or stabilize the neurologic symptomatology. A few paraneoplastic syndromes sometimes reverse spontaneously.[6a]

These distinctions are not absolute; some disorders initially classified as paraneoplastic syndromes were reclassified once another cause was identified. For example, progressive multifocal leukoencephalopathy (PML) was once considered a typical remote effect of cancer on the nervous system, but is now recognized to be an opportunistic viral infection (see Chapter 10). Other poorly understood disorders currently classified as paraneoplastic syndromes may eventually be reassigned as having a nutritional or infectious origin.

For the clinician, paraneoplastic syndromes affecting the nervous system can be divided into two large groups. The first group, *classic* paraneoplastic syndromes, are disorders that, when present, strongly suggest an underlying cancer. These include the Lambert–Eaton myasthenic syndrome (LEMS),[7] opsoclonus/myoclonus in adults and children,[8,9] and subacute cerebellar degeneration[10,11] (Table 15–1).

The second group of disorders consists of clinical syndromes sometimes associated with cancer but more often appearing in the absence of a neoplasm. These include inflammatory myositis (dermatomyositis, polymyositis),[12,13] particularly in children and young adults, amyotrophic lateral sclerosis (ALS),[14,15] and sensorimotor polyneuropathy.[16] Although some of these patients may have an associated cancer, an extensive search for a neoplasm often proves fruitless.

GENERAL CONSIDERATIONS

Classification

Paraneoplastic neurologic syndromes (PNS) can affect any portion of the nervous system, from the cerebral cortex through the brainstem and spinal cord to peripheral nerves, neuromuscular junction, and muscle (Table 15–1). Clinical symptomatology can reflect pathologic damage to a single specific structure in the nervous system (e.g., to cerebellar Purkinje cells as in paraneoplastic cerebellar degeneration [PCD] or to the cholinergic synapse as in LEMS) or can reflect damage to multiple areas of the nervous system simultaneously, as in encephalomyelitis associated with cancer.[17] In some clinically florid paraneoplastic syndromes, such as opsoclonus/myoclonus, the nervous system may be free of any identifiable pathologic change.[18]

Table 15–1. Paraneoplastic Syndromes of the Nervous System

Brain and Cranial Nerves

  Encephalomyelitis°
  Limbic encephalitis°
  Brainstem encephalitis
  Cerebellar degeneration°
  Opsoclonus–myoclonus°
  Visual syndromes
    Cancer-associated retinopathy
    Melanoma-associated retinopathy
    Optic neuropathy
  Chorea
  Parkinsonism

Spinal Cord

  Necrotizing myelopathy
  Inflammatory myelitis
  Motor neuron disease (ALS)
  Subacute motor neuronopathy°
  Stiff-person syndrome

Dorsal Root Ganglia

  Sensory neuronopathy°

Peripheral Nerves

  Autonomic neuropathy
    Chronic gastrointestinal
      pseudo-obstruction°
  Acute sensorimotor neuropathy
    Polyradiculopathy (Guillain–Barré)
    Brachial neuropathy
  Chronic sensorimotor neuropathy
  Vasculitic neuropathy
  Neuromyotonia

Neuromuscular Junction

  Lambert–Eaton myasthenic syndrome°
  Myasthenia gravis

Muscle

  Polymyositis
  Dermatomyositis°
  Necrotizing myopathy
  Myotonia

° Classic paraneoplastic syndromes.

Frequency

Several studies have addressed the frequency of paraneoplastic syndromes (Table 15–2).[16,19–23] Data from these studies depend on the definitions of the syndromes, the rigor used to exclude other causes of neurologic dysfunction, and the care with which the neurologic evaluation was performed. For example, typical LEMS occurs in approximately 3% of patients with small cell lung cancer (SCLC)[24] (the cancer that, along with ovarian cancer and thymoma, is believed to have the highest frequency of paraneoplastic disorders), but less well-defined neuromuscular dysfunction, including subjective or objective muscle weakness, affects 44% of patients with SCLC.[20] Croft and Wilkinson[19] reported that 16% of 250 patients with lung cancer had neuromyopathy, an illness characterized by proximal muscular weakness with one or more diminished or absent deep tendon reflexes. Approximately 10% of patients with malignant monoclonal gammopathy have an associated peripheral neuropathy,[25] as do approximately 50% of patients with osteosclerotic myeloma.[26] Antoine et al.[16] examined 422 consecutive patients with peripheral neuropathy; 26 (6%) had cancer. The likelihood of cancer was 9% in those over the age of 50 and 47% in those who suffered from a sensory neuropathy. On the other hand, Hawley et al.[27] found that most peripheral neuropathy in patients with SCLC could be related to weight loss and nutritional disturbances and thus was not truly paraneoplastic, as defined for this chapter. Croft and Wilkinson[19] found that 6.6% of 1476 patients with any type of cancer had a neuromyopathy as assessed by physical examination; however, Lipton and associates[28,29] found abnormalities of peripheral nerve function by quantitative thermal threshold testing in 43% of 29 patients with cancer, and Gomm et al.[21] found myopathic changes on muscle biopsy in 33 of 100 patients with lung cancer. Croft and Wilkinson[19] found only 2 instances of cerebellar degeneration in 319 patients with lung cancer and only 3 instances in 1476 patients with any type of cancer. Erlington and coworkers[20] found ataxia or nystagmus in 5 of 150 patients with SCLC, whereas Wessel et al.[30] reported cerebellar signs by posturographic analysis in 13 of 50 patients with lung cancer.[30] Although

Table 15–2. **Frequency of Paraneoplastic Syndromes in Several Studies**

No. of Patients with Cancer Examined	Type of Cancer	Type of Examination	Neurologic Problem Found	Percentage of Patients with Cancer and Neurologic Signs	Reference
1465	Any	Clinical	Neuromyopathy	6.6	19
1465	Any	Clinical	Cerebellar	0.2	19
150	SCLC	Clinical, EMG	LEMS	2.0	20
150	SCLC	Clinical	Weakness	44.0	20
171	Any	Sensory, clinical	Sensory neuropathy	12.0	28
50	Lung	Postural testing	Cerebellar	26.0	30
100	Lung	Muscle biopsy	Neuromuscular	33.0	21
641	SCLC	Clinical, EMG	LEMS	0.3	31
3843	Lung	Clinical	Encephalomyelitis	0.36	3
3843	Lung	Clinical	Peripheral neuropathy	0.7	3
908	Ovary	Clinical	Cerebellar degeneration	0.1	22

SCLC, small cell lung cancer; EMG, electromyogram; LEMS, Lambert–Eaton myasthenic syndrome.

such conflicting figures make it difficult to estimate the incidence of true neurologic paraneoplastic syndromes, our experience indicates that clinically significant classic paraneoplastic syndromes probably occur in fewer than 0.1% of patients with cancer.

The low incidence of classic neurologic paraneoplastic syndromes should not lead the physician to believe that patients with cancer do not endure disabling symptoms that may be paraneoplastic and related to nervous system dysfunction. As Table 15–3 indicates, almost one-half of patients with SCLC suffer symptoms that appear to involve the neuromuscular system. Table 4–1 (see Chapter 4) lists various symptoms in 922 patients with advanced cancer. Symptoms such as fatigue, anorexia, drowsiness, and concentration difficulties may very well be neurogenic in origin and possibly are paraneoplastic.

Because of the low incidence of classic paraneoplastic syndromes in patients with cancer, the oncologist is unlikely to encounter even one patient each year with a paraneoplastic syndrome. Thus, if a patient with cancer complains of neurologic symptomatology, other causes should be assiduously sought. On the other hand, the situation is quite different in a patient without known cancer who develops neurologic symptoms. In this situation, if the cause of the neurologic disability is not immediately obvious, and particularly if the patient presents with one of the classic paraneoplastic syndromes (see Table 15–1), the likelihood that the patient has cancer is considerable. Table 15–4 gives estimates of the likelihood that a patient presenting with a specific neurologic symptom has a cancer as the underlying cause. With the exception of LEMS, the data are not well established but represent our own estimates or those of others. Nevertheless, these estimates should encourage the physician to search extensively for cancer in patients with appropriate neurologic syndromes.

Importance

Although neurologic paraneoplastic syndromes are rare, their recognition by the physician is important for several reasons. First, in about two-thirds of patients, neurologic symptoms precede and prompt the diagnosis of systemic

Table 15–3. Neuromuscular and Somatic Clinical Features of 150 Patients with Small Cell Lung Cancer

Clinical Features	Percentage of Total
Symptoms	
Anorexia	53
Weight loss	51
Erectile impotence	44
Dry mouth	41
Weakness	31°
Sphincter disturbance	24
Sweating change	21
Visual change	6
Physical Signs	
Inability to rise from squatting position	21°
Sensory change	16
Brisk reflexes	13
Absent reflexes	10
Diminished reflexes	9
Weakness[†]	8°
Ataxia or nystagmus	5
Post-tetanic potentiation	3

From Ref. 20 with permission.
°One or more measures of weakness were present in 44% of subjects.
[†]Excluding inability to rise from squatting position.

Table 15–4. Estimated Likelihood that a Given Neurologic Disorder is a Paraneoplastic Syndrome

Syndrome	Percentage
Lambert–Eaton myasthenic syndrome	60
Subacute cerebellar degeneration	50
Subacute sensory neuronopathy	20
Opsoclonus/myoclonus (children)°	40
Opsoclonus/myoclonus	20
Myasthenia gravis	15
Sensorimotor peripheral neuropathy	10
Encephalomyelitis	10
Dermatomyositis	10

°All other percentages refer to adults.

cancer. Because specific paraneoplastic syndromes are associated with particular cancers, the diagnosis of a paraneoplastic syndrome can direct the search for the underlying malignancy to particular organs, such as the lung for LEMS and the ovaries for PCD. Second, in patients known to have cancer, paraneoplastic syndromes may be confused with the metastatic or the non-metastatic complications described in the previous chapter. Because, with some exceptions, there is no effective treatment for most paraneoplastic syndromes, other potentially reversible causes of neurologic dysfunction must be carefully excluded before a diagnosis of a paraneoplastic syndrome is made. Third, in many cases, the neurologic syndrome has its onset when the cancer is small and curable. Fourth, if the diagnosis is made early, before symptoms are disabling, treatment of the cancer may stabilize or reverse the neurologic illness. Finally, the identification of antibodies (see following text) associated with several paraneoplastic syndromes gives clues to the underlying etiology of these rare disorders and contributes to our understanding of the interactions of systemic cancers, immunologic mechanisms, and the nervous system.

Pathogenesis

Although the etiology of most neurologic paraneoplastic syndromes is unknown, several potential mechanisms have been proposed (Table 15–5).

In 1888, Oppenheim[32] proposed that some neurologic disorders associated with cancer were caused by a toxic substance released by the tumor. That tumors can secrete substances that interfere with central nervous system (CNS) function is now well established. The examples in Table 15–5 are peptide hormones that cause secondary neurologic dysfunction.[33–35] Cytokines such as tumor necrosis factor or IL-1 and IL-6, secreted by the tumor or by immune cells reacting to the tumor, may cause the cachexia[36] (see Fig. 4–1), fatigability, asthenia, muscle catabolism, and general sensation of weakness that can affect most patients with cancer during some stage (see Table 4–1).[37]

In 1948, Denny-Brown,[38] when describing probably the first cases of anti-Hu positive

Table 15–5. **Possible Pathogenesis of Paraneoplastic Syndromes**

Hypothesis	Example
Toxin secreted by tumor	ACTH → Cushing syndrome
	PTHRP → hypercalcemia
Competition for essential substrate	Carcinoid tumors compete with brain for tryptophan → pellagra-like syndrome
Opportunistic infection	Papovavirus → progressive multifocal leukoencephalopathy
Autoimmune process	Lambert–Eaton myasthenic syndrome

ACTH, adrenocorticotropic hormone; PTHRP, parathyroid hormone-related protein.

subacute sensory neuronopathy (see p 597), noted a similarity between the dorsal root ganglionitis in his patients and that seen in swine deprived of pantothenic acid. He suggested that the malignancy and the nervous system were competing for a vital nutrient. Metastatic carcinoid tumors appear to cause neurologic symptoms by such a mechanism,[39,40] but no evidence is available to show that small and occult cancers, such as those usually encountered in most paraneoplastic disorders, deprive the nervous system of any essential substrate.

Opportunistic viral infections involving the CNS complicate the clinical course of many patients immunosuppressed by tumor such as lymphomas or by chemotherapy (see Chapter 12). PML, originally classified as a remote effect,[2] is one such infection (see Chapter 10). Subacute motor neuronopathy (see following text) may be another. However, most paraneoplastic syndromes affect patients who are not immunosuppressed, making opportunistic infection unlikely unless these patients suffer unrecognized isolated immune dysfunction. The absence of common opportunistic infections such as PML or Herpes zoster also indicates that most patients with paraneoplastic syndromes are immunocompetent.

The current concept of the pathogenesis of paraneoplastic syndromes is that the illness is immune mediated (Fig. 15–1): The tumor ectopically expresses an antigen (onconeural

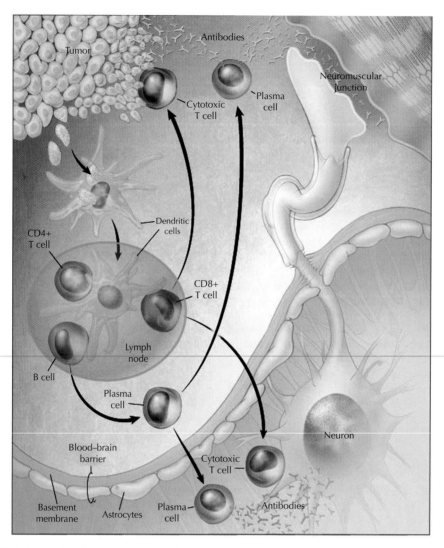

Figure 15–1. Proposed pathogenesis of paraneoplastic neurologic disorders. A tumor not involving the nervous system expresses a neuronal protein that the immune system recognizes as nonself. Apoptotic tumor cells are phagocytized by dendritic cells that migrate to lymph nodes where they activate antigen specific CD4+, CD8+, and B-cells. The B-cells mature into plasma cells that produce antibodies against the tumor antigen. The antibodies or the cytotoxic CD8+ T-cells (or both) slow the growth of the tumor, but they also react with portions of the nervous system expressing the antigen present in the tumor. In this illustration, antibodies react with voltage-gated calcium channels at the neuromuscular junction (Lambert–Eaton syndrome [see also Fig. 15–7]). Also in this illustration, immune cells cross the blood–brain barrier and attack neurons expressing the antigen they share with the tumor. (From Ref. 41 with permission.)

antigen) that normally is expressed exclusively in the nervous system,[41–44] or occasionally in the nervous system and testes, an organ that, like the brain, is an immunologically privileged site.[45–47] The onconeural antigen in the tumor cell is recognized by the immune system as foreign, leading to an immune attack.[48,48a] The immune attack may control the growth of the tumor and, in rare instances, obliterate it.[4–6] In

patients with paraneoplastic syndromes, the tumors, although identical in histologic type to tumors of patients without paraneoplastic syndromes, are more likely to be heavily infiltrated with inflammatory cells, including T-cells, B-cells, and plasma cells,[49] supporting the concept of an immune reaction. In addition, no matter how small the primary, lymph nodes are usually also involved.

The exact mechanism by which the immune reaction damages neural structures is not established for most paraneoplastic syndromes. The two paraneoplastic syndromes whose immune mechanism has been clearly established to meet the Drachman criteria for an antibody-mediated autoimmune disease[50] (LEMS and myasthenia gravis [see following text]) are primarily B-cell (i.e., antibody) mediated, but with a T-cell component. Paraneoplastic stiff-person syndrome may perhaps be caused by amphiphysin antibodies,[51] cancer-associated retinopathy (CAR) by antibodies to recovern,[52,53] and paraneoplastic autonomic neuropathy by antibodies to ganglionic acetylcholine receptor antibodies.[54] However, increasing evidence, particularly concerning paraneoplastic syndromes of the CNS, suggests a major T-cell component.[55-58] In Anti-Yo positive PCD (see p 570), cDR2 antigen–specific CD8 T-cells can be found in both tumor and brain.[59] Similarly, in CAR, antigen-specific T-cells have been found in patients,[60,61] and induction of recovern-specific CD8 T-cells and antibodies in mice has been reported to lead to abnormal electroretinograms (ERGs).[61]

Diagnosis

Specific diagnostic criteria have been defined,[62] dividing suspected paraneoplastic syndromes into either definite or possible categories (Table 15–6). A crucial criterion in diagnosis is that alternative causes that might explain the clinical symptoms must be excluded.

CLINICAL APPROACH

A clinician sees a patient with a possible paraneoplastic syndrome in one of two settings (Table 15–7):

First, a patient not known to have cancer presents with neurologic symptoms. If the patient has a classical paraneoplastic syndrome, or if the patient has a subacutely developing neurological disability that is not explained by the examination or conventional laboratory testing, a paraneoplastic syndrome should be suspected. Depending on the site of the neurologic disability, magnetic resonance imaging (MRI), cerebrospinal fluid (CSF) examination, electrodiagnostic tests, and measurement in the serum of paraneoplastic antibodies should be undertaken. The presence of paraneoplastic antibodies (Table 15–8) in the serum strongly indicates a paraneoplastic syndrome; however, the yield is small. A recent report describing a search for anti-neuronal antibodies in the sera of 60,000 patients screened for paraneoplastic syndromes revealed positive results in only 553 (<1%).[63] The presence of a classical paraneoplastic syndrome with a paraneoplastic antibody in the serum mandates a careful search for

Table 15–6. **Diagnostic Criteria for Paraneoplastic Neurological Syndromes (PNS)**

Definite PNS

1. A classical syndrome and cancer that develops within five years of the diagnosis of the neurological disorder.

2. A non-classical syndrome that resolves or significantly improves after cancer treatment without concomitant immunotherapy provided that the syndrome is not susceptible to spontaneous remission.

3. A non-classical syndrome with onconeural antibodies (well-characterized or not) and cancer that develops within five years of the diagnosis of the neurological disorder.

4. A neurological syndrome (classical or not) with well characterized onconeural antibodies (anti-Hu, Yo, CV2, Ri, Ma-2, or amphiphysin), and no cancer.

Possible PNS

1. A classical syndrome, no onconeural antibodies, no cancer but at high risk to have an underlying tumor.

2. A neurological syndrome (classical or not) with partially characterized onconeural antibodies and no cancer.

3. A non-classical syndrome, no onconeural antibodies, and cancer present within two years of diagnosis.

From Ref. 62 with permission.
PNS, paraneoplastic neurological syndromes.

Table 15–7. Approach to the Patient with a Suspected Paraneoplastic Disorder

Known Cancer	No Known Cancer
1. Search for metastases: MRI of involved site, CSF cytology	1. Search for cancer: Body CT or PET mammograms, serum cancer markers (e.g., CEA)
2. Search for non-metastatic disorder (see Chapters 9–14)	2. CSF for cells, IgG, oligoclonal bands, cytologic examination
3. CSF for cells and IgG	3. Serum and CSF for autoantibodies
4. Serum and CSF for autoantibodies	4. Follow and search again for cancer if numbers 2 and 3 are positive

CSF, cerebrospinal fluid; CEA, carcinoembryonic antigen; IgG, immunoglobulin G.

an underlying cancer, the specific cancer often being predicted by the nature of the antibody (e.g., an anti-Hu antibody indicates SCLC, an anti-Yo antibody indicates breast or ovarian cancer). Both body CT and 2-fluorodeoxy-D-glucose positron emission tomography (FDG-PET) scanning are mandated. PET scanning is probably the more sensitive, but doing both PET and CT increases sensitivity.[64,65] If no tumor is found, the patient should be followed carefully and studies repeated at intervals for the appearance of a cancer. Failure to find a cancer does not mean that the disorder is not paraneoplastic. The tumor may be too small to detect[66] or, on rare occasions, the tumor may disappear as a paraneoplastic syndrome develops.[5,6] Mediastinal biopsy may reveal only necrosis or inflammatory nodules where a SCLC had once been.[67] If the patient has neither a classical syndrome nor a paraneoplastic antibody, initial imaging is still indicated, but the patient can probably be studied at less frequent intervals.

Second, in a patient with known cancer, a careful search should be undertaken for non-paraneoplastic causes of nervous system dysfunction. Metastases are the most common cause of neurologic dysfunction and a careful search for metastatic tumor deposits in the brain, spinal cord, or leptomeninges should be made by appropriate imaging and CSF examination. However, if the patient has classical symptomatology, and particularly if paraneoplastic antibodies are present, symptoms are probably paraneoplastic.

Prognosis

Neurologists have long believed that patients with paraneoplastic syndromes fare better with respect to their cancer than patients with identical cancers without paraneoplastic syndromes. However, because the paraneoplastic syndrome is often the presenting complaint, the cancer is discovered earlier than it would have been otherwise, thus leading to a longer survival. One way of addressing this issue was to examine the 15% to 20% of patients with SCLC who have low titers of the anti-Hu antibody in their serum but no paraneoplastic syndrome. The assumption is that they have mounted an immune response to the cancer, but the immune response was insufficient to cause nervous system symptoms. In a study of 170 SCLC patients, low-titer anti-Hu antibodies were detected in 36 patients who did not have a paraneoplastic syndrome; the low-titer anti-Hu antibodies were associated with SCLC that was smaller, had a complete response to chemotherapy and a longer survival.[68] However, another study of 52 consecutive patients with SCLC found no difference in survival between those with the anti-Hu antibody (17%) and those whose serum did not harbor an antibody,[69] but a greater number of antibody-positive patients had limited disease at the time of diagnosis. A recent study of 200 patients with SCLC identified low-titer anti-Hu antibodies in 51. The presence of the antibodies did not correlate with the extent of disease or survival.[70] However, the technique to measure low titer antibodies was new, nonstandardized, and of unknown specificity.[71] Thus, the issue is unresolved.

A better prognosis with respect to the tumor has also been described in LEMS,[72] but surprisingly not in patients with SCLC but without LEMS who also harbored P/Q voltage-gated calcium channel (VGCC) antibodies.[73] Most patients with anti-Yo positive PCD have no known cancer at the time of diagnosis, and of those with known cancer, most have limited oncologic disease.[66] However, it was not possible to demonstrate that PCD patients have a better tumor prognosis in one small study.[74] The difficulty in demonstrating improved cancer prognosis in patients with paraneoplastic syndromes is compounded by the fact that the paraneoplastic syndrome itself may cause an early death or the cancer may not be treated equally vigorously in all patients with disabling neurologic symptoms.

The prognosis for resolution of neurologic signs in paraneoplastic syndromes varies with the nature of the syndrome. In those syndromes where damage occurs at a site in the nervous system that can regenerate, as for example the synapse (e.g., LEMS, myasthenia gravis), or the myelin sheath, as for example in chronic inflammatory demyelinating neuropathy (paraneoplastic CIDP), the neurologic disability may resolve with treatment of the underlying tumor, with immunosuppression, or sometimes spontaneously.[5] In those patients in whom there is destruction of neurons (e.g., PCD), treatment of the underlying tumor or immunosuppression may prevent progression but the syndrome, unless mild, usually does not improve substantially.

Treatment

There have been two general approaches to treatment[75] (Table 15–9). The most effective appears to be treatment of the underlying tumor, which often prevents progression of the disease[76] and, in some instances, leads to improvement or complete amelioration of the neurologic symptoms.[77] Effective treatment of the tumor should remove the inciting antigen and thus temper the immune response causing the neurologic damage.

Because paraneoplastic syndromes are believed to be immune mediated, the second approach to treatment is to suppress the immune response. It is important to note that suppression of the immune response does not appear to worsen the outcome of the tumor.[76] Several different forms of immunosuppression, including corticosteroids, plasma exchange, IVIg, tacrolimus, and rituximab and alemtuzumab, have been used to treat paraneoplastic syndromes.[55,78–82] Immunosuppression is clearly effective in some paraneoplastic syndromes (Table 15–9). In other paraneoplastic syndromes, especially those affecting the CNS,

the outcome is less clear.[76] However, the earlier in the course of the neurologic syndrome that immunosuppressive treatment or treatment of the cancer is begun, the more likely the patient is to stabilize or improve.[83] The clinical response of each paraneoplastic syndrome is discussed in the sections that follow.

SPECIFIC SYNDROMES

Paraneoplastic Cerebellar Degeneration

PCD is the most easily recognized and best characterized of the CNS paraneoplastic disorders. The disorder is uncommon, only several hundred patients having been described in the literature. In one report of 1476 patients with several types of carcinoma, only 3 (0.2%) had PCD.[19] PCD was first described in 1919,[84] when a patient with a rapidly developing cerebellar syndrome died and was found at autopsy to have loss of Purkinje cells throughout the cerebellum as well as pelvic cancer (probably ovarian carcinoma). However, no connection was made between the cancer and the cerebellar degeneration until the same authors suggested the relationship in 1938.[85] PCD can be associated with any cancer, but the most common culprits are lung cancer (particularly SCLC),[86] gynecologic cancers (especially ovarian cancer),[66] and lymphomas (particularly Hodgkin disease) (Fig. 15–2).[87,88] Neurologic symptoms send most patients to the physician before the cancer has been discovered. However, the offending cancer is usually found within months to a year after the onset of neurologic symptoms. In occasional patients, the cancer may elude detection for 2 to 5 years,[89] sometimes being found only at autopsy.

Symptoms often begin abruptly with dizziness, nausea, and vomiting and often diplopia, followed later by ataxia of gait and extremities, dysarthria, and dysphagia. The onset may be so abrupt as to simulate a stroke.[90] One of our patients, a middle-aged woman, went to bed one night feeling well. When she awoke the following morning, she fell while getting out of bed and was never able to walk again. A clinical diagnosis of brainstem infarct was made, although the MRI was normal. Eighteen months later, she was still bedridden and, while being bathed, the nurse encountered a lymph

Table 15–8. Antineuronal Antibody–Associated Paraneoplastic Disorders

Antibody	Neuronal Paraneoplastic Reactivity	Protein Antigens (mol wt)	Cloned Genes	Tumor	Symptoms
Well-Characterized Antibody Markers of PND and Tumor					
Anti-Hu	Nucleus > cytoplasm	35–40 kDa	HuD	SCLC, neuroblastoma, prostate	PEM, PSN, autonomic dysfunction
Anti-Yo	Cytoplasm, Purkinje cells	34, ~52 kDa (cdr2)	CDR34, CDR62	Ovary, breast, lung	PCD
Anti-Ri	Nucleus > cytoplasm	55, 80 kDa	Nova	Breast, Gyn, lung, bladder	Ataxia/opsoclonus
Anti-CRMP5 (anti-CV2)	Cytoplasm oligodendrocytes, neurons	66 kDa	CRMP5	SCLC, thymoma	PEM, PCD, chorea, optic, sensory neuropathy
Anti-amphiphysin	Pre-synaptic neurons	128 kDa	Amphiphysin	Breast, SCLC	Stiff-person syndrome
Anti-Ma2	Neurons (submnucleus)	41.5 kDa	Ma2	Testis	Limbic, brainstem encephalitis
Anti-recoverin	Photoreceptor, ganglion cells	23, 65, 145, 205 kDa	Recoverin	SCLC	CAR
Anti-Titin	Muscle	3000 kDa	Titin	Thymoma	Myasthenia gravis
Anti-NMDAR	Neurons, hippocampus	~150 kDa	NR1/NR2	Ovarian, teratoma	Limbic encephalitis
Well-Characterized Antibody Markers of Neurological Dysfunction that May not Predict the Presence of Cancer					
Anti-VGCC	Pre-synaptic NMJ	VGCC	P/Q VGCC	SCLC	LEMS
Anti-nAChR	Postsynaptic	64 kDa	Neuronal AChR nicotinic	SCLC	Autoimmune neuropathy

Partially Characterized Antibody Markers of PND and Cancer

Anti-Tr	Purkinje cytoplasm	?	?	Hodgkin	PCD
Anti-PCA-2	Purkinje cytoplasm, other neuras	280 kDa	?	SCLC	PCD
Anti-Ma1	Neurons (subnucleus)	40 kDa	Ma1	Lung, others	Brainstem, PCD
ANNA 3	Nuclei Purkinje cells	170 kDa	?	Lung cancer	Sensory neuronopathy, PEM
Anti-mGluR1	Purkinje cells, olfactory neurons, hippocampus	150 kDa	Glutamate receptor	Metabotropic Hodgkin	PCD
Anti-Zic4	Nuclei of cerebellar	55,50,48, and 35–38 kDa	Zic4	SCLC	PCD
Anti-bipolar	Bipolar retinal cells	?	Melanoma	MAR	Retinal cells
Anti-PKCS	Purkinje cells	80 kDa	PKCγ	NSCLC	PCD
Anti-gephyrin	Neuron (synapse)	97 kDa	Gephyrin	Unknown primary (1 case)	stiff-person syndrone

PND, paraneoplastic neurologic syndrome; SCLC, small cell lung cancer; PEM, paraneoplastic encephalomyelitis; PSN, paraneoplastic sensory neuronopathy; PCD, paraneoplastic cerebellar degeneration; CNS, central nervous system; Gyn, gynecological cancer; CRMP, collapsin response-mediating protein 5; CAR, cancer associated retinopathy; NMDAR, N-N-methyl-D-aspartate receptor; NMJ, neuromuscular junction; VGCC, voltage-gated calcium channel; MAR, melanoma associated retinopathy; PKCγ, protein kinase C gamma; AChR, acetylcholine receptor.

Table 15–9. **Treatment of Paraneoplastic Neurologic Syndromes (PNS)**

PNS that Usually Respond to Treatment

LEMS	Tumor Rx, plasma exchange, IVIg, 3,4, diaminopyridine, immunosuppressants°
Myasthenia gravis	Tumor Rx, plasma exchange, IVIg, immunosuppressants°
Dermatomyositis	Immunosuppressants°, IVIg
Opsoclonus/Myoclonus (pediatrics)	Tumor Rx, steroids, ACTH, IVIg
Neuropathy (osteosclerotic, myeloma)	Radiation, chemotherapy

PNS that May Respond to Treatment

Vasculitis (nerve/muscle)	Steroids, cyclophosphamide
Opsoclonus/Myoclonus (adults)	Steroids, tumor Rx, protein A column, clonazepam, diazepam, baclofen
Paraneoplastic cerebellar degeneration (Hodgkin)	Tumor Rx
Opsoclonus/ataxia (anti-Ri)	Steroids, cyclophosphamide
Guillain–Barré (Hodgkin)	Tumor Rx, plasma exchange, IVIg
Stiff-person	Tumor Rx, steroids, diazepam, baclofen, IVIg
Neuromyotonia	Plasma exchange
Demyelinating polyneuropathy	Immunosuppressants
Necrotizing myelopathy	Steroids

PNS that Usually do not Respond to Treatment°°
Paraneoplastic cerebellar degeneration
   SCLC (irrespective of anti-Hu)
   Anti-Yo antibodies (cancer of ovary, breast)
Paraneoplastic encephalomyelitis/sensory
  neuronopathy
   Limbic encephalopathy
   Cerebellar degeneration
   Brainstem encephalopathy
   Myelopathy
   Sensory neuronopathy
   Autonomic dysfunction (central or peripheral)
Cancer-associated retinopathy
Melanoma-associated retinopathy

° Corticosteroids, cyclophosphamide, cyclosporin, etc.
°° Although treatment usually does not improve neurologic symptoms, it may prevent progression. The tumor should be treated aggressively in all cases.

node in her axilla that led to the diagnosis of breast cancer.

Nystagmus (particularly downbeat nystagmus) with oscillopsia is common. The cerebellar signs usually progress over a few weeks to months and then stabilize, although slow progression over many months can also occur. At the time of stabilization, most patients cannot walk without support; many cannot sit unsupported. Handwriting is often impossible and independent eating is difficult. Speech may be understood only with great effort. In many patients, particularly those with gynecologic cancers and Hodgkin lymphoma, the symptoms are limited to the cerebellum but in others, particularly those with SCLC, symptoms are usually more widespread[91] (see section on Subacute Sensory Neuronopathy/Encephalomyelitis). In patients with SCLC, LEMS may accompany the cerebellar degeneration.[91] Antibodies

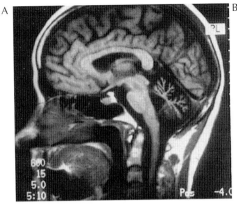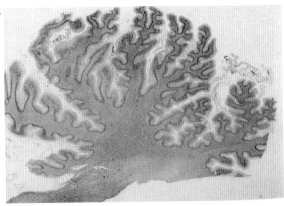

Figure 15–2. Paraneoplastic cerebellar degeneration (PCD) associated with Hodgkin disease. **A**: An MR scan of a patient with PCD. The scan shows prominent cerebellar sulci and an enlarged fourth ventricle indicating cerebellar atrophy. The cerebral hemispheres are normal. **B**: A mid-sagittal section of cerebellum from a 28-year-old man who died of Hodgkin disease 6 years after he developed PCD; note the prominent atrophy.

against VGCC, characteristic of LEMS (see section on LEMS), can also occur in patients with lung cancer and PCD who do not have symptoms of LEMS.[92]

Early in the course of the illness, the CSF may show a pleocytosis (10–50 lymphocytes) with a slightly elevated protein and increased IgG; later the CSF becomes acellular. Initially, the MRI is usually normal although we have encountered patients with cerebellar swelling and enhancement of cerebellar folia (Fig. 15–3). Later in the illness, cerebellar atrophy may be evident.

Several different paraneoplastic antibodies have been reported in patients with PCD (Table 15–10).

In one study of 50 PCD patients, 19 were associated with the anti-Yo antibody, 16 with the anti-Hu antibody, 7 with the anti-Tr antibody, and 6 with the anti-Ri antibody.[10] Occasional patients with typical PCD have no identifiable antibodies in their serum.

Anti-Yo antibodies almost invariably occur in women who have breast or gynecologic cancers.[66] However, on rare occasions other tumors, including cancers in men, cause Anti-Yo positive PCD. The anti-Yo antibody recognizes a major protein antigen, cDR2, that is expressed in the cytoplasm of Purkinje cells of the cerebellum. All patients with anti-Yo PCD who have ovarian or breast cancer express the cDR2 Purkinje cell antigen in their tumor cells. Approximately 60% of patients with ovarian cancer who do not have PCD express cDR2 in their tumors.[93]

The antibody is found at higher titer in the spinal fluid than in the serum, suggesting intrathecal synthesis from B-cells that have crossed into the brain.[94] In addition to the antibody response, cytotoxic T-cells against cDR2 have also been described.[55,59,95]

The clinical disorder is usually rapid in onset (in some of our patients overnight). It generally stabilizes with the patient substantially disabled often with a pure cerebellar syndrome. Anti-Yo PCD is rarely directly responsible for the patient's death although many patients have difficulty swallowing and are subject to aspiration pneumonia. Occasional patients are only mildly affected and some appear to recover function after treatment of the underlying tumor or with immunotherapy (IVIg) or other immunosuppressive agents.[96] Early treatment, before Purkinje cell destruction is complete, appears to be essential.[96] Many patients with anti-Yo–positive PCD live for years, often substantially disabled by the cerebellar degeneration, suggesting that the tumor runs an indolent course. However, one report indicates that patients with anti-Yo cerebellar degeneration associated with gynecologic cancers do not live longer than those who have the same cancer without the paraneoplastic syndrome.[74]

The tumors are often small. FDG-PET may be the only diagnostic test to identify the lesion.[81] In one study of 55 patients with anti-Yo PCD, the tumor could be found in only 52, and in that group was discovered only at exploratory laparotomy in 4 (8%).[66]

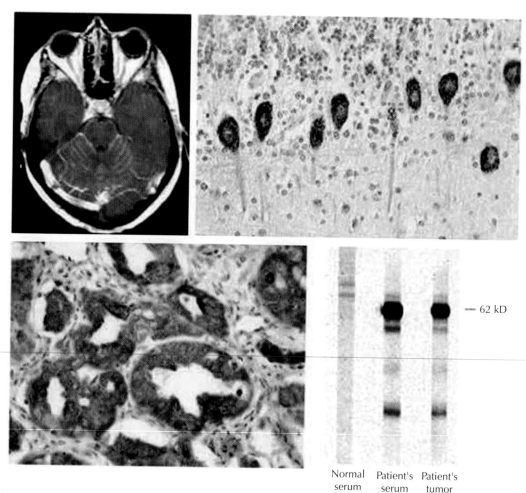

Figure 15–3. Anti-Yo–positive paraneoplastic cerebellar degeneration. The MRI is that of a woman who developed the subacute onset of a pure cerebellar syndrome that incapacitated her. The MRI demonstrates enhancement of the cerebellar folia, representing an acute inflammatory reaction that cleared within a few weeks. Antibodies in her serum reacted with Purkinje cells of the cerebellum but not with other neurons in the central nervous system. Approximately 2 months later, an ovarian carcinoma was discovered. Antibodies in her serum reacted with her ovarian cancer cells. The Western blot against cerebellar Purkinje cells and tumor revealed bands at 62 and 34 kDa. The patient's cerebellar syndrome stabilized, but she made no neurologic recovery during the 3 years she lived before dying of recurrent cancer. (From Ref. 41 with permission.)

Anti-Hu antibodies cause PCD as part of the encephalomyelitis syndrome (see following text). In some patients, cerebellar symptoms outweigh other symptoms[91,97]; in other patients, sensory neuronopathy is prominent so that it is difficult to discern whether the ataxia is due to sensory loss or cerebellar dysfunction. As with the anti-Yo antibody, symptoms are usually sudden in onset and pathological examination reveals substantial loss of Purkinje cells (Fig. 15–4). More inflammatory infiltrates are found in this disorder than in the anti-Yo syndrome, but that may be because patients survive for a shorter period of time, although one report seems to suggest otherwise.[10]

Anti-Tr antibodies occur in patients with PCD related to Hodgkin disease. This disorder often begins after the diagnosis of Hodgkin disease and sometimes when the patients are in remission.[88,87] Patients are generally less severely affected than are patients with the anti-Yo or anti-Hu antibody. The prognosis for life and successful treatment of the underlying tumor is good.

Other less common antibodies may also be associated with PCD (Table 15–10). These

Table 15–10. **Antibodies in Paraneoplastic Cerebellar Degeneration**

Antibody	Usual Cancer	Clinical Findings
Anti-Yo	Ovary, breast	Subacute severe PCD
Anti-Hu	SCLC	Part of PEM/SN
Anti-Tr	Hodgkin disease	Less severe, may remit
Anti–P/Q VGCC	SCLC	Absent DTRs
Anti-Ri	Breast	Opsoclonus/truncal ataxia
Anti-Ma1	Breast, colon	Brainstem
Anti-CRMP5 (CV2)	SCLC, thymoma	Optic neuritis, PEM
Anti-mGluR1	Hodgkin	PCD
Anti-Zic4	SCLC	PEM/LEMS
Anti-PCA2	SCLC	PEM
Anti-ANNA-3	SCLC	PEM/neuropathy
Anti-CARP VIII	Melanoma	PCD
Anti-Proteosome	Ovary, breast	PCD

PCD, paraneoplastic cerebellar degeneration; SCLC, small cell lung cancer; PEM/SN, paraneoplastic encephalomyelitis/subacute sensory neuronopathy; VGCC, voltage-gated calcium channel; DTR, deep tendon reflexes; LEMS, Lambert–Eaton myasthenic syndrome.

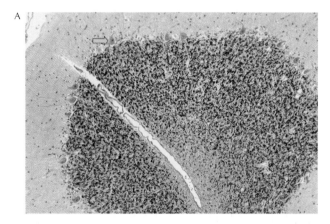

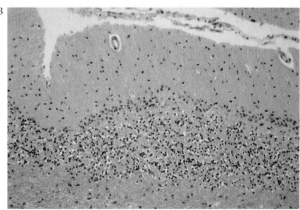

Figure 15–4. Pathologic changes in paraneoplastic cerebellar degeneration. **A**: A section of cerebellum taken from a patient who died without neurologic disease. Note the abundant neurons in the granular cell layer and the large number of Purkinje cells (*arrow*). **B**: A section of cerebellum taken from a patient with paraneoplastic cerebellar degeneration. Note the absence of Purkinje cells and the thinning of the granular cell layer.

include anti-Ma1, an antigen restricted to brain and testis,[47] anti-CRMP5 (anti-CV2) antibodies directed against an antigen in some glial cells and against peripheral nerve antigens,[98] Zic4 antibodies in patients with SCLC,[99] anti-mGluR1 antibodies associated with Hodgkin disease,[100] protein kinase Cγ antibodies,[101] and others.

DIAGNOSIS

In a middle-aged patient, particularly a woman who has the rapid onset of cerebellar dysfunction whether alone or in combination with other signs of encephalomyelitis, a paraneoplastic syndrome remains the major diagnostic consideration. Most other cerebellar degenerative diseases are gradual in onset and evolve much more slowly. Consideration, however, must be given to Creutzfeldt–Jacob disease,[102] celiac disease,[103] vitamin E deficiency, and HIV-associated cerebellar degeneration[104] (Table 15–11).

The presence of an appropriate autoantibody establishes a diagnosis of PCD, but even in those patients without an antibody, a careful search for an underlying neoplasm should be carried out at onset and, if negative, repeated in 3 to 6 months.

A subset of patients with SCLC and PCD also have LEMS.[91] The cerebellar symptoms may be predominant, but in addition to the cerebellar signs, deep tendon reflexes are diminished or absent (see LEMS). Antibodies against P/Q VGCC identify LEMS that responds to treatment, although the PCD usually does not improve. Cerebellar tissue is rich in P/Q VGCC protein, perhaps explaining the concurrence of the two syndromes.

PATHOGENESIS

The pathogenesis of PCD is unknown. Anti-Yo IgG injected into the CSF of experimental animals reacts with Purkinje cells of the cerebellum, but does not reproduce the neurologic disorder[105]; non-Yo anti-rabbit gammaglobulin injected into rabbits also reacts with Purkinje cells.[106] When anti-Yo IgG is presented to neurons in culture, the antibody does not kill neurons but does induce the expression of adhesion molecules and also appears to accelerate neuronal differentiation.[107] A single report indicates that anti-mGluR1 antibodies reduce activity of Purkinje cell cerebellar neurons in vitro and disturb compensatory eye movements in vivo.[108] Thus, for the most part, PCD does not appear to be strictly antibody mediated. However, in one study of cerebellar tissue from patients with PCD and LEMS, cerebellar P/Q type VGCC were reduced and the ratio of antibody VGCC complexes to total VGCC was increased, suggesting that the antibody may have been responsible for the reduction and perhaps the symptoms.[109]

Table 15–11. **Differential Diagnosis of Paraneoplastic Cerebellar Degeneration**

Differential Diagnosis	Additional Considerations in Patients Known to Have Cancer
Alcohol-related	Cerebellar metastasis
Vitamin deficiency (thiamine, vitamin E)	Chemotherapy toxicity (5-fluorouracil, cytarabine)
Toxins (anticonvulsants, other)	
Infectious or postinfectious cerebellitis	
Miller–Fisher syndrome	
GAD-associated cerebellar ataxia	
Idiopathic	
Creutzfeldt–Jacob disease	
HIV	
Celiac disease	

GAD, glutamic acid decarboxylase; HIV, human immunodeficiency virus.

In cerebral sections studied at autopsy from patients with anti-Yo PCD, one usually sees only loss of Purkinje cells without evidence of inflammation in the Purkinje cell layer. cDR2-specific T-cells have been found in the blood of PCD patients,[55,57,95] along with activated T-cells in the CSF,[55] suggesting the possibility that the destruction of Purkinje cells could be T-cell mediated, with the T-cells disappearing along with the Purkinje cells late in the course of the disease.

PATHOLOGY

The CNS may appear grossly normal when examined at autopsy, but usually the cerebellum is atrophic with abnormally widened sulci and small gyri. Microscopically, the PCD hallmark is an extensive and often complete loss of Purkinje cells of the cerebellar cortex[110,111] (Fig. 15–4). The degenerating Purkinje cells may have swellings, called torpedoes, along their axons. Other pathologic features seen include thinning of the molecular and granular layers of the cerebellar cortex, often without marked cell loss, and proliferation of Bergmann astrocytes. The deep cerebellar nuclei are usually well preserved, although rarefaction of white matter may surround the nuclei, corresponding to the loss of Purkinje cell axons. Basket cells and tangential fibers are usually intact. Lymphocytic infiltrates, if present in the cerebellum, are usually found in the leptomeninges, in the dentate nucleus, and in surrounding white matter, but only rarely in the Purkinje cell layer.

In many patients, the disorder is noninflammatory, with all pathologic changes restricted to the Purkinje cell layer of the cerebellum. Pathologic changes, sometimes seen outside the cerebellum, differ substantially from patient to patient. They may include dorsal column and pyramidal tract degeneration of the spinal cord; degeneration of the basal ganglia, specifically the pallidum; loss of peripheral nerve fibers; and inflammatory infiltrates in the brainstem, spinal cord, and cerebral cortex.

TREATMENT

Treatment of PCD is usually unsatisfactory, perhaps because death of Purkinje cells, the common pathologic finding, prevents neurologic recovery. However, several credible reports indicate that treatment of the tumor or immunosuppression either stabilizes or improves PCD associated with different antibodies.[10,78,96] Therapy aimed at reducing CNS T-cells has been tried; its efficacy has not been established.[55] In general, however, the anti-Hu and anti-Yo syndromes rarely improve, whereas improvement has been reported in some patients with anti-Tr or anti-CRMP5 antibodies.

Subacute Sensory Neuronopathy/Encephalomyelitis

Henson et al.[17] introduced the term *encephalomyelitis with carcinoma* to describe patients with cancer associated with clinical signs of damage to more than one area of the nervous system and with postmortem findings of inflammation within the brain, brainstem, spinal cord, dorsal root ganglia, and nerve roots (Fig. 15–5). Most of their patients had SCLC. Encephalomyelitis can affect any portion of the nervous system, including the hippocampus, hypothalamus, brainstem, cerebellum, spinal cord, dorsal root ganglia, autonomic ganglia, myenteric plexus, peripheral nerve or muscle, either as an isolated clinical finding or any combination. Even in patients with unifocal symptoms and signs, inflammatory changes and cellular destruction are more widespread than clinical symptoms suggested. The clinical features of paraneoplastic encephalomyelitis associated with the anti-Hu antibody have been described by Graus et al.[97,112] In about a third of patients, the disorder presents as a unifocal syndrome, usually either sensory neuronopathy or limbic encephalopathy. In the remainder, although one set of symptoms may predominate (e.g. cerebellar dysfunction), examination discloses findings involving other systems (e.g., dysautonomia, memory loss, sensory changes).

HIPPOCAMPUS (LIMBIC ENCEPHALITIS)

Paraneoplastic limbic encephalitis usually begins rapidly with changes in mood and personality worsening over days to weeks.[113] Accompanying the mood changes is severe impairment of recent memory with relatively preserved remote memory. Patients are often agitated and confused, with hallucinations and generalized or complex partial seizures.[113–115]

A

B

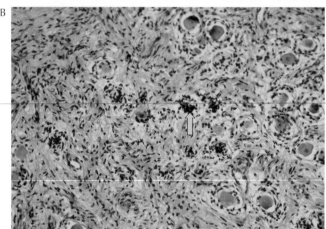

Figure 15–5. Pathologic changes in subacute sensory neuronopathy. **A**: A photomicrograph of a dorsal root ganglion taken from a patient who died of cancer but did not have a paraneoplastic syndrome. Note the abundant large neurons. **B**: A photomicrograph of a patient with paraneoplastic subacute sensory neuronopathy. Only a few neurons remain. Many dying neurons are surrounded by inflammatory cells (*arrow*).

The CSF is typically inflammatory, at least early in the course of the disease. The MRI may be normal although abnormalities, particularly hyperintensity on the T2-weighted or FLAIR image, sometimes with T1 contrast enhancement, are common especially in the hippocampus (Fig. 15–6).[113] Electroencephalographic findings include focal or generalized slowing and occasional epileptiform activity, particularly in the temporal areas.[114] About two-thirds of patients have clinical symptoms outside the limbic system.[113] The neoplasms associated with limbic encephalitis include SCLC, testicular cancer, thymoma, breast cancer, and Hodgkin lymphoma. Neurologic symptoms usually precede discovery of the tumor.

A number of different antibodies have been associated with limbic encephalopathy. The most common is probably the anti-Hu antibody, although anti-Ma1/2, CRMP5,

amphiphysin, and others have been reported[115a] (Table 15–8). An identical clinical syndrome related to abnormalities of voltage-gated potassium channels can occur as an autoimmune symptom in patients without cancer.[116] Women with ovarian teratomas may develop a reversible limbic encephalopathy characterized by behavioral abnormalities, seizures, memory deficits, central hypoventilation, and disorders of consciousness associated with antibodies against the N-methyl-D-aspartate (NMDA) receptor.[117,117a]

Pathologic changes may be restricted to the limbic system and insular cortex and consist of loss of neurons with reactive gliosis, perivascular lymphocytic cuffing, and microglial proliferation.

The differential diagnosis includes viral encephalitis, particularly Herpes simplex encephalitis, and non-paraneoplastic limbic encephalopathy related to voltage-gated potassium channels.[116]

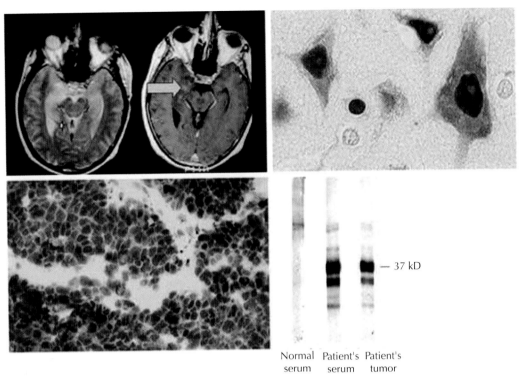

Figure 15–6. Anti-Hu–positive paraneoplastic encephalomyelitis. The woman whose MRI is depicted in the upper left developed, over a 2-month period, severe memory loss without other cognitive abnormalities. Complex partial seizures subsequently developed. The MRI revealed hyperintensity in the medial temporal lobes bilaterally with a small area of contrast enhancement (*arrow*). Her serum contained an antibody that reacted with the nuclei of all neurons in the central and peripheral nervous system and with all small cell lung cancers. The Western blot illustrates that the patient's serum reacted with a 37-kDa antigen found in nervous system tissue and small cell lung cancers. (From Ref. 41 with permission.)

Polymerase chain reaction analysis of CSF establishes the diagnosis of viral infection, and identification of voltage-gated potassium channel antibodies suggests that the patient does not have cancer.

Most paraneoplastic patients with limbic encephalitis do not respond well to treatment, although there are some cases of either spontaneous resolution or resolution in association with treatment of the underlying tumor, particularly Hodgkin disease[118–120] and ovarian teratoma.[117]

HYPOTHALAMUS

Hypothalamic dysfunction includes manifestations such as somnolence, hyperthermia, endocrine abnormalities, hyperhidrosis, and myokymia. Hypothalamic dysfunction occurs in patients with Ma-2 positive testicular cancer[115a,121,122] and CRMP5 (CV2) positive thymomas. Hypothalamic dysfunction

rarely occurs in isolation; it is usually associated with limbic encephalitis or brainstem encephalitis.[113]

BRAINSTEM

Paraneoplastic brainstem encephalitis is characterized by subacutely developing brainstem or basal ganglia signs, usually occurring as part of the more diffuse syndrome of encephalomyelitis, although sometimes presenting clinically as a dominant or isolated finding.[123,124] Upper brainstem and basal ganglia abnormalities may cause Parkinsonism,[125–127] chorea,[128] dystonia, bradykinesia, myoclonus, daytime sleepiness,[125] and supranuclear eye movement abnormalities.[122,129–131] Lower brainstem involvement is more likely to be associated with diplopia, vertigo, opsoclonus, oscillopsia, nystagmus, dysarthria, dysphagia, hypoventilation, hearing loss, and facial numbness.[132–135] Several different tumors can cause brainstem encephalitis, but the most characteristic are SCLC with the anti-Hu syndrome,[97] testicular cancer with the anti-Ma2 antibody,[115a,122,136] and thymomas with CRMP5 antibodies.[129]

CEREBELLUM

The clinical findings of PCD (see preceding text) may occur as the first or only manifestation of a more diffuse encephalomyelitis. The patient with encephalomyelitis is more likely to have evidence of other peripheral nervous system and CNS involvement, especially sensory neuronopathy (see following text), and the pathological changes are likely to be inflammatory rather than bland. In one of our patients with cerebellar signs associated with SCLC, infiltrates of T-cells were found in the Purkinje cell layer of the cerebellum, despite the absence of Purkinje cells. Prominent cerebellar signs occur in approximately 15% of patients with the anti-Hu syndrome.[97]

SPINAL CORD

Myelitis may be the most prominent finding in a patient with encephalomyelitis.[137] The disorder may appear as a transverse or a partial transverse myelopathy; in some instances, only motor neuron symptoms are present, resembling amyotrophic lateral sclerosis. Patients may harbor anti-Hu,[138] anti-CRMP-5,[98] amphiphysin,[139] or other[140] antibodies. They present with progressive weakness, sometimes with lower motor neuron signs including fasciculations, but unlike true motor neuron disease, sensory loss and autonomic dysfunction including incontinence and postural hypotension are often present. In a few patients with anti-Hu encephalomyelitis, the presenting symptoms are indistinguishable from those of motor neuron disease; however, the evolution is somewhat faster.[138] The CSF shows inflammatory changes and the pathology shows an intense inflammatory reaction and loss of neurons in anterior and posterior horns with secondary nerve root degeneration.

DORSAL ROOT GANGLIA

In 1948, Denny-Brown published a seminal paper describing two patients with a rapidly developing and incapacitating sensory neuronopathy.[38] This pathology included loss of dorsal root ganglion cells and inflammatory infiltrates in the dorsal root ganglia. These were probably the first two patients with what we would now recognize as an anti-Hu–positive sensory neuronopathy. One of the patients had myositis as well.

Sensory neuronopathy typically affects patients with SCLC and the anti-Hu antibody; the disorder has been described with other tumors as well, including Hodgkin disease.[141] Approximately 60% of patients with the anti-Hu paraneoplastic syndrome have sensory neuronopathy as the primary symptom.

The onset is usually rapid, with the patient complaining of sensory loss and paresthesias. Symptoms may begin distally in all four extremities, but unlike dying back neuropathies, can begin in other areas of the body, including the trunk and neck; the sensory loss is often asymmetrical. Examination reveals sensory loss to all modalities, distinguishing it from cisplatin neuropathy that affects proprioception and spares small fiber modalities including pin and temperature. The patients usually become so ataxic from the proprioceptive loss that, within weeks, they are unable to walk and may not be able to use their hands effectively. Deep tendon reflexes are lost, but motor function is usually preserved.

In patients with pure sensory neuronopathy, the CSF may be bland and intrathecal antibody synthesis is usually absent. When the sensory changes are associated with encephalomyelitis, the CSF is inflammatory and antibody synthesis is present.[142] Electrodiagnostic studies reveal that sensory nerve action potentials have low amplitude or are absent, while motor nerve action potentials are usually normal. Electrical evidence of degeneration is absent.[38]

At times, pathologic changes may be limited to the dorsal root ganglia showing inflammatory infiltrates and neuronal loss, but inflammatory changes are usually more widespread.

AUTONOMIC SYSTEM

Paraneoplastic autonomic neuropathy[143] can occur as an isolated phenomenon[144] or as part of a more widespread encephalomyelitis, particularly with sensory neuronopathy.[97] Usually associated with SCLC and the anti-Hu antibody, it can occur with other cancers.[144,145] Patients generally present with the subacute onset of postural hypotension, intestinal immobility, pupillary abnormalities,[146] and a neurogenic bladder.[144] In several patients, involvement of the myenteric plexus causes severe constipation as the predominant symptom preceding both the identification of cancer and the development of other neurologic abnormalities.[147] Autonomic neuropathy also occurs with an antibody against ganglionic acetylcholine receptor (anti-nAchR). Immunization with the protein produces an autonomic neuropathy,[148] and passive transfer of antibodies also causes the neuropathy, indicating a B-cell–mediated disease.

ANTIBODIES

Anti-Hu antibody–positive paraneoplastic encephalomyelitis is usually associated with SCLC. Other antibodies include CRMP5 (thymoma), Ma1 (a variety of tumors), and Ma2 (testicular cancer). Anti-Hu encephalomyelitis is more likely to be characterized by sensory neuronopathy and anti–Ma2 encephalomyelitis by brainstem and diencephalic involvement. CRMP5 may be associated with optic neuritis (see following text).

TREATMENT

Paraneoplastic encephalomyelitis is poorly responsive to treatment. Ma2-associated encephalomyelitis is more responsive than the other types, but in all, improvement may occur if caught early.

Opsoclonus/Myoclonus (OM)

Opsoclonus,[149] a disorder of saccadic stability, consists of involuntary arrhythmic multidirectional high amplitude conjugate saccades without an interval between them.[150] The disorder is often associated with diffuse or focal body myoclonus and truncal titubation with or without ataxia or other cerebellar signs. OM occurs in children either as a self-limited illness, probably the result of a viral infection in the brainstem, or as a complication of neuroblastoma. In approximately 40% of children, OM is paraneoplastic,[151] but the incidence of OM in neuroblastoma is under 2%.[152,153] However, given the known tendency of neuroblastoma to differentiate and spontaneously disappear, many of the cases reported as OM without tumor may have been paraneoplastic. The peak age of onset is approximately 18 months; girls are more affected than boys and the neurologic signs usually precede identification of the tumor in at least 50% of patients. Ataxia, irritability, insomnia, and dementia often accompany the opsoclonus. A viral-like prodrome, particularly with irritability, precedes the onset of paraneoplastic opsoclonus/myoclonus (POM) in 50% of children.[151]

Imaging studies are usually normal, although an occasional MRI is hyperintense in the brainstem or cerebellum on the T2 or FLAIR images. The CSF may show a mild pleocytosis, but most cell counts were within normal limits. However, there is a four to seven times higher percentage of B-cells in the CSF than in pediatric controls.[154] Treatment with immunotherapy, chemotherapy, or tumor resection does not normalize the B-cell percentages. Because the disorder responds to ACTH and high-dose dexamethasone pulses (20 mg/m^2 on days 1–3/28 day cycle)[155] and B-cells in the CSF are present, the pathophysiology of childhood OM is believed to be B-cell mediated. Rituximab, an anti-B cell antibody, is also effective in some patients.[156]

A number of autoantibodies have been described in neuroblastoma-related opsoclonus, but none consistently.[153,157] Affected children have a high incidence of antibodies that react with CNS antigens, but no particular antibody is characteristic of the syndrome. Recently, an autoantibody that binds the surface of cerebellar granule cells and is cytotoxic to neuroblastoma cell lines has been identified in pediatric patients with OM with or without neuroblastoma but not in controls.[158,159] Seven of the children had other antibodies that also reacted with cerebellar structures. These included two with the anti-Hu antibody.[153]

In adults, OM is paraneoplastic in approximately 20% of patients.[8] It often develops in association with truncal ataxia; other symptoms of brainstem or cerebellar dysfunction may also be present. Imaging is usually normal, but there may be hyperintensity in the pons on T2-weighted images. The most commonly associated antibody is the anti-Ri antibody occurring in women with breast cancer and gynecologic tumors as well as others.[160–162] The anti-Ri antibody identifies a protein called Nova,[163] an RNA-binding protein that regulates messenger RNAs encoding synaptic proteins, including the glycine and GABA$_A$ receptors, and inhibitory transmitter receptors in the brainstem.[164–166]

As indicated in the preceding text, children with OM may respond to treatment with ACTH, high-dose dexamethasone, rituximab, plasma exchange, or intravenous immunoglobulins and may improve after chemotherapy. However, many patients relapse and are left with significant cognitive deficits.[151] In adults, immunosuppressive agents[167] sometimes seem to be effective, although spontaneous remissions have also been described, making it difficult to interpret results. However, the syndrome probably responds less well to immunotherapy unless the tumor is also treated.[168]

Spinal Cord Syndromes

Paraneoplastic myelopathies may occur as part of a more widespread paraneoplastic encephalomyelitis (particularly the anti-Hu syndrome [see preceding text]) or as an isolated phenomenon.[137] The myelopathy may affect all of the structures within a segment of the spinal cord (transverse myelopathy) or may select a specific area of the spinal cord such as anterior horn cells. The different myelopathies may have different pathologies, some inflammatory, some necrotic, some apoptotic. Each has a presumably different pathogenetic mechanism. Taken together, paraneoplastic myelopathies are rare.

MYELITIS

Inflammatory paraneoplastic myelitis is usually a part of the encephalomyelitis syndrome and is associated with several different antibodies.

NECROTIZING MYELOPATHY

Although there had been several previous descriptions, subacute necrotic myelopathy was clearly recognized as a paraneoplastic syndrome by Mancall and Rosales in 1964.[169] This rare syndrome occurs with leukemia, lymphoma,[170–172] and lung or other cancers and may precede or follow the diagnosis of cancer.[173–175] Patients typically present with rapidly ascending flaccid paraplegia; back or radicular pain may precede the onset of neurologic dysfunction and the process may lead to respiratory failure and death. The CSF is usually inflammatory, at least early in the course and protein is elevated. The MRI may be normal or may show spinal cord swelling or sometimes contrast enhancement.[176] Treatment is usually unsatisfactory although one patient has been reported to respond to intrathecal dexamethasone.[177] Pathological examination reveals widespread spinal cord necrosis involving all components of the cord but with some white matter predominance. Inflammatory cells are usually absent. Unlike many other paraneoplastic syndromes, patients often die as a result of the spinal cord lesion rather than the underlying tumor.

The pathogenesis is not known. A T-cell–mediated immune reaction has been postulated based on an autopsy report of T-cells in the spinal fluid of a patient dying of acute necrotizing opticomyelopathy associated with thyroid cancer.[178]

Devic disease (neuromyelitis optica) is a nonparaneoplastic, autoimmune inflammatory disease that involves spinal cord and optic nerves. It is often confused clinically with multiple sclerosis (MS) but, unlike MS, individual attacks have a poor prognosis. A serum autoantibody against the aquaporin-4 water channel distinguishes it from MS.[179,180] Pathologic changes resemble paraneoplastic necrotizing myelopathy, and the presence in some patients with paraneoplastic myelopathy of pathologic changes in the optic nerves and chiasm suggest there may be a relationship. However, the antibody described by Antoine et al.[176] is not the same as the aquaporin-4 antibody.

The differential diagnosis includes viral infection, especially Herpes simplex type-2, Herpes zoster, and non-neoplastic necrotic myelopathy.[181]

MOTOR NEURON SYNDROMES

This term refers to a progressive degenerative disease largely restricted to motor neurons.[182] Motor neuron disease, particularly ALS, is fairly common. Epidemiologic studies of ALS do not show an increased incidence in patients with cancer when compared to the general population because the numbers of paraneoplastic patients are so small.[183] Nevertheless, there are isolated case reports of patients with cancer and clinically typical ALS whose neurological improvement was dramatic after treatment of the underlying tumor, suggesting that

ALS can be paraneoplastic.[184] Paraneoplastic motor neuron disease may be clinically indistinguishable from the non-neoplastic variety, but its course is usually more rapid.[185] In addition, a few patients have shown amelioration of their symptoms after treatment of the underlying cancer.[15,184]

The clinical picture is characterized by weakness, atrophy, and fasciculations, usually asymmetric, with a predilection for small muscles of the hand or bulbar muscles, reflex hyperactivity, and extensor plantar responses. Spasticity may also be present; bowel and bladder function are usually spared, as is the ocular motor system. A few patients with paraneoplastic motor neuron disease have had antibodies that react with nervous system elements.[186,187]

In one report, four patients with primary lateral sclerosis and one with ALS predominantly showing upper motor neuron signs were found to have breast cancer.[138] In three patients there was a close temporal relationship between the development of the neurologic syndrome and the cancer. Whether this was a coincidence or not is unclear.

The relationship between motor neuron disease and lymphoma may be more clear-cut than that with solid tumors. A series of articles, mostly from a single group, have reported a relationship between lymphoma and motor neuron disease, largely of the lower motor neuron type. In one series, asymptomatic lymphoma was found in 2 of 37 patients with ALS undergoing routine bone marrow examination.[188] One patient had a monoclonal protein. A patient with autopsy-proved ALS and Waldenström macroglobulinemia associated with antibodies to sulfated glucuronic acid paragloboside has also been reported.[189] Others have not seen a similar relationship between lymphoma and ALS.[190] A few reports also describe ALS in association with plasma cell dyscrasias.[191,192]

A strictly lower motor neuron syndrome, named *subacute motor neuronopathy*,[193] was characterized by subacutely developing lower motor neuron weakness without upper motor neuron signs.[194,195] However, in these reports, electrodiagnostic studies looking for conduction blocks that would characterize an autoimmune motor neuronopathy (see section on Peripheral Neuropathy) were not done. The disorder was thought to be viral by Walton et al. because of possible viral particles in anterior horn cells at autopsy.[195] Autopsy in a few patients clearly showed lower motor neuron involvement in the spinal cord. Patients reported by Schold et al.[193] improved spontaneously, whereas only one of nine patients reported by Younger et al.[196] improved and that patient had evidence of a proximal conduction block in nerve conduction studies. The other eight patients, all of whom had evidence of upper motor neuron dysfunction as well, did not improve. In another report, a patient with breast cancer and a lower motor neuron syndrome was shown to have antibodies in the serum that reacted with axons of motor neurons. The antigen was beta IV spectrin.[197]

STIFF-PERSON SYNDROME

Originally called the *stiff-man* syndrome, although it affects more women than men, the syndrome is characterized by muscle stiffness and rigidity that is usually painful.[198–200] The muscles involved almost always include paraspinal and abdominal muscles and usually those of the lower extremities as well. In some instances, the illness can be restricted to a single extremity. Sustained muscle contraction results in an abnormal posture such as an exaggerated lumbar lordosis. In addition to the chronic contractions, severe muscle spasms may be precipitated by voluntary movements, unexpected environmental stimuli, and emotional upset. The arms are less commonly involved and, although stiffness of the neck and face can occur, it is rarely severe. On examination the muscles feel hard and may be difficult to move passively. An unexpected external stimulus or a voluntary movement may precipitate an observable spasm. The electromyogram is characterized by sustained continuous motor unit activity that disappears during sleep and general anesthesia.

The non-paraneoplastic disorder is associated with diabetes and antibodies to glutamic acid decarboxylase. The paraneoplastic disorder is associated with antibodies against amphiphysin,[139] a synaptic protein that plays a role in vesicle endocytosis. (Amphiphysin antibodies have also been detected in patients with paraneoplastic myelopathy.[139]) Some patients with paraneoplastic stiff-person syndrome harbor GAD antibodies[201–203] with or without anti-amphiphysin antibodies. Anti-Ri antibodies were reported in one patient with stiff-person syndrome and metastatic adenocarcinoma.[204] A single patient with antibodies to gephyrin, a post-synaptic tubulin binding protein that binds inhibitory glycine and $GABA_A$ receptors, has also been reported.[205] The common tumors include breast cancer, SCLC, Hodgkin disease,[206] and colon cancer.

The pathogenesis is not completely understood but is believed to be B-cell mediated. A recent study of a patient with breast cancer and stiff-person syndrome demonstrated that injection of the patient's IgG, obtained by plasma exchange, into experimental animals resulted in dose-dependent stiffness with spasms. The animals were prepared by injection of encephalogenic T-helper lymphocytes to induce a leaky blood–brain barrier. Control IgG did not have the same effect, and IgG binding was demonstrated in the CNS of rats showing the disorder. The IgG contained amphiphysin antibodies but whether these were the culprits was not established.[51]

Benzodiazepines and baclofen may relieve symptoms. Treatment of the tumor and immunosuppression with corticosteroids, IVIg, and other immunosuppressive agents are also effective.

Cranial Nerves

RETINOPATHY

Carcinoma-Associated Retinopathy

Paraneoplastic syndromes can affect either the retina or the optic nerves.[207–212] Although all paraneoplastic visual disorders are rare, cancer associated retinopathy (CAR) is probably the most common. This disorder is characterized by photoreceptor degeneration of cones and rods[213] with relative sparing of the inner retina. CAR precedes the discovery of cancer in approximately 50% of patients and complicates a variety of tumors, most commonly SCLC and gynecologic tumors.[209] The onset is usually rapid, over days or weeks, with patients complaining of painless visual loss in both eyes often preceded by night blindness and accompanied by photopsias (shimmering or flickering lights) and peripheral ring-like scotomata. A single patient with CAR and lung cancer had slowly progressive visual loss over 10 years.[213a] CAR usually affects both rods and cones but rarely cones are affected in isolation[214] (see following text). The abnormalities that define loss of cones include photosensitivity, loss of

visual acuity with light-induced glare, loss of color vision, central scotomata, and abnormal phototopic electroretinograms (ERGs). The abnormalities that accompany rod dysfunction include night blindness, prolonged dark adaptation, and peripheral or ring-like scotomata. At onset, the retina may appear normal but later arterioles appear attenuated, retinal pigment epithelium becomes thin and mottled, and the optic disc may be slightly pale. Occasionally, white cells are found in the vitreous and in the CSF as well, suggesting an inflammatory response. The ERG is severely abnormal, demonstrating abnormalities of photoreceptors under both phototopic and scotopic conditions.

On histological examination, the retina shows loss of both the inner and outer segments of the photoreceptors and loss of the outer nuclear layer; the inner nuclear layer is generally preserved. Inflammatory infiltrates may or may not be present.

Several autoantibodies have been identified in patients with CAR,[215] the most common targeting recoverin.[216] Recoverin is a 23-kDa protein that modulates dark and light adaptation through calcium-dependent regulation of rhodopsin phosphorylation in photoreceptor cells. The antigen is found in lung cancer cell lines,[217] and antibodies have been identified in patients with lung cancer but without retinopathy.[216] Recoverin antibodies have been found in the aqueous, suggesting that they can cross the blood–retinal barrier and gain access to photoreceptors.[218] Visual loss associated with this antibody affects both rods and cones, is usually profound, and occurs before the discovery of the cancer.

Other anti-retinal antibodies have also been reported in patients with CAR. Anti-enolase antibodies[219–221] react with a 46-kDa protein in retinal bipolar cells, Müller cells, and ganglion cells but not photoreceptor cells. Anti-enolase antibodies are found to the same extent in patients with paraneoplastic retinopathy and in patients with non-paraneoplastic autoimmune retinopathy.[221] Visual loss associated with this antibody mainly affects cones, may be unilateral or asymmetric, and is usually less profound than anti-recoverin CAR. The disorder often follows discovery of the cancer.

The diagnosis of CAR should be suspected in any patient not known to have a family history of retinitis pigmentosa who presents with the subacute onset of episodes of transient visual loss or fixed visual loss accompanied by night blindness and photopsias. The presence of retinal arteriolar attenuation adds to the diagnosis; ERG studies indicating a photoreceptor disorder mandate measurement of autoantibodies and a search for cancer.

The pathogenesis of CAR is believed to be B-cell mediated. Some antibodies induce apoptosis of photoreceptor and bipolar cells in vivo.[220,222,223] Whether T-cells play a role is unclear. However, recoverin-specific cytotoxic T-lymphocytes have been identified in the peripheral blood of an experimental mouse model vaccinated with recoverin. The vaccination produced autoimmune retinal dysfunction as well as tumor regression in a recoverin-expressing tumor.[224] The antibody itself is thought to enter photoreceptor cells and inhibit rhodopsin phosphorylation in a calcium-dependent manner.[225]

A variety of immunosuppressive treatments have been tried. These include intravenous corticosteroids, plasma exchange, IVIg, and alemtuzumab[82] as well as treatment of the underlying malignancy.[209,226] Success using IVIg has been reported.[227] It is not clear whether the presence of the retinopathy or of anti-retinal autoantibodies affects the prognosis of the cancer.

Melanoma-Associated Retinopathy

Unlike anti-recoverin–associated CAR, patients with melanoma-associated retinopathy (MAR) are already known to have cancer, that is, cutaneous melanoma. A few cases identical to MAR in patients with colon cancer have also been reported.[207] The onset is acute, with the patient complaining of photopsias, night blindness, and mild peripheral visual field depression. Marked visual loss and central scotomata are uncommon. The fundus may appear normal, but retinal arterial attenuation may also be seen. ERGs show marked reduction of the dark-adapted and light-adapted B-waves with preservation of A-waves, suggesting bipolar and Muller cell dysfunction.

Anti-retinal antibodies also occur in patients with melanoma but no ocular symptoms.[228] In one series, 53 of 77 serum samples reacted by immunohistochemistry with retina. Three staining patterns were identified: inner nuclear layer only, inner and outer nuclear layer, both nuclear layers in combination with photoreceptor cells and/or the nerve fiber layer. Patients with advanced stages of melanoma were more likely to have antibodies.

Treatment usually has little effect on the illness. Corticosteroids have been reported to improve vision in some patients.[229] Treatment of MAR with immunosuppression IVIg and plasma exchange all appear to have some effect.[209]

MAR may be associated with prolonged survival of patients with metastatic melanoma, but conversely, some suggest that immunosuppression in patients with established melanoma may lead to spread of the tumor[230]; firm evidence is lacking.

Optic Neuropathy

Paraneoplastic optic neuropathy occurring in the absence of other neurologic symptoms is extremely uncommon. Cross et al.[98] described 16 patients with optic neuritis (5 of whom also had retinitis), all of whom harbored antibodies to CRMP5, a collapsin response mediator protein. The patients had subacute visual loss. Optic discs were swollen and visual field defects were present. All of the patients had a variety of other neurologic symptoms, including cognitive changes, ataxia, sensory neuropathy, and myelopathy. Three patients resembled Devic disease at onset of symptoms.[98] A variety of tumors were associated with the syndrome, the most common being SCLC; one patient had thyroid cancer (see also Ref. 178). Other reports have described this disorder in association with multiple myeloma,[231] nasopharyngeal carcinoma, neuroblastoma, and lymphomas.[209]

Peripheral Nerve Syndromes

The exact incidence of paraneoplastic peripheral neuropathy is unknown. In one series of 422 consecutive patients with peripheral neuropathy referred to a neurological referral center, 26 were believed to have a paraneoplastic peripheral neuropathy.[16] Seven patients had onconeuronal antibodies. The overall incidence of presumed paraneoplastic peripheral neuropathy was 6% to 8%, varying with the type of neuropathy. For pure sensory neuronopathy, the incidence was 47%, 1.7% for Guillain–Barré syndrome, 10.5% for chronic inflammatory demyelinating polyneuropathy (CIDP), 9% for mononeuritis multiplex, and

10.3% for axonal polyneuropathy of otherwise undetermined cause. Table 15–12 lists the wide range of peripheral neuropathies that can occur as paraneoplastic syndromes (see also Ref. 232).

In a patient without known cancer who presents with signs of a peripheral neuropathy, the first step is to determine the site of the neuropathy. A pure sensory neuronopathy, that is, pathology in the dorsal root ganglion, strongly suggests a paraneoplastic syndrome associated with the anti-Hu antibody (see section on Subacute Sensory Neuronopathy/Encephalomyelitis). A pure motor neuropathy subacutely developing could be the Guillain–Barré syndrome associated with Hodgkin disease[233] or a multifocal motor neuropathy with conduction block associated with plasma cell dyscrasias.[25] An autonomic neuropathy is sometimes associated with the anti-Hu syndrome (see preceding text). Mononeuritis multiplex suggests a vasculitis, possibly paraneoplastic in origin.[234]

The next step is to define the pathology of the neuropathy by electrodiagnostic studies. A demyelinating neuropathy (slow conduction velocities or conduction block) may be associated with monoclonal gammopathies, whereas

Table 15–12. **Paraneoplastic Peripheral Neuropathies**

Sensory
  Dorsal root ganglionitis
  Axonal
    Inflammatory
    Vasculitis
  Demyelination
    Acute (Guillain–Barré syndrome)
    Subacute (CIDP)
Sensorimotor
  Axonal
    Vasculitis
    Inflammation
  Demyelinating
Motor
  Demyelinating
Focal
  Mononeuritis multiplex
  Brachial neuropathy

CIDP, chronic inflammatory demyelinating polyneuropathy.

most other cancers cause axonal neuropathies. If only some nerves are involved (mononeuritis multiplex), the disorder might be a paraneoplastic microvasculitis. Comorbid illnesses such as diabetes and vitamin B_{12} deficiency that cause neuropathy must be excluded.

In the patient with known cancer, it is important to consider all of the chemotherapeutic agents the patient may have received as a possible cause of neuropathy (see Chapter 12).

SENSORY NEUROPATHY

Most paraneoplastic pure sensory neuropathies (actually neuronopathies) are a result of damage to dorsal root ganglia. The sensory disorder may occur in isolation or be part of the subacute sensory neuronopathy/encephalomyelitis syndrome, described in detail in the preceding text. Some patients with paraneoplastic sensory neuronopathy suffer from SCLC and the anti-Hu syndrome. In a study of 363 patients with peripheral sensory neuropathy,[235] 6 patients had known cancer without chemotherapy and 4 had a monoclonal gammopathy. Of the 53 patients with sensory neuropathy of unknown cause, a follow-up of 51 revealed cancer in 18; the cancer was found 3 to 72 months after the onset of the neuropathy. Only one patient was anti-Hu positive. Whether the lesions were in dorsal root ganglion or peripheral nerves was not determined.

Antoine et al.[16] studied 422 consecutive patients with peripheral neuropathy, 26 of whom had cancer. Six patients were anti-Hu positive, and one was anti-CV2 (CRMP5) positive. The anti-Hu patients had predominantly sensory changes. Of the remaining 19 patients, only 2 had a pure sensory neuropathy. Symptoms in both were distal and thought to be axonal in origin. A few other case reports describe paraneoplastic sensory neuropathy, some of which are axonal,[236] and some demyelinating,[237] in others,[238] the site of the lesion was not determined. Severe sensory demyelinating neuropathy characterized largely by sensory ataxia has been reported with monoclonal gammopathies, particularly IgM.[239]

The pathogenesis of paraneoplastic sensory neuropathies is variable. In some patients with monoclonal gammopathies, anti-MAG antibodies appear to bind to the nerve and elicit an inflammatory response that causes demyelination. In patients with antibody-positive peripheral neuropathies, both the antibody and T-cells may play a role, but the exact pathogenic mechanism has not been determined.

Treatment is generally unsatisfactory. A few patients respond to immune suppression with either plasma exchange or IVIg, and some patients report resolution of their symptoms with treatment of the underlying neoplasm. However, in the majority of instances, the neuropathy runs a course independent of the neoplasm and does not respond to immunosuppression.

SENSORIMOTOR NEUROPATHY

Most paraneoplastic peripheral neuropathies are sensorimotor and axonal. Approximately 10% to 15% of patients with cancer develop clinically apparent sensorimotor neuropathy. In most patients the neuropathy is mild and there is always some question about its relationship to the cancer. If one examines cancer patients electrophysiologically, as many as 35% to 50% have abnormalities suggestive of peripheral neuropathy.[240] Many of these neuropathies are probably nutritional or toxic. The typical paraneoplastic sensorimotor peripheral neuropathy is symmetrical in distribution (save for vasculitic neuropathies) and usually disabling. When the patient harbors an antibody, the neuropathy is usually associated with SCLC or thymoma.[241] Demyelinating neuropathy resembling CIDP has been reported in patients with melanoma,[242] lymphoma,[190] myeloma,[243] and a variety of solid tumors.[244]

Acute demyelinating polyneuropathy, the Guillain–Barré syndrome, has been reported with an increased incidence in Hodgkin disease[233] and may occasionally complicate other cancers. Possibly also demyelinating is a paraneoplastic brachial plexopathy associated with Hodgkin disease.[245] The pathogenesis of chronic demyelinating polyneuropathy is probably similar to those without cancer although the inciting antigen can be different.[246] Antibodies, cytokines, complement, inflammatory mediators, macrophages, and autoreactive T-cells are all believed to play a role. Because these disorders are clearly immune mediated and antibodies probably play a major role, IVIg or plasma exchange is sometimes effective.[247] Prednisone, azathioprine, cyclophosphamide, cyclosporin, and mycophenolate also may have some effect. Recent studies suggest that rituximab may also be effective.[248,249]

MOTOR NEUROPATHY

As indicated in the paragraphs on demyelinating neuropathies, the Guillain–Barré syndrome presenting as a pure motor neuropathy has been reported at an increased incidence in patients with Hodgkin disease and occasionally other cancers, although its exact relationship to the other cancers is unclear. A single case of multifocal motor neuropathy with conduction block has been reported in a patient with non–Hodgkin lymphoma.[250]

AUTONOMIC NEUROPATHY

Paraneoplastic disorders of the autonomic nervous system usually arise in the setting of the subacute sensory neuronopathy/encephalomyelitis syndrome (see preceding text).[144] However, autonomic symptoms may predominate or rarely be the only symptoms or signs of a paraneoplastic neuropathy.[143] The most common symptom is pseudo-obstruction of the bowel,[147] but anhydrosis, orthostatic hypotension, hypoventilation, sleep apnea, and cardiac arrhythmias can also present either alone or, more commonly, as part of a more widespread autonomic neuropathy. Most autonomic neuropathies are associated with SCLC and the anti-Hu syndrome but other tumors have also been described. Some patients have antibodies against ganglionic acetylcholine receptors.[145] Immunization of experimental animals with nicotinic acetylcholine receptor induces autonomic dysfunction, including intestinal pseudo-obstruction.[148] Although treatment of the tumor or immunosuppression may halt progression of the disease, patients usually do not improve substantially.[251,252]

PLASMA CELL DYSCRASIAS

Table 15–13 lists the peripheral neuropathies associated with plasma cell dyscrasias. Sensorimotor peripheral neuropathy, either demyelinating or axonal, can occur with monoclonal gammopathy of undetermined significance (MGUS), primary amyloidosis, multiple myeloma, and other disorders. Although MGUS is benign at diagnosis, malignancy often develops.

In one study of 104 patients with neuropathy associated with MGUS, 22% had a malignancy including lymphoma, multiple myeloma,

Table 15–13. **Plasma Cell Dyscrasias and Peripheral Neuropathy**

Monoclonal gammopathy of uncertain significance
Primary systemic amyloidosis
Multiple myeloma
  Osteolytic with amyloidosis
  Osteolytic without amyloidosis
  Osteosclerotic
Waldenström macroglobulinemia
Cryoglobulinemia
Gamma heavy-chain disease
Monoclonal gammopathy with solid tumors
Monoclonal gammopathy with benign lymph node
  hyperplasia
POEMS syndrome

POEMS, polyneuropathy, organomegaly, endocrinopathy, monoclonal protein, skin changes

plasmacytoma and POEMS (polyneuropathy, organomegaly, endocrinopathy, monoclonal protein, skin changes).[253] Factors suggesting malignancy include weight loss, progression of the neuropathy, a monoclonal (M) protein greater than 1 gram/liter, unexplained fever, and night sweats.[254] In approximately two-thirds of patients, the neuropathy responds to treatment of the tumor.[253]

VASCULITIC NEUROPATHY

This uncommon paraneoplastic syndrome presents clinically as a mononeuritis multiplex, not as diffuse symmetrical nerve dysfunction typical of other peripheral neuropathies.[255–257] One paper reports that among 151 patients with biopsy-proved vasculitic neuropathy collected over 25 years, only 2 (1.3%) had the disorder as a paraneoplastic syndrome.[234] However, another study reported that 15% of patients with vasculitic neuropathy developed various malignancies within 2 years of neuropathy onset.[256] The disorder sometimes appears as part of the anti-Hu syndrome in SCLC, but also in prostate and uterine cancer; Hodgkin and non–Hodgkin lymphoma have also been reported as causes. The diagnosis can be considered when the neuropathy is axonal but asymmetric, but can be proved only by biopsy. It is important to make the diagnosis because treatment is often effective. Treatment can include immunosuppression with corticosteroids and

cyclophosphamide. Treatment of the underlying tumor may also be effective.

Neuromuscular Junction

LAMBERT–EATON MYASTHENIC SYNDROME

LEMS is the most common of the classical paraneoplastic syndromes. The prevalence in patients with SCLC varies from 0% to 6%.[24,258] In approximately 40% to 60% of patients with LEMS, cancer is the underlying cause. In the others (more commonly women) there is no cancer, but the antibody is the same. In patients with LEMS and cancer, SCLC is the underlying tumor in the majority of cases although other neoplasms are occasional culprits.[259,260] In patients without cancer, LEMS is associated with familial autoimmune diseases with a preponderance of maternal inheritance.[261] In these patients, the HLA-B8 haplotype is over-represented[262]; the HLA-B8 haplotype in LEMS patients predicts that SCLC will not be present.[263]

At presentation, patients generally complain of fatigue and muscle weakness.[264] The weakness characteristically begins and is most prominent in the proximal muscles of the legs. The patient first notes difficulty climbing stairs and getting out of low chairs or off toilet seats without pushing down with the hands. Proximal arm muscles may also be affected; patients have difficulty with heavy lifting, for example, opening windows. Muscle weakness may fluctuate, becoming first better and then worse as patients exercise. Cranial nerve functions may also be involved but are much less severe than those seen in myasthenia gravis. The common cranial nerve symptoms are ptosis and diplopia; dysarthria and dysphagia may also be present.[265] When oculo-bulbar symptoms are prominent, one must take care to exclude myasthenia gravis as a comorbid illness.[266] Autonomic dysfunction includes impotence in men and dry mouth with a metallic taste in both sexes; orthostatic hypotension is a less common symptom. The clinical symptoms of LEMS usually precede diagnosis of the cancer.

On examination the patient's complaints often seem out of proportion to the degree of weakness found by the examiner. However, careful examination will reveal proximal muscle weakness in both the shoulder and pelvic girdle and diminished-to-absent deep tendon reflexes and sometimes sluggish pupillary reflexes.[146] In some patients, brief exercise of a muscle will increase its strength transiently and cause return of the deep tendon reflex associated with that muscle.

Laboratory tests include EMG and antibody measurement. Electrodiagnostic studies of neuromuscular transmission demonstrate a compound muscle action potential that is often less than 10% of normal (Fig. 15–7). However, when stimulated at frequencies from 22 to 50 Hz, the compound muscle action potential increases in size. An increase of at least 100% is considered typical of LEMS.[267,268]

All patients with paraneoplastic LEMS harbor antibodies against P/Q VGCC in their serum. Some patients with SCLC without clinical or electrophysiological evidence of LEMS will also harbor those antibodies, although at a lower titer.[269]

The diagnosis of LEMS is often difficult to make clinically. Patients complain of fatigability and weakness that seem out of proportion to what the examiner finds. To complicate matters, patients with cancer but without LEMS often have evidence of proximal weakness such as inability to rise from a squatting position and complaints of fatigability and erectile impotence[258] (Table 15–3). Once the clinician has convinced himself that the symptoms represent true neurologic dysfunction, the differential diagnosis includes myasthenia gravis, steroid-induced myopathy (sometimes from ectopic production of steroids by SCLC),[35] paraneoplastic polymyositis, or because of the diminution or absence of reflexes, peripheral neuropathy. All of these can be excluded by careful neurologic examination and electrodiagnostic and antibody testing.

LEMS itself is not a lethal disease. In fact, even undetected, it rarely incapacitates the patient. The presence of LEMS has been reported to confer an increased survival in patients with SCLC when compared to patients with the same tumor without LEMS.[72] Curiously, the presence of antibodies without LEMS does not confer an increased survival.[73]

LEMS is believed to be largely a B-cell–mediated disease (Fig. 15–8). All SCLCs express P/Q-type VGCC and a subset of such patients develop an immune response characterized by the development of anti–P/Q-VGCC antibodies. In one recent series, 10 of 148

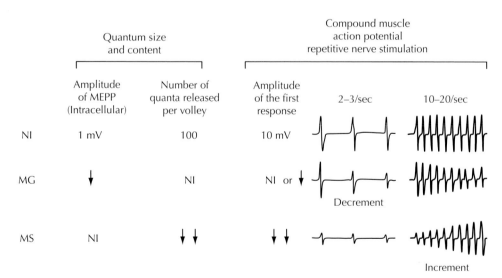

Figure 15–7. Typical changes in size and content as determined by intracellular recordings in myasthenia gravis (MG) and Lambert–Eaton myasthenic syndrome (MS) (*left*) and decrement (MG) or increment (MS) of the compound muscle action potential to repetitive nerve stimulation (*right*). The amplitude of the compound muscle action potential decreases as a result of dropout of individual muscle fibers according to the size of the end-plate potential (EPP), and the amplitude increases with recruitment of additional fibers. The amplitude of the miniature endplate potential (MEPP) is decreased in myasthenia gravis but normal in MS. NL = normal. (From Ref. 271 with permission.)

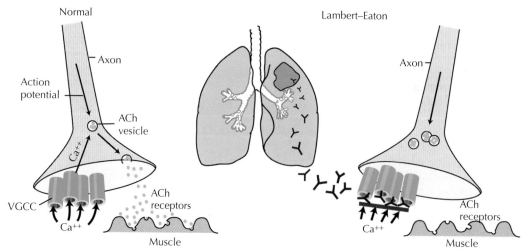

Figure 15–8. Pathogenesis of the Lambert–Eaton myasthenic syndrome. The portion of the cartoon on the left illustrates normal neuromuscular transmission. When an action potential reaches the presynaptic neuromuscular junction, it promotes opening of voltage-gated calcium channels. The channels allow entry of calcium that in turn promotes the release of acetylcholine from synaptic vesicles. When acetylcholine contacts receptors on the muscle, it causes the muscles to contract. All small cell lung cancers (center) contain voltage-gated calcium channels. Some patients develop an immune response to those channels and produce antibodies. The antibodies circulate in the serum and reach voltage-gated calcium channels at the presynaptic cholinergic junction, cross-linking them. As a result, when an action potential reaches the presynaptic synapse, calcium cannot enter and acetylcholine is not released.

consecutive patients (7%) with SCLC harbored VGCC antibodies.[73] Only four of the patients had LEMS. One patient with antibodies had PCD, a paraneoplastic disorder also associated with P/Q-VGCC antibodies.[86] Antibodies against VGCC bind to the channels at the active site of the presynaptic cholinergic junction, cross-linking these channels to prevent entry of calcium into the terminal axon when an action potential reaches the presynaptic synapse. Calcium is necessary for the release of acetylcholine, and its action is mediated through synaptogamin-1, a protein enriched in synaptic vesicles that may also be targeted in some patients with LEMS.[270]

For patients with SCLC, treatment of the tumor takes precedence and is often effective in ameliorating the neurologic symptoms.[272] Because neurologic symptoms facilitate an early diagnosis of cancer, and because the immune response probably helps control tumor growth, treatment of the tumor often yields long survival.

At the same time treatment of the tumor is undertaken, the patient, if the LEMS is symptomatic, can be treated with 3-4-diaminopyridine from 10 to 20 mg four times a day. The drug has the effect of facilitating release of acetylcholine at the presynaptic junction and usually produces marked improvement in the patient's autonomic and muscular symptoms.[273] If treatment of the tumor and 3-4-diaminopyridine are not effective, or if the drug is not available, immunosuppression with plasma exchange or IVIg ameliorates symptoms.[269,272]

MYASTHENIA GRAVIS

Approximately 15% of patients with myasthenia gravis are found to have a thymoma.[274] Conversely, approximately 30% of patients with thymoma develop myasthenia gravis. Myasthenia gravis is more common than LEMS. An epidemiologic study uncovered 32 patients with myasthenia gravis and thymomas while, at the same time, only 10 patients with LEMS and SCLC were found in the same population.[274] A few patients with lymphoma, both thymic and extrathymic, have been reported to develop myasthenia gravis.[275] In one series, 15% of patients with myasthenia developed extrathymic malignant tumors of various origins, suggesting that the immune disorder predisposed patients to cancer.[276]

The clinical symptoms include muscle weakness and fatigability. Muscle weakness is more likely to involve cranial nerves (particularly ptosis, diplopia, and dysphagia) and respiratory muscles. The weakness fluctuates, getting worse when the muscles are used, and usually resolving with rest. Limb weakness is substantially less common, approximately 12%.[277]

Patients are usually better or well on awakening or after rest, and worse later in the day. The diagnosis can usually be made clinically by demonstrating weakness of cranial nerve musculature that increases on fatigue and is relieved by rest.[278] Asking the patient to hold upward gaze for as long as possible will often result in the development of severe ptosis that disappears after eye closure and a short period of rest. Diplopia may also occur during the sustained upward gaze. Electrodiagnostic tests give a decrementing compound muscle action potential with repetitive stimulation mimicking the muscle weakness that develops with exercise (Fig. 15–7). Single fiber recordings of muscle show excessive jitter.

Virtually all patients with myasthenia gravis and thymoma have antibodies against the acetylcholine receptor and many have additional antibodies against muscle proteins, such as titan and ryanodine receptor.[279] Antibodies, when present, help with the diagnosis but do not distinguish paraneoplastic from nonparaneoplastic myasthenia gravis. Intravenous edrophonium, an anticholinesterase with a short duration of action, will usually improve weak muscles within 30 seconds and disappear within 5 minutes.[278]

As with LEMS, myasthenia gravis is predominantly a B-cell–mediated disease. The pathogenesis of thymoma-induced myasthenia gravis differs slightly from myasthenia gravis in the absence of thymoma.[280] Myasthenia gravis–associated thymomas contain autoreactive T-cells with specificity for the acetylcholine receptor alpha and epsilon subunits. The T-cells appear to be generated by intratumoral non-toleragenic thymopoesis in patients without thymoma.[280]

Thymectomy (whether or not the patient has a thymoma) although not yet proved by controlled studies, is often prescribed as treatment along with immunosuppression. The role of thymectomy has been documented only in retrospective studies. There is controversy over whether transcervical thymectomy yields equivalent results to the sternal splitting procedure.[281] Immunosuppressants, including azathioprine, cyclosporin, cyclophosphamide, tacrolimus, and mycophenolate, can be used either alone or in combination with corticosteroids. Both plasma exchange and IVIg are effective as short-term therapy. Corticosteroids, which in high doses may initially exacerbate the disease, also give relief.[282] The prognosis is good when the correct diagnosis is made and appropriate treatment is undertaken.

NEUROMYOTONIA

Several disorders characterized by hyperexcitability of terminal arborizations of peripheral nerves occur in patients with paraneoplastic syndromes. These include neuromyotonia, the cramp-fasciculation syndrome, and Morvan syndrome.[283,284] Neuromyotonia occurs when single motor unit potentials fire spontaneously at 150 to 300 Hz, leading to chronic contraction of muscles that can be either focal or generalized and sometimes associated with pain and hyperhidrosis. The disorder is autoimmune and related to antibodies against voltage-gated potassium channels.[285] When the disorder is paraneoplastic, antibodies against amphiphysin may also be found.[139] In its mildest form it is called myokymia (repetitive and recurrent firing of a motor unit potential at 2 to 60 Hz), clinically seen as undulating muscle twitching. Patients may also demonstrate pseudomyotonia (failure of muscles to relax after contraction), and intestinal pseudo-obstruction has also been reported.[283] In some patients, CNS symptoms consisting of memory loss, hallucinations, insomnia, and changes in mood (Morvan syndrome) also appear related to the presence of antibodies against voltage-gated potassium channels.[285]

Creatine kinase is elevated in about half the patients. The diagnosis can be made electrophysiologically when needle electromyography reveals spontaneous firing of single or multiple motor units at a rate of 150 to 300 Hz. These units fire at irregular intervals and may persist during sleep, general anesthesia, and even when peripheral nerves are blocked at proximal sites. Symptoms are relieved by blocking the neuromuscular junction, indicating that the action potentials arise from terminal arborizations of the motor nerve. When these disorders are paraneoplastic, thymoma is the most common cancer, but they have also been reported with SCLC[286] and Hodgkin disease.[287]

The pathophysiology of the signs and symptoms of both the central and peripheral nervous system symptoms is unknown. The fact that symptoms are relieved by plasma exchange suggests that the disorder, or at least the major part of it, is antibody mediated. This view is supported by an experimental study in which IgG voltage-gated potassium channel antibodies added to a neuroblastoma cell line reduced potassium currents, suggesting that cross-linking of the channel by antibody (as in LEMS) may be an important mechanism of symptom production. F(ab)$_2$ fragments were also effective.[288]

Some patients respond well to anticonvulsants such as phenytoin or carbamazepine or to muscle relaxants such as diazepam. However, the treatment of choice is immunosuppression with plasma exchange or IVIg; one report suggests that plasma exchange is superior to IVIg.[289]

Muscle

INFLAMMATORY MYOPATHY

The inflammatory myopathies (Table 15–14), including dermatomyositis, polymyositis, and inclusion body myositis, are three different but sometimes overlapping diseases.[290] Recent studies unequivocally established dermatomyositis

Table 15–14. **Proposed Classification of Idiopathic Inflammatory Myopathies**

Dermatomyositis
  Juvenile
  Adult
Polymyositis
  T-cell mediated (α/β, γ/δ)
  Eosinophilic
  Granulomatous
Overlap syndromes
  With polymyositis
  With dermatomyositis
With inclusion body myositis
Cancer-associated myositis
Inclusion body myositis
Other forms
  Focal: orbital myositis; localized nodular myositis; inflammatory pseudotumor
  Diffuse: macrophagic myofasciitis; necrotizing myopathy with pipe-stem capillaries; infantile myositis

Modified from Ref. 290.

as a paraneoplastic syndrome.[12] In one study, the incidence of cancer in 392 patients with dermatomyositis was 15% with a relative risk of 2.4 in men and 3.4 in women. In the same study, among 396 patients with polymyositis, the incidence was 9% with a relative risk of 1.7 in men and 1.8 in women.[291] A second study of 537 cases with biopsy-proved myositis found cancer in 116 patients (22%). Fifty-eight of 321 patients with polymyositis (18%) had cancer, for a standard incidence ratio (SIR) of 2.0. Thirty-six of 85 patients with dermatomyositis (42%) had cancer for an SIR of 6.2, and 12 of 52 patients with inclusion body myositis (23%) yielded an SIR of 2.4%.[292] In a Scottish population-based cohort study,[12] the SIR was 7.7 for dermatomyositis and 2.1 for polymyositis.

The disorders are associated with a wide range of cancer, particularly non–Hodgkin lymphoma and ovarian and lung cancer.[293] Most of the cancers are diagnosed near the time the myopathy is discovered. The likelihood of developing cancer decreases with time after the discovery of the myopathy. There are case reports of inclusion body myositis associated with cancer.[294]

In both polymyositis and dermatomyositis, muscle weakness usually evolves over several weeks or months, although it may be acute. The muscle weakness is usually proximal and typically symmetric. Spontaneous pain or muscle tenderness occurs in one-quarter to one-half of patients. The pain, however, is usually not severe. Weakness of pharyngeal or upper esophageal muscles causes dysphagia, nasal regurgitation, and aspiration. Muscle atrophy is uncommon and deep tendon reflexes are usually preserved until very late in the course of the disease.

Cutaneous changes distinguish dermatomyositis from polymyositis.[290] The most common cutaneous signs consist of a diffuse erythematous rash over the chest and shoulders in a "V" shaped distribution sometimes aggravated by sun exposure. A heliotrope rash, a red-violet eruption over the upper eyelids sometimes with swelling, is present in only a minority of patients, but is quite specific for dermatomyositis. Symmetric scaling and erythematous eruption over the extensor surfaces of the metacarpal phalangeal joints and intraphalangeal joints are called Grotton sign. These may also occur over the extensor surfaces of the elbow. Erythema, abnormal nail bed capillary loops, and often painful roughening and cracking of the skin, so-called mechanics hands, can also be observed.

Dermatomyositis and polymyositis differ from inclusion body myositis in that inclusion body myositis is usually more insidious in onset, more slowly progressive, and the weakness is more likely to be distal.

The laboratory evaluation consists of measurement of creatinine kinase, often grossly elevated (more than 10 times normal) in polymyositis and dermatomyositis but less so in inclusion body myositis. Electromyography reveals the typical small amplitude, short duration muscle potentials with a complete interference pattern. Fibrillations and positive sharp waves are present in some patients. A muscle biopsy is the definitive test for diagnosis. It should be taken from a muscle that is modestly to moderately weak, not one with severe weakness and from a muscle contralateral to one found abnormal by EMG so as to avoid any artifact caused by the EMG needle. Sometimes MRI with increased T2 signal in muscle identifies a muscle suitable for biopsy.

The pathogenesis of all these disorders appears to be T-cell–mediated.[295] In dermatomyositis, complement is activated when membrane attack complexes are deposited on the microvasculature, resulting in capillary damage, microinfarcts, and inflammation. In polymyositis and inclusion body myopathy, there appears to be antigen-directed, MHC-1-restricted, cytotoxicity mediated by CD8 killer T-cells against the muscle itself. Different patterns of macrophage differentiation have been described in polymyositis and dermatomyositis.[296] In polymyositis, the early activation markers MRP14 and 27E10 stained most macrophages that were located in the endomysium. In dermatomyositis, macrophages expressed delayed activation marker 25F9 and were found mainly in the perimysium. The inciting antigen is not known in any of these disorders.[297]

Most patients with dermatomyositis and polymyositis respond to immunosuppression. IVIg, mycophenolate, and cyclosporin are all effective agents, as are corticosteroids, azathioprine, and methotrexate.[290]

NECROTIZING MYOPATHY

Necrotizing myopathy is an extremely rare muscle disorder characterized by rapidly developing, painful, predominantly proximal weakness that may mimic polymyositis.[298,299] The creatine kinase (CK) is elevated; a muscle biopsy may reveal little or no inflammation but numerous necrotic muscle fibers, usually in groups. Some necrotic fibers are infiltrated with macrophages. This disorder is associated with a wide variety of tumors, including SCLC, breast cancer, and gastrointestinal cancers. Intravenous immunoglobulin has led to improvement in some patients[298] but the pathogenesis is unknown. Serum antibody binding to muscle protein was described in one patient.[300]

REFERENCES

1. Darnell RB, Posner JB. Paraneoplastic syndromes affecting the nervous system. *Semin Oncol* 2006;33:270–298.
2. Brain WR, Norris FHE. *The Remote Effects of Cancer on the Nervous System*. New York: Grune and Stratton, 1965.
3. Henson RA, Urich H. *Cancer and the Nervous System: The Neurological Manifestations of Systemic Malignant Disease*. London: Blackwell Scientific, 1982.
4. Horino T, Takao T, Yamamoto M, et al. Spontaneous remission of small cell lung cancer: a case report and review in the literature. *Lung Cancer* 2006;53:249–252.
5. Byrne T, Mason WP, Posner JB, et al. Spontaneous neurological improvement in anti-Hu associated encephalomyelitis. *J Neurol Neurosurg Psychiatry* 1997;62:276–278.
6. Darnell RB, DeAngelis LM. Regression of small-cell lung carcinoma in patients with paraneoplastic neuronal antibodies. *Lancet* 1993;341:21–22.
6a. Pruss H, Voltz R, Gelderblom H, et al. Spontaneous remission of anti-Ma associated paraneoplastic mesodiencephalic and brainstem encephalitis. *J Neurol* 255:292–4.
7. Mareska M, Gutmann L. Lambert-Eaton myasthenic syndrome. *Semin Neurol* 2004;24:149–153.
8. Digre KB. Opsoclonus in adults. Report of three cases and review of the literature. *Arch Neurol* 1986;43:1165–1175.
9. Pranzatelli MR, Tate ED, Travelstead AL, et al. Immunologic and clinical responses to rituximab in a child with opsoclonus-myoclonus syndrome. *Pediatrics* 2005;115:115–119.
10. Shams'ili S, Grefkens J, de Leeuw B, et al. Paraneoplastic cerebellar degeneration associated with antineuronal antibodies: analysis of 50 patients. *Brain* 2003;126:1409–1418.
11. Rojas-Marcos I, Rousseau A, Keime-Guibert F, et al. Spectrum of paraneoplastic neurologic disorders in women with breast and gynecologic cancer. *Medicine (Baltimore)* 2003;82:216–223.
12. Stockton D, Doherty VR, Brewster DH. Risk of cancer in patients with dermatomyositis or polymyositis, and follow-up implications: a Scottish population-based cohort study. *Br J Cancer* 2001;85:41–45.

13. Levine SM. Cancer and myositis: new insights into an old association. *Curr Opin Rheumatol* 2006;18:620–624.

14. Nobile-Orazio E, Carpo M, Meucci N. Are there immunologically treatable motor neuron diseases? *Amyotroph Lateral Scler Other Motor Neuron Disord* 2001;2 (Suppl 1):S23–S30.

15. Rosenfeld MR, Posner JB. Paraneoplastic motor neuron disease. In: Rowland LP, ed. *Advances in Neurology, Vol. 56: Amyotrophic Lateral Sclerosis and Other Motor Neuron Diseases*. New York: Raven, 1991. pp. 445–459.

16. Antoine JC, Mosnier JF, Absi L, et al. Carcinoma associated paraneoplastic peripheral neuropathies in patients with and without anti-onconeural antibodies. *J Neurol Neurosurg Psychiatry* 1999;67:7–14.

17. Henson RA, Hoffman HL, Urich H. Encephalomyelitis with carcinoma. *Brain* 1965;88:449–464.

18. Anderson NE, Budde-Steffen C, Rosenblum MK, et al. Opsoclonus, myoclonus, ataxia, and encephalopathy in adults with cancer: a distinct paraneoplastic syndrome. *Medicine* 1988;67:100–109.

19. Croft PB, Wilkinson M. The incidence of carcinomatous neuromyopathy in patients with various types of carcinomas. *Brain* 1965;88:427–434.

20. Erlington GM, Murray NM, Spiro SG, et al. Neurological paraneoplastic syndromes in patients with small cell lung cancer. A prospective survey of 150 patients. *J Neurol Neurosurg Psychiatry* 1991;54:764–767.

21. Gomm SA, Thatcher N, Barber PV, et al. A clinicopathological study of the paraneoplastic neuromuscular syndromes associated with lung cancer. *Q J Med* 1990;278:577–595.

22. Hudson CN, Curling M, Potsides P, et al. Paraneoplastic syndromes in patients with ovarian neoplasia. *J Royal Soc Med* 1993;86:202–204.

23. Naschitz JE, Abrahamson J, Yeshurun D. Clinical significance of paraneoplastic syndrome. *Oncology* 1989;46:40–44.

24. Seute T, Leffers P, ten Velde GP, et al. Neurologic disorders in 432 consecutive patients with small cell lung carcinoma. *Cancer* 2004;100:801–806.

25. Drappatz J, Batchelor T. Neurologic complications of plasma cell disorders. *Clin Lymphoma* 2004;5:163–171.

26. Kelly JJ, Jr, Kyle RA, Miles JM, et al. Osteosclerotic myeloma and peripheral neuropathy. *Neurology* 1983;33:202–210.

27. Hawley RJ, Cohen MH, Saini N, et al. The carcinomatous neuromyopathy of oat cell lung cancer. *Ann Neurol* 1980;7:65–72.

28. Lipton RB, Galer BS, Dutcher JP, et al. Quantitative sensory testing demonstrates that subclinical sensory neuropathy is prevalent in patients with cancer. *Arch Neurol* 1987;44:944–946.

29. Lipton RB, Galer BS, Dutcher JP, et al. Large and small fibre type sensory dysfunction in patients with cancer. *J Neurol Neurosurg Psychiatry* 1991;54:706–709.

30. Wessel K, Diener HC, Dichgans J, et al. Cerebellar dysfunction in patients with bronchogenic carcinoma: clinical and posturographic findings. *J Neurol* 1988;235:290–296.

31. Sculier J-P, Feld R, Evans WK, et al. Neurologic disorders in patients with small cell lung cancer. *Cancer* 1987;60:2275–2283.

32. Oppenheim H. Uber Hirnsymptome bei Carcinomatose ohne nachweisbare. Veranderungen im Gehirn. *Charite-Ann (Berlin)* 1888;13:335–344.

33. Matsen SL, Yeo CJ, Hruban RH, et al. Hypercalcemia and pancreatic endocrine neoplasia with elevated PTH-rP: report of two new cases and subject review. *J Gastrointest Surg* 2005;9:270–279.

34. Terzolo M, Reimondo G, Ali A, et al. Ectopic ACTH syndrome: molecular bases and clinical heterogeneity. *Ann Oncol* 2001;2:S83–S87.

35. Ilias I, Torpy DJ, Pacak K, et al. Cushing's syndrome due to ectopic corticotropin secretion: twenty years' experience at the National Institutes of Health. *J Clin Endocrinol Metab* 2005;90:4955–4962.

36. Cleeland CS, Bennett GJ, Dantzer R, et al. Are the symptoms of cancer and cancer treatment due to a shared biologic mechanism? A cytokine-immunologic model of cancer symptoms. *Cancer* 2003;97:2919–2925.

37. Walsh D, Rybicki L. Symptom clustering in advanced cancer. *Support Care Cancer* 2006;14(8):831–836.

38. Denny-Brown D. Primary sensory neuropathy with muscular changes associated with carcinoma. *J Neurol Neurosurg Psychiatry* 1948;11:73–87.

39. Modlin IM, I. A 5-decade analysis of 13,715 carcinoid tumors. *Cancer* 2003;97:934–959.

40. Robertson RG, Geiger WJ, Davis NB. Carcinoid tumors. *Am Fam Physician* 2006;74:429–434.

41. Darnell RB, Posner JB. Paraneoplastic syndromes involving the nervous system. *N Engl J Med* 2003;349:1543–1554.

42. Dalmau J, Gultekin HS, Posner JB. Paraneoplastic neurologic syndromes: pathogenesis and physiopathology. *Brain Pathology* 1999;9:275–284.

43. Furneaux HM, Rosenblum MK, Dalmau J, et al. Selective expression of Purkinje-cell antigens in tumor tissue from patients with paraneoplastic cerebellar degeneration. *N Engl J Med* 1990;322:1844–1851.

44. Dalmau J, Furneaux HM, Cordon-Cardo C, et al. The expression of the Hu (paraneoplastic encephalomyelitis/sensory neuronopathy) antigen in human normal and tumor tissues. *Am J Pathol* 1992;141:881–886.

45. Rosenfeld MR, Eichen JG, Wade DF, et al. Molecular and clinical diversity in paraneoplastic immunity to Ma proteins. *Ann Neurol* 2001;50:339–348.

46. Corradi JP, Yang CW, Darnell JC, et al. A post-transcriptional regulatory mechanism restricts expression of the paraneoplastic cerebellar degeneration antigen cdr2 to immune privileged tissues. *J Neurosci* 1997;17:1406–1415.

47. Dalmau J, Gultekin SH, Voltz R, et al. Ma1, a novel neuronal and testis specific protein, is recognized by the serum of patients with paraneoplastic neurologic disorders. *Brain* 1999;122:27–39.

48. Carpentier AF, Voltz R, Dechamps T, et al. Absence of HuD gene mutations in paraneoplastic small cell lung cancer tissue. *Neurology* 1998;50:1919.

48a. Savage PA, Vosseller K, Kang C, et al. Recognition of a ubiquitous self antigen by prostate cancer-infiltrating CD8+ T lymphocytes. *Science* 2008;319:164–165.

49. Hetzel DJ, Stanhope CR, O'Neill BP, et al. Gynecologic cancer in patients with subacute cerebellar degeneration predicted by anti-Purkinje cell antibodies and limited in metastatic volume. *Mayo Clin Proc* 1990;65:1558–1563.

50. Drachman DB. How to recognize an antibody-mediated autoimmune disease: criteria. *Res Publ Assoc Res Nerv Ment Dis* 1990;68:183–186.

51. Sommer C, Weishaupt A, Brinkhoff J, et al. Paraneoplastic stiff-person syndrome: passive transfer to rats by means of IgG antibodies to amphiphysin. *Lancet* 2005;365:1406–1411.

52. Adamus G, Machnicki M, Seigel GM. Apoptotic retinal cell death induced by antirecoverin autoantibodies of cancer-associated retinopathy. *Invest Ophthalmol Vis Sci* 1997;38:283–291.

53. Ohguro H, Ogawa K, Maeda T, et al. Cancer-associated retinopathy induced by both anti-recoverin and anti-hsc70 antibodies in vivo. *Invest Ophthalmol Vis Sci* 1999;40:3160–3167.

54. Vernino S, Low PA, Fealey RD, et al. Autoantibodies to ganglionic acetylcholine receptors in autoimmune autonomic neuropathies. *N Engl J Med* 2000;343:847–855.

55. Albert ML, Austin LM, Darnell RB. Detection and treatment of activated T cells in the cerebrospinal fluid of patients with paraneoplastic cerebellar degeneration. *Ann Neurol* 2000;47:9–17.

56. Rousseau A, Benyahia B, Dalmau J, et al. T cell response to Hu-D peptides in patients with anti-Hu syndrome. *J Neuro-Oncol* 2005;71:231–236.

57. Albert ML, Darnell RB. Paraneoplastic neurological degenerations: keys to tumour immunity. *Nat Rev Cancer* 2004;4:36–44.

58. Voltz R, Dalmau J, Posner JB, et al. T-cell receptor analysis in anti-Hu associated paraneoplastic encephalomyelitis. *Neurology* 1998;51:1146–1150.

59. Albert ML, Darnell JC, Bender A, et al. Tumor-specific killer cells in paraneoplastic cerebellar degeneration. *Nat Med* 1998;4:1321–1324.

60. Maeda A. Identification of human antitumor cytotoxic T lymphocytes epitopes of recoverin, a cancer-associated retinopathy antigen, possibly related with a better prognosis in a paraneoplastic syndrome. *Eur J Immunol* 2001;31:563–572.

61. Maeda A, Ohguro H, Nageta Y, et al. Vaccination with recoverin, a cancer-associated retinopathy antigen, induces autoimmune retinal dysfunction and tumor cell regression in mice. *Eur J Immunol* 2002;32:2300–2307.

62. Graus F, Delattre JY, Antoine JC, et al. Recommended diagnostic criteria for paraneoplastic neurological syndromes. *J Neurol Neurosurg Psychiatry* 2004;75:1135–1140.

63. Pittock SJ, Kryzer TJ, Lennon VA. Paraneoplastic antibodies coexist and predict cancer, not neurological syndrome. *Ann Neurol* 2004;56:715–719.

64. Rees JH, Hain SF, Johnson MR, et al. The role of [(18) F]fluoro-2-deoxyglucose-PET scanning in the diagnosis of paraneoplastic neurological disorders. *Brain* 2001;124:2223–2231.

65. Linke R, Schroeder M, Helmberger T, et al. Antibody-positive paraneoplastic neurologic syndromes: value of CT and PET for tumor diagnosis. *Neurology* 2004;63:282–286.

66. Peterson K, Rosenblum MK, Kotanides H, et al. Paraneoplastic cerebellar degeneration. I. A clinical analysis of 55 anti-Yo antibody positive patients. *Neurology* 1992;42:1931–1937.

67. Brierley JB, Corsellis JAN, Hierons R, et al. Subacute encephalitis of later adult life. Mainly affecting the limbic areas. *Brain* 1960;83:357–368.

68. Graus F, Dalmau J, René R, et al. Anti-Hu antibodies in patients with small-cell lung cancer: Association with complete response to therapy and improved survival. *J Clin Oncol* 1997;15:2866–2872.

69. Verschuuren JJ, Perquin M, Ten Velde G, et al. Anti-Hu antibody titre and brain metastases before and after treatment for small cell lung cancer. *J Neurol Neurosurg Psychiatry* 1999;67:353–357.

70. Monstad SE, Drivsholm L, Storstein A, et al. Hu and voltage-gated calcium channel (VGCC) antibodies related to the prognosis of small-cell lung cancer. *J Clin Oncol* 2004;22:795–800.

71. Darnell R. Tumor immunity in small-cell lung cancer. *J Clin Oncol* 2004;22:762–764.

72. Maddison P, Newsom-Davis J, Mills KR, et al. Favourable prognosis in Lambert-Eaton myasthenic syndrome and small-cell lung carcinoma [letter]. *Lancet* 1999;353:117–118.

73. Wirtz PW, Lang B, Graus F, et al. P/Q-type calcium channel antibodies, Lambert-Eaton myasthenic syndrome and small cell lung cancer. *J Neuroimmunol* 2005;164:161–165.

74. Rojas I, Graus F, Keime-Guibert F, et al. Long-term clinical outcome of paraneoplastic cerebellar degeneration and anti-Yo antibodies. *Neurology* 2000;55:713–715.

75. Dropcho EJ. Immunotherapy for paraneoplastic neurological disorders. *Expert Opin Biol Ther* 2005;5:1339–1348.

76. Keime-Guibert F, Graus F, Broet P, et al. Clinical outcome of patients with anti-Hu-associated encephalomyelitis after treatment of the tumor. *Neurology* 1999;53:1719–1723.

77. Douglas CA, Ellershaw J. Anti-Hu antibodies may indicate a positive response to chemotherapy in paraneoplastic syndrome secondary to small cell lung cancer. *Palliat Med* 2003;17:638–639.

78. Keime-Guibert F, Graus F, Fleury A, et al. Treatment of paraneoplastic neurological syndromes with antineuronal antibodies (Anti-Hu, Anti-Yo) with a combination of immunoglobulins, cyclophosphamide, and methylprednisolone. *J Neurol Neurosurg Psychiatry* 2000;68:479–482.

79. Vernino S, O'Neill BP, Marks RS, et al. Immunomodulatory treatment trial for paraneoplastic neurological disorders. *Neuro-Oncology* 2004;6:55–62.

80. Rosenfeld MR, Dalmau J. Current therapies for paraneoplastic neurologic syndromes. *Curr Treat Options Neurol* 2003;5:69–77.

81. Marchand V, Graveleau J, Lanctin-Garcia C, et al. A rare gynecological case of paraneoplastic cerebellar degeneration discovered by FDG-PET. *Gynecol Oncol* 2007;105:545–547.

82. Espandar L. Successful treatment of cancer-associated retinopathy with alemtuzumab. *J Neuro-Oncol* 2007;83:295–302.

83. Sillevis-Smitt P, Grefkens J, De Leeuw B, et al. Survival and outcome in 73 anti-Hu positive patients with paraneoplastic encephalomyelitis/sensory neuronopathy. *J Neurol* 2002;249:745–753.

84. Brouwer B. Beitrag zur Kenntnis der chronischen diffusen Kleinhirnerkrankungen. *Neurol Zentralbl* 1919;38:674–682.

85. Brouwer B, Biemond A. Les affections parenchymateuses du cervelet et leur signification du point de vue de l'anatomie et la physiologie de cet organe. *J Belg Neurol Psychiatrie* 1938;38:691–757.

86. Clouston PD, Saper CB, Arbizu T, et al. Paraneoplastic cerebellar degeneration. III. Cerebellar degeneration, cancer and the Lambert-Eaton myasthenic syndrome. *Neurology* 1992;42:1944–1950.

87. Hammack J, Kotanides H, Rosenblum MK, et al. Paraneoplastic cerebellar degeneration: II. Clinical and immunologic findings in 21 patients with Hodgkin's disease. *Neurology* 1992;42:1938–1943.

88. Bernal F, Shams'ili S, Rojas I, et al. Anti-Tr antibodies as markers of paraneoplastic cerebellar degeneration and Hodgkin's disease. *Neurology* 2003; 60:230–234.

89. Matthew RM, Cohen AB, Galetta SL, et al. Paraneoplastic cerebellar degeneration: Yo-expressing tumor revealed after a 5-year follow-up with FDG-PET. *J Neurol Sci* 2006;250:153–155.

90. Bonakis A. Acute onset paraneoplastic cerebellar degeneration. *J Neuro-Oncol* 2007;84:329–330.

91. Mason WP, Graus F, Lang B, et al. Small-cell lung cancer, paraneoplastic cerebellar degeneration and the Lambert-Eaton myasthenic syndrome. *Brain* 1997;120:1279–1300.

92. Graus F, Lang B, Pozo-Rosich P, et al. P/Q type calcium-channel antibodies in paraneoplastic cerebellar degeneration with lung cancer. *Neurology* 2002;59:764–766.

93. Darnell JC, Albert ML, Darnell RB. Cdr2, a target antigen of naturally occurring human tumor immunity, is widely expressed in gynecological tumors. *Cancer Res* 2000;60:2136–2139.

94. Furneaux HM, Reich L, Posner JB. Autoantibody synthesis in the central nervous system of patients with paraneoplastic syndromes. *Neurology* 1990;40:1085–1091.

95. Santomasso BD, Roberts WK, Thomas A, et al. A T-cell receptor associated with naturally occurring human tumor immunity. *Proc Natl Acad Sci USA*, 2007; 104: 19073–19078.

96. Widdess-Walsh P, Tavee JO, Schuele S, et al. Response to intravenous immunoglobulin in anti-Yo associated paraneoplastic cerebellar degeneration: case report and review of the literature. *J Neuro-Oncol* 2003;63:187–190.

97. Graus F, Keime-Guibert F, Reñe R, et al. Anti-Hu-associated paraneoplastic encephalomyelitis: analysis of 200 patients. *Brain* 2001;124:1138–1148.

98. Cross SA, Salomao DR, Parisi JE, et al. Paraneoplastic autoimmune optic neuritis with retinitis defined by CRMP-5-IgG. *Ann Neurol* 2003;54:38–50.

99. Bataller L, Wade DF, Fuller GN, et al. Cerebellar degeneration and autoimmunity to zinc-finger proteins of the cerebellum. *Neurology* 2002; 59:1985–1987.

100. Sillevis-Smitt P, Kinoshita A, De Leeuw B, et al. Paraneoplastic cerebellar ataxia due to autoantibodies against a glutamate receptor. *N Engl J Med* 2000;342:21–27.

101. Sabater L. Protein kinase C-gamma autoimmunity in paraneoplastic cerebellar degeneration and non-small-cell lung cancer. *J Neurol Neurosurg Psychiatry* 2006;77:1359–1362.

102. Saiz A, Graus F, Dalmau J, et al. Detection of 14-3-3 brain protein in the cerebrospinal fluid of patients with paraneoplastic neurological disorders. *Ann Neurol* 1999;46:774–777.

103. Hadjivassiliou M, Grunewald R, Sharrack B, et al. Gluten ataxia in perspective: epidemiology, genetic susceptibility and clinical characteristics. *Brain* 2003;126:685–691.

104. Tagliati M, Simpson D, Morgello S, et al. Cerebellar degeneration associated with human immunodeficiency virus infection. *Neurology* 1998;50:244–251.

105. Graus F, Illa I, Agusti M, et al. Effect of intraventricular injection of an anti-Purkinje cell antibody (anti-Yo) in a guinea pig model. *J Neurol Sci* 1991;106:82–87.

106. Yoshimi K, Woo M, Son Y, et al. IgG-immunostaining in the intact rabbit brain: variable but significant staining of hippocampal and cerebellar neurons with anti-IgG. *Brain Res* 2002;956:53–66.

107. Tanaka K, Ding X, Tanaka M. Effects of antineuronal antibodies from patients with paraneoplastic neurological syndrome on primary-cultured neurons. *J Neurol Sci* 2004;217:25–30.

108. Coesmans M, Smitt PA, Linden DJ, et al. Mechanisms underlying cerebellar motor deficits due to mGluR1-autoantibodies. *Ann Neurol* 2003;53:325–336.

109. Fukuda T, Motomura M, Nakao Y, et al. Reduction of P/Q-type calcium channels in the postmortem cerebellum of paraneoplastic cerebellar degeneration with Lambert-Eaton myasthenic syndrome. *Ann Neurol* 2003;53:21–28.

110. Storstein A, Krossnes B, Vedeler CA. Autopsy findings in the nervous system and ovarian tumour of two patients with paraneoplastic cerebellar degeneration. *Acta Neurol Scand Suppl* 2006;183:69–70.

111. Verschuuren J, Chuang L, Rosenblum MK, et al. Inflammatory infiltrates and complete absence of Purkinje cells in anti-Yo associated paraneoplastic cerebellar degeneration. *Acta Neuropathol* 1996; 91:519–525.

112. Dalmau J, Graus F, Rosenblum MK, Posner JB. Anti-Hu-associated paraneoplastic encephalomyelitis/sensory neuronopathy. A clinical study of 71 patients. *Medicine* 1992;71:59–72.

113. Gultekin SH, Rosenfeld MR, Voltz R, et al. Paraneoplastic limbic encephalitis: neurological symptoms, immunological findings and tumour association in 50 patients. *Brain* 2000;123:1481–1494.

114. Lawn ND, Westmoreland BF, Kiely MJ, et al. Clinical, magnetic resonance imaging, and electroencephalographic findings in paraneoplastic limbic encephalitis. *Mayo Clin Proc* 2003;78:1363–1368.

115. Fadul CE, Stommel EW, Dragnev KH, et al. Focal paraneoplastic limbic encephalitis presenting as orgasmic epilepsy. *J Neuro-Oncol* 2005;72:195–198.

115a. Hoffmann LA, Jarius S, Pellkofer HL, et al. Anti-Ma and anti-Ta associated paraneoplastic neurological syndromes: Twenty-two newly diagnosed patients and review of previous cases. *J Neurol Neurosurg Psychiatry* 2008 Jan 25 [Epub ahead of print].

116. Vincent A, Buckley C, Schott JM, et al. Potassium channel antibody-associated encephalopathy: a potentially immunotherapy-responsive form of limbic encephalitis. *Brain* 2004;127:701–712.

117. Sansing LH. A patient with encephalitis associated with NMDA receptor antibodies. *Nature Clin Prac Neurol* 2007;3:291–296.

117a. Dalmau J, Gleichman AJ, Rossi JE, et al. The syndrome and target epitopes of anti-NMDAR encephalitis: A study of 67 patients. *Neurology* 2008;(Suppl 1)70:A111–112.

118. Carr I. The Ophelia syndrome: memory loss in Hodgkin's disease. *Lancet* 1982;1:844–845.

119. Ances BM, Vitaliani R, Taylor RA, et al. Treatment-responsive limbic encephalitis identified by neuropil antibodies: MRI and PET correlates. *Brain* 2005;128:1764–1777.

120. Alamowitch S, Graus F, Uchuya M, et al. Limbic encephalitis and small cell lung cancer—clinical and immunological features. *Brain* 1997;120:923–928.

121. Voltz R, Gultekin SH, Rosenfeld MR, et al. A serologic marker of paraneoplastic limbic and brain-stem encephalitis in patients with testicular cancer. *N Engl J Med* 1999;340:1788–1795.

122. Dalmau J, Graus F, Villarejo A, et al. Clinical analysis of anti-Ma2-associated encephalitis. *Brain* 2004;127:1831–1844.

123. Kastrup O, Meyring S, Diener HC. Atypical paraneoplastic brainstem encephalitis associated with anti-Ri-antibodies due to thymic carcinoma with possible clinical response to immunoglobulins. *Eur Neurol* 2001;45:285–287.

124. Muni RH, Wennberg R, Mikulis DJ, et al. Bilateral horizontal gaze palsy in presumed paraneoplastic brainstem encephalitis associated with a benign ovarian teratoma. *J Neuroophthalmol* 2004;24:114–118.

125. Compta Y. REM sleep behavior disorder and narcoleptic features in anti-Ma2-associated encephalitis. *Sleep* 2007;30:767–769.

126. Golbe LI, Miller DC, Duvoisin RC. Paraneoplastic degeneration of the substantia nigra with dystonia and Parkinsonism. *Movement Dis* 1989;4:147–152.

127. Fahn S, Brin MF, Dwork AJ, et al. Case 1, 1996: rapidly progressive parkinsonism, incontinence, impotency, and levodopa-induced moaning in a patient with multiple myeloma. *Mov Disord* 1996;11:298–310.

128. Yu Z. CRMP-5 neuronal autoantibody: marker of lung cancer and thymoma-related autoimmunity. *Ann Neurol* 2001;49:146–154.

129. Vernino S, Tuite P, Adler CH, et al. Paraneoplastic chorea associated with CRMP-5 neuronal antibody and lung carcinoma. *Ann Neurol* 2002;51:625–630.

130. Samii A, Dahlen DD, Spence AM, et al. Paraneoplastic movement disorder in a patient with non-Hodgkin's lymphoma and CRMP-5 autoantibody. *Mov Disord* 2003;18:1556–1558.

131. Tan JH. Paraneoplastic progressive supranuclear palsy syndrome in a patient with B-cell lymphoma. *Parkinsonism Rel Disord* 2005;11:187–191.

132. Leek S, Higgins MJ, Patel BM, et al. Paraneoplastic coma and acquired central alveolar hypoventilation as a manifestation of brainstem encephalitis in a patient with ANNA-1 antibody and small-cell lung cancer. *Neurocrit Care* 2006;4:137–139.

133. Tonomura Y, Kataoka H, Hara Y, et al. Clinical analysis of paraneoplastic encephalitis associated with ovarian teratoma. *J Neuro-Oncol* 2007;84:287–292.

134. Rajabally YA, Naz S, Farrell D, et al. Paraneoplastic brainstem encephalitis with tetraparesis in a patient with anti-Ri antibodies. *J Neurol* 2004; 251:1528–1529.

135. Corato M, Marinou-Aktipi K, Nano R, et al. Paraneoplastic brainstem encephalitis in a patient with malignant fibrous histiocytoma and atypical anti-neuronal antibodies. *J Neurol* 2004;251:1415–1417.

136. Barnett M, Prosser J, Sutton I, et al. Paraneoplastic brain stem encephalitis in a woman with anti-Ma2 antibody. *J Neurol Neurosurg Psychiatry* 2001;70:222–225.

137. Pittock SJ. Inflammatory transverse myelitis: evolving concepts. *Curr Opin Neurol* 2006;19:362–368.

138. Forsyth PA, Dalmau J, Graus F, et al. Motor neuron syndromes in cancer patients. *Ann Neurol* 1997;41:722–730.

139. Pittock SJ, Lucchinetti CF, Parisi JE, et al. Amphiphysin autoimmunity: paraneoplastic accompaniments. *Ann Neurol* 2005;58:96–107.

140. Babikian VL, Stefansson K, Dieperink ME, et al. Paraneoplastic myelopathy: antibodies against protein in normal spinal cord and underlying neoplasm. Letter to the editor. *Lancet* 1985;2:49–50.

141. Hughes RA, Britton T, Richards M. Effects of lymphoma on the peripheral nervous system. *J R Soc Med* 1994;87:526–530.

142. Vega F, Graus F, Chen QM, et al. Intrathecal synthesis of the anti-Hu antibody in patients with paraneoplastic encephalomyelitis or sensory neuronopathy: Clinical-immunologic correlation. *Neurology* 1994;44:2145–2147.

143. Lorusso L. Autonomic paraneoplastic neurological syndromes. *Autoimmunity Rev* 2007;6:162–168.

144. Veilleux M, Bernier JP, Lamarche JB. Paraneoplastic encephalomyelitis and subacute dysautonomia due to an occult atypical carcinoid tumour of the lung. *Can J Neurol Sci* 1990;17:324–328.

145. Vernino S, Adamski J, Kryzer TJ, et al. Neuronal nicotinic ACh receptor antibody in subacute autonomic neuropathy and cancer-related syndromes. *Neurology* 1998;50:1806–1813.

146. Bremner FD, Smith SE. Pupil abnormalities in selected autonomic neuropathies. *J Neuroophthalmol* 2006;26:209–219.

147. Lennon VA, Sas DF, Busk MF, et al. Enteric neuronal autoantibodies in pseudoobstruction with small-cell lung carcinoma. *Gastroenterology* 1991; 100:137–142.

148. Lennon VA, Ermilov LG, Szurszewski JH, et al. Immunization with neuronal nicotinic acetylcholine receptor induces neurological autoimmune disease. *J Clin Invest* 2003;111:907–913.

149. Wong A. An update on opsoclonus. *Curr Opin Neurol* 2007;20:25–31.

150. Anderson NE, Budde-Steffen C, Wiley RG, et al. A variant of the anti-Purkinje cell antibody in a patient with paraneoplastic cerebellar degeneration. *Neurology* 1988;38:1018–1026.

151. Tate ED, Allison TJ, Pranzatelli MR, et al. Neuro-epidemiologic trends in 105 US cases of pediatric opsoclonus-myoclonus syndrome. *J Pediatr Oncol Nurs* 2005;22:8–19.

152. Rudnick E, Khakoo Y, Antunes NL, et al. Opsoclonus-myoclonus-ataxia syndrome in neuroblastoma: clinical outcome and antineuronal antibodies-a report from the Children's Cancer Group Study. *Med Pediatr Oncol* 2001;36:612–622.

153. Antunes NL, Khakoo Y, Matthay KK, et al. Antineuronal antibodies in patients with neuroblastoma and paraneoplastic opsoclonus-myoclonus. *J Pediatr Hematol Oncol* 2000;22:315–320.

154. Pranzatelli MR, Travelstead AL, Tate ED, et al. CSF B-cell expansion in opsoclonus-myoclonus syndrome: a biomarker of disease activity. *Mov Disord* 2004;19:770–777.

155. Ertle F, Behnisch W, Al Mulla NA, et al. Treatment of neuroblastoma-related opsoclonus-myoclonus-ataxia syndrome with high-dose dexamethasone pulses. *Pediatr Blood Cancer.* 2008; 50:683–687.

156. Pranzatelli MR. Rituximab (anti-CD20) adjunctive therapy for opsoclonus-myoclonus syndrome. *J Pediatr Hematol Oncol* 2006;28:585–593.

157. Connolly AM, Pestronk A, Mehta S, et al. Serum autoantibodies in childhood opsoclonus-myoclonus syndrome: An analysis of antigenic targets in neural tissues. *J Pediatr* 1997;130:878–884.

158. Blaes F, Fuhlhuber V, Korfei M, et al. Surface-binding autoantibodies to cerebellar neurons in opsoclonus syndrome. *Ann Neurol* 2005;58:313–317.

159. Korfei M, Fuhlhuber V, Schmidt-Woll T, et al. Functional characterisation of autoantibodies from patients with pediatric opsoclonus-myoclonus-syndrome. *J Neuroimmunol* 2005;170:150–157.

160. Hormigo A, Dalmau J, Rosenblum MK, et al. Immunological and pathological study of anti-Ri-associated encephalopathy. *Ann Neurol* 1994; 36:896–902.

161. Luque FA, Furneaux HM, Ferziger R, et al. Anti-Ri: an antibody associated with paraneoplastic opsoclonus and breast cancer. *Ann Neurol* 1991;29:241–251.

162. Pittock SJ, Lucchinetti CF, Lennon VA. Antineuronal nuclear autoantibody type 2: paraneoplastic accompaniments. *Ann Neurol* 2003;53:580–587.

163. Buckanovich RJ, Posner JB, Darnell RB. Nova, a paraneoplastic Ri antigen is homologous to an RNA-binding protein and is specifically expressed in the developing motor system. *Neuron* 1993;11:1–20.

164. Jensen KB, Dredge BK, Stefani G, et al. Nova-1 regulates neuron-specific alternative splicing and is essential for neuronal viability. *Neuron* 2000;25:359–371.

165. Ule J, Jensen KB, Ruggiu M, et al. CLIP identifies Nova-regulated RNA networks in the brain. *Science* 2003;302:1212–1215.

166. Ule J, Ule A, Spencer J, et al. Nova regulates brain-specific splicing to shape the synapse. *Nat Genet* 2005;37:844–852.

167. Cher LM, Hochberg FH, Teruya J, et al. Therapy for paraneoplastic neurologic syndromes in six patients with protein A column immunoadsorption. *Cancer* 1995;75:1678–1683.

168. Bataller L, Graus F, Saiz A, et al. Clinical outcome in adult onset idiopathic or paraneoplastic opsoclonus-myoclonus. *Brain* 2001;124:437–443.

169. Mancall EL, Rosales RK. Necrotizing myelopathy associated with visceral carcinoma. *Brain* 1964;87:639–656.

170. Hughes M, Ahern V, Kefford R, et al. Paraneoplastic myelopathy at diagnosis in a patient with pathologic stage 1A Hodgkin disease. *Cancer* 1992; 70:1598–1600.

171. Dansey RD, Hammond-Tooke GD, Lai K, et al. Subacute myelopathy: an unusual paraneoplastic complication of Hodgkin's disease. *Med Pediatr Oncol* 1989;16:284–286.

172. Grignani G, Gobbi PG, Piccolo G, et al. Progressive necrotic myelopathy as a paraneoplastic syndrome: report of a case and some pathogenetic considerations. *J Intern Med* 1992;231:81–85.

173. Ojeda VJ. Necrotizing myelopathy associated with malignancy. A clinicopathologic study of two cases and literature review. *Cancer* 1984;53:1115–1123.

174. Martin Escudero JC, Aparicio BM, Borrego PH, et al. [Necrotizing myelopathy associated with neoplasia. A clinico-pathological study of 2 cases

and a review of the literature]. *An Med Interna* 1991;8:497–500.

175. Kida E, Barcikowska M, Michalska T, et al. Peripheral nervous system alterations in small cell lung cancer. Clinico-pathological study. *Neuropatol Pol* 1992;30:43–56.

176. Antoine JC, Camdessanché JP, Absi L, et al. Devic disease and thymoma with anti-central nervous system and antithymus antibodies. *Neurology* 2004;62:978–980.

177. Handforth A, Nag S, Sharp D, et al. Paraneoplastic subacute necrotic myelopathy. *Can J Neurol Sci* 1983;10:204–207.

178. Kuroda Y, Miyahara M, Sakemi T, et al. Autopsy report of acute necrotizing opticomyelopathy associated with thyroid cancer. *J Neurol Sci* 1993;120:29–32.

179. Lennon VA, Wingerchuk DM, Kryzer TJ, et al. A serum autoantibody marker of neuromyelitis optica: distinction from multiple sclerosis. *Lancet* 2004;364:2106–2112.

180. Lennon VA, Kryzer TJ, Pittock SJ, et al. IgG marker of optic-spinal multiple sclerosis binds to the aquaporin-4 water channel. *J Exp Med* 2005; 202:473–477.

181. Katz JD, Ropper AH. Progressive necrotic myelopathy. *Arch Neurol* 2000;57:355–361.

182. Rowland LP, Shneider NA. Amyotrophic lateral sclerosis. *N Engl J Med* 2001;344:1688–1700.

183. Norris FH, Denys EH, Sang K, et al. Population study of amyotrophic lateral sclerosis. *Ann Neurol* 1989;26:139–140.

184. Mitchell DM, Olczak SA. Remission of a syndrome indistinguishable from motor neurone disease after resection of bronchial carcinoma. *BMJ* 1979;2:176–177.

185. Brain L, Croft PB, Wilkinson M. Motor neurone disease as a manifestation of neoplasm (with a note on the course of classical motor neurone disease). *Brain* 1965;88:479–500.

186. Couratier P, Yi FH, Preud'homme JL, et al. Serum autoantibodies to neurofilament proteins in sporadic amyotrophic lateral sclerosis. *J Neurol Sci* 1998;154:137–145.

187. Ferracci F, Fassetta G, Butler MH, et al. A novel anti-neuronal antibody in a motor neuron syndrome associated with breast cancer. *Neurology* 1999;53:852–855.

188. Rowland LP, Sherman WH, Latov N, et al. Amyotrophic lateral sclerosis and lymphoma: bone marrow examination and other diagnostic tests. *Neurology* 1992;42:1101–1102.

189. Rowland LP, Sherman WL, Hays AP, et al. Autopsy-proven amyotrophic lateral sclerosis, Waldenstrom's macroglobulinemia, and antibodies to sulfated glucuronic acid paragloboside. *Neurology* 1995; 45:827–829.

190. Kelly JJ, Karcher DS. Lymphoma and peripheral neuropathy: a clinical review. *Muscle Nerve* 2005;31:301–313.

191. Gericke CA, Zschenderlein R, Ludolph AC. Amytrophic lateral sclerosis associated with multiple myeloma, endocrinopathy and skin changes suggestive of a POEMS syndrome variant. *J Neurol Sci* 1995;129:58–60.

192. Gordon PH, Rowland LP, Younger DS, et al. Lymphoproliferative disorders and motor neuron disease: an update. *Neurology* 1997;48:1671–1678.

193. Schold SC, Cho ES, Somasundaram M, et al. Subacute motor neuronopathy: a remote effect of lymphoma. *Ann Neurol* 1979;5:271–287.

194. Rowland LP, Schneck SA. Neuromuscular disorders associated with malignant neoplastic disease. *J Chron Dis* 1963;16:777–795.

195. Walton JN, Tomlinson BE, Pearce GW. Subacute "poliomyelitis" and Hodgkin's disease. *J Neurol Sci* 1968;6:435–445.

196. Younger DS, Rowland LP, Latov N, et al. Motor neuron disease and amyotrophic lateral sclerosis: relation of high CSF protein content to paraproteinemia and clinical syndromes. *Neurology* 1990;40:595–599.

197. Berghs S, Ferracci F, Maksimova E, et al. Autoimmunity to *b*IV spectrin in paraneoplastic lower motor neuron syndrome. *Proc Natl Acad Sci USA* 2001;98:6945–6950.

198. Murinson BB. Stiff-person syndrome. *Neurologist* 2004;10:131–137.

199. Lockman J. Stiff-person syndrome. *Curr Treat Options Neurol* 2007;9:234–240.

200. Espay AJ. Rigidity and spasms from autoimmune encephalomyelopathies: stiff-person syndrome. *Muscle Nerve* 2006;34:677–690.

201. Darnell RB, Victor J, Rubin M, et al. A novel anti-neuronal antibody in stiff-man syndrome. *Neurology* 1993;43:114–120.

202. Schiff D, Dalmau J, Myers DJ. Anti-GAD antibody positive stiff-limb syndrome in multiple myeloma. *J Neuro-Oncol* 2003;65:173–175.

203. McHugh JC. GAD antibody positive paraneoplastic stiff person syndrome in a patient with renal cell carcinoma. *Mov Disord* 2007;22:1343–1346.

204. McCabe DJH, Turner NC, Chao D, et al. Paraneoplastic "stiff person syndrome" with metastatic adenocarcinoma and anti-Ri antibodies. *Neurology* 2004;62:1402–1404.

205. Butler MH, Hayashi A, Ohkoshi N, et al. Autoimmunity to gephyrin in Stiff-Man syndrome. *Neuron* 2000;26:307–312.

206. Werbrouck B, Meire V, De Bleecker JL. Multiple neurological syndromes during Hodgkin lymphoma remission. *Acta Neurol Belg* 2005;105:48–50.

207. Damek DM. Paraneoplastic retinopathy/optic neuropathy. *Curr Treat Options Neurol* 2005;7:57–67.

208. Ling CP, Pavesio C. Paraneoplastic syndromes associated with visual loss. *Curr Opin Ophthalmol* 2003;14:426–432.

209. Chan JW. Paraneoplastic retinopathies and optic neuropathies. *Surv Ophthalmol* 2003;48:12–38.

210. Cornblath WT. Paraneoplastic disorders of ophthalmic interest. *Ophthalmol Clin N Am* 2004; 17:447–454.

211. Khan N, Huang JJ, Foster CS. Cancer associated retinopathy (CAR): an autoimmune-mediated paraneoplastic syndrome. *Semin Ophthalmol* 2006; 21:135–141.

212. Thirkill CE. Cancer-induced, immune-mediated ocular degenerations. *Ocul Immunol Inflamm* 2005; 13:119–131.

213. Sawyer RA, Selhorst JB, Zimmerman LE, et al. Blindness caused by photoreceptor degeneration as a remote effect of cancer. *Am J Ophthal* 1976; 81:606–613.

213a. Saito W, Kase S, Ohguro H, et al. Slowly progressive cancer-associated retinopathy. *Arch Ophthalmol* 2007;125:1431–1433.

214. Jacobson DM, Thirkill CE. Paraneoplastic cone dysfunction: an unusual visual remote effect of cancer. *Arch Ophthalmol* 1995;113:1580–1582.

215. Savchenko MS, Bazhin AV, Shifrina ON, et al. Antirecoverin autoantibodies in the patient with non-small cell lung cancer but without cancer-associated retinopathy. *Lung Cancer* 2003;41: 363–367.

216. Bazhin AV. Recoverin as a cancer-retina antigen. *Cancer Immunol Immunother* 2007;56:110–116.

217. Bazhin AV, Savchenko MS, Belousov EV, et al. Stimu- lation of the aberrant expression of a paraneoplastic antigen, recoverin, in small cell lung cancer cell lines. *Lung Cancer* 2004;45:299–305.

218. Ohguro H, Maruyama I, Nakazawa M, et al. Antirecoverin antibody in the aqueous humor of a patient with cancer-associated retinopathy. *Am J Ophthalmol* 2002;134:605–607.

219. Weleber RG, Watzke RC, Shults WT, et al. Clinical and electrophysiologic characterization of paraneoplastic and autoimmune retinopathies associated with antienolase antibodies. *Am J Ophthalmol* 2005;139:780–794.

220. Adamus G, Amundson D, Seigel GM, et al. Antienolase-alpha autoantibodies in cancer-associated retinopathy: epitope mapping and cytotoxicity on retinal cells. *J Autoimmun* 1998;11:671–677.

221. Adamus G, Ren G, Weleber RG. Autoantibodies against retinal proteins in paraneoplastic and autoimmune retinopathy. *BMC Ophthalmol* 2004;4:5.

222. Adamus G, Machnicki M, Elerding H, et al. Antibodies to recoverin induce apoptosis of photoreceptor and bipolar cells in vivo. *J Autoimmun* 1998;11:523–533.

223. Shiraga S, Adamus G. Mechanism of CAR syndrome: anti-recoverin antibodies are the inducers of retinal cell apoptotic death via the caspase 9- and caspase 3-dependent pathway. *J Neuroimmunol* 2002;132:72–82.

224. Maeda A, Maeda T, Ohguro H, et al. Vaccination with recoverin, a cancer-associated retinopathy antigen, induces autoimmune retinal dysfunction and tumor cell regression in mice. *Eur J Immunol* 2002;32:2300–2307.

225. Ohguro H, Nakazawa M. Pathological roles of recoverin in cancer-associated retinopathy. *Adv Exp Med Biol* 2002;514:109–124.

226. Guy J, Aptsiauri N. Reversal of carcinoma associated retinopathy (CAR) induced blindness with IV Ig. *Neurology* 1997;48:A329–A330. Abstract.

227. Guy J, Aptsiauri N. Treatment of paraneoplastic visual loss with intravenous immunoglobulin—report of 3 cases. *Arch Ophthalmol* 1999;117:471–477.

228. Ladewig G, Reinhold U, Thirkill CE, et al. Incidence of antiretinal antibodies in melanoma: screening of 77 serum samples from 51 patients with American Joint Committee on Cancer stage I-IV. *Br J Dermatol* 2005;152:931–938.

229. Jacobzone C, Cochard-Marianowski C, Kupfer I, et al. Corticosteroid treatment for melanoma-associated retinopathy—effect on visual acuity and electrophysiologic findings. *Arch Dermatol* 2004; 140:1258–1261.

230. Chan C, O'Day J. Melanoma-associated retinopathy: does autoimmunity prolong survival? *Clin Experiment Ophthalmol* 2001;29:235–238.

231. Lieberman FS, Odel J, Hirsh J, et al. Bilateral optic neuropathy with IgGkappa multiple myeloma improved after myeloablative chemotherapy. *Neurology* 1999;52:414–416.

232. Antoine JC, Camdessanche JP. Peripheral nervous system involvement in patients with cancer. *Lancet Neurol* 2007;6:75–86.

233. Lisak RP, Mitchell M, Zweiman B, et al. Guillain-Barre syndrome and Hodgkin's disease: Three cases with immunological studies. *Ann Neurol* 1977;1:72–78.

234. Oh SJ. Paraneoplastic vasculitis of the peripheral nervous system. *Neurol Clin* 1997;15:849–863.

235. Camerlingo M, Nemni R, Ferraro B, et al. Malignancy and sensory neuropathy of unexplained cause. *Arch Neurol* 1998;55:981–984.

236. Oh BC, Lim YM, Kwon YM, et al. A case of Hodgkin's lymphoma associated with sensory neuropathy. *J Kor Med Sci* 2004;19:130–133.

237. Vallat JM, Leboutet MJ, Hugon J, et al. Acute pure sensory paraneoplastic neuropathy with perivascular endoneurial inflammation: ultrastructural study of capillary walls. *Neurology* 1986;36:1395–1399.

238. Iwahashi T, Inoue A, Koh CS, et al. A study on a new antineural antibody in a case of paraneoplastic sensory neuropathy associated with breast carcinoma. *J Neurol Neurosurg Psychiatry* 1997;63:516–519.

239. Lopate G, Choksi R, Pestronk A. Severe sensory ataxia and demyelinating polyneuropathy with IgM anti-GM2 and GalNAc-GD1A antibodies. *Muscle Nerve* 2002;25:828–836.

240. Teravainen H, Larsen A. Some features of the neuromuscular complications of pulmonary carcinoma. *Ann Neurol* 1977;6:495–502.

241. Antoine JC, Honnorat J, Camdessanché JP, et al. Paraneoplastic anti-CV2 antibodies react with peripheral nerve and are associated with a mixed axonal and demyelinating peripheral neuropathy. *Ann Neurol* 2001;49:214–221.

242. Bird SJ, Brown MJ, Shy ME, et al. Chronic inflammatory demyelinating polyneuropathy associated with malignant melanoma. *Neurology* 1996;46:822–824.

243. Nickel J, Neuen-Jacob E, Saleh A, et al. CIDP and isolated osteosclerotic myeloma. *Neurology* 2004;63:2439.

244. Sugai F, Abe K, Fujimoto T, et al. Chronic inflammatory demyelinating polyneuropathy accompanied by hepatocellular carcinoma. *Intern Med* 1997; 36:53–55.

245. Lachance DH, O'Neill BP, Harper CM, Jr, et al. Paraneoplastic brachial plexopathy in a patient with Hodgkin's disease. *Mayo Clin Proc* 1991; 66:97–101.

246. Koller H, Schroeter M, Kieseier BC, et al. Chronic inflammatory demyelinating polyneuropathy—update on pathogenesis, diagnostic criteria and therapy. *Curr Opin Neurol* 2005;18:273–278.

247. Finsterer J. Treatment of immune-mediated, dysimmune neuropathies. *Acta Neurol Scand* 2005; 112:115–125.

248. Goldfarb AR, Weimer LH, Brannagan TH, III. Rituximab treatment of an IgM monoclonal autonomic and sensory neuropathy. *Muscle Nerve* 2005; 31:510–515.

249. Weide R, Heymanns J, Köppler H. The polyneuropathy associated with Waldenstrom's macroglobulinaemia can be treated effectively with chemotherapy and the anti-CD20 monoclonal antibody rituximab. *Br J Haematol* 2000;109:838–841.

250. Garcia-Moreno JM, Castilla JM, Garcia-Escudero A, et al. [Multifocal motor neuropathy with conduction blocks and prurigo nodularis. A paraneoplastic syndrome in a patient with non-Hodgkin B-cell lymphoma?]. *Neurologia* 2004;19:220–224.

251. Liang BC, Albers JW, Sima AAF, et al. Paraneoplastic pseudo-obstruction, mononeuropathy multiplex, and sensory neuronopathy. *Muscle Nerve* 1994; 17:91–96.

252. Park DM, Johnson RH, Crean GP, et al. Orthostatic hypotension in bronchial carcinoma. *Br Med J* 1972;3:510–511.

253. Eurelings M, Notermans NC, Van de Donk NWCJ, et al. Risk factors for hematological malignancy in polyneuropathy associated with monoclonal gammopathy. *Muscle Nerve* 2001;24:1295–1302.

254. Eurelings M, Lokhorst HM, Kalmijn S, et al. Malignant transformation in polyneuropathy associated with monoclonal gammopathy. *Neurology* 2005;64:2079–2084.

255. Burns TM. Vasculitic neuropathies. *Neurol Clin* 2007;25:89–113.

256. Zivkovic SA. Vasculitic neuropathy—electrodiagnostic findings and association with malignancies. *Acta Neurol Scand* 2007;115:432–436.

257. Gorson KC. Vasculitic neuropathies: an update. *Neurologist* 2007;13:12–19.

258. Elrington GM, Murray NM, Spiro SG, et al. Neurological paraneoplastic syndromes in patients with small cell lung cancer. A prospective survey of 150 patients. *J Neurol Neurosurg Psychiatry* 1991;54:764–767.

259. Tyagi A, Connolly S, Hutchinson M. Lambert-Eaton myaesthenic syndrome: a possible association with Hodgkin's lymphoma. *Ir Med J* 2001;94:18–19.

260. Burns TM, Juel VC, Sanders DB, et al. Neuroendocrine lung tumors and disorders of the neuromuscular junction. *Neurology* 1999;52:1490–1491.

261. Wirtz PW, Bradshaw J, Wintzen AR, et al. Associated autoimmune diseases in patients with the Lambert-Eaton myasthenic syndrome and their families. *J Neurol* 2004;251:1255–1259.

262. Wirtz PW, Willcox N, van der Slik AR, et al. HLA and smoking in prediction and prognosis of small cell lung cancer in autoimmune Lambert-Eaton myasthenic syndrome. *J Neuroimmunol* 2005;159:230–237.

263. Wirtz PW, Willcox N, Roep BO, et al. HLA-B8 in patients with the Lambert-Eaton myasthenic syndrome reduces likelihood of associated small cell lung carcinoma. *Ann N Y Acad Sci* 2003; 998:200–201.

264. Sanders DB. Lambert-Eaton myasthenic syndrome: diagnosis and treatment. *Ann N Y Acad Sci* 2003;998:500–508.

265. Burns TM, Russell JA, Lachance DH, et al. Oculobulbar involvement is typical with Lambert-Eaton myasthenic syndrome. *Ann Neurol* 2003; 53:270–273.

266. Toyka KV, Schneider-Gold C. Oculomotor signs in Lambert-Eaton myasthenic syndrome-coincidence with myasthenia gravis. *Ann Neurol* 2003;54:135–136.

267. AAEM Quality Assurance Committee, American Association of Electrodiagnostic Medicine. Practice parameter for repetitive nerve stimulation and single fiber EMG evaluation of adults with suspected myasthenia gravis or Lambert-Eaton myasthenic

syndrome: summary statement. *Muscle Nerve* 2001;24:1236–1238.

268. AAEM Quality Assurance Committee. Literature review of the usefulness of repetitive nerve stimulation and single fiber EMG in the electrodiagnostic evaluation of patients with suspected myasthenia gravis or Lambert-Eaton myasthenic syndrome. *Muscle Nerve* 2001;24:1239–1247.

269. Tim RW, Massey JM, Sanders DB. Lambert-Eaton myasthenic syndrome: electrodiagnostic findings and response to treatment. *Neurology* 2000;54:2176–2178.

270. Takamori M. Lambert-Eaton myasthenic syndrome as an autoimmune calcium channelopathy. *Biochem Biophys Res Commun* 2004;322:1347–1351.

271. Kimura J. *Electrodiagnosis in Diseases of Nerve and Muscle*. 2nd ed. Philadelphia: FA Davis, 1989.

272. Newsom-Davis J. A treatment algorithm for Lambert-Eaton myasthenic syndrome. *Ann N Y Acad Sci* 1998;841:817–822.

273. Sanders DB, Massey JM, Sanders LL, et al. A randomized trial of 3,4-diaminopyridine in Lambert-Eaton myasthenic syndrome. *Neurology* 2000; 54:603–607.

274. Wirtz PW, Nijnuis MG, Sotodeh M, et al. The epidemiology of myasthenia gravis, Lambert-Eaton myasthenic syndrome and their associated tumours in the northern part of the province of South Holland. *J Neurol* 2003;250:698–701.

275. Abrey LE. Association of myasthenia gravis with extrathymic Hodgkin's lymphoma: complete resolution of myasthenic symptoms following antineoplastic therapy. *Neurology* 1995;45:1019.

276. Levin N, Abramsky O, Lossos A, et al. Extrathymic malignancies in patients with myasthenia gravis. *J Neurol Sci* 2005;237:39–43.

277. Wirtz PW, Sotodeh M, Nijnuis M, et al. Difference in distribution of muscle weakness between myasthenia gravis and the Lambert-Eaton myasthenic syndrome. *J Neurol Neurosurg Psychiatry* 2002; 73:766–768.

278. Scherer K, Bedlack RS, Simel DL. Does this patient have myasthenia gravis? *JAMA* 2005;293:1906–1914.

279. Romi F, Skeie GO, Aarli JA, et al. Muscle autoantibodies in subgroups of myasthenia gravis patients. *J Neurol* 2000;247:369–375.

280. Marx A, Muller-Hermelink HK, Strobel P. The role of thymomas in the development of myasthenia gravis. *Ann N Y Acad Sci* 2003;998:223–236.

281. Jaretzki A, III, Aarli JA, Kaminski HJ, et al. Thymectomy for myasthenia gravis: evaluation requires controlled prospective studies. *Ann Thorac Surg* 2003;76:1–3.

282. Schwendimann RN, Burton E, Minagar A. Management of myasthenia gravis. *Am J Ther* 2005; 12:262–268.

283. Viallard JF, Vincent A, Moreau JF, et al. Thymoma-associated neuromyotonia with antibodies against voltage-gated potassium channels presenting as chronic intestinal pseudo-obstruction. *Eur Neurol* 2005;53:60–63.

284. Diaz-Manera J. Antibodies to AChR, MuSK and VGKC in a patient with myasthenia gravis and Morvan's syndrome. *Nat Clin Prac Neurol* 2007; 3:405–410.

285. Newsom-Davis J, Buckley C, Clover L, et al. Autoimmune disorders of neuronal potassium channels. *Ann N Y Acad Sci* 2003;998:202–210.

286. Walsh JC. Neuromyotonia: an unusual presentaton of intrathoracic malignancy. *J Neurol Neurosurg Psychiatry* 1976;39:1086–1091.

287. Caress JB, Abend WK, Preston DC, et al. A case of Hodgkin's lymphoma producing neuromyotonia. *Neurology* 1997;49:258–259.

288. Tomimitsu H, Arimura K, Nagado T, et al. Mechanism of action of voltage-gated K^+ channel antibodies in acquired neuromyotonia. *Ann Neurol* 2004;56:440–444.

289. van den Berg JS, van Engelen BG, Boerman RH, et al. Acquired neuromyotonia: superiority of plasma exchange over high-dose intravenous human immunoglobulin. *J Neurol* 1999;246:623–625.

290. Mastaglia FL, Garlepp MJ, Phillips BA, et al. Inflammatory myopathies: clinical, diagnostic and therapeutic aspects. *Muscle Nerve* 2003;27:407–425.

291. Sigurgeirsson B, Lindelof B, Edhag O, et al. Risk of cancer in patients with dermatomyositis or polymyositis. A population-based study. *N Engl J Med* 1992;326:363–367.

292. Buchbinder R, Forbes A, Hall S, et al. Incidence of malignant disease in biopsy-proven inflammatory myopathy. A population-based cohort study. *Ann Intern Med* 2001;134:1087–1095.

293. Hill CL, Zhang Y, Sigurgeirsson B, et al. Frequency of specific cancer types in dermatomyositis and polymyositis: a population-based study. *Lancet* 2001;357:96–100.

294. Alexandrescu DT, Bhagwati NS, Fomberstein B, et al. Steroid-responsive inclusion body myositis associated with endometrial cancer. *Clin Exp Rheumatol* 2005;23(1):93–96.

295. Dalakas MC. Inflammatory disorders of muscle: progress in polymyositis, dermatomyositis and inclusion body myositis. *Curr Opin Neurol* 2004;17:561–567.

296. Rostasy KM, Piepkorn M, Goebel HH, et al. Monocyte/macrophage differentiation in dermatomyositis and polymyositis. *Muscle Nerve* 2004; 30:225–230.

297. Greenberg SA. Proposed immunologic models of the inflammatory myopathies and potential therapeutic implications. *Neurology*. 2007;69:2008–2019.298.

298. Sampson JB. Paraneoplastic myopathy: response to intravenous immunoglobulin. *Neuromusc Disord* 2007;17:404–408.

299. Levin MI, Mozaffar T, Al-Lozi MT, et al. Paraneoplastic necrotizing myopathy—clinical and pathologic features. *Neurology* 1998;50:764–767.

300. Ueyama H, Kumamoto T, Araki S. Circulating autoantibody to muscle protein in a patient with paraneoplastic myositis and colon cancer. *Eur Neurol* 1992;32:281–284.

Index